AF829265

tional Series on

LLIGENT SYSTEMS, CONTROL AND AUTOMATION:
NCE AND ENGINEERING

JME 34

or S. G. Tzafestas, National Technical University of Athens, Greece

Complex Dynamics

Inter
INT
SCI

VOL

Edito

Profe

Complex Dynamics

Advanced System Dynamics in Complex Variables

by

VLADIMIR G. IVANCEVIC
Defence Science and Technology Organisation, Adelaide, SA, Australia

and

TIJANA T. IVANCEVIC
The University of Adelaide, SA, Australia

Springer

A C.I.P. Catalogue record for this book is available from the Library of Congress.

ISBN 978-1-4020-6411-1 (HB)
ISBN 978-1-4020-6412-8 (e-book)

Published by Springer,
P.O. Box 17, 3300 AA Dordrecht, The Netherlands.

www.springer.com

Printed on acid-free paper

Dedicated to Nitya, Atma and Kali

Preface

Complex Dynamics: Advanced System Dynamics in Complex Variables is a graduate–level monographic textbook. It is designed as a comprehensive introduction into methods and techniques of modern complex–valued nonlinear dynamics with its various physical and non–physical applications. This book is a *complex–valued* continuation of our previous two monographs, *Geometrical Dynamics of Complex Systems* and *High–Dimensional Chaotic and Attractor Systems*, Volumes 31 and 32 in the *Springer* book series *Intelligent Systems, Control and Automation: Science and Engineering*, where we had developed the most powerful mathematical machinery to deal with high–dimensional nonlinear, attractor and chaotic *real–valued* dynamics. The present monograph is devoted to understanding, prediction and control of both low– and high–dimensional, as well as both continuous– and discrete–time, nonlinear systems dynamics in complex variables. Its objective is to provide a serious reader with a serious scientific tool that will enable him/her to actually *perform* a competitive research in modern complex-valued nonlinear dynamics.

This book has seven Chapters. The first, introductory Chapter explains 'in plain English' the objective of the book and provides the preliminaries in complex numbers and variables; it also gives a soft introduction to quantum dynamics. The second Chapter develops low–dimensional dynamics in the complex plane, theoretical and computational, continuous– and discrete–time. The third Chapter presents a modern introduction to quantum dynamics, mainly following Dirac's notation. The fourth Chapter develops geometrical machinery of complex manifolds, essential for the further text. The fifth Chapter develops high–dimensional complex continuous dynamics, which takes place on complex manifolds. The sixth Chapter develops the formalism of complex path integrals, which extends the continuous dynamics to the general high–dimensional dynamics, which can be both discrete and stochastic. In the last, seventh Chapter, all previously developed methods are employed to present the 'Holy Grail' of modern physical and cosmological science, the search for the 'theory of everything' and the 'true' cosmological dynamics.

Our approach to complex dynamics is somewhat similar to the approach to mathematical physics used at the beginning of the 20th Century by the two leading mathematicians: David Hilbert and John von Neumann – the approach of combining mathematical rigor with conceptual clarity, or *geometrical intuition* that underpins the rigor.

Note that Einstein's summation convention over repeated indices is used throughout the text (in accordance with our Geometrical Dynamics book). For its comprehensive reading the only necessary prerequisite is advanced engineering mathematics (namely, strong calculus with some complex variables and linear algebra), although some acquaintance with the two previous books mentioned above would certainly be an advantage.

The book contains both an extensive Index (which allows easy connections between related topics) and a number of cited references related to modern complex dynamics.

The intended audience includes (but is not restricted to): mechatronics, control, robotics and signal/image–processing engineers; theoretical and mathematical physicists; applied and pure mathematicians; computer and neural scientists; mathematically strong chemists, biologists, psychologists, economists and sociologists – both in academia and industry.

Adelaide,
May 2007

V. Ivancevic, Defence Science & Technology Organisation, Australia, e-mail: Vladimir.Ivancevic@dsto.defence.gov.au

T. Ivancevic, School of Mathematics, The University of Adelaide, e-mail: Tijana.Ivancevic@adelaide.edu.au

Acknowledgments

The authors wish to thank Land Operations Division, Defence Science & Technology Organisation, Australia, for the support in Human Biodynamics Engine (HBE) and Human–Robot Teaming (HRT), as well as for all the HBE– and HRT–related text in this monograph.

We also express our gratitude to *Springer* book series *Intelligent Systems, Control and Automation: Science and Engineering* and especially to the Editor, Professor Spyros Tzafestas.

Glossary of Frequently Used Symbols

General

– 'iff' means 'if and only if';
– 'r.h.s' means 'right hand side'; 'l.h.s' means 'l.h.s.';
– ODE means ordinary differential equation, while PDE means partial differential equation;
– *Einstein's summation convention over repeated indices* (not necessarily one up and one down) *is assumed in the whole text,* unless explicitly stated otherwise.

Sets

$\mathbb{N}$ – natural numbers; $\mathbb{Z}$ – integers; $\mathbb{R}$ – real numbers;
$\mathbb{C}$ – complex numbers; $\mathbb{H}$ – quaternions;
$\mathbb{K}$ – number field of real numbers, complex numbers, or quaternions.

Maps

$f : A \to B$ – a function, (or map) between sets $A \equiv \operatorname{Dom} f$ and $B \equiv \operatorname{Cod} f$;

$$\begin{aligned} \operatorname{Ker} f &= f^{-1}(e_B) - \text{a kernel of } f; \\ \operatorname{Im} f &= f(A) - \text{an image of } f; \\ \operatorname{Coker} f &= \operatorname{Cod} f / \operatorname{Im} f - \text{a cokernel of } f; \\ \operatorname{Coim} f &= \operatorname{Dom} f / \operatorname{Ker} f - \text{a coimage of } f; \end{aligned}$$

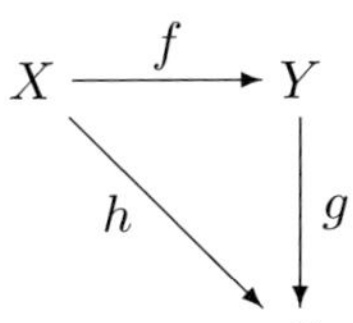

– a commutative diagram, requiring $h = g \circ f$.

Derivatives

$C^\infty(A, B)$ – set of $k-$times differentiable functions between sets A to B;
$C^\infty(A, B)$ – set of *smooth* functions between sets A to B;
$C^0(A, B)$ – set of *continuous* functions between sets A to B;
$f'(x) = \frac{df(x)}{dx}$ – derivative of f with respect to x;
$\dot{x}$ – total time derivative of x;
$\partial_t \equiv \frac{\partial}{\partial t}$ – partial time derivative;
$\partial_{x^i} \equiv \partial_i \equiv \frac{\partial}{\partial x^i}$ – partial coordinate derivative;

$\dot{f} = \partial_t f + \partial_{x^i} f \, \dot{x}^i$ – total time derivative of the scalar field $f = f(t, x^i)$;
$u_t \equiv \partial_t u$, $u_x \equiv \partial_x u$, $u_{xx} \equiv \partial_{x^2} u$ – only in partial differential equations;
$L_{x^i} \equiv \partial_{x^i} L$, $L_{\dot{x}^i} \equiv \partial_{\dot{x}^i} L$ – coordinate and velocity partial derivatives of the Lagrangian function;
d – exterior derivative;
d^n – coboundary operator;
∂_n – boundary operator;
$\nabla = \nabla(g)$ – affine Levi–Civita connection on a smooth manifold M with Riemannian metric tensor $g = g_{ij}$;
Γ^i_{jk} – Christoffel symbols of the affine connection ∇;
$\nabla_X T$ – covariant derivative of the tensor–field T with respect to the vector–field X, defined by means of Γ^i_{jk};
$T_{;x^i} \equiv T_{|x^i}$ – covariant derivative of the tensor–field T with respect to the coordinate basis $\{x^i\}$;
$\dot{\bar{T}} \equiv \frac{DT}{dt} \equiv \frac{\nabla T}{dt}$ – absolute (intrinsic, or Bianchi) derivative of the tensor–field T upon the parameter t; e.g., acceleration vector is the *absolute time derivative* of the velocity vector, $a^i = \dot{\bar{v}}^i \equiv \frac{Dv^i}{dt}$; note that in general, $a^i \neq \dot{v}^i$ – this is crucial for *proper definition of Newtonian force* (see Appendix);
$\mathcal{L}_X T$ – Lie derivative of the tensor–field T in direction of the vector–field X;
$[X, Y]$ – Lie bracket (commutator) of two vector–fields X and Y;
$[F, G]$, or $\{F, G\}$ – Poisson bracket, or Lie–Poisson bracket, of two functions F and G.

Smooth Manifolds, Fibre Bundles and Jet Spaces

Unless otherwise specified, all *manifolds* $M, N, ...$ are assumed C^∞–smooth, real, finite–dimensional, Hausdorff, paracompact, connected and without boundary,[1] while all maps are assumed smooth (C^∞). We use the symbols $\otimes$, $\vee$, $\wedge$ and $\oplus$ for the tensor, symmetrized and exterior products, as well as the Whitney sum[2], respectively, while $\rfloor$ denotes the interior product (contraction) of (multi)vectors and p–forms, and $\hookrightarrow$ denotes a manifold imbedding (i.e., both a submanifold and a topological subspace of the codomain manifold). The symbols ∂^A_B denote partial derivatives with respect to coordinates possessing multi–indices ^B_A (e.g., $\partial_\alpha = \partial/\partial x^\alpha$);
TM – tangent bundle of the manifold M;
$\pi_M : TM \to M$ – natural projection;
T^*M – cotangent bundle of the manifold M;
$\pi : Y \to X$ – fibre bundle;
(E, π, M) – vector bundle with total space E, base M and projection π;

[1] The only 1D manifolds obeying these conditions are the real line $\mathbb{R}$ and the circle S^1.

[2] *Whitney sum* $\oplus$ is an analog of the direct (Cartesian) product for vector bundles. Given two vector bundles Y and Y' over the same base X, their Cartesian product is a vector bundle over $X \times X$. The *diagonal map* induces a vector bundle over X called the Whitney sum of these vector bundles and denoted by $Y \oplus Y'$.

(Y, π, X, V) – fibre bundle with total space Y, base X, projection π and standard fibre V;
$J^k(M, N)$ – space of k–jets of smooth functions between manifolds M and N;
$J^k(X, Y)$ – k–jet space of a fibre bundle $Y \to X$; in particular, in mechanics we have a 1–jet space $J^1(\mathbb{R}, Q)$, with 1–jet coordinate maps $j^1_t s : t \mapsto (t, x^i, \dot{x}^i)$, as well as a 2–jet space $J^2(\mathbb{R}, Q)$, with 2–jet coordinate maps $j^2_t s : t \mapsto (t, x^i, \dot{x}^i, \ddot{x}^i)$;
$j^k_x s$ – k–jets of sections $s^i : X \to Y$ of a fibre bundle $Y \to X$;
We use the following kinds of *manifold maps*: immersion, imbedding, submersion, and projection. A map $f : M \to M'$ is called the *immersion* if the tangent map Tf at every point $x \in M$ is an injection (i.e., '1–1' map). When f is both an immersion and an injection, its image is said to be a submanifold of M'. A submanifold which also is a topological subspace is called imbedded submanifold. A map $f : M \to M'$ is called *submersion* if the tangent map Tf at every point $x \in M$ is a surjection (i.e., 'onto' map). If f is both a submersion and a surjection, it is called *projection* or *fibre bundle.*

Lie and (Co)Homology Groups

G – usually a general Lie group;
$GL(n)$ – general linear group with real coefficients in dimension n;
$SO(n)$ – group of rotations in dimension n;
T^n – toral (Abelian) group in dimension n;
$Sp(n)$ – symplectic group in dimension n;
$T(n)$ – group of translations in dimension n;
$SE(n)$ – Euclidean group in dimension n;
$H_n(M) = \operatorname{Ker} \partial_n / \operatorname{Im} \partial_{n-1}$ – nth homology group of the manifold M;
$H^n(M) = \operatorname{Ker} d^n / \operatorname{Im} d^{n+1}$ – nth cohomology group of the manifold M.

Other Spaces and Operators

$i \equiv \mathrm{i} \equiv \sqrt{-1}$ – imaginary unit;
$C^\infty(M)$ – space of k–differentiable functions on the manifold M;
$\Omega^k(M)$ – space of k–forms on the manifold M;
$\mathfrak{g}$ – Lie algebra of a Lie group G, i.e., the tangent space of G at its identity element;
$Ad(g)$ – adjoint endomorphism; recall that *adjoint representation* of a Lie group G is the linearized version of the action of G on itself by conjugation, i.e., for each $g \in G$, the inner automorphism $x \mapsto gxg^{-1}$ gives a linear transformation $Ad(g) : \mathfrak{g} \to \mathfrak{g}$, from the Lie algebra $\mathfrak{g}$ of G to itself;
nD space (group, system) means n–dimensional space (group, system), for $n \in \mathbb{N}$;
$\rhd$ – semidirect (noncommutative) product; e.g., $SE(3) = SO(3) \rhd \mathbb{R}^3$;

$\oint\!\!\!\!\!\!\int$ – Feynman path integral symbol, denoting integration over continuous spectrum of smooth paths and summation over discrete spectrum of Markov chains; e.g., $\oint\!\!\!\!\!\!\int \mathcal{D}[x]\, \mathrm{e}^{\mathrm{i}S[x]}$ denotes the path integral (i.e., sum–over–histories) over all possible paths $x^i = x^i(t)$ defined by the Hamiltonian action, $S[x] = \frac{1}{2}\int_{t_0}^{t_1} g_{ij}\, \dot{x}^i \dot{x}^j\, dt$, while $\oint\!\!\!\!\!\!\int \mathcal{D}[\Phi]\, \mathrm{e}^{\mathrm{i}S[\Phi]}$ denotes the path integral over all possible fields $\Phi^i = \Phi^i(x)$ defined by some field action $S[\Phi]$. In a similar way, we will define the path integral over all possible geometries and/or topologies.

Categories

$\mathcal{S}$ – all sets as objects and all functions between them as morphisms;
$\mathcal{V}$ – all vector spaces as objects and all linear maps between them as morphisms;
$\mathcal{B}$ – Banach spaces over $\mathbb{R}$ as objects and bounded linear maps between them as morphisms;
$\mathcal{G}$ – all groups as objects, all homomorphisms between them as morphisms;
$\mathcal{A}$ – Abelian groups as objects, homomorphisms between them as morphisms;
$\mathcal{T}$ – all topological spaces as objects, all continuous functions between them as morphisms;
$\mathcal{M}$ – all smooth manifolds as objects, all smooth maps between them as morphisms;
$\mathcal{LG}$ – all Lie groups as objects, all smooth homomorphisms between them as morphisms;
$\mathcal{LAL}$ – all Lie algebras (over a given field $\mathbb{K}$) as objects, all smooth homomorphisms between them as morphisms;
$\mathcal{TB}$ – all tangent bundles as objects, all smooth tangent maps between them as morphisms;
$\mathcal{T}^*\mathcal{B}$ – all cotangent bundles as objects, all smooth cotangent maps between them as morphisms;
$\mathcal{VB}$ – all smooth vector bundles as objects, all smooth homomorphisms between them as morphisms;
$\mathcal{FB}$ – all smooth fibre bundles as objects, all smooth homomorphisms between them as morphisms;
Symplec – all symplectic manifolds (i.e., physical phase–spaces), all symplectic maps (i.e., canonical transformations) between them as morphisms;
Hilbert – all Hilbert spaces and all unitary operators as morphisms.

Contents

1

Introduction

1.1 Why Complex Dynamics ?

Recall from [II06b, II07] that *dynamics* represents *a general scientific and engineering tool for prediction, or forecasting the future.* As a *predictive tool*, dynamics consists of two essential components:

(i) Newton–Maxwell like *dynamical laws* that govern *regularities in time*; and

(ii) *initial conditions* that govern how things started out (and therefore often specify *regularities in space*).

The first name associated with this so–called *Newtonian determinism* was Pierre-Simon Laplace around 1820. His famous statement reads [Lap51]:

> An intelligence knowing, at any given instant of time, all forces acting in nature, as well as the momentary positions of all things of which the universe consists, would be able to comprehend the motions of the largest bodies of the world and those of the smallest atoms in one single formula. ... To it, nothing would be uncertain, both future and past would be present before its eyes.

Both parts of dynamics are needed to make any predictions. For example, Newton's dynamical laws by themselves *do not* predict the trajectory of a tennis ball we might throw. To predict where it goes, we must also specify the position from which we throw it, the direction, and how fast. Technically, we must specify the ball's *initial conditions*.

Historically, most of classical dynamics has been developed in terms of real numbers and functions. On the other hand, complex dynamics is a predictive tool developed in terms of complex numbers and functions. A 19th Century scientist would naturally ask: Why complex dynamics? Isn't the world determined by real numbers? However, already 19th Century electrical engineers have realized the utility of using complex representation of trigonometric functions and series. They developed the so–called *complex–impendence* and *phasor–notation* methods for electric circuits,[1] as well as *frequency domain* methods in control theory,[2] and *complex Fourier methods* in signal/image analysis.[3] Later, in 1920s, with the advent of quantum mechanics,[4] scientists learned to use complex matrices and operators, with real eigen–values corre-

---

[1] In electrical engineering, when analyzing AC circuitry, the values for the electrical voltage (and current) are expressed as imaginary or complex numbers known as *phasors*. These are real voltages that can cause damage/harm to either humans or equipment even if their values contain no 'real part'. The study of AC (alternating current) entails introduction to electricity governed by trigonometric (i.e., oscillating) functions. From calculus, one knows that differentiating or integrating either $+/-\sin(t)$ or $+/-\cos(t)$ four times (with respect to t) results in the original function $+/-\sin(t)$ or $+/-\cos(t)$. From complex algebra, one knows that multiplying the imaginary unit quantity i, defined by $i^2 = -1$, by itself four times will result in the number 1 (identity), as $i^4 = i^3 i = (-i)i = -(i^2) = -(-1) = 1$. Thus, calculus can be represented by the algebraic properties of the imaginary unit quantity. Specifically, *Euler's formula*, which states that, for any real number x,

$$\mathrm{e}^{ix} = \cos x + i \sin x,$$

is used extensively to express signals (e.g., electromagnetic) that vary periodically over time as a combination of sine and cosine functions. Euler's formula accomplishes this more conveniently via an expression of exponential functions with imaginary exponents.

[2] Recall that in control theory, systems are often transformed from the time domain to the frequency domain using the *Laplace transform*. The system's poles and zeros are then analyzed in the *complex–plane*. The *root locus*, *Nyquist plot*, and *Nichols plot* techniques all make use of the complex–plane. For example, in the root locus method, it is especially important whether the poles and zeros are in the left or right half planes, i.e., have real part greater than or less than zero. If a system has poles that are: (i) in the right half plane, it will be unstable, (ii) all in the left half plane, it will be stable, (iii) in the imaginary axis, it will be marginally stable. If a system has zeros in the right half plane, it is a non–minimum–phase system.

[3] Recall that complex numbers of the form $z = x + iy$ (where x is the real part and y is the imaginary part) are used in *signal analysis* and other fields as a convenient description for periodically varying signals. The absolute value $|z|$ is interpreted as the *amplitude* and the argument $arg(z)$ as the *phase* of a sine wave of given frequency. If *Fourier analysis* is employed to write a given real–valued signal as a sum of periodic functions, these periodic functions are often written as the real part of complex–valued functions of the form

$$f(t) = z\mathrm{e}^{i\omega t},$$

where ω represents the *angular frequency* and the complex number z encodes the phase and amplitude.

[4] In quantum mechanics, the underlying theory is built on (infinite–dimensional) Hilbert spaces over the set of complex numbers $\mathbb{C}$.

sponding to measured observables. Our real scientific world is not any more restricted to the domain of real numbers.[5]

Today, we are using systems of differential equations of motion in the *complex–plane* (together with complex initial conditions), for low–resolution modelling of motion of groups of UGVs (or, soldiers). Similarly, we can model geographic motion of military forces by large systems of differential equations of motion taking place in the *Riemann sphere* (using complex initial conditions on the sphere). From these two basic examples, a generalization to the dynamics on high–dimensional complex manifolds, is natural.

In this book we discuss complex dynamical systems, both low–dimensional (flows in the complex–plane and Riemann sphere) and high–dimensional (flows in complex manifolds).

It is well–known that the real numbers have the advantage of being more directly tuned to describing real–life systems. However, complex numbers offer additional regularity, and besides real systems usually 'complexify' in a way that makes phenomena more clear; for example, periodic points disappear under parameter changes in the real case, but remain in the complex case.

Historically, complex dynamics in one complex dimension arose in the end of the 19th century as an outgrowth of studies of Newton's method and the 3–body problem in celestial mechanics (see [Ale94] for a historical treatment). In general, complex dynamics the study of dynamical systems for which the phase space is a complex manifold. More precisely, complex analytic dynamics specifies that it is analytic functions whose dynamics it is to study.

1.2 Preliminaries: Basics of Complex Numbers and Variables

In this section we briefly review the quintessence of complex numbers and functions. For more more detailed exposition, see any textbook in complex analysis (one of the oldest textbooks still in use is [Cop35]).

1.2.1 Complex Numbers and Vectors

For a *complex number*[6] $z = a + bi \in \mathbb{C}$, (with imaginary unit, $i = \sqrt{-1}$), its *complex–conjugate* is $\bar{z} \equiv z^* = \overline{a + bi} = a - bi \in \mathbb{C}$ (see Figure 1.1). Then

[5] For example, in *fluid dynamics*, complex functions are used to describe potential flow in 2D, while certain *fractals* (e.g., Mandelbrot and Julia sets) are plotted in the complex–plane; there are even complex neural networks (see e.g., [BP02].)

[6] Recall that the earliest fleeting reference to square roots of negative numbers occurred in the work of the Greek mathematician and inventor Heron of Alexandria in the 1st century AD, when he considered the volume of an impossible frustum of a pyramid. They became more prominent when in the 16th century closed formulas for the roots of third and fourth degree polynomials were discovered by Italian mathematicians (most notably Tartaglia and Cardano). It was soon

$z\bar{z} = a^2 + b^2 \in \mathbb{C}$; its *absolute value* or *modules* is $|z| = |a + bi| = \sqrt{z\bar{z}} = \sqrt{a^2 + b^2}$. z is *real*, $z \in \mathbb{R}$, iff $\bar{z} = z$.

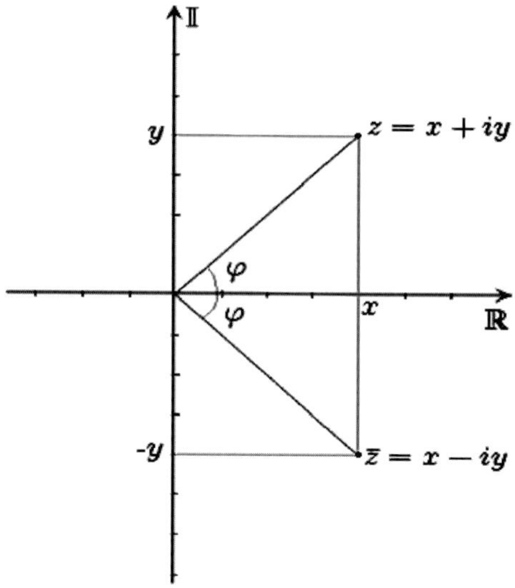

Fig. 1.1. Geometric representation of a complex number $z = x+iy$ and its conjugate $\bar{z} = x-iy$ as position vectors in the *complex–plane* (also called the *Argand diagram*).

The Cartesian coordinates of the complex number $z = x + iy$ are the real part x and the imaginary part y, while the polar coordinates are $r = |z|$, called the absolute value or modulus, and $\varphi = arg(z)$, called the complex argument of z. These numbers are connected by

---

realized that these formulas, even if one was only interested in real solutions, sometimes required the manipulation of square roots of negative numbers. This was doubly unsettling since not even negative numbers were considered to be on firm ground at the time. The term 'imaginary' for these quantities was coined by René Descartes in the 17th century and was meant to be derogatory. The 18th century saw the labors of Abraham de Moivre and Leonhard Euler. To De Moivre is due (1730) the well-known formula which bears his name, de Moivre's formula:

$$(\cos x + \mathrm{i} \sin x)^n = \cos nx + \mathrm{i} \sin nx,$$

and to Euler (1748), the Euler's formula given above. The existence of complex numbers was not completely accepted until the geometrical interpretation had been described by Caspar Wessel in 1799; it was rediscovered several years later and popularized by Carl Friedrich Gauss, and as a result the theory of complex numbers received a notable expansion. The general acceptance of the theory is not a little due to the labors of Augustin Louis Cauchy and Niels Henrik Abel.

$$x = r\cos\varphi, \qquad r = \sqrt{x^2 + y^2},$$
$$y = r\sin\varphi, \qquad \frac{y}{x} = \tan\varphi.$$

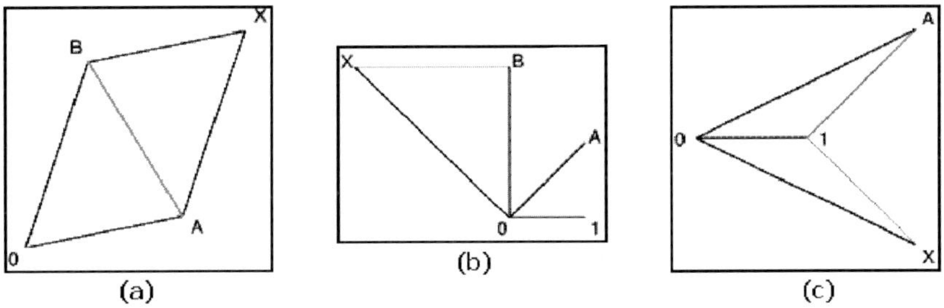

Fig. 1.2. Geometric interpretation of the operations on complex numbers. (a) Addition: $X = A+B$ (the sum of two points A and B is the point $X = A+B$ such that the triangles with vertices $0, A, B$ and X, B, A are *congruent*); (b) Multiplication: $X = AB$ (the product of two points A and B is the point $X = AB$ such that the triangles with vertices $0, 1, A$, and $0, B, X$ are *similar*); and (c) Conjugation: $X = A^*$ (the complex conjugate of a point A is a point $X = A^*$ such that the triangles with vertices $0, 1, A$ and $0, 1, X$ are mirror image of each other).

Using *Euler's formula*

$$\mathrm{e}^{i\varphi} = \cos\varphi + i\sin\varphi,$$

the *polar* and *exponential* forms of a complex number $z \in \mathbb{C}$ are given by

$$z = r(\cos\varphi + i\sin\varphi) = r\,cis\,\varphi = r\,\mathrm{e}^{i\varphi},$$

where $r = |z| = \sqrt{a^2 + b^2}$ and φ (*argument*, or *amplitude*) are polar coordinates, giving also

$$\cos\varphi = \frac{1}{2}(\mathrm{e}^{i\varphi} + \mathrm{e}^{-i\varphi}) = \frac{1}{2}\left(z + \frac{1}{z}\right),$$
$$\sin\varphi = \frac{1}{2i}(\mathrm{e}^{i\varphi} + \mathrm{e}^{-i\varphi}) = \frac{1}{2i}\left(z - \frac{1}{z}\right).$$

Product of two complex numbers is now given as (see Figure 1.2)

$$z_1 z_2 = r_1 r_2\left[\cos(\varphi_1 + \varphi_2) + i\sin(\varphi_1 + \varphi_2)\right] = r_1 r_2\, cis(\varphi_1 + \varphi_2) = r_1 r_2\,\mathrm{e}^{i(\varphi_1 + \varphi_2)},$$

there *quotient* is

$$\frac{z_1}{z_2} = \frac{r_1}{r_2}\left[\cos(\varphi_1 - \varphi_2) + i\sin(\varphi_1 - \varphi_2)\right] = \frac{r_1}{r_2}\, cis(\varphi_1 - \varphi_2) = \frac{r_1}{r_2}\,\mathrm{e}^{i(\varphi_1 - \varphi_2)},$$

the nth power *De Moivre's Theorem* holds (with $n \in \mathbb{N}$)

$$z^n = [r(\cos\varphi + i\sin\varphi)]^n = [r(\cos n\varphi + i\sin n\varphi)] = r^n\, cis(n\varphi) = r^n\,\mathrm{e}^{in\varphi},$$

while the nth root is (with $n, k \in \mathbb{N}$)

$$z^{1/n} = \left[r(\cos\frac{\varphi + 2k\pi}{n} + i\sin\frac{\varphi + 2k\pi}{n})\right]^{1/n} = r^{1/n}\, cis\frac{\varphi + 2k\pi}{n} = r^{1/n}\,\mathrm{e}^{i\frac{\varphi+2k\pi}{n}},$$

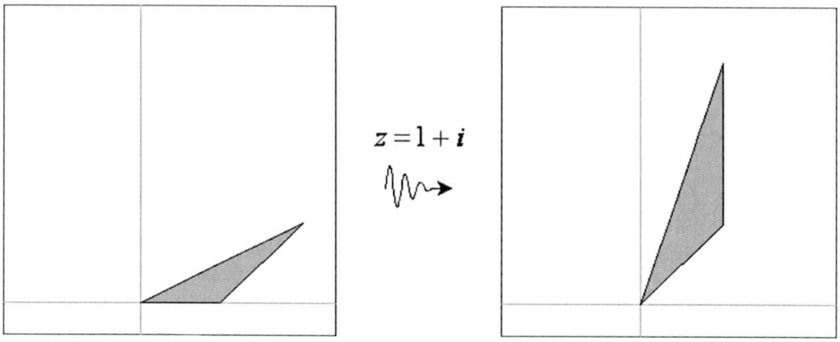

Fig. 1.3. Multiplication by a complex number $z = 1 + i$.

Transformations of the Complex Plane

A sample multiplication by a complex number is shown in Figure 1.3, while a sample transformation by a complex function is shown in Figure 1.4. Conformal maps by several complex functions are shown in Figure 1.5.

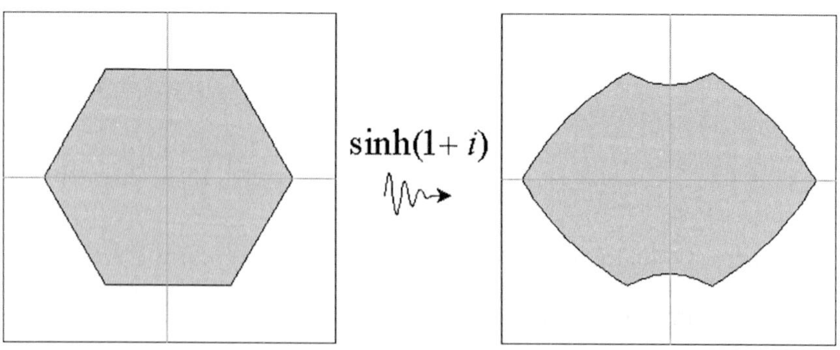

Fig. 1.4. Transformation by a complex function $w(z) = \sinh(1 + i)$.

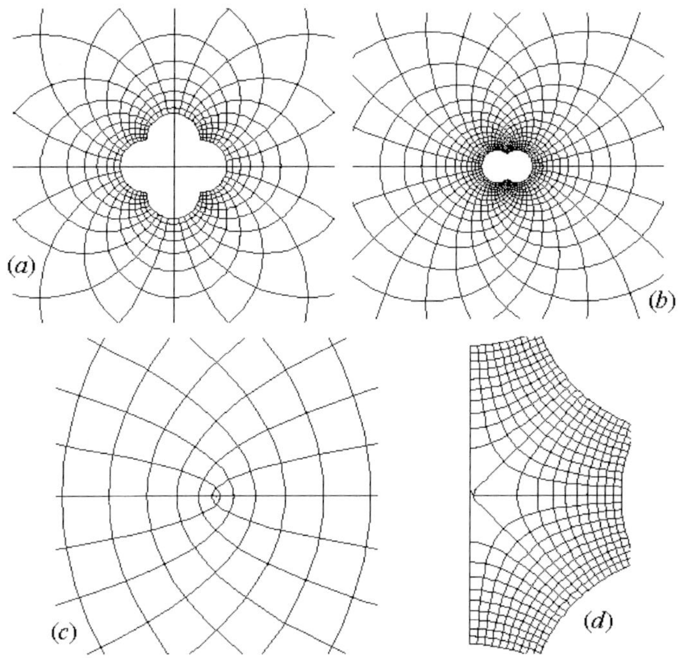

Fig. 1.5. Conformal maps by complex functions: (a) $w(z) = 1/z$; (b) $w(z) = 1/z^2$; (c) $w(z) = z^2$; and (d) $w(z) = \sqrt{z}$.

Complex n–Space

The elements of $\mathbb{C}^n$ are $n-$vectors. For any two $n-$vectors $x, y \in \mathbb{C}^n$ their *inner product* is defined as

$$x \cdot y = \langle x|y \rangle = \sum_{i=1}^{n} x_i y_i.$$

The *norm* of an $n-$vector $x \in \mathbb{C}^n$ is

$$\|x\| = \sqrt{x \cdot x} = \sqrt{\langle x|x \rangle}.$$

The space $\mathbb{C}^n$ with operations of vector addition, scalar multiplication, and inner product, is called *complex Euclidean n–space.*

M. Eastwood and R. Penrose (see [EP00]) developed a method for *drawing with complex numbers* in an ordinary Euclidean 3D space $\mathbb{R}^3$. They showed how the algebra of complex numbers can be used in an elegant way to represent the images of ordinary 3D figures, orthographically projected to the plane $\mathbb{R}^2 = \mathbb{C}$. For inspiration, see [HC99].

Quaternions and Rotations

Recall from topology that the set of Hamilton's *quaternions* $\mathbb{H}$ represents an extension of the set of complex numbers $\mathbb{C}$. Quaternions are widely used to represent rotations[7]. Instead of one imaginary unit $i = \sqrt{-1}$, we have three different numbers that are all square roots of -1 – labelled i, j, and k, respectively,

$$i \cdot i = -1, \qquad j \cdot j = -1, \qquad k \cdot k = -1.$$

When we multiply two quaternions, they behave similarly to cross products of the unit basis vectors,

$$i \cdot j = -j \cdot i = k, \qquad j \cdot k = -k \cdot j = i, \qquad k \cdot i = -i \cdot k = j.$$

The conjugate and magnitude of a quaternion are found in much the same way as complex conjugate and magnitude. If a quaternion q has length 1, we say that q is a *unit quaternion*

$$\begin{aligned} q &= w + xi + yj + zk, \\ q' &= w - xi - yj - zk, \\ |q| &= \sqrt{q \cdot q'} = \sqrt{w^2 + x^2 + y^2 + z^2}, \\ \text{unit quaternions:} \quad & |q| = 1 \Rightarrow q^{-1} = q', \\ \text{quaternions are associative:} \quad & (q_1 \cdot q_2) \cdot q_3 = q_1 \cdot (q_2 \cdot q_3), \\ \text{quaternions are } \textit{not} \text{ commutative:} \quad & q_1 \cdot q_2 \neq q_2 \cdot q_1. \end{aligned}$$

We can represent a quaternion in several ways: (i) as a linear combination of 1, i, j, and k, (ii) as a vector of the four coefficients in this linear combination, or (iii) as a scalar for the coefficient of 1 and a vector for the coefficients of the imaginary terms.

$$\begin{aligned} q &= w + xi + yj + zk = [x\ y\ z\ w] = (s, v), \\ s &= w, \qquad v = [x\ y\ z]. \end{aligned}$$

We can write the product of two quaternions in terms of the (s, v) representation using standard vector products in the following way:

$$\begin{aligned} q_1 &= (s_1, v_1), \qquad q_2 = (s_2, v_2), \\ q_1 \cdot q_2 &= (s_1 s_2 - v_1 \cdot v_2\ ,\ s_1 v_2 + s_2 v_1 + v_1 \times v_2). \end{aligned}$$

---

[7] Quaternions are superior to Euler angles in representing rotations, as they do not 'flip' at the angle of $\pm\pi/2$ (the well–known singularity of Euler angles).

Representing Rotations with Quaternions

We will compute a rotation about the unit vector, **u** by an angle θ. The quaternion that computes this rotation is

$$q = (s, v) = \left(\cos\frac{\theta}{2}\ ,\ u \sin\frac{\theta}{2} \right).$$

We will represent a point p in 3D space by the quaternion $P = (0, p)$. We compute the desired rotation about that point by

$$P = (0, p), \qquad P_{\text{rotated}} = q \cdot P \cdot q^{-1}.$$

Now, the quaternion P_{rotated} should be $(0, p_{\text{rotated}})$. Actually, we could put any value into the scalar part of P, i.e., $P = (c, p)$ and after performing the quaternion multiplication, we should get back $P_{\text{rotated}} = (c, p_{\text{rotated}})$.

You may want to confirm that q is a *unit quaternion*, since that will allow us to use the fact that the inverse of q is q' if q is a unit quaternion.

Concatenating Rotations

Suppose we want to perform two rotations on an object. This may come up in a manipulation interface where each movement of the mouse adds another rotation to the current object pose. This is very easy and numerically stable with a quaternion representation.

Suppose q_1 and q_2 are unit quaternions representing two rotations. We want to perform q_1 first and then q_2. To do this, we apply q_2 to the result of q_1, regroup the product using associativity, and find that the composite rotation is represented by the quaternion $q_2 \cdot q_1$.

$$q_2 \cdot (q_1 \cdot P \cdot q_1^{-1}) \cdot q_2^{-1} = (q_2 \cdot q_1) \cdot P \cdot (q_1^{-1} \cdot q_2^{-1}) = (q_2 \cdot q_1) \cdot P \cdot (q_2 \cdot q_1)^{-1}.$$

Therefore, the only time we need to compute the matrix is when we want to transform the object. For other operations we need only look at the quaternions. A matrix product requires many more operations than a quaternion product so we can save a lot of time and preserve more numerical accuracy with quaternions than with matrices.

Matrix Representation for Quaternion Multiplication

We can use the rules above to compute the product of two quaternions.

$$\begin{aligned}
q_1 &= w_1 + x_1 i + y_1 j + z_1 k, \\
q_2 &= w_2 + x_2 i + y_2 j + z_2 k, \\
q_1 \cdot q_2 &= (w_1 w_2 - x_1 x_2 - y_1 y_2 - z_1 z_2) \\
&\quad + (w_1 x_2 + x_1 w_2 + y_1 z_2 - z_1 y_2) i \\
&\quad + (w_1 y_2 - x_1 z_2 + y_1 w_2 + z_1 x_2) j \\
&\quad + (w_1 z_2 + x_1 y_2 - y_1 x_2 + z_1 w_2) k.
\end{aligned}$$

If we examine each term in this product, we can see that each term depends linearly on the coefficients for q_1. Also each term depends linearly on the coefficients for q_2.

So, we can write the product of two quaternions in terms of a matrix multiplication.

When the matrix $L_{\text{row}}(q_1)$ multiplies a row vector q_2, the result is a row vector representation for $q_1 \cdot q_2$. When the matrix $R_{\text{row}}(q_2)$ multiplies a row vector q_1, the result is also a row vector representation for $q_1 \cdot q_2$.

$$q_1 \cdot q_2 = q_2 L_{\text{row}}(q_1) = [\, x_2\ y_2\ z_2\ w_2 \,] \begin{bmatrix} w_1 & z_1 & -y_1 & -x_1 \\ -z_1 & w_1 & x_1 & -y_1 \\ y_1 & -x_1 & w_1 & -z_1 \\ x_1 & y_1 & z_1 & w_1 \end{bmatrix},$$

$$q_1 R_{\text{row}}(q_2) = [\, x_1\ y_1\ z_1\ w_1 \,] \begin{bmatrix} w_2 & -z_2 & y_2 & -x_2 \\ z_2 & w_2 & -x_2 & -y_2 \\ -y_2 & x_2 & w_2 & -z_2 \\ x_2 & y_2 & z_2 & w_2 \end{bmatrix}.$$

Computing Rotation Matrices from Quaternions.

Now we have all the tools we need to use quaternions to generate a rotation matrix for the given rotation. We have a matrix form for left–multiplication by q

$$P \cdot L_{\text{row}}(q) = [\, x_p\ y_p\ z_p\ 0 \,] \begin{bmatrix} w_q & z_q & -y_q & -x_q \\ -z_q & w_q & x_q & -y_q \\ y_q & -x_q & w_q & -z_q \\ x_q & y_q & z_q & w_q \end{bmatrix},$$

and a matrix form for right–multiplication by q^{-1}.

$$q^{-1} = q' = [\, -x_q\ -y_q\ -z_q\ w_q \,],$$

$$P \cdot R_{\text{row}}(q^{-1}) = [\, x_p\ y_p\ z_p\ 0 \,] \begin{bmatrix} w_q & z_q & -y_q & x_q \\ -z_q & w_q & x_q & y_q \\ y_q & -x_q & w_q & z_q \\ -x_q & -y_q & -z_q & w_q \end{bmatrix}.$$

The resulting rotation matrix is the product of these two matrices,

$$\begin{aligned} Q_{\text{row}} &= R_{\text{row}}(q^{-1}) \cdot L_{\text{row}}(q) \\ &= \begin{bmatrix} w_q & z_q & -y_q & x_q \\ -z_q & w_q & x_q & y_q \\ y_q & -x_q & w_q & z_q \\ -x_q & -y_q & -z_q & w_q \end{bmatrix} \cdot \begin{bmatrix} w_q & z_q & -y_q & -x_q \\ -z_q & w_q & x_q & -y_q \\ y_q & -x_q & w_q & -z_q \\ x_q & y_q & z_q & w_q \end{bmatrix} \end{aligned}$$

$$= \begin{bmatrix} w^2 + x^2 - y^2 - z^2 & 2xy + 2wz & 2xz - 2wy & 0 \\ 2xy - 2wz & w^2 - x^2 + y^2 - z^2 & 2yz + 2wx & 0 \\ 2xz + 2wy & 2yz - 2wx & w^2 - x^2 - y^2 + z^2 & 0 \\ 0 & 0 & 0 & w^2 + x^2 + y^2 + z^2 \end{bmatrix}$$

Although matrices do not generally commute (in general $AB \neq BA$), because these matrices represent left and right multiplication and quaternion multiplication is associative, these particular matrices do commute. So, we could write $Q_{\text{row}} = L_{\text{row}}(q) \cdot R_{\text{row}}(q^{-1})$ instead of $Q_{\text{row}} = R_{\text{row}}(q^{-1}) \cdot L_{\text{row}}(q)$ and we would get the same result. So using this matrix, we could compute P_{rotated} another way:

$$P_{\text{rotated}} = P\ Q_{\text{row}}.$$

1.2.2 Complex Functions

Now we return to complex variable theory. If to each of a set of complex numbers which a variable z may assume there corresponds one or more values of a variable w, then w is called a *function of the complex variable* z, written $w = f(z)$. A function is *single–valued* if for each value of z there corresponds only one value of w; otherwise it is *multiple–valued* or *many–valued.* In general we can write $w = f(z) = u(x, y) + iv(x, y)$, where u and v are real functions of x and y (called the *real and imaginary parts* of w, respectively).

Definitions of limits and continuity for functions of a complex variable are analogous to those for a real variable. Thus, $f(z)$ is said to have the *limit* l as z approaches z_0 if, given any $\epsilon > 0$, there exists a $\delta > 0$, such that $|f(z) - l| < \epsilon$ whenever $0 < |z - z_0| < \delta$. Similarly, $f(z)$ is said to be *continuous* at z_0 if, given any $\epsilon > 0$, there exists a $\delta > 0$, such that $|f(z) - f(z_0)| < \epsilon$ whenever $0 < |z - z_0| < \delta$; alternatively, $f(z)$ is continuous at z_0 if $\lim_{z \to z_0} f(z) = f(z_0)$.

If $f(z)$ is single–valued in some region of the z plane, the *derivative* of $f(z)$ is defined as

$$f'(z) = \lim_{\Delta z \to 0} \frac{f(z + \Delta z) - f(z)}{\Delta z}, \tag{1.1}$$

provided the limit exists independent of the manner in which $\Delta z \to 0$. If the limit (1.1) exists for $z = z_0$, then $f(z)$ is called *differentiable* at z_0. If the limit exists for all z such that $|z - z_0| < \delta$ for some $\delta > 0$, then $f(z)$ is called *holomorphic function*, or *analytic in a region* $\mathcal{R}$ in the complex–plane $\mathbb{C} \approx \mathbb{R}^2$. In order to be analytic, $f(z)$ must be single–valued and continuous. The converse is not necessarily true.

A *necessary condition* that $w = f(z) = u(x, y) + iv(x, y)$ be holomorphic (or, analytic) in a region $\mathcal{R} \in \mathbb{C}$ is that u and v satisfy the *Cauchy–Riemann equations*

$$\frac{\partial u}{\partial x} = \frac{\partial v}{\partial y}, \qquad \frac{\partial u}{\partial y} = -\frac{\partial v}{\partial x}. \tag{1.2}$$

If the partial derivatives in (1.2) are continuous in $\mathcal{R} \in \mathbb{C}$, the equations are also *sufficient conditions* that $f(z)$ be analytic in $\mathcal{R} \in \mathbb{C}$.

If the second derivatives of u and v with respect to x and y exist and are continuous, we find by differentiating (1.2) that the real and imaginary parts satisfy 2D *Laplace equation*

$$\frac{\partial^2 u}{\partial x^2} + \frac{\partial^2 u}{\partial y^2} = 0, \qquad \frac{\partial^2 v}{\partial x^2} + \frac{\partial^2 v}{\partial y^2} = 0.$$

Functions satisfying Laplace equation* are called *harmonic functions*.

A holomorphic function $w = f(z)$ gives a *surjective mapping* (or, transform) of its *domain of definition* in the complex z–plane *onto* its *range of values* in the complex w–plane (both planes are in $\mathbb{C}$). This mapping is *conformal*, i.e., the angle between two curves in the z plane intersecting at $z = z_0$, has the same magnitude and orientation as the angle between the images of the two curves, so long as $f'(z_0) \neq 0$. In other words, the mapping defined by analytic function $f(z)$ is conformal, except at *critical points* at which the derivative $f'(z) = 0$ (the *conformal property of analytic functions*).

If $f(z)$ is defined, single–valued and continuous in a region $\mathcal{R} \subset \mathbb{C}$, we define the *integral* of $f(z)$ along some path $c \in \mathcal{R}$ from point z_1 to point z_2, where $z_1 = x_1 + iy_1$, $z_2 = x_2 + iy_2$, as

$$\begin{aligned} \int_c f(z)\,dz &= \int_{(x_1,y_1)}^{(x_2,y_2)} (u + iv)(dx + idy) \\ &= \int_{(x_1,y_1)}^{(x_2,y_2)} u\,dx - v\,dy + i \int_{(x_1,y_1)}^{(x_2,y_2)} v\,dx + u\,dy. \end{aligned} \tag{1.3}$$

With this definition the integral of a function of a complex variable can be made to depend on line integrals of functions of real variables. It is equivalent to the definition based on the limit of a sum.

Let c be a simple closed curve in a region $\mathcal{R} \subset \mathbb{C}$. If $f(z)$ is analytic in $\mathcal{R}$ as well as on c, then we have the *Cauchy's Theorem*

$$\oint_c f(z)\,dz = 0. \tag{1.4}$$

Expressed in another way, (1.4) is equivalent to the statement that $\int_{z_1}^{z_2} f(z)\,dz$ has a value *independent of the path* joining z_1 and z_2. Such integrals can be evaluated as $F(z_2) - F(z_1)$ where $F'(z) = f(z)$.

If $f(z)$ is analytic within and on a simple closed curve c and a is any point interior to c, then

$$f(a) = \frac{1}{2\pi i} \oint_c \frac{f(z)}{z - a}\,dz, \tag{1.5}$$

where c is traversed in the positive (counterclockwise) sense. Similarly, the nth derivative of $f(z)$ at $z = a$ is given by

$$f^{(n)}(a) = \frac{n!}{2\pi i} \oint_c \frac{f(z)}{(z - a)^{n+1}}\,dz. \tag{1.6}$$

These are the *Cauchy's integral formulas*. They are quite remarkable because they show that if the function $f(z)$ is known *on* the closed curve c then it is also known *within* c, and the various derivatives at points within c can be calculated. Thus if a function of a complex variable has a first derivative, it has all higher derivatives as well, which is not necessarily true for functions of real variables.

Let $f(z)$ be analytic inside and on a circle having its center at $z = a$. Then for all points z in the circle we have the *Taylor series* representation of $f(z)$ given by

$$f(z) = f(a) + f'(a)(z-a) + \frac{f''(a)}{2!}(z-a)^2 + \frac{f'''(a)}{3!}(z-a)^3 + ... \quad (1.7)$$

A singular point of a function $f(z)$ is a value of z at which $f(z)$ fails to be analytic. If $f(z)$ is analytic everywhere in some region $\mathcal{R} \subset \mathbb{C}$ except at an interior point $z = a$, we call $z = a$ an *isolated singularity* of $f(z)$.

If $f(z) = \frac{\phi(z)}{(z-a)^n}$, $\phi(a) \neq 0$, where $\phi(z)$ is analytic everywhere in $\mathcal{R}$, with $n \in \mathbb{N}$, then $f(z)$ has an isolated singularity at $z = a$ which is called a *pole of order* n. If $n = 1$, the pole is called a *simple pole*; if $n = 2$ it is called a *double pole*, etc.

If $f(z)$ has a pole of order n at $z = a$ but is analytic at every other point inside and on a circle $c \subset \mathbb{C}$ with a center at a, then $(z-a)^n f(z)$ is analytic at all points inside and on c and has a Taylor series about $z = a$ so that

$$f(z) = \frac{a_{-n}}{(z-a)^n} + \frac{a_{-n+1}}{(z-a)^{n-1}} + ... + \frac{a_{-1}}{z-a} + a_0 + a_1(z-a) + a_2(z-a)^2 + ... \quad (1.8)$$

This is called a *Laurent series* for $f(z)$. The part $a_0 + a_1(z-a) + a_2(z-a)^2 + ...$ is called the *analytic part*, while the remainder consisting of inverse powers of $z - a$ is called the *principal part*. More generally, we refer to the series $\sum_{k=-\infty}^{\infty} a_k(z-a)^k$ as a Laurent series where the terms with $k < 0$ constitute the principal part. A function which is analytic in a region bounded by two concentric circles having center at $z = a$ can always be expanded into such Laurent series.

The coefficients in (1.8) can be obtained in the customary manner by writing the coefficients for the Taylor series corresponding to $(z-a)^n f(z)$. Specially, the coefficient a_{-1}, called the *residue* of $f(z)$ at the pole $z = a$, written $\texttt{Res}_{z=a} f(z)$, is very important. It can be found from the formula

$$\texttt{Res}_{z=a} f(z) = \frac{1}{(n-1)!} \lim_{z \to a} \frac{d^{n-1}}{dz^{n-1}}[(z-a)^n f(z)],$$

where n is the order of the pole. For simple poles the calculation of the residue is simple

$$\texttt{Res}_{z=a} f(z) = \lim_{z \to a}(z-a) f(z).$$

Caushy's residue Theorem: If $f(z)$ is analytic within and on the boundary c of a region $\mathcal{R} \subset \mathbb{C}$ except at a finite number of poles $a, b, c, ... \in \mathcal{R}$, having residues $a_{-1}, b_{-1}, c_{-1}, ...$ respectively, then

$$\oint_c f(z)\,dz = 2\pi i(a_{-1} + b_{-1} + c_{-1} + ...) \tag{1.9}$$

$$= 2\pi i \sum_{i=1}^{k} \mathtt{Res}_{z=z_i}\, f(z),$$

i.e., the integral of $f(z)$ equals $2\pi i$ times the sum of residues of $f(z)$ at the poles enclosed by c. Cauchy's Theorem and integral formulas are special cases of this result. It is used for evaluation of various definite integrals of both real and complex functions.

For example, (1.9) is used in *inverse Laplace transform*.

If $F(s) = L\{f(t)\} = \int_0^\infty \mathrm{e}^{-st} f(t)\,dt$, then $L^{-1}\{F(s)\}$ is given by

$$f(t) = L^{-1}\{F(s)\} = \frac{1}{2\pi i}\oint_c \mathrm{e}^{st} F(s)\,ds$$

$$= \sum \mathtt{Res}\,[\mathrm{e}^{st} F(s)]\; at\, poles\, of\; F(s)$$

where $c \subset \mathbb{C}$ is the so–called *Bromwich contour*.

1.2.3 Unit Circle and Riemann Sphere

Circle in the Complex Plane

Recall that in a high–school introduction of the sine and cosine functions (i.e., the *unit trigonometric circle*), an often used example is that of a rotating rod of length r with one end fixed; the position of the rod is plotted, its hight and horizontal distance from the centre against the angle through which the rod had rotated. Let us consider this problem with, in addition, specifying that the circular motion is at constant angular velocity ω. That is, the time derivative $\dot{\theta}$ of the angle θ between the rod and the $x-$axis (positive side), is given by

$$\dot{\theta} = \omega, \qquad \text{therefore} \qquad \theta(t) = \omega\, t + \theta(0).$$

If we start with the rod horizontal, then $\theta(0) = 0$ and we have $\theta = \omega\, t$. Thus, the $(x, y)-$position of the tip of the road is of length r is given as a function of time by

$$x = r\cos(\omega\, t), \qquad y = r\sin(\omega\, t). \tag{1.10}$$

Next, suppose that we want to find the acceleration of the distant end of the rod. Differentiating with respect to time gives the components of velocity in both directions as

$$\dot{x} = -r\omega \sin(\omega t), \qquad \dot{y} = r\omega \cos(\omega t),$$

while another differentiation gives the components of acceleration in both directions as

$$\ddot{x} = -r\omega^2 \cos(\omega t), \qquad \ddot{y} = -r\omega^2 \sin(\omega t),$$

and substituting (1.10), we finally get the expression for both x and y accelerations,

$$\ddot{x} = -\omega^2 x, \qquad \ddot{y} = -\omega^2 y. \tag{1.11}$$

Now, we can represent the motion, both in the $x-$direction and in the $y-$direction, by using a complex number $z = x + iy$, to represent a rotating vector with $(x, y)-$components, so instead of the two real second–order ODEs (1.11), we get one complex second–order ODE

$$\ddot{z} = -\omega^2 z. \tag{1.12}$$

We know that one solution[8] of (1.12), with initial condition $|z| = r$ when $t = 0$, is given by the *circle in the complex–plane* (see Figure 1.6 for the special case of the *unit circle*), that is

$$z = r\cos(\omega t) + i\, r \sin(\omega t),$$

with the corresponding complex velocity given by

$$\dot{z} = -r\omega \sin(\omega t) + i\, r\omega \cos(\omega t),$$

and at $t = 0$,

$$\dot{z} = i\, r\omega.$$

Riemann Sphere and Riemann Surfaces

Recall that the *Riemann sphere* (named after Bernhard Riemann, the father of differential geometry) is the unique way of viewing the *extended complex–plane* (the complex–plane plus a point at infinity) so that it looks exactly the same at the point infinity as at any complex number. The main application is to deal with extended complex functions (which may be defined at the point infinity and/or take the value infinity, in addition to complex numbers) in the

[8] Unfortunately, there is at least one other solution, given by the case when rod travels clockwise rather than anticlockwise, that is

$$z = r\cos(-\omega t) + i\, r \sin(-\omega t).$$

However, we can pin–down the solution to the anticlockwise direction of rotation by using the fact that we have initially defined the angular velocity by $\dot{\theta} = \omega$.

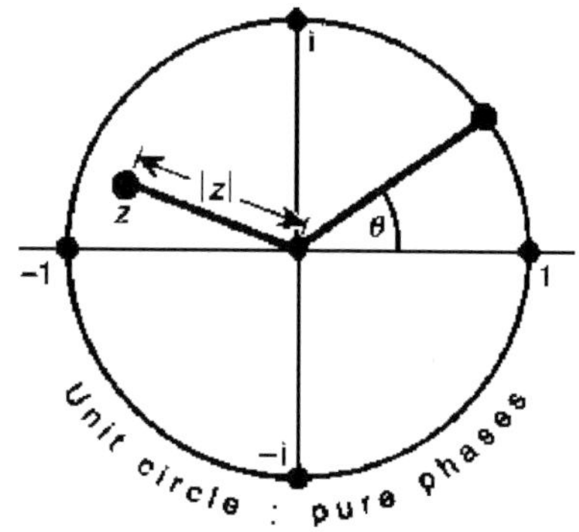

Fig. 1.6. The unit circle consists of *pure–phase* (or *unit modulus*) complex numbers, having the form: $z = \cos\theta + i\sin\theta = e^{i\theta}$, with θ real, i.e., $|z| \equiv r = 1$ (modified and adapted from [Pen94]).

same way at the point infinity as at any complex number, specifically with respect to continuity and differentiability.

From the geometrical view of the plane that deals with points, lines, circles and angles but not distances, the Riemann sphere is created by adding a point at infinity through which all lines cross, with parallel lines being tangent there and all other lines crossing at the same angle as they do at an existing point. This geometry is realized as a 2D sphere formed from the extended complex–plane using the stereographic projection, where lines in the complex–plane become circles through infinity. Angles in the Riemann sphere are identical to the corresponding angles in the complex–plane (and the same is true at infinity with the natural choice of the angle between two lines at infinity). Topologically, the Riemann sphere is the *one–point compactification* of the complex–plane. This gives it the topology of a 2D sphere, preserving the topology of the complex–plane. The Riemann sphere can be conveniently identified with a geometrical 2D sphere, in which lines become circles through infinity.

Define $\widehat{\mathbb{C}} = \mathbb{C} \cup \{\infty\}$ (i.e., the extended complex–plane: the complex numbers joined with the point at infinity). The Riemann sphere is based on the transformation from $\widehat{\mathbb{C}}$ to $\widehat{\mathbb{C}}$ in the form

$$w = f(z) = \frac{1}{z},$$

where $w, z \in \widehat{\mathbb{C}}$ and $\frac{1}{0} = \infty$. We visualize the Riemann sphere as a sphere in Euclidean 3–space $\mathbb{R}^3$ (see Figure 1.7). Every point on the sphere has both a $z-$value and $w-$value, related by the above transformation; i.e., $f(z)$ transforms the sphere onto itself.

To establish the correspondence between points in the extended complex–plane and the Riemann sphere, we first place the z plane tangent to the

Fig. 1.7. An outline of the Riemann sphere in Euclidean 3–space $\mathbb{R}^3$.

sphere's north pole. We then use *stereographic projection* from the south pole of the sphere (see Figure 1.8). This is done by drawing a line from the south pole that intersects both the sphere and the complex–plane; a unique, 1–1 correspondence is then established between points on the complex–plane and points on the Riemann sphere.

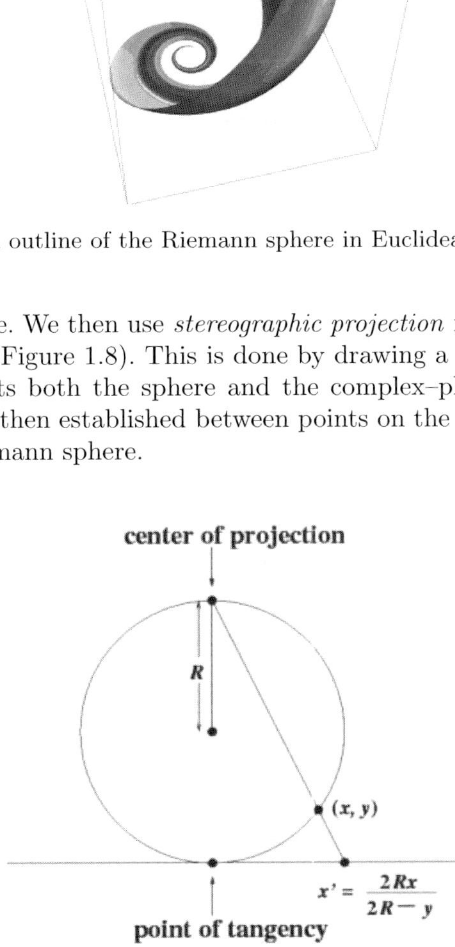

Fig. 1.8. Stereographic projection: the 1–1 correspondence between a sphere (represented by a circle) and the extended complex–plane (represented by a line).

In order to complete this 1–1 correspondence for the extended complex–plane, we define the south pole to be $z = \infty$ (note that the north pole is

$z = 0$). The correspondence between the w−plane and the Riemann sphere is done in much the same way, simply 'upside down'. That is, the w−plane is tangent to the south pole and oriented oppositely to the z−plane, such that $w = \{1, i, -1, -i\}$ matches to $z = \{1, -i, -1, i\}$. We then perform the stereographic projection from the north pole, and similarly define the north pole to be $w = \infty$. Now, every point on the sphere has both a z and w coordinate, related by the transformation: $w = f(z) = \frac{1}{z}$.

An alternate version of the stereographic projection places the planes at the equator, but preserves their opposite orientation. Thus, the planes are not geometrically distinct (see Figure 1.9).

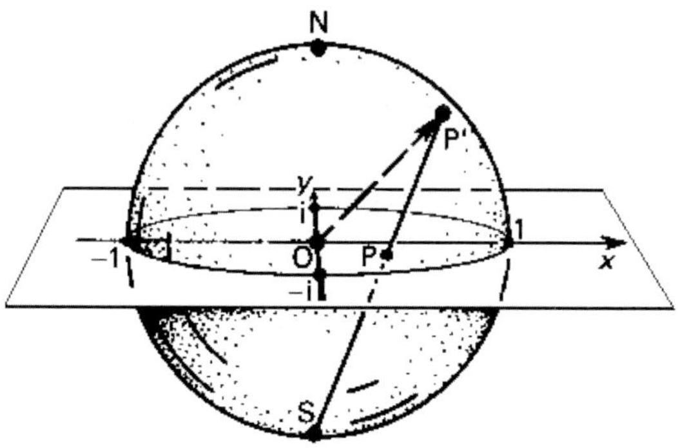

Fig. 1.9. The Riemann sphere. The point P, representing $u = z/w$ on the complex–plane, is projected from the south pole S to a point P' on the sphere. The direction OP', from the sphere's center O, is the direction of the spin axis for the superposed state of two spin–$\frac{1}{2}$ particles (modified and adapted from [Pen67, Pen94, Pen97]).

The so–called *Möbius transformations*, which send $\widehat{\mathbb{C}}$ to $\widehat{\mathbb{C}}$, are the automorphisms of the Riemann sphere (i.e., the conformal bijections), of the form

$$t = f(z) = \frac{az + b}{cz + d},$$

where $t, z \in \widehat{\mathbb{C}}$; $a, b, c, d \in \mathbb{C}$, and $ad - bc \neq 0$. They map the Riemann sphere to itself, preserving angles and orientation. This can be seen directly as they may be expressed as a composition of maps of the form:

$$z \to z + z_0, \qquad z \to ze^{i\theta}, \qquad z \to z + z_0, \qquad z \to \frac{1}{z},$$

(where r, θ are real numbers and z_0 is a complex number). These are respectively called: *elementary dilations*, *rotations*, *translations* and *complex inversion* (a composition of an inversion in the unit circle and a reflection in the real line), each of which is conformal on the complex–plane. Using the map $z \to \frac{1}{z}$, allows us to check that this is also true at infinity. Conversely, every everywhere–conformal bijection of the Riemann sphere is a Möbius transformation.

The 2–sphere admits a unique *complex structure* turning it into a *Riemann surface* (i.e., a 1D complex manifold). The Riemann sphere can be characterized as the unique simply–connected, compact Riemann surface, and may be taken to have the complex–plane as a complex submanifold. In all of these viewpoints, the point at infinity acquires an identical role to any point in the complex–plane. For example, the Riemann surface corresponding to the function $w = \sqrt{z}$ is given in Figure 1.10.

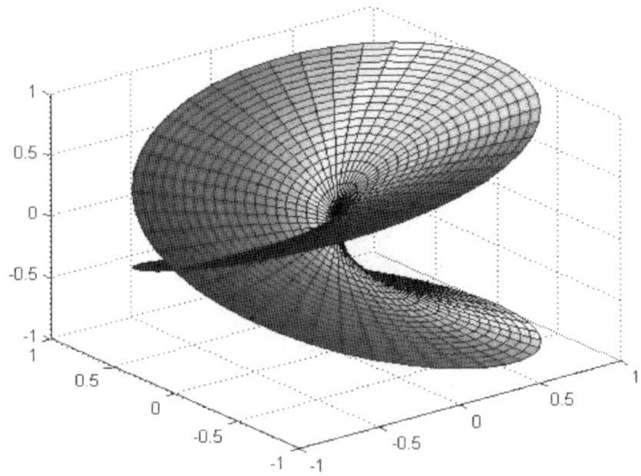

Fig. 1.10. Riemann surface corresponding to the function $w = \sqrt{z}$.

The complex manifold structure on the Riemann sphere is specified by an *atlas* with two charts defined by:

$$f : \widehat{\mathbb{C}} \setminus \{\infty\} \longrightarrow \mathbb{C}, \ f(z) = z, \qquad g : \widehat{\mathbb{C}} \setminus \{0\} \longrightarrow \mathbb{C}, \ g(z) = \frac{1}{z} \text{ and } g(\infty) = 0.$$

The overlap of these two charts is all points except 0 and ∞. On this overlap the transition function is given by $z \longrightarrow 1/z$, which is clearly holomorphic and so defines a complex structure.

The Riemann sphere has the same topology as S^2, that is, the sphere of radius 1 centered at the origin in the Euclidean space $\mathbb{R}^3$. A homeomorphism between them is given by the stereographic projection tangent to the south pole onto the complex–plane. Labelling the points in S^2 by (x_1, x_2, x_3) where $x_1^2 + x_2^2 + x_3^2 = 1$, the homeomorphism, given by $(x_1, x_2, x_3) \longrightarrow \frac{x_1 - ix_2}{1 - x_3}$, maps the south pole to the origin of the complex–plane and the north pole to ∞. In terms of standard spherical coordinates, this map can be given as $(\theta, \phi) \longrightarrow \mathrm{e}^{-i\phi} \cot \frac{\theta}{2}$. One can also use the stereographic projection tangent to the north pole given by $(x_1, x_2, x_3) \longrightarrow \frac{x_1 + ix_2}{1 + x_3}$, or in spherical coordinates $(\theta, \phi) \longrightarrow \mathrm{e}^{i\phi} \tan \frac{\theta}{2}$, which maps the north pole to the origin and the south pole to ∞.

Riemann surfaces can be thought of as 'deformed versions' of the complex–plane: locally near every point they look like patches of the complex–plane, but the global topology can be quite different. For example, they can look like a sphere or a torus or a couple of sheets glued together. The main point of Riemann surfaces is that holomorphic functions may be defined between them. Riemann surfaces are nowadays considered the natural setting for studying the global behavior of these functions, especially multi–valued functions such as the square root or the logarithm. Every Riemann surface is a 2D real analytic manifold (see Chapter 4), but it contains more structure (specifically a complex structure) which is needed for the unambiguous definition of holomorphic functions. A 2D real manifold can be turned into a Riemann surface (usually in several inequivalent ways) iff it is orientable; so the sphere and torus admit complex structures, but the Möbius strip, Klein bottle and projective plane do not.

Formally, let X be a *Hausdorff space*. A *homeomorphism* from an open subset $U \subset X$ to a subset of $\mathbb{C}$ is called a *chart*. Two charts f and g whose domains intersect are said to be compatible if the maps $f \circ g^-1$ and $g \circ f^-1$ are holomorphic over their domains. If A is a collection of compatible charts and if any $x \in X$ is in the domain of some $f \in A$, then we say that A is an *atlas*. When we endow X with an atlas A, we say that (X, A) is a Riemann surface. If the atlas is understood, we simply say that X is a Riemann surface.

Now, a function $f : M \longrightarrow N$ between two Riemann surfaces M and N is called *holomorphic* if for every chart g in the atlas of M and every chart h in the atlas of N, the map $h \circ f \circ g^-1$ is holomorphic (as a function from $\mathbb{C}$ to $\mathbb{C}$) wherever it is defined. The composition of two holomorphic maps is holomorphic. The two Riemann surfaces M and N are called *conformally equivalent* if there exists a bijective holomorphic function from M to N whose inverse is also holomorphic. Two conformally equivalent Riemann surfaces are for all practical purposes identical. Every *simply connected Riemann surface* is conformally equivalent to $\mathbb{C}$, or to the Riemann sphere $\mathbb{C} \cup \infty$, or to the open disk $\{z \in \mathbb{C} : |z| < 1\}$. This statement is known as the *uniformization Theorem*.

Every connected Riemann surface can be turned into a complete 2D real *Riemannian manifold* (see Chapter 4) with constant curvature –1, 0 or 1. This Riemann structure is unique up to scalings of the metric. The Riemann surfaces with curvature –1 are called *hyperbolic*; the open disk with the Poincarè–metric of constant curvature –1 is the canonical local model. Examples are all surfaces with genus $g > 1$. The Riemann surfaces with curvature 0 are called *parabolic*; $\mathbb{C}$ and the 2–torus are typical parabolic Riemann surfaces. Finally, the surfaces with curvature +1 are known as *elliptic*; the Riemann sphere $\mathbb{C} \cup \infty$ is the only example.

1.3 Soft Introduction to Quantum Dynamics

In this section we give a soft introduction to quantum dynamics (based on two 'classical' experiments), to be used in the following chapters. Recall that according to quantum mechanics, *light* consists of particles called *photons*, and the Figure 1.11 shows a photon source which we assume emits photons one at a time. There are two slits, A and B, and a screen behind them. The photons arrive at the screen as individual events, where they are detected separately, just as if they were ordinary particles. The curious quantum behavior arise in the following way [Pen97]. If only slit A were open and the other closed, there would be many places on the screen which the photon could reach. If we now close the slit A and open the slit B, we may again find that the photon could reach the same spot on the screen. However, if we open *both slits*, and if we have chosen the point on the screen carefully, we may now find that the photon cannot reach that spot, even though it could have done so if either slit alone were open. Somehow, the two possible things which the photon *might* do cancel each other out. This type of behavior does not take place in classical physics. Either one thing happens or another thing happens – we do not get two possible things which might happen, somehow conspiring to cancel each other out.

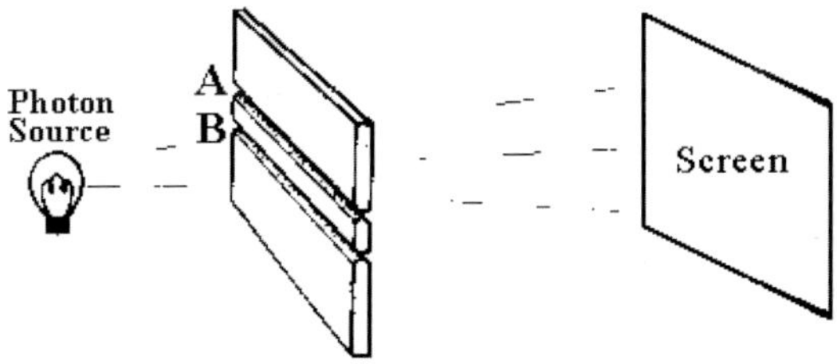

Fig. 1.11. The two–slit experiment, with individual photons of monochromatic light (modified and adapted from [Pen67, Pen97]).

The way we understand the outcome of this experiment in quantum mechanics is to say that when the photon is *en route* from the source to the screen, the state of the photon is not that of having gone through one slit or the other, but is some mysterious combination of the two, weighted by complex numbers [Pen97]. That is, we can write the state of the photon as a wave ψ–function,[9] which is the *linear superposition* of the two states, $|A>$ and $|B>$,[10] corresponding to the A–slot and B–slot alternatives,

[9] In the *Schrödinger picture*, the *unitary evolution* U of a quantum system is described by the *Schrödinger equation*, which provides the time rate of change of the quantum state or wave function $\psi = \psi(t)$.

[10] We are using here the standard Dirac 'bra–ket' notation for quantum states. Paul Dirac was one of the outstanding physicists of the 20th century. Among his

$$|\psi> = z_1|A> + z_2|B>,$$

where z_1 and z_2 are complex numbers (not both zero), while $|\cdot>$ denotes the *quantum state ket–vector*.

Now, in quantum mechanics, we are not so interested in the sizes of the complex numbers z_1 and z_2 themselves as we are in their ratio – it is only the ratio of these numbers which has direct physical meaning (as multiplying a quantum state with a nonzero complex number does not change the physical situation). Recall that the *Riemann sphere* (see Figure 1.9) is a way of representing complex numbers (plus ∞) and their ratios on a sphere on unit radius, whose equatorial plane is the complex–plane, whose center is the origin of that plane and the equator of this sphere is the unit circle in the complex–plane. We can project each point on the equatorial complex–plane onto the Riemann sphere, projecting from its south pole S, which corresponds to the *point at infinity* in the complex–plane. To represent a particular complex ratio, say $u = z/w$ (with $w \neq 0$), we take the stereographic projection from the sphere onto the plane.

The Riemann sphere plays a fundamental role in the quantum picture of two–state systems [Pen94]. If we have a spin–$\frac{1}{2}$ particle, such as an electron, a proton, or a neutron, then the various combinations of their spin states can be realised geometrically on the Riemann sphere. Spin–$\frac{1}{2}$ particles can have two spin states: (i) spin–up (with the rotation vector pointing upwards), and (ii) spin–down (with the rotation vector pointing downwards). The superposition of the two spin–states can be represented symbolically as

$$|\nearrow> = w|\uparrow> + z|\downarrow> .$$

Different combinations of these spin states give us rotation about some other axis and, if we want to know where that axis is, we take the ratio of complex numbers $u = z/w$. We place this new complex number u on the Riemann sphere and the direction of u from the center is the direction of the spin axis (see Figure 1.12).

More general quantum state vectors might have a form such as [Pen94]:

$$|\psi> = z_1|A_1> + z_2|A_2> + ... + z_n|A_n>,$$

where $z_1 ... z_n$ are complex numbers (not all zero) and the state vectors $|A_1>, ..., |A_n>$ might represent various possible locations for a particle (or perhaps some other property of a particle, such as its state of spin). Even more generally, infinite sums would be allowed for a wave ψ–function or quantum state vector.

achievements was a general formulation of quantum mechanics (having Heisenberg matrix mechanics and Shrödinger wave mechanics as special cases) and also its relativistic generalization involving the 'Dirac equation', which he discovered, for the electron. He had an unusual ability to 'smell out' the truth, judging his equations, to a large degree, by their aesthetic qualities!

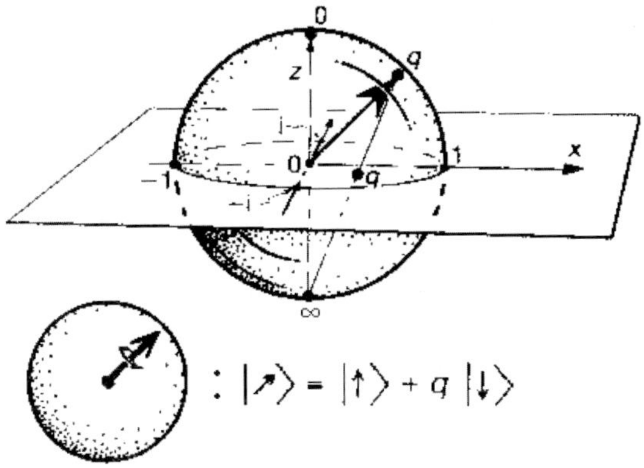

Fig. 1.12. The quantum Riemann sphere, represented as the space of physically distinct spin–states of a spin–$\frac{1}{2}$ particle (e.g., electron, proton, neutron): $|\nearrow>=|\uparrow> +q|\downarrow>$. The sphere is projected stereographically from its south pole (∞) to the complex–plane through its equator (modified and adapted from [Pen67]).

Now, the most basic feature of unitary quantum evolution U[11] is that it is *linear*. This means that, if we have two states, say $|\psi>$ and $|\phi>$, and if the Schrödinger equation would tell us that, after some time t, the states $|\psi>$ and $|\phi>$ would each individually evolve to new states $|\psi'>$ and $|\phi'>$, respectively then any linear superposition $z_1|\psi> + z_2|\phi>$, must evolve, after some time t, to the corresponding superposition $z_1|\psi'> + z_2|\phi'>$. Let us use the symbol $\rightsquigarrow$ to denote the evolution after time t, Then linearity asserts that if

$$|\psi>\rightsquigarrow|\psi'> \qquad \text{and} \qquad |\phi>\rightsquigarrow|\phi'>,$$

then the evolution

$$z_1|\psi> + z_2|\phi> \quad \rightsquigarrow \quad z_1|\psi'> + z_2|\phi'>$$

would also hold. This would consequently apply also to linear superpositions of more than two individual quantum states. For example, $z_1|\psi> + z_2|\phi> + z_3|\chi>$ would evolve, after time t, to $z_1|\psi'> + z_2|\phi'> + z_3|\chi'>$, if $|\psi>$

[11] Recall that unitary quantum evolution U is governed by the time–dependent *Schrödinger equation,*

$$i\hbar\,\partial_t|\psi(t)> \;=\; H|\psi(t)>,$$

where $\partial_t \equiv \partial/\partial t$, $\hbar$ is the *Planck's constant*, and H is the Hamiltonian (total energy) operator. Given the quantum state $|\psi(t)>$ at some initial time ($t=0$), we can integrate the Schrödinger equation to get the state at any subsequent time. In particular, if H is independent of time, then,

$$|\psi(t)> \;=\; \exp\left(-\frac{\mathrm{i}Ht}{\hbar}\right)|\psi(0)>.$$

$|\phi>$, and $|\chi>$ would each individually evolve to $|\psi'>, |\phi'>$, and $|\chi'>$, respectively. Thus, the evolution always proceeds as though each different component of a superposition were oblivious to the presence of the others.

As a second experiment, consider a situation in which light impinges on a half–silvered mirror, that is a semi-transparent mirror that reflects just half the light (composed of a *stream of photons*) falling upon it and transmits the remaining half [Pen94]. We might well have imagined that for a stream of photons impinging on our half–silvered mirror, half the photons would be reflected and half would be transmitted. Not so! Quantum theory tells us that, instead, each individual photon, as it impinges on the minor, is separately put into a superposed state of reflection and transmission. If the photon before its encounter with the minor is in state $|A>$, then afterwards it evolves according to U to become a state that can be written $|B> +i|C>$, where $|B>$ represents the state in which the photon is transmitted through the mirror and $|C>$ the state where the photon is reflected from it (see Figure 1.13). Let us write this as

$$|A> \rightsquigarrow |B> +i|C> .$$

The imaginary factor 'i' arises here because of a net phase shift by a quarter of a wavelength (see [KF86]), which occurs between the reflected and transmitted beams at such a mirror.

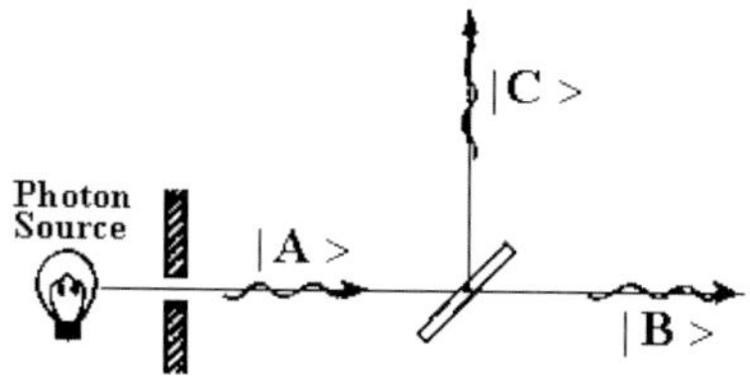

Fig. 1.13. A photon in state $|A>$ impinges on a half–silvered mirror and its state evolves according to U into a a superposition $|B> +i|C>$ (modified and adapted from [Pen94]).

Although, from the classical picture of a particle, we would have to imagine that $|B>$ and $|C>$ just represent alternative things that the photon *might* do, in quantum mechanics we have to try to believe that the photon is now actually doing *both things at once* in this strange, complex superposition. To see that it cannot just be a matter of classical probability–weighted alternatives, let us take this example a little further and try to bring the two parts of the photon state, i.e., the two photon beams, back together again [Pen94]. We can do this by first reflecting each beam with a fully silvered mirror. After reflection, the photon state $|B>$ would evolve according to U, into another state $i|D>$, whilst $|C>$ would evolve into $i|E>$,

$$|B> \rightsquigarrow i|D> \qquad \text{and} \qquad |C> \rightsquigarrow i|E> .$$

Thus the entire state $|B> +i|C>$ evolves by U into

$$|B> +i|C> \rightsquigarrow i|D> +i(i|E>) = i|D> -|E>,$$

since $i^2 = -1$. Now, suppose that these two beams come together at a fourth mirror, which is now half silvered (see Figure 1.14). The state $|D>$ evolves into a combination $|G> +i|F>$, where $|G>$ represents the transmitted state and $|F>$ the reflected one. Similarly, $|E>$ evolves into $|F> +i|G>$, since it is now the state $|F>$ that is the transmitted state and $|G>$ the reflected one,

$$|D> \rightsquigarrow |G> +i|F> \qquad \text{and} \qquad |E> \rightsquigarrow |F> +i|G> .$$

Our entire state $i|D> -|E>$ is now seen (because of the linearity of U) to evolve as:

$$\begin{aligned} i|D> -|E> &\rightsquigarrow i(|G> + + i|F>) - (|F> +i|G>) \\ &= i|G> -|F> -|F> -i|G> = -2|F> . \end{aligned}$$

As mentioned above, the multiplying factor -2 appearing here plays no physical role, thus we see that the possibility $|G>$ is *not* open to the photon; the two beams together combine to produce just a *single* possibility $|F>$. This curious outcome arises because *both* beams are present *simultaneously* in the physical state of the photon, between its encounters with the first and last mirrors. We say that the two beams *interfere* with one another.[12]

1.3.1 Complex Hilbert Space

Quantum Hilbert Space

The family of all possible states ($|\psi>, |\phi>$, etc.) of a quantum system confiture what is known as a *Hilbert space*. It is a *complex vector space*, which means that can perform the complex–number–weighted combinations that we considered before for quantum states. If $|\psi>$ and $|\phi>$ are both elements of the Hilbert space, then so also is $w|\psi> + z|\phi>$, for any pair of complex numbers w and z. Here, we even alow $w = z = 0$, to give the element $\mathbf{0}$ of the Hilbert space, which does not represent a possible physical state. We have the normal algebraic rules for a vector space:

[12] This is a property of single photons: each individual photon must be considered to feel out both routes that are open to it, but it remains one photon; it does not split into two photons in the intermediate stage, but its location undergoes the strange kind of complex–number–weighted *co–existence of alternatives* that is characteristic of quantum theory.

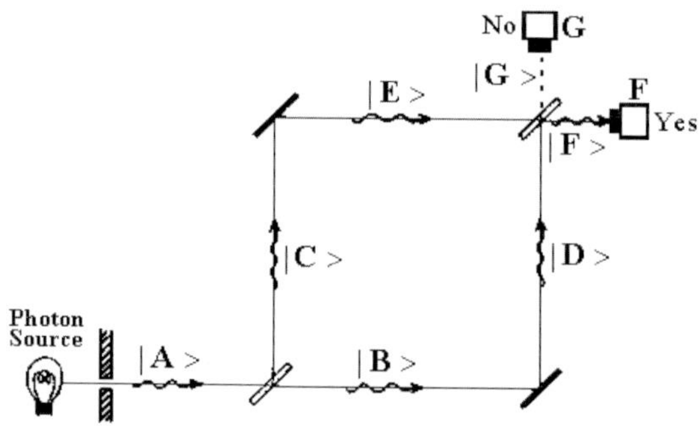

Fig. 1.14. *Mach–Zehnder interferometer*: the two parts of the photon state are brought together by two fully silvered mirrors (black), so as to encounter each other at a final half–silvered mirror (white). They interfere in such a way that the entire state emerges in state $|F>$, and the detector at G cannot receive the photon (modified and adapted from [Pen94]).

$$\begin{aligned}
&|\psi> + |\phi> = |\phi> + |\psi>, \\
&|\psi> + (|\phi> + |\chi>) = (|\psi> + |\phi>) + |\chi>, \\
&w(z|\psi>) = (wz)|\psi>, \\
&(w+z)|\psi> = w|\psi> + z|\psi>, \\
&z(|\psi> + |\phi>) = z|\psi> + z|\phi> \\
&0|\psi> = 0, \qquad z0 = 0.
\end{aligned}$$

A Hilbert space can sometimes have a finite number of dimensions, as in the case of the spin states of a particle. For spin $\frac{1}{2}$, the Hilbert space is just 2D, its elements being the complex linear combinations of the two states $|\uparrow>$ and $|\downarrow>$. For spin $\frac{1}{2}n$, the Hilbert space is $(n+1)$D. However, sometimes the Hilbert space can have an infinite number of dimensions, as e.g., the states of position or momentum of a particle. Here, each alternative position (or momentum) that the particle might have counts as providing a separate dimension for the Hilbert space. The general state describing the quantum location (or momentum) of the particle is a complex–number superposition of all these different individual positions (or momenta), which is the wave $\psi-$function for the particle.

Another property of the Hilbert space, crucial for quantum mechanics, is the *Hermitian inner (scalar) product*, which can be applied to any pair of Hilbert–space vectors to produce a single complex number. To understand how important the Hermitian inner product is for quantum mechanics, recall that the Dirac's 'bra–ket' notation is formulated on the its basis. If we have the two quantum states (i.e., Hilbert–space vectors) are $|\psi>$ and $|\phi>$, then their Hermitian scalar product is denoted $<\psi|\phi>$, and it satisfies a number of simple algebraic properties:

$$\overline{<\psi|\phi>} = <\phi|\psi>, \qquad \text{(bar denotes complex–conjugate)}$$
$$<\psi|(|\phi> + |\chi>) = <\psi|\phi> + <\psi|\chi>,$$
$$(z<\psi|)|\phi> = z<\psi|\phi>,$$
$$<\psi|\phi> \geq 0, \qquad <\psi|\phi> = 0 \quad \text{if} \quad |\psi> = 0.$$

For example, probability of finding a quantum particle at a given location is a *squared length* $|\psi|^2$ of a Hilbert–space position vector $|\psi>$, which is the scalar product $<\psi|\psi>$ of the vector $|\psi>$ with itself. A *normalized state* is given by a Hilbert–space vector whose squared length is *unity*.

The second important thing that the Hermitian scalar product gives us is the notion of *orthogonality* between Hilbert–space vectors, which occurs when the scalar product of the two vectors is *zero*. In ordinary terms, orthogonal states are things that are independent of one another. The importance of this concept for quantum physics is that the different alternative outcomes of any measurement are always orthogonal to each other. For example, states $|\uparrow>$ and $|\downarrow>$ are mutually orthogonal. Also, orthogonal are *all* different possible *positions* that a quantum particle might be located in [Pen94].

Formal Hilbert Space

A *norm* on a complex vector space $\mathcal{H}$ is a mapping from $\mathcal{H}$ into the complex numbers, $\|\cdot\| : \mathcal{H} \to \mathbb{C}$; $h \mapsto \|h\|$, such that the following set of norm–axioms hold:

(N1) $\|h\| \geq 0$ for all $h \in \mathcal{H}$ and $\|h\| = 0$ implies $h = 0$ (positive definiteness);

(N2) $\|\lambda\, h\| = |\lambda|\, \|h\|$ for all $h \in \mathcal{H}$ and $\lambda \in \mathbb{C}$ (homogeneity); and

(N3) $\|h_1 + h_2\| \leq \|h_1\| + \|h_2\|$ for all $h1, h2 \in \mathcal{H}$ (triangle inequality).

The pair $(\mathcal{H}, \|\cdot\|)$ is called a *normed space*.

A *Hermitian inner product* on a complex vector space $\mathcal{H}$ is a mapping $\langle\cdot,\cdot\rangle : \mathcal{H} \times \mathcal{H} \to \mathbb{C}$ such that the following set of inner–product–axioms hold:

(IP1) $\langle h\, h_1 + h_2 \rangle = \langle h\, h_1 + h\, h_2 \rangle$;

(IP2) $\langle \alpha\, h, h_1 \rangle = \alpha\, \langle h, h_1 \rangle$;

(IP3) $\langle h_1, h_2 \rangle = \overline{\langle h_1, h_2 \rangle}$ (so $\langle h, h \rangle$ is real);

(IP4) $\langle h, h \rangle \geq 0$ and $\langle h, h \rangle = 0$ provided $h = 0$.

These properties are to hold for all $h, h_1, h_2 \in \mathcal{H}$ and $\alpha \in \mathbb{C}$; $\bar{z}$ denotes the complex conjugate of the complex number z. (IP2) and (IP3) imply that $\langle \alpha\, h, h_1 \rangle = \bar{\alpha}\, \langle h_1, h_2 \rangle$. As is customary, for a complex number z we shall denote by $\mathrm{Re} z = \frac{z+z}{2}$, $\mathrm{Im} z = \frac{z-z}{2}$, $|z| = (z\bar{z})^{1/2}$ its real and imaginary parts and its absolute value.

The *standard inner product* on the product space $\mathbb{C}^n = \mathbb{C} \times \cdots \times \mathbb{C}$ is defined by $\langle z, w \rangle = \sum_{i=1}^n z_i w^i$, and axioms (IP1)–(IP4) are readily checked. Also $\mathbb{C}^n$ is a normed space with $\|z\|^2 = \sum_{i=1}^n |z_i|^2$.

The pair $(\mathcal{H}, \langle\cdot,\cdot\rangle)$ is called an *inner product space*.

In an inner product space $\mathcal{H}$, two vectors $h_1, h_2 \in \mathcal{H}$ are called orthogonal, and we write $h_1 \perp h_2$, provided $\langle h_1, h_2 \rangle = 0$. For a subset

In an inner product space $\mathcal{H}$, two vectors $h_1, h_2 \in \mathcal{H}$ are called orthogonal, and we write $h_1 \perp h_2$, provided $\langle h_1, h_2 \rangle = 0$. For a subset $A \subset \mathcal{H}$, the set $A^\perp$ defined by $A^\perp = \{h \in \mathcal{H} | \langle h, x \rangle = 0 \text{ for all } x \in A\}$ is called the orthogonal complement of A.

In an inner product space $\mathcal{H}$ the *Cauchy–Schwartz inequality* holds: $|\langle h_1, h_2 \rangle| \leq \langle h_1, h_2 \rangle^{1/2} \langle h_1, h_2 \rangle^{1/2}$. Here, equality holds provided h_1, h_2 are linearly dependent.

Let $(\mathcal{H}, \|\cdot\|)$ be an inner product space and set $\|h\| = \langle h, h \rangle^{1/2}$. Then $(\mathcal{H}, \|\cdot\|)$ is a normed space.

Let $(\mathcal{H}, \langle \cdot, \cdot \rangle)$ be an inner product space and $\|\cdot\|$ the corresponding norm. Then we have

1. *Polarization law*:
   $4 \langle h_1, h_2 \rangle = \|h_1 + h_2\|^2 - \|h_1 - h_2\|^2 + i \, \|h_1 + i\, h_2\|^2 - i \, \|h_1 - i\, h_2\|^2$, and
2. *Parallelogram law*:
   $2 \|h_1\|^2 + 2 \|h_2\|^2 = \|h_1 + h_2\|^2 - \|h_1 - h_2\|^2$.

Let $(\mathcal{H}, \|\cdot\|)$ be a normed space and define $d(h_1, h_2) = \|h_1 - h_2\|$. Then $(\mathcal{H}, d)$ is a metric space.

Let $(\mathcal{H}, \|\cdot\|)$ be a normed space. If the corresponding metric d is complete, we say $(\mathcal{H}, \|\cdot\|)$ is a Banach space. If $(\mathcal{H}, \|\cdot\|)$ is an inner product space whose corresponding metric is complete, we say $(\mathcal{H}, \|\cdot\|)$ is a *Hilbert space* (see, e.g., [AMR88]).

If $\mathcal{H}$ is a Hilbert space and F it closed subspace, then $\mathcal{H}$ *splits* into two mutually orthogonal subspaces, $\mathcal{H} = F \oplus F^\perp$, where $\oplus$ denotes the *orthogonal sum*. Thus every closed subspace of a Hilbert space splits.

Let $\mathcal{H}$ be a Hilbert space. A set $\{h_i\} i \in I$ is called orthonormal if $\langle h_i, h_j \rangle = \delta_{ij}$, the Kronecker delta. An orthonormal set $\{h_i\} i \in I$ is a *Hilbert basis* if $closure(\text{span}\{h_i\} i \in I) = \mathcal{H}$. Any Hilbert space has a Hilbert basis.

In the finite dimensional case equivalence and completeness are automatic. Let $\mathcal{H}$ be a finite–dimensional vector space. Then (i) there is a norm on $\mathcal{H}$; (ii) all norms on $\mathcal{H}$ are equivalent; (iii) all norms on $\mathcal{H}$ are complete.

Consider the space $L^2([a, b], \mathbb{C})$ of *square–Lebesgue–integrable* complex–valued functions defined on an interval $[a, b] \subset \mathbb{C}$, that is, functions f that satisfy $\int_a^b |f(x)|^2 \, dx < \infty$. It is a Banach space with the norm defined by $\|f\| = \left(\int_a^b |f(x)|^2 \, dx \right)^{1/2}$, and a Hilbert space with the inner product defined by $\langle f, g \rangle = \int_a^b f(x) \, \overline{g(x)} \, dx$.

Recall from elementary linear algebra that the dual space of a finite dimensional vector space of dimension n also has dimension n and so the space and its dual are isomorphic. It is also true for Hilbert space.

Riesz Representation Theorem. Let $\mathcal{H}$ be be a real (resp., complex) Hilbert space. The map $h \mapsto \langle \cdot, h \rangle$ is a linear (resp., antilinear) norm–preserving isomorphism of $\mathcal{H}$ with $\mathcal{H}*$; for short, $\mathcal{H} \cong \mathcal{H}*$. (A map $A : \mathcal{H} \to F$ between complex vector spaces is called *antilinear* if we have the identities $A(h+h\prime) = Ae + Ae\prime$, and $A(\alpha h) = \bar{\alpha} Ae$.)

Let $\mathcal{H}$ and F be Banach spaces. We say $\mathcal{H}$ and F are in *strong duality* if there is a non–degenerate continuous bilinear functional $\langle \cdot, \cdot \rangle : \mathcal{H} \times F \to \mathbb{R}$, also called a *pairing* of $\mathcal{H}$ with F. Now, let $\mathcal{H} = F$ and $\langle \cdot, \cdot \rangle : \mathcal{H} \times \mathcal{H} \to \mathbb{R}$ be an inner product on $\mathcal{H}$. If $\mathcal{H}$ is a Hilbert space, then $\langle \cdot, \cdot \rangle$ is a *strongly non–degenerate pairing* by the Riesz representation Theorem.

2

Nonlinear Dynamics in the Complex Plane

In this Chapter we start the low–dimensional nonlinear complex dynamics with both continuous–time and discrete–time systems in the complex–plane.

2.1 Complex Continuous Dynamics

Recall that a paradigm for continuous 2D dynamics is the so–called *complex velocity streamline*, formally given by $v(t) = v_1(t) + iv_2(t)$, representing a 2D–fluid flow in the complex–plane $\mathbb{C}^2$ is given in Figure 2.1. If the streamline is a closed curve in the complex–plane, then we have a *complex rotor*.

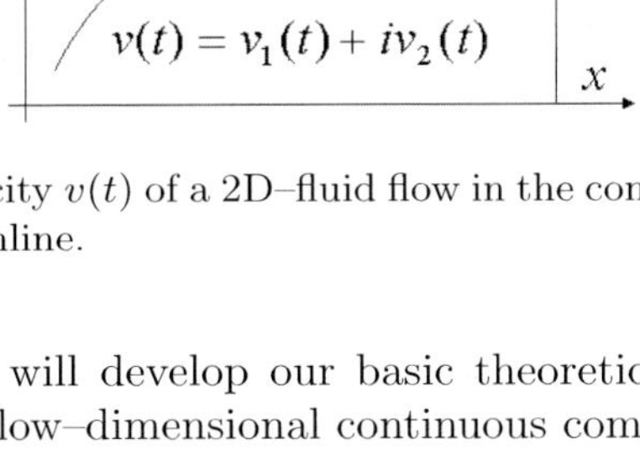

Fig. 2.1. Sketch of velocity $v(t)$ of a 2D–fluid flow in the complex–plane $\mathbb{C}^2$, showing its representative streamline.

In this section we will develop our basic theoretical and computational tools for dealing with low–dimensional continuous complex dynamics.

2.1.1 Complex Nonlinear ODEs

The study of ordinary differential equations (ODEs) in the complex variable and of complex function theory is a classical subject that has interested all the greatest mathematicians of the nineteenth century, such as Gauss, Cauchy, Abel, Jacobi, Einstein, Riemann, Weierstrass, Klein, Kowalesky and Poincaré. This subject has recently become very important in several areas of integrable systems and nonlinear physics. Its beauty is due to the fact that even though it requires several tools from different fields of mathematics such as Riemannian geometry, group theory, classical analysis, asymptotic analysis, Hamiltonian theory, it is conceptually rather simple.

We start–off by studying strongly–nonlinear ODEs, mainly following [Cve05]. Recall that the vibrations of a single mass system with 2 DOF are mostly described using a second–order ODE with a complex dependent variable. The differential equation is usually linear as is shown in the papers of [Dim59] and [Van88]. The solution of the differential equation clarifies the linear phenomena which occur in the system. If in the system some small nonlinearities exist they are introduced in the differential equation of motion as small nonlinear terms. Various methods for solving differential equations with complex dependent variable and small nonlinearity are introduced in [Cve92, Cve93, Mah98]. The solutions obtained describe the influence of small nonlinearities on the behavior of the system. As is known, in the real system both weak and also strong nonlinearities act. The motion of the system is described by a second-order strongly nonlinear complex differential equation. Some special cases of such differential equations are considered. The one–frequency solution of a special type of *Duffing equation* is obtained in [Cve92]. Besides the *Duffing type of nonlinearity* [Cve98], the *Lienard and Rayleigh systems* with complex functions are considered in [MMZ99]. The interaction between strong and weak nonlinearity in a system with complex dependent variable is also discussed in [Cve01]. An approximate analytic solution procedure is developed for analyzing such a system. The main disadvantage of the suggested procedures is that they do not give the general solution but are convenient only for some special cases of nonlinearities and corresponding special initial conditions.

In [Cve05] the initial conditions have been arbitrary but there has been a constraint to the differential equation: the coupling of the ODE has been due to the small nonlinearity. Separating the strong nonlinear term in the ODE with complex dependent variable into real and imaginary parts leads to functions that depend only on one real function and its time derivative. The real part depends only on a function $x(t)$ and its timede rivative $\dot{x}(t)$ and the imaginary part on a function $y(t)$ and its time derivative $\dot{y}(t)$: The mathematical model of the system is

$$\ddot{z} + f(z, \dot{z}) = \epsilon\varphi(z, \dot{z}, z^*, \dot{z}^*), \tag{2.1}$$

where $z = x + iy$ is a complex function, $z^* = x - iy$ is complex conjugate, i is the imaginary unit, x and y are real functions of time t, overdot denotes time derivative, $\epsilon << 1$ is a small parameter, $\varphi = \varphi_1 + i\varphi_2$ is the small nonlinear function, and

$$f(z, \dot{z}) = f_1 + if_2 \qquad \text{with} \qquad f_1 \equiv f_1(x, \dot{x}), \quad f_2 \equiv f_2(y, \dot{y}). \tag{2.2}$$

The arbitrary initial conditions are

$$z(0) = z_0, \qquad \dot{z}(0) = \dot{z}_0. \tag{2.3}$$

An approximate analytic solution procedure based on perturbation of the generating solution has been developed in [Cve05]. First, the closed form analytic solution of two independent 1 DOF systems, which are two decoupled strongly nonlinear second–order ODEs (2.1) is developed. The trial solution in the form of generating solution is formed and the differential equation of motion (2.1) is transformed into the system of four first–order ODEs. The solution of this system of ODEs represents the solution of (2.1). The suggested procedure is applied to system with strong cubic nonlinearity of Duffing type. The two–frequency solution is the *Jacobi elliptic function*. For the general case, an averaging procedure for solving such ODEs with small nonlinearity is developed. The method is used for calculation of the vibration properties of a rotor system with pure cubic nonlinearity and small nonlinearity of van der Pol type.

For the case when the small nonlinearity is neglected and $\epsilon = 0$, the ODE (2.1) transforms into

$$\ddot{z} + f(z, \dot{z}) = 0, \tag{2.4}$$

where $f(z, \dot{z}) = f_1(x, \dot{x}) + i f_2(y, \dot{y})$. It is a strongly nonlinear ODE. By separating the real and imaginary parts of the differential equation (2.4) and substituting $z = x + iy$ into (2.3) the following two independent ODEs are obtained:

$$\ddot{x} + f_1(x, \dot{x}) = 0, \qquad x(0) = x_0, \qquad \dot{x}(0) = \dot{x}_0, \tag{2.5}$$

$$\ddot{y} + f_2(y, \dot{y}) = 0, \qquad y(0) = y_0, \qquad \dot{y}(0) = \dot{y}_0. \tag{2.6}$$

The solutions of the equations are independent, and have the form

$$x = x(t, A, \alpha), \qquad y(t, B, \beta), \tag{2.7}$$

where $(A, \alpha), (B, \beta)$ are parameters which depend on the initial conditions (2.5–2.6). Specifically, A and B are the initial amplitudes, while α and β are the initial phase angles. In spite of the fact that (2.4) is a strongly nonlinear ODE, the generating solution is a superposition of solutions (2.7), i.e.,

$$z = x(t, A, \alpha) + iy(t, B, \beta). \tag{2.8}$$

It represents the closed form solution of the nonlinear complex ODE (2.4) which satisfies constraint (2.2).

Based on the generating solution the trial solution is formed. The following constraints are introduced:

(i) The trial solution has the form of the generating solution and it is

$$z = x(t, A(t), \alpha(t)) + iy(t, B(t), \beta(t)). \tag{2.9}$$

where $A(t), \alpha(t)$ and $B(t), \beta(t)$ are time–variable functions. Solution (2.9) has to satisfy the ODE (2.1) with the limitation (2.2).

(ii) The first time derivative of (2.9) has the form of the first time derivative of the generating solution (2.8) where A, α, B, β are supposed to have constant values

$$\dot{z} = \partial_t x(t, A(t), \alpha(t)) + i\partial_t y(t, B(t), \beta(t)). \tag{2.10}$$

The additional terms which exist due to the fact that $A(t), \alpha(t)$ and $B(t), \beta(t)$ are time–dependent give us a new relation

$$\left(\dot{A}\partial_A x + \dot{\alpha}\partial_\alpha x\right) + i\left(\dot{B}\partial_B y + \dot{\beta}\partial_\beta y\right) = 0. \tag{2.11}$$

(iii) The time derivative of (2.10) is

$$\ddot{z} = \partial_t \left(\partial_t x(t, A(t), \alpha(t)) + i\partial_t y(t, B(t), \beta(t))\right). \tag{2.12}$$

Substituting solution (2.9) the corresponding time derivatives (2.10) and (2.12) into (2.1) lead to

$$\left(\dot{A}\partial_A + \dot{\alpha}\partial_\alpha\right)\partial_t x + i\left(\dot{B}\partial_B + \dot{\beta}\partial_\beta\right)\partial_t y = \epsilon(\varphi_1 + i\varphi_2). \tag{2.13}$$

On separating the real and imaginary parts in (2.11) and (2.13), after some transformation, we get the following system of four first–order ODEs:

$$\begin{aligned}
\dot{A} &= \epsilon\,\mathrm{Re}(\varphi_1 + i\varphi_2)/\left(\partial_A - (\partial_A x/\partial_\alpha x)\partial_\alpha \partial_t x\right)\\
\dot{\alpha} &= -\epsilon\,\mathrm{Re}(\varphi_1 + i\varphi_2)/\left((\partial_\alpha x/\partial_A x)\,\partial_A - \partial_\alpha\right)\partial_t x\\
\dot{B} &= \epsilon\,\mathrm{Im}(\varphi_1 + i\varphi_2)/\left(\partial_B - (\partial_B y/\partial_\beta y)\partial_\beta \partial_t y\right)\\
\dot{\beta} &= -\epsilon\,\mathrm{Im}(\varphi_1 + i\varphi_2)/\left((\partial_\beta y/\partial_B y)\,\partial_B - \partial_\beta\right)\partial_t y.
\end{aligned} \tag{2.14}$$

By solving (2.14) the functions $A(t), \alpha(t)$ and $B(t), \beta(t)$ are determined, i.e., the exact solution (2.9). Unfortunately, obtaining the closed form solution of system (2.14) is usually impossible. Some approximation has to be introduced. As the motion is periodic it means that the solutions are also periodic functions. It is convenient to introduce the averaging of (2.14. Solutions of the averaged first–order ODEs represent the approximate solution of the original ODE (2.1).

For details how the above algorithm has been applied to the strongly nonlinear Duffing equation,

$$\ddot{z} + b_1 z + b_3\left(\left(\frac{z+z^*}{2}\right)^3 - \left(\frac{z-z^*}{2}\right)^3\right) = \epsilon(\varphi_1 + i\varphi_2),$$

and with a small nonlinearity of van der Pol type,

$$\ddot{z} + b_3\left(\left(\frac{z+z^*}{2}\right)^3 - \left(\frac{z-z^*}{2}\right)^3\right) = \epsilon\dot{z}[1 - zz^*], \tag{2.15}$$

describing the free vibrations of a symmetric nonlinear shaft–disc system (rotor) depicted in Figure 2.2, see [Cve05].

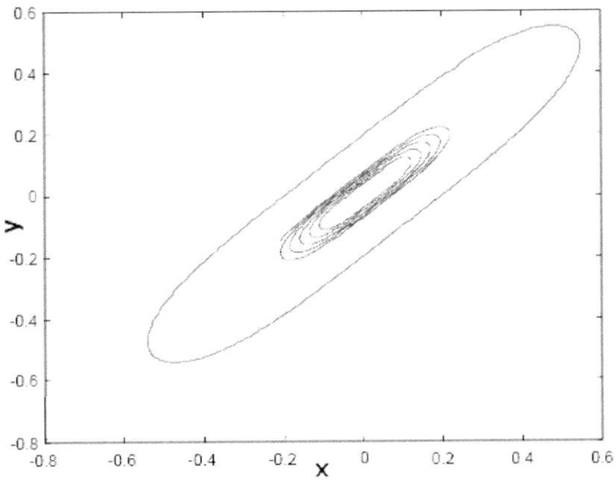

Fig. 2.2. The complex–plane diagram for the complex ODE (2.15): the rotor with linear properties (outer curve) and the strongly nonlinear rotor (inner curve) (modified and adapted from [Cve05]).

2.1.2 Numerical Integration of Complex ODEs

Complex Runge–Kutta–Nystrom Integrator

For numerical solution of systems of second–order complex–valued ODEs, we have designed the complex-valued Runge–Kutta–Nystrom integrator.

C–Code for Complex Runge–Kutta–Nystrom Integrator

In this subsection we give a C–code for complex Runge–Kutta–Nystrom integrator for numerical solution of systems of second–order complex–valued ODEs. The code has been used for simulating the collective motion of robot teams.

```
/* Complex Runge-Kutta-Nystrom integrator for systems of
second-order complex-valued ODEs. Each complex ODE defines the
motion of a single robot in the complex--plane. */
// Author: Dr. Vladimir Ivancevic, May 2007

#include <math.h> #include <stdio.h> #include <stdlib.h> #include
<complex.h>

static void ODE(int eq, double t, double _Complex z[], double
_Complex dz[], double _Complex ddz[]){ /* 2DOF Equation of motion
in the Complex-plane for one robot */
 ddz[0] = 0.2*ccos(dz[0]) - 0.5*csin(z[0])
        + 0.3*csin(2.*t) - 0.2*I*ccos(5.*t);
/* ddz[1] = ... for the second robot, etc */ }
```

```
static void RKN(int eq, int n, double h, double t, double _Complex
z[], double _Complex dz[], double _Complex ddz[]) {
 double _Complex k1,k2,k3,k4,K,L;//Runge-Kutta-Nystrom integrator
 double _Complex zmod[3];        // for systems of complex ODEs
 double _Complex dzmod[3];
 double _Complex ddzmod[3];

 zmod[eq] = z[eq];
 dzmod[eq] = dz[eq];
 ddzmod[eq] = ddz[eq];
 ODE(eq,t,zmod,dzmod,ddzmod);
 k1 = 0.5*h*ddzmod[eq];
 K = 0.5*h*(dz[eq] + 0.5*k1);\qquad \qquad \qquad

 zmod[eq] = z[eq]+ K;
 dzmod[eq] = dz[eq] + k1;
 ODE(eq,t+0.5*h,zmod,dzmod,ddzmod);
 k2 = 0.5*h*ddzmod[eq];\qquad

 zmod[eq] = z[eq]+ K;
 dzmod[eq] = dz[eq] + k2;
 ODE(eq,t+0.5*h,zmod,dzmod,ddzmod);
 k3 = 0.5*h*ddzmod[eq];
 L = h*(dz[eq] + k3);

 zmod[eq] = z[eq]+ L;
 dzmod[eq] = dz[eq] + 2*k3;
 ODE(eq,t+h,zmod,dzmod,ddzmod);
 k4 = 0.5*h*ddzmod[eq];

 dz[eq] = dz[eq]+(k1+2.0*k2+2.0*k3+k4)/3.0;
 z[eq] = z[eq]+h*(dz[eq]+(k1+k2+k3)/3.0);
}

int main() {
 /* declare variables */ // Do NOT use I as a loop variable!
 int n = 1; // Number of complex-valued ODEs
 double t, Tfin, h; // double times and step-size
 double _Complex z0[3]; /* Displacement */
 double _Complex dz0[3]; /* Velocity */
 double _Complex ddz0[3]; /* Acceleration */

 /* initialize variables */
 t = 0.0;
```

```
  Tfin = 10.0; /* initial and final times and time-step */
  h = 0.01;
  // Imaginary unit = I = (0.0F + 1.0iF);
  z0[0] = 0.1 + I*0.1; /* Initial 1.Displacement */
  dz0[0] = 0.1 + I*0.1; /* Initial 1.Velocity */

  /* output files */
  FILE *fp1;
  if ((fp1 = fopen("1.Compl.Displ.txt", "w")) == NULL)
  { printf("Error opening file\n"); exit(1); }

  FILE *fp2;
  if ((fp2 = fopen("1.Compl.Veloc.txt", "w")) == NULL)
  { printf("Error opening file\n"); exit(1); }

  /* time loop */
  while (t < Tfin)
  {
  RKN(0,1,h,t,z0,dz0,ddz0); // Integrate 1.ODE
  /* RKN(1,1,h,t,z0,dz0,ddz0); - Integrate 2.ODE, etc */
  // print t, real(z), imag(z)
  fprintf(fp1,"%lf\t%lf\t%lf\n",t,creal(z0[0]),cimag(z0[0]));
  // print t, real(dz), imag(dz)
  fprintf(fp2,"%lf\t%lf\t%lf\n",t,creal(dz0[0]),cimag(dz0[0]));
  t += h;
  }

  // close Output files
  fclose(fp1); fclose(fp2);
  return(0);
}
```

Sample outputs (2D robot trajectories) from the above code are given in Figures 2.3–2.4, showing the difference between a linear and a deformed (nonlinear) complex–valued ODEs.

Mathematica Code for Complex ODEs

A simple Mathematica code for numerical integration of the same complex ODE reads:

```
In[1]:=
Eq = Derivative[2][z][t] == 0.2*Cos[Derivative[1][z][t]] – 0.5*Sin[z[t]]
+ 0.3*Sin[2*t] – 0.2*I*Cos[5*t];
```

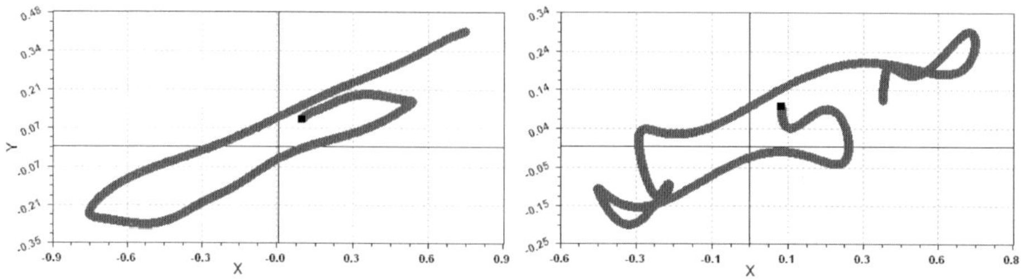

Fig. 2.3. Sample motion (displacement–left and velocity–right) of a single 2D robot in the complex–plane $\mathbb{C}^2$ defined by the complex–valued linear ODE: $\ddot{z} = 0.2\dot{z} - 0.5z + 0.3\sin(2t) - 0.2\mathrm{i}\cos(5t)$, with small initial conditions, $z(0) = \dot{z}(0) = 0.1 + 0.1\mathrm{i}$.

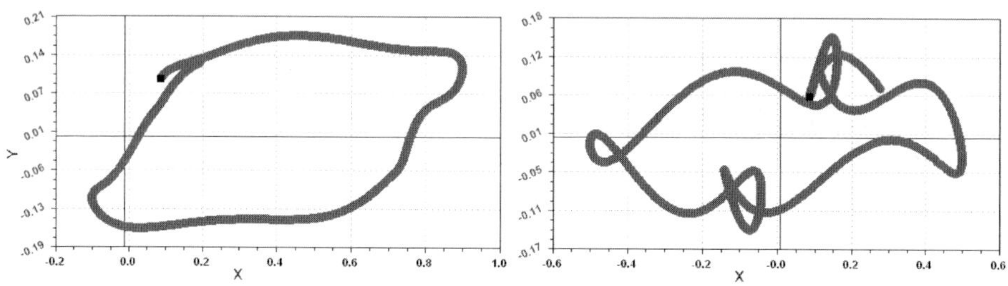

Fig. 2.4. Sample motion (displacement–left and velocity–right) of a single robot in $\mathbb{C}^2$ defined by the nonlinear (deformed) ODE: $\ddot{z} = 0.2\cos(\dot{z}) - 0.5\sin(z) + 0.3\sin(2t) - 0.2\mathrm{i}\cos(5t)$, with the same initial conditions as above.

```
In[2]:=
Init = {z[0] == Derivative[1][z][0] == 0.1 + 0.1*I}; Tfin = 10;

In[3]:=
sol = NDSolve[{Eq, Init}, z, {t, 0, Tfin}]

Out[3]=
{{z\[Rule]InterpolatingFunction[{{0.,10.}},<>]}}

In[4]:=
ParametricPlot[Evaluate[{Re[z[t]], Im[z[t]]} /. sol], {t, 0, Tfin}, Frame ->
True,
    GridLines -> Automatic, PlotStyle -> Thickness[0.01]];

In[5]:=
ParametricPlot[Evaluate[{Re[Derivative[1][z][t]], Im[Derivative[1][z][t]]} /. sol],
{t, 0, Tfin}, Frame ->True, GridLines -> Automatic, PlotStyle -> Thick-
ness[0.01]];
```

Sample outputs (2D robot trajectories) are given in Figures 2.5–2.6 (compare with the previous ones).

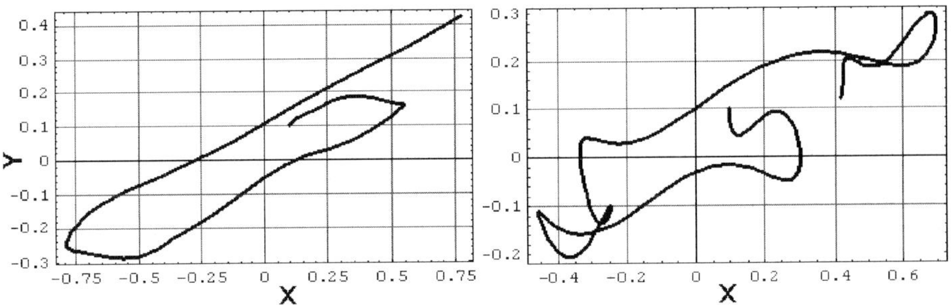

Fig. 2.5. Mathematica output in the complex–plane (displacement–left and velocity–right) for the complex-valued linear ODE: $\ddot{z} = 0.2\dot{z} - 0.5z + 0.3\sin(2t) - 0.2\mathrm{i}\cos(5t)$, with small initial conditions, $z(0) = \dot{z}(0) = 0.1 + 0.1\mathrm{i}$.

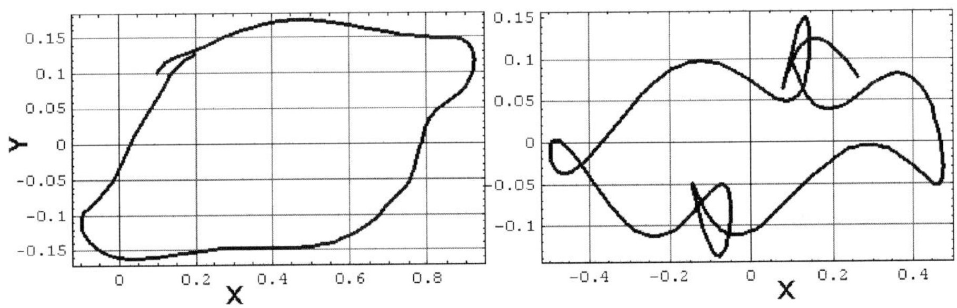

Fig. 2.6. Mathematica output in the complex–plane (displacement–left and velocity–right) for the nonlinear (deformed) ODE: $\ddot{z} = 0.2\cos(\dot{z}) - 0.5\sin(z) + 0.3\sin(2t) - 0.2\mathrm{i}\cos(5t)$, with the same initial conditions as above.

A Mathematica code for numerical integration of the system of 12 complex second–order ODEs, defining the motion of 12 robots in the complex–plane, reads:

```
In[28]:= n = 12;

In[29]:=
Eqs =Table[Derivative[2][Subscript[z, k]][t]== 0.2*Cos[Derivative[1][Subscript[z, k]][t]] – 0.5*Sin[Subscript[z, k][t]] + 0.3*Sin[2*t] – 0.2*I*Cos[5*t], k, n];
```

In[30]:=
Init = Table[Subscript[z, k][0] == Derivative[1][Subscript[z, k]][0] == k*(0.1 + 0.1*I), k, n];

In[31]:= Tfin = 7;

In[32]:=
sol = NDSolve[Eqs, Init, Table[Subscript[z, k], k, n], t, 0, Tfin];

(* Plot[Evaluate[Table[Re[Subscript[z, k][t]], k, n] /. sol], t, 0, Tfin, Frame -> True, GridLines -> Automatic, PlotStyle -> Thickness[0.005]];

Plot[Evaluate[Table[Im[Subscript[z, k][t]], k, n] /. sol], t, 0, Tfin, Frame -> True, GridLines -> Automatic, PlotStyle -> Thickness[0.005]]; *)

In[33]:=
Do[Subscript[g, k] = ParametricPlot[Evaluate[Re[Subscript[z, k][t]], Im[Subscript[z, k][t]] /. sol], t, 0, Tfin, PlotRange -> All, Frame -> True, GridLines -> Automatic, PlotStyle -> Thickness[0.007], DisplayFunction -> Identity], k, n];

In[34]:=
Show[Table[Subscript[g, k], k, n], DisplayFunction -> $DisplayFunction, Frame -> True];

In[35]:=
Do[Subscript[h, k] = ParametricPlot[Evaluate[Re[Derivative[1][Subscript[z, k]][t]], Im[Derivative[1][Subscript[z, k]][t]] /. sol], t, 0, Tfin, PlotRange -> All, Frame -> True, GridLines -> Automatic, PlotStyle -> Thickness[0.007], DisplayFunction -> Identity], k, n];

In[36]:=
Show[Table[Subscript[h, k], k, n], DisplayFunction -> $DisplayFunction, Frame -> True];

Sample output (2D robot trajectories) for 12 robots moving in the complex–plane are given in Figure 2.7.

2.1.3 Complex Hamiltonian Dynamics

Recall (see, e.g., [AM78, MR99, Wig90]) that classical *Hamiltonian equations*

$$\dot{q} = \partial H/\partial p, \quad \dot{p} = -\partial H/\partial q,$$

may be written in complex notation, by setting $z = q + ip, \quad z \in \mathbb{C}, i = \sqrt{-1}$, as

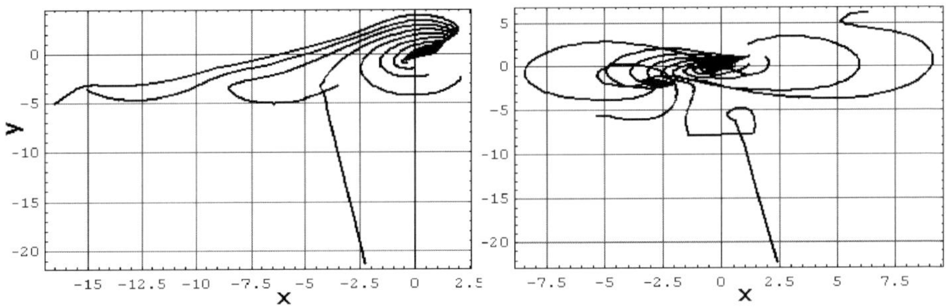

Fig. 2.7. Mathematica output in the complex–plane (displacement–left and velocity–right) for the system of 12 robots motion defined by a system of 12 nonlinear ODEs: $\ddot{z}_k = 0.2\cos(\dot{z}_k) - 0.5\sin(z_k) + 0.3\sin(2t) - 0.2\mathrm{i}\cos(5t)$, $(k = 1, ..., 12)$, with small initial conditions, $z_k(0) = \dot{z}_k(0) = k(0.1 + 0.1\mathrm{i})$.

$$\dot{z} = -2i\frac{\partial H}{\partial \bar{z}}. \tag{2.16}$$

Let U be an open set in the *complex phase–space manifold* M_C (i.e., manifold M modelled on $\mathbb{C}$, see next Chapter). A C^0 function $\gamma : [a,b] \to U \subset M_C$, $t \mapsto \gamma(t)$ represents a *solution curve* $\gamma(t) = q(t) + ip(t)$ of a complex Hamiltonian system (2.16). For instance, the curve $\gamma(\theta) = \cos\theta + i\sin\theta$, $0 \le \theta \le 2\pi$ is the unit circle. $\gamma(t)$ is a *parameterized curve.* We call $\gamma(a)$ the *beginning point*, and $\gamma(b)$ the *end point* of the curve. By a *point on the curve* we mean a point w such that $w = \gamma(t)$ for some $t \in [a,b]$.

The derivative $\dot{\gamma}(t)$ is defined in the usual way, namely

$$\dot{\gamma}(t) = \dot{q}(t) + i\dot{p}(t),$$

so that the usual rules for the derivative of a sum, product, quotient, and chain rule are valid. The *speed* is defined as usual to be $|\dot{\gamma}(t)|$. Also, if $f : U \to M_C$ represents a *holomorphic*, or *analytic* function, then the composite $f \circ \gamma$ is differentiable (as a function of the real variable t) and $(f \circ \gamma)'(t) = f'(\gamma(t))\,\dot{\gamma}(t)$.

A *path* represents a sequence of C^1–curves,

$$\gamma = \{\gamma_1, \gamma_2, \dots, \gamma_n\},$$

such that the end point of γ_j, $(j = 1, \dots, n)$ is equal to the beginning point of γ_{j+1}. If γ_j is defined on the interval $[a_j, b_j]$, this means that

$$\gamma_j(b_j) = \gamma_{j+1}(a_{j+1}).$$

We call $\gamma_1(a_1)$ the *beginning point* of γ_j, and $\gamma_n(b_n)$ the *end point* of γ_j. The path is said to *lie in an open set* $U \subset M_C$ if each curve γ_j lies in U, i.e., for each t, the point $\gamma_j(t)$ lies in U.

An open set U is *connected* if given two points α and β in U, there exists a path $\gamma = \gamma_1, \gamma_2, \dots, \gamma_n$ in U such that α is the beginning point of γ_1 and

β is the end point of γ_n; in other words, if there is a path γ in U which joins α to β. If U is a connected open set and f a holomorphic function on U such that $f' = 0$, then f is a constant. If g is a function on U such that $f' = g$, then f is called a *primitive* of g on U. Primitives can be either find out by integration or written down directly.

Let f be a C^0-function on an open set U, and suppose that γ is a curve in U, meaning that all values $\gamma(t)$ lie in U for $a \leq t \leq b$. The *integral of f along γ* is defined as

$$\int_\gamma f = \int_\gamma f(z) = \int_a^b f(\gamma(t))\,\dot{\gamma}(t)\,dt.$$

For example, let $f(z) = 1/z$, and $\gamma(\theta) = \mathrm{e}^{i\theta}$. Then $\dot{\gamma}(\theta) = i\mathrm{e}^{i\theta}$. We want to find the value of the integral of f over the circle, $\int_\gamma dz/z$, so $0 \leq \theta \leq 2\pi$. By definition, this integral is equal to $\int_0^{2\pi} i\mathrm{e}^{i\theta}/\mathrm{e}^{i\theta}\,d\theta = i\int_0^{2\pi} d\theta = 2\pi i$.

The *length* $L(\gamma)$ is defined to be the integral of the speed, $L(\gamma) = \int_a^b |\dot{\gamma}(t)|\,dt$.

If $\gamma = \gamma_1, \gamma_2, \ldots, \gamma_n$ is a path, then the integral of a C^0-function f on an open set U is defined as $\int_\gamma f = \sum_{i=1}^n \int_{\gamma_i} f$, i.e., the sum of the integrals of f over each curve γ_i ($i = 1, \ldots, n$ of the path γ. The *length of a path* is defined as $L(\gamma) = \sum_{i=1}^n L(\gamma_i)$.

Let f be continuous on an open set $U \subset M_C$, and suppose that f has a primitive g, that is, g is holomorphic and $g' = f$. Let α, β be two points in U, and let γ be a path in U joining α to β. Then $\int_\gamma f = g(\beta) - g(\alpha)$; this integral is independent of the path and depends only on the beginning and end point of the path.

A *closed path* is a path whose beginning point is equal to its end point. If f is a C^0-function on an open set $U \subset M_C$ admitting a holomorphic primitive g, and γ is any closed path in U, then $\int_\gamma f = 0$.

Let γ, η be two paths defined over the same interval $[a, b]$ in an open set $U \subset M_C$. Recall (see Introduction) that γ is *homotopic* to η if there exists a C^0-function $\psi : [a, b] \times [c, d] \to U$ defined on a rectangle $[a, b] \times [c, d] \subset U$, such that $\psi(t, c) = \gamma(t)$ and $\psi(t, d) = \eta(t)$ for all $t \in [a, b]$. For each number $s \in [c, d]$ we may view the function $|psi_s(t) = \psi(t, s)$ as a continuous curve defined on $[a, b]$, and we may view the family of continuous curves ψ_s as a *deformation* of the path γ to the path η. It is said that the homotopy ψ *leaves the end points fixed* if we have $\psi(a, s) = \gamma(a)$ and $\psi(b, s) = \gamma(b)$ for all values of $s \in [c, d]$. Similarly, when we speak of a homotopy of closed paths, we assume that each path ψ_s is a closed path.

Let γ, η be paths in an open set $U \subset M_C$ having the same beginning and end points. Assume that they are homotopic in U. Let f be holomorphic on U. Then $\int_\gamma f = \int_\eta f$. The same holds for closed homotopic paths in U. In particular, if γ is homotopic to a point in U, then $\int_\gamma f = 0$. Also, it is said that an open set $U \subset M_C$ is *simply connected* if it is connected and if every closed path in U is homotopic to a point.

In the previous example we found that

$$\frac{1}{2\pi I}\int_{\gamma}\frac{1}{z}\,dz = 1,$$

if γ is a circle around the origin, oriented counterclockwise. Now we define for any closed path γ its *winding number* with respect to a point α to be

$$W(\gamma,\alpha) = \frac{1}{2\pi i}\int_{\gamma}\frac{1}{z-\alpha}\,dz,$$

provided the path does not pass through α. If γ is a closed path, then $W(\gamma,\alpha)$ is an integer.

A closed path $\gamma \in U \subset M_C$ is *homologous to 0 in U* if

$$\int_{\gamma}\frac{1}{z-\alpha}\,dz = 0,$$

for every point α not in U, or in other words, $W(\gamma,\alpha) = 0$ for every such point.

Similarly, let γ,η be closed paths in an open set $U \subset M_C$. We say that they are *homologous in* U, and write $\gamma \sim \eta$, if $W(\gamma,\alpha) = W(\eta,\alpha)$ for every point α in the complement of U. We say that γ is *homologous to* 0 *in* U, and write $\gamma \sim 0$, if $W(\gamma,\alpha) = 0$ for every point α in the complement of U.

If γ and η are closed paths in U and are homotopic, then they are homologous. If γ and η are closed paths in U and are close together, then they are homologous.

Let $\gamma_1,\dots,\gamma_n$ be curves in an open set $U \subset M_C$, and let $m_1,\dots,m_n$ be integers. A formal sum $\gamma = m_1\gamma_1 + \dots + m_n\gamma_n = \sum_{i=1}^{n} m_i\gamma_i$ is called a *chain in* U. The chain is called *closed* if it is a finite sum of closed paths. If γ is the chain as above, then $\int_{\gamma} f = \sum_i m_i \int_{\gamma_i} f$. If γ and η are closed chains in U, then $W(\gamma+\eta,\alpha) = W(\gamma,\alpha) + W(\eta,\alpha)$. We say that γ and η are *homologous in* U, and write $\gamma \sim \eta$, if $W(\gamma,\alpha) = W(\eta,\alpha)$ for every point α in the complement of U. We say that γ is *homologous to* 0 *in* U, and write $\gamma \sim 0$, if $W(\gamma,\alpha) = 0$ for every point α in the complement of U.

Cauchy's Theorem states that if γ is a closed chain in an open set $U \subset M_C$, and γ is homologous to 0 in U, then $\int_{\gamma} f = 0$. If γ and η are closed chains in U, and $\gamma \sim \eta$ in U, then $\int_{\gamma} f = \int_{\eta} f$.

It follows from Cauchy's Theorem that if γ and η are homologous, then $\int_{\gamma} f = \int_{\eta} f$ for all holomorphic functions f on U [AM78, Wig90].

2.1.4 Dissipative Dynamics with Complex Hamiltonians

In this section, following [Raj07], we present dissipative oscillator dynamics with complex Hamiltonians. In many physical situations, loss of energy of the system under study to the outside environment cannot be ignored. Often, the long time behavior of the system is determined by this loss of energy, leading to interesting phenomena such as attractors. There is an extensive literature on dissipative systems at both the classical and quantum levels (see, e.g., the textbooks [Wei99, Sch01, SZ97]). Often the theory is based on an evolution equation of the density matrix of a 'small system' coupled to a *reservoir* with a large number of degrees of freedom, after the reservoir has been averaged out. In such approaches the system is described by a mixed state rather than a pure state: in quantum mechanics by a density instead of a wave00function and in classical mechanics by a density function rather than a point in the phase space.

There are other approaches that do deal with the evolution equations of a pure state. The canonical formulation of classical mechanics does not apply in a direct way to dissipative systems because the Hamiltonian usually has the meaning of energy and would be conserved. By redefining the Poisson brackets [Oku81], or by using time–dependent Hamiltonians [Sar98], it is possible to bring such systems within a canonical framework. Also, there are generalizations of the Poisson bracket that may not be anti–symmetric and/or may not satisfy the Jacobi identity [RMR04] which give dissipative equations.

We will follow another route, which turns out in many cases to be simpler than the above. It is suggested by the simplest example, that of the damped simple harmonic oscillator. As is well known, the effect of damping is to replace the natural frequency of oscillation by a complex number, the imaginary part of which determines the rate of exponential decay of energy. Any initial state will decay to the ground state (of zero energy) as time tends to infinity. The corresponding coordinates in phase space (normal modes) are complex as well. This suggests that the equations are of Hamiltonian form, but with a complex Hamiltonian.

It is not difficult to verify that this is true directly. The real part of the Hamiltonian is a harmonic oscillator, although with a shifted frequency; the imaginary part is its constant multiple. If we pass to the quantum theory in the usual way, we get a non–Hermitian Hamiltonian operator. Its eigenvalues are complex valued, except for the ground state which can be chosen to have a real eigenvalue. Thus all states except the ground state are unstable. Any state decays to its projection to the ground state as time tends to infinity. This is a reasonable quantum analogue of the classical decay of energy.

We will show that a wide class of dissipative systems can be brought to such a canonical form using a complex Hamiltonian. The usual equations of motion determined by a Hamiltonian and Poisson bracket are

$$\frac{d}{dt}\begin{pmatrix} p \\ x \end{pmatrix} = \begin{pmatrix} \{H,p\} \\ \{H,x\} \end{pmatrix}.$$

At first a complex function

$$\mathcal{H} = H_1 + iH_2$$

does not seem to make sense when put into the above formula:

$$\frac{d}{dt}\begin{pmatrix} p \\ x \end{pmatrix} = \begin{pmatrix} \{H_1, p\} \\ \{H_1, x\} \end{pmatrix} + i \begin{pmatrix} \{H_2, p\} \\ \{H_2, x\} \end{pmatrix},$$

since the l.h.s. has real components. How can we make sense of multiplication by i and still get a vector with only real components? [Raj07]

Let us consider a complex number $z = x + iy$ as an ordered pair of real numbers (x, y). The effect of multiplying z by i is the linear transformation

$$\begin{pmatrix} x \\ y \end{pmatrix} \mapsto \begin{pmatrix} -y \\ x \end{pmatrix}$$

on its components. That is, multiplication by i is equivalent to the action by the matrix

$$J = \begin{pmatrix} 0 & -1 \\ 1 & 0 \end{pmatrix}.$$

Note that $J^2 = -1$. Geometrically, this corresponds to a rotation by ninety degrees.

Generalizing this, we can interpret multiplication by i of a vector field in phase space to mean the action by some matrix J satisfying

$$J^2 = -1. \tag{2.17}$$

Given such a matrix, we can define the equations of motion generated by a complex function $\mathcal{H} = H_1 + iH_2$ to be

$$\frac{d}{dt}\begin{pmatrix} p \\ x \end{pmatrix} = \begin{pmatrix} \{H_1, p\} \\ \{H_1, x\} \end{pmatrix} + J \begin{pmatrix} \{H_2, p\} \\ \{H_2, x\} \end{pmatrix}.$$

Our point is that the infinitesimal time evolution of a wide class of mechanical systems is of this type for an appropriate choice of $\{,\}$, J, H_1 and H_2.

In most cases there is a complex coordinate system in which J reduces to a simple multiplication by i; e.g., on the plane this is just $z = x + iy$. For such a coordinate system to exist the tensor field has to satisfy certain integrability conditions in addition to (2.17) above. These conditions are automatically satisfied if the matrix elements of J are constants.

What would be the advantage of fitting dissipative systems into such a complex canonical formalism? A practical advantage is that they can lead to better numerical approximations, generalizing the symplectic integrators widely used in Hamiltonian systems: these integrators preserve the geometric structure of the underlying physical system. Another is that it allows us to use ideas from Hamiltonian mechanics to study structures unique to dissipative systems such as strange attractors.

Instead we will look into the canonical quantization of dissipative systems. The usual correspondence principle leads to a non-Hermitian Hamiltonian. As in the elementary example of the damped simple harmonic oscillator, the eigenvalues are complex. The excited states are unstable and decay to the ground state. Non-Hermitian Hamiltonians have arisen already in several dissipative systems in condensed matter physics [NS98] and in particle physics [MR91]. The Wigner–Weisskopf approximation provides a physical justification for using a non–Hermitian Hamiltonian. A dissipative system is modelled by coupling it to some other 'external' degrees of freedom so that the total Hamiltonian is Hermitian and is conserved. In second order perturbation theory we can eliminate the external degrees of freedom to get an effective Hamiltonian that is non–Hermitian [Raj07].

1D Dissipative Harmonic Oscillator

We start by recalling the most elementary example of a classical 1D dissipative harmonic oscillator (DSHO), described by the familiar linear ODE

$$\ddot{x} + 2\gamma\dot{x} + \omega^2 x = 0, \qquad \gamma > 0. \tag{2.18}$$

We will consider the under–damped case $\gamma < \omega$ so that the system is still oscillatory.

We can write these equations in phase space

$$\dot{x} = p, \qquad \dot{p} = -2\gamma p - \omega^2 x.$$

The energy $H = \frac{1}{2}[p^2 + \omega^2 x^2]$ decreases monotonically along the trajectory:

$$\dot{H} = p\dot{p} + \omega^2 x\dot{x} = -2\gamma p^2 \leq 0.$$

The only trajectory which conserves energy is the one with $p = 0$, which must have $x = 0$ as well to satisfy the equations of motion.

These equations can be brought to diagonal form by a linear transformation [Raj07]:

$$z = A\left[-i(p + \gamma x) + \omega_1 x\right], \qquad \dot{z} = [-\gamma + i\omega_1]z$$

where $\omega_1 = \sqrt{\omega^2 - \gamma^2}$. The constant A that can be chosen later for convenience. These complex coordinates are the natural variables (normal modes) of the system.

Complex Hamiltonian

We can think of the above DSHO (2.18) as a generalized Hamiltonian system with a complex–valued Hamiltonian.

The Poisson bracket $\{p, x\} = 1$ becomes, in terms of the variable z,

$$\{z^*, z\} = 2i\omega_1|A|^2$$

So, if we choose $A = \frac{1}{\sqrt{2\omega_1}}$, we get

$$\{z^*, z\} = i.$$

Therefore, the complex function

$$\mathcal{H} = (\omega_1 + i\gamma)zz^*$$

satisfies [Raj07]

$$\dot{z} = \{\mathcal{H}, z\}, \qquad \dot{z}^* = \{\mathcal{H}^*, z^*\}.$$

Clearly, the limit $\gamma \to 0$ this $\mathcal{H}$ tends to the usual Hamiltonian $H = \omega zz^*$. Thus, on any analytic function ψ, we will have

$$\dot{\psi} = \{\mathcal{H}, \psi\} = [\omega_1 + i\gamma]z\,\partial_z\psi.$$

Quantization

By the usual rules of canonical quantization, the quantum theory is given by turning $\mathcal{H}$ into a non-Hermitian operator by replacing $z \mapsto a^\dagger$, $z^* \mapsto \hbar a$ and

$$[a, a^\dagger] = 1, \qquad a^\dagger = z, \qquad a = \partial_z, \qquad \mathcal{H} = \hbar(\omega_1 + i\gamma)a^\dagger a.$$

The effective Hamiltonian $\mathcal{H} = H_1 + iH_2$ is normal (i.e., its Hermitian and anti-Hermitian parts commute, $[H_1, H_2] = 0$) so it is still meaningful to speak of eigenvectors of $\mathcal{H}$. The eigenvalues are complex

$$(\omega_1 + i\gamma)n, \qquad (n = 0, 1, 2, \cdots).$$

The higher excited states are more and more unstable. But the ground state is stable, as its eigenvalue is zero.

Thus a generic state

$$\psi = \sum_{n=0}^{\infty} \psi_n |n>$$

will evolve in time as

$$\psi(t) = \sum_{n=0}^{\infty} \psi_n \mathrm{e}^{i\hbar[\omega + i\gamma]nt} |n> .$$

Unless ψ happens to be orthogonal to the ground state $|0>$, the wave–function will tend to the ground state as time tends to infinity; final state will be the projection of the initial state to the ground state. This is the quantum analogue of the classical fact that the system will decay to the minimum energy state as time goes to infinity. All this sounds physically reasonable [Raj07].

The Schrödinger Representation

In the Schrödinger representation,this amounts to ∂_{x^2}

$$a = \frac{1}{\sqrt{2\hbar\omega_1}}\left[\omega_1 x + \hbar\partial_x\right], \qquad a^\dagger = \frac{1}{\sqrt{2\hbar\omega_1}}\left[\omega_1 x - \hbar\partial_x\right],$$
$$\hat{\mathcal{H}} = \left(1 + i\frac{\gamma}{\omega_1}\right)\left[-\frac{\hbar^2}{2}\partial_{x^2} + \frac{1}{2}\omega_1^2 x^2 - \frac{1}{2}\hbar\omega_1\right].$$

Thus, the operator representing momentum p is

$$\hat{p} = -i\hbar\partial_x - \gamma x,$$

which includes a subtle correction dependent on the friction.

The time evolution operator can be chosen to be [Raj07]

$$\hat{\mathcal{H}}_{\mathrm{Schr}} = \hat{H} + \hat{H}_{\mathrm{diss}}, \qquad \text{where}$$
$$\hat{H} = -\frac{\hbar^2}{2}\partial_{x^2} + \frac{1}{2}\omega^2 x^2$$

is the usual harmonic oscillator Hamiltonian and

$$\hat{H}_{\mathrm{diss}} = -\frac{1}{2}\gamma^2 x^2 + i\frac{\gamma}{\omega_1}\left[-\frac{\hbar^2}{2}\partial_{x^2} + \frac{1}{2}\omega_1^2 x^2 - \frac{1}{2}\hbar\omega_1\right].$$

This is slightly different from the operator $\hat{\mathcal{H}}$ above, because the ground state energy is not fixed to be zero. The constant in H_{diss} has been chosen so that this state has zero imaginary part for its eigenvalue.

Dissipative 1-DOF System

We will now generalize to a nonlinear 1D oscillator with friction [Raj07]:

$$\dot{p} = -\partial_x V - 2\gamma p, \qquad \dot{x} = p, \quad \gamma > 0.$$

The DSHO is the special case $V(x) = \frac{1}{2}\omega^2 x^2$. The idea is that we lose energy whenever the system is moving, at a rate proportional to its velocity. It again follows that

$$\dot{H} = -2\gamma p^2 \leq 0,$$

where $H = \frac{1}{2}p^2 + V$. These equations can be written as

$$\dot{\xi}^i = \{H, \xi^i\} - \gamma^{ij}\partial_j H,$$

where $\gamma^{ij} = \begin{pmatrix} 2\gamma & 0 \\ 0 & 0 \end{pmatrix}$ is a positive but degenerate matrix.

Complex Effective Hamiltonian

In the case of the DSHO, we saw that it is the combination

$$p_1 = p + \gamma x.$$

rather than p that appears naturally (e.g., in the normal coordinate z). In terms of (p_1, x) the equations above take the form

$$\dot{p}_1 = -\partial_x V_1 - \gamma p_1, \qquad \dot{x} = p_1 - \gamma x, \qquad \text{where}$$
$$V_1(x) = V(x) - \frac{1}{2}\gamma^2 x^2, \qquad \text{i.e.,}$$
$$\frac{d}{dt}\begin{pmatrix} p_1 \\ x \end{pmatrix} = \begin{pmatrix} \{H_1, p\} \\ \{H_1, x\} \end{pmatrix} - \gamma \begin{pmatrix} \{H_2, p\} \\ \{H_2, x\} \end{pmatrix}.$$

Now, we would like to see if we can write these as canonical equations of motion with a complex effective Hamiltonian. Suppose we define [Raj07]

$$H_2 = \frac{\gamma}{2\omega_2}[p^2 + \omega_2^2 x^2], \qquad J = \begin{pmatrix} 1 & -\omega_2 \\ \frac{1}{\omega_2} & 0 \end{pmatrix},$$

for some constant ω_2 which we will choose later. Note that J is a complex structure; i.e.,

$$J^2 = -\begin{pmatrix} 1 & 0 \\ 0 & 1 \end{pmatrix}.$$

Then the equations of motion take the form

$$\frac{d}{dt}\begin{pmatrix} p_1 \\ x \end{pmatrix} = \begin{pmatrix} \{H_1, p\} \\ \{H_1, x\} \end{pmatrix} + J \begin{pmatrix} \{H_2, p\} \\ \{H_2, x\} \end{pmatrix}.$$

In terms of the complex coordinate

$$z = \frac{1}{\sqrt{2\omega_2}}[\omega_2 x - i p_1]$$

this becomes

$$\dot{z} = \{H_1 + iH_2, z\}.$$

Thus the nonlinear oscillator also can be written as a canonical system with a complex valued Hamiltonian $\mathcal{H} = H_1 + iH_2$, with $\{H_1, H_2\} \neq 0$ in general. But there are many ways to do this, parametrized by ω_2. The natural choice is

$$\omega_2 = \sqrt{V''(x_0) - \gamma^2},$$

where x_0 is the equilibrium point at which $V'(x_0) = 0$. Then, in the neighborhood of the equilibrium point the complex structure reduces to that of the DSHO.

Quantization of a Dissipative 1D System

We can quantize the above system by applying the usual rules of canonical quantization to the complex effective Hamiltonian $H_1 + iH_2$ to get the operator:

$$\hat{\mathcal{H}} = -\frac{\hbar^2}{2}\partial_{x^2} + V(x) - \frac{1}{2}\gamma^2 x^2 + i\frac{\gamma}{\omega_2}\left[-\frac{\hbar^2}{2}\partial_{x^2} + \frac{1}{2}\omega_2^2 x^2 + c\right].$$

The *Schrödinger equation*

$$-i\hbar\dot{\psi} = \hat{\mathcal{H}}\psi$$

then determines the time evolution of the dissipative system.

The anti–Hermitian part of the Hamiltonian is bounded below so the imaginary part of the eigenvalues will be bounded below. We can choose the real constant c such that the eigenvalue with the smallest imaginary part is actually real. Then the generic state will evolve to this 'ground state'. We have the following remarks [Raj07]:

1. Our model of dissipation amounts to adding a term
$$\hat{H}_{\text{diss}} = -\frac{1}{2}\gamma^2 x^2 + i\frac{\gamma}{\omega_2}\left[-\frac{\hbar^2}{2}\partial_{x^2} + \frac{1}{2}\omega_2^2 x^2\right] + c$$
to the usual conservative Hamiltonian
$$\hat{H} = -\frac{\hbar^2}{2}\partial_{x^2} + V(x).$$
The dissipative term we add is very close to being anti–Hermitian; i.e., except for the term $-\frac{1}{2}\gamma^2 x^2$ which is second order in the dissipation.
2. The Schrödinger equation for time reversed QM (evolving to the past rather than future) is $\mathcal{H}^\dagger$. The eigenvalues of $\mathcal{H}^\dagger$ are the complex conjugates of those of $\mathcal{H}$, but the eigenfunctions may not be the same as those of $\mathcal{H}$ since $[\mathcal{H}, \mathcal{H}^\dagger] \neq 0$ in general; i.e., the operator $\mathcal{H}$ may not be normal. (For the DSHO the commutator vanished so the issue did not arise.) Moreover the eigenfunctions of $\mathcal{H}$ corresponding to distinct eigenvalues need not be orthogonal; they would still be linearly independent, of course.
There are examples of such non–normal Hamiltonians in nature when time–reversal invariance is violated (e.g., $K\bar{K}$ oscillation) [MR91]. But in ordinary quantum mechanics such loss of time reversal invariance might be unsettling. Carl Bender [BB98] has suggested that for such non–Hermitian Hamiltonians, real eigenvalues and time reversal invariant dynamics can be recovered by modifying the inner product in the Hilbert space. This is the quantum counterpart to the reformulation of classical DSHO as a conservative canonical system by modifying the Poisson bracket and Hamiltonian [Oku81].
However we note that in the presence of dissipation, the classical equations of motion are not time reversal invariant either: energy would grow rather than dissipate. So we should not expect quantum mechanics of dissipative systems to be time reversal invariant either. The appropriate symmetry is
$$[\mathcal{H}(\gamma)]^\dagger = \mathcal{H}(-\gamma)$$
which is satisfied in the present case.

3. Note that within the present model, the anti–Hermitian part of the Hamiltonian is a sort of harmonic oscillator even if the Hermitian part has nonlinear classical dynamics. This is because we chose a particularly simple form of dissipation, $\dot{p} \sim -\gamma p$. If we had chosen a more complicated (e.g., nonlinear) form of dissipation, the anti–Hermitian part would be more complicated. It is common to model a dissipative system by coupling it to a thermal bath of oscillators. The dissipation is determined by the spectral density of the frequencies of these oscillators. Each choice of spectral density leads to a different dissipation term and, in the present description, to a different anti-Hermitian part for the Hamiltonian. But if the dissipative force is small it is reasonable to expect that it is linear in the velocity.

Tunnelling in a Dissipative System

Perhaps the most interesting question about quantum dissipative systems is *how dissipation affects tunnelling?* [Raj07]. The standard WKB approximation method adapts easily to the present case.

For illustrative purposes, it suffices to consider a one-dimensional system with a potential

$$V(x) = \frac{1}{2}\omega^2 x^2 \left(1 - \frac{x}{a}\right).$$

We ask for the tunnelling probability amplitude from the origin $x = 0$ to the point $x = a$ in a long time. In the absence of dissipation this is given in the WKB approximation by

$$\mathrm{e}^{-\frac{1}{\hbar}\int_0^a \sqrt{2V(x)}dx}.$$

The integral in the exponent is the minimum of the imaginary time action

$$\int_0^\infty [\frac{1}{2}\dot{x}^2 + V(x(\tau))]\, d\tau.$$

among all paths satisfying the boundary conditions $x(0) = 0$ and $x(\infty) = a$. This minimizing path is called the instanton.

If we apply the WKB approximation to the Schrdinger equation we get

$$\psi = \mathrm{e}^{-\frac{\phi}{\hbar}}, \quad -\frac{1}{2}(\partial_x\phi)^2 + V_1(x) + i\frac{\gamma}{\omega_2}\left[-\frac{1}{2}(\partial_x\phi)^2 + \frac{1}{2}\omega_2^2 x\right] \approx 0.$$

Solving this,

$$\phi(x) = \int_0^x \sqrt{\frac{2V_1(x) + i\gamma\omega_2 x^2}{1 + i\frac{\gamma}{\omega_2}}}dx.$$

The tunnelling probability is given by

$$\mathrm{e}^{-2\mathrm{Re}\ \phi(b)},$$

where b is the point of escape from the potential barrier; it might depend on the dissipation.

It looks most natural to choose $\omega_2 = \omega_1 = \sqrt{\omega^2 - \gamma^2}$ as in the case of the DSHO. Then,

$$\phi(x) = \omega_1 \int_0^x \sqrt{1 - \frac{\omega^2}{\omega_1(\omega_1 + i\gamma)} \frac{x}{a}} x\, dx.$$

Now,

$$\int_0^b x\sqrt{b - x} dx = \frac{4}{15} b^{\frac{5}{2}}.$$

There are small discrepancies (up to higher order terms in γ) depending on whether we think of the point of escape as a, or $\frac{\omega_1^2}{\omega^2} a$ or the complex number $\frac{\omega_1(\omega_1+i\gamma)}{\omega^2} a$ which is the zero of the integrand. The last choice gives the simplest answer for the tunnelling probability

$$e^{-\frac{8}{15}\frac{a^2\omega^2}{\hbar\omega}\frac{(\omega^2-\gamma^2)^{\frac{3}{2}}(\omega^2-2\gamma^2)}{\omega^5}}.$$

This differs from the results of Caldeira and Legget [CL85]: the tunnelling probability is enhanced by dissipation. There could still be systems in nature that are described by the present model.

Multi–Dimensional Dissipative Systems

Now we will generalize to a multi–dimensional dynamical system while also allowing for a certain kind of non-linearity in the frictional force.

Suppose the Hamiltonian is the sum of a kinetic and a potential energy,

$$H = \frac{1}{2} p_a p_a + V(x).$$

With the usual Poisson brackets

$$\{p_a, x^b\} = \delta_a^b, \qquad \{p_a, p_b\} = 0 = \{x^a, x^b\}$$

the conservative equations of motion are

$$\dot{p}_a = -\partial_a V, \qquad \dot{x}^a = p^a$$

We now assume that the frictional force is of the form $-2\gamma_{ab}\dot{x}^b$ for some positive non–degenerate tensor γ_{ab} that might depend on x. The idea is that the system loses energy when parts of it move relative to other parts or relative to some medium in which the system is immersed. Then the equations of motion become

$$\dot{p}_a = -\partial_a V - 2\gamma_{ab} p_b, \qquad \dot{x}^a = p^a.$$

Here we raise and lower indices using the flat Euclidean metric δ_{ab}. So

$$\dot{H} = -2\gamma^{ab} p_a p_b \leq 0.$$

If the dissipation tensor happens to be the Hessian of some convex function

$$\gamma_{ab} = \partial_a \partial_b W,$$

it is possible to write these as Hamilton's equations with a complex Hamiltonian.

Then, we would have

$$\frac{d}{dt}\partial_a W = \gamma_{ab} p^b, \qquad \gamma_{ab}\partial_b W = \partial_a(\frac{1}{2}\partial_b W \partial_b W).$$

Using these identities we can rewrite the equations of motion in the new variables [Raj07]:

$$\tilde{p}_a = p_a + \partial_a W, \qquad \text{as}$$
$$\frac{d\tilde{p}_a}{dt} = -\partial_a[V - \frac{1}{2}(\partial W)^2] - \gamma_{ab}\tilde{p}_b, \qquad \dot{x}_a = \tilde{p}_a - \partial_a W.$$

This motivates us to define

$$H_1 = \frac{1}{2}\tilde{p}^2 + V - \frac{1}{2}(\partial W)^2, \qquad H_2 = \frac{1}{\omega_2}[\frac{1}{2}\gamma_{ab}\tilde{p}_a\tilde{p}_b + \omega_2^2 W], \qquad J = \begin{pmatrix} 1 & -\omega_2 \\ \frac{1}{\omega_2} & 0 \end{pmatrix},$$

for some positive constant ω_2. Note that J is a complex structure, $J^2 = -1$. Then we have

$$\frac{d}{dt}\begin{pmatrix} \tilde{p} \\ x \end{pmatrix} = \begin{pmatrix} \{H_1, \tilde{p}\} \\ \{H_1, x\} \end{pmatrix} + J\begin{pmatrix} \{H_2, \tilde{p}\} \\ \{H_2, x\} \end{pmatrix}.$$

In terms of the complex variable

$$z_a = \frac{1}{\sqrt{2\omega_2}}[\omega_2 x_a - i\tilde{p}_a],$$

this is, as before,

$$\dot{z}_a = \{H_1 + iH_2, z_a\}.$$

Thus once the dissipation tensor is of the form $\gamma_{ab} = \partial_a \partial_b W$ the whole framework generalizes easily. The effect of dissipation is to add the term

$$H_{\text{diss}} = -\frac{1}{2}(\partial W)^2 + \frac{i}{\omega_2}[\frac{1}{2}\gamma_{ab}p_a p_b + \omega_2^2 W]$$

to the Hamiltonian. The classical equations turn out to be independent of the choice of ω_2, but the quantum theory will depend on its choice.

Quantization

We can then quantize this operator in the Schrödinger picture:

$$\hat{\mathcal{H}} = \hat{H} + \hat{H}_{\text{diss}},$$

where [Raj07][1]

$$\hat{H}\psi = -\frac{1}{2}\nabla^2\psi + V\psi, \quad H_{\text{diss}}\psi = -\frac{1}{2}(\nabla W)^2\psi + \frac{i}{\omega_2}\left[-\frac{1}{2}\partial_a\left(\gamma^{ab}\partial_b\psi\right) + \omega_2^2 W\right].$$

The operator $\partial_a\left(\gamma^{ab}\partial_b\psi\right)$ is thus a kind of 'mixed' Laplacian that uses γ_{ab} as well as, implicitly, the Euclidean metric. Since $\partial_a\gamma^{bc} \neq 0$ in general, the ordering of factors is important. with the order we chose,$\partial_a\left(\gamma^{ab}\partial_b\psi\right)$ is Hermitian and positive.

Geometrical Dissipative Dynamics

We will now further generalize to the case of dissipative dynamics on a cotangent bundle T^*Q.

The Hamiltonian is again the sum of kinetic and potential energies [Raj07]

$$H = \frac{1}{2}g^{ab}p_a p_b + V(x).$$

except that we now allow the tensor g^{ab} not to be constant. We will use g_{ab} (the inverse of g^{ab}) as the Riemann metric, used to define covariant derivative ∇_a and to raise and lower indices.

The conservative equations of motion are

$$\dot{p}_a + \Gamma^a_{bc}p^b p^c = -g^{ab}\partial_b V, \qquad \dot{x}^a = p^a$$

where Γ^a_{bc} is the usual Christoffel symbol.

With friction added,

$$\dot{p}_a + \Gamma^a_{bc}p_b p^c = -g^{ab}\partial_b V - 2\gamma^{ab}p_b, \qquad \dot{x}^a = p^a$$

where γ^{ab} is the dissipation tensor, assumed to be positive and non-degenerate. Again, if the dissipation tensor is the Hessian of a convex function

$$\gamma_{ab} \equiv g_{ac}g_{bd}\gamma^{bd} = \nabla_a\partial_b W,$$

[1] A technical point to note here is that there are two competing metrics in the story. The kinetic energy is $\frac{1}{2}p_a p_a$, determined by the Euclidean metric. But the quadratic term in the dissipative part of the Hamiltonian is determined by some other tensor γ_{ab}. We have chosen to use the Euclidean metric δ_{ab} to define derivatives and to raise and lower indices. Thus, $\gamma^{ab} = \delta^{ac}\delta^{bd}\gamma_{cd}$ and not the inverse of γ_{ab}.

there are simplifications because

$$\frac{d}{dt}(\nabla^a W) + \Gamma^a_{bc}\nabla^b W p^c = \gamma_{ab}p^b$$

We define again

$$\tilde{p}_a = p_a + \partial_a W, \qquad \text{to get}$$
$$\frac{d}{dt}\begin{pmatrix}\tilde{p}\\ x\end{pmatrix} = \begin{pmatrix}\{H_1,\tilde{p}\}\\ \{H_1,x\}\end{pmatrix} + J\begin{pmatrix}\{H_2,\tilde{p}\}\\ \{H_2,x\}\end{pmatrix}, \qquad \text{where}$$

$$H_1 = \frac{1}{2}g^{ab}\tilde{p}_a\tilde{p}_b + V - (\nabla W)^2, \quad H_2 = \frac{1}{2\omega_2}\gamma^{ab}\tilde{p}_a\tilde{p}_b + \omega_2 W, \quad J = \begin{pmatrix}1 & -\omega_2\\ \frac{1}{\omega_2} & 0\end{pmatrix}.$$

Since again $J^2 = -1$, it is an almost complex structure; but it may not be integrable in general. Every tangent bundle has a natural almost complex structure; J is simply its translation to the cotangent bundle T^*Q using the metric g_{ab} which identifies the tangent and cotangent bundles.

Because J may not be integrable, we are not able to rewrite this in terms of a complex coordinate z in general. Nevertheless, we can think of the above equations as a generalization of Hamilton's equations to a complex Hamiltonian $H_1 + iH_2$.

Clearly, it is possible to quantize these system by applying the correspondence principle. Since the ideas are not very different, we will not work out the details. The Hamiltonian is:

$$\hat{\mathcal{H}} = \hat{H} + \hat{H}_{\text{diss}}, \qquad \text{where}$$
$$\hat{H}\psi = -\frac{1}{2}\nabla^2\psi + V\psi, \qquad \text{and}$$
$$H_{\text{diss}}\psi = -\frac{1}{2}g^{ab}\partial_a W\partial_b W\psi + \frac{i}{\omega_2}\left[-\frac{1}{2}\nabla_a\left(\gamma^{ab}\partial_b\psi\right) + \omega_2^2 W\right] + ic.$$

The imaginary constant ic is chosen such that the ground state is stable.

2.1.5 Classical Trajectories for Complex Hamiltonians

In this section, following [BCD06], we study complex non–Hermitian quantum–mechanical Hamiltonians whose spectra are real and which exhibit unitary time evolution. A particularly interesting class of such Hamiltonians is [DDT01, BBJ02]

$$H = p^2 + x^2(ix)^\varepsilon \qquad (\varepsilon \geq 0). \tag{2.19}$$

We will study the nature of the underlying classical theory described by this Hamiltonian. This question was addressed in several studies [BBM99, Nan04]. These papers presented numerical studies of the classical trajectories, that is, the position $x(t)$ of a particle of a given energy as a function of time. Some interesting features of these trajectories were discovered:

- While $x(t)$ for a Hermitian Hamiltonian is a real function, a complex Hamiltonian typically generates complex classical trajectories. Thus, even if the classical particle is initially on the real–x axis, it is subject to complex forces and thus it will move off the real axis and travel through the complex–plane $\mathbb{C}$.
- For the Hamiltonian in (2.19) the classical domain is a multi–sheeted Riemann surface when ε is non–integer. In this case, the classical trajectory may visit more than one sheet of the Riemann surface. Indeed, in [BBM99] classical trajectories that visit three sheets of the Riemann surface were displayed.
- As $\varepsilon \geq 0$, the $\mathcal{PT}$ symmetry of H in (2.19) is unbroken [BBJ02] and, as a result, the classical orbits are closed periodic paths in the complex–plane. When ε is negative, the classical trajectories are open (and nonperiodic).
- The classical trajectories manifest the $\mathcal{PT}$ symmetry of the Hamiltonian. Under parity reflection $\mathcal{P}$ the position of the particle changes sign: $\mathcal{P}: x(t) \to -x(t)$. Under time reversal $\mathcal{T}$ the sign of both t and i are reversed, so $\mathcal{T}: x(t) \to x^*(-t)$. Thus, under combined $\mathcal{PT}$ reflection the classical trajectory is replaced by its mirror image with respect to the imaginary axis on the principal sheet of the Riemann surface.

Although these features of classical non–Hermitian $\mathcal{PT}$–symmetric Hamiltonians were already known, we show in this section that the structure of the complex trajectories is much richer and more elaborate than was previously noticed. One can find trajectories that visit huge numbers of sheets of the Riemann surface and exhibit fine structure that is exquisitely sensitive to the initial condition $x(0)$ and to the value of ε. Small variations in $x(0)$ and ε give rise to dramatic changes in the topology of the classical orbits and to the size of the period.

We study the dependence on initial conditions of classical orbits governed by (2.19). To construct the classical trajectories, we first note that the value of the Hamiltonian in (2.19) is a constant of the motion. Without loss of generality, this constant (the energy E) may be chosen to be 1.[2] As $p(t)$ is the time derivative of $x(t)$, the trajectory $x(t)$ satisfies a first–order ODE whose solution is determined by the initial condition $x(0)$ and the sign of $\dot{x}(0)$.

Let us begin by examining the harmonic oscillator, which is obtained by setting $\varepsilon = 0$ in (2.19). For the harmonic oscillator the turning points (the solutions to the equation $x^2 = 1$) lie at $x = \pm 1$. If we chose $x(0)$ to lie between these turning points,

$$-1 \leq x(0) \leq 1, \tag{2.20}$$

then the classical trajectory oscillates between the turning points with period π. This orbit is shown in Figure 2.8 as the solid horizontal line joining the turning points.

[2] If E were not 1, we could then rescale x and t to make $E = 1$.

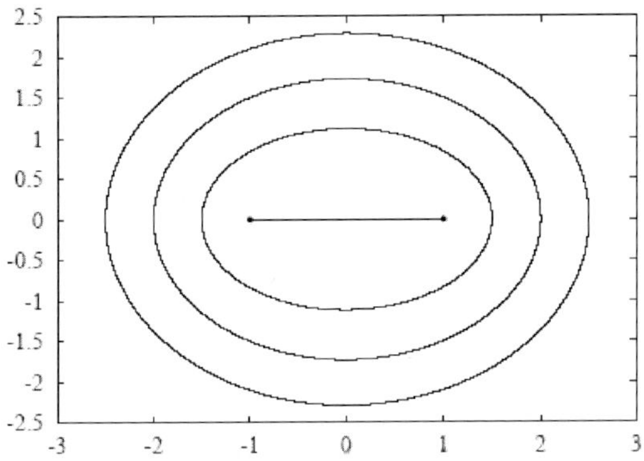

Fig. 2.8. Classical trajectories in the complex$-x$ plane for the harmonic oscillator whose Hamiltonian is $H = p^2 + x^2$. These trajectories represent the possible paths of a particle whose energy is $E = 1$. The trajectories are nested ellipses with foci located at the turning points at $x = \pm 1$. The real line segment (degenerate ellipse) connecting the turning points is the usual periodic classical solution to the harmonic oscillator. All closed paths have the same period π by virtue of *Cauchy's integral Theorem* (modified and adapted from [BCD06]).

However, while the harmonic–oscillator Hamiltonian is Hermitian, it can still have complex classical trajectories. To get one of these trajectories, we choose an initial condition that does not lie between the turning points and thus does not satisfy (2.20). The resulting trajectories are ellipses in the complex–plane (see Figure 2.8). The foci of these ellipses are the turning points [BBM99]. Note that for each of these closed orbits the period is always π; this is a consequence of the Cauchy integral Theorem applied to the integral that represents the period.

As ε increases from 0, the pair of turning points at $x = \pm 1$ moves downward into the complex–x plane. These turning points are determined by the equation

$$1 + (ix)^{2+\varepsilon} = 0. \tag{2.21}$$

When ε is non–integer, this equation has many solutions, all having absolute value 1. These solutions have the form

$$x = \exp\left(i\pi\frac{4N - 4 - \varepsilon}{4 + 2\epsilon}\right), \tag{2.22}$$

where N is an integer. These turning points occur in $\mathcal{PT}$–symmetric pairs (that is, pairs that are reflected through the imaginary axis) corresponding to the N values $(N = 1,\ N = 0)$, $(N = 2,\ N = -1)$, $(N = 3,\ N = -2)$, $(N = 4,\ N = -3)$, and so on. We label these pairs by the integer n ($n = 0,\ 1,\ 2,\ 3,\ \ldots$) so that the nth pair corresponds to $(N = n + 1,\ N = -n)$. Note that the pair of turning points at $\varepsilon = 0$ deforms continuously into the

$n = 0$ pair of turning points when $\varepsilon \neq 0$. For the case $\varepsilon = \pi - 2$ these turning points are shown in Figure 2.9 as dots.

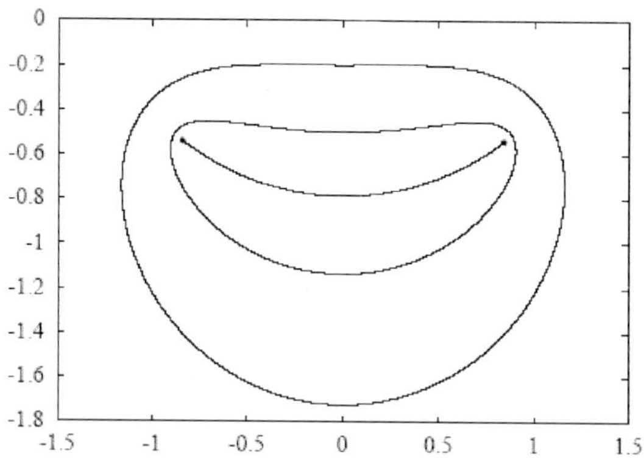

Fig. 2.9. Classical trajectories in the complex$-x$ plane for the complex oscillator whose Hamiltonian is $H = p^2 - (ix)^\pi$, which is (2.19) with $\varepsilon = \pi -$. As in Figure 2.8 the trajectories represent the possible paths of a particle whose energy is $E = 1$. The trajectories are deformed versions of the ellipses in Figure 2.8. By virtue of Cauchy's integral Theorem, all of the closed trajectories have the same period T as given in (2.23) (modified and adapted from [BCD06]).

In Figure 2.9 three closed classical trajectories are shown. First, there is the path connecting the $n = 0$ turning points, which is a deformed version of the straight line in Figure 2.8. Two other trajectories that enclose these two turning points are also indicated. These closed orbits are deformations of the ellipses shown in Figure 2.8. Furthermore, as in the $\varepsilon = 0$ case, the Cauchy integral Theorem implies that the period T for each of these orbits is the same. The general formula for the period of a closed orbit whose topology is like that of the orbits shown in Figure 2.9 is

$$T = 2\sqrt{\pi}\frac{\Gamma\left(\frac{3+\varepsilon}{2+\varepsilon}\right)}{\Gamma\left(\frac{4+\varepsilon}{4+2\epsilon}\right)}\cos\left(\frac{\epsilon\pi}{4+2\epsilon}\right). \qquad (2.23)$$

This formula is given in [BBM99] and is valid for all $\varepsilon \geq 0$. For the case of the closed orbits shown in Figure 2.9, we find that $T = 2.33276$.

The derivation of (2.23) is straightforward. The period T is given by a closed contour integral along the trajectory in the complex$-x$ plane. This trajectory encloses the square–root branch cut that joins the turning points. This contour can be deformed into a pair of rays that run from one turning point to the origin and then from the origin to the other turning point. The integral along each ray is easily evaluated as a beta function, which is then written in terms of gamma functions.

The key difference between classical paths for $\varepsilon > 0$ and for $\varepsilon < 0$ is that in the former case all the paths are closed orbits and in the latter case the paths are open orbits. In Figure 2.10 we consider the case $\varepsilon = -0.2$ and display two paths that begin on the negative imaginary axis. One path evolves forward in time and the other path evolves backward in time. Each path spirals outward and eventually moves off to infinity. Note that the pair of paths is a $\mathcal{PT}$–symmetric structure. Note also that the paths do not cross because they are on different sheets of the *Riemann surface*. The function $(ix)^{0.2}$ requires a branch cut, and we take this branch cut to lie along the positive imaginary axis. The forward–evolving path leaves the principal sheet (sheet 0) of the Riemann surface and crosses the branch cut in the positive sense and continues on sheet 1. The reverse path crosses the branch cut in the negative sense and continues on sheet -1. Figure 2.10 shows the projection of the classical orbit onto the principal sheet.

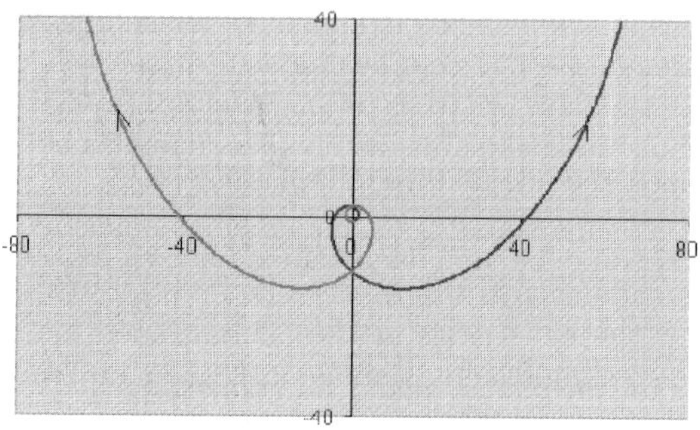

Fig. 2.10. Classical trajectories in the complex$-x$ plane for the Hamiltonian in (2.19) with $\varepsilon = -0.2$. These trajectories begin on the negative imaginary axis very close to the origin. One trajectory evolves forward in time and the other goes backward in time. The trajectories are open orbits and show the particle spiraling off to infinity. The trajectories begin on the principal sheet of the Riemann surface; as they cross the branch cut on the positive imaginary axis, they visit the higher and lower sheets of the surface. Note that the trajectories do not cross because they lie on different sheets (modified and adapted from [BCD06]).

Let us now examine closed orbits having a more complicated topological structure than the orbits shown in Figure 2.9. For the rest of this subsection we fix $\varepsilon = \pi - 2$ and study the effect of varying the initial conditions. It is not difficult to find an initial condition for which the classical trajectory crosses the branch cut on the positive imaginary axis and leaves the principal sheet of the Riemann surface. In Figure 2.11 we show such a trajectory. This trajectory visits three sheets of the Riemann surface, the principal sheet (sheet 0) on which the trajectory is shown as a solid line, and sheets ± 1 on which the trajectory is shown as a dashed line. On the Riemann surface the resulting trajectory is $\mathcal{PT}$–symmetric (left–right symmetric).

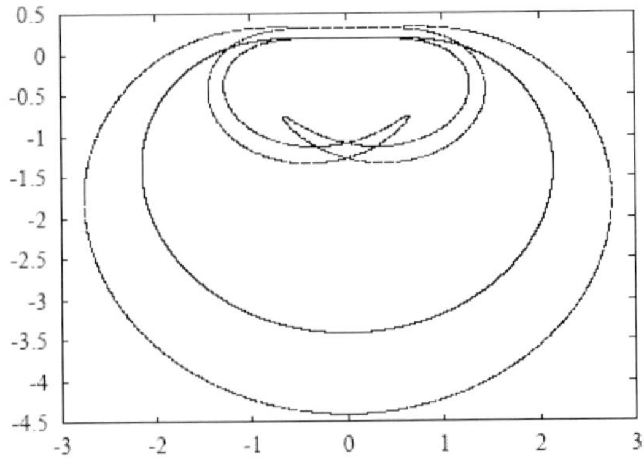

Fig. 2.11. A classical trajectory in the complex$-x$ plane for the Hamiltonian $H = p^2 - (ix)^\pi$, which is obtained by setting $\varepsilon = \pi - 2$ in (2.19). The initial condition is chosen so that the path crosses the branch cut on the positive imaginary axis and leaves the principal sheet of the Riemann surface. On the principal sheet the trajectory is indicated by a solid line. The classical particle visits two other sheets of the Riemann surface on which the trajectory is indicated by a dashed line. Note that the closed orbit is $\mathcal{PT}$–symmetric (has left–right symmetry) and that the period is $T = 11.8036$ (modified and adapted from [BCD06]).

The period of the orbit in Figure 2.11 is $T = 11.8036$, which is roughly five times longer than the periods of the orbits shown in Figure 2.9. This is because the orbit is topologically more complicated and encloses branch cuts joining three pairs rather than one pair of complex turning points.[3]

The closed orbit shown in Figure 2.11 only visits three sheets of the Riemann surface. It is possible to find initial conditions that generate trajectories that visit many sheets repeatedly. In Figure 2.12 we have plotted a classical trajectory starting at $x(0) = -7.1i$. This trajectory visits 11 sheets of the Riemann surface and its period is $T = 255.3$. The structure of this orbit near the origin is complicated and therefore a magnified version is shown in Figure 2.13.

As Figures 2.12 and 2.13 are so complicated, it is useful to give a more understandable representation of the classical orbit in which we plot the complex phase (argument) of $x(t)$ as a function of t. In particular, a characteristic feature of *long orbits* is the persistent oscillation in the classical path which makes huge numbers of U–turns in portions of the complex–plane. These U–turns focus about one of the many complex turning points and illustrate in a rather dramatic fashion the complex nature of the classical turning point.[4] For more details on complex Hamiltonians, see [BCD06].

[3] The period of the orbit is roughly proportional to the number of times that the orbit crosses the imaginary axis.

[4] The behavior of real trajectories is much simpler. When a real trajectory encounters a turning point on the real axis it merely stops and reverses direction.

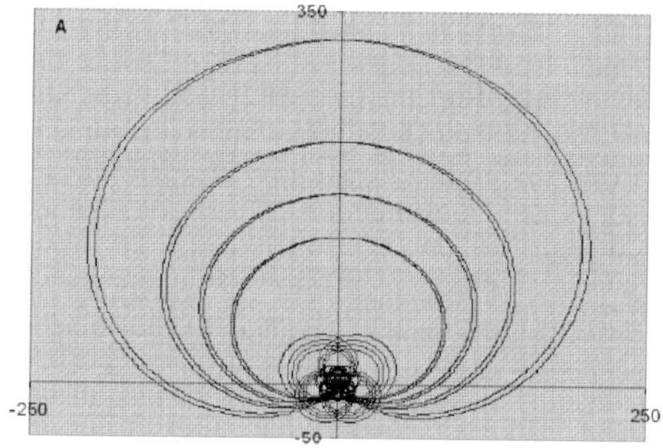

Fig. 2.12. A classical trajectory in the complex$-x$ plane for the complex Hamiltonian $H = p^2 - (ix)^\pi$. This complicated trajectory begins at $x(0) = -7.1i$ and visits 11 sheets of the Riemann surface. Its period is approximately $T = 255.3$. This figure displays the projection of the trajectory onto the principal sheet of the Riemann surface. Note that this trajectory does not cross itself (modified and adapted from [BCD06]).

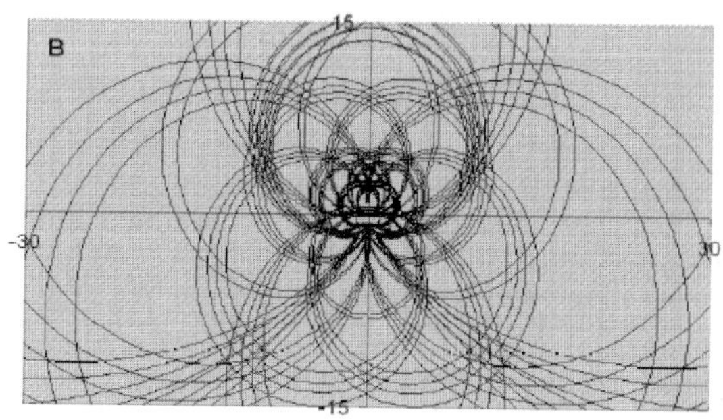

Fig. 2.13. An enlargement of the classical trajectory $x(t)$ in Figure 2.12 showing the detail near the origin in the complex$-x$ plane. This classical path never crosses itself – the apparent self–intersections are paths that lie on different sheets of the Riemann surface (modified and adapted from [BCD06]).

2.2 Complex Chaotic Dynamics: Discrete and Symbolic

In this section we present several models of both real and complex chaotic dynamics. For more details on this topic, see [II07].

A theory analogous to both the theories of polynomial–like maps and Smale's real horseshoes has been developed in [Obe87, Obe00] for the study of the dynamics of maps of two complex variables. In partial analogy with polynomials in a single variable there are the Hénon maps in two variables as well as higher dimensional analogues. From *polynomial–like maps*, *Hénon–like maps* and *quasi–Hénon–like maps* are defined following this analogy. A special form of the latter is the complex horseshoe. The major results about the real horseshoes of Smale remain true in the complex setting. In particular: (i) trapping fields of cones(which are sectors in the real case) in the tangent spaces can be defined and used to find horseshoes, (ii) the dynamics of a horseshoe is that of the two–sided shift on the symbol space on some number of symbols which depends on the type of the horseshoe, and (iii) transverse intersections of the stable and unstable manifolds of a hyperbolic periodic point guarantee the existence of horseshoes.

Recall that the study of the subject of the dynamics of complex analytic functions of one variable goes back to the early 1900's with the publication of several papers by Pierre Fatou and Gaston Julia around 1920, including the long memoirs [Fat19, Jul18]. Compared with the theory of complex analytic functions of a single variable, the theory of complex analytic maps of several variables is quite different. In particular, the theory of omitted values, including normal families, is not paralleled in the several complex variables case. This difference was really exposed by Fatou himself and L. Bieberbach in the 1920's in [Fat22, Bla84]. They showed the existence of the *Fatou–Bieberbach domains*: open subsets of $\mathbb{C}^n$ whose complements have nonempty interior and yet are the images of $\mathbb{C}^n$ under an injective analytic map. This is contrary to the one variable case where the image of every non-constant analytic function on $\mathbb{C}$ omits at most a single point. [Obe87, Obe00] attempted to understand these Fatou–Bieberbach domains, which arose naturally as the basins of attractive fixed–points of analytic automorphisms of $\mathbb{C}^n$; the basins were the image of the map conjugating the given automorphism to its linear part at the given fixed–point. For example, consider the map

$$F : \begin{pmatrix} x \\ y \end{pmatrix} \mapsto \begin{pmatrix} x^2 + 9/32 - y/8 \\ x \end{pmatrix},$$

which has two fixed–points, of which $(3/8, 3/8)$ is attractive with its linear part having resonant eigenvalues $1/4$ and $1/2$ (that is, $1/4 = (1/2)^2$). Moreover, none of the points in the region $\{\,(x,y)\mid |y| < 4|x|^2/3, \quad |x| > 4\,\}$, remain bounded under iteration of F. So the basin of $(3/8, 3/8)$ is not all of $\mathbb{C}^2$.

2.2.1 Basic Fractals and Biomorphs

Mandelbrot and Julia Sets

Mandelbrot and Julia sets are celebrated *fractals* (see Figure 2.14), defined either by a quadratic *conformal z–map* [Man80a, Man80b]

$$z_{n+1} = z_n^2 + c,$$

or by a real (x, y)–map

$$x_{n+1} = \sqrt{x_n} - \sqrt{y_n} + c_1, \qquad y_{n+1} = 2\,x_n\,y_n + c_2,$$

where c, c_1 and c_2 are parameters. For almost every c, this conformal transformation generates a fractal (probably, only for $c = -2$ it is not a fractal). Julia set J_c with $c \ll 1$, the *capacity dimension* is

$$d_{cap} = 1 + \frac{|c|^2}{4 \ln 2} + O(|c|^3).$$

The set of all points for which J_c is connected is the Mandelbrot set.[5]

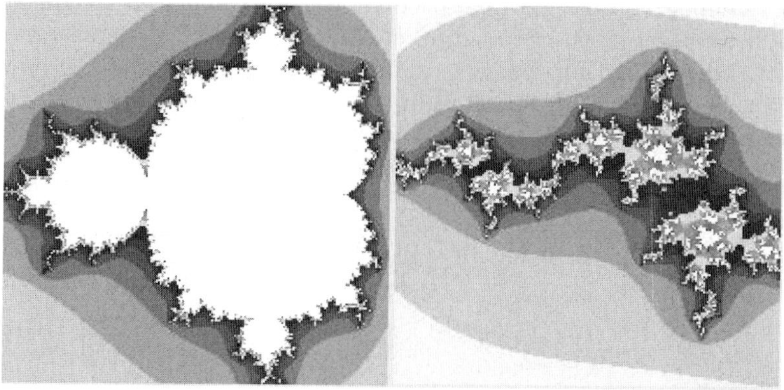

Fig. 2.14. The celebrated conformal Mandelbrot (left) and Julia (right) sets in the complex–plane, simulated using *Dynamics Solver*^{TM}.

Biomorphic Systems

Closely related to the Mandelbrot and Julia sets are *biomorphic systems*, which look like one–celled organisms. The term '*biomorph*' was proposed by C. Pickover from IBM [Pic86, Pic87]. Pickover's biomorphs inhabit the complex–plane like the the Mandelbrot and Julia sets and exhibit a *protozoan morphology*. Biomorphs began for Pickover as a 'bug' in a program intended to probe the fractal properties of various formulas. He accidentally used an OR logical operator instead of an AND operator in the conditional test for the size of z's real and imaginary parts. The cilia that project from the biomorphs are a

[5] The Mandelbrot set has its place in complex–valued dynamics, a field first investigated by the French mathematicians Pierre Fatou [Fat19, Fat22] and Gaston Julia [Jul18] at the beginning of the 20th century. For general families of holomorphic functions, the boundary of the Mandelbrot set generalizes to the bifurcation locus, which is a natural object to study even when the connectedness locus is not useful. A related *Mandelbar set* was encountered by mathematician John Milnor in his study of parameter slices of real cubic polynomials; it is not locally connected; this property is inherited by the connectedness locus of real cubic polynomials.

consequence of this 'error'. Each biomorph is generated by multiple iterations of a particular conformal map,

$$z_{n+1} = f(z_n, c),$$

where c is a parameter. Each iteration takes the output of the previous operations as the input of the next iteration. To generate a biomorph, one first needs to lay out a grid of points on a rectangle in the complex–plane [And01]. The coordinate of each point constitutes the real and imaginary parts of an initial value, z_0, for the iterative process. Each point is also assigned a pixel on the computer screen. Depending on the outcome of a simple test on the 'size' of the real and imaginary parts of the final value, the pixel is colored either black or white. The biomorphs presented in Figure 2.15 are generated using the following conformal functions:

1. $f(z,c) = z^3$,
2. $f(z,c) = z^3 + c$, $\quad c = 10$,
3. $f(z,c) = z^3 + c$, $\quad c = 10 - 10i$,
4. $f(z,c) = z^5 + c$, $\quad c = 0.77 - 0.77i$,
5. $f(z,c) = z^3 + \sin z + c$, $\quad c = 1 - i$,
6. $f(z,c) = z^6 + \sin z + c$, $\quad c = 0.5 - 0.5i$,
7. $f(z,c) = z^2 \sin z + c$, $\quad c = 0.78 - 0.78i$,
8. $f(z,c) = z^c$, $\quad c = 5 - i$,
9. $f(z,c) = |z|^c \sin z$, $\quad c = 4$,
10. $f(z,c) = |z|^c \cos z + c$, $\quad c = 3 + 3i$,
11. $f(z,c) = |z|^c(\cos z + z) + c$, $\quad c = 3 + 2i$.

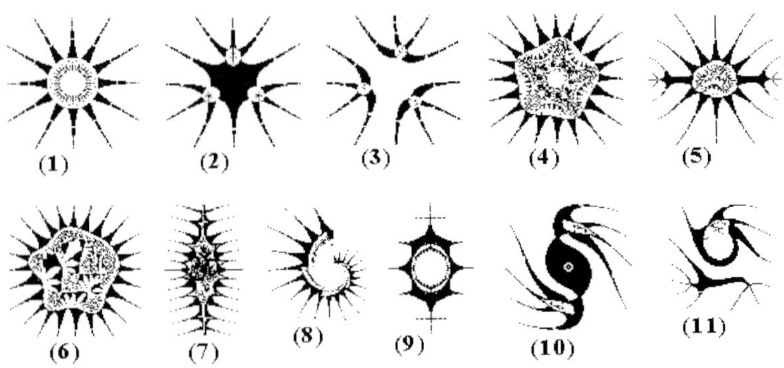

Fig. 2.15. *Pickover's biomorphs* (see text for details).

2.2.2 Mandelbrot Set

Recall from the previous subsection that the *Mandelbrot set*[6] M is defined as the *connectedness locus* of the family of complex quadratic polynomials,

$$f_c : \mathbb{C} \longrightarrow \mathbb{C}, \qquad z \mapsto z^2 + c.$$

That is, the Mandelbrot set is the subset of the complex–plane consisting of those parameters c for which the *Julia set* of f_c is connected. An equivalent way of defining is as the set of parameters for which the critical point does not tend to infinity. That is, $f_c^n(0) \not\to \infty$, where f_c^n is the n–fold composition of f_c with itself. The Mandelbrot set is generally considered to be a fractal. However, only the boundary of it is technically a fractal.

The Mandelbrot set M is a compact set, contained in the closed disk of radius 2 around the origin (see Figure 2.16). More precisely, if c belongs to M, then $|f^n(c)| \leq 2$ for all $n \geq 0$. The intersection of M with the real axis

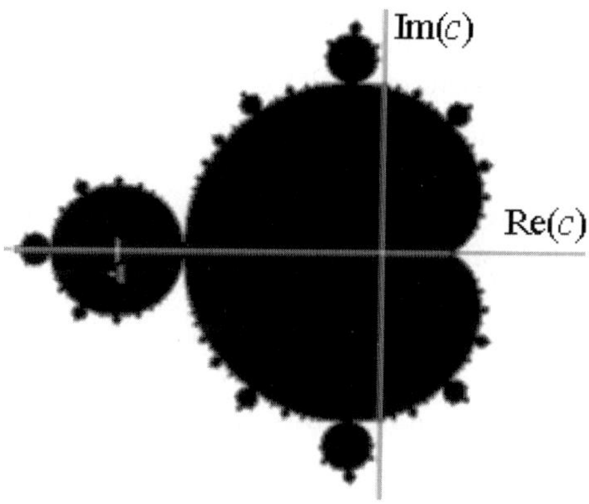

Fig. 2.16. The Mandelbrot set in a complex–plane.

is precisely the interval $[-2, 0.25]$. The parameters along this interval can be put in 1–1 correspondence with those of the real *logistic–map family*

$$z \mapsto \lambda z(z-1), \quad \lambda \in [1, 4].$$

[6] Benoit Mandelbrot studied the parameter space of quadratic polynomials in the 1980 article [Man80a]. The mathematical study of the Mandelbrot set really began with work by the mathematicians A. Douady and J.H. Hubbard [DH85], who established many fundamental properties of M, and named the set in honor of Mandelbrot. The Mandelbrot set has become popular far outside of mathematics both for its aesthetic appeal and its complicated structure, arising from a simple definition. This is largely due to the efforts of Mandelbrot (and others), who worked hard to communicate this area of mathematics to the general public.

Douady and Hubbard have shown in [DH85] that the Mandelbrot set is connected. In fact, they constructed an explicit conformal isomorphism between the complement of the Mandelbrot set and the complement of the closed unit disk. The dynamical formula for the uniformization of the complement of the Mandelbrot set, arising from Douady and Hubbard's proof of the connectedness of M, gives rise to *external rays* of the Mandelbrot set, which can be used to study the Mandelbrot set in combinatorial terms.

The boundary of the Mandelbrot set is exactly the bifurcation locus of the quadratic family; that is, the set of parameters c for which the dynamics changes abruptly under small changes of c. It can be constructed as the limit set of a sequence of plane algebraic *Mandelbrot curves*, of the general type known as *polynomial lemniscates*. The Mandelbrot curves are defined by setting

$$p_0 = z, \qquad p_n = p_{n-1}^2 + z,$$

and then interpreting the set of points $|p_n(z)| = 1$ in the complex–plane as a curve in the real Cartesian plane of degree 2^{n+1} in x and y.

Upon looking at a picture of the Mandelbrot set (Figure 2.16), one immediately notices the large cardioid–shaped region in the center. This *main cardioid* is the region of parameters c for which f_c has an *attracting fixed–point*. It consists of all parameters of the form $c = \frac{1-(\mu-1)^2}{4}$, for some μ in the open unit disk.

To the left of the main cardioid, attached to it at the point $c = -3/4$, a circular–shaped bulb is visible. This bulb consists of those parameters c for which f_c has an attracting cycle of period 2. This set of parameters is an actual circle, namely that of radius 1/4 around -1.

There are many other bulbs attached to the main cardioid: for every rational number p/q, with p and q coprime, there is such a bulb attached at the parameter,

$$c_{\frac{p}{q}} = \frac{1 - \left(e^{2\pi i \frac{p}{q}} - 1\right)^2}{4}.$$

This bulb is called the p/q–bulb of the Mandelbrot set. It consists of parameters which have an attracting cycle of period q and combinatorial rotation number p/q. More precisely, the q–periodic Fatou components containing the attracting cycle all touch at a common α–fixed–point. If we label these components $U_0, \dots, U_{q-1}$ in counterclockwise orientation, then f_c maps the component U_j to the component $U_{j+p \,(\mathrm{mod}\, q)}$.

The change of behavior occurring at $c_{\frac{p}{q}}$ is a *bifurcation*: the attracting fixed–point 'collides' with a repelling period q–cycle. As we pass through the bifurcation parameter into the q–bulb, the attracting fixed–point turns into a repelling α–fixed–point, and the period q–cycle becomes attracting.

All the above bulbs were interior components of the Mandelbrot set in which the maps f_c have an attracting periodic cycle. Such components are called *hyperbolic components* (see Figure 2.17). It has been conjectured that these are the only interior regions of M. This problem, known as *density of hyperbolicity*, may be the most important open problem in the field of complex dynamics. Hypothetical non–hyperbolic components of the Mandelbrot set are often referred to as 'queer' components.

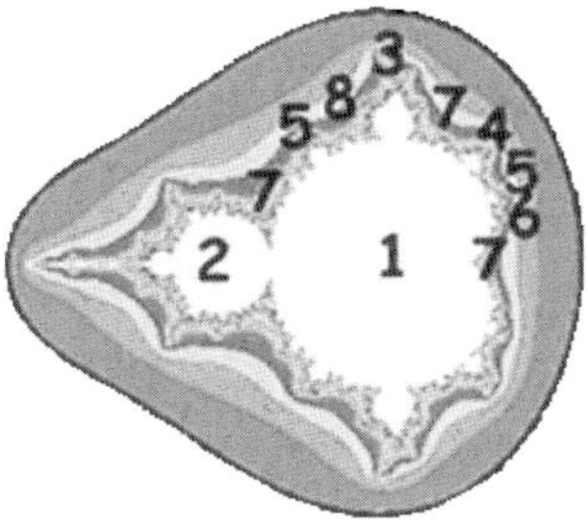

Fig. 2.17. Periods of hyperbolic components of the Mandelbrot set in a complex–plane.

The *Hausdorff dimension* of the boundary of the Mandelbrot set equals 2 (see [Shi98]). It is not known whether the boundary of the Mandelbrot set has positive planar *Lebesgue measure*.

Sometimes the connectedness loci of families other than the quadratic family are also referred to as the Mandelbrot sets of these families. The connectedness loci of the unicritical polynomial families $f_c = z^d + c$ for $d > 2$ are often called *Multibrot sets*. For general families of holomorphic functions, the boundary of the Mandelbrot set generalizes to the bifurcation locus, which is a natural object to study even when the connectedness locus is not useful. It is also possible to consider similar constructions in the study of non–analytic mappings. Of particular interest is the *tricorn* (also sometimes called the Mandelbar set), the connectedness locus of the anti–holomorphic family: $z \mapsto \bar{z}^2 + c$. The tricorn was encountered by John Milnor in his study of parameter slices of real cubic polynomials. It is not locally connected. This property is inherited by the connectedness locus of real cubic polynomials.

2.2.3 Hénon Maps

Real Hénon Maps

Recall that the famous *Hénon map* [Hen69] is a discrete–time dynamical system that is an extension of the *logistic map*

$$x_{t+1} = r\, x_t\, (1 - x_t), \tag{2.24}$$

(where r is the *Malthusian parameter* that varies between 0 and 4, and the initial value of the population $x_0 = x(0)$ is restricted to be between 0 and 1)

– and exhibits a chaotic behavior. The map was introduced by M. Hénon as a simplified model of the *Poincaré section* of the celebrated *Lorenz system*

$$\dot{x} = a(y - x), \qquad \dot{y} = bx - y - xz, \qquad \dot{z} = xy - cz, \qquad (2.25)$$

where x, y and z are dynamical variables, constituting the 3D *phase–space* of the *Lorenz system*; and a, b and c are the parameters of the system. The Hénon map is 2D–map which takes a point (x, y) in the plane and maps it to a new point defined by equations

$$x_{n+1} = y_n + 1 - ax_n^2, \qquad y_{n+1} = bx_n,$$

The map depends on two parameters, a and b, which for the canonical Hénon map have values of $a = 1.4$ and $b = 0.3$ (see Figure 2.18). For the canonical values the Hénon map is chaotic. For other values of a and b the map may be chaotic, intermittent, or converge to a periodic orbit. An overview of the

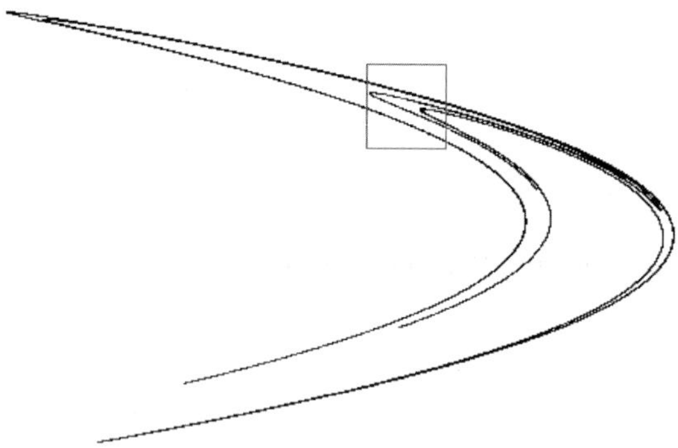

Fig. 2.18. *Hénon strange attractor* (see text for explanation), simulated using *Dynamics Solver*^{TM}.

type of behavior of the map at different parameter values may be obtained from its orbit (or, bifurcation) diagram (see Figure 2.19). For the canonical map, an initial point of the plane will either approach a set of points known as the *Hénon strange attractor*, or diverge to infinity. The Hénon attractor is a fractal, smooth in one direction and a Cantor set in another. Numerical estimates yield a correlation dimension of 1.42 ± 0.02 (Grassberger, 1983) and a Hausdorff dimension of 1.261 ± 0.003 (Russel 1980) for the Hénon attractor. As a dynamical system, the canonical Hénon map is interesting because, unlike the logistic map, its orbits defy a simple description. The Hénon map maps two points into themselves: these are the invariant points. For the canonical values of a and b, one of these points is on the attractor: $x = 0.631354477...$ and $y = 0.189406343...$ This point is unstable. Points close to this fixed–point and along the slope 1.924 will approach the fixed–point and points along the

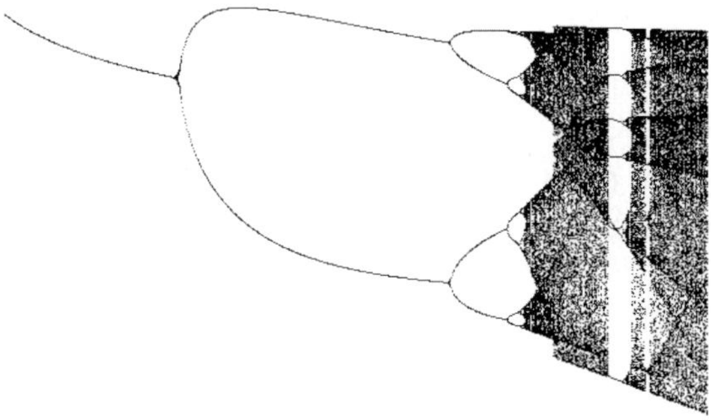

Fig. 2.19. Bifurcation diagram of the *Hénon strange attractor*, simulated using *Dynamics Solver*^{TM}.

slope –0.156 will move away from the fixed–point. These slopes arise from the linearizations of the *stable manifold* and *unstable manifold* of the fixed–point. The unstable manifold of the fixed–point in the attractor is contained in the strange attractor of the Hénon map. The Hénon map does not have a strange attractor for all values of the parameters a and b. For example, by keeping b fixed at 0.3 the bifurcation diagram shows that for a = 1.25 the Hénon map has a stable periodic orbit as an attractor. Cvitanovic *et al.* [CGP88] showed how the structure of the Hénon strange attractor could be understood in terms of unstable periodic orbits within the attractor.

For the (slightly modified) Hénon map: $x_{n+1} = ay_n + 1 - x_n^2,\ y_{n+1} = bx_n$, there are three *basins of attraction* (see Figure 2.20).

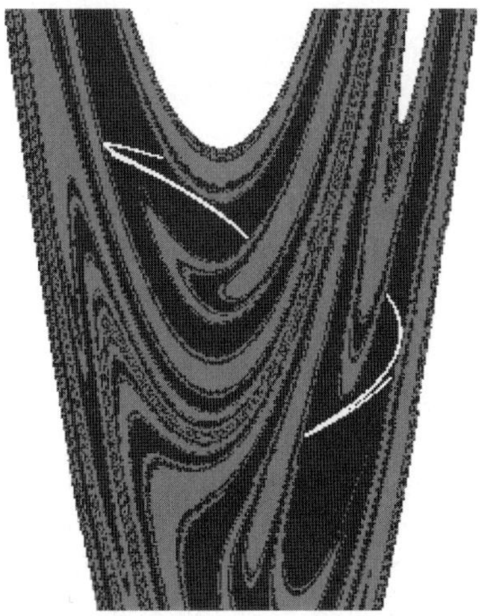

Fig. 2.20. Three basins of attraction for the Hénon map $x_{n+1} = ay_n + 1 - x_n^2$, $y_{n+1} = bx_n$, with $a = 0.475$.

The *generalized Hénon map* is a 3D–system (see Figure 2.21)

$$x_{n+1} = a\,x_n - z\,(y_n - x_n^2)), \qquad y_{n+1} = z\,x_n + a\,(y_n - x_n^2)), \qquad z_{n+1} = z_n,$$

where $a = 0.24$ is a parameter. It is an *area–preserving map*, and simulates the *Poincaré map* of period orbits in *Hamiltonian systems*. Repeated random initial conditions are used in the simulation and their gray–scale color is selected at random.

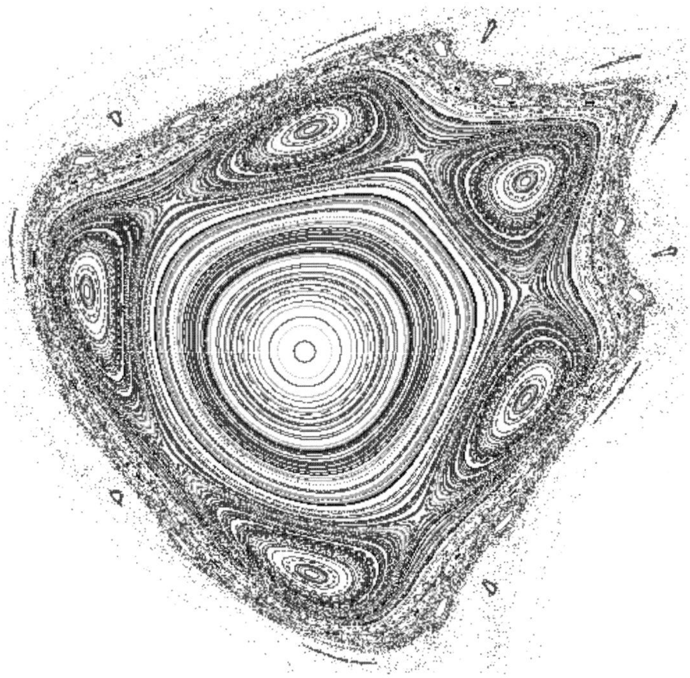

Fig. 2.21. Phase–plot of the *area–preserving generalized Hénon map*, simulated using *Dynamics Solver*^{TM}.

Complex Hénon Maps

In the time between Fatou and the present, most of the attention of those studying dynamical systems has been limited to maps in the real. This is somewhat surprising for two reasons. First, small perturbations of the coefficients of polynomial terms of real maps are liable to have large effects. For example, the number and periods of the periodic cycles may change. In the complex, the behavior is more uniform. Second, the major tools of complex analysis do not apply. These include the theory of normal families and the naturally contracting Poincaré metric together with the contracting map fixed–point Theorem. Recently, the maps $F_{a,c} : \mathbb{R}^2 \longrightarrow \mathbb{R}^2$, of the type

$$F_{a,c} : \begin{pmatrix} x \\ y \end{pmatrix} \mapsto \begin{pmatrix} x^2 + c - ay \\ x \end{pmatrix}, \qquad \text{with} \qquad a \neq 0,$$

have received much attention. Hénon first studied these maps numerically and they have become known as the *Hénon maps* [Hen69]. This family contains, up to conjugation, most of the most interesting of the simplest nonlinear polynomial maps of two variables. However, Hénon maps are still rather poorly understood and indeed the original question concerning the existence of a strange attractor for any values of the parameters is still unresolved today [Obe87, Obe00].

Despite the differences between the real and complex theories and the one variable and several variable theories, much of the development of the subject of complex analytic dynamics in several variables has been conceived through analogy. Recently, Hénon maps have started to be examined in the complex, that is, with both the variables and the parameters being complex. Also, there exist analogous maps of higher degree, called the *generalized Hénon maps*, of the form

$$\begin{pmatrix} x \\ y \end{pmatrix} \mapsto \begin{pmatrix} p(x) - ay \\ x \end{pmatrix},$$

where p is a polynomial of degree at least two and $a \neq 0$. Note that these are always invertible with inverses given by

$$\begin{pmatrix} x \\ y \end{pmatrix} \mapsto \begin{pmatrix} y \\ (p(y) - x)/a \end{pmatrix}.$$

For polynomials p of degree d, these maps are called *Hénon maps of degree d*.

Inspired by the definition of the *polynomial–like maps* [DH85], which was designed to capture the topological essence of polynomials on some disc or, more generally, on some open subset of $\mathbb{C}$ isomorphic to a disc, the *Hénon–like maps* have been defined in [Obe87, Obe00]. A polynomial–like map of degree d is a triple (U, U', f), where U and U' are open subsets of $\mathbb{C}$ isomorphic to discs, with U' relatively compact in U, and $f : U' \longrightarrow U$ analytic and proper of degree d. Note that it is convenient to think of polynomial–like of degree d as meaning an analytic map $f : U \longrightarrow \mathbb{C}$ such that $f(\partial U) \subset \mathbb{C} \setminus \overline{U}$ and $f|_{\partial U}$ of degree d (see Figure 2.22, which gives examples of the behavior, pictured in $\mathbb{R}^2$, that should be captured by the definition of Hénon–like maps of degree 2; in each case, the crescent-shaped region is the image of the square with A' the image of A, etc). It seems clear that the behaviors described by (a) and (b) versus (c) and (d) in Figure 2.22 must be described differently, albeit analogously, trading 'horizontal' for 'vertical' [Obe87, Obe00]. In the following, d will always be an arbitrary fixed integer greater than one. Let $\pi_1, \pi_2 : \mathbb{C}^2 \longrightarrow \mathbb{C}$ be the projections onto the first and second coordinates,

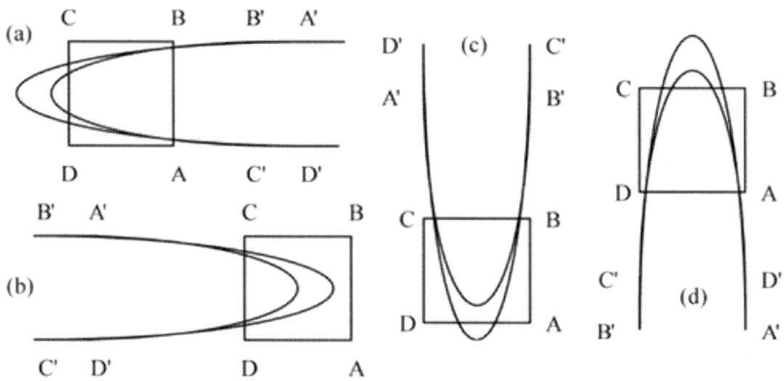

Fig. 2.22. The Hénon–like map of degree 2 (adapted from [Obe87, Obe00]).

respectively. We will consider a bidisc $B = D_1 \times D_2 \subset \mathbb{C}^2$, where $D_1, D_2 \subset \mathbb{C}$ are discs. Vertical and horizontal 'slices' of B are denoted respectively by

$$V_x = \{x\} \times D_2 \qquad \text{and} \qquad H_y = D_1 \times \{y\}, \qquad \text{for all} \quad x \in D_1, y \in D_2.$$

We will be considering maps of the bi–disc, $F : \overline{B} \to \mathbb{C}^2$, together with a map denoted by $F^{-1} : \overline{B} \to \mathbb{C}^2$, which is the inverse of F, where that makes sense. Now, for each $(x, y) \in B$, we can define

$$F_{1,y} = \pi_1 \circ F \circ (\mathrm{Id} \times y) : D_1 \to \mathbb{C}, \qquad F_{2,x}^{-1} = \pi_2 \circ F^{-1} \circ (x \times \mathrm{Id}) : D_2 \to \mathbb{C},$$
$$F_{2,x} = \pi_2 \circ F \circ (x \times \mathrm{Id}) : D_2 \to \mathbb{C}, \qquad F_{1,y}^{-1} = \pi_1 \circ F^{-1} \circ (\mathrm{Id} \times y) : D_1 \to \mathbb{C}.$$

The map $F : \overline{B} \to \mathbb{C}^2$ is a *Hénon-like map of degree d* if there exists a map $G : \overline{B} \to \mathbb{C}^2$ such that: (i) both F and G are injective and continuous on $\overline{B}$ and analytic on B, (ii) $F \circ G = \mathrm{Id}$ and $G \circ F = \mathrm{Id}$, where each makes sense; hence, we can rename G as F^{-1}, (iii) For all $x \in D_1$ and $y \in D_2$, either (i) $F_{1,y}$ and $F_{2,x}^{-1}$ are polynomial–like of degree d, or (ii) $F_{2,x}$ and $F_{1,y}^{-1}$ are polynomial-like of degree d. Depending on whether F satisfies condition (i) or (ii), we call it *horizontal* or *vertical* [Obe87, Obe00].

Now, let $\partial B_V = \partial \overline{D}_1 \times \overline{D}_2$ and $\partial B_H = \overline{D}_1 \times \partial \overline{D}_2$ be the 'vertical and horizontal boundaries'. If $F : \overline{B} \to \mathbb{C}^2$ is a Hénon–like map, then either

$$F(\partial B_V) \subset \mathbb{C}^2 \setminus \overline{B} \quad \text{and} \quad F^{-1}(\partial B_H) \subset \mathbb{C}^2 \setminus \overline{B} \quad \text{or}$$
$$F^{-1}(\partial B_V) \subset \mathbb{C}^2 \setminus \overline{B} \quad \text{and} \quad F(\partial B_H) \subset \mathbb{C}^2 \setminus \overline{B}.$$

This follows from the fact that the boundary of a polynomial-like map is mapped outside of the closure of its domain. Note that these are equivalent to

$$F(\partial B_V) \cap \overline{B} = \emptyset \quad \text{and} \quad F^{-1}(\partial B_H) \cap \overline{B} = \emptyset \quad \text{or}$$
$$F^{-1}(\partial B_V) \cap \overline{B} = \emptyset \quad \text{and} \quad F(\partial B_H) \cap \overline{B} = \emptyset.$$

The class of polynomial–like maps is stable under small perturbations [DH85] and the same is true for Hénon–like maps. More precisely, suppose $F : \overline{B} \to \mathbb{C}^2$ is Hénon–like of degree d. Let $H : \overline{B} \to \mathbb{C}^2$ be injective, continuous on $\overline{B}$, and analytic on B. If $\|H\|$ is sufficiently small, then $F + H$ is also Hénon–like of degree d.

For example, a simple computation shows that if F is the Hénon map (of degree 2) with parameters a and c, and D_R is the disc of radius R, with

$$R > (1/2)\left(1 + |a| + \sqrt{(1 + |a|)^2 + 4|c|}\right),$$

then $F : \overline{D}_R^2 \to \mathbb{C}^2$ is a horizontal Hénon–like map of degree 2. This R is exactly what is required so that $F(\partial\overline{D}_R \times \overline{D}_R) \cap \overline{D}_R{}^2 = \emptyset$ and $F^{-1}(\overline{D}_R \times \partial\overline{D}_R) \cap \overline{D}_R{}^2 = \emptyset$. Of course, their inverses are vertical Hénon–like maps.

More generally, consider the maps $G : \mathbb{C}^2 \to \mathbb{C}^2$ of the form

$$G : \begin{pmatrix} x \\ y \end{pmatrix} \mapsto \begin{pmatrix} x^d + c - ay \\ x \end{pmatrix},$$

with $a \neq 0$ and $d \geq 2$. When $d = 2$, we are back to the previous example. The lower bound on R came from solving the inequality

$$R^d - (1 + |a|)R - 4|c| > 0$$

for R. Note that when $d = 2$ we already had $R > 1$. Therefore, the same lower bound will work here as well. Of course, better lower bounds can be found.

Analogous to the invariant sets defined for Hénon maps, we can define the following sets for Hénon–like maps:

$$K_+ = \{\, z \in B | F^{\circ n}(z) \in B \quad \text{for all } n > 0\,\},$$
$$K_- = \{\, z \in B | F^{\circ -n}(z) \in B \quad \text{for all } n > 0\,\},$$
$$J_\pm = \partial K_\pm, \qquad K = K_+ \cap K_-, \qquad J = J_+ \cap J_-.$$

For every d, all Hénon–like maps of degree d have the same number of periodic cycles, counted with multiplicity, as a polynomial of degree d [Obe87, Obe00].

2.2.4 Smale Horseshoes

Real Horseshoes

Recall that the *Smale horseshoe map* (see Figure 2.23) *is any member of a class of chaotic maps of the square into itself.* This topological transformation provided a basis for understanding the chaotic properties of dynamical systems. Its basis are simple: A space is stretched in one direction, squeezed in another, and then folded. When the process is repeated, it produces something like a many–layered pastry dough, in which a pair of points that end up close together may have begun far apart, while two initially nearby points can end completely far apart.[7]

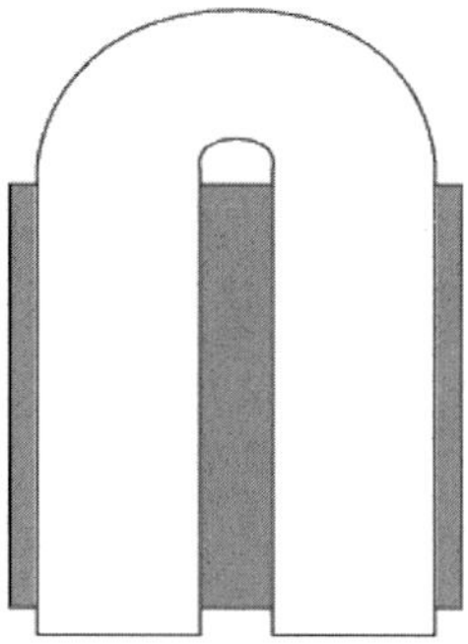

Fig. 2.23. The *Smale horseshoe map* consists of a sequence of operations on the unit square. First, stretch in the $y-$direction by more than a factor of two, then squeeze (compress) in the $x-$direction by more than a factor of two. Finally, fold the resulting rectangle and fit it back onto the square, overlapping at the top and bottom, and not quite reaching the ends to the left and right (and with a gap in the middle), as illustrated in the diagram. The shape of the stretched and folded map gives the horseshoe map its name. Note that it is vital to the construction process for the map to overlap and leave the middle and vertical edges of the initial unit square uncovered.

The horseshoe map was introduced by Smale while studying the behavior of the orbits of the *relaxation Van der Pol oscillator*. The action of the map is defined geometrically by squishing the square, then stretching the result into a long strip, and finally folding the strip into the shape of a horseshoe.

Most points eventually leave the square under the action of the map f. They go to the side caps where they will, under iteration, converge to a *fixed–point* in one of the caps. The points that remain in the square under repeated iteration form a *fractal set* and are part of the *invariant set* of the map f (see Figure 2.24).

The *stretching, folding* and *squeezing* of the horseshoe map are the essential elements that must be present in any chaotic system. In the horseshoe map the squeezing and stretching are uniform. They compensate each other so that the area of the square does not change. The folding is done neatly, so that the orbits that remain forever in the square can be simply described.

Repeating this generates the horseshoe attractor. If one looks at a cross section of the final structure, it is seen to correspond to a *Cantor set*.

---

[7] Originally, Smale had hoped to explain all dynamical systems in terms of *stretching* and *squeezing* – with no folding, at least no folding that would drastically undermine a system's stability. But *folding* turned out to be necessary, and folding allowed sharp changes in dynamical behavior (see [Gle87]).

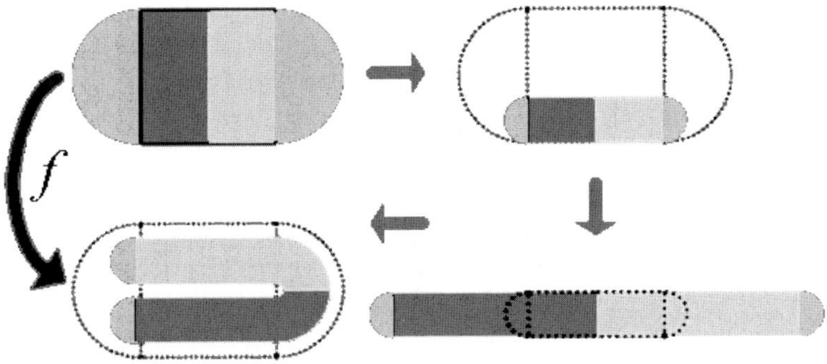

Fig. 2.24. The *Smale horseshoe map* f, defined by *stretching, folding* and *squeezing* of the system's phase–space.

The Smale horseshoe map is the set of basic topological operations for constructing an attractor consist of stretching (which gives sensitivity to initial conditions) and folding (which gives the attraction). Since *trajectories in phase–space cannot cross*, the repeated stretching and folding operations result in an object of great topological complexity. For any horseshoe map we have:

- There is an infinite number of periodic orbits;
- Periodic orbits of arbitrarily long period exist;
- The number or periodic orbits grows exponentially with the period; and
- Close to any point of the fractal invariant set there is a point of a periodic orbit.

More precisely, the horseshoe map f is a *diffeomorphism* defined from a region S of the plane into itself. The region S is a square capped by two semi–disks. The action of f is defined through the composition of three geometrically defined transformations. First the square is contracted along the vertical direction by a factor $a < 1/2$. The caps are contracted so as to remain semi-disks attached to the resulting rectangle. Contracting by a factor smaller than one half assures that there will be a gap between the branches of the horseshoe. Next the rectangle is stretched by a factor of $1/a$; the caps remain unchanged. Finally the resulting strip is folded into a horseshoe–shape and placed back into S.

The interesting part of the dynamics is the image of the square into itself. Once that part is defined, the map can be extended to a diffeomorphism by defining its action on the caps. The caps are made to contract and eventually map inside one of the caps (the left one in the figure). The extension of f to the caps adds a fixed–point to the *non–wandering set* of the map. To keep the class of horseshoe maps simple, the curved region of the horseshoe should not map back into the square.

The horseshoe map is one–to–one (1–1, or injection): any point in the domain has a unique image, even though not all points of the domain are the image of a point. The inverse of the horseshoe map, denoted by f^{-1}, cannot have as its domain the entire region S, instead it must be restricted to the image of S under f, that is, the domain of f^{-1} is $f(S)$.

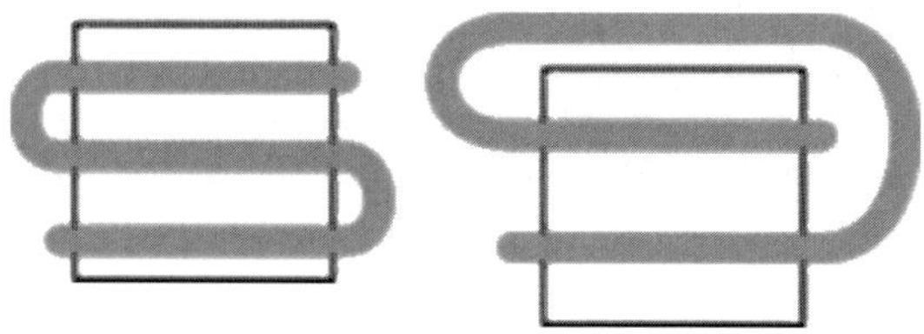

Fig. 2.25. Other types of horseshoe maps can be made by folding the contracted and stretched square in different ways.

By folding the contracted and stretched square in different ways, other types of horseshoe maps are possible (see Figure 2.25). The contracted square cannot overlap itself to assure that it remains 1–1. When the action on the square is extended to a diffeomorphism, the extension cannot always be done on the plane. For example, the map on the right needs to be extended to a diffeomorphism of the sphere by using a 'cap' that wraps around the equator.

The horseshoe map is an Axiom A diffeomorphism that serves as a model for the general behavior at a transverse *homoclinic point*, where the *stable and unstable manifolds* of a periodic point intersect.

The horseshoe map was designed by Smale to reproduce the chaotic dynamics of a *flow* in the neighborhood of a given periodic *orbit*. The neighborhood is chosen to be a small disk perpendicular to the orbit. As the system evolves, points in this disk remain close to the given periodic orbit, tracing out orbits that eventually intersect the disk once again. Other orbits diverge.

The behavior of all the orbits in the disk can be determined by considering what happens to the disk. The intersection of the disk with the given periodic orbit comes back to itself every period of the orbit and so do points in its neighborhood. When this neighborhood returns, its shape is transformed. Among the points back inside the disk are some points that will leave the disk neighborhood and others that will continue to return. The set of points that never leaves the neighborhood of the given periodic orbit form a fractal.

A symbolic name can be given to all the orbits that remain in the neighborhood. The initial neighborhood disk can be divided into a small number of regions. Knowing the sequence in which the orbit visits these regions allows the orbit to be pinpointed exactly. The visitation sequence of the orbits provide the so–called *symbolic dynamics*[8]

[8] Symbolic dynamics is the practice of modelling a dynamical system by a space consisting of infinite sequences of abstract symbols, each sequence corresponding

It is possible to describe the behavior of all initial conditions of the horseshoe map. An initial point $u_0 = x, y$ gets mapped into the point $u_1 = f(u_0)$. Its iterate is the point $u_2 = f(u_1) = f^2(u_0)$, and repeated iteration generates the orbit $u_0, u_1, u_2, \ldots$ Under repeated iteration of the horseshoe map, most orbits end up at the fixed–point in the left cap. This is because the horseshoe maps the left cap into itself by an *affine transformation*, which has exactly one fixed–point. Any orbit that lands on the left cap never leaves it and converges to the fixed–point in the left cap under iteration. Points in the right cap get mapped into the left cap on the next iteration, and most points in the square get mapped into the caps. Under iteration, most points will be part of orbits that converge to the fixed–point in the left cap, but some points of the square never leave.

Under forward iterations of the horseshoe map, the original square gets mapped into a series of horizontal strips. The points in these horizontal strips come from vertical strips in the original square. Let S_0 be the original square, map it forward n times, and consider only the points that fall back into the square S_0, which is a set of horizontal stripes $H_n = f^n(S_0) \cap S_0$. The points in the horizontal stripes came from the vertical stripes $V_n = f^{-n}(H_n)$, which are the horizontal strips H_n mapped backwards n times. That is, a point in V_n will, under n iterations of the horseshoe map, end up in the set H_n of vertical strips (see Figure 2.26).

Now, if a point is to remain indefinitely in the square, then it must belong to an *invariant set* Λ that maps to itself. Whether this set is empty or not has to be determined. The vertical strips V_1 map into the horizontal strips H_1, but not all points of V_1 map back into V_1. Only the points in the intersection of V_1 and H_1 may belong to Λ, as can be checked by following points outside the intersection for one more iteration. The intersection of the horizontal and vertical stripes, $H_n \cap V_n$, are squares that converge in the limit $n \to \infty$ to the invariant set Λ (see Figure 2.27).

The structure of invariant set Λ can be better understood by introducing a system of labels for all the intersections, namely a *symbolic dynamics*. The intersection $H_n \cap V_n$ is contained in V_1. So any point that is in Λ under iteration must land in the left vertical strip A of V_1, or on the right vertical strip B. The lower horizontal strip of H_1 is the image of A and the upper horizontal strip is the image of B, so $H_1 = f(A) \cap f(B)$. The strips A and B can be used to label the four squares in the intersection of V_1 and H_1 (see Figure 2.28) as:

---

to a state of the system, and a shift operator corresponding to the dynamics. Symbolic dynamics originated as a method to study general dynamical systems, now though, its techniques and ideas have found significant applications in data storage and transmission, linear algebra, the motions of the planets and many other areas. The distinct feature in symbolic dynamics is that time is measured in discrete intervals. So at each time interval the system is in a particular state. Each state is associated with a symbol and the evolution of the system is described by an infinite sequence of symbols (see text below).

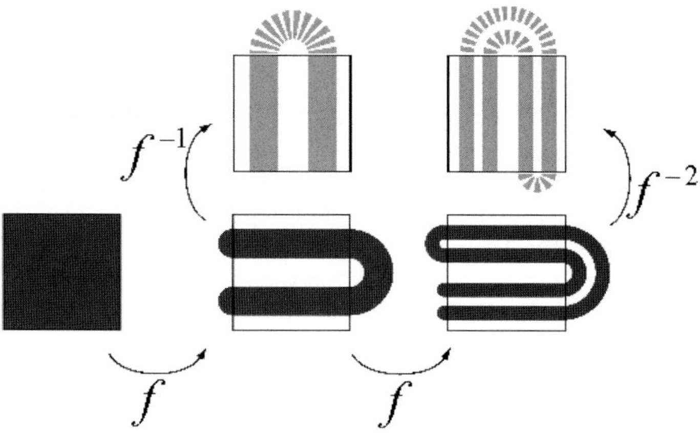

Fig. 2.26. Iterated horseshoe map: pre–images of the square region.

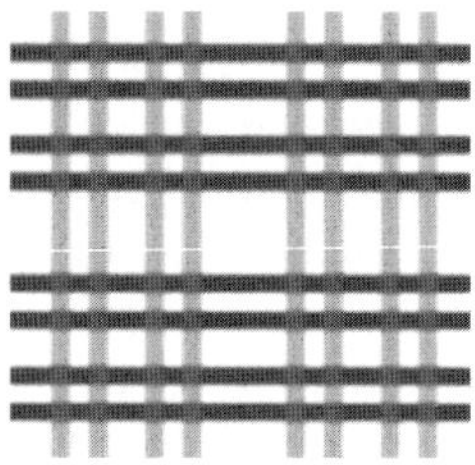

Fig. 2.27. Intersections that converge to the invariant set Λ.

$$\Lambda_{A\bullet A} = f(A) \cap A, \qquad \Lambda_{A\bullet B} = f(A) \cap B,$$
$$\Lambda_{B\bullet A} = f(B) \cap A, \qquad \Lambda_{B\bullet B} = f(B) \cap B.$$

The set $\Lambda_{B\bullet A}$ consist of points from strip A that were in strip B in the previous iteration. A dot is used to separate the region the point of an orbit is in from the region the point came from.

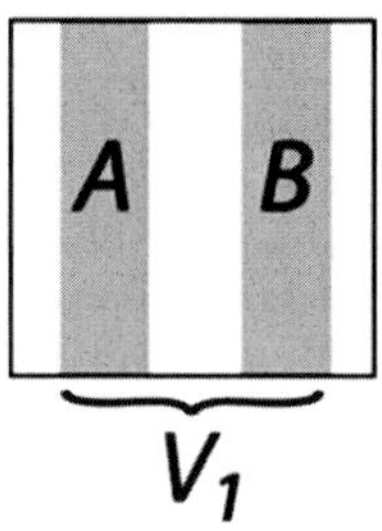

Fig. 2.28. The basic domains of the horseshoe map in symbolic dynamics.

This notation can be extended to higher iterates of the horseshoe map. The vertical strips can be named according to the sequence of visits to strip A or strip B. For example, the set $ABB \subset V_3$ consists of the points from A that will all land in B in one iteration and remain in B in the iteration after that:

$$ABB = \{x \in A | f(x) \in B \text{ and } f^2(x) \in B\}.$$

Working backwards from that trajectory determines a small region, the set ABB, within V_3.

The horizontal strips are named from their vertical strip pre–images. In this notation, the intersection of V_2 and H_2 consists of 16 squares, one of which is

$$\Lambda_{AB\bullet BB} = f^2(AB) \cap BB.$$

All the points in $\Lambda_{AB\bullet BB}$ are in B and will continue to be in B for at least one more iteration. Their previous trajectory before landing in BB was A followed by B.

Any one of the intersections $\Lambda_{P\bullet F}$ of a horizontal strip with a vertical strip, where P and F are sequences of As and Bs, is an affine transformation of a small region in V_1. If P has k symbols in it, and if $f^{-k}(\Lambda_{P\bullet F})$ and $\Lambda_{P\bullet F}$ intersect, then the region $\Lambda_{P\bullet F}$ will have a *fixed–point*. This happens when the sequence P is the same as F. For example, $\Lambda_{ABAB\bullet ABAB} \subset V_4 \cap H_4$ has at least one fixed–point. This point is also the same as the fixed–point in $\Lambda_{AB\bullet AB}$. By including more and more ABs in the P and F part of the label of intersection, the area of the intersection can be made as small as needed. It converges to a point that is part of a *periodic orbit of the horseshoe map*. The periodic orbit can be labelled by the simplest sequence of As and Bs that labels one of the regions the periodic orbit visits. For every sequence of As and Bs there is a periodic orbit.

The Smale horseshoe map is the same topological structure as the *homoclinic tangle*. To dynamically introduce homoclinic tangles, let us consider a classical engineering problem of *escape from a potential well.* Namely, if we have a motion, $x = x(t)$, of a damped particle in a well with potential energy $V = x^2/2 - x^3/3$ (see Figure 2.29) excited by a periodic driving force, $F\cos(wt)$ (with the period $T = 2\pi/w$), we are dealing with a nonlinear dynamical system given by [TS01]

$$\ddot{x} + a\dot{x} + x - x^2 = F\cos(wt). \tag{2.26}$$

Now, if the driving is switched off, i.e., $F = 0$, we have an autonomous 2D–system with the phase–portrait (and the safe basin of attraction) given in Figure 2.29 (below). The grey area of escape starts over the hilltop to infinity. Once we start driving, the system (2.26) becomes 3–dimensional, with its 3D phase–space. We need to see the basin in a *stroboscopic section* (see Figure 2.30). The hill–top solution still has an inset and and outset. As the driving increases, the inset and outset get tangled. They intersect one another an

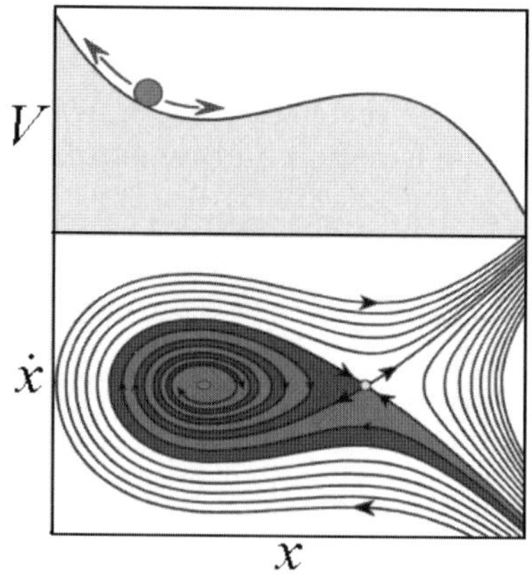

Fig. 2.29. Motion of a damped particle in a potential well, driven by a periodic force $F\cos(wt)$,. Up: potential $(x - V)$–plot, with $V = x^2/2 - x^3/3$; down: the corresponding phase $(x - \dot{x})$–portrait, showing the safe basin of attraction – if the driving is switched off $(F = 0)$.

infinite number of times. The boundary of the safe basin becomes fractal. As the driving increases even more, the so–called fractal–fingers created by the homoclinic tangling, make a sudden incursion into the safe basin. At that point, the integrity of the in–well motions is lost [TS01].

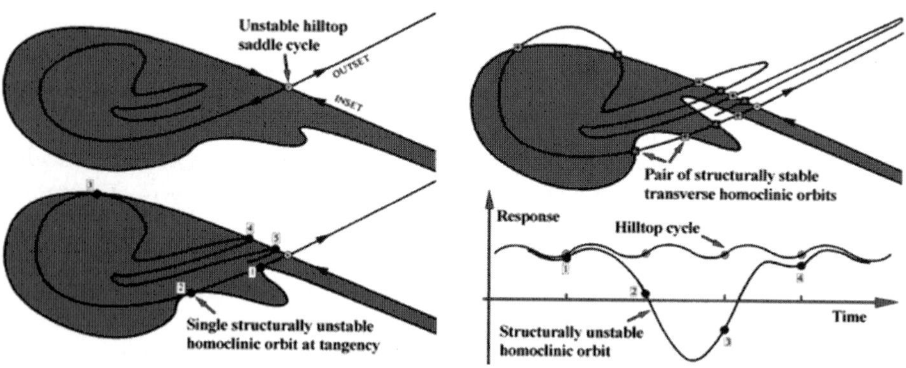

Fig. 2.30. Dynamics of a homoclinic tangle. The hill–top solution of a damped particle in a potential well driven by a periodic force. As the driving increases, the inset and outset get tangled.

Now, topologically speaking (referring to the Figure 2.31), let X be the point of intersection, with X' ahead of X on one manifold and ahead of X'' of the other. The map of each of these points TX' and TX'' must be ahead of the map of X, TX. The only way this can happen is if the manifold loops back and crosses itself at a new *homoclinic point*, i.e., a point where a stable and an unstable separatrix (invariant manifold) from the same fixed–point or same family intersect. Another loop must be formed, with T^2X another homoclinic point. Since T^2X is closer to the hyperbolic point than TX, the distance between T^2X and TX is less than that between X and TX. Area preservation requires the area to remain the same, so each new curve (which is closer than the previous one) must extend further. In effect, the loops become longer and thinner. The network of curves leading to a dense area of homoclinic points is known as a homoclinic tangle or tendril. Homoclinic points appear where chaotic regions touch in a *hyperbolic fixed–point*.

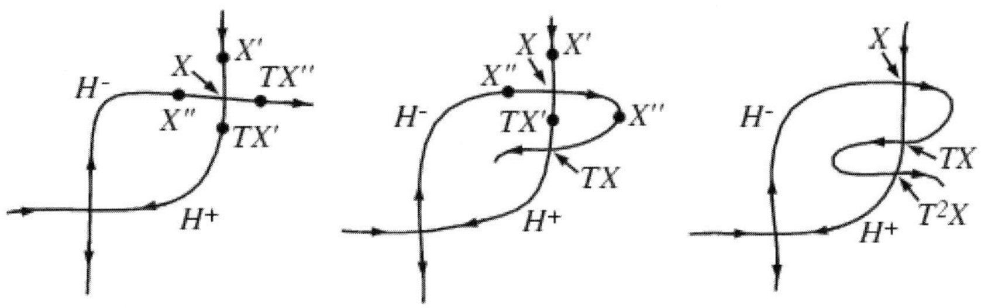

Fig. 2.31. More on homoclinic tangle (see text for explanation).

Complex Horseshoes

Here, following [Obe87, Obe00], we define and analyze complex analogs of Smale horseshoes. Using a criterion analogous to the one given by Moser [Mos73] in the real case, we will show that many Hénon maps are complex horseshoes. In particular, actual Hénon maps (of degree 2) are complex horseshoes when $|c|$ is sufficiently large.

Now, recall that in Figure 2.22 above, only (a) and (c) appear to be horseshoes. Basically, we would like to say that a horizontal Hénon–like map F of degree d is a complex horseshoe of degree d if the projections

$$\pi_1 : \bigcap_{0\leq m\leq n} F^{\circ m}(B) \to \mathbb{C} \qquad \text{and} \qquad \pi_2 : \bigcap_{0\leq m\leq n} F^{\circ -m}(B) \to \mathbb{C}$$

are *trivial fibrations* with fibers disjoint unions of d^n discs (see [II06b]). However, this is not general enough for our purpose here, so we give a definition with weaker conditions, which encompasses the Hénon-like maps defined above. Instead of requiring B to be an actual bi–disc, B may be an embedded bi–disc. More precisely, letting $D \subset \mathbb{C}$ be the open unit disc, assume that there is an embedding, $\varphi : \overline{D}^2 \to \mathbb{C}^2$, which is analytic on D^2 and such that $B = \varphi(D^2)$ and, naturally, $\overline{B} = \varphi(\overline{D}^2)$. By $\partial B_H = \varphi(\overline{D} \times \partial\overline{D})$

and $\partial B_V = \varphi(\partial\overline{D} \times \overline{D})$ we denote the horizontal and vertical boundaries of B. Also, define horizontal and vertical slices $H_y = \varphi(D \times \{y\})$ and $V_x = \varphi(\{x\} \times D)$ for all $x, y \in D$.

Consider the maps $F : \overline{B} \to \mathbb{C}^2$ which are injective and continuous on $\overline{B}$ and analytic on B, and such that either

$$\begin{array}{llll} F(\overline{B}) \cap \partial B_H = \emptyset & \text{and} & \overline{B} \cap F(\partial B_V) = \emptyset, & \text{or} \\ \overline{B} \cap F(\partial B_H) = \emptyset & \text{and} & F(\overline{B}) \cap \partial B_V = \emptyset. & \end{array}$$

Under these conditions, for all $y \in D$,

$$\pi_1 \circ \varphi^{-1} : F(H_y) \cap B \to D$$

is a proper map [Obe87, Obe00]. Such a proper map has a degree and since the degree is integer–valued and continuous in y, this defines a constant, the degree of such a map F. Now a class of maps generalizing the Hénon–like maps can be defined as follows.

$$F : \overline{B} \to \mathbb{C}^2$$

is a *quasi–Hénon–like map of degree* d if there exists a map $G : \overline{B} \to \mathbb{C}^2$ such that: (i) both F and G are injective and continuous on $\overline{B}$ and analytic on B, (ii) $F \circ G = \text{Id}$ and $G \circ F = \text{Id}$, where each makes sense; therefore, we can rename G as F^{-1}, and (iii) We have either

$$\begin{array}{llll} F(\overline{B}) \cap \partial B_H = \emptyset & \text{and} & \overline{B} \cap F(\partial B_V) = \emptyset, & \text{or} \\ \cap F(\partial B_H) = \emptyset & \text{and} & F(\overline{B}) \cap \partial B_V = \emptyset. & \end{array}$$

The degree is $d \geq 2$. Moreover, call F either *horizontal* or *vertical* according to whether it satisfies (a) or (b), respectively. Also, Hénon–like maps of degree d are quasi–Hénon–like of degree d.

Now, using the notion of quasi–Hénon–like maps, complex horseshoes may be defined as follows [Obe87, Obe00]. A *complex horseshoe of degree* d is a quasi–Hénon–like map of degree d, $F : \overline{B} \to \mathbb{C}^2$, such that, for all integers $n > 0$, depending on if F is horizontal or vertical, then either the projections

$$\pi_1 \circ \varphi^{-1} : \bigcap_{0 \leq m \leq n} F^{\circ m}(B) \to \mathbb{C} \quad \text{and} \quad \pi_2 \circ \varphi^{-1} : \bigcap_{0 \leq m \leq n} F^{\circ -m}(B) \to \mathbb{C}$$

or

$$\pi_2 \circ \varphi^{-1} : \bigcap_{0 \leq m \leq n} F^{\circ m}(B) \to \mathbb{C} \quad \text{and} \quad \pi_1 \circ \varphi^{-1} : \bigcap_{0 \leq m \leq n} F^{\circ -m}(B) \to \mathbb{C},$$

respectively, are trivial fibrations with fibers disjoint unions of d^n discs.

In the context of Hénon–like maps, the following results will show the close relation between complex horseshoe maps of degree d and polynomial-like maps of degree d whose critical points escape immediately. The following definition, borrowed from Moser [Mos73], is the key tool in this study of complex horseshoes [Ale68]. Let M be a differentiable manifold, $U \subset M$ an open subset, and $f : U \to M$ a differentiable map. A field of cones $\mathcal{C} = (C_x \subset T_x M)_{x \in U}$ on U is an f*-trapping field* if: (i) $C_{\mathbf{x}}$ depends continuously on x, and (ii) whenever $x \in U$ and $f(x) \in U$, then $d_x f(C_x) \subset C_{f(x)}$.

Now we consider the connection between trapping fields of cones and complex horseshoes. Let $F : \overline{B} \to \mathbb{C}^2$ be a quasi–Hénon–like map of degree d. The following are equivalent: (i) $F : \overline{B} \to \mathbb{C}^2$ is a complex horseshoe of degree d, (ii) there exist continuous, positive functions $\alpha(\mathbf{z})$ and $\beta(\mathbf{z})$ on B such that the field of cones $C_z = \{ (\xi_1, \xi_2) : |\xi_2| < \alpha(z)|\xi_1| \}$ is F–trapping and the field of cones $C'_z = \{ (\xi_1, \xi_2) : |\xi_1| < \beta(z)|\xi_2| \}$ is F^{-1}–trapping, and (iii) $F(\overline{B}) \cap \overline{B}$ and $F^{-1}(\overline{B}) \cap \overline{B}$ both have d connected components. Note that (ii) $\Rightarrow$ (i) is borrowed from Moser and that the implication (i) $\Rightarrow$ (ii) is what the contractive nature of complex analytic maps gives us for free. (iii) arises naturally in the proof of (ii) $\Rightarrow$ (i). When considering a map which is actually Hénon–like, consideration of the critical points of $F_{1,y}$ and $F_{2,x}^{-1}$ (or $F_{2,x}$ and $F_{1,y}^{-1}$) in light of the equivalences above yields the following. Let $F : \overline{D_1} \times \overline{D_2} \to \mathbb{C}^2$ be a Hénon–like map of degree d. The following are equivalent: (i) $F : \overline{D_1} \times \overline{D_2} \to \mathbb{C}^2$ is a complex horseshoe of degree d, and (ii) for all $(x, y) \in \overline{D_1} \times \overline{D_2}$, the critical values of the polynomial–like maps $F_{1,y}$ and $F_{2,x}^{-1}$ (or $F_{2,x}$ and $F_{1,y}^{-1}$) lie outside of $\overline{D_1}$ and $\overline{D_2}$, respectively. The diameters of the discs in the fibers above tend to 0 with n [Obe87, Obe00].

This criterion can be used to show that for each a there exists $r(a)$ such that if $|c| > r(a)$, then the Hénon map $F_{a,c}$ is a complex horseshoe. Of course, in the real locus this was known [DN79, New80], except that then c has to be taken very negative. When c is large and positive, all the 'horseshoe behavior' is complex. More precisely, For each $a \neq 0$ and each c such that $|c| > \left(5/4 + \sqrt{5}/2\right)(1 + |a|)^2$, there exists an R such that $F_{a,c} : \overline{D_R}^2 \to \mathbb{C}^2$ is a complex horseshoe. The key to showing that the former field of cones is F–trapping is the observation that $F(x, y) \in \overline{D_R}^2$ implies that $|x^2 + c - ay| \leq R$ which implies that

$$|x|^2 \geq |c| - R(1 + |a|).$$

This result says essentially everything about Hénon maps in the parameter range to which it applies [Obe87, Obe00].

Now, suppose that $F : \mathbb{C}^2 \to \mathbb{C}^2$ is an analytic map. Recall that a point q is a *homoclinic point* of F if there exists a positive integer k such that

$$\lim_{n \to \infty} F^{\circ kn}(q) = \lim_{n \to \infty} F^{\circ -kn}(q) = p, \tag{2.27}$$

where the limits exist. Note that the limit point, p, in (2.27) is a hyperbolic periodic point of F of period least k satisfying (2.27). To rephrase this definition, q is in both the *stable and unstable manifolds* of F at p, which are denoted by W^s and W^u, respectively. We call a homoclinic point *transversal*, if these

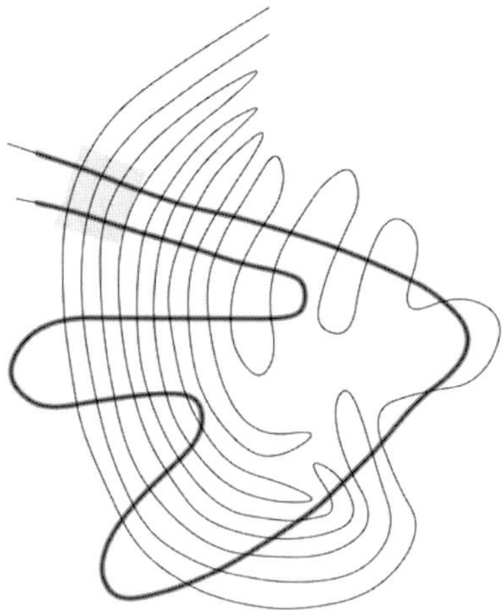

Fig. 2.32. Smale horseshoes from transverse homoclinic points.

invariant manifolds intersect transversally. Also, note that the invariant manifolds W^s and W^u tend to intersect in lots of points in $\mathbb{C}^2$ unless F is linear or affine. This is quite different from the case of these invariant manifolds in $\mathbb{R}^2$. These intersections are almost always transversal. In the real domain, Smale showed that in a neighborhood of a transversal homoclinic point there exist horseshoes (see Figure 2.32). Here we give an analogous result in the case of complex horseshoes: for every positive integer $d \geq 2$, there exists an embedded bi–disc, B_d, centered at p and a positive integer $N = N(d)$ such that $F^{\circ kN} : B_d \to \mathbb{C}^2$ is a complex horseshoe of degree d. There exist $D_s \subset W^s$ and $D_u \subset W^u$ isomorphic to discs with $D_s, D_u \subset U$ and a positive integer n such that $F^{\circ n}(D_u)$ intersects D_s in exactly d points and the following conditions hold

$$F^{\circ n}(D_u) \cap \partial\overline{D_s} = \emptyset \qquad \text{and} \qquad F^{\circ n}(\partial\overline{D_u}) \cap D_s = \emptyset.$$

Note that the set $F^{\circ n}(D_u) \cap D_s$ must consist of finitely many points for all positive integers n. If $F^{\circ n}(D_u) \cap \partial\overline{D_s} \neq \emptyset$, then take D_s to be slightly smaller. If $F^{\circ n}(\partial\overline{D_u}) \cap D_s \neq \emptyset$, then take D_u to be slightly smaller. In either case, make the adjustments to keep the same number of points in $F^{\circ n}(D_u) \cap D_s$. If there are more than d points in $F^{\circ n}(D_u) \cap D_s$, then deform D_u slightly, making it smaller, to exclude some points from $F^{\circ n}(D_u) \cap D_s$. This can be done in an orderly way. In particular, there is a natural metric on D_s – –-the Poincaré metric of a disc and consider the point x of $F^{\circ n}(D_u) \cap D_s$ which is furthest from p in this metric (or one such point if more than one has this property). Now deform D_u by taking out $F^{\circ -n}(x)$ and staying clear of the other preimages of points in $F^{\circ n}(D_u) \cap D_s$ [Obe87, Obe00].

There exists a nonnegative integer m such that, if $D_{u,m} = F^{\circ -m}(D_u)$ and $U_m = D_{u,m} \times D_s$, then $F^{\circ n+m}|_{U_m}$ is a Hénon–like map of degree d.

3

Complex Quantum Dynamics

In this Chapter we present the essence of quantum dynamics in a complex Hilbert space, mainly using the quantum formalism of P.A.M. Dirac.

3.1 Non–Relativistic Quantum Mechanics

Recall that *Heisenberg*, with his discovery of quantum mechanics (1925; see [Cas92]), introduced a *new outlook* on the nature of physical theory. Previously, it was always considered essential that there should be a detailed description of what is taking place in natural phenomena, and one used this description to calculate results comparable with experiment. Heisenberg put forward the view that it is sufficient to have a mathematical scheme from which one can calculate in a consistent manner the results of all experiments. That is, a detailed description in the traditional sense is unnecessary and may very well be impossible to establish [Dir28a, Dir28b, Dir26e].

Heisenberg's method focuses attention on the quantities which enter into experimental results. It was first applied to the *spectral theory*, for which these quantities are the energy levels of the atomic system and certain probability coefficients, which determine the probability of a radiative transition taking place from one level to another. The method sets up equations connecting these quantities and allows one to calculate them, but does not go beyond this. It does not provide any description of radiative transition processes. It does not even allow one to deduce how the results of a calculation are to be used, but requires one to assume *Einstein's laws of radiation* (the laws which tell how the probability of a radiative transition process depends on the intensity of the incident radiation), and to assume that certain quantities determined by the calculation are the coefficients appearing in the laws.

Shortly after Heisenberg's discovery, *Schrödinger* set up independently another form of quantum mechanics (1926; see [Moo89]), which also enables one to calculate energy levels and probability coefficients and gives results agreeing with those of Heisenberg, but which introduces an important new feature.

It connects together, in one calculation, a *set of probability coefficients* that act together under certain conditions in Nature; e.g., the set of probability coefficients referring to *transitions* from one particular initial state to any final state. In this respect, *Schrödinger's method* is to be contrasted with Heisenberg's method, which connects together in one calculation all the probability coefficients for a dynamical system, i.e., *the probability coefficients from all initial states to all final states.*

This feature of Schrödinger's method gives it two important advantages [Dir25, Dir26e]. First, as a consequence of its enabling one to get fewer results at a time, it makes the computation much simpler. Secondly, it supplies, in a certain sense, a description of what is taking place in Nature, since a calculation leading to results that come into play together under certain conditions in Nature will be in close correspondence with the physical process that is taking place under those conditions, various points in the calculation having their counterparts in the physical process. A description in this limited sense seems to be the most that is possible for atomic processes. It implies a much less complete connection between the mathematics and the physics than one has in classical mechanics, and one might be disinclined to call it a description at all, but one may at least consider it as an appropriate generalization of what one usually means by a description. On account of Schrödinger's method allowing a description in this new sense while Heisenberg's allows none, Schrödinger's method introduces an outlook on the nature of physical theory intermediate between Heisenberg's and the old classical (Newton–Maxwellian) one.

When Heisenberg's and Schrödinger's theories were developed it was soon found by Dirac that they both rested on the same mathematical formalism and differed only with regard to the method of physical interpretation (see [Dir49]). *Dirac's formalism* is a generalization of the Hamiltonian form of classical Newtonian dynamics, involving linear operators instead of ordinary algebraic variables, and is so natural and beautiful as to make one feel sure of its correctness as the foundation of the theory. The question of its interpretation, however, which involved unifying Heisenberg's and Schrödinger's ideas into a satisfactory comprehensive scheme, was not so easily settled.

The situation of a formalism (in this case, Dirac's) becoming established before one is clear about its interpretation should not be considered as surprising, but rather as a natural consequence of the drastic alterations which the development of physics had required in some of the basic physical concepts. This made it an easier matter to discover the mathematical formalism needed for a fundamental physical theory than its interpretation, since the number of things one had to choose between in discovering the formalism was very limited, the number of fundamental ideas in pure mathematics being not very great, while with the interpretation most unexpected things might turn up.

The best way of seeking the interpretation in such cases is probably from a discussion of simple examples. This way was used for the theory of quantum mechanics and led eventually to a satisfactory interpretation applicable to all phenomena for which relativistic effects are negligible. This interpretation is more closely connected with Schrödinger's method than Heisenberg's, as one would expect on account of the former affording in some sense a description of Nature, and is centered round a *Schrödinger's wave ψ–function*, which is one of the things that can be operated on by the linear operators which the dynamical variables have become. The correspondence which the existence of a description implies between the mathematics and the physics makes a wave ψ–function correspond to a state of motion of the atomic system, in such a way that, for example, a calculation which gives the transition probabilities from a particular initial state to any final state would be based on that wave ψ–function which represents the motion ensuing from this initial state. A wave ψ–function is a complex function $\psi = \psi(q_1, q_2, ..., q_n, t)$ of all the coordinates $q_1, q_2, ..., q_n, t$ of the system and of the time t, and it receives the interpretation that the square of its modulus, $|\psi(q_1, q_2, ..., q_n, t)|^2$, is the *probability*, for the state of motion it corresponds to, of the coordinates having values in the neighborhood of $q_1, q_2, ..., q_n$, per unit volume of coordinate space (or, configuration space), at the time t.

A wave ψ–function can be transformed so as to refer to other dynamical variables, for example, the momenta $p_1, p_2, ..., p_n$, when it is said to be in another representation. The square of its modulus $|\psi(p_1, p_2, ..., p_n, t)|^2$ is then the *probability*, per unit volume of momentum space (or, phase–space), of the momenta having values in the neighborhood of $p_1, p_2, ..., p_n$ at the time t. A wave ψ–function itself never has an interpretation, but only the square of its modulus, and the need for distinguishing between two wave functions having the same squares of their moduli arises only because, if they are transformed to a different representation, the squares of their moduli will in general become different. This brings out the *incompleteness of description*, which is possible with quantum mechanics [Dir28a, Dir28b, Dir26e, Dir49].

One may make a slight modification in the wave functions in any representation by introducing a weight factor λ and arranging for the probability to be $\lambda|\psi|^2$ instead of $|\psi|^2$. The weight factor may be any positive function of the variables occurring in the wave ψ–function.

Wave functions have to satisfy a certain *wave equation*, namely, the equation

$$i\hbar\, \partial_t \psi = H\psi, \tag{3.1}$$

where $\partial_t \equiv \partial/\partial_t$, $i = \sqrt{-1}$, $\hbar$ is the *Planck's constant*, and H is a *Hermitian (self–adjoint) linear operator* representing the Hamiltonian of the system (expressed in the representation concerned). The wave equation (3.1) is a generalization of the *Hamilton–Jacobi equation* of classical mechanics. If S is a solution of the latter equation, then

$$\psi = \mathrm{e}^{iS/\hbar} \tag{3.2}$$

will give a first approximation to a solution of the former.

An important property of the wave equation (3.1) is that it yields the *probability conservation law*: the total probability of the variables occurring in the wave ψ–function having any value is constant. The wave ψ–function should be *normalized* so as to make this probability initially unity and then it always remains unity. This conservation law is a mathematical consequence of the wave equation being linear in the operator ∂_t and of H being a self–adjoint operator.

The wave equation is linear and homogeneous in the wave ψ–function and so are the transformation equations. In consequence, one can add together two ψ's and get a third. The correspondence between ψ's and states of motion now allows one to infer that there is a relationship between the states of motion, such that one can add or superpose two states to get a third. This relationship constitutes the *Principle of superposition of states*, one of the general principles governing the interpretation of quantum mechanics.

Another of these principles is *Heisenberg's Principle of indeterminacy*. This is a consequence of the transformation laws connecting $\psi(q)$ and $\psi(p)$, which show that each of these functions is the *Fourier transform* of the other, apart from numerical coefficients, so that one meets the same limitations in giving values to a q and p as in giving values to the position and frequency of a train of waves [Dir26e, Dir49]. These general principles serve to bring out the departures needed from ordinary classical (Newton–Maxwellian) ideas. They are of so drastic and unexpected a nature that it is not to be wondered at that they were discovered only indirectly, as consequences of a previously established mathematical scheme, instead of being built up directly from experimental facts.

3.1.1 Dirac's Canonical Quantization

To make a leap into the quantum realm, recall that classical state–space for the biodynamic system of n point–particles is its $6ND$ phase–space $\mathcal{P}$, including all position and momentum vectors, $\mathbf{r}_i = (x, y, z)_i$ and $\mathbf{p}_i = (p_x, p_y, p_z)_i$, respectively, for $i = 1, ..., n$.

The *quantization* is performed as a *linear representation* of the real Lie algebra $\mathcal{L}_P$ of the phase–space $\mathcal{P}$, defined by the Poisson bracket $\{f, g\}$ of classical variables f, g – into the corresponding real Lie algebra $\mathcal{L}_H$ of the Hilbert space $\mathcal{H}$, defined by the commutator $[\hat{f}, \hat{g}]$ of skew–Hermitian operators $\hat{f}, \hat{g}$. This sounds like a functor, however it is not; as J. Baez says, 'First quantization is a mystery, but second quantization is a functor'. Mathematically, if quantization were *natural* it would be a functor from the category **Symplec**, whose objects are symplectic manifolds (i.e., phase–spaces) and whose morphisms are symplectic maps (i.e., canonical transformations) to the category **Hilbert**, whose objects are Hilbert spaces and whose morphisms are unitary operators.

Historically first, the so–called *canonical quantization* is based on the so–called *Dirac rules for quantization*. It is applied to 'simple' systems: finite number of degrees–of–freedom and 'flat' classical phase–spaces (an open set of $\mathbb{R}^{2n}$). Canonical quantization includes the following data [Dir49]:

1. *Classical description.* The system is described by the *Hamiltonian* or *canonical formalism*: its classical phase–space is locally coordinated by a set of *canonical coordinates* (q^j, p_j), the *position* and *momentum* coordinates. Classical observables are real functions $f(q^j, p_j)$. Eventually, a Lie group G of symmetries acts on the system.
2. *Quantum description.* The quantum phase–space is a complex Hilbert space $\mathcal{H}$. Quantum observables are Hermitian (i.e., self–adjoint) operators acting on $\mathcal{H}$. (The Hilbert space is complex in order to take into account the interference phenomena of wave functions representing the quantum states. The operators are self–adjoint in order to assure their eigenvalues are real.) The symmetries of the system are realized by a group of unitary operators $U_G(\mathcal{H})$.
3. *Quantization method.* As a Hilbert space we take the space of square integrable complex functions of the configuration space; that is, functions depending only on the position coordinates, $\psi(q^j)$. The quantum operator associated with $f(q^j, p_j)$ is obtained by replacing p_j by $-i\hbar\frac{\partial}{\partial q^j}$, and hence we have the correspondence $f(q^j, p_j) \mapsto \hat{f}(q^j, -i\hbar\frac{\partial}{\partial q^j})$. In this way, the classical commutation rules between the canonical coordinates are assured to have a quantum counterpart: the commutation rules between the quantum operators of position and momentum (which are related to the 'uncertainty principle' of quantum mechanics).

3.1.2 Quantum States and Operators

Quantum systems have two modes of evolution in time. The first, governed by standard, *time–dependent Schrödinger equation*:

$$i\hbar\,\partial_t\,|\psi\rangle = \hat{H}\,|\psi\rangle\,, \tag{3.3}$$

describes the time evolution of quantum systems when they are undisturbed by measurements. 'Measurements' are defined as *interactions* of the quantum system with its classical environment. As long as the system is sufficiently isolated from the environment, it follows Schrödinger equation. If an interaction with the environment takes place, i.e., a measurement is performed, the system abruptly *decoheres* i.e., collapses or reduces to one of its classically allowed states.

A *time–dependent state of a quantum system* is determined by a normalized, complex, *wave psi–function* $\psi = \psi(t)$. In Dirac's words, this is a unit *ket* vector $|\psi\rangle$, which is an element of the *Hilbert space* $L^2(\psi)$ with a coordinate basis (q^i). The state ket–vector $|\psi(t)\rangle$ is subject to action of the Hermitian operators, obtained by the procedure of *quantization* of classical biodynamic quantities, and whose real eigenvalues are being measured.

Quantum superposition is a generalization of the algebraic principle of linear combination of vectors. The Hilbert space has a set of states $|\varphi_i\rangle$ (where the index i runs over the degrees–of–freedom of the system) that form a basis and the most general state of such a system can be written as $|\psi\rangle = \sum_i c_i$ $|\varphi_i\rangle$. The system is said to be in a state $|\psi(t)\rangle$, describing the motion of the *de Broglie waves* (named after *Nobel Laureate, Prince Louis V.P.R. de Broglie*), which is a linear superposition of the basis states $|\varphi_i\rangle$ with weighting coefficients c_i that can in general be complex. At the microscopic or quantum level, the state of the system is described by the wave function $|\psi\rangle$, which in general appears as a linear superposition of all basis states. This can be interpreted as the system being in all these states at once. The coefficients c_i are called the *probability amplitudes* and $|c_i|^2$ gives the probability that $|\psi\rangle$ will collapse into state $|\varphi\rangle$ when it decoheres (interacts with the environment). By simple normalization we have the constraint that $\sum_i |c_i|^2 = 1$. This emphasizes the fact that the wavefunction describes a *real, physical system*, which must be in one of its allowable classical states and therefore by summing over all the possibilities, weighted by their corresponding probabilities, one must get unity. In other words, we have the *normalization condition* for the psi–function, determining the unit length of the state ket–vector

$$\langle\psi(t)|\psi(t)\rangle = \int \psi^*\psi \, dV = \int |\psi|^2 \, dV = 1,$$

where $\psi^* = \langle\psi(t)|$ denotes the *bra* vector, the complex–conjugate to the ket $\psi = |\psi(t)\rangle$, and $\langle\psi(t)|\psi(t)\rangle$ is their scalar product, i.e., Dirac *bracket*. For this reason the scene of quantum mechanics is the functional space of square–integrable complex psi–functions, i.e., the Hilbert space $L^2(\psi)$.

When the system is in the state $|\psi(t)\rangle$, the average value $\langle f\rangle$ of any physical observable f is equal to

$$\langle f\rangle = \langle\psi(t)|\,\hat{f}\,|\psi(t)\rangle,$$

where $\hat{f}$ is the Hermitian operator corresponding to f.

A quantum system is *coherent* if it is in a linear superposition of its basis states. If a measurement is performed on the system and this means that the system must somehow interact with its environment, the superposition is destroyed and the system is observed to be in only one basis state, as required classically. This process is called *reduction* or *collapse* of the wavefunction or simply *decoherence* and is governed by the form of the wavefunction $|\psi\rangle$.

Entanglement on the other hand, is a purely quantum phenomenon and has no classical analogue. It accounts for the ability of quantum systems to exhibit correlations in counterintuitive 'action–at–a–distance' ways. Entanglement is what makes all the difference in the operation of quantum computers versus classical ones. Entanglement gives 'special powers' to quantum computers because it gives quantum states the potential to exhibit and maintain correlations that cannot be accounted for classically. Correlations between bits are what make information encoding possible in classical computers. For instance, we can require two bits to have the same value thus encoding a relationship. If we are to subsequently change the encoded information, we must change the correlated bits in tandem by explicitly accessing each bit. Since quantum bits exist as superpositions, *correlations* between them also exist in superposition. When the superposition is destroyed (e.g., one qubit is measured), the correct correlations are *instantaneously* 'communicated' between the qubits and this communication allows *many qubits* to be accessed *at once*, preserving their correlations, something that is absolutely impossible classically.

More precisely, the *first quantization* is a *linear representation* of all classical dynamical variables (like coordinate, momentum, energy, or angular momentum) by linear *Hermitian* operators acting on the associated Hilbert state–space $L^2(\psi)$, which has the following properties [Dir49]:

1. Linearity:
$$\alpha f + \beta g \rightarrow \alpha \hat{f} + \beta \hat{g},$$
for all constants $\alpha, \beta \in \mathbb{C}$;
2. A 'dynamical' variable, equal to unity everywhere in the phase–space, corresponds to unit operator: $1 \rightarrow \hat{I}$; and
3. *Classical Poisson brackets*
$$\{f, g\} = \frac{\partial f}{\partial q^i}\frac{\partial g}{\partial p_i} - \frac{\partial f}{\partial p_i}\frac{\partial g}{\partial q^i}$$
quantize to the corresponding *commutators*
$$\{f, g\} \rightarrow -i\hbar[\hat{f}, \hat{g}], \qquad [\hat{f}, \hat{g}] = \hat{f}\hat{g} - \hat{g}\hat{f}.$$

Like Poisson bracket, commutator is bilinear and skew–symmetric operation, satisfying Jacobi identity. For Hermitian operators $\hat{f}, \hat{g}$ their commutator $[\hat{f}, \hat{g}]$ is anti–Hermitian; for this reason i is required in $\{f, g\} \rightarrow -i\hbar[\hat{f}, \hat{g}]$.

Property (2) is introduced for the following reason. In Hamiltonian mechanics each dynamical variable f generates some transformations in the phase–space via Poisson brackets. In quantum mechanics it generates transformations in the state–space by direct application to a state, i.e.,

$$\dot{u} = \{u, f\}, \qquad \partial_t|\psi\rangle = \frac{i}{\hbar}\hat{f}|\psi\rangle. \tag{3.4}$$

Exponent of anti–Hermitian operator is unitary. Due to this fact, transformations, generated by Hermitian operators

$$\hat{U} = \exp\frac{i\hat{f}t}{\hbar},$$

are unitary. They are *motions* – scalar product preserving transformations in the Hilbert state–space $L^2(\psi)$. For this property i is needed in (3.4).

Due to property (2), the transformations, generated by classical variables and quantum operators, have the same algebra.

For example, the quantization of energy E gives:

$$E \to \hat{E} = i\hbar\,\partial_t.$$

The relations between operators must be similar to the relations between the relevant physical quantities observed in classical mechanics.

For example, the quantization of the classical equation $E = H$, where

$$H = H(p_i, q^i) = T + U$$

denotes the Hamilton's function of the total system energy (the sum of the kinetic energy T and potential energy U), gives the Schrödinger equation of motion of the state ket–vector $|\psi(t)\rangle$ in the Hilbert state–space $L^2(\psi)$

$$i\hbar\,\partial_t|\psi(t)\rangle = \hat{H}\,|\psi(t)\rangle.$$

In the simplest case of a single particle in the potential field U, the operator of the total system energy – Hamiltonian is given by:

$$\hat{H} = -\frac{\hbar^2}{2m}\nabla^2 + U,$$

where m denotes the mass of the particle and ∇ is the classical gradient operator. So the first term on the r.h.s denotes the kinetic energy of the system, and therefore the momentum operator must be given by:

$$\hat{p} = -i\hbar\nabla.$$

Now, for each pair of states $|\varphi\rangle, |\psi\rangle$ their scalar product $\langle\varphi|\psi\rangle$ is introduced, which is [Nik95]:

1. Linear (for right multiplier):

$$\langle\varphi|\alpha_1\psi_1 + \alpha_2\psi_2\rangle = \alpha_1\langle\varphi|\psi_1\rangle + \alpha_2\langle\varphi|\psi_2\rangle;$$

2. In transposition transforms to complex conjugated:

$$\langle\varphi|\psi\rangle = \overline{\langle\psi|\varphi\rangle};$$

this implies that it is 'anti–linear' for left multiplier:

$$\langle\alpha_1\varphi_1 + \alpha_2\varphi_2\rangle = \bar{\alpha}_1\langle\varphi_1|\psi\rangle + \bar{\alpha}_2\langle\varphi_2|\psi\rangle);$$

3. Additionally it is often required, that the scalar product should be positively defined:

$$\text{for all } |\psi\rangle, \quad \langle\psi|\psi\rangle \geq 0 \qquad \text{and} \qquad \langle\psi|\psi\rangle = 0 \quad \text{iff} \quad |\psi\rangle = 0.$$

Complex conjugation of classical variables is represented as Hermitian conjugation of operators. We remind some definitions:

– two operators $\hat{f}, \hat{f}^+$ are called Hermitian conjugated (or adjoint), if

$$\langle\varphi|\hat{f}\psi\rangle = \langle\hat{f}^+\varphi|\psi\rangle \qquad (\text{for all } \varphi, \psi).$$

This scalar product is also denoted by $\langle\varphi|\hat{f}|\psi\rangle$ and called a matrix element of an operator.

– operator is Hermitian (self–adjoint) if $\hat{f}^+ = \hat{f}$ and anti–Hermitian if $\hat{f}^+ = -\hat{f}$;

– operator is unitary, if $\hat{U}^+ = \hat{U}^{-1}$; such operators preserve the scalar product:

$$\langle\hat{U}\varphi|\hat{U}\psi\rangle = \langle\varphi|\hat{U}^+\hat{U}|\psi\rangle = \langle\varphi|\psi\rangle.$$

Real classical variables should be represented by Hermitian operators; complex conjugated classical variables $(a, \bar{a})$ correspond to Hermitian conjugated operators $(\hat{a}, \hat{a}^+)$.

Multiplication of a state by complex numbers does not change the state physically.

Any Hermitian operator in Hilbert space has only real eigenvalues:

$$\hat{f}|\psi_i\rangle = f_i|\psi_i\rangle, \qquad (\text{for all } \ f_i \in \mathbb{R}).$$

Eigenvectors $|\psi_i\rangle$ form complete orthonormal basis (eigenvectors with different eigenvalues are automatically orthogonal; in the case of multiple eigenvalues one can form orthogonal combinations; then they can be normalized).

If the two operators $\hat{f}$ and $\hat{g}$ commute, i.e., $[\hat{f}, \hat{g}] = 0$ (see Heisenberg picture below), than the corresponding quantities can simultaneously have definite values. If the two operators do not commute, i.e., $[\hat{f}, \hat{g}] \neq 0$, the quantities corresponding to these operators cannot have definite values simultaneously, i.e., the general *Heisenberg's uncertainty relation* is valid:

$$(\Delta\hat{f})^2 \cdot (\Delta\hat{g})^2 \geq \frac{\hbar}{4}[\hat{f}, \hat{g}]^2,$$

where Δ denotes the deviation of an individual measurement from the mean value of the distribution. The well–known particular cases are ordinary uncertainty relations for coordinate–momentum $(q - p)$, and energy–time $(E - t)$:

$$\Delta q \cdot \Delta p_q \geq \frac{\hbar}{2}, \qquad \text{and} \qquad \Delta E \cdot \Delta t \geq \frac{\hbar}{2}.$$

For example, the rules of commutation, analogous to the classical ones written by the Poisson's brackets, are postulated for canonically–conjugate coordinate and momentum operators:

$$[\hat{q}^i, \hat{q}^j] = 0, \qquad [\hat{p}_i, \hat{p}_j] = 0, \qquad [\hat{q}^i, \hat{p}_j] = i\hbar\delta^i_j\hat{I},$$

where δ^i_j is the Cronecker's symbol. By applying the commutation rules to the system Hamiltonian $\hat{H} = \hat{H}(\hat{p}_i, \hat{q}^i)$, the *quantum Hamilton's equations* are obtained:

$$\frac{d(\hat{p}_i)}{dt} = -\frac{\partial \hat{H}}{\partial \hat{q}^i}, \qquad \text{and} \qquad \frac{d(\hat{q}^i)}{dt} = \frac{\partial \hat{H}}{\partial \hat{p}_i}.$$

A quantum state can be observed either in the *coordinate* $q-$*representation*, or in the *momentum* $p-$*representation*. In the $q-$representation, operators of coordinate and momentum have respective forms: $\hat{q} = q$, and $\hat{p}_q = -i\hbar\frac{\partial}{\partial q}$, while in the p–representation, they have respective forms: $\hat{q} = i\hbar\frac{\partial}{\partial p_q}$, and $\hat{p}_q = p_q$. The forms of the state vector $|\psi(t)\rangle$ in these two representations are mathematically related by a *Fourier–transform pair* (within the Planck constant).

3.1.3 Quantum Pictures

In the $q-$representation the quantum state is usually determined, i.e., the first quantization is performed, in one of the three *quantum pictures* (see e.g., [Dir49]):

1. *Schrödinger picture*,
2. *Heisenberg picture*, and
3. *Dirac interaction picture*.

These three pictures mutually differ in the time–dependence, i.e., time–evolution of the state vector $|\psi(t)\rangle$ and the Hilbert coordinate basis (q^i) together with the system operators.

1. In the *Schrödinger* (S) *picture*, under the action of the *evolution operator* $\hat{S}(t)$ the state–vector $|\psi(t)\rangle$ rotates:

$$|\psi(t)\rangle = \hat{S}(t)\,|\psi(0)\rangle,$$

and the coordinate basis (q^i) is fixed, so the operators are constant in time:

$$\hat{F}(t) = \hat{F}(0) = \hat{F},$$

and the system evolution is determined by the Schrödinger wave equation:

$$i\hbar\,\partial_t|\psi^S(t)\rangle = \hat{H}^S\,|\psi^S(t)\rangle.$$

If the Hamiltonian does not explicitly depend on time, $\hat{H}(t) = \hat{H}$, which is the case with the absence of variables of macroscopic fields, the state vector $|\psi(t)\rangle$ can be presented in the form:

$$|\psi(t)\rangle = \exp(-i\frac{E}{\hbar}t)\,|\psi\rangle,$$

satisfying the time–independent Schrödinger equation

$$\hat{H}\,|\psi\rangle = E\,|\psi\rangle,$$

which gives the eigenvalues E_m and eigenfunctions $|\psi_m\rangle$ of the Hamiltonian $\hat{H}$.

2. In the *Heisenberg* (H) *picture*, under the action of the evolution operator $\hat{S}(t)$, the coordinate basis (q^i) rotates, so the operators of physical variables evolve in time by the similarity transformation:

$$\hat{F}(t) = \hat{S}^{-1}(t)\,\hat{F}(0)\,\hat{S}(t),$$

while the state vector $|\psi(t)\rangle$ is constant in time:

$$|\psi(t)\rangle = |\psi(0)\rangle = |\psi\rangle,$$

and the system evolution is determined by the *Heisenberg equation of motion*:

$$i\hbar\,\partial_t\hat{F}^H(t) = [\hat{F}^H(t), \hat{H}^H(t)],$$

where $\hat{F}(t)$ denotes arbitrary Hermitian operator of the system, while the commutator, i.e., Poisson quantum bracket, is given by:

$$[\hat{F}(t), \hat{H}(t)] = \hat{F}(t)\,\hat{H}(t) - \hat{H}(t)\,\hat{F}(t) = \hat{\imath}K.$$

In both Schrödinger and Heisenberg picture the evolution operator $\hat{S}(t)$ itself is determined by the Schrödinger–like equation:

$$i\hbar\,\partial_t\hat{S}(t) = \hat{H}\,\hat{S}(t),$$

with the initial condition $\hat{S}(0) = \hat{I}$. It determines the Lie group of transformations of the Hilbert space $L^2(\psi)$ in itself, the Hamiltonian of the system being the generator of the group.

3. In the *Dirac interaction* (I) *picture* both the state vector $|\psi(t)\rangle$ and coordinate basis (q^i) rotate; therefore the system evolution is determined by both the Schrödinger wave equation and the Heisenberg equation of motion:

$$i\hbar\,\partial_t|\psi^I(t)\rangle = \hat{H}^I\,|\psi^I(t)\rangle, \qquad \text{and} \qquad i\hbar\,\partial_t\hat{F}^I(t) = [\hat{F}^I(t), \hat{H}^O(t)].$$

Here: $\hat{H} = \hat{H}^0 + \hat{H}^I$, where $\hat{H}^0$ corresponds to the Hamiltonian of the free fields and $\hat{H}^I$ corresponds to the Hamiltonian of the interaction.

Finally, we can show that the stationary Schrödinger equation

$$\hat{H}\,\psi = \hat{E}\,\psi$$

can be obtained from the condition for the minimum of the *quantum action*:

$$\delta S = 0.$$

The quantum action is usually defined by the integral:

$$S = \langle\psi(t)|\,\hat{H}\,|\psi(t)\rangle = \int \psi^*\hat{H}\psi\;dV,$$

with the additional normalization condition for the unit–probability of the psi–function:

$$\langle\psi(t)|\psi(t)\rangle = \int \psi^*\psi \, dV = 1.$$

When the functions ψ and ψ^* are considered to be formally independent and only one of them, say ψ^* is varied, we can write the condition for an extreme of the action:

$$\delta S = \int \delta\psi^* \hat{H}\psi \, dV - E\int \delta\psi^*\psi \, dV = \int \delta\psi^*(\hat{H}\psi - E\psi)\, dV = 0,$$

where E is a Lagrangian multiplier. Owing to the arbitrariness of $\delta\psi^*$, the Schrödinger equation $\hat{H}\psi - \hat{E}\psi = 0$ must hold.

3.1.4 Spectrum of a Quantum Operator

To recapitulate, each *state* of a system is represented by a *state vector* $|\psi\rangle$ with a unit–norm, $\langle\psi|\psi\rangle = 1$, in a complex Hilbert space $\mathcal{H}$, and vice versa. Each system *observable* is represented by a Hermitian operator $\hat{A}$ in a Hilbert space $\mathcal{H}$, and vice versa. A Hermitian operator $\hat{A}$ in a Hilbert space $\mathcal{H}$ has its domain $\mathcal{D}_{\hat{A}} \subset \mathcal{H}$ which must be dense in $\mathcal{H}$, and for any two state vectors $|\psi\rangle, |\varphi\rangle \in \mathcal{D}_{\hat{A}}$ holds $\langle\hat{A}\psi|\varphi\rangle = \langle\psi|\hat{A}\varphi\rangle$ (see, e.g., [Mes00]).

Discrete Spectrum. A Hermitian operator $\hat{A}$ in a *finite–dimensional Hilbert space* $\mathcal{H}_d$ has a *discrete spectrum* $\{a_i,\ a \in \mathbb{R},\ i \in \mathbb{N}\}$, defined as a set of *discrete eigenvalues* a_i, for which the *characteristic equation*

$$\hat{A}|\psi\rangle = a|\psi\rangle \tag{3.5}$$

has the solution eigenvectors $|\psi_a\rangle \neq 0 \in \mathcal{D}_{\hat{A}} \subset \mathcal{H}_d$. For each particular eigenvalue a of a Hermitian operator $\hat{A}$ there is a corresponding *discrete characteristic projector* $\hat{\pi}_a = |\psi_a\rangle\langle\psi_a|$ (i.e., the projector to the eigensubspace of $\hat{A}$ composed of all discrete eigenvectors $|\psi_a\rangle$ corresponding to a).

Now, the *discrete spectral form* of a Hermitian operator $\hat{A}$ is defined as

$$\hat{A} = a_i\hat{\pi}_i = \sum_i a_i\,|i\rangle\langle i|, \qquad \text{for all } i \in \mathbb{N} \tag{3.6}$$

where a_i are different eigenvalues and $\hat{\pi}_i$ are the corresponding projectors subject to

$$\sum_i \hat{\pi}_i = \hat{I}, \qquad \hat{\pi}_i\hat{\pi}_j = \delta_{ij}\hat{\pi}_j,$$

where $\hat{I}$ is identity operator in $\mathcal{H}_d$.

A Hermitian operator $\hat{A}$ defines, with its characteristic projectors $\hat{\pi}_i$, the *spectral measure* of any interval on the real axis $\mathbb{R}$; for example, for a closed interval $[a, b] \subset \mathbb{R}$ holds

$$\hat{\pi}_{[a,b]}(\hat{A}) = \sum_{a_i \in [a,b]} \hat{\pi}_i, \tag{3.7}$$

and analogously for other intervals, $(a, b], [a, b), (a, b) \subset \mathbb{R}$; if $a_i \in [a, b] = \emptyset$ then $\hat{\pi}_{[a,b]}(\hat{A}) = 0$, by definition.

Now, let us suppose that we measure an observable $\hat{A}$ of a system in state $|\psi\rangle$. The *probability* P to get a result within the a'priori given interval $[a, b] \subset \mathbb{R}$ is given by its spectral measure

$$P([a, b], \hat{A}, \psi) = \langle\psi|\hat{\pi}_{[a,b]}(\hat{A})|\psi\rangle. \tag{3.8}$$

As a consequence, the probability to get a discrete eigenvalue a_i as a result of measurement of an observable $\hat{A}$ equals its *expected value*

$$P(a_i, \hat{A}, \psi) = \langle\psi|\hat{\pi}_i|\psi\rangle = \langle\hat{\pi}_i\rangle,$$

where $\langle\hat{B}\rangle$ in general denotes the average value of an operator $\hat{B}$. Also, the probability to get a result a which is not a discrete eigenvalue of an observable $\hat{A}$ in a state $|\psi\rangle$ equals zero.

Continuous Spectrum. A Hermitian operator $\hat{A}$ in an *infinite–dimensional Hilbert space* $\mathcal{H}_c$ (the so–called *rigged Hilbert space*) has both a discrete spectrum $\{a_i,\ a \in \mathbb{R},\ i \in \mathbb{N}\}$ and a *continuous spectrum* $[c, d] \subset \mathbb{R}$. In other words, $\hat{A}$ has both a discrete sub–basis $\{|i\rangle : i \in \mathbb{N}\}$ and a continuous sub–basis $\{|s\rangle : s \in [c, d] \subset \mathbb{R}\}$. In this case s is called the *continuous eigenvalue* of $\hat{A}$. The corresponding characteristic equation is

$$\hat{A}|\psi\rangle = s|\psi\rangle. \tag{3.9}$$

Equation (3.9) has the solution eigenvectors $|\psi_s\rangle \neq 0 \in \mathcal{D}_{\hat{A}} \subset \mathcal{H}_c$, given by the *Lebesgue integral*

$$|\psi_s\rangle = \int_a^b \psi(s)\, |s\rangle\, ds, \qquad c \leq a < b \leq d,$$

where $\psi(s) = \langle s|\psi\rangle$ are continuous, *square integrable Fourier coefficients*,

$$\int_a^b |\psi(s)|^{\,2}\, ds < +\infty,$$

while the continuous eigenvectors $|\psi_s\rangle$ are orthonormal,

$$\psi(t) = \langle t|\psi_s\rangle = \int_c^d \psi(s)\, \delta(s - t)\, ds, \tag{3.10}$$

i.e., normed on the Dirac δ–function, with

$$\langle t|s\rangle = \delta(s - t), \qquad s, t \in [c, d].$$

The corresponding *continuous projectors* $\hat{\pi}^c_{[a,b]}(\hat{A})$ are defined as Lebesgue integrals

$$\hat{\pi}^c_{[a,b]}(\hat{A}) = \int_a^b |s\rangle\, ds\, \langle s| = |s\rangle\, \langle s|, \qquad -c \leq a < b \leq d. \tag{3.11}$$

In this case, projecting any vector $|\psi\rangle \in \mathcal{H}_c$ using $\hat{\pi}^c_{[a,b]}(\hat{A})$ is given by

$$\hat{\pi}^c_{[a,b]}(\hat{A})\, |\psi\rangle = \left(\int_a^b |s\rangle\, ds\, \langle s|\right) |\psi\rangle = \int_a^b \psi(s)\, |s\rangle\, ds.$$

Now, the *continuous spectral form* of a Hermitian operator $\hat{A}$ is defined as

$$\hat{A} = \int_c^d |s\rangle\, s\, ds\, \langle s|\,.$$

Total Spectrum. The *total Hilbert state–space* of the system is equal to the *orthogonal sum* of its *discrete* and *continuous subspaces*,

$$\mathcal{H} = \mathcal{H}_d \oplus \mathcal{H}_c. \tag{3.12}$$

The corresponding discrete and continuous projectors are mutually complementary,

$$\hat{\pi}_{a_i}(\hat{A}) + \hat{\pi}^c_{[c,d]}(\hat{A}) = \hat{I}.$$

Using the *closure property*

$$\sum_i |i\rangle\langle i| + \int_a^b |s\rangle\, ds\, \langle s| = \hat{I},$$

the *total spectral form* of a Hermitian operator $\hat{A} \in \mathcal{H}$ is given by

$$\hat{A} = \sum_i a_i\, |i\rangle\, \langle i| + \int_c^d |s\rangle\, s\, ds\, \langle s|\,, \tag{3.13}$$

while an arbitrary vector $|\psi\rangle \in \mathcal{H}$ is equal to

$$|\psi\rangle = \sum_i \psi_i |i\rangle + \int_c^d \psi(s)\, |s\rangle\, ds.$$

Here, $\psi_i = \langle i|\psi\rangle$ are *discrete Fourier coefficients*, while $\psi(s) = \langle s|\psi\rangle$ are continuous, square integrable, Fourier coefficients,

$$\int_a^b |\psi(s)|^2\, ds < +\infty.$$

Using both discrete and continuous Fourier coefficients, ψ_i and $\psi(s)$, the *total inner product* of $\mathcal{H}$ is defined as

$$\langle\varphi|\psi\rangle = \bar{\varphi}_i\psi_i + \int_c^d \bar{\varphi}(s)\,\psi(s)\;ds, \tag{3.14}$$

while the norm is

$$\langle\psi|\psi\rangle = \bar{\psi}_i\psi_i + \int_c^d \bar{\psi}(s)\,\psi(s)\;ds.$$

The *total spectral measure* is now given as

$$\hat{\pi}_{[a,b]}(\hat{A}) = \sum_i \hat{\pi}_i + \int_a^b |s\rangle\, ds\, \langle s|\,,$$

so the probability P to get a measurement result within the a'priori given interval $[a,b] \in \mathbb{R} \subset \mathcal{H}$ is given by

$$P([a,b],\hat{A},\psi) = \sum_i \langle\psi|\hat{\pi}_i|\psi\rangle + \int_a^b |\psi(s)|^{\,2}\, ds, \tag{3.15}$$

where $|\psi(s)|^{\,2} = \langle\psi|s\rangle\,\langle s|\psi\rangle$ is called the *probability density*. From this the expectation value of an observable $\hat{A}$ is equal to

$$\langle\hat{A}\rangle = \sum_i a_i\langle\psi|\hat{\pi}_i|\psi\rangle + \int_a^b s\;|\psi(s)|^{\,2}\, ds = \langle\psi|\hat{A}|\psi\rangle,$$

3.1.5 General Representation Model

In quantum mechanics the total spectral form of the complete observable is given by relation (3.13). We can split this total spectral form into:

1. *Pure discrete spectral form*,

$$\hat{A} = \sum_i a_i\,|i\rangle\,\langle i|,$$

   with its discrete eigenbasis $\{|i\rangle\ :\ i \in \mathbb{N}\}$, which is orthonormal ($\langle i|j\rangle = \delta_{ij}$) and closed ($\sum_i |i\rangle\,\langle i| = \hat{I}$); and
2. *Pure continuous spectral form*,

$$\hat{B} = \int_c^d |s\rangle\, s\, ds\, \langle s|\,,$$

   with its continuous eigenbasis $\{|s\rangle : s \in [c,d] \subset \mathbb{R}\}$, which is orthonormal ($\langle s|t\rangle = \delta(s-t)$) and closed ($\int_c^d |s\rangle\, ds\, \langle s| = \hat{I}$).

The completeness property of each basis means that any vector $|\psi\rangle \in \mathcal{H}$ can be expanded/developed along the components of the corresponding basis. In case of the discrete basis we have

$$|\psi\rangle = \hat{I}\,|\psi\rangle = \sum_i |i\rangle\,\langle i|\,\psi\rangle = \sum_i \psi_i |i\rangle,$$

with discrete Fourier coefficients of the development $\psi_i = \langle i|\,\psi\rangle$.

In case of the continuous basis we have

$$|\psi\rangle = \hat{I}\,|\psi\rangle = \int_c^d |s\rangle\, ds\, \langle s|\psi\rangle = \int_c^d \psi(s)\,|s\rangle\, ds.$$

with continuous Fourier coefficients of the two development $\psi(s) = \langle s|\psi\rangle$, which are square integrable, $\int_a^b |\psi\,(s)|\ ^2 ds < +\infty$.

3.1.6 Direct Product Space

Let $\mathcal{H}_1, \mathcal{H}_2, ..., \mathcal{H}_n$ and $\mathcal{H}$ be $n+1$ given Hilbert spaces such that dimension of $\mathcal{H}$ equals the product of dimensions of $\mathcal{H}_i$, ($i = 1, ..., n$ in this section). We say that the *composite Hilbert space* $\mathcal{H}$ is defined as a direct product of the *factor Hilbert spaces* $\mathcal{H}_i$ and write

$$\mathcal{H} = \mathcal{H}_1 \otimes \mathcal{H}_2 \otimes ... \otimes \mathcal{H}_n$$

if there exists a one–to–one mapping of the set of all *uncorrelated vectors* $\{|\psi_1\rangle, |\psi_2\rangle, ..., |\psi_n\rangle\}$, $|\psi_i\rangle \in \mathcal{H}_i$, with zero inner product (i.e., $\langle\psi_i|\psi_j\rangle = 0$, for $i \neq j$) – onto their direct product $|\psi_1\rangle \times |\psi_2\rangle \times ... \times |\psi_n\rangle$, so that the following conditions are satisfied:

1. Linearity per each factor:

$$\left(\sum_{j_1=1}^{J_1} b_{j_1}|\psi_{j_1}\rangle\right) \times \left(\sum_{j_2=1}^{J_2} b_{j_2}|\psi_{j_2}\rangle\right) \times ... \times \left(\sum_{j_n=1}^{J_n} b_{j_n}|\psi_{j_n}\rangle\right)$$
$$= \sum_{j_1=1}^{J_1}\sum_{j_2=1}^{J_2} ... \sum_{j_n=1}^{J_n} b_{j_1} b_{j_2} ... b_{j_n} |\psi_{j_1}\rangle \times |\psi_{j_2}\rangle \times ... \times |\psi_{j_n}\rangle.$$

2. Multiplicativity of scalar products of uncorrelated vectors $|\psi_i\rangle, |\varphi_i\rangle \in \mathcal{H}_i$:

$$(|\psi_1\rangle \times |\psi_2\rangle \times ... \times |\psi_n\rangle\,,\ |\varphi_1\rangle \times |\varphi_2\rangle \times ... \times |\varphi_n\rangle)$$
$$= \langle\psi_1|\varphi_1\rangle \times \langle\psi_2|\varphi_2\rangle \times ... \times \langle\psi_n|\varphi_n\rangle.$$

3. Uncorrelated vectors generate the whole composite space $\mathcal{H}$, which means that in a general case a vector in $\mathcal{H}$ equals the limit of linear combinations of uncorrelated vectors, i.e.,

$$|\psi\rangle = \lim_{K\to\infty}\sum_{k=1}^{K} b_k|\psi_1^k\rangle\times|\psi_2^k\rangle\times...\times|\psi_n^k\rangle.$$

Let $\{|k_i\rangle\}$ represent arbitrary bases in the factor spaces $\mathcal{H}_i$. They induce the basis $\{|k_1\rangle\times|k_2\rangle\times...\times|k_n\rangle\}$ in the composite space $\mathcal{H}$.

Let $\hat{A}_i$ be arbitrary operators (either all linear or all antilinear) in the factor spaces $\mathcal{H}_i$. Their direct product, $\hat{A}_1\otimes\hat{A}_2\otimes...\otimes\hat{A}_n$ acts on the uncorrelated vectors

$$\left(\hat{A}_1\otimes\hat{A}_2\otimes...\otimes\hat{A}_n\right)(|\psi_1\rangle\times|\psi_2\rangle\times...\times|\psi_n\rangle)$$
$$=\left(\hat{A}_1|\psi_1\rangle\right)\times\left(\hat{A}_2|\psi_2\rangle\right)\times...\times\left(\hat{A}_n|\psi_n\rangle\right)$$

3.1.7 State–Space for n Quantum Particles

Classical state–space for the system of n particles is its $6N$D phase–space $\mathcal{P}$, including all position and momentum vectors, $\mathbf{r}_i = (x,y,z)_i$ and $\mathbf{p}_i = (p_x,p_y,p_z)_i$ respectively, for $i = 1,...,n$.

The *quantization* is performed as a *linear representation* of the real Lie algebra $\mathcal{L}_P$ of the phase–space $\mathcal{P}$, defined by the Poisson bracket $\{A,B\}$ of classical variables A,B – into the corresponding real Lie algebra $\mathcal{L}_H$ of the Hilbert space $\mathcal{H}$, defined by the commutator $[\hat{A},\hat{B}]$ of skew–Hermitian operators $\hat{A},\hat{B}$.

We start with the *Hilbert space* $\mathcal{H}_x$ for a single 1D quantum particle, which is composed of all vectors $|\psi_x\rangle$ of the form

$$|\psi_x\rangle = \int_{-\infty}^{+\infty}\psi(x)\,|x\rangle\,dx,$$

where $\psi(x) = \langle x|\psi\rangle$ are square integrable Fourier coefficients,

$$\int_{-\infty}^{+\infty}|\psi(x)|^2\,dx < +\infty.$$

The position and momentum Hermitian operators, $\hat{x}$ and $\hat{p}$, respectively, act on the vectors $|\psi_x\rangle\in\mathcal{H}_x$ in the following way:

$$\hat{x}|\psi_x\rangle = \int_{-\infty}^{+\infty}\hat{x}\,\psi(x)\,|x\rangle\,dx, \qquad \int_{-\infty}^{+\infty}|x\,\psi(x)|^2\,dx < +\infty,$$
$$\hat{p}|\psi_x\rangle = \int_{-\infty}^{+\infty}-i\hbar\frac{\partial}{\partial\hat{x}}\psi(x)\,|x\rangle\,dx, \qquad \int_{-\infty}^{+\infty}\left|-i\hbar\frac{\partial}{\partial x}\psi(x)\right|^2 dx < +\infty.$$

The *orbit Hilbert space* $\mathcal{H}_1^o$ for a single 3D quantum particle with the full set of compatible observable $\hat{\mathbf{r}} = (\hat{x},\hat{y},\hat{z})$, $\hat{\mathbf{p}} = (\hat{p}_x,\hat{p}_y,\hat{p}_z)$, is defined as

$$\mathcal{H}_1^o = \mathcal{H}_x\otimes\mathcal{H}_y\otimes\mathcal{H}_z,$$

where $\hat{\mathbf{r}}$ has the common generalized eigenvectors of the form

$$|\hat{\mathbf{r}}\rangle = |x\rangle \times |y\rangle \times |z\rangle \,.$$

$\mathcal{H}_1^o$ is composed of all vectors $|\psi_r\rangle$ of the form

$$|\psi_r\rangle = \int_{\mathcal{H}^o} \psi(\mathbf{r})\, |\mathbf{r}\rangle\, d\mathbf{r} = \int_{-\infty}^{+\infty}\int_{-\infty}^{+\infty}\int_{-\infty}^{+\infty} \psi(x,y,z)\, |x\rangle\times|y\rangle\times|z\rangle\, dxdydz,$$

where $\psi(\mathbf{r}) = \langle \mathbf{r}|\psi_r\rangle$ are square integrable Fourier coefficients,

$$\int_{-\infty}^{+\infty} |\psi(\mathbf{r})|^{\,2}\, d\mathbf{r} < +\infty.$$

The position and momentum operators, $\hat{\mathbf{r}}$ and $\hat{\mathbf{p}}$, respectively, act on the vectors $|\psi_r\rangle \in \mathcal{H}_1^o$ in the following way:

$$\hat{\mathbf{r}}|\psi_r\rangle = \int_{\mathcal{H}_1^o} \hat{\mathbf{r}}\,\psi(\mathbf{r})\, |\mathbf{r}\rangle\, d\mathbf{r}, \qquad \int_{\mathcal{H}_1^o} |\mathbf{r}\,\psi(\mathbf{r})|^{\,2}\, d\mathbf{r} < +\infty,$$

$$\hat{\mathbf{p}}|\psi_r\rangle = \int_{\mathcal{H}_1^o} -i\hbar\frac{\partial}{\partial\hat{\mathbf{r}}}\psi(\mathbf{r})\, |\mathbf{r}\rangle\, d\mathbf{r}, \qquad \int_{\mathcal{H}_1^o} \left|-i\hbar\frac{\partial}{\partial\mathbf{r}}\psi(\mathbf{r})\right|^2 d\mathbf{r} < +\infty.$$

Now, if we have a system of n 3D particles, let $\mathcal{H}_i^o$ denote the orbit Hilbert space of the ith particle. Then the composite orbit state–space $\mathcal{H}_n^o$ of the whole system is defined as a direct product

$$\mathcal{H}_n^o = \mathcal{H}_1^o \otimes \mathcal{H}_2^o \otimes \ldots \otimes \mathcal{H}_n^o.$$

$\mathcal{H}_n^o$ is composed of all vectors

$$|\psi_r^n\rangle = \int_{\mathcal{H}_n^o} \psi(\mathbf{r}_1,\mathbf{r}_2,\ldots,\mathbf{r}_n)\, |\mathbf{r}_1\rangle\times|\mathbf{r}_2\rangle\times\ldots\times|\mathbf{r}_n\rangle\, d\mathbf{r}_1 d\mathbf{r}_2\ldots d\mathbf{r}_n$$

where $\psi(\mathbf{r}_1,\mathbf{r}_2,\ldots,\mathbf{r}_n) = \langle \mathbf{r}_1,\mathbf{r}_2,\ldots,\mathbf{r}_n|\psi_r^n\rangle$ are square integrable Fourier coefficients

$$\int_{\mathcal{H}_n^o} |\psi(\mathbf{r}_1,\mathbf{r}_2,\ldots,\mathbf{r}_n)|^2\, d\mathbf{r}_1 d\mathbf{r}_2\ldots d\mathbf{r}_n < +\infty,$$

The position and momentum operators $\hat{\mathbf{r}}_i$ and $\hat{\mathbf{p}}_i$ act on the vectors $|\psi_r^n\rangle \in \mathcal{H}_n^o$ in the following way:

$$\hat{\mathbf{r}}_i|\psi_r^n\rangle = \int_{\mathcal{H}_n^o} \{\hat{\mathbf{r}}_i\}\,\psi(\mathbf{r}_1,\mathbf{r}_2,\ldots,\mathbf{r}_n)\, |\mathbf{r}_1\rangle\times|\mathbf{r}_2\rangle\times\ldots\times|\mathbf{r}_n\rangle\, d\mathbf{r}_1 d\mathbf{r}_2\ldots d\mathbf{r}_n,$$

$$\hat{\mathbf{p}}_i|\psi_r^n\rangle = \int_{\mathcal{H}_n^o} \left\{-i\hbar\frac{\partial}{\partial\hat{\mathbf{r}}_i}\right\}\psi(\mathbf{r}_1,\mathbf{r}_2,\ldots,\mathbf{r}_n)\, |\mathbf{r}_1\rangle\times|\mathbf{r}_2\rangle\times\ldots\times|\mathbf{r}_n\rangle\, d\mathbf{r}_1 d\mathbf{r}_2\ldots d\mathbf{r}_n,$$

with the square integrable Fourier coefficients

$$\int_{\mathcal{H}_n^o} |\{\hat{\mathbf{r}}_i\}\,\psi(\mathbf{r}_1,\mathbf{r}_2,...,\mathbf{r}_n)|^2\; d\mathbf{r}_1 d\mathbf{r}_2...d\mathbf{r}_n < +\infty,$$

$$\int_{\mathcal{H}_n^o} \left|\left\{-i\hbar\frac{\partial}{\partial \mathbf{r}_i}\right\}\psi(\mathbf{r}_1,\mathbf{r}_2,...,\mathbf{r}_n)\right|^2\; d\mathbf{r}_1 d\mathbf{r}_2...d\mathbf{r}_n < +\infty,$$

respectively. In general, any set of vector Hermitian operators $\{\hat{\mathbf{A}}_i\}$ corresponding to all the particles, act on the vectors $|\psi_r^n\rangle \in \mathcal{H}_n^o$ in the following way:

$$\hat{\mathbf{A}}_i|\psi_r^n\rangle = \int_{\mathcal{H}_n^o} \{\hat{\mathbf{A}}_i\}\psi(\mathbf{r}_1,\mathbf{r}_2,...,\mathbf{r}_n)\;|\mathbf{r}_1\rangle\times|\mathbf{r}_2\rangle\times...\times|\mathbf{r}_n\rangle\; d\mathbf{r}_1 d\mathbf{r}_2...d\mathbf{r}_n,$$

with the square integrable Fourier coefficients

$$\int_{\mathcal{H}_n^o} \left|\left\{\hat{\mathbf{A}}_i\right\}\psi(\mathbf{r}_1,\mathbf{r}_2,...,\mathbf{r}_n)\right|^2\; d\mathbf{r}_1 d\mathbf{r}_2...d\mathbf{r}_n < +\infty.$$

3.2 Relativistic Quantum Mechanics and Electrodynamics

3.2.1 Difficulties of the Relativistic Quantum Mechanics

The theory outlined above is not in agreement with the Einstein's restricted *Principle of relativity*, as is at once evident from the special role played by the time t. Thus, while it works very well in the non–relativistic region of low velocities, where it appears to be in complete agreement with experiment, it can be considered only as an approximation, and one must face the task of extending it to make it conform to restricted relativity.[1] One should be prepared for possible further alterations being needed in basic physical concepts, and hence one should follow the route of first setting up the mathematical formalism and then seeking its physical interpretation.

Setting up the mathematical formalism is a fairly straightforward matter. One must first put classical Newtonian mechanics into *relativistic Hamiltonian form*. One must take into account that the various particles comprising the dynamical system interact through the medium of the electromagnetic field, and one must use Lorentz's equations of motion for them, including the damping terms which express the reaction of radiation. This is done in subsection 3.2.4 below, where, with the help of the *Dirac's electrodynamic action principle*, the equations of motion are obtained in the Hamiltonian form (3.62) with the Hamiltonians F_i, one for each particle, given by (3.61). This

[1] General relativity (i.e., gravitation theory) need not be considered here, since gravitational effects are negligible in purely atomic theory. We will consider them later in the book.

Hamiltonian formulation may now be made into a quantum theory by following rules which have become standardized from the non–relativistic quantum mechanics. The resulting formalism appears to be quite satisfactory mathematically, but when one proceeds to consider its physical interpretation one meets with serious difficulties [Dir26c, Dir26e, Dir32, Cha48].

Take an elementary example, that of a *free particle without spin*, moving in the absence of any field. The classical Hamiltonian for this system is the left–hand side of the equation

$$p_0^2 - p_1^2 - p_2^2 - p_3^2 - m^2 = 0, \tag{3.16}$$

where p_0 is the energy and p_1, p_2, p_3 the momentum of the particle, the velocity of light being taken as unity. Passing over to quantum theory by the standard rules, one gets from this Hamiltonian the so–called *Klein–Gordon equation*

$$(\hbar^2 \Box + m^2)\psi = 0, \tag{3.17}$$

where $\Box$ is the Dalambertian wave operator,

$$\Box \equiv \frac{\partial^2}{\partial x_0^2} - \frac{\partial^2}{\partial x_1^2} - \frac{\partial^2}{\partial x_2^2} - \frac{\partial^2}{\partial x_3^2}.$$

The wave function ψ here is a scalar, involving the coordinates x_1, x_2, x_3 and the time $t = x_0$ on the same footing, and so it is suitable for a relativistic theory.

If one now tries to use the old interpretation that $|\psi|^2$ is the probability per unit volume of the particle being in the neighborhood of the point $x = x_1, x_2, x_3$ at the time x_0, one immediately gets into conflict with relativity, since this probability ought to transform under *Lorentz transformations* like the time–component of a 4–vector, while $|\psi|^2$ is a scalar. Also the conservation law for total probability would no longer hold, the usual proof of it failing on account of the wave equation (3.17) not being linear in $\partial_{x_0} \equiv \partial/\partial_{x_0}$.

An important step forward was taken by [Gor26] and [Kle27], who proposed that instead of $|\psi|^2$ one should use the expression

$$\frac{1}{4\pi i}[\psi\, \partial_{x_0}\bar{\psi} - \bar{\psi}\, \partial_{x_0}\psi], \tag{3.18}$$

where $\bar{\psi} = \bar{\psi}(x_0, x_1, x_2, x_3)$ is the complex–conjugate wave ψ–function.

The expression (3.18) is the time component of a 4–vector. Further, it is easily verified that the divergence of this 4–vector vanishes, which gives the *conservation law* in relativistic form. Thus, (3.18) is evidently the correct mathematical form to use.

However, this form leads to trouble on the physical side, since, although it is real, it is not positive definite like $|\psi|^2$. Its employment would result in one having at times a negative probability for the particle being in a certain place.

This is not the only physical difficulty. Let us consider the energy and momentum of the particle, and take for simplicity a state for which these variables have definite values. The corresponding wave ψ−function will be of the form of plane waves,

$$\psi = \exp[-i(p_0x_0 - p_1x_1 - p_2x_2 - p_3x_3)/\hbar].$$

In order that the wave equation (3.17) may be satisfied, the energy and momentum values p_0, p_1, p_2, p_3 here must satisfy the classical equation (3.16). This equation allows of negative values for the energy p_0 as well as positive ones and is, in fact, symmetrical between positive and negative energies. The negative energies occur also in the classical theory, but do not then cause trouble, since a particle started off in a positive–energy state can never make a transition to a negative–energy one. In the quantum theory, however, such transitions are possible and do in general take place under the action of perturbing forces [Dir26c, Dir26e, Dir32].

The wave ψ−function may be transformed to the momentum and energy variables. The Klein–Gordon expression (3.18) then goes over into

$$|\psi(p_0, p_1, p_2, p_3)|^2 p_0^{-1} dp_1 dp_2 dp_3, \tag{3.19}$$

as the probability of the momentum having a value within the small domain $dp_1dp_2dp_3$ about the value p_1, p_2, p_3, with the energy having the value p_0, which must be connected with p_1, p_2, p_3 by (3.16). The weight factor p_0^{-1} appears in (3.19) and makes it *Lorentz invariant*, since $\psi(p)$ is a scalar (it is defined in terms of $\psi(x)$ to make it so), and the differential element $p_0^{-1}dp_1dp_2dp_3$ is also Lorentz invariant. This weight factor may be positive or negative, and makes the probability positive or negative accordingly. Thus the two undesirable things, negative energy and negative probability, always occur together.

Let us pass on to another simple example, that of a *free particle with spin half a quantum*. The wave equation is of the same form (3.17) as before, but the wave ψ−function is no longer a scalar. It must have two components, or four if there is a field present, and the way they transform under Lorentz transformations is given by the general connection between the theory of angular momentum in quantum mechanics and group theory. The expression $\sum |\psi(x)|^2$, summed for the components of ψ, turns out to be the time component of a 4–vector, and further the divergence of this 4–vector vanishes. Thus it is satisfactory to use this expression as the probability per unit volume of the particle being at any place at any time. One does not now have any negative probabilities in the theory. However, the negative energies remain, as in the case of no spin.

We may go on and consider particles of higher spin. The general result is that there are always states of negative energy as well as those of positive energy. For particles whose spin is an integral number of quanta, the negative–energy states occur with a negative probability and the positive–energy ones with a positive probability, while for particles whose spin is a half–odd integral number of quanta, all states occur with a positive probability [Dir26e, Dir32].

Negative energies and probabilities should not be considered as nonsense. They are well–defined concepts mathematically, like a negative sum of money, since the equations which express the important properties of energies and probabilities can still be used when they are negative. Thus negative energies and probabilities should be considered simply as things which do not appear in experimental results. The physical interpretation of relativistic quantum mechanics that one gets by a natural development of the non–relativistic theory involves these things and is thus in contradiction with experiment. We therefore have to consider ways of modifying or supplementing this interpretation.

3.2.2 Particles of Half–Odd Integral Spin

Let us first consider particles with a half–odd integral spin, for which there is only the negative–energy difficulty to be removed. The chief particle of this kind for which a relativistic theory is needed is the electron, with spin half a quantum. Now electrons, and also, it is believed, all particles with a half–odd integral spin, satisfy the *Pauli's Exclusion Principle*, according to which not more than one of them can be in any quantum state.[2] With this principle there are only two alternatives for a state, either it is unoccupied or it is occupied by one particle, and a symmetry appears with respect to these two alternatives.

Dirac proposed a way of dealing with the negative–energy difficulty for electrons, based on a theory in which nearly all their negative–energy states are occupied (see [Dir36]). An unoccupied negative–energy state now appears as a 'hole' in the distribution of occupied negative–energy states and thus has a deficiency of negative energy, i.e., a positive energy. From the wave equation one finds that a hole moves in the way one would expect a positively charged electron to move. It becomes reasonable to identify the holes with the recently discovered positrons, and thus to get an interpretation of the theory involving positrons together with electrons. An electron jumping from a positive– to a negative–energy state in the theory is now interpreted as an annihilation of an electron and a positron, and one jumping from a negative– to a positive–energy state as a creation of an electron and a positron.

The theory involves an infinite density of electrons everywhere. It becomes necessary to assume that the distribution of electrons for which all positive–energy states are unoccupied and all negative–energy states occupied, what one may call the *vacuum distribution*, as it corresponds to the absence of all electrons and positrons in the interpretation, is completely unobservable. Only departures from this distribution are observable and contribute to the electric density and current which give rise to electromagnetic field in accordance with Maxwell's equations.

[2] This principle is obtained in quantum mechanics from the requirement that wave functions shall be antisymmetric in all the particles.

The above theory does provide a way out from the negative–energy difficulty, but it is not altogether satisfactory. The infinite number of electrons that it involves requires one to deal with wave functions of very great complexity and leads to such complicated mathematics that one cannot solve even the simplest problems accurately, but must resort to crude and unreliable approximations. Such a theory is a most inconvenient one to have to work with, and on general philosophical grounds one feels that it must be wrong [Dir26c, Dir26e, Dir36].

Let us see whether one can modify the theory so as to make it possible to work out simple examples accurately, while retaining the basic idea of identifying unoccupied negative–energy states with positrons. The simple calculations that one can make involve simple wave functions, referring to only one or two electrons, and thus referring to nearly all the negative–energy states being unoccupied. The calculations therefore apply to a world almost saturated with positrons, i.e., having nearly every quantum state for a positron occupied. Such a world, of course, differs very much from the actual world. One can now calculate the probability of any kind of collision process occurring in this hypothetical world (in so far as electrons and positrons are concerned). One can deduce the probability coefficient for the process, i.e., the probability per unit number of incident particles or per unit intensity of the beam of incident particles, for each of the various kinds of incident particle taking part in the process. For this purpose one must use the laws of statistical mechanics, which tell how the probability of a collision process depends on the number of incident particles, paying due attention to the modified form of these laws arising from the Pauli's exclusion principle.

Let us now assume that probability coefficients so calculated for the hypothetical world are the same as those of the actual world. This single assumption provides a general physical interpretation for the formalism, enabling one to calculate collision probabilities in the actual world. It does not provide a complete physical theory, since it enables one to calculate only those experimental results that are reducible to collision probabilities, and some branches of physics, e.g., the structure of solids, do not seem to be so reducible. However, collision probabilities are the things for which a relativistic theory is at present most needed, and one may hope in the future to find ways of extending the scope of the theory to make it include the whole of physics.

Comparing the new theory with the old, one may say that the new assumption, identifying collision probability coefficients in the actual world with those in a certain hypothetical world, replaces the old assumption about the non–observability of the vacuum distribution of negative–energy electrons. The approximations needed for working out simple examples in the old theory are equivalent in their mathematical effect to making the new assumption; e.g., these approximations include the neglect of the Coulomb interaction between electron and positron in the calculation of the prob ability of pair creation and annihilation, and this interaction cannot appear in the new theory, since the calculation there is concerned with a one–electron system. Thus the new theory may be considered as a precise formulation of the old theory together with some general approximations needed for applying it.

The new theory for dealing with the negative–energy states of the electron may be applied to any kind of elementary particle with spin half a quantum, and probably also to particles with other half–odd integral spin values, provided, of course, they satisfy Pauli's exclusion principle. It may thus be applied to protons and neutrons. It requires for each particle the possibility of existence of an antiparticle of the opposite charge, if the original particle is charged. If the original particle is uncharged, one can arrange for the antiparticle to be identical with the original [Dir26c, Dir26e, Dir36].

3.2.3 Particles of Integral Spin

Most of the elementary particles of physics have half–odd integral spin, but there is the important exception of the photon (or, light–quantum), with spin one quantum, and there is the cosmic–ray particle, the meson, also probably with spin one quantum. All these kinds of particle, it is believed, satisfy the *Bose–Einstein statistics*, a statistics which allows any number of particles to be in the same quantum state with the same a priori probability.[3] For these kinds of particles the previous method of dealing with the negative–energy states is therefore no longer applicable, and there is the further difficulty of the negative probabilities.

When dealing with particles satisfying the Bose–Einstein statistics, it is useful to consider the operators corresponding to the absorption of a particle from a given state or the emission into a given state. These operators can be treated as dynamical variables, although they do not have any analogues in classical mechanics. If one works out their equations of motion and transformation equations, one finds a remarkable correspondence. The absorption operators from a set of independent states have the same equations of motion and transformation equations as the wave $\psi-$function representing a single particle, and similarly for the emission operators and the conjugate complex wave $\bar{\psi}-$function. Thus one can pass from a one–particle theory to a many–particle theory by making the ψ and $\bar{\psi}$ describing the one particle into *absorption and emission operators* (or anihilation and creation operators), which must satisfy the appropriate commutation relations. Such a passage is called *second quantization*.

One can get over the difficulties of negative energies and negative probabilities for Bose–Einstein particles by abandoning the attempt to get a satisfactory theory of a single particle and passing on to consider the problem of many particles, using a method given by Pauli and Weisskopf [PE34] for electrons having no spin and satisfying the Bose–Einstein statistics.[4] The method

[3] This statistics is obtained in quantum mechanics from the requirement that wave functions shall be symmetric in all the particles.

[4] Such electrons are not known experimentally, but there is no known theoretical reason why they should not exist.

of Pauli and Wiesskopf is to work entirely with positive–energy states. The operators of absorption from and emission into negative–energy states, arising in the application of second quantization to the one–electron theory, are replaced by the operators of emission into and absorption from positive–energy states of electrons with the opposite charge, respectively. This replacement does not disturb the laws of conservation of charge, energy and momentum. The resulting theory involves spinless electrons of both kinds of charge together, and leads to pair creation and annihilation, as with ordinary electrons and positrons [Dir26e, Dir26c].

The method of Pauli and Wiesskopf may be applied in a degenerate form to photons and leads to the quantum electrodynamics of Heisenberg and Pauli [HP29a, HP29b]. To take into account that photons have no charge, one must start with a one–particle theory in which the wave functions are real, so that $\bar{\psi} = \psi$. The part of the wave $\psi-$function referring to positive–energy states is then made into the absorption operators from positive–energy states, and the part referring to negative–energy states into the emission operators into positive energy states. The resulting scheme of operators, involving only positive energy photon states, may then be put into correspondence with classical electrodynamics, according to the usual laws governing the correspondence between quantum and classical theory.

It would seem that in this way the difficulties of negative energies and probabilities for Bose–Einstein particles can be overcome, but a new difficulty appears. When one tries to solve the wave equation (or the wave equations if there are several particles with their respective Hamiltonians) one gets divergent integrals in the solution, of the form, in the case of photons,

$$\int_0^\infty f(v)dv, \qquad f(v) \sim v^n \quad \text{for large } v, \tag{3.20}$$

v being the frequency of a photon. The values $1, 0$ and -1 for n are the chief ones occurring in simple examples. Thus the wave equation really has no solutions and the method fails [Dir26c, Dir26e].

Dirac had made a detailed study of the divergent integrals occurring in quantum electrodynamics and had shown [Dir36] with even values of n can be eliminated by introducing into the equations a certain limiting process, which one can justify by showing that a corresponding limiting process is needed in classical electrodynamics to get the equations of motion into Hamiltonian form (which appears according to the Dirac's electrodynamic action principle, see subsection 3.2.4 below). The divergent integrals with odd values of n remain, however, and indicate something more fundamentally wrong with the theory.

Divergent integrals are a general feature of quantum field theories, and it has usually been supposed that they should be avoided by altering the forces or the laws of interaction between the elementary particles at small distances, so as to get the integrals cut off for some high value of v. However, one can easily see that this is wrong, in the case of electrodynamics at any rate, by referring to the corresponding classical theory. The wave $\psi-$function should have its analogue in the solution of the Hamilton–Jacobi equation, in accordance with equation (3.2), but already when one tries to solve the Hamilton–Jacobi equation of classical electrodynamics corresponding to the wave equation of Heisenberg and Pauli's quantum electrodynamics, one meets with divergent integrals. Now the classical equations of motion concerned, namely, Lorentz's equations including radiation damping, have definite solutions when treated by straightforward methods and if, on trying to get these solutions by a Hamilton–Jacobi method, one meets with divergent integrals, it means simply that the Hamilton–Jacobi method is an unsuitable one, and not that one should try to alter the physical laws of interaction to get the integrals to converge. The correspondence between the quantum and classical theories is so close that one can infer that the corresponding divergent integrals in the quantum theory must also be due to an unsuitable mathematical method.

The appearance of divergent integrals with odd $n-$values in Heisenberg and Pauli's form of quantum electrodynamics may be ascribed to the asymmetrical treatment of positive– and negative–energy photon states. If instead of using Pauli and Weisskopf's method one keeps to plain second quantization, one can build up a form of quantum electrodynamics symmetrical between positive– and negative–energy photon states [Dir26e, Dir36]. The new theory leads to similar equations as the old one, but with integrals of the type

$$\int_{-\infty}^{\infty} f(v)dv, \tag{3.21}$$

instead of (3.20), and since $f(v)$ is always a rational algebraic function, and it is reasonable on physical grounds to approach the upper and lower limits of integration in (3.21) at the same rate, the divergencies with odd $n-$values all cancel out.

Dirac had shown that the new form of quantum electrodynamics also corresponds to classical electrodynamics in accordance with the usual laws, with the exception that operators corresponding to real dynamical variables in the classical theory are no longer always self-adjoint. This exception is not important, as it rather stands apart from the general mathematical connection between quantum and classical theory. The Hamilton–Jacobi equation corresponding to the wave equation of the new quantum electrodynamics differs from that of the old one only through being expressed in terms of a different set of coordinates, but the new Hamilton–Jacobi equation can be solved without divergent integrals and is connected with a satisfactory action principle [Dir32, Dir26e, Dir36]. Thus the correspondence with classical theory of the new form of quantum electrodynamics is more far–reaching than that of the old form, which provides a strong reason for preferring the new form. It now becomes necessary to find some new physical interpretation to avoid the difficulties of negative energies and probabilities.

Let us consider in more detail the relation between the *two forms of quantum electrodynamics*. In either form the electromagnetic potentials $\mathbf{A}$ at two points $\mathbf{x}$' and $\mathbf{x}$" must satisfy the *commutation relations*

$$[A_\mu(\mathbf{x'}), A_\nu(\mathbf{x''})] = g_{\mu\nu}\Delta(\mathbf{x'} - \mathbf{x''}), \tag{3.22}$$

obtained from analogy with the classical theory, Δ being the four–dimensional Lorentz–invariant function introduced by Jordan and Pauli (1928), which has a singularity on the light–cone and vanishes everywhere else. In the quantum electrodynamics of Heisenberg and Pauli the **A**'s are operators referring to the absorption and emission of photons into positive energy states. Let us call such operators $\mathbf{A}^1$. One could introduce a similar set of operators referring to the absorption and emission of photons into negative–energy states. Let us call these operators $\mathbf{A}^2$. They satisfy the same commutation relations (3.22) and commute with the $\mathbf{A}^1$'s. One can now introduce a third set of operators

$$\mathbf{A}^3 = \frac{\sqrt{2}}{2}(\mathbf{A}^1 + \mathbf{A}^2),$$

which operate on wave functions referring to photons in both positive– and negative–energy states, and which satisfy the same commutation relations (3.22). The use of this third set gives the new form of quantum electrodynamics arising from plain second quantization.

The three sets of **A**'s may be expressed in terms of their Fourier components as [Dir26e, Dir32, Dir36]

$$\mathbf{A}^1(\mathbf{x}) = \int [\mathbf{R}_k \mathrm{e}^{i(k,x)} + \bar{\mathbf{R}}_k \mathrm{e}^{-i(kx)}] k_0^{-1} dk_1 dk_2 dk_3, \quad \text{with} \quad k_0 = \sqrt{k_1^2 + k_2^2 + k_3^2}, \tag{3.23}$$

where $\int$ denotes the tripple integral, $\mathbf{R}_k$ is the emission operator and $\bar{\mathbf{R}}_k$ is the absorption operator,

$$\mathbf{A}^1(\mathbf{x}) = \int [\mathbf{R}_k \mathrm{e}^{i(k,x)} + \bar{\mathbf{R}}_k \mathrm{e}^{-i(kx)}] k_0^{-1} dk_1 dk_2 dk_3, \ \text{with} \ k_0 = -\sqrt{k_1^2 + k_2^2 + k_3^2}, \tag{3.24}$$

$$\mathbf{A}^3(\mathbf{x}) = \frac{\sqrt{2}}{2} \sum_{k_0 = \pm\sqrt{k_1^2+k_2^2+k_3^2}} \int [\mathbf{R}_k \mathrm{e}^{i(k,x)} + \bar{\mathbf{R}}_k \mathrm{e}^{-i(kx)}] k_0^{-1} dk_1 dk_2 dk_3. \tag{3.25}$$

Since the three sets of **A**'s all satisfy the same commutation relations, they must correspond merely to three different representations of the same dynamical variables, and the passage from one to another must be a transformation of the linear type usual in quantum mechanics. Thus, after obtaining the divergency–free solution of the wave equation in the representation corresponding to $\mathbf{A}^3$, one could apply a transformation to get the solution in the $\mathbf{A}^1$ representation. However, the transformation would then introduce the same divergent integrals as appear with the direct solution of the wave equation in the $\mathbf{A}^1$ representation, so one would not get any further this way [Dir36].

In working with the $\mathbf{A}^3$ representation one has redundant dynamical variables. It is as though, in dealing with a system of one degree of freedom with the variables q, p, one decided to treat it as a system of two degrees–of–freedom by putting

$$q = \frac{\sqrt{2}}{2}(q_1 + q_2) \qquad \text{and} \qquad p = \frac{\sqrt{2}}{2}(p_1 + p_2).$$

This would be quite a correct procedure, but would introduce an unnecessary complication. In the case of quantum electrodynamics, the complication is a necessary one, to avoid the divergent integrals. Let us put

$$\mathcal{B}(\mathbf{x}) = \frac{\sqrt{2}}{2}[\mathbf{A}^1(\mathbf{x}) - \mathbf{A}^2(\mathbf{x})]. \tag{3.26}$$

Then the **B**'s commute with the $\mathbf{A}^3$'s, and thus with all the dynamical variables appearing in the Hamiltonian, so they are the redundant variables.

To determine the significance of redundant variables in quantum mechanics one may consider a general case, and work in a representation which separates the redundant variables from the non–redundant ones. One then sees immediately that a solution of the wave equation corresponds in general, not to a single state, but to a set of states with a certain probability for each, what in the classical theory is called a *Gibbs ensemble*. The. probabilities of the various states depend on the weights attached to the various eigenvalues of the redundant variables in the wave ψ–function, these weights being arbitrary, depending on the weight factor in the representation used. If one works in a representation which does not separate the redundant and non–redundant variables, as is the case in quantum electrodynamics with the representation corresponding to the use of $\mathbf{A}^3$, the general result that wave functions represent Gibbs ensembles and not single states must still be valid. Thus one can conclude that there are no solutions of the wave equation of quantum electrodynamics representing single states, but only solutions representing Gibbs ensembles. The problem remains of interpreting the negative energies and probabilities occurring with these Gibbs ensembles.

For any **x**, $\mathcal{B}(\mathbf{x})$ commutes with the Hamiltonian and is a constant of the motion. We may give it any value we like, subject to not contradicting the commutation relations. Instead of $\mathcal{B}(\mathbf{x})$ it is more convenient to work with the potential field, $\mathbf{B}(\mathbf{x})$ say, obtained from $\mathcal{B}(\mathbf{x})$ by changing the sign of all the Fourier components containing $\mathrm{e}^{ik_0x_0}$ with negative values of k_0. From (3.26), (3.23) and (3.24), we have [Dir26e, Dir36]

$$\mathbf{B}(\mathbf{x}) = \frac{\sqrt{2}}{2} \sum_{k_0=\pm\sqrt{k_1^2+k_2^2+k_3^2}} \int [\mathbf{R}_k \mathrm{e}^{i(k,x)} - \bar{\mathbf{R}}_k \mathrm{e}^{-i(kx)}] k_0^{-1} dk_1 dk_2 dk_3. \tag{3.27}$$

Let us now take **B** equal to the initial value of $\mathbf{A}^3$, a proceeding which does not contradict the commutation relations since its consequences are self–consistent. Then for the initial wave ψ–function we have

$$[\mathbf{B}(\mathbf{x}) - \mathbf{A}^3(\mathbf{x})]\psi = 0,$$

or, from (3.25) and (3.27),

$$\bar{\mathbf{R}}_k\psi = 0, \tag{3.28}$$

with k_0 either positive or negative. Thus any absorption operator applied to the initial wave ψ–function gives the result zero, which means that the corresponding state is one with no photons present.

The following natural interpretation for the wave ψ–function at some later time now appears. That part of it corresponding to no photons present may be supposed to give (through the square of its modulus) the probability of no change having taken place in the field of photons; that part corresponding to one positive–energy photon present may be supposed to give the probability of a photon having been emitted; that corresponding to one negative–energy photon present may be supposed to give the probability of a photon having been absorbed; and so on for the parts corresponding to two or more photons present. 'The various parts of the wave ψ–function which referred to the existence' of positive– and negative–energy photons in the old interpretation now refer to the emissions and absorptions of photons. This disposes of the negative–energy difficulty in a satisfactory way, conforming to the laws of conservation of energy and momentum. It is possible only because of the redundant variables enabling one to arrange that the initial wave ψ–function shall correspond in its entirety to no emissions or absorptions having taken place.

The interpretation is not yet complete, because the theory at present would give a negative probability for a process involving the absorption of a photon, or the absorption of any odd number of photons. To find the origin of these negative probabilities, one must study the probability distribution of the photons initially present in the Gibbs ensemble, which one can do by transforming to the representation corresponding to the $\mathbf{A}^1$ potentials. It is true that one cannot apply this transformation to a solution of the wave equation without getting divergent integrals, as has already been mentioned, but one can apply it to the initial wave ψ–function, which is of a specially simple form in the photon variables. In [Dir32, Dir26e, Dir36] it is found that the probability of there being n photons initially in any photon state is $P_n = \pm 2$, according to whether n is even or odd. Strictly, to make $\sum_{n=0}^{\infty} P_n$ converge to the limit unity, one must consider P_n as a limit,

$$P_n = 2(\epsilon - 1)^n, \tag{3.29}$$

with ϵ a small positive quantity tending to zero.

Probabilities 2 and -2 are, clearly, not physically understandable, but one can use them mathematically in accordance with the rules for working with a Gibbs ensemble. One can suppose a hypothetical mathematical world with the initial probability distribution (3.29) for the photons, and one can work out the probabilities of radiative transition processes occurring in this world. One can deduce the corresponding probability coefficients, i.e., the probabilities per unit intensity of each beam of incident radiation concerned, by using Einstein's laws of radiation. For example, for a process involving the absorption of a photon, if the probability coefficient is B, the probability of the process is

$$\sum_{n=0}^{\infty} nP_n B = -\frac{1}{2}B, \tag{3.30}$$

and for a process involving the emission of a photon, if the probability coefficient is A, the probability of the process is

$$\sum_{n=0}^{\infty} (n+1)P_n A = \frac{1}{2}A. \tag{3.31}$$

Now the probability of an absorption process, as calculated from the theory, is negative, and that for an emission process is positive, so that, equating these calculated probabilities to (3.30) and (3.31) respectively, one obtains positive values for both B and A. Generally, it is easily verified that any radiative transition probability coefficient obtained by this method is positive.

It now becomes reasonable to assume that these probability coefficients obtained for a hypothetical world are the same as those of the actual world. One gets in this way a general physical interpretation for the quantum theory of photons. When applied to elementary examples, it gives the same results as Heisenberg and Pauli's quantum electrodynamics with neglect of the divergent integrals, since the extra factor $\sqrt{2}/2$ occurring in the matrix elements of the present theory owing to the $\sqrt{2}/2$ in the right–hand side of (3.25) compensates the factor $1/2$ in the right–hand side of (3.30) or (3.31). The present general method of physical interpretation is probably applicable to any kind of particle with an integral spin [Dir32, Dir26e, Dir36, Cha48].

Therefore, it appears that, whether one is dealing with particles of integral spin or of half-odd integral spin, one is led to a similar conclusion, namely, that the mathematical methods at present in use in quantum mechanics are capable of direct interpretation only in terms of a hypothetical world differing very markedly from the actual one. These mathematical methods can be made into a physical theory by the assumption that results about collision processes are the same for the hypothetical world as the actual one. One thus gets back to Heisenberg's view about physical theory, that all it does is to provide a consistent means of calculating experimental results. The limited kind of description of Nature which Schrödinger's method provides in the non–relativistic case is possible relativistically only for the hypothetical world, and even then is rather more indefinite (e.g., the principle of superposition of states no longer applies), because of the need to use a Gibbs ensemble for describing the photon distribution.

To have a description of Nature is philosophically satisfying, though not logically necessary, and it is somewhat strange that the attempt to get such a description should meet with a partial success, namely, in the non–relativistic domain, but yet should fail completely in the later development. It seems to suggest that the present mathematical methods are not final. Any improvement in them would have to be of a very drastic character, because the source of all the trouble, the symmetry between positive and negative energies arising from the association of energies with the Fourier components of functions of the time, is a fundamental feature of them [Dir32, Dir26e, Dir36, Cha48].

3.2.4 Dirac's Electrodynamics Action Principle

There are various forms which the action principle of classical electrodynamics may take, but most of them involve awkward conditions concerning the singularities of the field where the charged particles are situated and are not suitable for a subsequent passage to quantum mechanics.

Fokker [Fok29] set up a form of action principle which does not refer to the singularities of the field and which appears to be the best starting point for getting a quantum theory. *Fokker's action integral* may conveniently be written with the help of the $\delta-$function as

$$S = S_1 + S_2, \qquad \text{where}$$

$$S_1 = \sum_i m_i \int ds_i \qquad \text{and} \tag{3.32}$$

$$S_2 = \sum_i \sum_{j \neq i} e_i e_j \int \int \delta(\mathbf{z}_i - \mathbf{z}_j)^2 (\mathbf{v}_i, \mathbf{v}_j) ds_i ds_j \tag{3.33}$$

Here, the scalar product notation is used as

$$(\mathbf{a}, \mathbf{b}) = a^\mu b_\mu = a_0 b_0 - a_1 b_1 - a_2 b_2 - a_3 b_3,$$

and m_i and e_i are the mass and charge of the ith particle, the $4-$vector $\mathbf{z}_i$ gives the four coordinates of the point on the world–line of the ith particle whose proper–time is s_i, and $\mathbf{v}_i$ is the velocity $4-$vector of the ith particle satisfying

$$\mathbf{v}_i = \frac{d\mathbf{z}_i}{ds_i}, \tag{3.34}$$

$$\mathbf{v}_i^2 = 1. \tag{3.35}$$

The integrals in (3.32–3.33) are taken along the world–lines of the particles, and the occurrence of the $\delta-$function $\delta(\mathbf{z}_i - \mathbf{z}_j)^2$ in S_2 ensures that the only values for $\mathbf{z}_i$ and $\mathbf{z}_j$ contributing to the double integral are those for which $(\mathbf{z}_i - \mathbf{z}_j)^2 = 0$, which means that each of the points $\mathbf{z}_i, \mathbf{z}_j$ is on the past or future light–cone from the other.

The action integral as it stands is not a general one covering all possible states of motion. To make it general one must, as has been pointed out by the Dirac (1938), add to it a term of the form

$$S_3 = \sum_i e_i \int M_\mu(\mathbf{z}_i) \mathbf{v}_i^\mu ds_i. \tag{3.36}$$

The 4−vector potential $M_\mu(\mathbf{x})$ may be left for the present an arbitrary function of the field point $\mathbf{x}$.

For the purpose of deducing the equations of motion, one may take the limits of integration in the various integrals to be $-\infty$ and ∞, as was done by Fokker, but in order to introduce momenta and get the equations into Hamiltonian form one must take finite limits. Let us therefore suppose that each s_i goes from s_i^0 to s_i', and let the corresponding $\mathbf{z}_i$ and $\mathbf{v}_i$ be $\mathbf{z}_i^0$, $\mathbf{z}_i'$ and $\mathbf{v}_i^0$, $\mathbf{v}_i'$. It is desirable to restrict the initial values s_i^0 so that the points $\mathbf{z}_i^0$ all lie outside each other's light–cones, and similarly with the final values s_i'. Thus

$$(\mathbf{z}_i^0 - \mathbf{z}_j^0)^2 < 0, \qquad (\mathbf{z}_i' - \mathbf{z}_j')^2 < 0, \qquad (i \neq j). \tag{3.37}$$

Now, before making variations in S, one should replace S_1, by

$$S_1' = \sum_i m_i \int \sqrt{\mathbf{v}_i^2 ds_i}, \tag{3.38}$$

so as to make S homogeneous of degree zero in the differential elements ds_i, $\mathbf{v}_i$ counting as being of degree -1 [Dir26e]. The expression for S is then valid with s_i any parameter on the world–line of the ith particle, so that $\mathbf{v}_i$ defined by (3.34) does not necessarily satisfy (3.35).

Let us now make variations $\partial \mathbf{z}_i(s_i)$ in the world–lines of the particles, $\partial \mathbf{M}(\mathbf{x})$ in the field function $\mathbf{M}(\mathbf{x})$, and Ds_i' in the final values of the s_i, so that the end–points of the world–lines are changed by

$$D\mathbf{z}_i' = \partial \mathbf{z}_i' + \mathbf{v}_i' Ds_i', \tag{3.39}$$

$\partial \mathbf{z}_i'$ being written for $\partial \mathbf{z}_i(s_i')$. The initial values of the s_i and the initial points of the world–lines we suppose for simplicity to be fixed, since variations in them would give rise to terms of the same form as those arising from variations in the final values and would not lead to anything new.

Varying S_1' given by (3.38) and using (3.35), after the variation process, one gets with the help of (3.39),

$$S_1' = \sum_i m_i [- \int_{s_i^0}^{s_i'} \left(\frac{d\mathbf{v}_i}{ds_i}, \partial \mathbf{z}_i \right) ds_i + (\mathbf{v}_i', D\mathbf{z}_i')]. \tag{3.40}$$

To get the variation in S_2 given by (3.33) one may, owing to the symmetry between i and j in the double sum, vary only quantities involving i and multiply by 2. The result is [Dir36]

$$\begin{aligned}\partial S_2 = \sum_i \sum_{j \neq i} e_i e_j \{ &\int_{s_i^0}^{s_i'} \int_{s_j^0}^{s_j'} [\frac{\partial \delta (\mathbf{z}_i - \mathbf{z}_j)^2}{\partial \mathbf{z}_i} (\mathbf{v}_i, \mathbf{v}_j) - \frac{d}{ds_i} [\delta (\mathbf{z}_i - \mathbf{z}_j)^2 \mathbf{v}_j]] \partial \mathbf{z}_i ds_i ds_j \\ &+ \int_{s_i^0}^{s_i'} \delta (\mathbf{z}_i' - \mathbf{z}_j)^2 (\mathbf{v}_j', D\mathbf{z}_i') ds_j \}.\end{aligned} \tag{3.41}$$

Finally, in varying S_3 given by (3.36), one has to take into account that the total variation in $\mathbf{M}$ at a point $\mathbf{z}_i(s_i)$ on the ith world–line, let us call it $D\mathbf{M}(\mathbf{z}_i)$, consists of two parts, a part $\partial\mathbf{M}(\mathbf{z}_i)$ arising from the variation in the function $\mathbf{M}(\mathbf{x})$ and equal to the value of $\partial\mathbf{M}(\mathbf{x})$ at the point $\mathbf{x} = \mathbf{z}_i$, and a part arising from the variation in $\mathbf{z}_i$, equal to $\partial\mathbf{M}/\partial\mathbf{x}$, at the point $\mathbf{x} = \mathbf{z}_i$ multiplied into $\partial\mathbf{z}_i$; thus

$$D\mathbf{M}(\mathbf{z}_i) = \partial\mathbf{M}(\mathbf{z}_i) + (\partial\mathbf{M}/\partial\mathbf{x})_{\mathbf{z}_i}\partial\mathbf{z}_i. \tag{3.42}$$

The variation in S_3 is now [Dir36]

$$\partial S_3 = \sum_i e_i\{\int_{s_i^0}^{s_i'} \left[\partial M^\mu(\mathbf{z}_i)v_{\mu i} + \left(\frac{\partial M^\mu}{\partial x_\nu}\right)_{\mathbf{z}_i} v_{\mu i}\partial z_{\nu i} - \frac{dM^\mu(\mathbf{z}_i)}{ds_i}\partial z_{\mu i}\right] ds_i$$
$$+ M^\mu(\mathbf{z}_i')Dz_{\mu i}'\}. \tag{3.43}$$

The total variation in S is given by the sum of the three expressions (3.40), (3.41) and (3.43).

By equating to zero the total coefficient of $\partial z_{\mu i}$, one gets the equation of motion of the ith particle. It is

$$-m_i\frac{dv_i^\mu}{ds_i} + e_i\sum_{j\neq i} e_j \int_{s_j^0}^{s_j'} \left[\frac{\partial\delta(\mathbf{z}_i - \mathbf{z}_j)^2}{\partial\mathbf{z}_i}(\mathbf{v}_i, \mathbf{v}_j) - \frac{d}{ds_i}[\delta(\mathbf{z}_i - \mathbf{z}_j)^2\mathbf{v}_j]\right] ds_j$$
$$+ e_i\left[\left(\frac{\partial M^\mu}{\partial x_\nu}\right)_{\mathbf{z}_i} v_{\mu i} - \frac{dM^\mu(\mathbf{z}_i)}{ds_i}\right] = 0.$$

Introducing the field function

$$A_i^\mu(\mathbf{x}) = M^\mu(\mathbf{x}) + \sum_{j\neq i} e_j \int_{s_j^0}^{s_j'} \partial\delta(\mathbf{x} - \mathbf{z}_j)^2 v_j^\mu ds_j, \tag{3.44}$$

the above equation of motion may be written

$$m_i\frac{dv_i^\mu}{ds_i} = e_i\left[\left(\frac{\partial A_i^\mu}{\partial x_\nu}\right)_{\mathbf{z}_i} v_{\mu i} - \frac{dA^\mu(\mathbf{z}_i)}{ds_i}\right] = e_i\left[\frac{\partial A_i^\mu}{\partial x_\nu} - \frac{\partial A_i^\nu}{\partial x_\mu}\right]_{\mathbf{z}_i} v_{\mu i}. \tag{3.45}$$

It is the correct *Lorentz equation of motion* of the ith particle, provided A_i^μ is connected with the ingoing and outgoing fields and the retarded and advanced fields of the other particles by the relation, given by Dirac (1938),

$$A_i^\mu = \frac{1}{2}[A_{\text{in}}^\mu + A_{\text{out}}^\mu] + \frac{1}{2}\sum_{j\neq i}[A_{j\text{ret}}^\mu + A_{j\text{adv}}^\mu], \qquad \text{or}$$

$$A_i^\mu(\mathbf{x}) = \frac{1}{2}[A_{\text{in}}^\mu(\mathbf{x}) + A_{\text{out}}^\mu(\mathbf{x})] + \sum_{j\neq i} e_j \int_{-\infty}^{\infty} \delta(\mathbf{x} - \mathbf{z}_j)^2 v_j^\mu ds_j. \tag{3.46}$$

According to (3.44) this requires (in Dirac's notation for integrals)

$$M^{\mu}(\mathbf{x}) = \frac{1}{2}[A^{\mu}_{\text{in}}(\mathbf{x}) + A^{\mu}_{\text{out}}(\mathbf{x})] + \sum_j e_j \left[\int_{-\infty}^{s_j^0} + \int_{s_j'}^{\infty} \right] \delta(\mathbf{x} - \mathbf{z}_j)^2 v_j^{\mu} ds_j, \quad (3.47)$$

Note that we are summing here over all values of j [Dir36, Dir26e], as we are dealing with a space–time region which lies inside the future light–cone from $\mathbf{z}_i^0$ and inside the past light–cone from $\mathbf{z}_i'$. By assuming that (3.47) holds throughout space–time, one gets an expression for $M^{\mu}(\mathbf{x})$ independent of i, so that the equations of motion of all the particles follow from the same Fokker's action integral.

One can now pass to the Hamiltonian formulation of the equations of motion. For each point in space–time $\mathbf{x}$, $M^{\mu}(\mathbf{x})$ may be counted as a coordinate, depending on the proper–times s_i'[5], and will have a conjugate momentum, say $K_{\mu}(\mathbf{x})$. These momenta, together with the particle momenta p_i^{μ}, are defined, as in the general theory [Wei36], by the coefficients of $\partial M^{\mu}(\mathbf{x})$ and $Dz'_{\mu i}$ in the expression for ∂S, so that we have

$$\partial S = \sum_i p_i^{\mu} Dz'_{\mu i} + \int_{-\infty}^{\infty} K_{\mu}(\mathbf{x}) \partial M^{\mu}(\mathbf{x}) dx_0 dx_1 dx_2 dx_3, \quad (3.48)$$

where the integral sign denotes the quadruple space–time integral. Comparing (3.48) with the sum of (3.40), (3.41) and (3.43), one gets [Dir26e, Dir36]

$$K_{\mu}(\mathbf{x}) = \sum_i e_i \int_{s_i^0}^{s_i'} \delta(x_0 - z_{0i}) \delta(x_1 - z_{1i}) \delta(x_2 - z_{2i}) \delta(x_3 - z_{3i}) v_{\mu i} ds_i \quad (3.49)$$

$$\text{and} \quad p_i^{\mu} = m_i v_i'^{\mu} + e_i [M^{\mu}(\mathbf{z}_i') + \frac{1}{2} \sum_j e_j \int_{s_j^0}^{s_j'} \Delta(\mathbf{z}_i' - \mathbf{z}_j + \boldsymbol{\lambda}) v_j^{\mu} ds_j], \quad (3.50)$$

where $\boldsymbol{\lambda}$ is a small 4–vector whose direction is within the future light–cone (so that $\boldsymbol{\lambda}^2 > 0, \lambda_0 > 0$), $\Delta(\mathbf{y})$ denotes the Jordan and Pauli (1928) Δ–function of any 4–vector $\mathbf{y}$, satisfying the 4D wave equation (here $\Box$ is the *Dalambertian wave operator*, $\Box = \partial_t^2 - \partial_x^2 + \partial_y^2 + \partial_z^2$)

$$\Box \Delta(\mathbf{y}) = 0 \qquad \text{which implies} \qquad \Box M^{\mu}(\mathbf{y}) = 0,$$

and related to the corresponding δ–function by

$$\Delta(\mathbf{y}) = \pm 2\delta(\mathbf{y}^2).$$

The momenta satisfy the *Poisson bracket* commutation relationships

$$[p_{\mu i}, z_{vj}] = g_{\mu v} \delta_{ij}, \quad (3.51)$$
$$[K_{\mu}(\mathbf{x}), M_{\nu}(\mathbf{x}')] = g_{\mu v} \delta(x_0 - x_0') \delta(x_1 - x_1') \delta(x_2 - x_2') \delta(x_3 - x_3'), \quad (3.52)$$

[5] It also depends on the proper–times s_i^0, but this does not concern us here.

so that the Poisson bracket of any two momenta or of any two coordinates vanishes. Instead of $K_\mu(\mathbf{x})$ it is more convenient to work with the momentum field–function $N_\mu(\mathbf{x})$ defined by [Dir58, Dir29]

$$N_\mu(\mathbf{x}) = \frac{1}{2}\int_{-\infty}^{\infty} \Delta(\mathbf{x}-\mathbf{x}')K_\mu(\mathbf{x}')dx'_0 dx'_1 dx'_2 dx'_3, \tag{3.53}$$

and satisfying $\quad \Box N_\mu(\mathbf{x}) = 0. \quad$ (3.54)

Instead of (3.52) one has

$$[N_\mu(\mathbf{x}), M_\nu(\mathbf{x}')] = \frac{1}{2} g_{\mu\upsilon}\Delta(\mathbf{x}-\mathbf{x}'). \tag{3.55}$$

From (3.53) and (3.49) one gets

$$N_\mu(\mathbf{x}) = \frac{1}{2}\sum_i e_i \int_{s_i^0}^{s_i'} \Delta(\mathbf{x}-\mathbf{z}_i)v_{\mu\upsilon} ds_i, \tag{3.56}$$

so that (3.50) may be written

$$\begin{aligned} p_i^\mu &= m_i v_i'^\mu + e_i[M^\mu(\mathbf{z}_i') + N^\mu(\mathbf{z}_i' + \boldsymbol{\lambda})] \\ &= m_i v_i'^\mu + e_i A^\mu(\mathbf{z}_i'), \end{aligned} \tag{3.57}$$

where $\quad A^\mu(\mathbf{x}) = M^\mu(\mathbf{x}) + N^\mu(\mathbf{x}+\boldsymbol{\lambda}). \quad$ (3.58)

From (3.54) the potentials $A_\mu(\mathbf{x})$ satisfy

$$\Box A_\mu(\mathbf{x}) = 0, \tag{3.59}$$

showing that they can be resolved into waves travelling with the velocity of light, and from (3.55) it follows

$$[A_\mu(\mathbf{x}), A_\nu(\mathbf{x}')] = \frac{1}{2} g_{\mu\upsilon}[\Delta(\mathbf{x}-\mathbf{x}'+\boldsymbol{\lambda}) + \Delta(\mathbf{x}-\mathbf{x}'-\boldsymbol{\lambda})]. \tag{3.60}$$

From (3.35) and (3.57) it follows

$$F_i \equiv [p_i - e_i \mathbf{A}(\mathbf{z}_i')]^2 - m_i^2 = 0. \tag{3.61}$$

There is one of these equations for each particle. The expressions F_i may be used as Hamiltonians to determine how any dynamical variable ξ varies with the proper–times s_i', in accordance with the equations [Dir58, Dir26e, Dir49]

$$\kappa_i \frac{d\xi}{ds_i'} = [\xi, F_i], \tag{3.62}$$

were ξ is any function of the coordinates and momenta of the particles and of the fields $\mathbf{M}, \mathbf{K}, \mathbf{N}, \mathbf{A}$, and the κ's are multiplying factors not depending on ξ. Taking $\xi = z'_{\mu i}$, one finds that

$$\kappa_i = -2m_i,$$

to get agreement with (3.57). Taking $\xi = p_i^\mu$ gives one back the equation of motion (3.45) with the $\boldsymbol{\lambda}$ refinement. Taking $\xi = M_\mu(\mathbf{x})$, one gets from (3.58) and (3.55),

$$\frac{M_\mu(\mathbf{x})}{ds_i'} = e_i v_i'^\nu [M_\mu(\mathbf{x}), A_\nu(\mathbf{z}_i')] = \frac{1}{2} e_i v_{\mu i}' \Delta(\mathbf{x} - \mathbf{z}_i' - \boldsymbol{\lambda}).$$

This equation of motion for the field quantities $M_\mu(\mathbf{x})$ does not follow from the variation principle, as it involves only coordinates and velocities and not accelerations, and it has to be imposed as an extra condition in the variational method.

The above Hamiltonian formulation of the equations of classical electrodynamics may be taken over into the quantum theory in the usual way, by making the momenta into operators satisfying commutation relations corresponding to the Poisson bracket relations (3.51), (3.52). Equation (3.60) in the limit $\boldsymbol{\lambda} \to O$ goes over into the quantum equation (3.22). The Hamiltonians (3.61) provide the wave equations

$$F_i \psi = 0,$$

in which the wave $\psi-$function is a function of the coordinates $\mathbf{z}_i'$ of all the particles and of the field variables $M_\mu(\mathbf{x})$. One can apply the theory to spinning electrons instead of spinless particles, by modifying the Hamiltonians F_i in the appropriate way. For more details, see [Dir58, Dir26e, Dir49].

4

Complex Manifolds

In this Chapter we develop the concept of a complex smooth manifold, which is the essential tool in high–dimensional nonlinear complex–valued dynamics.

4.1 Smooth Manifolds

4.1.1 Intuition and Definition of a Smooth Manifold

Intuition Behind a Smooth Manifold

As we have already got the initial feeling, in the heart of geometrical dynamics is the concept of a *manifold* (see, e.g., [Rha84]). To get some *dynamical intuition* behind this concept, let us consider a simple 3DOF mechanical system determined by three generalized coordinates, $q^i = \{q^1, q^2, q^3\}$. There is a unique way to represent this system as a 3D manifold, such that to each point of the manifold there corresponds a definite configuration of the mechanical system with coordinates q^i; therefore, we have a geometrical representation of the configurations of our mechanical system, called the *configuration manifold*. If the mechanical system moves in any way, its coordinates are given as the functions of the time. Thus, the motion is given by equations of the form: $q^i = q^i(t)$. As t varies (i.e., $t \in \mathbb{R}$), we observe that the system's *representative point* in the configuration manifold describes a *curve* and $q^i = q^i(t)$ are the equations of this curve.

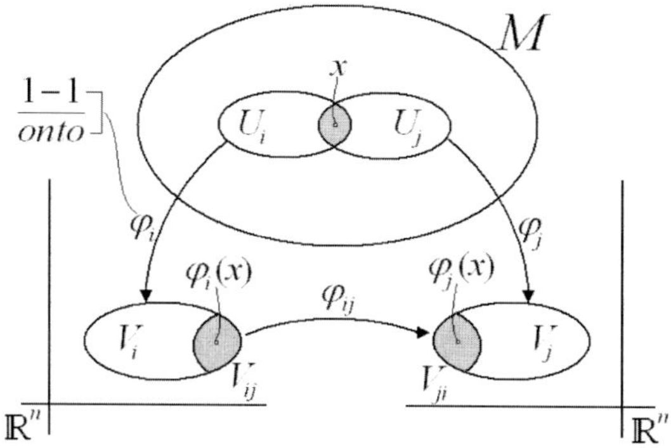

Fig. 4.1. An intuitive geometrical picture behind the manifold concept (see text).

On the other hand, to get some *geometrical intuition* behind the concept of a manifold, consider a set M (see Figure 4.1) which is a *candidate* for a manifold. Any point $x \in M$[1] has its *Euclidean chart*, given by a 1–1 and *onto* map $\varphi_i : M \to \mathbb{R}^n$, with its *Euclidean image* $V_i = \varphi_i(U_i)$. More precisely, a chart φ_i is defined by

$$\varphi_i : M \supset U_i \ni x \mapsto \varphi_i(x) \in V_i \subset \mathbb{R}^n,$$

where $U_i \subset M$ and $V_i \subset \mathbb{R}^n$ are open sets (see [Arn78, Rha84]).

Clearly, any point $x \in M$ can have several different charts (see Figure 4.1). Consider a case of two charts, $\varphi_i, \varphi_j : M \to \mathbb{R}^n$, having in their images two open sets, $V_{ij} = \varphi_i(U_i \cap U_j)$ and $V_{ji} = \varphi_j(U_i \cap U_j)$. Then we have *transition functions* φ_{ij} between them,

$$\varphi_{ij} = \varphi_j \circ \varphi_i^{-1} : V_{ij} \to V_{ji}, \qquad \text{locally given by} \qquad \varphi_{ij}(x) = \varphi_j(\varphi_i^{-1}(x)).$$

If transition functions φ_{ij} exist, then we say that two charts, φ_i and φ_j are *compatible*. Transition functions represent a general (nonlinear) *transformations of coordinates*, which are the core of classical *tensor calculus* (Appendix).

A set of compatible charts $\varphi_i : M \to \mathbb{R}^n$, such that each point $x \in M$ has its Euclidean image in at least one chart, is called an *atlas*. Two atlases are *equivalent* iff all their charts are compatible (i.e., transition functions exist between them), so their union is also an atlas. A *manifold structure* is a class of equivalent atlases.

Finally, as charts $\varphi_i : M \to \mathbb{R}^n$ were supposed to be 1-1 and onto maps, they can be either *homeomorphisms*, in which case we have a *topological* (C^0) manifold, or *diffeomorphisms*, in which case we have a *smooth* (C^∞) manifold.

[1] Note that sometimes we will denote the point in a manifold M by m, and sometimes by x (thus implicitly assuming the existence of coordinates $x = (x^i)$).

Slightly more precisely, a topological (respectively smooth) manifold is a separable space M which is locally homeomorphic (resp. diffeomorphic) to Euclidean space $\mathbb{R}^n$, having the following properties (reflected in Figure 4.1):

1. M is a *Hausdorff space*: For every pair of points $x_1, x_2 \in M$, there are disjoint open subsets $U_1, U_2 \subset M$ such that $x_1 \in U_1$ and $x_2 \in U_2$.
2. M is *second–countable space*: There exists a countable basis for the topology of M.
3. M *is locally Euclidean of dimension* n: Every point of M has a neighborhood that is homeomorphic (resp. diffeomorphic) to an open subset of $\mathbb{R}^n$.

This implies that for any point $x \in M$ there is a homeomorphism (resp. diffeomorphism) $\varphi : U \to \varphi(U) \subseteq \mathbb{R}^n$, where U is an open neighborhood of x in M and $\varphi(U)$ is an open subset in $\mathbb{R}^n$. The pair (U, φ) is called a *coordinate chart* at a point $x \in M$, etc.

Definition of a Smooth Manifold

Given a chart (U, φ), we call the set U a *coordinate domain*, or a coordinate neighborhood of each of its points. If in addition $\varphi(U)$ is an open ball in $\mathbb{R}^n$, then U is called a *coordinate ball*. The map φ is called a (*local*) *coordinate map*, and the component functions $(x^1, ..., x^n)$ of φ, defined by $\varphi(m) = (x^1(m), ..., x^n(m))$, are called *local coordinates* on U.

Two charts (U_1, φ_1) and (U_2, φ_2) such that $U_1 \cap U_2 \neq \varnothing$ are called *compatible* if $\varphi_1(U_1 \cap U_2)$ and $\varphi_2(U_2 \cap U_1)$ are open subsets of $\mathbb{R}^n$. A family $(U_\alpha, \varphi_\alpha)_{\alpha \in A}$ of compatible charts on M such that the U_α form a *covering* of M is called an *atlas*. The maps $\varphi_{\alpha\beta} = \varphi_\beta \circ \varphi_\alpha^{-1} : \varphi_\alpha(U_{\alpha\beta}) \to \varphi_\beta(U_{\alpha\beta})$ are called the *transition maps*, for the atlas $(U_\alpha, \varphi_\alpha)_{\alpha \in A}$, where $U_{\alpha\beta} = U_\alpha \cap U_\beta$, so that we have a commutative triangle:

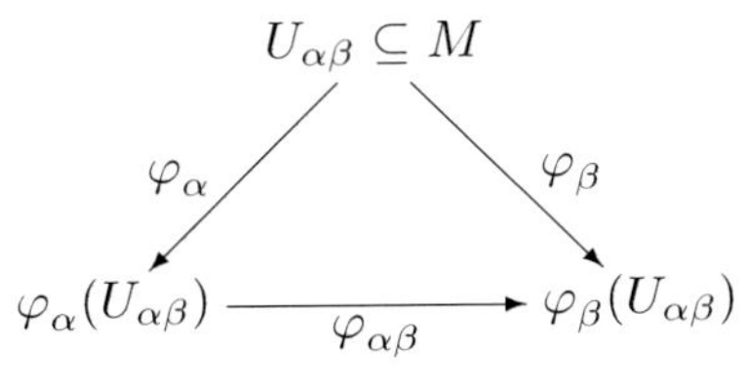

An atlas $(U_\alpha, \varphi_\alpha)_{\alpha \in A}$ for a manifold M is said to be a $C^\infty-$*atlas*, if all transition maps $\varphi_{\alpha\beta} : \varphi_\alpha(U_{\alpha\beta}) \to \varphi_\beta(U_{\alpha\beta})$ are of class C^∞. Two C^∞ atlases are called $C^\infty-$*equivalent*, if their union is again a $C^\infty-$atlas for M. An equivalence class of $C^\infty-$atlases is called a $C^\infty-$*structure* on M. In other words, a smooth structure on M is a *maximal* smooth atlas on M, i.e., such an atlas that is not contained in any strictly larger smooth atlas. By a $C^\infty-$*manifold* M, we mean a topological manifold together with a $C^\infty-$structure and a chart on M will be a chart belonging to some atlas of the $C^\infty-$structure. Smooth manifold means $C^\infty-$manifold, and the word '*smooth*' is used synonymously for C^∞ [Rha84].

Sometimes the terms 'local coordinate system' or 'parametrization' are used instead of charts. That M is not defined with any particular atlas, but with an equivalence class of atlases, is a mathematical formulation of the *general covariance* principle. Every suitable coordinate system is equally good. A Euclidean chart may well suffice for an open subset of $\mathbb{R}^n$, but this coordinate system is not to be preferred to the others, which may require many charts (as with polar coordinates), but are more convenient in other respects.

For example, the atlas of an $n-$sphere S^n has two charts. If $N = (1, 0, ..., 0)$ and $S = (-1, ..., 0, 0)$ are the north and south poles of S^n respectively, then the two charts are given by the stereographic projections from N and S:

$$\varphi_1 : S^n \backslash \{N\} \to \mathbb{R}^n, \varphi_1(x^1, ..., x^{n+1}) = (x^2/(1 - x^1), \dots, x^{n+1}/(1 - x^1)), \quad \text{and}$$
$$\varphi_2 : S^n \backslash \{S\} \to \mathbb{R}^n, \varphi_2(x^1, ..., x^{n+1}) = (x^2/(1 + x^1), \dots, x^{n+1}/(1 + x^1)),$$

while the overlap map $\varphi_2 \circ \varphi_1^{-1} : \mathbb{R}^n \backslash \{0\} \to \mathbb{R}^n \backslash \{0\}$ is given by the diffeomorphism $(\varphi_2 \circ \varphi_1^{-1})(z) = z/||z||^2$, for z in $\mathbb{R}^n \backslash \{0\}$, from $\mathbb{R}^n \backslash \{0\}$ to itself.

Various *additional structures* can be imposed on $\mathbb{R}^n$, and the corresponding manifold M will inherit them through its covering by charts. For example, if a covering by charts takes their values in a *Banach space* E, then E is called the *model space* and M is referred to as a $C^\infty-$*Banach manifold* modelled on E. Similarly, if a covering by charts takes their values in a *Hilbert space* $\mathcal{H}$, then $\mathcal{H}$ is called the *model space* and M is referred to as a $C^\infty-$*Hilbert manifold* modelled on $\mathcal{H}$. If not otherwise specified, we will consider M to be an Euclidean manifold, with its covering by charts taking their values in $\mathbb{R}^n$.

For a Hausdorff $C^\infty-$manifold the following properties are equivalent [KMS93]: (i) it is paracompact; (ii) it is metrizable; (iii) it admits a Riemannian metric;[2] (iv) each connected component is separable.

Smooth Maps Between Manifolds

A map $\varphi : M \to N$ between two manifolds M and N, with $M \ni m \mapsto \varphi(m) \in N$, is called a *smooth map*, or $C^\infty-$map, if we have the following charting:

[2] Recall the corresponding properties of a *Euclidean metric* d. For any three points $x, y, z \in \mathbb{R}^n$, the following axioms are valid:

$$M_1 : d(x, y) > 0, \quad \text{for } x \neq y; \qquad \text{and} \qquad d(x, y) = 0, \quad \text{for } x = y;$$
$$M_2 : d(x, y) = d(y, x); \qquad M_3 : d(x, y) \leq d(x, z) + d(z, y).$$

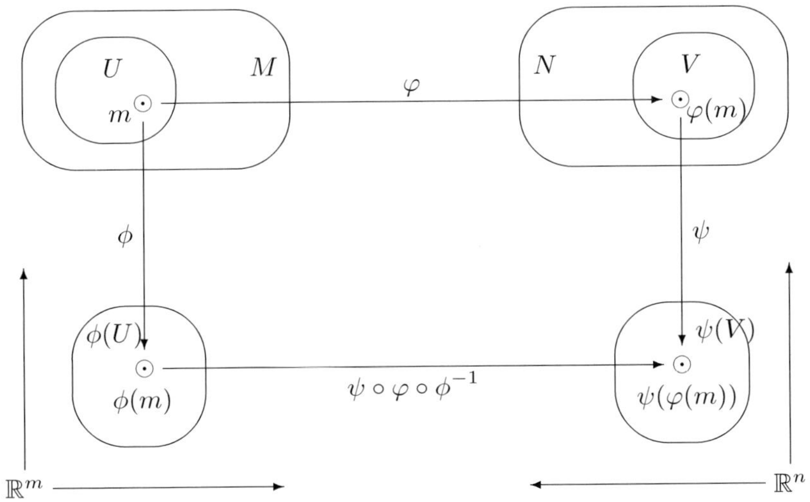

This means that for each $m \in M$ and each chart (V, ψ) on N with $\varphi(m) \in V$ there is a chart (U, ϕ) on M with $m \in U, \varphi(U) \subseteq V$, and $\Phi = \psi \circ \varphi \circ \phi^{-1}$ is C^∞, that is, the following diagram commutes:

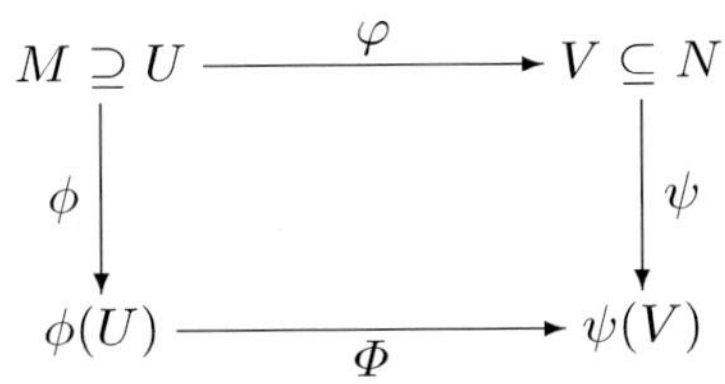

Let M and N be smooth manifolds and let $\varphi : M \to N$ be a smooth map. The map φ is called a *covering*, or equivalently, M is said to *cover* N, if φ is surjective and each point $n \in N$ admits an open neighborhood V such that $\varphi^{-1}(V)$ is a union of disjoint open sets, each diffeomorphic via φ to V.

A $C^\infty-$map $\varphi : M \to N$ is called a $C^\infty-$*diffeomorphism* if φ is a bijection, $\varphi^{-1} : N \to M$ exists and is also C^∞. Two manifolds are called diffeomorphic if there exists a diffeomorphism between them. All smooth manifolds and smooth maps between them form the category $\mathcal{M}$.

Intuition Behind Topological Invariants of Manifolds

Now, restricting to the topology of nD compact (i.e., closed and bounded) and connected manifolds, the only cases in which we have a complete understanding of topology are $n = 0, 1, 2$. The only compact and connected 0D manifold is a point. A 1D compact and connected manifold can either be a line element or a circle, and it is intuitively clear (and can easily be proven) that these two spaces are topologically different. In 2D, there is already an infinite number of different topologies: a 2D compact and connected surface can have an arbitrary number of handles and boundaries, and can either be orientable or non–orientable (see Figure 4.2). Again, it is intuitively quite clear that two surfaces are not homeomorphic if they differ in one of these respects. On the other hand, it can be proven that any two surfaces for which these data are the same can be continuously mapped to one another, and hence this gives a complete classification of the possible topologies of such surfaces.

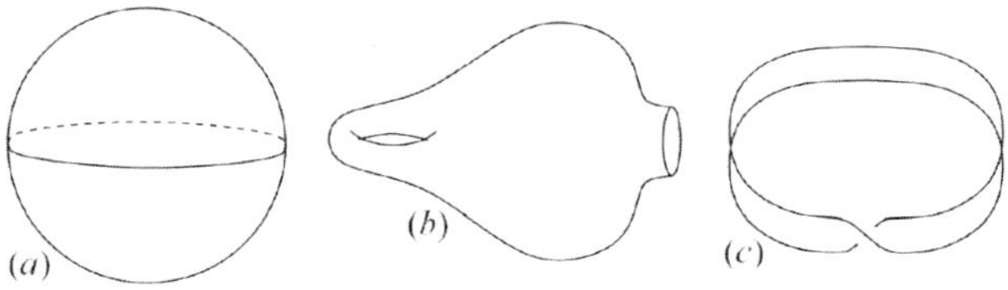

Fig. 4.2. Three examples of 2D manifolds: (*a*) The sphere S^2 is an orientable manifold without handles or boundaries. (*b*) An orientable manifold with one boundary and one handle. (*c*) The Möbius strip is an unorientable manifold with one boundary and no handles.

A quantity such as the number of boundaries of a surface is called a *topological invariant.* A topological invariant is a number, or more generally any type of structure, which one can associate to a topological space, and which does not change under continuous mappings. Topological invariants can be used to distinguish between topological spaces: if two surfaces have a different number of boundaries, they can certainly not be topologically equivalent. On the other hand, the knowledge of a topological invariant is in general not enough to decide whether two spaces are homeomorphic: a torus and a sphere have the same number of boundaries (zero), but are clearly not homeomorphic. Only when one has some *complete set* of topological invariants, such as the number of handles and boundaries in the 2D case, is it possible to determine whether or not two topological spaces are homeomorphic. In more than 2D, many topological invariants are known, but for no dimension larger than two has a complete set of topological invariants been found. In 3D, it is generally believed that a finite number of countable invariants would suffice for compact manifolds, but this is not rigorously proven, and in particular there is at present no generally accepted construction of a complete set. A very interesting and intimately related problem is the famous *Poincaré conjecture*, stating that if a 3D manifold has a certain set of topological invariants called its 'homotopy groups' equal to those of the 3–sphere S^3, it is actually homeomorphic to the three-sphere. In four or more dimensions, a complete set of topological invariants would consist of an uncountably infinite number of invariants! A general classification of topologies is therefore very hard to get, but even without such a general classification, each new invariant that can be constructed gives us a lot of interesting new information. For this reason, the construction of topological invariants of manifolds is one of the most important issues in topology.

4.1.2 (Co)Tangent Bundles of a Smooth Manifold

Intuition Behind a Tangent Bundle

In mechanics, to each nD *configuration manifold* M there is associated its $2n$D *velocity phase–space manifold*, denoted by TM and called the tangent bundle of M (see Figure 4.3). The original smooth manifold M is called the *base* of TM. There is an onto map $\pi : TM \longrightarrow M$, called the *projection*. Above each point $x \in M$ there is a *tangent space* $T_xM = \pi^{-1}(x)$ to M at x, which is called a *fibre*. The fibre $T_xM \subset TM$ is the subset of TM, such that the total tangent bundle, $TM = \bigsqcup_{m\in M} T_xM$, is a *disjoint union* of tangent spaces T_xM to M for all points $x \in M$. From dynamical perspective, the most important quantity in the tangent bundle concept is the smooth map $v : M \longrightarrow TM$, which is an inverse to the projection π, i.e, $\pi \circ v = \mathrm{Id}_M$, $\pi(v(x)) = x$. It is called the *velocity vector–field*. Its graph $(x, v(x))$ represents the *cross–section* of the tangent bundle TM. This explains the dynamical term velocity phase–space, given to the tangent bundle TM of the manifold M.

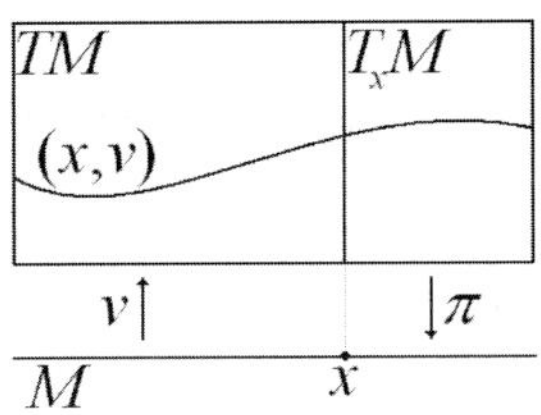

Fig. 4.3. A sketch of a tangent bundle TM of a smooth manifold M (see text for explanation).

Definition of a Tangent Bundle

Recall that if $[a, b]$ is a closed interval, a C^0–map $\gamma : [a, b] \rightarrow M$ is said to be *differentiable* at the endpoint a if there is a chart (U, ϕ) at $\gamma(a)$ such that the following limit exists and is finite [AMR88]:

$$\frac{d}{dt}(\phi \circ \gamma)(a) \equiv (\phi \circ \gamma)'(a) = \lim_{t\rightarrow a} \frac{(\phi \circ \gamma)(t) - (\phi \circ \gamma)(a)}{t-a}. \tag{4.1}$$

Generalizing (4.1), we get the notion of the *curve on a manifold*. For a smooth manifold M and a point $m \in M$ a curve at m is a C^0–map $\gamma : I \rightarrow M$ from an interval $I \subset \mathbb{R}$ into M with $0 \in I$ and $\gamma(0) = m$.

Two curves γ_1 and γ_2 passing though a point $m \in U$ are *tangent at* m with respect to the chart (U, ϕ) if $(\phi \circ \gamma_1)'(0) = (\phi \circ \gamma_2)'(0)$. Thus, two curves are tangent if they have identical tangent vectors (same direction and speed) in a local chart on a manifold.

For a smooth manifold M and a point $m \in M$, the *tangent space* $T_m M$ to M at m is the *set of equivalence classes* of curves at m:

$$T_m M = \{[\gamma]_m : \gamma \text{ is a curve at a point } m \in M\}.$$

A C^∞–map $\varphi : M \ni m \mapsto \varphi(m) \in N$ between two manifolds M and N induces a linear map $T_m\varphi : T_m M \to T_{\varphi(m)} N$ for each point $m \in M$, called a *tangent map*, if we have:

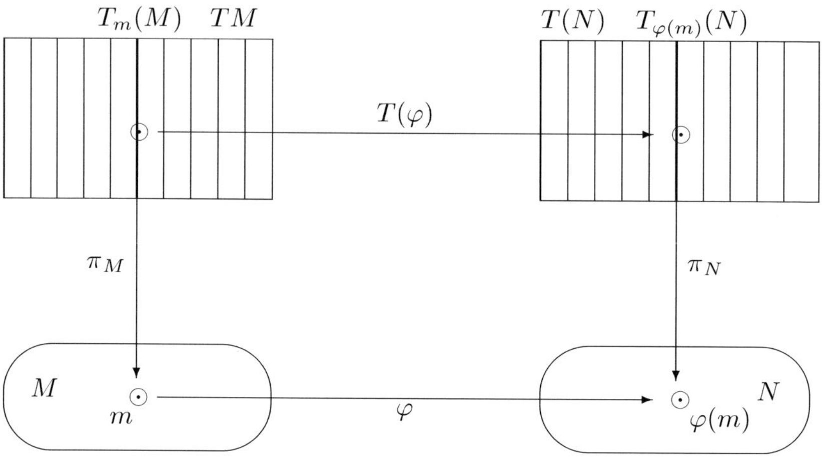

i.e., the following diagram commutes:

$$\begin{array}{ccc} T_m M & \xrightarrow{\;T_m\varphi\;} & T_{\varphi(m)} N \\ \downarrow{\scriptstyle \pi_M} & & \downarrow{\scriptstyle \pi_N} \\ M \ni m & \xrightarrow[\;\varphi\;]{} & \varphi(m) \in N \end{array}$$

with the *natural projection* $\pi_M : TM \to M$, given by $\pi_M(T_m M) = m$, that takes a tangent vector v to the point $m \in M$ at which the vector v is attached i.e., $v \in T_m M$.

For an nD smooth manifold M, its nD *tangent bundle* TM is the disjoint union of all its tangent spaces $T_m M$ at all points $m \in M$, $TM = \bigsqcup_{m \in M} T_m M$.

To define the smooth structure on TM, we need to specify how to construct local coordinates on TM. To do this, let $(x^1(m), ..., x^n(m))$ be local coordinates of a point m on M and let $(v^1(m), ..., v^n(m))$ be components of a tangent vector in this coordinate system. Then the $2n$ numbers $(x^1(m), ..., x^n(m), v^1(m), ..., v^n(m))$ give a *local coordinate system* on TM.

$TM = \bigsqcup_{m \in M} T_m M$ defines a family of vector spaces parameterized by M.
The inverse image $\pi_M^{-1}(m)$ of a point $m \in M$ under the natural projection π_M is the tangent space $T_m M$. This space is called the *fibre* of the tangent bundle over the point $m \in M$ [Ste72].

A C^∞–map $\varphi : M \to N$ between two manifolds M and N induces a linear *tangent map* $T\varphi : TM \to TN$ between their tangent bundles, i.e., the following diagram commutes:

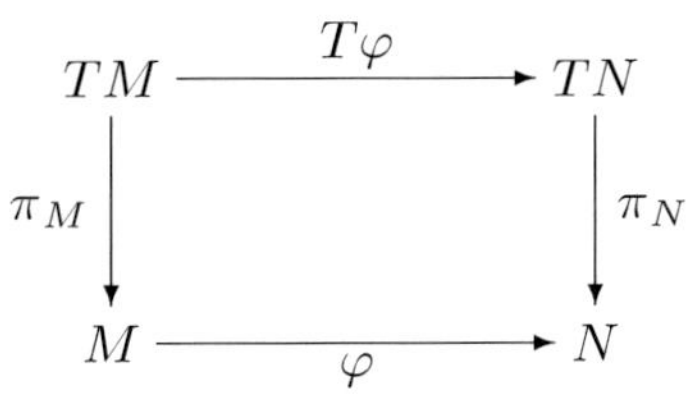

All tangent bundles and their tangent maps form the category $\mathcal{TB}$. The category $\mathcal{TB}$ is the natural framework for *Lagrangian dynamics*.

Now, we can formulate the *global version of the chain rule*. If $\varphi : M \to N$ and $\psi : N \to P$ are two smooth maps, then we have $T(\psi \circ \varphi) = T\psi \circ T\varphi$ (see [KMS93]). In other words, we have a functor $T : \mathcal{M} \Rightarrow \mathcal{TB}$ from the category $\mathcal{M}$ of smooth manifolds to the category $\mathcal{TB}$ of their tangent bundles:

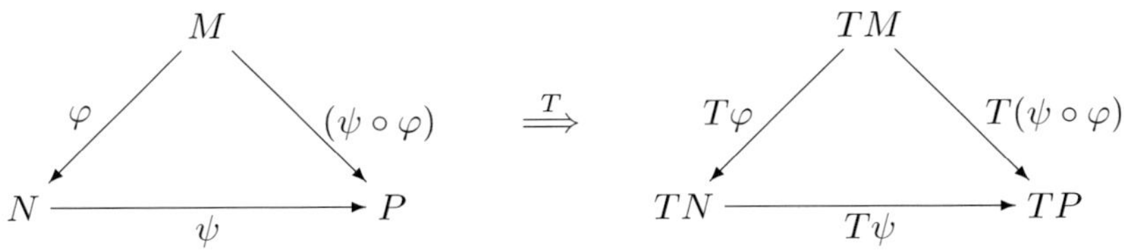

Definition of a Cotangent Bundle

A *dual* notion to the tangent space $T_m M$ to a smooth manifold M at a point m is its *cotangent space* $T_m^* M$ at the same point m. Similarly to the tangent bundle, for a smooth manifold M of dimension n, its *cotangent bundle* T^*M is the disjoint union of all its cotangent spaces $T_m^* M$ at all points $m \in M$, i.e., $T^*M = \bigsqcup_{m \in M} T_m^* M$. Therefore, the cotangent bundle of an n–manifold M is the vector bundle $T^*M = (TM)^*$, the (real) dual of the tangent bundle TM.

If M is an n–manifold, then T^*M is a $2n$–manifold. To define the smooth structure on T^*M, we need to specify how to construct local coordinates on T^*M. To do this, let $(x^1(m), ..., x^n(m))$ be local coordinates of a point m on M and let $(p_1(m), ..., p_n(m))$ be components of a covector in this coordinate system. Then the $2n$ numbers $(x^1(m), ..., x^n(m), \, p_1(m), ..., p_n(m))$ give a local coordinate system on T^*M. This is the basic idea one uses to prove that indeed T^*M is a $2n$–manifold.

$T^*M = \bigsqcup_{m\in M} T^*_m M$ defines a family of vector spaces parameterized by M, with the *conatural projection*, $\pi^*_M : T^*M \to M$, given by $\pi^*_M(T^*_m M) = m$, that takes a covector p to the point $m \in M$ at which the covector p is attached i.e., $p \in T^*_m M$. The inverse image $\pi^{-1}_M(m)$ of a point $m \in M$ under the conatural projection π^*_M is the cotangent space $T^*_m M$. This space is called the *fibre* of the cotangent bundle over the point $m \in M$.

In a similar way, a C^∞−map $\varphi : M \to N$ between two manifolds M and N induces a linear *cotangent map* $T^*\varphi : T^*M \to T^*N$ between their cotangent bundles, i.e., the following diagram commutes:

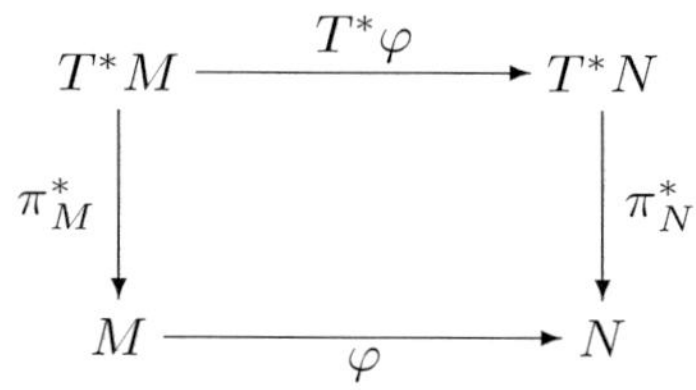

All cotangent bundles and their cotangent maps form the category $\mathcal{T}^*\mathcal{B}$. The category $\mathcal{T}^*\mathcal{B}$ is the natural stage for *Hamiltonian dynamics*.

Now, we can formulate the *dual version of the global chain rule.* If $\varphi : M \to N$ and $\psi : N \to P$ are two smooth maps, then we have $T^*(\psi\circ\varphi) = T^*\psi\circ T^*\varphi$. In other words, we have a cofunctor $T^* : \mathcal{M} \Rightarrow \mathcal{T}^*\mathcal{B}$ from the category $\mathcal{M}$ of smooth manifolds to the category $\mathcal{T}^*\mathcal{B}$ of their cotangent bundles:

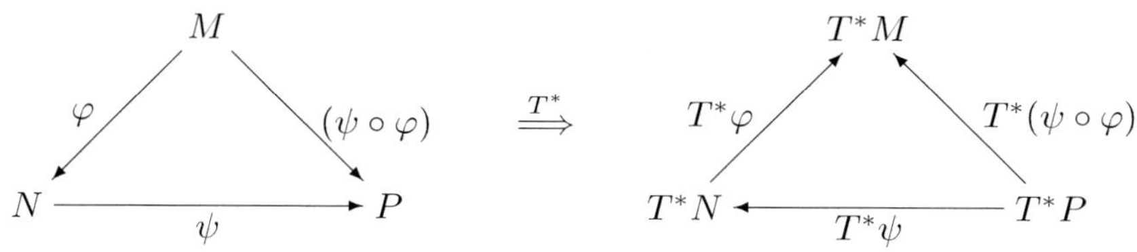

Tensor Fields and Bundles of a Smooth Manifold

A *tensor bundle* $\mathcal{T}$ associated to a smooth n−manifold M is defined as a *tensor product* of tangent and cotangent bundles:

$$\mathcal{T} = \bigotimes^{q} T^*M \otimes \bigotimes^{p} TM = \overbrace{TM \otimes ... \otimes TM}^{p \text{ times}} \otimes \overbrace{T^*M \otimes ... \otimes T^*M}^{q \text{ times}}.$$

Tensor bundles are special case of more general fibre bundles (see [II06b]).

A *tensor–field* of type (p, q) (see Appendix) on a smooth n−manifold M is defined as a *smooth section* $\tau : M \to \mathcal{T}$ of the tensor bundle $\mathcal{T}$. The coefficients of the tensor–field τ are smooth (C^∞) functions with p indices up and q indices down. The classical position of indices can be explained in modern terms as follows. If (U, ϕ) is a chart at a point $m \in M$ with local coordinates $(x^1, ..., x^n)$, we have the *holonomous frame field*

$$\partial_{x^{i_1}} \otimes \partial_{x^{i_2}} \otimes ... \otimes \partial_{x^{i_p}} \otimes dx^{j_1} \otimes dx^{j_2} ... \otimes dx^{j_q},$$

for $i \in \{1, ..., n\}^p$, $j = \{1, ..., n\}^q$, over U of this tensor bundle, and for any (p, q)–tensor–field τ we have

$$\tau|U = \tau^{i_1...i_p}_{j_1...j_q} \partial_{x^{i_1}} \otimes \partial_{x^{i_2}} \otimes ... \otimes \partial_{x^{i_p}} \otimes dx^{j_1} \otimes dx^{j_2} ... \otimes dx^{j_q}.$$

For such tensor–fields the *Lie derivative* along any vector–field is defined (see subsection 4.1.3 below), and it is a *derivation* (i.e., both linearity and Leibniz rules hold) with respect to the tensor product. Tensor bundle $\mathcal{T}$ admits many natural transformations (see [KMS93]). For example, a 'contraction' like the trace $T^*M \otimes TM = L(TM, TM) \to M \times \mathbb{R}$, but applied just to one specified factor of type T^*M and another one of type TM, is a natural transformation. And any 'permutation of the same kind of factors' is a natural transformation.

The tangent bundle $\pi_M : TM \to M$ of a manifold M (introduced above) is a special tensor bundle over M such that, given an atlas $\{(U_\alpha, \varphi_\alpha)\}$ of M, TM has the *holonomic atlas*

$$\Psi = \{(U_\alpha, \varphi_\alpha = T\varphi_\alpha)\}.$$

The associated linear bundle coordinates are the induced coordinates $(\dot{x}^\lambda)$ at a point $m \in M$ with respect to the *holonomic frames* $\{\partial_\lambda\}$ in tangent spaces $T_m M$. Their transition functions read (see Appendix)

$$\dot{x}'^\lambda = \frac{\partial x'^\lambda}{\partial x^\mu} \dot{x}^\mu.$$

Technically, the tangent bundle TM is a tensor bundle with the structure Lie group $GL(\dim M, \mathbb{R})$ (see section 4.1.3 below).

Recall that the cotangent bundle of M is the dual T^*M of TM. It is equipped with the induced coordinates $(\dot{x}_\lambda)$ at a point $m \in M$ with respect to *holonomic coframes* $\{dx^\lambda\}$ dual of $\{\partial_\lambda\}$. Their transition functions read

$$\dot{x}'_\lambda = \frac{\partial x'^\mu}{\partial x^\lambda} \dot{x}_\mu.$$

The Pull–Back and Push–Forward

In this subsection we define two important operations, following [AMR88], which will be used in the further text.

Let $\varphi : M \to N$ be a C^∞ map of manifolds and $f \in C^\infty(N, \mathbb{R})$. Define the *pull–back* of f by φ by

$$\varphi^* f = f \circ \varphi \in C^\infty(M, \mathbb{R}).$$

If f is a C^∞ diffeomorphism and $X \in \mathcal{X}^k(M)$, the *push–forward* of X by φ is defined by

$$\varphi_* X = T\varphi \circ X \circ \varphi^{-1} \in \mathcal{X}^k(N).$$

If x^i are local coordinates on M and y^j local coordinates on N, the preceding formula gives the components of $\varphi_* X$ by

$$(\varphi_* X)^j(y) = \frac{\partial \varphi^j}{\partial x^i}(x)\, X^i(x), \qquad \text{where} \quad y = \varphi(x).$$

We can interchange pull–back and push–forward by changing φ to φ^{-1}, that is, defining φ_* (resp. φ^*) by $\varphi_* = (\varphi^{-1})^*$ (resp. $\varphi^* = (\varphi^{-1})_*$). Thus the *push–forward* of a function f on M is $\varphi_* f = f \circ \varphi^{-1}$ and the *pull–back* of a vector–field Y on N is $\varphi^* Y = (T\varphi)^{-1} \circ Y \circ \varphi$.

Notice that φ must be a diffeomorphism in order that the pull–back and push–forward operations make sense, the only exception being pull–back of functions. Thus vector–fields can only be pulled back and pushed forward by diffeomorphisms. However, even when φ is not a diffeomorphism we can talk about *$\varphi-$related vector–fields* as follows.

Let $\varphi : M \to N$ be a C^∞ map of manifolds. The vector–fields $X \in \mathcal{X}^{k-1}(M)$ and Y$\in \mathcal{X}^{k-1}(N)$ are called *$\varphi-$related*, denoted $X \sim_\varphi Y$, if $T\varphi \circ X = Y \circ \varphi$.

Note that if φ is diffeomorphism and X and Y are $\varphi-$related, then $Y = \varphi_* X$. However, in general, X can be $\varphi-$related to more than one vector–field on N. $\varphi-$relatedness means that the following diagram commutes:

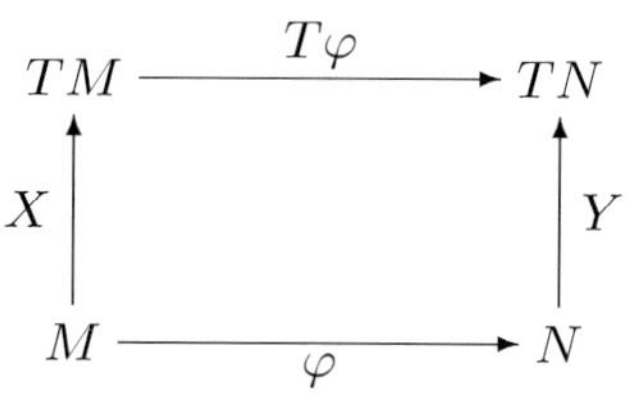

The behavior of flows under these operations is as follows: Let $\varphi : M \to N$ be a $C^\infty-$map of manifolds, $X \in \mathcal{X}^k(M)$ and $Y \in \mathcal{X}^k(N)$. Let F_t and G_t denote the flows of X and Y respectively. Then $X \sim_\varphi Y$ iff $\varphi \circ F_t = G_t \circ \varphi$. In particular, if φ is a diffeomorphism, then the equality $Y = \varphi_* X$ holds iff the flow of Y is $\varphi \circ F_t \circ \varphi^{-1}$ (This is called the push–forward of F_t by φ since it is the natural way to construct a diffeomorphism on N out of one on M). In particular, $(F_t)_* X = X$. Therefore, the flow of the push–forward of a vector–field is the push–forward of its flow.

Dynamical Evolution and Flow

As a motivational example, consider a mechanical system that is capable of assuming various states described by points in a set U. For example, U might be $\mathbb{R}^3 \times \mathbb{R}^3$ and a state might be the positions and momenta (x^i, p_i) of a particle moving under the influence of the central force field, with $i = 1, 2, 3$.

As time passes, the state evolves. If the state is $\gamma_0 \in U$ at time s and this changes to γ at a later time t, we set

$$F_{t,s}(\gamma_0) = \gamma,$$

and call $F_{t,s}$ the *evolution operator*; it maps a state at time s to what the state would be at time t; that is, after time $t - s$. has elapsed. Determinism is expressed by the *Chapman–Kolmogorov law* [AMR88]:

$$F_{\tau,t} \circ F_{t,s} = F_{\tau,s}, \qquad F_{t,t} = \text{identity}. \tag{4.2}$$

The evolution laws are called *time independent*, or *autonomous*, when $F_{t,s}$ depends only on $t - s$. In this case the preceding law (4.2) becomes the *group property*:

$$F_t \circ F_s = F_{t+s}, \qquad F_0 = \text{identity}. \tag{4.3}$$

We call such an F_t a *flow* and $F_{t,s}$ a *time–dependent flow*, or an evolution operator. If the system is irreversible, that is, defined only for $t \geq s$, we speak of a *semi–flow* [AMR88].

Usually, instead of $F_{t,s}$ the *laws of motion* are given in the form of ODEs that we must solve to find the flow. These equations of motion have the form:

$$\dot{\gamma} = X(\gamma), \qquad \gamma(0) = \gamma_0,$$

where X is a (possibly time–dependent) vector–field on U.

As a continuation of the previous example, consider the motion of a particle of mass m under the influence of the central force field (like gravity, or Coulombic potential) F^i $(i = 1, 2, 3)$, described by the *Newtonian equation of motion*:

$$m\ddot{x}^i = F^i(x). \tag{4.4}$$

By introducing momenta $p_i = m\dot{x}^i$, equation (4.4) splits into two Hamiltonian equations:

$$\dot{x}^i = p_i/m, \qquad \dot{p}_i = F_i(x). \tag{4.5}$$

Note that in Euclidean space we can freely interchange subscripts and superscripts. However, in general case of a Riemannian manifold, $p_i = mg_{ij}\dot{x}^j$ and (4.5) properly reads

$$\dot{x}^i = g^{ij}p_j/m, \qquad \dot{p}_i = F_i(x). \tag{4.6}$$

The phase–space here is the *Riemannian manifold* $(\mathbb{R}^3\backslash\{0\}) \times \mathbb{R}^3$, that is, the cotangent bundle of $\mathbb{R}^3\backslash\{0\}$, which is itself a smooth manifold for the central force field. The r.h.s of equations (5.137) define a Hamiltonian vector–field on this 6D manifold by

$$X(x, p) = \left((x^i, p_i), (p_i/m, F_i(x))\right). \tag{4.7}$$

Integration of equations (5.137) produces trajectories (in this particular case, planar conic sections). These trajectories comprise the flow F_t of the vector–field $X(x, p)$ defined in (4.7).

Vector–Fields and Their Flows

Vector–Fields on M

A *vector–field* X on U, where U is an open chart in $n-$manifold M, is a *smooth function* from U to M assigning to each point $m \in U$ a vector at that point, i.e., $X(m) = (m, X(m))$. If $X(m)$ is tangent to M for each $m \in M$, X is said to be a *tangent vector–field* on M. If $X(m)$ is orthogonal to M (i.e., $X(p) \in M_m^{\perp}$) for each $X(m) \in M$, X is said to be a *normal vector–field* on M.

In other words, let M be a $C^\infty-$manifold. A $C^\infty-$vector–field on M is a $C^\infty-$section of the tangent bundle TM of M. Thus a vector–field X on a manifold M is a $C^\infty-$map $X : M \to TM$ such that $X(m) \in T_m M$ for all points $m \in M$,and $\pi_M \circ X = Id_M$. Therefore, a vector–field assigns to each point m of M a vector based (i.e., bound) at that point. The set of all C^∞ vector–fields on M is denoted by $\mathcal{X}^k(M)$.

A vector–field $X \in \mathcal{X}^k(M)$ represents a field of direction indicators [Thi79]: to every point m of M it assigns a vector in the tangent space $T_m M$ at that point. If X is a vector–field on M and (U, ϕ) is a chart on M and $m \in U$, then we have $X(m) = X(m)\,\phi^i \frac{\partial}{\partial \phi^i}$. Following [KMS93], we write $X|_U = X\,\phi^i \frac{\partial}{\partial \phi^i}$.

Let M be a connected $n-$manifold, and let $f : U \to \mathbb{R}$ (U an open set in M) and $c \in \mathbb{R}$ be such that $M = f^{-1}(c)$ (i.e., M is the *level set* of the function f at *height* c) and $\nabla f(m) \neq 0$ for all $m \in M$. Then there exist on M exactly two smooth unit normal vector–fields $N_{1,2}(m) = \pm \frac{\nabla f(m)}{|\nabla f(m)|}$ (here $|X| = (X \cdot X)^{1/2}$ denotes the norm or length of a vector X, and $(\cdot)$ denotes the scalar product on M) for all $m \in M$, called *orientations* on M.

Let $\varphi : M \to N$ be a smooth map. Recall that two vector–fields $X \in \mathcal{X}^k(M)$ and $Y \in \mathcal{X}(N)$ are called $\varphi-$related, if $T\varphi \circ X = Y \circ \varphi$ holds, i.e., if the following diagram commutes:

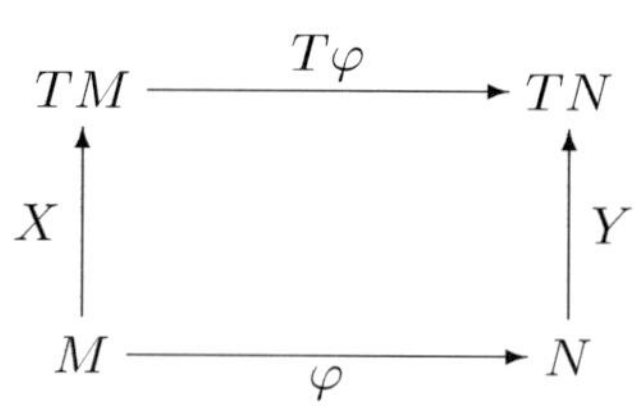

In particular, a diffeomorphism $\varphi : M \to N$ induces a linear map between vector–fields on two manifolds, $\varphi^* : \mathcal{X}^k(M) \to \mathcal{X}(N)$, such that $\varphi^* X = T\varphi \circ X \circ \varphi^{-1} : N \to TN$, i.e., the following diagram commutes:

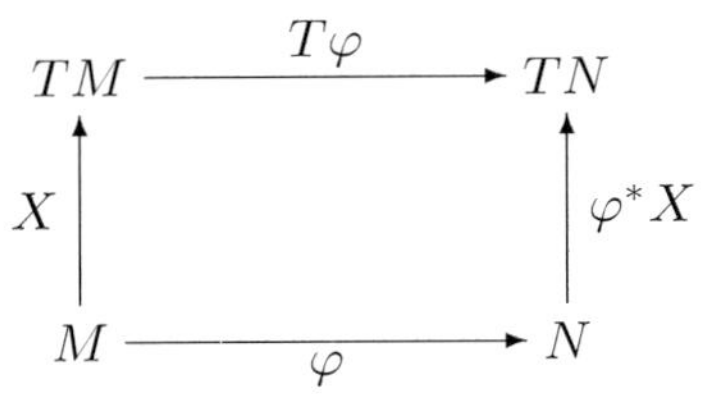

The correspondences $M \to TM$ and $\varphi \to T\varphi$ obviously define a functor $T : \mathcal{M} \Rightarrow \mathcal{M}$ from the category of smooth manifolds to itself. T is closely related to the tangent bundle functor (4.1.2).

A C^∞ *time–dependent vector–field* is a C^∞–map $X : \mathbb{R} \times M \to TM$ such that $X(t,m) \in T_m M$ for all $(t,m) \in \mathbb{R} \times M$, i.e., $X_t(m) = X(t,m)$.

Integral Curves as Dynamical Trajectories

Recall (4.1.2) that a curve γ at a point m of an n–manifold M is a C^0–map from an open interval $I \subset \mathbb{R}$ into M such that $0 \in I$ and $\gamma(0) = m$. For such a curve we may assign a tangent vector at each point $\gamma(t)$, $t \in I$, by $\dot{\gamma}(t) = T_t\gamma(1)$.

Let X be a smooth tangent vector–field on the smooth n–manifold M, and let $m \in M$. Then there exists an open interval $I \subset \mathbb{R}$ containing 0 and a parameterized curve $\gamma : I \to M$ such that:

1. $\gamma(0) = m$;
2. $\dot{\gamma}(t) = X(\gamma(t))$ for all $t \in I$; and
3. If $\beta : \tilde{I} \to M$ is any other parameterized curve in M satisfying (1) and (2), then $\tilde{I} \subset I$ and $\beta(t) = \gamma(t)$ for all $t \in \tilde{I}$.

A parameterized curve $\gamma : I \to M$ satisfying condition (2) is called an *integral curve* of the tangent vector–field X. The unique γ satisfying conditions (1)–(3) is the *maximal integral curve* of X through $m \in M$.

In other words, let $\gamma : I \to M$, $t \mapsto \gamma(t)$ be a smooth curve in a manifold M defined on an interval $I \subseteq \mathbb{R}$. $\dot{\gamma}(t) = \frac{d}{dt}\gamma(t)$ defines a smooth vector–field along γ since we have $\pi_M \circ \dot{\gamma} = \gamma$. Curve γ is called an *integral curve* or *flow line* of a vector–field $X \in \mathcal{X}^k(M)$ if the tangent vector determined by γ equals X at every point $m \in M$, i.e.,

$$\dot{\gamma} = X \circ \gamma,$$

or, if the following diagram commutes:

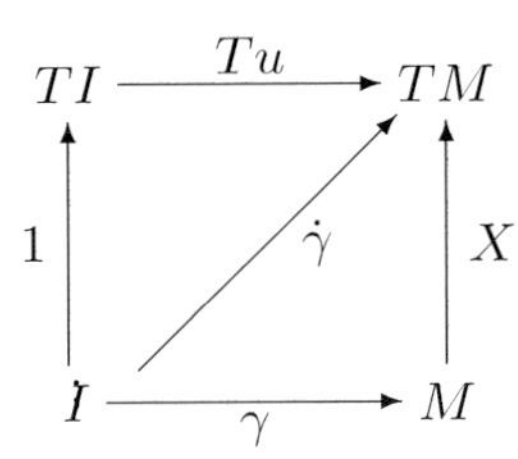

On a chart (U, ϕ) with coordinates $\phi(m) = \left(x^1(m), ..., x^n(m)\right)$, for which $\varphi \circ \gamma : t \mapsto \gamma_i(t)$ and $T\varphi \circ X \circ \varphi^{-1} : x^i \mapsto \left(x^i, X_i(m)\right)$, this is written

$$\dot{\gamma}_i(t) = X_i(\gamma(t)), \text{ for all } t \in I \subseteq \mathbb{R}, \tag{4.8}$$

which is an ordinary differential equation of first–order in n dimensions.

The *velocity* $\dot{\gamma}$ of the parameterized curve $\gamma(t)$ is a vector–field along γ defined by

$$\dot{\gamma}(t) = (\gamma(t), \dot{x}^1(t), \ldots \dot{x}^n(t)).$$

Its length $|\dot{\gamma}| : I \to \mathbb{R}$, defined by $|\dot{\gamma}|(t) = |\dot{\gamma}(t)|$ for all $t \in I$, is a function along α. $|\dot{\gamma}|$ is called *speed* of γ [Arn89].

Each vector–field X along γ is of the form $X(t) = (\gamma(t), X_1(t), \ldots, X_n(t))$, where each component X_i is a function along γ. X is *smooth* if each $X_i : I \to M$ is smooth. The *derivative* of a smooth vector–field X along a curve $\gamma(t)$ is the vector–field $\dot{X}$ along γ defined by

$$\dot{X}(t) = (\gamma(t), \dot{X}_1(t), \ldots \dot{X}_n(t)).$$

$\dot{X}(t)$ measures the *rate of change of the vector part* $(X_1(t), \ldots X_n(t))$ of $X(t)$ *along* γ. Thus, the *acceleration* $\ddot{\gamma}(t)$ of a parameterized curve $\gamma(t)$ is the vector–field along γ get by differentiating the velocity field $\dot{\gamma}(t)$.

Differentiation of vector–fields along parameterized curves has the following properties. For X and Y smooth vector–fields on M along the parameterized curve $\gamma : I \to M$ and f a smooth function along γ, we have:

1. $\frac{d}{dt}(X + Y) = \dot{X} + \dot{Y}$;
2. $\frac{d}{dt}(fX) = \dot{f}X + f\dot{X}$; and
3. $\frac{d}{dt}(X \cdot Y) = \dot{X}Y + X\dot{Y}$.

A *geodesic* in M is a parameterized curve $\gamma : I \to M$ whose acceleration $\ddot{\gamma}$ is everywhere orthogonal to M; that is, $\ddot{\gamma}(t) \in M^{\perp}_{\alpha(t)}$ for all $t \in I \subset \mathbb{R}$. Thus a geodesic is a curve in M which always goes 'straight ahead' in the surface. Its acceleration serves only to keep it in the surface. It has no component of acceleration tangent to the surface. Therefore, it also has a constant speed $\dot{\gamma}(t)$.

Let $v \in M_m$ be a vector on M. Then there exists an open interval $I \subset \mathbb{R}$ containing 0 and a geodesic $\gamma : I \to M$ such that:

1. $\gamma(0) = m$ and $\dot{\gamma}(0) = v$; and
2. If $\beta : \tilde{I} \to M$ is any other geodesic in M with $\beta(0) = m$ and $\dot{\beta}(0) = v$, then $\tilde{I} \subset I$ and $\beta(t) = \gamma(t)$ for all $t \in \tilde{I}$.

The geodesic γ is now called the *maximal geodesic* in M passing through m with initial velocity v.

By definition, a parameterized curve $\gamma : I \to M$ is a geodesic of M iff its acceleration is everywhere perpendicular to M, i.e., iff $\ddot{\gamma}(t)$ is a multiple of the orientation $N(\gamma(t))$ for all $t \in I$, i.e., $\ddot{\gamma}(t) = g(t)\, N(\gamma(t))$, where $g : I \to \mathbb{R}$. Taking the scalar product of both sides of this equation with $N(\gamma(t))$ we find $g = -\dot{\gamma}\dot{N}(\gamma(t))$. Thus $\gamma : I \to M$ is geodesic iff it satisfies the differential equation

$$\ddot{\gamma}(t) + \dot{N}(\gamma(t))\, N(\gamma(t)) = 0.$$

This vector equation represents the system of second–order component ODEs

$$\ddot{x}^i + N_i(x+1, \dots, x^n) \frac{\partial N_j}{\partial x^k}(x+1, \dots, x^n)\, \dot{x}^j \dot{x}^k = 0.$$

The substitution $u^i = \dot{x}^i$ reduces this second–order differential system (in n variables x^i) to the first–order differential system

$$\dot{x}^i = u^i, \qquad \dot{u}^i = -N_i(x+1, \dots, x^n) \frac{\partial N_j}{\partial x^k}(x+1, \dots, x^n)\, \dot{x}^j \dot{x}^k$$

(in $2n$ variables x^i and u^i). This first–order system is just the differential equation for the integral curves of the vector–field X in $U \times \mathbb{R}$ (U open chart in M), in which case X is called a *geodesic spray*.

Now, when an integral curve $\gamma(t)$ is the path a mechanical system Ξ follows, i.e., the solution of the equations of motion, it is called a *trajectory*. In this case the parameter t represents time, so that (4.8) describes motion of the system Ξ on its configuration manifold M.

If $X_i(m)$ is C^0 the existence of a local solution is guaranteed, and a *Lipschitz condition* would imply that it is unique. Therefore, exactly one integral curve passes through every point, and different integral curves can never cross. As $X \in \mathcal{X}^k(M)$ is C^∞, the following statement about the solution with arbitrary initial conditions holds [Thi79, Arn89]:

Theorem. Given a vector–field $X \in \mathcal{X}(M)$, for all points $p \in M$, there exist $\eta > 0$, a neighborhood V of p, and a function $\gamma : (-\eta, \eta) \times V \to M$, $(t, x^i(0)) \mapsto \gamma(t, x^i(0))$ such that

$$\dot{\gamma} = X \circ \gamma, \qquad \gamma(0, x^i(0)) = x^i(0) \qquad \text{for all } \ x^i(0) \in V \subseteq M.$$

For all $|t| < \eta$, the map $x^i(0) \mapsto \gamma(t, x^i(0))$ is a diffeomorphism f_t^X between V and some open set of M. For proof, see [Die69], I, 10.7.4 and 10.8.

This Theorem states that trajectories that are near neighbors cannot suddenly be separated. There is a well–known estimate (see [Die69], I, 10.5) according to which points cannot diverge faster than exponentially in time if the derivative of X is uniformly bounded.

An integral curve $\gamma(t)$ is said to be *maximal* if it is not a restriction of an integral curve defined on a larger interval $I \subseteq \mathbb{R}$. It follows from the existence and uniqueness theorems for ODEs with smooth r.h.s and from elementary properties of Hausdorff spaces that for any point $m \in M$ there exists a maximal integral curve γ_m of X, passing for $t = 0$ through point m, i.e., $\gamma(0) = m$.

Theorem (Local Existence, Uniqueness, and Smoothness) [AMR88]. Let E be a Banach space, $U \subset E$ be open, and suppose $X : U \subset E \to E$ is of class C^{∞}, $k \geq 1$. Then

1. For each $x_0 \in U$, there is a curve $\gamma : I \to U$ at x_0 such that $\dot{\gamma}(t) = X(\gamma(t))$ for all $t \in I$.

2. Any two such curves are equal on the intersection of their domains.

3. There is a neighborhood U_0 of the point $x_0 \in U$, a real number $a > 0$, and a C^{∞} map $F : U_0 \times I \to E$, where I is the open interval $]-a, a[$, such that the curve $\gamma_u : I \to E$, defined by $\gamma_u(t) = F(u,t)$ is a curve at $u \in E$ satisfying the ODEs $\dot{\gamma}_u(t) = X(\gamma_u(t))$ for all $t \in I$.

Proposition (Global Uniqueness). Suppose γ_1 and γ_2 are two integral curves of a vector–field X at a point $m \in M$. Then $\gamma_1 = \gamma_2$ on the intersection of their domains [AMR88].

If for every point $m \in M$ the curve γ_m is defined on the entire real axis $\mathbb{R}$, then the vector–field X is said to be *complete*.

The *support of a vector–field* X defined on a manifold M is defined to be the closure of the set $\{m \in M | X(m) = 0\}$. A C^{∞} vector–field with compact support on a manifold M is complete. In particular, a C^{∞} vector–field on a compact manifold is complete. Completeness corresponds to well–defined dynamics persisting eternally.

Now, following [AMR88], for the *derivative* of a C^{∞} function $f : E \to \mathbb{R}$ in the direction X we use the notation $X[f] = df \cdot X$, where df stands for the *derivative map*. In standard coordinates on $\mathbb{R}^n$ this is a standard *gradient*

$$df(x) = \nabla f = (\partial_{x^1} f, ..., \partial_{x^n} f), \qquad \text{and} \qquad X[f] = X^i \partial_{x^i} f.$$

Let F_t be the flow of X. Then $f(F_t(x)) = f(F_s(x))$ if $t \geq s$.

For example, Newtonian equations for a moving particle of mass m in a potential field V in $\mathbb{R}^n$ are given by $\ddot{q}^i(t) = -(1/m)\nabla V(q^i(t))$, for a smooth function $V : \mathbb{R}^n \to \mathbb{R}$. If there are constants $a, b \in \mathbb{R}$, $b \geq 0$ such that $(1/m)V(q^i) \geq a - b\left\|q^i\right\|^2$, then *every solution exists for all time.* To show this, rewrite the second–order equations as a first–order system $\dot{q}^i = (1/m)\,p_i$, $\dot{p}_i = -V(q^i)$ and note that the energy $E(q^i, p_i) = (1/2m)\left\| p_i\right\|^2 + V(q)$ is a *first integral* of the motion. Thus, for any solution $(q^i(t), p_i(t))$ we have $E(q^i(t), p_i(t)) = E(q^i(0), p_i(0)) = V(q(0))$.

Let X_t be a C^{∞} time–dependent vector–field on an $n-$manifold M, $k \geq 1$, and let m_0 be an *equilibrium* of X_t, that is, $X_t(m_0) = 0$ for all t. Then for any T there exists a neighborhood V of m_0 such that any $m \in V$ has integral curve existing for time $t \in [-T, T]$.

Dynamical Flows on M

Recall (4.1.2) that the *flow* F_t of a C^{∞} vector–field $X \in \mathcal{X}^k(M)$ is the *one–parameter group of diffeomorphisms* $F_t : M \to M$ such that $t \mapsto F_t(m)$ is the integral curve of X with initial condition m for all $m \in M$ and $t \in I \subseteq \mathbb{R}$. The flow $F_t(m)$ is C^{∞} by induction on k. It is defined as [AMR88]:

$$\frac{d}{dt}F_t(x) = X(F_t(x)).$$

Existence and uniqueness theorems for ODEs guarantee that F_t is smooth in m and t. From uniqueness, we get the *flow property*:

$$F_{t+s} = F_t \circ F_s$$

along with the initial conditions $F_0 =$ identity. The flow property generalizes the situation where $M = V$ is a linear space, $X(x) = A\,x$ for a (bounded) linear operator A, and where $F_t(x) = \mathrm{e}^{tA}x$ – to the nonlinear case. Therefore, the flow $F_t(m)$ can be defined as a *formal exponential*

$$F_t(m) = \exp(t\,X) = (I + t\,X + \frac{t^2}{2}X^2 + ...) = \sum_{k=0}^{\infty}\frac{X^k t^k}{k!}.$$

recall that a *time–dependent vector–field* is a map $X : M \times \mathbb{R} \to TM$ such that $X(m,t) \in T_mM$ for each point $m \in M$ and $t \in \mathbb{R}$. An *integral curve* of X is a curve $\gamma(t)$ in M such that

$$\dot{\gamma}(t) = X\left(\gamma\left(t\right), t\right), \qquad \text{for all } \ t \in I \subseteq \mathbb{R}.$$

In this case, the flow is the one–parameter group of diffeomorphisms $F_{t,s} : M \to M$ such that $t \mapsto F_{t,s}\left(m\right)$ is the integral curve $\gamma(t)$ with initial condition $\gamma(s) = m$ at $t = s$. Again, the existence and uniqueness Theorem from ODE–theory applies here, and in particular, uniqueness gives the time–dependent flow property, i.e., the *Chapman–Kolmogorov law*

$$F_{t,r} = F_{t,s} \circ F_{s,r}.$$

If X happens to be time independent, the two notions of flows are related by $F_{t,s} = F_{t-s}$ (see [MR99]).

Categories of ODEs

Ordinary differential equations are naturally organized into their categories (see [Koc81]). First order ODEs are organized into a category ODE_1. A first–order ODE on a manifold–like object M is a vector–field $X : M \to TM$, and a *morphism of vector–fields* $(M_1, X_1) \to (M_2, X_2)$ is a map $f : M_1 \to M_2$ such that the following diagram commutes

$$\begin{array}{ccc} TM_1 & \xrightarrow{\;Tf\;} & TM_2 \\ {\scriptstyle X_1}\uparrow & & \uparrow{\scriptstyle X_2} \\ M_1 & \xrightarrow[\;f\;]{} & M_2 \end{array}$$

A *global solution* of the differential equation (M, X), or a *flow line* of a vector–field X, is a morphism from $\left(\mathbb{R}, \frac{\partial}{\partial x}\right)$ to (M, X).

Similarly, second–order ODEs are organized into a category ODE_2. A second–order ODE on M is usually constructed as a vector–field on TM, $\xi : TM \to TTM$, and a morphism of vector–fields $(M_1, \xi_1) \to (M_2, \xi_2)$ is a map $f : M_1 \to M_2$ such that the following diagram commutes

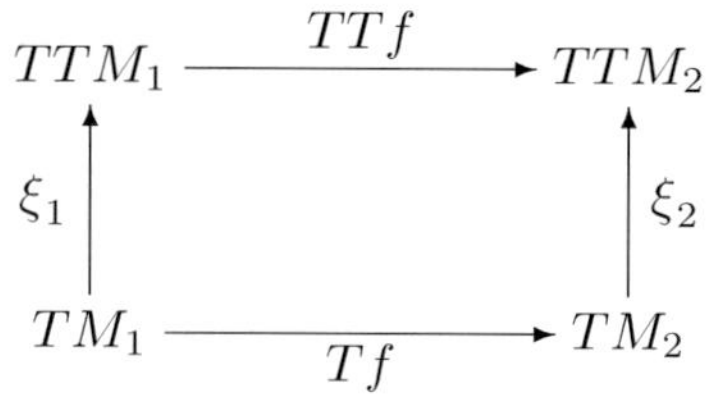

Unlike solutions for first–order ODEs, solutions for second–order ODEs are not in general homomorphisms from $\mathbb{R}$, unless the second–order ODE is a *spray* [KR03].

Differential Forms on Smooth Manifolds

Recall that *exterior differential forms* are a special kind of antisymmetrical covariant tensors, that formally occur as *integrands* under ordinary integral signs in $\mathbb{R}^3$. To give a more precise exposition, here we start with 1–forms, which are dual to vector–fields, and after that introduce general k–forms.

1–Forms on M

Dual to the notion of a C^∞ vector–field X on an n–manifold M is a C^∞ *covector–field*, or a C^∞ 1–form α, which is defined as a C^∞–section of the cotangent bundle T^*M, i.e., $\alpha : M \to T^*M$ is smooth and $\pi_M^* \circ X = Id_M$. We denote the set of all C^∞ 1–forms by $\Omega^1(M)$. A basic example of a 1–form is the differential *df* of a real–valued function $f \in C^\infty(M, \mathbb{R})$. With point wise addition and scalar multiplication $\Omega^1(M)$ becomes a vector space.

In other words, a C^∞ 1–form α on a C^∞ manifold M is a real–valued function on the set of all tangent vectors to M, i.e., $\alpha : TM \to \mathbb{R}$ with the following properties:

1. α is linear on the tangent space $T_m M$ for each $m \in M$;
2. For any C^∞ vector–field $X \in \mathcal{X}^k(M)$, the function $f : M \to \mathbb{R}$ is C^∞.

Given a 1–form α, for each point $m \in M$ the map $\alpha(m) : T_m M \to \mathbb{R}$ is an element of the dual space $T_m^* M$. Therefore, the space of 1–forms $\Omega^1(M)$ is dual to the space of vector–fields $\mathcal{X}^k(M)$.

In particular, the *coordinate* 1–*forms* $dx^1, ..., dx^n$ are locally defined at any point $m \in M$ by the property that for any vector–field $X = \left(X^1, ..., X^n\right) \in \mathcal{X}^k(M)$,

$$dx^i(X) = X^i.$$

The dx^i's form a *basis* for the 1–forms at any point $m \in M$, with local coordinates $(x^1, ..., x^n)$, so any 1–form α may be expressed in the form

$$\alpha = f_i(m)\, dx^i.$$

If a vector–field X on M has the form $X(m) = \left(X^1(m), ..., X^n(m)\right)$, then at any point $m \in M$,

$$\alpha_m(X) = f_i(m)\, X^i(m),$$

where $f \in C^\infty(M, \mathbb{R})$.

Suppose we have a 1D closed curve $\gamma = \gamma(t)$ inside a smooth manifold M. Using a simplified 'physical' notation, a 1–form $\alpha(x)$ defined at a point $x \in M$, given by

$$\alpha(x) = \alpha_i(x)\, dx^i, \tag{4.9}$$

can be unambiguously integrated over a curve $\gamma \in M$, as follows. Parameterize γ by a parameter t, so that its coordinates are given by $x^i(t)$. At time t, the velocity $\dot{x} = \dot{x}(t)$ is a tangent vector to M at $x(t)$. One can insert this tangent vector into the linear map $\alpha(x)$ to get a real number. By definition, inserting the vector $\dot{x}(t)$ into the linear map dx^i gives the component $\dot{x}^i = \dot{x}^i(t)$. Doing this for every t, we can then integrate over t,

$$\int \left(\alpha_i(x(t))\dot{x}^i\right) dt. \tag{4.10}$$

Note that this expression is independent of the parametrization in terms of t. Moreover, from the way that tangent vectors transform, one can deduce how the linear maps dx^i should transform, and from this how the coefficients $\alpha_i(x)$ should transform. Doing this, one sees that the above expression is also invariant under changes of coordinates on M. Therefore, a 1–form can be unambiguously integrated over a curve in M. We write such an integral as

$$\int_\gamma \alpha_i(x)\, dx^i, \qquad \text{or, even shorter, as} \qquad \int_\gamma \alpha.$$

Clearly, when M is itself a 1D manifold, (4.10) gives precisely the ordinary integration of a function $\alpha(x)$ over x, so the above notation is indeed natural.

The 1–forms on M are part of an algebra, called the *exterior algebra*, or *Grassmann algebra* on M. The multiplication $\wedge$ in this algebra is called *wedge product* (see (4.12) below), and it is skew–symmetric,

$$dx^i \wedge dx^j = -dx^j \wedge dx^i.$$

One consequence of this is that $dx^i \wedge dx^i = 0$.

k–Forms on M

A differential form, or an exterior form α of degree k, or a k–form for short, is a section of the vector bundle $\Lambda^k T^*M$, i.e., $\alpha : M \to \Lambda^k T^*M$. In other words, $\alpha(m) : T_m M \times ... \times T_m M \to \mathbb{R}$ (with k factors $T_m M$) is a function that assigns to each point $m \in M$ a skew–symmetric k–multilinear map on the tangent space $T_m M$ to M at m. Without the skew–symmetry assumption, α would be called a $(0, k)$–tensor–field. The space of all k–forms is denoted by $\Omega^k(M)$. It may also be viewed as the space of all skew symmetric $(0, k)$–tensor–fields, the space of all maps

$$\Phi : \mathcal{X}^k(M) \times ... \times \mathcal{X}^k(M) \to C^\infty(M, \mathbb{R}),$$

which are k–linear and skew–symmetric (see (4.12) below). We put $\Omega^k(M) = C^\infty(M, \mathbb{R})$.

In particular, a 2–form ω on an n–manifold M is a section of the vector bundle $\Lambda^2 T^*M$. If (U, ϕ) is a chart at a point $m \in M$ with local coordinates $(x^1, ..., x^n)$ let $\{e_1, ..., e_n\} = \{\partial_{x^1}, ..., \partial_{x^n}\}$ – be the corresponding basis for $T_m M$, and let $\{\mathrm{e}^1, ..., \mathrm{e}^n\} = \{dx^1, ..., dx^n\}$ – be the dual basis for $T^*_m M$. Then at each point $m \in M$, we can write a 2–form ω as

$$\omega_m(v, u) = \omega_{ij}(m)\, v^i u^j, \quad \text{where} \quad \omega_{ij}(m) = \omega_m(\partial_{x^i}, \partial_{x^j}).$$

Similarly to the case of a 1–form α (4.9), one would like to define a 2–form ω as something which can naturally be integrated over a 2D surface Σ within a smooth manifold M. At a specific point $x \in M$, the tangent plane to such a surface is spanned by a pair of tangent vectors, $(\dot{x}^1, \dot{x}^2)$. So, to generalize the construction of a 1–form, we should give a bilinear map from such a pair to $\mathbb{R}$. The most general form of such a map is

$$\omega_{ij}(x)\, dx^i \otimes dx^j, \tag{4.11}$$

where the tensor product of two cotangent vectors acts on a pair of vectors as,

$$dx^i \otimes dx^j\, (\dot{x}^1, \dot{x}^2) = dx^i(\dot{x}^1)\, dx^j(\dot{x}^2).$$

On the r.h.s. of this equation, one multiplies two ordinary numbers got by letting the linear map dx^i act on $\dot{x}^1$, and dx^j on $\dot{x}^2$.

However, the bilinear map (4.11) is slightly too general to give a good integration procedure. The reason is that we would like the integral to change sign if we change the orientation of integration, just like in the 1D case. In 2D, changing the orientation means exchanging $\dot{x}^1$ and $\dot{x}^2$, so we want our bilinear map to be antisymmetric under this exchange. This is achieved by defining a 2–form to be

$$\omega = \omega_{ij}(x)\, \left(dx^i \otimes dx^j - dx^j \otimes dx^i\right) \equiv \omega_{ij}(x)\, dx^i \wedge dx^j$$

We now see why a2–form corresponds to an *antisymmetric* tensor field: the symmetric part of ω_{ij} would give a vanishing contribution to ω. Now, parameterizing a surface Σ in M with two coordinates t_1 and t_2, and reasoning exactly like we did in the case of a 1–form, one can show that the integration of a 2–form over such a surface is indeed well–defined, and independent of the parametrization of both Σ and M.

If each summand of a differential form $\alpha \in \Omega^k(M)$ contains k basis 1–forms dx^i's, the form is called a k–form. Functions $f \in C^\infty(M, \mathbb{R})$ are considered to be 0–forms, and any form on an n–manifold M of degree $k > n$ must be zero due to the skew–symmetry.

Any k–form $\alpha \in \Omega^k(M)$ may be expressed in the form

$$\alpha = f_I\, dx^{i_1} \wedge ... \wedge dx^{i_k} = f_I\, dx^I,$$

where I is a *multiindex* $I = (i_1, ..., i_k)$ of length k, and $\wedge$ is the wedge product which is associative, bilinear and anticommutative.

Just as 1–forms act on vector–fields to give real–valued functions, so k–forms act on k–tuples of vector–fields to give real–valued functions.

The *wedge product* of two differential forms, a k–form $\alpha \in \Omega^k(M)$ and an l–form $\beta \in \Omega^l(M)$ is a $(k+l)$–form $\alpha \wedge \beta$ defined as:

$$\alpha \wedge \beta = \frac{(k+l)!}{k!l!}\mathbf{A}(\alpha \otimes \beta), \tag{4.12}$$

where $\mathbf{A} : \Omega^k(M) \to \Omega^k(M)$, $\mathbf{A}\tau(e_1, ..., e_k) = \frac{1}{k!}\sum_{\sigma \in S_k}(\operatorname{sign}\sigma)\,\tau(e_{\sigma(1)}, ..., e_{\sigma(k)})$, where S_k is the permutation group on k elements consisting of all bijections $\sigma : \{1, ..., k\} \to \{1, ..., k\}$.

For any k–form $\alpha \in \Omega^k(M)$ and l–form $\beta \in \Omega^l(M)$, the wedge product is defined fiberwise, i.e., $(\alpha \wedge \beta)_m = \alpha_x \wedge \beta_m$ for each point $m \in M$. It is also associative, i.e., $(\alpha \wedge \beta) \wedge \gamma = \alpha \wedge (\beta \wedge \gamma)$, and graded commutative, i.e., $\alpha \wedge \beta = (-1)^{kl}\beta \wedge \alpha$. These properties are proved in multilinear algebra. So $M \Longrightarrow \Omega^k(M)$ is a contravariant functor from the category $\mathcal{M}$ into the category of real graded commutative algebras [KMS93].

Let M be an n–manifold, $X \in \mathcal{X}^k(M)$, and $\alpha \in \Omega^{k+1}(M)$. The *interior product*, or *contraction*, $i_X\alpha = X\rfloor\alpha \in \Omega^k(M)$ of X and α (with *insertion operator* i_X) is defined as

$$i_X\alpha(X^1, ..., X^k) = \alpha(X, X^1, ..., X^k).$$

Insertion operator i_X of a vector–field $X \in \mathcal{X}^k(M)$ is natural with respect to the pull–back F^* of a diffeomorphism $F : M \to N$ between two manifolds, i.e., the following diagram commutes:

$$\begin{array}{ccc}
\Omega^k(N) & \xrightarrow{\;F^*\;} & \Omega^k(M) \\
\big\downarrow i_X & & \big\downarrow i_{F^*X} \\
\Omega^{k-1}(N) & \xrightarrow[\;F^*\;]{} & \Omega^{k-1}(M)
\end{array}$$

Similarly, insertion operator i_X of a vector–field $X \in \mathcal{Y}^k(M)$ is natural with respect to the push–forward F_* of a diffeomorphism $F : M \to N$, i.e., the following diagram commutes:

$$\begin{array}{ccc} \Omega^k(M) & \xrightarrow{F_*} & \Omega^k(N) \\ {\scriptstyle i_Y}\downarrow & & \downarrow{\scriptstyle i_{F_* Y}} \\ \Omega^{k-1}(M) & \xrightarrow[F_*]{} & \Omega^{k-1}(N) \end{array}$$

In case of *Riemannian manifolds* there is another exterior operation. Let M be a smooth $n-$manifold with Riemannian metric $g = \langle , \rangle$ and the corresponding volume element μ. The *Hodge star operator* $* : \Omega^k(M) \to \Omega^{n-k}(M)$ on M is defined as

$$\alpha \wedge *\beta = \langle \alpha, \beta \rangle \, \mu \quad \text{for } \alpha, \beta \in \Omega^k(M).$$

The Hodge star operator satisfies the following properties for $\alpha, \beta \in \Omega^k(M)$ [AMR88]:

1. $\alpha \wedge *\beta = \langle \alpha, \beta \rangle \, \mu = \beta \wedge *\alpha$;
2. $*1 = \mu, \quad *\mu = (-1)^{Ind(g)}$;
3. $* * \alpha = (-1)^{Ind(g)}(-1)^{k(n-k)}\alpha$;
4. $\langle \alpha, \beta \rangle = (-1)^{Ind(g)} \langle *\alpha, *\beta \rangle$, where $Ind(g)$ is the *index* of the metric g.

Exterior Differential Systems

Here we give an informal introduction to *exterior differential systems* (EDS, for short), which are expressions involving differential forms related to any manifold M. Central in the language of EDS is the notion of *coframing*, which is a real finite–dimensional smooth manifold M with a given global *cobasis* and coordinates, but without requirement for a proper topological and differential structures. For example, $M = \mathbb{R}^3$ is a coframing with cobasis $\{dx, dy, dz\}$ and coordinates $\{x, y, z\}$. In addition to the cobasis and coordinates, a coframing can be given structure equations (4.1.4) and restrictions. For example, $M = \mathbb{R}^2 \backslash \{0\}$ is a coframing with cobasis $\{\mathrm{e}^1, \mathrm{e}^2\}$, a single coordinate $\{r\}$, structure equations $\{dr = \mathrm{e}^1,\ d\mathrm{e}^1 = 0,\ d\mathrm{e}^2 = \mathrm{e}^1 \wedge \mathrm{e}^2/r\}$ and restrictions $\{r \neq 0\}$.

A *system* S on M in EDS terminology is a list of expressions including differential forms (e.g., $S = \{dz - ydx\}$).

Now, a simple EDS is a triple (S, Ω, M), where S is a system on M, and Ω is an *independence condition*: either a decomposable $k-$form or a system of $k-$forms on M. An EDS is a list of simple EDS objects where the various coframings are all disjoint.

An *integral element* of an exterior system (S, Ω, M) is a subspace $P \subset T_m M$ of the tangent space at some point $m \in M$ such that all forms in S vanish when evaluated on vectors from P. Alternatively, an integral element

$P \subset T_m M$ can be represented by its annihilator $P^\perp \subset T_m^* M$, comprising those 1–forms at m which annul every vector in P. For example, with $M = \mathbb{R}^3 = \{(x, y, z)\}$, $S = \{dx \wedge dz\}$ and $\Omega = \{dx, dz\}$, the integral element $P = \{\partial_x + \partial_z, \partial_y\}$ is equally determined by its annihilator $P^\perp = \{dz - dx\}$. Again, for $S = \{dz - y dx\}$ and $\Omega = \{dx\}$, the integral element $P = \{\partial_x + y\partial_z\}$ can be specified as $\{dy\}$.

Exterior Derivative on a Smooth Manifold

The *exterior derivative* is an operation that takes k–forms to $(k+1)$–forms on a smooth manifold M. It defines a unique family of maps $d : \Omega^k(U) \to \Omega^{k+1}(U)$, U open in M, such that (see [AMR88]):

1. d is a $\wedge$–*antiderivation*; that is, d is $\mathbb{R}$–linear and for two forms $\alpha \in \Omega^k(U)$, $\beta \in \Omega^l(U)$,
$$d(\alpha \wedge \beta) = d\alpha \wedge \beta + (-1)^k \alpha \wedge d\beta.$$
2. If $f \in C^\infty(U, \mathbb{R})$ is a function on M, then $df = \frac{\partial f}{\partial x^i} dx^i : M \to T^* M$ is the differential of f, such that $df(X) = i_X df = \mathcal{L}_X f - di_X f = \mathcal{L}_X f = X[f]$ for any $X \in \mathcal{X}^k(M)$; here, $\mathcal{L}_X$ denotes the Lie derivative (see below).
3. $d^2 = d \circ d = 0$ (that is, $d^{k+1}(U) \circ d^k(U) = 0$).
4. d is natural with respect to restrictions $|U$; that is, if $U \subset V \subset M$ are open and $\alpha \in \Omega^k(V)$, then $d(\alpha|U) = (d\alpha)|U$, or the following diagram commutes:
$$\begin{array}{ccc} \Omega^k(V) & \xrightarrow{\;|U\;} & \Omega^k(U) \\ {\scriptstyle d}\downarrow & & \downarrow{\scriptstyle d} \\ \Omega^{k+1}(V) & \xrightarrow[\;|U\;]{} & \Omega^{k+1}(U) \end{array}$$
5. d is natural with respect to the Lie derivative $\mathcal{L}_X$ along any vector–field $X \in \mathcal{X}^k(M)$; that is, for $\omega \in \Omega^k(M)$ we have $\mathcal{L}_X \omega \in \Omega^k(M)$ and $d\mathcal{L}_X \omega = \mathcal{L}_X d\omega$, or the following diagram commutes:
$$\begin{array}{ccc} \Omega^k(M) & \xrightarrow{\;\mathcal{L}_X\;} & \Omega^k(M) \\ {\scriptstyle d}\downarrow & & \downarrow{\scriptstyle d} \\ \Omega^{k+1}(M) & \xrightarrow[\;\mathcal{L}_X\;]{} & \Omega^{k+1}(M) \end{array}$$
6. Let $\varphi : M \to N$ be a C^∞ map of manifolds. Then $\varphi^* : \Omega^k(N) \to \Omega^k(M)$ is a homomorphism of differential algebras (with $\wedge$ and d) and d is natural with respect to $\varphi^* = F^*$; that is, $\varphi^* d\omega = d\varphi^*\omega$, or the following diagram commutes:

$$\begin{CD} \Omega^k(N) @>{\varphi^*}>> \Omega^k(M) \\ @V{d}VV @VV{d}V \\ \Omega^{k+1}(N) @>>{\varphi^*}> \Omega^{k+1}(M) \end{CD}$$

7. Analogously, d is natural with respect to diffeomorphism $\varphi_* = (F^*)^{-1}$; that is, $\varphi_* d\omega = d\varphi_* \omega$, or the following diagram commutes:

$$\begin{CD} \Omega^k(N) @>{\varphi_*}>> \Omega^k(M) \\ @V{d}VV @VV{d}V \\ \Omega^{k+1}(N) @>>{\varphi_*}> \Omega^{k+1}(M) \end{CD}$$

8. $\mathcal{L}_X = i_X \circ d + d \circ i_X$ for any $X \in \mathcal{X}^k(M)$ (the Cartan 'magic' formula).
9. $\mathcal{L}_X \circ d = d \circ \mathcal{L}_X$, i.e., $[\mathcal{L}_X, d] = 0$ for any $X \in \mathcal{X}^k(M)$.
10. $[\mathcal{L}_X, i_Y] = i_{[x,y]}$; in particular, $i_X \circ \mathcal{L}_X = \mathcal{L}_X \circ i_X$ for all $X, Y \in \mathcal{X}^k(M)$.

Given a k–form $\alpha = f_I\, dx^I \in \Omega^k(M)$, the exterior derivative is defined in local coordinates $(x^1, ..., x^n)$ of a point $m \in M$ as

$$d\alpha = d\left(f_I\, dx^I\right) = \frac{\partial f_I}{\partial x^{i_k}} dx^{i_k} \wedge dx^I = df_I \wedge dx^{i_1} \wedge ... \wedge dx^{i_k}.$$

In particular, the exterior derivative of a function $f \in C^\infty(M, \mathbb{R})$ is a 1–form $df \in \Omega^1(M)$, with the property that for any $m \in M$, and $X \in \mathcal{X}^k(M)$,

$$df_m(X) = X(f),$$

i.e., $df_m(X)$ is a Lie derivative of f at m in the direction of X. Therefore, in local coordinates $(x^1, ..., x^n)$ of a point $m \in M$ we have

$$df = \frac{\partial f}{\partial x^i} dx^i.$$

For any two functions $f, g \in C^\infty(M, \mathbb{R})$, exterior derivative obeys the *Leibniz rule*:

$$d(fg) = g\, df + f\, dg,$$

and the *chain rule*:

$$d\,(g(f)) = g'(f)\, df.$$

A k–form $\alpha \in \Omega^k(M)$ is called *closed form* if $d\alpha = 0$, and it is called *exact form* if there exists a $(k-1)$–form $\beta \in \Omega^{k-1}(M)$ such that $\alpha = d\beta$. Since $d^2 = 0$, every exact form is closed. The converse is only partially true (Poincaré Lemma): every closed form is *locally exact*. This means that given a closed $k-$form $\alpha \in \Omega^k(M)$ on an open set $U \subset M$, any point $m \in U$ has a neighborhood on which there exists a $(k-1)-$form $\beta \in \Omega^{k-1}(U)$ such that $d\beta = \alpha|_U$.

The Poincaré lemma is a generalization and unification of two well–known facts in vector calculus:

1. If $\operatorname{curl} F = 0$, then locally $F = \operatorname{grad} f$;
2. If $\operatorname{div} F = 0$, then locally $F = \operatorname{curl} G$.

Poincaré lemma for contractible manifolds: Any closed form on a smoothly contractible manifold is exact.

Intuition Behind Cohomology

The simple formula $d^2 = 0$ leads to the important topological notion of *cohomology*. Let us try to solve the equation $d\omega = 0$ for a $p-$form ω. A trivial solution is $\omega = 0$. From the above formula, we can actually find a much larger class of trivial solutions: $\omega = d\alpha$ for a $(p-1)-$form α. More generally, if ω is any solution to $d\omega = 0$, then so is $\omega + d\alpha$. We want to consider these two solutions as equivalent:

$$\omega \sim \omega + \omega' \qquad \text{if} \qquad \omega' \in \mathrm{Im}\ d,$$

where Im d is the image of d, that is, the collection of all $p-$forms of the form $d\alpha$.[3] The set of all $p-$forms which satisfy $d\omega = 0$ is called the kernel of d, denoted $\operatorname{Ker} d$, so we are interested in $\operatorname{Ker} d$ up to the *equivalence classes* defined by adding elements of Im d. (Again, strictly speaking, $\operatorname{Ker} d$ consists of $q-$forms for several values of q, so we should restrict it to the $p-$forms for our particular choice of p.) This set of equivalence classes is called $H^p(M)$, the $p-$th *de Rham cohomology group* of M,

$$H^p(M) = \frac{\operatorname{Ker} d}{\mathrm{Im}\ d}.$$

Clearly, $\operatorname{Ker} d$ is a group under addition: if two forms $\omega^{(1)}$ and $\omega^{(2)}$ satisfy $d\omega^{(1)} = d\omega^{(2)} = 0$, then so does $\omega^{(1)} + \omega^{(2)}$. Moreover, if we change $\omega^{(i)}$ by adding some $d\alpha^{(i)}$, the result of the addition will still be in the same cohomology class, since it differs from $\omega^{(1)} + \omega^{(2)}$ by $d(\alpha^{(1)} + \alpha^{(2)})$. Therefore, we can view this addition really as an addition of cohomology classes: $H^p(M)$ is itself an additive group. Also note that if $\omega^{(3)}$ and $\omega^{(4)}$ are in the same cohomology class (that is, their difference is of the form $d\alpha^{(3)}$), then so are $c\omega^{(3)}$ and $c\omega^{(4)}$ for any constant factor c. In other words, we can multiply a cohomology class by a constant to get another cohomology class: cohomology classes actually form a *vector space*.

[3] To be precise, the image of d contains $q-$forms for any $0 < q \leq n$, so we should restrict this image to the $p-$forms for the p we are interested in.

Intuition Behind Homology

Another operator similar to the exterior derivative d is the *boundary operator* δ, which maps compact submanifolds of a smooth manifold M to their boundary. Here, $\delta C = 0$ means that a submanifold C of M has no boundary, and $C = \delta U$ means that C is itself the boundary of some submanifold U. It is intuitively clear, and not very hard to prove, that $\delta^2 = 0$: the boundary of a compact submanifold does not have a boundary itself. That the objects on which δ acts are independent of its coordinates is also clear. So is the grading of the objects: the degree p is the dimension of the submanifold C.[4] What is less clear is that the collection of submanifolds actually forms a vector space, but one can always *define* this vector space to consist of formal linear combinations of submanifolds, and this is precisely how one proceeds. The pD elements of this vector space are called *$p-$chains.* One should think of $-C$ as C with its orientation reversed, and of the sum of two disjoint sets, $C^1 + C^2$, as their union. The equivalence classes constructed from δ are called *homology* classes.

For example, in Figure 4.4, C^1 and C^2 both satisfy $\delta C = 0$, so they are elements of $\mathrm{Ker}\,\delta$. Moreover, it is clear that neither of them separately can be viewed as the boundary of another submanifold, so they are not in the trivial homology class Im δ. However, the boundary of U is $C^1 - C^2$.[5] This can be written as $C^1 - C^2 = \delta U$, or equivalently $C^1 = C^2 + \delta U$, showing that C^1 and C^2 are in the same homology class.

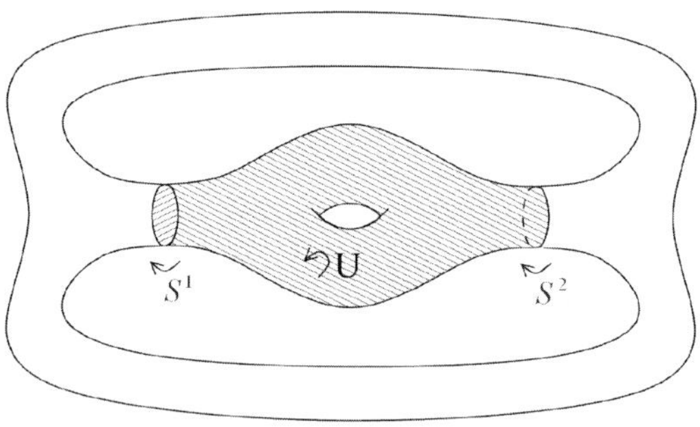

Fig. 4.4. The 1D submanifolds S^1 and S^2 represent the same homology class, since their difference is the boundary of U.

[4] Note that here we have an example of an operator that maps objects of degree p to objects of degree $p-1$ instead of $p+1$.

[5] The minus sign in front of C^2 is a result of the fact that C^2 itself actually has the wrong orientation to be considered a boundary of U.

The cohomology groups for the δ–operator are called *homology groups*, and denoted by $H_p(M)$, with a lower index.[6] The p–chains C that satisfy $\delta C = 0$ are called *p–cycles*. Again, the $H_p(M)$ only exist for $0 \leq p \leq n$.

There is an interesting relation between cohomology and homology groups. Note that we can construct a bilinear map from $H^p(M) \times H_p(M) \to \mathbb{R}$ by

$$([\omega], [C]) \mapsto \int_C \omega, \tag{4.13}$$

where $[\omega]$ denotes the cohomology class of a p–form ω, and $[\Sigma]$ the homology class of a p–cycle Σ. Using Stokes' Theorem, it can be seen that the result does not depend on the representatives for either ω or C

$$\begin{aligned} \int_{C+\delta U} \omega + d\alpha &= \int_C \omega + \int_C d\alpha + \int_{\delta U} \omega + d\alpha \\ &= \int_C \omega + \int_{\delta C} \alpha + \int_U d(\omega + d\alpha) = \int_C \omega, \end{aligned}$$

where we used that by the definition of (co)homology classes, $\delta C = 0$ and $d\omega = 0$. As a result, the above map is indeed well–defined on homology and cohomology classes. A very important Theorem by de Rham says that this map is nondegenerate [Rha84]. This means that if we take some $[\omega]$ and we know the result of the map (4.13) for all $[C]$, this uniquely determines $[\omega]$, and similarly if we start by picking an $[C]$. This in particular means that the vector space $H^p(M)$ is the dual vector space of $H_p(M)$.

The de Rham Complex and Homotopy Operators on M

After an intuitive introduction of (co)homology ideas, we now turn to their proper definitions. Given a smooth manifold M, let $\Omega^p(M)$ denote the space of all smooth p–forms on M. The differential d, mapping p–forms to $(p+1)$–forms, serves to define the *de Rham complex* on M

$$0 \to \Omega^0(M) \xrightarrow{d^0} \Omega^1(M) \xrightarrow{d^1} \dots \xrightarrow{d^{n-1}} \Omega^n(M) \to 0. \tag{4.14}$$

Recall that in general, a *complex* is defined as a sequence of vector spaces, and linear maps between successive spaces, with the property that the composition of any pair of successive maps is identically 0. In the case of the de Rham complex (4.14), this requirement is a restatement of the closure property for the exterior differential: $d \circ d = 0$.

In particular, for $n = 3$, the de Rham complex on a manifold M reads

6 Historically, as can be seen from the terminology, homology came first and cohomology was related to it in the way we will discuss below. However, since the cohomology groups have a more natural additive structure, it is the name 'cohomology' which is actually used for generalizations.

$$0 \to \Omega^0(M) \xrightarrow{d^0} \Omega^1(M) \xrightarrow{d^1} \Omega^2(M) \xrightarrow{d^2} \Omega^3(M) \to 0. \quad (4.15)$$

If $\omega \equiv f(x, y, z) \in \Omega^0(M)$, then

$$d^0\omega \equiv d^0 f = \frac{\partial f}{\partial x}dx + \frac{\partial f}{\partial y}dy + \frac{\partial f}{\partial z}dz = \operatorname{grad}\omega.$$

If $\omega \equiv f dx + g dy + h dz \in \Omega^1(M)$, then

$$d^1\omega \equiv \left(\frac{\partial g}{\partial x} - \frac{\partial f}{\partial y}\right) dx\wedge dy + \left(\frac{\partial h}{\partial y} - \frac{\partial g}{\partial z}\right) dy\wedge dz + \left(\frac{\partial f}{\partial z} - \frac{\partial h}{\partial x}\right) dz\wedge dx = \operatorname{curl}\omega.$$

If $\omega \equiv F dy \wedge dz + G dz \wedge dx + H dx \wedge dy \in \Omega^2(M)$, then

$$d^2\omega \equiv \frac{\partial F}{\partial x} + \frac{\partial G}{\partial y} + \frac{\partial H}{\partial z} = \operatorname{div}\omega.$$

Therefore, the de Rham complex (4.15) can be written as

$$0 \to \Omega^0(M) \xrightarrow{\text{grad}} \to \Omega^1(M) \xrightarrow{\text{curl}} \Omega^2(M) \xrightarrow{\text{div}} \Omega^3(M) \to 0.$$

Using the closure property for the exterior differential, $d \circ d = 0$, we get the standard identities from vector calculus

$$\operatorname{curl} \cdot \operatorname{grad} = 0 \qquad \text{and} \qquad \operatorname{div} \cdot \operatorname{curl} = 0.$$

The definition of the complex requires that the kernel of one of the linear maps contains the image of the preceding map. The complex is *exact* if this containment is *equality*. In the case of the de Rham complex (4.14), exactness means that a closed p–form ω, meaning that $d\omega = 0$, is necessarily an exact p–form, meaning that there exists a $(p-1)$–form θ such that $\omega = d\theta$. (For $p = 0$, it says that a smooth function f is closed, $df = 0$, iff it is constant). Clearly, any exact form is closed, but the converse need not hold. Thus the de Rham complex is *not* in general exact. The celebrated *de Rham Theorem* states that the extent to which this complex fails to be exact measures purely topological information about the manifold M, its cohomology group.

On the local side, for special types of domains in Euclidean space $\mathbb{R}^m$, there is only trivial topology and we do have exactness of the de Rham complex (4.14). This result, known as the *Poincaré lemma*, holds for *star–shaped* domains $M \subset \mathbb{R}^m$: Let $M \subset \mathbb{R}^m$ be a star–shaped domain. Then the de Rham complex over M is exact.

The key to the proof of exactness of the de Rham complex lies in the construction of suitable *homotopy operators*. By definition, these are linear operators $h : \Omega^p \to \Omega^{p-1}$, taking differential p–forms into $(p-1)$–forms, and satisfying the basic identity [Olv86]

$$\omega = dh(\omega) + h(d\omega), \tag{4.16}$$

for all p–forms $\omega \in \Omega^p$. The discovery of such a set of operators immediately implies exactness of the complex. For if ω is closed, $d\omega = 0$, then (4.16) reduces to $\omega = d\theta$ where $\theta = h(\omega)$, so ω is exact.

Stokes Theorem and de Rham Cohomology of M

Stokes Theorem states that if α is an $(n-1)-$form on an orientable $n-$manifold M, then the integral of $d\alpha$ over M equals the integral of α over ∂M, the boundary of M. The classical theorems of Gauss, Green, and Stokes are special cases of this result.

A *manifold with boundary* is a set M together with an atlas of charts (U, ϕ) with boundary on M. Define (see [AMR88]) the *interior* and *boundary* of M respectively as

$$\text{Int}\, M = \bigcup_U \phi^{-1}\left(\text{Int}\left(\phi(U)\right)\right) \qquad \text{and} \qquad \partial M = \bigcup_U \phi^{-1}\left(\partial\left(\phi(U)\right)\right).$$

If M is a manifold with boundary, then its interior $\text{Int}\, M$ and its boundary ∂M are smooth manifolds without boundary. Moreover, if $f : M \to N$ is a diffeomorphism, N being another manifold with boundary, then f induces, by restriction, two diffeomorphisms

$$\text{Int}\, f : \text{Int}\, M \to \text{Int}\, N, \qquad \text{and} \qquad \partial f : \partial M \to \partial N.$$

If $n = \dim M$, then $\dim(\text{Int}\, M) = n$ and $\dim(\partial M) = n - 1$.

To integrate a differential $n-$form over an $n-$manifold M, M must be oriented. If $\text{Int}\, M$ is oriented, we want to choose an *orientation* on ∂M compatible with it. As for manifolds without boundary a volume form on an $n-$manifold with boundary M is a nowhere vanishing $n-$form on M. Fix an orientation on $\mathbb{R}^n$. Then a chart (U, ϕ) is called *positively oriented* if the map $T_m\phi : T_mM \to \mathbb{R}^n$ is orientation preserving for all $m \in U$.

Let M be a compact, oriented kD smooth manifold with boundary ∂M. Let α be a smooth $(k-1)-$form on M. Then the classical *Stokes formula* holds

$$\int_M d\alpha = \int_{\partial M} \alpha.$$

If $\partial M = \emptyset$ then $\int_M d\alpha = 0$.

The quotient space

$$H^k(M) = \frac{\text{Ker}\left(d : \Omega^k(M) \to \Omega^{k+1}(M)\right)}{\text{Im}\left(d : \Omega^{k-1}(M) \to \Omega^k(M)\right)}$$

represents the kth *de Rham cohomology group* of a manifold M. recall that the *de Rham Theorem* states that these Abelian groups are isomorphic to the so–called singular cohomology groups of M defined in algebraic topology in terms of simplices and that depend only on the topological structure of M and not on its differentiable structure. The isomorphism is provided by integration; the fact that the integration map drops to the preceding quotient is guaranteed by Stokes' Theorem.

The exterior derivative commutes with the pull–back of differential forms. That means that the vector bundle $\Lambda^k T^*M$ is in fact the value of a functor, which associates a bundle over M to each manifold M and a vector bundle homomorphism over φ to each (local) diffeomorphism φ between manifolds of the same dimension. This is a simple example of the concept of a natural bundle. The fact that the exterior derivative d transforms sections of $\Lambda^k T^* M$ into sections of $\Lambda^{k+1} T^* M$ for every manifold M can be expressed by saying that d is an operator from $\Lambda^k T^* M$ into $\Lambda^{k+1} T^* M$. That the exterior derivative d commutes with (local) diffeomorphisms now means, that d is a natural operator from the functor $\Lambda^k T^*$ into functor $\Lambda^{k+1} T^*$. If $k > 0$, one can show that d is the unique natural operator between these two natural bundles up to a constant. So even linearity is a consequence of naturality [KMS93].

Euler–Poincaré Characteristics of M

The *Euler–Poincaré characteristics* of a manifold M equals the sum of its *Betti numbers*

$$\chi(M) = \sum_{p=0}^{n} (-1)^p \, b_p.$$

In case of $2n$D oriented compact Riemannian manifold M (*Gauss–Bonnet Theorem*) its Euler–Poincaré characteristics is equal

$$\chi(M) = \int_M \gamma,$$

where γ is a closed $2n$ form on M, given by

$$\gamma = \frac{(-1)^n}{(4\pi)^n n!} \epsilon^{1\ldots 2n}_{i_1 \ldots i_{2n}} \Omega^{i_1}_{i_2} \wedge \Omega^{i_{2n-1}}_{i_{2n}},$$

where Ω^i_j is the curvature 2−form of a Riemannian connection on M.

Poincaré–Hopf Theorem: The Euler–Poincaré characteristics $\chi(M)$ of a compact manifold M equals the sum of indices of zeros of any vector–field on M which has only isolated zeros.

Duality of Chains and Forms on M

In topology of finite–dimensional smooth (i.e., C^{p+1} with $p \geq 0$) manifolds, a fundamental notion is the *duality* between p−chains C and p−forms (i.e., p−cochains) ω on the smooth manifold M, or domains of integration and integrands – as an integral on M represents a bilinear functional (see [BM82, DP97])

$$\int_C \omega \equiv \langle C, \omega \rangle \,, \tag{4.17}$$

where the integral is called the *period* of ω. Period depends only on the cohomology class of ω and the homology class of C. A closed form (cocycle) is exact (coboundary) if all its periods vanish, i.e., $d\omega = 0$ implies $\omega = d\theta$. The duality (4.17) is based on the classical Stokes formula

$$\int_C d\omega = \int_{\partial C} \omega.$$

This is written in terms of scalar products on M as

$$\langle C, d\omega \rangle = \langle \partial C, \omega \rangle \,,$$

where ∂C is the boundary of the $p-$chain C oriented coherently with C. While the boundary operator ∂ is a global operator, the coboundary operator, that is, the exterior derivative d, is local, and thus more suitable for applications. The main property of the exterior differential,

$$d^2 = 0 \qquad \text{implies} \qquad \partial^2 = 0,$$

can be easily proved by the use of Stokes' formula

$$\left\langle \partial^2 C, \omega \right\rangle = \langle \partial C, d\omega \rangle = \left\langle C, d^2\omega \right\rangle = 0.$$

The analysis of p–chains and p–forms on the finite–dimensional smooth manifold M is usually performed in (co)homology categories (see [DP97, Die88]) related to M.

Let $\mathcal{M}^\bullet$ denote the category of cochains, (i.e., p–forms) on the smooth manifold M. When $\mathcal{C} = \mathcal{M}^\bullet$, we have the category $\mathcal{S}^\bullet(\mathcal{M}^\bullet)$ of generalized cochain complexes $A^\bullet$ in $\mathcal{M}^\bullet$, and if $A' = 0$ for $n < 0$ we have a subcategory $\mathcal{S}^\bullet_{\mathcal{DR}}(\mathcal{M}^\bullet)$ of the de Rham differential complexes in $\mathcal{M}^\bullet$

$$\begin{aligned} A^\bullet_{DR} : 0 \to \Omega^0(M) &\xrightarrow{\quad d \quad} \Omega^1(M) \xrightarrow{\quad d \quad} \Omega^2(M) \cdots \\ \cdots &\xrightarrow{\quad d \quad} \Omega^n(M) \xrightarrow{\quad d \quad} \cdots . \end{aligned} \tag{4.18}$$

Here $A' = \Omega^n(M)$ is the vector space over $\mathbb{R}$ of all p–forms ω on M (for $p = 0$ the smooth functions on M) and $d_n = d : \Omega^{n-1}(M) \to \Omega^n(M)$ is the exterior differential. A form $\omega \in \Omega^n(M)$ such that $d\omega = 0$ is a closed form or n–cocycle. A form $\omega \in \Omega^n(M)$ such that $\omega = d\theta$, where $\theta \in \Omega^{n-1}(M)$, is an exact form or n–coboundary. Let $Z^n(M) = \mathrm{Ker}(d)$ (resp. $B^n(M) = \mathrm{Im}(d)$) denote a real vector space of cocycles (resp. coboundaries) of degree n. Since $d_{n+1}\, d_n = d^2 = 0$, we have $B^n(M) \subset Z^n(M)$. The quotient vector space

$$H^n_{DR}(M) = \mathrm{Ker}(d)/\,\mathrm{Im}(d) = Z^n(M)/B^n(M)$$

is the de Rham cohomology group. The elements of $H^n_{DR}(M)$ represent equivalence sets of cocycles. Two cocycles ω_1, ω_2 belong to the same equivalence set, or are cohomologous (written $\omega_1 \sim \omega_2$) iff they differ by a coboundary $\omega_1 - \omega_2 = d\theta$. The de Rham cohomology class of any form $\omega \in \Omega^n(M)$ is $[\omega] \in H^n_{DR}(M)$. The de Rham differential complex (4.18) can be considered as a system of second–order ODEs $d^2\theta = 0$, $\theta \in \Omega^{n-1}(M)$ having a solution represented by $Z^n(M) = \mathrm{Ker}(d)$.

Analogously let $\mathcal{M}_\bullet$ denote the category of chains on the smooth manifold M. When $\mathcal{C} = \mathcal{M}_\bullet$, we have the category $\mathcal{S}_\bullet(\mathcal{M}_\bullet)$ of generalized chain complexes $A_\bullet$ in $\mathcal{M}_\bullet$, and if $A_n = 0$ for $n < 0$ we have a subcategory $\mathcal{S}_\bullet^C(\mathcal{M}_\bullet)$ of chain complexes in $\mathcal{M}_\bullet$

$$A_\bullet : 0 \leftarrow C^0(M) \xleftarrow{\partial} C^1(M) \xleftarrow{\partial} C^2(M) \cdots \xleftarrow{\partial} C^n(M) \xleftarrow{\partial} \cdots .$$

Here $A_n = C^n(M)$ is the vector space over $\mathbb{R}$ of all finite chains C on the manifold M and $\partial_n = \partial : C^{n+1}(M) \to C^n(M)$. A finite chain C such that $\partial C = 0$ is an $n-$cycle. A finite chain C such that $C = \partial B$ is an $n-$boundary. Let $Z_n(M) = \mathrm{Ker}(\partial)$ (resp. $B_n(M) = \mathrm{Im}(\partial)$) denote a real vector space of cycles (resp. boundaries) of degree n. Since $\partial_{n+1}\partial_n = \partial^2 = 0$, we have $B_n(M) \subset Z_n(M)$. The quotient vector space

$$H_n^C(M) = \mathrm{Ker}(\partial)/\,\mathrm{Im}(\partial) = Z_n(M)/B_n(M)$$

is the $n-$homology group. The elements of $H_n^C(M)$ are equivalence sets of cycles. Two cycles C_1, C_2 belong to the same equivalence set, or are homologous (written $C_1 \sim C_2$), iff they differ by a boundary $C_1 - C_2 = \partial B$). The homology class of a finite chain $C \in C^n(M)$ is $[C] \in H_n^C(M)$.

The dimension of the $n-$cohomology (resp. $n-$homology) group equals the nth Betti number b^n (resp. b_n) of the manifold M. *Poincaré lemma* says that on an open set $U \in M$ diffeomorphic to $\mathbb{R}^N$, all closed forms (cycles) of degree $p \geq 1$ are exact (boundaries). That is, the Betti numbers satisfy $b^p = 0$ (resp. $b_p = 0$) for $p = 1, \ldots, n$.

The *de Rham Theorem* states the following. The map $\Phi\colon H_n \times H^n \to \mathbb{R}$ given by $([C], [\omega]) \to \langle C, \omega\rangle$ for $C \in Z_n, \omega \in Z^n$ is a bilinear nondegenerate map which establishes the duality of the groups (vector spaces) H_n and H^n and the equality $b_n = b^n$.

Hodge Star Operator and Harmonic Forms

As the configuration manifold M is an oriented ND Riemannian manifold, we may select an orientation on all tangent spaces T_mM and all cotangent spaces T_m^*M, with the local coordinates $x^i = (q^i, p_i)$ at a point $m \in M$, in a consistent manner. The simplest way to do that is to choose the Euclidean orthonormal basis $\partial_1, ..., \partial_N$ of $\mathbb{R}^N$ as being positive.

Since the manifold M carries a Riemannian structure $g = \langle , \rangle$, we have a scalar product on each T_m^*M. So, we can define (as above) the linear *Hodge star operator*

$$* : \Lambda^p(T_m^*M) \to \Lambda^{N-p}(T_m^*M),$$

which is a base point preserving operator

$$* : \Omega^p(M) \to \Omega^{N-p}(M), \qquad (\Omega^p(M) = \Gamma(\Lambda^p(M)))$$

(here $\Lambda^p(V)$ denotes the p-fold exterior product of any vector space V, $\Omega^p(M)$ is a space of all $p-$forms on M, and $\Gamma(E)$ denotes the space of sections of the vector bundle E). Also,

$$** = (-1)^{p(N-p)} : \Lambda^p(T_x^* M) \to \Lambda^p(T_m^* M).$$

As the metric on $T_m^* M$ is given by $g^{ij}(x) = (g_{ij}(x))^{-1}$, we have the *volume form* defined in local coordinates as

$$*(1) = \sqrt{\det(g_{ij})} dx^1 \wedge ... \wedge dx^n, \qquad \text{and} \qquad \mathrm{vol}(M) = \int_M *(1).$$

For any to $p-$forms $\alpha, \beta \in \Omega^p(M)$ with compact support, we define the (bilinear and positive definite) L^2-product as

$$(\alpha, \beta) = \int_M \langle \alpha, \beta \rangle * (1) = \int_M \alpha \wedge *\beta.$$

We can extend the product $(\cdot, \cdot)$ to $L^2(\Omega^p(M))$; it remains bilinear and positive definite, because as usual, in the definition of L^2, functions that differ only on a set of measure zero are identified.

Using the Hodge star operator $*$, we can introduce the *codifferential* operator δ, which is formally adjoint to the exterior derivative $d : \Omega^p(M) \to \Omega^{p+1}(M)$ on $\oplus_{p=0}^N \Omega^p(M)$ w.r.t. $(\cdot, \cdot)$. This means that for $\alpha \in \Omega^{p-1}(M), \beta \in \Omega^p(M)$

$$(d\alpha, \beta) = (\alpha, \delta\beta).$$

Therefore, we have $\delta : \Omega^p(M) \to \Omega^{p-1}(M)$ and

$$\delta = (-1)^{N(p+1)+1} * d * .$$

Now, the *Laplace–Beltrami operator* (or, *Hodge Laplacian*, see [Gri83b, Voi02]), Δ on $\Omega^p(M)$, is defined by relation similar to (4.16) above

$$\Delta = d\delta + \delta d : \Omega^p(M) \to \Omega^p(M) \tag{4.19}$$

and an exterior differential form $\alpha \in \Omega^p(M)$ is called *harmonic* if $\Delta\alpha = 0$.

Let M be a compact, oriented Riemannian manifold, E a vector bundle with a bundle metric $\langle \cdot, \cdot \rangle$ over M,

$$D = d + A : \Omega^{p-1}(Ad_E) \to \Omega^p(Ad_E), \qquad \text{with} \quad A \in \Omega^1(Ad_E)$$

– a tensorial and $\mathbb{R}-$linear metric connection on E with curvature $F_D \in \Omega^2(Ad_E)$ (Here by $\Omega^p(Ad_E)$ we denote the space of those elements of $\Omega^p(End_E)$ for which the endomorphism of each fibre is skew symmetric; End_E denotes the space of linear endomorphisms of the fibers of E).

4.1.3 Lie Derivatives, Lie Groups and Lie Algebras

Lie Derivatives on Smooth Manifolds

Lie derivative is popularly called '*fisherman's derivative*'. In continuum mechanics it is called *Liouville operator*. This is a central differential operator in modern differential geometry and its physical and control applications.

Lie Derivative Operating on Functions

To define how vector–fields operate on functions on an $m-$manifold M, we will use the *Lie derivative.* Let $f : M \to \mathbb{R}$ so $Tf : TM \to T\mathbb{R} = \mathbb{R} \times \mathbb{R}$. Following [AMR88] we write Tf acting on a vector $v \in T_m M$ in the form

$$Tf \cdot v = (f(m), df(m) \cdot v) .$$

This defines, for each point $m \in M$, the element $df(m) \in T_m^* M$. Thus df is a section of the cotangent bundle $T^* M$, i.e., a $1-$form. The $1-$form $df : M \to T^* M$ defined this way is called the *differential* of f. If f is C^∞, then df is C^{k-1}.

If $\phi : U \subset M \to V \subset E$ is a local chart for M, then the local representative of $f \in C^\infty(M, \mathbb{R})$ is the map $f : V \to \mathbb{R}$ defined by $f = f \circ \phi^{-1}$. The local representative of Tf is the tangent map for local manifolds,

$$Tf(x, v) = (f(x), Df(x) \cdot v) .$$

Thus the local representative of df is the derivative of the local representative of f. In particular, if $(x^1, ..., x^n)$ are local coordinates on M, then the local components of df are

$$(df)^i = \partial_{x^i} f.$$

The introduction of df leads to the following definition of the Lie derivative. The *directional* or *Lie derivative* $\mathcal{L}_X : C^\infty(M, \mathbb{R}) \to C^{k-1}(M, \mathbb{R})$ of a function $f \in C^\infty(M, \mathbb{R})$ along a vector–field X is defined by

$$\mathcal{L}_X f(m) = X[f](m) = df(m) \cdot X(m),$$

for any $m \in M$. Denote by $X[f] = df(X)$ the map $M \ni m \mapsto X[f](m) \in \mathbb{R}$. If f is $F-$valued, the same definition is used, but now $X[f]$ is $F-$valued.

If a local chart (U, ϕ) on an $n-$manifold M has local coordinates $(x^1, ..., x^n)$, the local representative of $X[f]$ is given by the function

$$\mathcal{L}_X f = X[f] = X^i \, \partial_{x^i} f.$$

Evidently if f is C^∞ and X is C^{k-1} then $X[f]$ is C^{k-1}.

Let $\varphi : M \to N$ be a diffeomorphism. Then $\mathcal{L}_X$ is natural with respect to push–forward by φ. That is, for each $f \in C^\infty(M, \mathbb{R})$,

$$\mathcal{L}_{\varphi_* X}(\varphi_* f) = \varphi_* \mathcal{L}_X f,$$

i.e., the following diagram commutes:

$$\begin{array}{ccc}
C^\infty(M, \mathbb{R}) & \xrightarrow{\varphi_*} & C^\infty(N, \mathbb{R}) \\
\mathcal{L}_X \downarrow & & \downarrow \mathcal{L}_{\varphi_* X} \\
C^\infty(M, \mathbb{R}) & \xrightarrow[\varphi_*]{} & C^\infty(N, \mathbb{R})
\end{array}$$

Also, $\mathcal{L}_X$ is natural with respect to restrictions. That is, for U open in M and $f \in C^\infty(M, \mathbb{R})$,

$$\mathcal{L}_{X|U}(f|U) = (\mathcal{L}_X f)|U,$$

where $|U : C^\infty(M, \mathbb{R}) \to C^\infty(U, \mathbb{R})$ denotes restriction to U, i.e., the following diagram commutes:

$$\begin{CD} C^\infty(M,\mathbb{R}) @>{|U}>> C^\infty(U,\mathbb{R}) \\ @V{\mathcal{L}_X}VV @VV{\mathcal{L}_{X|U}}V \\ C^\infty(M,\mathbb{R}) @>>{|U}> C^\infty(U,\mathbb{R}) \end{CD}$$

Since $\varphi^* = (\varphi^{-1})_*$ the Lie derivative is also natural with respect to pull–back by φ. This has a generalization to φ–related vector–fields as follows: Let $\varphi : M \to N$ be a C^∞–map, $X \in \mathcal{X}^{k-1}(M)$ and $Y \in \mathcal{X}^{k-1}(N)$, $k \geq 1$. If $X \sim_\varphi Y$, then

$$\mathcal{L}_X(\varphi^* f) = \varphi^* \mathcal{L}_Y f$$

for all $f \in C^\infty(N, \mathbb{R})$, i.e., the following diagram commutes:

$$\begin{CD} C^\infty(N,\mathbb{R}) @>{\varphi^*}>> C^\infty(M,\mathbb{R}) \\ @V{\mathcal{L}_Y}VV @VV{\mathcal{L}_X}V \\ C^\infty(N,\mathbb{R}) @>>{\varphi^*}> C^\infty(M,\mathbb{R}) \end{CD}$$

The Lie derivative map $\mathcal{L}_X : C^\infty(M, \mathbb{R}) \to C^{k-1}(M, \mathbb{R})$ is a *derivation*, i.e., for two functions $f, g \in C^\infty(M, \mathbb{R})$ the *Leibniz rule* is satisfied

$$\mathcal{L}_X(fg) = g\mathcal{L}_X f + f\mathcal{L}_X g;$$

Also, Lie derivative of a constant function is zero, $\mathcal{L}_X(\text{const}) = 0$.

The connection between the Lie derivative $\mathcal{L}_X f$ of a function $f \in C^\infty(M, \mathbb{R})$ and the flow F_t of a vector–field $X \in \mathcal{X}^{k-1}(M)$ is given as:

$$\frac{d}{dt}(F_t^* f) = F_t^* (\mathcal{L}_X f).$$

Lie Derivative of Vector Fields

If $X, Y \in \mathcal{X}^k(M)$, $k \geq 1$ are two vector–fields on M, then

$$[\mathcal{L}_X, \mathcal{L}_Y] = \mathcal{L}_X \circ \mathcal{L}_Y - \mathcal{L}_Y \circ \mathcal{L}_X$$

is a derivation map from $C^{k+1}(M, \mathbb{R})$ to $C^{k-1}(M, \mathbb{R})$. Then there is a unique vector–field, $[X, Y] \in \mathcal{X}^k(M)$ of X and Y such that $\mathcal{L}_{[X,Y]} = [\mathcal{L}_X, \mathcal{L}_Y]$ and

$[X,Y](f) = X(Y(f)) - Y(X(f))$ holds for all functions $f \in C^\infty(M,\mathbb{R})$. This vector–field is also denoted $\mathcal{L}_X Y$ and is called the Lie derivative of Y with respect to X, or the *Lie bracket* of X and Y. In a local chart (U,ϕ) at a point $m \in M$ with coordinates $(x^1,...,x^n)$, for $X|_U = X^i\partial_{x^i}$ and $Y|_U = Y^i\partial_{x^i}$ we have

$$\left[X^i\partial_{x^i}, Y^j\partial_{x^j}\right] = \left(X^i\left(\partial_{x^i}Y^j\right) - Y^i\left(\partial_{x^i}X^j\right)\right)\partial_{x^j},$$

since second partials commute. If, also X has flow F_t, then [AMR88]

$$\frac{d}{dt}\left(F_t^* Y\right) = F_t^*\left(\mathcal{L}_X Y\right).$$

In particular, if $t = 0$, this formula becomes

$$\frac{d}{dt}|_{t=0}\left(F_t^* Y\right) = \mathcal{L}_X Y.$$

Then the unique C^{k-1} vector–field $\mathcal{L}_X Y = [X,Y]$ on M defined by

$$[X,Y] = \frac{d}{dt}|_{t=0}\left(F_t^* Y\right),$$

is called the Lie derivative of Y with respect to X, or the Lie bracket of X and Y, and can be interpreted as the leading order term that results from the sequence of flows

$$F_t^{-Y} \circ F_t^{-X} \circ F_t^{Y} \circ F_t^{-X}(m) = \epsilon^2[X,Y](m) + \mathcal{O}(\epsilon^3), \tag{4.20}$$

for some real $\epsilon > 0$. Therefore a Lie bracket can be interpreted as a 'new direction' in which the system can flow, by executing the sequence of flows (4.20).

Lie bracket satisfies the following property:

$$[X,Y][f] = X[Y[f]] - Y[X[f]],$$

for all $f \in C^{k+1}(U,\mathbb{R})$, where U is open in M.

An important relationship between flows of vector–fields is given by the *Campbell–Baker–Hausdorff formula*:

$$F_t^Y \circ F_t^X = F_t^{X+Y+\frac{1}{2}[X,Y]+\frac{1}{12}([X,[X,Y]]-[Y,[X,Y]])+\dots} \tag{4.21}$$

Essentially, if given the composition of multiple flows along multiple vector–fields, this formula gives the one flow along one vector–field which results in the same net flow. One way to prove the Campbell–Baker–Hausdorff formula (4.21) is to expand the product of two formal exponentials and equate terms in the resulting formal power series.

Lie bracket is the $\mathbb{R}$–bilinear map $[,] : \mathcal{X}^k(M) \times \mathcal{X}^k(M) \to \mathcal{X}^k(M)$ with the following properties:

1. $[X, Y] = -[Y, X]$, i.e., $\mathcal{L}_X Y = -\mathcal{L}_Y X$ for all $X, Y \in \mathcal{X}^k(M)$ – skew–symmetry;
2. $[X, X] = 0$ for all $X \in \mathcal{X}^k(M)$;
3. $[X, [Y, Z]] + [Y, [Z, X]] + [Z, [X, Y]] = 0$ for all $X, Y, Z \in \mathcal{X}^k(M)$ – the Jacobi identity;
4. $[fX, Y] = f[X, Y] - (Yf)X$, i.e., $\mathcal{L}_{fX}(Y) = f(\mathcal{L}_X Y) - (\mathcal{L}_Y f)X$ for all $X, Y \in \mathcal{X}^k(M)$ and $f \in C^\infty(M, \mathbb{R})$;
5. $[X, fY] = f[X, Y] + (Xf)Y$, i.e., $\mathcal{L}_X(fY) = f(\mathcal{L}_X Y) + (\mathcal{L}_X f)Y$ for all $X, Y \in \mathcal{X}^k(M)$ and $f \in C^\infty(M, \mathbb{R})$;
6. $[\mathcal{L}_X, \mathcal{L}_Y] = \mathcal{L}_{[x,y]}$ for all $X, Y \in \mathcal{X}^k(M)$.

The pair $(\mathcal{X}^k(M), [,])$ is the prototype of a *Lie algebra* [KMS93]. In more general case of a *general linear Lie algebra* $\mathfrak{gl}(n)$, which is the Lie algebra associated to the Lie group $GL(n)$, Lie bracket is given by a *matrix commutator*

$$[A, B] = AB - BA,$$

for any two matrices $A, B \in \mathfrak{gl}(n)$.

Let $\varphi : M \to N$ be a diffeomorphism. Then $\mathcal{L}_X : \mathcal{X}^k(M) \to \mathcal{X}^k(M)$ is natural with respect to push–forward by φ. That is, for each $f \in C^\infty(M, \mathbb{R})$,

$$\mathcal{L}_{\varphi_* X}(\varphi_* Y) = \varphi_* \mathcal{L}_X Y,$$

i.e., the following diagram commutes:

$$\begin{CD}
\mathcal{X}^k(M) @>{\varphi_*}>> \mathcal{X}^k(N) \\
@V{\mathcal{L}_X}VV @VV{\mathcal{L}_{\varphi_* X}}V \\
\mathcal{X}^k(M) @>>{\varphi_*}> \mathcal{X}^k(N)
\end{CD}$$

Also, $\mathcal{L}_X$ is natural with respect to restrictions. That is, for U open in M and $f \in C^\infty(M, \mathbb{R})$,

$$[X|U, Y|U] = [X, Y]|U,$$

where $U : C^\infty(M, \mathbb{R}) \to C^\infty(U, \mathbb{R})$ denotes restriction to U, i.e., the following diagram commutes [AMR88]:

$$\begin{CD}
\mathcal{X}^k(M) @>{|U}>> \mathcal{X}^k(U) \\
@V{\mathcal{L}_X}VV @VV{\mathcal{L}_{X|U}}V \\
\mathcal{X}^k(M) @>>{|U}> \mathcal{X}^k(U)
\end{CD}$$

If a local chart (U, ϕ) on an $n-$manifold M has local coordinates $(x^1, ..., x^n)$, then the local components of a Lie bracket are

$$[X,Y]^j = X^i\,\partial_{x^i} Y^j - Y^i\,\partial_{x^i} X^j,$$

that is, $[X,Y] = (X\cdot\nabla)Y - (Y\cdot\nabla)X$.

Let $\varphi : M \to N$ be a C^∞–map, $X \in \mathcal{X}^{k-1}(M)$ and $Y \in \mathcal{X}^{k-1}(N)$, $k \geq 1$. Then $X \sim_\varphi Y$, iff

$$(Y[f])\circ\varphi = X[f\circ\varphi]$$

for all $f \in C^\infty(V,\mathbb{R})$, where V is open in N.

For every $X \in \mathcal{X}^k(M)$, the operator $\mathcal{L}_X$ is a derivation on $\left(C^\infty(M,\mathbb{R}), \mathcal{X}^k(M)\right)$, i.e., $\mathcal{L}_X$ is $\mathbb{R}$–linear.

For any two vector–fields $X \in \mathcal{X}^k(M)$ and $Y \in \mathcal{X}^k(N)$, $k \geq 1$ with flows F_t and G_t, respectively, if $[X,Y] = 0$ then $F_t^*Y = Y$ and $G_t^*X = X$.

Derivative of the Evolution Operator

Recall that the time–dependent flow or evolution operator $F_{t,s}$ of a vector–field $X \in \mathcal{X}^k(M)$ is defined by the requirement that $t \mapsto F_{t,s}(m)$ be the integral curve of X starting at a point $m \in M$ at time $t = s$, i.e.,

$$\frac{d}{dt}F_{t,s}(m) = X\left(t, F_{t,s}(m)\right) \qquad \text{and} \qquad F_{t,t}(m) = m.$$

By uniqueness of integral curves we have $F_{t,s}\circ F_{s,r} = F_{t,r}$ (replacing the flow property $F_{t+s} = F_t + F_s$) and $F_{t,t} =$ identity.

Let $X_t \in \mathcal{X}^k(M)$, $k \geq 1$ for each t and suppose $X(t,m)$ is continuous in $(t,m) \in \mathbb{R}\times M$. Then $F_{t,s}$ is of class C^∞ and for $f \in C^{k+1}(M,\mathbb{R})$ [AMR88], and $Y \in \mathcal{X}^k(M)$, we have

1. $\frac{d}{dt}F_{t,s}^*\,f = F_{t,s}^*\left(\mathcal{L}_{X_t} f\right)$, and
2. $\frac{d}{dt}F_{t,s}^*\,f = F_{t,s}^*([X_t,Y]) = F_{t,s}^*\left(\mathcal{L}_{X_t} Y\right)$.

From the above Theorem, the following identity holds:

$$\frac{d}{dt}F_{t,s}^*\,f = -X_t\left[F_{t,s}^*\,f\right].$$

Lie Derivative of Differential Forms

Since $\mathcal{F} : M \Longrightarrow \Lambda^k T^*M$ is a vector bundle functor on $\mathcal{M}$, the Lie derivative of a k–form $\alpha \in \Omega^k(M)$ along a vector–field $X \in \mathcal{X}^k(M)$ is defined by

$$\mathcal{L}_X\alpha = \frac{d}{dt}\,|_{t=0}\; F_t^*\alpha.$$

It has the following properties:

1. $\mathcal{L}_X(\alpha\wedge\beta) = \mathcal{L}_X\,\alpha\wedge\beta + \alpha\wedge\mathcal{L}_X\,\beta$, so $\mathcal{L}_X$ is a derivation.
2. $[\mathcal{L}_X, \mathcal{L}_Y]\;\alpha = \mathcal{L}_{[X,Y]}\,\alpha$.
3. $\frac{d}{dt}F_t^*\alpha = F_t^*\mathcal{L}_X\alpha = \mathcal{L}_X\left(F_t^*\alpha\right)$.

Formula (3) holds also for time–dependent vector–fields in the sense that $\frac{d}{dt}F^*_{t,s}\alpha = F^*_{t,s}\mathcal{L}_X\alpha = \mathcal{L}_X\left(F^*_{t,s}\alpha\right)$ and in the expression $\mathcal{L}_X\alpha$ the vector–field X is evaluated at time t.

The famous *Cartan magic formula* (see [MR99]) states: the Lie derivative of a $k-$form $\alpha \in \Omega^k(M)$ along a vector–field $X \in \mathcal{X}^k(M)$ on a smooth manifold M is defined as

$$\mathcal{L}_X\alpha = di_X\alpha + i_Xd\alpha = d(X\rfloor\alpha) + X\rfloor d\alpha.$$

Also, the following identities hold [MR99, KMS93]:

1. $\mathcal{L}_{fX}\alpha = f\mathcal{L}_X\alpha + df \wedge i_x\alpha$.
2. $\mathcal{L}_{[X,Y]}\alpha = \mathcal{L}_X\mathcal{L}_Y\alpha - \mathcal{L}_Y\mathcal{L}_X\alpha$.
3. $i_{[X,Y]}\alpha = \mathcal{L}_X i_Y\alpha - i_Y\mathcal{L}_X\alpha$.
4. $\mathcal{L}_X d\alpha = d\mathcal{L}_X\alpha$, i.e., $[\mathcal{L}_X, d] = 0$.
5. $\mathcal{L}_X i_X\alpha = i_X\mathcal{L}_X\alpha$, i.e., $[\mathcal{L}_X, i_X] = 0$.
6. $\mathcal{L}_X(\alpha\wedge\beta) = \mathcal{L}_X\alpha\wedge\beta + \alpha\wedge\mathcal{L}_X\beta$.

Lie Derivative of Various Tensor Fields

In this subsection, we use local coordinates x^i $(i = 1, ..., n)$ on a biomechanical $n-$manifold M, to calculate the Lie derivative $\mathcal{L}_{X^i}$ with respect to a generic vector–field X^i. (As always, $\partial_{x^i} \equiv \frac{\partial}{\partial x^i}$).

Lie Derivative of a Scalar Field

Given the scalar field ϕ, its Lie derivative $\mathcal{L}_{X^i}\phi$ is given as

$$\mathcal{L}_{X^i}\phi = X^i\partial_{x^i}\phi = X^1\partial_{x^1}\phi + X^2\partial_{x^2}\phi + ... + X^n\partial_{x^n}\phi.$$

Lie Derivative of Vector and Covector–Fields

Given a contravariant vector–field V^i, its Lie derivative $\mathcal{L}_{X^i}V^i$ is given as

$$\mathcal{L}_{X^i}V^i = X^k\partial_{x^k}V^i - V^k\partial_{x^k}X^i \equiv [X^i, V^i] - \text{the Lie bracket.}$$

Given a covariant vector–field (i.e., a one–form) ω_i, its Lie derivative $\mathcal{L}_{X^i}\omega_i$ is given as

$$\mathcal{L}_{X^i}\omega_i = X^k\partial_{x^k}\omega_i + \omega_k\partial_{x^i}X^k.$$

Lie Derivative of a Second–Order Tensor–Field

Given a $(2,0)$ tensor–field S^{ij}, its Lie derivative $\mathcal{L}_{X^i}S^{ij}$ is given as

$$\mathcal{L}_{X^i}S^{ij} = X^i\partial_{x^i}S^{ij} - S^{ij}\partial_{x^i}X^i - S^{ii}\partial_{x^i}X^j.$$

Given a $(1,1)$ tensor–field S^i_j, its Lie derivative $\mathcal{L}_{X^i}S^i_j$ is given as

$$\mathcal{L}_{X^i}S^i_j = X^i\partial_{x^i}S^i_j - S^i_j\partial_{x^i}X^i + S^i_i\partial_{x^j}X^i.$$

Given a $(0,2)$ tensor–field S_{ij}, its Lie derivative $\mathcal{L}_{X^i}S_{ij}$ is given as

$$\mathcal{L}_{X^i}S_{ij} = X^i\partial_{x^i}S_{ij} + S_{ij}\partial_{x^i}X^i + S_{ii}\partial_{x^j}X^i.$$

Lie Derivative of a Third–Order Tensor–Field

Given a $(3,0)$ tensor–field T^{ijk}, its Lie derivative $\mathcal{L}_{X^i}T^{ijk}$ is given as

$$\mathcal{L}_{X^i}T^{ijk} = X^i\partial_{x^i}T^{ijk} - T^{ijk}\partial_{x^i}X^i - T^{iik}\partial_{x^i}X^j - T^{iji}\partial_{x^i}X^k.$$

Given a $(2,1)$ tensor–field T_k^{ij}, its Lie derivative $\mathcal{L}_{X^i}T_k^{ij}$ is given as

$$\mathcal{L}_{X^i}T_k^{ij} = X^i\partial_{x^i}T_k^{ij} - T_k^{ij}\partial_{x^i}X^i + T_i^{ij}\partial_{x^k}X^i - T_k^{ii}\partial_{x^i}X^j.$$

Given a $(1,2)$ tensor–field T_{jk}^i, its Lie derivative $\mathcal{L}_{X^i}T_{jk}^i$ is given as

$$\mathcal{L}_{X^i}T_{jk}^i = X^i\partial_{x^i}T_{jk}^i - T_{jk}^i\partial_{x^i}X^i + T_{ik}^i\partial_{x^j}X^i + T_{ji}^i\partial_{x^k}X^i.$$

Given a $(0,3)$ tensor–field T_{ijk}, its Lie derivative $\mathcal{L}_{X^i}T_{ijk}$ is given as

$$\mathcal{L}_{X^i}T_{ijk} = X^i\partial_{x^i}T_{ijk} + T_{ijk}\partial_{x^i}X^i + T_{iik}\partial_{x^j}X^i + T_{iji}\partial_{x^k}X^i.$$

Lie Derivative of a Fourth–Order Tensor–Field

Given a $(4,0)$ tensor–field R^{ijkl}, its Lie derivative $\mathcal{L}_{X^i}R^{ijkl}$ is given as

$$\mathcal{L}_{X^i}R^{ijkl} = X^i\partial_{x^i}R^{ijkl} - R^{ijkl}\partial_{x^i}X^i - R^{iikl}\partial_{x^i}X^j - R^{ijil}\partial_{x^i}X^k - R^{ijki}\partial_{x^i}X^l.$$

Given a $(3,1)$ tensor–field R_l^{ijk}, its Lie derivative $\mathcal{L}_{X^i}R_l^{ijk}$ is given as

$$\mathcal{L}_{X^i}R_l^{ijk} = X^i\partial_{x^i}R_l^{ijk} - R_l^{ijk}\partial_{x^i}X^i + R_i^{ijk}\partial_{x^l}X^i - R_l^{iik}\partial_{x^i}X^j - R_l^{iji}\partial_{x^i}X^k.$$

Given a $(2,2)$ tensor–field R_{kl}^{ij}, its Lie derivative $\mathcal{L}_{X^i}R_{kl}^{ij}$ is given as

$$\mathcal{L}_{X^i}R_{kl}^{ij} = X^i\partial_{x^i}R_{kl}^{ij} - R_{kl}^{ij}\partial_{x^i}X^i + R_{il}^{ij}\partial_{x^k}X^i + R_{ki}^{ij}\partial_{x^l}X^i - R_{kl}^{ii}\partial_{x^i}X^j.$$

Given a $(1,3)$ tensor–field R_{jkl}^i, its Lie derivative $\mathcal{L}_{X^i}R_{jkl}^i$ is given as

$$\mathcal{L}_{X^i}R_{jkl}^i = X^i\partial_{x^i}R_{jkl}^i - R_{jkl}^i\partial_{x^i}X^i + R_{ikl}^i\partial_{x^j}X^i + R_{jil}^i\partial_{x^k}X^i + R_{jki}^i\partial_{x^l}X^i.$$

Given a $(0,4)$ tensor–field R_{ijkl}, its Lie derivative $\mathcal{L}_{X^i}R_{ijkl}$ is given as

$$\mathcal{L}_{X^i}R_{ijkl} = X^i\partial_{x^i}R_{ijkl} + R_{ijkl}\partial_{x^i}X^i + R_{iikl}\partial_{x^j}X^i + R_{ijil}\partial_{x^k}X^i + R_{ijki}\partial_{x^l}X^i.$$

Finally, recall that a *spinor* is a two–component complex column vector. Physically, spinors can describe both bosons and fermions, while tensors can describe only bosons. The Lie derivative of a spinor ϕ is defined by

$$\mathcal{L}_X\phi(x) = \lim_{t\to 0}\frac{\bar{\phi}_t(x) - \phi(x)}{t},$$

where $\bar{\phi}_t$ is the image of ϕ by a one–parameter group of isometries with X its generator. For a vector–field X^a and a covariant derivative ∇_a, the Lie derivative of ϕ is given explicitly by

$$\mathcal{L}_X\phi = X^a\nabla_a\phi - \frac{1}{8}\left(\nabla_a X_b - \nabla_b X_a\right)\gamma^a\gamma^b\phi,$$

where γ^a and γ^b are *Dirac matrices* (see, e.g., [BM00]).

The Lie Derivative and Lie Bracket in Control Theory

Recall (see (4.1.3) above) that given a scalar function $h(x)$ and a vector–field $f(x)$, we define a new scalar function, $\mathcal{L}_f h = \nabla h f$, which is the Lie derivative of h w.r.t. f, i.e., the directional derivative of h along the direction of the vector f. Repeated Lie derivatives can be defined recursively:

$$\mathcal{L}_f^0 h = h, \qquad \mathcal{L}_f^i h = \mathcal{L}_f\left(\mathcal{L}_f^{i-1} h\right) = \nabla\left(\mathcal{L}_f^{i-1} h\right) f, \qquad (\text{for } i = 1, 2, ...)$$

Or given another vector–field, g, then $\mathcal{L}_g\mathcal{L}_f h(x)$ is defined as

$$\mathcal{L}_g\mathcal{L}_f h = \nabla\left(\mathcal{L}_f h\right) g.$$

For example, if we have a control system

$$\dot{x} = f(x), \qquad y = h(x),$$

with the state $x = x(t)$ and the output y, then the derivatives of the output are:

$$\dot{y} = \frac{\partial h}{\partial x}\dot{x} = \mathcal{L}_f h, \qquad \text{and} \qquad \ddot{y} = \frac{\partial L_f h}{\partial x}\dot{x} = \mathcal{L}_f^2 h.$$

Also, recall that the curvature of two vector–fields, g_1, g_2, gives a non–zero Lie bracket, $[g_1, g_2]$ ((4.1.3) see Figure 4.5). Lie bracket motions can generate new directions in which the system can move.

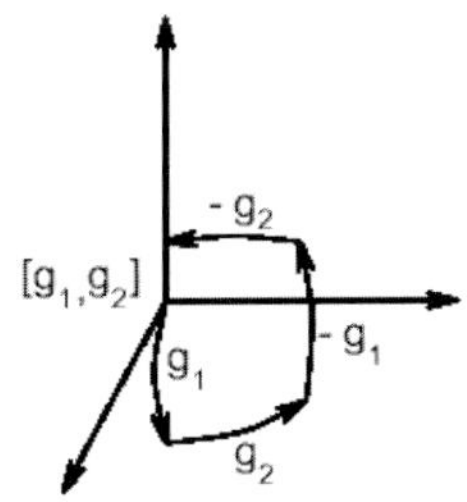

Fig. 4.5. The so–called 'Lie bracket motion' is possible by appropriately modulating the control inputs (see text for explanation).

In general, the Lie bracket of two vector–fields, $f(x)$ and $g(x)$, is defined by

$$[f, g] = Ad_f g = \nabla g f - \nabla f g = \frac{\partial g}{\partial x} f - \frac{\partial f}{\partial x} g,$$

where $\nabla f = \partial f / \partial x$ is the Jacobian matrix. We can define Lie brackets recursively,

$$Ad_f^0 g = g, \qquad Ad_f^i g = [f, Ad_f^{i-1} g], \qquad (\text{for } i = 1, 2, ...)$$

Lie brackets have the properties of bilinearity, skew–commutativity and Jacobi identity.

For example, if

$$f = \begin{pmatrix} \cos x_2 \\ x_1 \end{pmatrix}, \qquad g = \begin{pmatrix} x_1 \\ 1 \end{pmatrix},$$

then we have

$$[f, g] = \begin{pmatrix} 1 & 0 \\ 0 & 0 \end{pmatrix} \begin{pmatrix} \cos x_2 \\ x_1 \end{pmatrix} - \begin{pmatrix} 0 & -\sin x_2 \\ 1 & 0 \end{pmatrix} \begin{pmatrix} x_1 \\ 1 \end{pmatrix} = \begin{pmatrix} \cos x_2 + \sin x_2 \\ -x_1 \end{pmatrix}.$$

Now, recall that nonlinear MIMO–systems are generally described by differential equations of the form (see [Isi89, NS90, SI89]):

$$\dot{x} = f(x) + g_i(x)\, u^i, \qquad (i = 1, ..., n), \tag{4.22}$$

defined on a smooth $n-$manifold M, where $x \in M$ represents the state of the control system, $f(x)$ and $g_i(x)$ are vector–fields on M and the u^i are control inputs, which belong to a set of *admissible controls*, $u^i \in U$. The system (4.22) is called *driftless*, or *kinematic*, or *control linear* if $f(x)$ is identically zero; otherwise, it is called a *system with drift*, and the vector–field $f(x)$ is called the *drift term*. The flow $\phi_t^g(x_0)$ represents the solution of the differential equation $\dot{x} = g(x)$ at time t starting from x_0. Geometrical way to understand the *controllability* of the system (4.22) is to understand the geometry of the vector–fields $f(x)$ and $g_i(x)$.

Example: Car–Parking Using Lie Brackets

In this popular example, the driver has two different transformations at his disposal. He/she can turn the steering wheel, or he/she can drive the car forward or back. Here, we specify the state of a car by four coordinates: the (x, y) coordinates of the center of the rear axle, the direction θ of the car, and the angle ϕ between the front wheels and the direction of the car. L is the constant length of the car. Therefore, the configuration manifold of the car is 4D, $M = (x, y, \theta, \phi)$.

Using (4.22), the driftless car kinematics can be defined as:

$$\dot{x} = g_1(x)\, u_1 + g_2(x)\, u_2, \tag{4.23}$$

with two vector–fields $g_1, g_2 \in \mathcal{X}^k(M)$.

The infinitesimal transformations will be the vector–fields

$$g_1(x) \equiv \text{DRIVE} = \cos\theta \frac{\partial}{\partial x} + \sin\theta \frac{\partial}{\partial y} + \frac{\tan\phi}{L} \frac{\partial}{\partial \theta} \equiv \begin{pmatrix} \cos\theta \\ \sin\theta \\ \frac{1}{L}\tan\phi \\ 0 \end{pmatrix},$$

and $$g_2(x) \equiv \text{STEER} = \frac{\partial}{\partial \phi} \equiv \begin{pmatrix} 0 \\ 0 \\ 0 \\ 1 \end{pmatrix}.$$

Now, STEER and DRIVE do not commute; otherwise we could do all your steering at home before driving of on a trip. Therefore, we have a Lie bracket

$$[g_2, g_1] \equiv [\text{STEER}, \text{DRIVE}] = \frac{1}{L\cos^2\phi}\frac{\partial}{\partial\theta} \equiv \text{ROTATE}.$$

The operation $[g_2, g_1] \equiv$ ROTATE $\equiv$ [STEER,DRIVE] is the infinitesimal version of the sequence of transformations: steer, drive, steer back, and drive back, i.e.,

$$\{\text{STEER}, \text{DRIVE}, \text{STEER}^{-1}, \text{DRIVE}^{-1}\}.$$

Now, ROTATE can get us out of some parking spaces, but not tight ones: we may not have enough room to ROTATE out. The usual tight parking space restricts the DRIVE transformation, but not STEER. A truly tight parking space restricts STEER as well by putting your front wheels against the curb.

Fortunately, there is still another commutator available:

$$[g_1, [g_2, g_1]] \equiv [\text{DRIVE}, [\text{STEER}, \text{DRIVE}]] = [[g_1, g_2], g_1] \equiv$$
$$[\text{DRIVE}, \text{ROTATE}] = \frac{1}{L\cos^2\phi}\left(\sin\theta\frac{\partial}{\partial x} - \cos\theta\frac{\partial}{\partial y}\right) \equiv \text{SLIDE}.$$

The operation $[[g_1, g_2], g_1] \equiv$ SLIDE $\equiv$ [DRIVE,ROTATE] is a displacement at right angles to the car, and can get us out of any parking place. We just need to remember to steer, drive, steer back, drive some more, steer, drive back, steer back, and drive back:

$$\{\text{STEER}, \text{DRIVE}, \text{STEER}^{-1}, \text{DRIVE}, \text{STEER}, \text{DRIVE}^{-1}, \text{STEER}^{-1}, \text{DRIVE}^{-1}\}.$$

We have to reverse steer in the middle of the parking place. This is not intuitive, and no doubt is part of the problem with parallel parking.

Thus from only two controls u_1 and u_2 we can form the vector–fields DRIVE $\equiv g_1$, STEER $\equiv g_2$, ROTATE $\equiv [g_2, g_1]$, and SLIDE $\equiv [[g_1, g_2], g_1]$, allowing us to move anywhere in the configuration manifold M. The car kinematics $\dot{x} = g_1 u_1 + g_2 u_2$ is thus expanded as:

$$\begin{pmatrix} \dot{x} \\ \dot{y} \\ \dot{\theta} \\ \dot{\phi} \end{pmatrix} = \text{DRIVE} \cdot u_1 + \text{STEER} \cdot u_2 \equiv \begin{pmatrix} \cos\theta \\ \sin\theta \\ \frac{1}{L}\tan\phi \\ 0 \end{pmatrix} \cdot u_1 + \begin{pmatrix} 0 \\ 0 \\ 0 \\ 1 \end{pmatrix} \cdot u_2 \,.$$

The *parking Theorem* says: One can get out of any parking lot that is larger than the car.

Lie Algebras

Recall from Introduction that an *algebra* A is a vector space with a product. The product must have the property that

$$a(uv) = (au)v = u(av),$$

for every $a \in \mathbb{R}$ and $u, v \in A$. A map $\phi : A \to A'$ between algebras is called an *algebra homomorphism* if $\phi(u \cdot v) = \phi(u) \cdot \phi(v)$. A vector subspace $\mathfrak{J}$ of an algebra A is called a *left ideal* (resp. *right ideal*) if it is closed under algebra multiplication and if $u \in A$ and $i \in \mathfrak{J}$ implies that $ui \in \mathfrak{J}$ (resp. $iu \in \mathfrak{J}$). A subspace $\mathfrak{J}$ is said to be a *two–sided ideal* if it is both a left and right ideal. An ideal may not be an algebra itself, but the quotient of an algebra by a two–sided ideal inherits an algebra structure from A.

A *Lie algebra* is an algebra A where the multiplication, i.e., the *Lie bracket* $(u, v) \mapsto [u, v]$, has the following properties:

LA 1. $[u, u] = 0$ for every $u \in A$, and

LA 2. $[u, [v, w]] + [w, [u, v]] + [v, w, u]] = 0$ for all $u, v, w \in A$.

The condition LA 2 is usually called *Jacobi identity*. A subspace $E \subset A$ of a Lie algebra is called a *Lie subalgebra* if $[u, v] \in E$ for every $u, v \in E$. A map $\phi : A \to A'$ between Lie algebras is called a *Lie algebra homomorphism* if $\phi([u, v]) = [\phi(u), \phi(v)]$ for each $u, v \in A$.

All Lie algebras (over a given field $\mathbb{K}$) and all smooth homomorphisms between them form the category $\mathcal{LAL}$, which is itself a complete subcategory of the category $\mathcal{AL}$ of all algebras and their homomorphisms.

Lie Groups and Associated Lie Algebras

In the middle of the 19th Century S. Lie made a far reaching discovery that techniques designed to solve particular unrelated types of ODEs, such as separable, homogeneous and exact equations, were in fact all special cases of a general form of integration procedure based on the invariance of the differential equation under a continuous group of symmetries. Roughly speaking a symmetry group of a system of differential equations is a group that transforms solutions of the system to other solutions. Once the symmetry group has been identified a number of techniques to solve and classify these differential equations becomes possible. In the classical framework of Lie, these groups were local groups and arose locally as groups of transformations on some Euclidean space. The passage from the local Lie group to the present day definition using manifolds was accomplished by E. Cartan at the end of the 19th Century, whose work is a striking synthesis of Lie theory, classical geometry, differential geometry and topology.

These continuous groups, which originally appeared as symmetry groups of differential equations, have over the years had a profound impact on diverse areas such as algebraic topology, differential geometry, numerical analysis, control theory, classical mechanics, quantum mechanics etc. They are now universally known as Lie groups.

Definition of a Lie Group

A *Lie group* is a smooth (Banach) manifold M that has at the same time a group $G-$structure consistent with its manifold $M-$structure in the sense that *group multiplication*

$$\mu : G \times G \to G, \qquad (g, h) \mapsto gh \tag{4.24}$$

and the *group inversion*

$$\nu : G \to G, \qquad g \mapsto g^{-1} \tag{4.25}$$

are $C^\infty-$maps [Che55, AMR88, MR99, Put93]. A point $e \in G$ is called the *group identity element*.

For example, any nD Banach vector space V is an Abelian Lie group with group operations $\mu : V \times V \to V$, $\mu(x, y) = x + y$, and $\nu : V \to V$, $\nu(x) = -x$. The identity is just the zero vector. We call such a Lie group a *vector group*.

Let G and H be two Lie groups. A map $G \to H$ is said to be a *morphism* of Lie groups (or their *smooth homomorphism*) if it is their homomorphism as abstract groups and their smooth map as manifolds [Pos86].

All Lie groups and all their morphisms form the category $\mathcal{LG}$ (more precisely, there is a countable family of categories $\mathcal{LG}$ depending on C^k-smoothness of the corresponding manifolds).

Similarly, a group G which is at the same time a topological space is said to be a *topological group* if maps (4.24–4.25) are continuous, i.e., C^0-maps for it. The homomorphism $G \to H$ of topological groups is said to be continuous if it is a continuous map. Topological groups and their continuous homomorphisms form the category $\mathcal{TG}$.

A topological group (as well as a smooth manifold) is not necessarily Hausdorff. A topological group G is Hausdorff iff its identity is closed. As a corollary we have that every Lie group is a Hausdorff topological group (see [Pos86]).

For every g in a Lie group G, the two maps,

$$\begin{aligned} L_g &: G \to G, \qquad h \mapsto gh, \qquad \text{and} \\ R_h &: G \to G, \qquad g \mapsto gh, \end{aligned}$$

are called *left* and *right translation* maps. Since $L_g \circ L_h = L_{gh}$, and $R_g \circ R_h = R_{gh}$, it follows that $(L_g)^{-1} = L_{g^{-1}}$ and $(R_g)^{-1} = R_{g^{-1}}$, so both L_g and R_g are diffeomorphisms. Moreover $L_g \circ R_h = R_h \circ L_g$, i.e., left and right translation commute.

A vector–field X on G is called *left–invariant vector–field* if for every $g \in G$, $L_g^* X = X$, that is, if $(T_h L_g)X(h) = X(gh)$ for all $h \in G$, i.e., the following diagram commutes:

$$
\begin{array}{ccc}
TG & \xrightarrow{TL_g} & TG \\
X \uparrow & & \uparrow X \\
G & \xrightarrow[L_g]{} & G
\end{array}
$$

The correspondences $G \to TG$ and $L_g \to TL_g$ obviously define a functor $\mathcal{F} : \mathcal{LG} \Rightarrow \mathcal{LG}$ from the category G of Lie groups to itself. $\mathcal{F}$ is a special case of the *vector bundle functor*.

Let $\mathcal{X}_L(G)$ denote the set of left–invariant vector–fields on G; it is a Lie subalgebra of $\mathcal{X}(G)$, the set of all vector–fields on G, since $L_g^*[X,Y] = [L_g^* X, L_g^* Y] = [X,Y]$, so the Lie bracket $[X,Y] \in \mathcal{X}_L(G)$.

Let e be the identity element of G. Then for each ξ on the tangent space T_eG we define a vector–field X_ξ on G by

$$X_\xi(g) = T_e L_g(\xi).$$

$\mathcal{X}_L(G)$ and T_eG are isomorphic as vector spaces. Define the Lie bracket on T_eG by

$$[\xi, \eta] = [X_\xi, X_\eta](e),$$

for all $\xi, \eta \in T_eG$. This makes T_eG into a Lie algebra. Also, by construction, we have

$$[X_\xi, X_\eta] = X_{[\xi,\eta]},$$

this defines a bracket in T_eG via *left extension*. The vector space T_eG with the above algebra structure is called the Lie algebra of the Lie group G and is denoted $\mathfrak{g}$.

For example, let V be a nD vector space. Then $T_eV \simeq V$ and the left–invariant vector–field defined by $\xi \in T_eV$ is the constant vector–field $X_\xi(\eta) = \xi$, for all $\eta \in V$. The Lie algebra of V is V itself.

Since any two elements of an Abelian Lie group G commute, it follows that all adjoint operators Ad_g, $g \in G$, equal the identity. Therefore, the Lie algebra g is Abelian; that is, $[\xi, \eta] = 0$ for all $\xi, \eta \in \mathfrak{g}$ [MR99].

Recall (4.1.3) that Lie algebras and their smooth homomorphisms form the category $\mathcal{LAL}$. We can now introduce the fundamental *Lie functor*, $\mathcal{F} : \mathcal{LG} \Rightarrow \mathcal{LAL}$, from the category of Lie groups to the category of Lie algebras [Pos86].

Let X_ξ be a left–invariant vector–field on G corresponding to ξ in $\mathfrak{g}$. Then there is a unique integral curve $\gamma_\xi : \mathbb{R} \to G$ of X_ξ starting at e, i.e.,

$$\dot{\gamma}_\xi(t) = X_\xi\left(\gamma_\xi(t)\right), \qquad \gamma_\xi(0) = e.$$

$\gamma_\xi(t)$ is a smooth *one–parameter subgroup* of G, i.e.,

$$\gamma_\xi(t+s) = \gamma_\xi(t) \cdot \gamma_\xi(s),$$

since, as functions of t both sides equal $\gamma_\xi(s)$ at $t = 0$ and both satisfy differential equation

$$\dot{\gamma}(t) = X_\xi\left(\gamma_\xi(t)\right)$$

by left invariance of X_ξ, so they are equal. Left invariance can be also used to show that $\gamma_\xi(t)$ is defined for all $t \in \mathbb{R}$. Moreover, if $\phi : \mathbb{R} \to G$ is a one–parameter subgroup of G, i.e., a *smooth homomorphism* of the additive group $\mathbb{R}$ into G, then $\phi = \gamma_\xi$ with $\xi = \dot{\phi}(0)$, since taking derivative at $s = 0$ in the relation

$$\phi(t+s) = \phi(t) \cdot \phi(s) \qquad \text{gives} \qquad \dot{\phi}(t) = X_{\dot{\phi}(0)}\left(\phi(t)\right),$$

so $\phi = \gamma_\xi$ since both equal e at $t = 0$. Therefore, all one–parameter subgroups of G are of the form $\gamma_\xi(t)$ for some $\xi \in \mathfrak{g}$.

The map $\exp : \mathfrak{g} \to G$, given by

$$\exp(\xi) = \gamma_\xi(1), \qquad \exp(0) = e, \tag{4.26}$$

is called the *exponential map* of the Lie algebra $\mathfrak{g}$ of G into G. exp is a C^∞–map, similar to the projection π of tangent and cotangent bundles; exp is locally a diffeomorphism from a neighborhood of zero in $\mathfrak{g}$ onto a neighborhood of e in G; if $f : G \to H$ is a smooth homomorphism of Lie groups, then

$$f \circ \exp_G = \exp_H \circ T_e f\,.$$

Also, in this case (see [Che55, MR99, Pos86])

$$\exp(s\xi) = \gamma_\xi(s).$$

Indeed, for fixed $s \in \mathbb{R}$, the curve $t \mapsto \gamma_\xi(ts)$, which at $t = 0$ passes through e, satisfies the differential equation

$$\frac{d}{dt}\gamma_\xi(ts) = sX_\xi\left(\gamma_\xi(ts)\right) = X_{s\xi}\left(\gamma_\xi(ts)\right).$$

Since $\gamma_{s\xi}(t)$ satisfies the same differential equation and passes through e at $t = 0$, it follows that $\gamma_{s\xi}(t) = \gamma_\xi(st)$. Putting $t = 1$ induces $\exp(s\xi) = \gamma_\xi(s)$ [MR99].

Hence exp maps the line $s\xi$ in $\mathfrak{g}$ onto the one–parameter subgroup $\gamma_\xi(s)$ of G, which is tangent to ξ at e. It follows from left invariance that the flow F_t^ξ of X satisfies $F_t^\xi(g) = g\exp(s\xi)$.

Globally, the exponential map exp, as given by (4.26), is a natural operation, i.e., for any morphism $\varphi : G \to H$ of Lie groups G and H and a Lie functor $\mathcal{F}$, the following diagram commutes [Pos86]:

$$\begin{array}{ccc}
\mathcal{F}(G) & \xrightarrow{\mathcal{F}(\varphi)} & \mathcal{F}(H) \\
\big\downarrow{\scriptstyle \exp} & & \big\downarrow{\scriptstyle \exp} \\
G & \xrightarrow[\varphi]{} & H
\end{array}$$

Let G_1 and G_2 be Lie groups with Lie algebras $\mathfrak{g}_1$ and $\mathfrak{g}_2$. Then $G_1 \times G_2$ is a Lie group with Lie algebra $\mathfrak{g}_1 \times \mathfrak{g}_2$, and the exponential map is given by [MR99].

$$\exp : \mathfrak{g}_1 \times \mathfrak{g}_2 \to G_1 \times G_2, \qquad (\xi_1, \xi_2) \mapsto (\exp_1(\xi_1), \exp_2(\xi_2)).$$

For example, in case of a nD vector space, or infinite–dimensional Banach space, the exponential map is the identity.

The unit circle in the complex–plane $S^1 = \{z \in \mathbb{C} : |z| = 1\}$ is an Abelian Lie group under multiplication. The tangent space $T_e S^1$ is the imaginary axis, and we identify $\mathbb{R}$ with $T_e S^1$ by $t \mapsto 2\pi i t$. With this identification, the exponential map $\exp : \mathbb{R} \to S^1$ is given by $\exp(t) = \mathrm{e}^{2\pi i t}$.

The nD torus $T^n = S^1 \times \cdots \times S^1$ (n times) is an Abelian Lie group. The exponential map $\exp : \mathbb{R}^n \to T^n$ is given by

$$\exp(t_1, ..., t_n) = (\mathrm{e}^{2\pi i t_1}, ..., \mathrm{e}^{2\pi i t_n}).$$

Since $S^1 = \mathbb{R}/\mathbb{Z}$, it follows that

$$T^n = \mathbb{R}^n/\mathbb{Z}^n,$$

the projection $\mathbb{R}^n \to T^n$ being given by the exp map (see [MR99, Pos86]).

For every $g \in G$, the map

$$Ad_g = T_e \left(R_{g^{-1}} \circ L_g \right) : \mathfrak{g} \to \mathfrak{g}$$

is called the *adjoint map* (or *operator*) associated with g.

For each $\xi \in \mathfrak{g}$ and $g \in G$ we have

$$\exp\left(Ad_g \xi \right) = g \left(\exp \xi \right) g^{-1}.$$

The relation between the adjoint map and the Lie bracket is the following: For all $\xi, \eta \in \mathfrak{g}$ we have

$$\left.\frac{d}{dt}\right|_{t=0} Ad_{\exp(t\xi)} \eta = [\xi, \eta].$$

A Lie subgroup H of G is a subgroup H of G which is also a submanifold of G. Then $\mathfrak{h}$ is a Lie subalgebra of $\mathfrak{g}$ and moreover $\mathfrak{h} = \{\xi \in \mathfrak{g} | \exp(t\xi) \in H,$ for all $t \in \mathbb{R}\}$.

Recall that one can characterize *Lebesgue measure* up to a multiplicative constant on $\mathbb{R}^n$ by its invariance under translations. Similarly, on a locally compact group there is a unique (up to a nonzero multiplicative constant) left–invariant measure, called *Haar measure*. For Lie groups the existence of such measures is especially simple [MR99]: Let G be a Lie group. Then there is a volume form $Ub5$, unique up to nonzero multiplicative constants, that is left–invariant. If G is compact, $Ub5$ is right invariant as well.

Actions of Lie Groups on Smooth Manifolds

Let M be a smooth manifold. An *action of a Lie group* G (with the unit element e) on M is a smooth map $\phi : G \times M \to M$, such that for all $x \in M$ and $g, h \in G$, (i) $\phi(e, x) = x$ and (ii) $\phi(g, \phi(h, x)) = \phi(gh, x)$. In other words, letting $\phi_g : x \in M \mapsto \phi_g(x) = \phi(g, x) \in M$, we have (i') $\phi_e = id_M$ and (ii') $\phi_g \circ \phi_h = \phi_{gh}$. ϕ_g is a diffeomorphism, since $(\phi_g)^{-1} = \phi_{g^{-1}}$. We say that the map $g \in G \mapsto \phi_g \in Diff(M)$ is a homomorphism of G into the group of diffeomorphisms of M. In case that M is a vector space and each ϕ_g is a linear operator, the function of G on M is called a representation of G on M [Put93]

An action ϕ of G on M is said to be *transitive group action*, if for every $x, y \in M$, there is $g \in G$ such that $\phi(g, x) = y$; *effective group action*, if $\phi_g = id_M$ implies $g = e$, that is $g \mapsto \phi_g$ is 1–1; and *free group action*, if for each $x \in M$, $g \mapsto \phi_g(x)$ is 1–1.

For example,

1. $G = \mathbb{R}$ acts on $M = \mathbb{R}$ by translations; explicitly,
$$\phi : G \times M \to M, \qquad \phi(s, x) = x + s.$$
Then for $x \in \mathbb{R}$, $O_x = \mathbb{R}$. Hence M/G is a single point, and the action is transitive and free.
2. A complete flow ϕ_t of a vector–field X on M gives an action of $\mathbb{R}$ on M, namely
$$(t, x) \in \mathbb{R} \times M \mapsto \phi_t(x) \in M.$$
3. Left translation $L_g : G \to G$ defines an effective action of G on itself. It is also transitive.
4. The coadjoint action of G on $\mathfrak{g}^*$ is given by
$$Ad^* : (g, \alpha) \in G \times \mathfrak{g}^* \mapsto Ad^*_{g^{-1}}(\alpha) = \left(T_e(R_{g^{-1}} \circ L_g)\right)^* \alpha \in \mathfrak{g}^*.$$

Let ϕ be an action of G on M. For $x \in M$ the *orbit* of x is defined by
$$O_x = \{\phi_g(x) | g \in G\} \subset M$$
and the *isotropy group* of ϕ at x is given by
$$G_x = \{g \in G | \phi(g, x) = x\} \subset G.$$

An action ϕ of G on a manifold M defines an equivalence relation on M by the relation belonging to the same orbit; explicitly, for $x, y \in M$, we write $x \sim y$ if there exists a $g \in G$ such that $\phi(g, x) = y$, that is, if $y \in O_x$. The set of all orbits M/G is called the *group orbit space*.

For example, let $M = \mathbb{R}^2 \backslash \{0\}$, $G = SO(2)$, the group of rotations in plane, and the action of G on M given by

$$\left(\begin{bmatrix} \cos\theta & -\sin\theta \\ \sin\theta & \cos\theta \end{bmatrix}, (x,y)\right) \longmapsto (x\cos\theta - y\sin\theta,\ x\sin\theta + y\cos\theta).$$

The action is always free and effective, and the orbits are concentric circles, thus the orbit space is $M/G \simeq \mathbb{R}_+^*$.

A crucial concept in mechanics is the infinitesimal description of an action. Let $\phi : G \times M \to M$ be an action of a Lie group G on a smooth manifold M. For each $\xi \in \mathfrak{g}$,

$$\phi_\xi : \mathbb{R} \times M \to M, \qquad \phi_\xi(t,x) = \phi\,(\exp(t\xi), x)$$

is an $\mathbb{R}$–action on M. Therefore, $\phi_{\exp(t\xi)} : M \to M$ is a flow on M; the corresponding vector–field on M, given by

$$\xi_M(x) = \frac{d}{dt}\bigg|_{t=0} \phi_{\exp(t\xi)}(x)$$

is called the infinitesimal generator of the action, corresponding to ξ in $\mathfrak{g}$.

The tangent space at x to an orbit O_x is given by

$$T_x O_x = \{\xi_M(x) | \xi \in \mathfrak{g}\}.$$

Let $\phi : G \times M \to M$ be a smooth G-–action. For all $g \in G$, all $\xi, \eta \in \mathfrak{g}$ and all $\alpha, \beta \in \mathbb{R}$, we have:

$(Ad_g \xi)_M = \phi_{g^{-1}}^* \xi_M$, $[\xi_M, \eta_M] = -\,[\xi, \eta]_M$, and $(\alpha\xi + \beta\eta)_M = \alpha\xi_M + \beta\eta_M$.

Let M be a smooth manifold, G a Lie group and $\phi : G \times M \to M$ a G–action on M. We say that a smooth map $f : M \to M$ is with respect to this action if for all $g \in G$,

$$f \circ \phi_g = \phi_g \circ f.$$

Let $f : M \to M$ be an equivariant smooth map. Then for any $\xi \in \mathfrak{g}$ we have

$$Tf \circ \xi_M = \xi_M \circ f.$$

Basic Dynamical Groups

Here we give the first two examples of Lie groups, namely Galilei group and general linear group. Further examples will be given in association with particular dynamical systems.

Galilei Group

The *Galilei group* is the group of transformations in space and time that connect those Cartesian systems that are termed 'inertial frames' in Newtonian mechanics. The most general relationship between two such frames is the following. The origin of the time scale in the inertial frame S' may be shifted compared with that in S; the orientation of the Cartesian axes in S' may be different from that in S; the origin O of the Cartesian frame in S' may be moving relative to the origin O in S at a uniform velocity. The transition from S to S' involves ten parameters; thus the Galilei group is a ten parameter group. The basic assumption inherent in Galilei–Newtonian relativity is that there is an absolute time scale, so that the only way in which the time variables used by two different 'inertial observers' could possibly differ is that the zero of time for one of them may be shifted relative to the zero of time for the other.

Galilei space–time structure involves the following three elements:

1. *World*, as a 4D affine space A^4. The points of A^4 are called *world points* or *events*. The parallel transitions of the world A^4 form a linear (i.e., Euclidean) space $\mathbb{R}^4$.
2. *Time*, as a linear map $t : \mathbb{R}^4 \to \mathbb{R}$ of the linear space of the world parallel transitions onto the real 'time axes'. Time interval from the event $a \in A^4$ to $b \in A^4$ is called the number $t(b-a)$; if $t(b-a) = 0$ then the events a and b are called synchronous. The set of all mutually synchronous events consists a 3D affine space A^3, being a subspace of the world A^4. The kernel of the mapping t consists of the parallel transitions of A^4 translating arbitrary (and every) event to the synchronous one; it is a linear 3D subspace $\mathbb{R}^3$ of the space $\mathbb{R}^4$.
3. *Distance* (*metric*) between the synchronous events,
$$\rho(a,b) = \| a - b \|, \qquad \text{for all} \quad a,b \in A^3,$$
given by the scalar product in $\mathbb{R}^3$. The distance transforms arbitrary space of synchronous events into the well known 3D Euclidean space E^3.

The space A^4, with the Galilei space–time structure on it, is called Galilei space. Galilei group is the group of all possible transformations of the Galilei space, preserving its structure. The elements of the Galilei group are called Galilei transformations. Therefore, Galilei transformations are affine transformations of the world A^4 preserving the time intervals and distances between the synchronous events.

The direct product $\mathbb{R} \times \mathbb{R}^3$, of the time axes with the 3D linear space R3 with a fixed Euclidean structure, has a natural Galilei structure. It is called Galilei coordinate system.

General Linear Group

The group of linear isomorphisms of $\mathbb{R}^n$ to $\mathbb{R}^n$ is a Lie group of dimension n^2, called the *general linear group* and denoted $Gl(n, \mathbb{R})$. It is a smooth manifold, since it is a subset of the vector space $L(\mathbb{R}^n, \mathbb{R}^n)$ of all linear maps of $\mathbb{R}^n$ to $\mathbb{R}^n$, as $Gl(n, \mathbb{R})$ is the inverse image of $\mathbb{R}\backslash\{0\}$ under the continuous map $A \mapsto \det A$ of $L(\mathbb{R}^n, \mathbb{R}^n)$ to $\mathbb{R}$. The group operation is composition

$$(A,B) \in Gl(n,\mathbb{R}) \times Gl(n,\mathbb{R}) \mapsto A \circ B \in Gl(n,\mathbb{R})$$

and the inverse map is

$$A \in Gl(n,\mathbb{R}) \mapsto A^{-1} \in Gl(n,\mathbb{R}).$$

If we choose a basis in $\mathbb{R}^n$, we can represent each element $A \in Gl(n,\mathbb{R})$ by an invertible $(n \times n)$–matrix. The group operation is then matrix multiplication and the inversion is matrix inversion. The identity is the identity matrix I_n. The group operations are smooth since the formulas for the product and inverse of matrices are smooth in the matrix components.

The Lie algebra of $Gl(n,\mathbb{R})$ is $\mathfrak{gl}(n)$, the vector space $L(\mathbb{R}^n,\mathbb{R}^n)$ of all linear transformations of $\mathbb{R}^n$, with the commutator bracket

$$[A,B] = AB - BA.$$

For every $A \in L(\mathbb{R}^n,\mathbb{R}^n)$,

$$\gamma_A : t \in \mathbb{R} \mapsto \gamma_A(t) = \sum_{i=0}^{\infty} \frac{t^i}{i!} A^i \in Gl(n,\mathbb{R})$$

is a one–parameter subgroup of $Gl(n,\mathbb{R})$, because

$$\gamma_A(0) = I, \qquad \text{and} \qquad \dot{\gamma}_A(t) = \sum_{i=0}^{\infty} \frac{t^{i-1}}{(i-1)!} A^i = \gamma_A(t)\, A.$$

Hence γ_A is an integral curve of the left–invariant vector–field X_A. Therefore, the exponential map is given by

$$\exp : A \in L(\mathbb{R}^n,\mathbb{R}^n) \mapsto \exp(A) \equiv \mathrm{e}^A = \gamma_A(1) = \sum_{i=0}^{\infty} \frac{A^i}{i!} \in Gl(n,\mathbb{R}).$$

For each $A \in Gl(n,\mathbb{R})$ the corresponding adjoint map

$$Ad_A : L(\mathbb{R}^n,\mathbb{R}^n) \to L(\mathbb{R}^n,\mathbb{R}^n)$$

is given by

$$Ad_A B = A \cdot B \cdot A^{-1}.$$

Classical Lie Theory

In this section we present the basics of *classical theory of Lie groups* and their Lie algebras, as developed mainly by Sophus Lie, Elie Cartan, Felix Klein, Wilhelm Killing and Hermann Weyl. For more comprehensive treatment see e.g., [Che55, Hel01].

Basic Tables of Lie Groups and Their Lie Algebras

One classifies Lie groups regarding their algebraic properties (simple, semisimple, solvable, nilpotent, Abelian), their connectedness (connected or simply connected) and their compactness (see Tables A.1–A.3). This is the content of the *Hilbert 5th problem*.

Some real Lie groups and their Lie algebras:

Lie group	**Description**	**Remarks**	**Lie algb.**	**Description**	**dim $/\mathbb{R}$**
$\mathbb{R}^n$	Euclidean space with addition	Abelian, simply connected, not compact	$\mathbb{R}^n$	the Lie bracket is zero	n
$\mathbb{R}^\times$	nonzero real numbers with multiplication	Abelian, not connected, not compact	$\mathbb{R}$	the Lie bracket is zero	1
$\mathbb{R}^{>0}$	positive real numbers with multiplication	Abelian, simply connected, not compact	$\mathbb{R}$	the Lie bracket is zero	1
$S^1 = \mathbb{R}/\mathbb{Z}$	complex numbers of absolute value 1, with multiplication	Abelian, connected, not simply connected, compact	$\mathbb{R}$	the Lie bracket is zero	1
$\mathbb{H}^\times$	non–zero quaternions with multiplication	simply connected, not compact	$\mathbb{H}$	quaternions, with Lie bracket the commutator	4
S^3	quaternions of absolute value 1, with multiplication; a 3–sphere	simply connected, compact, simple and semi–simple, isomorphic to $SU(2)$, $SO(3)$ and to $Spin(3)$	$\mathbb{R}^3$	real 3–vectors, with Lie bracket the cross product; isomorphic to $\mathfrak{su}(2)$ and to $\mathfrak{so}(3)$	3
$GL(n,\mathbb{R})$	general linear group: invertible n–by–n real matrices	not connected, not compact	$\mathrm{M}(n,\mathbb{R})$	n–by–n matrices, with Lie bracket the commutator	n^2
$GL^+(n,\mathbb{R})$	n–by–n real matrices with positive determinant	simply connected, not compact	$\mathrm{M}(n,\mathbb{R})$	n–by–n matrices, with Lie bracket the commutator	n^2

Classical real Lie groups and their Lie algebras:

Lie group	Description	Remarks	Lie algb.	Description	dim /$\mathbb{R}$
$SL(n,\mathbb{R})$	special linear group: real matrices with determinant 1	simply connected, not compact if $n > 1$	$\mathfrak{sl}(n,\mathbb{R})$	square matrices with trace 0, with Lie bracket the commutator	$n^2 - 1$
$O(n,\mathbb{R})$	orthogonal group: real orthogonal matrices	not connected, compact	$\mathfrak{so}(n,\mathbb{R})$	skew–symmetric square real matrices, with Lie bracket the commutator; $\mathfrak{so}(3,\mathbb{R})$ is isomorphic to $\mathfrak{su}(2)$ and to $\mathbb{R}^3$ with the cross product	$n(n-1)/2$
$SO(n,\mathbb{R})$	special orthogonal group: real orthogonal matrices with determinant 1	connected, compact, for $n \geq 2$: not simply connected, for $n = 3$ and $n \geq 5$: simple and semisimple	$\mathfrak{so}(n,\mathbb{R})$	skew–symmetric square real matrices, with Lie bracket the commutator	$n(n-1)/2$
$Spin(n)$	spinor group	simply connected, compact, for $n = 3$ and $n \geq 5$: simple and semisimple	$\mathfrak{so}(n,\mathbb{R})$	skew–symmetric square real matrices, with Lie bracket the commutator	$n(n-1)/2$
$U(n)$	unitary group: complex unitary n–by–n matrices	isomorphic to S^1 for $n = 1$, not simply connected, compact	$\mathfrak{u}(n)$	square complex matrices A satisfying $A = -A^*$, with Lie bracket the commutator	n^2
$SU(n)$	special unitary group: complex unitary n–by–n matrices with determinant 1	simply connected, compact, for $n \geq 2$: simple and semisimple	$\mathfrak{su}(n)$	square complex matrices A with trace 0 satisfying $A = -A^*$, with Lie bracket the commutator	$n^2 - 1$

Basic complex Lie groups and their Lie algebras:[7]

Lie group	Description	Remarks	Lie algb.	Description	dim /$\mathbb{C}$
$\mathbb{C}^n$	group operation is addition	Abelian, simply connected, not compact	$\mathbb{C}^n$	the Lie bracket is zero	n
$\mathbb{C}^\times$	nonzero complex numbers with multiplication	Abelian, not simply connected, not compact	$\mathbb{C}$	the Lie bracket is zero	1
$GL(n, \mathbb{C})$	general linear group: invertible n-by-n complex matrices	simply connected, not compact, for $n = 1$: isomorphic to $\mathbb{C}^\times$	$M(n, \mathbb{C})$	n-by-n matrices, with Lie bracket the commutator	n^2
$SL(n, \mathbb{C})$	special linear group: complex matrices with determinant 1	simple, semisimple, simply connected, for $n \geq 2$: not compact	$\mathfrak{sl}(n, \mathbb{C})$	square matrices with trace 0, with Lie bracket the commutator	$n^2 - 1$
$O(n, \mathbb{C})$	orthogonal group: complex orthogonal matrices	not connected, for $n \geq 2$: not compact	$\mathfrak{so}(n, \mathbb{C})$	skew-symmetric square complex matrices, with Lie bracket the commutator	$n(n-1)/2$
$SO(n, \mathbb{C})$	special orthogonal group: complex orthogonal matrices with determinant 1	for $n \geq 2$: not compact, not simply connected, for $n = 3$ and $n \geq 5$: simple and semisimple	$\mathfrak{so}(n, \mathbb{C})$	skew-symmetric square complex matrices, with Lie bracket the commutator	$n(n-1)/2$

Representations of Lie groups

The idea of a *representation of a Lie group* plays an important role in the study of continuous symmetry (see, e.g., [Hel01]). A great deal is known about such representations, a basic tool in their study being the use of the corresponding 'infinitesimal' representations of Lie algebras.

Formally, a representation of a Lie group G on a vector space V (over a field K) is a group homomorphism $G \to Aut(V)$ from G to the automorphism

---

[7] The dimensions given are dimensions over $\mathbb{C}$. Note that every complex Lie group/algebra can also be viewed as a real Lie group/algebra of twice the dimension.

group of V. If a basis for the vector space V is chosen, the representation can be expressed as a homomorphism into $GL(n, K)$. This is known as a *matrix representation*.

On the Lie algebra level, there is a corresponding linear map from the Lie algebra of G to $End(V)$ preserving the Lie bracket $[\cdot, \cdot]$.

If the homomorphism is in fact an monomorphism, the representation is said to be *faithful*.

A unitary representation is defined in the same way, except that G maps to unitary matrices; the Lie algebra will then map to skew–Hermitian matrices.

Now, if G is a semisimple group, its finite–dimensional representations can be decomposed as direct sums of irreducible representations. The irreducibles are indexed by highest weight; the allowable (*dominant*) highest weights satisfy a suitable positivity condition. In particular, there exists a set of *fundamental weights*, indexed by the vertices of the *Dynkin diagram* of G (see below), such that dominant weights are simply non–negative integer linear combinations of the fundamental weights.

If G is a commutative compact Lie group, then its irreducible representations are simply the continuous characters of G. A *quotient representation* is a quotient module of the group ring.

Root Systems and Dynkin Diagrams

A *root system* is a special configuration in Euclidean space that has turned out to be fundamental in Lie theory as well as in its applications. Also, the classification scheme for root systems, by *Dynkin diagrams*, occurs in parts of mathematics with no overt connection to Lie groups (such as singularity theory, see e.g., [Hel01]).

Definitions

Formally, a *root system* is a finite set Φ of non–zero vectors (*roots*) spanning a finite–dimensional Euclidean space V and satisfying the following properties:

1. The only scalar multiples of a root α in V which belong to Φ are α itself and $-\alpha$.
2. For every root α in V, the set Φ is symmetric under reflection through the hyperplane of vectors perpendicular to α.
3. If α and β are vectors in Φ, the projection of 2β onto the line through α is an integer multiple of α.

The *rank* of a root system Φ is the dimension of V. Two root systems may be combined by regarding the Euclidean spaces they span as mutually orthogonal subspaces of a common Euclidean space. A root system which does not arise from such a combination, such as the systems A_2, B_2, and G_2 in Figure 4.6, is said to be *irreducible*.

Two irreducible root systems (V_1, Φ_1) and (V_2, Φ_2) are considered to be the same if there is an invertible linear transformation $V_1 \to V_2$ which preserves distance up to a scale factor and which sends Φ_1 to Φ_2.

The group of isometries of V generated by reflections through hyperplanes associated to the roots of Φ is called the Weyl group of Φ as it acts faithfully on the finite set Φ, the Weyl group is always finite.

Classification

It is not too difficult to classify the root systems of rank 2 (see Figure 4.6).

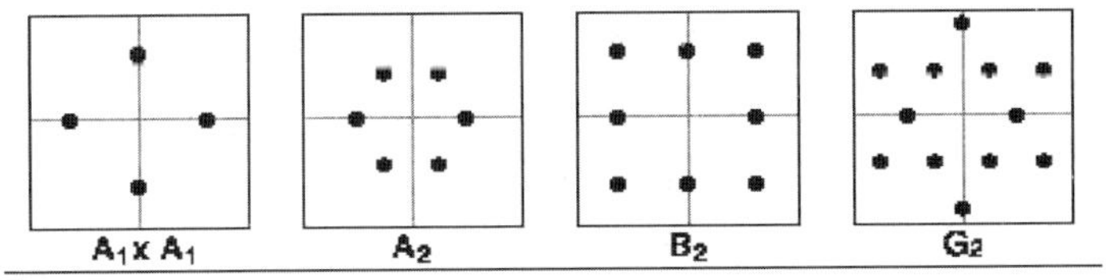

Fig. 4.6. Classification of root systems of rank 2.

Whenever Φ is a root system in V and W is a subspace of V spanned by $\Psi = \Phi \cap W$, then Ψ is a root system in W. Thus, our exhaustive list of root systems of rank 2 shows the geometric possibilities for any two roots in a root system. In particular, two such roots meet at an angle of 0, 30, 45, 60, 90, 120, 135, 150, or 180 degrees.

In general, irreducible root systems are specified by a family (indicated by a letter A to G) and the rank (indicated by a subscript n). There are four *infinite families*:

- A_n $(n \geq 1)$, which corresponds to the special unitary group, $SU(n+1)$;
- B_n $(n \geq 2)$, which corresponds to the special orthogonal group, $SO(2n+1)$;
- C_n $(n \geq 3)$, which corresponds to the symplectic group, $Sp(2n)$;
- D_n $(n \geq 4)$, which corresponds to the special orthogonal group, $SO(2n)$,

as well as five *exceptional cases*: E_6, E_7, E_8, F_4, G_2.

Dynkin Diagrams

A Dynkin diagram is a graph with a few different kinds of possible edges (see Figure 4.7). The connected components of the graph correspond to the irreducible subalgebras of $\mathfrak{g}$. So a simple Lie algebra's Dynkin diagram has only one component. The rules are restrictive. In fact, there are only certain possibilities for each component, corresponding to the classification of semi–simple Lie algebras (see, e.g., [CCN85]).

The *roots* of a complex Lie algebra form a lattice of rank k in a *Cartan subalgebra* $\mathfrak{h} \subset \mathfrak{g}$, where k is the Lie algebra rank of $\mathfrak{g}$. Hence, the *root lattice*

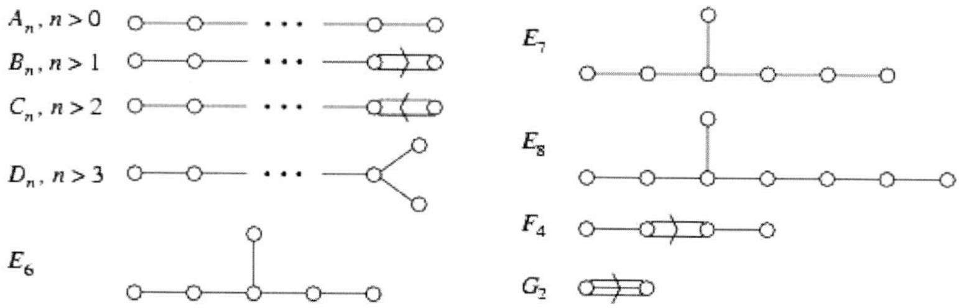

Fig. 4.7. The problem of classifying irreducible root systems reduces to the problem of classifying connected Dynkin diagrams.

can be considered a lattice in $\mathbb{R}^k$. A vertex, or node, in the Dynkin diagram is drawn for each *Lie algebra simple root*, which corresponds to a generator of the root lattice. Between two nodes α and β, an edge is drawn if the simple roots are not perpendicular. One line is drawn if the angle between them is $2\pi/3$, two lines if the angle is $3\pi/4$, and three lines are drawn if the angle is $5\pi/6$. There are no other possible angles between Lie algebra simple roots. Alternatively, the number of lines N between the simple roots α and β is given by

$$N = A_{\alpha\beta} A_{\beta\alpha} = \frac{2\langle\alpha,\beta\rangle}{|\alpha|^2} \frac{2\langle\beta,\alpha\rangle}{|\beta|^2} = 4\cos^2\theta,$$

where $A_{\alpha\beta} = \frac{2\langle\alpha,\beta\rangle}{|\alpha|^2}$ is an entry in the *Cartan matrix* $(A_{\alpha\beta})$ (for details on Cartan matrix see, e.g., [Hel01]). In a Dynkin diagram, an arrow is drawn from the longer root to the shorter root (when the angle is $3\pi/4$ or $5\pi/6$).

Here are some properties of *admissible Dynkin diagrams*:

1. A diagram obtained by removing a node from an admissible diagram is admissible.
2. An admissible diagram has no loops.
3. No node has more than three lines attached to it.
4. A sequence of nodes with only two single lines can be collapsed to give an admissible diagram.
5. The only connected diagram with a triple line has two nodes.

A *Coxeter–Dynkin diagram*, also called a *Coxeter graph*, is the same as a Dynkin diagram, but without the arrows. The Coxeter diagram is sufficient to characterize the algebra, as can be seen by enumerating connected diagrams.

The simplest way to recover a simple Lie algebra from its Dynkin diagram is to first reconstruct its Cartan matrix (A_{ij}). The ith node and jth node are connected by $A_{ij}A_{ji}$ lines. Since $A_{ij} = 0$ iff $A_{ji} = 0$, and otherwise $A_{ij} \in \{-3,-2,-1\}$, it is easy to find A_{ij} and A_{ji}, up to order, from their product. The arrow in the diagram indicates which is larger. For example, if node 1 and node 2 have two lines between them, from node 1 to node 2, then $A_{12} = -1$ and $A_{21} = -2$.

However, it is worth pointing out that each simple Lie algebra can be constructed concretely. For instance, the infinite families A_n, B_n, C_n, and D_n correspond to the special linear Lie algebra $\mathfrak{gl}(n+1,\mathbb{C})$, the odd orthogonal Lie algebra $\mathfrak{so}(2n+1,\mathbb{C})$, the symplectic Lie algebra $\mathfrak{sp}(2n,\mathbb{C})$, and the even orthogonal Lie algebra $\mathfrak{so}(2n,\mathbb{C})$. The other simple Lie algebras are called *exceptional Lie algebras*, and have constructions related to the *octonions*.

To prove this classification Theorem, one uses the angles between pairs of roots to encode the root system in a much simpler combinatorial object, the Dynkin diagram. The Dynkin diagrams can then be classified according to the scheme given above.

To every root system is associated a corresponding Dynkin diagram. Otherwise, the Dynkin diagram can be extracted from the root system by choosing a *base*, that is a subset Δ of Φ which is a basis of V with the special property that every vector in Φ when written in the basis Δ has either all coefficients ≥ 0 or else all ≤ 0.

The vertices of the Dynkin diagram correspond to vectors in Δ. An edge is drawn between each non–orthogonal pair of vectors; it is a double edge if they make an angle of 135 degrees, and a triple edge if they make an angle of 150 degrees. In addition, double and triple edges are marked with an angle sign pointing toward the shorter vector.

Although a given root system has more than one base, the Weyl group acts transitively on the set of bases. Therefore, the root system determines the Dynkin diagram. Given two root systems with the same Dynkin diagram, we can match up roots, starting with the roots in the base, and show that the systems are in fact the same.

Thus the problem of classifying root systems reduces to the problem of classifying possible Dynkin diagrams, and the problem of classifying irreducible root systems reduces to the problem of classifying connected Dynkin diagrams. Dynkin diagrams encode the inner product on E in terms of the basis Δ, and the condition that this inner product must be positive definite turns out to be all that is needed to get the desired classification (see Figure 4.7).

In detail, the individual root systems can be realized case–by–case, as in the following paragraphs:

$\mathbf{A}_n$. Let V be the subspace of $\mathbb{R}^{n+1}$ for which the coordinates sum to 0, and let Φ be the set of vectors in V of length $\sqrt{2}$ and with integer coordinates in $\mathbb{R}^{n+1}$. Such a vector must have all but two coordinates equal to 0, one coordinate equal to 1, and one equal to -1, so there are n^2+n roots in all.

$\mathbf{B}_n$. Let $V=\mathbb{R}^n$, and let Φ consist of all integer vectors in V of length 1 or $\sqrt{2}$. The total number of roots is $2n^2$.

$\mathbf{C}_n$: Let $V=\mathbb{R}^n$, and let Φ consist of all integer vectors in V of $\sqrt{2}$ together with all vectors of the form 2λ, where λ is an integer vector of length 1. The total number of roots is $2n^2$. The total number of roots is $2n^2$.

$\mathbf{D}_n$. Let $V=\mathbb{R}^n$, and let Φ consist of all integer vectors in V of length $\sqrt{2}$. The total number of roots is $2n(n-1)$.

$\mathbf{E}_n$. For V_8, let $V = \mathbb{R}^8$, and let E_8 denote the set of vectors α of length $\sqrt{2}$ such that the coordinates of 2α are all integers and are either all even or all odd. Then E_7 can be constructed as the intersection of E_8 with the hyperplane of vectors perpendicular to a fixed root α in E_8, and E_6 can be constructed as the intersection of E_8 with two such hyperplanes corresponding to roots α and β which are neither orthogonal to one another nor scalar multiples of one another. The root systems E_6, E_7, and E_8 have 72, 126, and 240 roots respectively.

$\mathbf{F}_4$. For F_4, let $V = \mathbb{R}^4$, and let Φ denote the set of vectors α of length 1 or $\sqrt{2}$ such that the coordinates of 2α are all integers and are either all even or all odd. There are 48 roots in this system.

$\mathbf{G}_2$. There are 12 roots in G_2, which form the vertices of a *hexagram.*

Root Systems and Lie Theory

Irreducible root systems classify a number of related objects in Lie theory, notably:

1. Simple complex Lie algebras;
2. Simple complex Lie groups;
3. Simply connected complex Lie groups which are simple modulo centers; and
4. Simple compact Lie groups.

In each case, the roots are non–zero weights of the adjoint representation.

A root system can also be said to describe a *plant's root* and associated systems.

Simple and Semisimple Lie Groups and Algebras

A *simple Lie group* is a Lie group which is also a simple group. These groups, and groups closely related to them, include many of the so–called *classical groups* of geometry, which lie behind projective geometry and other geometries derived from it by the *Erlangen programme* of Felix Klein. They also include some *exceptional groups*, that were first discovered by those pursuing the classification of simple Lie groups. The exceptional groups account for many special examples and configurations in other branches of mathematics. In particular the classification of finite simple groups depended on a thorough prior knowledge of the 'exceptional' possibilities.

The complete listing of the simple Lie groups is the basis for the theory of the semisimple Lie groups and reductive groups, and their representation theory. This has turned out not only to be a major extension of the theory of compact Lie groups (and their representation theory), but to be of basic significance in mathematical physics.

Such groups are classified using the prior classification of the complex simple Lie algebras. It has been shown that a simple Lie group has a simple

Lie algebra that will occur on the list given there, once it is complexified (that is, made into a complex vector space rather than a real one). This reduces the classification to two further matters.

The groups $SO(p, q, \mathbb{R})$ and $SO(p+q, \mathbb{R})$, for example, give rise to different real Lie algebras, but having the same Dynkin diagram. In general there may be different *real forms* of the same complex Lie algebra.

Secondly, the Lie algebra only determines uniquely the simply connected (universal) cover G^* of the component containing the identity of a Lie group G. It may well happen that G^* is not actually a simple group, for example having a non–trivial center. We have therefore to worry about the global topology, by computing the fundamental group of G (an Abelian group: a Lie group is an H–space). This was done by Elie Cartan.

For an example, take the special orthogonal groups in even dimension. With $-I$ a scalar matrix in the center, these are not actually simple groups; and having a two–fold spin cover, they aren't simply–connected either. They lie 'between' G^* and G, in the notation above.

Recall that a *semisimple module* is a module in which each submodule is a direct summand. In particular, a *semisimple representation* is completely reducible, i.e., is a direct sum of irreducible representations (under a descending chain condition). Similarly, one speaks of an Abelian category as being semisimple when every object has the corresponding property. Also, a semisimple ring is one that is semisimple as a module over itself.

A *semisimple matrix* is diagonalizable over any algebraically closed field containing its entries. In practice this means that it has a diagonal matrix as its Jordan normal form.

A *Lie algebra* $\mathfrak{g}$ is called *semisimple* when it is a direct sum of *simple Lie algebras*, i.e., non–trivial Lie algebras $\mathfrak{L}$ whose only ideals are $\{0\}$ and $\mathfrak{L}$ itself. An equivalent condition is that the *Killing form*

$$\mathcal{B}(X, Y) = \mathrm{Tr}(Ad(X)\, Ad(Y))$$

is non–degenerate [Sch96]. The following properties can be proved equivalent for a finite–dimensional algebra $\mathfrak{L}$ over a field of characteristic 0:

1. $\mathfrak{L}$ is semisimple.
2. $\mathfrak{L}$ has no nonzero Abelian ideal.
3. $\mathfrak{L}$ has zero radical (the radical is the biggest solvable ideal).
4. Every representation of $\mathfrak{L}$ is fully reducible, i.e., is a sum of irreducible representations.
5. $\mathfrak{L}$ is a (finite) direct product of simple Lie algebras (a Lie algebra is called simple if it is not Abelian and has no nonzero ideal).

A *connected Lie group* is called *semisimple* when its Lie algebra is semisimple; and the same holds for algebraic groups. Every finite dimensional representation of a semisimple Lie algebra, Lie group, or algebraic group in characteristic 0 is semisimple, i.e., completely reducible, but the converse is not true. Moreover, in characteristic $p > 0$, semisimple Lie groups and Lie algebras have finite dimensional representations which are not semisimple. An element of a semisimple Lie group or Lie algebra is itself semisimple if its image in every finite–dimensional representation is semisimple in the sense of matrices.

Every semisimple Lie algebra $\mathfrak{g}$ can be classified by its Dynkin diagram [Hel01].

4.1.4 Riemannian, Finsler and Symplectic Manifolds

Riemannian Manifolds

Local Riemannian Geometry

An important class of problems in Riemannian geometry is to understand the interaction between the curvature and topology on a smooth manifold (see [CC99]). A prime example of this interaction is the *Gauss–Bonnet formula* on a closed surface M^2, which says

$$\int_M K\, dA = 2\pi\, \chi(M), \tag{4.27}$$

where dA is the area element of a metric g on M, K is the Gaussian curvature of g, and $\chi(M)$ is the *Euler characteristic* of M.

To study the geometry of a smooth manifold we need an additional structure: the *Riemannian metric tensor*. The metric is an inner product on each of the tangent spaces and tells us how to measure angles and distances infinitesimally. In local coordinates $(x^1, x^2, \cdots, x^n)$, the metric g is given by $g_{ij}(x)\, dx^i \otimes dx^j$, where $(g_{ij}(x))$ is a positive definite symmetric matrix at each point x. For a smooth manifold one can differentiate functions. A Riemannian metric defines a natural way of differentiating vector–fields: *covariant differentiation.* In Euclidean space, one can change the order of differentiation. On a Riemannian manifold the commutator of twice covariant differentiating vector–fields is in general nonzero and is called the *Riemann curvature tensor*, which is a 4–tensor–field on the manifold.

For surfaces, the Riemann curvature tensor is equivalent to the *Gaussian curvature* K, a scalar function. In dimensions 3 or more, the Riemann curvature tensor is inherently a tensor–field. In local coordinates, it is denoted by R_{ijkl}, which is anti-symmetric in i and k and in j and l, and symmetric in the pairs $\{ij\}$ and $\{kl\}$. Thus, it can be considered as a bilinear form on 2–forms which is called the *curvature operator.* We now describe heuristically the various curvatures associated to the Riemann curvature tensor. Given a point $x \in M^n$ and 2-plane Π in the tangent space of M at x, we can define a surface S in M to be the union of all geodesics passing through x and tangent to Π. In a neighborhood of x, S is a smooth 2D submanifold of M. We define the *sectional curvature* $K(\Pi)$ of the 2–plane to be the Gauss curvature of S at x:

$$K(\Pi) = K_S(x).$$

Thus the sectional curvature K of a Riemannian manifold associates to each 2-plane in a tangent space a real number. Given a line L in a tangent space, we can average the sectional curvatures of all planes through L to get the *Ricci tensor* $Rc(L)$. Likewise, given a point $x \in M$, we can average the Ricci curvatures of all lines in the tangent space of x to get the *scalar curvature* $R(x)$. In local coordinates, the Ricci tensor is given by $R_{ik} = g^{jl} R_{ijkl}$ and the scalar curvature is given by $R = g^{ik} R_{ik}$, where $(g^{ij}) = (g_{ij})^{-1}$ is the inverse of the metric tensor (g_{ij}).

Riemannian Metric on M

Riemann in 1854 observed that around each point $m \in M$ one n pick a *special* coordinate system $(x^1, \dots, x^n)$ such that there is a symmetric $(0,2)$–tensor–field $g_{ij}(m)$ called the *metric tensor* defined as

$$g_{ij}(m) = g(\partial_{x^i}, \partial_{x^j}) = \delta_{ij}, \qquad \partial_{x^k} g_{ij}(m) = 0.$$

Thus the metric, at the specified point $m \in M$, in the coordinates $(x^1, \dots, x^n)$ looks like the Euclidean metric on $\mathbb{R}^n$. We emphasize that these conditions only hold at the specified point $m \in M$. When passing to different points it is necessary to pick different coordinates. If a curve γ passes through m, say, $\gamma(0) = m$, then the acceleration at 0 is defined by firstly, writing the curve out in our special coordinates

$$\gamma(t) = (\gamma^1(t), \dots, \gamma^n(t)),$$

secondly, defining the tangent, *velocity* vector–field, as

$$\dot{\gamma} = \dot{\gamma}^i(t) \cdot \partial_{x^i},$$

and finally, the *acceleration* vector–field as

$$\ddot{\gamma}(0) = \ddot{\gamma}^i(0) \cdot \partial_{x^i}.$$

Here, the background idea is that we have a *connection* [Pet99, Pet98].

Recall that a *connection* on a smooth manifold M tells us how to *parallel transport a vector* at a point $x \in M$ to a vector at a point $x' \in M$ along a curve $\gamma \in M$. Roughly, to parallel transport vectors along curves, it is enough if we can define parallel transport under an infinitesimal displacement: given a vector X at x, we would like to define its parallel transported version $\tilde{X}$ after an infinitesimal displacement by ϵv, where v is a tangent vector to M at x.

More precisely, a vector–field X along a parameterized curve $\alpha : I \to M$ in M is *tangent to M along α* if $X(t) \in M_{\alpha(t)}$ for all for $t \in I \subset \mathbb{R}$. However, the derivative $\dot{X}$ of such a vector–field is, in general, not tangent to M. We can, nevertheless, get a vector–field tangent to M by projecting $\dot{X}(t)$ orthogonally onto $M_{\alpha(t)}$ for each $t \in I$. This process of differentiating and then projecting onto the tangent space to M defines an operation with the same properties as differentiation, except that now differentiation of vector–fields tangent to M induces vector–fields tangent to M. This operation is called *covariant differentiation*.

Let $\gamma : I \to M$ be a parameterized curve in M, and let X be a smooth vector–field tangent to M along α. The *absolute covariant derivative* of X is the vector–field $\dot{\bar{X}}$ tangent to M along α, defined by $\dot{\bar{X}} = \dot{X}(t) - [\dot{X}(t) \cdot N(\alpha(t))]\, N(\alpha(t))$, where N is an orientation on M. Note that $\dot{\bar{X}}$ is independent of the choice of N since replacing N by $-N$ has no effect on the above formula.

Lie bracket (4.1.3) defines a *symmetric affine connection* ∇ on any manifold M:

$$[X,Y] = \nabla_X Y - \nabla_Y X.$$

In case of a Riemannian manifold M, the connection ∇ is also compatible with the Riemannian metrics g on M and is called the *Levi–Civita connection* on TM.

For a function $f \in C^\infty(M,\mathbb{R})$ and a vector a vector–field $X \in \mathcal{X}^k(M)$ we always have the Lie derivative (4.1.3)

$$\mathcal{L}_X f = \nabla_X f = df(X).$$

But there is no natural definition for $\nabla_X Y$, where $Y \in \mathcal{X}^k(M)$, unless one also has a Riemannian metric. Given the tangent field $\dot{\gamma}$, the acceleration can then be computed by using a Leibniz rule on the r.h.s, if we can make sense of the derivative of ∂_{x^i} in the direction of $\dot{\gamma}$. This is exactly what the *covariant derivative* $\nabla_X Y$ does. If $Y \in T_m M$ then we can write $Y = a^i \partial_{x^i}$, and therefore

$$\nabla_X Y = \mathcal{L}_X a^i \partial_{x^i}. \tag{4.28}$$

Since there are several ways of choosing these coordinates, one must check that the definition does not depend on the choice. Note that for two vector–fields we define $(\nabla_Y X)(m) = \nabla_{Y(m)} X$. In the end we get a *connection*

$$\nabla : \mathcal{X}^k(M) \times \mathcal{X}^k(M) \to \mathcal{X}^k(M),$$

which satisfies (for all $f \in C^\infty(M,\mathbb{R})$ and $X, Y, Z \in \mathcal{X}^k(M)$):

1. $Y \to \nabla_Y X$ is tensorial, i.e., linear and $\nabla_{fY} X = f \nabla_Y X$.
2. $X \to \nabla_Y X$ is linear.
3. $\nabla_X(fY) = (\nabla_X f) Y(m) + f(m) \nabla_X Y$.
4. $\nabla_X Y - \nabla_Y X = [X,Y]$.
5. $\mathcal{L}_X g(Z,Y) = g(\nabla_X Z, Y) + g(Z, \nabla_X Y)$.

A semicolon is commonly used to denote covariant differentiation with respect to a natural basis vector. If $X = \partial_{x^i}$, then the components of $\nabla_X Y$ in (4.28) are denoted

$$Y^k_{;i} = \partial_{x^i} Y^k + \Gamma^k_{ij} Y^j, \tag{4.29}$$

where Γ_{ij}^{k} are *Christoffel symbols* defined in (4.30) below. Similar relations hold for higher–order tensor–fields (with as many terms with Christoffel symbols as is the tensor valence).

Therefore, no matter which coordinates we use, we can now define the acceleration of a curve in the following way:

$$\begin{aligned}\gamma(t) &= (\gamma^1(t), \dots, \gamma^n(t)),\\ \dot{\gamma}(t) &= \dot{\gamma}^i(t)\partial_{x^i},\\ \ddot{\gamma}(t) &= \ddot{\gamma}^i(t)\partial_{x^i} + \dot{\gamma}^i(t)\nabla_{\dot{\gamma}(t)}\partial_{x^i}.\end{aligned}$$

We call γ a *geodesic* if $\gamma(t) = 0$. This is a second–order nonlinear ODE in a fixed coordinate system $(x^1, \dots, x^n)$ at the specified point $m \in M$. Thus we see that given any tangent vector $X \in T_mM$, there is a unique geodesic $\gamma_X(t)$ with $\dot{\gamma}_X(0) = X$. If the manifold M is closed, the geodesic must exist for all time, but in case the manifold M is open this might not be so. To see this, take as M any open subset of Euclidean space with the induced metric.

Given an arbitrary vector–field $Y(t)$ along γ, i.e., $Y(t) \in T_{\gamma(t)}M$ for all t, we can also define the derivative $\dot{Y} \equiv \frac{dY}{dt}$ in the direction of $\dot{\gamma}$ by writing

$$\begin{aligned}Y(t) &= a^i(t)\partial_{x^i},\\ \dot{Y}(t) &= \dot{a}^i(t)\partial_{x^i} + a^i(t)\nabla_{\dot{\gamma}(t)}\partial_{x^i}.\end{aligned}$$

Here the derivative of the tangent field $\dot{\gamma}$ is the acceleration γ. The field Y is said to be *parallel* iff $\dot{Y} = 0$. The equation for a field to be parallel is a first–order linear ODE, so we see that for any $X \in T_{\gamma(t_0)}M$ there is a unique parallel field $Y(t)$ defined on the entire domain of γ with the property that $Y(t_0) = X$. Given two such parallel fields $Y, Z \in \mathcal{X}^k(M)$, we have that

$$\dot{g}(Y, Z) = D_{\dot{\gamma}}g(Y, Z) = g(\dot{Y}, Z) + g(Y, \dot{Z}) = 0.$$

Thus X and Y are both of constant length and form constant angles along γ. Hence, 'parallel translation' along a curve defines an orthogonal transformation between the tangent spaces to the manifold along the curve. However, in contrast to Euclidean space, this parallel translation will depend on the choice of curve.

An infinitesimal distance between the two nearby local points m and n on M is defined by an *arc–element*

$$ds^2 = g_{ij}\, dx^i dx^j,$$

and realized by the curves $x^i(s)$ of shortest distance, called *geodesics*, addressed by the *Hilbert 4th problem*. In local coordinates $(x^1(s), ..., x^n(s))$ at a point $m \in M$, the geodesic defining equation is a second–order ODE,

$$\ddot{x}^i + \Gamma_{jk}^i\, \dot{x}^j\, \dot{x}^k = 0,$$

where the overdot denotes the derivative with respect to the affine parameter s, $\dot{x}^i(s) = \frac{d}{ds}x^i(s)$ is the tangent vector to the base geodesic, while the *Christoffel symbols* $\Gamma^i_{jk} = \Gamma^i_{jk}(m)$ of the *affine Levi–Civita connection* ∇ at the point $m \in M$ are defined, in a holonomic coordinate basis e_i as

$$\Gamma^k_{ij} = g^{kl}\Gamma_{ijl}, \qquad \text{with} \qquad g^{ij} = (g_{ij})^{-1} \qquad \text{and} \tag{4.30}$$
$$\Gamma_{ijk} = \frac{1}{2}(\partial_{x^k} g_{ij} + \partial_{x^j} g_{ki} - \partial_{x^i} g_{jk}).$$

Note that the Christoffel symbols (4.30) do not transform as tensors on the tangent bundle. They are the components of an object on the second tangent bundle, a *spray*. However, they do transform as tensors on the *jet space* (see [II06b]).

In *nonholonomic coordinates*, (4.30) takes the extended form

$$\Gamma^i_{kl} = \frac{1}{2}g^{im}\left(\partial_{x^l} g_{mk} + \partial_{x^k}\partial g_{ml} - \partial_{x^m}\partial g_{kl} + c_{mkl} + c_{mlk} - c_{klm}\right),$$

where $c_{klm} = g_{mp}c^p_{kl}$ are the commutation coefficients of the basis, i.e., $[e_k, e_l] = c^m_{kl}e_m$.

The *torsion* tensor–field T of the connection ∇ is the function $T : \mathcal{X}^k(M) \times \mathcal{X}^k(M) \to \mathcal{X}^k(M)$ given by

$$T(X,Y) = \nabla_X Y - \nabla_Y X - [X,Y].$$

From the skew symmetry ($[X,Y] = -[Y,X]$) of the Lie bracket, follows the skew symmetry ($T(X,Y) = -T(Y,X)$) of the torsion tensor. The mapping T is said to be $f-$bilinear since it is linear in both arguments and also satisfies $T(fX,Y) = fT(X,Y)$ for smooth functions f. Since $[\partial_{x^i}, \partial_{x^j}] = 0$ for all $1 \leq i,j \leq n$, it follows that

$$T(\partial_{x^i}, \partial_{x^j}) = (\Gamma^k_{ij} - \Gamma^k_{ji})\partial_{x^k}.$$

Consequently, torsion T is a $(1,2)$ tensor–field, locally given by

$$T = T^k_{i\,j}\, dx^i \otimes \partial_{x^k} \otimes dx^j,$$

where the torsion components $T^k_{i\,j}$ are given by

$$T^k_{i\,j} = \Gamma^k_{ij} - \Gamma^k_{ji}.$$

Therefore, the torsion tensor gives a measure of the nonsymmetry of the connection coefficients. Hence, $T = 0$ if and only if these coefficients are symmetric in their subscripts. A connection ∇ with $T = 0$ is said to be *torsion free* or *symmetric*.

The connection also enables us to define many other classical concepts from calculus in the setting of Riemannian manifolds. Suppose we have a function $f \in C^\infty(M, \mathbb{R})$. If the manifold is not equipped with a Riemannian metric, then we have the differential of f defined by $df(X) = \mathcal{L}_X f$, which is a 1–form. The dual concept, the *gradient* of f, is supposed to be a vector–field. But we need a metric g to define it. Namely, ∇f is defined by the relationship

$$g(\nabla f, X) = df(X).$$

Having defined the gradient of a function on a Riemannian manifold, we can then use the connection to define the *Hessian* as the linear map

$$\nabla^2 f : TM \to TM, \qquad \nabla^2 f(X) = \nabla_X \nabla f.$$

The corresponding bilinear map is then defined as

$$\nabla^2 f(X, Y) = g(\nabla^2 f(X), Y).$$

One can check that this is a symmetric bilinear form. The *Laplacian* of f, Δf, is now defined as the trace of the Hessian

$$\Delta f = \mathrm{Tr}(\nabla^2 f(X)) = \mathrm{Tr}(\nabla_X \nabla f),$$

which is a linear map. It is also called the *Laplace–Beltrami operator*, since Beltrami first considered this operator on Riemannian manifolds.

Riemannian metric has the following mechanical interpretation. Let M be a closed Riemannian manifold with the *mechanical metric* $g = g_{ij} v^i v^j \equiv \langle v, v\rangle$, with $v^i = \dot{x}^i$. Consider the *Lagrangian function*

$$L : TM \to \mathbb{R}, \qquad (x, v) \mapsto \frac{1}{2}\langle v, v\rangle - U(x) \tag{4.31}$$

where $U(x)$ is a smooth function on M called the *potential.* On a fixed level of energy E, bigger than the maximum of U, the Lagrangian flow generated by (4.31) is conjugate to the geodesic flow with metric $\bar{g} = 2(e - U(x))\langle v, v\rangle$. Moreover, the reduced action of the Lagrangian is the distance for $g = \langle v, v\rangle$ [Arn89, AMR88]. Both of these statements are known as the *Maupertius action principle*.

Geodesics on M

For a $C^\infty, k \geq 2$ curve $\gamma : I \to M$, we define its *length* on I as

$$L(\gamma, I) = \int_I |\dot{\gamma}|\, dt = \int_I \sqrt{g(\dot{\gamma}, \dot{\gamma})} dt.$$

This length is independent of our parametrization of the curve γ. Thus the curve γ can be reparameterized, in such a way that it has unit velocity. The *distance between two points* m_1 and m_2 on M, $d(m_1, m_2)$, can now be defined as the infimum of the lengths of all curves from m_1 to m_2, i.e.,

$$L(\gamma, I) \to \min.$$

This means that the distance measures the shortest way one can travel from m_1 to m_2.

If we take a variation $V(s,t) : (-\varepsilon,\varepsilon) \times [0,\ell] \to M$ of a smooth curve $\gamma(t) = V(0,t)$ parameterized by arc–length L and of length ℓ, then the first derivative of the arc–length function

$$L(s) = \int_0^\ell |\dot{V}|\, dt, \qquad \text{is given by}$$

$$\frac{dL(0)}{ds} \equiv \dot{L}(0) = g(\dot{\gamma}, X)|_0^\ell - \int_0^\ell g(\gamma, X)\, dt, \tag{4.32}$$

where $X(t) = \frac{\partial V}{\partial s}(0,t)$ is the so–called *variation vector–field*. Equation (4.32) is called the *first variation formula*. Given any vector–field X along γ, one can produce a variation whose variational field is X. If the variation fixes the endpoints, $X(a) = X(b) = 0$, then the second term in the formula drops out, and we note that the length of γ can always be decreased as long as the acceleration of γ is not everywhere zero. Thus the *Euler–Lagrangian equations* for the arc–length functional are the equations for a curve to be a *geodesic*.

In local coordinates $x^i \in U$, where U is an open subset in the Riemannian manifold M, the geodesics are defined by the *geodesic equation*

$$\ddot{x}^i + \Gamma^i_{jk}\dot{x}^j\dot{x}^k = 0, \tag{4.33}$$

where overdot means derivative upon the line parameter s, while Γ^i_{jk} are Christoffel symbols of the affine Levi–Civita connection ∇ on M. From (4.33) it follows that the linear *connection homotopy*,

$$\bar{\Gamma}^i_{jk} = s\Gamma^i_{jk} + (1-s)\Gamma^i_{jk}, \qquad (0 \le s \le 1),$$

determines the same geodesics as the original Γ^i_{jk}.

Riemannian Curvature on M

The *Riemann curvature tensor* is a rather ominous tensor of type $(1,3)$; i.e., it has three vector variables and its value is a vector as well. It is defined through the Lie bracket (4.1.3) as

$$R(X,Y)Z = \left(\nabla_{[X,Y]} - [\nabla_X, \nabla_Y]\right) Z = \nabla_{[X,Y]}Z - \nabla_X\nabla_Y Z + \nabla_Y\nabla_X Z.$$

This turns out to be a vector valued $(1,3)$–tensor–field in the three variables $X, Y, Z \in \mathcal{X}^k(M)$. We can then create a $(0,4)$–tensor,

$$R(X,Y,Z,W) = g\left(\nabla_{[X,Y]}Z - \nabla_X\nabla_Y Z + \nabla_Y\nabla_X Z, W\right).$$

Clearly this tensor is skew–symmetric in X and Y, and also in Z and $W \in \mathcal{X}^k(M)$. This was already known to Riemann, but there are some further, more subtle properties that were discovered a little later by Bianchi. The *Bianchi symmetry condition* reads

$$R(X, Y, Z, W) = R(Z, W, X, Y).$$

Thus the Riemann curvature tensor is a symmetric *curvature operator*

$$\mathfrak{R} : \Lambda^2 TM \to \Lambda^2 TM.$$

The *Ricci tensor* is the $(1, 1)-$ or $(0, 2)-$tensor defined by

$$\mathrm{Ric}(X) = R(\partial_{x^i}, X)\partial_{x^i}, \qquad \mathrm{Ric}(X, Y) = g(R(\partial_{x^i}, X)\partial_{x^i}, Y),$$

for any orthonormal basis (∂_{x^i}). In other words, the Ricci curvature is a trace of the curvature tensor. Similarly one can define the *scalar curvature* as the trace

$$\mathrm{scal}(m) = \mathrm{Tr}\,(\mathrm{Ric}) = \mathrm{Ric}(\partial_{x^i}, \partial_{x^i}).$$

When the Riemannian manifold has dimension 2, all of these curvatures are essentially the same. Since $\dim \Lambda^2 TM = 1$ and is spanned by $X \wedge Y$ where $X, Y \in \mathcal{X}^k(M)$ form an orthonormal basis for $T_m M$, we see that the curvature tensor depends only on the scalar value

$$K(m) = R(X, Y, X, Y),$$

which also turns out to be the *Gaussian curvature*. The Ricci tensor is a *homothety*

$$\mathrm{Ric}(X) = K(m)X, \qquad \mathrm{Ric}(Y) = K(m)Y,$$

and the scalar curvature is twice the Gauss curvature. In dimension 3 there are also some redundancies as $\dim TM = \dim \Lambda^2 TM = 3$. In particular, the Ricci tensor and the curvature tensor contain the same amount of information.

The *sectional curvature* is a kind of generalization of the Gauss curvature whose importance Riemann was already aware of. Given a 2$-$plane $\pi \subset T_m M$ spanned by an orthonormal basis $X, Y \in \mathcal{X}^k(M)$ it is defined as

$$\sec(\pi) = R(X, Y, X, Y).$$

The remarkable observation by Riemann was that the *curvature operator is a homothety*, i.e., looks like $\mathfrak{R} = kI$ on $\Lambda^2 T_m M$ iff all sectional curvatures of planes in $T_m M$ are equal to k. This result is not completely trivial, as the sectional curvature is not the entire quadratic form associated to the symmetric operator $\mathfrak{R}$. In fact, it is not true that $\sec \geq 0$ implies that the curvature operator is nonnegative in the sense that all its eigenvalues are nonnegative. What Riemann did was to show that our special coordinates $(x^1, \ldots, x^n)$ at m can be chosen to be *normal* at m, i.e., satisfy the condition

$$x^i = \delta^i_j x^j, \qquad (\delta^i_j x^j = g_{ij})$$

on a neighborhood of m. One can show that such coordinates are actually exponential coordinates together with a choice of an orthonormal basis for T_mM so as to identify T_mM with $\mathbb{R}^n$. In these coordinates one can then expand the metric as follows:

$$g_{ij} = \delta_{ij} - \frac{1}{3}R_{ikjl}x^k x^l + O\left(r^3\right).$$

Now the equations $x^i = g_{ij}x^j$ evidently give conditions on the curvatures R_{ijkl} at m.

If $\Gamma^i_{jk}(m) = 0$, the manifold M is flat at the point m. This means that the (1,3) curvature tensor, defined locally at $m \in M$ as

$$R^l_{ijk} = \partial_{x^j}\Gamma^l_{ik} - \partial_{x^k}\Gamma^l_{ij} + \Gamma^l_{rj}\Gamma^r_{ik} - \Gamma^l_{rk}\Gamma^r_{ij},$$

also vanishes at that point, i.e., $R^l_{ijk}(m) = 0$.

Now, the rate of change of a vector–field A^k on the manifold M along the curve $x^i(s)$ is properly defined by the *absolute covariant derivative*

$$\frac{D}{ds}A^k = \dot{x}^i\,\nabla_i\,A^k = \dot{x}^i\left(\partial_{x^i}A^k + \Gamma^k_{ij}\,A^j\right) = \dot{A}^k + \Gamma^k_{ij}\,\dot{x}^i A^j.$$

By applying this result to itself, we can get an expression for the second covariant derivative of the vector–field A^k along the curve $x^i(s)$:

$$\frac{D^2}{ds^2}A^k = \frac{d}{ds}\left(\dot{A}^k + \Gamma^k_{ij}\,\dot{x}^i A^j\right) + \Gamma^k_{ij}\,\dot{x}^i(\dot{A}^j + \Gamma^j_{mn}\,\dot{x}^m\,A^n).$$

In the local coordinates $(x^1(s), ..., x^n(s))$ at a point $m \in M$, if $\delta x^i = \delta x^i(s)$ denotes the *geodesic deviation*, i.e., the infinitesimal vector describing perpendicular separation between the two neighboring geodesics, passing through two neighboring points $m, n \in M$, then the *Jacobi equation of geodesic deviation* on the manifold M holds:

$$\frac{D^2\delta x^i}{ds^2} + R^i_{jkl}\,\dot{x}^j\,\delta x^k\,\dot{x}^l = 0. \tag{4.34}$$

This equation describes the *relative acceleration* between two infinitesimally close facial geodesics, which is proportional to the facial curvature (measured by the Riemann tensor R^i_{jkl} at a point $m \in M$), and to the geodesic deviation δx^i. Solutions of equation (4.34) are called *Jacobi fields*.

In particular, if the manifold M is a 2D–surface in $\mathbb{R}^3$, the Riemann curvature tensor simplifies into

$$R^i_{jmn} = \frac{1}{2}R\,g^{ik}(g_{km}\,g_{jn} - g_{kn}\,g_{jm}),$$

where R denotes the *scalar Gaussian curvature*. Consequently the equation of geodesic deviation (4.34) also simplifies into

$$\frac{D^2}{ds^2}\delta x^i + \frac{R}{2}\delta x^i - \frac{R}{2}\dot{x}^i(g_{jk}\,\dot{x}^j\,\delta x^k) = 0. \tag{4.35}$$

This simplifies even more if we work in a locally Cartesian coordinate system; in this case the covariant derivative $\frac{D^2}{Ds^2}$ reduces to an ordinary derivative $\frac{d^2}{ds^2}$ and the metric tensor g_{ij} reduces to identity matrix I_{ij}, so our 2D equation of geodesic deviation (4.35) reduces into a simple second–order ODE in just two coordinates x^i $(i = 1, 2)$

$$\ddot{x}^i + \frac{R}{2}\delta x^i - \frac{R}{2}\dot{x}^i(I_{jk}\,\dot{x}^j\,\delta x^k) = 0.$$

Global Riemannian Geometry

The Second Variation Formula

Cartan also establishes another important property of manifolds with nonpositive curvature. First he observes that all spaces of constant zero curvature have torsion–free fundamental groups. This is because any isometry of finite order on Euclidean space must have a fixed point (the center of mass of any orbit is necessarily a fixed point). Then he notices that one can geometrically describe the L^∞ center of mass of finitely many points $\{m_1, \dots, m_k\}$ in Euclidean space as the unique minimum for the strictly convex function

$$x \longrightarrow \max_{i=1,\cdots,k} \frac{1}{2}\left\{(d\,(m_i, x))^2\right\}.$$

In other words, the center of mass is the center of the ball of smallest radius containing $\{m_1, \dots, m_k\}$. Now Cartan's observation from above was that the exponential map is expanding and globally distance nondecreasing as a map:

$$(T_m M,\ \text{Euclidean metric}) \to (T_m M,\ \text{with pull–back metric}).$$

Thus distance functions are convex in nonpositive curvature as well as in Euclidean space. Hence the above argument can in fact be used to conclude that any Riemannian manifold of nonpositive curvature must also have torsion free fundamental group.

Now, let us set up the *second variation formula* and explain how it is used. We have already seen the first variation formula and how it can be used to characterize geodesics. Now suppose that we have a unit speed geodesic $\gamma(t)$ parameterized on $[0, \ell]$ and consider a variation $V(s, t)$, where $V(0, t) = \gamma(t)$. Synge then shows that $(\ddot{L} \equiv \frac{d^2 L}{ds^2})$

$$\ddot{L}(0) = \int_0^\ell \{g(\dot{X}, \dot{X}) - (g(\dot{X}, \dot{\gamma}))^2 - g(R(X, \dot{\gamma})X, \dot{\gamma})\}dt + g(\dot{\gamma}, A)|_0^\ell,$$

where $X(t) = \frac{\partial V}{\partial s}(0, t)$ is the variational vector–field, $\dot{X} = \nabla_{\dot{\gamma}} X$, and $A(t) = \nabla_{\frac{\partial V}{\partial s}} X$. In the special case where the variation fixes the endpoints, i.e., $s \to$

$V(s,a)$ and $s \to V(s,b)$ are constant, the term with A in it falls out. We can also assume that the variation is perpendicular to the geodesic and then drop the term $g\left(\dot{X},\dot{\gamma}\right)$. Thus, we arrive at the following simple form:

$$\ddot{L}(0) = \int_0^\ell \{g(\dot{X},\dot{X}) - g\left(R\left(X,\dot{\gamma}\right)X,\dot{\gamma}\right)\}dt = \int_0^\ell \{|\dot{X}|^2 - \sec(\dot{\gamma},X)\,|X|^2\}dt.$$

Therefore, if the sectional curvature is nonpositive, we immediately observe that any geodesic locally minimizes length (that is, among close–by curves), even if it does not minimize globally (for instance γ could be a closed geodesic). On the other hand, in positive curvature we can see that if a geodesic is too long, then it cannot minimize even locally. The motivation for this result comes from the unit sphere, where we can consider geodesics of length $> \pi$. Globally, we know that it would be shorter to go in the opposite direction. However, if we consider a variation of γ where the variational field looks like $X = \sin\left(t \cdot \frac{\pi}{\ell}\right)E$ and E is a unit length parallel field along γ which is also perpendicular to γ, then we get

$$\begin{aligned}
\ddot{L}(0) &= \int_0^\ell \left\{\left|\dot{X}\right|^2 - \sec\left(\dot{\gamma},X\right)|X|^2\right\}dt \\
&= \int_0^\ell \left\{\left(\frac{\pi}{\ell}\right)^2 \cdot \cos^2\left(t\cdot\frac{\pi}{\ell}\right) - \sec\left(\dot{\gamma},X\right)\sin^2\left(t\cdot\frac{\pi}{\ell}\right)\right\}dt \\
&= \int_0^\ell \left(\left(\frac{\pi}{\ell}\right)^2 \cdot \cos^2\left(t\cdot\frac{\pi}{\ell}\right) - \sin^2\left(t\cdot\frac{\pi}{\ell}\right)\right)dt = -\frac{1}{2\ell}\left(\ell^2 - \pi^2\right),
\end{aligned}$$

which is negative if the length ℓ of the geodesic is greater than π. Therefore, the variation gives a family of curves that are both close to and shorter than γ. In the general case, we can then observe that if $\sec \geq 1$, then for the same type of variation we get

$$\ddot{L}(0) \leq -\frac{1}{2\ell}\left(\ell^2 - \pi^2\right).$$

Thus we can conclude that, if the space is complete, then the diameter must be $\leq \pi$ because in this case any two points are joined by a segment, which cannot minimize if it has length $> \pi$. With some minor modifications one can now conclude that any complete Riemannian manifold (M,g) with $\sec \geq k^2 > 0$ must satisfy $\operatorname{diam}(M,g) \leq \pi \cdot k^{-1}$. In particular, M must be compact. Since the universal covering of M satisfies the same curvature hypothesis, the conclusion must also hold for this space; hence M must have compact universal covering space and finite fundamental group.

In odd dimensions all spaces of constant positive curvature must be orientable, as orientation reversing orthogonal transformation on odd–dimensional spheres have fixed points. This can now be generalized to manifolds of varying positive curvature. Synge did it in the following way: Suppose M is not simply–connected (or not orientable), and use this to find a shortest closed geodesic in a free homotopy class of curves (that reverses orientation). Now consider parallel translation around this geodesic. As the tangent field to the geodesic is itself a parallel field, we see that parallel translation preserves the orthogonal complement to the geodesic. This complement is now odd dimensional (even dimensional), and by assumption parallel translation preserves (reverses) the orientation; thus it must have a fixed point. In other words, there must exist a closed parallel field X perpendicular to the closed geodesic γ. We can now use the above second variation formula

$$\ddot{L}(0) = \int_0^\ell \{|\dot{X}|^2 - |X|^2 \sec(\dot{\gamma}, X)\}dt + g(\dot{\gamma}, A)|_0^\ell = -\int_0^\ell |X|^2 \sec(\dot{\gamma}, X)\, dt.$$

Here the boundary term drops out because the variation closes up at the endpoints, and $\dot{X} = 0$ since we used a parallel field. In case the sectional curvature is always positive we then see that the above quantity is negative. But this means that the closed geodesic has nearby closed curves which are shorter. However, this is in contradiction with the fact that the geodesic was constructed as a length minimizing curve in a free homotopy class.

In 1941 Myers generalized the diameter bound to the situation where one only has a lower bound for the Ricci curvature. The idea is that $\operatorname{Ric}(\dot{\gamma}, \dot{\gamma}) = \sum_{i=1}^{n-1} \sec(E_i, \dot{\gamma})$ for any set of vector–fields E_i along γ such that $\dot{\gamma}, E_1, \ldots, E_{n-1}$ forms an orthonormal frame. Now assume that the fields are parallel and consider the $n-1$ variations coming from the variational vector–fields $\sin\left(t \cdot \frac{\pi}{\ell}\right) E_i$. Adding up the contributions from the variational formula applied to these fields then induces

$$\begin{aligned}\sum_{i=1}^{n-1} \ddot{L}(0) &= \sum_{i=1}^{n-1} \int_0^\ell \left\{\left(\frac{\pi}{\ell}\right)^2 \cdot \cos^2\left(t \cdot \frac{\pi}{\ell}\right) - \sec(\dot{\gamma}, E_i) \sin^2\left(t \cdot \frac{\pi}{\ell}\right)\right\} dt \\ &= \int_0^\ell \left\{(n-1)\left(\frac{\pi}{\ell}\right)^2 \cdot \cos^2\left(t \cdot \frac{\pi}{\ell}\right) - \operatorname{Ric}(\dot{\gamma}, \dot{\gamma}) \sin^2\left(t \cdot \frac{\pi}{\ell}\right)\right\} dt.\end{aligned}$$

Therefore, if $\operatorname{Ric}(\dot{\gamma}, \dot{\gamma}) \geq (n-1)\, k^2$ (this is the Ricci curvature of S_k^n), then

$$\begin{aligned}\sum_{i=1}^{n-1} \ddot{L}(0) &\leq (n-1) \int_0^\ell \left\{\left(\frac{\pi}{\ell}\right)^2 \cdot \cos^2\left(t \cdot \frac{\pi}{\ell}\right) - k^2 \sin^2\left(t \cdot \frac{\pi}{\ell}\right)\right\} dt \\ &= -(n-1)\frac{1}{2\ell}\left(\ell^2 k^2 - \pi^2\right),\end{aligned}$$

which is negative when $\ell > \pi \cdot k^{-1}$ (the diameter of S_k^n). Thus at least one of the contributions $\frac{d^2 L_i}{ds^2}(0)$ must be negative as well, implying that the geodesic cannot be a segment in this situation.

Gauss–Bonnet Formula

In 1926 Hopf proved that in fact there is a *Gauss–Bonnet formula* for all even–dimensional hypersurfaces $H^{2n} \subset \mathbb{R}^{2n+1}$. The idea is that the determinant of the differential of the Gauss map $G : H^{2n} \to S^{2n}$ is the Gaussian curvature of the hypersurface. Moreover, this is an intrinsically computable quantity. If we integrate this over the hypersurface, we get,

$$\frac{1}{\mathrm{vol}\, S^{2n}} \int_H \det(DG) = \deg(G)\,,$$

where $\deg(G)$ is the *Brouwer degree* of the *Gauss map*. Note that this can also be done for odd–dimensional surfaces, in particular curves, but in this case the degree of the Gauss map will depend on the embedding or immersion of the hypersurface. Instead one gets the so–called winding number. Hopf then showed, as Dyck had earlier done for surfaces, that $\deg(G)$ is always half the *Euler characteristic* of H, thus yielding

$$\frac{2}{\mathrm{vol}\, S^{2n}} \int_H \det(DG) = \chi(H)\,. \tag{4.36}$$

Since the l.h.s of this formula is in fact intrinsic, it is natural to conjecture that such a formula should hold for all manifolds.

Ricci Flow on M

Ricci flow, or the *parabolic Einstein equation*,[8] was introduced by R. Hamilton in 1982 [Ham82] in the form

[8] Recall that the *Einstein field equations* (EFE) are a set of ten equations in Einstein's general relativity theory in which the fundamental force of gravitation is described as a curved space–time caused by matter and energy. The EFE can be written in a covariant (tensor) form as

$$R_{\mu\nu} - \frac{1}{2} R\, g_{\mu\nu} = \kappa T_{\mu\nu} = \frac{8\pi G}{c^4} T_{\mu\nu}$$

where $R_{\mu\nu}$ is the *Ricci tensor*, R is the scalar curvature, $g_{\mu\nu}$ is the metric tensor and $T_{\mu\nu}$ is the *stress–energy tensor*. The constant κ (kappa) is called the *Einstein gravitation constant*, where G is the gravitational constant and c is the speed of light. The EFE is a tensor equation relating a set of symmetric 4×4 tensors. When fully written out, the EFE are a system of 10 coupled, nonlinear, hyperbolic–elliptic PDEs. One can write the EFE in a more compact form by defining the *Einstein tensor*

$$G_{\mu\nu} = R_{\mu\nu} - \frac{1}{2} R g_{\mu\nu},$$

which is a symmetric second–rank tensor that is a function of the metric $g_{\mu\nu}$. Working in geometrized (normal) units where $G = c = 1$, the EFE can be written as

$$G_{\mu\nu} = 8\pi T_{\mu\nu}.$$

The expression on the left represents the curvature of space–time as determined by the metric and the expression on the right represents the matter/energy content of space–time. The EFE can then be interpreted as a set of equations dictating how the curvature of space–time is related to the matter/energy content of the

$$\partial_t g_{ij} = -2R_{ij}. \tag{4.37}$$

Now, because of the minus sign in the front of the Ricci tensor R_{ij} in this equation, the solution metric g_{ij} to the Ricci flow shrinks in positive Ricci curvature direction while it expands in the negative Ricci curvature direction. For example, on the 2−sphere S^2, any metric of positive Gaussian curvature will shrink to a point in finite time. Since the Ricci flow (4.37) does not preserve volume in general, one often considers the *normalized* Ricci flow defined by

$$\partial_t g_{ij} = -2R_{ij} + \frac{2}{n} r g_{ij}, \tag{4.38}$$

where $r = \int R dV / \int dV$ is the average scalar curvature. Under this normalized flow, which is equivalent to the (unnormalized) Ricci flow (4.37) by reparameterizing in time t and scaling the metric in space by a function of t, the volume of the solution metric is constant in time. Also that Einstein metrics (i.e., $R_{ij} = cg_{ij}$) are fixed points of (4.38).

Hamilton [Ham82] showed that on a closed Riemannian 3−manifold M^3 with initial metric of positive Ricci curvature, the solution $g(t)$ to the normalized Ricci flow (4.38) exists for all time and the metrics $g(t)$ converge exponentially fast, as time t tends to the infinity, to a constant positive sectional curvature metric g_∞ on M^3.

Since the Ricci flow lies in the realm of parabolic partial differential equations, where the prototype is the heat equation, here is a brief review of the *heat equation* [CC99].

Let (M^n, g) be a Riemannian manifold. Given a C^2 function $u : M \to \mathbb{R}$, its Laplacian is defined in local coordinates $\{x^i\}$ to be

$$\Delta u = \operatorname{Tr}\left(\nabla^2 u\right) = g^{ij} \nabla_i \nabla_j u,$$

where $\nabla_i = \nabla_{\partial_{x^i}}$ is its associated covariant derivative (Levi–Civita connection). We say that a C^2 function $u : M^n \times [0, T) \to \mathbb{R}$, where $T \in (0, \infty]$, is a solution to the heat equation if

---

universe. An important consequence of the EFE is the local *conservation of energy and momentum*; this result arises by using the differential Bianchi identity to get

$$G^{\mu\nu}_{;\nu} = 0,$$

which, by using the EFE, results in

$$T^{\mu\nu}_{;\nu} = 0,$$

(semicolon denotes the covariant derivative), which expresses the local conservation of stress-energy. This conservation law is a physical requirement. In designing the field equations, Einstein aimed at finding equations which automatically satisfied this conservation condition. For more details on EFE, general relativity and classical cosmology, see [MTW73].

$$\partial_t u = \Delta u.$$

One of the most important properties satisfied by the heat equation is the maximum principle, which says that for any smooth solution to the heat equation, whatever pointwise bounds hold at $t = 0$ also hold for $t > 0$. Let $u : M^n \times [0, T) \to \mathbb{R}$ be a C^2 solution to the heat equation on a complete Riemannian manifold. If $C_1 \le u(x, 0) \le C_2$ for all $x \in M$, for some constants $C_1, C_2 \in \mathbb{R}$, then $C_1 \le u(x, t) \le C_2$ for all $x \in M$ and $t \in [0, T)$ [CC99].

Now, given a smooth manifold M, a one–parameter family of metrics $g(t)$, where $t \in [0, T)$ for some $T > 0$, is a solution to the Ricci flow if (4.37) is valid at all $x \in M$ and $t \in [0, T)$. The minus sign in the equation (4.37) makes the Ricci flow a *forward* heat equation [CC99] (with the normalization factor 2).

In local geodesic coordinates $\{x^i\}$, we have [CC99]

$$g_{ij}(x) = \delta_{ij} - \frac{1}{3} R_{ipjq} x^p x^q + O\left(|x|^3\right), \qquad \text{therefore,} \qquad \Delta g_{ij}(0) = -\frac{1}{3} R_{ij},$$

where Δ is the standard Euclidean Laplacian. Hence the Ricci flow is like the heat equation for a Riemannian metric

$$\partial_t g_{ij} = 6 \Delta g_{ij}.$$

The practical study of the Ricci flow is made possible by the following short–time existence result: Given any smooth compact Riemannian manifold (M, g_o), there exists a unique smooth solution $g(t)$ to the Ricci flow defined on some time interval $t \in [0, \epsilon)$ such that $g(0) = g_o$ [CC99].

Now, given that short–time existence holds for any smooth initial metric, one of the main problems concerning the Ricci flow is to determine under what conditions the solution to the normalized equation exists for all time and converges to a constant curvature metric. Results in this direction have been established under various curvature assumptions, most of them being some sort of positive curvature. Since the Ricci flow (4.37) does not preserve volume in general, one often considers, as we mentioned in the Introduction, the normalized Ricci flow (4.38). Under this flow, the volume of the solution $g(t)$ is independent of time.

To study the long–time existence of the normalized Ricci flow, it is important to know what kind of curvature conditions are preserved under the equation. In general, the Ricci flow tends to preserve some kind of positivity of curvatures. For example, positive scalar curvature is preserved in all dimensions. This follows from applying the maximum principle to the evolution equation for scalar curvature R, which is

$$\partial_t R = \Delta R + 2 |R_{ij}|^2 .$$

In dimension 3, positive Ricci curvature is preserved under the Ricci flow. This is a special feature of dimension 3 and is related to the fact that the Riemann curvature tensor may be recovered algebraically from the Ricci tensor and the metric in dimension 3. Positivity of sectional curvature is not preserved in general. However, the stronger condition of positive curvature operator is preserved under the Ricci flow.

Structure Equations on M

Let $\{X_a\}_{a=1}^m$, $\{Y_i\}_{i=1}^n$ be local orthonormal framings on M, N respectively and $\{e_i\}_{i=1}^n$ be the induced framing on E defined by $e_i = Y_i \circ \phi$, then there exist smooth local coframings $\{\omega_a\}_{a=1}^m$, $\{\eta_i\}_{i=1}^n$ and $\{\phi^*\eta_i\}_{i=1}^n$ on TM, TN and E respectively such that (locally)

$$g = \sum_{a=1}^m \omega_a^2 \qquad \text{and} \qquad h = \sum_{i=1}^n \eta_i^2.$$

The corresponding *first structure equations* are [Mus99]:

$$\begin{aligned} d\omega_a &= \omega_b \wedge \omega_{ba}, & \omega_{ab} &= -\omega_{ba}, \\ d\eta_i &= \eta_j \wedge \eta_{ji}, & \eta_{ij} &= -\eta_{ji}, \\ d(\phi^*\eta_i) &= \phi^*\eta_j \wedge \phi^*\eta_{ji}, & \phi^*\eta_{ij} &= -\phi^*\eta_{ji}, \end{aligned}$$

where the unique 1–forms ω_{ab}, η_{ij}, $\phi^*\eta_{ij}$ are the respective connection forms. The *second structure equations* are

$$d\omega_{ab} = \omega_{ac} \wedge \omega_{cb} + \Omega_{ab}^M, \qquad d\eta_{ij} = \eta_{ik} \wedge \eta_{kj} + \Omega_{ij}^N,$$
$$d(\phi^*\eta_{ij}) = \phi^*\eta_{ik} \wedge \phi^*\eta_{kj} + \phi^*\Omega_{ij}^N,$$

where the curvature 2–forms are given by

$$\Omega_{ab}^M = -\frac{1}{2} R_{abcd}^M \omega_c \wedge \omega_d \qquad \text{and} \qquad \Omega_{ij}^N = -\frac{1}{2} R_{ijkl}^N \eta_k \wedge \eta_l.$$

The pull back map ϕ^* and the push forward map ϕ_* can be written as [Mus99]

$$\phi^*\eta_i = f_{ia}\omega_a$$

for unique functions f_{ia} on $U \subset M$, so that

$$\phi_* = e_i \otimes \phi^*\eta_i = f_{ia} e_i \otimes \omega_a.$$

Note that ϕ_* is a section of the vector bundle $\phi^{-1}TN \otimes T^*M$.

The covariant differential operators are represented as

$$\nabla^M X_a = \omega_{ab} \otimes X_b, \qquad \nabla^N Y_i = \eta_{ij} \otimes Y_j, \qquad \nabla^* \omega_a = -\omega_{ca} \otimes \omega_c,$$

where ∇^* is the dual connection on the cotangent bundle T^*M.

Furthermore, the induced connection ∇^ϕ on E is

$$\nabla^\phi e_i = \big(\eta_{ij}(Y_k) \circ \phi\big)\, e_j \otimes f_{ka}\omega_a.$$

The components of the Ricci tensor and scalar curvature are defined respectively by

$$R^M_{ab} = R^M_{acbc} \qquad \text{and} \qquad R^M = R^M_{aa}.$$

Given a function $f : M \to$, there exist unique functions $f_{cb} = f_{bc}$ such that

$$df_c - f_b\omega_{cb} = f_{cb}\omega_b\,, \tag{4.39}$$

where $f_c = df(X_c)$ for a local orthonormal frame $\{X_c\}_{c=1}^m$. To prove this we take the exterior derivative of $df = \sum_{c=1}^m f_c\omega_c$ and using structure equations, we have

$$0 = [df_c \wedge \omega_c + f_{bc}\omega_b \wedge \omega_{bc}] = [(df_c - f_b\omega_{cb}) \wedge \omega_c]\,.$$

Hence by *Cartan's lemma* (cf. [Wil93]), there exist unique functions $f_{cb} = f_{bc}$ such that

$$df_c - f_b\omega_{cb} = f_{cb}\omega_b.$$

The Laplacian of a function f on M is given by

$$\Delta f = -\operatorname{Tr}(\nabla df),$$

that is, negative of the usual Laplacian on functions.

Basics of Morse and (Co)Bordism Theories

Morse Theory on Smooth Manifolds

At the same time the variational formulae were discovered, a related technique, called *Morse theory*, was introduced into Riemannian geometry. This theory was developed by Morse, first for functions on manifolds in 1925, and then in 1934, for the loop space. The latter theory, as we shall see, sets up a very nice connection between the first and second variation formulae from the previous section and the topology of M. It is this relationship that we shall explore at a general level here. In section 5 we shall then see how this theory was applied in various specific settings.

If we have a *proper function* $f : M \to \mathbb{R}$, then its Hessian (as a quadratic form) is in fact well defined at its *critical points* without specifying an underlying Riemannian metric. The *nullity* of f at a *critical point* is defined as the dimension of the *kernel* of $\nabla^2 f$, while the *index* is the number of negative eigenvalues counted with multiplicity. A function is said to be a *Morse function* if the nullity at any of its critical points is zero. Note that this guarantees in particular that all critical points are isolated. The first fundamental Theorem of Morse theory is that one can determine the topological structure of a manifold from a Morse function. More specifically, if one can order the critical points $x_1, \ldots, x_k$ so that $f(x_1) < \cdots < f(x_k)$ and the index of x_i is denoted λ_i, then M has the structure of a CW complex with a cell of dimension λ_i for each i. Note that in case M is closed then x_1 must be a minimum and so

$\lambda_1 = 0$, while x_k is a maximum and $\lambda_k = n$. The classical example of Milnor of this Theorem in action is a torus in 3–space and f the height function.

We are now left with the problem of trying to find appropriate Morse functions. While there are always plenty of such functions, there does not seem to be a natural way of finding one. However, there are natural choices for Morse functions on the loop space to a Riemannian manifold. This is, somewhat inconveniently, infinite–dimensional. Still, one can develop Morse theory as above for suitable functions, and moreover *the loop space of a manifold determines the topology of the underlying manifold.*

If $m, p \in M$, then we denote by Ω_{mp} the space of all C^∞ paths from m to p. The first observation about this space is that

$$\pi_{i+1}(M) = \pi_i(\Omega_{mp}).$$

To see this, just fix a path from m to q and then join this path to every curve in Ω_{mp}. In this way Ω_{mp} is identified with Ω_m, *the space of loops fixed at m.* For this space the above relationship between the homotopy groups is almost self-evident.

On the space Ω_{mp} we have two naturally defined functions, the *arc–length* and *energy functionals*:

$$L(\gamma, I) = \int_I |\dot{\gamma}|\, dt, \quad \text{and} \quad E(\gamma, I) = \frac{1}{2}\int_I |\dot{\gamma}|^2\, dt.$$

While the energy functional is easier to work with, it is the arc–length functional that we are really interested in. In order to make things work out nicely for the arc–length functional, it is convenient to parameterize all curves on $[0, 1]$ and proportionally to arc–length. We shall think of Ω_{mp} as an *infinite–dimensional manifold.* For each curve $\gamma \in \Omega_{mp}$ the natural choice for the tangent space consists of the vector–fields along γ which vanish at the endpoints of γ. This is because these vector–fields are exactly the variational fields for curves through γ in Ω_{mp}, i.e., fixed endpoint variations of γ. An inner product on the tangent space is then naturally defined by

$$(X, Y) = \int_0^1 g(X, Y)\, dt.$$

Now the first variation formula for arc–length tells us that the gradient for L at γ is $-\nabla_{\dot{\gamma}}\dot{\gamma}$. Actually this cannot be quite right, as $-\nabla_{\dot{\gamma}}\dot{\gamma}$ does not vanish at the endpoints. The real gradient is gotten in the same way we find the gradient for a function on a surface in space, namely, by projecting it down into the correct tangent space. In any case we note that the critical points for L are exactly the geodesics from m to p. The second variation formula tells us that the Hessian of L at these critical points is given by

$$\nabla^2 L(X) = \ddot{X} + R(X, \dot{\gamma})\,\dot{\gamma},$$

at least for vector–fields X which are perpendicular to γ. Again we ignore the fact that we have the same trouble with endpoint conditions as above. We now need to impose the Morse condition that this Hessian is not allowed to have any kernel. The vector–fields J for which $\ddot{J} + R(J, \dot{\gamma})\dot{\gamma} = 0$ are called *Jacobi fields*. Thus we have to Figure out whether there are any Jacobi fields which vanish at the endpoints of γ. The first observation is that Jacobi fields must always come from geodesic variations. The Jacobi fields which vanish at m can therefore be found using the exponential map $\exp_m$. If the Jacobi field also has to vanish at p, then p must be a critical value for $\exp_m$. Now Sard's Theorem asserts that the set of critical values has measure zero. For given $m \in M$ it will therefore be true that the arc–length functional on Ω_{mp} is a Morse function for almost all $p \in M$. Note that it may not be possible to choose $p = m$, the simplest example being the standard sphere. We are now left with trying to decide what the *index* should be. This is the dimension of the largest subspace on which the Hessian is negative definite. It turns out that this index can also be computed using Jacobi fields and is in fact always finite. Thus one can calculate the topology of Ω_{mp}, and hence M, by finding all the geodesics from m to p and then computing their index.

In geometrical situations it is often unrealistic to suppose that one can calculate the index precisely, but as we shall see it is often possible to given lower bounds for the index. As an example, note that if M is not simply–connected, then Ω_{mp} is not connected. Each curve of minimal length in the path components is a geodesic from m to p which is a local minimum for the arc–length functional. Such geodesics evidently have index zero. In particular, if one can show that all geodesics, except for the minimal ones from m to p, have index > 0, then the manifold must be simply–connected.

(Co)Bordism Theory on Smooth Manifolds

(Co)bordism appeared as a revival of Poincaré's unsuccessful 1895 attempts to define homology using only manifolds. Smooth manifolds (without boundary) are again considered as 'negligible' when they are *boundaries* of smooth manifolds–with–boundary. But there is a big difference, which keeps definition of 'addition' of manifolds from running into the difficulties encountered by Poincaré; it is now the disjoint union. The (unoriented) *(co)bordism relation* between two compact smooth manifolds M_1, M_2 of same dimension n means that their disjoint union $\partial W = M_1 \cup M_2$ is the boundary ∂W of an $(n+1)$D smooth manifold–with–boundary W. This is an *equivalence relation*, and the classes for that relation of nD manifolds form a *commutative group* $\mathfrak{N}_n$ in which every element has order 2. The direct sum $\mathfrak{N}_\bullet = \oplus_{n\geq 0}\mathfrak{N}_n$ is a ring for the multiplication of classes deduced from the Cartesian product of manifolds.

More precisely, a manifold M is said to be a (co)bordism from A to B if exists a diffeomorphism from a disjoint sum, $\varphi \in \mathrm{diff}(A^* \cup B, \partial M)$. Two (co)bordisms $M(\varphi)$ and $M'(\varphi')$ are equivalent if there is a $\Phi \in \mathrm{diff}(M, M')$

such that $\varphi' = \Phi \circ \varphi$. The equivalence class of (co)bordisms is denoted by $M(A,B) \in Cob(A,B)$ [Sto68].

Composition c_{Cob} of (co)bordisms comes from gluing of manifolds [BD95]. Let $\varphi' \in \mathrm{diff}(C^* \cup D, \partial N)$. One can glue (co)bordism M with N by identifying B with C^*, $(\varphi')^{-1} \circ \varphi \in \mathrm{diff}(B, C^*)$. We get the glued (co)bordism $(M \circ N)(A,D)$ and a semigroup operation,

$$c(A,B,D) : Cob(A,B) \times Cob(B,D) \longrightarrow Cob(A,D).$$

A *surgery* is an operation of cutting a manifold M and gluing to cylinders. A surgery gives new (co)bordism: from $M(A,B)$ into $N(A,B)$. The disjoint sum of $M(A,B)$ with $N(C,D)$ is a (co)bordism $(M \cup N)(A \cup C, B \cup D)$. We got a 2–graph of (co)bordism Cob with $Cob_0 = Man_d$, $Cob_1 = Man_{d+1}$, whose 2–cells from Cob_2 are surgery operations.

There is an $n-$category of (co)bordisms $\mathcal{BO}$ [Lei03] with:

- 0–cells: 0–manifolds, where 'manifold' means 'compact, smooth, oriented manifold'. A typical 0–cell is • • • •.
- 1–cells: 1–manifolds with corners, i.e., (co)bordisms between 0–manifolds, such as 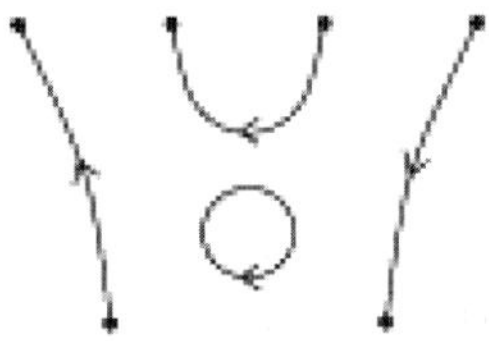 (this being a 1–cell from the 4–point manifold to the 2–point 0–manifold).
- 2–cells: 2–manifolds with corners, such as 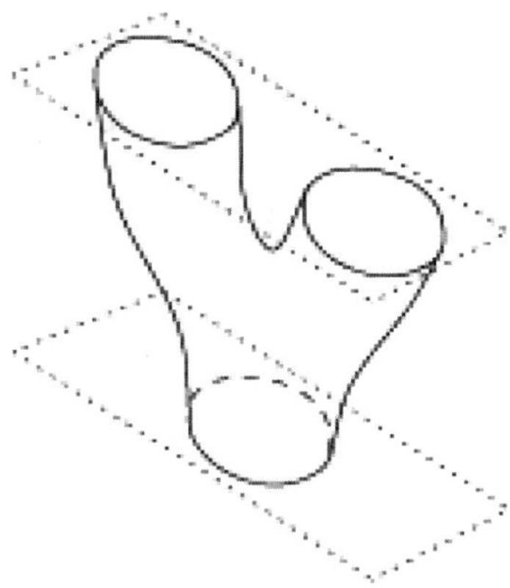
- 3–cells, 4–cells,... are defined similarly;
- Composition is gluing of manifolds.

The (co)bordisms theme was taken a step further by [BD95], when when they started a programme to understand the subtle relations between certain TMFT models for manifolds of different dimensions, frequently referred to as the dimensional ladder. This programme is based on higher–dimensional algebra, a generalization of the theory of categories and functors to $n-$categories and $n-$functors. In this framework a topological quantum field theory (TMFT) becomes an $n-$functor from the $n-$category $\mathcal{BO}$ of $n-$cobordisms to the $n-$category of $n-$Hilbert spaces.

Finsler Manifolds

Recall that Finsler geometry is such a generalization of Riemannian geometry, that is closely related to multivariable calculus of variations.

Definition of a Finsler Manifold

Let M be a real, smooth, connected, finite–dimensional manifold. The pair (M, F) is called a *Finsler manifold* iff there exists a *fundamental function* $F : TM \to \mathbb{R}$ that satisfies the following set of axioms (see, e.g., [UN99]):

F1 $F(x, y) > 0$ for all $x \in M,\ y \neq 0$.
F2 $F(x, \lambda y) = |\lambda| F(x, y)$ for all $\lambda \in \mathbb{R}$, $(x, y) \in TM$.
F3 the *fundamental metric tensor* g_{ij} on M, given by

$$g_{ij}(x,y) = \frac{1}{2}\frac{\partial^2 F^2}{\partial y^i \partial y^j},$$

is *positive definite.*
F4 F is *smooth* (C^∞) at every point $(x, y) \in TM$ with $y \neq 0$ and *continuous* (C^0) at every $(x, 0) \in TM$. Then, the *absolute Finsler energy function* is given by

$$F^2(x,y) = g_{ij}(x,y)y^i y^j.$$

Let $c = c(t) : [a, b] \to M$ be a smooth regular curve on M. For any two vector–fields $X(t) = X^i(t)\, \frac{\partial}{\partial x^i}\big|_{c(t)}$ and $Y(t) = Y^i(t)\, \frac{\partial}{\partial x^i}\big|_{c(t)}$ along the curve $c = c(t)$, we introduce the *scalar (inner) product* [Che96]

$$g(X,Y)(c) = g_{ij}(c,\dot{c})X^i Y^j$$

along the curve c.

In particular, if $X = Y$ then we have $\|X\| = \sqrt{g(X, X)}$. The vector–fields X and Y are *orthogonal* along the curve c, denoted by $X \bot Y$, iff $g(X, Y) = 0$.

Let $C\Gamma(N) = (L^i_{jk}, N^i_j, C^i_{jk})$ be the *Cartan canonical $N-$linear metric connection* determined by the metric tensor $g_{ij}(x, y)$. The coefficients of this connection are expressed by [UN99]

$$L^i_{jk} = \frac{1}{2}g^{im}\left(\frac{\delta g_{mk}}{\delta x^j} + \frac{\delta g_{jm}}{\delta x^k} - \frac{\delta g_{jk}}{\delta x^m}\right), \quad C^i_{jk} = \frac{1}{2}g^{im}\left(\frac{\partial g_{mk}}{\partial y^j} + \frac{\partial g_{jm}}{\partial y^k} - \frac{\partial g_{jk}}{\partial y^m}\right),$$

$$N^i_j = \frac{1}{2}\frac{\partial}{\partial y^j}\left(\Gamma^i_{kl}y^k y^l\right) = \frac{1}{2}\frac{\partial \Gamma^i_{00}}{\partial y^j}, \qquad \Gamma^i_{jk} = \frac{1}{2}g^{im}\left(\frac{\partial g_{mk}}{\partial x^j} + \frac{\partial g_{jm}}{\partial x^k} - \frac{\partial g_{jk}}{\partial x^m}\right),$$

$$\text{where} \quad \frac{\delta}{\delta x^i} = \frac{\partial}{\partial x^i} + N^j_i\frac{\partial}{\partial y^j}.$$

Let X be a vector–field along the curve c expressed locally by $X(t) = X^i(t)\, \frac{\partial}{\partial x^i}\big|_{c(t)}$. Using the Cartan $N-$linear connection, we define the *covariant derivative* $\frac{\nabla X}{dt}$ of $X(t)$ *along the curve* $c(t)$, by [UN99]

$$\frac{\nabla X}{dt} = \{\dot{X}^i + X^m[L^i_{mk}(c,\dot{c})\dot{c}^k + C^i_{mk}(c,\dot{c})\frac{\delta}{\delta t}(\dot{c}^k)]\}\left.\frac{\partial}{\partial x^i}\right|_{c(t)}.$$

Since $\frac{\delta}{\delta t}(\dot{c}^k) = \ddot{c}^k + N^k_l(c,\dot{c})\dot{c}^l,$

we have
$$\frac{\nabla X}{dt} = \{\dot{X}^i + X^m[\Gamma^i_{mk}(c,\dot{c})\dot{c}^k + C^i_{mk}(c,\dot{c})\ddot{c}^k]\}\left.\frac{\partial}{\partial x^i}\right|_{c(t)}, \qquad (4.40)$$

where $\Gamma^i_{mk}(c,\dot{c}) = L^i_{mk}(c,\dot{c}) + C^i_{ml}(c,\dot{c})N^l_k(c,\dot{c}).$

In particular, c is a *geodesic* iff $\frac{\nabla \dot{c}}{dt} = 0.$

Since $C\Gamma(N)$ is a metric connection, we have

$$\frac{d}{dt}\left[g(X,Y)\right] = g\left(\frac{\nabla X}{dt}, Y\right) + g\left(X, \frac{\nabla Y}{dt}\right).$$

Energy Functional, Variations and Extrema

Let $x_0, x_1 \in M$ be two points not necessarily distinct. We introduce the $\Omega-$set on M, as

$$\Omega = \{c : [0,1] \to M \mid c \text{ is piecewise } C^\infty \text{ regular curve}, c(0) = x_0, c(1) = x_1\}.$$

For every $p \in \mathbb{R} - \{0\}$, we can define the $p-$*energy functional* on M [UN99]

$$E_p : \Omega \to \mathbb{R}_+, \qquad \text{as}$$

$$E_p(c) = \int_0^1 [g_{ij}(c,\dot{c})\dot{c}^i\dot{c}^j]^{p/2} dt = \int_0^1 [g(\dot{c},\dot{c})]^{p/2} dt = \int_0^1 \|\dot{c}\|^p dt.$$

In particular, for $p = 1$ we get the *length functional*

$$L(c) = \int_0^1 \|\dot{c}\| dt,$$

and for $p = 2$ we get the *energy functional*

$$E(c) = \int_0^1 \|\dot{c}\|^2 dt.$$

Also, for any naturally parametrized curve (i.e., $\|\dot{c}\| = \text{const}$) we have

$$E_p(c) = (L(c))^p = (E(c))^{p/2}.$$

Note that the $p-$energy of a curve is dependent of parametrization if $p \neq 1$.

For every curve $c \in \Omega$, we define the *tangent space* $T_c\Omega$ as

$$T_c\Omega = \{X : [0,1] \to TM \mid X \text{ is continuous, piecewise } C^\infty, X(t) \in T_{c(t)}M,$$
$$\text{for all } t \in [0,1],\ X(0) = X(1) = 0\}.$$

Let $(c_s)_{s\in(-\epsilon,\epsilon)} \subset \Omega$ be a *one–parameter variation* of the curve $c \in \Omega$. We define

$$X(t) = \frac{dc_s}{ds}(0,t) \in T_c\Omega.$$

Using the equality

$$g\left(\frac{\nabla \dot{c}_s}{\partial s}, \dot{c}_s\right) = g\left(\frac{\nabla}{\partial t}\left(\frac{\partial c_s}{\partial s}\right), \dot{c}_s\right),$$

we can prove the following Theorem: The first variation of the $p-$energy is

$$\frac{1}{p}\frac{dE_p(c_s)}{ds}(0) = -\sum_t g(X, \Delta_t(\|\dot{c}\|^{p-2}\dot{c}))$$
$$- \int_0^1 \|\dot{c}\|^{p-4} g\left(X, \|\dot{c}\|^2\frac{\nabla \dot{c}}{dt} + (p-2)g\left(\frac{\nabla \dot{c}}{dt}, \dot{c}\right)\dot{c}\right) dt,$$

where $\Delta_t(\|\dot{c}\|^{p-2}\dot{c}) = (\|\dot{c}\|^{p-2}\dot{c})_{t^+} - (\|\dot{c}\|^{p-2}\dot{c})_{t^-}$ represents the jump of $\|\dot{c}\|^{p-2}\dot{c}$ at the discontinuity point $t \in (0,1)$ [UN99].

The curve c is a *critical point* of E_p iff c is a geodesic.

In particular, for $p = 1$ the curve c is a *reparametrized geodesic*.

Now, let $c \in \Omega$ be a critical point for E_p (i.e., the curve c is a geodesic). Let $(c_{s_1s_2})_{s_1,s_2\in(-\epsilon,\epsilon)} \subset \Omega$ be a *two–parameter variation* of c. Using the notations:

$$X(t) = \frac{\partial c_{s_1s_2}}{\partial s_1}(0,0,t) \in T_c\Omega, \qquad Y(t) = \frac{\partial c_{s_1s_2}}{\partial s_2}(0,0,t) \in T_c\Omega,$$

$$\|\dot{c}\| = v = \text{constant}, \qquad \text{and} \qquad I_p(X,Y) = \frac{\partial^2 E_p(c_{s_1s_2})}{\partial s_1 \partial s_2}(0,0),$$

we get the following Theorem: The second variation of the $p-$energy is [UN99]

$$\frac{1}{pv^{p-4}} I_p(X,Y) = -\sum_t g\left(Y, v^2\Delta_t\left(\frac{\nabla X}{dt}\right) + (p-2)g\left(\Delta_t\left(\frac{\nabla X}{dt}\right), \dot{c}\right)\dot{c}\right)$$
$$-\int_0^1 g\left(Y, v^2\left[\frac{\nabla}{dt}\frac{\nabla X}{dt} + R^2(X,\dot{c})\dot{c}\right] + (p-2)g\left(\left[\frac{\nabla}{dt}\frac{\nabla X}{dt} + R^2(X,\dot{c})\dot{c}\right], \dot{c}\right)\dot{c}\right) dt,$$

where $\Delta_t\left(\frac{\nabla X}{dt}\right) = \left(\frac{\nabla X}{dt}\right)_{t^+} - \left(\frac{\nabla X}{dt}\right)_{t^-}$ represents the jump of $\frac{\nabla X}{dt}$ at the discontinuity point $t \in (0,1)$; also, if $R^l_{ijk}(c,\dot{c})$ represents the components of the *Finsler curvature tensor*, then

$$R^2(X,\dot{c})\dot{c} = R^l_{ijk}(c,\dot{c})\dot{c}^i\dot{c}^j X^k \frac{\partial}{\partial x^l} = R^l_{jk}(c,\dot{c})\dot{c}^j X^k \frac{\partial}{\partial x^l}.$$

In particular, we have

$$R^i_{jk} = \frac{\delta N^i_j}{\delta x^k} - \frac{\delta N^i_k}{\delta x^j}, \quad \text{and} \quad R^i_{hjk} = \frac{\delta L^i_{hj}}{\delta x^k} - \frac{\delta L^i_{hk}}{\delta x^j} + L^s_{hj}L^i_{sk} - L^s_{hk}L^i_{sj} + C^i_{hs}R^s_{jk}.$$

Moreover, using the *Ricci identities* for the *deflection tensors*, we also have

$$R^i_{jk} = R^i_{mjk}y^m = R^i_{0jk}.$$

$I_p(X,Y) = 0$ (for all $Y \in T_c\Omega$) iff X is a *Jacobi field*, i.e.,

$$\frac{\nabla}{dt}\frac{\nabla X}{dt} + R^2(X,\dot{c})\dot{c} = 0.$$

In these conditions we have the following definition: A point $c(b)$ $(0 \leq a < b < 1)$ of a geodesic $c \in \Omega$ is called a *conjugate point* of a point $c(a)$ along the curve $c(t)$, if there exists a non–zero Jacobi field which vanishes at $t \in \{a,\ b\}$.

Now, integrating by parts and using the property of metric connection, we find

$$\frac{1}{pv^{p-4}} I_p(X,Y) = \int_0^1 v^2 \left[g\left(\frac{\nabla X}{dt}, \frac{\nabla Y}{dt}\right) - R^2(X,\dot{c},Y,\dot{c}) \right] + (p-2) g\left(\dot{c}, \frac{\nabla X}{dt}\right) g\left(\dot{c}, \frac{\nabla Y}{dt}\right) dt,$$

where $R^2(X,\dot{c},Y,\dot{c}) = g(R^2(Y,\dot{c})\dot{c},X) = R_{0i0j}(c(t),\dot{c}(t))X^iY^j$.

Let $R_{ijk} = g_{jm}R^m_{ik}$. In any Finsler space the following identity is satisfied,

$$R_{ijk} + R_{jki} + R_{kij} = 0,$$

get by the *Bianchi identities*. As $R_{0i0j} = R_{i0j} = R_{j0i} = R_{0j0i}$ we get $R^2(X,\dot{c},Y,\dot{c}) = R^2(Y,\dot{c},X,\dot{c})$.

The quadratic form associated to the *Hessian of the p–energy* is given by

$$I_p(X) = I_p(X,X) = \int_0^1 v^2 \left[\left\| \frac{\nabla X}{dt} \right\|^2 - R^2(X,\dot{c},X,\dot{c}) \right] + (p-2) \left[g\left(\dot{c}, \frac{\nabla X}{dt}\right) \right]^2 dt.$$

Let

$$T^{\perp}_c\Omega = \{X \in T_c\Omega \mid g(X,\dot{c}) = 0\}, \quad \text{and}$$
$$T^{'}_c\Omega = \{X \in T_c\Omega \mid X = f\dot{c}, \text{ where } f:[0,1] \to R \text{ is continuous,}$$

piecewise C^∞, $f(0) = f(1) = 0\}$.

Let c be a geodesic and $p \in R - \{0,1\}$. Then $I_p(T'_c\Omega) \geq 0$ for $p \in (-\infty,0) \cup (1,\infty)$, and $I_p(T'_c\Omega) \leq 0$ for $p \in (0,1)$. Moreover, in both cases: $I_p(X) = 0$ iff $X = 0$. To prove it, let $X = f\dot{c} \in T'_c\Omega$. Then we have [UN99]

$$\frac{1}{v^{p-4}} I_p(X) = p \int_0^1 \left\{ v^2 \left[g(f'\dot{c}, f'\dot{c}) - R^2(f\dot{c}, \dot{c}, f\dot{c}, \dot{c}) \right] + (p-2) \left[g(\dot{c}, f'\dot{c}) \right]^2 \right\} dt$$
$$= p \int_0^1 \left[v^4 (f')^2 + (p-2) v^4 (f')^2 \right] dt = \int_0^1 p(p-1) v^4 (f')^2 dt.$$

Moreover, if $I_p(X) = 0$, then $f' = 0$, which means that f is constant. The conditions $f(0) = f(1) = 0$ imply that $f = 0$.

As $I_p(T'_c\Omega)$ is positive definite for $p \in (-\infty, 0) \cup (1, \infty)$ and negative definite for $p \in (0,1)$, it is sufficient to study the behavior of I_p restricted to $T_c^{\perp}\Omega$. Since $X \perp \dot{c}$ and the curve c is a geodesic it follows

$$g\left(\dot{c}, \frac{\nabla X}{dt} \right) = 0.$$

Hence, for all $X \in T_c^{\perp}\Omega$, we have

$$\frac{1}{p v^{p-2}} I_p(X) = \int_0^1 \left[\left\| \frac{\nabla X}{dt} \right\|^2 - R^2(X, \dot{c}, X, \dot{c}) \right] dt = I(X).$$

Constant Curvature Finsler Manifolds

We assume the Finsler space (M,F) is complete, of dimension $n \geq 3$ and of constant curvature $K \in R$. Hence, we have

$$H_{ijkl} = K(g_{ik} g_{jl} - g_{il} g_{jk}),$$

where H_{ijkl} are the components of the $h-$*curvature tensor* H of the *Berwald connection* $B\Gamma$. It follows that

$$R_{ijk} = KF \left(g_{ik} \frac{y_j}{F} - g_{ij} \frac{y_k}{F} \right),$$

where $y_j = g_{jk} y^k$. We also have

$$R_{i0k} = R_{ijk} y^j = K(g_{ik} F^2 - y_i y_k).$$

Hence, along the geodesic $c \in \Omega$, we get

$$R^2(X, \dot{c})\dot{c} = K\{ \|\dot{c}\|^2 X - g(X, \dot{c})\dot{c} \}.$$

This equality is also true in the case of constant $h-$curvature for the Cartan canonical connection. Following [Mat82] we have:
(i) If $K \leq 0$, then the geodesic c has no conjugate points to $x_0 = c(0)$.
(ii) If $K \geq 0$ and the geodesic c has conjugate points to $x_0 = c(0)$, then the number of conjugate points is finite, according to the *Morse index Theorem for Finsler manifolds.* Moreover, in the case (ii), choosing an orthonormal frame of vector–fields $\{E_i\}_{i=\overline{1,n-1}} \in T_c^{\perp}\Omega$ parallel–propagated along the geodesic c,

we can build a basis $\{U_i, V_i\}_{i=\overline{1,n-1}}$ in the set of Jacobi fields orthogonal to $\dot{c}$, defining

$$U_i(t) = \sin(\sqrt{K}vt)E_i, \quad \text{and} \quad V_i(t) = \cos(\sqrt{K}vt)E_i,$$

where $v = \|\dot{c}\| = \text{const}$. In conclusion, the distance between two consecutive conjugate points is $\pi/\sqrt{K}$. In these conditions we can prove the following Theorem [UN99]: Let (M, F) be a Finsler space, as above, and let $c = c_p \in \Omega$ be a global extremum point for the $p-$energy functional E_p, where p is a number in $R - \{0, 1\}$. In these conditions we have:
(i) If $p \in (-\infty, 0)$, then c has conjugate points, $K > 0$ and

$$\left[\frac{(m(c)+1)\pi}{\sqrt{K}}\right]^p \leq E_p(c) \leq \left[\frac{m(c)\pi}{\sqrt{K}}\right]^p,$$

where $m(c)$ is the maximal number of conjugate points to $x_0 = c(0)$ along the geodesic c.
(ii) If $p \in (0, 1)$, then c has conjugate points, $K > 0$ and

$$\left[\frac{m(c)\pi}{\sqrt{K}}\right]^p \leq E_p(c) \leq \left[\frac{(m(c)+1)\pi}{\sqrt{K}}\right]^p.$$

(iii) If $p \in (1, \infty)$, then c is a minimal geodesic (i.e., it minimizes the length functional). If we denote $m = \sup\{m(c) \mid c \in \Omega,\ c-\text{geodesic}\} \in N$, we get the following corollary: If there is $c \in \Omega$ a global extremum point for the $p-$energy functional E_p, where $p \in (-\infty, 0) \cup (0, 1)$, we must have $m < \infty$ and $m(c) = m$.

For example, in the case of Riemannian unit sphere $S^n \subset R^{n+1}$, $n \geq 2$, it is well known that the geodesics are precisely the great circles, that is the intersections of S^n with the hyperplanes trough the center of S^n. Moreover, two arbitrary points on S^n are conjugate along a geodesic γ if they are antipodal points. In these conditions, for any two points x_0 and x_1 on the sphere S^n, there is no geodesic trough these points which has a finite maximal number of conjugate points, because we can surround the sphere infinite times. Hence, for the unit sphere S^n, we have $m = \infty$. In conclusion, in the case $p \in (-\infty, 0) \cup (0, 1)$, the $p-$energy functional on the sphere has no global extremum points [UN99].

Symplectic Manifolds

Symplectic Algebra

Symplectic algebra works in the category of symplectic vector spaces V_i and linear symplectic mappings $t \in L(V_i, V_j)$ [Put93].

Let V be a nD real vector space and $L^2(V, \mathbb{R})$ the space of all bilinear maps from $V \times V$ to $\mathbb{R}$. We say that a *bilinear map* $\omega \in L^2(V, \mathbb{R})$ is *nondegenerate*, i.e., if $\omega(v_1, v_2) = 0$ for all $v_2 \in V$ implies $v_1 = 0$.

If $\{e_1, ..., e_n\}$ is a basis of V and $\{\mathrm{e}^1, ..., \mathrm{e}^n\}$ is the dual basis, $\omega_{ij} = \omega(e_i, e_j)$ is the matrix of ω. A bilinear map $\omega \in L^2(V, \mathbb{R})$ is nondegenerate iff its matrix ω_{ij} is nonsingular. The transpose ω^t of ω is defined by $\omega^t(e_i, e_j) = \omega(e_j, e_i)$. ω is *symmetric* if $\omega^t = \omega$, and *skew–symmetric* if $\omega^t = -\omega$.

Let $A^2(V)$ denote the space of skew–symmetric bilinear maps on V. An element $\omega \in A^2(V)$ is called a 2−form on V. If $\omega \in A^2(V)$ is nondegenerate then in the basis $\{e_1, ..., e_n\}$ its matrix $\omega(e_i, e_j)$ has the form $J = \begin{pmatrix} 0 & I_n \\ -I_n & 0 \end{pmatrix}$.

A *symplectic form* on a real vector space V of dimension $2n$ is a nondegenerate 2−form $\omega \in A^2(V)$. The pair (V, ω) is called a symplectic vector space. If (V_1, ω_1) and (V_2, ω_2) are symplectic vector spaces, a linear map $t \in L(V_1, V_2)$ is a *symplectomorphism* (i.e., a symplectic mapping) iff $t^*\omega_2 = \omega_1$. If (V, ω) is a symplectic vector space, we have an orientation Ω_ω on V given by

$$\Omega_\omega = \frac{(-1)^{\frac{n(n-1)}{2}}}{n!}\omega^n.$$

Let (V, ω) be a $2n$D symplectic vector space and $t \in L(V, V)$ a symplectomorphism. Then t is volume preserving, i.e., $t^*(\Omega_\omega) = \Omega_\omega$, and $\det_{\Omega_\omega}(t) = 1$.

The set of all symplectomorphisms $t : V \to V$ of a $2n$D symplectic vector space (V, ω) forms a group under composition, called the *symplectic group*, denoted by $Sp(V, \omega)$.

In matrix notation, there is a basis of V in which the matrix of ω is $J = \begin{pmatrix} 0 & I_n \\ -I_n & 0 \end{pmatrix}$, such that $J^{-1} = J^t = -J$, and $J^2 = -I$. For $t \in L(V, V)$ with matrix $T = [T^i_j]$ relative to this basis, the condition $t \in Sp(V, \omega)$, i.e., $t^*\omega = \omega$, becomes

$$T^t J T = J.$$

In general, by definition a matrix $A \in M_{2n \times 2n}(\mathbb{R})$ is symplectic iff $A^t J A = J$.

Let (V, ω) be a symplectic vector space, $t \in Sp(V, \omega)$ and $\lambda \in \mathbb{C}$ an eigenvalue of t. Then λ^{-1}, $\bar{\lambda}$ and $\bar{\lambda}^{-1}$ are eigenvalues of t.

Symplectic Geometry

Symplectic geometry is a globalization of symplectic algebra [Put93]; it works in the category **Symplec** of symplectic manifolds M and symplectic diffeomorphisms f. The phase–space of a conservative dynamical system is a symplectic manifold, and its time evolution is a one–parameter family of symplectic diffeomorphisms.

A *symplectic form* or a symplectic structure on a smooth (*i.e.*, C^∞) manifold M is a nondegenerate closed 2−form ω on M, i.e., for each $x \in M$ $\omega(x)$ is nondegenerate, and $d\omega = 0$. A *symplectic manifold* is a pair (M, ω) where M is a smooth $2n$D manifold and ω is a symplectic form on it. If (M_1, ω_1) and (M_2, ω_2) are symplectic manifolds then a smooth map $f : M_1 \to M_2$ is called *symplectic map* or *canonical transformation* if $f^*\omega_2 = \omega_1$.

For example, any symplectic vector space (V, ω) is also a symplectic manifold; the requirement $d\omega = 0$ is automatically satisfied since ω is a constant map. Also, any orientable, compact surface Σ is a symplectic manifold; any nonvanishing 2−form (volume element) ω on Σ is a symplectic form on Σ.

If (M, ω) is a symplectic manifold then it is orientable with the standard volume form

$$\Omega_\omega = \frac{(-1)^{\frac{n(n-1)}{2}}}{n!}\omega^n,$$

If $f : M \to M$ is a symplectic map, then f is volume preserving, $\det_{\Omega_\omega}(f) = 1$ and f is a local diffeomorphism.

In general, if (M, ω) is a $2n$D compact symplectic manifold then ω^n is a volume element on M, so the *de Rham cohomology class* $[\omega^n] \in H^{2n}(M, \mathbb{R})$ is nonzero. Since $[\omega^n] = [\omega]^n$, $[\omega] \in H^2(M, \mathbb{R})$ and all of its powers through the nth must be nonzero as well. The existence of such an element of $H^2(M, \mathbb{R})$ is a *necessary condition* for the compact manifold to admit a symplectic structure.

However, if M is a $2n$D compact manifold without boundary, then there does not exist any exact symplectic structure, $\omega = d\theta$ on M, as its total volume is zero (by Stokes' Theorem),

$$\int_M \Omega_\omega = \frac{(-1)^{\frac{n(n-1)}{2}}}{n!}\int_M \omega^n = \frac{(-1)^{\frac{n(n-1)}{2}}}{n!}\int_M d(\theta \wedge \omega^{n-1}) = 0.$$

For example, spheres S^{2n} do not admit a symplectic structure for $n \geq 2$, since the second de Rham group vanishes, i.e., $H^2(S^{2n}, \mathbb{R}) = 0$. This argument applies to any compact manifold without boundary and having $H^2(M, \mathbb{R}) = 0$.

In mechanics, the phase–space is the cotangent bundle T^*M of a configuration space M. There is a natural symplectic structure on T^*M that is usually defined as follows. Let M be a smooth nD manifold and pick local coordinates $\{dq^1, ..., dq^n\}$. Then $\{dq^1, ..., dq^n\}$ defines a basis of the tangent space T_q^*M, and by writing $\theta \in T_q^*M$ as $\theta = p_i dq^i$ we get local coordinates $\{q^1, ..., q^n, p_1, ..., p_n\}$ on T^*M. Define the canonical symplectic form ω on T^*M by

$$\omega = dp_i \wedge dq^i.$$

This 2−form ω is obviously independent of the choice of coordinates $\{q^1, ..., q^n\}$ and independent of the base point $\{q^1, ..., q^n, p_1, ..., p_n\} \in T_q^*M$; therefore, it is locally constant, and so $d\omega = 0$.

The canonical 1−form θ on T^*M is the unique 1−form with the property that, for any 1−form β which is a section of T^*M we have $\beta^*\theta = \theta$.

Let $f : M \to M$ be a diffeomorphism. Then T^*f preserves the canonical 1−form θ on T^*M, i.e., $(T^*f)^*\theta = \theta$. Thus T^*f is symplectic diffeomorphism.

If (M, ω) is a $2n$D symplectic manifold then about each point $x \in M$ there are local coordinates $\{q^1, ..., q^n, p_1, ..., p_n\}$ such that $\omega = dp_i \wedge dq^i$. These coordinates are called canonical or symplectic. By the Darboux Theorem, ω is constant in this local chart, i.e., $d\omega = 0$.

Momentum Map and Symplectic Reduction

Let (M, ω) be a connected symplectic manifold and $\phi : G \times M \to M$ a symplectic action of the Lie group G on M, that is, for each $g \in G$ the map $\phi_g : M \to M$ is a symplectic diffeomorphism. If for each $\xi \in \mathfrak{g}$ there exists a globally defined function $\hat{J}(\xi) : M \to \mathbb{R}$ such that $\xi_M = X_{\hat{J}(\xi)}$, then the map $J : M \to \mathfrak{g}^*$, given by

$$J : x \in M \mapsto J(x) \in \mathfrak{g}^*, \qquad J(x)(\xi) = \hat{J}(\xi)(x)$$

is called the *momentum map* for ϕ [MR99, Put93].

Since ϕ is symplectic, $\phi_{\exp(t\xi)}$ is a one–parameter family of canonical transformations, i.e., $\phi^*_{\exp(t\xi)}\omega = \omega$, hence ξ_M is locally Hamiltonian and not generally Hamiltonian. That is why not every symplectic action has a momentum map. $\phi : G \times M \to M$ is Hamiltonian iff $\hat{J} : \mathfrak{g} \to C^\infty(M, \mathbb{R})$ is a Lie algebra homomorphism.

Let $H : M \to \mathbb{R}$ be G–-invariant, that is $H\left(\phi_g(x)\right) = H(x)$ for all $x \in M$ and $g \in G$. Then $\hat{J}(\xi)$ is a constant of motion for dynamics generated by H.

Let ϕ be a symplectic action of G on (M, ω) with the momentum map J. Suppose $H : M \to \mathbb{R}$ is G–-invariant under this action. Then the Noether's Theorem states that J is a constant of motion of H, i.e., $J \circ \phi_t = J$, where ϕ_t is the flow of X_H.

A *Hamiltonian action* is a symplectic action with an Ad^*–equivariant momentum map J, i.e.,

$$J\left(\phi_g(x)\right) = Ad^*_{g^{-1}}\left(J(x)\right),$$

for all $x \in M$ and $g \in G$.

Let ϕ be a symplectic action of a Lie group G on (M, ω). Assume that the symplectic form ω on M is exact, i.e., $\omega = d\theta$, and that the action ϕ of G on M leaves the one form $\theta \in M$ invariant. Then $J : M \to \mathfrak{g}^*$ given by $(J(x))(\xi) = \left(i_{\xi_M}\theta\right)(x)$ is an Ad^*–equivariant momentum map of the action.

In particular, in the case of the cotangent bundle $(M = T^*M,\ \omega = d\theta)$ of a mechanical configuration manifold M, we can *lift* up an action ϕ of a Lie group G on M to get an action of G on T^*M. To perform this lift, let G act on M by transformations $\phi_g : M \to M$ and define the *lifted action* to the cotangent bundle by $(\phi_g)_* : T^*M \to T^*M$ by pushing forward one forms, $(\phi_g)_*(\alpha) \cdot v = \alpha\left(T\phi_g^{-1}v\right)$,where $\alpha \in T^*_qM$ and $v \in T_{\phi_g(q)}M$. The lifted action $(\phi_g)_*$ preserves the canonical one form θ on T^*M and the momentum map for $(\phi_g)_*$ is given by

$$J : T^*M \to \mathfrak{g}^*, \qquad J\left(\alpha_q\right)(\xi) = \alpha_q\left(\xi_M(q)\right).$$

For example, let $M = \mathbb{R}^n$, $G = \mathbb{R}^n$ and let G act on $\mathbb{R}^n$ by translations:

$$\phi : (t, q) \in \mathbb{R}^n \times \mathbb{R}^n \mapsto t + q \in \mathbb{R}^n.$$

Then $\mathfrak{g} = \mathbb{R}^n$ and for each $\xi \in \mathfrak{g}$ we have $\xi_{\mathbb{R}^n}(q) = \xi$.

In case of the group of rotations in $\mathbb{R}^3$, $M = \mathbb{R}^3$, $G = SO(3)$ and let G act on $\mathbb{R}^3$ by $\phi(A, q) = A \cdot q$. Then $\mathfrak{g} \simeq \mathbb{R}^3$ and for each $\xi \in \mathfrak{g}$ we have $\xi_{\mathbb{R}^3}(q) = \xi \times q$.

Let G act transitively on (M, ω) by a Hamiltonian action. Then $J(M) = \{Ad^*_{g^{-1}}(J(x)) \,|g \in G\}$ is a coadjoint orbit.

Now, let (M, ω) be a symplectic manifold, G a Lie group and $\phi : G \times M \to M$ a Hamiltonian action of G on M with Ad^*–equivariant momentum map $J : M \to \mathfrak{g}^*$. Let $\mu \in \mathfrak{g}^*$ be a regular value of J; then $J^{-1}(\mu)$ is a submanifold of M such that $\dim\left(J^{-1}(\mu)\right) = \dim(M) - \dim(G)$. Let $G_\mu = \{g \in G | Ad^*_g \mu = \mu\}$ be the isotropy subgroup of μ for the coadjoint action. By Ad^*–equivariance, if $x \in J^{-1}(\mu)$ then $\phi_g(x) = J^{-1}(\mu)$ for all $g \in G$, i.e., $J^{-1}(\mu)$ is invariant under the induced G_μ–action and we can form the quotient space $M_\mu = J^{-1}(\mu)/G_\mu$, called the *reduced phase–space* at $\mu \in \mathfrak{g}^*$.

Let (M, ω) be a symplectic $2n$D manifold and let $f_1, ..., f_k$ be k functions in involution, i.e., $\{f_i, f_j\}_\omega = 0$, $i = 1, ..., k$. Because the flow of X_{f_i} and X_{f_j} commute, we can use them to define a symplectic action of $G = \mathbb{R}^k$ on M. Here $\mu \in \mathbb{R}^k$ is in the range space of $f_1 \times ... \times f_k$ and $J = f_1 \times ... \times f_k$ is the momentum map of this action. Assume that $\{df_1, ..., df_k\}$ are independent at each point, so μ is a regular value for J. Since G is Abelian, $G_\mu = G$ so we get a symplectic manifold $J^{-1}(\mu)/G$ of dimension $2n - 2k$. If $k = n$ we have *integrable systems.*

For example, let $G = SO(3)$ and $(M, \omega) = \left(\mathbb{R}^6, \sum_{i=1}^3 dp_i \wedge dq^i\right)$, and the action of G on $\mathbb{R}^6$ is given by $\phi : (R, (q, p)) \mapsto (R_q, R_p)$. Then the momentum map is the well known angular momentum and for each $\mu \in \mathfrak{g}^* \simeq \mathbb{R}^3 \mu \neq 0$, $G_\mu \simeq S^1$ and the reduced phase–space (M_μ, ω_μ) is $(T^*\mathbb{R}, \omega = dp_i \wedge dq^i)$, so that $\dim(M_\mu) = \dim(M) - \dim(G) - \dim(G_\mu)$. This reduction is in celestial mechanics called by Jacobi 'the elimination of the nodes'.

The equations of motion: $\dot{f} = \{f, H\}_\omega$ on M reduce to the equations of motion: $\dot{f}_\mu = \{f_\mu, H_\mu\}_{\omega_\mu}$on M_μ (see [MR99]).

4.1.5 Hamilton–Poisson Geometry and Human Biodynamics

Now, instead of using symplectic structures arising in Hamiltonian biodynamics, we propose the more general *Poisson manifold* $(\mathbf{g}^*, \{F, G\})$. Here $\mathbf{g}^*$ is a chosen Lie algebra with a $(\pm)$ *Lie–Poisson bracket* $\{F, G\}_\pm(\mu))$ and carries an abstract *Poisson evolution equation* $\dot{F} = \{F, H\}$. This approach is well–defined in both the finite– and the infinite–dimensional case. It is equivalent to the strong symplectic approach when this exists and offers a viable formulation for Poisson manifolds which are not symplectic (for technical details, see see [Wei90, AMR88, MR99, Put93, IP01a]).

Let E_1 and E_2 be Banach spaces. A continuous bilinear functional $<,>$: $E_1 \times E_2 \longrightarrow \mathbb{R}$ is nondegenerate if $< x, y > = 0$ implies $x = 0$ and $y = 0$ for all $x \in E_1$ and $y \in E_2$. We say E_1 and E_2 are in *duality* if there is a nondegenerate bilinear functional $<,>: E_1 \times E_2 \longrightarrow \mathbb{R}$. This functional is also referred to as an L^2-*pairing* of E_1 with E_2.

Recall that a *Lie algebra* consists of a vector space $\mathbf{g}$ (usually a Banach space) carrying a bilinear skew–symmetric operation $[,] : \mathbf{g} \times \mathbf{g} \to \mathbf{g}$, called the *commutator* or Lie bracket. This represents a pairing $[\xi, \eta] = \xi\eta - \eta\xi$ of elements $\xi, \eta \in \mathbf{g}$ and satisfies *Jacobi identity*

$$[[\xi, \eta], \mu] + [[\eta, \mu], \xi] + [[\mu, \xi], \eta] = 0.$$

Let $\mathbf{g}$ be a (finite– or infinite–dimensional) Lie algebra and $\mathbf{g}^*$ its dual Lie algebra, that is, the vector space L^2 paired with $\mathbf{g}$ *via* the inner product $<,>: \mathbf{g}^* \times \mathbf{g} \to \mathbb{R}$. If $\mathbf{g}$ is finite–dimensional, this pairing reduces to the usual action (interior product) of forms on vectors. The standard way of describing any finite–dimensional Lie algebra $\mathbf{g}$ is to provide its n^3 *structural constants* γ_{ij}^k, defined by $[\xi_i, \xi_j] = \gamma_{ij}^k \xi_k$, in some basis ξ_i, $(i = 1, \dots, n)$

For any two smooth functions $F : \mathbf{g}^* \to \mathbb{R}$, we define the *Fréchet derivative* D on the space $L(\mathbf{g}^*, \mathbb{R})$ of all linear diffeomorphisms from $\mathbf{g}^*$ to $\mathbb{R}$ as a map $DF : \mathbf{g}^* \to L(\mathbf{g}^*, \mathbb{R})$; $\mu \mapsto DF(\mu)$. Further, we define the *functional derivative* $\delta F/\delta\mu \in \mathbf{g}$ by

$$DF(\mu) \cdot \delta\mu = < \delta\mu, \frac{\delta F}{\delta\mu} >$$

with arbitrary 'variations' $\delta\mu \in \mathbf{g}^*$.

For any two smooth functions $F, G : \mathbf{g}^* \to \mathbb{R}$, we define the $(\pm)$ *Lie–Poisson bracket* by

$$\{F, G\}_{\pm}(\mu) = \pm < \mu, \left[\frac{\delta F}{\delta\mu}, \frac{\delta G}{\delta\mu}\right] > . \tag{3.1}$$

Here $\mu \in \mathbf{g}^*$, $[\xi, \mu]$ is the Lie bracket in $\mathbf{g}$ and $\delta F/\delta\mu$, $\delta G/\delta\mu \in \mathbf{g}$ are the functional derivatives of F and G.

The $(\pm)$ Lie–Poisson bracket (3.1) is clearly a bilinear and skew–symmetric operation. It also satisfies the Jacobi identity

$$\{\{F, G\}, H\}_{\pm}(\mu) + \{\{G, H\}, F\}_{\pm}(\mu) + \{\{H, F\}, G\}_{\pm}(\mu) = 0$$

thus confirming that $\mathbf{g}^*$ is a Lie algebra, as well as Leibniz' rule

$$\{FG, H\}_{\pm}(\mu) = F\{G, H\}_{\pm}(\mu) + G\{F, H\}_{\pm}(\mu). \tag{4.41}$$

If $\mathbf{g}$ is a finite–dimensional phase–space manifold with structure constants γ_{ij}^k, the $(\pm)$ Lie–Poisson bracket (4.41) becomes

$$\{F, G\}_{\pm}(\mu) = \pm\mu_k \gamma_{ij}^k \frac{\delta F}{\delta\mu_i} \frac{\delta G}{\delta\mu_j}. \tag{4.42}$$

The $(\pm)$ Lie–Poisson bracket represents a Lie–algebra generalization of the classical finite–dimensional Poisson bracket $[F, G] = \omega(X_f, X_g)$ on the symplectic phase–space manifold (P, ω) for any two real–valued smooth functions $F, G : P \longrightarrow \mathbb{R}$.

As in the classical case, any two smooth functions $F, G : \mathbf{g}^* \longrightarrow \mathbb{R}$ are *in involution* if $\{F, G\}_\pm(\mu) = 0$.

The Lie–Poisson Theorem states that a Lie algebra $\mathbf{g}^*$ with a $\pm$ Lie–Poisson bracket $\{F, G\}_\pm(\mu)$ represents a Poisson manifold $(\mathbf{g}^*, \{F, G\}_\pm(\mu))$.

Given a smooth Hamiltonian function $H : \mathbf{g}^* \to \mathbb{R}$ on the Poisson manifold $(\mathbf{g}^*, \{F, G\}_\pm(\mu))$, the time evolution of any smooth function $F : \mathbf{g}^* \to \mathbb{R}$ is given by the abstract *Poisson evolution equation*

$$\dot{F} = \{F, H\}. \tag{4.43}$$

Hamilton–Poisson Biodynamic Systems

Let $(P, \{\})$ be a Poisson manifold and $H \in C^k(P, \mathbb{R})$ a smooth real valued function on P. The vector–field X_H defined by

$$X_H(F) = \{F, H\},$$

is the Hamiltonian vector–field with energy function H. The triple $(P, \{\}, H)$ we call the *Hamilton–Poisson biodynamic system* (HPBS) [MR99, Put93, IP01a]. The map $F \mapsto \{F, H\}$ is a derivation on the space $C^k(P, \mathbb{R})$, hence it defines a vector–field on P. The map $F \in C^k(P, \mathbb{R}) \mapsto X_F \in \mathcal{X}(P)$ is a Lie algebra anti–homomorphism, i.e., $[X_F, X_g] = -X_{\{F,g\}}$.

Let $(P, \{\}, H)$ be a HPBS and ϕ_t the flow of X_H. Then for all $F \in C^k(P, \mathbb{R})$ we have the *conservation of energy*:

$$H \circ \phi_t = H,$$

and the *equations of motion in Poisson bracket form*,

$$\frac{d}{dt}(F \circ \phi_t) = \{F, H\} \circ \phi_t = \{F \circ \phi_t, H\},$$

that is, the above Poisson evolution equation (4.43) holds. Now, the function F is constant along the integral curves of the Hamiltonian vector–field X_H iff

$$\{F, H\} = 0.$$

ϕ_t preserves the Poisson structure.

Next we present two main examples of HPBS.

'Ball–and–Socket' Joint Dynamics in Euler Vector Form.

The dynamics of human body–segments, classically modelled *via* Lagrangian formalism (see [Hat77b, Iva91]), may be also prescribed by Euler's equations of rigid body dynamics. The equations of motion for a free rigid body, described by an observer fixed on the moving body, are usually given by *Euler's vector equation*

$$\dot{p} = p \times w. \tag{4.44}$$

Here $p, w \in \mathbb{R}^3$, $p_i = I_i w_i$ and I_i $(i = 1, 2, 3)$ are the principal moments of inertia, the coordinate system in the segment is chosen so that the axes are principal axes, w is the angular velocity of the body and p is the corresponding angular momentum.

The kinetic energy of the segment is the Hamiltonian function $H : \mathbb{R}^3 \to \mathbb{R}$ given by [IP01a]

$$H(p) = \frac{1}{2} p \cdot w$$

and is a conserved quantity for (4.44).

The vector space $\mathbb{R}^3$ is a Lie algebra with respect to the bracket operation given by the usual cross product. The space $\mathbb{R}^3$ is paired with itself *via* the usual dot product. So if $F : \mathbb{R}^3 \to \mathbb{R}$, then $\delta F/\delta p = \nabla F(p)$ and the (−) Lie–Poisson bracket $\{F, G\}_-(p)$ is given *via* (4.42) by the triple product

$$\{F, G\}_-(p) = -p \cdot (\nabla F(p) \times \nabla G(p)).$$

Euler's vector equation (4.44) represents a generalized Hamiltonian system in $\mathbb{R}^3$ relative to the Hamiltonian function $H(p)$ and the (−) Lie–Poisson bracket $\{F, G\}_-(p)$. Thus the Poisson manifold $(\mathbb{R}^3, \{F, G\}_-(p))$ is defined and the abstract Poisson equation is equivalent to Euler's equation (4.44) for a body segment and associated joint.

Solitary Model of Muscular Contraction.

The basis of the molecular model of muscular contraction is oscillations of Amid I peptide groups with associated dipole electric momentum inside a spiral structure of myosin filament molecules (see [Dav81, Dav91]).

There is a simultaneous resonant interaction and strain interaction generating a collective interaction directed along the axis of the spiral. The resonance excitation jumping from one peptide group to another can be represented as an exciton, the local molecule strain caused by the static effect of excitation as a phonon and the resultant collective interaction as a *soliton*.

The simplest model of Davydov's solitary particle–waves is given by the *nonlinear Schrödinger equation* [IP01a]

$$i\partial_t \psi = -\partial_{x^2} \psi + 2\chi |\psi|^2 \psi \tag{4.45}$$

for $-\infty < x < +\infty$. Here $\psi(x, t)$ is a smooth complex–valued wave function with initial condition $\psi(x, t)|_{t=0} = \psi(x)$ and χ is a nonlinear parameter. In the linear limit ($\chi = 0$) (4.45) becomes the ordinary Schrödinger equation for the wave function of the free 1D particle with mass $m = 1/2$ (see section 4.3 below).

We may define the infinite–dimensional phase–space manifold $\mathcal{P} = \{(\psi, \bar{\psi}) \in S(\mathbb{R}, \mathbf{C})\}$, where $S(\mathbb{R}, \mathbf{C})$ is the Schwartz space of rapidly–decreasing complex–valued functions defined on $\mathbb{R}$). We define also the algebra $\boldsymbol{\chi}(\mathcal{P})$ of observables on $\mathcal{P}$ consisting of real–analytic functional derivatives $\delta F/\delta\psi$, $\delta F/\delta\bar{\psi} \in S(\mathbb{R}, \mathbf{C})$.

The Hamiltonian function $H : \mathcal{P} \longrightarrow \mathbb{R}$ is given by

$$H(\psi) = \int_{-\infty}^{+\infty} \left(\left| \frac{\partial \psi}{\partial x} \right|^2 + \chi |\psi|^4 \right) dx$$

and is equal to the total energy of the soliton. It is a conserved quantity for (4.3) (see [Sei95]).

The Poisson bracket on $\boldsymbol{\chi}(\mathcal{P})$ represents a direct generalization of the classical finite–dimensional Poisson bracket

$$\{F, G\}_+(\psi) = i \int_{-\infty}^{+\infty} \left(\frac{\delta F}{\delta \psi} \frac{\delta G}{\delta \bar{\psi}} - \frac{\delta F}{\delta \bar{\psi}} \frac{\delta G}{\delta \psi} \right) dx. \qquad (4.46)$$

It manifestly exhibits skew–symmetry and satisfies Jacobi identity. The functionals are given by $\delta F/\delta\psi = -i\{F, \bar{\psi}\}$ and $\delta F/\delta\bar{\psi} = i\{F, \psi\}$. Therefore the algebra of observables $\boldsymbol{\chi}(\mathcal{P})$ represents the Lie algebra and the Poisson bracket is the (+) Lie–Poisson bracket $\{F, G\}_+(\psi)$.

The nonlinear Schrödinger equation (4.45) for the solitary particle–wave is a Hamiltonian system on the Lie algebra $\boldsymbol{\chi}(\mathcal{P})$ relative to the (+) Lie–Poisson bracket $\{F, G\}_+(\psi)$ and Hamiltonian function $H(\psi)$. Therefore the Poisson manifold $(\boldsymbol{\chi}(\mathcal{P}), \{F, G\}_+(\psi))$ is defined and the abstract Poisson evolution equation (4.43), which holds for any smooth function $F : \boldsymbol{\chi}(\mathcal{P}) \to \mathbb{R}$, is equivalent to equation (4.45).

A more subtle model of soliton dynamics is provided by the *Korteveg–De Vries equation* [IP01a]

$$f_t - 6 f f_x + f_{xxx} = 0, \qquad (f_x = \partial_x f) \qquad (4.47)$$

where $x \in \mathbb{R}$ and f is a real–valued smooth function defined on $\mathbb{R}$. This equation is related to the ordinary Schrödinger equation by the inverse scattering method [Sei95, IP01a].

We may define the infinite–dimensional phase–space manifold $\mathcal{V} = \{f \in S(\mathbb{R})\}$, where $S(\mathbb{R})$ is the Schwartz space of rapidly–decreasing real–valued functions $\mathbb{R}$). We define further $\boldsymbol{\chi}(\mathcal{V})$ to be the algebra of observables consisting of functional derivatives $\delta F/\delta f \in S(\mathbb{R})$.

The Hamiltonian $H : \mathcal{V} \to \mathbb{R}$ is given by

$$H(f) = \int_{-\infty}^{+\infty} \left(f^3 + \frac{1}{2} f_x^2 \right) dx$$

and provides the total energy of the soliton. It is a conserved quantity for (4.47) (see [Sei95]).

As a real–valued analogue to (4.46), the (+) Lie–Poisson bracket on $\chi(\mathcal{V})$ is given *via* (4.41) by

$$\{F,G\}_+(f) = \int_{-\infty}^{+\infty} \frac{\delta F}{\delta f}\frac{d}{dx}\frac{\delta G}{\delta f}dx.$$

Again it possesses skew–symmetry and satisfies Jacobi identity. The functionals are given by $\delta F/\delta f = \{F,f\}$.

The Korteveg–De Vries equation (KdV1), describing the behavior of the molecular solitary particle–wave, is a Hamiltonian system on the Lie algebra $\chi(\mathcal{V})$ relative to the (+) Lie–Poisson bracket $\{F,G\}_+(f)$ and the Hamiltonian function $H(f)$. Therefore, the Poisson manifold $(\chi(\mathcal{V}),\{F,G\}_+(f))$ is defined and the abstract Poisson evolution equation (4.43), which holds for any smooth function $F:\chi(\mathcal{V})\to\mathbb{R}$, is equivalent to (4.47).

Finally, it is clear that the two solitary equations, (4.47) and (4.45), have a quantum–mechanical origin. By the use of the first quantization method, every classical biodynamic observable F is represented in the Hilbert space $L^2(\psi)$ of square–integrable complex ψ–functions by a Hermitian (self–adjoint) linear operator $\hat{F}$ with real eigenvalues. The classical Poisson bracket $\{F,G\} = K$ corresponds to the *quantum commutator* $[\hat{F},\hat{G}] = i\hbar\hat{K}$. Therefore the classical evolution equation (4.43) corresponds, in the *Heisenberg picture*, to the *quantum evolution equation*

$$i\hbar\dot{\hat{F}} = [\hat{F},\hat{H}],$$

for any representative operator $\hat{F}$ and quantum Hamiltonian $\hat{H}$. By Ehrenfest's Theorem, this equation is also valid for expectation values $<\cdot>$ of observables, that is,

$$i\hbar < \dot{\hat{F}} > = < [\hat{F},\hat{H}] > .$$

4.2 Complex Manifolds

Just as a smooth manifold has enough structure to define the notion of differentiable functions, a *complex manifold* is one with enough structure to define the notion of holomorphic (or, analytic) functions $f:X\to\mathbb{C}$. Namely, if we demand that the transition functions $\phi_j\circ\phi_i^{-1}$ in the charts U_i on M (see Figure 4.8) satisfy the *Cauchy–Riemann equations*

$$\partial_x u = \partial_y v, \qquad \partial_y u = -\partial_x v,$$

then the analytic properties of f can be studied using its coordinate representative $f\circ\phi_i^{-1}$ with assurance that the conclusions drawn are patch independent. Introducing local complex coordinates in the charts U_i on M, the ϕ_i

can be expressed as maps from U_i to an open set in $\mathbb{C}^{\frac{n}{2}}$, with $\phi_j \circ \phi_i^{-1}$ being a holomorphic map from $\mathbb{C}^{\frac{n}{2}}$ to $\mathbb{C}^{\frac{n}{2}}$. Clearly, n must be even for this to make sense. In local complex coordinates, we recall that a function $h : \mathbb{C}^{\frac{n}{2}} \to \mathbb{C}^{\frac{n}{2}}$ is *holomorphic* if $h(z^1, \bar{z}^1, ..., z^{\frac{n}{2}}, \bar{z}^{\frac{n}{2}})$ is actually independent of all the $\bar{z}^j$.

In a given patch on any even–dimensional manifold, we can always introduce local complex coordinates by, for instance, forming the combinations $z^j = x^j + ix^{\frac{n}{2}+j}$, where the x^j are local real coordinates on M. The real test is whether the transition functions from one patch to another—when expressed in terms of the local complex coordinates—are holomorphic maps. If they are, we say that M is a complex manifold of complex dimension $d = n/2$. The local complex coordinates with holomorphic transition functions give M with a *complex structure* (see [Gre96]).

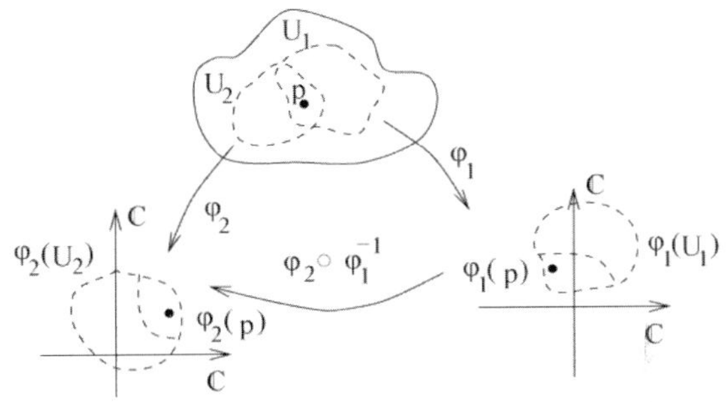

Fig. 4.8. The charts for a complex manifold M have complex coordinates (see text for explanation).

Given a smooth manifold with even real dimension n, it can be a difficult question to determine whether or not a complex structure exists. On the other hand, if some smooth manifold M does admit a complex structure, we are not able to decide whether it is unique, i.e., there may be numerous inequivalent ways of defining complex coordinates on M.

Now, in the same way as a *homeomorphism* defines an equivalence between topological manifolds, and a *diffeomorphism* defines an equivalence between smooth manifolds, a *biholomorphism* defines an equivalence between complex manifolds. If M and N are complex manifolds, we consider them to be equivalent if there is a map $\phi : M \to N$ which in addition to being a diffeomorphism, is also a holomorphic map. That is, when expressed in terms of the complex structures on M and N respectively, ϕ is holomorphic. It is not hard to show that this necessarily implies that ϕ^{-1} is holomorphic as well and hence ϕ is known as a biholomorphism. Such a map allows us to identify the complex structures on M and N and hence they are isomorphic as complex manifolds.

These definitions are important because there are pairs of smooth manifolds M and N which are homeomorphic but not diffeomorphic, as well as, there are complex manifolds M and N which are diffeomorphic but not biholomorphic. This means that if one simply ignored the fact that M and N admit local complex coordinates (with holomorphic transition functions), and one only worked in real coordinates, there would be no distinction between M and N. The difference between them only arises from the way in which complex coordinates have been laid down upon them.

Again, recall that a *tangent space* to a manifold M *at a point* p is the closest flat approximation to M at that point. A convenient basis for the tangent space of M at p consists of the n linearly independent partial derivatives,

$$T_pM : \{\partial_{x^1}|_p, ..., \partial_{x^n}|_p\}. \tag{4.48}$$

A vector $v \in T_pM$ can then be expressed as $v = v^\alpha \partial_{x^\alpha}|_p$.

Also, a convenient basis for the dual, *cotangent space* T_p^*M, is the basis of one–forms, which is dual to (4.48) and usually denoted by

$$T_p^*M : \{dx^1|_p, ..., dx^n|_p\}, \tag{4.49}$$

where, by definition, $dx^i : T_pM \to \mathbb{R}$ is a linear map with $dx_p^i(\partial_{x^j}|_p) = \delta_j^i$.

Now, if M is a complex manifold of complex dimension $d = n/2$, there is a notion of the *complexified tangent space* of M, denoted by $T_pM^{\mathbb{C}}$, which is the same as the real tangent space T_pM except that we allow complex coefficients to be used in the vector space manipulations. This is often denoted by writing $T_pM^{\mathbb{C}} = T_pM \otimes \mathbb{C}$. We can still take our basis to be as in (4.48) with an arbitrary vector $v \in T_pM^{\mathbb{C}}$ being expressed as $v = v^\alpha \frac{\partial}{\partial x^\alpha}|_p$, where the v^α can now be complex numbers. In fact, it is convenient to rearrange the basis vectors in (4.48) to more directly reflect the underlying complex structure. Specifically, we take the following linear combinations of basis vectors in (4.48) to be our new basis vectors:

$$\begin{aligned} T_pM^{\mathbb{C}} \quad &: \quad \{(\partial_{x^1} + \mathrm{i}\partial_{x^{d+1}})|_p, ..., \\ &(\partial_{x^d} + \mathrm{i}\partial_{x^{2D}})|_p, (\partial_{x^1} - \mathrm{i}\partial_{x^{d+1}})|_p, ..., (\partial_{x^d} - \mathrm{i}\partial_{x^{2D}})|_p\}. \end{aligned} \tag{4.50}$$

In terms of complex coordinates we can write the basis (4.50) as

$$T_pM^{\mathbb{C}} : \{\partial_{z^1}|_p, ..., \partial_{z^d}|_p, \partial_{\bar{z}^1}|_p, ..., \partial_{\bar{z}^d}|_p\}.$$

From the point of view of real vector spaces, $\partial_{x^j}|_p$ and $i\partial_{x^j}|_p$ would be considered linearly independent and hence $T_pM^{\mathbb{C}}$ has real dimension $4D$.

In exact analogy with the real case, we can define the dual to $T_pM^{\mathbb{C}}$, which we denote by $T_p^*M^{\mathbb{C}} = T_p^*M \otimes \mathbb{C}$, with the one–forms basis

$$T_p^*M^{\mathbb{C}} : \{dz^1|_p, ..., dz^d|_p, d\bar{z}^1|_p, ..., d\bar{z}^d|_p\}.$$

For certain types of complex manifolds M, it is worthwhile to refine the definition of the complexified tangent and cotangent spaces, which pulls apart the holomorphic and anti–holomorphic directions in each of these two vector spaces. That is, we can write

$$T_pM^{\mathbb{C}} = T_pM^{(1,0)} \oplus T_pM^{(0,1)},$$

where $T_pM^{(1,0)}$ is the vector space spanned by $\{\partial_{z^1}|_p, ..., \partial_{z^d}|_p\}$ and $T_pM^{(0,1)}$ is the vector space spanned by $\{\partial_{\bar{z}^1}|_p, ..., \partial_{\bar{z}^d}|_p\}$. Similarly, we can write

$$T_p^*M^{\mathbb{C}} = T_p^*M^{(1,0)} \oplus T_p^*M^{(0,1)},$$

where $T_p^*M^{(1,0)}$ is the vector space spanned by $\{dz^1|_p, ..., dz^d|_p\}$ and $T_p^*M^{(0,1)}$ is the vector space spanned by $\{d\bar{z}^1|_p, ..., d\bar{z}^d|_p\}$. We call $T_pM^{(1,0)}$ the *holomorphic tangent space*; it has complex dimension d and we call $T_p^*M^{1,0}$ the *holomorphic cotangent space*. It also has complex dimension d. Their complements are known as the *anti–holomorphic* tangent and cotangent spaces respectively [Gre96].

Now, a *complex vector bundle* is a vector bundle $\pi : E \to M$ whose fiber bundle $\pi^{-1}(x)$ is a complex vector space. It is not necessarily a complex manifold, even if its base manifold M is a complex manifold. If a complex vector bundle also has the structure of a complex manifold, and is holomorphic, then it is called a holomorphic vector bundle.

4.2.1 Complex Metrics: Hermitian and Kähler

If M is a complex manifold, there is a natural extension of the metric g to a map

$$g : T_pM^{\mathbb{C}} \times T_pM^{\mathbb{C}} \to \mathbb{C},$$

defined in the following way. Let r, s, u, v be four vectors in the tangent space T_pM to a complex manifold M. Using them, we can construct, for example, two vectors $w_{(1)} = r + is$ and $w_{(2)} = u + iv$ which lie in $T_pM^{\mathbb{C}}$. Then, we evaluate g on $w_{(1)}$ and $w_{(2)}$ by linearity:

$$g(w_{(1)}, w_{(2)}) = g(r + is, u + iv) = g(r, u) - g(s, v) + i\left[g(r, v) + g(s, u)\right].$$

We can define components of this extension of the original metric (which we have called by the same symbol) with respect to complex coordinates in the usual way: $g_{ij} = g(\frac{\partial}{\partial z^i}, \frac{\partial}{\partial z^j})$, $g_{i\bar{j}} = g(\frac{\partial}{\partial z^i}, \frac{\partial}{\partial \bar{z}^j})$ and so forth. The reality of our original metric g and its symmetry implies that in complex coordinates we have $g_{ij} = g_{ji}$, $g_{i\bar{j}} = g_{\bar{j}i}$ and $\overline{g_{ij}} = g_{\bar{i}\bar{j}}$, $\overline{g_{i\bar{j}}} = g_{\bar{i}j}$.

Now, recall that a *Hermitian metric* on a complex vector bundle assigns a *Hermitian inner product* to every fiber bundle. The basic example is the trivial bundle $\pi : U \times \mathbb{C}^2 \to U$, where U is an open set in $\mathbb{R}^n$. Then a positive definite Hermitian matrix H defines a Hermitian metric by

$$\langle v, w\rangle = v^T H \bar{w},$$

where $\bar{w}$ is the complex conjugate of w. By a partition of unity, any complex vector bundle has a Hermitian metric.

In local coordinates of a complex manifold M, a metric g is Hermitian if $g_{ij} = g_{\bar{\imath}\bar{\jmath}} = 0$. In this case, only the mixed type components of g are nonzero and hence it can be written as

$$g = g_{i\bar{\jmath}}\, dz^i \otimes d\bar{z}^{\bar{\jmath}} + g_{\bar{\imath}j}\, d\bar{z}^{\bar{\imath}} \otimes dz^j \ .$$

With a little bit of algebra one can work out the constraint this implies for the original metric written in real coordinates. Formally, if J is a *complex structure* acting on the real tangent space T_pM, i.e.,

$$J : T_pM \to T_pM \qquad \text{with} \qquad J^2 = -I,$$

then the *Hermiticity condition* on g is $g(Jv_{(1)}, Jv_{(2)}) = g(v_{(1)}, v_{(2)})$.

On a holomorphic vector bundle with a Hermitian metric h, there is a unique connection compatible with h and the complex structure. Namely, it must be $\nabla = \partial + \bar{\partial}$.

In the special case of a complex manifold, the complexified tangent bundle $TM \otimes \mathbb{C}$ may have a Hermitian metric, in which case its real part is a Riemannian metric and its imaginary part is a nondegenerate alternating multilinear form ω. When ω is closed, i.e., in this case a symplectic form, then ω is called the *Kähler form*. Formally, given a Hermitian metric g on M, we can build a form in $\Omega^{1,1}(M)$ — that is, a form of type $(1,1)$ in the following way:

$$\omega = ig_{i\bar{\jmath}}\, dz^i \otimes d\bar{z}^{\bar{\jmath}} - ig_{\bar{\jmath}i}\, d\bar{z}^{\bar{\jmath}} \otimes dz^i .$$

By the symmetry of g, we can write this as

$$\omega = ig_{i\bar{\jmath}}\, dz^i \wedge d\bar{z}^{\bar{\jmath}} .$$

Now, if ω is closed, that is, if $dJ = 0$, then ω is called a *Kähler form* and M is called a *Kähler manifold*. At first sight, this *Kählerity condition* might not seem too restrictive. However, it leads to remarkable simplifications in the resulting differential geometry on M.

A *Kähler structure* on a complex manifold M combines a Riemannian metric on the underlying real manifold with the complex structure. Such a structure brings together geometry and complex analysis, and the main examples come from algebraic geometry. When M has n complex dimensions, then it has $2n$ real dimensions. A Kähler structure is related to the unitary group $U(n)$, which embeds in $SO(2n)$ as the orthogonal matrices that preserve the *almost complex structure* (multiplication by i). In a coordinate chart, the complex structure of M defines a multiplication by i and the metric defines orthogonality for tangent vectors. On a *Kähler manifold*, these two notions (and their derivatives) are related.

A Kähler manifold is a complex manifold for which the exterior derivative of the fundamental form ω associated with the given *Hermitian metric* vanishes, so $d\omega = 0$. In other words, it is a complex manifold with a *Kähler structure.* It has a *Kähler form*, so it is also a symplectic manifold. It has a *Kähler metric*, so it is also a Riemannian manifold.

The simplest example of a Kähler manifold is a *Riemann surface*, which is a complex manifold of dimension 1. In this case, the imaginary part of any Hermitian metric must be a closed form since all 2−forms are closed on a real 2D manifold.

In other words, a *Kähler form* is a closed two–form ω on a complex manifold M which is also the negative imaginary part of a Hermitian metric $h = g - iw$. In this case, M is called a Kähler manifold and g, the real part of the Hermitian metric, is called a Kähler metric. The Kähler form combines the metric and the complex structure, $g(M,Y) = \omega(M, JY)$,where ω is the *almost complex structure* induced by multiplication by i. Since the Kähler form comes from a Hermitian metric, it is preserved by ω, since $h(M,Y) = h(JX, JY)$. The equation $d\omega = 0$ implies that the metric and the complex structure are related. It gives M a Kähler structure, and has many implications.

In particular, on $\mathbb{C}^2$, the Kähler form can be written as

$$\omega = -\frac{i}{2}\left(dz_1 \wedge \overline{dz_1} + dz_2 \wedge \overline{dz_2}\right) = dx_1 \wedge dy_1 + dx_2 \wedge dy_2,$$

where $z_n = x_n + i\, y_n$. In general, the Kähler form can be written in coordinates

$$\omega = g_{ij}\, dz_i \wedge \overline{dz_j},$$

where g_{ij} is a Hermitian metric, the real part of which is the Kähler metric. Locally, a Kähler form can be written as $i\partial\bar{\partial}f$, where f is a function called a *Kähler potential.* The Kähler form is a real $(1,1)$−complex form. The Kähler potential is a real–valued function f on a Kähler manifold for which the Kähler form ω can be written as $\omega = i\partial\bar{\partial}f$, where,

$$\partial = \partial_{z_k} dz_k \qquad \text{and} \qquad \bar{\partial} = \partial_{\bar{z}_k} d\bar{z}_k.$$

In local coordinates, the fact that $dJ = 0$ for a Kähler manifold M implies

$$dJ = (\partial + \bar{\partial}) i g_{i\bar{\jmath}}\, dz^i \wedge d\bar{z}^{\bar{\jmath}} = 0.$$

This implies that

$$\partial_{z^l} g_{i\bar{\jmath}} = \partial_{z^i} g_{l\bar{\jmath}} \tag{4.51}$$

and similary with z and $\bar{z}$ interchanged. From this we see that locally we can express $g_{i\bar{\jmath}}$ as

$$g_{i\bar{\jmath}} = \frac{\partial^2 \phi}{\partial z^i \partial \bar{z}^{\bar{\jmath}}}.$$

That is, $\omega = i\partial\bar{\partial}\phi$, where ϕ is a locally defined function in the patch whose local coordinates we are using, which is known as the *Kähler potential.*

If ω on M is a Kähler form, the conditions (4.51) imply that there are numerous cancellations in (4.51). so that the only nonzero *Christoffel symbols* (of the standard *Levi–Civita connection*) in *complex coordinates* are those of the form Γ^l_{jk} and $\Gamma^{\bar{l}}_{\bar{j}\bar{k}}$, with all indices holomorphic or anti–holomorphic. Specifically,

$$\Gamma^l_{jk} = g^{l\bar{s}}\,\partial_{z^j} g_{k\bar{s}} \qquad \text{and} \qquad \Gamma^{\bar{l}}_{\bar{j}\bar{k}} = g^{\bar{l}s}\,\partial_{\bar{z}^j} g_{\bar{k}s}.$$

The curvature tensor also greatly simplifies. The only non–zero components of the *Riemann tensor*, when written in complex coordinates, have the form $R_{i\bar{j}k\bar{l}}$ (up to index permutations consistent with symmetries of the curvature tensor). And we have

$$R_{i\bar{j}k\bar{l}} = g_{i\bar{s}}\,\partial_{z^k}\Gamma^{\bar{s}}_{\bar{j}\bar{l}},$$

as well as the *Ricci tensor*

$$R_{\bar{\imath}j} = R^{\bar{k}}_{\ \bar{\imath}\bar{k}j} = -\partial_{z^j}\Gamma^{\bar{k}}_{\bar{\imath}\bar{k}}.$$

Since the Kähler form ω is closed, it represents a cohomology class in the *de Rham cohomology*. On a compact manifold, it cannot be exact because $\omega^n/n! \neq 0$ is the volume form determined by the metric. In the special case of a projective variety, the Kähler form represents an integral cohomology class. That is, it integrates to an integer on any 1D submanifold, i.e., an algebraic curve. The *Kodaira Embedding Theorem* says that if the Kähler form represents an integral cohomology class on a compact manifold, then it must be a projective variety. There exist Kähler forms which are not projective algebraic, but it is an open question whether or not any Kähler manifold can be deformed to a projective variety (in the compact case).

A Kähler form satisfies *Wirtinger's inequality*,

$$|\omega(M,Y)| \leq |M \wedge Y|,$$

where the r.h.s is the volume of the parallelogram formed by the tangent vectors M and Y. Corresponding inequalities hold for the *exterior powers* of ω. Equality holds iff M and Y form a complex subspace. Therefore, there is a *calibration form*, and the complex submanifolds of a Kähler manifold are *calibrated submanifolds*. In particular, the complex submanifolds are locally volume minimizing in a Kähler manifold. For example, the graph of a holomorphic function is a locally area–minimizing surface in $\mathbb{C}^2 = \mathbb{R}^4$.

Kähler identities is a collection of identities which hold on a Kähler manifold, also called the *Hodge identities*. Let ω be a Kähler form, $d = \partial + \bar{\partial}$ be the exterior derivative, $[A, B] = AB - BA$ be the commutator of two differential operators, and A^* denote the formal adjoint of A. The following operators also act on differential forms α on a Kähler manifold:

$$L(\alpha) = \alpha \wedge \omega, \qquad \Lambda(\alpha) = L^*(\alpha) = \alpha \rfloor \omega, \qquad d_c = -JdJ,$$

where J is the almost complex structure, $J = -I$, and $\rfloor$ denotes the interior product. Then we have

$$[L,\bar\partial]=[L,\partial]=0,\qquad [\Lambda,\bar\partial^*]=[\Lambda,\partial^*]=0,$$
$$[L,\bar\partial^*]=-i\partial,\qquad [L,\partial^*]=i\bar\partial,\qquad [\Lambda,\bar\partial]=-i\partial^*,\qquad [\Lambda,\partial]=-i\bar\partial.$$

These identities have many implications. For example, the two operators

$$\Delta_d=dd^*+d^*d\qquad\text{and}\qquad \Delta_{\bar\partial}=\bar\partial\bar\partial^*+\bar\partial^*\bar\partial$$

(called Laplacians because they are elliptic Laplacian–like operators) satisfy

$$\Delta_d=2\Delta_{\bar\partial}.$$

At this point, assume that M is also a compact manifold. Along with Hodge's Theorem, this equality of Laplacians proves the *Hodge decomposition*. The operators L and Λ commute with these Laplacians. By *Hodge's Theorem*, they act on cohomology, which is represented by *harmonic forms*. Moreover, defining

$$H=[L,\Lambda]=\sum(p+q-n)\,\Pi^{p,q},$$

where $\Pi^{p,q}$ is projection onto the $(p,q)-$Dolbeault cohomology, they satisfy

$$[L,\Lambda]=H,\qquad [H,L]=-2L,\qquad [H,\Lambda]=2L.$$

In other words, these operators give a group representation of the special linear Lie algebra $\mathfrak{sl}_2(\mathbb{C})$ on the complex cohomology of a compact Kähler manifold (*Lefschetz Theorem*).

4.2.2 Dolbeault Cohomology and Hodge Numbers

A generalization of the real–valued de Rham cohomology to complex manifolds is called the *Dolbeault cohomology*. On complex mD manifolds, we have local coordinates z^i and $\bar z^i$. One can now study $(p,q)-$forms, which are forms containing p factors of dz^i and q factors of $d\bar z^j$:

$$\omega=\omega_{i_1\dots i_p,j_1\dots j_q}(z,\bar z)\,dz^{i_1}\wedge\cdots\wedge dz^{i_p}\wedge d\bar z^{j_1}\wedge\cdots\wedge d\bar z^{j_q}.$$

Moreover, one can introduce *two* exterior derivative operators ∂ and $\bar\partial$, where ∂ is defined by

$$\partial\omega\equiv\frac{\partial\omega_{i_1\dots i_p,j_1\dots j_q}}{\partial z^k}\,dz^k\wedge dz^{i_1}\wedge\cdots\wedge dz^{i_p}\wedge d\bar z^{j_1}\wedge\cdots\wedge d\bar z^{j_q},$$

and $\bar\partial$ is defined similarly by differentiating with respect to $\bar z^k$ and adding a factor of $d\bar z^k$. Again, both of these operators square to zero. We can now construct two cohomologies – one for each of these operators – but as we will see, in the cases that we are interested in, the information contained in them is the same. Conventionally, one uses the cohomology defined by the $\bar\partial-$operator.

For complex manifolds, the *Hodge Theorem* also holds: *each cohomology class $H^{p,q}(M)$ contains a unique harmonic form.* Here, a harmonic form ω_h is a form for which the *complex Laplacian*

$$\Delta_{\bar{\partial}} = \bar{\partial}\bar{\partial}^* + \bar{\partial}^*\bar{\partial}$$

has a zero eigenvalue: $\Delta_{\bar{\partial}}\omega_h = 0$. In general, this operator does not equal the ordinary Laplacian, but one can prove that in the case where M is a *Kähler manifold*,

$$\Delta = 2\Delta_{\bar{\partial}} = 2\Delta_{\partial}.$$

In other words, on a Kähler manifold the notion of a harmonic form is the same, independently of which exterior derivative one uses. As a first consequence, we find that the vector spaces $H^{p,q}_{\partial}(M)$ and $H^{p,q}_{\bar{\partial}}(M)$ both equal the vector space of harmonic $(p,q)-$forms, so the two cohomologies are indeed equal. Moreover, every $(p,q)-$form is a $(p+q)-$form in the de Rham cohomology, and by the above result we see that a *harmonic* $(p,q)-$form can also be viewed as a de Rham harmonic $(p+q)-$form. Conversely, any de Rham $p-$form can be written as a sum of Dolbeault forms:

$$\omega_p = \omega_{p,0} + \omega_{p-1,1} + \ldots + \omega_{0,p}. \tag{4.52}$$

Acting on this with the Laplacian, one sees that for a harmonic $p-$form,

$$\Delta\omega_p = \Delta_{\bar{\partial}}\omega_p = \Delta_{\bar{\partial}}\omega_{p,0} + \Delta_{\bar{\partial}}\omega_{p-1,1} + \ldots + \Delta_{\bar{\partial}}\omega_{0,p} = 0.$$

Since $\Delta_{\bar{\partial}}$ does not change the degree of a form, $\Delta_{\bar{\partial}}\omega_{p_1,p_2}$ is also a $(p_1,p_2)-$form. Therefore, the r.h.s. can only vanish if each term vanishes separately, so all the terms on the r.h.s. of (4.52) must be harmonic forms. Summarizing, we have shown that the vector space of harmonic de Rham $p-$forms is a direct sum of the vector spaces of harmonic Dolbeault $(p_1,p_2)-$forms with $p_1+p_2=p$. Since the harmonic forms represent the cohomology classes in a 1–1 way, we find the important result that for Kähler manifolds,

$$H^p(M) = H^{p,0}(M) \oplus H^{p-1,1}(M) \oplus \cdots \oplus H^{0,p}(M).$$

That is, the Dolbeault cohomology can be viewed as a refinement of the de Rham cohomology. In particular, we have

$$b^p = h^{p,0} + h^{p-1,1} + \ldots + h^{0,p},$$

where $h^{p,q} = \dim H^{p,q}(M)$ are called the *Hodge numbers* of M.

The Hodge numbers of a Kähler manifold give us several topological invariants, but not all of them are independent. In particular, the following two relations hold:

$$h^{p,q} = h^{q,p}, \qquad h^{p,q} = h^{m-p,m-q}. \tag{4.53}$$

The first relation immediately follows if we realize that $\omega \mapsto \overline{\omega}$ maps $\partial-$harmonic $(p,q)-$forms to $\bar{\partial}-$harmonic $(q,p)-$forms, and hence can be viewed as an invertible map between the two respective cohomologies. As we have seen, the $\partial-$cohomology and the $\bar{\partial}-$cohomology coincide on a Kähler manifold, so the first of the above two equations follows.

The second relation can be proved using the map

$$(\alpha, \omega) \mapsto \int_M \alpha \wedge \omega$$

from $H^{p,q} \times H^{m-p,m-q}$ to $\mathbb{C}$. It can be shown that this map is nondegenerate, and hence that $H^{p,q}$ and $H^{m-p,m-q}$ can be viewed as dual vector spaces. In particular, it follows that these vector spaces have the same dimension, which is the statement in the second line of (4.53).

Note that the last argument also holds for de Rham cohomology, in which case we find the relation $b^p = b^{n-p}$ between the Betti numbers. We also know that $H^{n-p}(M)$ is dual to $H_{n-p}(M)$, so combining these statements we find an identification between the vector spaces $H^p(M)$ and $H_{n-p}(M)$. Recall that this identification between $p-$form cohomology classes and $(n-p)-$cycle homology classes represents the *Poincaré duality*. Intuitively, take a certain $(n-p)-$cycle Σ representing a homology class in H_{n-p}. One can now try to define a 'delta function' $\delta(\Sigma)$ which is localized on this cycle. Locally, Σ can be parameterized by setting p coordinates equal to zero, so $\delta(\Sigma)$ is a 'pD delta function' – that is, it is an object which is naturally integrated over pD submanifolds: a $p-$form. This intuition can be made precise, and one can indeed view the cohomology class of the resulting 'delta–function' $p-$form as the Poincar é dual to Σ.

Going back to the relations (4.53), we see that the Hodge numbers of a Kähler manifold can be nicely written in a so–called *Hodge–diamond* form:

$$\begin{array}{ccccc}
 & & h^{0,0} & & \\
 & h^{1,0} & & h^{0,1} & \\
 \cdot^{\cdot^{\cdot}} & & & & \ddots \\
h^{m,0} & & \cdots & & h^{0,m} \\
 \ddots & & & & \cdot^{\cdot^{\cdot}} \\
 & h^{m,m-1} & & h^{m-1,m} & \\
 & & h^{m,m} & &
\end{array}$$

The integers in this diamond are symmetrical under the reflection in its horizontal and vertical axes.

4.3 Basics of Kähler Geometry

Let M be an nD compact Kähler manifold. A Kähler metric can be given by its *Kähler form* ω on M. In local complex coordinates $z_1, \cdots, z_n$, this ω is of the form [CT02]

$$\omega = \mathrm{i}\, g_{i\bar{j}} d\, z^i \wedge d\, z^{\bar{j}},$$

where $\{g_{i\bar{j}}\}$ is a positive definite Hermitian matrix function. The Kähler condition requires that ω is a closed positive $(1,1)-$form. In other words, the following holds

$$\partial_{z^j} g_{i\bar{k}} = \partial_{z^i} g_{j\bar{k}} \qquad \text{and} \qquad \partial_{z^{\bar{j}}} g_{k\bar{i}} = \partial_{z^{\bar{i}}} g_{k\bar{j}}, \qquad \text{for all } \ i, j, k = 1, 2, \cdots, n.$$

The Kähler metric corresponding to ω is given by

$$\mathrm{i}\, g_{\alpha\overline{\beta}}\, d\, z^\alpha \ \otimes d\, z^{\overline{\beta}}.$$

For simplicity, in the following, we will often denote by ω the corresponding *Kähler metric*. The *Kähler class* of ω is its cohomology class $[\omega]$ in $H^2(M, \mathbb{R})$. By the *Hodge Theorem*, any other Kähler metric in the same Kähler class is of the form

$$\omega_\varphi = \omega + \mathrm{i}\, \partial^2_{z^i z^{\bar{j}}} \varphi\, d\, z_i \wedge d\, z_{\bar{j}} > 0$$

for some real value function φ on M. The functional *space of Kähler potentials* is given by

$$\mathcal{P}(M, \omega) = \{\varphi \ : \ \omega_\varphi = \omega + \mathrm{i}\, \partial\overline{\partial}\varphi > 0 \ \text{ on } M\}.$$

Given a Kähler metric ω, its *volume form* is [CT02]

$$\omega^n = \mathrm{i}^n \det\left(g_{i\bar{j}}\right) d\, z^1 \wedge d\, z^{\bar{1}} \wedge \cdots \wedge d\, z^n \wedge d\, z^{\bar{n}}.$$

Its *Christoffel symbols* are given by

$$\Gamma^k_{i\, j} = g^{k\bar{l}} \partial_{z^j} g_{i\bar{l}} \quad \text{and} \quad \Gamma^{\bar{k}}_{\bar{i}\, \bar{j}} = g^{\bar{k}l} \partial_{z^{\bar{j}}} g_{l\bar{i}}, \qquad \text{for all } \ i, j, k = 1, 2, \cdots, n.$$

The *bisectional curvature tensor* is

$$R_{i\bar{j}k\bar{l}} = -\partial^2_{z^k z^{\bar{l}}} g_{i\bar{j}} + g^{p\bar{q}} \partial_{z^k} g_{i\bar{q}} \partial_{z^{\bar{l}}} g_{p\bar{j}}, \qquad \text{for all } i, j, k, l = 1, 2, \cdots, n.$$

We say that ω is of nonnegative bisectional curvature if

$$R_{i\bar{j}k\bar{l}} v^j v^{\bar{i}} w^l w^{\bar{k}} \geq 0$$

for all non–zero vectors v and w in the holomorphic tangent bundle of M. The bisectional curvature and the curvature tensor can be mutually determined by each other [CT02]. The *Ricci curvature* of ω is locally given by

$$R_{i\bar{j}} = -\partial^2_{z_i \bar{z}_j} \log\det(g_{k\bar{l}}).$$

So its *Ricci curvature form* is

$$\mathrm{Ric}(\omega) = \mathrm{i}\, R_{i\bar{j}}(\omega) d\, z^i \wedge d\, z^{\bar{j}} = -\mathrm{i}\, \partial\overline{\partial} \log\det(g_{k\bar{l}}).$$

It is a real, closed $(1,1)-$form. Recall that $[\omega]$ is a canonical Kähler class if this Ricci form is cohomologous to $\lambda\, \omega$, for some constant λ.

4.3.1 The Kähler Ricci Flow

Now we assume that the first *Chern class* $c_1(M)$ is positive. The *Ricci flow* (see for instance [Ham82] and [Ham86]) on a Kähler manifold M is of the form

$$\partial_t g_{i\bar{j}} = g_{i\bar{j}} - R_{i\bar{j}}, \qquad \text{for all } i,\ j = 1, 2, \cdots, n. \tag{4.54}$$

If we choose the initial Kähler metric ω with $c_1(M)$ as its Kähler class. Then the flow (4.54) preserves the Kähler class $[\omega]$. It follows that on the level of Kähler potentials, the Ricci flow becomes

$$\partial_t \varphi = \log \frac{{\omega_\varphi}^n}{\omega^n} + \varphi - h_\omega, \tag{4.55}$$

where h_ω is defined by

$$\mathrm{Ric}(\omega) - \omega = \mathrm{i}\,\partial\overline{\partial} h_\omega, \text{ and } \int_M (\mathrm{e}^{h_\omega} - 1)\omega^n = 0.$$

As usual, the flow (4.55) is referred as the *Kähler Ricci flow* on M. Differentiating on both sides of equation (4.55) on t, we get

$$\partial_t \partial_t \varphi = \triangle_\varphi \partial_t \varphi + \partial_t \varphi,$$

where $\triangle_\varphi$ is the Laplacian operator of the metric ω_φ. Then it follows from R. Hamilton's *Maximum Principle for tensors*: along the Kähler Ricci flow (4.54) $|\frac{\partial\varphi}{\partial t}|$ grows at most exponentially. In particular, the C^0-norm of φ can be bounded by a constant depending only t. Using this fact and following Yau's calculation in [Yau78], one can prove that for any initial metric with Kähler class $c_1(M)$, the evolution equation (4.55) has a global solution for all time $0 \le t < \infty$ [Cao85].

Now, for any $k = 0, 1, \cdots, n$, we can define a functional E_k^0 on $\mathcal{P}(M, \omega)$ by [CT02]

$$E_{k,\omega}^0(\varphi) = \frac{1}{V} \int_M \left(\log \frac{{\omega_\varphi}^n}{\omega^n} - h_\omega\right) \left(\sum_{i=0}^k \mathrm{Ric}(\omega_\varphi)^i \wedge \omega^{k-i}\right) \wedge {\omega_\varphi}^{n-k} + c_k,$$

$$\text{where} \qquad c_k = \frac{1}{V} \int_M h_\omega \left(\sum_{i=0}^k \mathrm{Ric}(\omega)^i \wedge \omega^{k-i}\right) \wedge \omega^{n-k}.$$

For each $k = 0, 1, 2, \cdots, n-1$, we will define $J_{k,\omega}$ as follows: let $\varphi(t)$ ($t \in [0,1]$) be a path from 0 to φ in $\mathcal{P}(M, \omega)$; then we define

$$J_{k,\omega}(\varphi) = -\frac{n-k}{V} \int_0^1 \int_M \partial_t \varphi \,\left({\omega_\varphi}^{k+1} - \omega^{k+1}\right) \wedge {\omega_\varphi}^{n-k-1} \wedge dt.$$

Put $J_n = 0$ for convenience in notations.

In a non canonical Kähler class, we need to modify the definition slightly since h_ω is not defined there. For any $k = 0, 1, \cdots, n,$ we define [CT02]

$$\begin{aligned} E_{k,\omega}(\varphi) = \tfrac{1}{V} \int_M \log \tfrac{{\omega_\varphi}^n}{\omega^n} \left(\textstyle\sum_{i=0}^k \operatorname{Ric}(\omega_\varphi)^i \wedge \operatorname{Ric}(\omega)^{k-i} \right) \wedge {\omega_\varphi}^{n-k} \\ - \tfrac{n-k}{V} \int_M \varphi \left(\operatorname{Ric}(\omega)^{k+1} - \omega^{k+1} \right) \wedge \omega^{n-k-1} - J_{k,\omega}(\varphi). \end{aligned}$$

The second integral on the right hand side is to offset the change from ω to $Ric(\omega)$ in the first term. The derivative of this functional is exactly the same as in the canonical Kähler class. In other words, the Euler–Lagrange equation is not changed.

Direct computations lead to the following result: for any $k = 0, 1, \cdots, n,$ we have

$$\begin{aligned} \partial_t E_k &= \frac{k+1}{V} \int_M \Delta_\varphi \left(\partial_t \varphi\right) \operatorname{Ric}(\omega_\varphi)^k \wedge {\omega_\varphi}^{n-k} \\ &\quad - \frac{n-k}{V} \int_M \partial_t \varphi \left(\operatorname{Ric}(\omega_\varphi)^{k+1} - {\omega_\varphi}^{k+1} \right) \wedge {\omega_\varphi}^{n-k-1}. \end{aligned} \tag{4.56}$$

Here $\{\varphi(t)\}$ is any path in $\mathcal{P}(M, \omega)$.

Along the Kähler Ricci flow, we have [CT02]

$$\partial_t E_k \leq -\frac{k+1}{V} \int_M (R(\omega_\varphi) - r) \operatorname{Ric}(\omega_\varphi)^k \wedge {\omega_\varphi}^{n-k}. \tag{4.57}$$

When $k = 0, 1,$ we have

$$\begin{aligned} \partial_t E_0 &= -\frac{n\,\mathrm{i}}{V} \int_M \partial(\partial_t \varphi) \wedge \overline{\partial}(\partial_t \varphi) {\omega_\varphi}^{n-1} \leq 0, \\ \partial_t E_1 &\leq -\frac{2}{V} \int_M (R(\omega_\varphi) - r)^2 {\omega_\varphi}^n \leq 0. \end{aligned}$$

In particular, both E_0 and E_1 are decreasing along the Kähler Ricci flow.

We then prove that the derivatives of these functionals along a holomorphic automorphisms give rise to holomorphic invariants. For any holomorphic vector field X, and for any Kähler metric ω, there exists a potential function θ_X such that

$$L_X \omega = \mathrm{i}\, \partial \overline{\partial} \theta_X.$$

Here L_X denotes the *Lie derivative* along a vector field X and θ_X is defined up to the addition of any constant. Now we define $\Im_k(X, \omega)$ for each $k = 0, 1, \cdots, n$ by

$$\begin{aligned} \Im_k(X, \omega) &= (n-k) \textstyle\int_M \theta_X\, \omega^n \\ &+ \textstyle\int_M \left((k+1) \Delta(\omega) \theta_X \operatorname{Ric}(\omega)^k \wedge \omega^{n-k} - (n-k)\, \theta_X \operatorname{Ric}(\omega)^{k+1} \wedge \omega^{n-k-1} \right). \end{aligned}$$

Clearly, the integral is unchanged if we replace θ_X by $\theta_X + c$ for any constant c.

The next result assures that the above integral gives rise to a holomorphic invariant. The integral $\Im_k(X,\omega)$ is independent of choices of Kähler metrics in the Kähler class $[\omega]$, that is, $\Im_k(X,\omega) = \Im_k(X,\omega')$ so long as the Kähler forms ω and ω' represent the same Kähler class. Hence, the integral $\Im_k(X,\omega)$ is a holomorphic invariant, which will be denoted by $\Im_k(X,[\omega])$.

The above invariants $\Im_k(X,c_1(M))$ all vanish for any holomorphic vector fields X on a compact *Kähler–Einstein manifold* [CT02]. In particular, these invariants all vanish on P^n.

For any Kähler Einstein manifold, $E_k(k=0,1,\cdots,n)$ is invariant under actions of holomorphic automorphisms.

One crucial step in [CT02] is to modify the Kähler–Einstein metric so that the evolved Kähler form is centrally positioned with respect to this new Kähler Einstein metric. For the convenience of a reader, we include the definition of 'centrally positioned' here. Any Kähler form ω_φ is called centrally positioned with respect to some Kähler–Einstein metric

$$\omega_\rho = \omega + \mathrm{i}\,\partial\bar{\partial}\rho$$

if it satisfies the following:

$$\int_M (\varphi-\rho)\,\theta\,\omega_\rho{}^n = 0, \qquad \text{for all } \theta \in \Lambda_1(\omega_\rho). \tag{4.58}$$

Let $\varphi(t)$ be the evolved Kähler potentials. For any $t>0$, there always exists an automorphism $\sigma(t)\in \mathrm{Aut}(M)$ such that $\omega_{\varphi(t)}$ is centrally positioned with respect to $\omega_{\rho(t)}$. Here

$$\sigma(t)^*\omega_1 = \omega_{\rho(t)} = \omega + \mathrm{i}\,\partial\bar{\partial}\rho(t).$$

It was proved in [CT02] that the existence of at least one Kähler–Einstein metric $\omega_{\rho(t)}$ such that $\omega_{\varphi(t)}$ is centrally positioned with respect to $\omega_{\rho(t)}$. As a matter of fact, such a Kähler–Einstein metric is unique. However, a priori we do not know if the curve $\rho(t)$ is differentiable or not. On a Kähler–Einstein manifold, the $K-$energy ν_ω is uniformly bounded from above and below along the Kähler Ricci flow. Moreover, there exists a uniform constant C such that

$$\begin{aligned} |J_{k,\omega_{\rho(t)}}(\varphi(t)-\rho(t))| &\le \{\nu_\omega(\varphi(t))+C\}^{\frac{1}{\delta}},\\ \log\frac{\omega_\varphi{}^n}{\omega_{\rho(t)}{}^n} &\ge -4C''\,\mathrm{e}^{2(\nu_\omega(\varphi(t))+C)^{\frac{1}{\delta}}+C')},\\ E_k(\varphi(t)) &\ge -\mathrm{e}^{c\left(1+\max\{0,\nu_\omega(\varphi(t))\}+(\nu_\omega(\varphi(t))+C)^{\frac{1}{\delta}}\right)}, \end{aligned}$$

where c, C, C' and C'' are some uniform constants. And $\rho(t)$ is defined in the preceding proposition.

The energy functional $E_k(k=0,1,\cdots,n)$ has a uniform lower bound from below along the Kähler Ricci flow. For each $k=0,1,\cdots,n$, there exists a uniform constant C such that the following holds (for any $T\le\infty$) along the Kähler Ricci flow [CT02]:

$$\int_0^T \frac{k+1}{V} \int_M \left(R(\omega_{\varphi(t)}) - r\right) \, \mathrm{Ric}(\omega_{\varphi(t)})^k \wedge \omega_{\varphi(t)}{}^{n-k} \, dt \le C.$$

When $k = 1$, we have

$$\int_0^\infty \frac{1}{V} \int_M (R(\omega_{\varphi(t)}) - r)^2 \, \omega_{\varphi(t)}{}^n \, dt \le C < \infty.$$

4.3.2 Kähler Orbifolds

Now, following [CT01], we will define *Kähler orbifolds* and subsequently derive the associated Kähler Ricci flow.

Let M be a connected analytic space. An $n-$dimensional *complex orbifold structure* on M is given by the following data: for any point $p \in M$, there are neighborhoods U_p and their $n-$dimensional uniformization systems (V_p, G_p, π_p) such that for any $q \in U_p$, (V_p, G_p, π_p) and (V_q, G_q, π_q) are equivalent at q. A point $p \in M$ is called regular if there exists a uniformization system (V_p, G_p, π_p) over $U_p \ni p$ such that G_p is trivial; otherwise it is called singular. The set of regular points is denoted by M_{reg}, the set of singular points by M_{sing}; and $M = M_{reg} \bigcup M_{sing}$.

In order to introduce metric in a Kähler orbifold, we need first introduce some differential structure to it. Let U be uniformized by (V, G, π) and U' be uniformized by (V', G', π'), and $f : U \to U'$ be a continuous map. A C^l lifting, $0 \le l \le \infty$, of f is a C^l map $\tilde{f} : V \to V'$ and a homomorphism $\lambda : G \to G'$ such that $\pi' \circ \tilde{f} = f \circ \pi$, and $\lambda(g) \cdot \tilde{f}(x) = \tilde{f}(g \cdot x)$ for any $x \in V$. Two liftings on two neighborhood of p are equivalent if they induced an isomorphic lifting in a smaller neighborhood. Now we define a C^l map between two Kähler orbifolds.

A C^l map $f (0 \le l \le \infty)$ between orbifolds M_1 and M_2 consists of the following data [CT01]:

1. f is a continuous map from M_1 to M_2, and is a C^l map when restricted to the regular part of the orbifold.
2. For any $p \in M_1$, and $q = f(p) \in M_2$, consider the local open set $U_{1,p} \subset M_1$ and $U_{2,q} \subset M_2$. Suppose $U_{1,p}$ is uniformized by $(V_{1,p}, G_{1,p}, \pi_{1,p})$ and $U_{2,q}$ is uniformized by $(V_{2,q}, G_{2,q}, \pi_{2,q})$. For simplicity, suppose $U_{1,p} = f^{-1}(U_{2,p})$. Then $f : U_{1,p} \to U_{2,q}$ admits a C^l lifting $\tilde{f}$: $(V_{1,p}, G_{1,p}, \pi_{1,p}) \to (V_{2,q}, G_{2,q}, \pi_{2,q})$.
3. For any $p \in M_1$, and any two open neighborhood U_1 and U_2, then any lifting induces from the two uniformization system near p must be equivalent at p.

Next we define orbifold vector bundles over a complex orbifold. As before, we begin with local uniformization systems for orbifold vector bundles. Given an analytic space U which is uniformized by (V, G, π) and a complex analytic space E with a surjective holomorphic map $pr : E \to U$, a uniformization system of rank k complex vector bundle for E over U consists of the following data.

1. A uniformization system (V, G, π) of U.
2. A uniformization system $(V \times {}^k, G, \tilde{\pi})$ for E. The action of G on $V \times {}^k$ is an extension of the action of G on V given by $g(x, v) = (g \cdot x, \rho(x, g) \cdot v)$, where $\rho : V \times G \to GL({}^k)$ is a holomorphic map satisfying

$$\rho(g \cdot x, h) \circ \rho(x, g) = \rho(x, h \circ g), \qquad \text{for all } g, h \in G,\ x \in V.$$

3. The natural projection map $\tilde{pr} : V \times {}^k \to V$ satisfies

$$\pi \circ \tilde{pr} = pr \circ \tilde{\pi}.$$

We can similarly define isomorphisms between two uniformization systems of *orbifold vector bundles* for E over U. The only additional requirement is that the diffeomorphism between $V \times {}^k$ are linear on each fiber of $\tilde{pr}$: $V \times {}^k \to V$. Moreover, we can also define the equivalence relation between two uniformization systems of complex vector bundles at any specific point. Here is the definition of orbifold vector bundles over complex orbifolds [CT01]:

Let M be a complex orbifold and E a complex vector space with a surjective holomorphic map $pr : E \to M$. A rank k complex orbifold vector bundle structure on E over M consists of the following data: for each point $p \in M$, there is a unformized neighborhood U_p and a uniformization system of a rank k complex vector bundle for $pr^{-1}(U_p)$ over U_p such that for any $q \in U_p$, the rank k complex orbifold vector bundles over U_p and U_q are isomorphic in a smaller open subset $U_p \bigcap U_q$. Two orbifold vector bundles $pr_1 : E_1 \to M$ and $pr_2 : E_2 \to M$ are isomorphic if there is a holomorphic map $\tilde{\psi} : E_1 \to E_2$ given by $\tilde{\psi}_p : (V_{1,p} \times {}^k, G_{1,p}, \tilde{\pi}_{1,p}) \to (V_{2,p} \times {}^k, G_{2,p}, \tilde{\pi}_{2,p})$ which induces an isomorphism between $(V_{1,p}, G_{1,p}, \tilde{\pi}_{1,p})$ and $(V_{2,p}, G_{2,p}, \tilde{\pi}_{2,p})$, and is a linear isomorphism between the fibers of $\tilde{pr}_{1,p}$ and $\tilde{pr}_{2,p}$.

For a complex orbifold, one can define the tangent bundle, the cotangent bundle, and various exterior or tensor powers of these bundles. All the differential geometric quantities such as cohomology class, connections, metrics, and curvatures can be introduced on the complex orbifold.

The following gives a definition of what a smooth Kähler metric or a Kähler form on the complex orbifold is [CT01]:

For any point $p \in M$, let U_p be uniformized by (V_p, G_p, π_p). A Kähler metric g (resp. a Kähler form ω) on a complex orbifold M is a smooth metric on M_{reg} such that for any $p \in M$, ${\pi_p}^* g$ (resp. Kähler form ${\pi_p}^* \omega$) can extends to a smooth Kähler metric (resp. smooth Kähler form) on V_p.

A function f is called a smooth function on an orbifold M if for any $p \in M$, $f \circ \pi_p$ is a smooth function on V_p.

Similarly, one can define any tensor to be smooth on M if its pre-image on each local uniformization system is smooth. Clearly, the curvature tensor and the Ricci tensor of any smooth metric on orbifolds, as well as their derivatives, are smooth tensors. A complex orbifold admits a Kähler metric is called a Kähler orbifold.

A curve $c(t)$ on Kähler orbifold M is called geodesic if near any point p on it, $c(t) \bigcap U_p$ can be lifted to a geodesic on V_p and at least one preimage of $c(t)$ is smooth in V_p. Here U_p is any open connected neighborhood of p over which (V_p, G_p, π_p) is a uniformization system.

Under this definition, we have the following result: Any minimizing geodesic between two regular points never pass any singular point of the Kähler orbifold.

4.3.3 Kähler Ricci Flow on Kähler–Einstein Orbifolds

A *Kähler–Einstein orbifold metric* is a metric on orbifold such that the Ricci curvature is proportional to the metric. A Kähler orbifold with a Kähler–Einstein metric is called a Kähler–Einstein orbifold [CT01].

Let M be any Kähler Einstein orbifold. If there is another Kähler metric in the same cohomology class which has non–negative bisectional curvature and positive at least at one point, then the Kähler–Ricci flow converges to a Kähler–Einstein metric with positive bi–setcional curvature.

We want to generalize the above Kähler–manifold results to the orbifold case. Note that the analysis for Kähler orbifolds is exactly the same as that for Kähler manifolds [DT92]. We use the Kähler form ω as a smooth Kähler form on the orbifold M. Locally on M_{reg}, it can be written as

$$\omega = \mathrm{i}\, g_{i\bar{j}} d\, z^i \wedge d\, z^{\bar{j}},$$

where $\{g_{i\bar{j}}\}$ is a positive definite Hermitian matrix function. Denote by $\mathcal{B}$ the set of all real valued smooth functions on M in the orbifold sense. Then the Kähler class $[\omega]$ consists of all Kähler form which can be expressed as

$$\omega_\varphi = \omega + \mathrm{i}\,\partial\bar{\partial}\varphi > 0$$

on M for some $\varphi \in \mathcal{B}$. In other words, the space of all Kähler potentials in this Kähler class is

$$\mathcal{H} = \{\varphi \in \mathcal{B} : \omega_\varphi = \omega + \mathrm{i}\,\partial\bar{\partial}\varphi > 0\}.$$

The Ricci form for ω is:

$$Ric(\omega) = -\mathrm{i}\,\partial\bar{\partial}\log\omega^n.$$

As in the case of smooth manifolds, $[\omega]$ is the canonical Kähler class if ω and the Ricci form $Ric(\omega)$ is in the same cohomology class after proper re-scalling. In the canonical Kähler class, consider the Kähler Ricci flow

$$\partial_t \varphi = \log \frac{\omega_\varphi{}^n}{\omega^n} + \varphi - h_\omega,$$

where h_ω is defined above. Clearly, this flow preserves the structure of Kähler orbifold, in particular, preserves the Kähler class $[\omega]$. Here we emphasize the following three characteristics of Kähler manifolds [CT01]:

1. The preservation of positive bisectional curvature under the Kähler Ricci flow.
2. The introduction of a set of new functionals E_k and new invariants $\Im_k (k = 0, 1, \cdots, n)$.
3. The uniform estimate on the diameter; consequently, the uniform control on the Sobolev constant and the Poincare constant.

To extend these to the case of Kähler orbifolds, we really need to make sure that the following tools for geometric analysis hold in the orbifold case:

1. Maximum principle for smooth functions and tensors on Kähler orbifold;
2. Integration by parts for smooth functions/tensors in the orbifold case;
3. The second variation formula for any smooth geodesics.
   For more details, see [CT01].

4.3.4 Induced Evolution Equations

The Kähler Ricci flow induces an evolution equation on the bisectional curvature [CT02]

$$\begin{aligned}\partial_t R_{i\bar{j}k\bar{l}} = \triangle R_{i\bar{j}k\bar{l}} + R_{i\bar{j}p\bar{q}} R_{q\bar{p}k\bar{l}} - R_{i\bar{p}k\bar{q}} R_{p\bar{j}q\bar{l}} + R_{i\bar{l}p\bar{q}} R_{q\bar{p}k\bar{j}} + R_{i\bar{j}k\bar{l}} \\ -\tfrac{1}{2}\left(R_{i\bar{p}} R_{p\bar{j}k\bar{l}} + R_{p\bar{j}} R_{i\bar{p}k\bar{l}} + R_{k\bar{p}} R_{i\bar{j}p\bar{l}} + R_{p\bar{l}} R_{i\bar{j}k\bar{p}} \right).\end{aligned}$$

Similarly, we have an evolution equation for the Ricci tensor and the scalar curvature

$$\begin{aligned}\partial_t R_{i\bar{j}} &= \triangle R_{i\bar{j}} + R_{l\bar{k}} R_{i\bar{j}k\bar{l}} - R_{i\bar{k}} R_{k\bar{j}}, \qquad \text{and} \\ \partial_t R &= \triangle R + |\mathrm{Ric}|^2 - R.\end{aligned}$$

4.4 Conformal Killing–Riemannian Geometry

In this subsection we present some basic facts from *conformal Killing–Riemannian geometry*. In mechanics it is well–known that *symmetries of Lagrangian* or *Hamiltonian* result in *conservation laws*, that are used to deduce constants of motion for the trajectories (geodesics) on the configuration manifold M. The same constants of motion are get using geometrical language, where a *Killing vector–field* is the standard tool for the description of symmetry [MTW73]. A Killing vector–field ξ^i is a vector–field on a Riemannian manifold M with metrics g, which in coordinates $x^j \in M$ satisfies the *Killing equation*

$$\xi^{i;j} + \xi^{j;i} = \xi^{(i;j)} = 0, \qquad \text{or} \qquad \mathcal{L}_{\xi^i} g_{ij} = 0, \tag{4.59}$$

where semicolon denotes the covariant derivative on M, the indexed bracket denotes the tensor symmetry, and $\mathcal{L}$ is the Lie derivative.

The *conformal Killing vector–fields* are, by definition, infinitesimal conformal symmetries i.e., the flow of such vector–fields preserves the conformal class of the metric. The number of linearly–independent conformal Killing fields measures the degree of *conformal symmetry of the manifold.* This number is bounded by $\frac{1}{2}(n+1)(n+2)$, where n is the dimension of the manifold. It is the maximal one if the manifold is conformally flat [Bau00].

Now, to properly initialize our conformal geometry, recall that *conformal twistor spinor–fields* φ were introduced by R. Penrose into physics (see [Pen67, PR86]) as solutions of the *conformally covariant twistor equation*

$$\nabla^S_X \varphi + \frac{1}{n} X \cdot D\varphi = 0,$$

for each vector–fields X on a Riemannian manifold (M,g), where D is the *Dirac operator*. Each twistor spinor–field φ on (M,g) defines a *conformal vector–field* V_φ on M by

$$g(V_\varphi, X) = i^{k+1} \langle X \cdot \varphi, \varphi \rangle.$$

Also, each twistor spinor–field φ that satisfies the *Dirac equation* on (M,g),

$$D\varphi = \mu\varphi,$$

is called a *Killing spinor–field*. Each twistor spinor–field without zeros on (M,g) can be transformed by a conformal change of the metric g into a Killing spinor–field [Bau00].

4.4.1 Conformal Killing Vector–Fields and Forms on M

The space of all conformal Killing vector–fields forms the Lie algebra of the *isometry group* of a Riemannian manifold (M,g) and the number of linearly independent Killing vector–fields measures the *degree of symmetry* of M. It is known that this number is bounded from above by the dimension of the isometry group of the standard sphere and, on compact manifolds, equality is attained if and only if the manifold M is isometric to the standard sphere or the real projective space. Slightly more generally one can consider *conformal vector–fields*, i.e., vector–fields with a flow preserving a given conformal class of metrics. There are several geometrical conditions which force a conformal vector–field to be Killing [Sem02].

A natural generalization of conformal vector–fields are the *conformal Killing forms* [Yan52], also called *twistor forms* [MS03]. These are p–forms α satisfying for any vector–field X on the manifold M the *Killing–Yano equation*

$$\nabla_X \alpha \;-\; \tfrac{1}{p+1}\, X \rfloor d\alpha \;+\; \tfrac{1}{n-p+1}\, X^* \wedge d^*\alpha \;=\; 0, \qquad (4.60)$$

where n is the dimension of the manifold (M, g), ∇ denotes the covariant derivative of the Levi–Civita connection on M, X^* is 1–form dual to X and $\rfloor$ is the operation dual to the wedge product on M. It is easy to see that a conformal Killing 1–form is dual to a conformal vector–field. *Coclosed* conformal Killing p–forms are called *Killing forms.* For $p = 1$ they are dual to Killing vector–fields.

Let α be a Killing p–form and let γ be a geodesic on (M, g), i.e., $\nabla_{\dot\gamma}\dot\gamma = 0$. Then

$$\nabla_{\dot\gamma}\,(\dot\gamma \rfloor \alpha) \;=\; (\nabla_{\dot\gamma}\,\dot\gamma) \rfloor \alpha \;+\; \dot\gamma \rfloor \nabla_{\dot\gamma}\,\alpha \;=\; 0,$$

i.e., $\dot\gamma \rfloor \alpha$ is a $(p-1)$–form parallel along the geodesic γ and in particular its length is constant along γ.

The l.h.s of equation (4.60) defines a first–order elliptic differential operator T, the so–caled *twistor operator.* Equivalently one can describe a conformal Killing form as a form in the kernel of twistor operator T. From this point of view conformal Killing forms are similar to Penrose's *twistor spinors* in *Lorentzian spin geometry.* One shared property is the conformal invariance of the defining equation. In particular, any form which is parallel for some metric g, and thus a Killing form for trivial reasons, induces non–parallel conformal Killing forms for metrics conformally equivalent to g (by a non–trivial change of the metric) [Sem02].

4.4.2 Conformal Killing Tensors and Laplacian Symmetry on M

In an nD Riemannian manifold (M, g), a *Killing tensor–field* (of order 2) is a symmetric tensor K^{ab} satisfying (generalizing (4.59))

$$K^{(ab;c)} = 0. \qquad (4.61)$$

A *conformal Killing tensor–field* (of order 2) is a symmetric tensor Q^{ab} satisfying

$$Q^{(ab;c)} = q^{(a} g^{bc)}, \qquad \text{with} \qquad q^a = (Q^{,a} + 2Q_d^{a;d})/(n+2), \qquad (4.62)$$

where comma denotes partial derivative and $Q = Q_d^d$. When the associated *conformal vector* q^a is nonzero, the conformal Killing tensor will be called *proper* and otherwise it is a (ordinary) Killing tensor. If q^a is a Killing vector, Q^{ab} is referred to as a *homothetic Killing tensor.* If the associated conformal vector $q^a = q^{,a}$ is the gradient of some scalar field q, then Q^{ab} is called a *gradient conformal Killing tensor.* For each gradient conformal Killing tensor Q^{ab} there is an associated Killing tensor K^{ab} given by

$$K^{ab} = Q^{ab} - qg^{ab}, \qquad (4.63)$$

which is defined only up to the addition of a constant multiple of the inverse metric tensor g^{ab}.

Some authors define a conformal Killing tensor as a *trace–free* tensor P^{ab} satisfying $P^{(ab;c)} = p^{(a}g^{bc)}$. Note that there is no contradiction between the two definitions: if P^{ab} is a trace–free conformal Killing tensor then for any scalar field λ, $P^{ab} + \lambda g^{ab}$ is a conformal Killing tensor and conversely if Q^{ab} is a conformal Killing tensor, its trace–free part $Q^{ab} - \frac{1}{n}Qg^{ab}$ is a trace–free Killing tensor [REB03].

Killing tensor–fields are of importance owing to their connection with quadratic first integrals of the geodesic equations: if p^a is tangent to an affinely parameterized geodesic (i.e., $p^a{}_{;b}p^b = 0$) it is easy to see that $K_{ab}p^ap^b$ is constant along the geodesic. For conformal Killing tensors $Q_{ab}p^ap^b$ is constant along *null* geodesics and here, only the trace–free part of Q_{ab} contributes to the constants of motion. Both Killing tensors and conformal Killing tensors are also of importance in connection with the separability of the Hamiltonian–Jacobi equations [CH64] (as well as other PDEs).

A Killing tensor is said to be *reducible* if it can be written as a constant linear combination of the metric and symmetrized products of Killing vectors,

$$K_{ab} = a^0 g_{ab} + a^{IJ}\xi_{I(a}\xi_{|J|b)}, \tag{4.64}$$

where ξ_I for $I = 1 \dots N$ are the Killing vectors admitted by the manifold (M, g) and a^0 and a^{IJ} for $1 \le I \le J \le N$ are constants. Generally one is interested only in Killing tensors which are not reducible since the quadratic constant of motion associated with a reducible Killing tensor is a constant linear combination of p^ap_a and of pairwise products of the linear constants of motion $\xi_{Ia}p^a$ [REB03].

More generally, any linear differential operator on a Riemannian manifold (M, g) may be written in the form [EG91, Eas02]

$$\mathcal{D} = V^{bc\cdots d}\nabla_b\nabla_c \cdots \nabla_d + \text{ lower order terms},$$

where $V^{bc\cdots d}$ is symmetric in its indices, and $\nabla_a = \partial/\partial x^a$ (differentiation in coordinates). This tensor is called the symbol of $\mathcal{D}$. We shall write $\phi^{(ab\cdots c)}$ for the symmetric part of $\phi^{ab\cdots c}$.

Now, a *conformal Killing tensor* on (M, g) is a symmetric trace–free tensor field, with s indices, satisfying

$$\text{the trace–free part of } \nabla^{(a}V^{bc\cdots d)} = 0, \tag{4.65}$$

or, equivalently,

$$\nabla^{(a}V^{bc\cdots d)} = g^{(ab}T^{c\cdots d)}, \tag{4.66}$$

for some tensor field $T^{c\cdots d}$ or, equivalently,

$$\nabla^{(a}V^{bc\cdots d)} = \tfrac{s}{n+2s-2}g^{(ab}\nabla_e V^{c\cdots d)e}, \tag{4.67}$$

where $\nabla^a = g^{ab}\nabla_b$ (the standard convention of raising and lowering indices with the metric tensor g_{ab}). When $s = 1$, these equations define a *conformal Killing vector.*

M. Eastwood proved the following Theorem: any symmetry $\mathcal{D}$ of the Laplacian $\Delta = \nabla^a \nabla_a$ on a Riemannian manifold (M, g) is canonically equivalent to one whose symbol is a conformal Killing tensor [EG91, Eas02].

4.5 Stringy Manifolds

4.5.1 Calabi–Yau Manifolds

Fundamental geometrical objects in modern superstring theory are the so–called *Calabi–Yau manifolds* [Cal57, Yau78]. A Calabi–Yau manifold is a compact *Ricci–flat Kähler manifold* with a vanishing first Chern class. A Calabi–Yau manifold of complex dimension n is also called a Calabi–Yau $n-$fold, which is a manifold with an $SU(n)$ holonomyi.e., it admits a global nowhere vanishing holomorphic $(n, 0)-$form.

For example, in one complex dimension, the only examples are family of *tori*. Note that the Ricci–flat metric on the torus is actually a flat metric, so that the holonomy is the trivial group SU(1). In particular, 1D *Calabi–Yau manifolds* are also called elliptic curves.

In two complex dimensions, the torus T^4 and the $K3$ surfaces[9] are the only examples. T^4 is sometimes excluded from the classification of being a Calabi–Yau, as its holonomy (again the trivial group) is a proper subgroup of $SU(2)$, instead of being isomorphic to $SU(2)$. On the other hand, the holonomy group of a $K3$ surface is the full $SU(2)$ group, so it may properly be called a Calabi–Yau in 2D.

In three complex dimensions, classification of the possible Calabi–Yau manifolds is an open problem. One example of a 3D Calabi–Yau is the quintic threefold in CP^4.

In string theory, the term *compactification* refers to 'curling up' the extra dimensions (6 in the superstring theory), usually on Calabi–Yau spaces or on orbifolds. The mechanism behind this type of compactification is described by the Kaluza–Klein theory.

In the most conventional superstring models, 10 conjectural dimensions in string theory are supposed to come as 4 of which we are aware, carrying some kind of fibration with fiber dimension 6. Compactification on Calabi–Yau $n-$folds are important because they leave some of the original supersymmetry unbroken. More precisely, compactification on a Calabi–Yau $3-$fold (with real dimension 6) leaves one quarter of the original supersymmetry unbroken.

[9] Recall that $K3$ surfaces are compact, complex, simply–connected surfaces, with trivial canonical line bundle, named after three algebraic geometers, Kummer, Kähler and Kodaira. Otherwise, they are *hyperkähler manifolds* of real dimension 4 with $SU(2)$ holonomy.

4.5.2 Orbifolds

Recall that in topology, an *orbifold* is a generalization of a manifold, a topological space (called the underlying space) with an orbifold structure. The underlying space locally looks like a quotient of a Euclidean space under the action of a finite group of isometries.

The formal orbifold definition goes along the same lines as a definition of manifold, but instead of taking domains in $\mathbb{R}^n$ as the target spaces of charts one should take domains of finite quotients of $\mathbb{R}^n$. A (topological) orbifold O, is a Hausdorff topological space X with a countable base, called the *underlying space*, with an *orbifold structure*, which is defined by orbifold atlas, given as follows.

An *orbifold chart* is an open subset $U \subset X$ together with open set $V \subset \mathbb{R}^n$ and a continuous map $\varphi : U \to V$ which satisfy the following property: there is a finite group Γ acting by linear transformations on V and a homeomorphism $\theta : U \to V/\Gamma$ such that $\varphi = \theta \circ \pi$, where π denotes the projection $V \to V/\Gamma$. A collection of orbifold charts, $\{\varphi_i = U_i \to V_i\}$, is called the *orbifold atlas* if it satisfies the following properties:

(i) $\cup_i U_i = X$;

(ii) if $\varphi_i(x) = \varphi_j(y)$ then there is a neighborhood $x \in V_x \subset V_i$ and $y \in V_y \subset V_j$ as well as a homeomorphism $\psi : V_x \to V_y$ such that $\varphi_i = \varphi_j \circ \psi$.

The orbifold atlas defines the orbifold structure completely and we regard two orbifold atlases of X to give the same orbifold structure if they can be combined to give a larger orbifold atlas. One can add differentiability conditions on the gluing map in the above definition and get a definition of *smooth* (C^∞) *orbifolds* in the same way as it was done for manifolds.

The main example of underlying space is a quotient space of a manifold under the action of a finite group of diffeomorphisms, in particular manifold with boundary carries natural orbifold structure, since it is $\mathbb{Z}_2-$factor of its double. A factor space of a manifold along a smooth S^1-action without fixed points cares an orbifold structure. The orbifold structure gives a natural stratification by open manifolds on its underlying space, where one strata corresponds to a set of singular points of the same type.

Note that one topological space can carry many different orbifold structures. For example, consider the orbifold O associated with a factor space of a $2-$sphere S^2 along a rotation by π. It is homeomorphic to S^2, but the natural orbifold structure is different. It is possible to adopt most of the characteristics of manifolds to orbifolds and these characteristics are usually different from the correspondent characteristics of the underlying space. In the above example, its orbifold fundamental group of O is $\mathbb{Z}_2$ and its *orbifold Euler characteristic* is 1.

Manifold orbifolding denotes an operation of wrapping, or folding in the case of mirrors, to superimpose all equivalent points on the original manifold– to get a new one.

In string theory, the word *orbifold* has a new flavor. In physics, the notion of an orbifold usually describes an object that can be globally written as a coset M/G where M is a manifold (or a theory) and G is a group of its isometries (or symmetries). In string theory, these symmetries do not need to have a geometric interpretation. The so–called *orbifolding* is a general procedure of string theory to derive a new string theory from an old string theory in which the elements of the group G have been identified with the identity. Such a procedure reduces the number of string states because the states must be invariant under G, but it also increases the number of states because of the extra twisted sectors. The result is usually a new, perfectly smooth string theory.

4.5.3 Mirror Symmetry

The so–called *mirror symmetry* is a surprising relation that can exist between two Calabi–Yau manifolds. It happens, usually for two such 6D manifolds, that the shapes may look very different geometrically, but nevertheless they are equivalent if they are employed as hidden dimensions of a (super)string theory. More specifically, mirror symmetry relates two manifolds M and W whose *Hodge numbers* $h^{1,1} = \dim H^{1,1}$ and $h^{1,2} = \dim H^{1,2}$ are swapped; string theory compactified on these two manifolds leads to identical physical phenomena (see [Gre00]).

[Str90] showed that mirror symmetry is a special example of the so–called $T-$duality: the Calabi–Yau manifold may be written as a fiber bundle whose fiber is a 3D torus $T^3 = S^1 \times S^1 \times S^1$. The simultaneous action of $T-$duality on all three dimensions of this torus is equivalent to mirror symmetry.

Mirror symmetry allowed the physicists to calculate many quantities that seemed virtually incalculable before, by invoking the 'mirror' description of a given physical situation, which can be often much easier. Mirror symmetry has also become a very powerful tool in mathematics, and although mathematicians have proved many rigorous theorems based on the physicists' intuition, a full mathematical understanding of the phenomenon of mirror symmetry is still lacking.

4.5.4 String Theory in 'Plain English'

With modern (super)string theory,[10] scientists might be on the verge of fulfilling Einstein's dream: formulating the sought for 'theory of everything', which

[10] Recall that 'superstring' means 'supersymmetric string'. The *supersymmetry* (often abbreviated SUSY) is a hypothetical symmetry that relates bosons (particles that transmit forces) and fermions (particles of matter). In supersymmetric theories, every fundamental fermion has a bosonic 'super–partner' and vice versa.

would unite our understanding of the *four fundamental forces* of Nature[11] into a single equation (like, e.g., Newton, or Einstein, or Schrödinger equation) and explaining the basic nature of matter and energy.

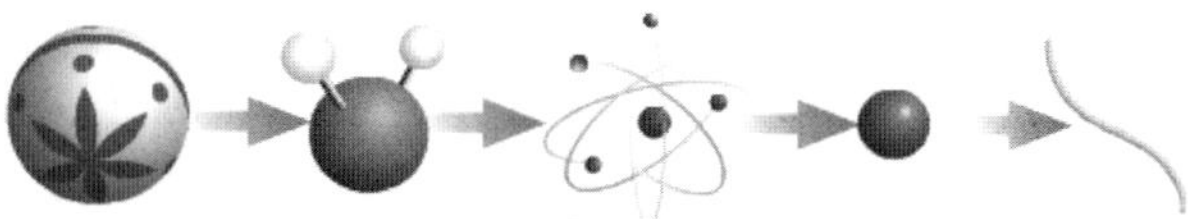

Fig. 4.9. All particles and forces of Nature are supposed to be manifestations of different resonances of tiny 1D strings vibrating in a 10D hyper–space: (a) An ordinary matter; (b) A molecule; (c) An atom (around ten billionths of a centimeter in diameter; (d) A subatomic particle (e.g., proton – around 100.000 times smaller than an atom); (e) A super–string (around 10^{20} times smaller than a proton).

In simplest terms, string theory states that all particles and forces of Nature are manifestations of different resonances of tiny 1–dimensional strings (rather than the zero–dimensional points (particles) that are the basis of the *Standard Model of particle physics*),[12] vibrating in 10 dimensions (see Figure 4.9).

[11] Recall that the *four fundamental forces* are: (i) *Gravity* (it describes the attractive force of matter; it is the same force that holds planets and moons in their orbits and keeps our feet on the ground; it is the weakest force of the four by many orders of magnitude); (ii) *Electromagnetism* (it describes how electric and magnetic fields work together; it also makes objects solid; once believed to be two separate forces, could be described by a relatively simple set of *Maxwell equations*); (iii) *Strong nuclear force* (it is responsible for holding the nucleus of atoms together; without it, protons would repel one another so no elements other than hydrogen, which has only one proton, would be able to form); (iv) *Weak nuclear force* (it explains beta decay and the associated radioactivity; it also describes how elementary particles can change into other particles with different energies and masses).

[12] Recall that the Standard Model of particle physics is a theory which describes three of the four known fundamental interactions between the elementary particles that make up all matter. It is a quantum field theory developed between 1970 and 1973 which is consistent with both quantum mechanics and special relativity. To date, almost all experimental tests of the three forces described by the Standard Model have agreed with its predictions. However, the Standard Model falls short of being a complete theory of fundamental interactions, primarily because of its lack of inclusion of gravity, the fourth known fundamental interaction. The matter particles described by the Standard Model all have an intrinsic spin whose value is determined to be 1/2, making them fermions. For this reason, they follow the

Recall that the *Standard Model* is a theory which describes the strong, weak, and electromagnetic fundamental forces, as well as the fundamental particles that make up all matter. Developed between 1970 and 1973, it is a *quantum field theory*, and consistent with both quantum mechanics and special relativity. The Standard Model contains both *fermionic and bosonic fundamental particles.* Fermions are particles which possess half–integer *spin*, obey the *Fermi–Dirac statistics* and also the *Pauli exclusion principle*, which states that no fermions can share the same quantum state. On the other hand, bosons possess integer spin, obey the *Bose–Einstein statistics*, and do not obey the Pauli exclusion principle. In the Standard Model, the theory of the *electro–weak interaction* (which describes the weak and electromagnetic interactions) is combined with the theory of *quantum chromodynamics*. All of these theories are *gauge theories*,[13] meaning that they model the forces

---

Pauli Exclusion Principle. Apart from their antiparticle partners, a total of twelve different matter particles are known as of early 2007. Six of these are classified as *quarks* (up, down, strange, charm, top and bottom), and the other six as leptons (electron, muon, tau, and their corresponding neutrinos). All particles in the Standard Model have an intrinsic spin, allowing us to roughly visualize each particle as a miniature top spinning in space.

[13] Recall that the familiar *Maxwell gauge field theory*(or, in the non–Abelian case, *Yang–Mills gauge field theory*) is defined in terms of the fundamental *gauge field* (which geometrically represents a *connection*) $A_\mu = (A_0, \mathbf{A})$, that is $\mu = 0, 3$. Here A_0 is the scalar potential and $\mathbf{A}$ is the vector potential. The *Maxwell Lagrangian*

$$\mathcal{L}_{\mathrm{M}} = -\frac{1}{4} F_{\mu\nu} F^{\mu\nu} - A_\mu J^\mu \tag{4.68}$$

is expressed in terms of the field strength tensor (curvature) $F_{\mu\nu} = \partial_\mu A_\nu - \partial_\nu A_\mu$, and a matter current J^μ that is conserved: $\partial_\mu J^\mu = 0$. This Maxwell Lagrangian is manifestly invariant under the gauge transformation $A_\mu \to A_\mu + \partial_\mu \Lambda$; and, correspondingly, the classical Euler-Lagrange equations of motion

$$\partial_\mu F^{\mu\nu} = J^\nu \tag{4.69}$$

are gauge invariant. Observe that current conservation $\partial_\nu J^\nu = 0$ follows from the antisymmetry of $F_{\mu\nu}$.

Note that this Maxwell theory could easily be defined in *any* space–time dimension d simply by taking the range of the space–time index μ on the gauge field A_μ to be $\mu = 0, 1, 2, \ldots, (d-1)$ in dD space–time. The field strength tensor is still the antisymmetric tensor $F_{\mu\nu} = \partial_\mu A_\nu - \partial_\nu A_\mu$, and the Maxwell Lagrangian (4.68) and the field equations of motion (4.69) do not change their form. The only real difference is that the number of independent fields contained in the field strength tensor $F_{\mu\nu}$ is different in different dimensions. (Since $F_{\mu\nu}$ can be regarded as a $d \times d$ antisymmetric matrix, the number of fields is equal to

between fermions by coupling them to bosons which mediate the forces. The Lagrangian of each set of mediating bosons is invariant under a transformation called a *gauge transformation*, so these mediating bosons are referred to as gauge bosons. There are twelve different 'flavors' of fermions in the Standard

---

$\frac{1}{2}d(d-1)$.) So at this level, planar (2+1)D Maxwell theory is quite similar to the familiar (3+1)D Maxwell theory. The main difference is simply that the magnetic field is a (pseudo–) scalar $B = \epsilon^{ij}\partial_i A_j$ in (2 + 1)D, rather than a (pseudo–) vector $\mathbf{B} = \nabla \times \mathbf{A}$ in (3 + 1)D. This is just because in (2 + 1)D the vector potential $\mathbf{A}$ is a 2D vector, and the curl in 2D produces a scalar. On the other hand, the electric field $\mathbf{E} = -\nabla A_0 - \dot{\mathbf{A}}$ is a 2D vector. So the antisymmetric 3×3 field–strength tensor has three nonzero field components: two for the electric field $\mathbf{E}$ and one for the magnetic field B.

The real novelty of (2 + 1)D is that, instead of considering this 'reduced' form of Maxwell theory, we can also define a completely different type of gauge theory: a *Chern–Simons gauge theory*. It satisfies the usual *criteria for a sensible gauge theory*: it is Lorentz invariant, gauge invariant, and local. The *Chern–Simons Lagrangian* is (see, e.g., [Dun99])

$$\mathcal{L}_{\mathrm{CS}} = \frac{\kappa}{2}\epsilon^{\mu\nu\rho}A_\mu\partial_\nu A_\rho - A_\mu J^\mu. \tag{4.70}$$

Two things are important about this Chern–Simons Lagrangian. First, it does not *look* gauge invariant, because it involves the gauge field A_μ itself, rather than just the (manifestly gauge invariant) field strength $F_{\mu\nu}$. Nevertheless, under a gauge transformation, the Chern–Simons Lagrangian changes by a total space–time derivative

$$\delta\mathcal{L}_{\mathrm{CS}} = \frac{\kappa}{2}\partial_\mu\left(\lambda\,\epsilon^{\mu\nu\rho}\partial_\nu A_\rho\right). \tag{4.71}$$

Therefore, if we can neglect boundary terms then the corresponding *Chern–Simons action*,

$$S_{\mathrm{CS}} = \int d^3x\,\mathcal{L}_{\mathrm{CS}},$$

is gauge invariant. This is reflected in the fact that the classical Euler–Lagrange equations

$$\frac{\kappa}{2}\epsilon^{\mu\nu\rho}F_{\nu\rho} = J^\mu, \qquad \text{or equivalently} \qquad F_{\mu\nu} = \frac{1}{\kappa}\epsilon_{\mu\nu\rho}J^\rho, \tag{4.72}$$

are clearly gauge invariant. Note that the *Bianchi identity*, $\epsilon^{\mu\nu\rho}\partial_\mu F_{\nu\rho} = 0$, is compatible with the *current conservation*: $\partial_\mu J^\mu = 0$, which follows from the Noether Theorem. A second important feature of the Chern–Simons Lagrangian (4.70) is that it is *first-order* in space–time derivatives. This makes the canonical

Model. The proton, neutron are made up of two of these: the up–quark and down–quark,[14] bound together by the strong nuclear force. Together with the electron (bound to the nucleus in atoms by the electromagnetic force), those fermions constitute the vast majority of everyday matter. To date, almost all experimental tests of the three forces described by the Standard Model have agreed with its predictions. However, the Standard Model is not a complete theory of fundamental interactions, primarily because it does not describe the gravitational force.

For this reason, string theories are able to avoid problems associated with the presence of point–like particles in a physical theory. The basic idea is that the fundamental constituents of Nature are *strings of energy* of the *Planck length* (around 10^{-35} m), which *vibrate at specific resonant frequencies*

---

structure of these theories significantly different from that of Maxwell theory. A related property is that the Chern–Simons Lagrangian is particular to (2+1)D, in the sense that we cannot write down such a term in (3+1)D – the indices simply do not match up. Actually, it is possible to write down a 'Chern–Simons theory' in any odd space–time dimension (for example, the Chern–Simons Lagrangian in 5D space–time is $\mathcal{L} = \epsilon^{\mu\nu\rho\sigma\tau} A_\mu \partial_\nu A_\rho \partial_\sigma A_\tau$), but it is only in $(2+1)$D that the Lagrangian is quadratic in the gauge field.

Recently, increasingly popular has become *Seiberg–Witten gauge theory*. It refers to a set of calculations that determine the low–energy physics, namely the *moduli space* and the masses of electrically and magnetically charged supersymmetric particles as a function of the moduli space. This is possible and nontrivial in gauge theory with $N = 2$ *extended supersymmetry*, by combining the fact that various parameters of the Lagrangian are holomorphic functions (a consequence of supersymmetry) and the known behavior of the theory in the classical limit. The extended supersymmetry is *supersymmetry* whose infinitesimal generators Q_i^α carry not only a spinor index α, but also an additional index $i = 1, 2...$ The more extended supersymmetry is, the more it constrains physical observables and parameters. Only the minimal (un–extended) supersymmetry is a realistic conjecture for particle physics, but extended supersymmetry is very important for analysis of mathematical properties of *quantum field theory* and *superstring theory*.

[14] Recall that in particle physics, *quarks* are one of the two basic constituents of matter (the other are the *leptons*). Quarks are the only fundamental particles that interact through all four of the fundamental forces. The word was borrowed by M. Gell–Mann from the book Finnegans Wake by James Joyce. Quarks come in six flavors, and their names (up, down, strange, charm, bottom, and top) were also chosen arbitrarily based on the need to name them something that could be easily remembered and used. Antiparticles of quarks are called antiquarks. Isolated quarks are never found naturally; they are almost always found in groups of two (*mesons*) or groups of three (*baryons*) called *hadrons*.

(modes). Another key claim of the theory is that no measurable differences can be detected between strings that wrap around dimensions smaller than themselves and those that move along larger dimensions (i.e., physical processes in a dimension of size R match those in a dimension of size 1/R). Singularities are avoided because the observed consequences of 'big crunches' never reach zero size. In fact, should the universe begin a 'Big–Crunch' sort of process, string theory dictates that the universe could never be smaller than the size of a string, at which point it would actually begin expanding.

Recently, physicists have been exploring the possibility that the strings are actually membranes, that is strings with 2 or more dimensions (membranes are refereed to as $p-$branes, where p is the number of dimensions, see Figure 4.10). Every $p-$brane sweeps out a $(p+1)-$dimensional *world–volume* as it propagates through space–time. A special class of $p-$branes are the so–called *D–branes*, named for the mathematician J. Dirichlet.[15] D–branes are typically classified by their dimension, which is indicated by a number written after the D: a D0–brane is a single point, a D1–brane is a line (sometimes called a 'D-string'), a D2–brane is a plane, and a D25–brane fills the highest–dimensional space considered in old bosonic string theory.[16]

---

[15] Recall that *Dirichlet boundary conditions* have long been used in the study of fluids and potential theory, where they involve specifying some quantity all along a boundary. In *fluid dynamics*, fixing a Dirichlet boundary condition could mean assigning a known fluid velocity to all points on a surface; when studying *electrostatics*, one may establish Dirichlet boundary conditions by fixing the voltage to known values at particular locations, like the surfaces of conductors. In either case, the locations at which values are specified is called a *D–brane*. These constructions take on special importance in string theory, because open strings must have their endpoints attached to D–branes.

[16] The central idea of the so–called *brane–world scenario* is that our visible 3D universe is entirely restricted to a D3–brane embedded in a higher–dimensional space–time, called the *bulk*. The additional dimensions may be taken to be compact, in which case the observed universe contains the extra dimensions, and then no reference to the bulk is appropriate in this context. In the bulk model, other branes may be moving through this bulk. Interactions with the bulk, and possibly with other branes, can influence our brane and thus introduce effects not seen in more standard cosmological models. As one of its attractive features, the model can 'explain' the weakness of gravity relative to the other fundamental forces of nature. In the brane picture, the other three forces (electromagnetism and the weak and strong nuclear forces) are localized on the brane, but gravity has no such constraint and so much of its attractive power 'leaks' into the bulk. As a consequence, the force of gravity should appear significantly stronger on small

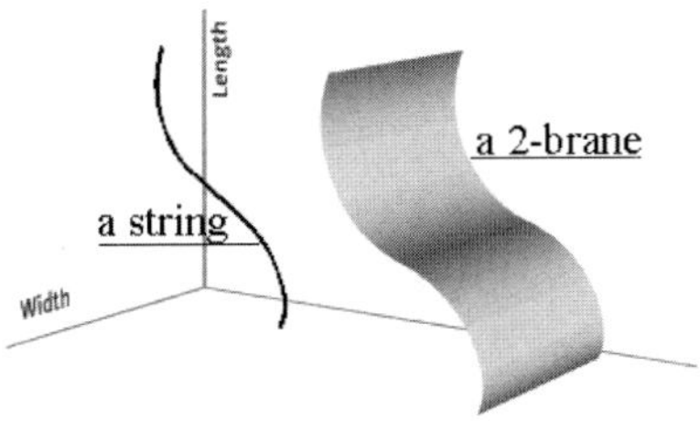

Fig. 4.10. Visualizing strings and p−branes.

According to superstring theory, all the different types of elementary particles can be derived from only five types of interactions between just two different states of strings, *open* and *closed*: (i) an open string can split to create two smaller open strings (see Figure 4.11); (ii) a closed string can split to create two smaller closed strings; (iii) an open string can form both a new open and a new closed string; (iv) two open strings can collide and create two new open strings; (v) an open string can join its ends to become a closed string. All the forces and particles of Nature are just *different modes of vibrating strings* (somewhat like vibrating strings on string instruments to produce a music: different strings have different frequencies that sound as different notes and combining several strings gives chords). For example, gravity is caused by the lowest vibratory mode of a circular string. Higher frequencies and different interactions of superstrings create different forms of matter and energy.

String theory is a possible solution of the *core quantum gravity problem*, and in addition to gravity it can naturally describe interactions similar to electromagnetism and the other forces of nature. Superstring theories include *fermions*, the building blocks of matter, and incorporate the so–called *supersymmetry*.[17] It is not yet known whether string theory will be able to describe

(sub–millimeter) scales, where less gravitational force has 'leaked'. Various experiments are currently underway to test this. For example, in a particle accelerator, if a graviton were to be discovered and then observed to suddenly disappear, it might be assumed that the graviton 'leaked' into the bulk.

[17] In a world based on *supersymmetry*, when a particle moves in space, it also can vibrate in the new fermionic dimensions. This new kind of vibration produces a 'cousin' or 'superpartner' for every elementary particle that has the same electric charge but differs in other properties such as *spin*. Supersymmetric theories make detailed predictions about how superpartners will behave. To confirm supersymmetry, scientists would like to produce and study the new supersymmetric particles. The crucial step is building a particle accelerator that achieves high enough energies. At present, the highest–energy particle accelerator is the Tevatron at Fermilab near Chicago. There, protons and antiprotons collide with

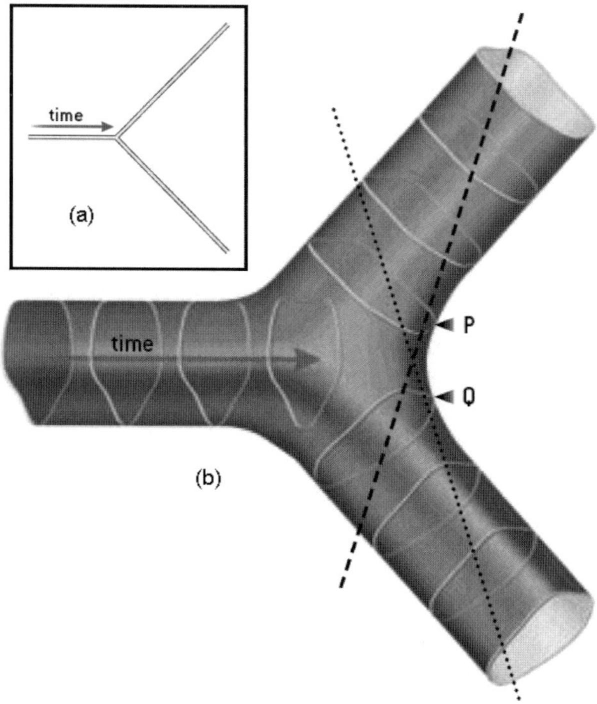

Fig. 4.11. An elementary particle split (a) and string split (b). When a single elementary particle splits in two particles, it occurs at a definite moment in space–time. On the other hand, when a string splits into two strings, different observers will disagree about when and where this occurred. A relativistic observer who considers the dotted line to be a surface of constant time believes the string broke at the space–time point P while another observer who considers the dashed line to be a surface of constant time believes the string broke at Q.

a universe with the precise collection of forces and matter that is observed, nor how much freedom to choose those details that the theory will allow. String theory as a whole has not yet made falsifiable predictions that would allow it

---

an energy nearly 2,000 times the rest energy of an individual proton (given by Einsteins well–known formula $E = mc^2$). Earlier in this decade, physicists capitalized on Tevatron's unsurpassed energy in their discovery of the top quark, the heaviest known elementary particle. After a shutdown of several years, the Tevatron resumed operation in 2001 with even more intense particle beams. In 2007, the available energies will make a 'quantum jump' when the European Laboratory for Particle Physics, or CERN (located near Geneva, Switzerland) turns on the Large Hadron Collider (LHC). The LHC should reach energies 15,000 times the proton rest energy. The LHC is a multi–billion dollar international project, funded mainly by European countries with substantial contributions from the United States, Japan, and other countries.

to be experimentally tested, though various special corners of the theory are accessible to planned observations and experiments. Work on string theory has led to advances in both mathematics (mainly in differential and algebraic geometry) and physics (supersymmetric gauge theories).[18]

Historically, string theory was originally invented to explain peculiarities of hadron (subatomic particle which experiences the strong nuclear force) behavior. In particle–accelerator experiments, physicists observed that the spin of a hadron is never larger than a certain multiple of the square of its energy. No simple model of the hadron, such as picturing it as a set of smaller particles held together by spring–like forces, was able to explain these relationships. In 1968, theoretical physicist G. Veneziano was trying to understand the strong nuclear force when he made a startling discovery. He found that a 200–year–old *Euler beta function* perfectly matched modern data on the strong force. Veneziano applied the Euler beta function to the strong force, but no one could explain why it worked.

In 1970, Y. Nambu, H.B. Nielsen, and L. Susskind presented a physical explanation for Euler's strictly theoretical formula. By representing nuclear forces as vibrating, 1D strings, these physicists showed how Euler's function

[18] Recall that *gauge theories* are a class of physical theories based on the idea that symmetry transformations can be performed locally as well as globally. *Yang–Mills theory* is a particular example of gauge theories with non–Abelian symmetry groups specified by the *Yang–Mills action*. For example, the Yang–Mills action for the $O(n)$ gauge theory for a set of n non–interacting scalar fields φ_i, with equal masses m is

$$S = \int (\sum_{i=1}^{n} \frac{1}{2}\partial_\mu \varphi_i \partial^\mu \varphi_i - \frac{1}{2} m^2 \varphi_i^2)\, d^4x.$$

Other gauge theories with a non–Abelian gauge symmetry also exist, e.g., the Chern–Simons model. Most physical theories are described by Lagrangians which are invariant under certain transformations, when the transformations are identically performed at every space–time point-they have global symmetries. Gauge theory extends this idea by requiring that the Lagrangians must possess local symmetries as well-it should be possible to perform these symmetry transformations in a particular region of space–time without affecting what happens in another region. This requirement is a generalized version of the equivalence principle of general relativity. Gauge symmetries reflect a redundancy in the description of a system. The importance of gauge theories for physics stems from the tremendous success of the mathematical formalism in providing a unified framework to describe the quantum field theories of electromagnetism, the weak force and the strong force. This theory, known as the Standard Model (see footnote 5), accurately describes experimental predictions regarding three of the four fundamental forces of nature, and is a gauge theory with the gauge group $SU(3)\times SU(2)\times U(1)$. Modern theories like string theory, as well as some formulations of general relativity, are, in one way or another, gauge theories. Sometimes, the term gauge symmetry is used in a more general sense to include any local symmetry, like for example, *diffeomorphisms*.

accurately described those forces. But even after physicists understood the physical explanation for Veneziano's insight, the string description of the strong force made many predictions that directly contradicted experimental findings. The scientific community soon lost interest in string theory, and the Standard Model, with its particles and fields, remained un–threatened.

Then, in 1974, J. Schwarz and J. Scherk studied the messenger–like patterns of string vibration and found that their properties exactly matched those of the gravitational force's hypothetical messenger particle - the *graviton*. They argued that string theory had failed to catch on because physicists had underestimated its scope. This led to the development of *bosonic string theory*, which is still the version first taught to many students. The original need for a viable theory of hadrons has been fulfilled by *quantum chromodynamics* (QCD), the theory of Gell–Mann's *quarks* and their interactions. It is now hoped that string theory (or some descendant of it) will provide a fundamental understanding of the quarks themselves.

Bosonic string theory is formulated in terms of the so–called *Polyakov action*, a mathematical quantity which can be used to predict how strings move through space and time. By applying the ideas of quantum mechanics to the Polyakov action - a procedure known as *quantization* - one can deduce that each string can vibrate in many different ways, and that each vibrational state appears to be a different particle. The mass the particle has, and the fashion with which it can interact, are determined by the way the string vibrates - in essence, by the 'note' which the string sounds. The scale of notes, each corresponding to a different kind of particle, is termed the *spectrum* of the theory. These early models included both open strings, which have two distinct endpoints, and closed strings, where the endpoints are joined to make a complete loop. The two types of string behave in slightly different ways, yielding two spectra. Not all modern string theories use both types; some incorporate only the closed variety. However, the bosonic theory has problems. Most importantly, the theory has a fundamental instability, believed to result in the decay of space-time itself. Additionally, as the name implies, the spectrum of particles contains only bosons, particles like the photon which obey particular rules of behavior. While bosons are a critical ingredient of the Universe, they are not its only constituents. Investigating how a string theory may include fermions in its spectrum led to supersymmetry, a mathematical relation between bosons and fermions which is now an independent area of study. String theories which include fermionic vibrations are now known as superstring theories; several different kinds have been described.

Roughly between 1984 and 1986, physicists realized that string theory could describe all elementary particles and interactions between them, and hundreds of them started to work on string theory as the most promising idea to unify theories of physics. This so–called *first superstring revolution* was started by a discovery of *anomaly* cancellation in type I string theory by M. Green and J. Schwarz in 1984. The anomaly is cancelled due to the

Green–Schwarz mechanism. Several other ground–breaking discoveries, such as the heterotic string, were made in 1985.

Contemporary String Theories		
Type	**Dim**	**Details**
Bosonic	26	Only *bosons*, no *fermions* means only forces, no matter, with both open and closed strings; major flaw: a *particle* with imaginary *mass*, called the *tachyon*, representing an instability in the theory
I	10	*Supersymmetry* between forces and matter, with both open and closed strings, no *tachyon*, group symmetry is $SO(32)$
IIA	10	*Supersymmetry* between forces and matter, with closed strings and open strings bound to *D–branes*, no *tachyon*, massless *fermions* spin both ways (nonchiral)
IIB	10	*Supersymmetry* between forces and matter, with closed strings and open strings bound to *D–branes*, no *tachyon*, massless *fermions* only spin one way (chiral)
HO	10	*Supersymmetry* between forces and matter, with closed strings only, no *tachyon*, *heterotic*, meaning right moving and left moving strings differ, group symmetry is $SO(32)$
HE	10	*Supersymmetry* between forces and matter, with closed strings only, no *tachyon*, *heterotic*, meaning right moving and left moving strings differ, group symmetry is $E_8 \times E_8$

Note that in the type IIA and type IIB string theories closed strings are allowed to move everywhere throughout the 10D space-time (called the *bulk*), while open strings have their ends attached to *D–branes*, which are membranes of lower dimensionality (their dimension is odd - 1,3,5,7 or 9 – in type IIA and even – 0,2,4,6 or 8 – in type IIB, including the time direction).

While understanding the details of string and superstring theories requires considerable geometrical sophistication, some qualitative properties of quantum strings can be understood in a fairly intuitive fashion. For example, quantum strings have tension, much like regular strings made of twine; this tension is considered a fundamental parameter of the theory. The tension of a quantum string is closely related to its size. Consider a closed loop of string, left to move through space without external forces. Its tension will tend to contract it into a smaller and smaller loop. Classical intuition suggests that it might shrink to a single point, but this would violate *Heisenberg's uncertainty principle*. The characteristic size of the string loop will be a balance between the tension force, acting to make it small, and the uncertainty effect, which keeps it 'stretched'. Consequently, the minimum size of a string must be related to the string tension.

Before the 1990s, string theorists believed that there were five distinct superstring theories: *type I*, *types IIA* and *IIB*, and the two *heterotic string* theories ($SO(32)$ and $E_8 \times E_8$). The thinking was that out of these five candidate theories, only one was the actual correct *theory of everything*, and that theory was the theory whose low energy limit, with ten dimensions space–time *compactified* down to four, matched the physics observed in our world today. But now it is known that this naïve picture was wrong, and that the five superstring theories are connected to one another as if they are each a special case of some more fundamental theory, of which there is only one. These theories are related by transformations that are called dualities. If two theories are related by a duality transformation, it means that the first theory can be transformed in some way so that it ends up looking just like the second theory. The two theories are then said to be dual to one another under that kind of transformation. Put differently, the two theories are two different mathematical descriptions of the same phenomena. These dualities link quantities that were also thought to be separate. Large and small distance scales, strong and weak coupling strengths – these quantities have always marked very distinct limits of behavior of a physical system, in both classical *field theory* and quantum *particle physics*. But strings can obscure the difference between large and small, strong and weak, and this is how these five very different theories end up being related.

This type of duality is called *T–duality*. T–duality relates type IIA superstring theory to type IIB superstring theory. That means if we take type IIA and Type IIB theory and 'compactify' them both on a circle, then switching the momentum and winding modes, and switching the distance scale, changes one theory into the other. The same is also true for the two heterotic theories. T–duality also relates type I superstring theory to both type IIA and type IIB superstring theories with certain boundary conditions (termed 'orientifold'). Formally, the location of the string on the circle is described by two fields living on it, one which is left-moving and another which is right–moving. The movement of the string center (and hence its momentum) is related to the sum of the fields, while the string stretch (and hence its winding number) is related to their difference. T-duality can be formally described by taking the left-moving field to minus itself, so that the sum and the difference are interchanged, leading to switching of momentum and winding.

On the other hand, every force has a *coupling constant*, which is a measure of its strength, and determines the chances of one particle to emit or receive another particle. For electromagnetism, the coupling constant is proportional to the square of the electric charge. When physicists study the quantum behavior of electromagnetism, they can't solve the whole theory exactly, because every particle may emit and receive many other particles, which may also do the same, endlessly. So events of emission and reception are considered as perturbations and are dealt with by a series of approximations, first assuming there is only one such event, then correcting the result for allowing two such events, etc (this method is called Perturbation theory. This is a reasonable approximation only if the coupling constant is small, which is the case for electromagnetism. But if the coupling constant gets large, that method of calculation breaks down, and the little pieces become worthless as an approximation to the real physics. This can also happen in string theory. String theories have a *string coupling constant*. But unlike in particle theories, the string coupling constant is not just a number, but depends on one of the oscillation modes of the string, called the *dilaton*. Exchanging the dilaton field with minus itself exchanges a very large coupling constant with a very small one. This symmetry is called *S–duality*. If two string theories are related by S–duality, then one theory with a strong coupling constant is the same as the other theory with weak coupling constant. The theory with strong coupling cannot be understood by means of perturbation theory, but the theory with weak coupling can. So if the two theories are related by S-duality, then we just need to understand the weak theory, and that is equivalent to understanding the strong theory. Superstring theories related by S–duality are: type I superstring theory with heterotic $SO(32)$ superstring theory, and type IIB theory with itself.

Around 1995, Ed Witten and others found strong evidence that the different superstring theories were different limits of a new 11D theory called M–theory. With the discovery of M–theory, an extra dimension appeared and the fundamental string of string theory became a 2-dimensional membrane called an M2–brane (or supermembrane). Its magnetic dual is an M5–brane. The various branes of string theory are thought to be related to these higher dimensional M–branes wrapped on various cycles. These discoveries sparked the so–called *second superstring revolution*.

One intriguing feature of string theory is that it *predicts* the number of dimensions which the universe should possess. Nothing in Maxwell's theory of electromagnetism, or Einstein's theory of relativity, makes this kind of prediction; these theories require physicists to insert the number of dimensions 'by hand'. The first person to add a fifth dimension to Einstein's four space–time dimensions was German mathematician T. Kaluza in 1919. The reason for the un–observability of the fifth dimension (its compactness) was suggested by Swedish physicist O. Klein in 1926. Today, this is called the 5D *Kaluza–Klein theory*.

Instead, string theory allows one to compute the number of space–time dimensions from first principles. Technically, this happens because for a different number of dimensions, the theory has a gauge anomaly. This can be understood by noting that in a consistent theory which includes a photon (technically, a particle carrying a force related to an unbroken gauge symmetry), it must be massless. The mass of the photon which is predicted by string theory depends on the energy of the string mode which represents the photon. This energy includes a contribution from the *Casimir effect*, namely from *quantum fluctuations in the string*. The size of this contribution depends on the number of dimensions since for a larger number of dimensions, there are more possible fluctuations in the string position. Therefore, the photon will be massless – and the theory consistent – only for a particular number of dimensions.

The only problem is that when the calculation is done, the universe's dimensionality is not four as one may expect (three axes of space and one of time), but 26. More precisely, bosonic string theories are 26D, while superstring and M–theories turn out to involve 10 and 11 dimensions, respectively. In bosonic string theories, the 26 dimensions come from the *Polyakov equation*. However, these results appear to contradict the observed four dimensional space–time.

Fig. 4.12. Calabi–Yau manifold – a 3D projection created using Mathematica^{TM}.

Two different ways have been proposed to solve this apparent contradiction. The first is to *compactify the extra dimensions*; i.e., the 6 or 7 extra dimensions are so small as to be undetectable in our phenomenal experience. The 6D model's resolution is achieved with the so–called *Calabi–Yau manifolds* (see Figure 4.12). In 7D, they are termed G_2-manifolds. Essentially these extra dimensions are compactified by causing them to loop back upon themselves. A standard analogy for this is to consider multidimensional space as a garden hose. If the hose is viewed from a sufficient distance, it appears to have only one dimension, its length. Indeed, think of a ball small enough to enter the hose but not too small. Throwing such a ball inside the hose, the ball would move more or less in one dimension; in any experiment we make by throwing such balls in the hose, the only important movement will be one-dimensional, that is, along the hose. However, as one approaches the hose, one discovers that it contains a second dimension, its circumference. Thus, a ant crawling inside it would move in two dimensions (and a fly flying in it would move in three dimensions). This 'extra dimension' is only visible within a relatively close range to the hose, or if one 'throws in' small enough objects. Similarly, the extra compact dimensions are only visible at extremely small distances, or by experimenting with particles with extremely small wave lengths (of the order of the compact dimension's radius), which in quantum mechanics means very high energies. Another possibility is that we are stuck in a 3+1 dimensional (i.e., three spatial dimensions plus one time dimension) subspace of the full universe. This subspace is supposed to be a D–brane, hence this is known as a *brane–world theory*. In either case, gravity acting in the hidden dimensions affects other non–gravitational forces such as electromagnetism. In principle, therefore, it is possible to deduce the nature of those extra dimensions by requiring consistency with the Standard Model, but this is not yet a practical possibility. It is also be possible to extract information regarding the hidden dimensions by precision tests of gravity, but so far these have only put upper limitations on the size of such hidden dimensions.

For popular expose on string theory, see [Wit02, Gre00], while the main textbook is still [GSW87].

5

Nonlinear Dynamics on Complex Manifolds

In this Chapter we develop high–dimensional nonlinear complex–valued dynamics on complex manifolds.

5.1 Gauge Theories

Recall that most physical theories are described by Lagrangians which are invariant under certain transformations, when the transformations are identically performed at every space–time point, i.e., they have global symmetries. Gauge theory extends this idea by requiring that the Lagrangians must possess local symmetries as well, that is it should be possible to perform these symmetry transformations in a particular region of space-time without affecting what happens in another region. This requirement is a generalized version of the equivalence principle of general relativity. Gauge 'symmetries' reflect a redundancy in the description of a system. Sometimes, the term *gauge symmetry* is used in a more general sense to include any local symmetry, like for example, *diffeomorphisms*.

Yang–Mills theories are a particular example of gauge theories with non-Abelian symmetry groups specified by the *Yang–Mills action*. Other gauge theories with a non-Abelian gauge symmetry also exist, e.g., the *Chern–Simons theory* (see below).

The importance of *gauge theories* for physics stems from the tremendous success of the mathematical formalism in providing a unified framework to describe the quantum field theories of electromagnetism, the weak force and the strong force. This theory, known as the *Standard Model*, accurately describes experimental predictions regarding three of the four fundamental forces of nature, and is a gauge theory with the gauge group $SU(3) \times SU(2) \times U(1)$. Modern theories like *string theory*, as well as some formulations of general relativity, are, in one way or another, gauge theories.

The earliest physical theory which had a gauge symmetry was *Maxwell's electrodynamics*. However, the importance of this symmetry remained unnoticed in the earliest formulations. After Einstein's development of general relativity, H. Weyl, in an attempt to unify general relativity and electromagnetism, conjectured that invariance under the change of scale (or *gauge*) might also be a local symmetry of the theory of general relativity. After the development of quantum mechanics, Weyl, V. Fock and F. London realized that the idea, with some modifications (replacing the scale factor with a complex–valued quantity, and turning the scale transformation into a change of phase, that is a $U(1)$–gauge symmetry) provided a neat explanation for the effect of an electromagnetic field on the *wave function* of a charged quantum–mechanical particle. This was the first gauge theory, popularised by W. Pauli in the 1940s.

In the 1950s, attempting to resolve some of the great confusion in elementary particle physics, C. Yang and R. Mills introduced *non–Abelian gauge theories* as models to understand the strong interaction holding together nucleons in atomic nuclei. Generalizing the gauge invariance of electromagnetism, they attempted to construct a theory based on the action of the (non–Abelian) $SU(2)$–symmetry group on the isospin doublet of protons and neutrons, similar to the action of the $U(1)$–group on the spinor fields of quantum electrodynamics. In particle physics the emphasis was on using *quantized gauge theories*.

This idea later found application in the quantum field theory of the weak force, and its unification with electromagnetism in the electroweak theory. Gauge theories became even more attractive when it was realized that non–Abelian gauge theories reproduced a feature called *asymptotic freedom*, that was believed to be an important characteristic of strong interactions, thereby motivating the search for a gauge theory of the strong force. This theory, now known as *quantum chromodynamics*, is a gauge theory with the action of the $SU(3)$–group on the *color triplet* of *quarks*. The Standard Model unifies the description of electromagnetism, weak interactions and strong interactions in the language of gauge theory.

In the seventies, M. Atiyah began a program of studying the mathematics of solutions to the classical *Yang–Mills equations*. In 1983, Atiyah's student S. Donaldson built on this work to show that the differentiable classification of smooth 4–manifolds is very different from their classification up to homeomorphism. M. Freedman used Donaldson's work to exhibit exotic differentiable structures on Euclidean 4D space $\mathbb{R}^4$. This led to an increasing interest in gauge theory for its own sake, independent of its successes in fundamental physics. In 1994, E. Witten and N. Seiberg invented gauge–theoretic techniques based on *supersymmetry*[1] which enabled the calculation of certain

[1] Recall that in particle physics, supersymmetry (often abbreviated SUSY) is a symmetry that interchanges *bosons* and *fermions*. In supersymmetric theories, every fundamental fermion has a bosonic superpartner and vice versa. A supersymmetric quantum field theory tames quantum mechanical dynamics and sometimes allows the theory to be solved. If supersymmetry is applied to the Standard Model of particle physics, the hierarchy problem can be solved. The minimal

sutopological invariants. These contributions to mathematics from gauge theory have led to a renewed interest in this area.

The definition of electrical ground in an electric circuit is an example of a gauge symmetry; when the electric potentials across all points in a circuit are raised by the same amount, the circuit would still operate identically; as the potential differences (voltages) in the circuit are unchanged. A common illustration of this fact is the sight of a bird perched on a high voltage power line without electrocution, as the bird is insulated from the ground. This is called a *global gauge symmetry*. The absolute value of the potential is immaterial; what matters to circuit operation is the potential differences across the components of the circuit. The definition of the ground point is arbitrary, but once that point is set, then that definition must be followed globally. In contrast, if some symmetry could be defined arbitrarily from one position to the next, that would be a *local gauge symmetry*.

5.1.1 Classical Gauge Theory

Scalar $O(n)$ Gauge Theory

Consider a set of n non–interacting scalar fields, with equal masses m. This system is described by an action which is the sum of the (usual) action for each scalar field ϕ_i,

$$\mathcal{S}=\int \mathrm{d}^4x\sum_{i=1}^{n}\left[\frac{1}{2}\partial_\mu\varphi_i\partial^\mu\varphi_i-\frac{1}{2}m^2\varphi_i^2\right].$$

---

persymmetric Standard Model is one of the best studied candidates for physics beyond the Standard Model.

Traditional symmetries in physics are generated by objects that transform under the tensor representations of the *Poincaré group* and *internal symmetries*. Supersymmetries, on the other hand, are generated by objects that transform under the *spinor representations*. According to the spin–statistics Theorem, bosonic fields commute while fermionic fields anticommute. In order to combine the two kinds of fields into a single algebra requires the introduction of a Z_2-grading under which the bosons are the even elements and the fermions are the odd elements. Such an algebra is called a *Lie superalgebra*.

The simplest supersymmetric extension of the *Poincaré algebra* contains two *Weyl spinors* with the following anti–commutation relation:

$$\{Q_\alpha,\bar{Q}_{\dot\beta}\}=2(\sigma^\mu)_{\alpha\dot\beta}P_\mu$$

and all other anti–commutation relations between the Qs and Ps vanish. In the above expression, $P_\mu=-i\partial_\mu$ denote the generators of translation and σ^μ are the *Pauli matrices*.

There are representations of a *Lie superalgebra* that are analogous to representations of a Lie algebra. Each Lie algebra has an associated Lie group and a Lie superalgebra can sometimes be extended into representations of a *Lie supergroup*.

By introducing vector of fields $\Phi = (\varphi_1, \varphi_2, \dots, \varphi_n)^T$, the Lagrangian density can be compactly written as

$$\mathcal{L} = \frac{1}{2}(\partial_\mu \Phi)^T \partial^\mu \Phi - \frac{1}{2} m^2 \Phi^T \Phi.$$

It is now transparent that the Lagrangian is invariant under the transformation $\Phi \mapsto G\Phi$,whenever G is a constant matrix belonging to the orthogonal group $O(n)$. This is the global symmetry of this particular Lagrangian, and the symmetry group is often called the *gauge group*. Incidentally, *Noether's Theorem* implies that invariance under this group of transformations leads to the conservation of the current,

$$J_\mu^a = i\partial_\mu \Phi^T T^a \Phi,$$

where the T^a matrices are generators of the $SO(n)-$group. There is one conserved current for every generator.

Now, demanding that this Lagrangian should have local $O(n)-$invariance requires that the G matrices (which were earlier constant) should be allowed to become functions of the space–time coordinates x^μ. Unfortunately, the G matrices do not 'pass through' the derivatives. When $G = G(x)$, we get

$$\partial_\mu (G\Phi)^T \partial^\mu (G\Phi) \neq \partial_\mu \Phi^T \partial^\mu \Phi.$$

This suggests defining the *gauge–covariant derivative* D with the property

$$D_\mu (G(x)\Phi(x)) = G(x) D_\mu \Phi.$$

It can be checked that such a covariant derivative is

$$D_\mu = \partial_\mu + g A_\mu(x),$$

where the *gauge field* $A(x)$ is defined to have the *transformation law*

$$A_\mu(x) \mapsto G(x) A_\mu(x) G^{-1}(x) - \frac{1}{g} \partial_\mu G(x) G^{-1}(x),$$

and g is the coupling constant, a quantity defining the strength of an interaction. The gauge field $A(x)$ is an element of the Lie algebra, and can therefore be expanded as

$$A_\mu(x) = \sum_a A_\mu^a(x) T^a.$$

Therefore, there are as many gauge fields as there are generators of the Lie algebra.

Finally, we now have a *locally gauge invariant Lagrangian*

$$\mathcal{L}_{loc} = \frac{1}{2}(D_\mu \Phi)^T D^\mu \Phi - \frac{1}{2} m^2 \Phi^T \Phi.$$

The difference between this Lagrangian and the original *globally gauge-invariant Lagrangian* is seen to be the *interaction Lagrangian*,

$$\mathcal{L}_{int} = \frac{g}{2}\Phi^T A_\mu^T \partial^\mu \Phi + \frac{g}{2}(\partial_\mu \Phi)^T A^\mu \Phi + \frac{g^2}{2}(A_\mu \Phi)^T A^\mu \Phi.$$

This term introduces interactions between the n scalar fields just as a consequence of the demand for local gauge invariance. In the quantized version of this classical field theory, the quanta of the gauge field $A(x)$ are called *gauge bosons*. The interpretation of the interaction Lagrangian in quantum field theory is of scalar bosons interacting by the exchange of these gauge bosons.

Yang–Mills Lagrangian

Our picture of classical gauge theory is almost complete except for the fact that to define the covariant derivatives D, one needs to know the value of the gauge field $A(x)$ at all space–time points. Instead of manually specifying the values of this field, it can be given as the solution to a field equation. Further requiring that the Lagrangian which generates this field equation is locally gauge invariant as well, one possible form for the gauge field Lagrangian is (conventionally) written as

$$\mathcal{L}_{gf} = -\frac{1}{4}\operatorname{Tr}(F^{\mu\nu}F_{\mu\nu}) \qquad \text{with} \qquad F_{\mu\nu} = [D_\mu, D_\nu]$$

and the trace being taken over the vector space of the fields. This is called the *Yang–Mills action*.[2] The complete Lagrangian for the $O(n)$ gauge theory is now[3]

$$\mathcal{L} = \mathcal{L}_{loc} + \mathcal{L}_{gf} = \mathcal{L}_{glob} + \mathcal{L}_{int} + \mathcal{L}_{gf}.$$

QED Lagrangian

As a simple application of the above formalism, consider the case of *quantum electrodynamics* (QED) with only the electron field. Recall that QED is a *relativistic quantum field theory* of electromagnetism. QED mathematically describes all phenomena involving electrically charged particles interacting by means of exchange of photons, whether the interaction is between light and matter or between two charged particles. It has been called 'the jewel of

[2] In this Lagrangian there is not a field Φ whose transformation counterweights the one of A. Invariance of this term under gauge transformations is a particular case of *a priori* classical (or geometrical) symmetry. This symmetry must be restricted in order to perform quantization, the procedure called *gauge fixing*, but even after restriction, gauge transformations are possible.

[3] Other gauge invariant actions also exist (e.g., nonlinear electrodynamics, Born–Infeld action, Chern–Simons model, etc.).

physics' for its extremely accurate predictions of quantities like the anomalous magnetic moment of the electron, and the *Lamb shift* of the energy levels of hydrogen.

Recall that in classical physics, due to interference, light is observed to take the stationary path between two points; but how does light know where it's going? That is, if the start and end points are known, the path that will take the shortest time can be calculated. However, when light is first emitted, the end point is not known, so how is it that light always takes the quickest path? In some interpretations, it is suggested that according to QED light does not have to — it simply goes over every possible path, and the observer (at a particular location) simply detects the mathematical result of all wave functions added up (as a sum of all line integrals – histories). For other interpretations, paths are viewed as non physical, mathematical constructs that are equivalent to other, possibly infinite, sets of mathematical expansions. According to R. Feynman, light can go slower or faster than c, but will travel at speed c on average.

Physically, QED describes charged particles (and their antiparticles) interacting with each other by the exchange of photons. The magnitude of these interactions can be computed using perturbation theory; these rather complex formulas have a remarkable pictorial representation as Feynman diagrams. QED was the theory to which Feynman diagrams were first applied. These diagrams were invented on the basis of Lagrangian mechanics. Using a Feynman diagram, one decides every possible path between the start and end points. Each path is assigned a *complex–valued probability amplitude*, and the actual amplitude we observe is the sum of all amplitudes over all possible paths. Obviously, among all possible paths the ones with stationary phase contribute most (due to lack of destructive interference with some neighboring counter–phase paths); this results in the stationary classical path between the two points.

The bare–bones action which generates the electron field's *Dirac equation* is

$$\mathcal{S} = \int \bar{\psi}(i\hbar c\,\gamma^{\mu}\partial_{\mu} - mc^2)\psi\,\mathrm{d}^4x.$$

The global symmetry for this system is

$$\psi \mapsto \mathrm{e}^{i\theta}\psi.$$

The gauge group here is $U(1)$, that is just the phase angle of the field, with a constant θ. Localising this symmetry implies the replacement of θ by $\theta(x)$. An appropriate covariant derivative is then

$$D_{\mu} = \partial_{\mu} - i\frac{e}{\hbar}A_{\mu}.$$

Identifying the 'charge' e with the usual electric charge (this is the origin of the usage of the term in gauge theories), and the gauge field $A(x)$ with the four–vector potential of electromagnetic field results in an interaction Lagrangian

$$\mathcal{L}_{int} = \frac{e}{\hbar}\bar{\psi}(x)\gamma^\mu\psi(x)A_\mu(x) = J^\mu(x)A_\mu(x).$$

where $J^\mu(x)$ is the usual four–vector electric current density. The gauge principle is therefore seen to introduce the so–called *minimal coupling* of the electromagnetic field to the electron field in a natural fashion. Adding a Lagrangian for the gauge field $A_\mu(x)$ in terms of the field strength tensor exactly as in electrodynamics, one gets the Lagrangian which is used as the starting point in *quantum electrodynamics*

$$\mathcal{L}_{\mathrm{QED}} = \bar{\psi}(i\hbar c\,\gamma^\mu D_\mu - mc^2)\psi - \frac{1}{4\mu_0}F_{\mu\nu}F^{\mu\nu}.$$

In the so–called *normal units*, this simplifies into

$$\mathcal{L}_{\mathrm{QED}} = \bar{\psi}(i\gamma^\mu D_\mu - m)\psi - \frac{1}{4}F_{\mu\nu}F^{\mu\nu},$$

where γ_μ are *Dirac gamma matrices*

$$\gamma^\mu = (\gamma^0, \gamma^1, \gamma^2, \gamma^3), \qquad \text{and} \qquad \gamma_\mu = (\gamma^0, -\gamma^1, -\gamma^2, -\gamma^3). \qquad \text{with}$$

$$\gamma^0 = \begin{pmatrix} 1 & 0 & 0 & 0 \\ 0 & 1 & 0 & 0 \\ 0 & 0 & -1 & 0 \\ 0 & 0 & 0 & -1 \end{pmatrix}, \qquad \gamma^1 = \begin{pmatrix} 0 & 0 & 0 & 1 \\ 0 & 0 & 1 & 0 \\ 0 & -1 & 0 & 0 \\ -1 & 0 & 0 & 0 \end{pmatrix},$$

$$\gamma^2 = \begin{pmatrix} 0 & 0 & 0 & -i \\ 0 & 0 & i & 0 \\ 0 & i & 0 & 0 \\ -i & 0 & 0 & 0 \end{pmatrix}, \qquad \gamma^3 = \begin{pmatrix} 0 & 0 & 1 & 0 \\ 0 & 0 & 0 & -1 \\ -1 & 0 & 0 & 0 \\ 0 & 1 & 0 & 0 \end{pmatrix};$$

ψ and its Dirac adjoint $\bar{\psi}$ are the wave–fields representing electrically charged particles, specifically electron and positron fields represented as *Dirac spinors*; $D_\mu = \partial_\mu + ieA_\mu$ is the gauge covariant derivative, with e the coupling strength (equal to the elementary charge), A_μ is the covariant vector potential of the electromagnetic field and $F_{\mu\nu} = \partial_\mu A_\nu - \partial_\nu A_\mu$ is the electromagnetic field tensor. By replacing the definition of D into the Lagrangian we get (skipping the subscript $_{\mathrm{QED}}$ for simplicity)

$$\mathcal{L} = i\bar{\psi}\gamma^\mu\partial_\mu\psi - e\bar{\psi}\gamma_\mu A^\mu\psi - m\bar{\psi}\psi - \frac{1}{4}F_{\mu\nu}F^{\mu\nu}.$$

To get the field equations for QED, we need to plug this Lagrangian into the *Euler–Lagrange equation* of motion for a field,

$$\partial_\mu\left(\frac{\partial\mathcal{L}}{\partial(\partial_\mu\psi)}\right) - \frac{\partial\mathcal{L}}{\partial\psi} = 0.$$

The two terms from this Lagrangian are then

$$\partial_\mu \left(\frac{\partial \mathcal{L}}{\partial(\partial_\mu \psi)} \right) = \partial_\mu \left(i\bar{\psi}\gamma^\mu \right)$$
$$\frac{\partial \mathcal{L}}{\partial \psi} = -e\bar{\psi}\gamma_\mu A^\mu - m\bar{\psi}$$

Putting these two terms back into the Euler–Lagrange equation results in

$$i\partial_\mu \bar{\psi}\gamma^\mu + e\bar{\psi}\gamma_\mu A^\mu + m\bar{\psi} = 0$$

and the complex–conjugate

$$i\gamma^\mu \partial_\mu \psi - e\gamma_\mu A^\mu \psi - m\psi = 0.$$

If we bring the middle term to the right-hand side, we finally get

$$i\gamma^\mu \partial_\mu \psi - m\psi = e\gamma_\mu A^\mu \psi.$$

The right hand side of of this equation is the interaction with the electromagnetic field, while the left hand side is the celebrated *Dirac equation*,[4]

$$i\gamma^\mu \partial_\mu \psi - m\psi = 0,$$

which in our original units reads

[4] Dirac equation was originally invented to describe the electron; however, the equation also applies to quarks, which are also elementary spin–$\frac{1}{2}$ particles. A *modified Dirac equation* can be used to approximately describe protons and neutrons, which are not elementary particles (they are made up of quarks), but have a net spin of $\frac{1}{2}$. Another modification of the Dirac equation, called the *Majorana equation*, is thought to describe neutrinos, which are also spin–$\frac{1}{2}$ particles.

The Dirac equation describes the *probability amplitudes* for a single electron. This is a single–particle theory; in other words, it does not account for the creation and destruction of the particles. It gives a good prediction of the magnetic moment of the electron and explains much of the fine structure observed in atomic spectral lines. It also explains the spin of the electron. Two of the four solutions of the equation correspond to the two spin states of the electron. The other two solutions make the peculiar prediction that there exist an infinite set of quantum states in which the electron possesses negative energy. This strange result led Dirac to predict, via a remarkable hypothesis known as 'hole theory', the existence of positrons, particles behaving like positively–charged electrons.

Despite these successes, Dirac's theory is flawed by its neglect of the possibility of creating and destroying particles, one of the basic consequences of relativity. This difficulty is resolved by reformulating it as a *quantum field theory*. Adding a quantized electromagnetic field to this theory leads to the theory of quantum electrodynamics (QED). Moreover the equation cannot fully account for particles of negative energy but is restricted to positive energy particles. A similar equation for spin 3/2 particles is called the *Rarita–Schwinger equation*.

$$i\hbar\gamma^\mu\partial_\mu\psi - mc\psi = 0,$$

Similarly, from the complex–conjugate relation above we get the Dirac equation for anti–particles,

$$i\hbar\partial_\mu\bar{\psi}\gamma^\mu + mc\bar{\psi} = 0.$$

Geometrical Gauge Formalism

Gauge theories are usually discussed in the language of *differential geometry*. Technically, a gauge is a choice of a local section of some *principal bundle*. A gauge transformation is a transformation between two such sections. If we have a principal bundle P whose base space is space or space–time and structure group is a Lie group, then the sections of P form a principal homogeneous space of the group of gauge transformations. We can define a *gauge connection* on this principal bundle, yielding a covariant derivative D in each associated vector bundle. If we choose a local frame X (a local basis of sections) then we can represent D by the connection form A, as

$$DX = dX + AX,$$

where d is the *exterior derivative* and A is a Lie algebra–valued 1–form which is called the *gauge potential* and which is not an intrinsic but a *frame–dependent quantity*. From this connection form we can construct the curvature form F, a Lie algebra–valued 2–form which is an *intrinsic quantity*, by

$$F = dA + A \wedge A,$$

where $\wedge$ is the *wedge product*. The *Yang–Mills action* is now given by

$$\frac{1}{4g^2}\int \mathrm{Tr}[*F \wedge F],$$

where $*$ is the *Hodge–star* dual, while the integral is defined as in differential geometry.

5.2 Monopoles

Recall that *monopoles* are solutions of a first order PDE called the *Bogomolny equation*. They can be thought of as approximated by static, magnetic particles in $\mathbb{R}^3$. As a background material, see [AH88]; as a survey, see [Sut97]. In this section we mainly follow [Mur01], using the notation from [AH88].

5.2.1 Monopoles in $\mathbb{R}^3$

We start with $\mathfrak{su}(2)$, the Lie algebra of all skew–hermitian 2×2 matrices. Let A be a one–form with values in $\mathfrak{su}(2)$, so that

$$A = A_i dx^i,$$

and each A_i is a function

$$A_i \colon \mathbb{R}^3 \to \mathfrak{su}(2).$$

Similarly, the Higgs field Φ is a function

$$\Phi \colon \mathbb{R}^3 \to \mathfrak{su}(2).$$

The one–form A can be thought of as the connection one–form for a connection

$$\nabla = d + A$$

on a trivial $SU(2)$ bundle on $\mathbb{R}^3$ [Mur01]. The curvature of such a connection is the two–form

$$F_A = \frac{1}{2} F_{ij} dx^i \wedge dx^j, \qquad \text{where}$$
$$F_{ij} = [\nabla_i, \nabla_j] = \partial_i A_j - \partial_j A_i + [A_i, A_j].$$

The connection A can be used to covariantly differentiate the Higgs field Φ to get

$$\nabla_A \Phi = (\partial_i \Phi + [A_i, \Phi])\, dx^i.$$

A monopole is a pair (A, Φ) satisfying the *Bogomolny equations* and some particular boundary conditions. The Bogomolny equations read [Mur01]:

$$F_A = *\nabla_A \Phi, \qquad (5.1)$$

where $*$ is the *Hodge star (duality) operator*, mapping one–forms into two–forms, as

$$*dx^1 = dx^2 \wedge dx^3, \qquad *dx^2 = dx^3 \wedge dx^1, \qquad *dx^3 = dx^1 \wedge dx^2.$$

If A and B are elements of $\mathfrak{su}(2)$, let $\langle A, B \rangle$ denote the *invariant form* as

$$\langle A, B \rangle = -\mathrm{tr}(AB^t).$$

Then the *energy density* of a pair (A, Φ) is defined by

$$e(A, \Phi) = \frac{1}{2} |F_A|^2 + \frac{1}{2} |\nabla_A \Phi|^2, \qquad \text{where} \qquad (5.2)$$
$$|F_A|^2 = \sum_{i<j} \langle F_{ij}, F_{ij} \rangle, \qquad \text{and}$$
$$|\nabla_A \Phi|^2 = \frac{1}{2} \langle \nabla_i \Phi, \nabla_i \Phi \rangle.$$

The *Yang–Mills–Higgs action* $\mathcal{L}_{YMH}$ of a pair (A, Φ) is the integral of the energy density over Euclidean three space,

$$\mathcal{L}_{YMH}(A, \Phi) = \int_{\mathbb{R}^3} e(A, \Phi) d^3x. \tag{5.3}$$

If B_R is a ball of radius R, integrating by parts shows that

$$\int_{B_R} e(A, \Phi)\, d^3x = \int_{B_R} \frac{1}{2} |F_A \pm *\nabla_A \Phi|^2 d^3x \mp \int_{S^2_R} \langle F_A, \Phi \rangle,$$

where S^2_R is the sphere of radius R. If the limits of all these integrals exist, as $R \to \infty$ we get

$$\mathcal{L}_{YMH}(A, \Phi) = \int_{\mathbb{R}^3} \frac{1}{2} |F_A \pm *\nabla_A \Phi|^2 d^3x \mp \lim_{R \to \infty} \int_{S^2_R} \langle F_A, \Phi \rangle.$$

From this expression we can deduce that the minima of the Yang–Mills–Higgs functional are solutions of the Bogomolny equations (5.1) or the anti–Bogomolny equations

$$F_A = - * \nabla_A \Phi.$$

As changing Φ to $-\Phi$ changes a solution of the Bogomolny equations to a solution of the anti–Bogomolny equations we can focus our attention on the former [Mur01].

The Bogomolny equations are invariant under *gauge transformations*, i.e., replacing (A, Φ) by

$$(gAg^{-1} + gd(g^{-1}), g\Phi g^{-1}), \qquad \text{where} \qquad g \colon \mathbb{R}^3 \to SU(2).$$

The energy density (5.2) is also invariant under gauge transformations. When we talk about a monopole we are really talking about an equivalence class of (A, Φ) under gauge transformations.

The boundary conditions imposed on a monopole are primarily that the energy density (5.2) should have finite integral, i.e., the action (5.3) is finite.[5] From these boundary conditions we can deduce that, after a suitable gauge transformation, we can arrange for the Higgs field to have a limiting value at infinity

$$\Phi^\infty(u) = \lim_{t \to \infty} \Phi(tu), \qquad \text{where} \quad u \in S^2.$$

The boundary conditions can be used to show that the eigenvalues of the Higgs field at infinity are independent of the direction $u \in S^2$. In the case of $SU(2)$ this is equivalent to the fact that $|\Phi(u)|^2$ is constant for all u.

It is easy to show that if $c > 0$ and (A, Φ) solves the Bogomolny equations, then

[5] There are also some purely technical conditions; however, it is believed that these can all be deduced from finiteness of the action and the Bogomolny equations.

$$(\hat{A}(x), \hat{\Phi}(x)) = (cA(x/c), c\Phi(x/c))$$

also solves the Bogomolny equations. So we may as well normalize the Higgs field so that $|\Phi(u)|^2 = 1$ for all directions u. Because the Lie algebra $\mathfrak{su}(2)$ is three dimensional the Higgs field at infinity is a map

$$\Phi^\infty : S^2 \to S^2 \subset \mathfrak{su}(2).$$

The space of all continuous maps $S^2 \to S^2$ breaks up into connected components labelled by a winding number $k \in$ just as for maps $S^1 \to S^1$.

Because of this boundary condition we can arrange by a gauge transformation for any Higgs field to satisfy the relation [Mur01]

$$\lim_{t\to\infty} \Phi(0,0,t) = \frac{1}{\sqrt{2}} \begin{pmatrix} i & 0 \\ 0 & -i \end{pmatrix}.$$

We call such a Higgs field *framed*. We define the *moduli space* $\mathcal{M}_k$ of all monopoles of charge k to be the space of all (A, Φ) solving the Bogomolny equations and satisfying the appropriate boundary conditions, with the Higgs field framed and modulo the action of gauge transformations satisfying

$$\lim_{t\to\infty} g(0,0,t) = \begin{pmatrix} 1 & 0 \\ 0 & 1 \end{pmatrix}. \tag{5.4}$$

If $k \leq 0$ then $\mathcal{M}_k = \emptyset$, otherwise $\mathcal{M}_k$ is a smooth manifold of dimension $4k$.

Notice that the Bogomolny equations are translation invariant. Moreover, because of the way we have defined the framing, the group of all diagonal matrices (a copy of the circle group S^1) acts by constant gauge transformations on $\mathcal{M}_k$. Hence $\mathbb{R}^3 \times S^1$ acts on $\mathcal{M}_k$.

If $k = 1$ there is, up to this action of $\mathbb{R}^3 \times S^1$, a unique monopole, the so–called Bogomolny–Prasad–Sommerfield (BPS) monopole, given by

$$\Phi(x) = \left(\frac{1}{r} - \frac{1}{\tanh r}\right)\frac{e}{r},$$
$$A(x) = \left(\frac{1}{\sinh r} - \frac{1}{r}\right)\frac{[e, de]}{r},$$

where $r = |x|$ and $e(x) = x^i \mathrm{e}^i$ for an orthonormal basis e^1, e^2, e^3 of $\mathfrak{su}(2)$.

The other $k = 1$ monopoles are obtained by acting by $\mathbb{R}^3 \times S^1$ so that $\mathcal{M}_1 = \mathbb{R}^3 \times S^1$. It is easy to calculate the energy density of the BPS monopole as follows. For any monopole there is a useful formula

$$e(A, \Phi) = \sum_{i=1}^{3} \frac{\partial^2}{(\partial x^i)^2} \langle \Phi, \Phi \rangle,$$

which can be proved using the Bogomolny equations and the Bianchi identity [Mur01]. From this we have for the BPS monopole,

$$e(A,\Phi) = \frac{6}{\tanh^4 r} - \frac{8}{\tanh^2 r} + 2 + \frac{2}{r^4} - \frac{8}{r\tanh^3 r} + \frac{8}{r\tanh r},$$

where $r = |x|$. Clearly, the energy density is spherically symmetric and concentrated around the origin in $\mathbb{R}^3$. We can think of the monopole as a particle located at the origin.

The BPS monopole was discovered by Prasad and Sommerfield in 1975 [PS75]. For some time this was the only monopole known and it was unclear whether higher charge monopoles existed. In 1977 Manton [Man77] showed that to first order the forces between two monopoles, due the Higgs field and the connection, cancelled. This lead to the conjecture that stable higher charge monopoles would exist. In 1979 Weinberg [Wei79] calculated that the dimension of the moduli space of monopoles would be $4k$ if it was non-empty. Finally in 1981 Taubes [JT80] showed that the moduli space was non–empty.

It is important to note that when the charge is greater than one we cannot associate to every monopole (A,Φ) a collection of k points which are the locations of the k particles we think of as its constituents. We can however associate sensibly to a k monopole a *center of mass* or location [HMM95].

The analysis of monopoles directly in terms of the connection and Higgs field on $\mathbb{R}^3$, for example the definition of the location of a monopole, while possible, is difficult. Part of this difficulty stems from the infinite dimensional symmetry group of gauge transformations. Research on monopoles has focused on various transformations which are designed to construct some other mathematical data equivalent to the monopole. Study of these data then, hopefully, sheds light on the original monopole. This process is particularly useful if the object produced is an invariant of the monopole, something which does not change under gauge transformation.

5.2.2 Spectral Curve

Let $\gamma(t)$ be an oriented line in $\mathbb{R}^3$. This can be put in the form [Mur01]

$$\gamma(t) = v + tu,$$

with vectors u and v determined uniquely by the requirement that $|u| = 1$, $\langle u, v\rangle = 0$ and u points in the direction of the orientation. Along the line γ we have the *Hitchin differential equation* [Hit83]

$$\left(\partial_t + u^i A_i(\gamma(t)) - i\Phi(\gamma(t))\right) s(t) = 0. \tag{5.5}$$

This is an ODE, so it has a 2D space of solutions E_γ.

Notice that by a gauge transformation we can arrange, for any given line γ, that $u^i A_i = 0$, so we can essentially disregard this term. The boundary conditions can be used to show that we can expand the Higgs field along any line as

$$\Phi(\gamma(t)) = \begin{pmatrix} i & 0 \\ 0 & -i \end{pmatrix} - \frac{1}{2t}\begin{pmatrix} ik & 0 \\ 0 & -ik \end{pmatrix} + O\left(\frac{1}{t^2}\right),$$

where k denotes the *monopole charge*. We want to consider the Hitchin equation as a perturbation of a modified Hitchin equation,

$$\left(\partial_t + \begin{pmatrix} 1 & 0 \\ 0 & -1 \end{pmatrix} - \frac{1}{2t}\begin{pmatrix} k & 0 \\ 0 & -k \end{pmatrix}\right) s = 0. \tag{5.6}$$

The modified equation is the Hitchin equation with the $o(1/t^2)$ term in the Higgs field expansion set to zero. We use this to study the behavior of solutions of the Hitchin equation. The solutions of (5.6) are given by [Mur01]

$$s_1(t) = \begin{pmatrix} t^{k/2}\mathrm{e}^{-t} \\ 0 \end{pmatrix}, \qquad s_2(t) = \begin{pmatrix} 0 \\ t^{-k/2}\mathrm{e}^{t} \end{pmatrix}.$$

Asymptotic analysis of ordinary differential equations shows that for any line there are solutions $s_1(t)$ and $s_2(t)$ of the Hitchin equation (5.5) which behave asymptotically like the solutions to the modified Hitchin equation 5.6, that is they satisfy

$$\lim_{t\to\infty} t^{-k/2}\mathrm{e}^{t} s_1(t) = \begin{pmatrix} 1 \\ 0 \end{pmatrix}, \qquad \lim_{t\to\infty} t^{k/2}\mathrm{e}^{-t} s_2(t) = \begin{pmatrix} 0 \\ 1 \end{pmatrix}.$$

Similarly there are solutions that decay and blow up exponentially as $t \to -\infty$. For the modified Hitchin equation a solution that blows up (decays) at one end of the line decays (blows up) at the other end. In general this will not be true of solutions of the Hitchin equation. In particular asymptotic analysis tells us that there will be a ball in $\mathbb{R}^3$ of radius $R > 0$ with the property that if a line lies outside the ball then the solutions $s_1(t)$ and $s_2(t)$ behave like the solutions to the modified Hitchin equation, that is $s_1(t)$ decays as $t \to -\infty$ and $s_2(t)$ blows up as $t \to -\infty$.

We expect then that lines which *do not* exhibit this behavior are somehow close to the monopole. We call a line γ a *spectral line* if there is a solution to the Hitchin equation which decays at both ends. We call the set of all spectral lines the *spectral curve* of the monopole. It is easy to see that being a spectral line for a monopole is independent of gauge transformations so the spectral curve is an invariant of the monopole.

It is not difficult to show that for the BPS monopole located at the point $x \in \mathbb{R}^3$ the spectral lines are exactly the lines passing through x. Note that this is a two-dimensional set, indeed a copy of S^2. This is more generally true: the spectral curve is always a two-dimensional family of lines. To say more about the structure of the spectral curve we need to consider the set of all oriented lines in $\mathbb{R}^3$.

The importance of the spectral curve is the following *Hitchin spectral Theorem*: If monopoles (A, Φ) and (A', Φ') have spectral curves S and S' and $S = S'$ then (A, Φ) is an unframed gauge transform of (A', Φ').

5.2.3 Twistor Theory of Monopoles

As we have discussed above, each oriented line in $\mathbb{R}^3$ is determined uniquely by vectors u and v, satisfying $|u| = 1$, $\langle u, v\rangle = 0$. It follows that the set of all oriented lines is the tangent bundle TS^2 to the 2–sphere S^2, given by [Mur01]

$$TS^2 = \{(u, v) : |u| = 1, \quad \langle u, v\rangle = 0\}.$$

This is often called the *mini–twistor space* of $\mathbb{R}^3$. Mini–twistor space is naturally a complex smooth manifold and we can introduce local coordinates (η, ζ) on the open subset where $u \neq (0, 0, 1)$ by letting

$$\zeta = \frac{u^1 + iu^2}{1 - u^3} \qquad \text{and} \qquad \eta = (v^1 + iv^2) + 2v^3\zeta + (-v^1 + iv^2)\zeta^2. \tag{5.7}$$

The relationship between mini–twistor space and $\mathbb{R}^3$ is summarized by the equation [Mur01]

$$\eta = (x^1 + ix^2) + 2x^3\zeta + (-x^1 + ix^2)\zeta^2. \tag{5.8}$$

If we hold (η, ζ) fixed then the (x^1, x^2, x^3) satisfying (5.8) define a line in $\mathbb{R}^3$. On the other hand if we hold (x^1, x^2, x^3) fixed then the (η, ζ) satisfying (5.8) parameterize the set of all lines through the point $x = (x^1, x^2, x^3)$ which is a copy of S^2 inside TS^2.

Mini–twistor space has an *involution*

$$\tau\colon TS^2 \to TS^2,$$

which sends each oriented line to the same line with opposite orientation. In local coordinates this is given by

$$\tau(\eta, \zeta) = \left(-\frac{\bar{\eta}}{\bar{\zeta}^2}, -\frac{1}{\bar{\zeta}}\right).$$

As τ is similar to a conjugation it is called the *real structure*. The set of all lines through the point x is real in the sense that it is fixed by the real structure.

We can now state the basic result concerning the spectral curve. It is a subset of TS^2 defined by an equation of the form [Mur01]

$$p(\eta, \zeta) = \eta^k + a_1(\zeta)\eta^{k-1} + \cdots + a_k(\zeta) = 0, \tag{5.9}$$

where each of the $a_i(\zeta)$ is a polynomial of degree $2i$.

Note that by no means every such curve is the spectral curve of a monopole. One constraint is immediate from our definition. If a line is a spectral line then so also is the line with the opposite orientation. So the spectral curve is *real*, that is, fixed by the real structure. But more is true. The family of real curves defined by equations of the form (5.9) is $(k-1)^2 - 1$ real dimensional whereas the moduli space of monopoles is $4k$ dimensional. So there have to be further constraints on the spectral curve. In particular a certain holomorphic line bundle must be trivial when restricted to the spectral curve. It is possible to say quite precisely what the other constraints are and hence to say, in principle, which spectral curves give rise to monopoles [Hit82].

Spectral curves can be used to deduce a number of useful facts about monopoles. It is easy to show that the only real curves of the type (5.9) for $k = 1$ are those of the form (5.8) for some point (x^1, x^2, x^3). Hence the BPS monopoles are the only charge one monopoles.

The coefficient $a_1(\zeta)$ in (5.9) defines a real curve and hence has the form

$$a_1(\zeta) = (x^1 + ix^2) + 2x^3\zeta + (-x^1 + ix^2)\zeta^2,$$

for some point $(x^1, x^2, x^3) \in \mathbb{R}^3$. This point is called the center of the monopole. If $(u, v_1), \ldots, (u, v_k)$ is a collection of k parallel lines let us define their average to be the line $(u, (1/k)\sum_{i=1}^{k} v_i)$. Notice from definition of η (5.7) that if these lines have complex coordinates $(\eta_1, \zeta), \ldots, (\eta_k, \zeta)$ then their average has complex coordinates $((1/k)\sum_{i=1}^{k} \eta_i, \zeta)$. If we fix a particular direction in $\mathbb{R}^3$, that is fix a ζ and look for all the spectral lines in that direction we are finding all the η satisfying a degree k polynomial and hence there are generically k of them. If we take the average of all these lines then it will pass through the monopole center. This gives us a way of defining the center entirely in $\mathbb{R}^3$. Take the average of the spectral lines in each direction in $\mathbb{R}^3$, the resulting family of lines will (nearly) all intersect in a single point, that point is the center of the monopole.

The definition of the spectral curve of a monopole clearly preserves the action of the rotations and translations of $\mathbb{R}^3$ and this gives us a way of looking for monopoles with particular symmetries. We look first for spectral curves with these symmetries. This approach can be used to show that the only spherically symmetric monopole is the BPS monopole at the origin. It was used by Hitchin [Hit82] to classify the axially symmetric monopoles and more recently [HMM95] to find monopoles with symmetry groups those of the regular solids.

The various properties of the spectral curve such as the form of equation (5.9) are proved by using *Hitchin's twistor transform*. Hitchin [Hit83] introduces the vector space E_γ of all solutions to his equation (5.5). This is a two-dimensional space and the collection of them all defines a complex vector bundle E over the mini–twistor space TS^2. Hitchin shows that the Bogomolny equations imply that E is a holomorphic vector bundle and moreover the monopole can be recovered from knowing E. The boundary conditions of the monopole then enter by noting that there are two distinguished holomorphic sub–bundles E^+ and E^- of solutions to Hitchin's equation which decay at $+$ and $-$ infinity. The spectral curve is the set where these line bundles coincide. Algebraic geometry can then be used to prove the Hitchin spectral Theorem and that the spectral curve satisfies an equation of the form (5.9). Various constraints on the spectral curve also follow from the twistor theory. The precise constraints that a curve must satisfy to be the spectral curve are given in [Hit82].

5.2.4 Nahm Transform and Nahm Equations

An alternative point of view on monopoles comes via Nahm's adaption of the Atiyah, Drinfeld, Hitchin, Manin construction of instantons [Nah82]. Nahm considers a *Dirac operator* D_x on $\mathbb{R}^3$ coupled to the monopole. In more detail let σ^i be an orthonormal basis for the Lie algebra $\mathfrak{su}(2)$. This particular $\mathfrak{su}(2)$ should be regarded as different to the monopole $\mathfrak{su}(2)$ in which the connection and Higgs field take values. It is, in fact, the Lie algebra of the *spin group* of the *group of rotations* of $\mathbb{R}^3$. The Dirac operator D_x is defined by

$$D_x = \sigma^i \nabla_i - (\Phi + ix)$$

and acts on $\mathbb{C}^2 \otimes \mathbb{C}^2$–valued functions on $\mathbb{R}^3$. The first $\mathbb{C}^2$ is the *spin–space* on which the σ^i act and the second is the space on which the A_i and Φ act. Here x is any real number. We also define the adjoint

$$D_x^* = \sigma^i \nabla_i + (\Phi + ix).$$

If we compute the composition the Bogomolny equations show us that [Mur01]

$$D_x D_x^* = \sum_{i=1}^{3} \nabla_i \nabla_i - (\Phi + ix)(\Phi + ix)$$

which is a positive operator and hence has no L^2 kernel. From this we conclude that D_x^* has no L^2 kernel. An L^2 *index Theorem of Callias* shows that D_x has index k if $-1 < x < 1$ and 0 otherwise. Hence it follows that D_x has a k–dimensional L^2 kernel N_x if $-1 < x < 1$. The point of view we wish to adopt is that N_x is a kD vector bundle over the interval $(-1, 1)$.

We are interested in sections of this vector bundle, that is functions

$$\psi\colon (-1,1) \times \mathbb{R}^3 \to \mathbb{C}^2 \otimes \mathbb{C}^2, \qquad \text{such that}$$
$$D_x \psi(x, x) = 0 \qquad \text{for every} \quad x \in (-1, 1).$$

Choose k of these $\psi^1, \ldots, \psi^k$ so that for each x they span N_x. Moreover choose them so that they are orthonormal

$$\int_{\mathbb{R}^3} (\psi^i, \psi^j)\, d^3x = \delta^{ij}, \qquad \text{and satisfy}$$
$$\int_{\mathbb{R}^3} \left(\psi^i, \frac{\partial \psi^j}{\partial z} \right) d^3x = 0, \qquad \text{for all} \quad i, j = 1 \ldots, k.$$

Notice that there is no obstruction to satisfying these extra conditions. We can use *Gramm–Schmidt orthogonalization* for the first and the second is just solving an ordinary differential equation on $(-1, 1)$.

Now we define three $k \times k$ matrix functions of x by [Mur01]

$$T_a^{ij}(x) = \int_{\mathbb{R}^3} (\psi^i, x^a \psi^j)\, d^3x, \qquad \text{for } i,j = 1\ldots,k \text{ and } a = 1,2,3.$$

The remarkable thing about this *Nahm transformation* is that these matrix–valued functions satisfy some simple ODEs, called *Nahm equations*

$$\frac{dT_1}{dx} = [T_2, T_3], \qquad \frac{dT_2}{dx} = [T_3, T_1], \qquad \frac{dT_3}{dx} = [T_1, T_2]. \tag{5.10}$$

It is possible to cast the Nahm equations into Lax form and solve them (by the Krichever method). Define

$$\begin{aligned} A(\zeta) &= (T_1 + iT_2) + 2T_3\zeta + (-T_1 + iT_2)\zeta^2, \qquad \text{and} \\ A_+(\zeta) &= iT_3 - (iT_1 + T_2)\zeta. \end{aligned}$$

Then the Nahm equations (5.10) are equivalent to

$$\frac{dA}{dx} = [A_+, A],$$

which is in Lax form. Now consider the curve S_x in $\mathbb{C} \times \mathbb{C}$ defined by

$$\det(\eta - A(\zeta)) = 0. \tag{5.11}$$

Then we have the following result: the curve S_x is independent of x. For proof, see [Mur01].

If we identify the (η, ζ) in (5.11) with the coordinates (5.7) on mini–twistor space, we realise S_x as a curve in mini–twistor space. It is a remarkable fact [HM88] that the curve $S = S_x$ defined via Nahm's transform is the same as the spectral curve of the monopole defined before.

Standard methods from integrable systems can be used to solve the Nahm equations using the curve S and some additional structure. Indeed in [Hit82] Hitchin uses this approach to determine exactly which spectral curves correspond to monopoles.

One of the important properties of the Nahm equations is that it is straightforward to define a monopole from a solution of the Nahm equations plus some boundary conditions. Given such a solution and a point (x^1, x^2, x^3) in $\mathbb{R}^3$ we define

$$x = T_i\sigma^i \qquad \text{and} \qquad x = x^i\sigma^i.$$

Now define E_x to be the L^2 kernel of the 2D operator

$$D_x = \partial_x - T - x.$$

We define the connection and Higgs field by choosing an orthonormal basis (v_1, v_2) for E_x and letting

$$\begin{aligned} \Phi(v_a) &= \sum_{b=1}^{2} \int_{-1}^{1} (v_b, x v_a)\, dx, \qquad \text{and} \\ A_i(v_a) &= \sum_{b=1}^{2} \int_{-1}^{1} \left(v_b, \frac{dv_a}{dx^i}\right) dx. \end{aligned}$$

This (A, Φ) define a monopole if the T_i satisfy the Nahm equations with the appropriate boundary conditions. Moreover if the solution of the Nahm equations came from a monopole this construction returns the same monopole. For more details see [Nah82], as well as section 6.2 below.

5.3 Hermitian Geometry and Complex Relativity

5.3.1 About Space–Time Complexification

The idea of complexifying space–time in general relativity was put forward in the early sixties. It appeared in different but related lines of research. These include complexifying the 4D manifold and equipping it with a *holomorphic metric*, asymptotically complex null surfaces and Penrose's twistor theory [Syn61], [New61], [Pen67], [PR86], [Fla76], [Fla80]. More recently, Witten [Wit88c] considered string propagation on complexified space–time where he presented some evidence that the imaginary part of the complex coordinates enters in the study of the high–energy behavior of scattering amplitudes [GM87]. In this string picture it is assumed that the imaginary parts of the coordinates are small at low–energies. At a fundamental level the complex coordinates X^μ, $\mu = 1, \cdots, n$ with complex conjugates $\overline{X^\mu} \equiv X^{\overline{\mu}}$ are described by the topological σ model action [Wit88c]

$$I = \int d\sigma d\overline{\sigma} g_{\mu\overline{\nu}} \left(X(\sigma, \overline{\sigma}), \overline{X}(\sigma, \overline{\sigma})\right) \partial_{\overline{\sigma}} X^\mu \partial_\sigma X^{\overline{\nu}},$$

where the world–sheet coordinates are denoted by σ and $\overline{\sigma}$, and where the background metric for the complex nD manifold M is Hermitian so that

$$\overline{g_{\mu\overline{\nu}}} = g_{\nu\overline{\mu}}, \qquad g_{\mu\nu} = g_{\overline{\mu}\,\overline{\nu}} = 0.$$

Decomposing the metric into real and imaginary components

$$g_{\mu\overline{\nu}} = G_{\mu\nu} + iB_{\mu\nu},$$

the hermiticity condition implies that $G_{\mu\nu}$ is symmetric and $B_{\mu\nu}$ is antisymmetric. The low–energy effective string action is given by the *Einstein–Hilbert action* coupled to the field strength of the antisymmetric tensor. This can be related to the invariance of the sigma model under complex transformations

$$X^\mu \to X^\mu + \zeta^\mu(X), \qquad X^{\overline{\mu}} \to X^{\overline{\mu}} + \zeta^{\overline{\mu}}(\overline{X}).$$

A related phenomena was observed in noncommutative geometry [AC94] where the space–time coordinates are deformed and become noncommuting [CDS98],

$$[x^\mu, x^\nu] = i\theta^{\mu\nu}.$$

Furthermore, It was found that in the *effective action of open–string theory*, the inverse of the combinations $(G_{\mu\nu} + B_{\mu\nu})^{-1}$ does appear [SW99]. This was taken as a motivation to study the dynamics of a complex Hermitian metric on a real manifold [AHC01], considered first by Einstein and Strauss [ES46]. In [AHC01] it was shown that the invariant action constructed have the required behavior for the propagation of the fields $G_{\mu\nu}$ and $B_{\mu\nu}$ at the linearized level, but problems do arise when nonlinear interactions are taken into account. This is due to the fact that there is no gauge symmetry to prevent the ghost components of $B_{\mu\nu}$ from propagating. It is then important to address the question of whether it is possible to have consistent interactions in which the field $B_{\mu\nu}$ appears explicitly in analogy with $G_{\mu\nu}$ and not only through the combination of derivatives

$$H_{\mu\nu\rho} = \partial_\mu B_{\nu\rho} + \partial_\nu B_{\rho\mu} + \partial_\rho B_{\mu\nu}.$$

This suggests that the gauge parameters for the transformation

$$B_{\mu\nu} \to B_{\mu\nu} + \partial_\mu \Lambda_\nu - \partial_\nu \Lambda_\mu$$

that keep $H_{\mu\nu\rho}$ invariant must be combined with the diffeomorphism parameters on the real manifold. For this to happen there must be diffeomorphism invariance of the Hermitian manifold M of complex dimensions n, with complex coordinates $z^\mu = x^\mu + iy^\mu$, $\mu = 1, \cdots, n$. The line element is then given by [Gol56]

$$ds^2 = 2g_{\mu\overline{\nu}} dz^\mu d\overline{z}^{\overline{\nu}},$$

where we have denoted $\overline{z^\mu} = z^{\overline{\mu}}$. The metric preserves its form under infinitesimal transformations

$$z^\mu \to z^\mu - \zeta^\mu(z), \qquad z^{\overline{\mu}} \to z^{\overline{\mu}} - \zeta^{\overline{\mu}}(\overline{z}),$$

as can be seen from the transformations

$$\begin{aligned} 0 = \delta g_{\mu\nu} &= \partial_\mu \zeta^{\overline{\lambda}} g_{\overline{\lambda}\nu} + \partial_\nu \zeta^{\overline{\lambda}} g_{\mu\overline{\lambda}}, \\ 0 = \delta g_{\overline{\mu}\,\overline{\nu}} &= \partial_{\overline{\mu}} \zeta^{\lambda} g_{\lambda\overline{\nu}} + \partial_{\overline{\nu}} \zeta^{\lambda} g_{\overline{\mu}\lambda}, \\ \delta g_{\mu\overline{\nu}} &= \partial_\mu \zeta^\lambda g_{\lambda\overline{\nu}} + \partial_{\overline{\nu}} \zeta^{\overline{\lambda}} g_{\mu\overline{\lambda}} + \zeta^\lambda \partial_\lambda g_{\mu\overline{\nu}} + \zeta^{\overline{\lambda}} \partial_{\overline{\lambda}} g_{\mu\overline{\nu}}. \end{aligned}$$

It is instructive to express these transformations in terms of the fields $G_{\mu\nu}(x,y)$ and $B_{\mu\nu}(x,y)$ by writing

$$\zeta^\mu(z) = \alpha^\mu(x,y) + i\beta^\mu(x,y), \qquad \zeta^{\overline{\mu}}(\overline{z}) = \alpha^\mu(x,y) - i\beta^\mu(x,y).$$

The holomorphicity conditions on ζ^μ and $\zeta^{\overline{\mu}}$ imply the relations

$$\partial_{y^\mu} \beta^\nu = \partial_{x^\mu} \alpha^\nu, \qquad \partial_{y^\mu} \alpha^\nu = -\partial_{x^\mu} \beta^\nu.$$

The transformations of $G_{\mu\nu}(x,y)$ and $B_{\mu\nu}(x,y)$ are then given by

$$\delta G_{\mu\nu}(x,y) = \partial_{x^\mu}\alpha^\lambda G_{\lambda\nu} + \partial_{x^\nu}\alpha^\lambda G_{\mu\lambda} + \alpha^\lambda\partial_{x^\lambda}G_{\mu\nu} - \partial_{x^\mu}\beta^\lambda B_{\lambda\nu} + \partial_{x^\nu}\beta^\lambda B_{\mu\lambda} + \beta^\lambda\partial_{x^\lambda}G_{\mu\nu},$$

$$\delta B_{\mu\nu}(x,y) = \partial_{x^\mu}\beta^\lambda G_{\lambda\nu} - \partial_{x^\nu}\beta^\lambda G_{\mu\lambda} + \alpha^\lambda\partial_{x^\lambda}B_{\mu\nu} + \partial_{x^\mu}\alpha^\lambda B_{\lambda\nu} + \partial_{x^\nu}\alpha^\lambda B_{\mu\lambda} + \beta^\lambda\partial_{x^\lambda}B_{\mu\nu}.$$

One readily recognizes that in the vicinity of small y^μ the fields $G_{\mu\nu}(x,0)$ and $B_{\mu\nu}(x,0)$ transform as symmetric and antisymmetric tensors with gauge parameters $\alpha^\mu(x)$ and $\beta^\mu(x)$ where

$$\alpha^\mu(x,y) = \alpha^\mu(x) - \partial_{x^\nu}\beta^\mu(x)y^\nu + O(y^2),$$
$$\beta^\mu(x,y) = \beta^\mu(x) + \partial_{x^\nu}\alpha^\mu(x)y^\nu + O(y^2),$$

as implied by the holomorphicity conditions.

In this section, following [Cha05], we will investigate the dynamics of the Hermitian metric $g_{\mu\overline{\nu}}$ on a complex space–time with complex dimensions four, such that in the limit of vanishing imaginary values of the coordinates, the action reduces to that of a symmetric metric $G_{\mu\nu}$ and an antisymmetric field $B_{\mu\nu}$.

5.3.2 Hermitian Geometry

The Hermitian manifold M of complex dimensions n is defined as a Riemannian manifold with real dimensions $2n$ with Riemannian metric g_{ij} and complex coordinates $z^i = \left\{z^\mu, z^{\overline{\mu}}\right\}$, where Latin indices $i,j,k,\cdots$, run over the range $1,2,\cdots,n,\overline{1},\overline{2},\cdots,\overline{n}$. The *invariant line element* is then [Yan65]

$$ds^2 = g_{ij}dz^i dz^j,$$

where the metric g_{ij} is hybrid

$$g_{ij} = \begin{pmatrix} 0 & g_{\mu\overline{\nu}} \\ g_{\nu\overline{\mu}} & 0 \end{pmatrix}.$$

It has also an integrable complex structure F_i^j satisfying

$$F_i^k F_k^j = -\delta_i^j,$$

and with a vanishing *Nijenhuis tensor*

$$N_{ji}{}^h = F_j^t\left(\partial_t F_i^h - \partial_i F_t^h\right) - F_i^t\left(\partial_t F_j^h - \partial_j F_t^h\right).$$

The complex structure has components [Cha05]

$$F_i^j = \begin{pmatrix} i\delta_\mu^\nu & 0 \\ 0 & -i\delta_{\overline{\mu}}^{\overline{\nu}} \end{pmatrix}.$$

The affine connection with torsion Γ^h_{ij} is introduced so that the following two conditions are satisfied

$$\nabla_k g_{ij} = \partial_k g_{ij} - \Gamma^h_{ik} g_{hj} - \Gamma^h_{jk} g_{ih} = 0,$$
$$\nabla_k F^j_i = \partial_k F^j_i - \Gamma^h_{ik} F^j_h + \Gamma^j_{hk} F^h_i = 0.$$

These conditions do not determine the affine connection uniquely and there exists several possibilities used in the literature. We shall adopt the *Chern connection*, which is the one most commonly used. It is defined by prescribing that the linear differential forms

$$\omega^i{}_j = \Gamma^i_{jk} dz^k,$$

be such that ω^μ_ν and $\omega^{\overline{\mu}}_{\overline{\nu}}$ are given by [Gol56]

$$\omega^\mu_\nu = \Gamma^\mu_{\nu\rho} dz^\rho, \qquad \overline{\omega^\mu_\nu} = \omega^{\overline{\mu}}_{\overline{\nu}} = \Gamma^{\overline{\mu}}_{\overline{\nu}\overline{\rho}} dz^{\overline{\rho}}.$$

For ω^μ_ν to have a metrical connection the differential of the metric tensor g must be given by

$$dg_{\mu\overline{\nu}} = \omega^\rho{}_\mu g_{\rho\overline{\nu}} + \omega^{\overline{\rho}}{}_{\overline{\nu}} g_{\mu\overline{\rho}},$$

from which we get

$$\partial_\lambda g_{\mu\overline{\nu}} dz^\lambda + \partial_{\overline{\lambda}} g_{\mu\overline{\nu}} dz^{\overline{\lambda}} = \Gamma^\rho_{\mu\lambda} g_{\rho\overline{\nu}} dz^\lambda + \Gamma^{\overline{\rho}}_{\overline{\nu}\overline{\lambda}} g_{\mu\overline{\rho}} dz^{\overline{\lambda}},$$
$$\text{so that} \qquad \Gamma^\rho_{\mu\lambda} = g^{\overline{\nu}\rho} \partial_\lambda g_{\mu\overline{\nu}}, \qquad \Gamma^{\overline{\rho}}_{\overline{\nu}\overline{\lambda}} = g^{\overline{\rho}\mu} \partial_{\overline{\lambda}} g_{\mu\overline{\nu}},$$

where the inverse metric $g^{\overline{\nu}\mu}$ is defined by

$$g^{\overline{\nu}\mu} g_{\mu\overline{\kappa}} = \delta^{\overline{\nu}}_{\overline{\kappa}}.$$

The condition $\nabla_k F^j_i = 0$ is then automatically satisfied and the connection is metric. The torsion forms are defined by [Cha05]

$$\Theta^\mu \equiv -\frac{1}{2} T_{\nu\rho}{}^\mu dz^\nu \wedge dz^\rho = \omega^\mu{}_\nu dz^\nu = -\Gamma^\mu_{\nu\rho} dz^\nu \wedge dz^\rho,$$

which implies that

$$T_{\nu\rho}{}^\mu = \Gamma^\mu_{\nu\rho} - \Gamma^\mu_{\rho\nu} = g^{\overline{\sigma}\mu} \left(\partial_\rho g_{\nu\overline{\sigma}} - \partial_\nu g_{\rho\overline{\sigma}} \right).$$

The torsion form is related to the differential of the Hermitian form

$$F = \frac{1}{2} F_{ij} dz^i \wedge dz^j, \qquad \text{where} \qquad F_{ij} = F^k_i g_{kj} = -F_{ji},$$

is antisymmetric and satisfy

$$F_{\mu\nu} = 0 = F_{\overline{\mu}\,\overline{\nu}}, \qquad F_{\mu\overline{\nu}} = i g_{\mu\overline{\nu}} = -F_{\overline{\nu}\mu}, \qquad \text{so that}$$
$$F = i g_{\mu\overline{\nu}} dz^\mu \wedge dz^{\overline{\nu}}.$$

The differential of F is then

$$dF = \frac{1}{6}F_{ijk}dz^i \wedge dz^j \wedge dz^k, \qquad \text{so that}$$
$$F_{ijk} = \partial_i F_{jk} + \partial_j F_{ki} + \partial_k F_{ij}.$$

The only non–vanishing components of this tensor are

$$F_{\mu\nu\overline{\rho}} = i\left(\partial_\mu g_{\nu\overline{\rho}} - \partial_\nu g_{\mu\overline{\rho}}\right) = -iT_{\mu\nu}{}^{\sigma} g_{\sigma\overline{\rho}} = -iT_{\mu\nu\overline{\rho}},$$
$$F_{\overline{\mu}\,\overline{\nu}\rho} = -i\left(\partial_{\overline{\mu}} g_{\rho\overline{\nu}} - \partial_{\overline{\nu}} g_{\rho\overline{\mu}}\right) = iT_{\overline{\mu}\,\overline{\nu}}{}^{\overline{\sigma}} g_{\rho\overline{\sigma}} = iT_{\overline{\mu}\,\overline{\nu}\rho}.$$

The curvature tensor of the metric connection is constructed in the usual manner

$$\Omega^i{}_j = d\omega^i{}_j - \omega^i{}_k \wedge \omega^k{}_j,$$

with the only non–vanishing components $\Omega^\nu{}_\mu$ and $\Omega^{\overline{\nu}}_{\overline{\mu}}$. These are given by [Cha05]

$$\Omega^\nu{}_\mu = -R^\nu{}_{\mu\kappa\lambda} dz^\kappa \wedge dz^\lambda - R^\nu{}_{\mu\kappa\overline{\lambda}} dz^\kappa \wedge dz^{\overline{\lambda}}$$
$$= \left(\partial_\kappa \Gamma^\nu_{\mu\lambda} - \Gamma^\rho_{\mu\kappa}\Gamma^\nu_{\rho\lambda}\right) dz^\kappa \wedge dz^\lambda - \partial_{\overline{\lambda}}\Gamma^\nu_{\mu\kappa} dz^\kappa \wedge dz^{\overline{\lambda}}.$$

Comparing both sides we get

$$R^\nu{}_{\mu\kappa\lambda} = \partial_\lambda \Gamma^\nu_{\mu\kappa} - \partial_\kappa \Gamma^\nu_{\mu\lambda} + \Gamma^\rho_{\mu\kappa}\Gamma^\nu_{\rho\lambda} - \Gamma^\rho_{\mu\lambda}\Gamma^\nu_{\rho\kappa}, \qquad R^\nu{}_{\mu\kappa\overline{\lambda}} = \partial_{\overline{\lambda}}\Gamma^\nu_{\mu\kappa}.$$

One can easily show that

$$R^\nu{}_{\mu\kappa\lambda} = 0, \qquad R^\nu{}_{\mu\kappa\overline{\lambda}} = g^{\overline{\rho}\nu}\partial_\kappa \partial_{\overline{\lambda}} g_{\mu\overline{\rho}} + \partial_{\overline{\lambda}} g^{\overline{\rho}\nu}\partial_\kappa g_{\mu\overline{\rho}}.$$

Transvecting the last relation with $g_{\nu\overline{\sigma}}$ we get

$$-R_{\mu\overline{\sigma}\kappa\overline{\lambda}} = \partial_\kappa \partial_{\overline{\lambda}} g_{\mu\overline{\sigma}} + g_{\nu\overline{\sigma}}\partial_{\overline{\lambda}} g^{\overline{\rho}\nu}\partial_\kappa g_{\mu\overline{\rho}}.$$

Therefore the only non-vanishing covariant components of the curvature tensor are

$$R_{\mu\overline{\nu}\kappa\overline{\lambda}}, \quad R_{\mu\overline{\nu}\,\overline{\kappa}\lambda}, \quad R_{\overline{\mu}\nu\kappa\overline{\lambda}}, \quad R_{\overline{\mu}\nu\overline{\kappa}\lambda},$$

which are related by

$$R_{\mu\overline{\nu}\kappa\overline{\lambda}} = -R_{\overline{\nu}\mu\kappa\overline{\lambda}} = -R_{\mu\overline{\nu}\overline{\lambda}\kappa},$$

and satisfy the *first Bianchi identity* [Gol56]

$$R^\nu{}_{\mu\kappa\overline{\lambda}} - R^\nu{}_{\kappa\mu\overline{\lambda}} = \nabla_{\overline{\lambda}} T_{\mu\kappa}{}^{\nu}.$$

The *second Bianchi identity* is given by

$$\nabla_\rho R_{\mu\overline{\nu}\kappa\overline{\lambda}} - \nabla_\kappa R_{\mu\overline{\nu}\rho\overline{\lambda}} = R_{\mu\overline{\nu}\sigma\overline{\lambda}} T_{\rho\kappa}{}^{\sigma},$$

together with the conjugate relations. There are three possible contractions for the curvature tensor which are called the Ricci tensors

$$R_{\mu\bar{\nu}} = -g^{\bar{\lambda}\kappa} R_{\mu\bar{\lambda}\kappa\bar{\nu}}, \quad S_{\mu\bar{\nu}} = -g^{\bar{\lambda}\kappa} R_{\mu\bar{\nu}\kappa\bar{\lambda}}, \quad T_{\mu\bar{\nu}} = -g^{\bar{\lambda}\kappa} R_{\kappa\bar{\lambda}\mu\bar{\nu}}.$$

Upon further contraction these result in two possible curvature scalars

$$R = g^{\bar{\nu}\mu} R_{\mu\bar{\nu}}, \quad S = g^{\bar{\nu}\mu} S_{\mu\bar{\nu}} = g^{\bar{\nu}\mu} T_{\mu\bar{\nu}}.$$

Note that when the torsion tensors vanishes, the manifold M becomes Kähler. We shall not impose the Kähler condition as we are interested in Hermitian non–Kählerian geometry. We note that it is also possible to consider the *Levi–Civita connection* $\overset{\circ}{\Gamma}{}^k_{ij}$ and the associated Riemann curvature ${K_{kij}}^h$ where [Cha05]

$$\begin{aligned}\overset{\circ}{\Gamma}{}^k_{ij} &= \frac{1}{2} g^{kl} \left(\partial_i g_{lj} + \partial_j g_{il} - \partial_l g_{ij} \right), \\ {K_{kij}}^h &= \partial_k \overset{\circ}{\Gamma}{}^h_{ij} - \partial_i \overset{\circ}{\Gamma}{}^h_{kj} + \overset{\circ}{\Gamma}{}^h_{kt} \overset{\circ}{\Gamma}{}^t_{ij} - \overset{\circ}{\Gamma}{}^h_{it} \overset{\circ}{\Gamma}{}^t_{kj}.\end{aligned}$$

The relation between the Chern connection and the Levi–Civita connection is given by

$$\Gamma^k_{ij} = \overset{\circ}{\Gamma}{}^k_{ij} + \frac{1}{2} \left({T_{ij}}^k - T^k{}_{ij} - T^k{}_{ji} \right).$$

The Ricci tensor and curvature scalar are

$$K_{ij} = {K_{tij}}^t \qquad \text{and} \qquad K = g^{ij} K_{ij}.$$

Moreover,

$$H_{kj} = {K_{kji}}^t F^i_t \qquad \text{and} \qquad H = g^{kj} H_{kj}.$$

The two scalar curvatures K and H are related by [Gau84]

$$\begin{aligned}K - H = {} & \overset{\circ}{\nabla}{}^h F^{ij} \overset{\circ}{\nabla}_j F_{ih} - \overset{\circ}{\nabla}{}^k F_{ki} \overset{\circ}{\nabla}_h F^{hi} \\ & -2 F^{ji} \overset{\circ}{\nabla}_j \overset{\circ}{\nabla}{}^k F_{ki}.\end{aligned}$$

There are also relations between curvatures of the Chern connection and those of the Levi–Civita connection, mainly [Gau84]

$$\frac{1}{2} K = S - \nabla^\mu T_\mu - \nabla^{\bar{\mu}} T_{\bar{\mu}} - T_\mu T_{\bar{\nu}} g^{\bar{\nu}\mu},$$

where $T_\mu = {T_{\mu\nu}}^\nu$. There are two natural conditions that can be imposed on the torsion. The first is $T_\mu = 0$ which results in a *semi–Kähler manifold*. The other is when the torsion is complex analytic so that $\nabla_{\bar{\lambda}} {T_{\mu\kappa}}^\nu = 0$ implying that the curvature tensor has the same symmetry properties as in the Kähler case. In this work we shall not impose any conditions on the torsion tensor.

5.3.3 Invariant Action

We now specialize to the realistic case of a complexified four dimensional space–time. To construct invariants up to second order in derivatives we write the following possible terms [Cha05]

$$I = \int_{M^4} d^4z d^4\overline{z} g \left(aR + bS + c\, T_{\mu\nu\overline{\kappa}} T_{\overline{\rho}\overline{\sigma}\lambda} g^{\overline{\rho}\mu} g^{\overline{\sigma}\nu} g^{\overline{\kappa}\lambda} + d T_{\mu\nu\overline{\kappa}} T_{\overline{\rho}\overline{\sigma}\lambda} g^{\overline{\rho}\mu} g^{\overline{\sigma}\lambda} g^{\overline{\kappa}\nu} + e\right).$$

The density factor is

$$|\det g_{ij}|^{\frac{1}{2}} = \det g_{\mu\overline{\nu}} \equiv g.$$

We shall set the cosmological term to zero $(e = 0)$. The above action can equivalently be written in terms of the Riemannian metric g_{ij} in the form

$$I = \int_M d^4z d^4\overline{z} \,|\det g_{ij}|^{\frac{1}{2}} \left(a'K + b'H + c' F_{ijk}F^{ijk} + d' F_i F^i\right),$$

where $F_i = F_{ijk}F^{jk}$ and a', b', c', d' are parameters linearly related to the parameters a, b, c, d. We shall now impose the requirement that the linearized action, in the limit of $y \to 0$ gives the correct kinetic terms for $G_{\mu\nu}(x)$ and $B_{\mu\nu}(x)$. Therefore writing

$$G_{\mu\nu}(x,y) = \eta_{\mu\nu} + h_{\mu\nu}(x), \qquad B_{\mu\nu}(x,y) = B_{\mu\nu}(x),$$

and keeping only quadratic terms in the action, we get, after integrating by parts, the quadratic $h_{\mu\nu}$ terms [Cha05],

$$I = \int d^4x d^4y \left(2c\partial_{x^\kappa} h_{\mu\nu} \partial_{x^\kappa} h_{\mu\nu} + (a - 2c + d)\, \partial_{x^\nu} h_{\mu\nu} \partial_{x^\lambda} h^{\mu\lambda} + \right.$$
$$\left. - (a - b + 2d)\, \partial_{x^\lambda} h_{\mu\nu} \partial_{x^\nu} h_\lambda{}^\lambda + (d - b)\, \partial_{x^\mu} h_\nu{}^\nu \partial^{x\mu} h_\lambda{}^\lambda \right).$$

Comparing with the linearized Einstein action we get the following conditions

$$2c = 1, \quad a - 2c + d = -2, \quad -a + b - 2d = 2, \quad d - b = -1,$$

which are equivalent to

$$b = -a, \quad c = \frac{1}{2}, \quad d = -1 - a.$$

With this choice of coefficients, the quadratic B contributions simplify to

$$\int d^4x d^4y \left(\partial_{x^\mu} B_{\nu\rho} \partial_{x^\mu} B^{\nu\rho} - 2\partial_{x^\mu} B_{\mu\lambda} \partial_{x^\nu} B^{\nu\lambda}\right),$$

which is identical to the term

$$\frac{1}{3}\int d^4xd^4yH_{\mu\nu\rho}H^{\mu\nu\rho},$$

where $H_{\mu\nu\rho} = \partial^x_\mu B_{\nu\rho} + \partial^x_\nu B_{\rho\mu} + \partial^x_\rho B_{\mu\nu}$. The action can then be regrouped into the form [Cha05]

$$I = \int_M d^4zd^4\overline{z}g\left(a\left(R - S - T_{\mu\nu\overline{\kappa}}T_{\overline{\rho}\,\overline{\sigma}\lambda}g^{\overline{\rho}\mu}g^{\overline{\sigma}\lambda}g^{\overline{\kappa}\nu}\right)\right.$$
$$\left.+\frac{1}{2}T_{\mu\nu\overline{\kappa}}T_{\overline{\rho}\,\overline{\sigma}\lambda}\left(g^{\overline{\rho}\mu}g^{\overline{\sigma}\nu}g^{\overline{\kappa}\lambda} - 2g^{\overline{\rho}\mu}g^{\overline{\sigma}\lambda}g^{\overline{\kappa}\nu}\right)\right).$$

Using the first Bianchi identity we have

$$\int_M d^4zd^4\overline{z}g\left(R - S\right) = \int_M d^4zd^4\overline{z}gg^{\overline{\lambda}\mu}\partial_{\overline{\lambda}}T_{\mu\nu}{}^{\nu}$$
$$= \int_M d^4zd^4\overline{z}gT_{\mu\nu\overline{\kappa}}T_{\overline{\rho}\,\overline{\sigma}\lambda}g^{\overline{\rho}\mu}g^{\overline{\sigma}\lambda}g^{\overline{\kappa}\nu},$$

where we have integrated by parts and ignored a surface term. This imply that the group of terms with coefficient a drop out, and the action becomes unique:

$$I = \frac{1}{2}\int_M d^4zd^4\overline{z}gT_{\mu\nu\overline{\kappa}}T_{\overline{\rho}\,\overline{\sigma}\lambda}\left(g^{\overline{\rho}\mu}g^{\overline{\sigma}\nu}g^{\overline{\kappa}\lambda} - 2g^{\overline{\rho}\mu}g^{\overline{\sigma}\lambda}g^{\overline{\kappa}\nu}\right).$$

Substituting for the torsion tensor in terms of the metric $g_{\mu\overline{\nu}}$, the above action reduces to

$$I = \frac{1}{2}\int_M d^4zd^4\overline{z}gX^{\overline{\kappa}\overline{\lambda}\overline{\sigma}\mu\nu\rho}\partial_\nu g_{\mu\overline{\sigma}}\partial_{\overline{\lambda}}g_{\rho\overline{\kappa}}, \qquad \text{where}$$

$$X^{\overline{\kappa}\overline{\lambda}\overline{\sigma}\mu\nu\rho} = g^{\overline{\sigma}\rho}\left(g^{\overline{\kappa}\mu}g^{\overline{\lambda}\nu} - g^{\overline{\kappa}\nu}g^{\overline{\lambda}\mu}\right) + g^{\overline{\sigma}\mu}\left(g^{\overline{\kappa}\nu}g^{\overline{\lambda}\rho} - g^{\overline{\kappa}\rho}g^{\overline{\lambda}\nu}\right)$$
$$+g^{\overline{\sigma}\nu}\left(g^{\overline{\kappa}\rho}g^{\overline{\lambda}\mu} - g^{\overline{\kappa}\mu}g^{\overline{\lambda}\rho}\right),$$

which is completely antisymmetric in the indices $\mu\nu\rho$ and in $\overline{\kappa}\overline{\lambda}\overline{\sigma}$

$$X^{\overline{\kappa}\overline{\lambda}\overline{\sigma}\mu\nu\rho} = X^{\left[\overline{\kappa}\overline{\lambda}\overline{\sigma}\right][\mu\nu\rho]}.$$

This is remarkable because the simple requirement that the linearized action for $G_{\mu\nu}$ should be recovered determines the action uniquely. This form of the action is valid in all complex dimensions n, however, when $n = 4$, we can write

$$X^{\overline{\kappa}\overline{\lambda}\overline{\sigma}\mu\nu\rho} = -\frac{1}{g}\epsilon^{\overline{\kappa}\overline{\lambda}\overline{\sigma}\,\overline{\eta}}\epsilon^{\mu\nu\rho\tau}g_{\tau\overline{\eta}},$$

and the action takes the very simple form [Cha05]

$$I = -\frac{1}{2} \int_M d^4 z d^4 \overline{z} \epsilon^{\overline{\kappa}\overline{\lambda}\overline{\sigma}\,\overline{\eta}} \epsilon^{\mu\nu\rho\tau} g_{\tau\overline{\eta}} \partial_\mu g_{\nu\overline{\sigma}} \partial_{\overline{\kappa}} g_{\rho\overline{\lambda}}.$$

The above expression has the advantage that the action is a function of the metric $g_{\mu\overline{\nu}}$ and there is no need to introduce the inverse metric $g^{\overline{\nu}\mu}$. This suggests that the action could be expressed in terms of the *Kähler form* F. Indeed, we can write

$$I = \frac{i}{2} \int_M F \wedge \partial F \wedge \overline{\partial} F.$$

The equations of motion are given by

$$\epsilon^{\overline{\kappa}\overline{\lambda}\overline{\sigma}\,\overline{\eta}} \epsilon^{\mu\nu\rho\tau} \left(g_{\nu\overline{\sigma}} \partial_\mu \partial_{\overline{\kappa}} g_{\rho\overline{\lambda}} + \frac{1}{2} \partial_\mu g_{\nu\overline{\sigma}} \partial_{\overline{\kappa}} g_{\rho\overline{\lambda}} \right) = 0.$$

Notice that the above equations are trivially satisfied when the metric $g_{\mu\overline{\nu}}$ is Kähler

$$\partial_\mu g_{\nu\overline{\rho}} = \partial_\nu g_{\mu\overline{\rho}}, \qquad \partial_{\overline{\sigma}} g_{\nu\overline{\rho}} = \partial_{\overline{\rho}} g_{\nu\overline{\sigma}},$$

where these conditions are locally equivalent to $g_{\mu\overline{\nu}} = \partial_\mu \partial_{\overline{\nu}} K$ for some scalar function K.

5.4 Gradient Kähler Ricci Solitons

In this section, following [Bry04], we study the local and global geometry of *gradient Kähler Ricci solitons*, that is *Kähler metrics* g on a complex $n-$manifold M that admit a *Ricci potential*, i.e., a function f such that

$$\mathrm{Ric}(g) = \nabla^2 f,$$

where ∇ denotes the *Levi–Civita connection* of M. These metrics arise as limiting metrics in the study of the *Ricci flow*

$$g_t = -2\,\mathrm{Ric}(g),$$

applied to Kähler metrics. Under the Ricci flow, a gradient Kähler Ricci soliton g_0 evolves by flowing under the vector–field ∇f. In particular, if the flow of ∇f is complete, then the Ricci flow with initial value g_0 exists for all time. The reader who wants more background on these metrics might consult the references and survey articles [Cao97, Che02].

5.4.1 Introduction

Unless the metric g admits flat factors, the *Ricci potential equation* $\mathrm{Ric}(g) = \nabla^2 f$ determines f up to an additive constant and it does no harm to fix a choice of f for the discussion. For simplicity, it does no harm to assume that g has no (local) flat factors and so this will frequently be done. Also, the *Ricci–flat* case (aka the *Calabi–Yau* case), in which $\mathrm{Ric}(g) = 0$, is a special case that is usually treated by different methods, so it is here assumed that $\mathrm{Ric}(g) \neq 0$.

The Associated Holomorphic Vector–Field Z

One of the earliest observations [Cao94] made about gradient Kähler Ricci solitons is that the vector–field ∇f is the real part of a holomorphic vector–field and that, moreover, $J(\nabla f)$ is a *Killing field* for g. In this section, we will formulate the *holomorphic vector–field associated* to g as

$$Z = \frac{1}{2}\left(\nabla f - iJ(\nabla f)\right).$$

The Holomorphic Volume Form Υ

In the Ricci–flat case, at least when M is simply connected, it is well–known that there is a g–parallel holomorphic volume form Υ, i.e., one which satisfies the condition that $i^{n^2} 2^{-n}\,\Upsilon\overline{\Upsilon}$ is the real volume form determined by g and the J–orientation. For any gradient Kähler Ricci soliton g with Ricci potential f defined on a simply connected M, there is a holomorphic volume form Υ (unique up to a constant multiple of modulus 1) such that $i^{n^2} 2^{-n\,-f}\,\Upsilon\overline{\Upsilon}$ is the real volume form determined by g and the J–orientation. Clearly, Υ is not g–parallel (unless g is Ricci–flat) but satisfies [Bry04]

$$\nabla \Upsilon = \frac{1}{2}\,\partial f \otimes \Upsilon.$$

This leads to a notion of *special coordinate charts* for (g, f) i.e., coordinate charts (U, z) such that the associated coordinate volume form $dz = dz^1 \wedge \cdots \wedge dz^n$ is the restriction of Υ to U. In such coordinate charts, several of the usual formulae simplify for gradient Kähler Ricci solitons.

The Υ–Divergence of Z

Given a vector–field and and volume form, the divergence of the vector–field with respect to the volume form is well defined. It turns out to be useful to consider this quantity for Z and Υ. The divergence in this case is the (necessarily holomorphic) function h that satisfies

$$\mathcal{L}_Z \Upsilon = h\,\Upsilon,$$

where $\mathcal{L}_Z$ denotes the *Lie derivative* along the vector–field Z.

By general principles, the scalar function h must be expressible in terms of the first and second derivatives of f. Explicit computation yields

$$2h = \mathrm{Tr}_g(\nabla^2 f) + |\nabla f|^2 = R(g) + |\nabla f|^2, \qquad \text{where}$$
$$R(g) = \mathrm{Tr}_g\,(\mathrm{Ric}(g))$$

is the scalar curvature of g. In particular, h is real–valued and therefore constant. Now, the constancy of $R(g) + |\nabla f|^2$ had been noted and utilized by Hamilton and Cao [CR00]. However, its interpretation as a holomorphic divergence seems to be due to Bryant [Bry04].

Generality

An interesting question is: How many gradient Kähler Ricci solitons are there? Clearly, this rather vague question can be sharpened in several ways. The point of view adopted in this section is to start with a complex $n-$manifold M already endowed with a holomorphic volume form Υ and a holomorphic vector–field Z and ask how many gradient Kähler solitons on M there might be (locally or globally) that have Z and Υ as their associated holomorphic data. An obvious necessary condition is that the divergence h of Z with respect to Υ must be a real constant.

Nonsingular Extension

Away from the singularities (i.e., zeroes) of Z, this divergence condition turns out to be locally sufficient. More precisely, if $H \subset M$ is an embedded complex hypersurface that is transverse at each of its points to Z, and g_0 and f_0 are, respectively, a real–analytic Kähler metric and function on H, then there is an open neighborhood U of H in M on which there exists a gradient Kähler Ricci soliton g with potential f whose associated holomorphic quantities are Z and Υ and such that g and f pull back to H to become g_0 and f_0. The pair (g, f) is essentially uniquely specified by these conditions. The real–analyticity of the 'initial data' g_0 and f_0 is necessary in order for an extension to exist since any gradient Kähler Ricci soliton is real–analytic anyway. Roughly speaking, this result shows that, away from singular points of Z, the local solitons g with associated holomorphic data (Z, Υ) depend on two arbitrary (real–analytic) functions of $2n{-}2$ variables [Bry04].

Singular Existence

The existence of (local) gradient Kähler solitons in a neighborhood of a singularity p of Z is both more subtle and more interesting. Even if $\mathcal{L}_Z\Upsilon$, the divergence of Z with respect to Υ, is a real constant, it is not true in general that a gradient Kähler Ricci solition with Z and Υ as associated holomorphic data exists in a neighborhood of such a p. A necessary condition is that there exist $p-$centered holomorphic coordinates $z = (z^i)$ on a $p-$neighborhood $U \subset M$ and real numbers $h_1, \ldots, h_n$ such that

$$Z = h_1 \, z^1 \, \partial_{z^1} + \cdots + h_n \, z^n \, \partial_{z^n} \qquad \text{on} \quad U.$$

In other words, Z must be holomorphically linearizable, with real eigenvalues. In such a case, if $\mathcal{L}_Z \Upsilon = h\Upsilon$ where h is a constant, then $h = h_1 + \cdots + h_n$. Moreover, in this case one can always choose $Z-$linearizing coordinates as above so that $\Upsilon = dz^1 \wedge \cdots \wedge dz^n$. Thus, the possible local singular pairs (Z, Υ) that can be associated to a gradient Kähler Ricci soliton are, up to biholomorphism, parametrized by n real constants. Using this normal form, one then observes that, by taking products of solitons of dimension 1, any set of real constants $(h_1, \ldots, h_n)$ can occur. Since, for any gradient Kähler Ricci soliton g with associated holomorphic data (Z, Υ), the following formula holds [Bry04]

$$\mathrm{Ric}(g) = \mathcal{L}_{\mathrm{Re}(Z)} \, g,$$

it follows that if g is such a Kähler Ricci soliton defined on a neighborhood of a point p with $Z(p) = 0$, then $h_1, \ldots, h_n$ are the eigenvalues (each of even multiplicity) of $\mathrm{Ric}(g)$ with respect to g at p.

However, this does not fully answer the question of how 'general' the solitons are in a neighborhood of such a p. In fact, this very subtly depends on the numbers h_i. For example, if the $h_i \in \mathbb{R}$ are linearly independent over $\mathbb{Q}$, then any gradient Kähler Ricci soliton g with associated data (Z, Υ) defined on a neighborhood of p must be invariant under the compact $n-$torus action generated by the closure of the flow of the imaginary part of Z. This puts severe restrictions on the possibilities for such solitons.

The Positive Case

An interesting special case is this: The case where g is complete, the Ricci curvature is positive, and the scalar curvature $R(g)$ attains its maximum at some (necessarily unique) point $p \in M$. This case has been studied before by Cao and Hamilton [CR00], who proved that this point p is a minimum of the Ricci potential f, that f is a proper plurisubharmonic exhaustion function on M (which is therefore Stein), and that, moreover, the Killing field $J(\nabla f)$ has a periodic orbit on 'many' of its level sets.

For simplicity, the Ricci potential f is normalized so that $f(p) = 0$, so that f is positive away from p. Under these assumptions there exist global $Z-$linearizing coordinates

$$z = (z^i) : M \to \mathbb{C}^n,$$

so that M is biholomorphic to $\mathbb{C}^n$ (which generalizes an earlier result of Chau and Tam [CT03]). Moreover, as a consequence, it follows that every positive level set of f has at least n periodic orbits, a considerable sharpening of Cao and Hamilton's original results.

This global coordinate system has several other applications. For example, we show that there is a Kähler potential ϕ for g that is invariant under the flow of $J(\nabla f)$ and that this potential is unique up to an additive constant (which can be normalized away by requiring that $\phi(p) = 0$).

As another application, we show how to normalize the choice of $Z-$linearizing holomorphic coordinates up to an ambiguity that lies in a compact subgroup of $\mathrm{U}(n)$. This makes the function $|z|$ well–defined on M, so it is available for estimates.

The Toric Case

This section studies the geometry of the reduced equation in the case when a gradient Käher Ricci soliton g defined on a neighborhood of $0 \in \mathbb{C}^n$ has toric symmetry, i.e., is invariant under the action of $\mathbb{T}^n$, the diagonal subgroup of $\mathrm{U}(n)$. This may seem specialized, but, for example, if the associated holomorphic vector–field is Z_{h} where $\mathrm{h} = (h_1, \dots, h_n)$ and the real numbers $h_1, \dots, h_n$ have the 'generic' property of being linearly independent over $\mathbb{Q}$, then g has toric symmetry. Thus, metrics with toric symmetry are the rule when Z has a 'generic' singularity.

We first derive the equation satisfied by the reduced potential, which turns out to be a singular *Monge–Ampére equation*.[6] Nevertheless, this singular equation has good regularity and its singular initial value problem is well–posed in the sense of [GH96].

As a consequence, it follows that, for any $\mathrm{h} \in \mathbb{R}^n$, any real–analytic $\mathbb{T}^{n-1}-$invariant Kähler metric on a neighborhood of $0 \in \mathbb{C}^{n-1}$ is the restriction to $\mathbb{C}^{n-1}$ of an essentially unique toric gradient Kähler Ricci soliton on an open subset of $\mathbb{C}^n$ with associated holomorphic vector–field $Z = Z_{\mathrm{h}}$ and associated holomorphic volume form $\Upsilon = z$. In particular, it follows that, in a sense made precise in that section, the toric gradient Kähler Ricci solitons on $\mathbb{C}^n$ depend on one 'arbitrary' real–analytic function of $(n-1)$ real variables.

Next, we show that the reduced singular Monge–Ampère equation is of Euler–Lagrange type, at least, away from its singular locus, and discuss some of its conservation laws via an application of Noether's Theorem (this is in contrast to the unreduced soliton equation, which is not variational).

5.4.2 Associated Holomorphic Quantities

In this subsection, constructions of some holomorphic quantities associated to a gradient Kähler Ricci soliton g on a complex $n-$manifold M^n with Ricci potential f is described [Bry04].

[6] The singularities are related to the singular orbits of the $\mathbb{T}^n-$action.

Preliminaries

In order to avoid confusion because of various different conventions in the literature, we will collect the notations, conventions, and normalizations to be used in this section.

Tensors and inner products

Factors of 2 are sometimes troubling and confusing in Kähler geometry. For a and b in a vector space V, we will use Bryant's conventions [Bry04]

$$a \circ b = \frac{1}{2}(a \otimes b + b \otimes a) \qquad \text{and} \qquad a \wedge b = a \otimes b - b \otimes a;$$

in particular, $a \otimes b = a \circ b + \frac{1}{2}\, a \wedge b.$

A real–valued *inner product* $\langle , \rangle$ on a real vector space V can be extended to the complex vector space

$$V^{\mathbb{C}} = \mathbb{C} \otimes V$$

in several different ways. A natural way is to extend it as an *Hermitian form*, i.e., so that

$$\langle v_1 + v_2, w_1 + w_2 \rangle = (\langle v_1, w_1 \rangle + \langle v_2, w_2 \rangle) + (\langle v_2, w_1 \rangle - \langle v_1, w_2 \rangle)$$

and that is the convention adopted here.

If the real vector space V has a *complex structure* $J : V \to V$, then $V^{\mathbb{C}} = V^{1,0} \oplus V^{0,1}$ where $V^{1,0}$ is the (+)eigenspace of J extended complex linearly to $V^{\mathbb{C}}$ while $V^{0,1}$ is the (−)eigenspace of J. It is common practice to identify $v \in V$ with $v^{1,0} = v - Jv \in V^{1,0}$, but some care must be taken with this.

For example, an inner product $\langle , \rangle$ on V is *compatible* with J if $\langle Jv, Jw \rangle = \langle v, w \rangle$ for all $v, w \in V$. Note the identity

$$\langle v^{1,0}, v^{1,0} \rangle = 2\langle v, v \rangle.$$

For any J−compatible inner product $\langle , \rangle$ on V (or equivalently, quadratic form) there is an *associated* 2−form η defined by

$$\eta(v, w) = \langle Jv, w \rangle.$$

Coordinate expressions and the Ricci form

Let $z = (z^i) : U \to \mathbb{C}^n$ be a holomorphic coordinate chart on an open set $U \subset M$. The metric g restricted to U can be expressed in the form

$$g = g_{i\bar{j}}\, dz^i \circ d\bar{z}^j$$

for some functions $g_{i\bar{j}} = \overline{g_{j\bar{i}}}$ on U. The associated *Kähler form* Ω then has the coordinate expression

$$\Omega = \frac{i}{2}\, g_{i\bar{\jmath}}\, dz^i \wedge d\bar{z}^j.$$

Note that [Bry04]

$$g_{i\bar{\jmath}}\, dz^i \otimes d\bar{z}^j = g - 2i\Omega.$$

The Ricci tensor $\mathrm{Ric}(g)$ is $J-$compatible since g is Kähler, and hence has a coordinate expression

$$\mathrm{Ric}(g) = R_{j\bar{k}}\, dz^j \circ d\bar{z}^k,$$

where $R_{j\bar{k}} = \overline{R_{k\bar{\jmath}}}$. Its associated $2-$form ρ is computed by the formula

$$\rho = \frac{i}{2}\, R_{i\bar{\jmath}}\, dz^i \wedge d\bar{z}^j = -\, i\partial\overline{\partial}\, G, \qquad \text{where} \tag{5.12}$$

$$G = \log\det\left(g_{i\bar{\jmath}}\right). \tag{5.13}$$

While ρ is independent of the coordinate chart used to compute it, the function G does depend on the coordinate chart.

The scalar curvature

$$R(g) = \mathrm{Tr}_g(\mathrm{Ric}(g))$$

has the coordinate expression

$$R(g) = 2g^{i\bar{\jmath}} R_{i\bar{\jmath}}$$

and satisfies

$$R(g)\, \Omega^n = 2n\, \rho \wedge \Omega^{n-1}.$$

The gradient Kähler Ricci soliton condition

The following equivalent formulation of the gradient Kähler Ricci soliton condition is well–known: A real–valued function f on M satisfies [Bry04]

$$\mathrm{Ric}(g) = D^2 f \qquad \text{iff} \qquad \rho = i\partial\overline{\partial}\, f \qquad \text{and} \qquad D^{0,2} f = 0.$$

This latter condition is equivalent to the condition that the $g-$gradient of f be the real part of a holomorphic vector–field on M.

The Associated Holomorphic Volume Form

In this subsection, given a gradient Kähler Ricci soliton g with Ricci potential f on a simply–connected complex $n-$manifold M, a holomorphic volume form on M (unique up to a complex multiple of modulus 1) is constructed.

Existence of Special Coordinates

The following result shows that there are coordinate systems in which the Ricci potential is more closely tied to the local coordinate quantities. If g is a gradient Kähler Ricci soliton on M with Ricci potential f, then M has an atlas of holomorphic charts (U, z) satisfying [Bry04]

$$\log \det (g_{i\bar{j}}) = -f. \tag{5.14}$$

Let g be a gradient Kähler Ricci soliton on M with Ricci potential f. A coordinate chart (U, z) for which (5.14) is said to be *special* for (g, f). A coordinate chart (U, z) is special for (g, f) iff the volume form of g satisfies

$$\mathrm{dvol}_g = \frac{1}{n!}\,\Omega^n = \left(\frac{i^n}{2}\right)^n \mathrm{e}^{-f}\, dz\wedge d\bar{z}.$$

Let M be a simply connected complex $n-$manifold endowed with a gradient Kähler Ricci soliton g with associated Kähler form Ω and a choice of Ricci potential f. Then there exists a holomorphic volume form Υ on M, unique up to multiplication by a complex number of modulus 1, with the property that

$$\mathrm{dvol}_g = \frac{1}{n!}\,\Omega^n = \left(\frac{i^n}{2}\right)^n \mathrm{e}^{-f}\, \Upsilon\wedge\overline{\Upsilon}. \tag{5.15}$$

Given a gradient Kähler Ricci soliton g with Ricci potential f, a holomorphic volume form Υ satisfying (5.15) is said to be *associated* to the pair (g, f).[7] If Υ is associated to (g, f), then, for any real constants $\lambda > 0$ and c, the $n-$form $\lambda^n \mathrm{e}^c\, \Upsilon$ is associated to $(\lambda^2\, g,\ f{+}2c)$.

The Holomorphic Flow

Write the $g-$gradient of f as $Z + \bar{Z}$ where Z is of type $(1, 0)$. Thus,

$$Z = \frac{1}{2}\left(\nabla f - iJ(\nabla f)\right).$$

The Infinitesimal Symmetry

By the standard *Kähler identities*, Z is the unique vector–field of type $(1, 0)$ satisfying [Bry04]

$$\bar{\partial} f = -\,iZ \lrcorner\, \Omega\,. \tag{5.16}$$

If we write

$$Z = X - i\,Y = X - i\,JX,$$

[7] Note that scaling a gradient Kähler Ricci soliton g by a constant produces another gradient Kähler Ricci soliton and adding a constant to f will produce another Ricci potential for g.

it follows that, in addition to X being the one–half the gradient of f, the vector–field $Y = JX$ is Ω–Hamiltonian. Thus, the flow of Y preserves Ω. Since Z is holomorphic the flow of Y also preserves the complex structure on M. Hence, Y must be a Killing vector–field for the metric g. Thus, a gradient Kähler Ricci soliton that is not Ricci–flat always has a nontrivial infinitesimal symmetry. The singular locus of Z is a disjoint union of non-singular complex submanifolds of M, each of which is totally geodesic in the metric g.

Z in Special Coordinates

Assume (U, z) is a special local coordinate system. Since

$$\bar{\partial} G = g^{i\bar{\jmath}}\, \partial_{\bar{z}^k}\, g_{i\bar{\jmath}}\, d\bar{z}^k = -\bar{\partial} f\,,$$

the formula for Z in special coordinates is

$$Z = Z^\ell \partial_{z^\ell} = -\left(2 g^{\ell\bar{k}} g^{i\bar{\jmath}} \partial_{\bar{z}^k} g_{i\bar{\jmath}}\right) \partial_{z^\ell}\,. \qquad (5.17)$$

Thus, the equations for a gradient Kähler Ricci soliton in special coordinates are that the functions Z^ℓ defined by (5.17) be holomorphic. In fact, the expression in (5.17) can be simplified, since the closure of Ω is equivalent to the equations [Bry04]

$$\partial_{\bar{z}^k} g_{i\bar{\jmath}} = \partial_{\bar{z}^j} g_{i\bar{k}}. \qquad \text{Thus,} \qquad (5.18)$$

$$Z^\ell = -2\, g^{\ell\bar{k}} g^{i\bar{\jmath}}\, \partial_{\bar{z}^k} g_{i\bar{\jmath}} = -2\, g^{i\bar{\jmath}} g^{\ell\bar{k}}\, \partial_{\bar{z}^j} g_{i\bar{k}} = 2\, g^{i\bar{\jmath}} g_{i\bar{k}}\, \partial_{\bar{z}^j} g^{\ell\bar{k}} = 2\, \partial_{\bar{z}^j} g^{\ell\bar{\jmath}}, \quad (5.19)$$

where we have used the identity $g^{i\bar{\jmath}} g_{i\bar{k}} = \delta^{\bar{\jmath}}_{\bar{k}}$ and the identity $g_{i\bar{k}} g^{\ell\bar{k}} = \delta^\ell_i$ and its derivatives.

The Υ–Divergence of Z

Since Z is holomorphic, the Lie derivative of Υ with respect to Z must be of the form $h\Upsilon$ where h is a holomorphic function on M (usually called the divergence of Z with respect to Υ).

Replacing Υ by $\lambda\Upsilon$ for any $\lambda \in \mathbb{C}^*$ will not affect the definition of h, so the function h is intrinsic to the geometry of the soliton. On general principle, it must be computable in terms of the first and second covariant derivatives of f, which leads to the following interpretation of a result of Cao and Hamilton: The holomorphic function h is real–valued (and therefore constant). Moreover,

$$2h = R(g) + 2|Z|^2, \qquad (5.20)$$

where $R(g)$ is the scalar curvature of g and $|Z|^2$ is the squared g–norm of Z.

Since $\rho = i\partial\bar{\partial} f$, it follows that [Bry04]

$$\frac{i}{2} R_{i\bar{\jmath}}\, dz^i \wedge d\bar{z}^j = \rho = i\partial\bar{\partial} f = \partial\,(Z \lrcorner\, \Omega)$$
$$= \frac{i}{2} \left(g_{l\bar{\jmath}} \partial_{z^i} Z^l + Z^l \partial_{z^i} g_{l\bar{\jmath}}\right) dz^i \wedge d\bar{z}^j.$$

In particular,

$$R(g) = 2g^{i\bar{j}}R_{i\bar{j}} = 2h - 2|Z|^2.$$

It was Cao and Hamilton [CR00] who first observed that the quantity $R(g) + |\nabla f|^2$ is constant for a (steady) gradient Kähler Ricci soliton. Since

$$Z = \frac{1}{2}\left(\nabla f - iJ(\nabla f)\right), \qquad \text{one has} \qquad 2|Z|^2 = |\nabla f|^2,$$

so their expression is the right hand side of (5.20). The interpretation of $R(g) + |\nabla f|^2$ as the Υ–divergence of Z seems to be new. In a sense, this constancy can be regarded as a sort of conservation law for the Ricci flow. Note that, since $\Delta f = R(g)$, this relation is equivalent to the equation $\Delta_g(\mathrm{e}^f) = 2h\mathrm{e}^f$.

Examples

The associated holomorphic objects constructed so far make it possible to simplify somewhat the usual treatment of the known explicit examples [Bry04].

Suppose that M is a Riemann surface. Then Υ is a nowhere vanishing 1-form on M and Z is a holomorphic vector–field on M that satisfies $\Upsilon(Z) = h\Upsilon$, where h is a constant. There are essentially two cases to consider.

First, suppose that $h = 0$. Then $\Upsilon(Z)$ is a constant, say $\Upsilon(Z) = c$.

If $c = 0$, then Z is identically zero, and, from (5.19) it follows that, in special coordinates $z = (z^1)$ the real–valued function $g^{1\bar{1}}$ is constant. In particular, in special coordinates $g = g_{1\bar{1}}|z^1|^2$, so g is flat.

If $c \neq 0$, then Z is nowhere vanishing and, after adjusting Υ and the special coordinate system by a constant multiple, it can be assumed that $c = 2$, i.e., that $\Upsilon = dz^1$ and $Z = 2\,\partial_{z^1}$. Then (5.19) implies that $g^{1\bar{1}} = z^1 + \bar{z}^1 + C$ for some constant C. By adding a constant to z^1, it can be assumed that $C = 0$, so it follows that, in this coordinate system

$$g = \frac{|z^1|^2}{(z^1 + \bar{z}^1)}. \tag{5.21}$$

Since M is supposed to be simply connected, one can take z^1 to be globally defined. Thus M is immersed into the right half–plane in in such a way that g is the pullback of the conformal metric defined by (5.21). Clearly, this metric is not complete, even on the entire right half–plane.

Second, assume that h is not zero. Then $\Upsilon(Z)$ is a holomorphic function on M that has nowhere vanishing differential. Write $\Upsilon(Z) = hz^1$ for some (globally defined) holomorphic immersion $z^1 : M \to \mathbb{C}$. Then, by construction, $\Upsilon = dz^1$ and $Z = hz^1\partial_{z^1}$. By (5.19), it follows that

$$g^{1\bar{1}} = \frac{1}{2}(c + h\,|z^1|^2)$$

for some constant c, so $z^1(M) \subset \mathbb{C}$ must lie in the open set U in the w–plane on which $c+h|w|^2 > 0$. In fact, g must be the pullback under $z^1 : M \to U \subset \mathbb{C}$ of the metric

$$\frac{2\,|w|^2}{c + h\,|w|^2}. \tag{5.22}$$

This metric on the domain $U \subset \mathbb{C}$ is not complete unless both c and h are nonnegative and it is flat unless both c and h are positive. In this latter case, this metric is simply *Hamilton's cigar soliton* [Ham88].

Consequently, in dimension 1, the only complete gradient Kähler Ricci solitons are either flat or one of Hamilton's 'cigar' solitons (which are all homothetic to a single example).[8]

Now, by taking products of the 1D examples, one can construct a family of complete examples on $\mathbb{C}^n$: Let $h_1, \dots, h_n$ and $c_1, \dots, c_n$ be positive real numbers and consider the metric on $\mathbb{C}^n$ defined by

$$g = \frac{2\,|w^k|^2}{(c_k + h_k\,|w^k|^2)}. \tag{5.23}$$

Clearly, this is a gradient Kähler Ricci soliton, with associated holomorphic volume form and vector–field

$$\Upsilon = dw^1 \wedge \cdots \wedge dw^n, \qquad Z = h_k\,w^k\,\partial_{w^k}.$$

The Ricci curvature is

$$\mathrm{Ric}(g) = \frac{2c_k h_k\,|w^k|^2}{(c_k + h_k\,|w^k|^2)^2}.$$

Although these product examples are trivial generalizations of Hamilton's cigar soliton, they is useful in observations to be made below.

Also, note that, even if the h_k are not positive, as long as the c_k are positive, the formula (5.23) defines a not-necessarily-complete gradient Kähler Ricci soliton on the polycylinder defined by the inequalities $c_k + h_k|w^k|^2 > 0$.

One more case of an easily constructed example is the gradient Kähler Ricci soliton metric g on $\mathbb{C}^n$ that is invariant under $\mathrm{U}(n)$, discovered by Cao [Cao94]. The form of this metric can be derived as follows:

---

8 Note that, under the Ricci flow

$$g_t = -2\,\mathrm{Ric}(g),$$

the metric (5.22) evolves as

$$g(t) = \frac{2\,|w|^2}{{}^{2ht}c + h\,|w|^2} = \frac{2\,|({}^{-ht}w)|^2}{c + h\,|{}^{-ht}w|^2} = \Phi(-t)^*(g_0),$$
$$\text{where} \qquad \Phi(t)(w) = \mathrm{e}^{ht}w$$

is the flow of twice the real part of $Z = hw\,\partial_w$.

Suppose that such a metric g is given on $\mathbb{C}^n$. (One could do this analysis on any $\mathrm{U}(n)$–invariant domain in $\mathbb{C}^n$, and Cao does this, but we will not pursue this more general case further here.) The group $\mathrm{U}(n)$ must preserve the associated holomorphic volume form Υ up to a constant multiple and this implies that Υ must be a constant multiple of the standard volume form $dz^1 \wedge \cdots \wedge dz^n$. Since Υ is only determined up to a constant multiple anyway, there is no loss of generality in assuming that $\Upsilon = dz^1 \wedge \cdots \wedge dz^n$. Furthermore, the vector–field Z must also be invariant under $\mathrm{U}(n)$, which implies that Z must be a multiple of the radial vector–field. Since $d(Z \lrcorner \Upsilon) = h\,\Upsilon$ where h is real, it follows that

$$Z = h\ z^k\, \partial_{z^k}\,.$$

Now, the condition that g be rotationally invariant with associated Kähler form closed implies that

$$g_{i\bar{j}} = a(r)\delta_{ij} + a'(r)\,\bar{z}^i\,z^j \tag{5.24}$$

for some function a of $r = |z^1|^2+\cdots+|z^n|^2$ that satisfies $ra'(r) + a(r) > 0$ and $a(r) > 0$ (when $n > 1$). Thus

$$G = \log\left(a(r)^{n-1}(ra'(r){+}a(r))\right)$$

in this coordinate system. Now, the identity $G = -f$, the equation (5.16), and the above formula for the coefficients of Ω combine to yield

$$\bar{\partial}G = i\,Z \lrcorner\,\Omega = -\frac{h}{2}\,(ra'(r){+}a(r))\,\bar{\partial}r = -\frac{h}{2}\bar{\partial}\,(ra(r))\,.$$

Supposing that $n > 1$ (since the $n = 1$ case has already been treated), it follows that $G + \frac{h}{2}ra(r)$ must be constant, i.e., that

$$a(r)^{n-1}\,(ra(r))'\,\mathrm{e}^{(h/2)ra(r)} = a(0)^n. \tag{5.25}$$

Upon scaling Υ by a constant, it can be assumed that $a(0) = 1$, so assume this from now on. Also, one can assume that h is nonzero since, otherwise, the solution that is smooth at $r = 0$ is simply $a(r) \equiv a(0) = 1$, which gives the flat metric.

The ODE (5.25) for a is singular at $r = 0$, so the existence of a smooth solution near $r = 0$ is not immediately apparent. Fortunately, (5.25) can be integrated by quadrature: Set $b(r) = (h/2)ra(r)$ and note that (5.25) can be written in terms of b as

$$b(r)^{n-1}\mathrm{e}^{b(r)}\,b'(r) = (h/2)^n r^{n-1}. \tag{5.26}$$

Integrating both sides from 0 to $r > 0$ yields an equation of the form

$$(-1)^n(n-1)!\,\mathrm{e}^{b(r)}\left(\mathrm{e}^{-b(r)} - \sum_{k=0}^{n-1}\frac{(-b(r))^k}{k!}\right) = \left(\frac{h}{2}\right)^n\frac{r^n}{n}. \tag{5.27}$$

Set

$$F(b) = (-1)^n (n-1)!\, \mathrm{e}^b \left(\mathrm{e}^{-b} - \sum_{k=0}^{n-1} \frac{(-b)^k}{k!} \right) \simeq \mathrm{e}^b \left(\frac{b^n}{n} - \frac{b^{n+1}}{n(n+1)} + \cdots \right).$$

Now, F has a power series of the form

$$F(b) = \frac{1}{n}\, b^n (1 + \frac{n}{n+1} b + \cdots),$$

so F can be written in the form $F(b) = \frac{1}{n} f(b)^n$ for an analytic function of the form $f(b) = b(1 + \frac{1}{n+1} b + \cdots)$. The analytic function f is easily seen to satisfy $f'(b) > 0$ for all b and to satisfy the limits

$$\lim_{b \to +\infty} f(b) = \infty \qquad \text{and} \qquad \lim_{b \to -\infty} f(b) = -\sqrt[n]{n!}\,.$$

In particular, f maps diffeomorphically onto $\left(-\sqrt[n]{n!}, \infty \right)$ and is smoothly invertible. Clearly, $f(0) = 0$.

Since (5.27) is equivalent to

$$f\left(b(r)\right)^n = \left(\frac{h}{2}\, r \right)^n,$$

when $h > 0$ it can be solved for $r \geq 0$ by setting $b(r) = f^{-1}\left(\frac{h}{2}\, r \right)$, yielding a unique real–analytic solution with a power series of the form

$$b(r) = \frac{h}{2}\, r - \frac{h^2}{4(n+1)}\, r^2 + \cdots\,.$$

Consequently, when $h > 0$, the solution b is defined for all $r \geq 0$ and is positive and strictly increasing on the half–line $r \geq 0$. In particular, the function

$$a(r) = \frac{2}{h}\, \frac{b(r)}{r} = 1 - \frac{h}{2(n+1)}\, r + \cdots\,.$$

is a positive real–analytic solution of (5.25) that is defined on the range $0 \leq r < \infty$ and satisfies

$$r a'(r) + a(r) = b'(r) > 0$$

on this range, so that the expression (5.24) defines a gradient Kähler Ricci soliton on $\mathbb{C}^n$.

An ODE–analysis of this solution [Cao94] shows that when $h > 0$ the resulting metric is complete on $\mathbb{C}^n$ and has positive sectional curvature. When $h < 0$, the solution $b(r)$ only exists for $r < -\frac{2}{h}\sqrt[n]{n!}$. It is not difficult to see that the corresponding gradient Kähler Ricci soliton on a bounded ball in $\mathbb{C}^n$ is inextendible and incomplete.

5.4.3 Potentials and Local Generality

In this subsection, the question of 'how many' gradient Kähler Ricci soliton metrics could give rise to specified holomorphic data (Υ, Z) on a complex manifold M is considered. While this question is not easy to answer globally, it is not so difficult to answer locally. Thus, throughout this section, assume that a complex n–manifold M is specified, together with a nonvanishing holomorphic volume form Υ on M and a holomorphic vector–field Z on M such that [Bry04]

$$d\,(Z \lrcorner \Upsilon) = h\,\Upsilon$$

for some real constant h.

Local Potentials

Suppose that $U \subset M$ is an open subset on which there exists a function ϕ such that

$$\Omega = \frac{i}{2}\,\partial\bar{\partial}\phi$$

is a positive definite $(1,1)$–form whose associated Kähler metric g is a gradient Ricci soliton with associated holomorphic data Υ and Z and Ricci potential f.

By (5.16), we have [Bry04]

$$\begin{aligned} 2\bar{\partial} f &= -2iZ \lrcorner \Omega = Z \lrcorner (\partial\bar{\partial}\phi) = -Z \lrcorner (\bar{\partial}\partial\phi) \\ &= -Z \lrcorner\, (d(\partial\phi)) = -\mathcal{L}_Z(\partial\phi) + d(\partial\phi(Z)) \\ &= \bar{\partial}\,(\partial\phi(Z)) - (\mathcal{L}_Z(\partial\phi)) - \partial\,(\mathcal{L}_Z(\phi))\,. \end{aligned}$$

By decomposition into type, it follows that

$$\bar{\partial}\,(2f - \partial\phi(Z)) = 0. \tag{5.28}$$

Consequently,

$$F = 2f - \partial\phi(Z) = 2f - \phi(Z)$$

is a holomorphic function on U.

Nonsingular Extension Problems

Suppose now that $p \in U$ is not a singular point of Z. Then, by shrinking U if necessary, F can be written in the form

$$F = dH(Z)$$

for some holomorphic function H on the p-neighborhood U. Replacing ϕ by $\phi + H + \bar{H}$, gives a new potential for Ω that satisfies the stronger condition

$$\partial\phi(Z) = \phi(Z) = 2f. \tag{5.29}$$

This function ϕ is unique up to the addition of the real part of a holomorphic function that is constant on the orbits of Z. Clearly, (5.29) implies that $\phi(Y) = 0$, i.e., that ϕ is invariant under the flow of Y, the imaginary part of Z.

Local Reduction to Equations

In local coordinates $z = (z^i)$ for which $\Upsilon = dz^1 \wedge \cdots \wedge dz^n$, one has $f = -G$ so ϕ satisfies the *Monge–Ampére equation* [Bry04]

$$\det \left(\frac{\partial^2 \phi}{\partial z^i \partial \bar{z}^j} \right) e^{\frac{1}{2}\phi(X)} = 1 \qquad (5.30)$$

as well as the equation

$$\phi(Y) = 0. \qquad (5.31)$$

Conversely, if ϕ is a strictly pseudo–convex function defined on a $p-$neighborhood U that satisfies both (5.30) and (5.31), then the Kähler metric g whose associated Kähler form is $\Omega = \frac{i}{2}\partial\bar{\partial}\phi$ – is a gradient Kähler Ricci soliton on U with associated holomorphic form Υ and holomorphic vector–field Z.

Note that, because (5.30) is a real–analytic elliptic equation for the strictly pseudo–convex function ϕ, it follows by elliptic regularity that ϕ (and hence g) is real–analytic as well.

Now, (5.30) and (5.31) are two PDE for ϕ, the first of second order and the second of first order. While this is an overdetermined system, it is not difficult to show that it is involutive in Cartan's sense.

In fact, an analysis along the lines of exterior differential systems leads to the following result as a proper formulation of a 'Cauchy problem' for gradient Kähler Ricci solitons in the nonsingular case:

Let M^n be a complex $n-$manifold endowed with a holomorphic volume form Υ and a nonzero vector–field Z satisfying $d(Z \lrcorner \Upsilon) = h\Upsilon$ for some real constant h. Let $H^{n-1} \subset M$ be any embedded complex hypersurface that is transverse to Z, let Ω_0 be any real–analytic Kähler form on H, and let f_0 be any real–analytic function on H. Then there is an open $H-$neighborhood $U \subset M$ on which there exists a gradient Kähler Ricci soliton g with associated Kähler form Ω, holomorphic volume form Υ, holomorphic vector–field Z, and Ricci potential f that satisfy

$$H^*\Omega = \Omega_0, \qquad \text{and} \qquad H^*f = f_0.$$

Moreover, g is locally unique in the sense that any other gradient Kähler Ricci soliton $\tilde{g}$ defined on an open $H-$neighborhood $\tilde{U} \subset M$ satisfying these initial conditions agrees with g on some open neighborhood of H in $U \cap \tilde{U}$.

This Theorem essentially says that the local gradient Kähler Ricci solitons depend on two real–analytic functions of $2n-2$ variables, namely the potential functions ψ_0 (which is assumed to be strictly pseudo–convex but otherwise arbitrary) and f_0 (which is arbitrary). There is, of course, some ambiguity in the choice of the holomorphic coordinates z^i, but this ambiguity turns out to depend on essentially $n-2$ holomorphic functions of $n-1$ holomorphic variables, which is negligible when compared with two arbitrary (real–analytic) functions of $2n-2$ real variables.

Finally, consider the initial value problem for a function ϕ on a neighborhood of R in U given by the real–analytic PDE [Bry04]

$$\det\left(\frac{\partial^2\phi}{\partial z^i\partial\bar{z}^j}\right) e^{\frac{1}{2}\phi(X)} = 1 \tag{5.32}$$

subject to the real–analytic initial conditions

$$\begin{aligned} \phi(z) &= \psi_1(z) \\ \mathsf{L}_X(\phi)(z) &= 2f_1(z) \end{aligned} \qquad \text{for all } z \in R \subset U. \tag{5.33}$$

It is easy to check that (5.32) and (5.33) constitutes a non–characteristic Cauchy problem. Hence, by the Cauchy–Kovalewska Theorem, there exists an open neighborhood $W \subset U$ containing R on which there exists a solution ϕ to this problem.

Near Singular Points of Z

The situation near a singular point of Z is considerably more delicate and interesting.

Linear Parts and Linearizability

Recall that, at a point $p \in M$ where Z vanishes, there is a well–defined linear map

$$Z'_p : T_pM \to T_pM,$$

often called 'the linear part of Z at p', defined by setting $Z'_p(v) = w$ if $w = [V, Z](p)$ for some (and hence any) holomorphic vector–field V defined near p and satisfying $V(p) = v \in T_pM$. In local coordinates $z = (z^i)$ centered on p, if

$$Z = Z^j(z)\,\partial_{z^j},$$

where, by assumption $Z^j(0) = 0$ for $1 \leq j \leq n$, then [Bry04]

$$Z'_p\left(\partial_{z^l}(p)\right) = \partial_{z^l} Z^j(0)\,\partial_{z^j}(p).$$

The linear map $Z'_p : T_pM \to T_pM$ has a *Jordan normal form* and this is an important invariant of the singularity. In particular, the set of eigenvalues of Z'_p is well–defined.

Let Z be the holomorphic vector–field associated to a gradient Kähler Ricci soliton g on M. At any singular point of Z, the linear part Z'_p is diagonalizable, with all eigenvalues real. A holomorphic vector–field Z on M is said to be *linearizable* near a singular point p if there exist p–centered coordinates $w = (w^i)$ on an open p–neighborhood W and constants a^i_j such that, on W, one has

$$Z = a^i_j w^j\,\partial_{w^i}\,.$$

The coordinates $w = (w^i)$ are said to be *linearizing* or *Poincaré coordinates* for Z near p.

Not every holomorphic vector–field is linearizable near its singular points, even if the linear part at such a point has all of its eigenvalues nonzero and distinct. For example, the vector–field

$$Z = z^1\partial_{z^1} + \left(2z^2 + (z^1)^2\right)\partial_{z^2}$$

on $\mathbb{C}^2$ is not linearizable at the origin, even though its linear part there is diagonalizable with eigenvalues 1 and 2. This nonlinearizability is perhaps most easily seen as follows: The flow $\Phi(t)$ of the vector–field Z is

$$\Phi(t)(z^1, z^2) = \left({}^t z^1, {}^{2t}(z^2 + (z^1)^2 t)\right).$$

In particular $\Phi(t+2\pi) \neq \Phi(t)$, which would be true if Z were holomorphically conjugate to the linear vector–field

$$Z'_{(0,0)} = z^1\partial_{z^1} + 2z^2\partial_{z^2}.$$

This phenomenon, however, does not happen for singular points of holomorphic vector fields associated to a gradient Kähler Ricci soliton: Let Z be a nonzero holomorphic vector–field on the complex $n-$manifold M that is associated to a gradient Kahler Ricci soliton g. Then Z is linearizable at each of its singular points. Moreover, the linear part of Z at a singular point is diagonalizable and has all its eigenvalues real. Clearly, the exponential map $\exp_p : T_pM \to M$ of g also intertwines the flow of Y'_p on T_pM with the flow of Y on M, but the exponential map is not generally holomorphic and so cannot be used to linearize Z holomorphically.

Recall that, for a holomorphic vector–field $Z = X - iY$, the two real vector fields X and Y have commuting flows and that, moreover, the following identity holds:[9]

$$\exp_{(a+b)Z} = \exp_{2aX} \circ \exp_{2bY}.$$

Let g be a gradient Kähler Ricci soliton on M and let Z be its associated holomorphic vector–field. Let $p \in M$ be a singular point of Z and let $\lambda \in \mathbb{R}^*$ be a nonzero eigenvalue of Z'_p of multiplicity $k \geq 1$. Then there exists a $k-$dimensional complex submanifold $N_\lambda \subset M$ that passes through p, to which Z is everywhere tangent, and on which Y is periodic of period $4\pi/|\lambda|$.[10]

[9] The factors of 2 are neglected in some references.

[10] The reader should be careful not to confuse the submanifolds N_λ with the images under the exponential mapping of the eigenspaces of Z'_p acting on T_pM. Indeed, the N_λ need not be unique. For example, for the linear vector field

$$Z = z^1\partial_{z^1} + 2z^2\partial_{z^2}$$

on $\mathbb{C}^2$, each of the parabolas $z^2 - c(z^1)^2 = 0$ for $c \in \mathbb{C}$ is tangent to Z and the imaginary part of Z has period 4π on all of $\mathbb{C}^2$, so each could be regarded as N_1.

On the other hand, the line $z^1 = 0$ is the only curve that could be regarded as N_2, since this is the union of the 2π–periodic points of Y. As shown above, diagonalizability with real eigenvalues is sufficient for a linear vector–field to be the linear part of a vector–field associated to a (locally defined) gradient Kähler Ricci soliton.

Prescribed Eigenvalues

Let $\mathrm{h} = (h_1, \dots, h_n) \in \mathbb{R}^n$ be a nonzero real vector and define [Bry04]

$$\Lambda_{\mathrm{h}} = \{\mathrm{k} \in \mathbb{Z}^n \ : \ \mathrm{k} \cdot \mathrm{h} = 0\} = \mathbb{Z}^n \cap \mathrm{h}^\perp \subset \mathbb{R}^n.$$

Then Λ_{h} is a free Abelian group of rank $n - k$ for some $1 \leq k \leq n$. The number k is the dimension over $\mathbb{Q}$ of the $\mathbb{Q}$–span of the numbers $h_1, \dots, h_n$ in $\mathbb{R}$. Let $\Lambda_{\mathrm{h}}^+ \subset \Lambda_{\mathrm{h}}$ consist of the $\mathrm{k} \in \Lambda_{\mathrm{h}}$ such that $\mathrm{k} = (k_1, \dots, k_n)$ with each k_i nonnegative.

Consider the linear holomorphic vector–field

$$Z_{\mathrm{h}} = h_j z^j \partial_{z^j} \qquad \text{on } \mathbb{C}^n. \tag{5.34}$$

Let $Z_{\mathrm{h}} = X_{\mathrm{h}} - iY_{\mathrm{h}}$ be the decomposition into real and imaginary parts.

Normalizing Volume Forms

In addition to knowing that Z can be linearized near a singular point, it is useful to know that this can be done in such a way that it simplifies the coordinate expression for Υ as well: Set $h = h_1 + \cdots + h_n$ and let Υ be a non–vanishing holomorphic n–form defined on an open neighborhood U of the origin in $\mathbb{C}^n$ that satisfies $d(Z_{\mathrm{h}} \lrcorner \Upsilon) = h\Upsilon$. Then there exist Z_{h}–linearizing coordinates $w = (w^i)$ near the origin in $\mathbb{C}^n$ such that, on the domain of these coordinates $\Upsilon = dw^1 \wedge \cdots \wedge dw^n$.

Let Z and Υ be a holomorphic vector–field and volume form, respectively on a complex n–manifold M. Let $p \in M$ be a singular point of Z. If there exists a gradient Kähler Ricci soliton g with Ricci potential f on a neighborhood of p whose associated holomorphic vector–field and volume form are Z and Υ, respectively, then there exists an $\mathrm{h} \in \mathbb{R}^n$ and a p–centered holomorphic chart $z = (z^i) : U \to \mathbb{C}^n$ such that, on U,

$$Z = h_i\, z^i \partial_{z^i} \qquad \text{and} \qquad \Upsilon = dz = dz^1 \wedge \cdots \wedge dz^n.$$

Local Solitons near a Singular Point

In view of the above statement, questions about the local existence and generality of gradient Kähler Ricci solitons with prescribed Z and Υ near a singular point of Z can be reduced by a holomorphic change of variables to the study of solitons on an open neighborhood of $0 \in \mathbb{C}^n$ with $Z = Z_{\mathrm{h}}$ for some $\mathrm{h} \neq 0$ and $\Upsilon = dz = dz^1 \wedge \cdots \wedge dz^n$.

Let ϕ be a strictly pseudo–convex function defined on a $\mathbb{T}_{\mathrm{h}}$–invariant, contractible neighborhood of $0 \in \mathbb{C}^n$ that satisfies [Bry04]

$$\det\left(\frac{\partial^2\phi}{\partial z^i \partial \bar{z}^j}\right) e^{\frac{1}{2}d\phi(X_{\mathrm{h}})} = 1 \qquad \text{and} \tag{5.35}$$

$$d\phi(Y_{\mathrm{h}}) = 0. \tag{5.36}$$

Then $\Omega = \frac{i}{2}\,\partial\bar{\partial}\phi$ is the associated Kähler form of a gradient Kähler Ricci soliton with Ricci potential $f = \frac{1}{2}d\phi(X_{\mathrm{h}})$ whose associated holomorphic vector–field and volume form are Z_{h} and $dz^1 \wedge \cdots \wedge dz^n$, respectively.

Conversely, if g is a gradient Kähler Ricci soliton defined on a $\mathbb{T}_{\mathrm{h}}$–invariant, contractible neighborhood of $0 \in \mathbb{C}^n$ and f is a Ricci potential for g that satisfies $f(0) = 0$ such that the associated holomorphic vector–field and volume form are Z_{h} and $dz^1 \wedge \cdots \wedge dz^n$, respectively, then g has a Kähler potential ϕ that satisfies (5.35) and (5.36).

The equation (5.35) is a $\mathbb{T}_{\mathrm{h}}$–invariant real–analytic Monge–Ampère equation whose linearization at a strictly pseudo–convex solution ϕ is given by

$$\Delta u + 2\,\mathcal{L}_{X_{\mathrm{h}}} u = 0, \tag{5.37}$$

where Δ is the Laplacian with respect to the metric g associated to $\Omega = \frac{i}{2}\,\partial\bar{\partial}\phi$. Clearly, this is an elliptic equation. It follows by elliptic regularity that any gradient Kähler Ricci soliton is real–analytic, even in the neighborhood of singular points of Z. We can see now that, for any h, there is a sufficiently small ball around the origin on which there is at least one strictly pseudo–convex solution ϕ to (5.35).

A Boundary–Value Formulation

Suppose now that ϕ is a strictly pseudo–convex solution of (5.35) defined on a $\mathbb{T}_{\mathrm{h}}$–invariant bounded neighborhood $D \subset \mathbb{C}^n$ of $0 \in \mathbb{C}^n$ with smooth boundary ∂D. Let g be the corresponding gradient Kähler Ricci soliton. Any solution u of (5.37) in D that vanishes on the boundary will also satisfy [Bry04]

$$0 = \int_D |\nabla u|^2 + \frac{1}{2}R(g)\,u^2 \;d\mathrm{vol}_g\,,$$

as follows by integration by parts using the identities $\rho = \mathcal{L}_{X_{\mathrm{h}}}\Omega$ and $d\mathrm{vol}_g = \frac{1}{n!}\Omega^n$.[11]

In particular, by shrinking D if necessary, it can be assumed that any solution u to (5.37) in D that vanishes on ∂D must vanish on D.

[11] Note that the metric g does not always uniquely determine ϕ by the construction given above since one can add to ϕ the real part of any $\mathbb{T}_{\mathrm{h}}$–invariant holomorphic function that vanishes at $0 \in \mathbb{C}^n$ (depending on h, there may or may not be any nonconstant $\mathbb{T}_{\mathrm{h}}$–invariant holomorphic functions on a neighborhood of $0 \in \mathbb{C}^n$). However, this ambiguity is relatively small.

It then follows, by the implicit function Theorem, that any $\mathbb{T}_h$−invariant function ψ on ∂D that is sufficiently close (in the appropriate norm) to ϕ on ∂D is the boundary value of a unique pseudo-convex solution $\tilde{\phi}$ of (5.35) that is near ϕ on D. The uniqueness then implies that $\tilde{\phi}$ must also be $\mathbb{T}_h$−invariant and so must, in particular, satisfy (5.36).

Thus, local gradient Kähler Ricci solitons near $0 \in \mathbb{C}^n$ with prescribed holomorphic data $(Z, \Upsilon) = (Z_h, dz)$ do exist and have a 'degree of generality' that depends on the number k. The most constraints appear when k reaches its maximum value n and the least when k reaches its minimum value 1.

Finally, Cao and Hamilton [CR00] proved the following useful result: The scalar curvature $R(g)$ has only one critical point and it is both a local maximum and the unique critical point of f, which is a strictly convex proper function on M. As Cao and Hamilton remark, since $\rho = i\partial\bar{\partial}f$ is the Ricci form of g, which is positive, this shows that f is a strictly plurisubharmonic proper exhaustion function on M. This implies that M is *Stein* and that M is diffeomorphic to $\mathbb{R}^{2n}$.

The following result, also known to Cao and Hamilton, gives constraints on the rate of growth of the Ricci potential. Let p be the critical point of $R(g)$ and let f be the Ricci potential, normalized so that $f(p) = 0$. There exist positive constants c_1 and c_2 such that, for all $x \in M$,

$$\sqrt{1 + (c_1\, d(x,p))^2} - 1 \leq f(x) \leq c_2\, d(x,p).$$

For any vector $v \in TM$, one has

$$\mathrm{Ric}(g)(v,v) \leq \lambda_{\max}(g)\,|v|^2$$

where $\lambda_{\max}(g) : M \to \mathbb{R}$ is the maximum eigenvalue function for $\mathrm{Ric}(g)$. Since g is Kähler, the eigenvalues of $\mathrm{Ric}(g)$ occur in pairs and, since $\mathrm{Ric}(g) > 0$, it follows that $\lambda_{\max}(g) \leq \frac{1}{2}R(g)$. In particular, one has the more explicit inequality

$$\mathrm{Ric}(g)(v,v) \leq \frac{1}{2}R(g)\,|v|^2 \leq \frac{1}{2}\left(2h - |\nabla f|^2\right)|v|^2. \tag{5.38}$$

Now let $\gamma : (0,\infty) \to M$ be the arc–length parametrization of a non-constant integral curve of ∇f, such that p is the limit of $\gamma(s)$ as $s \to 0^+$. Thus,

$$|\nabla f(\gamma(s))|\,\gamma'(s) = \nabla f(\gamma(s)) \qquad \text{for all} \quad s > 0.$$

Let $\phi(s) = f(\gamma(s))$. One then computes via the Chain Rule that

$$\phi'(s) = |\nabla f(\gamma(s))| \leq \sqrt{2h},$$

and hence that

$$\phi''(s) = \mathrm{Ric}(g)\left(\frac{\nabla f(\gamma(s))}{|\nabla f(\gamma(s))|}, \frac{\nabla f(\gamma(s))}{|\nabla f(\gamma(s))|}\right). \tag{5.39}$$

By the positivity of Ric(g) and (5.38), this implies

$$0 < \phi''(s) \leq \frac{1}{2}\left(2h - (\phi'(s))^2\right).$$

Moreover, it is clear that, as $s \to 0^+$, the quantity on the right hand side of (5.39) has $\lambda_{\min}(g)(0) > 0$ as a lower bound for its infimum limit. Thus, the infimum limit of $\phi''(s)$ as $s \to 0^+$ is positive.

From these relations, several conclusions can be drawn. The function ϕ is increasing and strictly convex up on $(0, \infty)$. On the other hand, since ϕ' is bounded above, it follows that ϕ grows at most linearly. Moreover, there must be a sequence of distances $s_k \to \infty$ such that $\phi''(s_k) \to 0$. Since, by (5.39)

$$\phi''(s_k) \geq \lambda_{\min}(g)\left(\gamma(s_k)\right),$$

it follows that $\lambda_{\min}(g)\left(\gamma(s_k)\right) \to 0$ as $k \to \infty$. For more details on gradient Kähler Ricci solitons, see [Bry04]

5.5 Monge–Ampère Equations

In this section, following [Ban07], we review the so–called *Monge–Ampère PDEs* in the framework of *generalized complex geometry*. Recall that a general approach to the study of nonlinear PDEs, which goes back to Sophus Lie, is to see a $k-$order equation on an nD manifold M^n as a closed subset in the manifold of $k-$jets J^kM (see [II06b]). In particular, a second–order differential equation lives in the space J^2M. Nevertheless, as it was noticed by Lychagin in his seminal paper [Lyc79], it is sometimes possible to decrease one dimension and to work on the contact space J^1M. The idea is to define for any differential form $\omega \in \Omega^n(J^1M)$, a second order differential operator $\Delta_\omega : C^\infty(M) \to \Omega^n(M)$ acting according to the rule

$$\Delta_\omega(f) = j_1(f)^*\omega,$$

where $j_1(f) : M \to J^1M$ is the section corresponding to the function f.

The differential equations of the form $\Delta_\omega = 0$ are said to be of *Monge–Ampère type* because of their 'Hessian–like' nonlinearity. Despite its very simple description, this classical class of differential equations attends much interest due to its appearance in different problems of geometry or mathematical physics. We refer to [KLR03] for a complete exposition of the theory and for numerous examples.

A *Monge–Ampére equation* $\Delta_\omega = 0$ is said to be symplectic if the *Monge–Ampère operator* Δ_ω is invariant with respect to the Reeb vector field. In other words, the $n-$form ω lives actually on the cotangent bundle T^*M, and symplectic geometry takes place of contact geometry. The Monge–Ampère operator is then defined by

$$\Delta_\omega(f) = (df)^*\omega.$$

This partial case is in some sense quite generic because of the beautiful result of Lychagin which says that any Monge–Ampère equation admitting a contact symmetry is equivalent (by a Legendre transform on J^1M) to a symplectic one.

We are interested here in symplectic Monge–Ampère equations in two variables. These equations read [Ban07]:

$$A\frac{\partial^2 f}{\partial q_1^2} + 2B\frac{\partial^2 f}{\partial q_1 \partial q_2} + C\frac{\partial^2 f}{\partial q_2^2} + D\left(\frac{\partial^2 f}{\partial q_1^2}\frac{\partial^2 f}{\partial q_2^2} - \left(\frac{\partial^2 f}{\partial q_1 \partial q_2}\right)^2\right) + E = 0, \quad (5.40)$$

with A, B, C, D and E smooth functions of $(q, \frac{\partial f}{\partial q})$. These equations correspond to 2–form ω on $T^*\mathbb{R}^2$, or equivalently to tensors on $T^*\mathbb{R}^2$ using the correspondence

$$\omega(\cdot,\cdot) = \Omega(A\cdot,\cdot),$$

Ω being the symplectic form on T^*M. In the non-degenerate case, the traceless part of this tensor A defines either an almost complex structure or an almost product structure and it is integrable if and only the corresponding Monge–Ampère equation is equivalent to the Laplace equation or the wave equation. This elegant result of [LR93] is quite frustrating: which kind of integrable geometry could we define for more general Monge–Ampère equations [Ban07]?

It has been noticed in [Cra04] that such a pair of forms (ω, Ω) defines an almost generalized complex structure, a very rich concept defined recently by Hitchin ([Hit03]) and developed by Gualtieri ([Gua04a]), which interpolates between complex and symplectic geometry. It is easy to see that this almost generalized complex structure is integrable for a very large class of $2D$–Monge–Ampère equations, the equations of *divergent type*. This observation is the starting point for the approach proposed in this section: the aim is to present these differential equations as 'generalized Laplace equations'.

In the first part, we write down this correspondence between Monge–Ampère equations in two variable and 4D generalized complex geometry.

In the second part we study the $\overline{\partial}$–operator associated with a Monge–Ampère equation of divergent type and we show how the corresponding conservation laws and generating functions can be seen as 'holomorphic objects'.

5.5.1 Monge–Ampère Equations and Hitchin Pairs

In what follows M is the smooth symplectic space $T^*\mathbb{R}^2$ endowed with the canonical symplectic form Ω. Our point of view is local (in particulary we do not make any distinction between closed and exact forms) but most of the results presented here have a global version [Ban07].

A primitive 2–form is a differential form $\omega \in \Omega^2(M)$ such that $\omega \wedge \Omega = 0$. We denote by $\perp : \Omega^k(M) \to \Omega^{k-2}(M)$ the operator $\theta \mapsto \iota_{X_\Omega}(\theta)$, the bivector X_Ω being the bivector dual to Ω. It is straightforward to check that in dimension 4, a 2–form ω is primitive iff $\perp\omega = 0$.

Monge–Ampère Operators

Let ω be a 2–form on M. A 2D submanifold L is a generalized solution of the equation $\Delta_\omega = 0$ if it is bi–Lagrangian with respect to Ω and ω.[12] Consider the $2D$–Laplace equation [Ban07]

$$f_{q_1 q_1} + f_{q_2 q_2} = 0.$$

It corresponds to the form

$$\omega = dq_1 \wedge dp_2 - dq_2 \wedge dp_1,$$

while the symplectic form is

$$\Omega = dq_1 \wedge dp_1 + dq_2 \wedge dp_2.$$

Introducing the *complex coordinates*

$$z_1 = q_1 + iq_2 \qquad \text{and} \qquad z_2 = p_2 + ip_1, \qquad \text{we get}$$
$$\omega + i\Omega = dz_1 \wedge dz_2.$$

Generalized solution of the $2D$ Laplace equation appears then as a family of complex curves in $\mathbb{C}^2$.

The following so called Hodge–Lepage–Lychagin Theorem [Lyc79] establishes the 1–1 correspondence between Monge–Ampère operators and primitive 2–forms:

1. Any 2–form ω admits the unique decomposition $\omega = \omega_0 + \lambda\omega$, with ω_0 primitive.
2. If two primitive forms vanish on the same Lagrangian subspaces, then they are proportional.

A Monge–Ampère operator Δ_ω is therefore uniquely defined by the primitive part ω_0 of ω, since $\lambda\Omega$ vanish on any Lagrangian submanifold. The function λ can be arbitrarily chosen.

Let $\omega = \omega_0 + \lambda\Omega$ be a 2–form. We define the tensor A by $\omega = \Omega(A\cdot, \cdot)$. One has $A = A_0 + \lambda Id$ and

$$A_0^2 = -\mathrm{pf}(\omega_0)Id,$$

where the function $\mathrm{pf}(\omega_0)$ is the Pfaffian of ω_0 defined by

$$\omega_0 \wedge \omega_0 = \mathrm{pf}(\omega_0)\Omega \wedge \Omega.$$

[12] A Lagrangian submanifold of $T^*\mathbb{R}^2$ which projects isomorphically on $\mathbb{R}^2$ is a graph of a closed 1–form $df : \mathbb{R}^2 \to T^*\mathbb{R}^2$. A generalized solution can be thought as a smooth patching of classical solutions of the Monge–Ampère equation $\Delta_\omega = 0$ on $\mathbb{R}^2$.

Therefore,

$$A^2 = 2\lambda A - (\lambda^2 + \mathrm{pf}(\omega_0))Id.$$

The equation $\Delta_\omega = 0$ is said to be elliptic if $\mathrm{pf}(\omega_0) > 0$, hyperbolic if $\mathrm{pf}(\omega_0) < 0$, parabolic if $\mathrm{pf}(\omega_0) = 0$. In the elliptic/hyperbolic case, one can define the tensor

$$J_0 = \frac{A_0}{\sqrt{|\mathrm{pf}(\omega_0)|}}$$

which is either an almost complex structure or an almost product structure.

The following assertions are equivalent [LR93]:

1. The tensor J_0 is integrable.
2. The form $\omega_0/\sqrt{|\mathrm{pf}(\omega_0)|}$ is closed.
3. The Monge–Ampère equation $\Delta_\omega = 0$ is equivalent (with respect to the action of local symplectomorphisms) to the elliptic *Laplace equation*

$$f_{q_1q_1} + f_{q_2q_2} = 0,$$

or the hyperbolic *wave equation*

$$f_{q_1q_1} - f_{q_2q_2} = 0.$$

Let us introduce now the *Euler operator* and the notion of *Monge–Ampère equation of divergent type* [Lyc79]. The Euler operator is the second–order differential operator

$$\mathcal{E} : \Omega^2(M) \to \Omega^2(M) \qquad \text{defined by} \qquad \mathcal{E}(\omega) = d\bot d\omega.$$

A Monge–Ampère equation $\Delta_\omega = 0$ is said to be of divergent type if $\mathcal{E}(\omega) = 0$.

For example, the *Born–Infeld equation* is

$$(1 - f_t)^2 f_{xx} + 2f_t f_x f_{tx} - (1 + f_x^2) f_{tt} = 0.$$

The corresponding primitive form is

$$\omega_0 = (1 - p_1^2)dq_1 \wedge dp_2 + p_1p_2(dq_1 \wedge dp_1) + (1 + p_2^2)dq_2 \wedge dp_1.$$

with $q_1 = t$ and $q_2 = x$. A direct computation gives

$$d\omega_0 = 3(p_1dp_2 - p_2dp_1) \wedge \Omega,$$

and then the Born–Infeld equation is not of divergent type.

For example, the *Tricomi equation* is

$$v_{xx}xv_{yy} + \alpha v_x + \beta v_y + \gamma(x, y).$$

The corresponding primitive form is

$$\omega_0 = (\alpha p_1 + \beta p_2 + \gamma(q))dq_1 \wedge dq_2 + dq_1 \wedge dp_2 - q_2dq_2 \wedge dp_1,$$

with $x = q_1$ and $y = q_2$. Since

$$d\omega_0 = (-\alpha dq_2 + \beta dq_1) \wedge \Omega,$$

we conclude that the Tricomi equation is of divergent type [Ban07].

A Monge–Ampère equation $\Delta_\omega = 0$ is said to be of divergent type iff there exists a function μ on M such that the form $\omega + \mu\Omega$ is closed. Since the exterior product by Ω is an isomorphism from $\Omega^1(M)$ to $\Omega^3(M)$, for any 2–form ω, there exists a 1–form α_ω such that

$$d\omega = \alpha_\omega \wedge \Omega.$$

Since $\perp(\alpha_\omega \wedge \Omega) = \alpha_\omega$ we deduce that

$$\begin{aligned} \mathcal{E}(\omega) = 0 \qquad &\text{iff} \qquad d\alpha_\omega = 0, \qquad \text{that is} \\ d(\omega + \mu\Omega) = 0 \qquad &\text{with} \qquad d\mu = -\alpha_\omega. \end{aligned}$$

Hence, if $\Delta_\omega = 0$ is of divergent type, one can choose ω being closed. The point is that it is not primitive in general.

Hitchin Pairs

The natural indefinite interior product on $TM \oplus T^*M$ is [Ban07]

$$(X + \xi, Y + \eta) = \frac{1}{2}(\xi(Y) + \eta(X)),$$

and the *Courant bracket* on sections of $TM \oplus T^*M$ is

$$[X + \xi, Y + \eta] = [X, Y] + L_X\eta - L_Y\xi - \frac{1}{2}d(\iota_X\eta - \iota_Y\xi),$$

where L_X denotes the *Lie derivative* in the direction of the vector–field X. According to [Hit03], an almost generalized complex structure is a bundle map

$$\begin{aligned} \mathbb{J} : TM \oplus T^*M \to TM \oplus T^*M, \qquad &\text{satisfying} \\ \mathbb{J}^2 = -1 \qquad \text{and} \qquad &(\mathbb{J}\cdot, \cdot) = -(\cdot, \mathbb{J}\cdot). \end{aligned}$$

Such an almost generalized complex structure is said to be integrable if the spaces of sections of its two eigenspaces are closed under the Courant bracket.

The standard examples are:

$$\mathbb{J}_1 = \begin{pmatrix} J & 0 \\ 0 & -J^* \end{pmatrix} \qquad \text{and} \qquad \mathbb{J}_2 = \begin{pmatrix} 0 & \Omega^{-1} \\ -\Omega & 0 \end{pmatrix}$$

with J a complex structure and Ω a symplectic form.

Let Ω be a symplectic form and ω any 2–form. Define the tensor A by $\omega = \Omega(A\cdot, \cdot)$ and the form $\tilde{\omega}$ by $\tilde{\omega} = -\Omega(1 + A^2\cdot, \cdot)$. The almost generalized complex structure [Cra04]

$$\mathbb{J} = \begin{pmatrix} A & \Omega^{-1} \\ \tilde{\omega} & -A^* \end{pmatrix} \tag{5.41}$$

is integrable iff ω is closed. Such a pair (ω, Ω) with $d\omega = 0$ is called a *Hitchin pair*.

We get then immediately the following result [Ban07]: To any 2D symplectic Monge–Ampère equation of divergent type $\Delta_\omega = 0$ corresponds a Hitchin pair (ω, Ω) and therefore a 4D generalized complex structure. Let $L^2 \subset M^4$ be a 2D submanifold. Let $TM_L \subset TM$ be its tangent bundle and $TM_L^0 \subset T^*M$ its annihilator. L is a generalized complex submanifold (according to the terminology of [Gua04a]) or a generalized Lagrangian submanifold (according to the terminology of [BB04]) if $TM_L \oplus TM_L^0$ is closed under $\mathbb{J}$. When $\mathbb{J}$ is defined by (5.41), this is equivalent to saying that L is Lagrangian with respect to Ω and closed under A, that is, L is a generalized solution of $\Delta_\omega = 0$.

Systems of First–Order PDEs

On $2n$D manifold, a generalized complex structure is defined by

$$\mathbb{J} = \begin{pmatrix} A & \pi \\ \sigma & -A^* \end{pmatrix},$$

with different relations detailed in [Cra04] between the tensor A, the bivector π and the 2–form σ (the most outstanding being $[\pi, \pi] = 0$, that is π is a *Poisson bivector*). In [Cra04], a generalized complex structures is said to be non–degenerate if the Poisson bivector π is non–degenerate, that is, if the two eigenspaces $E = \ker(\mathbb{J} - i)$ and $\overline{E} = \ker(\mathbb{J} + i)$ are transverse to T^*M. This leads to our symplectic form $\Omega = \pi^{-1}$ and to our 2–form $\omega = \Omega(A\cdot, \cdot)$.

One could also take the dual point of view and study generalized complex structure transverse to TM. In this situation, the eigenspace E writes as

$$E = \{\xi + \iota_\xi P, \xi \in T^*M \otimes \mathbb{C}\},$$

with $P = \pi + i\Pi$ a complex bivector. This space defines a generalized complex structure iff it is a Dirac subbundle of $(TM \oplus T^*M) \otimes \mathbb{C}$ and if it is transverse to its conjugate $\overline{E}$. According to the *Maurer–Cartan type* equation described in the famous paper [LWX97], the first condition is

$$[\pi + i\Pi, \pi + i\Pi] = 0.$$

The second condition says that Π is non–degenerate.

Hence, we get some analog of the Crainic's result [Ban07]: A Hitchin pair of bivectors is a pair consisting of two bivectors π and Π, Π being non-degenerate, and satisfying

$$[\Pi,\Pi] = [\pi,\pi] = 0. \tag{5.42}$$

There is a 1–1 correspondence between the generalized complex structure

$$\mathbb{J} = \begin{pmatrix} A & \pi_A \\ \sigma & -A^* \end{pmatrix},$$

with σ non degenerate and Hitchin pairs of bivectors (π,Π). In this correspondence, we have

$$\begin{cases} \sigma = \Pi^{-1}, \\ A = \pi \circ \Pi^{-1}, \\ \pi_A = -(1+A^2)\Pi. \end{cases}$$

For example, if $\pi + i\Pi$ is non-degenerate, it defines a 2−form $\omega + i\Omega$ which is necessarily closed (this is the complex version of the classical result which says that a non-degenerate Poisson bivector is actually symplectic). We find again an Hitchin pair. So new examples occur only in the degenerate case. Note that $\pi + i\Pi = (A+i)\Pi$, so $\det(\pi + i\Pi) = 0$ iff -i is an eigenvalue for A. In dimension 4, this implies that $A^2 = -1$ but this is not any more true in greater dimensions (see for example the classification of pair of 2−forms on 6-dimensional manifolds in [LR93]). Nevertheless, the case $A^2 = -1$ is interesting by itself. It corresponds to generalized complex structure of the form

$$\mathbb{J} = \begin{pmatrix} J & 0 \\ \sigma & -J^* \end{pmatrix}$$

with J an integrable complex structure and σ a 2−form satisfying $J^*\sigma = -\sigma$ and

$$d\sigma_J = d\sigma(J\cdot,\cdot,\cdot) + d\sigma(\cdot,J\cdot,\cdot) + d\sigma(\cdot,\cdot,J\cdot).$$

where $\sigma_J = \sigma(J\cdot,\cdot)$ (see [Cra04]). Or equivalently $\sigma + i\sigma_J$ is a $(2,0)$−form satisfying

$$\partial(\sigma + i\sigma_J) = 0.$$

One typical example of such geometry is the so called *HyperKähler geometry with torsion* which is an elegant generalization of HyperKähler geometry ([GP00]). Unlike the HyperKäler case, such geometry is always generated by potentials ([BS04]).

Let us consider now an Hitchin pair of bivectors (π,Π) in dimension 4. Since Π is non-degenerate, it defines two 2−forms ω and Ω, which are not necessarily closed, and related by the tensor A. A generalized Lagrangian surface is a surface closed under A, or equivalently, bi–Lagrangian: $\omega|_L = \Omega|_L = 0$. Locally, L is defined by two functions u and v satisfying a first–order system

$$\begin{cases} a + b\frac{\partial u}{\partial x} + c\frac{\partial u}{\partial y} + d\frac{\partial v}{\partial x} + e\frac{\partial v}{\partial y} + f\det J_{u,v} \\ A + B\frac{\partial u}{\partial x} + C\frac{\partial u}{\partial y} + D\frac{\partial v}{\partial x} + E\frac{\partial v}{\partial y} + E\det J_{u,v} \end{cases} \quad \text{with} \quad J_{u,v} = \begin{pmatrix} \frac{\partial u}{\partial x} & \frac{\partial u}{\partial y} \\ \frac{\partial v}{\partial x} & \frac{\partial v}{\partial y} \end{pmatrix}$$

Such a system generalizes both Monge–Ampère equations and Cauchy-Riemann systems and is called Jacobi-system (see [KLR03]). With the help of Hitchin's formalism, we understand now the integrability condition (5.42) as a 'divergent type' condition for Jacobi equations [Ban07].

5.5.2 The $\overline{\partial}-$Operator

Let us fix now a $2D-$ symplectic Monge–Ampère equation of divergent type $\Delta_\omega = 0$, the $2-$form $\omega = \omega_0 + \lambda\Omega$ being closed. We still denote by $A = A_0 + \lambda$ the associated tensor. For any $1-$form α, the following relation holds [Ban07]:

$$\alpha \wedge \omega - B^* \alpha \wedge \Omega = 0, \tag{5.43}$$

with $B = \lambda - A_0$. Let $\alpha = \iota_X \Omega$ be a 1-form. Since ω_0 is primitive, we get

$$0 = \iota_X(\omega_0 \wedge \Omega) = (\iota_X \omega_0) \wedge \Omega + (\iota_X \Omega) \wedge \omega_0 = A_0^* \alpha \wedge \Omega + \alpha \wedge \omega_0.$$

Therefore,

$$\alpha \wedge \omega = \alpha \wedge \omega_0 + \lambda \alpha \wedge \Omega = (-A_0 + \lambda)^* \alpha \wedge \Omega.$$

Now, by $\mathbb{J}$ we denote the generalized complex structure associated with the Hitchin pair (ω, Ω). Also,

$$\Theta = \omega - i\Omega \qquad \text{and} \qquad \Phi = \exp(\Theta) = 1 + \Theta + \frac{\Theta^2}{2}.$$

Decomposition of Differential Forms

Using the tensor $\mathbb{J}$, Gualtieri defines a decomposition [Gua04a]

$$\Lambda^*(T^*M) \otimes \mathbb{C} = U_2 \oplus U_{-1} \oplus U_0 \oplus U_1 \oplus U_2,$$

which generalizes the Dolbeault decomposition for a complex structure.

Let us introduce some notations to understand this decomposition. The space $TM \oplus T^*M$ acts on $\Lambda^*(T^*M)$ by

$$\rho(X + \xi)(\theta) = \iota_X \theta + \xi \wedge \theta,$$

and this action extends to an isomorphism (the standard spin representation) between the Clifford algebra $CL(TM \oplus T^*M)$ and the space of linear endomorphisms $End(\Lambda^*(T^*M))$.

With these notations, the eigenspace $E = \ker(\mathbb{J} - i)$ is also defined by

$$E = \{X + \xi \in TM \oplus T^*M, \rho(X + \xi)(\Phi) = 0\},$$

The space U_k is defined by[13]

[13] $\mathbb{J}$ identified with the $2-$form $(\mathbb{J}\cdot, \cdot)$ lives in $\Lambda^2(TM \oplus T^*M) \subset CL(TM \oplus T^*M)$. We get then an infinitesimal action of $\mathbb{J}$ on $\Lambda^*(T^*M)$.

$$U_k = \rho\left(\Lambda^{2-k}\overline{E}\right)(\Phi).$$

U_k is the $ik-$eigenspace of $\mathbb{J}$.

We see then immediately that $U_{-k} = \overline{U_k}$, since $\mathbb{J}$ is a real tensor. We have the following results [Ban07]:

1. $U_2 = \mathbb{C}\Phi$.
2. $U_1 = \left\{\alpha \wedge \Phi, \alpha \in \Lambda^1(T^*M) \otimes \mathbb{C}\right\}$.
3. $U_0 = \left\{(\theta - \frac{i}{2}\perp\theta) \wedge \Phi, \theta \in \Lambda^2(T^*M) \otimes \mathbb{C}\right\}$.

The next proposition describes the space $U_0^{\mathbb{R}}$ of *real* forms in U_0. It is a direct consequence of the proposition above. Let Λ_0^2 be the space of (real) primitive 2–forms. Then [Ban07]

$$U_0^{\mathbb{R}} = \left\{[\theta + a(i\Omega + 1)] \wedge \Phi,\ \theta \in \Lambda_0^2 \text{ and } a \in \mathbb{R}\right\}.$$

We have actually

$$(\Lambda^1 \oplus \Lambda^3) \otimes \mathbb{C} = U_{-1} \oplus U_1 \qquad \text{and}$$
$$(\Lambda^0 \oplus \Lambda^2 \oplus \Lambda^4) \otimes \mathbb{C} = U_{-2} \oplus U_0 \oplus U_2.$$

For example, the decomposition of a 1–form $\alpha \in \Lambda^1(T^*M)$ is

$$\alpha = \frac{\alpha - iB\alpha}{2} \wedge \Phi + \frac{\alpha + iB\alpha}{2} \wedge \overline{\Phi}.$$

This decomposition is a point–wise decomposition. Denote now by $\mathcal{U}_k$ the space of smooth sections of the bundle U_k. The Gualtieri decomposition now reads[14]

$$\Omega^*(M) \otimes \mathbb{C} = \mathcal{U}_{-2} \oplus \mathcal{U}_{-1} \oplus \mathcal{U}_0 \oplus \mathcal{U}_1 \oplus \mathcal{U}_2.$$

The operator $\overline{\partial} : \mathcal{U}_k \to \mathcal{U}_{k+1}$ is simply $\overline{\partial} = \pi_{k+1} \circ d$.

The next Theorem is completely analogous to the corresponding statement involving an almost complex structure and the Dolbeault operator $\overline{\partial}$. The almost generalized complex structure $\mathbb{J}$ is integrable iff [Gua04a]

$$d = \partial + \overline{\partial}.$$

For example, let $\alpha \in \Omega^1(M)$ be a 1–form. From $d(\alpha \wedge \Phi) = d\alpha \wedge \Phi$ we get

$$\begin{cases} \overline{\partial}(\alpha \wedge \Phi) = \frac{i}{2}(\perp d\alpha)\Phi \\ \partial(\alpha \wedge \Phi) = (d\alpha - \frac{i}{2}\perp d\alpha) \wedge \Phi. \end{cases}$$

It is worth mentioning that one can also define the real differential operator $d^{\mathbb{J}} = [d, \mathbb{J}]$, or equivalently [Cav05]

$$d^{\mathbb{J}} = -i(\partial - \overline{\partial}).$$

[14] Conservations laws are actually well defined up to closed forms.

Cavalcanty establishes in [Cav05], for the particular case $\omega = 0$, an isomorphism

$$\begin{aligned} \Xi : \Omega^*(M) \otimes \mathbb{C} \to \Omega^*(M) \otimes \mathbb{C}, \qquad & \text{satisfying} \\ \Xi(d\theta) = \partial \Xi(\theta), \quad \Xi(\delta\theta) = \overline{\partial} \Xi(\theta), & \end{aligned}$$

with $\delta = [d, \perp]$ the symplectic codifferential. Since $d\delta$ is the Euler operator, Monge–Ampère equations of divergent type write as $\Delta_\omega = 0$ with $\Xi(\omega)$ pluriharmonic on the generalized complex manifold $\big(M^4, \exp(i\Omega)\big)$.

Conservation Laws and Generating Functions

Recall that the notion of conservation laws is a natural generalization to partial differential equations of the notion of first integrals. A 1−form α is a conservation law for the equation $\Delta_\omega = 0$ if the restriction of α to any generalized solution is closed.

For example, let us consider the Laplace equation and the complex structure J associated with. The 2−form $d\alpha$ vanish on any complex curve iff $[d\alpha]_{1,1} = 0$, that is [Ban07]

$$\begin{aligned} \overline{\partial}\alpha_{1,0} + \partial\alpha_{0,1} = 0, \qquad & \text{or equivalently} \\ \overline{\partial}\alpha_{1,0} = \overline{\partial}\partial\psi, & \end{aligned}$$

for some real function ψ. Here $\overline{\partial}$ is the usual *Dolbeault operator* defined by the integrable complex structure J. We deduce that

$$\alpha - d\psi = \beta_{1,0} + \beta_{0,1} \qquad \text{with} \qquad \beta_{1,0} = \alpha_{1,0} - \partial\psi$$

is a holomorphic $(1,0)$−form. Hence, the conservation laws of the $2D$ Laplace equation are (up to exact forms) real parts of $(1,0)$−holomorphic forms.

According to the Hodge–Lepage–Lychagin Theorem, α is a conservation law iff there exist two functions f and g such that $d\alpha = f\omega + g\Omega$. The function f is called a generating function of the Monge–Ampère equation $\Delta_\omega = 0$. By analogy with the Laplace equation, we will say that the function g is the conjugate function to the generating function f. A function f is a generating function iff [Ban07]

$$dBdf = 0.$$

f is a generating function iff there exists a function g such that

$$0 = d(f\omega + g\Omega) = df \wedge \omega + dg \wedge \Omega = (dg + Bdf) \wedge \Omega,$$

and therefore g exists iff $dBdf = 0$.

If f is a generating function and g is its conjugate then for any $c \in \mathbb{C}$, $L_c = (f + ig)^{-1}(c)$ is a generalized solution of the Monge–Ampère equation $\Delta_\omega = 0$. The tangent space TM_aL_c is generated by the hamiltonian vector fields X_f and X_g. Since

$$\Omega(BX_f, Y) = \Omega(X_f, BY) = df(BY) = Bdf(Y) = dg(Y),$$

we deduce that $X_g = BX_f$ and therefore L_c is closed under $B = \lambda - A_0$. L_c is then closed under A_0 and so bi–Lagrangian with respect to Ω and ω.

For example, a generating function of the $2D$ Laplace equation satisfies $dJdf = 0$, and hence it is the real part of a holomorphic function.

The above lemma has a nice interpretation in the Hitchin/Gualtieri formalism: A function f is a generating function of the Monge–Ampère equation $\Delta_\omega = 0$ iff f is a pluriharmonic function on the generalized complex manifold $(M^4, \exp(\omega - i\Omega))$, that is

$$\partial\overline{\partial} f = 0.$$

The spaces U_1 and U_{-1} are respectively the i and $-i$ eigenspaces for the infinitesimal action of $\mathbb{J}$. So, we have [Ban07]

$$\begin{aligned} \mathbb{J}df &= \mathbb{J}\left(\frac{df - iBdf}{2} \wedge \Phi + \frac{df + iBdf}{2} \wedge \overline{\Phi}\right) \\ &= i\left(\frac{df - iBdf}{2} \wedge \Phi - \frac{df + iBdf}{2} \wedge \overline{\Phi}\right) \\ &= Bdf + (B^2 + 1)df \wedge \Omega. \end{aligned}$$

Moreover,

$$d\left((B^2 + 1)df \wedge \Omega\right) = d(B^2 df \wedge \Omega) = d(Bdf \wedge \omega) = (dBdf) \wedge \omega.$$

So, we can deduce that

$$d\mathbb{J}df = 0 \qquad \text{iff} \qquad dBdf = 0.$$

Since $d\mathbb{J}df = 2i\partial\overline{\partial} f$, the proposition is proved.

Decompose the function f as $f = f_{-2} + f_0 + f_2$. Since $\partial f_{-2} = 0$ and $\overline{\partial} f_2 = 0$, f is pluriharmonic iff f_0 is so. Assume that the $\partial\overline{\partial}$–lemma holds (see [Cav05] and [Gua04b]). Then there exists $\psi \in \mathcal{U}_1$ such that

$$\overline{\partial} f_0 = \overline{\partial}\partial\psi.$$

Define then $G_0 \in \mathcal{U}_0$ by $G_0 = i(\partial\psi - \overline{\partial\psi})$. We get

$$\overline{\partial}(f_0 + iG_0) = 0$$

and f_0 appears as the real part of an 'holomorphic object'. Nevertheless, this assumption is not really clear. Does the $\partial\overline{\partial}$-lemma always hold locally?

The following proposition gives an alternative 'holomorphic object' when the closed form ω is primitive (that is $\lambda = 0$) [Ban07]. Assume that the closed form ω is primitive and consider the real forms $U = \omega\wedge\Phi$ and $V = (i\Omega+1)\wedge\Phi$. A function f is a generating function of the Monge–Ampère equation $\Delta_\omega = 0$ with conjugate function g iff

$$\overline{\partial}(fU - igV) = 0.$$

For example, the $2D$ *Von Karman equation* is

$$v_x v_{xx} - v_{yy} = 0.$$

The corresponding primitive form is

$$\omega = p_1 dq_2 \wedge dp_1 + dq_1 \wedge dp_2,$$

which is obviously closed. The form U and V are

$$\begin{cases} U = p_1 dq_2 \wedge dp_1 + dq_1 \wedge dp_2 + 2p_1 dq_1 \wedge dq_2 \wedge dp_1 \wedge dp_2, \\ V = 1 + p_1 dq_2 \wedge dp_1 + dq_1 \wedge dp_2 + (p_1 - 1) dq_1 \wedge dq_2 \wedge dp_1 \wedge dp_2. \end{cases}$$

Generalized Kähler Partners

Gualtieri has also introduced the notion of *generalized Kähler structure*. This is a pair of commuting generalized complex structure such that the symmetric product $(\mathbb{J}_1\mathbb{J}_2)$ is definite positive. The remarkable fact in this theory is that such a structure gives for free two integrable complex structures and a compatible metric [Gua04a]. This theory has been used to construct explicit examples of bi–Hermtian structures on 4D compact manifolds [Hit05]. The idea is that the $+1-$eigenspace V_+ of $\mathbb{J}_1\mathbb{J}_2$ is closed under $\mathbb{J}_1$ and $\mathbb{J}_2$ and that the restriction of $(\cdot,\cdot)$ to it is definite positive. The complex structures and the metric come then from the natural isomorphism $V_+ \rightarrow TM$.

From the point of view of [Ban07], this approach gives us the possibility to associate to a given partial differential equation, natural integrable complex structures and inner products. Nevertheless, at least for hyperbolic equations, such inner product should have a signature, and we have may be to a relax a little bit the definition of generalized Kähler structure: Let $\Delta_\omega = 0$ be a $2D$ symplectic Monge–Ampère equation of divergent type and let $\mathbb{J}$ be the generalized complex structure associated with. We will say that this Monge–Ampère equation admits a generalized Kähler partner if there exists a generalized complex structure $\mathbb{K}$ commuting with $\mathbb{J}$ such that the two eigenspaces of $\mathbb{J}\mathbb{K}$ are transverse to TM and T^*M.

Note that a powerful tool has been done in [Hit05] to construct such structures: Let $\exp\beta_1$ and $\exp\beta_2$ be two complex closed form defining generalized complex structure $\mathbb{J}_1$ and $\mathbb{J}_2$ on 4D manifold. Suppose that

$$(\beta_1 - \beta_2)^2 = 0 = (\beta_1 - \overline{\beta_2})^2,$$

then $\mathbb{J}_1$ and $\mathbb{J}_2$ commute.

Let us see now on a particular case how one can use this tool. Consider an elliptic Monge–Ampère equation $\Delta_\omega = 0$ with $d\omega = 0$ and $\Omega \wedge \omega = 0$. Moreover, assume that there exists a closed $2-$form Θ such that

$$\Omega \wedge \Theta = \omega \wedge \Theta = 0 \qquad \text{and}$$
$$4\omega = \Omega^2 + \Theta^2.$$

Note that $\exp(\omega - i\Omega)$ and $\exp(-\omega - i\Theta)$ satisfy the conditions of the above lemma. We suppose also that $\Theta^2 = \lambda^2\Omega$ with λ a non vanishing function. This implies that $\omega^2 = \mu^2\Omega^2$ with

$$\mu = \frac{\sqrt{1+\lambda^2}}{2}.$$

The triple (ω, Ω, Θ) defines a metric G and an almost hypercomplex structure (I, J, K) such that

$$\omega = \mu G(I\cdot, \cdot), \quad \Omega = G(J\cdot, \cdot), \quad \Theta = \lambda G(K\cdot, \cdot).$$

Define now the two almost complex structures

$$I_+ = \frac{K + \lambda J}{\mu}, \quad I_- = \frac{K - \lambda J}{\mu}.$$

From

$$\omega = \frac{\Omega + \Theta}{2}(I_-\cdot, \cdot) \qquad \text{and}$$
$$\omega = \frac{\Omega - \Theta}{2}(I_+\cdot, \cdot),$$

we deduce that I_+ and I_- are integrable [Ban07].

A function g is the conjugate of a generating function f of the Monge–Ampère equation $\Delta_\omega = 0$ iff

$$dI_+ dg = -dI_- dg.$$

f is a generating function with conjugate g iff

$$0 = df \wedge \omega + dg \wedge \Omega = (-\mu K df + dg) \wedge \Omega,$$

that is iff $d\frac{K}{\mu}dg = 0$.

For example, consider again the Von Karman equation

$$v_x v_{xx} - v_{yy} = 0.$$

with corresponding primitive and closed form

$$\omega = p_1 dq_2 \wedge dp_1 + dq_1 \wedge dp_2.$$

Define then Θ by

$$\Theta = dp_1 \wedge dp_2 + (1 + 4p_1) dq_1 \wedge dq_2.$$

With the triple (ω, Ω, Θ) we construct I_+ and I_- defined by [Ban07]

$$I_+ = \frac{1}{2}\begin{pmatrix} 0 & -1 & 1 & 0 \\ -1/p_1 & 0 & 0 & -1/p_1 \\ -(1+4p_1)/p_1 & 0 & 0 & -1/p_1 \\ 0 & 1+4p_1 & -1 & 0 \end{pmatrix},$$

$$I_- = \frac{1}{2}\begin{pmatrix} 0 & -1 & -1 & 0 \\ -1/p_1 & 0 & 0 & 1/p_1 \\ (1+4p_1)/p_1 & 0 & 0 & -1/p_1 \\ 0 & -(1+4p_1) & -1 & 0 \end{pmatrix}.$$

It is worth mentioning that I_+ and I_- are well defined for all $p_1 \neq 0$. But the metric G is definite positive only for $p_1 < -\frac{1}{4}$.

It would be very interesting to understand the behavior of generating functions and generalized solution of this kind of Monge–Ampère equations with respect to the Gualtieri metric. In particulary, Gualtieri has introduced a *scheming generalized Laplacian* $dd^* + d^*d$ [Gua04b] and to know if generating functions (which are pluriharmonic as we have seen above) are actually harmonic would give important informations on the global nature of the solutions. For more technical details, see [Ban07].

5.6 Quantum Mechanics Viewed as a Complex Structure on a Classical Phase Space

5.6.1 Introduction

In this section, following [Isi04a], we interpret quantum as complex structure on a classical phase space. Complex differentiability on a given real manifold often admits several, mathematically nonequivalent definitions. This is of utmost importance for quantum mechanics when the latter is formulated with the aid of classical phase space. Roughly speaking, complex differentiability amounts to a declaration of what depends on z *vs.* what depends on $\bar{z}$, through the *Cauchy–Riemann equations*. Setting *Darboux coordinates*:

$$z = \frac{1}{\sqrt{2}}(q + \mathrm{i}p) \qquad \text{and} \qquad \bar{z} = \frac{1}{\sqrt{2}}(q - \mathrm{i}p)$$

on a classical phase space and proceeding to quantization, z and $\bar{z}$ respectively become *annihilation and creation operators* of the quantum theory. Clearly, having more than one possible definition of complex differentiability on classical phase space implies having more than one notion of what an elementary quantum is. This is precisely the concept of a *quantum duality* [Vaf97].

An important disclaimer is in order. The choice of a complex structure on classical phase space is not new to geometric quantization [Sni80]. Thus, e.g., in the particular case of *Chern–Simons theory* [AW91], special effort is devoted to proving the independence of the quantum theory with respect to the choice of a complex structure on classical phase space. In geometric quantization, independence of the quantum theory with respect to the complex structure can be formulated more or less as follows. There is a *bundle of Hilbert spaces of quantum states* over a certain base manifold. The latter is the space of all complex structures (satisfying certain natural requirements) that one can place on classical phase space. This bundle of Hilbert spaces admits a projectively flat connection. Being projectively flat, this connection allows for a canonical identification between the fibres corresponding to different choices of a complex structure.

Geometric quantization was firmly established already in the 1980's. It never faced the notion of duality, which arose during the second superstring revolution of the mid 1990's [Vaf97]. In this section we interpret dualities as the possibility of having different notions of what an elementary quantum is, depending on the choice of a complex structure on classical phase space. Conclusions presented in this section do not clash with geometric quantization [Isi04a]. Notwithstanding the canonical identification between different fibres alluded to above, our approach to varying the complex structure is entirely different. We perform a canonical quantization of a number of *Kähler manifolds* (classical phase spaces) whose (local) *Kähler potentials* are taken to be their classical *Hamiltonian functions*. Solving the corresponding time–independent *Schrödinger equations*, we observe that both the eigenvectors and the eigenvalues exhibit an unambiguous dependence on the complex structure chosen, despite the possibility of parallel–transporting them into those corresponding to another complex structure. As an example, what appears to be a semiclassical state (with respect to a certain complex structure) may well be mapped, by the parallel transport mentioned above, into a highly quantum excitation, as measured by a complex structure that is non–biholomorphic with the former. Indeed, non–holomorphic maps involving not just z but also $\bar{z}$, the corresponding quantum creation and annihilation operators are no longer kept separate. One can imagine a transformation mapping, e.g., a large number of creation operators (large quantum numbers: the semiclassical regime in terms of the original variables) into a large number of annihilation operators (small quantum numbers: the strong quantum regime in the new variables) [Isi04a].

Mathematical background pertaining to the topics analyzed here can be found in [KN96, GH94, BGM02]. References to the quantization of Kähler manifolds, from different standpoints, are *Berezin quantization* [Ber74] for the homogeneous case, reviewed in [Per86]; *Berezin–Toeplitz quantization* [Sch01] for the compact case, and *deformation quantization* [RT99] for the general case. From the point of view of geometric quantization, the independence of the quantum theory with respect to the complex structure has been analyzed in [AW91]; see also [Sni80]. Although related with our theme, these approaches are entirely different from the viewpoint taken, the techniques applied, and the goals achieved here. Related papers are also [CPP02, MT03, Isi03, BFM00, BH01, FA04].

Notations

Throughout this section, $\mathcal{C}$ will denote a real $2n$D Kähler manifold that will play the role of classical phase space. We will denote the corresponding symplectic form and complex structure by ω and $\mathcal{J}$, respectively. Since ω and $\mathcal{J}$ are compatible, holomorphic coordinates on $\mathcal{C}$ will also be Darboux coordinates, up to a possible conformal factor. We will normalize all symplectic forms such that the symplectic volume of $\mathcal{C}$ equals the dimension of the complex *Hilbert state–space* obtained by quantization of $\mathcal{C}$ [Isi04a]:

$$\int_{\mathcal{C}} \omega^n = \dim \mathcal{H}. \tag{5.44}$$

The linear $2n$D space $\mathbb{R}^{2n}$ has the Darboux coordinates q^k, p^k, $(k = 1, \dots, n)$, and the symplectic form

$$\omega = \mathrm{d}q^k \wedge \mathrm{d}p^k. \tag{5.45}$$

The corresponding quantum operators satisfy

$$[Q^j, P^k] = \mathrm{i}\delta^{jk}. \tag{5.46}$$

We endow $\mathbb{R}^{2n}$ with the *Euclidean metric*

$$g_{\mathrm{lin}} = \frac{1}{2}\sum_{k=1}^{n} \left((\mathrm{d}q^k)^2 + (\mathrm{d}p^k)^2\right). \tag{5.47}$$

Complex nD space $\mathbb{C}^n$ has the holomorphic coordinates

$$z^k = \frac{1}{\sqrt{2}}\left(q^k + \mathrm{i}p^k\right), \qquad (k = 1, \dots, n), \tag{5.48}$$

and is endowed with the same metric as $\mathbb{R}^{2n}$, now Hermitian instead of real bilinear,

$$g_{\mathrm{lin}} = \mathrm{d}\bar{z}^k \mathrm{d}z^k = ||\mathrm{d}z||^2. \tag{5.49}$$

The corresponding quantum operators

$$A^k = \frac{1}{\sqrt{2}}\left(Q^k + \mathrm{i}P^k\right), \qquad (A^k)^+ = \frac{1}{\sqrt{2}}\left(Q^k - \mathrm{i}P^k\right) \tag{5.50}$$

$$\text{satisfy} \qquad [A^j, (A^k)^+] = \delta^{jk}. \tag{5.51}$$

On a general Kähler phase space $\mathcal{C}$ other than $\mathbb{C}^n$, equations (5.45) to (5.51) will hold locally on every coordinate chart, possibly up to conformal factors.

5.6.2 Varying the Vacuum

Given a complex manifold $\mathcal{C}$, complex line bundles over it can be arranged into holomorphic equivalence classes, and the quotient set can be given a group structure. The result is the *Picard group* of $\mathcal{C}$, denoted $\mathrm{Pic}\,(\mathcal{C})$ [GH94]. The way in which $\mathrm{Pic}\,(\mathcal{C})$ enters the quantum theory is as follows [Isi04a].

Assume that $\mathcal{C}$ is covered by a collection of holomorphic coordinate charts $(\mathcal{U}_j, z^k_{(j)})$, where $k = 1, \ldots, n$ runs over the complex dimensions of $\mathcal{C}$ and j runs over all the charts in a holomorphic atlas. As in [Isi04a] we erect, on every chart $(\mathcal{U}_j, z^k_{(j)})$, a vector–space fibre that will play the role of the Hilbert state–space; this is done by identifying a vacuum state $|0(j)\rangle$ and acting on it with (products of powers of) the creation operators $(A^k(j))^+$. Assuming initially that the vacuum is nondegenerate, every choice of a set of holomorphic transition functions for the complex vector $|0(j)\rangle$ defines a complex line bundle on $\mathcal{C}$ as we cover the latter with coordinate charts $(\mathcal{U}_j, z^k_{(j)})$. Hence the vacuum state defines a class of holomorphic vector bundles over $\mathcal{C}$, i.e., an element of $\mathrm{Pic}\,(\mathcal{C})$. Conversely, every class in $\mathrm{Pic}\,(\mathcal{C})$ determines a holomorphic line bundle, whose fibrewise generator (a complex vector) we can take to be the vacuum state on each coordinate chart. The action of creation operators on this vector gives rise to excitations of the vacuum, i.e., quantum states.

Degenerate vacua can be treated similarly. Assume that we have a d–fold degenerate vacuum, spanned on the chart $(\mathcal{U}_j, z^k_{(j)})$ by the vectors $|0(j)_1\rangle, \ldots,$ $|0(j)_d\rangle$. By the assumption of degeneracy, they are all physically indistinguishable, so it makes sense to consider their wedge product $\wedge_{m=1}^d|0(j)_m\rangle$, which changes at most by a sign under permutations of its d factors. Taking this wedge product to be the fibrewise generator of a line bundle over $\mathcal{C}$, we determine a class in $\mathrm{Pic}\,(\mathcal{C})$. Now not every class in $\mathrm{Pic}\,(\mathcal{C})$ gives rise to a d–fold degenerate vacuum, as the transition functions must be such that a dth root must exist, so that this root also defines a set of transition functions for each and every one of the d vectors $|0(j)_m\rangle$. The assumption of degeneracy implies that the same set of transition functions must be valid for all $m = 1, \ldots, d$. This can be ensured by taking the parameter space for inequivalent d–fold degenerate vacua to be the dth power of the Picard group $\mathrm{Pic}\,(\mathcal{C})$. That is, we take the dth power of all classes in $\mathrm{Pic}\,(\mathcal{C})$, as their dth root is then (and only then) a true class. The resulting set provides the correct parameter space for degenerate vacua.

5.6.3 Kähler Manifolds as Classical Phase Spaces

The simplest Kähler manifold is the linear space $\mathbb{C}^n$. A possible Kähler potential is [Isi04a]

$$K_{\mathrm{lin}} = ||z||^2, \tag{5.52}$$

the Kähler symplectic form being

$$\omega_{\mathrm{lin}} = -\mathrm{i}\,\mathrm{d}\bar{z}^k \wedge \mathrm{d}z^k. \tag{5.53}$$

The symplectic volume of $\mathbb{C}^n$ is infinite,

$$\int_{\mathbb{C}^n} \omega_{\rm lin}^n = \infty. \tag{5.54}$$

$\mathbb{C}^n$ being contractible, it can be covered with a single coordinate chart (the z^k above), so all vector bundles over $\mathbb{C}^n$ are necessarily trivial. In particular its Picard group is trivial,

$$\mathrm{Pic}\,(\mathbb{C}^n) = 0. \tag{5.55}$$

We denote by $|0\rangle_{\rm lin}$ the fibrewise generator of the trivial complex line bundle over $\mathbb{C}^n$. Now the classical Hamiltonian function $H_{\rm lin}$ on $\mathbb{C}^n$ equals the Kähler potential (5.52), $H_{\rm lin} = K_{\rm lin}$. This is the dynamics of the nD linear harmonic oscillator, whose canonical equations of motion read

$$\dot{z}^k = -\mathrm{i}\,\frac{\partial K_{\rm lin}}{\partial \bar{z}^k} = -\mathrm{i}\,z^k. \tag{5.56}$$

The quantization of this dynamics is well known. The classical coordinates z^k and their complex conjugates $\bar{z}^k$ respectively give rise to annihilation and creation operators $A_{\rm lin}^k$ and $(A_{\rm lin}^k)^+$ acting on the vacuum $|0\rangle_{\rm lin}$. The quantum Hamiltonian operator is [Isi04a]

$$H_{\rm lin} = \sum_{k=1}^n \left((A_{\rm lin}^k)^+ A_{\rm lin}^k + \frac{1}{2} \right), \tag{5.57}$$

and its eigenvalue equation

$$\sum_{k=1}^n \left((A_{\rm lin}^k)^+ A_{\rm lin}^k + \frac{1}{2} \right) |m_1, \dots, m_n\rangle_{\rm lin} = E_n |m_1, \dots, m_n\rangle_{\rm lin} \tag{5.58}$$

is solved by the eigenvalues $E_n = \sum_{k=1}^n \left(m_k + \frac{1}{2} \right)$, with the eigenstates

$$|m_1, \dots, m_n\rangle_{\rm lin} = \frac{1}{\sqrt{m_1! \cdots m_n!}} \left((A_{\rm lin}^1)^+ \right)^{m_1} \cdots \left((A_{\rm lin}^n)^+ \right)^{m_n} |0\rangle_{\rm lin}, \tag{5.59}$$

which are excitations of the vacuum $|0\rangle_{\rm lin}$. The Hilbert space $\mathcal{H}_{\rm lin}$ is (the closure of) the linear span of all the states $|m_1, \dots, m_n\rangle_{\rm lin}$, where the occupation numbers m_k $k = 1, \dots, n$, run over all the nonnegative integers. Thus $\mathcal{H}_{\rm lin}$ is infinite–dimensional, in agreement with equations (5.44) and (5.54).

The previous results can be extended to more general Kähler manifolds. As before, let us consider a Kähler manifold $\mathcal{C}$ covered by a holomorphic atlas with coordinate charts $(\mathcal{U}_j, z_{(j)}^k)$. For simplicity we will drop the subindex j from our notations, bearing in mind, however, that we are working locally on the jth chart. On the latter, Kähler potentials $K(\bar{z}^k, z^k)$ are defined only up to gauge transformations

$$K(\bar{z}^k, z^k) \longrightarrow K(\bar{z}^k, z^k) + F(z^k) + G(\bar{z}^k), \tag{5.60}$$

where $F(z^k)$ is an arbitrary holomorphic function and $G(\bar{z}^k)$ an arbitrary antiholomorphic function on the given chart. Hence terms depending exclusively on z^k or exclusively on $\bar{z}^k$ can be gauged away. With this choice of gauge the Kähler potential on the given chart is unique, given that its overall normalization is fixed by (5.44). Such a potential always exists locally on a Kähler manifold $\mathcal{C}$; it is a real, smooth function that factorizes as the product of a holomorphic function $F(z^k)$ times an antiholomorphic function $G(\bar{z}^k)$,

$$K(z^k, \bar{z}^k) = F(z^k)G(\bar{z}^k), \tag{5.61}$$

or, more generally, as a sum of such terms. In general, however, no Kähler potential can be defined *globally* on $\mathcal{C}$, and cases like $\mathbb{C}^n$, where a global Kähler potential *does* exist, are rather exceptional. A nontrival de Rham cohomology group $H^2(\mathcal{C}, \mathbb{R})$ is an obstruction to the existence of a globally–defined Kähler potential [KN96, GH94].

Now let us assume that, on every coordinate chart $(\mathcal{U}_j, z^k_{(j)})$, a Kähler potential can be found that is a function of

$$||z||^2 = \sum_{k=1}^{n} \bar{z}^k z^k, \qquad \text{in the form of}$$

$$K(\bar{z}^k, z^k) = K_{\mathcal{C}}(||z||^2). \tag{5.62}$$

Now for the potential (5.61) to satisfy the requirement (5.62) it is necessary and sufficient that it be $U(n)$–invariant. Indeed, given $U \in U(n)$, consider the coordinates z^k as a column vector and their complex conjugates $\bar{z}^k$ as a row vector, with the z^k transforming into $U^k_m z^m$ and the $\bar{z}^k$ into $\bar{z}^m U^k_m$. Then $||z||^2$ (or functions thereof) is the unique $U(n)$–invariant one can build. This assumption rules out potentials like, e.g., $z^2\bar{z} + \bar{z}^2 z$, whose summands are *unbalanced*, so to speak, in their holomorphic/antiholomorphic dependence. This gives a Kähler metric

$$g_{km} = \frac{\partial^2 K_{\mathcal{C}}(||z||^2)}{\partial \bar{z}^k \partial z^m}. \tag{5.63}$$

Under the above assumptions, we define a dynamics on $\mathcal{C}$ by taking the classical Hamiltonian function on the coordinate chart $(\mathcal{U}_j, z^k_{(j)})$ equal to the Kähler potential on the same chart,

$$H_{\mathcal{C}} = K_{\mathcal{C}}. \tag{5.64}$$

Now (5.64) defines only a *local dynamics* on $\mathcal{C}$ since, as explained above, a general $\mathcal{C}$ admits no globally–defined potential. More importantly, implicitly contained in (5.64) is the following statement: the space of all solutions to Hamilton's classical equations of motion with respect to the Hamiltonian (5.64) is the manifold $\mathcal{C}$ itself. Thus our choice (5.64) makes sense, because classical phase space is in fact the space of all solutions to Hamilton's equations (modulo possible gauge symmetries). Finally, the extrema of $H_{\mathcal{C}}$ will always be minima, as follows from the positivity of the metric (5.63). Thus picking a Hamiltonian equal to the Kähler potential is physically sound.

In order to quantize the dynamics (5.64) on the coordinate chart $(\mathcal{U}_j, z^k_{(j)})$ we will conformally transform the Kähler metric (5.63) into the Euclidean metric $\mathrm{d}\bar{w}^k \mathrm{d}w^k$ by means of a coordinate transformation

$$z^k \to w^k(\bar{z}^m, z^m). \tag{5.65}$$

Next we will replace the classical function $||w||^2$ with the operator of (5.57),

$$||w||^2 \mapsto H_{\mathrm{lin}} = \sum_{k=1}^{n} \left((A^k_{\mathrm{lin}})^+ A^k_{\mathrm{lin}} + \frac{1}{2} \right). \tag{5.66}$$

This quantization prescription also carries a choice of operator ordering attached to it. Then the Kähler potential gives rise to a quantum Hamiltonian operator whose diagonalization, in principle, can be performed using equations (5.57)–(5.59). In this way we can erect, over each coordinate chart $(\mathcal{U}_j, z^k_{(j)})$ on $\mathcal{C}$, a vector–space fibre given by the Hilbert space of quantum states. However, in order to get the complete quantum theory, one also needs the following elements [Isi04a]:

(i) the precise conformal transformation (5.65);

(ii) a set of transition functions for patching together the Hilbert–space fibres across overlapping charts (when C is not contractible);

(iii) the Picard group $Pic(C)$ (when C is not contractible); AND

(iv) the symplectic volume of $\mathcal{C}$.

5.6.4 Complex–Structure Deformations

The above analysis assumes that a complex structure has been picked on $\mathcal{C}$ and kept fixed throughout. However one can also consider varying the complex structure on classical phase space.

Let us first consider $\mathbb{C}^n$. It has a moduli space of complex structures that are compatible with a given orientation [Vio83]. This moduli space is denoted $\mathcal{M}(\mathbb{C}^n)$; it is the symmetric space

$$\mathcal{M}(\mathbb{C}^n) = SO(2n)/U(n). \tag{5.67}$$

This is a compact space of real dimension $n(n-1)$. Here the embedding of $U(n)$ into $SO(2n)$ is given by [Isi04a]

$$A + \mathrm{i}B \longrightarrow \begin{pmatrix} A & B \\ -B & A \end{pmatrix}, \tag{5.68}$$

where $A + \mathrm{i}B \in U(n)$ with A, B real, $n \times n$ matrices [Hel01].

Let us see how the symmetric space (5.67) appears as a moduli space [Vio83, Hel01]. Consider the Euclidean metric g_{lin} of (5.47). Requiring rotations to preserve the orientation, the isometry group of g_{lin} is $SO(2n)$. In the complex coordinates of (5.48), $g_{\rm lin}$ becomes the Hermitian form (5.49), whose isometry group is $U(n)$. Notice that we no longer impose the condition of unit determinant, since $U(n) = SU(n) \times U(1)$ and $g_{\rm lin}$ is invariant under the $U(1)$ action

$$z^k \to e^{i\alpha} z^k, \qquad (k = 1, \dots, n) \text{ for all } \alpha \in \mathbb{R}.$$

Now every choice of orthogonal axes x^k, y^k in $\mathbb{R}^{2n}$, i.e., every element of $SO(2n)$, defines a complex structure on $\mathbb{R}^{2n}$ upon setting

$$w^k = \frac{1}{\sqrt{2}} \left(x^k + i y^k \right), \qquad (k = 1, \dots, n). \tag{5.69}$$

Generically the w^k are related non–biholomorphically with the z^k, because the orthogonal transformation

$$\begin{aligned} z^k &\longrightarrow w^k = R^k_m z^m + S^k_m \bar{z}^m \\ \bar{z}^k &\longrightarrow \bar{w}^k = \bar{R}^k_m \bar{z}^m + \bar{S}^k_m z^m , \end{aligned} \tag{5.70}$$

while satisfying the orthogonality conditions

$$R^k_m \bar{R}^k_n + S^k_n \bar{S}^k_m = \delta_{mn}, \qquad R^k_m \bar{S}^k_n = 0 = S^k_m \bar{R}^k_n, \tag{5.71}$$

need not satisfy the Cauchy–Riemann conditions

$$\frac{\partial \bar{w}^k}{\partial z^m} = \bar{S}^k_m = 0 = S^k_m = \frac{\partial w^k}{\partial \bar{z}^m}. \tag{5.72}$$

However, when (5.72) holds, the transformation (5.70) is not just orthogonal but also unitary. Therefore one must divide $SO(2n)$ by the action of the unitary group $U(n)$, in order to get the parameter space for rotations that truly correspond to inequivalent complex structures on $\mathbb{R}^{2n} \simeq \mathbb{C}^n$. Non–biholomorphic complex structures on $\mathbb{C}^n$ are 1–to–1 with rotations of $\mathbb{R}^{2n}$ that are *not* unitary transformations.

When $n = 1$ the moduli space (5.67) reduces to a point. Therefore on the complex–plane $\mathbb{C}$ there exists a unique complex structure, that we can identify as the one whose holomorphic atlas consists of the open set $\mathbb{C}$ endowed with the holomorphic coordinate $z = (q + ip)/\sqrt{2}$. Physically this corresponds to the 1D harmonic oscillator. Consider now n independent harmonic oscillators, where $\mathcal{C} = \mathbb{C}^n = \mathbb{C} \times \overset{n}{\dots} \times \mathbb{C}$. Although it is never explicitly stated, the complex structure on this product space is always understood to be the n–fold Cartesian product of the unique complex structure on $\mathbb{C}$. Obviously, removing the requirement of compatibility between the complex structure and the orientation chosen, we duplicate the number of complex structures.

Let us finally analyse the complex–structure moduli of projective and hyperbolic spaces (see appendix). For projective space we have

$$\mathbb{CP}^n = \mathbb{C}^n \cup \mathbb{CP}^{n-1}, \qquad \text{with} \quad \mathbb{CP}^{n-1} \quad \text{a hyperplane at infinity.}$$

It follows that $\mathcal{M}(\mathbb{CP}^n) = \mathcal{M}(\mathbb{C}^n)$. The case $n = 1$ is interesting because $\mathbb{CP}^1 = S^2$, the Riemann sphere. The latter can be regarded as the classical phase space of a spin–1/2 system, inasmuch as spin possesses a classical counterpart. Hence there are no complex–structure moduli on the Riemann sphere. For hyperbolic space we also have $\mathcal{M}(B^n) = \mathcal{M}(\mathbb{C}^n)$, since the complex structure on B^n is the one induced by $\mathbb{C}^n$. Grassmann manifolds $G_{r,r'}(\mathbb{C})$ and bounded domains $\mathrm{D}_{r,r'}(\mathbb{C})$ are natural generalizations of projective and hyperbolic space, respectively, so analogous conclusions apply to them.

5.6.5 Kähler Deformations

Next we study the dependence of the quantum theory on the Kähler moduli, while keeping the complex moduli fixed, in the cases when $\mathcal{C}$ is linear space $\mathbb{C}^n$, hyperbolic space B^n and projective space $\mathbb{CP}^n$. We will show that these 3 cases correspond to different approximation regimes of the Kähler class [Isi04a].

Let us first consider the restriction of the Kähler potential (5.52) to the unit ball B^n, and let us deform it by a polynomial of degree $N > 1$,

$$K_{\text{lin}} \to K_{(N)} = ||z||^2 + \frac{1}{2}||z||^4 + \frac{1}{3}||z||^6 + \ldots + \frac{1}{N}||z||^{2N}. \qquad (5.73)$$

This deformation gives rise to a new Kähler potential on B^n. Let $\omega_{(N)}$ denote the deformed symplectic form corresponding to $K_{(N)}$, and let $B^n_{(N)}$ denote the resulting manifold, with $B^n_{(N=1)} = B^n$. Any deformation of finite degree N increases the symplectic volume by a finite amount. This increase is positive because all summands in (5.73) are positive definite. Despite its increase, the symplectic volume of $B^n_{(N)}$ measured by the $2n$–form $\omega^n_{(N)}$ always remains finite:

$$\int_{B^n_{(N)}} \omega^n_{(N)} < \infty, \qquad (1 < N < \infty). \qquad (5.74)$$

For all finite $N > 1$, $K_{(N)}$ is a Kähler deformation of K_{lin} that increases the symplectic volume of B^n. Then equations (5.58), (5.59) allow one to diagonalize the Hamiltonian

$$H_{(N)} = \sum_{j=1}^{N} \frac{1}{j} \left(\sum_{k=1}^{n} \left((A^k_{\text{lin}})^+ A^k_{\text{lin}} + \frac{1}{2} \right) \right)^j. \qquad (5.75)$$

It has eigenstates $|m_1, \ldots, m_n\rangle_{\text{lin}}$ corresponding to the (nondegenerate) eigenvalues

$$\sum_{j=1}^{N} \frac{1}{j} \left(\sum_{k=1}^{n} \left(m_k + \frac{1}{2} \right) \right)^j, \qquad (5.76)$$

where the occupation numbers m_k do not run over all the nonnegative integers: they are limited by the constraint (5.74) to a finite range. Although the precise value of this range is immaterial for our purposes, let us say that it can be determined (up to irrelevant multiplicative factors) as the whole part of the integral (5.74); as such it is a positive, monotonically increasing function of N, divergent in the limit $N \to \infty$ where, thanks to the Taylor expansion ($x = ||z||^2$)

$$-\ln(1-x) = x + \frac{1}{2}x^2 + \frac{1}{3}x^3 + \frac{1}{4}x^4 + \ldots \qquad |x| < 1, \tag{5.77}$$

the above results are reproduced. In the limit $N \to \infty$ the manifold $B^n_{(N)}$ becomes the hyperbolic manifold $B^n_{\rm hyp}$, and equations (5.75), (5.76) become their hyperbolic partners (5.94), (5.95); the function $h_{\rm hyp}$ of (5.90) appears in the process. Thus the effect of the Kähler deformation (5.73) is that of enlarging the Hilbert space, allowing for excitations of the vacuum obtained by the action of more than just one creation operator $(A^k_{\rm lin})^+$. Analogous conclusions would hold if we considered arbitrary positive coefficients c_j multiplying the deformations $||z||^{2j}$ in (5.73), instead of $c_j = 1/j$.

Choices for the c_j not all positive, such as $c_j = (-1)^{j+1}/j$, lead to different deformations of the Kähler potential (5.52) on $\mathbb{C}^n$ [Isi04a]:

$$K_{\rm lin} \to K_{(N)} = ||z||^2 - \frac{1}{2}||z||^4 + \frac{1}{3}||z||^6 - \ldots + \frac{(-1)^{N+1}}{N}||z||^{2N}. \tag{5.78}$$

In the limit $N \to \infty$, apply the Taylor expansion ($x = ||z||^2$)

$$\ln(1+x) = x - \frac{1}{2}x^2 + \frac{1}{3}x^3 - \frac{1}{4}x^4 + \ldots \qquad |x| < 1, \tag{5.79}$$

initially only to the manifold B^n for convergence. Once the series (5.79) has been summed up, take the left–hand side as the Kähler potential on all of $\mathbb{C}^n$, and declare the latter to be just one of the $n+1$ coordinate charts on $\mathbb{CP}^n$ described below.

Conversely, taking due care of the domains for the coordinate charts, the hyperbolic and projective dynamics can be linearized, as per equations (5.77), (5.79), respectively, in order to yield the linear dynamics. In this way the effect of varying the Kähler moduli is to deform the symplectic volume of $\mathcal{C}$. By (5.44), this is reflected as a variation in the number of quantum states.

The moduli space of Kähler structures on $\mathbb{CP}^n$ is $\mathbb{R}^+$, i.e., the positive reals. All these Kähler deformations are compatible with the fixed complex structure.[15]

[15] Kähler moduli are associated with what in [Isi04a] it was called *representations* for $\mathbb{CP}^n$.

5.6.6 Dynamics on Kähler Spaces

Hyperbolic Space

Within $\mathbb{C}^n$ we have the unit ball

$$B^n = \{(z^1, \dots, z^n) \in \mathbb{C}^n : \|z\| < 1\}. \tag{5.80}$$

Consider the Kähler potential on B^n [Isi04a]

$$K_{\text{hyp}} = -\ln\left(1 - ||z||^2\right), \tag{5.81}$$

from which the hyperbolic symplectic form

$$\omega_{\text{hyp}} = \frac{-\mathrm{i}}{(1 - ||z||^2)^2} \mathrm{d}\bar{z}^k \wedge \mathrm{d}z^k \tag{5.82}$$

and the hyperbolic metric

$$g_{\text{hyp}} = \frac{1}{(1 - ||z||^2)^2} \mathrm{d}\bar{z}^k \mathrm{d}z^k \tag{5.83}$$

follow. Hyperbolic space is the Kähler manifold obtained by endowing the unit ball (5.80) with the Kähler potential (5.81). It has constant negative scalar curvature. The symplectic volume of B^n is infinite,

$$\int_{B^n} \omega_{\text{hyp}}^n = \infty. \tag{5.84}$$

B^n is contractible. Hence it can be covered with a single coordinate chart (the z^k above), and all vector bundles over B^n are trivial. In particular its Picard group is trivial,

$$\text{Pic}\,(B^n) = 0. \tag{5.85}$$

Let $|0\rangle_{\text{hyp}}$ denote the (fibrewise) generator of the trivial complex line bundle over B^n. We take the classical Hamiltonian function on B^n equal to the Kähler potential (5.81). Let π^{rs}_{hyp} denote the Poisson tensor corresponding to the symplectic form ω_{hyp} of (5.82). Now one can verify that the space of all solutions to Hamilton's equations [Isi04a]

$$\dot{z}^k = \left\{z^k, K_{\text{hyp}}\right\} = \pi^{rs}_{\text{hyp}} \frac{\partial z^k}{\partial z^r} \frac{\partial K_{\text{hyp}}}{\partial \bar{z}^s} = \pi^{ks}_{\text{hyp}} \frac{\partial K_{\text{hyp}}}{\partial \bar{z}^s} = -\mathrm{i}\, z^k \left(1 - ||z||^2\right) \tag{5.86}$$

is in fact B^n. On the latter manifold the Hamiltonian (5.81) is bounded from below, as expected physically. The right–hand side of the equations of motion (5.86) contains the square root of the conformal factor

$$f_{\text{hyp}} = \left(1 - ||z||^2\right)^2 \tag{5.87}$$

needed to transform the hyperbolic metric (5.83) into the Euclidean metric (5.49), i.e.,

$$g_{\rm lin} = f_{\rm hyp} g_{\rm hyp}. \tag{5.88}$$

The above conformal transformation to $g_{\rm lin} = {\rm d}\bar{w}^k {\rm d}w^k$ is induced by the change of variables (5.65) that solves the differential equations

$${\rm d}w^k = \frac{{\rm d}z^k}{1 - ||z||^2}, \qquad {\rm d}\bar{w}^k = \frac{{\rm d}\bar{z}^k}{1 - ||z||^2}. \tag{5.89}$$

The solution to (5.89) provides us with a positive function $h_{\rm hyp}(x)$ such that

$$||z||^2 = h_{\rm hyp}(||w||^2). \tag{5.90}$$

Now let hyperbolic creation and annihilation operators $(A^k_{\rm hyp})^+$ and $A^k_{\rm hyp}$ correspond to the coordinates $\bar{z}^k$ and z^k, respectively. Linear creation and annihilation operators $(A^k_{\rm lin})^+$ and $A^k_{\rm lin}$ respectively correspond to the coordinates $\bar{w}^k$ and w^k obtained as a solution to (5.89). The classical Hamiltonian (5.81)

$$H_{\rm hyp} = -\ln\left(1 - ||z||^2\right) \tag{5.91}$$

can be reexpressed as

$$H_{\rm hyp} = -\ln\left(1 - h_{\rm hyp}(||w||^2)\right). \tag{5.92}$$

Quantum–mechanically we make the replacement

$$||w||^2 \mapsto \sum_{k=1}^n \left((A^k_{\rm lin})^+ A^k_{\rm lin} + \frac{1}{2}\right), \tag{5.93}$$

so the quantum Hamiltonian operator is

$$H_{\rm hyp} = -\ln\left\{1 - h_{\rm hyp}\left(\sum_{k=1}^n\left((A^k_{\rm lin})^+ A^k_{\rm lin} + \frac{1}{2}\right)\right)\right\}. \tag{5.94}$$

Diagonalising first the argument of the logarithm as in equations (5.57), (5.58), (5.59), the eigenvalue equation for the hyperbolic Hamiltonian (5.94) reads

$$H_{\rm hyp}|m_1, \ldots, m_n\rangle_{\rm hyp} = -\ln\left\{1 - h_{\rm hyp}\left(\sum_{k=1}^n\left(m_k + \frac{1}{2}\right)\right)\right\}|m_1, \ldots, m_n\rangle_{\rm hyp}, \tag{5.95}$$

where $$|m_1, \ldots, m_n\rangle_{\rm hyp} = \frac{1}{\sqrt{m_1! \cdots m_n!}}\left((A^1_{\rm lin})^+\right)^{m_1} \cdots \left((A^n_{\rm lin})^+\right)^{m_n}|0\rangle_{\rm hyp}. \tag{5.96}$$

The occupation numbers m_k, $k = 1, \ldots, n$, run over all the nonnegative integers, and the Hilbert space $\mathcal{H}_{\rm hyp}$ is (the closure of) the linear span of all the states $|m_1, \ldots, m_n\rangle_{\rm hyp}$. Thus $\mathcal{H}_{\rm hyp}$ is infinite–dimensional, in agreement with equations (5.44), (5.84). One could also express quantum states on B^n as the result of the action of hyperbolic creation operators $(A^k_{\mathrm{hyp}})^+$ on the hyperbolic vacuum $|0\rangle_{\mathrm{hyp}}$, at the cost of losing the nice interpretation of (5.96), namely, that each integer power of a creation operator contributes to the state $|m_1, \dots, m_n\rangle_{\mathrm{hyp}}$ by one quantum.

When $x \to 0$ we have $h_{\mathrm{hyp}}(x) \simeq x$. This ensures that, in the limit of small quantum numbers, equations (5.94) and (5.95) correctly reduce to their expected limits (5.57) and (5.58). This makes sense as, in a neighborhood of the origin in B^n, the hyperbolic oscillator reduces to the linear oscillator, and curvature effects can be neglected. The limit of small quantum numbers is the strong quantum regime. On the contrary, in the limit of large quantum numbers, or classical limit, we have $||z|| \to 1$, $||w|| \to \infty$, so it must hold that $h_{\mathrm{hyp}}(x) \to 1$ as $x \to \infty$. Hence the effects of the negative curvature of B^n can no longer be neglected as we approach the boundary of hyperbolic space. The effects of classical curvature due to the logarithm in the Kähler potential (5.81) are suppressed, or smoothed out, quantum–mechanically.

Projective Space

Let $Z^1, \dots, Z^{n+1}$ denote homogeneous coordinates on $\mathbb{CP}^n$. The chart defined by $Z^j \neq 0$ covers one copy of the open set $\mathcal{U}_j = \mathbb{C}^n$. On the latter we have the holomorphic coordinates $z^k_{(j)} = Z^k/Z^j$, $(k = 1, \dots, n+1, k \neq j)$; there are $n+1$ such coordinate charts. $\mathbb{CP}^n$ is a Kähler manifold with respect to the *Fubini–Study metric*, with constant positive scalar curvature. On $(\mathcal{U}_j, z^k_{(j)})$ the Kähler potential reads, dropping the subindex j for simplicity [Isi04a],

$$K_{\mathrm{proj}} = \ln\left(1 + ||z||^2\right). \tag{5.97}$$

On the same chart, the projective symplectic form is

$$\omega_{\mathrm{proj}} = \frac{-\mathrm{i}}{(1 + ||z||^2)^2}\, \mathrm{d}\bar{z}^k \wedge \mathrm{d}z^k, \tag{5.98}$$

while the Fubini–Study metric reads

$$g_{\mathrm{proj}} = \frac{1}{(1 + ||z||^2)^2}\, \mathrm{d}\bar{z}^k \mathrm{d}z^k. \tag{5.99}$$

The Picard group is the additive group of integers [GH94],

$$\mathrm{Pic}\,(\mathbb{CP}^n) = \mathbb{Z}. \tag{5.100}$$

The class $l = 0$ corresponds to the trivial complex line bundle; all other classes $l \neq 0$ correspond to nontrivial line bundles. On $(\mathcal{U}_j, z^k_{(j)})$, we denote the (fibrewise) generator of the line bundle corresponding to the class l by $|0(j)\rangle^l_{\mathrm{proj}}$. For simplicity we will concentrate on the class $l = 1$; see [Isi04a] for the general case. Then the symplectic volume of $\mathbb{CP}^n$ can be normalized as

$$\int_{\mathbb{CP}^n} \omega_{\text{proj}}^n = n + 1. \tag{5.101}$$

As explained, we take the classical Hamiltonian function on the coordinate chart $(\mathcal{U}_j, z^k_{(j)})$ equal to the Kähler potential (5.97). Let π^{rs}_{proj} denote the Poisson tensor corresponding to the symplectic form (5.98). Now the space of all solutions to Hamilton's equations

$$\dot{z}^k = \{z^k, K_{\text{proj}}\} = \pi^{rs}_{\text{proj}} \frac{\partial z^k}{\partial z^r} \frac{\partial K_{\text{proj}}}{\partial \bar{z}^s} = \pi^{ks}_{\text{proj}} \frac{\partial K_{\text{proj}}}{\partial \bar{z}^s} = -\mathrm{i}\, z^k \left(1 + ||z||^2\right), \tag{5.102}$$

is in fact the whole coordinate chart $(\mathcal{U}_j, z^k_{(j)})$. The right–hand side of (5.102) contains the square root of the conformal factor

$$f_{\text{proj}} = \left(1 + ||z||^2\right)^2 \tag{5.103}$$

that transforms the Fubini–Study metric (5.99) into the Euclidean metric (5.49), i.e.,

$$g_{\text{lin}} = f_{\text{proj}} g_{\text{proj}}. \tag{5.104}$$

The above conformal transformation to

$$g_{\text{lin}} = \mathrm{d}\bar{w}^k \mathrm{d}w^k$$

is induced by the change of variables that, on the chart $(\mathcal{U}_j, z^k_{(j)})$, solves the differential equations

$$\mathrm{d}w^k = \frac{\mathrm{d}z^k}{1 + ||z||^2}, \qquad \mathrm{d}\bar{w}^k = \frac{\mathrm{d}\bar{z}^k}{1 + ||z||^2}. \tag{5.105}$$

Thus, e.g., when $n = 1$, this change of variables is given by the usual stereographic projection from the plane to the Riemann sphere. By the same reasoning as above, a positive function $h_{\text{proj}}(x)$ exists such that

$$||z||^2 = h_{\text{proj}}(||w||^2). \tag{5.106}$$

Moreover,

$$h_{\text{proj}}(x) \simeq x \qquad \text{when} \qquad x \to 0,$$

because the projective oscillator approaches the linear oscillator in this limit. This corresponds to dropping the logarithm in the Kähler potential (5.97).

On the coordinate chart under consideration, $\bar{z}^k$ and z^k respectively give rise to projective creation and annihilation operators $(A^k_{\text{proj}})^+$ and A^k_{proj} acting on the vacuum $|0\rangle^{(l=1)}_{\text{proj}}$. Linear creation and annihilation operators $(A^k_{\text{lin}})^+$ and A^k_{lin} correspond to the coordinates $\bar{w}^k$ and w^k, respectively. The classical Hamiltonian (5.97)

$$H_{\text{proj}} = \ln\left(1 + ||z||^2\right) \tag{5.107}$$

can be reexpressed as

$$H_{\text{proj}} = \ln\left(1 + h_{\text{proj}}(||w||^2)\right). \tag{5.108}$$

Now quantum–mechanically we apply the replacement

$$||w||^2 \mapsto \sum_{k=1}^{n}\left((A^k_{\text{lin}})^+ A^k_{\text{lin}} + \frac{1}{2}\right), \tag{5.109}$$

so the quantum Hamiltonian operator is, on the given chart [Isi04a],

$$H_{\text{proj}} = \ln\left\{1 + h_{\text{proj}}\left(\sum_{k=1}^{n}\left((A^k_{\text{lin}})^+ A^k_{\text{lin}} + \frac{1}{2}\right)\right)\right\}. \tag{5.110}$$

Proceeding as in previous sections we find

$$H_{\text{proj}}|m_1,\ldots,m_n\rangle^{(l=1)}_{\text{proj}} = \ln\left\{1 + h_{\text{proj}}\left(\sum_{k=1}^{n}\left(m_k + \frac{1}{2}\right)\right)\right\}|m_1,\ldots,m_n\rangle^{(l=1)}_{\text{proj}}, \tag{5.111}$$

where

$$|m_1,\ldots,m_n\rangle^{(l=1)}_{\text{proj}} = \frac{1}{\sqrt{m_1!\cdots m_n!}}\left((A^1_{\text{lin}})^+\right)^{m_1}\cdots\left((A^n_{\text{lin}})^+\right)^{m_n}|0\rangle^{(l=1)}_{\text{proj}}. \tag{5.112}$$

In agreement with equations (5.44), (5.101) there are $n+1$ states, as the Hilbert space $\mathcal{H}^{(l=1)}_{\text{proj}}$ over the given chart is the linear span of the vectors $|m_1,\ldots,m_n\rangle^{(l=1)}_{\text{proj}}$, where the occupation numbers are either all zero [for the vacuum $|0\rangle^{(l=1)}_{\text{proj}}$], or all are zero but one [for the pth excited state $(A^p_{\text{lin}})^+|0\rangle^{(l=1)}_{\text{proj}}$, $p = 1,\ldots,n$]. One could also express quantum states on $\mathbb{CP}^n$ as the result of the action of projective creation operators $(A^k_{\text{proj}})^+$ on the projective vacuum $|0\rangle^{(l=1)}_{\text{proj}}$. Transition functions for this bundle of Hilbert spaces over $\mathbb{CP}^n$ have been given in [Isi04a]. The corresponding bundles over $\mathbb{C}^n$ and B^n were trivial due to contractibility.

5.6.7 Interpretations

As we have seen above, complex–differentiable structures on classical phase spaces $\mathcal{C}$ have a twofold meaning. Geometrically they define complex differentiability, or analyticity, of functions on complex manifolds such as $\mathcal{C}$. Quantum–mechanically they define the notion of a quantum, i.e., an elementary excitation of the vacuum state. In this section we have elaborated on this latter meaning. The mathematical possibility of having two or more non–biholomorphic complex–differentiable structures on a given classical phase space leads to the physical notion of a quantum–mechanical duality, i.e., to the relativity of the notion of an elementary quantum. This relativity is understood as the dependence of the notion of a quantum on the choice of a complex–differentiable structure on $\mathcal{C}$. We have summarized this fact in the statement that a quantum is a complex–differentiable structure on classical phase space [Isi04a].

In this section we have proposed a solution to the problem suggested in [Vaf97], namely, how to implement duality transformations in the quantum mechanics of a finite number of degrees of freedom. We have first drawn attention to the key role that complex–differentiable structures on classical phase space play in the formulation of quantum mechanics, without resorting to geometric quantization. This raises the question, how does quantum mechanics vary with each choice of a complex structure on classical phase space? What does it mean, to have different possible quantum–mechanical descriptions of a given physics? We claim that there are at least three ways in which one can get different quantum–mechanical theories over a given classical phase space, thus giving rise to dualities:
(i) by varying the ground state, i.e., the vacuum;
(ii) by varying the type of excitations of the vacuum, i.e., the creation and annihilation operators;
(iii) by varying the number of excitations of the vacuum, i.e., the dimension of the Hilbert space of quantum states.

Each one of these variations, while referring to the quantum theory, concerns properties of classical phase space. Moreover, each one of these variations has its own parameter space. The parameter space for physically inequivalent vacua is the Picard group of classical phase space. The parameter space for physically inequivalent pairs of creation and annihilation operators is the moduli space of complex structures on classical phase space. The parameter space for the dimension of the Hilbert state–space is the moduli space of Kähler classes on classical phase space.

On $\mathbb{C}^n$ every complex structure induces a compatible symplectic structure, by taking the real and imaginary parts of $z^k = (q^k + \mathrm{i}p^k)/\sqrt{2}$ as Darboux coordinates. On C^n the converse is also true: Darboux coordinates can be arranged into the real and imaginary parts of holomorphic coordinates. Hence, on $\mathbb{C}^n$, there is a 1–to–1 correspondence between complex structures and symplectic structures, and a variation in one of them induces a corresponding variation in the other. Differences in the notion of an elementary quantum on $\mathbb{C}^n$ can therefore be traced back to different choices of the classical symplectic structure. Moreover, the Picard group of $\mathbb{C}^n$ being trivial, the corresponding vacuum is also unique (for each choice of a complex structure). Altogether there is no room for dualities on $\mathbb{C}^n$.

However, on other classical phase spaces (see appendix) there is room for independent variations of complex and symplectic structures, and/or for choosing physically nonequivalent vacua. We have shown that, on the unit ball $B^n \subset \mathbb{C}^n$, we can deform the symplectic structure while keeping the complex structure fixed. This is not quite a quantum–mechanical duality yet, as the quantum theory refers to a complex structure, but further examples can be manufactured. Thus, on the complex 1D torus one can vary the complex structure while keeping the symplectic structure fixed [Isi04b]. This is an example of a quantum–mechanical duality that passes completely unnoticed at the classical level, as it leaves the symplectic structure unchanged. On complex projective space $\mathbb{CP}^n$ there is a nontrivial Picard group, which allows for different vacua [Isi04a].

A duality arises as the possibility of having two or more, apparently different, quantum–mechanical descriptions of the same physics. Above we have enumerated three possible ways in which one can vary the description of a given physics. These facts imply that the concept of a quantum is not absolute, but relative to the quantum theory used to measure it [Vaf97]. That is, duality expresses the relativity of the concept of a quantum. In particular *classical* and *quantum*, for long known to be intimately related, are not necessarily always the same for all observers on phase space.

When $\mathcal{C}$ is not only complex but also Kähler, we have a natural arena for the study of quantum–mechanical dualities. A (local) classical Hamiltonian function can always be found, namely the Kähler potential, such that the corresponding canonical equations of motion have $\mathcal{C}$ as the space of all solutions. We have quantized this classical dynamics by means of a change of variables that essentially reduces the problem to a variant of the harmonic oscillator on Euclidean space $\mathbb{C}^n$ (itself the simplest Kähler manifold). Now Kähler spaces typically have complex–structure deformation moduli as well as Kähler–deformation moduli. We have argued that moving around in their respective moduli spaces, i.e., varying these moduli, we get different quantum–mechanical descriptions of a given physics. This is precisely the notion of a *quantum duality* [Vaf97]. For more details, see [Isi04a].

5.7 Geometric Quantization

5.7.1 Quantization of Ordinary Hamiltonian Mechanics

Recall from Chapter 4 that classical *Dirac quantization* states [Dir49]:

$$\{f, g\} = \frac{1}{i\hbar}[\hat{f}, \hat{g}],$$

which means that the quantum Poisson brackets (i.e., commutators) have the same values as the classical Poisson brackets. In other words, we can associate smooth functions defined on the symplectic phase–space manifold (M, ω) of the classical biodynamic system with operators on a Hilbert space $\mathcal{H}$ in such a way that the Poisson brackets correspond. Therefore, *there is a functor* from the category **Symplec** to the category **Hilbert**. This functor is called *prequantization.*[16]

Let us start with the simplest symplectic manifold ($M = T^*\mathbb{R}^n$, $\omega = dp_i \wedge dq^i$) and state the *Dirac problem*: A prequantization of ($T^*\mathbb{R}^n$, $\omega = dp_i \wedge dq^i$)

[16] We emphasize this fact because *there is no a quantization functor.*

is a map $\delta : f \mapsto \delta_f$, taking smooth functions $f \in C^\infty(T^*\mathbb{R}^n, \mathbb{R})$ to Hermitian operators δ_f on a Hilbert space $\mathcal{H}$, satisfying the *Dirac conditions*:

1. $\delta_{f+g} = \delta_f + \delta_g$, for each $f, g \in C^\infty(T^*\mathbb{R}^n, \mathbb{R})$;
2. $\delta_{\lambda f} = \lambda \delta_f$, for each $f \in C^\infty(T^*\mathbb{R}^n, \mathbb{R})$ and $\lambda \in \mathbb{R}$;
3. $\delta_{1_{\mathbb{R}^n}} = Id_{\mathcal{H}}$; and
4. $[\delta_f, \delta_g] = (\delta_f \circ \delta_g - \delta_g \circ \delta_f) = i\hbar \delta_{\{f,g\}_\omega}$, for each $f, g \in C^\infty(T^*\mathbb{R}^n, \mathbb{R})$;

The pair $(\mathcal{H}, \delta)$, where

$$\mathcal{H} = L^2(\mathbb{R}^n, \mathbb{C}); \qquad \delta : f \in C^\infty(T^*\mathbb{R}^n, \mathbb{R}) \mapsto \delta_f : \mathcal{H} \to \mathcal{H};$$
$$\delta_f = -i\hbar X_f - \theta(X_f) + f; \qquad \theta = p_i dq^i,$$

gives a prequantization of $(T^*\mathbb{R}^n, dp_i \wedge dq^i)$, or equivalently, the answer to the Dirac problem is affirmative [Put93].

Now, let $(M = T^*Q, \omega)$ be the cotangent bundle of an arbitrary manifold Q with its canonical symplectic structure $\omega = d\theta$. The prequantization of M is given by the pair $\left(L^2(M, \mathbb{C}), \delta^\theta\right)$, where for each $f \in C^\infty(M, \mathbb{R})$, the operator $\delta^\theta_f : L^2(M, \mathbb{C}) \to L^2(M, \mathbb{C})$ is given by

$$\delta^\theta_f = -i\hbar X_f - \theta(X_f) + f.$$

Here, *symplectic potential* θ is not uniquely determined by the condition $\omega = d\theta$; for instance $\theta' = \theta + du$ has the same property for any real function u on M. On the other hand, in the general case of an arbitrary symplectic manifold (M, ω) (not necessarily the cotangent bundle) we can find only locally a 1–form θ such that $\omega = d\theta$.

In general, a symplectic manifold $(M, \omega = d\theta)$ is quantizable (i.e., we can define the Hilbert representation space $\mathcal{H}$ and the prequantum operator δ_f in a globally consistent way) if ω defines an *integral cohomology class*. Now, by the *construction Theorem of a fiber bundle*, we can see that this condition on ω is also sufficient to guarantee the existence of a *complex line bundle* $L^\omega = (L, \pi, M)$ over M, which has $\exp(i\, u_{ji}/\hbar)$ as *gauge transformations* associated to an open cover $\mathcal{U} = \{U_i | i \in I\}$ of M such that θ_i is a *symplectic potential* defined on U_i (i.e., $d\theta_i = \omega$ and $\theta_i = \theta_i + d\, u_{ji}$ on $U_i \cap U_j$).

In particular, for exact symplectic structures ω (as in the case of cotangent bundles with their canonical symplectic structures) an integral cohomology condition is automatically satisfied, since then we have only one set $U_i = M$ and do not need any gauge transformations.

Now, for each vector–field $X \in M$ there exists an operator ∇^ω_X on the *space of sections* $\Gamma(L^\omega)$ of L^ω,

$$\nabla^\omega_X : \Gamma(L^\omega) \to \Gamma(L^\omega), \qquad \text{given by} \qquad \nabla^\omega_X f = X(f) - \frac{\mathrm{i}}{\hbar}\theta(X) f,$$

and it is easy to see that ∇^ω is a *connection* on L^ω whose curvature is $\omega / i\hbar$. In terms of this connection, the definition of δ_f becomes

$$\delta_f = -i\hbar\nabla^{\omega}_{X_f} + f.$$

The complex line bundle $L^\omega = (L, \pi, M)$ together with its compatible connection and Hermitian structure is usually called the *prequantum bundle* of the symplectic manifold (M, ω).

If (M, ω) is a quantizable manifold then the pair $(\mathcal{H}, \delta)$ defines its prequantization.

Quantization Examples

Each exact symplectic manifold $(M, \omega = d\theta)$ is quantizable, for the cohomology class defined by ω is zero. In particular, the cotangent bundle, with its canonical symplectic structure is always quantizable.

Let $(M, \omega = d\theta)$ be an exact symplectic manifold. Then it is quantizable with the prequantum bundle given by [Put93]:

$$\begin{aligned}
&L^\omega = (M \times \mathbb{C}, pr_1, M); \\
&\Gamma(L^\omega) \simeq C^\infty(M, \mathbb{C}); \qquad \nabla^{\omega}_X f = X(f) - \frac{\mathrm{i}}{\hbar}\theta(X) f; \\
&((x, z_1), (x, z_2))_x = \bar{z}_1 z_2; \qquad \delta_f = -i\hbar[X_f - \frac{i}{\hbar}\theta(X_f)] + f.
\end{aligned}$$

Let $(M, \omega) = (T^*\mathbb{R}, dp \wedge dq)$. It is quantizable with [Put93]:

$$\begin{aligned}
L^\omega &= (\mathbb{R}^2 \times \mathbb{C}, pr_1, \mathbb{R}^2); \qquad \Gamma(L^\omega) = C^\infty(\mathbb{R}^2, \mathbb{C}); \\
\nabla^{\omega}_X f &= X(f) - \frac{\mathrm{i}}{\hbar} p dq(X) f; \qquad ((x, z_1), (x, z_2))_x = \bar{z}_1 z_2; \\
\delta_f &= -i\hbar\left[\frac{\partial f}{\partial p}\frac{\partial}{\partial q} - \frac{\partial f}{\partial q}\frac{\partial}{\partial p}\right] - p\frac{\partial f}{\partial p} + f.
\end{aligned}$$

Therefore,

$$\delta_q = i\hbar\frac{\partial}{\partial p} + q, \qquad \delta_p = -i\hbar\frac{\partial}{\partial q},$$

which differs from the classical result of the *Schrödinger quantization*:

$$\delta_q = q, \qquad \delta_p = -i\hbar\frac{\partial}{\partial q}.$$

Let $\mathcal{H}$ be a complex Hilbert space and $U_t : \mathcal{H} \to \mathcal{H}$ a continuous one–parameter unitary group, i.e., a homomorphism $t \mapsto U_t$ from $\mathbb{R}$ to the group of unitary operators on $\mathcal{H}$ such that for each $x \in \mathcal{H}$ the map $t \mapsto U_t(x)$ is continuous. Then we have the self–adjoint generator A of U_t, defined by

$$Ax = \frac{1}{i}\frac{d}{dt}U_t(x) = \frac{1}{i}\lim_{h\to 0}\frac{U_h(x) - x}{h}.$$

Let $\left(\mathbb{R}^2, \omega = dp \wedge dq, H = \frac{1}{2}(p^2 + q^2\right)$ be the Hamiltonian structure of the 1D harmonic oscillator.

If we take $\theta = \frac{1}{2}(pdq - qdp)$ as the symplectic potential of ω, then the spectrum of the prequantum operator $\delta_H = i\hbar\left(q\frac{\partial}{\partial p} - p\frac{\partial}{\partial q}\right)$ is [Put93] $Spec(\delta_H) = \{..., -2\hbar, -\hbar, 0, \hbar, 2\hbar, ...\}$, where each eigenvalue occurs with infinite multiplicity.

Let $\mathfrak{g}$ be the vector space spanned by the prequantum operators $\delta_q, \delta_p, \delta_H$, given by

$$\delta_q = i\hbar\frac{\partial}{\partial p} + q, \qquad \delta_p = -i\hbar\frac{\partial}{\partial q}, \qquad \delta_H = i\hbar\left(q\frac{\partial}{\partial p} - p\frac{\partial}{\partial q}\right),$$

and Id. Then we have [Put93]:

1. $\mathfrak{g}$ is a Lie algebra called the *oscillator Lie algebra*, given by:

$$[\delta_p, \delta_q] = i\hbar\delta_{\{p,q\}_\omega} = i\hbar\, Id,$$
$$[\delta_H, \delta_q] = i\hbar\delta_{\{H,q\}_\omega} = -i\hbar\delta_p,$$
$$[\delta_H, \delta_p] = i\hbar\delta_{\{H,p\}_\omega} = i\hbar\delta_q,$$

2. $[\mathfrak{g}, \mathfrak{g}]$ is spanned by $\delta_q, \delta_p, \delta_H$ and Id, or equivalently, it is a *Heisenberg Lie algebra.*
3. The oscillator Lie algebra $\mathfrak{g}$ is *solvable.*

5.7.2 Quantization of Relativistic Hamiltonian Mechanics

Given a symplectic manifold (Z, Ω) and a Hamiltonian H on Z, a Dirac constraint system on a closed imbedded submanifold $i_N : N \longrightarrow Z$ of Z is defined as a Hamiltonian system on N admitting the pull–back presymplectic form $\Omega_N = i_N^*\Omega$ and the pull–back Hamiltonian $i_N^* H$ [GNH78, MS98, MR92]. Its solution is a vector–field γ on N which fulfils the equation

$$\gamma\rfloor\Omega_N + i_N^* dH = 0.$$

Let N be coisotropic. Then a solution exists if the Poisson bracket $\{H, f\}$ vanishes on N whenever f is a function vanishing on N. It is the Hamiltonian vector–field of H on Z restricted to N [Sar03].

Recall that a configuration space of non–relativistic time–dependent mechanics (henceforth NRM) of m degrees of freedom is an $(m+1)$D smooth fibre bundle $Q \longrightarrow \mathbb{R}$ over the time axis $\mathbb{R}$ [MS98, Sar98]. It is coordinated by $(q^\alpha) = (q^0, q^i)$, where $q^0 = t$ is the standard Cartesian coordinate on $\mathbb{R}$. Let T^*Q be the cotangent bundle of Q equipped with the induced coordinates $(q^\alpha, p_\alpha = \dot{q}_\alpha)$ with respect to the holonomic coframes $\{dq^\alpha\}$. The cotangent bundle T^*Q plays the role of a homogeneous momentum phase–space of NRM, admitting the canonical symplectic form

$$\Omega = dp_\alpha \wedge dq^\alpha. \tag{5.113}$$

Its momentum phase–space is the vertical cotangent bundle V^*Q of the configuration bundle $Q \longrightarrow \mathbb{R}$, coordinated by (q^α, q^i). A Hamiltonian $\mathcal{H}$ of NRM is defined as a section $p_0 = -\mathcal{H}$ of the fibre bundle $T^*Q \longrightarrow V^*Q$. Then the momentum phase–space of NRM can be identified with the image N of $\mathcal{H}$ in T^*Q which is the one–codimensional (consequently, coisotropic) imbedded submanifold given by the constraint

$$\mathcal{H}_T = p_0 + \mathcal{H}(q^\alpha, p_k) = 0.$$

Furthermore, a solution of a non–relativistic Hamiltonian system with a Hamiltonian $\mathcal{H}$ is the restriction γ to $N \cong V^*Q$ of the Hamiltonian vector–field of $\mathcal{H}_T$ on T^*Q. It obeys the equation $\gamma \rfloor \Omega_N = 0$ [MS98, Sar98]. Moreover, one can show that geometrical quantization of V^*Q is equivalent to geometrical quantization of the cotangent bundle T^*Q where the quantum constraint $\widehat{\mathcal{H}}_T \psi = 0$ on sections ψ of the quantum bundle serves as the Schrödinger equation [Sar03].

A configuration space of *relativistic mechanics* (henceforth RM) is an oriented pseudo–Riemannian manifold (Q, g), coordinated by (t, q^i). Its momentum phase–space is the cotangent bundle T^*Q provided with the symplectic form Ω (5.113). Note that one also considers another symplectic form $\Omega + F$ where F is the strength of an electromagnetic field [Sni80]. A relativistic Hamiltonian is defined as a smooth real function H on T^*Q [MS98, Sar98]. Then a relativistic Hamiltonian system is described as a Dirac constraint system on the subspace N of T^*Q given by the equation

$$H_T = g_{\mu\nu} \partial^\mu H \partial^\nu H - 1 = 0. \tag{5.114}$$

To perform geometrical quantization of NRM, we give geometrical quantization of the cotangent bundle T^*Q and characterize a quantum relativistic Hamiltonian system by the quantum constraint

$$\widehat{H}_T \psi = 0. \tag{5.115}$$

We choose the vertical polarization on T^*Q spanned by the tangent vectors ∂^α. The corresponding *quantum algebra* $\mathcal{A} \subset C^\infty(T^*Q)$ consists of affine functions of momenta

$$f = a^\alpha(q^\mu) p_\alpha + b(q^\mu) \tag{5.116}$$

on T^*Q. They are represented by the *Schrödinger operators*

$$\widehat{f} = -i a^\alpha \partial_\alpha - \frac{\mathrm{i}}{2} \partial_\alpha a^\alpha - \frac{i}{4} a^\alpha \partial_\alpha \ln(-g) + b, \qquad (g = \det(g_{\alpha\beta})) \tag{5.117}$$

in the space $\mathbb{C}^\infty(Q)$ of smooth complex functions on Q.

Note that the function H_T (5.114) need not belong to the quantum algebra $\mathcal{A}$. Nevertheless, one can show that, if H_T is a polynomial of momenta of degree k, it can be represented as a finite composition

$$H_T = \sum_i f_{1i} \cdots f_{ki} \tag{5.118}$$

of products of affine functions (5.116), i.e., as an element of the enveloping algebra $\overline{\mathcal{A}}$ of the Lie algebra $\mathcal{A}$ [GMS02b]. Then it is quantized

$$H_T \mapsto \widehat{H}_T = \sum_i \widehat{f}_{1i} \cdots \widehat{f}_{ki} \tag{5.119}$$

as an element of $\overline{\mathcal{A}}$. However, the representation (5.118) and, consequently, the quantization (5.119) fail to be unique.

The space of relativistic velocities of RM on Q is the tangent bundle TQ of Q equipped with the induced coordinates $(t, q^i, \dot{q}^\alpha)$ with respect to the holonomic frames $\{\partial_\alpha\}$. Relativistic motion is located in the subbundle W_g of hyperboloids [MS98, MS00b]

$$g_{\mu\nu}(q)\dot{q}^\mu \dot{q}^\nu - 1 = 0 \tag{5.120}$$

of TQ. It is described by a second–order dynamical equation

$$\ddot{q}^\alpha = \Xi^\alpha(q^\mu, \dot{q}^\mu) \tag{5.121}$$

on Q which preserves the subbundle (5.120), i.e.,

$$(\dot{q}^\alpha \partial_\alpha + \Xi^\alpha \dot{\partial}_\alpha)(g_{\mu\nu}\dot{q}^\mu \dot{q}^\nu - 1) = 0, \qquad (\dot{\partial}_\alpha = \partial/\partial \dot{q}^\alpha).$$

This condition holds if the r.h.s. of the equation (5.121) takes the form

$$\Xi^\alpha = \Gamma^\alpha_{\mu\nu}\dot{q}^\mu \dot{q}^\nu + F^\alpha,$$

where $\Gamma^\alpha_{\mu\nu}$ are Christoffel symbols of a metric g, while F^α obey the relation $g_{\mu\nu}F^\mu \dot{q}^\nu = 0$. In particular, if the dynamical equation (5.121) is a geodesic equation,

$$\ddot{q}^\alpha = K^\alpha_\mu \dot{q}^\mu$$

with respect to a (non-linear) connection on the tangent bundle $TQ \to Q$,

$$K = dq^\alpha \otimes (\partial_\alpha + K^\mu_\alpha \dot{\partial}_\mu),$$

this connections splits into the sum

$$K^\alpha_\mu = \Gamma^\alpha_{\mu\nu}\dot{q}^\nu + F^\alpha_\mu \tag{5.122}$$

of the Levi–Civita connection of g and a soldering form

$$F = g^{\lambda\nu}F_{\mu\nu}dq^\mu \otimes \dot{\partial}_\alpha, \qquad F_{\mu\nu} = -F_{\nu\mu}.$$

As was mentioned above, the momentum phase–space of RM on Q is the cotangent bundle T^*Q provided with the symplectic form Ω (5.113). Let H be a smooth real function on T^*Q such that the map

$$\widetilde{H} : T^*Q \longrightarrow TQ, \qquad \dot{q}^\mu = \partial^\mu H \tag{5.123}$$

is a bundle isomorphism. Then the inverse image $N = \widetilde{H}^{-1}(W_g)$ of the subbundle of hyperboloids W_g (5.120) is a one-codimensional (consequently, coisotropic) closed imbedded subbundle of T^*Q given by the constraint $H_T = 0$ (5.114). We say that H is a relativistic Hamiltonian if the Poisson bracket $\{H, H_T\}$ vanishes on N. This means that the Hamiltonian vector–field

$$\gamma = \partial^\alpha H \partial_\alpha - \partial_\alpha H \partial^\alpha \tag{5.124}$$

of H preserves the constraint N and, restricted to N, it obeys the Hamiltonian equation

$$\gamma \rfloor \Omega_N + i_N^* dH = 0 \tag{5.125}$$

of a Dirac constraint system on N with a Hamiltonian H.

The map (5.123) sends the vector–field γ (5.124) onto the vector–field

$$\gamma_T = \dot{q}^\alpha \partial_\alpha + (\partial^\mu H \partial^\alpha \partial_\mu H - \partial_\mu H \partial^\alpha \partial^\mu H) \dot{\partial}_\alpha$$

on TQ. This vector–field defines the second–order dynamical equation

$$\ddot{q}^\alpha = \partial^\mu H \partial^\alpha \partial_\mu H - \partial_\mu H \partial^\alpha \partial^\mu H \tag{5.126}$$

on Q which preserves the subbundle of hyperboloids (5.120).

The following is a basic example of relativistic Hamiltonian systems. Put

$$H = \frac{1}{2m} g^{\mu\nu} (p_\mu - b_\mu)(p_\nu - b_\nu),$$

where m is a constant and $b_\mu dq^\mu$ is a covector–field on Q. Then $H_T = 2m^{-1}H - 1$ and $\{H, H_T\} = 0$. The constraint $H_T = 0$ defines a closed imbedded one-codimensional subbundle N of T^*Q. The Hamiltonian equation (5.125) takes the form $\gamma \rfloor \Omega_N = 0$. Its solution (5.124) reads

$$\dot{q}^\alpha = \frac{1}{m} g^{\alpha\nu} (p_\nu - b_\nu),$$
$$\dot{p}_\alpha = -\frac{1}{2m} \partial_\alpha g^{\mu\nu} (p_\mu - b_\mu)(p_\nu - b_\nu) + \frac{1}{m} g^{\mu\nu} (p_\mu - b_\mu) \partial_\alpha b_\nu.$$

The corresponding second–order dynamical equation (5.126) on Q is

$$\ddot{q}^\alpha = \Gamma^\alpha_{\mu\nu} \dot{q}^\mu \dot{q}^\nu - \frac{1}{m} g^{\lambda\nu} F_{\mu\nu} \dot{q}^\mu, \tag{5.127}$$
$$\Gamma^\alpha_{\mu\nu} = -\frac{1}{2} g^{\lambda\beta} (\partial_\mu g_{\beta\nu} + \partial_\nu g_{\beta\mu} - \partial_\beta g_{\mu\nu}), \qquad F_{\mu\nu} = \partial_\mu b_\nu - \partial_\nu b_\mu.$$

It is a geodesic equation with respect to the affine connection

$$K^\alpha_\mu = \Gamma^\alpha_{\mu\nu}\dot{q}^\nu - \frac{1}{m}g^{\lambda\nu}F_{\mu\nu}$$

of type (5.122). For example, let g be a metric gravitational field and let $b_\mu = eA_\mu$, where A_μ is an electromagnetic potential whose gauge holds fixed. Then the equation (5.127) is the well–known equation of motion of a relativistic massive charge in the presence of these fields.

Let us now perform the quantization of RM, following the standard geometrical quantization of the cotangent bundle (see [Bla83, Sni80, Woo92]). As the canonical symplectic form Ω (5.113) on T^*Q is exact, the prequantum bundle is defined as a trivial complex line bundle C over T^*Q. Note that this bundle need no metaplectic correction since T^*X is with canonical coordinates for the symplectic form Ω. Thus, C is called the *quantum bundle*. Let its trivialization

$$C \cong T^*Q \times \mathbb{C} \tag{5.128}$$

hold fixed, and let (t, q^i, p_α, c), with $c \in \mathbb{C}$, be the associated bundle coordinates. Then one can treat sections of C (5.128) as smooth complex functions on T^*Q. Note that another trivialization of C leads to an equivalent quantization of T^*Q.

Recall that the Kostant–Souriau prequantization formula associates to each smooth real function $f \in C^\infty(T^*Q)$ on T^*Q the first–order differential operator

$$\widehat{f} = -i\nabla_{\vartheta_f} + f \tag{5.129}$$

on sections of C, where $\vartheta_f = \partial^\alpha f \partial_\alpha - \partial_\alpha f \partial^\alpha$ is the Hamiltonian vector–field of f and ∇ is the covariant differential with respect to a suitable $U(1)$-principal connection A on C. This connection preserves the *Hermitian metric* $g(c, c') = c\bar{c}'$ on C, and its curvature form obeys the prequantization condition $R = i\Omega$. For the sake of simplicity, let us assume that Q and, consequently, T^*Q is simply–connected. Then the connection A up to gauge transformations is

$$A = dp_\alpha \otimes \partial^\alpha + dq^\alpha \otimes (\partial_\alpha + icp_\alpha\partial_c), \tag{5.130}$$

and the prequantization operators (5.129) read

$$\widehat{f} = -i\vartheta_f + (f - p_\alpha\partial^\alpha f). \tag{5.131}$$

Let us choose the vertical polarization on T^*Q. It is the vertical tangent bundle VT^*Q of the fibration $\pi : T^*Q \to Q$. As was mentioned above, the corresponding quantum algebra $\mathcal{A} \subset C^\infty(T^*Q)$ consists of affine functions f (5.116) of momenta p_α. Its representation by operators (5.131) is defined in the space E of sections ρ of the quantum bundle C of compact support which obey the condition $\nabla_\vartheta \rho = 0$ for any vertical Hamiltonian vector–field ϑ on T^*Q. This condition takes the form

$$\partial_\alpha f \partial^\alpha \rho = 0, \qquad (f \in C^\infty(Q)).$$

It follows that elements of E are independent of momenta and, consequently, fail to be compactly supported, unless $\rho = 0$. This is the well–known problem of Schrödinger quantization which is solved as follows [Bla83, GMS02b].

Let $i_Q : Q \longrightarrow T^*Q$ be the canonical zero section of the cotangent bundle T^*Q. Let $C_Q = i_Q^* C$ be the pull–back of the bundle C (5.128) over Q. It is a trivial complex line bundle $C_Q = Q \times \mathbb{C}$ provided with the pull–back Hermitian metric $g(c, c') = c\bar{c}'$ and the pull–back

$$A_Q = i_Q^* A = dq^\alpha \otimes (\partial_\alpha + icp_\alpha \partial_c)$$

of the connection A (5.130) on C. Sections of C_Q are smooth complex functions on Q, but this bundle need metaplectic correction.

Let the cohomology group $H^2(Q; \mathbb{Z}_2)$ of Q be trivial. Then a metalinear bundle $\mathcal{D}$ of complex half-forms on Q is defined. It admits the canonical lift of any vector–field τ on Q such that the corresponding Lie derivative of its sections reads

$$\mathfrak{L}_\tau = \tau^\alpha \partial_\alpha + \frac{1}{2} \partial_\alpha \tau^\alpha.$$

Let us consider the tensor product $Y = C_Q \otimes \mathcal{D}$ over Q. Since the Hamiltonian vector–fields

$$\vartheta_f = a^\alpha \partial_\alpha - (p_\mu \partial_\alpha a^\mu + \partial_\alpha b) \partial^\alpha$$

of functions f (5.116) are projected onto Q, one can assign to each element f of the quantum algebra $\mathcal{A}$ the first–order differential operator

$$\widehat{f} = (-i\nabla_{\pi\vartheta_f} + f) \otimes \mathrm{Id} + \mathrm{Id} \otimes \mathfrak{L}_{\pi\vartheta_f} = -ia^\alpha \partial_\alpha - \frac{\mathrm{i}}{2} \partial_\alpha a^\alpha + b$$

on sections ρ_Q of Y. For the sake of simplicity, let us choose a trivial metalinear bundle $\mathcal{D} \to Q$ associated to the orientation of Q. Its sections can be written in the form $\rho_Q = (-g)^{1/4}\psi$, where ψ are smooth complex functions on Q. Then the quantum algebra $\mathcal{A}$ can be represented by the operators $\widehat{f}$ (5.117) in the space $\mathbb{C}^\infty(Q)$ of these functions. It can be justified that these operators obey the *Dirac condition*

$$[\widehat{f}, \widehat{f'}] = -i\widehat{\{f, f'\}}.$$

One usually considers the subspace $E_Q \subset \mathbb{C}^\infty(Q)$ of functions of compact support. It is a pre–Hilbert space with respect to the non–degenerate Hermitian form

$$\langle\psi|\psi'\rangle = \int_Q \psi\overline{\psi}'(-g)^{1/2}d^{m+1}q$$

Note that $\widehat{f}$ (5.117) are symmetric operators $\widehat{f} = \widehat{f}^*$ in E_Q, i.e., $\langle\widehat{f}\psi|\psi'\rangle = \langle\psi|\widehat{f}\psi'\rangle$. However, the space E_Q gets no physical meaning in RM.

As was mentioned above, the function H_T (5.114) need not belong to the quantum algebra $\mathcal{A}$, but a polynomial function H_T can be quantized as an element of the enveloping algebra $\overline{\mathcal{A}}$ by operators $\widehat{H}_T$ (5.119). Then the quantum constraint (5.115) serves as a relativistic quantum equation.

Let us again consider a massive relativistic charge whose relativistic Hamiltonian is

$$H = \frac{1}{2m}g^{\mu\nu}(p_\mu - eA_\mu)(p_\nu - eA_\nu).$$

It defines the constraint

$$H_T = \frac{1}{m^2}g^{\mu\nu}(p_\mu - eA_\mu)(p_\nu - eA_\nu) - 1 = 0. \tag{5.132}$$

Let us represent the function H_T (5.132) as symmetric product of affine functions of momenta,

$$H_T = \frac{(-g)^{-1/4}}{m}(p_\mu - eA_\mu)(-g)^{1/4}g^{\mu\nu}(-g)^{1/4}(p_\nu - eA_\nu)\frac{(-g)^{-1/4}}{m} - 1.$$

It is quantized by the rule (5.119), where

$$(-g)^{1/4}\circ\widehat{\partial}_\alpha\circ(-g)^{-1/4} = -i\partial_\alpha.$$

Then the well–known relativistic quantum equation

$$(-g)^{-1/2}[(\partial_\mu - ieA_\mu)g^{\mu\nu}(-g)^{1/2}(\partial_\nu - ieA_\nu) + m^2]\psi = 0. \tag{5.133}$$

is reproduced up to the factor $(-g)^{-1/2}$.

5.8 K–Theory and Complex Dynamics

Recall from the history of topology [Die88] that the 1930s were the decade of the development of the *cohomology theory*, as several research directions grew together and the *de Rham cohomology*, that was implicit in Poincaré's work, became the subject of definite theorems. The development of algebraic topology from 1940 to 1960 was very rapid, and the role of homology theory was often as 'baseline' theory, easy to compute and in terms of which topologists sought to calculate with other functors. The axiomatization of homology theory by Eilenberg and Steenrod (celebrated *Eilenberg–Steenrod Axioms*) revealed that what various candidate homology theories had in common was, roughly speaking, some *exact sequences* (in particular, the *Mayer–Vietoris Theorem* and the *Dimension Axiom* that calculated the homology of the point).

5.8.1 Topological K−Theory

Now, *K-theory* is an *extraordinary cohomology theory*, which consists of topological K−theory and algebraic K−theory. The *topological K-theory* was founded to study vector bundles on general topological spaces, by means of ideas now recognisee as (general) K−theory that were introduced by Alexander Grothendieck. The early work on topological K−theory was due to Michael Atiyah and Friedrich Hirzebruch.

Let X be a compact Hausdorff space and $k = \mathbb{R}$ or $k = \mathbb{C}$. Then $K_k(X)$ is the *Grothendieck group* of the *commutative monoid*[17] which elements are the isomorphism classes of finite dimensional k−vector bundles on X with the operation

$$[E \oplus F] := [E] \oplus [F]$$

for vector bundles E, F.[18] Usually, $K_k(X)$ is denoted $KO(X)$ in real case and $KU(X)$ in the complex case.

More precisely, the *stable equivalence*, i.e., the equivalence relation on bundles E and F on X of defining the same element in $K(X)$, occurs when there is a trivial bundle G, so that $E \oplus G \cong F \oplus G$. Under the *tensor product of vector bundles*, $K(X)$ then becomes a *commutative ring*. The rank of a vector bundle carries over to the K−group define the homomorphism: $K(X) \to \check{H}^0(X, \mathbb{Z})$, where $\check{H}^0(X, \mathbb{Z})$ is the 0−group of the *Chech cohomology* which is equal to group of locally constant functions with values in $\mathbb{Z}$.

The constant map $X \longrightarrow \{x_0\}$, $x_0 \in X$ defines the reduced K−group (of reduced homology)

$$\check{K}(X) = \mathrm{Coker}(K(X) \longrightarrow \{x_0\}).$$

In particular, when X is a connected space, then

$$\tilde{K}(X) \cong \mathrm{Ker}(K(X) \to \check{H}^0(X, \mathbb{Z}) = \mathbb{Z}).$$

[17] Recall that a monoid is an algebraic structure with a single, associative binary operation and an identity element; a monoid whose operation is commutative is called a commutative monoid (or, an Abelian monoid); e.g., every group is a monoid and every Abelian group a commutative monoid.

[18] The Grothendieck group construction in abstract algebra constructs an Abelian group from a commutative monoid 'in the best possible way'.

Bott Periodicity Theorem

An important property in the topological $K-$theory is the *Bott Periodicity Theorem* [Bot59][19], which can be formulated this way:

1. $K(X \times S^2) = K(X) \otimes K(S^2)$, and $K(S^2) = [H]/(H-1)^2$, where H is the class of the *tautological bundle* on the $S^2 = P^1$, i.e., the *Riemann sphere* as complex projective line;
2. $\tilde{K}^{n+2}(X) = \tilde{K}^n(X)$;
3. $\Omega^2\mathrm{BU} \simeq \mathrm{BU} \times \mathbb{Z}$.

In real $K-$theory there is a similar periodicity, but modulo 8.

5.8.2 Algebraic $K-$Theory

On the other hand, the so–called *algebraic K–theory* is an advanced part of *homological algebra* concerned with defining and applying a sequence $K_n(R)$ of functors from rings to Abelian groups, for $n = 0, 1, 2, ...$ Here, for traditional reasons, the cases of K_0 and K_1 are thought of in somewhat different terms from the higher algebraic $K-$groups K_n for $n \geq 2$. In fact K_0 generalizes the construction of the ideal class group, using projective modules; and K_1 as applied to a commutative ring is the unit group construction, which was generalized to all rings for the needs of topology (simple homotopy theory) by means of elementary matrix theory. Therefore the first two cases counted as relatively accessible; while after that the theory becomes quite noticeably deeper, and certainly quite hard to compute (even when R is the ring of integers).

---

[19] The Bott Periodicity Theorem is a result from homotopy theory discovered by Raoul Bott during the latter part of the 1950s, which proved to be of foundational significance for much further research, in particular in $K-$theory of stable complex vector bundles, as well as the stable homotopy groups of spheres. Bott periodicity can be formulated in numerous ways, with the periodicity in question always appearing as a period 2 phenomenon, with respect to dimension, for the theory associated to the unitary group. The context of Bott periodicity is that the homotopy groups of spheres, which would be expected to play the basic part in algebraic topology by analogy with homology theory, have proved elusive (and the theory is complicated). The subject of stable homotopy theory was conceived as a simplification, by introducing the suspension (smash product with a circle) operation, and seeing what (roughly speaking) remained of homotopy theory once one was allowed to suspend both sides of an equation, as many times as one wished. The stable theory was still hard to compute with, in practice. What Bott periodicity offered was an insight into some highly non-trivial spaces, with central status in topology because of the connection of their cohomology with characteristic classes, for which all the (unstable) homotopy groups could be calculated. These spaces are the (infinite, or stable) unitary, orthogonal and symplectic groups U, O and Sp.

Historically, the roots of the theory were in *topological K–theory* (based on vector bundle theory); and its motivation the conjecture of Serre[20] that now is the *Quillen–Suslin Theorem.*[21]

Applications of $K-$groups were found from 1960 onwards in *surgery theory* for manifolds, in particular; and numerous other connections with classical algebraic problems were found. A little later a branch of the theory for *operator algebras* was fruitfully developed. It also became clear that $K-$theory could play a role in *algebraic cycle theory* in algebraic geometry: here the higher $K-$groups become connected with the higher codimension phenomena, which are exactly those that are harder to access. The problem was that the definitions were lacking (or, too many and not obviously consistent). A definition of K_2 for fields by John Milnor, for example, gave an attractive theory that was too limited in scope, constructed as a quotient of the multiplicative group of the field tensored with itself, with some explicit relations imposed; and closely connected with central extensions [MS74)].

Eventually the foundational difficulties were resolved (leaving a deep and difficult theory), by a definition of D. Quillen:

$$K_n(R) = \pi_n(BGL(R)^+).$$

This is a very compressed piece of abstract mathematics. Here π_n is an nth homotopy group, $GL(R)$ is the direct limit of the general linear groups over R for the size of the matrix tending to infinity, B is the classifying space construction of homotopy theory, and the $^+$ is *Quillen's plus construction.*

5.8.3 Chern Classes and Chern Character

An important properties in K–theory are the *Chern classes* and *Chern character* [Che46]. The Chern classes are a particular type of *characteristic classes* (topological invariants, see [MS74)]). associated to complex vector bundles of a smooth manifold. Recall that a characteristic class is a way of associating to each principal bundle on a topological space X a cohomology class of X. The

[20] Jean–Pierre Serre used the analogy of vector bundles with projective modules to found in 1959 what became algebraic $K-$theory. He formulated the *Serre's Conjecture*, that projective modules over the ring of polynomials over a field are free modules; this resisted proof for 20 years.

[21] The Quillen–Suslin Theorem is a Theorem in abstract algebra about the relationship between free modules and projective modules. Projective modules are modules that are locally free. Not all projective modules are free, but in the mid–1950s, Jean–Pierre Serre found evidence that a limited converse might hold. He asked the question: Is every projective module over a polynomial ring over a field a free module? A more geometric variant of this question is whether every algebraic vector bundle on affine space is trivial. This was open until 1976, when Daniel Quillen and Andrei Suslin independently proved that the answer is yes. Quillen was awarded the Fields Medal in 1978 in part for his proof.

cohomology class measures the extent to which the bundle is 'twisted' – particularly, whether it possesses sections or not. In other words, characteristic classes are global invariants which measure the deviation of a local product structure from a global product structure. They are one of the unifying geometric concepts in algebraic topology, differential geometry and algebraic geometry.[22]

If we describe the same vector bundle on a manifold in two different ways, the Chern classes will be the same, i.e., if the Chern classes of a pair of vector bundles do not agree, then the vector bundles are different (the converse is not true, though). In topology, differential geometry, and algebraic geometry, it is often important to count how many linearly independent sections a vector bundle has. The Chern classes offer some information about this through, for

[22] Recall that characteristic classes are in an essential way phenomena of cohomology theory – they are contravariant functors, in the way that a section is a kind of function on a space, and to lead to a contradiction from the existence of a section we do need that variance. In fact cohomology theory grew up after homology and homotopy theory, which are both covariant theories based on mapping into a space; and characteristic class theory in its infancy in the 1930s (as part of obstruction theory) was one major reason why a 'dual' theory to homology was sought. The characteristic class approach to curvature invariants was a particular reason to make a theory, to prove a general *Gauss–Bonnet Theorem*.

When the theory was put on an organized basis around 1950 (with the definitions reduced to homotopy theory) it became clear that the most fundamental characteristic classes known at that time (the *Stiefel–Whitney class*, the *Chern class*, and the *Pontryagin class*) were reflections of the classical linear groups and their maximal torus structure. What is more, the Chern class itself was not so new, having been reflected in the Schubert calculus on Grassmannians, and the work of the Italian school of algebraic geometry. On the other hand there was now a framework which produced families of classes, whenever there was a vector bundle involved.

The prime mechanism then appeared to be this: Given a space X carrying a vector bundle, implied in the homotopy category a mapping from X to a classifying space BG, for the relevant linear group G. For the homotopy theory, the relevant information is carried by compact subgroups such as the orthogonal groups and unitary groups of G. Once the cohomology $H^*(BG)$ was calculated, once and for all, the contravariance property of cohomology meant that characteristic classes for the bundle would be defined in $H^*(X)$ in the same dimensions. For example, the Chern class is really one class with graded components in each even dimension.

This is still the classic explanation, though in a given geometric theory it is profitable to take extra structure into account. When cohomology became 'extra–ordinary' with the arrival of $K-$theory and Thom's cobordism theory from 1955 onwards, it was really only necessary to change the letter H everywhere to say what the characteristic classes were.

Characteristic classes were later found for foliations of manifolds; they have (in a modified sense, for foliations with some allowed singularities) a classifying space theory in homotopy theory.

instance, the *Riemann–Roch Theorem* and the *Atiyah–Singer Index Theorem*. Chern classes are also feasible to calculate in practice. In differential geometry (and some types of algebraic geometry), the Chern classes can be expressed as polynomials in the coefficients of the *curvature form*.

In particular, given a complex hermitian vector bundle V of complex rank n over a smooth manifold M, a representative of each Chern class (also called a *Chern form*) $c_k(V)$ of V are given as the coefficients of the characteristic polynomial

$$\det\left(\frac{\mathrm{i}t\Omega}{2\pi}+I\right)=c_k(V)t^k,$$

of the curvature form Ω of V, which is defined as

$$\Omega=d\omega+\frac{1}{2}[\omega,\omega],$$

with ω the connection form and d the exterior derivative, or via the same expression in which ω is a *gauge form* for the *gauge group* of V. The scalar t is used here only as an indeterminate to generate the sum from the determinant, and I denotes the $n\times n$ identity matrix. To say that the expression given is a *representative* of the Chern class indicates that 'class' here means up to addition of an exact differential form. That is, Chern classes are cohomology classes in the sense of *de Rham cohomology*. It can be shown that the cohomology class of the Chern forms do not depend on the choice of connection in V.

For example, let $\mathbb{C}P^1$ be the *Riemann sphere*: a 1D complex projective space. Suppose that z is a holomorphic local coordinate for the Riemann sphere. Let $V=T\mathbb{C}P^1$ be the bundle of complex tangent vectors having the form $a\partial/\partial z$ at each point, where a is a complex number. In the following we prove the complex version of the *Hairy Ball Theorem*: *V has no section which is everywhere nonzero.*

For this, we need the following fact: the first Chern class of a trivial bundle is zero, i.e., $c_1(\mathbb{C}P^1\times\mathbb{C})=0$. This is evinced by the fact that a trivial bundle always admits a flat metric. So, we will show that $c_1(V)\neq 0$.

Consider the Kähler metric

$$h=\frac{dzd\bar{z}}{(1+|z|^2)}..$$

One can show that the curvature 2–form is given by

$$\Omega=\frac{2dz\wedge d\bar{z}}{(1+|z|^2)^2}.$$

Furthermore, by the definition of the first Chern class

$$c_1=\frac{\mathrm{i}}{2\pi}\Omega..$$

We need to show that the cohomology class of this is non–zero. It suffices to compute its integral over the Riemann sphere:

$$\int c_1 = \frac{\mathrm{i}}{\pi}\int \frac{dz\wedge d\bar{z}}{(1+|z|^2)^2} = 2,$$

after switching to polar coordinates. By *Stokes Theorem*, an exact form would integrate to 0, so the cohomology class is nonzero. This proves that $T\mathbb{C}P^1$ is not a trivial vector bundle.

An important special case occurs when V is a *line bundle*. Then the only nontrivial Chern class is the *first Chern class*, which is an element of $H^2(X;Z)$–the second cohomology group of X. As it is the top Chern class, it equals the *Euler class* of the bundle. The first Chern class turns out to be a complete topological invariant with which to classify complex line bundles. That is, there is a bijection between the isomorphism classes of line bundles over X and the elements of $H^2(X;Z)$, which associates to a line bundle its first Chern class. Addition in the second cohomology group coincides with tensor product of complex line bundles.

In algebraic geometry, this classification of (isomorphism classes of) complex line bundles by the first Chern class is a crude approximation to the classification of (isomorphism classes of) holomorphic line bundles by linear equivalence classes of divisors. For complex vector bundles of dimension greater than one, the Chern classes are not a complete invariant.

The Chern classes can be used to construct a homomorphism of rings from the topological K–theory of a space to (the completion of) its rational cohomology. For line bundles V, the Chern character ch is defined by

$$\mathrm{ch}(V) = \exp[c_1(V)].$$

For sums of line bundles, the Chern character is defined by additivity. For arbitrary vector bundles, it is defined by pretending that the bundle is a sum of line bundles; more precisely, for sums of line bundles the Chern character can be expressed in terms of Chern classes, and we use the same formulas to define it on all vector bundles. For example, the first few terms are

$$\mathrm{ch}(V) = \dim(V) + c_1(V) + c_1(V)2/2 - c_2(V) + ...$$

If V is filtered by line bundles $L_1, ..., L_k$ having first Chern classes $x_1, ..., x_k$, respectively, then

$$\mathrm{ch}(V) = \mathrm{e}^{x_1} + \cdots + \mathrm{e}^{x_k}..$$

If a connection is used to define the Chern classes, then the explicit form of the Chern character is

$$\mathrm{ch}(V) = \mathrm{Tr}\left[\exp\left(\frac{\mathrm{i}\Omega}{2\pi}\right)\right],$$

where Ω is the curvature of the connection.

The Chern character is useful in part because it facilitates the computation of the Chern class of a tensor product. Specifically, it obeys the following identities:

$$\mathrm{ch}(V \oplus W) = \mathrm{ch}(V) + \mathrm{ch}(W), \qquad \mathrm{ch}(V \otimes W) = \mathrm{ch}(V)\mathrm{ch}(W).$$

Using the *Grothendieck Additivity Axiom* for Chern classes, the first of these identities can be generalized to state that ch is a homomorphism of Abelian groups from the $K-$theory $K(X)$ into the rational cohomology of X. The second identity establishes the fact that this homomorphism also respects products in $K(X)$, and so ch is a homomorphism of rings. The Chern character is used in the Hirzebruch–Riemann–Roch Theorem.

The so–called *twisted K–theory* a particular variant of $K-$theory, in which the twist is given by an integral 3D cohomology class. In physics, it has been conjectured to classify $D-$branes, Ramond–Ramond field strengths and in some cases even spinors in type II string theory.

5.8.4 Atiyah's View on *K*−Theory

According to Michael Atiyah [AA67, Ati00], *K–theory* may roughly be described as the study of additive (or, Abelian) invariants of large matrices. The key point is that, although matrix multiplication is not commutative, matrices which act in orthogonal subspaces do commute. Given 'enough room' we can put matrices A and B into the block form

$$\begin{pmatrix} A & 0 \\ 0 & 1 \end{pmatrix}, \quad \begin{pmatrix} 1 & 0 \\ 0 & B \end{pmatrix},$$

which obviously commute. Examples of Abelian invariants are traces and determinants.

The prime motivation for the birth of $K-$theory came from Hirzebruch's generalization of the classical *Riemann–Roch Theorem* (see [Hir66]). This concerns a complex projective algebraic manifold X and a holomorphic (or algebraic) vector bundle E over X. Then one has the sheaf cohomology groups $H^q(X, E)$, which are finite–dimensional vector spaces, and the corresponding *Euler characteristics*

$$\chi(X, E) = \sum_{q=0}^{n} (-1)^q \dim H^q(X, E),$$

where n is the complex dimension of X. One also has topological invariants of E and of the tangent bundle of X, namely their *Chern classes*. From these one defines a certain explicit polynomial $T(X, E)$ which by evaluation on X becomes a rational number. *Hirzebruch's Riemann–Roch Theorem* asserts the equality: $\chi(X, E) = T(X, E)$.

It is an important fact, easily proved, that both χ and T are additive for exact sequences of vector bundles:

$$0 \to E' \to E \to E'' \to 0,$$

$$\chi(X, E) = \chi(X, E') + \chi(X, E''), \qquad T(X, E) = T(X, E') + T(X, E'').$$

This was the starting point of the Grothendieck's generalization. Grothendieck defined an Abelian group $K(X)$ as the universal additive invariant of exact sequences of algebraic vector bundles over X, so that χ and T both gave homomorphisms of $K(X)$ into the integers (or rationals).

More precisely, Grothendieck defined two different K–groups, one arising from vector bundles (denoted by K^0) and the other using coherent sheaves (denoted by K_0). These are formally analogous to cohomology and homology respectively. Thus $K^0(X)$ is a ring (under tensor product) while $K_0(X)$ is a $K^0(X)$–module. Moreover, K^0 is contravariant while K_0 is covariant (using a generalization of χ). Finally, Grothendieck established the analogue of *Poincaré duality*. While $K^0(X)$ and $K_0(X)$ can be defined for an arbitrary projective variety X, singular or not, the natural map $K^0(X) \to K_0(X)$ is an isomorphism if X is non–singular.

The *Grothendieck's Riemann–Roch Theorem* concerns a morphism $f : X \to Y$ and compares the direct image of f in K–theory and cohomology. It reduces to the Hirzeburch's version when Y is a point.

Topological K–theory started with the famous *Bott Periodicity Theorem* [Bot59], concerning the homotopy of the large unitary groups $U(N)$ (for $N \to \infty$). Combining Bott's Theorem with the formalism of Grothendieck, Atiyah and Hirzebruch, in the late 1950's, developed a K–theory based on topological vector bundles over a compact space [AH61]. Here, in addition to a group $K^0(X)$, they also introduced an odd counterpart $K^1(X)$, defined as the group of homotopy classes of X into $U(N)$, for N large. Putting these together,

$$K^*(X) = K^0(X) \oplus K^1(X),$$

they obtained a periodic 'generalized cohomology theory'. Over the rationals, the *Chern character* gave an isomorphism:

$$\mathrm{ch} : K^*(X) \otimes Q \cong H^*(X, Q).$$

But, over the integers, K–theory is much more subtle and it has had many interesting topological applications. Most notable was the solution of the vector fields on spheres problem by Frank Adams, using real K–theory (based on the orthogonal groups) [Ada62].

An old generalisation of K–theory is related to *projective bundles* [AA67, Ati00]. Given a vector bundle V over a space X, we can form the bundle $P(V)$ whose fibre at $x \in X$ is the projective space $P(V_x)$. In terms of groups and principal bundles, this is the passage from $GL(n, C)$ to $PGL(n, C)$, or from $U(n)$ to $PU(n)$. We have two exact sequences of groups:

$$\begin{array}{l} 1 \to U(1) \to U(n) \quad \to PU(n) \to 1 \\ 1 \to Z_n \quad \to SU(n) \to PU(n) \to 1 \end{array}.$$

The first gives rise to an obstruction $\alpha \in H^3(X, Z)$ to lifting a projective bundle to a vector bundle, while the second gives an obstruction $\beta \in H^2(X, Z_n)$ to lifting a projective bundle to a special unitary bundle. They are related by

$$\alpha = \delta(\beta), \qquad \text{where} \qquad \delta : H^2(X, Z_n) \to H^3(X, Z)$$

is the *coboundary operator*. This shows that $n\alpha = 0$. In fact, one can show that any $\alpha \in H^3(X, Z)$ of order dividing n arises in this way.

Can we define an appropriate $K-$theory for projective bundles with $\alpha \neq 0$? The answer is yes. For each fixed α of finite order we can define an Abelian group $K_\alpha(X)$. Moreover this is a $K(X)$ module. We will now indicate how these *twisted K–groups* (i.e., twisted $K-$theory) can be defined.

Note first that, for any vector space V, $\mathrm{End}\, V = V \otimes V^*$ depends only $P(V)$. Hence, given a projective bundle P over X, we can define the associated bundle $\mathcal{E}(P)$ of endomorphism (matrix) algebras. The sections of $\mathcal{E}(P)$ form a non–commutative C^*-algebra and one can define its $K-$group by using finitely–generated projective modules. This $K-$group turns out to depend not on P but only on its obstruction class $\alpha \in H^3(X, Z)$ and so can be denoted by $K_\alpha(X)$.

In addition to the $K(X)-$module structure of $K_\alpha(X)$ there are multiplications

$$K_\alpha(X) \otimes K_\beta(X) \to K_{\alpha+\beta}(X).$$

Today, there are many variants and generalizations of $K-$theory, something which is not surprising given the universality of linear algebra and matrices [AA67, Ati00]. In each case there are specific features and techniques relevant to the particular area.

First, as already mentioned, is the *real* $K-$theory based on real vector bundles and the Bott periodicity theorems for the orthogonal groups: here the period is 8 rather than 2.

Next there is *equivariant theory* $K_G(X)$, where G is a a compact Lie group acting on the space X. If X is a point, we just get the representation or character ring $R(G)$ of the group G. In general $K_G(X)$ is a module over $R(G)$ and this can be exploited in terms of the fixed–point sets in X of elements of G.

If we pass from the space X to the ring $C(X)$ of continuous complex–valued functions on X then $K(X)$ can be defined purely algebraically in terms of finitely–general projective modules over X. This then lends itself to a major generalization if we replace $C(X)$, which is a commutative C^*-algebra, by a *non–commutative* C^*-algebra. This has become a rich theory linked to many basic ideas in functional analysis, in particular to the *von Neumann dimension theory*.

5.8.5 Atiyah–Singer Index Theorem

We shall recall here very briefly some essential results of *Atiyah–Singer Index Theory.* The reader who is not familiar with the topological and analytic properties of the index of elliptic operators is urged to gain some familiarity with the *Atiyah–Singer Index Theorem* [AS63, AS68][23] (for technical details, see also [BB04]).

A differential operator of order m, mapping the smooth sections of a vector bundle E over a compact manifold Y to those of another such bundle F, can be described in local coordinates and local trivializations of the bundles as

$$D = \sum_{|\alpha|\leq m} a_\alpha(x) D^\alpha,$$

with $\alpha = (\alpha_1, \ldots, \alpha_n)$. The coefficients $a_\alpha(x)$ are matrices of smooth functions that represent elements of $Hom(E, F)$ locally; and $D^\alpha = \frac{\partial}{\partial x_1^{\alpha_1}} \cdots \frac{\partial}{\partial x_n^{\alpha_n}}$.

The principal symbol associated to the operator D is the expression

$$\sigma_m(D)(x, p) = \sum_{|\alpha|=m} a_\alpha(x) p^\alpha.$$

Given the differential operator $D : \Gamma(Y, E) \to \Gamma(Y, F)$, the principal symbol with the local expression above defines a global map $\sigma_m : \pi^*(E) \to \pi^*(F)$, where $T^*Y \xrightarrow{\pi} Y$ is the cotangent bundle; that is, the variables (x, p) are local coordinates on T^*Y.

Consider bundles E_i, $i = 1 \ldots k$, over a compact nD manifold Y such that there is a complex $\Gamma(Y, E)$ formed by the spaces of sections $\Gamma(Y, E_i)$ and differential operators d_i of order m:

---

[23] In the geometry of manifolds and differential operators, the Atiyah—Singer Index Theorem is an important unifying result that connects topology and analysis. It deals with elliptic differential operators (such as the Laplacian) on compact manifolds. It finds numerous applications, including many in theoretical physics. When Michael Atiyah and Isadore Singer were awarded the Abel Prize by the Norwegian Academy of Science and Letters in 2004, the prize announcement explained the Atiyah—Singer Index Theorem in these words:

> "Scientists describe the world by measuring quantities and forces that vary over time and space. The rules of nature are often expressed by formulas, called differential equations, involving their rates of change. Such formulas may have an 'index', the number of solutions of the formulas minus the number of restrictions that they impose on the values of the quantities being computed. The Atiyah–Singer index Theorem calculated this number in terms of the geometry of the surrounding space. A simple case is illustrated by a famous paradoxical etching of M. C. Escher, 'Ascending and Descending', where the people, going uphill all the time, still manage to circle the castle courtyard. The index Theorem would have told them this was impossible."

$$0 \to \Gamma(Y, E_1) \xrightarrow{d_1} \cdots \xrightarrow{d_{k-1}} \Gamma(Y, E_k) \to 0.$$

Construct the principal symbols $\sigma_m(d_i)$; these determine an associated symbol complex

$$0 \to \pi^*(E_1) \xrightarrow{\sigma_m(d_1)} \cdots \xrightarrow{\sigma_m(d_{k-1})} \pi^*(E_k) \to 0.$$

The complex $\Gamma(Y, E)$ is elliptic iff the associated symbol complex is exact. In the case of just one operator, this means that $\sigma_m(d)$ is an isomorphism off the zero section.

Recall that the *Hodge Theorem* states that the cohomology of the complex $\Gamma(Y, E)$ coincides with the harmonic forms, i.e.,

$$H^i(E) = \frac{\mathrm{Ker}(d_i)}{\mathrm{Im}(d_{i-1})} \cong \mathrm{Ker}(\Delta_i), \qquad \text{where} \qquad \Delta_i = d_i^* d_i + d_{i-1} d_{i-1}^*.$$

Without loss of generality, by passing to the assembled complex

$$E^+ = E^1 \oplus E^3 \oplus \cdots, \qquad E^- = E^2 \oplus E^4 \oplus \cdots,$$

we can always think of one elliptic operator

$$D : \Gamma(Y, E^+) \to \Gamma(Y, E^-), \qquad D = \sum_i (d_{2i-1} + d_{2i}^*).$$

The Index Theorem states: Consider an elliptic complex over a compact, orientable, even dimensional manifold Y without boundary. The index of D, which is given by

$$\mathrm{Ind}(D) = \dim[\mathrm{Ker}(D)] - \dim[\mathrm{Coker}(D)] = \sum_i (-1)^i \dim[\mathrm{Ker}\,\Delta_i] = -\chi(E),$$

$\chi(E)$ being the Euler characteristic of the complex, can be expressed in terms of characteristic classes as:

$$\mathrm{Ind}(D) = (-1)^{n/2} \left\langle \frac{\mathrm{ch}(\sum_i (-1)^i [E_i])}{e(Y)} \mathrm{td}(TY_{\mathbf{C}}), [Y] \right\rangle.$$

Here, ch is the Chern character, e is the Euler class of the tangent bundle of Y, $\mathrm{td}(TY_{\mathbf{C}})$ is the Todd class of the complexified tangent bundle. The Atiyah–Singer Index Theorem, which computes the index of a family of elliptic differential operators, is naturally formulated in terms of $K-$theory and is an extension of the Riemann–Roch Theorem.

5.8.6 The Infinite–Order Case

Topological $K-$theory turned out to have a very natural link with the theory of operators in *quantum Hilbert space*. If $\mathcal{H}$ is an infinite–dimensional complex

Hilbert space and $\mathcal{B}(\mathcal{H})$ the space of bounded operators on $\mathcal{H}$ with the uniform norm, then one defines the subspace $\mathcal{F}(\mathcal{H}) \subset \mathcal{B}(\mathcal{H})$ of *Fredholm operators* T, by the requirement that both $\mathrm{Ker}(T)$ and $\mathrm{Coker}(T)$ have finite–dimensions. The Atiyah–Singer index is then defined by

$$\mathrm{Ind}(T) = \dim[\mathrm{Ker}(T)] - \dim[\mathrm{Coker}(T)],$$

and it has the key property that it is continuous, and therefore constant on each connected component of $\mathcal{F}(\mathcal{H})$. Moreover, $\mathrm{Ind} : \mathcal{F} \to Z$ identifies the components of $\mathcal{F}$.

This has a generalization to any compact space X. To any continuous map $X \to \mathcal{F}$ (i.e., a continuous family of Fredholm operators parametrized by X) one can assign an index in $K(X)$. Moreover one gets in this way an isomorphism $\mathrm{Ind} : [X, \mathcal{F}] \to K(X)$, where $[X, \mathcal{F}]$ denotes the set of homotopy classes of maps of X into $\mathcal{F}$. A notable example of a Fredholm operator is an elliptic differential operator on a compact manifold (these are turned into bounded operators by using appropriate *Sobolev norms*).

Now, in the quantum–physical situation, one meets infinite–order elements $\alpha \in H^3(X, Z)$ and the question arises of whether one can still define a 'twisted' group $K_\alpha(X)$. It turns out that it is possible to do this and one approach is being developed by [AS71].

Since an α of order n arises from an obstruction problem involving nD vector bundles, it is plausible that, for α of infinite order, we need to consider bundles of Hilbert spaces $\mathcal{H}$. But here we have to be careful not to confuse the 'small' unitary group

$$U(\infty) = \lim_{N\to\infty} U(N)$$

with the 'large' group $U(\mathcal{H})$ of all unitary operators in Hilbert space. The small unitary group has interesting homotopy groups given by Bott's periodicity Theorem, but $U(\mathcal{H})$ is contractible, by *Kuiper's Theorem*. This means that all Hilbert space bundles (with $U(\mathcal{H})$) as structure group) are trivial. This implies the following homotopy equivalences:

$$PU(\mathcal{H}) = U(\mathcal{H})/U(1) \sim CP_\infty = K(Z, 2), \qquad BPU(\mathcal{H}) \sim K(Z, 3),$$

where B denotes here the classifying space and on the right we have the *Eilenberg–MacLane spaces*. It follows that $P(\mathcal{H})$–bundles over X are classified completely by $H^3(X, Z)$. Thus, for each $\alpha \in H^3(X, Z)$, there is an essentially unique bundle P_α over X with fibre $P(\mathcal{H})$.

As in finite dimensions. $\mathcal{B}(\mathcal{H})$ depends only on $P(\mathcal{H})$, we can define a bundle $\mathcal{B}_\alpha$ of algebras over X. We now let $\mathcal{F}_\alpha \subset \mathcal{B}_\alpha$ be the corresponding bundle of Fredholm operators. Finally we define

$$K_\alpha(X) = \text{Homotopy classes of sections of } \mathcal{F}_\alpha.$$

This definition works for all α. If α is of finite order, then P_α contains a finite–dimensional sub–bundle, but if α is of infinite order this is not true. Thus we are essentially in an infinite–dimensional analytic situation.

To get the twisted odd groups we recall that $\mathcal{F}^1 \subset \mathcal{F}$, the space of self–adjoint Fredholm operators, is a classifying space for K^1 and so we take $\mathcal{F}^1_\alpha \subset \mathcal{F}_\alpha$ to define

$$K^1_\alpha(X) = \text{Homotopy classes of sections of } \mathcal{F}^1_\alpha.$$

One peculiar feature of the infinite–order case is that all sections of $\mathcal{F}_a$ lie in the zero–index component, or equivalently that the restriction map

$$K_\alpha(X) \to K_\alpha(\text{point})$$

is zero.

Now, what can we say about the relation between twisted $K-$groups and cohomology? Over the rationals, if α is of finite order, nothing much changes [AA67, Ati00]. In particular the *Chern character* induces an isomorphism. However if α is of infinite order something new happens. We can now consider the operation $u \to \alpha u$ on $H^*(X,Q)$ as a differential d_α ($\alpha^2 = 0$ since α has odd dimension). We can then form the cohomology with respect to this differential

$$\mathcal{H}_a = \text{Ker } d_\alpha / \operatorname{Im} d_\alpha.$$

One can then prove that there is an isomorphism

$$K^*_\alpha(X) \otimes Q \cong \mathcal{H}_\alpha.$$

In the usual Atiyah–Hirzebruch spectral sequence [AH61], relating $K-$theory to integral cohomology, all differentials are of finite order and so vanish over Q. In particular, $d_3 = Sq^3$, the *Steenrod operation*. However for K_α one finds

$$d_3 u = Sq^3 u + \alpha u$$

and this explains why an α of infinite–order gives the isomorphism above over the rationals.

Chern classes over the integers are a more delicate matter. One can proceed as follows. In $\mathcal{F}$ there are various subspaces $\mathcal{F}_{r,s}$ (of finite codimension) where

$$\dim[\text{Ker}] = r, \qquad \dim[\text{Coker}] = s,$$

and these lie in the component of $\text{Ind}(r-s)$.

Since the $\mathcal{F}_{r,s} \subset \mathcal{F}$ are invariant under the action of $U(\mathcal{H})$, it follows that they can be defined fibrewise and this shows that the classes $c_{r,s}$ can be defined for $K_\alpha(X)$. However the classes for $r = s$ (and so of index zero) are not sufficient to generate all Chern classes. It is a not unreasonable conjecture that the $c_{r,r}$ are the only integral characteristic classes for the twisted $K-$theories [AA67, Ati00].

While the use of Hilbert spaces H and the corresponding projective spaces $P(\mathcal{H})$ may not come naturally to a topologist, these are perfectly natural in physics. Recall that $P(\mathcal{H})$ is the *space of quantum states*. Bundles of such arise naturally in quantum field theory.

5.8.7 Twisted $K-$Theory and the Verlinde Algebra

Twistings of cohomology theories are most familiar for ordinary cohomology [Fre01, FHT03]. Let M be a smooth manifold. Then a flat real vector bundle $E \to M$ determines twisted real cohomology groups $H^\bullet(M; E)$. In differential geometry these cohomology groups are defined by extending the de Rham complex to forms with coefficients in E using the flat connection. The sorts of twistings of $K-$theory we consider are 1D, so analogous to the case when E is a line bundle. There are also 1D twistings of *integral* cohomology, determined by a *local system* $Z \to M$. This is a bundle of groups isomorphic to $\mathbb{Z}$, so is determined up to isomorphism by an element of $H^1(M; \mathrm{Aut}(\mathbb{Z})) \cong H^1(M; \mathbb{Z} \bmod 2)$, since the only nontrivial automorphism of $\mathbb{Z}$ is multiplication by -1. The twisted integral cohomology $H^\bullet(M; Z)$ may be thought of as sheaf cohomology, or defined using a cochain complex. We give a Čech description as follows. Let $\{U_i\}$ be an open covering of M and

$$g_{ij} : U_i \cap U_j \longrightarrow \{\pm 1\} \tag{5.134}$$

a cocycle defining the local system Z. Then an element of $H^q(M; Z)$ is represented by a collection of $q-$cochains $a_i \in Z^q(U_i)$ which satisfy

$$a_j = g_{ij}\, a_i \qquad \text{on} \quad U_{ij} = U_i \cap U_j. \tag{5.135}$$

We can use any model of co–chains, since the group $\mathrm{Aut}(\mathbb{Z}) \cong \{\pm 1\}$ always acts. In place of co–chains we represent integral cohomology classes by maps to an Eilenberg–MacLane space $K(\mathbb{Z}, q)$. The cohomology group is the set of homotopy classes of maps, but here we use honest maps as representatives. The group $\mathrm{Aut}(\mathbb{Z})$ acts on $K(\mathbb{Z}, q)$. One model of $K(\mathbb{Z}, 0)$ is the integers, with -1 acting by multiplication. The circle is a model for $K(\mathbb{Z}, 1)$, and -1 acts by reflection. Using the action of $\mathrm{Aut}(\mathbb{Z})$ on $K(\mathbb{Z}, q)$ and the cocycle (5.134) we build an associated bundle $\mathcal{H}^q \to M$ with fiber $K(\mathbb{Z}, q)$. Equation (5.135) says that twisted cohomology classes are represented by sections of $\mathcal{H}^q \to M$; the twisted cohomology group $H^q(M; Z)$ is the set of homotopy classes of sections of $\mathcal{H}^q \to M$.

Twistings may be defined for any generalized cohomology theory; our interest is in complex $K-$theory [Fre01, FHT03]. In homotopy theory one regards K as a marriage of a ring and a space (more precisely, spectrum), and it makes sense to ask for the units in K, denoted $GL_1(K)$. In the previous paragraph we used the units in integral cohomology, the group $\mathbb{Z}$ mod 2. For complex $K-$theory there is a richer group of units

$$GL_1(K) \sim \mathbb{Z}mod2 \times CP^{\infty} \times BSU. \tag{5.136}$$

In our problem the last factor does not enter and all the interest is in the first two, which we denote $GL_1(K)'$. As a first approximation, view K as the category of all finite dimensional $\mathbb{Z}$ mod 2–graded complex vector spaces. Then CP^{∞} is the subcategory of even complex lines, and it is a group under tensor product. It acts on K by tensor product as well. The nontrivial element of $\mathbb{Z}$ mod 2 in (5.136) acts on K by reversing the parity of the grading. This model is deficient since there is not an appropriate topology. One may consider instead complexes of complex vector spaces, or spaces of operators as we do below. Of course, there are good topological models of CP^{∞}, for example the space of all 1D subspaces of a fixed complex Hilbert space $\mathcal{H}$. For a manifold M the twistings of $K-$theory of interest are classified up to isomorphism by

$$H^1\left(M; GL_1(K)'\right) \cong H^1(M; \mathbb{Z}mod2) \times H^3(M; \mathbb{Z}).$$

In this section we will not encounter twistings from the first factor and will focus exclusively on the second. These twistings are represented by co–cycles g_{ij} with values in the space of lines, in other words by complex line bundles $L_{ij} \to U_{ij}$ which satisfy a cocycle condition. This is the data often given to define a *gerbe*.[24]

Now, let $X = G$ be a compact Lie group and, for simplicity, we shall assume that it is simply connected, though the theory works in the general case. We consider G as $G-$space, the group acting on itself by conjugation.

Since $H^3(G, Z) \cong Z$ we can construct twisted $K-$theories for each integer k. Moreover, we can also do this equivariantly, thus obtaining Abelian groups $K^*_{G,k}(G)$. These will all be $R(G)-$modules.

Now, the group multiplication map $\mu : G \times G \to G$ is compatible with conjugation and so is a $G-$map. In addition, to the pull back μ^*, we can also consider the push–forward μ_*. This depends on Poincaré duality for $K-$theory and it works also, when appropriately formulated, in the present context.

If $\dim(G)$ is even, this gives us a commutative multiplication on $K^0_{G,k}(G)$, while for $\dim(G)$ odd, our multiplication is on $K^1_{G,k}(G)$. In either case we get a ring.

The claim of [Fre01, FHT03] is that this ring (according to the parity of $\dim(G)$ is naturally isomorphic to the *Verlinde algebra* of G at level $k - h$ (where h is the Coxeter number). The Verlinde algebra is a key tool in certain

[24] Recall that a *gerbe* is a construct in homological algebra. It is defined as a *stack* over a topological space which is locally isomorphic to the *Picard groupoid* of that space. Recall that the Picard groupoid on an open set U is the category whose objects are *line bundles* on U and whose morphisms are isomorphisms. A stack refers to any category acting like a *moduli space* with a universal family (analogous to a classifying space) parameterizing a family of related mathematical objects such as schemes or topological spaces, especially when the members of these families have nontrivial automorphisms.

quantum field theories and it has been much studied by physicists, topologists, group theorists and algebraic geometers. The $K-$theory approach is totally new and much more direct than most other ways.

The Verlinde algebra is defined in the theory of loop groups. Let G be a compact Lie group. There is a version of the Theorem for *any* compact group G, but here for the most part we focus on connected, simply connected, and simple groups—$G = SU_2$ is the simplest example. In this case a central extension of the free loop group LG is determined by the *level*, which is a positive integer k. There is a finite set of equivalence classes of positive energy representations of this central extension; let $V_k(G)$ denote the free Abelian group they generate. One of the influences of 2D conformal field theory on the theory of loop groups is the construction of an algebra structure on $V_k(G)$, the *fusion product*. This is the *Verlinde algebra* [Ver88].

More precisely, let G act on itself by conjugation. Then with our assumptions the equivariant cohomology group $H^3_G(G)$ is free of rank one. Let $h(G)$ be the dual *Coxeter number* of G, and define $\zeta(k) \in H^3_G(G)$ to be $k + h(G)$ times a generator. We will see that elements of H^3 may be used to *twist* $K-$theory, and so elements of equivariant H^3 twist equivariant $K-$theory.

The *Freed–Hopkins–Teleman Theorem* [Fre01, FHT03] states: There is an isomorphism of algebras

$$V_k(G) \cong K_G^{\dim G + \zeta(k)}(G),$$

where the right hand side is the $\zeta(k)-$twisted equivariant $K-$theory in degree $\dim(G)$. The group structure on the right–hand side is induced from the multiplication map $G \times G \to G$.

For an arbitrary compact Lie group G the level k is replaced by a class in $H^4(BG; \mathbb{Z})$ and the dual Coxeter number $h(G)$ is pulled back from a universal class in $H^4(BSO; \mathbb{Z})$ via the adjoint representation. The twisting class is obtained from their sum by transgression.

5.8.8 Stringy and Brane Dynamics via $K-$Theory

In string theory, $K-$theory has been conjectured to classify the allowed *Ramond–Ramond field* strengths;[25] and also the charges of stable $D-$branes.

Classification of Ramond–Ramond Fluxes

In the classical limit of type II string theory, which is type II supergravity, the Ramond–Ramond (RR) field strengths are differential forms. In the quantum

[25] Recall that Ramond–Ramond (RR) fields are differential–form fields in the 10D space–time of type II supergravity theories, which are the classical limits of type II string theory. The ranks of the fields depend on which type II theory is considered. As Joe Polchinski argued in 1995, $D-$branes are the charged objects that act as sources for these fields, according to the rules of $p-$form electrodynamics. It has been conjectured that quantum RR fields are not differential forms, but instead are classified by twisted $K-$theory.

theory the well–definedness of the partition functions of $D-$branes implies that the RR–field strengths obey *Dirac quantization* conditions when space–time is compact, or when a spatial slice is compact and one considers only the (magnetic) components of the field strength which lie along the spatial directions. This led twentieth century physicists to classify RR field strengths using cohomology with integral coefficients.

However, some authors have argued that the cohomology of space–time with integral coefficients is too big. For example, in the presence of Neveu–Schwarz (NS) $H-$flux, or non–spin cycles, some RR–fluxes dictate the presence of $D-$branes. In the former case this is a consequence of the supergravity equation of motion which states that the product of a RR–flux with the NS 3–form is a $D-$brane charge density. Thus the set of topologically distinct RR–field strengths that can exist in brane–free configurations is only a subset of the cohomology with integral coefficients.

This subset is still too big, because some of these classes are related by large gauge transformations. In QED there are large gauge transformations which add integral multiples of 2π to *Wilson loops*.[26]

[26] Recall that in gauge theory, a *Wilson loop* (named after Ken Wilson) is a gauge–invariant observable obtained from the holonomy of the gauge connection around a given loop. In the classical theory, the collection of all Wilson loops contains sufficient information to reconstruct the gauge connection, up to gauge transformation. In *quantum field theory*, the definition of Wilson loops observables as *bona fide* operators on *Fock space* (actually, Haag's Theorem states that Fock space does not exist for interacting QFTs) is a mathematically delicate problem and requires regularization, usually by equipping each loop with a framing. The action of Wilson loop operators has the interpretation of creating an elementary excitation of the quantum field which is localized on the loop. In this way, Faraday's "flux tubes" become elementary excitations of the quantum electromagnetic field.

Wilson loops were introduced in the 1970s in an attempt at a non–perturbative formulation of quantum chromodynamics (QCD), or at least as a convenient collection of variables for dealing with the strongly–interacting regime of QCD. The problem of confinement, which Wilson loops were designed to solve, remains unsolved to this day. The fact that strongly–coupled quantum gauge field theories have elementary non–perturbative excitations which are loops motivated Alexander Polyakov to formulate the first string theories, which described the propagation of an elementary quantum loop in space–time.

Wilson loops played an important role in the formulation of *loop quantum gravity*, but there they are superseded by spin networks, a certain generalization of Wilson loops.

In particle physics and string theory, Wilson loops are often called *Wilson lines*, especially Wilson loops around non–contractible loops of a compact manifold.

A Wilson line W_C is a quantity defined by a *path–ordered exponential* of a *gauge field* A_μ

$$W_C = \mathrm{Tr}\, \mathcal{P} \exp \mathrm{i} \oint_C A_\mu dx^\mu .$$

The $p-$form potentials in type II supergravity theories also enjoy these large gauge transformations, but due to the presence of Chern–Simons terms in the supergravity actions these large gauge transformations transform not only the $p-$form potentials but also simultaneously the $(p+3)-$form field strengths. Thus to get the space of inequivalent field strengths from the forementioned subset of integral cohomology we must quotient by these large gauge transformations.

The Atiyah–Hirzebruch spectral sequence constructs twisted $K-$theory, with a twist given by the NS 3−form field strength, as a quotient of a subset of the cohomology with integral coefficients. In the classical limit, which corresponds to working with rational coefficients, this is precisely the quotient of a subset described above in supergravity. The quantum corrections come from torsion classes and contain mod 2 torsion corrections due to the Freed–Witten anomaly.

Thus twisted $K-$theory classifies the subset of RR–field strengths that can exist in the absence of $D-$branes quotiented by large gauge transformations.

Classification of $D-$Branes

Now, in many applications one wishes to add sources for the RR fields. These sources are called $D-$branes. As in classical electromagnetism, one may add sources by including a coupling $Cp\mathcal{J}_{10-p}$ of the $p-$form *potential* to a $(10-p)-$form *current* $\mathcal{J}_{10-p}$ in the Lagrangian (density). The usual convention in the string theory literature appears to be to not write this term explicitly in the action.

The current $\mathcal{J}_{10-p}$ modifies the equation of motion that comes from the variation of Cp. As is the case with magnetic monopoles in electromagnetism, this source also invalidates the dual *Bianchi identity* as it is a point at which

---

Here, C is a contour in space, $\mathcal{P}$ is the path–ordering operator, and the trace Tr guarantees that the operator is invariant under gauge transformations. Note that the quantity being traced over is an element of the *gauge Lie group* and the trace is really the character of this element with respect to an *irreducible representation*, which means there are infinitely many traces, one for each *irrep*.

Precisely because we're looking at the trace, it doesn't matter which point on the loop is chosen as the initial point. They all give the same value.

Actually, if A is viewed as a connection over a principal $G-$bundle, the equation above really ought to be 'read' as the *parallel transport* of the identity around the loop which would give an element of the Lie group G.

Note that a path–ordered exponential is a convenient shorthand notation common in physics which conceals a fair number of mathematical operations. A mathematician would refer to the path–ordered exponential of the connection as 'the holonomy of the connection' and characterize it by the parallel–transport differential equation that it satisfies.

In finite temperature QCD, the expectation value of the Wilson line distinguishes between the 'confined phase' and the 'deconfined phase' of the theory.

the dual field is not defined. In the modified equation of motion $\mathcal{J}_{p+2}$ appears on the left hand side of the equation of motion instead of zero. For simplicity, we will also interchange p and $7-p$, then the equation of motion in the presence of a source is

$$\mathcal{J}_{9-p} = d^2 C_{7-p} = dG_{8-p} = dF_{8-p} + H \wedge G_{p-1}.$$

The $(9-p)-$form $\mathcal{J}_{9-p}$ is the $Dp-$brane current, which means that it is *Poincaré dual* to the world–volume of a $(p+1)-$D extended object called a $Dp-$brane. The discrepancy of one in the naming scheme is historical and comes from the fact that one of the $p+1$ directions spanned by the $Dp-$brane is often time–like, leaving p spatial directions.

The above Bianchi identity is interpreted to mean that the $Dp-$brane is, in analogy with magnetic monopoles in electromagnetism, magnetically charged under the RR $p-$form $C7-p$. If instead one considers this Bianchi identity to be a field equation for $Cp+1$, then one says that the $Dp-$brane is electrically charged under the $(p+1)-$form $Cp+1$.

The above equation of motion implies that there are two ways to derive the $Dp-$brane charge from the ambient fluxes. First, one may integrate dG_{8-p} over a surface, which will give the $Dp-$brane charge intersected by that surface. The second method is related to the first by *Stoke's Theorem*. One may integrate G_{8-p} over a cycle, this will yield the $Dp-$brane charge linked by that cycle. The quantization of $Dp-$brane charge in the quantum theory then implies the quantization of the field strengths G, but not of the improved field strengths F.

It has been conjectured that twisted $K-$theory classifies classifies $D-$branes in noncompact space–times, intuitively in space–times in which we are not concerned about the flux sourced by the brane having nowhere to go. While the $K-$theory of a 10D space–time classifies $D-$branes as subsets of that space–time, if the space–time is the product of time and a fixed $9-$manifold then $K-$theory also classifies the conserved $D-$brane charges on each 9D spatial slice. While we were required to forget about RR potentials to get the $K-$theory classification of RR field strengths, we are required to forget about RR field strengths to get the $K-$theory classification of $D-$branes.

5.9 Self–Similar Liouville Neurodynamics

Recall that neurodynamics has its physical behavior both on the *macroscopic*, classical, *inter–neuronal level*, and on the *microscopic*, quantum, *intra–neuronal level*. On the *macroscopic* level, various models of neural networks (NNs, for short) have been proposed as goal–oriented models of the specific neural functions, like for instance, function–approximation, pattern–recognition, classification, or control (see, e.g., [Hay94]). In the physically–based, Hopfield–type models of NNs [Hop82, Hop84] the information is stored as a content–addressable memory in which synaptic strengths are modified after the Hebbian rule (see [Heb49]. Its retrieval is made when the network with the symmetric couplings works as the point–attractor with the fixed–points. Analysis of both *activation* and *learning dynamics* of Hopfield–Hebbian NNs using the techniques of statistical mechanics [DHS91], gives us with the most important information of storage capacity, role of noise and recall performance.

On the other hand, on the general *microscopic* intra–cellular level, energy transfer across the cells, without dissipation, had been first conjectured to occur in biological matter by [FK83]. The phenomenon conjectured by them was based on their 1D superconductivity model: in 1D electron systems with holes, the formation of *solitonic structures* due to electron–hole pairing results in the transfer of electric current without dissipation. In a similar manner, Frölich and Kremer conjectured that energy in biological matter could be transferred without dissipation, if appropriate solitonic structures are formed inside the cells. This idea has lead theorists to construct various models for the energy transfer across the cell, based on the formation of *kink* classical solutions (see [STZ93, SZT98].

The interior of living cells is structurally and dynamically organized by *cytoskeletons*, i.e., networks of protein polymers. Of these structures, *microtubules* (MTs, for short) appear to be the most fundamental (see [Dus84]). Their dynamics has been studied by a number of authors in connection with the mechanism responsible for dissipation–free energy transfer. Hameroff and Penrose [Ham87] have conjectured another fundamental role for the MTs, namely being responsible for *quantum computations* in the human neurons. [Pen67, Pen94, Pen97] further argued that the latter is associated with certain aspects of quantum theory that are believed to occur in the cytoskeleton MTs, in particular quantum superposition and subsequent collapse of the wave function of coherent MT networks. These ideas have been elaborated by [MN95a, MN95b] and [Nan95], based on the quantum–gravity EMN–language of [EMN92, EMN99] where MTs have been physically modelled as non-critical (SUSY) bosonic strings. It has been suggested that the neural MTs are the microsites for the emergence of stable, macroscopic quantum coherent states, identifiable with the *preconscious states*; stringy–quantum space–time effects trigger an organized collapse of the coherent states down to a specific or *conscious state*. More recently, [TVP99] have presented the evidence for biological self–organization and pattern formation during embryogenesis.

Now, we have two space–time biophysical scales of neurodynamics. Naturally the question arises: are these two scales somehow inter-related, is there a space–time self–similarity between them?

The purpose of this section is to prove the formal positive answer to the self–similarity question. We try to describe neurodynamics on both physical levels by the *unique form* of a single equation, namely *open Liouville equation*: NN–dynamics using its classical form, and MT–dynamics using its quantum form in the Heisenberg picture. If this formulation is consistent, that would prove the *existence* of the *formal neurobiological space–time self–similarity.*

Hamiltonian Framework

Suppose that on the macroscopic NN–level we have a conservative Hamiltonian system acting in a $2N$D symplectic phase–space $T^*Q = \{q^i(t), p_i(t)\}$, $(i = 1 \dots N)$ (which is the cotangent bundle of the NN–configuration manifold $Q = \{q^i\}$), with a Hamiltonian function $H = H(q^i, p_i, t) : T^*Q \times \mathbb{R} \to \mathbb{R}$. The conservative dynamics is defined by classical Hamiltonian canonical equations:

$$\dot{q}^i = \partial_p H, \qquad \dot{p}_i = -\partial_q H. \tag{5.137}$$

Recall that within the conservative Hamiltonian framework, we can apply the formalism of classical Poisson brackets: for any two functions $A = A(q^i, p_i, t)$ and $B = B(q^i, p_i, t)$ their Poisson bracket is defined as

$$[A, B] = \left(\frac{\partial A}{\partial q^i} \frac{\partial B}{\partial p_i} - \frac{\partial A}{\partial p_i} \frac{\partial B}{\partial q^i} \right).$$

Conservative Classical System

Any function $A(q^i, p_i, t)$ is called a *constant* (or integral) of motion of the conservative system (5.137) if

$$\dot{A} \equiv \partial_t A + [A, H] = 0, \qquad \text{which implies} \qquad \partial_t A = -[A, H]. \tag{5.138}$$

For example, if $A = \rho(q^i, p_i, t)$ is a *density function* of ensemble phase–points (or, a probability density to see a state $x(t) = (q^i(t), p_i(t))$ of *ensemble* at a moment t), then equation

$$\partial_t \rho = -[\rho, H] = -\mathrm{i}L\, \rho \tag{5.139}$$

represents the *Liouville Theorem*, where L denotes the (Hermitian) *Liouville operator*

$$\mathrm{i}L = [..., H] \equiv \left(\frac{\partial H}{\partial p_i} \frac{\partial}{\partial q^i} - \frac{\partial H}{\partial q^i} \frac{\partial}{\partial p_i} \right) = \mathrm{div}(\rho \dot{\mathbf{x}}),$$

which shows that the conservative Liouville equation (5.139) is actually equivalent to the mechanical *continuity equation*

$$\partial_t \rho + \mathrm{div}(\rho \dot{\mathbf{x}}) = 0. \tag{5.140}$$

Conservative Quantum System

We perform the formal quantization of the conservative equation (5.139) in the Heisenberg picture: all variables become Hermitian operators (denoted by '$\wedge$'), the symplectic phase–space $T^*Q = \{q^i, p_i\}$ becomes the Hilbert state–space $\mathcal{H} = \mathcal{H}_{\hat{q}^i} \otimes \mathcal{H}_{\hat{p}_i}$ (where $\mathcal{H}_{\hat{q}^i} = \mathcal{H}_{\hat{q}^1} \otimes ... \otimes \mathcal{H}_{\hat{q}^N}$ and $\mathcal{H}_{\hat{p}_i} = \mathcal{H}_{\hat{p}_1} \otimes ... \otimes$

$\mathcal{H}_{\hat{p}_N}$), the classical Poisson bracket $[\,,\,]$ becomes the quantum commutator $\{\,,\,\}$ multiplied by $-\mathrm{i}/\hbar$

$$[\,,\,] \longrightarrow -\mathrm{i}\{\,,\,\} \qquad (\hbar = 1 \text{ in normal units})\,. \tag{5.141}$$

In this way the classical Liouville equation (5.139) becomes the *quantum Liouville equation*

$$\partial_t \hat{\rho} = \mathrm{i}\{\hat{\rho}, \hat{H}\}\,, \tag{5.142}$$

where $\hat{H} = \hat{H}(\hat{q}^i, \hat{p}_i, t)$ is the Hamiltonian evolution operator, while

$$\hat{\rho} = P(a)|\Psi_a><\Psi_a|, \qquad \text{with} \qquad \mathrm{Tr}(\hat{\rho}) = 1,$$

denotes the von Neumann *density matrix operator*, where each quantum state $|\Psi_a>$ occurs with probability $P(a)$; $\hat{\rho} = \hat{\rho}(\hat{q}^i, \hat{p}_i, t)$ is closely related to another von Neumann concept: *entropy* $S = -\,\mathrm{Tr}(\hat{\rho}[\ln \hat{\rho}])$.

Open Classical System

We now move to the *open (nonconservative) system*: on the macroscopic NN–level the *opening operation* equals to the *adding* of a *covariant* vector of external (dissipative and/or motor) forces $F_i = F_i(q^i, p_i, t)$ to (the r.h.s of) the covariant Hamiltonian *force equation*, so that Hamiltonian equations get the *open (dissipative and/or forced) form*

$$\dot{q}^i = \frac{\partial H}{\partial p_i}, \qquad \dot{p}_i = F_i - \frac{\partial H}{\partial q^i}\,. \tag{5.143}$$

In the framework of the open Hamiltonian system (5.143), dynamics of any function $A(q^i, p_i, t)$ is defined by the *open evolution equation*:

$$\partial_t A = -[A, H] + \Phi,$$

where $\Phi = \Phi(F_i)$ represents the general form of the scalar force term.

In particular, if $A = \rho(q^i, p_i, t)$ represents the density function of ensemble phase–points, then its dynamics is given by the (dissipative/forced) *open Liouville equation*:

$$\partial_t \rho = -[\rho, H] + \Phi\,. \tag{5.144}$$

In particular, the scalar force term can be cast as a linear Poisson–bracket form

$$\Phi = F_i[A, q^i]\,, \qquad \text{with} \qquad [A, q^i] = -\frac{\partial A}{\partial p_i}\,. \tag{5.145}$$

Now, in a similar way as the conservative Liouville equation (5.139) resembles the continuity equation (5.140) from continuum dynamics, also the open Liouville equation (5.144) resembles the probabilistic *Fokker–Planck equation* from statistical mechanics. If we have a ND stochastic process $x(t) = (q^i(t), p_i(t))$ defined by the vector *Itô SDE*

$$dx(t) = f(x,t)\,dt + G(x,t)\,dW,$$

where f is a ND vector function, W is a KD Wiener process, and G is a $N \times K$D matrix valued function, then the corresponding probability density function $\rho = \rho(x,t|\dot{x},t')$ is defined by the ND Fokker–Planck equation (see, e.g., [Gar85])

$$\partial_t \rho = -\operatorname{div}[\rho\, f(x,t)] + \frac{1}{2}\frac{\partial^2}{\partial x_i \partial x_j}(Q_{ij}\,\rho), \tag{5.146}$$

where $Q_{ij} = \left(G(x,t)\,G^T(x,t)\right)_{ij}$. It is obvious that the Fokker–Planck equation (5.146) represents the particular, stochastic form of our general open Liouville equation (5.144), in which the scalar force term is given by the (second–derivative) noise term

$$\Phi = \frac{1}{2}\frac{\partial^2}{\partial x_i \partial x_j}(Q_{ij}\,\rho).$$

Equation (5.144) will represent the open classical model of our macroscopic NN–dynamics.

Continuous Neural Network Dynamics

The generalized NN–dynamics, including two special cases of *graded response neurons* (GRN) and *coupled neural oscillators* (CNO), can be presented in the form of a stochastic *Langevin rate equation*

$$\dot{\sigma}_i = f_i + \eta_i(t), \tag{5.147}$$

where $\sigma_i = \sigma_i(t)$ are the continual neuronal variables of ith neurons (representing either membrane action potentials in case of GRN, or oscillator phases in case of CNO); J_{ij} are individual synaptic weights; $f_i = f_i(\sigma_i, J_{ij})$ are the deterministic forces (given, in GRN–case, by $f_i = \sum_j J_{ij}\tanh[\gamma\sigma_j] - \sigma_i + \theta_i$, with $\gamma > 0$ and with the θ_i representing injected currents, and in CNO–case, by $f_i = \sum_j J_{ij}\sin(\sigma_j - \sigma_i) + \omega_i$, with ω_i representing the natural frequencies of the individual oscillators); the noise variables are given as $\eta_i(t) = \lim_{\Delta\to 0}\zeta_i(t)\sqrt{2T/\Delta}$ where $\zeta_i(t)$ denote uncorrelated Gaussian distributed random forces and the parameter T controls the amount of noise in the system, ranging from $T = 0$ (deterministic dynamics) to $T = \infty$ (completely random dynamics).

More convenient description of the neural random process (5.147) is provided by the Fokker–Planck equation describing the time evolution of the probability density $P(\sigma_i)$

$$\partial_t P(\sigma_i) = -\frac{\partial}{\partial \sigma_i}\,(f_i P(\sigma_i)) + T\frac{\partial^2}{\partial \sigma_i^2}P(\sigma_i). \tag{5.148}$$

Now, in the case of deterministic dynamics $T = 0$, equation (5.148) can be put into the form of the conservative Liouville equation (5.139), by making the substitutions: $P(\sigma_i) \to \rho$, $f_i = \dot{\sigma}_i$, and $[\rho, H] = \mathrm{div}(\rho\,\dot{\sigma}_i) \equiv \sum_i \frac{\partial}{\partial \sigma_i}(\rho\,\dot{\sigma}_i)$, where $H = H(\sigma_i, J_{ij})$. Further, we can formally identify the stochastic forces, i.e., the second–order noise–term $T\sum_i \frac{\partial^2}{\partial \sigma_i^2}\rho$ with $F^i[\rho, \sigma_i]$, to get the open Liouville equation (5.144).

Therefore, on the NN–level deterministic dynamics corresponds to the conservative system (5.139). Inclusion of stochastic forces corresponds to the system opening (5.144), implying the *macroscopic arrow of time.*

Open Quantum System

By formal quantization of equation (5.144) with the scalar force term defined by (5.145), in the same way as in the case of the conservative dynamics, we get the *quantum open Liouville equation*

$$\partial_t \hat{\rho} = \mathrm{i}\{\hat{\rho}, \hat{H}\} + \hat{\Phi}, \qquad \text{with} \qquad \hat{\Phi} = -\mathrm{i}\hat{F}_i\{\hat{\rho}, \hat{q}^i\}, \tag{5.149}$$

where $\hat{F}_i = \hat{F}_i(\hat{q}^i, \hat{p}_i, t)$ represents the covariant quantum operator of external friction forces in the Hilbert state–space $\mathcal{H} = \mathcal{H}_{\hat{q}^i} \otimes \mathcal{H}_{\hat{p}_i}$.

Equation (5.149) will represent the open quantum–friction model of our microscopic MT–dynamics. Its system–independent properties are [EMN92, EMN99, MN95a, MN95b, Nan95]:

1. Conservation of probability P
$$\partial_t P = \partial_t[\mathrm{Tr}(\hat{\rho})] = 0.$$
2. Conservation of energy E, on the average
$$\partial_t \langle\langle E \rangle\rangle \equiv \partial_t[\mathrm{Tr}(\hat{\rho}\, E)] = 0.$$
3. Monotonic increase in entropy
$$\partial_t S = \partial_t[-\,\mathrm{Tr}(\hat{\rho} \ln \hat{\rho})] \geq 0,$$
and thus automatically and naturally implies a *microscopic arrow of time*, so essential in realistic biophysics of neural processes.

Non–Critical Stringy MT–Dynamics

In EMN–language of non–critical (SUSY) bosonic strings, our MT–dynamics equation (5.149) reads

$$\partial_t \hat{\rho} = \mathrm{i}\{\hat{\rho}, \hat{H}\} - \mathrm{i}\hat{g}_{ij}\{\hat{\rho}, \hat{q}^i\}\hat{\dot{q}}^j\,, \tag{5.150}$$

where the target–space density matrix $\hat{\rho}(\hat{q}^i, \hat{p}_i)$ is viewed as a function of coordinates $\hat{q}^i$ that parameterize the couplings of the generalized $\sigma-$models on the bosonic string world–sheet, and their conjugate momenta $\hat{p}_i$, while $\hat{g}_{ij} = \hat{g}_{ij}(\hat{q}^i)$ is the quantum operator of the *positive definite metric* in the space of couplings. Therefore, the covariant quantum operator of external friction forces is in EMN–formulation given as $\hat{F}_i(\hat{q}^i, \hat{\dot{q}}^i) = \hat{g}_{ij}\,\hat{\dot{q}}^j$.

Equation (5.150) establishes the conditions under which a large–scale coherent state appearing in the MT–network, which can be considered responsible for loss–free energy transfer along the tubulins.

Equivalence of Neurodynamic Forms

It is obvious that both the macroscopic NN–equation (5.144) and the microscopic MT–equation (5.149) have the same open Liouville form, which implies the arrow of time. These proves the existence of the formal neuro–biological space–time self–similarity.

In this way, we have described neurodynamics of both NN and MT ensembles, belonging to completely different biophysical space–time scales, by the unique form of open Liouville equation, which implies the arrow of time. The existence of the formal *neuro–biological self–similarity* has been proved.

6

Path Integrals and Complex Dynamics

In this Chapter we develop the formalism of complex path integrals, the essential tool in highly–nonlinear high–dimensional complex dynamics.

6.1 Path Integrals: Sums Over Histories

Recall that in the core of modern physics there is a powerful *conceptual and computational tool*, the celebrated *Feynman path integral*. In the path–integral formalism, we first formulate the specific classical *action* of a new theory, and subsequently perform its quantization by means of the associated *amplitude*. This *action–amplitude picture* is the core structure in any new physical theory. Unlike mathematical manifolds, bundles and jets, the path integral is an invention of the physical mind of Richard (Dick) Feynman. Its *virtual paths* are in general neither deterministic not smooth, although they include bundles and jets of deterministic and smooth paths, as well as Markov chains. Yet, it is essentially a (broader) geometrical dynamics, with its Riemannian and symplectic versions, among many others. At the beginning, it worked only for *conservative* physical systems. Today it includes also *dissipative structures*, as well as various *sources and sinks*. Its smooth part reveals all celebrated equations of the 20th Century, both classical and quantum. It is the core of modern quantum gravity and string theory. It is arguably the most important construct of mathematical physics. At the edge of a new millennium, if you asked a typical theoretical physicist: what will be your main research tool in the new millennium, he/she would most probably say: path integral. And today, we see it moving out from physics, into the realm of social sciences. Finally, since Feynman's fairly intuitive invention of the path integral [Fey51], a lot of research has been done to make it mathematically rigorous (see e.g., [Loo99, Loo00, AFH86, Kla97, SK98a, Kla00]).

6.1.1 Intuition Behind a Path Integral

Classical Probability Concept

Recall that a *random variable* X is defined by its *distribution function* $f(x)$. Its *probabilistic description* is based on the following rules: (i) $P(X = x_i)$ is the probability that $X = x_i$; and (ii) $P(a \leq X \leq b)$ is the probability that X lies in a closed interval $[a, b]$. Its statistical description is based on: (i) μ_X or $E(X)$ is the mean or expectation of X; and (ii) σ_X is the standard deviation of X. There are two cases of random variables: discrete and continuous, each having its own probability (and statistics) theory.

Discrete Random Variable

A discrete random variable X has only a countable number of values $\{x_i\}$. Its distribution function $f(x_i)$ has the following properties:

$$P(X = x_i) = f(x_i), \qquad f(x_i) \geq 0, \qquad \sum_i f(x_i)\, dx = 1.$$

Statistical description of X is based on its discrete mean value μ_X and standard deviation σ_X, given respectively by

$$\mu_X = E(X) = \sum_i x_i f(x_i), \qquad \sigma_X = \sqrt{E(X^2) - \mu_X^2}.$$

Continuous Random Variable

Here $f(x)$ is a piecewise continuous function such that:

$$P(a \leq X \leq b) = \int_a^b f(x)\, dx, \qquad f(x) \geq 0, \qquad \int_{-\infty}^{\infty} f(x)\, dx = \int_{\mathbb{R}} f(x)\, dx = 1.$$

Statistical description of X is based on its continuous mean μ_X and standard deviation σ_X, given respectively by

$$\mu_X = E(X) = \int_{-\infty}^{\infty} x f(x)\, dx, \qquad \sigma_X = \sqrt{E(X^2) - \mu_X^2}.$$

Now, let us observe the similarity between the two descriptions. The same kind of similarity between discrete and continuous quantum spectrum stroke Dirac when he suggested the combined integral approach, that he denoted by $\sum\!\!\!\!\!\!\int$ – meaning 'both integral and sum at once': summing over discrete spectrum and integration over continuous spectrum.

To emphasize this similarity even further, as well as to set–up the stage for the path integral, recall the notion of a *cumulative distribution function* of a random variable X, that is a function $F : \mathbb{R} \longrightarrow \mathbb{R}$, defined by

$$F(a) = P(X) \leq a.$$

In particular, suppose that $f(x)$ is the distribution function of X. Then

$$F(x) = \sum_{x_i \leq x} f(x_i), \qquad \text{or} \qquad F(x) = \int_{-\infty}^{\infty} f(t)\, dt,$$

according to as x is a discrete or continuous random variable. In either case, $F(a) \leq F(b)$ whenever $a \leq b$. Also,

$$\lim_{x \longrightarrow -\infty} F(x) = 0 \qquad \text{and} \qquad \lim_{x \longrightarrow \infty} F(x) = 1,$$

that is, $F(x)$ is monotonic and its limit to the left is 0 and the limit to the right is 1. Furthermore, its cumulative probability is given by

$$P(a \leq X \leq b) = F(b) - F(a),$$

and the Fundamental Theorem of Calculus tells us that, in the continuum case,

$$f(x) = \partial_x F(x).$$

General Markov Stochastic Dynamics

Recall that *Markov stochastic process* is a random process characterized by a *lack of memory*, i.e., the statistical properties of the immediate future are uniquely determined by the present, regardless of the past [Gar85].

For example, a *random walk* is an example of the *Markov chain*, i.e., a discrete–time Markov process, such that the motion of the system in consideration is viewed as a sequence of states, in which the transition from one state to another depends only on the preceding one, or the probability of the system being in state k depends only on the previous state $k-1$. The property of a Markov chain of prime importance in biomechanics is the existence of an *invariant distribution of states*: we start with an initial state x_0 whose absolute probability is 1. Ultimately the states should be distributed according to a specified distribution.

Between the pure deterministic dynamics, in which all DOF of the system in consideration are explicitly taken into account, leading to classical dynamical equations, for example in Hamiltonian form (5.137), i.e.,

$$\dot{q}^i = \partial_{p_i} H, \qquad \dot{p}_i = -\partial_{q^i} H$$

– and pure stochastic dynamics (Markov process), there is so–called *hybrid dynamics*, particularly *Brownian dynamics*, in which some of DOF are represented only through their *stochastic influence* on others. As an example, suppose a system of particles interacts with a viscous medium. Instead of specifying a detailed interaction of each particle with the particles of the viscous medium, we represent the medium as a *stochastic force* acting on the particle. The stochastic force *reduces the dimensionally* of the dynamics.

Recall that the Brownian dynamics represents the phase–space trajectories of a collection of particles that individually obey *Langevin rate equations* in the field of force (i.e., the particles interact with each other via some deterministic force). For a free particle, the Langevin equation reads [Gar85]:

$$m\dot{v} = R(t) - \beta v,$$

where m denotes the mass of the particle and v its velocity. The right–hand side represent the coupling to a *heat bath*; the effect of the random force $R(t)$ is to heat the particle. To balance overheating (on the average), the particle is subjected to *friction* β. In humanoid dynamics this is performed with the Rayleigh–Van der Pol's *dissipation.* Formally, the solution to the Langevin equation can be written as

$$v(t) = v(0)\exp\left(-\frac{\beta}{m}t\right) + \frac{1}{m}\int_0^t \exp[-(t-\tau)\beta/m]\,R(\tau)\,d\tau,$$

where the integral on the right–hand side is a *stochastic integral* and the solution $v(t)$ is a random variable. The stochastic properties of the solution depend significantly on the stochastic properties of the random force $R(t)$. In the Brownian dynamics the random force $R(t)$ is Gaussian distributed. Then the problem boils down to finding the solution to the Langevin stochastic differential equation with the supplementary condition (mean zero and variance)

$$< R(t) >= 0, \qquad < R(t)\,R(0) >= 2\beta k_B T \delta(t),$$

where $< . >$ denotes the mean value, T is temperature, k_B-*equipartition* (i.e., uniform distribution of energy) coefficient, Dirac $\delta(t)-$function.

Algorithm for computer simulation of the Brownian dynamics (for a single particle) can be written as [Hee90]:

1. Assign an initial position and velocity.
2. Draw a random number from a Gaussian distribution with mean zero and variance.
3. Integrate the velocity to get v^{n+1}.
4. Add the random component to the velocity.

Another approach to taking account the coupling of the system to a heat bath is to subject the particles to collisions with *virtual particles* [Hee90]. Such collisions are imagined to affect only momenta of the particles, hence they affect the kinetic energy and introduce fluctuations in the total energy. Each stochastic collision is assumed to be an instantaneous event affecting only one particle.

The collision–coupling idea is incorporated into the Hamiltonian model of dynamics (5.137) by adding a stochastic force $R_i = R_i(t)$ to the $\dot{p}$ equation

$$\dot{q}^i = \partial_{p_i} H, \qquad \dot{p}_i = -\partial_{q^i} H + R_i(t).$$

On the other hand, the so–called *Ito stochastic integral* represents a kind of classical Riemann–Stieltjes integral from linear functional analysis, which is (in 1D case) for an arbitrary time–function $G(t)$ defined as the *mean square limit*

$$\int_{t_0}^{t} G(t)dW(t) = ms \lim_{n\to\infty} \{\sum_{i=1}^{n} G(t_{i-1}[W(t_i) - W(t_{i-1}]\}.$$

Now, the general ND Markov process can be defined by *Ito* stochastic differential equation (SDE),

$$\begin{aligned} dx_i(t) &= A_i[x^i(t), t]dt + B_{ij}[x^i(t), t]\, dW^j(t), \\ x^i(0) &= x_{i0}, \qquad (i, j = 1, \dots, N) \end{aligned}$$

or corresponding *Ito stochastic integral equation*

$$x^i(t) = x^i(0) + \int_0^t ds\, A_i[x^i(s), s] + \int_0^t dW^j(s)\, B_{ij}[x^i(s), s],$$

in which $x^i(t)$ is the variable of interest, the vector $A_i[x(t), t]$ denotes deterministic *drift*, the matrix $B_{ij}[x(t), t]$ represents continuous stochastic *diffusion fluctuations*, and $W^j(t)$ is an N-variable *Wiener process* (i.e., generalized Brownian motion) [Wie61], and $dW^j(t) = W^j(t + dt) - W^j(t)$.

Now, there are three well–known special cases of the *Chapman–Kolmogorov equation* (see [Gar85]):

1. When both $B_{ij}[x(t), t]$ and $W(t)$ are zero, i.e., in the case of pure deterministic motion, it reduces to the *Liouville equation*

$$\partial_t P(x', t'|x'', t'') = -\sum_i \frac{\partial}{\partial x^i} \{A_i[x(t), t]\, P(x', t'|x'', t'')\}.$$

2. When only $W(t)$ is zero, it reduces to the *Fokker–Planck equation*

$$\begin{aligned} \partial_t P(x', t'|x'', t'') &= -\sum_i \frac{\partial}{\partial x^i} \{A_i[x(t), t]\, P(x', t'|x'', t'')\} \\ &+ \frac{1}{2}\sum_{ij} \frac{\partial^2}{\partial x^i \partial x^j} \{B_{ij}[x(t), t]\, P(x', t'|x'', t'')\}. \end{aligned}$$

3. When both $A_i[x(t), t]$ and $B_{ij}[x(t), t]$ are zero, i.e., the state–space consists of integers only, it reduces to the *Master equation* of discontinuous jumps

$$\begin{aligned} &\partial_t P(x', t'|x'', t'') = \\ &\int dx\, \{W(x'|x'', t)\, P(x', t'|x'', t'') - W(x''|x', t)\, P(x', t'|x'', t'')\}. \end{aligned}$$

The *Markov assumption* can now be formulated in terms of the conditional probabilities $P(x^i, t_i)$: if the times t_i increase from right to left, the conditional probability is determined entirely by the knowledge of the most recent condition. Markov process is generated by a set of conditional probabilities whose probability–density $P = P(x', t'|x'', t'')$ evolution obeys the general *Chapman–Kolmogorov integro–differential equation*

$$\partial_t P = -\sum_i \frac{\partial}{\partial x^i}\{A_i[x(t), t]\,P\} \\ + \frac{1}{2}\sum_{ij}\frac{\partial^2}{\partial x^i \partial x^j}\{B_{ij}[x(t), t]\,P\} + \int dx\,\{W(x'|x'', t)\,P - W(x''|x', t)\,P\}$$

including *deterministic drift*, *diffusion fluctuations* and *discontinuous jumps* (given respectively in the first, second and third terms on the r.h.s.).

It is this general Chapman–Kolmogorov integro–differential equation, with its conditional probability density evolution, $P = P(x', t'|x'', t'')$, that we are going to model by various forms of the Feynman path integral, providing us with the physical insight behind the abstract (conditional) probability densities.

Quantum Probability Concept

An alternative concept of probability, the so–called *quantum probability*, is based on the following physical facts (elaborated in detail in this section):

1. *The time–dependent Schrödinger equation* represents a *complex–valued generalization* of the real–valued *Fokker–Planck equation* for describing the spatio–temporal *probability density function* for the system exhibiting *continuous–time Markov stochastic process.*
2. The *Feynman path integral* $\oint$ is a generalization of the time–dependent Schrödinger equation, including both continuous–time and discrete–time Markov stochastic processes.
3. Both Schrödinger equation and path integral give 'physical description' of any system they are modelling in terms of its physical energy, instead of an abstract probabilistic description of the Fokker–Planck equation.

Therefore, the *Feynman path integral* $\oint$, as a generalization of the time–dependent Schrödinger equation, gives a unique physical description for the general Markov stochastic process, in terms of the physically based generalized probability density functions, valid both for continuous–time and discrete–time Markov systems.

Basic consequence: a different way for calculating probabilities. The difference is rooted in the fact that *sum of squares is different from the square of sums*, as is explained in the following text.

Namely, in Dirac–Feynman quantum formalism, each possible route from the initial system state A to the final system state B is called a *history*. This history comprises any kind of a route (see Figure 6.1), ranging from continuous and smooth deterministic (mechanical–like) paths to completely discontinues and random Markov chains (see, e.g., [Gar85]). Each history (labelled by index i) is quantitatively described by a *complex number*[1] z_i called the 'individual transition amplitude'. Its absolute square, $|z_i|^2$, is called the *individual transition probability*. Now, the *total transition amplitude* is the sum of all individual transition amplitudes, $\sum_i z_i$, called the *sum–over–histories*. The absolute square of this sum–over–histories, $|\sum_i z_i|^2$, is the *total transition probability*.

In this way, the overall probability of the system's transition from some initial state A to some final state B is given *not* by adding up the probabilities for each history–route, but by 'head–to–tail' adding up the sequence of amplitudes making–up each route first (i.e., performing the sum–over–histories) – to get the total amplitude as a 'resultant vector', and then squaring the total amplitude to get the overall transition probability.

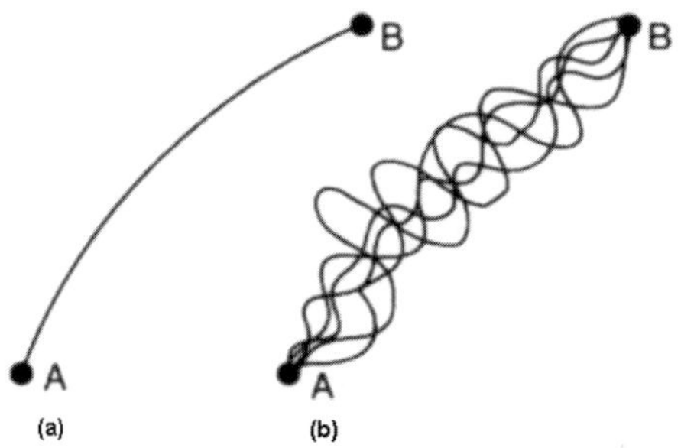

Fig. 6.1. Two ways of physical *transition* from an *initial state* A to the corresponding *final state* B. (a) Classical physics proposes *a single deterministic trajectory*, minimizing the total system's energy. (b) Quantum physics proposes *a family of Markov stochastic histories*, namely *all possible routes* from A to B, both continuous–time and discrete–time Markov chains, each giving an equal contribution to the total *transition probability*.

[1] Recall that a *complex number* $z = x + \mathrm{i}y$, where $\mathrm{i} = \sqrt{-1}$ is the *imaginary unit*, x is the *real part* and y is the *imaginary part*, can be represented also in its *polar form*, $z = r(\cos\theta + \mathrm{i}\sin\theta)$, where the radius vector in the complex–plane, $r = |z| = \sqrt{x^2 + y^2}$, is the *modulus* or *amplitude*, and angle θ is the *phase*; as well as in its exponential form $z = r\mathrm{e}^{\mathrm{i}\theta}$. In this way, complex numbers actually represent 2D vectors with usual vector 'head–to–tail' addition rule.

Quantum Coherent States

Recall that a *quantum coherent state* is a specific kind of quantum state of the quantum harmonic oscillator whose dynamics most closely resemble the oscillating behavior of a classical harmonic oscillator. It was the first example of quantum dynamics when Erwin Schrödinger derived it in 1926 while searching for solutions of the *Schrödinger equation* that satisfy the *correspondence principle*. The quantum harmonic oscillator and hence, the coherent state, arise in the quantum theory of a wide range of physical systems. For instance, a coherent state describes the oscillating motion of the particle in a quadratic potential well. In the quantum electrodynamics and other bosonic quantum field theories they were introduced by the 2005 Nobel Prize winning work of R. Glauber in 1963 [Gla63a, Gla63b]. Here the coherent state of a field describes an oscillating field, the closest quantum state to a classical sinusoidal wave such as a continuous laser wave.

In classical optics, light is thought of as electromagnetic waves radiating from a source. Specifically, coherent light is thought of as light that is emitted by many such sources that are in phase. For instance, a light bulb radiates light that is the result of waves being emitted at all the points along the filament. Such light is incoherent because the process is highly random in space and time. On the other hand, in a laser, light is emitted by a carefully controlled system in processes that are not random but interconnected by stimulation and the resulting light is highly ordered, or coherent. Therefore a coherent state corresponds closely to the quantum state of light emitted by an ideal laser. Semi–classically we describe such a state by an electric field oscillating as a stable wave. Contrary to the coherent state, which is the most wave–like quantum state, the *Fock state* (e.g., a single photon) is the most particle–like state. It is indivisible and contains only one quanta of energy. These two states are examples of the opposite extremes in the concept of *wave–particle duality*. A coherent state distributes its quantum–mechanical uncertainty equally, which means that the phase and amplitude uncertainty are approximately equal. Conversely, in a single–particle state the phase is completely uncertain.

Formally, the coherent state $|\alpha\rangle$ is defined to be the eigenstate of the annihilation operator a, i.e., $a|\alpha\rangle = \alpha|\alpha\rangle$. Note that since a is not Hermitian, $\alpha = |\alpha|\mathrm{e}^{i\theta}$ is complex. $|\alpha|$ and θ are called the *amplitude* and *phase* of the state.

Physically, $a|\alpha\rangle = \alpha|\alpha\rangle$ means that a coherent state is left unchanged by the detection (or annihilation) of a particle. Consequently, in a coherent state, one has exactly the same probability to detect a second particle. Note, this condition is necessary for the coherent state's *Poisson detection statistics*. Compare this to a single–particle's Fock state: Once one particle is detected, we have zero probability of detecting another.

Now, recall that a *Bose–Einstein condensate* (BEC) is a collection of boson atoms that are all in the same quantum state. An approximate theoretical description of its properties can be derived by assuming the BEC is in a coherent state. However, unlike photons, atoms interact with each other so it now appears that it is more likely to be one of the *squeezed coherent states* (see [BSM97]). In quantum field theory and string theory, a generalization of coherent states to the case of infinitely many degrees of freedom is used to define a *vacuum state* with a different vacuum expectation value from the original vacuum.

Dirac's $<\boldsymbol{bra}\,|\,\boldsymbol{ket}>$ Transition Amplitude

Now, we are ready to move–on into the realm of quantum mechanics. Recall that P. Dirac [Dir49] described behavior of quantum systems in terms of complex–valued *ket–vectors* $|A>$ living in the Hilbert space $\mathcal{H}$, and their duals, *bra–covectors* $<B|$ (i.e., 1–forms) living in the *dual* Hilbert space $\mathcal{H}^*$.[2] The *Hermitian inner product* of kets and bras, the *bra–ket* $<B|A>$, is a *complex number*, which is the evaluation of the ket $|A>$ by the bra $<B|$. This complex number, say $r\mathrm{e}^{\mathrm{i}\theta}$ represents the system's *transition amplitude*[3] from its *initial state* A to its *final state* B[4], i.e.,

$$Transition\,Amplitude =< B|A >= r\mathrm{e}^{\mathrm{i}\theta}.$$

That is, there is a process that can mediate a transition of a system from initial state A to the final state B and the amplitude for this transition equals

[2] Recall that a *norm* on a complex vector space $\mathcal{H}$ is a mapping from $\mathcal{H}$ into the complex numbers, $\|\cdot\| : \mathcal{H} \to \mathbb{C};\ h \mapsto \|h\|$, such that the following set of norm–axioms hold:
(N1) $\|h\| \geq 0$ for all $h \in \mathcal{H}$ and $\|h\| = 0$ implies $h = 0$ (positive definiteness);
(N2) $\|\lambda\, h\| = |\lambda|\, \|h\|$ for all $h \in \mathcal{H}$ and $\lambda \in \mathbb{C}$ (homogeneity); and
(N3) $\|h_1 + h_2\| \leq \|h_1\| + \|h_2\|$ for all $h_1, h_2 \in \mathcal{H}$ (triangle inequality). The pair $(\mathcal{H}, \|\cdot\|)$ is called a *normed space*.
A *Hermitian inner product* on a complex vector space $\mathcal{H}$ is a mapping $\langle\cdot,\cdot\rangle : \mathcal{H} \times \mathcal{H} \to \mathbb{C}$ such that the following set of inner–product–axioms hold:
(IP1) $\langle h\, h_1 + h_2\rangle = \langle h\, h_1 + h\, h_2\rangle$;
(IP2) $\langle \alpha\, h, h_1\rangle = \alpha\, \langle h, h_1\rangle$;
(IP3) $\langle h_1, h_2\rangle = \overline{\langle h_1, h_2\rangle}$ (so $\langle h, h\rangle$ is real);
(IP4) $\langle h, h\rangle \geq 0$ and $\langle h, h\rangle = 0$ provided $h = 0$.
The *standard inner product* on the product space $\mathbb{C}^n = \mathbb{C} \times \cdots \times \mathbb{C}$ is defined by $\langle z, w\rangle = \sum_{i=1}^n z_i w^i$, and axioms (IP1)–(IP4) are readily checked. Also $\mathbb{C}^n$ is a normed space with $\|z\|^2 = \sum_{i=1}^n |z_i|^2$. The pair $(\mathcal{H}, \langle\cdot,\cdot\rangle)$ is called an *inner product space*.
Let $(\mathcal{H}, \|\cdot\|)$ be a normed space. If the corresponding metric d is complete, we say $(\mathcal{H}, \|\cdot\|)$ is a Banach space. If $(\mathcal{H}, \|\cdot\|)$ is an inner product space whose corresponding metric is complete, we say $(\mathcal{H}, \|\cdot\|)$ is a *Hilbert space*.

[3] Transition amplitude is otherwise called *probability amplitude*, or just *amplitude*.

[4] Recall that in quantum mechanics, complex numbers are regarded as the *vacuum–state*, or the *ground–state*, and the entire amplitude $<b|a>$ is a *vacuum-to-vacuum amplitude* for a process that includes the *creation* of the state a, its *transition* to b, and the *annihilation* of b to the vacuum once more.

$< B|A >= re^{i\theta}$. The absolute square of the amplitude, $| < B|A > |^2$ represents the *transition probability*. Therefore, the probability of a transition event equals the absolute square of a complex number, i.e.,

$$Transition\, Probability = | < B|A > |^2 = |re^{i\theta}|^2.$$

These complex amplitudes obey the usual *laws of probability*: when a transition event can happen in alternative ways then we add the complex numbers,

$$< B_1|A_1 > + < B_2|A_2 >= r_1 e^{i\theta_1} + r_2 e^{i\theta_2},$$

and when it can happen only as a succession of intermediate steps then we multiply the complex numbers,

$$< B|A >=< B|c >< c|A >= (r_1 e^{i\theta_1})(r_2 e^{i\theta_2}) = r_1 r_2 e^{i(\theta_1+\theta_2)}.$$

In general,

1. The amplitude for n *mutually alternative processes* equals the *sum* $\sum_{k=1}^{n} r_k e^{i\theta_k}$ of the amplitudes for the alternatives; and
2. If transition from A to B occurs in a *sequence of* m *steps*, then the total transition amplitude equals the *product* $\prod_{j=1}^{m} r_j e^{i\theta_j}$ of the amplitudes of the steps.

Formally, we have the so–called *expansion principle*, including both products and sums,[5]

$$< B|A >= \sum_{i=1}^{n} < B|c_i >< c_i|A > . \tag{6.1}$$

Feynman's Sum–over–Histories

Now, *iterating* the Dirac's expansion principle (6.1) over a *complete set of all possible states* of the system, leads to the simplest form of the *Feynman path integral*, or, *sum–over–histories*. Imagine that the *initial* and *final* states, A and B, are points on the vertical lines $x = 0$ and $x = n + 1$, respectively, in the $x - y$ plane, and that $(c(k)_{i(k)}, k)$ is a given point on the line $x = k$ for $0 < i(k) < m$ (see Figure 6.2). Suppose that the sum of projectors for each

[5] In Dirac's language, the *completeness* of intermediate states becomes the statement that a certain sum of projectors is equal to the identity. Namely, suppose that $\sum_i |c_i >< c_i| = 1$ with $< c_i|c_i >= 1$ for each i. Then

$$< b|a >=< b||a >=< b| \sum_i |c_i >< c_i||a >= \sum_i < b|c_i >< c_i|a > .$$

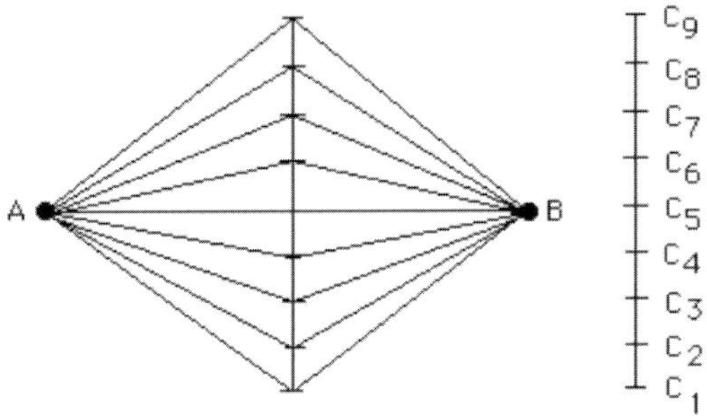

Fig. 6.2. Analysis of all possible routes from the source A to the detec or B is simplified to include only double straight lines (in a plane).

intermediate state is complete[6] Applying the completeness iteratively, we get the following expression for the transition amplitude:

$$< B|A >= \sum \sum ... \sum < B|c(1)_{i(1)} >< c(1)_{i(1)}|c(2)_{i(2)} > ... < c(n)_{i(n)}|A >,$$

where the sum is taken over all $i(k)$ ranging between 1 and m, and k ranging between 1 and n. Each term in this sum can be construed as a *combinatorial route* from A to B in the 2D space of the $x - y$ plane. Thus the transition amplitude for the system going from some initial state A to some final state B is seen as a summation of contributions from *all the routes* connecting A to B.

Feynman used this description to produce his celebrated *path integral* expression for a transition amplitude (see, e.g., [GS98, Sch81]). His path integral takes the form

$$Transition\,Amplitude =< B|A >= \int\!\!\!\!\!\!\sum \mathcal{D}[x]\, \mathrm{e}^{\mathrm{i}\mathcal{S}[x]}, \tag{6.2}$$

where the sum–integral $\int\!\!\!\!\!\!\sum$ is taken over all possible routes $x = x(t)$ from the initial point $A = A(t_{ini})$ to the final point $B = B(t_{fin})$, and $\mathcal{S} = \mathcal{S}[x]$ is the classical *action* for a particle to travel from A to B along a given extremal path x. In this way, Feynman took seriously Dirac's conjecture interpreting the exponential of the classical *action functional* ($\mathcal{D}\mathrm{e}^{\mathrm{i}\mathcal{S}}$), resembling a complex number ($r\mathrm{e}^{\mathrm{i}\theta}$), as an *elementary amplitude*. By integrating this elementary amplitude, $\mathcal{D}\mathrm{e}^{\mathrm{i}\mathcal{S}}$, over the infinitude of all possible histories, we get the total system's transition amplitude.[7]

[6] We assume that following sum is equal to one, for each k from 1 to $n - 1$:

$$|c(k)_1 >< c(k)_1| + ... + |c(k)_m >< c(k)_m| = 1.$$

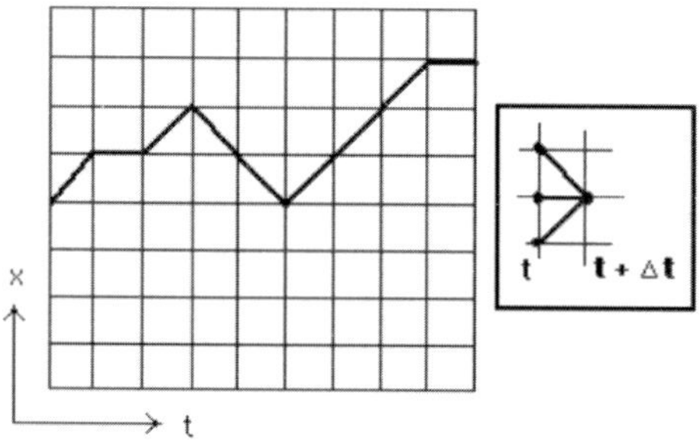

Fig. 6.3. Random walk (a particular case of Markov chain) on the $x-$axis.

The Basic Form of a Path Integral

In Feynman's version of non–relativistic quantum mechanics, the time evolution $\psi(x', t') \mapsto \psi(x'', t'')$ of the wave function $\psi = \psi(x, t)$ of the elementary 1D particle may be described by the integral equation [GS98]

$$\psi(x'', t'') = \int_{\mathbb{R}} K(x'', x'; t'', t')\, \psi(x', t'), \tag{6.3}$$

where the *propagator* or *Feynman kernel* $K = K(x'', x'; t'', t')$ is defined through a limiting procedure,

$$K(x'', x'; t'', t') = \lim_{\epsilon \to 0} A^{-N} \prod_{k=1}^{N-1} \int dx_k \; \mathrm{e}^{\mathrm{i} \sum_{j=0}^{N-1} \epsilon L(x_{j+1}, (x_{j+1} - x_j)/\epsilon)}. \tag{6.4}$$

---

[7] For the quantum physics associated with a classical (Newtonian) particle the action S is given by the integral along the given route from a to b of the difference $T - V$ where T is the classical kinetic energy and V is the classical potential energy of the particle.
The beauty of Feynman's approach to quantum physics is that it shows the relationship between the classical and the quantum in a particularly transparent manner. Classical motion corresponds to those regions where all nearby routes contribute constructively to the summation. This classical path occurs when the *variation of the action* is null. To ask for those paths where the variation of the action is zero is a problem in the calculus of variations, and it leads directly to Newton's equations of motion (derived using the Euler–Lagrangian equations). Thus with the appropriate choice of action, classical and quantum points of view are unified.
Also, a discretization of the Schrodinger equation

$$i\hbar \frac{d\psi}{dt} = -\frac{\hbar^2}{2m} \frac{d^2\psi}{dx^2} + V\psi,$$

leads to a sum–over–histories that has a discrete path integral as its solution. Therefore, the transition amplitude is equivalent to the wave ψ. The particle travelling on the $x-$axis is executing a one–step *random walk*, see Figure 6.3.

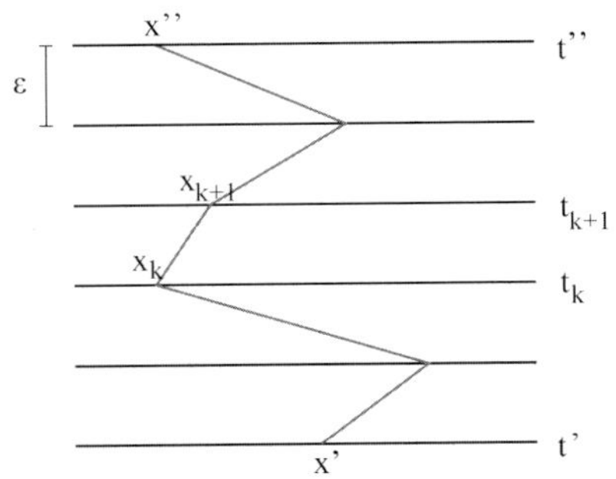

Fig. 6.4. A piecewise linear particle path contributing to the discrete Feynman propagator.

The time interval $t'' - t'$ has been discretized into N steps of length $\epsilon = (t'' - t')/N$, and the r.h.s. of (6.4) represents an integral over all piecewise linear paths $x(t)$ of a 'virtual' particle propagating from x' to x'', illustrated in Figure 6.4.

The prefactor A^{-N} is a normalization and L denotes the Lagrangian function of the particle. Knowing the propagator G is tantamount to having solved the quantum dynamics. This is the simplest instance of a path integral, and is often written schematically as

$$K(x', t'; x'', t'') = \sum\!\!\!\!\!\!\int \mathcal{D}[x(t)]\ \mathrm{e}^{\mathrm{i}S[x(t)]},$$

where $\mathcal{D}[x(t)]$ is a functional measure on the 'space of all paths', and the exponential weight depends on the classical action $S[x(t)]$ of a path. Recall also that this procedure can be defined in a mathematically clean way if we Wick–rotate the time variable t to imaginary values $t \mapsto \tau = it$, thereby making all integrals real [RS75].

Adaptive Path Integral

Now, we can extend the Feynman sum–over–histories (6.2), by adding the synaptic–like weights $w^i = w^i(t)$ into the measure $\mathcal{D}[x]$, to get the *adaptive path integral*:

$$Adaptive\,Transition\,Amplitude = < B|A >_w = \sum\!\!\!\!\!\!\int \mathcal{D}[w, x]\,\mathrm{e}^{\mathrm{i}S[x]}, \qquad (6.5)$$

where the *adaptive measure* $\mathcal{D}[w, x]$ is defined by the weighted product (of discrete time steps)

$$\mathcal{D}[w, x] = \lim_{n \longrightarrow \infty} \prod_{t=1}^{n} w^i(t)\, dx^i(t). \qquad (6.6)$$

In (6.6) the *synaptic weights* $w^i = w^i(t)$ are updated by the unsupervised *Hebbian–like learning* rule [Heb49]:

$$w^i(t+1) = w^i(t) + \frac{\sigma}{\eta}(w^i_d(t) - w^i_a(t)), \tag{6.7}$$

where $\sigma = \sigma(t)$, $\eta = \eta(t)$ represent local *signal* and *noise* amplitudes, respectively, while superscripts d and a denote *desired* and *achieved* system states, respectively. Theoretically, equations (6.5–6.7) define an ∞–*dimensional complex–valued neural network.*[8] Practically, in a computer simulation we can use $10^7 \leq n \leq 10^8$, approaching the number of neurons in the brain. Such equations are usually solved using Markov–Chain Monte–Carlo methods on parallel (cluster) computers (see, e.g., [WW83a, WW83b]).

6.1.2 Path Integral History

Extract from Feynman's Nobel Lecture

In his Nobel Lecture, December 11, 1965, Richard (Dick) Feynman said that he and his PhD supervisor, John Wheeler, had found the *action* $A = A[x; t_i, t_j]$, directly involving the *motions of the charges only,*[9]

$$\begin{aligned} A[x;t_i,t_j] &= m_i \int (\dot{x}^i_\mu \dot{x}^i_\mu)^{\frac{1}{2}}\, dt_i + \frac{1}{2} e_i e_j \int\!\!\int \delta(I^2_{ij})\, \dot{x}^i_\mu(t_i) \dot{x}^j_\mu(t_j)\, dt_i dt_j \\ &\quad \text{with } (i \neq j) \\ I^2_{ij} &= \left[x^i_\mu(t_i) - x^j_\mu(t_j)\right]\left[x^i_\mu(t_i) - x^j_\mu(t_j)\right], \end{aligned} \tag{6.8}$$

where $x^i_\mu = x^i_\mu(t_i)$ is the four–vector *position* of the ith particle as a function of the proper time t_i, while $\dot{x}^i_\mu(t_i) = dx^i_\mu(t_i)/dt_i$ is the *velocity* four–vector.

The first term in the action $A[x; t_i, t_j]$ (6.8) is the integral of the proper time t_i, the *ordinary action of relativistic mechanics of free particles of mass* m_i (summation over μ). The second term in the action $A[x; t_i, t_j]$ (6.8) represents the *electrical interaction of the charges.* It is summed over each pair of charges (the factor $\frac{1}{2}$ is to count each pair once, the term $i = j$ is omitted to avoid self–action). The *interaction is a double integral over a delta function of the square of space–time interval* I^2 *between two points on the paths.* Thus, interaction occurs only when this interval vanishes, that is, *along light cones* (see [WF49]).

Feynman comments here: "The fact that the interaction is exactly one–half advanced and half–retarded meant that we could write such a principle of

---

8 For details on complex–valued neural networks, see e.g., complex–domain extension of the standard backpropagation learning algorithm [GK92, BP02].

9 *Wheeler–Feynman Idea* [WF49] "The energy tensor can be regarded only as a provisional means of representing matter. In reality, *matter consists of electrically charged particles.*"

least action, whereas interaction via retarded waves alone cannot be written in such a way. So, all of classical electrodynamics was contained in this very simple form."

"...The problem is only to make a quantum theory, which has as its classical analog, this expression (6.8). Now, there is no unique way to make a quantum theory from classical mechanics, although all the textbooks make believe there is. What they would tell you to do, was find the momentum variables and replace them by $(\hbar/i)(\partial/\partial x)$, but I couldn't find a momentum variable, as there wasn't any."

"The character of quantum mechanics of the day was to write things in the famous *Hamiltonian way* (in the form of Schrödinger equation), which described how the wave function changes from instant to instant, and in terms of the Hamiltonian operator H. If the classical physics could be reduced to a Hamiltonian form, everything was all right. Now, least action does not imply a Hamiltonian form if the action is a function of anything more than positions and velocities at the same moment. If the action is of the form of the integral of the Lagrangian $L = L(\dot{x}, x)$, a function of the velocities and positions at the same time t,

$$S[x] = \int L(\dot{x}, x)\, dt, \tag{6.9}$$

then you can start with the Lagrangian L and then create a Hamiltonian H and work out the quantum mechanics, more or less uniquely. But the action $A[x; t_i, t_j]$ (6.8) involves the key variables, positions (and velocities), at two different times t_i and t_j and therefore, it was not obvious what to do to make the quantum–mechanical analogue..."

So, Feynman was looking for the action integral in quantum mechanics. He says: "...I simply turned to Professor Jehle and said, 'Listen, do you know any way of doing quantum mechanics, starting with action – where the action integral comes into the quantum mechanics?" 'No", he said, 'but Dirac has a paper in which the Lagrangian, at least, comes into quantum mechanics." What Dirac said was the following: There is in quantum mechanics a very important quantity which carries the wave function from one time to another, besides the differential equation but equivalent to it, a kind of a kernel, which we might call $K(x', x)$, which carries the wave function $\psi(x)$ known at time t, to the wave function $\psi(x')$ at time $t + \varepsilon$,

$$\psi(x', t + \varepsilon) = \int K(x', x)\, \psi(x, t)\, dx.$$

Dirac points out that this function K was analogous to the quantity in classical mechanics that you would calculate if you took the exponential of [$i\varepsilon$ multiplied by the Lagrangian $L(\dot{x}, x)$], imagining that these two positions x, x' corresponded to t and $t + \varepsilon$. In other words,

$$K(x', x) \quad \text{is analogous to} \quad \mathrm{e}^{\mathrm{i}\varepsilon L(\frac{x'-x}{\varepsilon}, x)/\hbar}.$$

So, Feynman continues: "What does he mean, they are analogous; what does that mean, *analogous*? What is the use of that?" Professor Jehle said, 'You Americans! You always want to find a use for everything!" I said that I thought that Dirac must mean that they were *equal.* 'No", he explained, 'he doesn't mean they are equal." 'Well", I said, 'Let's see what happens if we make them equal."

"So, I simply put them equal, taking the simplest example where the Lagrangian is

$$L = \frac{1}{2}M\dot{x}^2 - V(x),$$

but soon found I had to put a constant of proportionality N in, suitably adjusted. When I substituted for K to get

$$\psi(x', t+\varepsilon) = \int N \exp\left[\frac{\mathrm{i}\varepsilon}{\hbar}L(\frac{x'-x}{\varepsilon}, x)\right]\psi(x,t)\,dx \qquad (6.10)$$

and just calculated things out by Taylor series expansion, *out came the Schrödinger equation.* So, I turned to Professor Jehle, not really understanding, and said, 'Well, you see, Dirac meant that they were proportional." Professor Jehle's eyes were bugging out – he had taken out a little notebook and was rapidly copying it down from the blackboard, and said, 'No, no, this is an important discovery. You Americans are always trying to find out how something can be used. That's a good way to discover things!" So, I thought I was finding out what Dirac meant, but, as a matter of fact, had made the discovery that what Dirac thought was analogous, was, in fact, equal. I had then, at least, the connection between the Lagrangian and quantum mechanics, but still with wave functions and infinitesimal times."

"It must have been a day or so later when I was lying in bed thinking about these things, that I imagined what would happen if I wanted to calculate the wave function at a finite interval later. I would put one of these factors $\mathrm{e}^{\mathrm{i}\varepsilon L}$ in here, and that would give me the wave functions the next moment, $t+\varepsilon$, and then I could substitute that back into (6.10) to get another factor of $\mathrm{e}^{\mathrm{i}\varepsilon L}$ and give me the wave function the next moment, $t+2\varepsilon$, and so on and so on. In that way I found myself thinking of a large number of integrals, one after the other in sequence. In the integrand was the product of the exponentials, which was the exponential of the sum of terms like εL. Now, L is the Lagrangian and ε is like the time interval dt, so that if you took a sum of such terms, that's exactly like an integral. That's like Riemann's formula for the integral $\int L dt$, you just take the value at each point and add them together. We are to take the limit as $\varepsilon \to 0$. Therefore, the connection between the wave function of one instant and the wave function of another instant a finite time later could be get by an infinite number of integrals (because ε goes to zero), of exponential where S is the action expression (6.9). At last, I had succeeded in representing quantum mechanics directly in terms of the action $S[x]$."

Fully satisfied, Feynman comments: "This led later on to the idea of the *transition amplitude* for a path: that for each possible way that the particle can go from one point to another in space–time, there's an amplitude. That amplitude is **e** to the power of [i/$\hbar$ times the action $S[x]$ for the path], i.e., $e^{iS[x]/\hbar}$. Amplitudes from various paths superpose by addition. This then is another, a third way, of describing quantum mechanics, which looks quite different from that of Schrödinger or Heisenberg, but which is equivalent to them."

"...Now immediately after making a few checks on this thing, what we wanted to do, was to substitute the action $A[x; t_i, t_j]$ (6.8) for the other $S[x]$ (6.9). The first trouble was that I could not get the thing to work with the relativistic case of spin one–half. However, although I could deal with the matter only nonrelativistically, I could deal with the light or the photon interactions perfectly well by just putting the interaction terms of (6.8) into any action, replacing the mass terms by the non–relativistic $Ldt = \frac{1}{2}M\dot{x}^2 dt$,

$$A[x; t_i, t_j] = \frac{1}{2}\sum_i m_i \int (\dot{x}^i_\mu)^2 dt_i + \frac{1}{2}\sum_{i,j(i\neq j)} e_i e_j \int\int \delta(I^2_{ij})\, \dot{x}^i_\mu(t_i)\dot{x}^j_\mu(t_j)\, dt_i dt_j.$$

When the action has a delay, as it now had, and involved more than one time, I had to lose the idea of a wave function. That is, I could no longer describe the program as: given the amplitude for all positions at a certain time to calculate the amplitude at another time. However, that didn't cause very much trouble. It just meant developing a new idea. *Instead of wave functions we could talk about this: that if a source of a certain kind emits a particle, and a detector is there to receive it, we can give the amplitude that the source will emit and the detector receive,* $e^{iA[x;t_i,t_j]/\hbar}$. We do this without specifying the exact instant that the *source* emits or the exact instant that any *detector* receives, without trying to specify the state of anything at any particular time in between, but by just finding the *amplitude for the complete experiment.* And, then we could discuss how that amplitude would change if you had a scattering sample in between, as you rotated and changed angles, and so on, without really having any wave functions...It was also possible to discover what the old concepts of energy and momentum would mean with this generalized action. And, so I believed that I had a quantum theory of classical electrodynamics – or rather of this new classical electrodynamics described by the action $A[x; t_i, t_j]$ (6.8)..."

Lagrangian Path Integral

Dirac and Feynman first developed the lagrangian approach to functional integration. To review this approach, we start with the time–dependent *Schrödinger equation*

$$i\hbar\, \partial_t \psi(x,t) = -\partial_{x^2}\psi(x,t) + V(x)\,\psi(x,t)$$

appropriate to a particle of mass m moving in a potential $V(x)$, $x \in \mathbb{R}$. A solution to this equation can be written as an integral (see e.g., [Kla97, Kla00]),

$$\psi(x'',t'') = \int K(x'',t'';x',t')\,\psi(x',t')\,dx' ,$$

which represents the wave function $\psi(x'',t'')$ at time t'' as a linear superposition over the wave function $\psi(x',t')$ at the initial time t', $t' < t''$. The integral kernel $K(x'',t'';x',t')$ is known as the *propagator*, and according to Feynman [Fey48] it may be given by

$$K(x'',t'';x',t') = \mathcal{N} \int \mathcal{D}[x]\; \mathrm{e}^{(\mathrm{i}/\hbar)\int[(m/2)\,\dot{x}^2(t)-V(x(t))]\,dt},$$

which is a formal expression symbolizing an integral over a suitable set of paths. This integral is supposed to run over all continuous paths $x(t)$, $t' \leq t \leq t''$, where $x(t'') = x''$ and $x(t') = x'$ are fixed end points for all paths. Note that the integrand involves the *classical Lagrangian* for the system.

To overcome the convergence problems, Feynman adopted a *lattice regularization* as a procedure to yield well–defined integrals which was then followed by a limit as the lattice spacing goes to zero called the continuum limit. With $\varepsilon > 0$ denoting the lattice spacing, the details regarding the lattice regularization procedure are given by

$$K(x'',t'';x',t') = \lim_{\varepsilon\to 0}\,(m/2\pi\mathrm{i}\hbar\varepsilon)^{(N+1)/2} \int\cdots$$

$$\cdots\int \exp\{(\mathrm{i}/\hbar)\sum_{l=0}^{N}[(m/2\varepsilon)(x_{l+1}-x_l)^2 - \varepsilon\,V(x_l)\,]\}\prod_{l=1}^{N} dx_l\,,$$

where $x_{N+1} = x''$, $x_0 = x'$, and $\varepsilon \equiv (t''-t')/(N+1)$, $N \in \{1,2,3,\dots\}$. In this version, at least, we have an expression that has a reasonable chance of being well defined, provided, that one interprets the conditionally convergent integrals involved in an appropriate manner. One common and fully acceptable interpretation adds a convergence factor to the exponent of the preceding integral in the form $-(\varepsilon^2/2\hbar)\sum_{l=1}^{N} x_l^2$, which is a term that formally makes no contribution to the final result in the continuum limit save for ensuring that the integrals involved are now rendered absolutely convergent.

Hamiltonian Path Integral

It is necessary to retrace history at this point to recall the introduction of the *phase–space path integral* by Feynman [Fey51, GS98]. In Appendix B to this article, Feynman introduced a formal expression for the configuration or q–space propagator given by (see e.g., [Kla97, Kla00])

$$K(q'',t'';q',t') = \mathcal{M}\int \mathcal{D}[p]\,\mathcal{D}[q]\;\exp\{(\mathrm{i}/\hbar)\int[\,p\,\dot{q} - H(p,q)\,]\,dt\}.$$

In this equation one is instructed to integrate over all paths $q(t)$, $t' \leq t \leq t''$, with $q(t'') \equiv q''$ and $q(t') \equiv q'$ held fixed, as well as to integrate over all paths $p(t)$, $t' \leq t \leq t''$, without restriction.

It is widely appreciated that the phase–space path integral is more generally applicable than the original, Lagrangian, version of the path integral. For example, the original configuration space path integral is satisfactory for Lagrangians of the general form

$$L(x) = \frac{1}{2}\, m\dot{x}^2 + A(x)\,\dot{x} - V(x)\;,$$

but it is unsuitable, for example, for the case of a relativistic particle with the Lagrangian

$$L(x) = -m\, qrt1 - \dot{x}^2$$

expressed in units where the speed of light is unity. For such a system – as well as many more general expressions – the phase–space form of the path integral is to be preferred. In particular, for the relativistic free particle, the phase–space path integral

$$\mathcal{M}\int \mathcal{D}[p]\,\mathcal{D}[q]\;\exp\{(\mathrm{i}/\hbar)\textstyle\int[\,p\,\dot{q} - \;qrtp^2 + m^2\,]\,dt\},$$

is readily evaluated and induces the correct propagator.

Feynman–Kac Formula

Through his own research, M. Kac was fully aware of *Wiener's theory of Brownian motion* and the *associated diffusion equation* that describes the corresponding *distribution function.* Therefore, it is not surprising that he was well prepared to give a path integral expression in the sense of Feynman for an equation similar to the time–dependent Schrödinger equation save for a rotation of the time variable by $-\pi/2$ in the complex–plane, namely, by the change $t \longrightarrow -it$ (see e.g., [Kla97, Kla00]). In particular, Kac [Kac51] considered the equation

$$\partial_t\rho(x,t) = \partial_{x^2}\rho(x,t) - V(x)\;\rho(x,t). \tag{6.11}$$

This equation is analogous to Schrödinger equation but differs from it in certain details. Besides certain constants which are different, and the change $t \longrightarrow -it$, the nature of the dependent variable function $\rho(x,t)$ is quite different from the normal quantum mechanical wave function. For one thing, if the function ρ is initially real it will remain real as time proceeds. Less obvious is the fact that if $\rho(x,t) \geq 0$ for all x at some time t, then the function will continue to be nonnegative for all time t. Thus we can interpret $\rho(x,t)$ more like a probability density; in fact in the special case that $V(x) = 0$, then $\rho(x,t)$ is the probability density for a Brownian particle which underlies the *Wiener measure.* In this regard, ν is called the diffusion constant.

The fundamental solution of (6.11) with $V(x) = 0$ is readily given as

$$W(x,T;y,0) = \frac{1}{qrt2\pi\nu T}\;\exp\left(-\;\frac{(x-y)^2}{2\nu T}\right),$$

which describes the solution to the diffusion equation subject to the initial condition

$$\lim_{T\to 0+} W(x,T;y,0) = \delta(x-y)\ .$$

Moreover, it follows that the solution of the diffusion equation for a general initial condition is given by

$$\rho(x'',t'') = \int W(x'',t'';x',t')\,\rho(x',t')\,dx'\ .$$

Iteration of this equation N times, with $\epsilon = (t''-t')/(N+1)$, leads to the equation

$$\rho(x'',t'') = N' \int \cdots \int \mathrm{e}^{-(1/2\nu\epsilon)\sum_{l=0}^{N}(x_{l+1}-x_l)^2} \prod_{l=1}^{N} dx_l\ \rho(x',t')\,dx',$$

where $x_{N+1} \equiv x''$ and $x_0 \equiv x'$. This equation features the imaginary time propagator for a free particle of unit mass as given formally as

$$W(x'',t'';x',t') = \mathcal{N}\int \mathcal{D}[x]\ \mathrm{e}^{-(1/2\nu)\int \dot{x}^2\,dt},$$

where $\mathcal{N}$ denotes a formal normalization factor.

The similarity of this expression with the Feynman path integral [for $V(x)=0$] is clear, but there is a profound difference between these equations. In the former (Feynman) case the underlying measure is only *finitely additive*, while in the latter (Wiener) case the continuum limit actually defines a genuine measure, i.e., a *countably additive measure* on paths, which is a version of the famous *Wiener measure*. In particular,

$$W(x'',t'';x',t') = \int d\mu_W^\nu(x),$$

where μ_W^ν denotes a measure on continuous paths $x(t)$, $t' \le t \le t''$, for which $x(t'') \equiv x''$ and $x(t') \equiv x'$. Such a measure is said to be a *pinned* Wiener measure, since it specifies its path values at two time points, i.e., at $t=t'$ and at $t=t''>t'$.

We note that Brownian motion paths have the property that with probability one they are concentrated on continuous paths. However, it is also true that the time derivative of a Brownian path is almost nowhere defined, which means that, with probability one, $\dot{x}(t)=\pm\infty$ for all t.

When the potential $V(x)\neq 0$ the propagator associated with (6.11) is formally given by

$$W(x'',t'';x',t') = \mathcal{N}\int \mathcal{D}[x]\mathrm{e}^{-(1/2\nu)\int \dot{x}^2\,dt - \int V(x)\,dt},$$

an expression which is well defined if $V(x)\ge c$, $-\infty<c<\infty$. A mathematically improved expression makes use of the Wiener measure and reads

$$W(x'',t'';x',t') = \int \mathrm{e}^{-\int V(x(t))\,dt}\, d\mu_W^\nu(x).$$

This is an elegant relation in that it represents a solution to the differential equation (6.11) in the form of an integral over Brownian motion paths suitably weighted by the potential V. Incidentally, since the propagator is evidently a strictly positive function, it follows that the solution of the differential equation (6.11) is nonnegative for all time t provided it is nonnegative for any particular time value.

Itô Formula

Itô [Ito60] proposed another version of a *continuous–time regularization* that resolved some of the troublesome issues. In essence, the proposal of Itô takes the form given by

$$\lim_{\nu\to\infty} \mathcal{N}_\nu \int \mathcal{D}[x]\ \exp\{(\mathrm{i}/\hbar) \textstyle\int [\frac{1}{2}m\dot{x}^2 - V(x)]\,dt\} \exp\{-(1/2\nu) \textstyle\int [\ddot{x}^2 + \dot{x}^2]\,dt\}.$$

Note well the alternative form of the auxiliary factor introduced as a regulator. The additional term $\ddot{x}^2$, the square of the second derivative of x, acts to smooth out the paths sufficiently well so that in the case of (21) both $x(t)$ and $\dot{x}(t)$ are continuous functions, leaving $\ddot{x}(t)$ as the term which does not exist. However, since only x and $\dot{x}$ appear in the rest of the integrand, the indicated path integral can be well defined; this is already a positive contribution all by itself (see e.g., [Kla97, Kla00]).

6.1.3 Standard Path–Integral Quantization

Canonical versus Path–Integral Quantization

Recall that in the usual, *canonical formulation* of quantum mechanics, the system's phase–space coordinates, q, and momenta, p, are replaced by the corresponding Hermitian operators in the Hilbert space, with real measurable eigenvalues, which obey *Heisenberg commutation relations.*

The *path–integral quantization* is instead based directly on the notion of a propagator $K(q_f,t_f;q_i,t_i)$ which is defined such that (see [Ryd96, CL84, Gun03])

$$\psi(q_f,t_f) = \int K(q_f,t_f;q_i,t_i)\,\psi(q_i,t_i)\,dq_i, \tag{6.12}$$

i.e., the wave function $\psi(q_f,t_f)$ at final time t_f is given by a Huygens principle in terms of the wave function $\psi(q_i,t_i)$ at an initial time t_i, where we have to integrate over all the points q_i since all can, in principle, send out little wavelets that would influence the value of the wave function at q_f at the later time t_f. This equation is very general and is an expression of causality. We use the normal units with $\hbar = 1$.

According to the usual interpretation of quantum mechanics, $\psi(q_f, t_f)$ is the *probability amplitude* that the particle is at the point q_f and the time t_f, which means that $K(q_f, t_f; q_i, t_i)$ is the probability amplitude for a transition from q_i and t_i to q_f and t_f. The probability that the particle is observed at q_f at time t_f if it began at q_i at time t_i is

$$P(q_f, t_f; q_i, t_i) = |K(q_f, t_f; q_i, t_i)|^2 .$$

Let us now divide the time interval between t_i and t_f into two, with t as the intermediate time, and q the intermediate point in space. Repeated application of (6.12) gives

$$\psi(q_f, t_f) = \int\int K(q_f, t_f; q, t)\, dq\, K(q, t; q_i, t_i)\, \psi(q_i, t_i)\, dq_i,$$

from which it follows that

$$K(q_f, t_f; q_i, t_i) = \int dq\, K(q_f, t_f; q, t)\, K(q, t; q_i, t_i).$$

This equation says that the transition from (q_i, t_i) to (q_f, t_f) may be regarded as the result of the transition from (q_i, t_i) to all available intermediate points q followed by a transition from (q, t) to (q_f, t_f). This notion of *all possible paths* is crucial in the *path–integral formulation* of quantum mechanics.

Now, recall that the *state vector* $|\psi, t\rangle_S$ in the *Schrödinger picture* is related to that in the *Heisenberg picture* $|\psi\rangle_H$ by

$$|\psi, t\rangle_S = \mathrm{e}^{-\mathrm{i}Ht}\, |\psi\rangle_H ,$$

or, equivalently,

$$|\psi\rangle_H = \mathrm{e}^{\mathrm{i}Ht}\, |\psi, t\rangle_S .$$

We also define the vector

$$|q, t\rangle_H = \mathrm{e}^{\mathrm{i}Ht}\, |q\rangle_S ,$$

which is the Heisenberg version of the Schrödinger state $|q\rangle$. Then, we can equally well write

$$\psi(q, t) = \langle q, t\, |\psi\rangle_H . \qquad (6.13)$$

By completeness of states we can now write

$$\langle q_f, t_f\, |\psi\rangle_H = \int \langle q_f, t_f\, |q_i, t_i\rangle_H\, \langle q_i, t_i\, |\psi\rangle_H\, dq_i,$$

which with the definition of (6.13) becomes

$$\psi(q_f, t_f) = \int \langle q_f, t_f\, |q_i, t_i\rangle_H\, \psi(q_i, t_i)\, dq_i.$$

Comparing with (6.12), we get

$$K(q_f, t_f; q_i, t_i) = \langle q_f, t_f \,| q_i, t_i\rangle_H \,.$$

Now, let us calculate the *quantum–mechanics propagator*

$$\langle q', t' \,| q, t\rangle_H = \left\langle q' | \mathrm{e}^{-\mathrm{i}H(t-t')} \,| q \right\rangle$$

using the *path–integral formalism* that will incorporate the direct quantization of the coordinates, without Hilbert space and Hermitian operators.

The first step is to divide up the time interval into $n+1$ tiny pieces: $t_l = l\varepsilon + t$ with $t' = (n+1)\varepsilon + t$. Then, by completeness, we can write (dropping the Heisenberg picture index H from now on)

$$\begin{aligned}\langle q', t' \,| q, t\rangle &= \int dq_1(t_1) ... \int dq_n(t_n) \,\langle q', t' \,| q_n, t_n\rangle \\ &\times \langle q_n, t_n \,| q_{n-1}, t_{n-1}\rangle \dots \langle q_1, t_1 \,| q, t\rangle \,. \end{aligned} \tag{6.14}$$

The integral $\int dq_1(t_1)...dq_n(t_n)$ is an *integral over all possible paths*, which are not trajectories in the normal sense, since there is no requirement of continuity, but rather *Markov chains*.

Now, for small ε we can write

$$\langle q', \varepsilon \,| q, 0\rangle = \left\langle q' | \mathrm{e}^{-\mathrm{i}\varepsilon H(P,Q)} \,| q \right\rangle = \delta(q'-q) - \mathrm{i}\varepsilon \,\langle q' | H(P,Q) \,| q\rangle \,,$$

where $H(P,Q)$ is the Hamiltonian (e.g., $H(P,Q) = \frac{1}{2}P^2 + V(Q)$, where P, Q are the momentum and coordinate operators). Then we have (see [Ryd96, CL84, Gun03])

$$\langle q' | H(P,Q) \,| q\rangle = \int \frac{dp}{2\pi} \,\mathrm{e}^{\mathrm{i}p(q'-q)} H\left(p, \frac{1}{2}(q'+q)\right).$$

Putting this into our earlier form we get

$$\langle q', \varepsilon \,| q, 0\rangle \simeq \int \frac{dp}{2\pi} \exp\left[\mathrm{i}\left\{p(q'-q) - \varepsilon H\left(p, \frac{1}{2}(q'+q)\right)\right\}\right],$$

where the 0th order in $\varepsilon \to \delta(q'-q)$ and the 1st order in $\varepsilon \to -\mathrm{i}\varepsilon\,\langle q' | H(P,Q) | q\rangle$. If we now substitute many such forms into (6.14) we finally get

$$\langle q', t' \,| q, t\rangle = \lim_{n\to\infty} \int \prod_{i=1}^{n} dq_i \prod_{k=1}^{n+1} \frac{dp_k}{2\pi} \times \tag{6.15}$$

$$\times \exp\left\{\mathrm{i}\sum_{j=1}^{n+1} [p_j(q_j - q_{j-1})] - H\left(p_j, \frac{1}{2}(q_j + q_{j+1})\right)(t_j - t_{j-1})]\right\},$$

with $q_0 = q$ and $q_{n+1} = q'$. Roughly, the above formula says to *integrate over all possible momenta and coordinate values associated with a small interval*, weighted by something that is going to turn into the *exponential of the action* $\mathrm{e}^{\mathrm{i}S}$ in the limit where $\varepsilon \to 0$. It should be stressed that the different q_i and p_k integrals are independent, which implies that p_k for one interval can be completely different from the $p_{k'}$ for some other interval (including the neighboring intervals). In principle, the integral (6.15) should be defined by *analytic continuation into the complex–plane* of, for example, the p_k integrals.

Now, if we go to the differential limit where we call $t_j - t_{j-1} \equiv d\tau$ and write $\frac{(q_j - q_{j-1})}{(t_j - t_{j-1})} \equiv \dot{q}$, then the above formula takes the form

$$\langle q', t' \, | q, t \rangle = \int \mathcal{D}[p]\mathcal{D}[q] \exp\left\{ \mathrm{i} \int_t^{t'} [p\dot{q} - H(p,q)] \, d\tau \right\},$$

where we have used the shorthand notation

$$\int \mathcal{D}[p]\mathcal{D}[q] \equiv \int \prod_\tau \frac{dq(\tau)dp(\tau)}{2\pi}.$$

Note that the above integration is an integration over the p and q values at every time τ. This is what we call a *functional integral.* We can think of a given set of choices for all the $p(\tau)$ and $q(\tau)$ as defining a *path in the* 6*D phase–space.* The most important point of the above result is that we have get an expression for a *quantum–mechanical transition amplitude* in terms of an integral involving only pure complex numbers, without operators.

We can actually perform the above integral for Hamiltonians of the type $H = H(P, Q)$. We use square completion in the exponential for this, defining the integral in the complex p plane and continuing to the physical situation. In particular, we have

$$\int_{-\infty}^{\infty} \frac{dp}{2\pi} \exp\left\{ \mathrm{i}\varepsilon (p\dot{q} - \frac{1}{2}p^2] \right\} = \frac{1}{\sqrt{2\pi \mathrm{i}\varepsilon}} \exp\left[\frac{1}{2} \mathrm{i}\varepsilon \dot{q}^2 \right],$$

(see [Ryd96, CL84, Gun03]) which, substituting into (6.15) gives

$$\langle q', t' \, | q, t \rangle = \lim_{n\to\infty} \int \prod_i \frac{dq_i}{\sqrt{2\pi \mathrm{i}\varepsilon}} \exp\{\mathrm{i}\varepsilon \sum_{j=1}^{n+1} [\frac{1}{2}(\frac{q_j - q_{j-1}}{\varepsilon})^2 - V(\frac{q_j + q_{j+1}}{2})]\}.$$

This can be formally written as

$$\langle q', t' \, | q, t \rangle = \int \mathcal{D}[q] \, \mathrm{e}^{\mathrm{i}S[q]},$$

where

$$\int \mathcal{D}[q] \equiv \int \prod_i \frac{dq_i}{\sqrt{2\pi \mathrm{i}\varepsilon}},$$

while

$$S[q] = \int_t^{t'} L(q, \dot{q})\, d\tau$$

is the *standard action* with the *Lagrangian*

$$L = \frac{1}{2}\dot{q}^2 - V(q).$$

Generalization to many degrees of freedom is straightforward:

$$\langle q_1{}'...q_N{}', t'|q_1...q_N, t\rangle = \int \mathcal{D}[p]\mathcal{D}[q] \exp\left\{ \mathrm{i} \int_t^{t'} \left[\sum_{n=1}^{N} p_n \dot{q}_n - H(p_n, q_n) \right] d\tau \right\},$$

$$\text{with} \quad \int \mathcal{D}[p]\mathcal{D}[q] = \int \prod_{n=1}^{N} \frac{dq_n dp_n}{2\pi}.$$

Here, $q_n(t) = q_n$ and $q_n(t') = q_n{}'$ for all $n = 1, ..., N$, and we are allowing for the full Hamiltonian of the system to depend upon all the N momenta and coordinates collectively.

Elementary Applications

(i) Consider first

$$\begin{aligned} &\langle q', t'|Q(t_0)|q, t\rangle \\ &= \int \prod dq_i(t_i) \langle q', t'|q_n, t_n\rangle ... \langle q_{i0}, t_{i0}|Q(t_0)|q_{i-1}, t_{i-1}\rangle ... \langle q_1, t_1|q, t\rangle, \end{aligned}$$

where we choose one of the time interval ends to coincide with t_0, i.e., $t_{i0} = t_0$. If we operate $Q(t_0)$ to the left, then it is replaced by its eigenvalue $q_{i0} = q(t_0)$. Aside from this one addition, everything else is evaluated just as before and we will obviously get

$$\langle q', t'|Q(t_0)|q, t\rangle = \int \mathcal{D}[p]\mathcal{D}[q]\, q(t_0) \exp\left\{ \mathrm{i} \int_t^{t'} [p\dot{q} - H(p, q)] d\tau \right\}.$$

(ii) Next, suppose we want a *path–integral expression* for $\langle q', t'|Q(t_1)Q(t_2)|q, t\rangle$ in the case where $t_1 > t_2$. For this, we have to insert as intermediate states $|q_{i1}, t_{i1}\rangle \langle q_{i1}, t_{i1}|$ with $t_{i1} = t_1$ and $|q_{i2}, t_{i2}\rangle \langle q_{i2}, t_{i2}|$ with $t_{i2} = t_2$ and since we have ordered the times at which we do the insertions we must have the first insertion to the left of the 2nd insertion when $t_1 > t_2$. Once these insertions are done, we evaluate $\langle q_{i1}, t_{i1}|\, Q(t_1) = \langle q_{i1}, t_{i1}|\, q(t_1)$ and $\langle q_{i2}, t_{i2}|\, Q(t_2) = \langle q_{i2}, t_{i2}|\, q(t_2)$ and then proceed as before and get

$$\langle q', t'|Q(t_1)Q(t_2)|q, t\rangle = \int \mathcal{D}[p]\mathcal{D}[q]\, q(t_1)\, q(t_2) \exp\left\{ \mathrm{i} \int_t^{t'} [p\dot{q} - H(p, q)] d\tau \right\}.$$

Now, let us ask what the above integral is equal to if $t_2 > t_1$? It is obvious that what we get for the above integral is $\langle q', t'|Q(t_2)Q(t_1)|q, t\rangle$. Clearly, this generalizes to an arbitrary number of Q operators.

(iii) When we enter into quantum field theory, the Q's will be replaced by fields, since it is the fields that play the role of coordinates in the 2nd quantization conditions.

Sources

The *source* is represented by modifying the Lagrangian:

$$L \to L + J(t)q(t).$$

Let us define $|0, t\rangle^J$ as the ground state (vacuum) vector (in the moving frame, i.e., with the e^{iHt} included) in the presence of the source. The required *transition amplitude* is

$$Z[J] \propto \langle 0, +\infty|0, -\infty\rangle^J ,$$

where the source $J = J(t)$ plays a role analogous to that of an electromagnetic current, which acts as a source of the electromagnetic field. In other words, we can think of the scalar product $J_\mu A^\mu$, where J_μ is the current from a scalar (or Dirac) field acting as a source of the potential A^μ. In the same way, we can always define a current J that acts as the source for some arbitrary field ϕ. $Z[J]$ (otherwise denoted by $W[J]$) is a functional of the current J, defined as (see [Ryd96, CL84, Gun03])

$$Z[J] \propto \int \mathcal{D}[p]\mathcal{D}[q] \exp\left\{ \mathrm{i} \int_t^{t'} [p(\tau)\dot{q}(\tau) - H(p, q) + J(\tau)q(\tau)]d\tau \right\},$$

with the *normalization condition* $Z[J = 0] = 1$. Here, the argument of the exponential depends upon the functions $q(\tau)$ and $p(\tau)$ and we then integrate over all possible forms of these two functions. So the exponential is a functional that maps a choice for these two functions into a number. For example, for a quadratically completable $H(p, q)$, the p integral can be performed as a q integral

$$Z[J] \propto \int \mathcal{D}[q] \exp\left\{ \mathrm{i} \int_{-\infty}^{+\infty} \left(L + Jq + \frac{1}{2}\mathrm{i}\varepsilon q^2 \right) d\tau \right\},$$

where the addittion to H was chosen in the form of a *convergence factor* $-\frac{1}{2}\mathrm{i}\varepsilon q^2$.

Fields

Let us now treat the *abstract scalar field* $\phi(x)$ as a coordinate in the sense that we imagine dividing space up into many little cubes and the average value of the field $\phi(x)$ in that cube is treated as a coordinate for that little cube. Then, we go through the multi–coordinate analogue of the procedure we just considered above and take the continuum limit. The final result is

$$Z[J] \propto \int \mathcal{D}[\phi] \exp\left\{ \mathrm{i} \int d^4x \left(\mathcal{L}\left(\phi(x)\right) + J(x)\phi(x) + \frac{1}{2}\mathrm{i}\varepsilon\phi^2 \right) \right\},$$

where for $\mathcal{L}$ we would employ the *Klein–Gordon Lagrangian* form. In the above, the dx_0 integral is the same as $d\tau$, while the $d^3\mathbf{x}$ integral is summing over the sub–Lagrangians of all the different little cubes of space and then taking the continuum limit. $\mathcal{L}$ is the *Lagrangian density* describing the Lagrangian for each little cube after taking the many–cube limit (see [Ryd96, CL84, Gun03]) for the full derivation).

We can now introduce *interactions,* $\mathcal{L}_I$. Assuming the simple form of the Hamiltonian, we have

$$Z[J] \propto \int \mathcal{D}[\phi] \exp\left\{ \mathrm{i} \int d^4x \left(\mathcal{L}\left(\phi(x)\right) + \mathcal{L}_I\left(\phi(x)\right) + J(x)\phi(x) \right) \right\},$$

again using the normalization factor required for $Z[J=0]=1$.

For example of Klein Gordon theory, we would use

$$\mathcal{L} = \mathcal{L}_0 + \mathcal{L}_I, \qquad \mathcal{L}_0 \frac{1}{2}[\partial_\mu\phi\partial^\mu\phi - \mu^2\phi^2], \qquad \mathcal{L}_I = \mathcal{L}_I(\phi),$$

where $\partial_\mu \equiv \partial_{x^\mu}$ and we can freely manipulate indices, as we are working in Euclidean space $\mathbb{R}^3$. In order to define the above $Z[J]$, we have to include a convergence factor $\mathrm{i}\varepsilon\phi^2$,

$$\mathcal{L}_0 \to \frac{1}{2}[\partial_\mu\phi\partial^\mu\phi - \mu^2\phi^2 + \mathrm{i}\varepsilon\phi^2], \qquad \text{so that}$$

$$Z[J] \propto \int \mathcal{D}[\phi] \exp\{\mathrm{i} \int d^4x (\frac{1}{2}[\partial_\mu\phi\partial^\mu\phi - \mu^2\phi^2 + \mathrm{i}\varepsilon\phi^2] + \mathcal{L}_I(\phi(x)) + J(x)\phi(x))\}$$

is the appropriate *generating function* in the free field theory case.

Gauges

In the path integral approach to quantization of the *gauge theory,* we implement *gauge fixing* by restricting in some manner or other the path integral over gauge fields $\int \mathcal{D}[A_\mu]$. In other words we will write instead

$$Z[J] \propto \int \mathcal{D}[A_\mu]\,\delta\,(\text{some gauge fixing condition}) \exp\{\mathrm{i} \int d^4x \mathcal{L}\left(A_\mu\right)\}.$$

A common approach would be to start with the *gauge condition*

$$\mathcal{L} = -\frac{1}{4}F_{\mu\nu}F^{\mu\nu} - \frac{1}{2}(\partial^\mu A_\mu)^2$$

where the electrodynamic field tensor is given by $F_{\mu\nu} = \partial_\mu A_\nu - \partial_\nu A_\mu$, and calculate

$$Z[J] \propto \int \mathcal{D}[A_\mu] \exp\left\{ \mathrm{i} \int d^4x \, [\mathcal{L}(A_\mu(x)) + J_\mu(x) A^\mu(x)] \right\}$$

as the *generating function* for the *vacuum expectation* values of *time ordered products* of the A_μ fields. Note that J_μ should be conserved ($\partial^\mu J_\mu = 0$) in order for the full expression $\mathcal{L}(A_\mu) + J_\mu A^\mu$ to be *gauge–invariant* under the integral sign when $A_\mu \to A_\mu + \partial^\mu \Lambda$. For a proper approach, see [Ryd96, CL84, Gun03].

Riemannian–Symplectic Geometries

In this subsection, following [SK98b], we describe path integral quantization on Riemannian–symplectic manifolds. Let $\hat{q}^j$ be a set of Cartesian coordinate canonical operators satisfying the Heisenberg commutation relations $[\hat{q}^j, \hat{q}^k] = \mathrm{i}\omega^{jk}$. Here $\omega^{jk} = -\omega^{kj}$ is the canonical symplectic structure. We introduce the canonical coherent states as $|q\rangle \equiv \mathrm{e}^{\mathrm{i}q^j \omega_{jk} \hat{q}^k} |0\rangle$, where $\omega_{jn}\omega^{nk} = \delta_j^k$, and $|0\rangle$ is the ground state of a harmonic oscillator with unit angular frequency. Any state $|\psi\rangle$ is given as a function on phase–space in this representation by $\langle q|\psi\rangle = \psi(q)$. A general operator $\hat{A}$ can be represented in the form $\hat{A} = \int dq \, a(q) |q\rangle\langle q|$, where $a(q)$ is the lower symbol of the operator and dq is a properly normalized form of the Liouville measure. The function $A(q, q') = \langle q|\hat{A}|q'\rangle$ is the *kernel of the operator*.

The main object of the path integral formalism is the integral kernel of the evolution operator

$$K_t(q, q') = \langle q|\mathrm{e}^{-\mathrm{i}t\hat{H}}|q'\rangle = \int_{q(0)=q'}^{q(t)=q} \mathcal{D}[q] \; \mathrm{e}^{\mathrm{i}\int_0^t d\tau \left(\frac{1}{2} q^j \omega_{jk} \dot{q}^k - h\right)} \,. \tag{6.16}$$

Here $\hat{H}$ is the Hamiltonian, and $h(q)$ its symbol. The measure formally implies a sum over all phase-space paths pinned at the initial and final points, and a Wiener measure regularization implies the following replacement

$$\mathcal{D}[q] \to \mathcal{D}[\mu_\nu(q)] = \mathcal{D}[q] \, \mathrm{e}^{-\frac{1}{2\nu}\int_0^t d\tau \, \dot{q}^2} = N_\nu(t) \, d\mu_W^\nu(q) \,. \tag{6.17}$$

The factor $N_\nu(t)$ equals $2\pi \mathrm{e}^{\nu t/2}$ for every degree of freedom, $d\mu_W^\nu(q)$ stands for the Wiener measure, and ν denotes the diffusion constant. We denote by $K_t^\nu(q, q')$ the integral kernel of the evolution operator for a finite ν. The Wiener measure determines a stochastic process on the *flat* phase–space. The integral of the symplectic 1–form $\int q\omega dq$ is a stochastic integral that is interpreted in the Stratonovich sense. Under general coordinate transformations $q = q(\bar{q})$, the Wiener measure describes *the same* stochastic process on *flat* space in the curvilinear coordinates $dq^2 = d\sigma(\bar{q})^2$, so that the value of the integral is not changed apart from a possible phase term. After the calculation of the integral, the evolution operator kernel is get by taking the limit $\nu \to \infty$. The existence of this limit, and also the covariance under general phase-space coordinate transformations, can be proved through the *operator* formalism for the regularized kernel $K_t^\nu(q, q')$.

Note that the integral (6.16) with the Wiener measure inserted can be regarded as an ordinary Lagrangian path integral with a complex action, where the configuration space is the original phase–space and the Hamiltonian $h(q)$ serves as a potential. Making use of this observation it is not hard to derive the corresponding Schrödinger–like equation

$$\partial_t K_t^\nu(q, q') = \left[\frac{\nu}{2} \left(\partial_{q^j} + \frac{\mathrm{i}}{2} \omega_{jk} q^k \right)^2 - \mathrm{i} h(q) \right] K_t^\nu(q, q') \, , \qquad (6.18)$$

subject to the initial condition $K_{t=0}^\nu(q, q') = \delta(q - q'), 0 < \nu < \infty$. One can show that $\hat{K}_t^\nu \to \hat{K}_t$ as $\nu \to \infty$ for all $t > 0$. The covariance under general coordinate transformations follows from the covariance of the "kinetic" energy of the Schrödinger operator in (6.18): The Laplace operator is replaced by the Laplace–Beltrami operator in the new curvilinear coordinates $q = q(\bar{q})$, so the solution is not changed, but written in the new coordinates. This is similar to the covariance of the ordinary Schrödinger equation and the corresponding *Lagrangian* path integral relative to general coordinate transformations on the configuration space: The kinetic energy operator (the Laplace operator) in the ordinary Schrödinger equation gives a term *quadratic* in time derivatives in the path integral measure which is sufficient for the general coordinate covariance. We remark that the regularization procedure based on the modified Schrödinger equation (6.18) applies to far more general Hamiltonians than those quadratic in canonical momenta and leading to the conventional *Lagrangian* path integral.

6.1.4 Sum over Geometries and Topologies

Recall that the term *quantum gravity* (or *quantum geometrodynamics*, or *quantum geometry*), is usually understood as a consistent fundamental quantum description of gravitational space–time geometry whose classical limit is Einstein's general relativity. Among the possible ramifications of such a theory are a model for the structure of space–time near the Planck scale, a consistent calculational scheme to calculate gravitational effects at all energies, a description of quantum geometry near space–time singularities and a non–perturbative quantum description of 4D black holes. It might also help us in understanding cosmological issues about the beginning and end of the universe, i.e., the so–called *'big bang'* and *'Big–Crunch'* (see e.g., [Pen67, Pen94, Pen97]).

From what we know about the quantum dynamics of other fundamental interactions it seems eminently plausible that also the gravitational excitations should at very short scales be governed by quantum laws. Now, conventional *perturbative path integral expansions of gravity*, as well as perturbative expansion in the string coupling in the case of unified approaches, both have difficulty in finding any direct or indirect evidence for quantum gravitational effects, be they experimental or observational, which could give a feedback for model building. The outstanding problems mentioned above require a non–perturbative treatment; it is not sufficient to know the first few terms of a perturbation series. The real goal is to search for a *non–perturbative* definition of such a theory, where the initial input of any fixed 'background metric' is inessential (or even undesirable), and where 'space–time' is determined *dynamically*. Whether or not such an approach necessarily requires the inclusion of higher dimensions and fundamental supersymmetry is currently unknown (see [AK93, AL98, AJL00a, AJL00b, AJL01a, AJL01b, AJL01d, DL01]).

Such a non–perturbative viewpoint is very much in line with how one proceeds in classical geometrodynamics, where a metric space–time $(M, g_{\mu\nu})$ (+ matter) emerges only as a *solution* to the familiar *Einstein equation*

$$G_{\mu\nu}[g] \equiv R_{\mu\nu}[g] - \frac{1}{2}g_{\mu\nu}R[g] = -8\pi T_{\mu\nu}[\Phi], \tag{6.19}$$

which define the classical dynamics of fields $\Phi = \Phi^{\mu\nu}$ on the space $\mathcal{M}(M)$, the space of all metrics $g = g_{\mu\nu}$ on a given smooth manifold M. The analogous question we want to address in the quantum theory is: Can we get 'quantum space–time' as a solution to a set of non–perturbative quantum equations of motion on a suitable quantum analogue of $\mathcal{M}(M)$ or rather, of the space of geometries, $Geom(M) = \mathcal{M}(M)/Diff(M)$?

Now, this is not a completely straightforward task. Whichever way we want to proceed non–perturbatively, if we give up the privileged role of a flat, Minkowskian background space–time on which the quantization is to take place, we also have to abandon the central role usually played by the Poincaré group, and with it most standard quantum field–theoretic tools for regularization and renormalization. If one works in a continuum metric formulation of gravity, the symmetry group of the *Einstein–Hilbert action* is instead the group $Diff(M)$ of diffeomorphisms on M, which in terms of local charts are the smooth invertible coordinate transformations $x^\mu \mapsto y^\mu(x^\mu)$.

In the following, we will describe a non–perturbative path integral approach to quantum gravity, defined on the *space of all geometries*, without distinguishing any background metric structure [Lol01]. This is closely related in spirit with the canonical approach of loop quantum gravity [Rov98] and its more recent incarnations using so–called *spin networks* (see, e.g., [Ori01]). 'Non–perturbative' here means in a covariant context that the path sum or integral will have to be performed explicitly, and not just evaluated around its stationary points, which can only be achieved in an appropriate regularization. The method we will employ uses a discrete lattice regularization as an intermediate step in the construction of the quantum theory.

Simplicial Quantum Geometry

In this section we will explain how one may construct a theory of quantum gravity from a non–perturbative path integral, using the method of Lorentzian dynamical triangulations. The method is minimal in the sense of employing standard tools from quantum field theory and the theory of critical phenomena and adapting them to the case of *generally covariant systems*, without invoking any symmetries beyond those of the classical theory. At an intermediate stage of the construction, we use a regularization in terms of simplicial *Regge geometries*, that is, piecewise linear manifolds. In this approach, 'computing the path integral' amounts to a conceptually simple and geometrically transparent 'counting of geometries', with additional weight factors which are determined by the EH action. This is done first of all at a regularized level. Subsequently, one searches for interesting continuum limits of these discrete models which are possible candidates for theories of quantum gravity, a step that will always involve a renormalization. From the point of view of statistical mechanics, one may think of Lorentzian dynamical triangulations as a new class of statistical models of Lorentzian random surfaces in various dimensions, whose building blocks are flat simplices which carry a 'time arrow', and whose dynamics is entirely governed by their intrinsic geometric properties.

Before describing the details of the construction, it may be helpful to recall the path integral representation for a 1D non–relativistic particle (see previous subsection). The time evolution of the particle's wave function ψ may be described by the integral equation (6.3) above, where the *propagator*, or the *Feynman kernel* G, is defined through a limiting procedure (6.4). The time interval $t''-t'$ has been discretized into N steps of length $\epsilon=(t''-t')/N$, and the r.h.s. of (6.4) represents an integral over all piecewise linear paths $x(t)$ of a 'virtual' particle propagating from x' to x'', illustrated in Figure 6.4 above.

The prefactor A^{-N} is a normalization and L denotes the Lagrangian function of the particle. Knowing the propagator G is tantamount to having solved the quantum dynamics. This is the simplest instance of a path integral, and is often written schematically as

$$G(x',t';x'',t'')=\sum\!\!\!\!\!\!\int\mathcal{D}[x(t)]\ \mathrm{e}^{\mathrm{i}S[x(t)]}, \tag{6.20}$$

where $\mathcal{D}[x(t)]$ is a functional measure on the 'space of all paths', and the exponential weight depends on the classical action $S[x(t)]$ of a path. Recall also that this procedure can be defined in a mathematically clean way if we Wick–rotate the time variable t to imaginary values $t\mapsto\tau=\mathrm{i}t$, thereby making all integrals real [RS75].

Can a similar strategy work for the case of Einstein geometrodynamics? As an analogue of the particle's position we can take the geometry $[g_{ij}(x)]$ (i.e., an equivalence class of spatial metrics) of a constant–time slice. Can one then define a gravitational propagator

$$G([g'_{ij}],[g''_{ij}])=\sum\!\!\!\!\!\!\int_{Geom(M)}\mathcal{D}[g_{\mu\nu}]\ \mathrm{e}^{\mathrm{i}S^{\mathrm{EH}}[g_{\mu\nu}]} \tag{6.21}$$

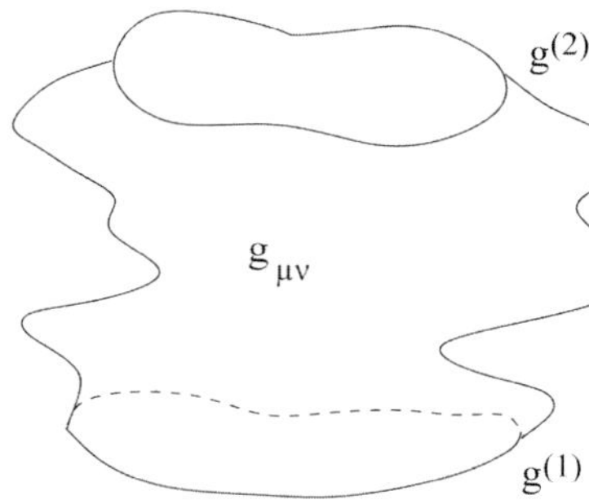

Fig. 6.5. The time–honoured way [HE79] of illustrating the gravitational path integral as the propagator from an initial to a final spatial boundary geometry.

from an initial geometry $[g']$ to a final geometry $[g'']$ (Figure 6.5) as a limit of some discrete construction analogous to that of the non-relativistic particle (6.4)? And crucially, what would be a suitable class of 'paths', that is, space–times $[g_{\mu\nu}]$ to sum over?

Now, to be able to perform the integration $\oint \mathcal{D}[g_{\mu\nu}]$ in a meaningful way, the strategy we will be following starts from a regularized version of the space $Geom(M)$ of all geometries. A regularized path integral $G(a)$ can be defined which depends on an ultraviolet cutoff a and is *convergent* in a non–trivial region of the space of coupling constants. Taking the continuum limit corresponds to letting $a \to 0$. The resulting continuum theory – if it can be shown to exist – is then investigated with regard to its geometric properties and in particular its semiclassical limit.

Discrete Gravitational Path Integrals

Trying to construct non–perturbative path integrals for gravity from sums over discretized geometries, using approach of *Lorentzian dynamical triangulations*, is not a new idea. Inspired by the successes of lattice gauge theory, attempts to describe quantum gravity by similar methods have been popular on and off since the late 70's. Initially the emphasis was on gauge–theoretic, first–order formulations of gravity, usually based on (compactified versions of) the Lorentz group, followed in the 80's by 'quantum Regge calculus', an attempt to represent the gravitational path integral as an integral over certain piecewise linear geometries (see [Wil97] and references therein), which had first made an appearance in approximate descriptions of *classical* solutions of the Einstein equations. A variant of this approach by the name of 'dynamical triangulation(s)' attracted a lot of interest during the 90's, partly because it had proved a powerful tool in describing 2D quantum gravity (see the textbook [ADJ97] and lecture notes [AJL00a] for more details).

The problem is that none of these attempts have so far come up with convincing evidence for the existence of an underlying continuum theory of 4D quantum gravity. This conclusion is drawn largely on the basis of numerical simulations, so it is by no means water–tight, although one can make an argument that the 'symptoms' of failure are related in the various approaches [Lol98]. What goes wrong generically seems to be a dominance in the continuum limit of highly degenerate geometries, whose precise form depends on the approach chosen. One would expect that non–smooth geometries play a decisive role, in the same way as it can be shown in the particle case that the support of the measure in the continuum limit is on a set of nowhere differentiable paths. However, what seems to happen in the case of the path integral for 4–geometries is that the structures get are *too* wild, in the sense of not generating, even at coarse–grained scales, an effective geometry whose dimension is anywhere near four.

The schematic phase diagram of Euclidean dynamical triangulations shown in Figure 6.6 gives an example of what can happen. The picture turns out to be essentially the same in both three and four dimensions: the model possesses infinite-volume limits everywhere along the critical line $k_3^{\rm crit}(k_0)$, which fixes the bare cosmological constant as a function of the inverse Newton constant $k_0 \sim G_N^{-1}$. Along this line, there is a critical point $k_0^{\rm crit}$ (which we now know to be of first–order in $d = 3, 4$) below which geometries generically have a very large effective or Hausdorff dimension.[10] Above $k_0^{\rm crit}$ we find the opposite phenomenon of 'polymerization': a typical element contributing to the state sum is a thin branched polymer, with one or more dimensions 'curled up' such that its effective dimension is around two.

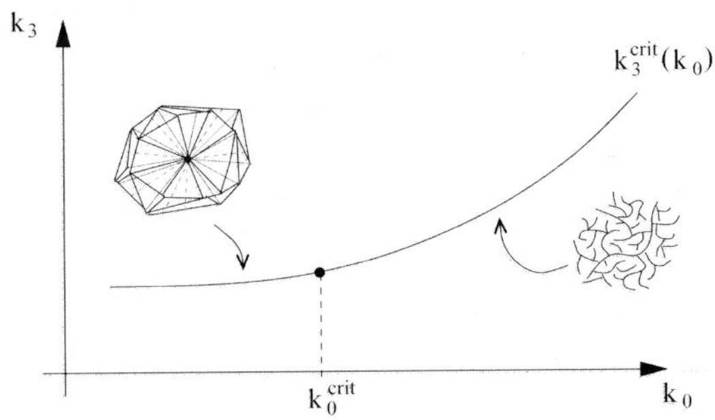

Fig. 6.6. The phase diagram of 3D and 4D Euclidean dynamical triangulations (adapted from [AJL00b, AJL01a]).

This problem has to do with the fact that the gravitational action is unbounded below, causing potential havoc in Euclidean versions of the path

[10] In terms of geometry, this means that there are a few vertices at which the entire space–time 'condenses' in the sense that almost every other vertex in the simplicial space–time is about one link-distance away from them.

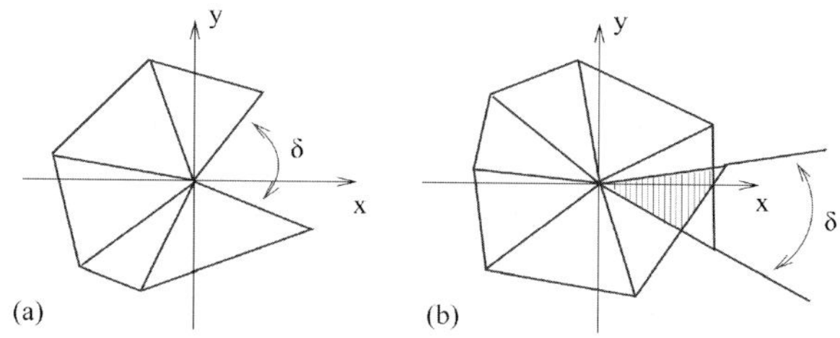

Fig. 6.7. Positive (a) and negative (b) space–like deficit angles δ (adapted from [Lol01, Lol98]).

integral. Namely, what all the above-mentioned approaches have in common is that they work from the outset with *Euclidean* geometries, and associated Boltzmann-type weights $\exp(-S^{\text{eu}})$ in the path integral. In other words, they integrate over 'space–times' which know nothing about time, light cones and causality. This is done mainly for technical reasons, since it is difficult to set up simulations with complex weights and since until recently a suitable Wick rotation was not known.

'Lorentzian dynamical triangulations', first proposed in [AL98] and further elaborated in [AJL00b, AJL01a] tries to establish a logical connection between the fact that non–perturbative path integrals were constructed for Euclidean instead of Lorentzian geometries and their apparent failure to lead to an interesting continuum theory.

Regge Calculus

The use of simplicial methods in general relativity goes back to the pioneering work of Regge [Reg61]. In classical applications one tries to approximate a classical space–time geometry by a triangulation, that is, a piecewise linear space get by gluing together flat simplicial building blocks, which in dimension d are dD generalizations of triangles. By 'flat' we mean that they are isometric to a subspace of dD Euclidean or Minkowski space. We will only be interested in gluings leading to genuine manifolds, which therefore look locally like an R^d. A nice feature of such simplicial manifolds is that their geometric properties are completely described by the discrete set $\{l_i^2\}$ of the squared lengths of their edges. Note that this amounts to a description of geometry *without the use of coordinates*. There is nothing to prevent us from re–introducing coordinate patches covering the piecewise linear manifold, for example, on each individual simplex, with suitable transition functions between patches. In such a coordinate system the metric tensor will then assume a definite form. However, for the purposes of formulating the path integral we will not be interested in doing this, but rather work with the edge lengths, which constitute a direct, regularized parametrization of the space $Geom(M)$ of geometries.

How precisely is the intrinsic geometry of a simplicial space, most importantly, its curvature, encoded in its edge lengths? A useful example to keep in mind is the case of dimension two, which can easily be visualized. A 2d piecewise linear space is a triangulation, and its scalar curvature $R(x)$ coincides with the *Gaussian curvature* (see subsection 4.1.4 above). One way of measuring this curvature is by parallel–transporting a vector around closed curves in the manifold. In our piecewise–flat manifold such a vector will always return to its original orientation *unless* it has surrounded lattice vertices v at which the surrounding angles did not add up to 2π, but $\sum_{i\supset v}\alpha_i = 2\pi - \delta$, for $\delta \neq 0$, see Figure 6.7. The so–called deficit angle δ is precisely the rotation angle picked up by the vector and is a direct measure for the scalar curvature at the vertex. The operational description to get the scalar curvature in higher dimensions is very similar, one basically has to sum in each point over the Gaussian curvatures of all 2D submanifolds. This explains why in Regge calculus the curvature part of the EH action is given by a sum over building blocks of dimension $(d-2)$ which are the objects dual to those local 2d submanifolds. More precisely, the continuum curvature and volume terms of the action become

$$\frac{1}{2}\int_{\mathcal{R}} d^d x\,\sqrt{|\det g|}^{(d)}R \longrightarrow \sum_{i\in\mathcal{R}} Vol(i^{th}\ (d-2)\text{–simplex})\ \delta_i \tag{6.22}$$

$$\int_{\mathcal{R}} d^d x\,\sqrt{|\det g|} \longrightarrow \sum_{i\in\mathcal{R}} Vol(i^{th}\ d\text{–simplex}) \tag{6.23}$$

in the simplicial discretization. It is then a simple exercise in trigonometry to express the volumes and angles appearing in these formulas as functions of the edge lengths l_i, both in the Euclidean and the Minkowskian case.

The approach of dynamical triangulations uses a certain class of such simplicial space–times as an explicit, regularized realization of the space $Geom(M)$. For a given volume N_d, this class consists of all gluings of manifold–type of a set of N_d simplicial building blocks of top–dimension d whose edge lengths are restricted to take either one or one out of two values. In the Euclidean case we set $l_i^2 = a^2$ for all i, and in the Lorentzian case we allow for both space- and time–like links with $l_i^2 \in \{-a^2, a^2\}$, where the geodesic distance a serves as a short-distance cutoff, which will be taken to zero later. Coming from the classical theory this may seem a grave restriction at first, but this is indeed not the case. Firstly, keep in mind that for the purposes of the quantum theory we want to sample the space of geometries 'ergodically' at a coarse-grained scale of order a. This should be contrasted with the classical theory where the objective is usually to approximate a given, *fixed* space–time to within a length scale a. In the latter case one typically requires a much finer topology on the space of metrics or geometries. It is also straightforward to see that no local curvature degrees of freedom are suppressed by fixing the edge lengths; deficit angles in all directions are still present, although they take on only a discretized set of values. In this sense, in dynamical triangulations all geometry is in the gluing of the fundamental building blocks. This is dual to how quantum Regge calculus is set up, where one usually fixes a triangulation T and then 'scans' the space of geometries by letting the l_i's run continuously over all values compatible with the triangular inequalities.

In a nutshell, Lorentzian dynamical triangulations give a definite meaning to the 'integral over geometries', namely, as a sum over inequivalent Lorentzian gluings T over any number N_d of d–simplices,

$$\int_{Geom(M)} \mathcal{D}[g_{\mu\nu}]\, \mathrm{e}^{\mathrm{i}S[g_{\mu\nu}]} \quad \overset{\text{LDT}}{\longrightarrow} \quad \sum_{T\in\mathcal{T}} \frac{1}{C_T}\, \mathrm{e}^{\mathrm{i}S^{\mathrm{Reg}}(T)}, \tag{6.24}$$

where the symmetry factor $C_T = |\mathrm{Aut}(T)|$ on the r.h.s. is the order of the automorphism group of the triangulation, consisting of all maps of T onto itself which preserve the connectivity of the simplicial lattice. We will specify below what precise class $\mathcal{T}$ of triangulations should appear in the summation.

It follows from the above that in this formulation all curvatures and volumes contributing to the *Regge simplicial action* come in discrete units. This can be illustrated by the case of a 2D triangulation with Euclidean signature, which according to the prescription of dynamical triangulations consists of *equilateral* triangles with squared edge lengths $+a^2$. All interior angles of such a triangle are equal to $\pi/3$, which implies that the deficit angle at any vertex v can take the values $2\pi - k_v\pi/3$, where k_v is the number of triangles meeting at v. As a consequence, the Einstein–Regge action S^{Reg} assumes the simple form

$$S^{\mathrm{Reg}}(T) = \kappa_{d-2}N_{d-2} - \kappa_d N_d, \tag{6.25}$$

where the coupling constants $\kappa_i = \kappa_i(\lambda, G_N)$ are simple functions of the bare cosmological and Newton constants in d dimensions. Substituting this into the path sum in (6.24) leads to

$$Z(\kappa_{d-2}, \kappa_d) = \sum_{N_d} \mathrm{e}^{-\mathrm{i}\kappa_d N_d} \sum_{N_{d-2}} \mathrm{e}^{\mathrm{i}\kappa_{d-2}N_{d-2}} \sum_{T|_{N_d, N_{d-2}}} \frac{1}{C_T}, \tag{6.26}$$

The point of taking separate sums over the numbers of d– and $(d-2)$–simplices in (6.26) is to make explicit that 'doing the sum' is tantamount to the combinatorial problem of *counting* triangulations of a given volume and number of simplices of codimension 2 (corresponding to the last summation in (6.26)).[11] It turns out that at least in two space–time dimensions the counting of geometries can be done completely explicitly, turning both Lorentzian and Euclidean quantum gravity into exactly soluble statistical models.

Lorentzian Path Integral

Now, the simplicial building blocks of the models are taken to be pieces of Minkowski space, and their edges have squared lengths $+a^2$ or $-a^2$. For example, the two types of 4–simplices that are used in Lorentzian dynamical

[11] The symmetry factor C_T is almost always equal to 1 for large triangulations.

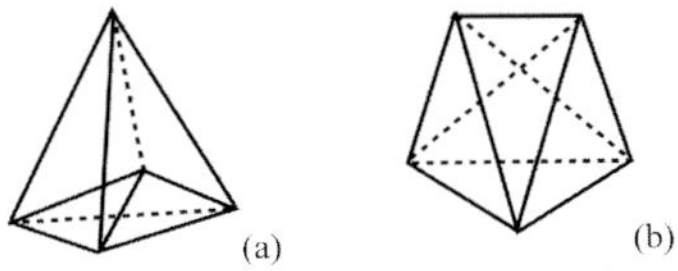

Fig. 6.8. Two types of Minkowskian 4–simplices in 4D (adapted from [Lol01, Lol98]).

triangulations in dimension four are shown in Figure 6.8. The first of them has four time–like and six space–like links (and therefore contains 4 time–like and 1 space–like tetrahedron), whereas the second one has six time–like and four space–like links (and contains 5 time–like tetrahedra). Since both are subspaces of flat space with signature $(- + ++)$, they possess well–defined light–cone structures everywhere [Lol01, Lol98].

In general, gluings between pairs of d–simplices are only possible when the metric properties of their $(d-1)$–faces match. Having local light cones implies causal relations between pairs of points in local neighborhoods. Creating closed time–like curves will be avoided by requiring that all space–times contributing to the path sum possess a global 'time' function t. In terms of the triangulation this means that the d–simplices are arranged such that their space–like links all lie in slices of constant integer t, and their time–like links interpolate between adjacent spatial slices t and $t+1$. Moreover, with respect to this time, we will not allow for any *spatial* topology changes[12].

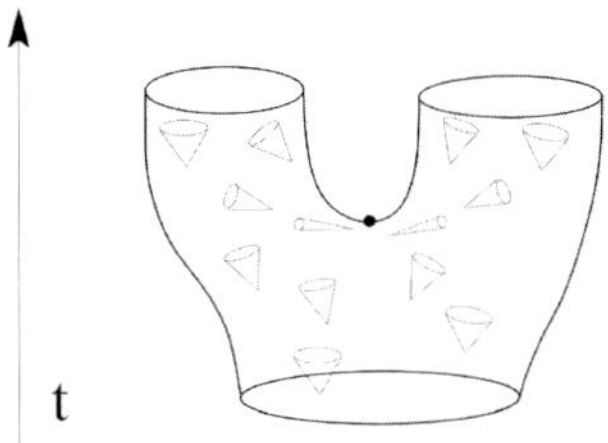

Fig. 6.9. At a branching point associated with a spatial topology change, light-cones get 'squeezed' [Lol01, Lol98].

This latter condition is always satisfied in classical applications, where 'trouser points' like the one depicted in Figure 6.9 are ruled out by the requirement of having a non–degenerate Lorentzian metric defined everywhere on M (it is geometrically obvious that the light cone and hence $g_{\mu\nu}$ must degenerate in at least one point along the 'crotch'). Another way of thinking

[12] Note that if we were in the continuum and had introduced coordinates on space–time, such a statement would actually be diffeomorphism–invariant.

about such configurations (and their time–reversed counterparts) is that the causal past (future) of an observer changes discontinuously as her world–line passes near the singular point (see [Dow02] and references therein for related discussions about the issue of topology change in quantum gravity).

There is no *a priori* reason in the quantum theory to not relax some of these classical causality constraints. After all, as we stressed right at the outset, path integral histories are not in general classical solutions, nor can we attribute any other direct physical meaning to them individually. It might well be that one can construct models whose path integral configurations violate causality in this strict sense, but where this notion is somehow recovered in the resulting continuum theory. What the approach of Lorentzian dynamical triangulations has demonstrated is that *imposing causality constraints will in general lead to a different continuum theory.* This is in contrast with the intuition one may have that 'including a few isolated singular points will not make any difference'. On the contrary, tampering with causality in this way is not innocent at all, as was already anticipated by Teitelboim many years ago [Tei83].

We want to point out that one cannot conclude from the above that spatial topology changes or even *fluctuations in the space–time topology* cannot be treated in the formulation of dynamical triangulations. However, if one insists on including geometries of variable topology in a Lorentzian discrete context, one has to come up with a prescription of how to weigh these singular points in the path integral, both before and after the Wick rotation [Das02]. Maybe this can be done along the lines suggested in [LS97]; this is clearly an interesting issue for further research.

Having said this, we next have to address the question of the *Wick rotation*, in other words, of how to get rid of the factor of i in the exponent of (6.26). Without it, this expression is an infinite sum (since the volume can become arbitrarily large) of complex terms whose convergence properties will be very difficult to establish. In this situation, a Wick rotation is simply a technical tool which – in the best of all worlds – enables us to perform the state sum and determine its continuum limit. The end result will have to be Wick–rotated back to Lorentzian signature.

Fortunately, Lorentzian dynamical triangulations come with a natural notion of Wick rotation, and the strategy we just outlined can be carried out explicitly in two space–time dimensions, leading to a unitary theory. In higher dimensions we do not yet have sufficient analytical control of the continuum theories to make specific statements about the *inverse* Wick rotation. Since we use the Wick rotation at an intermediate step, one can ask whether other Wick rotations would lead to the same result. Currently this is a somewhat academic question, since it is in practice difficult to find such alternatives. In fact, it is quite miraculous we have found a single prescription for Wick–rotating in our regularized setting, and it does not seem to have a direct continuum analogue (for more comments on this issue, see [DL01, Das02]).

Our Wick rotation W in any dimension is an injective map from Lorentzian– to Euclidean–signature simplicial space–times. Using the notation $\mathtt{T}$ for a simplicial manifold together with length assignments l_s^2 and l_t^2 to its space– and time–like links, it is defined by

$$\mathtt{T}^{\rm lor} = (T, \{l_s^2 = a^2, l_t^2 = -a^2\}) \stackrel{W}{\longmapsto} \mathtt{T}^{\rm eu} = (T, \{l_s^2 = a^2, l_t^2 = a^2\}). \quad (6.27)$$

Note that we have not touched the connectivity of the simplicial manifold T, but only its metric properties, by mapping all time–like links of T into space–like ones, resulting in a Euclidean 'space–time' of equilateral building blocks. It can be shown [AJL01a] that at the level of the corresponding weight factors in the path integral this Wick rotation[13] has precisely the desired effect of rotating to the exponentiated Regge action of the 'Euclideanized' geometry,

$$\mathrm{e}^{\mathrm{i}S(\mathcal{T}^{\rm lor})} \stackrel{W}{\longmapsto} \mathrm{e}^{-S(\mathcal{T}^{\rm eu})}. \quad (6.28)$$

The Euclideanized path sum after the Wick rotation has the form

$$\begin{aligned} Z^{\rm eu}(\kappa_{d-2}, \kappa_d) &= \sum_T \frac{1}{C_T}\, \mathrm{e}^{-\kappa_d N_d(T) + \kappa_{d-2} N_{d-2}(T)} \\ &= \sum_{N_d} \mathrm{e}^{-\kappa_d N_d} \sum_{T|_{N_d}} \frac{1}{C_T}\, \mathrm{e}^{\kappa_{d-2} N_{d-2}(T)} \\ &= \sum_{N_d} \mathrm{e}^{-\kappa_d N_d}\, \mathrm{e}^{\kappa_d^{\rm crit}(\kappa_{d-2}) N_d} \times \text{subleading}(N_d). \end{aligned} \quad (6.29)$$

In the last equality we have used that the number of Lorentzian triangulations of discrete volume N_d to leading order scales exponentially with N_d for large volumes. This can be shown explicitly in space–time dimension 2 and 3. For $d = 4$, there is strong (numerical) evidence for such an exponential bound for *Euclidean triangulations*, from which the desired result for the Lorentzian case follows (since W maps to a strict subset of all Euclidean simplicial manifolds).

From the functional form of the last line of (6.29) one can immediately read off some qualitative features of the phase diagram, an example of which appeared already earlier in Figure 6.6. Namely, the sum over geometries $Z^{\rm eu}$ converges for values $\kappa_d > \kappa_d^{\rm crit}$ of the bare cosmological constant, and diverges (i.e., is not defined) below this critical line. Generically, for all models of dynamical triangulations the infinite–volume limit is attained by approaching the critical line $\kappa_d^{\rm crit}(\kappa_{d-2})$ from above, ie. from inside the region of convergence of $Z^{\rm eu}$. In the process of taking $N_d \to \infty$ and the cutoff $a \to 0$, one gets a renormalized cosmological constant Λ through

[13] To get a genuine Wick rotation and not just a discrete map, one introduces a complex parameter α in $l_t^2 = -\alpha a^2$. The proper prescription leading to (6.28) is then an analytic continuation of α from 1 to -1 through the lower–half complex–plane.

$$(\kappa_d - \kappa_d^{\text{crit}}) = a^\mu \Lambda + O(a^{\mu+1}). \tag{6.30}$$

If the scaling is canonical (which means that the dimensionality of the renormalized coupling constant is the one expected from the classical theory), the exponent is given by $\mu = d$. Note that this construction requires a positive *bare* cosmological constant in order to make the state sum converge. Moreover, by virtue of relation (6.30) also the *renormalized* cosmological constant must be positive. Other than that, its numerical value is not determined by this argument, but by comparing observables of the theory which depend on Λ with actual physical measurements.[14] Another interesting observation is that the inclusion of a sum over topologies in the discretized sum (6.29) would lead to a super–exponential growth of at least $\propto N_d!$ of the number of triangulations with the volume N_d. Such a divergence of the path integral cannot be compensated by an additive renormalization of the cosmological constant of the kind outlined above.

There are ways in which one can sum divergent series of this type, for example, by performing a Borel sum. The problem with these stems from the fact that two different functions can share the same asymptotic expansion. Therefore, the series in itself is *not* sufficient to define the underlying theory uniquely. The non–uniqueness arises because of non–perturbative contributions to the path integral which are not represented in the perturbative expansion.[15] In order to fix these uniquely, an independent, non–perturbative definition of the theory is necessary. Unfortunately, for dynamically triangulated models of quantum gravity, no such definitions have been found so far. In the context of 2D (Euclidean) quantum gravity this difficulty is known as the 'absence of a physically motivated double-scaling limit' [AK93].

Lastly, getting an interesting continuum limit may or may not require an additional fine–tuning of the inverse gravitational coupling κ_{d-2}, depending on the dimension d. In four dimensions, one would expect to find a second-order transition along the critical line, corresponding to local gravitonic excitations. The situation in $d = 3$ is less clear, but results get so far indicate that no fine–tuning of Newton's constant is necessary [AJL01b, AJL01d].

Before delving into the details, let me summarize briefly the results that have been get so far in the approach of Lorentzian dynamical triangulations. At the regularized level, that is, in the presence of a finite cutoff a for the edge lengths and an infrared cutoff for large space–time volume, they are well–defined statistical models of Lorentzian random geometries in $d = 2, 3, 4$. In particular, they obey a suitable notion of reflection-positivity and possess self–adjoint Hamiltonians.

The crucial questions are then to what extent the underlying combinatorial problems of counting all dD geometries with certain causal properties can

[14] The non–negativity of the renormalized cosmological coupling may be taken as a first 'prediction' of our construction, which in the physical case of four dimensions is indeed in agreement with current observations.

[15] A field–theoretic example would be instantons and renormalons in QCD.

be solved, whether continuum theories with non–trivial dynamics exist and how their bare coupling constants get renormalized in the process. What we know about Lorentzian dynamical triangulations so far is that they lead to continuum theories of quantum gravity in dimension 2 and 3. In $d = 2$, there is a complete analytic solution, which is distinct from the continuum theory produced by Euclidean dynamical triangulations. Also the matter–coupled model has been studied. In $d = 3$, there are numerical and partial analytical results which show that both a continuum theory exists and that it again differs from its Euclidean counterpart. Work on a more complete analytic solution which would give details about the geometric properties of the quantum theory is under way. In $d = 4$, the first numerical simulations are currently being set up. The challenge here is to do this for sufficiently large lattices, to be able to perform meaningful measurements. So far, we cannot make any statements about the existence and properties of a continuum theory in this physically most interesting case.

6.2 Complex Dynamics of Quantum Fields

6.2.1 Topological Quantum Field Theory

Before we come to (super)strings, we give a brief on *topological quantum field theory* (TQFT), as developed by Ed Witten, from his original path integral point of view (see [Wit88b, LL98]). TQFT originated in 1982, when Witten rewrote classical *Morse theory* (see section 4.1.4 above) in Dick Feynman's language of quantum field theory [Wit82]. Witten's arguments made use of Feynman's path integrals and consequently, at first, they were regarded as mathematically non–rigorous. However, a few years later, A. Floer reformulated a rigorous Morse–Witten theory [Flo87] (that won a Fields medal for Witten). This trend in which some mathematical structure is first constructed by quantum field theory methods and then reformulated in a rigorous mathematical ground constitutes one of the tendencies in modern physics.

In TQFT our basic topological space is an nD Riemannian manifold M with a metric $g_{\mu\nu}$. Let us consider on it a set of fields $\{\phi_i\}$, and let $S[\phi_i]$ be a real functional of these fields which is regarded as the *action* of the theory. We consider 'operators', $O_\alpha(\phi_i)$, which are in general arbitrary functionals of the fields. In TQFT these functionals are real functionals labelled by some set of indices α carrying topological or group–theoretical data. The *vacuum expectation value* (VEV) of a product of these operators is defined as

$$\langle O_{\alpha_1} O_{\alpha_2} \cdots O_{\alpha_p} \rangle = \int [D\phi_i] O_{\alpha_1}(\phi_i) O_{\alpha_2}(\phi_i) \cdots O_{\alpha_p}(\phi_i) \exp\left(-S[\phi_i]\right).$$

A quantum field theory is considered *topological* if the following relation is satisfied:

$$\frac{\delta}{\delta g^{\mu\nu}}\langle O_{\alpha_1} O_{\alpha_2} \cdots O_{\alpha_p}\rangle = 0, \tag{6.31}$$

i.e., if the VEV of some set of selected operators is independent of the metric $g_{\mu\nu}$ on M. If such is the case those operators are called 'observables'.

There are two ways to guarantee, at least formally, that condition (6.31) is satisfied. The first one corresponds to the situation in which both, the action $S[\phi_i]$, as well as the operators O_{α_i} are *metric independent*. These TQFTs are called of *Schwarz* type. The most important representative is *Chern–Simons gauge theory*. The second one corresponds to the case in which there exist a *symmetry*, whose infinitesimal form is denoted by δ, satisfying the following properties:

$$\delta O_{\alpha_i} = 0, \qquad T_{\mu\nu} = \delta G_{\mu\nu}, \tag{6.32}$$

where $T_{\mu\nu}$ is the SEM–tensor of the theory, i.e.,

$$T_{\mu\nu}(\phi_i) = \frac{\delta}{\delta g^{\mu\nu}} S[\phi_i]. \tag{6.33}$$

The fact that δ in (6.32) is a symmetry of the theory implies that the transformations $\delta\phi_i$ of the fields are such that both $\delta A[\phi_i] = 0$ and $\delta O_{\alpha_i}(\phi_i) = 0$. Conditions (6.32) lead, at least formally, to the following relation for VEVs:

$$\frac{\delta}{\delta g^{\mu\nu}}\langle O_{\alpha_1} O_{\alpha_2} \cdots O_{\alpha_p}\rangle = -\int [D\phi_i] O_{\alpha_1}(\phi_i) O_{\alpha_2}(\phi_i) \cdots O_{\alpha_p}(\phi_i) T_{\mu\nu} \mathrm{e}^{-S[\phi_i]}$$
$$= -\int [D\phi_i] \delta\left(O_{\alpha_1}(\phi_i) O_{\alpha_2}(\phi_i) \cdots O_{\alpha_p}(\phi_i) G_{\mu\nu} \exp\left(-S[\phi_i]\right)\right) = 0, \tag{6.34}$$

which implies that the quantum field theory can be regarded as topological. This second type of TQFTs are called of *Witten type*. One of its main representatives is the theory related to Donaldson invariants, which is a twisted version of $N = 2$ supersymmetric *Yang–Mills gauge theory*. It is important to remark that the symmetry δ must be a scalar symmetry, i.e., that its symmetry parameter must be a scalar. The reason is that, being a global symmetry, this parameter must be covariantly constant and for arbitrary manifolds this property, if it is satisfied at all, implies strong restrictions unless the parameter is a scalar.

Most of the TQFTs of *cohomological type* satisfy the relation:

$$S[\phi_i] = \delta \Lambda(\phi_i), \tag{6.35}$$

for some functional $\Lambda(\phi_i)$. This has far–reaching consequences, for it means that the topological observables of the theory, in particular the partition function, (path integral) itself are independent of the value of the coupling constant. Indeed, let us consider for example the VEV:

$$\langle O_{\alpha_1} O_{\alpha_2} \cdots O_{\alpha_p} \rangle = \int [D\phi_i] O_{\alpha_1}(\phi_i) O_{\alpha_2}(\phi_i) \cdots O_{\alpha_p}(\phi_i)\, \mathrm{e}^{-\frac{1}{g^2} S[\phi_i]}. \qquad (6.36)$$

Under a change in the coupling constant, $1/g^2 \to 1/g^2 - \Delta$, one has (assuming that the observables do not depend on the coupling), up to first–order in Δ:

$$\begin{aligned} &\langle O_{\alpha_1} O_{\alpha_2} \cdots O_{\alpha_p} \rangle \longrightarrow \langle O_{\alpha_1} O_{\alpha_2} \cdots O_{\alpha_p} \rangle \\ &+ \Delta \int [D\phi_i] \delta \left[O_{\alpha_1}(\phi_i) O_{\alpha_2}(\phi_i) \cdots O_{\alpha_p}(\phi_i) \Lambda(\phi_i) \exp\left(-\frac{1}{g^2} S[\phi_i] \right) \right] \\ &= \langle O_{\alpha_1} O_{\alpha_2} \cdots O_{\alpha_p} \rangle. \end{aligned} \qquad (6.37)$$

Hence, observables can be computed either in the *weak coupling limit*, $g \to 0$, or in the *strong coupling limit*, $g \to \infty$.

So far we have presented a rather general definition of TQFT and made a series of elementary remarks. Now we will analyze some aspects of its structure. We begin pointing out that given a theory in which (6.32) holds one can build *correlators* which correspond to *topological invariants* (in the sense that they are invariant under deformations of the metric $g_{\mu\nu}$) just by considering the operators of the theory which are *invariant under the symmetry*. We will call these operators *observables*. In virtue of (6.34), if one of these operators can be written as a symmetry transformation of another operator, its presence in a correlation function will make it vanish. Thus we may identify operators satisfying (6.32) which differ by an operator which corresponds to a symmetry transformation of another operator. Let us denote the set of the resulting classes by $\{\Phi\}$. By restricting the analysis to the appropriate set of operators, one has that in fact,

$$\delta^2 = 0. \qquad (6.38)$$

Property (6.38) has consequences on the features of TQFT. First, the *symmetry must be odd* which implies the presence in the theory of *commuting* and *anticommuting fields*. For example, the tensor $G_{\mu\nu}$ in (6.32) must be anticommuting. This is the first appearance of an odd non–spinorial field in TQFT. Those kinds of objects are standard features of *cohomological* TQFTs. Second, if we denote by Q the operator which implements this symmetry, the observables of the theory can be described as the cohomology classes of Q:

$$\{\Phi\} = \frac{\mathrm{Ker}\, Q}{\mathrm{Im}\, Q}, \qquad Q^2 = 0. \qquad (6.39)$$

Equation (6.32) means that in addition to the *Poincaré group* the theory possesses a symmetry generated by an odd version of the Poincaré group. The corresponding odd generators are constructed out of the tensor $G_{\mu\nu}$ in much the same way as the ordinary Poincaré generators are built out of $T_{\mu\nu}$. For example, if P_μ represents the ordinary momentum operator, there exists a corresponding odd one G_μ such that

$$P_\mu = \{Q, G_\mu\}. \tag{6.40}$$

Now, let us discuss the structure of the Hilbert space of the theory in virtue of the symmetries that we have just described. The states of this space must correspond to representations of the algebra generated by the operators in the Poincaré groups and by Q. Furthermore, as follows from our analysis of operators leading to (6.39), if one is interested only in states $|\Psi\rangle$ leading to topological invariants one must consider states which satisfy

$$Q|\Psi\rangle = 0, \tag{6.41}$$

and two states which differ by a $Q-$exact state must be identified. The odd Poincaré group can be used to generate descendant states out of a state satisfying (6.41). The operators G_μ act non–trivially on the states and in fact, out of a state satisfying (6.41) we can build additional states using this generator. The simplest case consists of

$$\int_{\gamma_1} G_\mu |\Psi\rangle,$$

where γ_1 is a 1–cycle. One can verify using (6.32) that this new state satisfies (6.41):

$$Q\int_{\gamma_1} G_\mu|\Psi\rangle = \int_{\gamma_1} \{Q, G_\mu\}|\Psi\rangle = \int_{\gamma_1} P_\mu|\Psi\rangle = 0.$$

Similarly, one may construct other invariants tensoring n operators G_μ and integrating over $n-$cycles γ_n:

$$\int_{\gamma_n} G_{\mu_1} G_{\mu_2} ... G_{\mu_n} |\Psi\rangle. \tag{6.42}$$

Notice that since the operator G_μ is odd and its algebra is Poincaré–like the integrand in this expression is an exterior differential $n-$form. These states also satisfy condition (6.41). Therefore, starting from a state $|\Psi\rangle \in \ker Q$ we have built a set of partners or descendants giving rise to a topological multiplet. The members of a multiplet have well defined *ghost* number. If one assigns ghost number -1 to the operator G_μ the state in (6.42) has ghost number $-n$ plus the ghost number of $|\Psi\rangle$. Now, n is bounded by the dimension of the manifold X. Among the states constructed in this way there may be many which are related via another state which is $Q-$exact, i.e., which can be written as Q acting on some other state. Let us try to single out representatives at each level of ghost number in a given topological multiplet.

Consider an $(n-1)-$cycle which is the boundary of an nD surface, $\gamma_{n-1} = \partial S_n$. If one builds a state taking such a cycle one finds ($P_\mu = -\mathrm{i}\partial_\mu$),

$$\int_{\gamma_{n-1}} G_{\mu_1} G_{\mu_2} ... G_{\mu_{n-1}} |\Psi\rangle = \mathrm{i} \int_{S_n} P_{[\mu_1} G_{\mu_2} G_{\mu_3} ... G_{\mu_n]} |\Psi\rangle \qquad (6.43)$$

$$= \mathrm{i}Q \int_{S_n} G_{\mu_1} G_{\mu_2} ... G_{\mu_n} |\Psi\rangle,$$

i.e., it is $Q-$exact. The square–bracketed subscripts in (6.43) denote that all indices between them must by antisymmetrized. In (6.43) use has been made of (6.40). This result tells us that the representatives we are looking for are built out of the homology cycles of the manifold X. Given a manifold X, the homology cycles are equivalence classes among cycles, the equivalence relation being that two $n-$cycles are equivalent if they differ by a cycle which is the boundary of an $n+1$ surface. Thus, knowledge on the homology of the manifold on which the TQFT is defined allows us to classify the representatives among the operators (6.42). Let us assume that X has dimension d and that its homology cycles are γ_{i_n}, ($i_n = 1, ..., d_n$, $n = 0, ..., d$), where d_n is the dimension of the $n-$homology group, and d the dimension of X. Then, the non–trivial partners or descendants of a given $|\Psi\rangle$ *highest–ghost–number state* are labelled in the following way:

$$\int_{\gamma_{i_n}} G_{\mu_1} G_{\mu_2} ... G_{\mu_n} |\Psi\rangle, \qquad (i_n = 1, ..., d_n, \;\; n = 0, ..., d).$$

A similar construction to the one just described can be made for fields. Starting with a field $\phi(x)$ which satisfies,

$$[Q, \phi(x)] = 0, \qquad (6.44)$$

one can construct other fields using the operators G_μ. These fields, which we call *partners* are antisymmetric tensors defined as,

$$\phi^{(n)}_{\mu_1 \mu_2 ... \mu_n}(x) = \frac{1}{n!} [G_{\mu_1}, [G_{\mu_2} ... [G_{\mu_n}, \phi(x)\} ...\}\}, \qquad (n = 1, ..., d).$$

Using (6.40) and (6.44) one finds that these fields satisfy the so–called *topological descent equations*:

$$d\phi^{(n)} = \mathrm{i}[Q, \phi^{(n+1)}\},$$

where the subindices of the forms have been suppressed for simplicity, and the highest–ghost–number field $\phi(x)$ has been denoted as $\phi^{(0)}(x)$. These equations enclose all the relevant properties of the observables which are constructed out of them. They constitute a very useful tool to build the observables of the theory.

6.2.2 Seiberg–Witten Theory and TQFT

Recall that the field of low–dimensional geometry and topology [Ati88b] has undergone a dramatic phase of progress in the last decade of the 20th Century, prompted, to a large extend, by new ideas and discoveries in mathematical physics. The discovery of quantum groups [Dri86] in the study of the *Yang–Baxter equation* [Bax82] has reshaped the theory of knots and links [Jon85, RT91, ZGD91]; the study of conformal field theory and quantum *Chern–Simons theory* [Wit89] in physics had a profound impact on the theory of 3–manifolds; and most importantly, investigations of the classical Yang–Mills (YM) theory led to the creation of the *Donaldson theory* of 4–manifolds [FU84, Don87]. Witten [Wit94] discovered a new set of invariants of 4–manifolds in the study of the Seiberg–Witten (SW) monopole equations, which have their origin in supersymmetric gauge theory. The SW theory, while closely related to Donaldson theory, is much easier to handle. Using SW theory, proofs of many theorems in Donaldson theory have been simplified, and several important new results have also been obtained [Tau90, Tau94].

In [ZOC95] a topological quantum field theory was introduced which reproduces the SW invariants of 4–manifolds. A geometrical interpretation of the 3D quantum field theory was also given.

SW Invariants and Monopole Equations

Recall that the SW monopole equations are classical field theoretical equations involving a $U(1)$ gauge field and a complex Weyl spinor on a 4D manifold. Let X denote the 4–manifold, which is assumed to be oriented and closed. If X is spin, there exist positive and negative spin bundles $S^{\pm}$ of rank two. Introduce a complex line bundle $L \to X$. Let A be a connection on L and M be a section of the product bundle $S^{+} \otimes L$. Recall that the SW monopole equations read

$$\mathcal{F}^{+}_{kl} = -\frac{\mathrm{i}}{2}\bar{M}\Gamma_{kl}M, \qquad D_A M = 0, \tag{6.45}$$

where D_A is the twisted Dirac operator, $\Gamma_{ij} = \frac{1}{2}[\gamma_i, \gamma_j]$, and $\mathcal{F}^{+}$ represents the self–dual part of the curvature of L with connection A.

If X is not a spin manifold, then spin bundles do not exist. However, it is always possible to introduce the so called $Spin_c$ bundles $S^{\pm} \otimes L$, with L^2 being a line bundle. Then in this more general setting, the SW monopoles equations look formally the same as (6.45), but the M should be interpreted as a section of the the $Spin_C$ bundle $S^{+} \otimes L$.

Denote by $\mathcal{M}$ the moduli space of solutions of the SW monopole equations up to gauge transformations. Generically, this space is a manifold. Its virtual dimension is equal to the number of solutions of the following equations

$$(d\psi)^+_{kl} + \frac{\mathrm{i}}{2}\left(\bar{M}\Gamma_{kl}N + \bar{N}\Gamma_{kl}M\right) = 0, \qquad D_A N + \psi M = 0,$$
$$\nabla_k \psi^k + \frac{\mathrm{i}}{2}(\overline{N}M - \overline{M}N) = 0, \tag{6.46}$$

where A and M are a given solution of (6.45), $\psi \in \Omega^1(X)$ is a one form, $(d\psi)^+ \in \Omega^{2,+}(X)$ is the self dual part of the two form $d\psi$, and $N \in S^+ \otimes L$. The first two of the equations in (6.46) are the linearization of the monopole equations (6.45), while the last one is a gauge fixing condition. Though with a rather unusual form, it arises naturally from the dual operator governing gauge transformations

$$C : \Omega^0(X) \to \Omega^1(X) \oplus (S^+ \otimes L), \qquad \phi \mapsto (-d\phi, \mathrm{i}\phi M).$$
Let $$T : \Omega^1(X) \oplus (S^+ \otimes L) \to \Omega^0(X) \oplus \Omega^{2,+}(X) \oplus (S^- \otimes L),$$

be the operator governing equation (6.46), namely, the operator which allows us to rewrite (6.46) as $T(\psi, N) = 0$. Then T is an elliptic operator, the index $\mathrm{Ind}(T)$ of which yields the virtual dimension of $\mathcal{M}$. A straightforward application of the Atiyah–Singer index Theorem gives

$$\mathrm{Ind}(T) = -\frac{2\chi(X) + 3\sigma(X)}{4} + c_1(L)^2,$$

where $\chi(X)$ is the Euler character of X, $\sigma(X)$ its signature index and $c_1(L)^2$ is the square of the first Chern class of L evaluated on X in the standard way.

When $\mathrm{Ind}(T)$ equals zero, the moduli space generically consists of a finite number of points, $\mathcal{M} = \{p_t : t = 1, 2, ..., I\}$. Let ϵ_t denote the sign of the determinant of the operator T at p_t, which can be defined with mathematical rigor. Then the SW invariant of the 4–manifold X is defined by $\sum_1^I \epsilon_t$.

The fact that this is indeed an invariant(i.e., independent of the metric) of X is not very difficult to prove, and we refer to [Wit94] for details. As a matter of fact, the number of solutions of a system of equations weighted by the sign of the operator governing the equations(i.e., the analog of T) is a topological invariant in general [Wit94]. This point of view has been extensively explored by Vafa and Witten [VW94] within the framework of topological quantum field theory in connection with the so called S duality. Here we wish to explore the SW invariants following a similar line as that taken in [Wit88b, VW94].

Topological Lagrangian

Introduce a *Lie super–algebra* with an odd generator Q and two even generators U and δ obeying the following (anti)commutation relations [ZOC95]

$$[U,Q]=Q, \qquad [Q,Q]=2\delta, \qquad [Q,\delta]=0. \tag{6.47}$$

We will call U the ghost number operator, and Q the *BRST–operator*.

Let A be a connection of L and $M \in S^+ \otimes L$. We define the action of the super–algebra on these fields by requiring that δ coincide with a gauge transformation with a gauge parameter $\phi \in \Omega^0(X)$. The field multiplets associated with A and M furnishing representations of the super–algebra are (A,ψ,ϕ), and (M,N), where $\psi \in \Omega^1(X)$, $\phi \in \Omega^0(X)$, and N is a section of $S^+ \otimes L$. They transform under the action of the super–algebra according to

$$\begin{aligned} & & [Q,A_i]=\psi_i, \qquad & [Q,M]=N, \\ [Q,\psi_i] &= -\partial_i\phi, \qquad & [Q,N]=\mathrm{i}\phi M, \qquad & [Q,\phi]=0. \end{aligned}$$

We assume that both A and M have ghost number 0, and thus will be regarded as bosonic fields when we study their quantum field theory. The ghost numbers of other fields can be read off the above transformation rules. We have that ψ and N are of ghost number 1, thus are fermionic, and ϕ is of ghost number 2 and bosonic. Note that the multiplet (A,ψ,ϕ) is what one would get in the topological field theory for Donaldson invariants except that our gauge group is $U(1)$, while the existence of M and N is a new feature. Also note that both M and ψ have the wrong statistics.

In order to construct a quantum field theory which will reproduce the SW invariants as correlation functions, anti–ghosts and Lagrangian multipliers are also required. We introduce the anti–ghost multiplet $(\lambda,\eta) \in \Omega^0(X)$, such that

$$[U,\lambda]=-2\lambda, \qquad [Q,\lambda]=\eta, \qquad [Q,\eta]=0,$$

and the Lagrangian multipliers $(\chi,H) \in \Omega^{2,+}(X)$, and $(\mu,\nu) \in S^- \otimes L$ such that

$$\begin{aligned} [U,\chi] &= -\chi, \qquad & [Q,\chi] &= H, \qquad & [Q,H] &= 0; \\ [U,\mu] &= -\mu, \qquad & [Q,\mu] &= \nu, \qquad & [Q,\nu] &= \mathrm{i}\phi\mu. \end{aligned}$$

With the given fields, we construct the following functional which has ghost number -1:

$$\begin{aligned} V = \int_X \Big\{ & [\nabla_k\psi^k + \frac{\mathrm{i}}{2}(\overline{N}M - \overline{M}N)]\lambda - \chi^{kl}\left(H_{kl} - \mathcal{F}^+_{kl} - \frac{\mathrm{i}}{2}\bar{M}\Gamma_{kl}M\right) \\ & - \bar{\mu}\,(\nu - \mathrm{i}D_AM) - \overline{(\nu - \mathrm{i}D_AM)}\mu \Big\}, \end{aligned} \tag{6.48}$$

where the indices of the tensorial fields are raised and lowered by a given metric g on X, and the integration measure is the standard $\sqrt{g}d^4x$. Also, $\overline{M}$ and $\bar{\mu}$ etc. represent the Hermitian conjugate of the spinorial fields. In a formal language, $\overline{M} \in S^+ \otimes L^{-1}$ and $\bar{\mu},\bar{\nu},\overline{D_AM} \in S^- \otimes L^{-1}$. Following the standard procedure in constructing topological quantum field theory, we take the classical action of our theory to be [ZOC95]: $S = [Q, V]$, which has ghost number 0. One can easily show that S is also BRST invariant, i.e., $[Q, S] = 0$, thus it is invariant under the full super–algebra (6.47).

The bosonic Lagrangian multiplier fields H and ν do not have any dynamics, and so can be eliminated from the action by using their equations of motion

$$H_{kl} = \frac{1}{2}\left(\mathcal{F}^{+}_{kl} + \frac{\mathrm{i}}{2}\bar{M}\Gamma_{kl}M\right), \qquad \nu = \frac{1}{2}\mathrm{i}D_A M. \tag{6.49}$$

Then we arrive at the following expression for the action [ZOC95]

$$\begin{aligned} S = \int_X \Big\{ & [-\Delta\phi + \overline{M}M\phi - \mathrm{i}\overline{N}N]\lambda - [\nabla_k\psi^k + \frac{\mathrm{i}}{2}(\overline{N}M - \overline{M}N)]\eta + 2\mathrm{i}\phi\bar{\mu}\mu \\ & + \overline{(\mathrm{i}D_A N - \gamma.\psi M)}\mu - \bar{\mu}\,(\mathrm{i}D_A N - \gamma.\psi M) \\ & - \chi^{kl}\left[\left(\nabla_k\psi^l - \nabla_l\psi^k\right)^+ + \frac{\mathrm{i}}{2}\left(\bar{M}\Gamma_{kl}N + \bar{N}\Gamma_{kl}M\right)\right]\Big\} + S_0, \end{aligned} \tag{6.50}$$

where S_0 is given by

$$S_0 = \int_X \left\{\frac{1}{4}|\mathcal{F}^+ + \frac{\mathrm{i}}{2}\bar{M}\Gamma M|^2 + \frac{1}{2}|D_A M|^2\right\}.$$

It is interesting to observe that S_0 is nonnegative, and vanishes if and only if A and M satisfy the SW monopole equations. As pointed out in [Wit94], S_0 can be rewritten as

$$S_0 = \int_X \left\{\frac{1}{4}|\mathcal{F}^+|^2 + \frac{1}{4}|M|^4 + \frac{1}{8}R|M|^2 + g^{ij}\overline{D_i M}D_j M\right\},$$

where R is the scalar curvature of X associated with the metric g. If R is nonnegative over the entire X, then the only square integrable solution of the monopole equations (6.45) is A is a anti-self-dual connection and $M = 0$.

Quantum Field Theory

We will now investigate the quantum field theory defined by the classical action (6.50) with the path integral method. Let $\mathcal{F}$ collectively denote all the fields. The partition function of the theory is defined by [ZOC95]

$$Z = \int \mathcal{D}\mathcal{F}\exp(-\frac{1}{e^2}S),$$

where $e \in \mathbb{R}$ is the coupling constant. The integration measure $\mathcal{D}\mathcal{F}$ is defined on the space of all the fields. However, since S is invariant under the gauge transformations, we assume the integration over the gauge field to be performed over the gauge orbits of A. In other words, we fix a gauge for the A field using, say, a *Faddeev–Popov procedure*. This can be carried out in the standard manner, thus there is no need for us to spell out the details here. The integration measure $\mathcal{DF}$ can be shown to be invariant under the super charge Q. Also, it does not explicitly involve the metric g of X.

Let W be any operator in the theory. Its correlation function is defined by

$$Z[W] = \int \mathcal{DF} \exp(-\frac{1}{e^2} S) \, W.$$

It follows from the Q invariance of both the action S and the path integration measure that for any operator W,

$$Z[[Q, W]] = \int \mathcal{DF} \exp(-\frac{1}{e^2} S)[Q, W] = 0.$$

For the purpose of constructing topological invariants of the 4–manifold X, we are particularly interested in operators W which are BRST–closed,

$$[Q, W] = 0, \tag{6.51}$$

but not BRST–exact, i.e., can not be expressed as the (anti)–commutators of Q with other operators. For such a W, if its variation with respect to the metric g is $BRST$ exact,

$$\delta_g W = [Q, W'], \tag{6.52}$$

then its correlation function $Z[W]$ is a topological invariant of X (by that we really mean that it does not depend on the metric g):

$$\delta_g Z[W] = \int \mathcal{DF} \exp(-\frac{1}{e^2} S)[Q, W' - \frac{1}{e^2} \delta_g V.W] = 0.$$

In particular, the partition function Z itself is a topological invariant.

Another important property of the partition function is that it does not depend on the coupling constant e:

$$\frac{\partial Z}{\partial e^2} = \int \mathcal{DF} \, \frac{1}{e^4} \exp(-\frac{1}{e^2} S)[Q, V] = 0.$$

Therefore, Z can be computed *exactly* in the limit when the coupling constant goes to zero. Such a computation can be carried out in the standard way: Let A^o, M^o be a solution of the equations of motion of A and M arising from the action S. We expand the fields A and M around this classical configuration,

$$A = A^o + ea, \qquad M = M^o + em,$$

where a and m are the quantum fluctuations of A and M respectively. All the other fields do not acquire background components, thus are purely quantum mechanical. We scale them by the coupling constant e, by setting N to eN, ϕ to $e\phi$ etc.. To the order $o(1)$ in e^2, we have

$$Z = \sum_p \exp(-\frac{1}{e^2} S_{cl}^{(p)}) \int \mathcal{DF}' \exp(-S_q^{(p)}),$$

where $S_q^{(p)}$ is the quadratic part of the action in the quantum fields and depends on the gauge orbit of the classical configuration A^o, M^o, which we label by p. Explicitly [ZOC95],

$$\begin{aligned} S_q^{(p)} = \int_X \Big\{ & [-\Delta\phi + \overline{M}^o M^o \phi - \mathrm{i}\overline{N}N]\lambda - [\nabla_k \psi^k + \frac{\mathrm{i}}{2}(\overline{N}M^o - \overline{M}^o N)]\eta + 2\mathrm{i}\phi\bar{\mu}\mu \\ & + \overline{(\mathrm{i}D_{A^o}N - \gamma.\psi M^o)}\mu - \bar{\mu}\,(\mathrm{i}D_{A^o}N - \gamma.\psi M^o) \\ & - \chi^{kl}\left[\left(\nabla_k\psi^l - \nabla_l\psi^k\right)^+ + \frac{\mathrm{i}}{2}\left(\bar{M}^o\Gamma_{kl}N + \bar{N}\Gamma_{kl}M^o\right)\right] \\ & + \frac{1}{4}|f^+ + \frac{\mathrm{i}}{2}(\bar{m}\Gamma M^o + \bar{M}^o\Gamma m)|^2 + \frac{1}{2}|\mathrm{i}D_{A^o}m + \gamma.aM^o|^2 \Big\}, \end{aligned}$$

with f^+ the self–dual part of $f = da$. The classical part of the action is given by $S_{cl}^{(p)} = S_0|_{A=A^o, M=M^o}$.The integration measure $\mathcal{DF}'$ has exactly the same form as $\mathcal{DF}$ but with A replaced by a, and M by m, $\bar{M}$ by $\bar{m}$ respectively. Needless to say, the summation over p runs through all gauge classes of classical configurations.

Let us now examine further features of our quantum field theory. A gauge class of classical configurations may give a non–zero contribution to the partition function in the limit $\mathrm{e}^2 \to 0$ only if $S_{cl}^{(p)}$ vanishes, and this happens if and only if A^o and M^o satisfy (6.45). Therefore, the SW monopole equations are recovered from the quantum field theory.

The equations of motion of the fields ψ and N in the semi–classical approximation can be easily derived from the quadratic action $S_q^{(p)}$, solutions of which are the zero modes of the quantum fields ψ and N. The equations of motion read

$$\begin{aligned} (d\psi)_{kl}^+ + \frac{\mathrm{i}}{2}\left(\bar{M}^o\Gamma_{kl}N + \bar{N}\Gamma_{kl}M^o\right) = 0, \qquad & D_{A^o}N + \gamma.\psi M^0 = 0, \\ \nabla_k\psi^k + \frac{\mathrm{i}}{2}(\overline{N}M - \overline{M}N) = 0. & \end{aligned} \tag{6.53}$$

Note that they are exactly the same equations which we have already discussed in (6.46). The first two equations are the linearization of the monopole equations, while the last is a 'gauge fixing condition' for ψ. The dimension of the space of solutions of these equations is the virtual dimension of the moduli space $\mathcal{M}$. Thus, within the context of our quantum field theoretical model, the virtual dimension of $\mathcal{M}$ is identified with the number of the zero modes of the quantum fields ψ and N.

For simplicity we assume that there are no zero modes of ψ and N, i.e., the moduli space is zero–dimensional. Then no zero modes exist for the other two fermionic fields χ and μ. To compute the partition function in this case, we first observe that the quadratic action $S_q^{(p)}$ is invariant under the supersymmetry obtained by expanding Q to first order in the quantum fields around the monopole solution A^o, M^o (equations of motion for the nonpropagating fields H and ν should also be used.). This supersymmetry transforms the set of 8 real bosonic fields (each complex field is counted as two real ones; the a_i contribute 2 upon gauge fixing.) and the set of 16 fermionic fields to each other. Thus at a given monopole background we get [ZOC95]

$$\int \mathcal{D}\mathcal{F}' \exp(-S_q^{(p)}) = \frac{\text{Pfaff}(\nabla_{\mathcal{F}})}{|\text{Pfaff}(\nabla_{\mathcal{F}})|} = \epsilon^{(p)},$$

where $\epsilon^{(p)}$ is +1 or −1. In the above equation, $\nabla_{\mathcal{F}}$ is the skew symmetric first order differential operator defining the fermionic part of the action $S_q^{(p)}$, which can be read off from $S_q^{(p)}$ to be $\nabla_{\mathcal{F}} = \begin{pmatrix} 0 & T \\ -T^* & 0 \end{pmatrix}$. Therefore, $\epsilon^{(p)}$ is the sign of the determinant of the elliptic operator T at the monopole background A^o, M^o, and the partition function $Z = \sum_p \epsilon^{(p)}$ coincides with the SW invariant of the 4–manifold X.

When the dimension of the moduli space $\mathcal{M}$ is greater than zero, the partition function Z vanishes identically, due to integration over zero modes of the fermionic fields. In order to get any non trivial topological invariants for the underlying manifold X, we need to examine correlations functions of operators satisfying equations (6.51) and (6.52). A class of such operators can be constructed following the standard procedure [Wit94]. We define the following set of operators

$$\begin{aligned}
W_{k,0} &= \frac{\phi^k}{k!}, \qquad W_{k,1} = \psi W_{k-1,0}, \\
W_{k,2} &= \mathcal{F} W_{k-1,0} - \frac{1}{2}\psi \wedge \psi W_{k-2,0}, \\
W_{k,3} &= \mathcal{F} \wedge \psi W_{k-2,0} - \frac{1}{3!}\psi \wedge \psi \wedge \psi W_{k-3,0}, \\
W_{k,4} &= \frac{1}{2}\mathcal{F} \wedge F W_{k-2,0} - \frac{1}{2}\mathcal{F} \wedge \psi \wedge \psi W_{k-3,0} - \frac{1}{4!}\psi \wedge \psi \wedge \psi \wedge \psi W_{k-4,0}.
\end{aligned} \tag{6.54}$$

These operators are clearly independent of the metric g of X. Although they are not BRST invariant except for $W_{k,0}$, they obey the following equations [ZOC95]

$$\begin{aligned}
dW_{k,0} &= -[Q, W_{k,1}], \qquad dW_{k,1} = [Q, W_{k,2}], \\
dW_{k,2} &= -[Q, W_{k,3}], \qquad dW_{k,3} = [Q, W_{k,4}], \qquad dW_{k,4} = 0,
\end{aligned}$$

which allow us to construct BRST invariant operators from the the W's in the following way: Let X_i, $i = 1, 2, 3$, $X_4 = X$, be compact manifolds without boundary embedded in X. We assume that these submanifolds are homologically nontrivial. Define

$$\widehat{O}_{k,0} = W_{k,0}, \qquad \widehat{O}_{k,i} = \int_{X_i} W_{k,i}, \qquad (i = 1, 2, 3, 4). \tag{6.55}$$

As we have already pointed out, $\widehat{O}_{k,0}$ is BRST invariant. It follows from the descendent equations that

$$[Q, \widehat{O}_{k,i}] = \int_{X_i} [Q, W_{k,i}] = \int_{X_i} dW_{k,i-1} = 0.$$

Therefore the operators $\widehat{O}$ indeed have the properties (6.51) and (6.52). Also, for the boundary ∂K of an $i+1$D manifold K embedded in X, we have

$$\int_{\partial K} W_{k,i} = \int_K dW_{k,i} = [Q, \int_K W_{k,i+1}],$$

is BRST trivial. The correlation function of $\int_{\partial K} W_{k,i}$ with any BRST invariant operator is identically zero. This in particular shows that the $\widehat{O}$'s only depend on the homological classes of the submanifolds X_i.

Dimensional Reduction and 3D Field Theory

In this subsection we dimensionally reduce the quantum field theoretical model for the SW invariant from 4D to 3D, thus to get a new topological quantum field theory defined on 3−manifolds. Its partition function yields a 3−manifold invariant, which can be regarded as the SW version of Casson's invariant [AM90, Tau94].

We take the 4–manifold X to be of the form $Y \times [0,\ 1]$ with Y being a compact 3−manifold without boundary. The metric on X will be taken to be

$$(ds)^2 = (dt)^2 + g_{ij}(x)dx^i dx^j,$$

where the 'time' t−independent $g(x)$ is the Riemannian metric on Y. We assume that Y admits a spin structure which is compatible with the $Spin_c$ structure of X, i.e., if we think of Y as embedded in X, then this embedding induces maps from the $Spin_c$ bundles $S^{\pm} \otimes L$ of X to $\tilde{S} \otimes L$, where $\tilde{S}$ is a spin bundle and L is a line bundle over Y.

To perform the dimensional reduction, we impose the condition that all fields are t in dependent. This leads to the following action [ZOC95]

$$\begin{aligned} S = \int \sqrt{g} d^3x \Big\{ & [-\Delta\phi + \overline{M}M\phi - \mathrm{i}\overline{N}N]\lambda - [\nabla_k \psi^k + \frac{\mathrm{i}}{2}(\overline{N}M - \overline{M}N)]\eta + 2\mathrm{i}\phi\bar{\mu}\mu \\ & + \overline{[\mathrm{i}(D_A + b)N - (\sigma.\psi - \tau)M]}\mu - \bar{\mu}\,[\mathrm{i}(D_A + b)N - (\sigma.\psi - \tau)M] \\ & - 2\chi^k \left[-\partial_k\tau + *(\nabla\psi)_k - \bar{M}\sigma_k N - \bar{N}\sigma_k M\right] \\ & + \frac{1}{4}| * \mathcal{F} - \partial b - \bar{M}\sigma M|^2 + \frac{1}{2}|(D_A + b)M|^2 \Big\}, \end{aligned} \tag{6.56}$$

where the k is a 3D index, and σ_k are the Pauli matrices. The fields $b, \tau \in \Omega^0(Y)$ respectively arose from A_0 and ψ_0 of the 4D theory, while the meanings of the other fields are clear. The BRST symmetry in 4D carries over to the 3D theory. The BRST transformations rules for (A_i, ψ_i, ϕ), $i = 1, 2, 3$, (M, N), and (λ, η) are the same as before, but for the other fields, we have

$$[Q, b] = \tau, \qquad [Q, \tau] = 0,$$
$$[Q, \chi_k] = \frac{1}{2}\left(*\mathcal{F}_k - \partial_k b - \bar{M}\sigma_k M\right),$$
$$[Q, \mu] = \frac{1}{2}\mathrm{i}(D_A + b)M.$$

The action S is cohomological in the sense that $S = [Q, V_3]$, with V_3 being the dimensionally reduced version of V defined by (6.48), and $[Q, S] = 0$. Thus it gives rise to a topological field theory upon quantization. The partition function of the theory

$$Z = \int \mathcal{DF} \exp(-\frac{1}{\mathrm{e}^2} S),$$

can be computed exactly in the limit $\mathrm{e}^2 \to 0$, as it is coupling constant independent. We have, as before,

$$Z = \sum_p \exp(-\frac{1}{\mathrm{e}^2} S_{cl}^{(p)}) \int \mathcal{DF}' \exp(-S_q^{(p)}),$$

where $S_q^{(p)}$ is the quadratic part of S expanded around a classical configuration with the classical parts for the fields A, M, b being A^o, M^o, b^o, while those for all the other fields being zero. The classical action $S_{cl}^{(p)}$ is given by

$$S_{cl}^{(p)} = \int_Y \left\{ \frac{1}{4} | *\mathcal{F}^o - db^o - \bar{M}^o \sigma M^o |^2 + \frac{1}{2} |(D_{A^o} + b^o) M^o|^2 \right\},$$

which can be rewritten as [ZOC95]

$$S_{cl}^{(p)} = \int_Y \left\{ \frac{1}{4} | *\mathcal{F}^o - \bar{M}^o \sigma M^o |^2 + \frac{1}{2} |D_{A^o} M^o|^2 + \frac{1}{2} |db^o|^2 + \frac{1}{2} |b^o M^o|^2 \right\}.$$

In order for the classical configuration to have non–vanishing contributions to the partition function, all the terms in $S_{cl}^{(p)}$ should vanish separately. Therefore,

$$*\mathcal{F}^o - \bar{M}^o \sigma M^o = 0, \qquad D_{A^o} M^o = 0, \qquad b^o = 0, \tag{6.57}$$

where the last condition requires some explanation. When we have a trivial solution of the equations (6.57), it can be replaced by the less stringent condition $db^o = 0$. However, in a more rigorous treatment of the problem at hand, we in general perturb the equations (6.57), then the trivial solution does not arise.

Let us define an operator

$$\begin{aligned} \tilde{T}: \Omega^0(Y) \oplus \Omega^1(Y) \oplus (\tilde{S} \otimes L) &\to \Omega^0(Y) \oplus \Omega^1(Y) \oplus (\tilde{S} \otimes L), \\ (\tau, \psi, N) \mapsto (-d^*\psi + \frac{\mathrm{i}}{2}(\overline{N}M - \overline{M}N), \quad & *(d\psi) - d\tau - \bar{N}\sigma M - \overline{M}\sigma N, \\ \mathrm{i}D_A N - (\sigma.\psi - \tau)M), & \qquad (6.58) \end{aligned}$$

where the complex bundle $\tilde{S} \otimes L$ should be regarded as a real one with twice the rank. This operator is self–adjoint, and is also obviously elliptic. We will assume that it is Fredholm as well. In terms of $\tilde{T}$, the equations of motion of the fields χ^i and μ can be expressed as [ZOC95] $\tilde{T}^{(p)}(\tau, \psi, N) = 0$, where $\tilde{T}^{(p)}$ is the opeartor $\tilde{T}$ with the background fields (A^o, M^o) belonging to the gauge class p of classical configurations .

When the kernel of $\tilde{T}$ is zero, the partition function Z does not vanish identically. An easy computation leads to $Z = \sum_p \epsilon^{(p)}$, where the sum is over all gauge inequivalent solutions of (6.57), and $\epsilon^{(p)}$ is the sign of the determinant of $\tilde{T}^{(p)}$.

A rigorous definition of the sign of the $\det(\tilde{T})$ can be devised. However, if we are to compute only the absolute value of Z, then it is sufficient to know the sign of $\det(\tilde{T})$ relative to a fixed gauge class of classical configurations. This can be achieved using the *mod* $-$ 2 spectral flow of a family of Fredholm operators $\tilde{T}_t$ along a path of solutions of (6.57). More explicitly, let (A^o, M^o) belong to the gauge class of classical configurations p, and $(\tilde{A}^o, \tilde{M}^o)$ in $\tilde{p}$. We consider the solution of the SW equation on $X = Y \times [0, 1]$ with $A_0 = 0$ and also satisfying the following conditions

$$(A, M)|_{t=0} = (A^o, M^o), \qquad (A, M)|_{t=1} = (\tilde{A}^o, \tilde{M}^o).$$

Using this solution in $\tilde{T}$ results in a family of *Fredholm operators*, which has zero kernels at $t = 0$ and 1. The spectral flow of $\tilde{T}_t$, denoted by $q(p, \tilde{p})$, is defined to be the number of eigenvalues which cross zero with a positive slope minus the number which cross zero with a negative slope. This number is a well defined quantity, and is given by the index of the operator $\frac{\partial}{\partial t} - \tilde{T}_t$. In terms of the spectral flow, we have [ZOC95]

$$\frac{\det(\tilde{T}^{(p)})}{\det(\tilde{T}^{(\tilde{p})})} = (-1)^{q(p, \tilde{p})}.$$

Equations (6.57) can be derived from the functional

$$S_{c-s} = \frac{1}{2} \int_Y A \wedge \mathcal{F} + \mathrm{i} \int_Y \sqrt{g} d^3x \overline{M} D_A M.$$

(It is interesting to observe that this is almost the standard Lagrangian of a $U(1)$ Chern–Simons theory coupled to spinors, except that we have taken M to have bosonic statistics.) S_{c-s} is gauge invariant modulo a constant arising from the Chern–Simons term upon a gauge transformation. Therefore,

$(\frac{\delta S_{c-s}}{\delta A}, \frac{\delta S_{c-s}}{\delta \bar{M}})$ defines a vector field on the quotient space of all $U(1)$ connections $\mathcal{A}$ tensored with the $\tilde{S} \times L$ sections by the $U(1)$ gauge group G, i.e., $\mathcal{W} = (\mathcal{A} \times (\tilde{S} \otimes L))/G$. Solutions of (6.57) are zeros of this vector field, and $\tilde{T}^{(p)}$ is the Hessian at the point $p \in \mathcal{W}$. Thus the partition Z is nothing else but the Euler character of $\mathcal{W}$. This geometrical interpretation will be spelt out more explicitly in the next subsection by re–interpreting the theory using the *Mathai–Quillen formula* [MQ86].

Geometrical Interpretation

To elucidate the geometric meaning of the 3D theory obtained in the last section, we now cast it into the framework of Atiyah and Jeffrey [AJ90]. Let us briefly recall the geometric set up of the *Mathai–Quillen formula* as reformulated in [AJ90]. Let P be a Riemannian manifold of dimension $2m + \dim G$, and G be a compact Lie group acting on P by isometries. Then $P \to P/G$ is a principle bundle. Let V be a $2m$ dimensional real vector space, which furnishes a representation $G \to SO(2m)$. Form the associated vector bundle $P \times_G V$. Now the Thom form of $P \times_G V$ can be expressed [ZOC95]

$$\begin{aligned} U = \frac{\exp(-x^2)}{(2\pi)^{\dim G}\pi^m} \int \exp\left\{\frac{\mathrm{i}\chi\phi\chi}{4} + \mathrm{i}\chi dx - \mathrm{i}\langle\delta\nu, \lambda\rangle\right. \\ \left. - \langle\phi, R\lambda\rangle + \langle\nu, \eta\rangle\right\} \mathcal{D}\eta\mathcal{D}\chi\mathcal{D}\phi\mathcal{D}\lambda, \end{aligned} \tag{6.59}$$

where $x = (x^1, ..., x^{2m})$ is the coordinates of V, ϕ and λ are bosonic variables in the Lie algebra g of G, and η and χ are Grassmannian variables valued in the Lie algebra and the tangent space of the fiber respectively. In the above equation, C maps any $\eta \in g$ to the element of the vertical part of TP generated by η; ν is the g - valued one form on P defined by $\langle\nu(\alpha), \eta\rangle = \langle\alpha, C(\eta)\rangle$, for all vector fields α; and $R = C^*C$. Also, δ is the exterior derivative on P.

Now we choose a G invariant map $s : P \to V$, and pull back the Thom form U. Then the top form on P in s^*U is the Euler class. If $\{\delta p\}$ forms a basis of the cotangent space of P(note that ν and δs are one forms on P), we replace it by a set of Grassmannian variables $\{\psi\}$ in s^*U, then intergrate them away. We arrive at

$$\begin{aligned} \Upsilon = \frac{1}{(2\pi)^{\dim G}\pi^m} \int \exp\left\{-|s|^2 + \frac{\mathrm{i}\chi\phi\chi}{4} + \mathrm{i}\chi\delta s - \mathrm{i}\langle\delta\nu, \lambda\rangle\right. \\ \left. - \langle\phi, R\lambda\rangle + \langle\psi, C\eta\rangle\right\} \mathcal{D}\eta\mathcal{D}\chi\mathcal{D}\phi\mathcal{D}\lambda\mathcal{D}\psi, \end{aligned} \tag{6.60}$$

the precise relationship of which with the Euler character of $P \times_G V$ is

$$\int_P \Upsilon = \mathrm{Vol}(G)\chi(P\times_G).$$

It is rather obvious that the action S defined by (6.50) for the 4D theory can be interpreted as the exponent in the integrand of (6.60), if we identify

P with $\mathcal{A} \times \Gamma(W^+)$, and V with $\Omega^{2,+}(X) \times \Gamma(W^-)$, and set $s = (\mathcal{F}^+ + \frac{\mathrm{i}}{2}\bar{M}\Gamma M, D_A M)$. Here $\mathcal{A}$ is the space of all $U(1)$ connections of $\det(W^+)$, and $\Gamma(W^\pm)$ are the sections of $S^\pm \otimes L$ respectively.

For the 3D theory, we wish to show that the partition function yields the Euler number of $\mathcal{W}$. However, the tangent bundle of $\mathcal{W}$ cannot be regarded as an associated bundle with the principal bundle, for which for the formulae (6.59) or (6.60) can readily apply, some further work is required.

Let P be the principal bundle over P/G, V, V' be two orthogonal representions of G. Suppose there is an embedding from $P \times_G V'$ to $P \times_G V$ via a $G-$map $\gamma(p) : V' \to V$ for $p \in P$. Denote the resulting quotient bundle as E. In order to derive the Thom class for E, one needs to choose a section of E, or equivalently, a $G-$map $s : P \to V$ such that $s(p) \in (\mathrm{Im}\gamma(p))^\perp$. Then the Euler class of E can be expressed as $\pi_* \rho^* U$, where U is the Thom class of $P \times_G V$, ρ is a $G-$map: $P \times V' \to P \times V$ defined by

$$\rho(p, \tau) = (p, \gamma(p)\tau + s(p)),$$

and π_* is the integration along the fiber for the projection $\pi : P \times V' \to P/G$. Explicitly, [ZOC95]

$$\begin{aligned} \pi_* \rho^*(U) = \int \exp\{&-|\gamma(p)\tau + s(p)|^2 + \mathrm{i}\chi\phi\chi + \mathrm{i}\chi\delta(\gamma(p)\tau + s(p)) \\ &- \mathrm{i}\langle\delta\nu, \lambda\rangle - \langle\phi, R\lambda\rangle + \langle\nu, C\eta\rangle\ \}\mathcal{D}\chi\mathcal{D}\phi\mathcal{D}\tau\mathcal{D}\eta\mathcal{D}\lambda. \end{aligned} \tag{6.61}$$

Consider the exact sequence

$$0 \longrightarrow (\mathcal{A} \times \Gamma(W)) \times_G \Omega^0(Y) \stackrel{j}{\longrightarrow} (\mathcal{A} \times \Gamma(W)) \times_G (\Omega^1(Y) \times \Gamma(W)),$$

where $j_{(A,M)} : b \mapsto (-db, bM)$ (assuming that $M \neq 0$). Then the tangent bundle of $\mathcal{A} \times_G \Gamma(W)$ can be Regarded as the quotient bundle

$$(\mathcal{A} \times \Gamma(W)) \times_G (\Omega^1(Y) \times \Gamma(W))/\mathrm{Im}(j).$$

We define a vector field on $\mathcal{A} \times_G \Gamma(W)$ by

$$s(A, M) = (*\mathcal{F}_A - \bar{M}\sigma M, D_A M),$$

which lies in $\mathrm{Im}(j)^\perp$:

$$\int_Y (*\mathcal{F}_A - \bar{M}\sigma M) \wedge *(-db) + \int_Y \sqrt{g} d^3x \langle D_A M, bM\rangle = 0, \tag{6.62}$$

where we have used the short hand notation $\langle M_1, M_2\rangle = \frac{1}{2}(\overline{M}_1 M_2 + \overline{M}_2 M_1)$.

Formally applying the formula (6.61) to the present infinite–dimensional situation, we get the *Euler class* $\pi_*\rho^*(U)$ for the tangent bundle $T(\mathcal{A} \times_G \Gamma(W))$, where ρ is the $\mathcal{G}-$invariant map ρ is defined by

$$\rho : \quad \Omega^0(Y) \longrightarrow \Omega^1(Y) \times \Gamma(W), \qquad \rho(b) = (-db + *\mathcal{F}_A - \bar{M}\sigma M, (D_A + b)M),$$

π is the projection $(\mathcal{A} \times \Gamma(W)) \times_G \Omega^0(Y) \longrightarrow \mathcal{A} \times_G \Gamma(W)$, and π_* signifies the integration along the fiber. Also U is the Thom form of the bundle

$$(\mathcal{A} \times \Gamma(W)) \times_G (\Omega^1(Y) \times \Gamma(W)) \longrightarrow \mathcal{A} \times_G \Gamma(W).$$

To get a concrete feel about U, we need to explain the geometry of this bundle. The metric on Y and the Hermitian metric $\langle . \, , \, . \rangle$ on $\Gamma(W)$ naturally define a connection. The *Maurer–Cartan connection* on $\mathcal{A} \longrightarrow \mathcal{A}/G$ is flat while the Hermitian connection on has the curvature $\mathrm{i}\phi\mu \wedge \bar{\mu}$. This gives the expression of term $\mathrm{i}(\chi, \mu)\phi(\chi, \mu)$ in (6.60) in our case.

In our infinite–dimensional setting, the map C is given by

$$C: \quad \Omega^0(Y) \longrightarrow T_{(A,M)}(\mathcal{A} \times \Gamma(W)), \qquad C(\eta) = (-d\eta, \mathrm{i}\eta M),$$

and its dual is given by

$$C^*: \quad \Omega^1(Y) \times \Gamma(W) \longrightarrow \Omega^0(Y), \qquad C^*(\psi, N) = -d^*\psi + \langle N, \mathrm{i}M \rangle.$$

The one form $\langle \nu, \eta \rangle$ on $\mathcal{A} \times \Gamma(W)$ takes the value

$$\langle (\psi, N), C\eta \rangle = \langle -d^*\psi, \eta \rangle + \langle N, \mathrm{i}M \rangle \eta$$

on the vector field (ψ, N). We also easily get $R(\lambda) = -\Delta\lambda + \langle M, M \rangle \lambda$, where $\Delta = d^*d$. The $\langle \delta\nu, \lambda \rangle$ is a two form on $\mathcal{A} \times \Gamma(W)$ whose value on $(\psi_1, N_1), (\psi_2, N_2)$ is -$\langle N_1, N_2 \rangle \lambda$.

Combining all the information together, we arrive at the following formula,

$$\begin{aligned} \pi_*\rho^*(U) = \int \exp\Big\{ & -\frac{1}{2}|\rho|^2 + \mathrm{i}(\chi, \mu)\delta\rho + 2\mathrm{i}\phi\mu\bar{\mu} \\ & + \langle \Delta\phi, \lambda \rangle - \phi\lambda\langle M, M \rangle + \mathrm{i}\langle N, N \rangle \lambda \\ & + \langle \nu, \eta \rangle \Big\} \mathcal{D}\chi\mathcal{D}\phi\mathcal{D}\lambda\mathcal{D}\eta\mathcal{D}b. \end{aligned} \tag{6.63}$$

Note that the 1–form $\mathrm{i}(\chi, \mu)\delta\rho$ on $\mathcal{A} \times \Gamma(W) \times \Omega^0(Y)$ contacted with the vector field (ϕ, N, b) leads to

$$2\chi^k \left[-\partial_k\tau + *(\nabla\psi)_k - \bar{M}\sigma_k N - \bar{N}\sigma_k M \right] + 2\langle \mu, [\mathrm{i}(D_A + b)N - (\sigma.\psi - \tau)M] \rangle;$$

and the relation (6.62) gives $|\rho|^2 = |*\mathcal{F} - \bar{M}\sigma M|^2 + |db|^2 + |D_A M|^2 + b^2|M|^2$. Finally we get the *Euler character*

$$\pi_*\rho^*(U) = \int \exp(-S)\mathcal{D}\chi\mathcal{D}\phi\mathcal{D}\lambda\mathcal{D}\eta\mathcal{D}b, \tag{6.64}$$

where S is the action (6.56) of the 3D theory defined on the manifold Y.

Integrating (6.64) over $\mathcal{A} \times_G \Gamma(W)$ leads to the Euler number

$$\sum_{[(A,M)]: s(A,M)=0} \epsilon^{(A,M)},$$

which coincides with the partition function Z of our 3D theory.

6.2.3 TQFTs Associated with SW–Monopoles

Recall that TQFTs are often used to study topological nature of manifolds. In particular, 3D and 4D TQFTs are well developed. The most well–known 3D TQFT would be the Chern–Simons theory, whose partion function gives *Ray–Singer torsion* of 3–manifolds and the other topological invariants can be obtained as gauge invariant observables i.e., *Wilson loops*. The correlation functions can be identified with knot or link invariants e.g., Jones polynomal or its generalizations. On the other hand, in 4D, a twisted $N = 2$ supersymmetric YM theory developed by Witten [Wit88b] also has a nature of TQFT. This YM theory can be interpreted as *Donaldson theory* and the *correlation functions* are identified with *Donaldson polynomials*, which classify smooth structures of topological 4–manifolds. A new TQFT on 4–manifolds was discovered in SW studies of electric–magnetic duality of supersymmetric gauge theory. Seiberg and Witten [SW94a, SW94b] studied the electric–magnetic duality of $N = 2$ supersymmetric $SU(2)$ YM gauge theory, by using a version of *Montonen–Olive duality* and obtained exact solutions. According to this result, the exact *low energy effective action* can be determined by a certain elliptic curve with a parameter $u = \langle \mathrm{Tr}(\phi)^2 \rangle$, where ϕ is a complex scalar field in the adjoint representation of the gauge group, describing the quantum moduli space. For large u, the theory is weakly–coupled and semi–classical, but at $u = \pm\Lambda^2$ corresponding to strong coupling regime, where Λ is the dynamically generated mass scale, the elliptic curve becomes singular and the situation of the theory changes drastically. At these singular points, magnetically charged particles become massless. Witten showed that at $u = \pm\Lambda^2$ the TQFT was related to the moduli problem of counting the solution of the (Abelian) 'Seiberg–Witten monopole equations' [Wit94] and it gave a dual description for the $SU(2)$ Donaldson theory.

It turns out that in 3D a particular TQFT of *Bogomol'nyi monopoles* can be obtained from a dimensional reduction of Donaldson theory and the partition function of this theory gives the so–called *Casson invariant* of 3–manifolds [AJ90].

Ohta [Oht98] discussed TQFTs associated with the 3D version of both Abelian and non–Abelian SW–monopoles, by applying *Batalin–Vilkovisky quantization* procedure. In particular, Ohta constructed the topological actions, topological observables and BRST transformation rules.

In this subsection, mainly following [Oht98], we will discuss TQFTs associated with both Abelian and non–Abelian SW–monopoles. We will use the following notation.

Let X be a compact orientable Spin 4–manifold without boundary and $g_{\mu\nu}$ be its Riemannian metric tensor (with $g = \det\ g_{\mu\nu}$). Here we use x_μ as the local coordinates on X. γ_μ are Dirac's gamma matrices and $\sigma_{\mu\nu} = [\gamma_\mu, \gamma_\nu]/2$ with $\{\gamma_\mu, \gamma_\nu\} = g_{\mu\nu}$. M is a Weyl fermion and $\overline{M}$ is a complex conjugate of M. (We will suppress spinor indices.) The Lie algebra $\mathfrak{g}$ is defined by $[T^a, T^b] = \mathrm{i} f_{abc} T^c$, where T^a is a generator normalized as $\mathrm{Tr}(T^a T^b) =$

δ^{ab}. The symbol f_{abc} is a structure constant of $\mathfrak{g}$ and is antisymmetric in its indices. The Greek indices μ, ν, α etc run from 0 to 3. The Roman indices $a, b, c, \cdots$ are used for the Lie algebra indices running from 1 to dim $\mathfrak{g}$, whereas $i, j, k, \cdots$ are the indices for space coordinates. Space–time indices are raised and lowered with $g_{\mu\nu}$. The repeated indices are assumed to be summed. $\epsilon_{\mu\nu\rho\sigma}$ is an antisymmetric tensor with $\epsilon_{0123} = 1$. We often use the abbreviation of roman indices as $\theta = \theta^a T^a$ etc., in order to suppress the summation over Lie algebra indices.

Brief Review of TQFT

Firstly, we give a brief review of TQFT (compare with Witten's TQFT presented in subsection 6.2.1 above).

Let ϕ be any field content. For a local symmetry of ϕ, we can construct a nilpotent *BRST–operator* Q_B ($Q_B^2 = 0$). The variation of any functional $\mathcal{O}$ of ϕ is denoted by $\delta\mathcal{O} = \{Q_B, \mathcal{O}\}$, where the bracket $\{*, *\}$ represnts a graded commutator, that is, if $\mathcal{O}$ is bosonic, the bracket means a commutator $[*, *]$ and otherwise it is an anti–bracket. Now, we can give the definition of topological field theory, as given in [BBR91]:

A topological field theory consists of:

1. *a collection of Grassmann graded fields ϕ on an nD Riemannian manifold X with a metric g,*
2. *a nilpotent Grassmann odd operator Q,*
3. *physical states to be $Q-$cohomology classes,*
4. *an energy–momentum tensor $T_{\alpha\beta}$ which is $Q-$exact for some functional $V_{\alpha\beta}$ such as*
$$T_{\alpha\beta} = \{Q, V_{\alpha\beta}(\phi, g)\}.$$
In this definition, Q is often identified with Q_B and is in general independent of the metric. Now, recall that there are two broad types of TQFTs satisfying this definition and they are classified into Witten–type [Wit94] or Schwarz–type [Sch78)].

For *Witten*–type TQFT, the quantum action S_q which comprises the classical action, ghost and gauge fixing terms, can be represented by $S_q = \{Q_B, V\}$, for some function V of metric and fields and BRST charge Q_B. Under the metric variation δ_g of the partition function Z, it is easy to see that

$$\begin{aligned}\delta_g Z &= \int \mathcal{D}\phi \; \mathrm{e}^{-S_q} \left(-\frac{1}{2} \int_X d^n x \sqrt{g} \delta g^{\alpha\beta} T_{\alpha\beta} \right) \\ &= \int \mathcal{D}\phi \; \mathrm{e}^{-S_q} \{Q, \chi\} \;\equiv\; \langle \{Q, \chi\} \rangle = 0, \end{aligned} \tag{6.65}$$

$$\text{where} \qquad \chi = -\frac{1}{2} \int_X d^n x \sqrt{g} \delta g^{\alpha\beta} V_{\alpha\beta}.$$

The last equality in (6.65) follows from the BRST invariance of the vacuum and means that Z is independent of the local structure of X, that is, Z is a *topological invariant* of X.

In general, for Witten type theory, Q_B can be constructed by an introduction of a topological shift with other local gauge symmetry [Oht98]. For example, in order to get the topological YM theory on four manifold M^4, we introduce the shift in the gauge transformation for the gauge field A^a_μ such as $\delta A^a_\mu = D_\mu \theta^a + \epsilon^a_\mu$, where D_μ is a covariant derivative, θ^a and ϵ^a_μ are the (Lie algebra valued) usual gauge transformation parameter and topological shift parameter, respectively. In order to see the role of this shift, let us consider the first *Pontryagin class* on M^4 given by

$$S = \frac{1}{8} \int_{M^4} \epsilon^{\mu\nu\rho\sigma} F^a_{\mu\nu} F^a_{\rho\sigma} d^4x, \tag{6.66}$$

where $F^a_{\mu\nu}$ is a field strength of the gauge field. We can easily check the invariance of (6.66) under the action of δ. In this sense, (6.66) has a larger symmetry than the usual YM gauge symmetry. Taking this into account, we can construct the topological YM gauge theory. We can also consider similar 'topological' shifts for matter fields.

In addition, in general, Witten type topological field theory can be obtained from the quantization of some Langevin equations. This approach has been used for the construction of several topological field theories, e.g., supersymmetric quantum mechanics, topological sigma models or Donaldson theory (see [BBR91]).

On the other hand, *Schwarz–type TQFT* [Sch78)] begins with any metric independent classical action S_c as a starting point, but S_c is assumed not to be a total derivative. Then the quantum action (up to gauge fixing term) can be written by

$$S_q = S_c + \{Q, V(\phi, g)\}, \tag{6.67}$$

for some function V. For this quantum action, we can easily check the topological nature of the partition function, but note that the energy–momentum tensor contributes only from the second term in (6.67). One of the differences between Witten type and Schwarz type theories can be seen in this point. Namely, the energy–momentum tensor of the classical action for Schwarz type theory vanishes because it is derived as a result of metric variation.

Finally, we comment on the local symmetry of Schwarz type theory. Let us consider the Chern–Simons theory as an example. The classical action,

$$S_{CS} = \int_{M^3} d^3x \left(A \wedge dA + \frac{2}{3} A \wedge A \wedge A \right), \tag{6.68}$$

is a *topological invariant*, which gives the *second Chern class* of 3–manifold M^3. As is easy to find, S_{CS} is not invariant under the topological gauge transformation, although it is YM gauge invariant. Therefore the quantization is proceeded by the standard BRST method. This is a general feature of Schwarz–type TQFT.

Dimensional Reduction

First, let us recall the SW monopole equations in 4D. We assume that X has Spin structure. Then there exist rank two positive and negative spinor bundles $S^{\pm}$. For Abelian gauge theory, we introduce a complex line bundle L and a connection A_μ on L. The *Weyl spinor* $M(\overline{M})$ is a section of $S^+ \otimes L$ $(S^+ \otimes L^{-1})$, hence M satisfies the *positive chirality condition* $\gamma^5 M = M$. If X does not have Spin structure, we introduce *Spin*^{c} structure and *Spin*^{c} bundles $S^{\pm} \otimes L$, where L^2 is a line bundle. In this case, M should be interpreted as a section of $S^+ \otimes L$. Below, we assume Spin structure.

Recall that the 4D Abelian SW monopole equations are the following set of differential equations

$$F^+_{\mu\nu} + \frac{\mathrm{i}}{2}\overline{M}\sigma_{\mu\nu}M = 0, \qquad \mathrm{i}\gamma^\mu D_\mu M = 0, \tag{6.69}$$

where $F^+_{\mu\nu}$ is the self–dual part of the $U(1)$ curvature tensor

$$F_{\mu\nu} = \partial_\mu A_\nu - \partial_\nu A_\mu, \qquad F^+_{\mu\nu} = P^+_{\mu\nu\rho\sigma}F^{\rho\sigma}, \tag{6.70}$$

while $P^+_{\mu\nu\rho\sigma}$ is the self–dual projector defined by

$$P^+_{\mu\nu\rho\sigma} = \frac{1}{2}\left(\delta_{\mu\rho}\delta_{\nu\sigma} + \frac{\sqrt{g}}{2}\epsilon_{\mu\nu\rho\sigma}\right).$$

Note that the second term in the first equation of (6.69) is also self–dual. On the other hand, the second equation in (6.69) is a *twisted Dirac equation*, whose covariant derivative D_μ is given by

$$D_\mu = \partial_\mu + \omega_\mu - \mathrm{i}A_\mu, \qquad \text{where} \qquad \omega_\mu = \frac{1}{4}\omega^{\alpha\beta}_\mu[\gamma_\alpha, \gamma_\beta]$$

is the *spin connection 1–form* on X.

In order to perform a reduction to 3D, let us first assume that X is a *product manifold* of the form $X = Y \times [0, 1]$, where Y is a 3D compact manifold which has Spin structure. We may identify $t \in [0, 1]$ as a 'time' variable, or, we assume t as the zero–th coordinate of X, whereas x_i $(i = 1, 2, 3)$ are the coordinates on (space manifoild) Y. Then the metric is given by

$$ds^2 = dt^2 + g_{ij}dx^i dx^j.$$

The dimensional reduction is proceeded by assumnig that all fields are independent of t. (Below, we suppress the volume factor $\sqrt{g}$ of Y for simplicity.)

First, let us consider the Dirac equation. After the dimensional reduction, the Dirac equation will be

$$\gamma^i D_i M - \mathrm{i}\gamma^0 A_0 M = 0.$$

As for the first monopole equation, using (6.70) we find that

$$F_{i0} + \frac{1}{2}\epsilon_{i0jk}F^{jk} = -\mathrm{i}\overline{M}\sigma_{i0}M, \qquad F_{ij} + \epsilon_{ijk0}F^{k0} = -\mathrm{i}\overline{M}\sigma_{ij}M. \qquad (6.71)$$

Since the above two equations are dual each other, the first one, for instance, can be reduced to the second one by a contraction with the totally anti–symmetric tensor. Thus, it is sufficient to consider one of them. Here, we take the first equation in (6.71).

After the dimensional reduction, (6.71) will be

$$\partial_i A_0 - \frac{1}{2}\epsilon_{ijk}F^{jk} = -\mathrm{i}\overline{M}\sigma_{i0}M, \qquad (6.72)$$

where we have set $\epsilon_{ijk} \equiv \epsilon_{0ijk}$.

Therefore, the 3D version of the SW equations are given by

$$\partial_i b - \frac{1}{2}\epsilon_{ijk}F^{jk} + \mathrm{i}\overline{M}\sigma_{i0}M = 0, \qquad \mathrm{i}(\gamma^i D_i - \mathrm{i}\gamma^0 b)M = 0, \qquad (b \equiv A_0). \quad (6.73)$$

It is now easy to establish the non–Abelian 3D monopole equations as

$$\partial_i b^a + f_{abc}A_i^b b^c - \frac{1}{2}\epsilon_{ijk}F^{ajk} + \mathrm{i}\overline{M}\sigma_{i0}T^a M = 0, \qquad \mathrm{i}(\gamma^i D_i - \mathrm{i}\gamma^0 b)M = 0,$$

where we have abbreaviated $\overline{M}\sigma_{\mu\nu}T^a M \equiv \overline{M}^i\sigma_{\mu\nu}(T^a)_{ij}M^j$, subscripts of $(T^a)_{ij}$ run from 1 to dim $\mathfrak{g}$ and $b^a \equiv A_0^a$.

Next, let us find an action which produces (6.73). The simplest one is given by

$$S = \frac{1}{2}\int_Y \left[\left(\partial_i b - \frac{1}{2}\epsilon_{ijk}F^{jk} + \mathrm{i}\overline{M}\sigma_{i0}M\right)^2 + |\mathrm{i}(\gamma^i D_i - \mathrm{i}\gamma^0 b)M|^2\right] d^3x. \qquad (6.74)$$

Note that the minimum of (6.74) is given by (6.73). In this sense, the 3D monopole equations are not equations of motion but rather of constraints. Furthermore, there is a constraint for b. To see this, let us rewrite (6.74) as

$$S = \int_Y d^3x \left[\frac{1}{2}\left(\frac{1}{2}\epsilon_{ijk}F^{jk} - \mathrm{i}\overline{M}\sigma_{i0}M\right)^2 + \frac{1}{2}|\gamma^i D_i M|^2 + \frac{1}{4}(\partial_i b)^2 + \frac{1}{2}b^2|M|^2\right].$$

The minimum of this action is clearly given by the 3D monopole equations with $b = 0$, for non–trivial A_i and M. However, for trivial A_i and M, we may relax the condition $b = 0$ to $\partial_i b = 0$, i.e., b is in general a non–zero constant. This can be also seen from (6.72). Therefore, we get

$$\frac{1}{2}\epsilon_{ijk}F^{jk} - \mathrm{i}\overline{M}\sigma_{i0}M = 0, \qquad \mathrm{i}\gamma^i D_i M = 0,$$
$$\text{with} \qquad b = 0 \quad \text{or} \quad \partial_i b = 0, \qquad (6.75)$$

as an equivalent expression to (6.73), but we will rather use (6.73) for convenience. The Gaussian action will be used in the next subsection in order to construct a TQFT by Batalin–Vilkovisky quantization algorithm. The non–Abelian version of (6.74) and (6.75) would be obvious.

TQFTs of 3D Monopoles

In this subsection, we construct TQFTs associated with both the Abelian and non–Abelian 3D monopoles by Batalin–Vilkovisky quantization algorithm.

Abelian Case

A 3D action for the Abelian 3D monopoles was found by the direct dimensional reduction of the 4D one [ZOC95], but here we rather show that the 3D topological action can be directly constructed from the 3D monopole equations [Oht98].

Topological Bogomol'nyi Action

A *topological Bogomol'nyi action* was constructed by using *Batalin–Vilkovisky quantization* algorithm [BRT89], or quantization of a magnetic charge [BG88]. The former is based on the quantization of a certain Langevin equation ('Bogomol'nyi monopole equation') and the classical action is quadratic, but the latter is based on the 'quantization' of the pure topological invariant by using the Bogomol'nyi monopole equation as a gauge fixing condition.

In order to compare the action to be constructed with those of Bogomol'nyi monopoles [BRT89, BG88], we take Batalin–Vilkovisky procedure (see also [BBR91]).

In order to get the topological action associated with 3D monopoles, we introduce random Gaussian fields G_i and $\nu(\overline{\nu})$ and then start with the action

$$S_c = \frac{1}{2}\int_Y [(G_i - \partial_i b + \frac{1}{2}\epsilon_{ijk}F^{jk} - \mathrm{i}\overline{M}\sigma_{i0}M)^2 + |(\nu - \mathrm{i}\gamma^i D_i M - \gamma^0 bM)|^2] d^3x. \tag{6.76}$$

Note that G_i and $\nu(\overline{\nu})$ are also regarded as auxiliary fields. This action reduces to (6.74) in the gauge

$$G_i = 0, \qquad \nu = 0. \tag{6.77}$$

Firstly, note that (6.76) is invariant under the topological gauge transformation

$$\begin{aligned} \delta A_i &= \partial_i\theta + \epsilon_i, \qquad \delta b = \tau, \qquad \delta M = \mathrm{i}\theta M + \varphi, \\ \delta G_i &= \partial_i\tau - \epsilon_{ijk}\partial^j\epsilon^k + \mathrm{i}(\overline{\varphi}\sigma_{i0}M + \overline{M}\sigma_{i0}\varphi), \\ \delta\nu &= \mathrm{i}\theta\nu + \gamma^i\epsilon_i M + \mathrm{i}\gamma^i D_i\varphi + \gamma^0 b\varphi + \gamma^0\tau M, \end{aligned} \tag{6.78}$$

where θ is the parameter of gauge transformation, ϵ_i and $\tau \equiv \epsilon_4$ are parameters which represent the topological shifts and φ the shift on the spinor space. The brackets for indices means anti–symmetrization, i.e., $A_{[i}B_{j]} = A_iB_j - A_jB_i$.

Here, let us classify the gauge algebra (6.78). This is necessary to use Batalin–Vilkovisky algorithm. Let us recall that the local symmetry for fields ϕ_i can be written generally in the form $\delta\phi_i = R^i_\alpha(\phi)\epsilon^\alpha$, where the indices mean the label of fields and ϵ^α is a some local parameter. When $\delta\phi_i = 0$ for non–zero ϵ^α, this symmetry is called *first–stage reducible*. In the reducible theory, we can find zero–eigenvectors Z_a^α satisfying $R_\alpha^i Z_a^\alpha = 0$. Moreover, when the theory is *on–shell reducible*, we can find such eigenvectors by using equations of motion.

For the case at hand, under the identifications

$$\begin{aligned} &\theta = \Lambda, \qquad \epsilon_i = -\partial_i \Lambda, \qquad \varphi = -\mathrm{i}\Lambda M, \\ \text{and} \qquad &\tau = 0, \qquad \text{so that (6.78) will be} \end{aligned} \tag{6.79}$$

$$\begin{aligned} &\delta A_i = 0, \qquad \delta b = 0, \qquad \delta M = 0, \qquad \delta G_i = 0, \\ &\delta\nu = \mathrm{i}\Lambda(\nu - \mathrm{i}\gamma^i D_i M - \gamma^0 b M)|_{on-shell} = 0. \end{aligned} \tag{6.80}$$

Then for δA_i, for example, the R coefficients and the zero–eigenvectors are derived from

$$\begin{aligned} &\delta A_i = R_\theta^{A_i} Z_\Lambda^\theta + R_{\epsilon_j}^{A_i} Z_\Lambda^{\epsilon_j} = 0, \qquad \text{that is} \\ &R_\theta^{A_i} = \partial_i, \qquad R_{\epsilon_j}^{A_i} = \delta_{ij}, \qquad Z_\Lambda^\theta = 1, \qquad Z_\Lambda^{\epsilon_j} = -\partial_j. \end{aligned}$$

Obviously, similar relations hold for other fields. The reader may think that the choice (6.79) is not suitable as a first stage reducible theory, but note that the zero–eigenvectors appear on every point where the gauge equivalence and the topological shift happen to coincide. In this three dimensional theory, $b(A_0)$ is invariant for the usual infinitesimal gauge transformation because of its 'time' independence, so (6.79) means that the existence of the points on spinor space where the topological shift trivializes indicates the first stage reducibility.

If we carry out *BRST quantization* via *Faddeev–Popov procedure* in this situation, the Faddeev–Popov determinant will have zero modes. Therefore in order to fix the gauge further we need a ghost for ghost. This reflects on the second generation gauge invariance (6.80) realized on–shell. However, since b is irrelevant to Λ, the ghost for τ will not couple to the second generation ghost. With this in mind, we use Batalin–Vilkovisky algorithm in order to make BRST quantization (for details, see [Oht98] and references therein).

Let us assign new ghosts carrying opposite statistics to the local parameters. The assortment is given by

$$\theta \longrightarrow c, \qquad \epsilon_i \longrightarrow \psi_i, \qquad \tau \longrightarrow \xi, \qquad \varphi \longrightarrow N, \qquad \Lambda \longrightarrow \phi. \tag{6.81}$$

Ghosts in (6.81) are first generations, in particular, c is Faddeev–Popov ghost, whereas ϕ is a second generation ghost. Their Grassmann parity and ghost number (U number) are given by

$$\begin{matrix} c & \psi_i & \xi & N & \phi \\ 1^- & 1^- & 1^- & 1^- & 2^+, \end{matrix} \tag{6.82}$$

where the superscript of ghost number denotes the Grassmann parity. Note that the ghost number counts the degree of differential form on the moduli space $\mathcal{M}$ of the solution to the 3D monopole equations. The minimal set Φ_{min} of fields consists of

$$\begin{array}{ccccc} A_i & b & M & G_i & \nu \\ 0^+ & 0^+ & 0^+ & 0^+ & 0^+, \end{array} \qquad \text{and} \quad (6.82).$$

On the other hand, the set of anti–fields Φ^*_{min} carrying opposite statistics to Φ_{min} is given by

$$\begin{array}{ccccccccc} A^*_i & b^* & M^* & G^*_i & \nu^* & c^* & \psi^*_i & N^* & \phi^* \\ -1^- & -1^- & -1^- & -1^- & -1^- & -2^+ & -2^+ & -2^+ & -3^- \end{array}.$$

Next step is to find a solution to the master equation with Φ_{min} and Φ^*_{min}, given by

$$\frac{\partial_r S}{\partial \Phi^A}\frac{\partial_l S}{\partial \Phi^*_A} - \frac{\partial_r S}{\partial \Phi^*_A}\frac{\partial_l S}{\partial \Phi^A} = 0, \tag{6.83}$$

where $r(l)$ denotes right (left) derivative.

The general solution for the first stage reducible theory at hand can be expressed by

$$S = S_c + \Phi^*_i R^i_\alpha C^\alpha_1 + C^*_{1\alpha}(Z^\alpha_\beta C^\beta_2 + T^\alpha_{\beta\gamma} C^\gamma_1 C^\beta_1) + C^*_{2\gamma} A^\gamma_{\beta\alpha} C^\alpha_1 C^\beta_2 + \Phi^*_i \Phi^*_j B^{ji}_\alpha C^\alpha_2 + \cdots, \tag{6.84}$$

where $C^\alpha_1(C^\alpha_2)$ denotes generally the first (second) generation ghost and only relevant terms in our case are shown. We often use $\Phi^A_{min} = (\phi^i, C^\alpha_1, C^\beta_2)$, where ϕ^i denote generally the fields. In this expression, the indices should be interpreted as the label of fields. Do not confuse with space–time indices. The coefficients $Z^\alpha_\beta, T^\alpha_{\beta\gamma}$, etc can be directly determined from the master equation. In fact, it is known that these coefficients satisfy the following relations

$$R^i_\alpha Z^\alpha_\beta C^\beta_2 - 2\frac{\partial_r S_c}{\partial \phi^j} B^{ji}_\alpha C^\alpha_2 (-1)^{|i|} = 0, \quad \frac{\partial_r R^i_\alpha C^\alpha_1}{\partial \phi^j} R^j_\beta C^\beta_1 + R^i_\alpha T^\alpha_{\beta\gamma} C^\gamma_1 C^\beta_1 = 0,$$
$$\frac{\partial_r Z^\alpha_\beta C^\beta_2}{\partial \phi^j} R^j_\gamma C^\gamma_1 + 2T^\alpha_{\beta\gamma} C^\gamma_1 Z^\beta_\delta C^\delta_2 + Z^\alpha_\beta A^\beta_{\delta\gamma} C^\gamma_1 C^\delta_2 = 0, \tag{6.85}$$

where $|i|$ means the Grassmann parity of the ith field.

In these expansion coefficients, R^i_α and Z^α_β are related to the local symmetry (6.78). On the other hand, as $T^\alpha_{\beta\gamma}$ is related to the structure constant of a given Lie algebra for a gauge theory, it is generally called as structure function. Of course if the theory is Abelian, such structure function does not appear. However, for a theory coupled with matters, all of the structure functions do not always vanish, even if the gauge group is Abelian. At first sight, this seems to be strange, but the expansion (6.84) obviously detects the coupling of matter fields and ghosts. In fact, the appearance of this type of structure function is required in order to make the action to be constructed being full BRST invariant.

After some algebraic works, we will find the solution to be

$$S(\Phi_{min}, \Phi^*_{min}) = S_c + \int_Y \Delta S d^3x, \qquad \text{where}$$

$$\begin{aligned}\Delta S = & A^*_i(\partial^i c + \psi^i) + b^*\xi + M^*(\mathrm{i}cM + N) + \overline{M}^*(-\mathrm{i}c\overline{M} + \overline{N}) \\ & + G^*_i\left[\partial^i\xi - \epsilon^{ijk}\partial_j\psi_k + \mathrm{i}(\overline{N}\sigma^{i0}M + \overline{M}\sigma^{i0}N)\right] \\ & + \nu^*(\mathrm{i}c\nu + \mathrm{i}\gamma^i D_i N + \gamma^i\psi_i M + \gamma^0 bN + \gamma^0\xi M) \\ & + \overline{\nu}^*\overline{(\mathrm{i}c\nu + \mathrm{i}\gamma^i D_i N + \gamma^i\psi_i M + \gamma^0 bN + \gamma^0\xi M)} \\ & + c^*\phi - \psi^*_i\partial^i\phi - \mathrm{i}N^*(\phi M + cN) + \mathrm{i}\overline{N}^*\left(\phi\overline{M} + c\overline{N}\right) + 2\mathrm{i}\nu^*\overline{\nu}^*\phi.\end{aligned}$$

We augment Φ_{min} by new fields $\chi_i, d_i, \mu(\overline{\mu}), \zeta(\overline{\zeta}), \lambda, \rho, \eta, e$ and the corresponding anti–fields. Their ghost number and Grassmann patity are given by

$$\begin{array}{cccccccc} \chi_i & d_i & \mu & \zeta & \lambda & \rho & \eta & e \\ -1^- & 0^+ & -1^- & 0^+ & -2^+ & -1^- & -1^- & 0^+ \end{array},$$

$$\text{and} \quad \begin{array}{cccc} \chi^*_i & \mu^* & \lambda^* & \rho^* \\ 0^+ & 0^+ & 1^- & 0^+ \end{array}.$$

Then we look for the solution

$$S' = S(\Phi_{min}, \Phi^*_{min}) + \int_Y (\chi^{*i}d_i + \overline{\mu}^*\overline{\zeta} + \mu^*\zeta + \rho^* e + \lambda^*\eta)d^3x, \qquad (6.86)$$

where d_i, ζ, e, η, are Lagrange multiplier fields.

In order to get the quantum action we must fix the gauge. The best choice for the gauge fixing condition, which can reproduce the action obtained from the dimensional reduction of the 4D one, is found to be [Oht98]

$$\begin{aligned} G_i &= 0, \qquad \nu = 0, \qquad \partial^i A_i = 0, \\ & -\partial^i\psi_i + \frac{\mathrm{i}}{2}(\overline{N}M - \overline{M}N) = 0. \end{aligned}$$

Thus we can get the gauge fermion carrying the ghost number –1 and odd Grassmann parity,

$$\Psi = -\chi^i G_i - \overline{\mu}\nu - \mu\overline{\nu} + \rho\partial^i A_i - \lambda\left[-\partial^i\psi_i + \frac{\mathrm{i}}{2}(\overline{N}M - \overline{M}N)\right].$$

The quantum action S_q can be obtained by eliminating anti–fields and are restricted to lie on the gauge surface $\Phi^* = \frac{\partial_r\Psi}{\partial\Phi}$. Therefore the anti–fields will be

$$
\begin{aligned}
&G_i^* = -\chi_i, \qquad \chi_i^* = -G_i, \qquad \nu^* = -\overline{\mu}, \qquad \overline{\nu}^* = -\mu, \qquad \overline{\mu}^* = -\nu, \\
&\mu^* = -\overline{\nu}, \qquad M^* = -\frac{\mathrm{i}}{2}\lambda\overline{N}, \qquad \overline{M}^* = \frac{\mathrm{i}}{2}\lambda N, \qquad N^* = \frac{\mathrm{i}}{2}\lambda\overline{M}, \\
&\overline{N}^* = -\frac{\mathrm{i}}{2}\lambda M, \qquad \rho^* = \partial^i A_i, \qquad A_i^* = -\partial_i\rho, \qquad \psi_i^* = -\partial_i\lambda, \qquad (6.87) \\
&\lambda^* = -\left[-\partial^i\psi_i + \frac{\mathrm{i}}{2}(\overline{N}M - \overline{M}N)\right], \quad c^* = \phi^* = b^* = \zeta^*(\overline{\zeta}^*) = 0.
\end{aligned}
$$

Then the quantum action S_q is given by $S_q = S'(\Phi, \Phi^* = \partial_r\Psi/\partial\Phi)$. Substituting (6.87) into S_q, we find that

$$S_q = S_c + \int_Y \widetilde{\Delta}S d^3x, \qquad \text{where}$$

$$
\begin{aligned}
\widetilde{\Delta}S = &(-\triangle\phi + \phi\overline{M}M - \mathrm{i}\overline{N}N)\lambda - \left[-\partial^i\psi_i + \frac{\mathrm{i}}{2}(\overline{N}M - \overline{M}N)\right]\eta \\
&- \overline{\mu}(\mathrm{i}c\nu + \mathrm{i}\gamma^i D_i N + \gamma^i\psi_i M + \gamma^0 bN + \gamma^0\xi M) \\
&+ \overline{(\mathrm{i}c\nu + \mathrm{i}\gamma^i D_i N + \gamma^i\psi_i M + \gamma^0 bN + \gamma^0\xi M)}\mu + 2\mathrm{i}\phi\overline{\mu}\mu \\
&- \chi^i\left[\partial_i\xi - \epsilon_{ijk}\partial^j\psi^k + \mathrm{i}(\overline{N}\sigma_{i0}M + \overline{M}\sigma_{i0}N)\right] \\
&+ \rho(\triangle c + \partial^i\psi_i) - d^i G_i - \overline{\zeta}\nu - \overline{\nu}\zeta + e\partial^i A_i. \qquad (6.88)
\end{aligned}
$$

Using the condition (6.77) with $c = 0$, we can arrive at

$$S_q' = S_c|_{G_i=\nu(\overline{\nu})=0} + \int_Y \widetilde{\Delta}S|_{c=0} d^3x, \qquad \text{where} \qquad (6.89)$$

$$
\begin{aligned}
\widetilde{\Delta}S|_{c=0} = &(-\triangle\phi + \phi\overline{M}M - \mathrm{i}\overline{N}N)\lambda - \left[-\partial^i\psi_i + \frac{\mathrm{i}}{2}(\overline{N}M - \overline{M}N)\right]\eta \\
&- \overline{\mu}(\mathrm{i}\gamma^i D_i N + \gamma^i\psi_i M + \gamma^0 bN + \gamma^0\xi M) \\
&+ \overline{(\mathrm{i}\gamma^i D_i N + \gamma^i\psi_i M + \gamma^0 bN + \gamma^0\xi M)}\mu + 2\mathrm{i}\phi\overline{\mu}\mu \\
&- \chi^i\left[\partial_i\xi - \epsilon_{ijk}\partial^j\psi^k + \mathrm{i}(\overline{N}\sigma_{i0}M + \overline{M}\sigma_{i0}N)\right] + \rho\partial^i\psi_i + e\partial^i A_i.
\end{aligned}
$$

It is easy to find that (6.89) is consistent with the action found by the dimensional reduction of the 4D topological action [ZOC95].

BRST Transformation

The Batalin–Vilkovisky algorithm also facilitates to construct BRST transformation rule. The BRST transformation rule for a field Φ is defined by

$$\delta_B\Phi = \epsilon \left.\frac{\partial_r S'}{\partial\Phi^*}\right|_{\Phi^*=\frac{\partial_r\Psi}{\partial\Phi}}, \qquad (6.90)$$

where ϵ is a constant Grassmann odd parameter. With this definition for (7.105), we get

$$\begin{aligned} \delta_B A_i &= -\epsilon(\partial_i c + \psi_i), \qquad \delta_B b = -\epsilon\xi, \qquad \delta_B M = -\epsilon(\mathrm{i}cM + N), \\ \delta_B G_i &= -\epsilon\left[\partial_i\xi - \epsilon_{ijk}\partial^j\psi^k + \mathrm{i}(\overline{N}\sigma_{i0}M + \overline{M}\sigma_{i0}N)\right], \\ \delta_B\nu &= -\epsilon(\mathrm{i}c\nu + \mathrm{i}\gamma^i D_i N + \gamma^i\psi_i M + \gamma^0 bN + \gamma^0\xi M - \mathrm{i}\mu\phi), \\ \delta_B c &= \epsilon\phi, \qquad \delta_B\psi_i = -\epsilon\partial_i\phi, \qquad \delta_B\rho = \epsilon e, \qquad \delta_B\lambda = -\epsilon\eta, \\ \delta_B\mu &= \epsilon\zeta, \qquad \delta_B N = -\mathrm{i}\epsilon(\phi M + cN), \qquad \delta_B\chi_i = \epsilon d_i, \\ \delta_B\phi &= \delta_B\xi = \delta_B d_i = \delta_B e = \delta_B\zeta = \delta_B\eta = 0. \end{aligned} \tag{6.91}$$

It is clear at this stage that (6.91) has on-shell nilpotency, i.e., the quantum equation of motion for ν must be used in order to have $\delta_B^2 = 0$. This is due to the fact that the gauge algebra has on–shell reducibility. Accordingly, the Batalin–Vilkovisky algorithm gives a BRST invariant action and on–shell nilpotent BRST transformation. Note that the equations

$$\begin{aligned} \partial_i\xi - \epsilon_{ijk}\partial^j\psi^k + \mathrm{i}(\overline{N}\sigma_{i0}M + \overline{M}\sigma_{i0}N) = 0, \\ \mathrm{i}\gamma^i D_i N + \gamma^i\psi_i M + \gamma^0 bN + \gamma^0\xi M = 0 \end{aligned}$$

can be recognized as linearizations of the 3D monopole equations and the number of linearly independent solutions gives the dimension of the moduli space $\mathcal{M}$.

It is now easy to show that the global supersymmetry can be recovered from (6.91). In Witten type theory, Q_B can be interpreted as a supersymmetric BRST charge. We define the supersymmetry transformation as [Oht98]

$$\delta_S\Phi := \delta_B\Phi|_{c=0}.$$

Off–Shell Action

As was mentioned before, the quantum action of Witten type TQFT can be represented by BRST commutator with nilpotent BRST charge Q_B. However, since our BRST transformation rule is on-shell nilpotent, we should integrate out ν and G_i in order to get off–shell BRST transformation and off–shell quantum action.

For this purpose, let us consider the following terms in (7.105),

$$\frac{1}{2}(G_i - X_i)^2 + \frac{1}{2}|\nu - A|^2 - \mathrm{i}\overline{\mu}c\nu + \overline{\mathrm{i}c\nu}\mu - \overline{\zeta}\nu - \overline{\nu}\zeta - d^i G_i, \tag{6.92}$$
$$\text{where} \qquad X_i = \partial_i b - \frac{1}{2}\epsilon_{ijk}F^{jk} + \mathrm{i}\overline{M}\sigma_{i0}M, \; A = \mathrm{i}\gamma^i D_i M + \gamma^0 bM.$$

Here, let us define

$$\nu' = \nu - A, \qquad B = -\mathrm{i}c\mu - \zeta.$$

$\nu'(\overline{\nu}')$ and G_i can be integrated out and then (6.92) will be

$$-\frac{1}{2}d_i d^i - d_i X^i - 2|B|^2 + \overline{B}A + B\overline{A}.$$

Consequently, we get the off–shell quantum action

$$S_q = \left\{Q, \widetilde{\Psi}\right\}, \qquad \text{where} \tag{6.93}$$

$$\widetilde{\Psi} = -\chi^i \left(X_i + \frac{\alpha}{2}d_i\right) - \mu(\overline{\mathrm{i}\gamma^i D_i M + \gamma^0 bM - \beta B)}$$
$$-\overline{\mu}(\mathrm{i}\gamma^i D_i M + \gamma^0 bM - \beta B) + \rho\partial^i A_i - \lambda\left[-\partial^i\psi_i + \frac{\mathrm{i}}{2}(\overline{N}M - \overline{M}N)\right].$$

α and β are arbitrary gauge fixing paramaters. Convenience choice for them is $\alpha = \beta = 1$. The BRST transformation rule for X_i and B fields can be easily obtained, although we do not write down here.

Observables

We can now discuss the observables. For this purpose, let us define [BG88]

$$\mathcal{A} = A + c, \qquad \mathcal{F} = F + \psi - \phi, \qquad \mathcal{K} = db + \xi,$$

where we have introduced differential form notations, but their meanings would be obvious. A and c are considered as a $(1,0)$ and $(0,1)$ part of 1–form on $(Y, \mathcal{M})$. Similarly, F, ψ and ϕ are $(2,0), (1,1)$ and $(0,2)$ part of the 2–form $\mathcal{F}$, and db and ξ are $(1,0)$ and $(0,1)$ part of the 1–form $\mathcal{K}$. Thus $\mathcal{A}$ defines a connection 1–form on $(Y, \mathcal{M})$ and $\mathcal{F}$ is a curvature 2–form. Note that the exterior derivative d maps any (p_1, p_2)–form X_p of total degree $p = p_1 + p_2$ to $(p_1 + 1, p_2)$–form, but δ_B maps any (p_1, p_2)–form to $(p_1, p_2 + 1)$–form. Also note that $X_p X_q = (-1)^{pq} X_q X_p$. Then the action of δ_B is

$$(d + \delta_B)\mathcal{A} = \mathcal{F}, \qquad (d + \delta_B)b = \mathcal{K}. \tag{6.94}$$

$\mathcal{F}$ and $\mathcal{K}$ also satisfy the following Bianchi identities in Abelian theory:

$$(d + \delta_B)\mathcal{F} = 0, \qquad (d + \delta_B)\mathcal{K} = 0. \tag{6.95}$$

Equations (6.94) and (6.95) mean anti–commuting property between the BRST variation δ_B and the exterior differential d, i.e., $\{\delta_B, d\} = 0$.

The BRST transformation rule in geometric sector can be easily read from (6.91), i.e., $\delta_B A, \delta_B \psi, \delta_B c$ and $\delta_B \phi$. (6.95) implies

$$(d + \delta_B)\mathcal{F}^n = 0, \tag{6.96}$$

and expanding the above expression by ghost number and form degree, we get the following $(i, 2n - i)$–form $W_{n,i}$,

$$
\begin{aligned}
W_{n,0} &= \frac{\phi^n}{n!}, \qquad W_{n,1} = \frac{\phi^{n-1}}{(n-1)!}\psi, \\
W_{n,2} &= \frac{\phi^{n-2}}{2(n-2)!}\psi \wedge \psi - \frac{\phi^{n-1}}{(n-1)!}F, \\
W_{n,3} &= \frac{\phi^{n-3}}{6(n-3)!}\psi \wedge \psi \wedge \psi - \frac{\phi^{n-2}}{(n-2)!}F \wedge \psi,
\end{aligned}
\tag{6.97}
$$

$$
\begin{aligned}
\text{where} \quad 0 &= \delta_B W_{n,0}, \qquad dW_{n,0} = \delta_B W_{n,1}, \\
dW_{n,1} &= \delta_B W_{n,2}, \qquad dW_{n,2} = \delta_B W_{n,3}, \qquad dW_{n,3} = 0.
\end{aligned}
\tag{6.98}
$$

Picking a certain $k-$cycle γ as a representative and defining the integral

$$
W_{n,k}(\gamma) = \int_\gamma W_{n,k},
$$

we can easily prove that

$$
\delta_B W_{n,k}(\gamma) = -\int_\gamma dW_{n,k-1} = -\int_{\partial\gamma} W_{n,k-1} = 0,
$$

as a consequence of (6.98). Note that the last equality follows from the fact that the cycle γ is a simplex without boundary, i.e., $\partial\gamma = 0$. Therefore, $W_{n,k}(\gamma)$ indeed gives a topological invariant associated with $n-$th *Chern class* on $Y \times \mathcal{M}$.

On the other hand, since we have a scalar field b and its ghosts, we may construct topological observables associated with them. Therefore, the observables can be obtained from the ghost expansion of

$$
(d + \delta_B)\mathcal{F}^n \wedge \mathcal{K}^m = 0.
$$

Explicitly, for $m = 1$, for example, we get

$$
\begin{aligned}
0 &= \delta_B W_{n,1,0}, \qquad dW_{n,1,0} = \delta_B W_{n,1,1}, \qquad dW_{n,1,1} = \delta_B W_{n,1,2}, \\
dW_{n,1,2} &= \delta_B W_{n,1,3}, \qquad dW_{n,1,3} = 0, \qquad \text{where}
\end{aligned}
$$

$$
\begin{aligned}
W_{n,1,0} &= \frac{\phi^n}{n!}\xi, \qquad W_{n,1,1} = \frac{\phi^{n-1}}{(n-1)!}\psi\xi - \frac{\phi^n}{n!}db, \\
W_{n,1,2} &= \frac{\phi^{n-2}}{2(n-2)!}\psi \wedge \psi\xi - \frac{\phi^{n-1}}{(n-1)!}F\xi - \frac{\phi^{n-1}}{(n-1)!}\psi \wedge db, \\
W_{n,1,3} &= \frac{\phi^{n-3}}{6(n-3)!}\psi \wedge \psi \wedge \psi\xi + \frac{\phi^{n-1}}{(n-1)!}F \wedge db \\
&\quad + \frac{\phi^{n-2}}{2(n-2)!}(2\psi \wedge F\xi + \psi \wedge \psi \wedge db).
\end{aligned}
\tag{6.99}
$$

These relations correspond to the cocycles [BG88] in $U(1)$ case.

Next, let us look for the observables for matter sector. The BRST transformation rules in this sector is given by $\delta_B, \delta_B N, \delta_B c$ and $\delta_B \phi$. At first sight, the matter sector does not have any observable, but we can find the combined form

$$\widetilde{W} = \mathrm{i}\phi \overline{M} M + \overline{N} N \tag{6.100}$$

is an observable. However, unfortunately, as $\widetilde{W}$ is cohomologically trivial because $\delta_B \widetilde{W} = 0$ but $d\widetilde{W} \neq \delta_B \widetilde{W}'$ for some $\widetilde{W}'$. Accordingly, $\widetilde{W}$ does not give any new topological invariant [Oht98].

In topological Bogomol'nyi theory, there is a sequence of observables associated with a magnetic charge. For the Abelian case, it is given by

$$W = \int_Y F \wedge db. \tag{6.101}$$

As is pointed out for the case of Bogomol'nyi monopoles [BRT89], we can not get the observables related with this magnetic charge by the action of δ_B as well, but we can construct those observables by anti–BRST variation $\overline{\delta}_B$ which maps (m,n)–form to $(m,n-1)$–form. $\overline{\delta}_B$ can be obtained by a discrete symmetry which is realized as 'time reversal symmetry' in 4D. In our 3D theory, the discrete symmetry is given by

$$\begin{aligned}
&\phi \longrightarrow -\lambda, \qquad \lambda \longrightarrow -\phi, \qquad N \longrightarrow \mathrm{i}\sqrt{2}\mu, \qquad \mu \longrightarrow \frac{\mathrm{i}}{\sqrt{2}} N, \\
&\psi_i \longrightarrow \frac{\chi_i}{\sqrt{2}}, \qquad \chi_i \longrightarrow \sqrt{2}\psi_i, \qquad \eta \longrightarrow \sqrt{2}\xi, \qquad \xi \longrightarrow -\frac{\eta}{\sqrt{2}}
\end{aligned} \tag{6.102}$$

$$\text{with} \qquad b \longrightarrow -b, \tag{6.103}$$

where (6.103) represents an additional symmetry [BRT89]. Note that we must also change N and μ (and their conjugates). The positive chirality condition for M should be used in order to check the invariance of the action. In this way, we can get anti–BRST transformation rule by substituting (6.102) and (6.103) into (6.91) and then we can get the observables associated with the magnetic charge by using the action of this anti–BRST variation [BRT89].

The topological observables available in this theory are the same with those of topological Bogomol'nyi monopoles.

Finally, let us briefly comment on our three dimensional theory. First note that Lagrangian L and Hamiltonian H in dimensional reduction can be considered as equivalent. This is because the relation between them is defined by

$$H = p\dot{q} - L,$$

where q is any field, the overdot means time derivative and p is a canonical conjugate momentum of q, and the dimensional reduction requires the time independence of all fields, thus $H = -L$ in this sense. Though we have constructed the three dimensional action directly from the 3D monopole equations, our action may be interpreted essentially as the Hamiltonian of the four dimensional SW theory.

Non–Abelian Case

It is easy to extend the results obtained in the previous subsection to non–Abelian case. In this subsection, we summarize the results for the non–Abelian 3D monopoles (for details, see [Oht98] and references therein).

Non–Abelian Topological Action

With the auxiliary fields $G^a_{\mu\nu}$ and ν, we consider

$$S_c = \frac{1}{2}\int_Y d^3x \left[(G^a_i - K^a_i)^2 + |\nu - \mathrm{i}\gamma^i D_i M - \gamma^0 bM|^2\right], \quad (6.104)$$

$$\text{where} \qquad K^a_i = \partial_i b^a + f_{abc}A^b_i b^c - \frac{1}{2}\epsilon_{ijk}F^a_{jk} + \mathrm{i}\overline{M}\sigma_{i0}T^a M.$$

Note that the minimum of (6.104) with the gauge $G^a_i = \nu = 0$ are given by the non–Abelian 3D monopoles. We take the generator of Lie algebra in the fundamental representation, e.g., for $SU(n)$,

$$(T_a)_{ij}(T^a)_{kl} = \delta_{il}\delta_{jk} - \frac{1}{n}\delta_{ij}\delta_{kl}.$$

Extension to other Lie algebra and representation is straightforward.

The gauge transformation rule for (6.104) is given by

$$\begin{aligned}
\delta A^a_i &= \partial_i\theta^a + f_{abc}A^b_i\theta^c + \epsilon^a_i, \qquad \delta b^a = f_{abc}b^b\theta^c + \tau^a, \qquad \delta M = \mathrm{i}\theta M + \varphi,\\
\delta G^a_i &= f_{abc}G^b_i\theta^c + \left[-\epsilon_{ijk}(\partial^j\epsilon^{ak} + f_{abc}\epsilon^{jb}A^{ck})\right.\\
&\quad \left. + \partial_i\tau^a + f_{abc}(\epsilon^b_i b^c - \tau^b A^c_i) + \mathrm{i}(\overline{\varphi}\sigma_{i0}T^a M + \overline{M}\sigma_{i0}T^a\varphi)\right],\\
\delta\nu &= \mathrm{i}\gamma^i D_i\varphi + \gamma^i\epsilon_i M + \gamma^0 b\varphi + \gamma^0\tau M + \mathrm{i}\theta\nu.
\end{aligned} \quad (6.105)$$

Note that we have a G^a_i term in the transformation of G^a_i, while it did not appear in Abelian theory.

The gauge algebra (6.105) possesses on–shell zero modes as in the Abelian case. Setting

$$\theta^a = \Lambda^a, \quad \epsilon^a_i = -\partial_i\Lambda^a - f_{abc}A^b_i\Lambda^c, \quad \tau^a = -f_{abc}b^b\Lambda^c, \quad \varphi = -\mathrm{i}\Lambda M,$$

we can easily find that (6.105) closes, i.e.,

$$\begin{aligned}
\delta A^a_i &= 0, \qquad \delta b^a = 0, \qquad \delta M = 0,\\
\delta G^a_i &= f_{abc}\Lambda^c[G^b_i - K^b_i]|_{on-shell} = 0,\\
\delta\nu &= \mathrm{i}\Lambda[\nu - \mathrm{i}(\gamma^i D_i - \mathrm{i}\gamma^0 b)M]|_{on-shell} = 0,
\end{aligned} \quad (6.106)$$

when the equations of motion of G_i^a and ν are used. Note that we must use both equations of motion of G_i^a and ν in the non–Abelian case, while only 'ν' was needed for the Abelian theory. Furthermore, as φ is a parameter in the spinor space, φ is not $\mathfrak{g}$–valued, in other words, $\varphi \neq \varphi^a T^a$. (6.105) is first stage reducible.

The assortment of ghost fields, the minimal set Φ_{min} of the fields and the ghost number and the Grassmann parity, furthermore those for Φ^*_{min} would be obvious.

Then the solution to the master equation will be

$$S(\Phi_{min}, \Phi^*_{min}) = S_c + \int_Y \mathrm{Tr}\,(\Delta S_n)\, d^3x, \qquad \text{where}$$

$$\begin{aligned}
&\Delta S_n = \\
&A_i^*(D^i c + \psi^i) + b^*(\mathrm{i}[b,c] + \xi) + M^*(\mathrm{i}cM + N) + \overline{M}^*(-\mathrm{i}c\overline{M} + \overline{N}) + G_i^* \widetilde{G}^i - \\
&\mathrm{i}N^*(\phi M + cN) + \mathrm{i}\overline{N}^*(\phi\overline{M} + c\overline{N}) + \nu^*(\mathrm{i}c\nu + \mathrm{i}\gamma^i D_i N + \gamma^i \psi_i M + \gamma^0 bN + \gamma^0 \xi M) \\
&+\overline{\nu}^*\overline{(\mathrm{i}c\nu + \mathrm{i}\gamma^i D_i N + \gamma^i \psi_i M + \gamma^0 bN + \gamma^0 \xi M)} + 2\mathrm{i}\nu^* \overline{\nu}^* \phi + \psi_i^*(-D^i\phi - \mathrm{i}\{\psi^i, c\}) \\
&+c^*\left(\phi - \frac{\mathrm{i}}{2}\{c,c\}\right) - \mathrm{i}\phi^*[\phi, c] - \frac{\mathrm{i}}{2}\{G_i^*, G^{*i}\}\phi + \mathrm{i}\xi^*([b,\phi] - \{\xi, c\}).
\end{aligned}$$

Here

$$\widetilde{G}_i = \mathrm{i}[c, G_i] - \epsilon_{ijk} D^j \psi^k + D_i \xi + [\psi_i, \xi] + \mathrm{i}(\overline{N}\sigma_{i0} T_a T^a M + \overline{M}\sigma_{i0} T_a T^a N).$$

The equations

$$\begin{aligned}
-\epsilon_{ijk} D^j \psi^k + D_i \xi + [\psi_i, \xi] + \mathrm{i}(\overline{N}\sigma_{i0} T_a T^a M + \overline{M}\sigma_{i0} T_a T^a N) &= 0, \\
\mathrm{i}\gamma^i D_i N + \gamma^i \psi_i M + \gamma^0 bN + \gamma^0 \xi M &= 0,
\end{aligned}$$

can be seen as linearizations of non–Abelian 3D monopoles.

We augment Φ_{min} by new fields $\chi_i^a, d_i^a, \mu(\overline{\mu}), \zeta(\overline{\zeta}), \lambda, \rho, \eta, e$ and the corresponding anti–fields, but Lagrange multipliers fields $d_i^a, \zeta(\overline{\zeta}), e, \eta$, are assumed not to have anti–fields for simplicity and therefore their BRST transformation rules are set to zero. This simplification means that we do not take into account of BRST exact terms. In this sense, the result to be obtained will correspond to those of the dimensionally reduced version of the 4D theory up to these terms, i.e., topological numbers.

From the gauge fixing condition

$$G_i^a = 0, \qquad \nu = 0, \qquad \partial^i A_i = 0, \qquad -D^i \psi_i + \frac{\mathrm{i}}{2}(\overline{N}M - \overline{M}N) = 0,$$

the gauge fermion will be

$$\Psi = -\chi^i G_i - \overline{\mu}\nu - \mu\overline{\nu} + \rho\partial^i A_i - \lambda\left[-D^i\psi_i + \frac{\mathrm{i}}{2}(\overline{N}M - \overline{M}N)\right].$$

The anti–fields are then given by

$$
\begin{aligned}
&G_i^* = -\chi_i, \quad \chi_i^* = -G_i, \quad \nu^* = -\overline{\mu}, \quad \overline{\nu}^* = -\mu, \quad \overline{\mu}^* = -\nu, \quad \mu = -\overline{\nu}, \\
&M^* = -\frac{\mathrm{i}}{2}\lambda\overline{N}, \qquad \overline{M}^* = \frac{\mathrm{i}}{2}\lambda N, \qquad N^* = \frac{\mathrm{i}}{2}\lambda\overline{M}, \qquad \overline{N}^* = -\frac{\mathrm{i}}{2}\lambda M, \\
&\rho^* = \partial^i A_i, \quad A_i^* = -\partial_i\rho + \mathrm{i}[\lambda, \psi_i], \quad b^* = c^* = \xi^* = \phi^* = \zeta^*(\overline{\zeta}^*) = 0, \\
&\lambda^* = -\left[-D_i\psi^i + [b, \xi] + \frac{\mathrm{i}}{2}(\overline{N}M - \overline{M}N)\right], \quad \psi_i^* = -D_i\lambda.
\end{aligned}
$$

Therefore we find the quantum action

$$
S_q = S_c + \int_Y \mathrm{Tr}\left(\widetilde{\Delta} S_n\right) d^3x, \qquad \text{where} \tag{6.107}
$$

$$
\begin{aligned}
\widetilde{\Delta} S_n &= -\left[-D_i\psi^i + [b, \xi] + \frac{\mathrm{i}}{2}(\overline{N}M - \overline{M}N)\right]\eta - \lambda(D_iD^i\phi + \mathrm{i}D_i\{\psi^i, c\}), \\
&+\mathrm{i}\lambda\{\psi_i, D^ic + \psi^i\} + (\phi\overline{M}M - \mathrm{i}\overline{N}N)\lambda, \\
-\chi^i &\left[\mathrm{i}[c, G_i] + \epsilon_{ijk}D^j\psi^k + D_k\xi + [\psi_k, \xi] + \frac{\mathrm{i}}{2}(\overline{N}\sigma^{ij}T_aT^aM + \overline{M}\sigma^{ij}T_aT^aN)\right], \\
&-\overline{\mu}(\mathrm{i}\gamma^\mu D_\mu N + \gamma^\mu\psi_\mu M + \mathrm{i}c\nu) + \overline{(\mathrm{i}\gamma^i D_i N + \gamma^\mu\psi_\mu M + \mathrm{i}c\nu)}\mu, \\
&+2\mathrm{i}\phi\overline{\mu}\mu - \frac{\mathrm{i}}{2}\{\chi_i, \chi^i\}\phi + \rho(\partial_iD^ic + \partial_i\psi^i) - d^iG_i - \overline{\zeta}\nu - \overline{\nu}\zeta + e\partial^iA_i.,
\end{aligned} \tag{6.108}
$$

In this quantum action, setting

$$
M(\overline{M}) = N(\overline{N}) = \mu(\overline{\mu}) = \nu(\overline{\nu}) = 0,
$$

we can find that the resulting action coincides with that of Bogomol'nyi monopoles [BRT89].

Finally, in order to get the off–shell quantum action, both the auxiliary fields should be integrated out by the similar technique presented in Abelian case.

BRST transformation

The BRST transformation rule is given by

$$
\begin{aligned}
&\delta_B A_i = -\epsilon(D_ic + \psi_i), \qquad \delta_B b = -\epsilon(\mathrm{i}[c, b] + \xi), \qquad \delta_B\xi = \mathrm{i}\epsilon([b, \phi] - \{\xi, c\}), \\
&\delta_B M = -\epsilon(\mathrm{i}cM + N), \qquad \delta_B G_i = -\epsilon(\widetilde{G}_i - \mathrm{i}[\chi_i, \phi]), \\
&\delta_B\nu = -\epsilon(\mathrm{i}c\nu + \gamma^\mu D_\mu N + \gamma^\mu\psi_\mu M - \mathrm{i}\mu\phi), \qquad \delta_B c = \epsilon\left(\phi - \frac{\mathrm{i}}{2}\{c, c\}\right), \\
&\delta_B\psi_i = -\epsilon\left(D_i\phi + \mathrm{i}\{\psi_i, c\}\right), \qquad \delta_B\rho = \epsilon e, \qquad \delta_B\lambda = -\epsilon\eta, \\
&\delta_B\mu = \epsilon\zeta, \qquad \delta_B N = -\mathrm{i}\epsilon(\phi M + cN), \qquad \delta_B\chi_i = \epsilon d_i, \\
&\delta_B\phi = \mathrm{i}\epsilon[\phi, c], \qquad \delta_B d_i = \delta_B e = \delta_B\zeta = \delta_B\eta = 0.
\end{aligned} \tag{6.109}
$$

It is easy to get supersymmetry also in this case. However, as we have omitted the BRST exact terms, the supersymmetry in our construction does not detect them.

Observables

We have already constructed the topological observables for Abelian case. Also in non-Abelian case, the construction of observables is basically the same. But the relation (6.94) and (6.95) are required to modify

$$(d+\delta_B)\mathcal{A} - \frac{\mathrm{i}}{2}[\mathcal{A},\mathcal{A}] = \mathcal{F}, \qquad (d+\delta_B)b - \mathrm{i}i[\mathcal{A},b] = \mathcal{K}, \qquad \text{and} \quad (6.110)$$

$$(d+\delta_B)\mathcal{F} - \mathrm{i}[\mathcal{A},\mathcal{F}] = 0, \qquad (d+\delta_B)\mathcal{K} - \mathrm{i}[\mathcal{A},\mathcal{K}] = \mathrm{i}[\mathcal{F},b], \qquad (6.111)$$

respectively, where $[*,*]$ is a graded commutator. The observables in geometric and matter sector are the same as before, but we should replace db by $d_A b$ in (6.99) as well as (6.110) and (6.111), where d_A is a exterior covariant derivative and trace is required. In addition, the magnetic charge observables are again obtained by anti-BRST variation as outlined before.

The observables in geometric sector are those in (6.97) and follow the cohomological relation (6.98). In this way, the topological observables available in this three dimensional theory are precisely the Bogomol'nyi monopole cocycles [BG88].

6.3 Complex Stringy Dynamics

6.3.1 Stringy Actions and Amplitudes

Now we give a brief review of modern path–integral methods in superstring theory (mainly following [DEF99]). Recall that the fundamental quantities in quantum field theory (QFT) are the transition amplitudes $Amp : IN \Longrightarrow OUT$, describing processes in which a number IN of incoming particles scatter to produce a number OUT of outgoing particles. The square modulus of the transition amplitude yields the *probability* for this process to take place.

Strings

Recall that in string theory, *elementary particles* are not described as 0–dimensional *points*, but instead as 1D *strings*. If M_s and $M(\sim \mathbb{R}\times M_s)$ denote the 3D space and 4D space–time manifolds respectively, then we picture strings as in Figure 6.10.

While the point–particle sweeps out a 1D *world–line*, the string sweeps out a *world–sheet*, i.e., a 2D real surface. For a *free string*, the topology of the world–sheet is a *cylinder* (in the case of a closed string) or a *sheet* (for an open string).

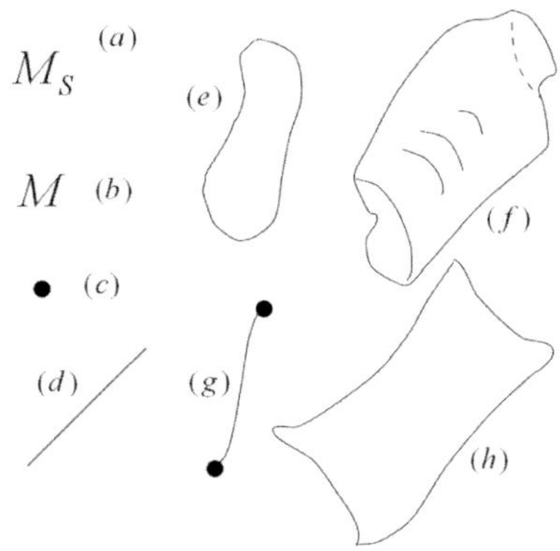

Fig. 6.10. Basic geometrical objects of string theory: (*a*) a space with fixed time; (*b*) a space–time picture; (*c*) a point–particle; (*d*) a world–line of a point–particle; (*e*) a closed string; (*f*) a world–sheet of a closed string; (*g*) an open string; (*h*) a world–sheet of an open string.

Roughly, different elementary particles correspond to different *vibration modes* of the string just as different minimal notes correspond to different vibrational modes of musical string instruments.

It turns out that the physical size of strings is set by *gravity*, more precisely the *Planck length* $\ell_P \sim 10^{-33}$ cm. This scale is so small that we effectively only see point–particles at our distance scales. Thus, for length scales much larger than ℓ_P, we expect to recover a QFT–description of point–particles, plus typical *string corrections* that represent physics at the Planck scale.

Interactions

While the string itself is an extended 1D object, the fundamental string interactions are *local*, just as for point–particles. The interaction takes place when strings overlap in space at the same time. In case of *closed string theories* the interactions have a form depicted in Figure 6.11, while in case of *open string theories* the interactions have a form depicted in Figure 6.12. Other interactions result from combining the interactions defined above.

In point–particle theories, the fundamental interactions are read off from the QFT–Lagrangian. An interaction occurs at a *geometrical point*, where the world–lines join and cease to be a manifold. In *Lorentz–invariant theories* (where manifold M is a flat Minkowski space–time), the interaction point is Lorentz–invariant. To specify how the point–particles interact, additional data must be supplied at the interaction point, giving rise to *many possible* distinct quantum field theories.

In string theory, the interaction point *depends* upon the Lorentz frame chosen to observe the process. In the Figure above, equal time slices are indicated from the point of view of two different Lorentz frames, schematically

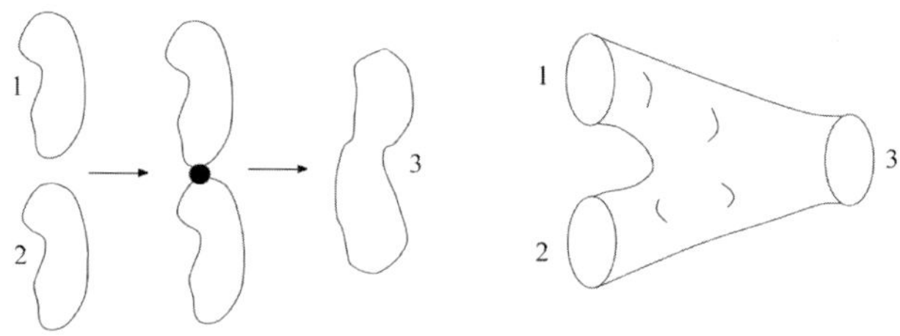

Fig. 6.11. Interactions in closed string theories (left 2D–picture and right 3D–picture).

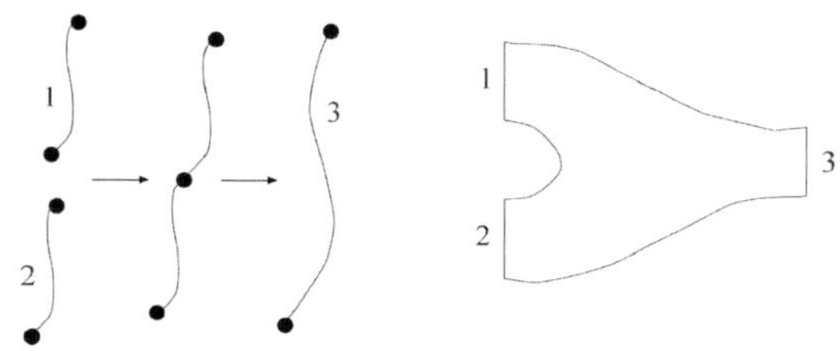

Fig. 6.12. Interactions in open string theories (left 2D–picture and right 3D–picture).

indicated by t and t'. The closed string interaction, as seen from frames t and t', occurs at times t_2 and t'_2 and at (distinct) points P and P' respectively.

Lorentz invariance of interaction forbids that any point on the world–sheet be singled out as interaction point. Instead, the interaction results purely from the joining and splitting of strings. While free closed strings are characterized by their topology being that of a cylinder, interacting strings are characterized by the fact that their associated world–sheet is connected to at least 3 strings, incoming and/or outgoing.

As a result, the free string determines the nature of the interactions completely, leaving only the string coupling constant undetermined.

The *orientation* is an additional structure of closed strings, dividing them into two categories: (i) *oriented strings*, in which all world–sheets are assumed to be orientable; and (ii) non–oriented strings, in which world–sheets are non–orientable, such as the Möbius strip, Klein bottle, etc.

Loop Expansion – Topology of Closed Surfaces

For simplicity, here we consider closed oriented strings only, so that the associated world–sheet is also oriented. A general string configuration describing the process in which M incoming strings interact and produce N outgoing strings looks at the topological level like a closed surface with $M + N = E$ *boundary components* and any number of *handles* (see Figure 6.13). This picture is a kind of topological generalization of nonlinear control MIMO–systems with M inputs, N outputs X states.

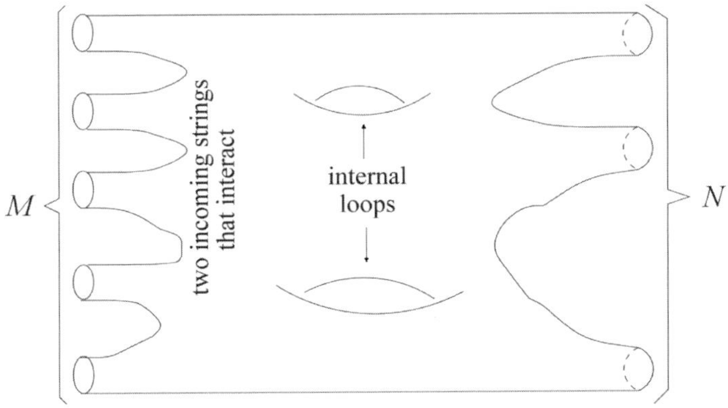

Fig. 6.13. Boundary components and handles of closed oriented system of M incoming strings, interacting through internal loops, to produce N outgoing strings. Note the striking similarity with MIMO–systems of *nonlinear control theory*, with M input processes and N output processes (see [II06b]).

The internal loops may arise when virtual particle pairs are produced, just as in quantum field theory. For example, a *Feynman diagram* in quantum field theory that involves a loop is shown in Figure 6.14 together with the corresponding string diagram.

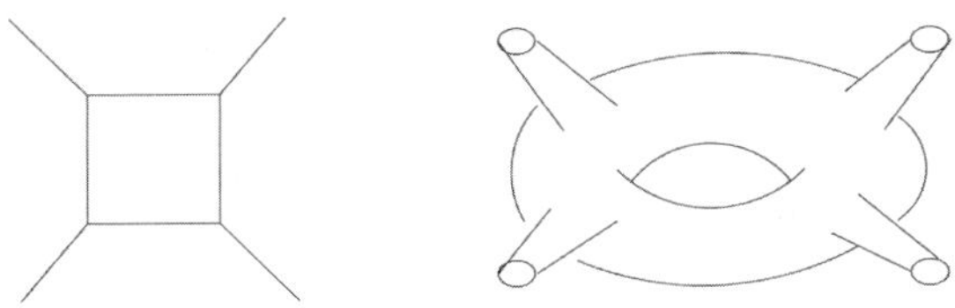

Fig. 6.14. A QFT Feynman diagram that involves an internal loop (left), with the corresponding string diagram (right).

Surfaces associated with closed oriented strings have two topological invariants: (i) the *number of boundary components* $E = M + N$ (which may be shrunk to punctures, under certain conditions), and (ii) the number h of *handles* on the surface, which equals the surface *genus*.

When $E = 0$, we just have the topological classification of compact oriented surfaces without boundary. Rendering $E > 0$ is achieved by removing E discs from the surface.

Recall that in QFT, an expansion in powers of Planck's constant $\hbar$ yields an expansion in the number of loops of the associated Feynman diagram, for a given number of external states:

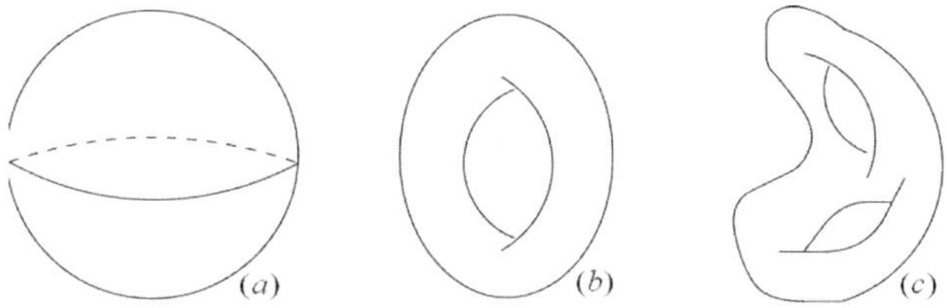

Fig. 6.15. Number h of handles on the surface of closed oriented strings, which equals the string–surface genus: (a) $h = 0$ for sphere S^2; (b) $h = 1$ for torus T^2; (c) $h = 2$ for string–surfaces with higher genus, etc.

$$\hbar^{E+h-1} = \begin{cases} \hbar & \text{for every propagator} \\ \hbar^{-1} & \text{for every vertex} \\ -1 & \text{for overall momentum conservation} \end{cases}$$

Thus, in string theory we expect that, for a given number of external strings E, the topological expansion genus by genus should correspond to a *loop* expansion as well.

Recall that in QFT, there are in general many Feynman diagrams that correspond to an amplitude with a given number of external particles and a given number of loops. For example, for $E = 4$ external particles and $h = 1$ loop in ϕ^3 theory are given in Figure 6.16, together with the same process in string theory (for closed oriented strings), where it is described by just a single diagram (right).

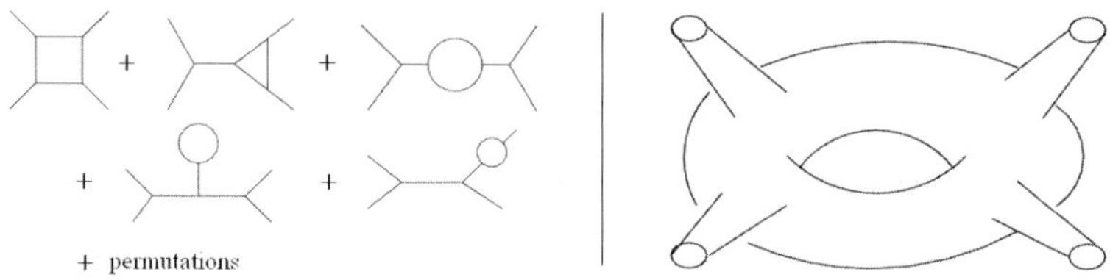

Fig. 6.16. Feynman QFT–diagrams for ϕ^3 theory with $E = 4$ external particles and $h = 1$ loop (left), and a single corresponding string diagram (right). In this way the usual Feynman diagrams of quantum field theory are generalized by arbitrary Riemannian surfaces.

Much of recent interest has been focused on the so–called $D-$branes. A $D-$*brane* is a submanifold of space–time with the property that strings can *end* or *begin* on it.

6.3.2 Transition Amplitudes for Strings

The only way we have today to define string theory is by giving a *rule* for the evaluation of transition amplitudes, order by order in the loop expansion, i.e., genus by genus. The rule is to assign a relative weight to a given configuration and then to sum over all configurations [DEF99]. To make this more precise, we first describe the system's configuration manifold M (see Figure 6.17).

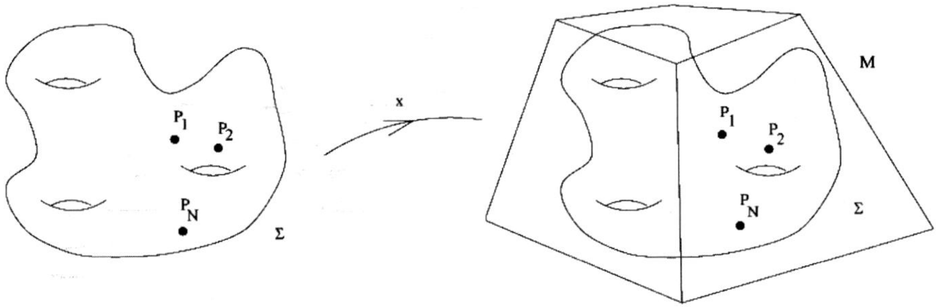

Fig. 6.17. The embedding map x from the reference surface Σ into the pseudo–Riemannian configuration manifold M (see text for explanation).

We assume that Σ and M are smooth manifolds, of dimensions 2 and n respectively, and that x is a continuous map from Σ to M. If ξ^m, (for $m = 1, 2$), are local coordinates on Σ and x^μ, $(\mu = 1, \ldots, n)$, are local coordinates on M then the map x may be described by functions $x^\mu(\xi^m)$ which are continuous.

To each system configuration we can associate a weight $\mathrm{e}^{-S[x,\Sigma,M]}$, (for $S \in \mathbb{C}$) and the transition amplitude Amp for specified external strings (incoming and outgoing) is get by summing over all surfaces Σ and all possible maps x,

$$Amp = \sum_{surfaces\ \Sigma} \sum_{x} \mathrm{e}^{-S[x,\Sigma,M]}.$$

We now need to specify each of these ingredients:

(1) We assume M to be an nD Riemannian manifold, with metric g. A special case is flat Euclidean space–time $\mathbb{R}^n$. The space–time metric is assumed *fixed.*

$$ds^2 = (dx, dx)_g = g_{\mu\nu}(x)dx^\mu \otimes dx^\nu.$$

(2) The metric g on M induces a metric on Σ: $\gamma = x^*(g)$,

$$\gamma = \gamma_{mn} d\xi^m \otimes d\xi^n, \qquad \gamma_{mn} = g_{\mu\nu} \frac{\partial x^\mu}{\partial \xi^m} \frac{\partial x^\nu}{\partial \xi^n}.$$

This metric is non–negative, but depends upon x. It is advantageous to introduce an intrinsic Riemannian metric g on Σ, independently of x; in local coordinates, we have

$$g = g_{mn}(\xi) d\xi^m \otimes d\xi^n.$$

A natural intrinsic candidate for S is the area of $x(\Sigma)$, which gives the so–called *Nambu–Goto action*[16]

$$Area\,(x\,(\Sigma)) = \int_\Sigma d\mu_\gamma = \int_\Sigma n^2\xi\sqrt{\det\,\gamma_{mn}}, \qquad (6.112)$$

which depends only upon g and x, but not on g [Got71]. However, the transition amplitudes derived from the Nambu–Goto action are *not* well–defined quantum–mechanically.

Otherwise, we can take as starting point the so–called *Polyakov action*[17]

$$S[x,g] = \kappa\int_\Sigma (dx, *dx)_g = \kappa\int_\Sigma d\mu_g g^{mn}\partial_m x^\mu \partial_n x^\nu g_{\mu\nu}(x), \qquad (6.113)$$

where κ is the *string tension* (a positive constant with dimension of inverse length square). The stationary points of S with respect to g are at $g^0 = e^\phi\gamma$ for some function ϕ on Σ, and thus $S[x,g^0] \sim Area\ (x\,(\Sigma))$.

The Polyakov action leads to *well–defined* transition amplitudes, get by integration over the space $Met(\Sigma)$ of all positive metrics on Σ for a given topology, as well as over the space of all maps $Map(\Sigma,M)$. We can define the path integral

$$Amp = \sum_{\substack{topologies \\ \Sigma}} \int_{Met(\Sigma)} \frac{1}{N(g)} \int_{Map(\Sigma,M)} \mathcal{D}[x]\,e^{-S[x,g,g]},$$

where N is a normalization factor, while the measures $\mathcal{D}[g]$ and $\mathcal{D}[x]$ are constructed from $Diff^+(\Sigma)$ and $Diff(M)$ invariant L^2 norms on Σ and M. For fixed metric g, the action S is well–known: its stationary points are the harmonic maps $x\colon\ \Sigma \to M$ (see, e.g., [EL78]). However, g here varies and in fact is to be integrated over. For a general metric g, the action S defines a *nonlinear sigma model*, which is renormalizable because the dimension of Σ is 2. It would not in general be renormalizable in dimension higher than 2, which is usually regarded as an argument against the existence of fundamental membrane theories (see [DEF99]).

The Nambu–Goto action (6.112) and Polyakov action (6.113) represent the core of the so–called *bosonic string theory*, the original version of string theory, developed in the late 1960s. Although it has many attractive features, it also predicts a particle called the *tachyon* possessing some unsettling properties, and it has no fermions. All of its particles are *bosons*, the matter particles. The

[16] Nambu–Goto action is the starting point of the analysis of string behavior, using the principles of ordinary Lagrangian mechanics. Just as the Lagrangian for a free point particle is proportional to its proper timei.e., the 'length' of its world–line, a relativistic string's Lagrangian is proportional to the area of the sheet which the string traces as it travels through space–time.

[17] The Polyakov action is the 2D action from *conformal field theory*, used in string theory to describe the world–sheet of a moving string.

physicists have also calculated that bosonic string theory requires 26 space–time dimensions: 25 spatial dimensions and one dimension of time. In the early 1970s, *supersymmetry* was discovered in the context of string theory, and a new version of string theory called *superstring theory* (i.e., supersymmetric string theory) became the real focus, as it includes also *fermions*, the force particles. Nevertheless, bosonic string theory remains a very useful 'toy model' to understand many general features of *perturbative string theory*.

6.3.3 Weyl Invariance and Vertex Operator Formulation

The action S is also invariant under *Weyl rescalings* of the metric g by a positive function on $\sigma : \Sigma \to \mathbb{R}$, given by $g \to e^{2\sigma} g$. In general, Weyl invariance of the full amplitude may be spoiled by anomalies. Assuming Weyl invariance of the full amplitude, the integral defining Amp may be *simplified* in two ways.

1) The integration over $Met(\Sigma)$ effectively collapses to an integration over the *moduli space of surfaces*, which is finite dimensional, for each genus h.

2) The boundary components of Σ — characterizing external string states — may be mapped to regular points on an underlying compact surface without boundary by conformal transformations. The *data*, such as momenta and other quantum numbers of the external states, are mapped into *vertex operators*. The amplitudes are now given by the path integral

$$Amp = \sum_{h=0}^{\infty} \int_{Met(\Sigma)} \mathcal{D}[g] \frac{1}{N(g)} \int_{Map(\Sigma,M)} \mathcal{D}[x] \, V_1 \ldots V_N \, e^{-S},$$

for suitable vertex operators $V_1, \ldots V_N$.

6.3.4 More General Actions

Generalizations of the action S given above are possible when M carries extra structure.

1) M carries a 2–form $B \in \Omega^{(2)}(M)$. The resulting contribution to the action is also that of a 'nonlinear sigma model'

$$S_B[x, B] = \int_\Sigma x^*(B) = \int_\Sigma dx^\mu \wedge dx^\nu B_{\mu\nu}(x)$$

2) M may carry a *dilaton field* $\Phi \in \Omega^{(0)}(M)$ so that

$$S_\Phi[x, \Phi] = \int_\Sigma d\mu_g R_g \Phi(x).$$

where R_g is the Gaussian curvature of Σ for the metric g.

3) There may be a *tachyon field* $T \in \Omega^{(0)}(M)$ contributing

$$S_T[x, T] = \int_\Sigma d\mu_g T(x).$$

6.3.5 Transition Amplitude for a Single Point Particle

The transition amplitude for a single point–particle could in fact be get in a way analogous to how we prescribed string amplitudes. Let space–time be again a Riemannian manifold M, with metric g. The prescription for the transition amplitude of a particle travelling from a point $y \in M$ to a point y' to M is expressible in terms of a sum over all (continuous) paths connecting y to y':

$$Amp(y, y') = \sum_{\substack{paths \\ joining\ y\ and\ y'}} \mathrm{e}^{-S[path]}.$$

Paths may be parametrized by maps from $C = [0, 1]$ into M with $x(0) = y$, $x(1) = y'$. A simple world–line action for a massless particle is get by introducing a metric g on $[0, 1]$

$$S[x, g] = \frac{1}{2} \int_C d\tau\, g(\tau)^{-1} \dot{x}^\mu \dot{x}^\nu g_{\mu\nu}(x),$$

which is invariant under $Diff^+(C)$ and $Diff(M)$.

Recall that the analogous prescription for the point–particle transition amplitude is the path integral

$$Amp(y, y') = \int_{Met(C)} \mathcal{D}[g] \frac{1}{N} \int_{Map(C,M)} \mathcal{D}[x]\, \mathrm{e}^{-S[x,g]}.$$

Note that for string theory, we had a prescription for transition amplitudes valid for all topologies of the world–sheet. For point–particles, there is only the topology of the interval C, and we can only describe a single point–particle, but not interactions with other point–particles. To put those in, we would have to supply additional information.

Finally, it is very instructive to work out the amplitude Amp by carrying out the integrations. The only $Diff^+(C)$ invariant of g is the length $L = \int_0^1 d\tau\, g(\tau)$; all else is generated by $Diff^+(C)$. Defining the normalization factor to be the volume of $Diff(C)$: $N = \mathrm{Vol}(Diff(C))$ we have $\mathcal{D}[g] = \mathcal{D}[v]\, dL$ and the transition amplitude becomes

$$Amp(y, y') = \int_0^\infty dL \int \mathcal{D}[x]\, \mathrm{e}^{-\frac{1}{2L} \int_0^1 d\tau (\dot{x}, \dot{x})_g} = \int_0^\infty dL\, \langle y' | \mathrm{e}^{-L\Delta} | y \rangle = \left\langle y' | \frac{1}{\Delta} | y \right\rangle.$$

Thus, the amplitude is just the Green function at (y, y') for the Laplacian Δ and corresponds to the propagation of a massless particle (see [DEF99]).

6.3.6 Witten's Open String Field Theory

Noncommutative nature of space–time has often appeared in non–perturbative aspects of string theory. It has been used in a formulation of interacting open string field theory by Ed Witten [Wit88c, Wit86a]. Witten has written a classical action of open string field theory in terms of noncommutative geometry, where the noncommutativity appears in a product of string fields. Later, the *Dirichlet branes* (or, D–branes) have been recognized as solitonic objects in superstring theory [Pol95]. Further, it has been found that the low energy behavior of the D–branes are well described by supersymmetric Yang–Mills theory (SYM) [Wit96]. In the situation of some D–branes coinciding, the space–time coordinates are promoted to matrices which appear as the fields in SYM. Then the size of the matrices corresponds to the number of the D–branes, so noncommutativity of the matrices is related to the noncommutative nature of space–time.

In this subsection, mainly following [Sug00], we review some basic properties of Witten's bosonic open string field theory [Wit88c] and its explicit construction based on a *Fock space* representation of string field functional and δ–function overlap vertices [GJ87a, GJ87b, CST86].

Witten introduced a beautiful formulation of open string field theory in terms of a noncommutative extension of differential geometry, where string fields, the BRST operator Q and the integration over the string configurations $\int$ in string field theory are analogs of differential forms, the exterior derivative d and the integration over the manifold $\int_M$ in the differential geometry, respectively. The ghost number assigned to the string field corresponds to the degree of the differential form. Also the (noncommutative) product between the string fields $*$ is interpreted as an analog of the wedge product $\wedge$ between the differential forms.

The axioms obeyed by the system of $\int$, $*$ and Q are

$$\int QA = 0, \qquad Q(A * B) = (QA) * B + (-1)^{n_A} A * (QB),$$

$$(A * B) * C = A * (B * C), \qquad \int A * B = (-1)^{n_A n_B} \int B * A,$$

where A, B and C are arbitrary string fields, whose ghost number is half–integer valued: The *ghost number* of A is defined by the integer n_A as $n_A + \frac{1}{2}$.

Then Witten discussed the following *string–field–theory action*

$$S = \frac{1}{G_s} \int \left(\frac{1}{2} \psi * Q\psi + \frac{1}{3} \psi * \psi * \psi \right), \tag{6.114}$$

where G_s is the open string coupling constant and ψ is the string field with the ghost number $-\frac{1}{2}$. The action is invariant under the gauge transformation

$$\delta\psi = Q\Lambda + \psi * \Lambda - \Lambda * \psi,$$

with the gauge parameter Λ of the ghost number $-\frac{3}{2}$.

Operator Formulation of String Field Theory

The objects defined above can be explicitly constructed by using the operator formulation, where the string field is represented as a Fock space, and the integration $\int$ as an inner product on the Fock space. It was considered by [GJ87a, GJ87b] in the case of the *Neumann boundary condition*. We will heavily use the notation of [GJ87a, GJ87b]. In the operator formulation, the action (6.114) is described as

$$S = \frac{1}{G_s}\left(\frac{1}{2}{}_{12}\langle V_2||\psi\rangle_1 Q|\psi\rangle_2 + \frac{1}{3}{}_{123}\langle V_3||\psi\rangle_1|\psi\rangle_2|\psi\rangle_3\right), \tag{6.115}$$

where the structure of the product $*$ in the kinetic and potential terms is encoded to that of the overlap vertices $\langle V_2|$ and $\langle V_3|$ respectively (here, subscripts put to vectors in the Fock space label the strings concerning the vertices).

As a preparation for giving the explicit form of the overlaps, let us consider open strings in 26-dimensional space–time with the constant metric G_{ij} in the Neumann boundary condition. The world sheet action is given by

$$S_{WS} = \frac{1}{4\pi\alpha'}\int d\tau \int_0^\pi d\sigma G_{ij}(\partial_\tau X^i \partial_\tau X^j - \partial_\sigma X^i \partial_\sigma X^j) + S_{gh}, \tag{6.116}$$

where S_{gh} is the action of the $bc-$ghosts:

$$S_{gh} = \frac{\mathrm{i}}{2\pi}\int d\tau \int_0^\pi d\sigma [c_+(\partial_\tau - \partial_\sigma)b_+ + c_-(\partial_\tau + \partial_\sigma)b_-]. \tag{6.117}$$

Under the Neumann boundary condition, the string coordinates have the standard mode expansions:

$$X^j(\tau,\sigma) = x^j + 2\alpha'\tau p^j + \mathrm{i}\sqrt{2\alpha'}\sum_{n\neq 0}\frac{1}{n}\alpha_n^j \mathrm{e}^{-in\tau}\cos(n\sigma), \tag{6.118}$$

also the mode expansions of the ghosts are given by

$$c_\pm(\tau,\sigma) = \sum_{n\in\mathbb{Z}} c_n \mathrm{e}^{-in(\tau\pm\sigma)} \equiv c(\tau,\sigma) \pm \mathrm{i}\pi_b(\tau,\sigma),$$
$$b_\pm(\tau,\sigma) = \sum_{n\in\mathbb{Z}} b_n \mathrm{e}^{-in(\tau\pm\sigma)} \equiv \pi_c(\tau,\sigma) \mp \mathrm{i}b(\tau,\sigma).$$

As a result of the quantization, the modes obey the commutation relatons:

$$[x^i, p^j] = \mathrm{i}G^{ij}, \qquad [\alpha_n^i, \alpha_m^j] = nG^{ij}\delta_{n+m,0}, \qquad \{b_n, c_m\} = \delta_{n+m,0},$$

while the other therms vanish.

The overlap $|V_N\rangle = |V_N\rangle^X |V_N\rangle^{gh}$, $(N = 1, 2, \cdots)$ is the state satisfying the continuity conditions for the string coordinates and the ghosts at the $N-$string vertex of the string field theory. The superscripts X and gh show the contribution of the sectors of the coordinates and the ghosts respectively. The continuity conditions for the coordinates are

$$(X^{(r)j}(\sigma) - X^{(r-1)j}(\pi-\sigma))|V_N\rangle^X = 0, \quad (P_i^{(r)}(\sigma) + P_i^{(r-1)}(\pi-\sigma))|V_N\rangle^X = 0, \tag{6.119}$$

for $0 \le \sigma \le \frac{\pi}{2}$ and $r = 1, \cdots, N$. Here $P_i(\sigma)$ is the momentum conjugate to the coordinate $X^j(\sigma)$ at $\tau = 0$, and the superscript (r) labels the string (r) meeting at the vertex. In the above formulas, we regard $r = 0$ as $r = N$ because of the cyclic property of the vertex. For the ghost sector, we impose the following conditions on the variables $c(\sigma)$, $b(\sigma)$ and their conjugate momenta $\pi_c(\sigma)$, $\pi_b(\sigma)$:

$$(\pi_c^{(r)}(\sigma) - \pi_c^{(r-1)}(\pi - \sigma))|V_N\rangle^{gh} = 0, \quad (b^{(r)}(\sigma) - b^{(r-1)}(\pi - \sigma))|V_N\rangle^{gh} = 0,$$
$$(c^{(r)}(\sigma) + c^{(r-1)}(\pi - \sigma))|V_N\rangle^{gh} = 0, \quad (\pi_b^{(r)}(\sigma) + \pi_b^{(r-1)}(\pi - \sigma))|V_N\rangle^{gh} = 0,$$

for $0 \le \sigma \le \frac{\pi}{2}$ and $r = 1, \cdots, N$.

Open Strings in Constant B–Field Background

We consider a constant background of the second–rank antisymmetric tensor field B_{ij} in addition to the constant metric g_{ij} where open strings propagate. Then the boundary condition at the end points of the open strings changes from the Neumann type, and thus the open string has a different mode expansion from the Neumann case (6.118). As a result, the end point is to be noncommutative, in the picture of the D–branes which implies noncommutativity of the world volume coordinates on the D–branes. Here we derive the mode–expanded form of the open string coordinates as a preparation for a calculation of the overlap vertices in the next section.

We start with the world sheet action

$$\begin{aligned} S_{WS}^B = \frac{1}{4\pi\alpha'} \int d\tau \int_0^\pi d\sigma [g_{ij}(\partial_\tau X^i \partial_\tau X^j - \partial_\sigma X^i \partial_\sigma X^j) \\ - 2\pi\alpha' B_{ij}(\partial_\tau X^i \partial_\sigma X^j - \partial_\sigma X^i \partial_\tau X^j)] + S_{gh}. \end{aligned} \tag{6.120}$$

Because the term proportional to B_{ij} can be written as a total derivative term, it does not affect the equation of motion but does the boundary condition, which requires

$$g_{ij}\partial_\sigma X^j - 2\pi\alpha' B_{ij}\partial_\tau X^j = 0 \tag{6.121}$$

on $\sigma = 0, \pi$. This can be rewritten to the convenient form

$$\begin{aligned} E_{ij}\partial_- X^j &= (E^T)_{ij}\partial_+ X^j, \\ \text{where} \qquad E_{ij} &\equiv g_{ij} + 2\pi\alpha' B_{ij}, \end{aligned} \tag{6.122}$$

and $\partial_\pm$ are derivatives with respect to the light cone variables $\sigma^\pm = \tau \pm \sigma$. We can easily see that $X^j(\tau, \sigma)$ satisfying the boundary condition (6.122) has the following mode expansion:

$$\begin{aligned} X^j(\tau, \sigma = \tilde{x}^j + \alpha' \left[(E^{-1})^{jk} g_{kl} p^l \sigma^- + (E^{-1T})^{jk} g_{kl} p^l \sigma^+ \right] \\ + \mathrm{i}\sqrt{\frac{\alpha'}{2}} \sum_{n \neq 0} \frac{1}{n} \left[(E^{-1})^{jk} g_{kl} \alpha_n^l \mathrm{e}^{-in\sigma^-} + (E^{-1T})^{jk} g_{kl} \alpha_n^l \mathrm{e}^{-in\sigma^+} \right]. \end{aligned} \tag{6.123}$$

We will get the commutators between the modes from the propagator of the open strings, which gives another derivation different from the method by [Ch99] based on the quantization via the Dirac bracket. When performing the *Wick rotation*: $\tau \to -\mathrm{i}\tau$ and mapping the world sheet to the upper half plane

$$z = \mathrm{e}^{\tau + \mathrm{i}\sigma}, \qquad \bar{z} = \mathrm{e}^{\tau - \mathrm{i}\sigma}(0 \le \sigma \le \pi),$$

the boundary condition (6.122) becomes

$$E_{ij}\partial_{\bar{z}}X^j = (E^T)_{ij}\partial_z X^j, \tag{6.124}$$

which is imposed on the real axis $z = \bar{z}$. The propagator $\langle X^i(z,\bar{z})X^j(z',\bar{z'})\rangle$ satisfying the boundary condition (6.124)is determined as

$$\begin{aligned}\langle X^i(z,\bar{z})X^j(z',\bar{z'})\rangle = -\alpha' \big[& g^{ij}\ln|z-z'| - g^{ij}\ln|z-\bar{z'}| \\ & + G^{ij}\ln|z-\bar{z'}|^2 + \frac{1}{2\pi\alpha'}\theta^{ij}\ln\frac{z-\bar{z'}}{\bar{z}-z'} + D^{ij},\end{aligned}$$

where G^{ij} and θ^{ij} are given by

$$G^{ij} = \frac{1}{2}(E^{T-1} + E^{-1})^{ij} = (E^{T-1}gE^{-1})^{ij} = (E^{-1}gE^{T-1})^{ij}, \tag{6.125}$$

$$\begin{aligned}\theta^{ij} &= 2\pi\alpha' \cdot \frac{1}{2}(E^{T-1} - E^{-1})^{ij} = (2\pi\alpha')^2(E^{T-1}BE^{-1})^{ij} \\ &= -(2\pi\alpha')^2(E^{-1}BE^{T-1})^{ij}.\end{aligned} \tag{6.126}$$

Also the constant D^{ij} remains unknown from the boundary condition alone. However it is an irrelevant parameter, so we can fix an appropriate value. The mode-expanded form (6.123) is mapped to

$$\begin{aligned}X^j(z,\bar{z}) = {} & \tilde{x}^j - \mathrm{i}\alpha'[(E^{-1})^{jk}p_k\ln\bar{z} + (E^{-1T})^{jk}p_k\ln z] \\ & + \mathrm{i}\sqrt{\frac{\alpha'}{2}}\sum_{n\neq 0}\frac{1}{n}\left[(E^{-1})^{jk}\alpha_{n,k}\bar{z}^{-n} + (E^{-1T})^{jk}\alpha_{n,k}z^{-n}\right].\end{aligned}$$

Note that the indices of p^l and α_n^l were lowered by the metric g_{ij} not G_{ij}. Recall the definition of the propagator

$$\langle X^i(z,\bar{z})X^j(z',\bar{z'})\rangle \equiv R(X^i(z,\bar{z})X^j(z',\bar{z'})) - N(X^i(z,\bar{z})X^j(z',\bar{z'})), \tag{6.127}$$

where R and N stand for the radial ordering and the normal ordering respectively. We take a prescription for the normal ordering which pushes p_i to the right and $\tilde{x}_j$ to the left with respect to the zero–modes p_i and $\tilde{x}_j$. It corresponds to considering the vacuum satisfying

$$p_j|0\rangle = \alpha_{n,j}|0\rangle = 0 \qquad (n > 0), \qquad \langle 0|\alpha_{n,j} = 0 \qquad (n < 0), \tag{6.128}$$

which is the standard prescription for calculating the propagator of the massless scalar field in 2D conformal field theory from the operator formalism.

Making use of (6.127), (6.128) and techniques of the contour integration, it is easy to get the commutators

$$[\alpha_{n,i}, \alpha_{m,j}] = n\delta_{n+m,0}G_{ij}, \qquad [\tilde{x}^i, p_j] = \mathrm{i}\delta^i_j,$$

where the first equation holds for all integers with $\alpha_{0,i} \equiv \sqrt{2\alpha'}p_i$. The constant D^{ij} is written as $\alpha' D^{ij} = -\langle 0|\tilde{x}^i \tilde{x}^j|0\rangle$. Let us fix D^{ij} as $\alpha' D^{ij} = -\frac{\mathrm{i}}{2}\theta^{ij}$, which is the convention taken in [SW99]. Then the coordinates $\tilde{x}^i$ become noncommutative:

$$[\tilde{x}^i, \tilde{x}^j] = \mathrm{i}\theta^{ij},$$

but the center of mass coordinates $x^i \equiv \tilde{x}^i + \frac{1}{2}\theta^{ij}p_j$ can be seen to commute each other.

Now we have the mode–expanded form of the string coordinates and the commutation relations between the modes, which are

$$\begin{aligned} X^j(\tau,\sigma) &= x^j + 2\alpha' \left(G^{jk}\tau + \frac{1}{2\pi\alpha'}\theta^{jk}(\sigma - \frac{\pi}{2}) \right) p_k \\ &\quad + \mathrm{i}\sqrt{2\alpha'} \sum_{n\neq 0} \frac{1}{n} \mathrm{e}^{-in\tau} \left[G^{jk}\cos(n\sigma) - \mathrm{i}\frac{1}{2\pi\alpha'}\theta^{jk}\sin(n\sigma) \right] \alpha_{n,k}, \end{aligned}$$

$$[\alpha_{n,i}, \alpha_{m,j}] = n\delta_{n+m,0}G_{ij}, \qquad [x^i, p_j] = \mathrm{i}\delta^i_j,$$

with all the other commutators vanishing.

Also, due to the formula

$$\sum_{n=1}^{\infty} \frac{2}{n}\sin(n(\sigma+\sigma')) = \begin{cases} \pi - \sigma - \sigma', & (\sigma+\sigma' \neq 0, 2\pi) \\ 0, & (\sigma+\sigma' = 0, 2\pi), \end{cases}$$

we can see by a direct calculation that the end points of the string become noncommutative

$$[X^i(\tau,\sigma), X^j(\tau,\sigma')] = \begin{cases} \mathrm{i}\theta^{ij}, & (\sigma = \sigma' = 0) \\ -\mathrm{i}\theta^{ij}, & (\sigma = \sigma' = \pi) \\ 0, & (\text{otherwise}). \end{cases}$$

On the other hand, it is noted that the conjugate momenta have the mode expansion identical with that in the Neumann case:

$$\begin{aligned} P_i(\tau,\sigma) &= \frac{1}{2\pi\alpha'}(g_{ij}\partial_\tau - 2\pi\alpha' B_{ij}\partial_\sigma)X^j(\tau,\sigma) \\ &= \frac{1}{\pi}p_i + \frac{1}{\pi\sqrt{2\alpha'}} \sum_{n\neq 0} \mathrm{e}^{-in\tau}\cos(n\sigma)\alpha_{n,i}. \end{aligned}$$

Note that the relations (6.125) and (6.126) are in the same form as a T–duality transformation, although the correspondence is a formal sense, because we are not considering any compactification of space–time. The generalized T–duality transformation, namely $O(D,D)$–transformation, is defined by

$$E' = (aE + b)(cE + d)^{-1}, \tag{6.129}$$

with a, b, c and d being $D \times D$ real matrices. (D is the dimension of space–time.) The matrix $h = \begin{pmatrix} a & b \\ c & d \end{pmatrix}$ is $O(D,D)$ matrix, which satisfies

$$h^T J h = J, \qquad \text{where} \qquad J = \begin{pmatrix} 0 & \mathbf{1}_D \\ \mathbf{1}_D & 0 \end{pmatrix}.$$

The relations (6.125) and (6.126) correspond to the case of the inversion $a = d = 0$, $b = c = \mathbf{1}_D$.

Construction of Overlap Vertices

Here we construct Witten's open string theory in the constant B–field background by obtaining the explicit formulas of the overlap vertices. As is understood from the fact that the action of the ghosts (6.117) contains no background fields, the ghost sector is not affected by turning on the B–field background. Thus we may consider the coordinate sector only. First, let us see the mode-expanded forms of the coordinates and the momenta at $\tau = 0$

$$\begin{aligned} X^j(\sigma) &= G^{jk} y_k + \frac{1}{\pi}\theta^{jk}(\sigma - \frac{\pi}{2}) p_k \\ &\quad + 2\sqrt{\alpha'} \sum_{n=1}^{\infty} \left[G^{jk} \cos(n\sigma) x_{n,k} + \frac{1}{2\pi\alpha'}\theta^{jk} \sin(n\sigma) \frac{1}{n} p_{n,k} \right], \\ P_i(\sigma) &= \frac{1}{\pi} p_i + \frac{1}{\pi\sqrt{\alpha'}} \sum_{n=1}^{\infty} \cos(n\sigma) p_{n,i}, \end{aligned}$$

where $x^j = G^{jk} y_k$, the coordinates and the momenta for the oscillator modes are

$$\begin{aligned} x_{n,k} &= \frac{\mathrm{i}}{2}\sqrt{\frac{2}{n}}(a_{n,k} - a^\dagger_{n,k}) = \frac{\mathrm{i}}{\sqrt{2n}}(\alpha_{n,k} - \alpha_{-n,k}), \\ p_{n,k} &= \sqrt{\frac{n}{2}}(a_{n,k} + a^\dagger_{n,k}) = \frac{1}{\sqrt{2}}(\alpha_{n,k} + \alpha_{-n,k}). \end{aligned}$$

The nonvanishing commutators are given by

$$[x_{n,k}, p_{m,l}] = \mathrm{i} G_{kl} \delta_{n,m}, \qquad [y_k, p_l] = \mathrm{i} G_{kl}. \tag{6.130}$$

We should note that the metric appearing in eqs. (6.130) is G_{ij}, instead of g_{ij}. So it can be seen that if we employ the variables with the lowered space–time indices y_k, p_k, $x_{n,k}$ and $p_{n,k}$, the metric used in the expression of the overlaps is G^{ij} not g^{ij}.

The continuity condition (6.119) is universal for any background, and the mode expansion of the momenta $P_i(\sigma)$'s is of the same form as in the Neumann case, thus the continuity conditions for the momenta in terms of the modes $p_{n,i}$ are identical with those in the Neumann case. Also, since $p_{n,i}$'s mutually commute, it is natural to find a solution of the continuity condition, assuming the following form for the overlap vertices:

$$|\hat{V}_N\rangle^X_{1\cdots N} = \exp\left[\frac{\mathrm{i}}{4\pi\alpha}\theta^{ij}\sum_{r,s=1}^{N} p^{(r)}_{n,i}Z^{rs}_{nm}p^{(s)}_{m,j}\right]|V_N\rangle^X_{1\cdots N}, \tag{6.131}$$

where $|\hat{V}_N\rangle^X_{1\cdots N}$ and $|V_N\rangle^X_{1\cdots N}$ are the overlaps in the background corresponding to the world sheet actions (6.120) and (6.116) respectively, the explicit form of the latter is given in appendix A. Clearly the expression (6.131) satisfies the continuity conditions for the modes of the momenta, and the coefficients Z^{rs}_{nm} are determined so that the continuity conditions for the coordinates are satisfied [Sug00].

- $|\hat{I}\rangle^X \equiv |\hat{V}_1\rangle^X$

For the $N = 1$ case, we consider the identity overlap $|\hat{I}\rangle^X \equiv |\hat{V}_1\rangle^X$. The continuity conditions for the momenta require that

$$P_i(\sigma) + P_i(\pi - \sigma) = \frac{2}{\pi}p_i + \frac{2}{\pi\sqrt{\alpha'}}\sum_{n=2,4,6,\cdots}\cos(n\sigma)p_{n,i}$$

should vanish for $0 \le \sigma \le \frac{\pi}{2}$, namely,

$$p_i = 0, \qquad p_{n,i} = 0 \qquad (n = 2,4,6,\cdots), \tag{6.132}$$

which is satisfied by the overlap in the Neumann case $|I\rangle$. In addition, the conditions for the coordinates are that

$$X^j(\sigma) - X^j(\pi - \sigma) = \frac{2}{\pi}\theta^{jk}(\sigma - \frac{\pi}{2})p_k \; + \tag{6.133}$$
$$4\sqrt{\alpha'}\sum_{n=1,3,5,\cdots} G^{jk}\cos(n\sigma)x_{n,k} + 4\sqrt{\alpha'}\sum_{n=2,4,6,\cdots}\frac{1}{2\pi\alpha'}\theta^{jk}\sin(n\sigma)\frac{1}{n}p_{n,k},$$

should vanish for $0 \le \sigma \le \frac{\pi}{2}$. The first and third lines in the r. h. s. can be put to zero by using (6.132). So what we have to consider is the remaining condition $x_{n,k} = 0$ for $n = 1,3,5,\cdots$, which however is nothing but the continuity condition for the coordinates in the Neumann case. It can be understood from the point that the second line in (6.133) does not depend on θ^{ij}. Thus it turns out that the continuity conditions in the case of the $B-$field turned on are satisfied by the identity overlap made in the Neumann case. The solution is [Sug00]

$$|\hat{I}\rangle^X = |I\rangle^X = \exp\left[-\frac{1}{2}G^{ij}\sum_{n=0}^{\infty}(-1)^n a_{n,i}^\dagger a_{n,j}^\dagger\right]|0\rangle, \tag{6.134}$$

where also the zero modes y_i and p_i are written by using the creation and annihilation operators $a_{0,i}^\dagger$ and $a_{0,i}$ as

$$y_i = \frac{\mathrm{i}}{2}\sqrt{2\alpha'}(a_{0,i} - a_{0,i}^\dagger), \qquad p_i = \frac{1}{\sqrt{2\alpha'}}(a_{0,i} + a_{0,i}^\dagger).$$

- $|\hat{V}_2\rangle_{12}^X$

For the $N = 2$ case, we are to do the same argument as in the $N = 1$ case. The continuity conditions mean that

$$P_i^{(1)}(\sigma) + P_i^{(2)}(\pi - \sigma) = \frac{1}{\pi}(p_i^{(1)} + p_i^{(2)}) + \frac{1}{\pi\sqrt{\alpha'}}\sum_{n=1}^{\infty}\cos(n\sigma)(p_{n,i}^{(1)} + (-1)^n p_{n,i}^{(2)}),$$

$$\begin{aligned} X^{(1)j}(\sigma) - X^{(2)j}(\pi - \sigma) &= G^{jk}(y_k^{(1)} - y_k^{(2)}) + \frac{1}{\pi}\theta^{jk}(\sigma - \frac{\pi}{2})(p_k^{(1)} + p_k^{(2)}) \\ &+ 2\sqrt{\alpha'}\sum_{n=1}^{\infty}\Big[G^{jk}\cos(n\sigma)(x_{n,k}^{(1)} - (-1)^n x_{n,k}^{(2)}) \\ &+ \frac{1}{2\pi\alpha'}\theta^{jk}\sin(n\sigma)\frac{1}{n}(p_{n,k}^{(1)} + (-1)^n p_{n,k}^{(2)})\Big] \end{aligned}$$

should be zero for $0 \le \sigma \le \pi$. It turns out again that the conditions for the modes are identical with those in the Neumann case:

$$p_i^{(1)} + p_i^{(2)} = 0, \qquad p_{n,i}^{(1)} + (-1)^n p_{n,i}^{(2)} = 0,$$
$$y_i^{(1)} - y_i^{(2)} = 0, \qquad x_{n,i}^{(1)} - (-1)^n x_{n,i}^{(2)} = 0,$$

for $n \ge 1$. Thus we have the solution [Sug00]

$$|\hat{V}_2\rangle_{12}^X = |V_2\rangle_{12}^X = \exp\left[-G^{ij}\sum_{n=0}^{\infty}(-1)^n a_{n,i}^{(1)\dagger} a_{n,j}^{(2)\dagger}\right]|0\rangle_{12}. \tag{6.135}$$

- $|\hat{V}_4\rangle^X_{1234}$

We find a solution of the continuity conditions (6.119) in the $N = 4$ case assuming the form

$$|\hat{V}_4\rangle^X_{1234} = \exp\left[\frac{\mathrm{i}}{4\pi\alpha}\theta^{ij}\sum_{r,s=1}^{4} p^{(r)}_{n,i} Z^{rs}_{nm} p^{(s)}_{m,j}\right] |V_4\rangle^X_{1234}. \tag{6.136}$$

When considering the continuity conditions, it is convenient to employ the Z_4–Fourier transformed variables:

$$\begin{aligned}
Q^j_1(\sigma) &= \frac{1}{2}[\mathrm{i}X^{(1)j}(\sigma) - X^{(2)j}(\sigma) - iX^{(3)j}(\sigma) + X^{(4)j}(\sigma)] \equiv Q^j(\sigma),\\
Q^j_2(\sigma) &= \frac{1}{2}[-X^{(1)j}(\sigma) + X^{(2)j}(\sigma) - X^{(3)j}(\sigma) + X^{(4)j}(\sigma)],\\
Q^j_3(\sigma) &= \frac{1}{2}[-\mathrm{i}X^{(1)j}(\sigma) - X^{(2)j}(\sigma) + iX^{(3)j}(\sigma) + X^{(4)j}(\sigma)] \equiv \bar{Q}^j(\sigma),\\
Q^j_4(\sigma) &= \frac{1}{2}[X^{(1)j}(\sigma) + X^{(2)j}(\sigma) + X^{(3)j}(\sigma) + X^{(4)j}(\sigma)].
\end{aligned}$$

For the momentum variables we also define the Z_4–Fourier transformed variables $P_{1,i}(\sigma)(\equiv P_i(\sigma))$, $P_{2,i}(\sigma)$, $P_{3,i}(\sigma)(\equiv \bar{P}_i(\sigma))$ and $P_{4,i}(\sigma)$ by the same combinations of $P^{(r)}_i(\sigma)$'s as the above. These variables have the following mode expansions

$$\begin{aligned}
P_{t,i}(\sigma) &= \frac{1}{\pi\sqrt{2\alpha'}}P_{t,0,i} + \frac{1}{\pi\sqrt{\alpha'}}\sum_{n=1}^{\infty}\cos(n\sigma)P_{t,n,i},\\
Q^j_t(\sigma) &= G^{jk}\sqrt{2\alpha'}Q_{t,0,k} + \frac{1}{\pi}\theta^{jk}(\sigma - \frac{\pi}{2})\frac{1}{\sqrt{2\alpha'}}P_{t,0,k}\\
&\quad + \sqrt{2\alpha'}\sum_{n=1}^{\infty}\left[G^{jk}\cos(n\sigma)Q_{t,n,k} + \frac{1}{2\pi\alpha'}\theta^{jk}\sin(n\sigma)\frac{1}{n}P_{t,n,k}\right],
\end{aligned} \tag{6.137}$$

where $t = 1, 2, 3, 4$. From now on, we frequently omit the subscript t for the $t = 1$ case, and at the same time we employ the expression with a bar instead of putting the subscript t for the $t = 3$ case.

Using those variables, the continuity conditions are written as

$$\begin{aligned}
&Q^j_4(\sigma) - Q^j_4(\pi - \sigma) = 0, \quad P_{4,i}(\sigma) + P_{4,i}(\pi - \sigma) = 0,\\
&Q^j_2(\sigma) + Q^j_2(\pi - \sigma) = 0, \quad P_{2,i}(\sigma) - P_{2,i}(\pi - \sigma) = 0,\\
&Q^j(\sigma) - \mathrm{i}Q^j(\pi - \sigma) = 0, \quad P_i(\sigma) + \mathrm{i}P_i(\pi - \sigma) = 0,\\
&\bar{Q}^j(\sigma) + \mathrm{i}\bar{Q}^j(\pi - \sigma) = 0, \quad \bar{P}_i(\sigma) - \mathrm{i}\bar{P}_i(\pi - \sigma) = 0
\end{aligned} \tag{6.138}$$

for $0 \leq \sigma \leq \frac{\pi}{2}$. In terms of the modes, the conditions for the sectors of $t = 2$ and 4 are identical with the Neumann case

$$(1-C)|Q_{4,k})|\hat{V}_4\rangle^X = (1+C)|P_{4,k})|\hat{V}_4\rangle^X = 0,$$
$$(1+C)|Q_{2,k})|\hat{V}_4\rangle^X = (1-C)|P_{2,k})|\hat{V}_4\rangle^X = 0,$$

which can be seen from the point that the conditions (6.138) for the sectors of $t = 2$ and 4 lead the same relations between the modes as those without the terms containing θ^{jk}. Here we adopted the vector notation for the modes

$$|Q_{t,k}) = \begin{bmatrix} Q_{t,0,k} \\ Q_{t,1,k} \\ \vdots \end{bmatrix}, \qquad |P_{t,k}) = \begin{bmatrix} P_{t,0,k} \\ P_{t,1,k} \\ \vdots \end{bmatrix},$$

and C is a matrix such that $(C)_{nm} = (-1)^n \delta_{nm}$ $(n, m \geq 0)$. Thus there is needed no correction containing θ^{ij} for the sectors of $t = 2$ and 4, so it is natural to assume the form of the phase factor in (6.136) as

$$\frac{1}{2}\theta^{ij} \sum_{r,s=1}^{4} (p_i^{(r)}|Z^{rs}|p_j^{(s)}) = \theta^{ij}(P_i|Z|\bar{P}_j) \tag{6.139}$$

with Z being anti–Hermitian.

Next let us consider the conditions for the sectors of $t = 1$ and 3. We rewrite the mode expansions of $Q^j(\sigma)$ and $\bar{Q}^j(\sigma)$ as [Sug00]

$$\begin{aligned} Q^j(\sigma) &= G^{jk}(\sqrt{2\alpha'}Q_{0,k} + 2\sqrt{\alpha'}\sum_{n=1}^{\infty}\cos(n\sigma)Q_{n,k}) \\ &+ \theta^{jk}\left[\int_{\pi/2}^{\sigma} d\sigma' P_i(\sigma') + \frac{1}{\pi\sqrt{\alpha'}}\sum_{n=1,3,5,\cdots}\frac{1}{n}(-1)^{(n-1)/2}P_{n,k}\right] \\ &\equiv \theta^{jk}\int_{\pi/2}^{\sigma} d\sigma' P_i(\sigma') + \Delta Q^j(\sigma), \end{aligned} \tag{6.140}$$

$$\begin{aligned} \bar{Q}^j(\sigma) &= G^{jk}(\sqrt{2\alpha'}\bar{Q}_{0,k} + 2\sqrt{\alpha'}\sum_{n=1}^{\infty}\cos(n\sigma)\bar{Q}_{n,k}) \\ &+ \theta^{jk}\left[\int_{\pi/2}^{\sigma} d\sigma' \bar{P}_i(\sigma') + \frac{1}{\pi\sqrt{\alpha'}}\sum_{n=1,3,5,\cdots}\frac{1}{n}(-1)^{(n-1)/2}\bar{P}_{n,k}\right] \\ &\equiv \theta^{jk}\int_{\pi/2}^{\sigma} d\sigma' \bar{P}_i(\sigma') + \Delta\bar{Q}^j(\sigma). \end{aligned} \tag{6.141}$$

Using the conditions for $P_i(\sigma)$ and $\bar{P}_i(\sigma)$ in (6.138), we can reduce the conditions for $Q^j(\sigma)$ and $\bar{Q}^j(\sigma)$ to those for $\Delta Q^j(\sigma)$ and $\Delta\bar{Q}^j(\sigma)$:

$$\Delta Q^j(\sigma) = \begin{cases} \mathrm{i}\Delta Q^j(\pi-\sigma) & (0 \leq \sigma \leq \frac{\pi}{2}) \\ -\mathrm{i}\Delta Q^j(\pi-\sigma) & (\frac{\pi}{2} \leq \sigma \leq \pi), \end{cases}$$

$$\Delta\bar{Q}^j(\sigma) = \begin{cases} -\mathrm{i}\Delta\bar{Q}^j(\pi-\sigma) & (0 \leq \sigma \leq \frac{\pi}{2}) \\ \mathrm{i}\Delta\bar{Q}^j(\pi-\sigma) & (\frac{\pi}{2} \leq \sigma \leq \pi). \end{cases}$$

These formulas are translated to the relations between the modes via the Fourier transformation. The result is expressed in the vector notation as

$$(1-X)|\mathcal{Q}_i)|\hat{V}_4\rangle^X = (1+X)|\overline{\mathcal{Q}}_i)|\hat{V}_4\rangle^X = 0, \tag{6.142}$$

where the vectors $|\mathcal{Q}_i)$ and $|\overline{\mathcal{Q}}_i)$ stand for

$$|\mathcal{Q}_i) = \begin{bmatrix} Q_{0,i} + \frac{i}{4\alpha'}G_{ik}\theta^{kj}\sum_{n=0}^{\infty} X_{0n}P_{n,j} \\ Q_{1,i} \\ Q_{2,i} \\ \vdots \end{bmatrix},$$

$$|\overline{\mathcal{Q}}_i) = \begin{bmatrix} \bar{Q}_{0,i} + \frac{i}{4\alpha'}G_{ik}\theta^{kj}\sum_{n=0}^{\infty} X_{0n}\bar{P}_{n,j} \\ \bar{Q}_{1,i} \\ \bar{Q}_{2,i} \\ \vdots \end{bmatrix}.$$

In (6.142), passing the vectors through the phase factor of the $|\hat{V}_4\rangle$ and using the continuity conditions in the Neumann case

$$\begin{aligned} (1+X)|P_i)|V_4\rangle^X &= (1-X)|\bar{P}_i)|V_4\rangle^X = 0, \\ (1-X)|Q_i)|V_4\rangle^X &= (1+X)|\bar{Q}_i)|V_4\rangle^X = 0, \end{aligned} \tag{6.143}$$

we get the equations, which the coefficients Z_{nm}'s should satisfy,

$$[(1-X)_{m0}\sum_{n=0}^{\infty}(\bar{Z}_{0n} + \mathrm{i}\frac{\pi}{2}\bar{X}_{0n})P_{n,j} + \sum_{n=1}^{\infty}(1-X)_{mn}\sum_{n'=0}^{\infty}\bar{Z}_{nn'}P_{n',j}]|V_4\rangle^X = 0$$

$$[(1+X)_{m0}\sum_{n=0}^{\infty}(Z_{0n} - \mathrm{i}\frac{\pi}{2}X_{0n})\bar{P}_{n,j} + \sum_{n=1}^{\infty}(1+X)_{mn}\sum_{n'=0}^{\infty}Z_{nn'}\bar{P}_{n',j}]|V_4\rangle^X = 0$$

for $m \geq 0$. Now all our remaining task is to solve these equations. It is easy to see that a solution of them is given by [Sug00]

$$\begin{aligned} Z_{mn} &= -\mathrm{i}\frac{\pi}{2}(1-X)_{mn} + \mathrm{i}\beta\frac{\pi}{2}C_{mn}, \quad (m, n \geq 0, \text{ except for } m = n = 0), \\ Z_{00} &= \mathrm{i}\beta\frac{\pi}{2}, \end{aligned}$$

if we pay attention to (6.143). Here β is an unknown real constant, which is not fixed by the continuity conditions alone. This ambiguity of the solution comes from the property of the matrix X: $XC = -CX$. However it will become clear that the term containing the constant β does not contribute to the vertex $|\hat{V}_4\rangle^X$.

Therefore, we have the expression of the phase (6.139)

$$\theta^{ij}(P_i|Z|\bar{P}_j) = \theta^{ij}[\mathrm{i}\frac{\pi}{2}P_{0,i}\bar{P}_{0,j} + \mathrm{i}\beta\frac{\pi}{2}\sum_{n=0}^{\infty}(-1)^n P_{n,i}\bar{P}_{n,j} - \mathrm{i}\frac{\pi}{2}\sum_{m,n=0}^{\infty} P_{m,i}(1-X)_{mn}\bar{P}_{n,j}].$$

Then recalling (6.143) again, the last term in the r. h. s. can be discarded. Also we can rewrite the term containing β

$$-\frac{\theta^{ij}}{4\alpha'}\beta(P_i|C|\bar{P}_j) = +\frac{\theta^{ij}}{4\alpha'}\beta(P_i|X^T CX|\bar{P}_j) = +\frac{\theta^{ij}}{4\alpha'}\beta(P_i|C|\bar{P}_j),$$

on $|V_4\rangle^X$. The above formula means that the term containing β can be set to zero on $|V_4\rangle^X$. After all, the form of the 4-string vertex becomes

$$|\hat{V}_4\rangle^X_{1234} = \exp\left[-\frac{\theta^{ij}}{4\alpha'}P_{0,i}\bar{P}_{0,j}\right]|V_4\rangle^X_{1234}.$$

Note that the phase factor has the cyclic symmetric form

$$-\frac{\theta^{ij}}{4\alpha'}P_{0,i}\bar{P}_{0,j} = \mathrm{i}\frac{\theta^{ij}}{8\alpha'}(p^{(1)}_{0.i}p^{(2)}_{0,j} + p^{(2)}_{0.i}p^{(3)}_{0,j} + p^{(3)}_{0.i}p^{(4)}_{0,j} + p^{(4)}_{0.i}p^{(1)}_{0,j}),$$

which is a property the vertices should have[18].

- $|\hat{V}_3\rangle^X_{123}$

We can get the 3-string overlap in the similar manner as in the 4-string case. First, we introduce the Z_3-Fourier transformed variables

$$Q^j_1(\sigma) = \frac{1}{\sqrt{3}}[eX^{(1)j}(\sigma) + \bar{e}X^{(2)j}(\sigma) + X^{(3)j}(\sigma)] \equiv Q^j(\sigma),$$
$$Q^j_2(\sigma) = \frac{1}{\sqrt{3}}[\bar{e}X^{(1)j}(\sigma) + eX^{(2)j}(\sigma) + X^{(3)j}(\sigma)] \equiv \bar{Q}^j(\sigma),$$
$$Q^j_3(\sigma) = \frac{1}{\sqrt{3}}[X^{(1)j}(\sigma) + X^{(2)j}(\sigma) + X^{(3)j}(\sigma)],$$

where $e \equiv \mathrm{e}^{\mathrm{i}2\pi/3}$, $\bar{e} \equiv \mathrm{e}^{-\mathrm{i}2\pi/3}$. The momenta $P_{1,i}(\sigma)(\equiv P_i(\sigma))$, $P_{2,i}(\sigma)(\equiv \bar{P}_i(\sigma))$ and $P_{3,i}(\sigma)$ are defined in the same way. The mode expansions take the same form as those in (6.137). In these variables, the continuity conditions require

$$\begin{aligned} Q^j(\sigma) - eQ^j(\pi-\sigma) &= 0, & P_i(\sigma) + eP_i(\pi-\sigma) &= 0, \\ \bar{Q}^j(\sigma) - \bar{e}\bar{Q}^j(\pi-\sigma) &= 0, & \bar{P}_i(\sigma) + \bar{e}\bar{P}_i(\pi-\sigma) &= 0, \\ Q^j_3(\sigma) - Q^j_3(\pi-\sigma) &= 0, & P_{3,i}(\sigma) + P_{3,i}(\pi-\sigma) &= 0 \end{aligned}$$

[18] Here the momentum $p^{(r)}_{0,i}$ is given by $p^{(r)}_{0,i} = \sqrt{2\alpha'}p^{(r)}_i$.

for $0 \leq \sigma \leq \frac{\pi}{2}$. The conditions imposed to the modes with respect to the $t = 3$ component are identical with those in the Neumann case

$$(1 + C)|P_{3,i})|\hat{V}_3\rangle^X = (1 - C)|Q_{3,i})|\hat{V}_3\rangle^X = 0.$$

Thus the $t = 3$ component does not couple with θ^{ij}, so we can find the solution by determining the single anti–Hermitian matrix Z in the phase factor whose form is assumed as [Sug00]

$$\frac{1}{2}\theta^{ij}\sum_{r,s=1}^{3}(p_i^{(r)}|Z^{rs}|p_j^{(s)}) = \theta^{ij}(P_i|Z|\bar{P}_j). \tag{6.144}$$

For the sectors of $t = 1$ and 2, the same argument goes on as in the 4-string case. $Q^j(\sigma)$ and $\bar{Q}^j(\sigma)$ have the mode expansions same as in eqs. (6.140) and (6.141). The conditions we have to consider are

$$\Delta Q^j(\sigma) = \begin{cases} e\Delta Q^j(\pi - \sigma), & (0 \leq \sigma \leq \frac{\pi}{2}) \\ \bar{e}\Delta Q^j(\pi - \sigma), & (\frac{\pi}{2} \leq \sigma \leq \pi), \end{cases}$$

$$\Delta \bar{Q}^j(\sigma) = \begin{cases} \bar{e}\Delta \bar{Q}^j(\pi - \sigma), & (0 \leq \sigma \leq \frac{\pi}{2}) \\ e\Delta \bar{Q}^j(\pi - \sigma), & (\frac{\pi}{2} \leq \sigma \leq \pi), \end{cases}$$

which are rewritten as the relations between the modes

$$(1 - Y)|\mathcal{Q}_i)|\hat{V}_3\rangle^X = (1 - Y^T)|\overline{\mathcal{Q}}_i)|\hat{V}_3\rangle^X = 0. \tag{6.145}$$

Recalling the conditions in the Neumann case

$$\begin{aligned} (1 + Y)|P_i)|V_3\rangle^X &= (1 + Y^T)|\bar{P}_i)|V_3\rangle^X = 0, \\ (1 - Y)|Q_i)|V_3\rangle^X &= (1 - Y^T)|\bar{Q}_i)|V_3\rangle^X = 0, \end{aligned} \tag{6.146}$$

we end up with the following equations

$$[(1 - Y)_{m0}\sum_{n=0}^{\infty}(\bar{Z}_{0n} + \frac{\pi}{2}\bar{X}_{0n})P_{n,j} + \sum_{n=1}^{\infty}(1 - Y)_{mn}\sum_{n'=0}^{\infty}\bar{Z}_{nn'}P_{n',j}]|V_3\rangle^X = 0,$$

$$[(1 - Y^T)_{m0}\sum_{n=0}^{\infty}(Z_{0n} - \mathrm{i}\frac{\pi}{2}X_{0n})\bar{P}_{n,j} + \sum_{n=1}^{\infty}(1 - Y^T)_{mn}\sum_{n'=0}^{\infty}Z_{nn'}\bar{P}_{n',j}]|V_3\rangle^X = 0$$

for $m \geq 0$. It can be easily found out that the expression

$$\begin{aligned} Z_{mn} &= -\mathrm{i}\frac{\pi}{\sqrt{3}}(1 + Y^T)_{mn} \quad (m, n \geq 0, \text{ except for } m = n = 0), \\ Z_{00} &= 0, \end{aligned}$$

satisfies the above equations. It should be noted that in this case, because of $CYC = \bar{Y} \neq -Y$, it does not contain any unknown constant differently from the 4–string case.

Owing to the condition (6.146) we can write the phase factor only in terms of the zero-modes. Finally we have [Sug00]

$$
\begin{aligned}
|\hat{V}_3\rangle_{123}^X &= \exp\left[-\frac{\theta^{ij}}{4\sqrt{3}\alpha'} P_{0,i}\bar{P}_{0,j}\right] |V_3\rangle_{123}^X \\
&= \exp\left[i\frac{\theta^{ij}}{12\alpha'}(p_{0,i}^{(1)}p_{0,j}^{(2)} + p_{0,i}^{(2)}p_{0,j}^{(3)} + p_{0,i}^{(3)}p_{0,j}^{(1)})\right] |V_3\rangle_{123}^X . \quad (6.147)
\end{aligned}
$$

It is not clear whether the solutions we have obtained here are unique or not. However we can show that the phase factors are consistent with the relations between the overlaps which they should satisfy,

$$
{}_3\langle\hat{I}|\hat{V}_3\rangle_{123} = |\hat{V}_2\rangle_{12}, \quad {}_4\langle\hat{I}|\hat{V}_4\rangle_{1234} = |\hat{V}_3\rangle_{123}, \quad {}_{34}\langle\hat{V}_2||\hat{V}_3\rangle_{123}|\hat{V}_3\rangle_{456} = |\hat{V}_4\rangle_{1256},
$$

by using the momentum conservation on the vertices $(p_i^{(1)}+\cdots+p_i^{(N)})|\hat{V}_N\rangle_{1\cdots N}^X = 0$. Furthermore we can see that the phase factors successfully reproduce the Moyal product structures of the correlators among vertex operators obtained in the perturbative approach to open string theory in the constant B–field background [SW99]. These facts convince us that the solutions obtained here are physically meaningful.

Transformation of String Fields

In the previous section, we have explicitly constructed the overlap vertices in the operator formulation under the constant B–field background. Then we have obtained the vertices with a new noncommutative structure of the Moyal type originating from the constant B–field, in addition to the ordinary product $*$ of string fields. Denoting the product with the new structure by $\star$, the action of the string field theory is written as

$$
\begin{aligned}
S_B &= \frac{1}{G_s}\int\left(\frac{1}{2}\psi\star Q\psi + \frac{1}{3}\psi\star\psi\star\psi\right) \\
&= \frac{1}{G_s}\left(\frac{1}{2}{}_{12}\langle\hat{V}_2||\psi\rangle_1 Q|\psi\rangle_2 + \frac{1}{3}{}_{123}\langle\hat{V}_3||\psi\rangle_1|\psi\rangle_2|\psi\rangle_3\right), \quad (6.148)
\end{aligned}
$$

where the BRST charge Q is constructed from the world sheet action (6.120). The theory (6.148) gives the noncommutative $U(1)$ Yang–Mills theory in the low energy region in the same sense as Witten's open string field theory in the case of the Neumann boundary condition leads to the ordinary $U(1)$ Yang–Mills theory in the low energy limit.[19]

[19] It can be explicitly seen by repeating a similar calculation as that carried out in [Dea90].

In [SW99] the authors argued that open string theory in the constant B–field background leads to either commutative or noncommutative Yang–Mills theories, corresponding to the different regularization scheme (the so–called Pauli–Villars regularization or the point–splitting regularization) in the world sheet formulation. They discussed a map between the gauge fields in the commutative and noncommutative Yang–Mills theories. In string field theory perspective, there also should be a certain transformation (hopefully simpler than the Yang–Mills case) from the string field ψ in (6.148) to a string field in a new string field theory which leads to the commutative Yang–Mills theory in the low energy limit.

Here we get the new string field theory by finding a unitary transformation which absorbs the noncommutative structure of the Moyal type in the product $\star$ into a redefinition of the string fields. There are used the two vertices $|\hat{V}_2\rangle$ and $|\hat{V}_3\rangle$ in the action (6.148). Recall that the 2–string vertex is in the same form as in the Neumann case and has no Moyal type noncommutative structure. First, we consider the phase factor of the 3–string vertex which multiplies in front of $|V_3\rangle$ (see (6.147)). Making use of the continuity conditions

$$P_{0,i} = -2\sum_{n=1}^{\infty} Y_{0n} P_{n,i}, \qquad \bar{P}_{0,i} = -2\sum_{n=1}^{\infty} \bar{Y}_{0n} \bar{P}_{n,i}, \tag{6.149}$$

it can be rewritten as [Sug00]

$$\begin{aligned}
-\frac{\theta^{ij}}{4\sqrt{3}\alpha'} P_{0,i}\bar{P}_{0,j} &= \frac{\theta^{ij}}{4\sqrt{3}\alpha'} \sum_{n=1}^{\infty} (P_{0,i}\bar{Y}_{0n}\bar{P}_{n,j} + P_{n,i}Y_{0n}\bar{P}_{0,j}) \\
&= -\frac{\theta^{ij}}{24\alpha'} \sum_{n=1}^{\infty} X_{0n}[(-p^{(2)}_{0,i} - p^{(3)}_{0,i} + 2p^{(1)}_{0,i})p^{(1)}_{n,j} \\
&+ (-p^{(3)}_{0,i} - p^{(1)}_{0,i} + 2p^{(2)}_{0,i})p^{(2)}_{n,j} + (-p^{(1)}_{0,i} - p^{(2)}_{0,i} + 2p^{(3)}_{0,i})p^{(3)}_{n,j}] \\
&= -\frac{\theta^{ij}}{8\alpha'} \sum_{r=1}^{3}\sum_{n=1}^{\infty} X_{0n} p^{(r)}_{0,i} p^{(r)}_{n,j},
\end{aligned}$$

where we used the property of the matrix Y: $Y_{0n} = -\bar{Y}_{0n} = \frac{\sqrt{3}}{2} X_{0n}$ for $n \geq 1$ and the momentum conservation on $|V_3\rangle$: $p^{(1)}_{0,i} + p^{(2)}_{0,i} + p^{(3)}_{0,i} = 0$. We manage to represent the phase factor of the Moyal type as a form factorized into the product of the unitary operators

$$\mathcal{U}_r = \exp\left(\frac{\theta^{ij}}{8\alpha'} \sum_{n=1,3,5,\cdots} X_{0n} p^{(r)}_{0,i} p^{(r)}_{n,j}\right). \tag{6.150}$$

Note that the unitary operator acts to a single string field. So the Moyal type noncommutativity can be absorbed by the unitary rotation of the string field

$${}_{123}\langle\hat{V}_3||\psi\rangle_1|\psi\rangle_2|\psi\rangle_3 = {}_{123}\langle V_3|\mathcal{U}_1\mathcal{U}_2\mathcal{U}_3|\psi\rangle_1|\psi\rangle_2|\psi\rangle_3 = {}_{123}\langle V_3||\tilde{\psi}\rangle_1|\tilde{\psi}\rangle_2|\tilde{\psi}\rangle_3, \tag{6.151}$$

with $|\tilde{\psi}\rangle_r = \mathcal{U}_r|\psi\rangle_r$. It should be remarked that this manipulation has been suceeded owing to the factorized expression of the phase factor, which originates from the continuity conditions relating the zero-modes to the nonzero-modes (6.149). It is a characteristic feature of string field theory that can not be found in any local field theories.

Next let us see the kinetic term. In doing so, it is judicious to write the kinetic term as follows:

$$ {}_{12}\langle\hat{V}_2||\psi\rangle_1(Q|\psi\rangle_2) = {}_{123}\langle\hat{V}_3||\psi\rangle_1(Q_L|I\rangle_2|\psi\rangle_3 + |\psi\rangle_2 Q_L|I\rangle_3), \qquad (6.152) $$

where Q_L is defined by integrating the BRST current $j_{BRST}(\sigma)$ with respect to σ over the left half region

$$ Q_L = \int_0^{\pi/2} d\sigma j_{BRST}(\sigma). $$

Equation (6.152) is also represented by the product $\star$ as

$$ \psi \star (Q\psi) = \psi \star [(Q_L I) \star \psi + \psi \star (Q_L I)]. \qquad (6.153) $$

Here, I stands for the identity element with respect to the $\star-$product, carrying the ghost number $-\frac{3}{2}$, which corresponds to $|I\rangle$ in the operator formulation. As is discussed by [HLR86], in order to show the relation (6.153) we need the formulas

$$ Q_R I = -Q_L I, \qquad (Q_R\psi) \star \xi = -(-1)^{n_\psi}\psi \star (Q_L\xi) \qquad (6.154) $$

for arbitrary string fields ψ and ξ, where Q_R is the integrated BRST current over the right half region of σ. n_ψ stands for the ghost number of the string field ψ minus $\frac{1}{2}$, and takes an integer value. The first formula means that the identity element is a physical quantity, also the second does the conservation of the BRST charge. By using these formulas, the first term in the bracket in r. h. s. of (6.153) becomes

$$ (Q_L I) \star \psi = -(Q_R I) \star \psi = I \star (Q_L\psi) = Q_L\psi. $$

Also, it turns out that the second term is equal to $Q_R\psi$. Combining these, we can see that (6.153) holds.

Further, we should remark that because the BRST current does not contain the center of mass coordinate x^j, it commute with the momentum p_i. From the continuity condition $p_i|I\rangle = 0$, it can be seen that $p_i Q_L|I\rangle = 0$. Expanding the exponential in the expression of the unitary operator (6.150) and passing the momentum $p_{0,i}$ to the right, we get

$$ \mathcal{U} Q_L|I\rangle = Q_L|I\rangle. \qquad (6.155) $$

Now we can write down the result of the kinetic term. As a result of the same manipulation as in eq. (6.151) and the use of eq. (6.155), we have[20]

$$\begin{aligned} {}_{12}\langle\hat{V}_2||\psi\rangle_1(Q|\psi\rangle_2) &= {}_{123}\langle\hat{V}_3||\psi\rangle_1(Q_L|I\rangle_2|\psi\rangle_3+|\psi\rangle_2 Q_L|I\rangle_3) \\ &= {}_{123}\langle V_3||\tilde{\psi}\rangle_1(Q_L|I\rangle_2|\tilde{\psi}\rangle_3+|\tilde{\psi}\rangle_2 Q_L|I\rangle_3) = {}_{12}\langle V_2||\tilde{\psi}\rangle_1(Q|\tilde{\psi}\rangle_2). \quad (6.156) \end{aligned}$$

Here we have a comment [Sug00]. If we considered the kinetic term itself without using (6.152), what would be going on? Let us see this. From the continuity conditions for $|\hat{V}_2\rangle_{12}^X = |V_2\rangle_{12}^X$:

$$p_{0,i}^{(1)}+p_{0,i}^{(2)}=0, \qquad p_{n,i}^{(1)}+(-1)^n p_{n,i}^{(2)}=0 \qquad (n=1,2,\cdots),$$

it could be shown that the 2–string overlap is invariant under the unitary rotation

$$\mathcal{U}_1\mathcal{U}_2|V_2\rangle_{12}=|V_2\rangle_{12}.$$

So we would find the expression for the kinetic term after the rotation

$${}_{12}\langle V_2||\psi\rangle_1 Q|\psi\rangle_2 = {}_{12}\langle V_2||\tilde{\psi}\rangle_1\tilde{Q}|\tilde{\psi}\rangle_2,$$

where $\tilde{Q}$ is the BRST charge similarity transformed by $\mathcal{U}$

$$\tilde{Q}=\mathcal{U}Q\mathcal{U}^\dagger. \qquad (6.157)$$

However, after some computations of the r. h. s. of (6.157), we could see that $\tilde{Q}$ has divergent term proportional to

$$\sum_{n=1,3,5,\cdots} 1$$

and thus it is not well–defined. It seems that this procedure is not correct and needs some suitable regularization, which preserves the conformal symmetry[21]. It is considered that the use of eq. (6.152) gives that kind of regularization, which will be justified at the end of the next section.

[20] Strictly speaking, in general this formula holds in the case that both of the string fields $|\psi\rangle$ and $|\tilde{\psi}\rangle$ belong to the Fock space which consists of states excited by *finite* number of creation operators. This point is subtle for giving a proof. However, for the infinitesimal θ case, by keeping arbitrary finite order terms in the expanded form of the exponential of $\mathcal{U}$, we can make the situation of both $|\psi\rangle$ and $|\tilde{\psi}\rangle$ being inside the Fock space, and thus clearly eq. (6.156) holds. From this fact, it is plausible to expect that eq. (6.156) is correct in the finite θ case.

[21] That divergence comes from the mid–point singularity of the string coordinates transformed by $\mathcal{U}$. In fact, after some calculations, we have

$$\mathcal{U}X^j(\sigma)\mathcal{U}^\dagger = X^j(\sigma) - \mathrm{i}\frac{\theta^{jk}}{4\sqrt{2\alpha'}}\sum_{n=1,3,5,\cdots} X_{n0}p_{n,k} - \frac{\theta^{jk}}{4}p_k\,\mathrm{sgn}\left(\sigma-\frac{\pi}{2}\right). \quad (6.158)$$

The last term leads to the mid-point sigularity in the energy–momentum tensor and the BRST charge Q. It seems that the use of (6.152) corresponds to taking the

Therefore, the string field theory action (6.148) with the Moyal type noncommutativity added to the ordinary noncommutativity is equivalently rewritten as the one with the ordinary noncommutativity alone [Sug00]:

$$S_B = \frac{1}{G_s}\int\left(\frac{1}{2}\tilde{\psi} * Q\tilde{\psi} + \frac{1}{3}\tilde{\psi} * \tilde{\psi} * \tilde{\psi}\right)$$
$$= \frac{1}{G_s}\left(\frac{1}{2}{}_{12}\langle V_2||\tilde{\psi}\rangle_1 Q|\tilde{\psi}\rangle_2 + \frac{1}{3}{}_{123}\langle V_3||\tilde{\psi}\rangle_1|\tilde{\psi}\rangle_2|\tilde{\psi}\rangle_3\right). \quad (6.159)$$

It is noted that the BRST charge Q, which is constructed from the world sheet action (6.120), has the same form as the one obtained from the action (6.116) with the relation (6.125). So all the $B-$dependence has been stuffed into the string fields except that existing in the metric G_{ij}. Furthermore, recalling that the relation between the metrics G^{ij} and g_{ij} is the same form as the T–duality inversion transformation, which was pointed out at the end of section 3, we can make the metric g_{ij} appear in the overlap vertices, instead of the metric G_{ij}. To do so, we consider the following transformation for the modes:

$$\hat{\alpha}^i_n = (E^{T-1})^{ik}\alpha_{n,k}, \qquad \hat{p}^i = (E^{T-1})^{ik}p_k, \qquad \hat{x}_i = E_{ik}x^k. \quad (6.160)$$

By this transformation, the commutators become

$$[\hat{\alpha}^i_n, \hat{\alpha}^j_m] = ng^{ij}\delta_{n+m,0}, \qquad [\hat{p}^i, \hat{x}_j] = -\mathrm{i}\delta^i_j,$$

and the bilinear form of the modes

$$G^{ij}\alpha_{n,i}\alpha_{m,j} = g_{ij}\hat{\alpha}^i_n\hat{\alpha}^j_m, \qquad G^{ij}p_i\alpha_{m,j} = g_{ij}\hat{p}^i\hat{\alpha}^j_m, \qquad G^{ij}p_ip_j = g_{ij}\hat{p}^i\hat{p}^j. \quad (6.161)$$

6.3.7 Topological Strings

The 2D field theories we have constructed are already very similar to string theories. However, one ingredient from string theory is missing: in string theory, the world–sheet theory does not only involve a path integral over the maps ϕ^i to the target space and their fermionic partners, but also a path integral over the world–sheet metric $h_{\alpha\beta}$. So far, we have set this metric to a fixed background value.

We have also encountered a drawback of our construction. Even though the theories we have found can give us some interesting 'semi–topological' information about the target spaces, one would like to be able to define general nonzero $n-$point functions at genus g instead of just the partition function

---

point splitting regularization with respect to the mid–point. Because of the discontinuity of the last term in (6.158), it is considered that the transformed string coordinates have no longer a good picture as a string. It could be understood from the point that the transformation $\mathcal{U}$ drives states around a perturbative vacuum to those around highly non–perturbative one like coherent states.

at genus one and the particular correlation functions we calculated at genus zero.

It turns out that these two remarks are intimately related. In this section we will go from topological field theory to topological string theory by introducing integrals over all metrics, and in doing so we will find interesting nonzero correlation functions at any genus (see [Von05]).

Coupling to Topological Gravity

In coupling an ordinary field theory to gravity, one has to perform the following three steps.

- First of all, one rewrites the Lagrangian of the theory in a covariant way by replacing all the flat metrics by the dynamical ones, introducing covariant derivatives and multiplying the measure by a factor of $\sqrt{\det h}$.
- Secondly, one introduces an Einstein–Hilbert term as the 'kinetic' term for the metric field, plus possibly extra terms and fields to preserve the symmetries of the original Lagrangian.
- Finally, one has to integrate the resulting theory over the space of all metrics.

Here we will not discuss the first two steps in this procedure. As we have seen in our discussion of topological field theories, the precise form of the Lagrangian only plays a comparatively minor role in determining the properties of the theory, and we can derive many results without actually considering a Lagrangian. Therefore, let us just state that it is possible to carry out the analog of the first two steps mentioned above, and construct a Lagrangian with a 'dynamical' metric which still possesses the topological Q−symmetry we have constructed. The reader who is interested in the details of this construction is referred to the paper [Wit90] and to the lecture notes [DVV91].

The third step, integrating over the space of all metrics, is the one we will be most interested in here. Naively, by the metric independence of our theories, integrating their partition functions over the space of all metrics, and then dividing the results by the volume of the topological 'gauge group', would be equivalent to multiplication by a factor of 1,

$$Z[h_0] \stackrel{?}{=} \frac{1}{G_{top}} \int \mathcal{D}[h]\, Z[h], \tag{6.162}$$

for any arbitrary background metric h_0. There are several reasons why this naive reasoning might go wrong:

- There may be metric configurations which cannot be reached from a given metric by continuous changes.
- There may be anomalies in the topological symmetry at the quantum level preventing the conclusion that all gauge fixed configurations are equivalent.

- The volume of G_{top} is infinite, so even if we could rigorously define a path integral the above multiplication and division would not be mathematically well–defined.

For these reasons, we should really be more careful and precisely define what we mean by the *'integral over the space of all metrics'*. Let us note the important fact that just like in ordinary string theory (and even before twisting), the 2D sigma models become conformal field theories when we include the metric in the Lagrangian. This means that we can borrow the technology from string theory to integrate over all conformally equivalent metrics. As is well known, and as we will discuss in more detail later, the conformal symmetry group is a huge group, and integrating over conformally equivalent metrics leaves only a nD integral over a set of world–sheet moduli. Therefore, our strategy will be to use the analogy to ordinary string theory to first do this integral over all conformally equivalent metrics, and then perform the integral over the remaining nD moduli space.

In integrating over conformally equivalent metrics, one usually has to worry about conformal anomalies. However, here a very important fact becomes our help. To understand this fact, it is useful to rewrite our twisting procedure in a somewhat different language (see [Von05]).

Let us consider the SEM–tensor $T_{\alpha\beta}$, which is the conserved Noether current with respect to global translations on $\mathbb{C}$. From conformal field theory, it is known that $T_{z\bar{z}} = T_{\bar{z}z} = 0$, and the fact that T is a conserved current, $\partial_\alpha T^\alpha{}_\beta = 0$, means that $T_{zz} \equiv T(z)$ and $T_{\bar{z}\bar{z}} \equiv \bar{T}(\bar{z})$ are (anti–)holomorphic in z. One can now expand $T(z)$ in Laurent modes,

$$T(z) = \sum L_m z^{-m-2}. \tag{6.163}$$

The L_m are called the *Virasoro generators*, and it is a well–known result from conformal field theory that in the quantum theory their commutation relations are

$$[L_m, L_n] = (m-n)L_{m+n} + \frac{c}{12}m(m^2-1)\delta_{m+n}.$$

The number c depends on the details of the theory under consideration, and it is called the central charge. When this central charge is nonzero, one runs into a technical problem. The reason for this is that the equation of motion for the metric field reads

$$\frac{\delta S}{\delta h^{\alpha\beta}} = T_{\alpha\beta} = 0.$$

In conformal field theory, one imposes this equation as a constraint in the quantum theory. That is, one requires that for physical states $|\psi\rangle$,

$$L_m|\psi\rangle = 0 \qquad (\text{for all } m \in \mathbb{Z}).$$

However, this is clearly incompatible with the above commutation relation unless $c = 0$. In string theory, this value for c can be achieved by taking the target space of the theory to be 10D. If $c \neq 0$ the quantum theory is problematic to define, and we speak of a 'conformal anomaly' [Von05].

The whole above story repeats itself for $\bar{T}(\bar{z})$ and its modes $\bar{L}_m$. At this point there is a crucial difference between open and closed strings. On an open string, left–moving and right–moving vibrations are related in such a way that they combine into standing waves. In our complex notation, 'left–moving' translates into 'z–dependent' (i.e., holomorphic), and 'right–moving' into '$\bar{z}$–dependent' (i.e., anti–holomorphic). Thus, on an open string all holomorphic quantities are related to their anti–holomorphic counterparts. In particular, $T(z)$ and $\bar{T}(\bar{z})$, and their modes L_m and $\bar{L}_m$, turn out to be complex conjugates. There is therefore only one independent algebra of Virasoro generators L_m.

On a *closed string* on the other hand, which is the situation we have been studying so far, left– and right–moving waves are completely independent. This means that all holomorphic and anti–holomorphic quantities, and in particular $T(z)$ and $\bar{T}(\bar{z})$, are independent. One therefore has two sets of Virasoro generators, L_m and $\bar{L}_m$.

Let us now analyze the problem of central charge in the twisted theories. To twist the theory, we have used the $U(1)$–symmetries. Any global $U(1)$–symmetry of our theory has a conserved current J_α. The fact that it is conserved again means that $J_z \equiv J(z)$ is holomorphic and $J_{\bar{z}} \equiv \bar{J}(\bar{z})$ is anti–holomorphic. Once again, on an open string J and $\bar{J}$ will be related, but in the closed string theory we are studying they will be independent functions. In particular, this means that we can view a global $U(1)$–symmetry as really consisting of two independent, left– and right–moving, $U(1)$–symmetries, with generators F_L and F_R.

Note that the sum of $U(1)$–symmetries $F_V + F_A$ only acts on objects with a + index. That is, it acts purely on left–moving quantities. Similarly, $F_V - F_A$ acts purely on right–moving quantities. From our discussion above, it is therefore natural to identify these two symmetries with the two components of a single global $U(1)$ symmetry:

$$F_V = \frac{1}{2}(F_L + F_R) \qquad F_A = \frac{1}{2}(F_L - F_R).$$

A more detailed construction shows that this can indeed be done.

Let us expand the left–moving conserved $U(1)$–current into Laurent modes,

$$J(z) = \sum J_m z^{-m-1}. \tag{6.164}$$

The commutation relations of these modes with one another and with the Virasoro modes can be calculated, either by writing down all of the modes in terms of the fields of the theory, or by using more abstract knowledge from the theory of superconformal symmetry algebras. In either case, one finds

$$[L_m, L_n] = (m-n)L_{m+n} + \frac{c}{12}m(m^2-1)\delta_{m+n}[L_m, J_n]$$
$$= -nJ_{m+n}[J_m, J_n] = \frac{c}{3}m\delta_{m+n}.$$

Note that the same central charge c appears in the $J-$ and in the $L-$commutators. This turns out to be crucial.

Following the standard Noether procedure, we can now construct a conserved charge by integrating the conserved current $J(z)$ over a space–like slice of the $z-$plane. In string theory, the physical time direction is the radial direction in the $z-$plane, so a space–like slice is just a curve around the origin. The integral is therefore calculated using the *Cauchy Theorem*,

$$F_L = \oint_{z=0} J(z)dz = 2\pi \mathrm{i} J_0.$$

In the quantum theory, it will be this operator that generates the $U(1)_L-$symmetry. Now recall that to twist the theory we want to introduce new Lorentz rotation generators,

$$M_A = M - F_V = M - \frac{1}{2}(F_L + F_R)M_B = M - F_A = M - \frac{1}{2}(F_L - F_R).$$

A well–known result from string theory (see [Von05]) is that the generator of Lorentz rotations is $M = 2\pi \mathrm{i}(L_0 - \bar{L}_0)$. Therefore, we find that the twisting procedure in this new language amounts to

$$A: L_{0,A} = L_0 - \frac{1}{2}J_0, \qquad \bar{L}_{0,A} = \bar{L}_0 + \frac{1}{2}\bar{J}_0,$$
$$B: L_{0,B} = L_0 - \frac{1}{2}J_0, \qquad \bar{L}_{0,B} = \bar{L}_0 - \frac{1}{2}\bar{J}_0.$$

Let us now focus on the left–moving sector; we see that for both twistings the new Lorentz rotation generator is the difference of L_0 and $\frac{1}{2}J_0$. The new Lorentz generator should also correspond to a conserved 2–tensor, and from (6.163) and (6.164) there is a very natural way to get such a current:

$$\tilde{T}(z) = T(z) + \frac{1}{2}\partial J(z), \tag{6.165}$$

which clearly satisfies $\bar{\partial}\tilde{T} = 0$ and

$$\tilde{L}_m = L_m - \frac{1}{2}(m+1)J_m, \tag{6.166}$$

so in particular we find that $\tilde{L}_0$ can serve as $L_{0,A}$ or $L_{0,B}$. We should apply the same procedure (with a minus sign in the $A-$model case) in the right–moving sector. Equations (6.165) and (6.166) tell us how to implement the twisting procedure not only on the conserved charges, but on the whole $N=2$

superconformal algebra – or at least on the part consisting of the $J-$ and $L-$modes, but a further investigation shows that this is the only part that changes. We have motivated, but not rigorously derived (6.165); for a complete justification the reader is referred to the original papers [LVW89] and [CV91].

Now, we come to the crucial point. The algebra that the new modes $\tilde{L}_m$ satisfy can be directly calculated from (6.165), and we find

$$[\tilde{L}_m, \tilde{L}_n] = (m-n)\tilde{L}_{m+n}.$$

That is, there is no central charge left. This means that we do not have any restriction on the dimension of the theory, and topological strings will actually be well–defined in target spaces of any dimension.

From this result, we see that we can integrate our partition function over conformally equivalent metrics without having to worry about the conformal anomaly represented by the nonzero central charge. After having integrated over this large part of the space of all metrics, it turns out that there is a nD integral left to do. In particular, it is known that one can always find a conformal transformation which in the neighborhood of a chosen point puts the metric in the form $h_{\alpha\beta} = \eta_{\alpha\beta}$, with η the usual flat metric with diagonal entries ±1. (Or, $+1$ in the Euclidean setting.) On the other hand, when one considers the global situation, it turns out that one cannot always enforce this gauge condition everywhere. For example, if the world–sheet is a torus, there is a left–over complex parameter τ that cannot be gauged away. The easiest way to visualize this parameter (see [Von05]) is by drawing the resulting torus in the complex–plane and rescaling it in such a way that one of its edges runs from 0 to 1; the other edge then runs from 0 to τ, see Figure 6.18. It seems intuitively clear that a conformal transformation – which should leave all angles fixed – will never deform τ, and even though intuition often fails when considering conformal mappings, in this case this can indeed be proven. Thus, τ is really a modular parameter which we need to integrate over. Another fairly intuitive result is that any locally flat torus can, after a rescaling, be drawn in this form, so τ indeed is the only modulus of the torus.

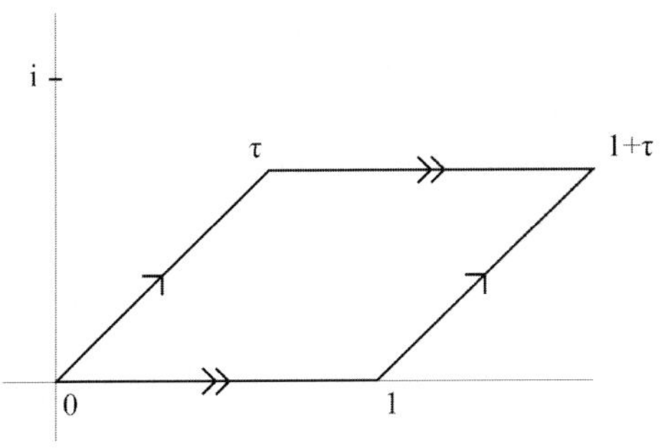

Fig. 6.18. The only modulus τ of a torus T^2.

More generally, one can show that a Riemann surface of genus g has $m_g = 3(g-1)$ complex modular parameters. As usual, this is the *virtual* dimension of the moduli space. If $g > 1$, one can show that this virtual dimension equals the actual dimension. For $g = 0$, the sphere, we have a negative virtual dimension $m_g = -3$, but the actual dimension is 0: there is always a flat metric on a surface which is topologically a sphere (just consider the sphere as a plane with a point added at infinity), and after having chosen this metric there are no remaining parameters such as τ in the torus case. For $g = 1$, the virtual dimension is $m_g = 0$, but as we have seen the actual dimension is 1.

We can explain these discrepancies using the fact that, after we have used the conformal invariance to fix the metric to be flat, the sphere and the torus have leftover symmetries. In the case of the sphere, it is well known in string theory that one can use these extra symmetries to fix the positions of three labelled points. In the case of the torus, after fixing the metric to be flat we still have rigid translations of the torus left, which we can use to fix the position of a single labelled point. To see how this leads to a difference between the virtual and the actual dimensions, let us for example consider tori with n labelled points on them. Since the virtual dimension of the moduli space of tori without labelled points is 0, the virtual dimension of the moduli space of tori with n labelled points is n. One may expect that at some point (and in fact, this happens already when $n = 1$), one reaches a sufficiently generic situation where the virtual dimension really is the actual dimension. However, even in this case we can fix one of the positions using the remaining conformal (translational) symmetry, so the positions of the points only represent $n - 1$ moduli. Hence, there must be an nth modulus of a different kind, which is exactly the shape parameter τ that we have encountered above. In the limiting case where $n = 0$, this parameter survives, thus causing the difference between the virtual and the real dimension of the moduli space.

For the sphere, the reasoning is somewhat more formal: we analogously expect to have three 'extra' moduli when $n = 0$. In fact, three extra parameters are present, but they do not show up as moduli. They must be viewed as the three parameters which need to be added to the problem to find a 0D moduli space.

Since the cases $g = 0, 1$ are thus somewhat special, let us begin by studying the theory on a Riemann surface with $g > 1$. To arrive at the topological string correlation functions, after gauge fixing we have to integrate over the remaining moduli space of complex dimension $3(g-1)$. To do this, we need to fix a measure on this moduli space. That is, given a set of $6(g-1)$ tangent vectors to the moduli space, we want to produce a number which represents the size of the volume element spanned by these vectors, see Figure 6.19. We should do this in a way which is invariant under coordinate redefinitions of both the moduli space and the world–sheet. Is there a 'natural' way to do this?

To answer this question, let us first ask how we can describe the tangent vectors to the moduli space (see [Von05]). In two dimensions, conformal

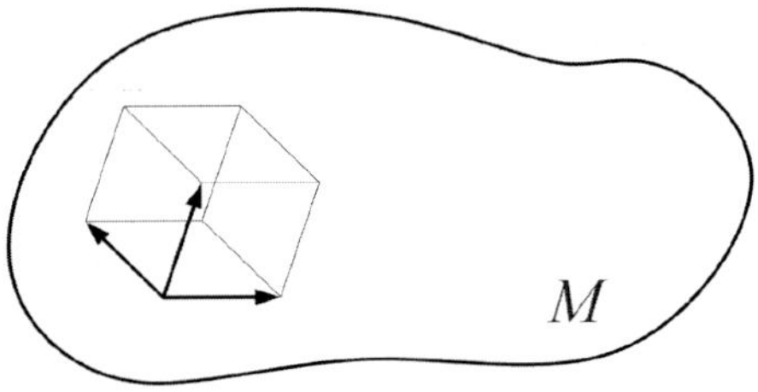

Fig. 6.19. A measure on the moduli space M assigns a number to every set of three tangent vectors. This number is interpreted as the volume of the element spanned by these vectors.

transformations are equivalent to holomorphic transformations: $z \mapsto f(z)$. It thus seems natural to assume that the moduli space we have left labels different complex structures on Σ, and indeed this can be shown to be the case. Therefore, a tangent vector to the moduli space is an infinitesimal change of complex structure, and these changes can be parameterized by holomorphic 1–forms with anti–holomorphic vector indices,

$$dz \mapsto dz + \epsilon\mu_{\bar{z}}^{z}(z)d\bar{z}.$$

The dimension counting above tells us that there are $3(g-1)$ independent $(\mu_i)_{\bar{z}}^{z}$, plus their $3(g-1)$ complex conjugates which change $d\bar{z}$. So the tangent space is spanned by these $\mu_i(z,\bar{z}), \bar{\mu}_i(z,\bar{z})$. How do we get a number out of a set of these objects? Since μ_i has a z and a $\bar{z}$ index, it seems natural to integrate it over Σ. However, the $z-$index is an upper index, so we need to lower it first with some tensor with two $z-$indices. It turns out that a good choice is to use the $Q-$partner G_{zz} of the SEM–tensor component T_{zz}, and thus to define the integration over moduli space as

$$\int_{M_g} \prod_{i=1}^{3g=3} \left(dm^i d\bar{m}^i \int_\Sigma G_{zz}(\mu_i)_{\bar{z}}^{z} \int_\Sigma G_{\bar{z}\bar{z}}(\bar{\mu}_i)_{z}^{\bar{z}} \right). \qquad (6.167)$$

Note that by construction, this integral is also invariant under a change of basis of the moduli space. There are several reasons why using G_{zz} is a natural choice. First of all, this choice is analogous to what one does in bosonic string theory. There, one integrates over the moduli space using exactly the same formula, but with G replaced by the conformal ghost b. This ghost is the BRST–partner of the SEM–tensor in exactly the same way as G is the $Q-$partner of T. Secondly, one can make the not unrelated observation that since $\{Q,G\} = T$, we can still use the standard arguments to show independence of the theory of the parameters in a Lagrangian of the form $L = \{Q,V\}$. The only difference is that now we also have to commute Q through G to make it act on the vacuum, but since $T_{\alpha\beta}$ itself is the derivative of the action with respect to the metric $h_{\alpha\beta}$, the terms we get in this way amount to integrating a total derivative over the moduli space. Therefore, apart from possible boundary terms these contributions vanish. Note that this reasoning also gives us an argument for using G_{zz} instead of T_{zz} (which is more or less the only other reasonable option) in (6.167): if we had chosen T_{zz} then all path integrals would have been over total derivatives on the moduli space, and apart from boundary contributions the whole theory would have become trivial.

If we consider the vector and axial charges of the full path integral measure, including the new path integral over the world–sheet metric h, we find a surprising result. Since the world–sheet metric does not transform under $R-$symmetry, naively one might expect that its measure does not either. However, this is clearly not correct since one should also take into account the explicit $G-$insertions in (6.167) that do transform under $R-$symmetry. From the $N = 2$ superconformal algebra (or, more down–to–earth, from expressing the operators in terms of the fields), it follows that the product of G and $\bar{G}$ has vector charge zero and axial charge 2. Therefore, the total vector charge of the measure remains zero, and the axial charge gets an extra contribution of $6(g-1)$, so we find a total axial $R-$charge of $6(g-1) - 2m(g-1)$. From this, we see that the case of complex target space dimension 3 is very special: here, the axial charge of the measure vanishes for any g, and hence the partition function is nonzero at every genus. If $m > 3$ and $g > 1$, the total axial charge of the measure is negative, and we have seen that we cannot cancel such a charge with local operators. Therefore, for these theories only the partition function at $g = 1$ and a specific set of correlation functions at genus zero give nonzero results. Moreover, for $m = 2$ and $m = 1$, the results can be shown to be trivial by other arguments. Therefore, a Calabi–Yau threefold is by far the most interesting target space for a topological string theory. It is a 'happy coincidence' (see [Von05]) that this is exactly the dimension we are most interested in from the string theory perspective.

Finally, let us come back to the special cases of genus 0 and 1. At genus zero, the Riemann surface has a single point as its moduli space, so there are no extra integrals or $G-$insertions to worry about. Therefore, we can copy the topological field theory result saying that we have to introduce local operators with total degree (m, m) in the theory. The only remnant of the fact that we are integrating over metrics is that we should also somehow fix the remaining three symmetries of the sphere. The most straightforward way to do this is to consider 3–point functions with insertions on three labelled points. As a gauge choice, we can then for example require these points to be at the points 0, 1 and ∞ in the compactified complex–plane. For example, in the $A-$model on a Calabi–Yau threefold, the 3–point function of three operators corresponding to $(1, 1)-$forms would thus give a nonzero result.

In the case of the torus, we have seen that there is one 'unexpected' modular parameter over which we have to integrate. This means we have to insert one $G-$ and one $\bar{G}-$operator in the measure, which spoils the absence of the axial anomaly we had for $g = 1$ in the topological field theory case. However, we also must fix the one remaining translational symmetry, which we can do by inserting a local operator at a labelled point. Thus, we can restore the axial R–charge to its zero value by choosing this to be an operator of degree $(1, 1)$.

Summarizing, we have found that in topological string theory on a target Calabi–Yau 3–fold, we have a non–vanishing 3–point function of total degree $(3, 3)$ at genus zero; a non–vanishing 1–point function of degree $(1, 1)$ at genus one, and a non–vanishing partition ('zero–point') function at all genera $g > 1$.

Nonlocal Operators

In one respect, what we have achieved is great progress: we can now for any genus define a nonzero partition function (or for low genus a correlation function) of the topological string theory. On the other hand, we would also like to define correlation functions of an arbitrary number of operators at these genera. As we have seen, the insertion of extra local operators in the correlation functions is not possible, since any such insertion will spoil our carefully constructed absence of R–symmetry anomalies. Therefore, we have to introduce nonlocal operators.

There is one class of nonlocal operators which immediately becomes mind. Before we saw, using the descent equations, that for every local operator we can define a corresponding 1–form and a 2–form operator. If we check the axial and vector charges of these operators, we find that if we start with an operator of degree $(1, 1)$, the 2–form operator we end up with actually has vanishing axial and vector charges. This has two important consequences. First of all, we can add the integral of this operator to our action [Von05],

$$S[t] = S_0 + t^a \int \mathcal{O}_a^{(2)},$$

without spoiling the axial and vector symmetry of the theory. Secondly, we can insert the integrated operator into correlation functions,

$$\langle \int \mathcal{O}_1^{(2)} \cdots \int \mathcal{O}_n^{(2)} \rangle$$

and still get a nonzero result by the vanishing of the axial and vector charges. These two statements are related: one obtains such correlators by differentiating $S[t]$ with respect to the appropriate t's, and then setting all $t^a = 0$.

A few remarks are in place here. First of all, recall that the integration over the insertion points of the operators can be viewed as part of the integration over the moduli space of Riemann surfaces, where now we label a certain number of points on the Riemann surface. From this point of view, the $g = 0, 1$ cases fit naturally into the same framework. We could unite the descendant fields into a world–sheet *super–field*,

$$\Phi_a = \mathcal{O}_a^{(0)} + \mathcal{O}_{a\alpha}^{(1)} \theta^\alpha + \mathcal{O}_{a\alpha\beta}^{(2)} \theta^\alpha \theta^\beta$$

where we formally replaced each dz and $d\bar{z}$ by corresponding fermionic coordinates θ^z and $\theta^{\bar{z}}$. Now, one can write the above correlators as integrals over n copies of this *super–space*,

$$\int \prod_{s=1}^{n} d^2 z_s d^2 \theta_s \, \langle \Phi_{a_1}(z_1, \theta_1) \cdots \Phi_{a_n}(z_n, \theta_n) \rangle$$

The integration prescription at genus 0 and 1 tells us to fix 3 and 1 points respectively, so we need to remove this number of super–space integrals. Then, integrating over the other super–space coordinates, the genus 0 correlators indeed become

$$\langle \mathcal{O}^{(0)}_{a_1} \mathcal{O}^{(0)}_{a_2} \mathcal{O}^{(0)}_{a_3} \int \mathcal{O}^{(2)}_{a_4} \cdots \int \mathcal{O}^{(2)}_{a_n} \rangle$$

From this prescription we note that these expressions are symmetric in the exchange of all a_i and a_j. In particular, this means that the genus zero 3–point functions at arbitrary t,

$$c_{abc}[t] = \langle \mathcal{O}^{(0)}_a \mathcal{O}^{(0)}_b \mathcal{O}^{(0)}_c \rangle [t]$$

have symmetric derivatives:

$$\frac{\partial c_{abc}}{\partial t^d} = \frac{\partial c_{abd}}{\partial t^c},$$

and similarly with permuted indices. These equations can be viewed as integrability conditions, and using the Poincaré lemma we see that they imply that

$$c_{ijk}[t] = \frac{\partial Z_0[t]}{\partial t^i \partial t^j \partial t^k}.$$

for some function $Z_0[t]$. Following the general philosophy that $n-$point functions are nth derivatives of the $t-$dependent partition function, we see that $Z_0[t]$ can be naturally thought of as the partition function at genus zero. Similarly, the partition function at genus 1 can be defined by integrating up the one-point functions once.

The quantities we have calculated above should be semi–topological invariants, meaning that they only depend on 'half' of the moduli (either the Kähler ones or the complex structure ones) of the target space. For example, in the $A-$model we find the Gromov–Witten invariants. In the $B-$model, it turns out that $F_0[t] = \ln Z_0[t]$ is actually a quantity we already knew: it is the prepotential of the Calabi–Yau manifold. A discussion of why this is the case can be found in the paper [BCO94]. The higher genus partition functions can be thought of as 'quantum corrections' to the prepotential.

Finally, there is a type of operator we have not discussed at all so far. Recall that in the topological string theory, the metric itself is now a dynamical field. We could not include the metric in our physical operators, since this would spoil the topological invariance. However, the metric is part of a

$Q-$multiplet, and the highest field in this multiplet is a scalar field which is usually labelled φ. (It should not be confused with the fields ϕ^i.) We can get more correlation functions by inserting operators φ^k and the operators related to them by the descent equations into the correlation functions. These operators are called 'gravitational descendants'. Even the case where the power is $k = 0$ is nontrivial; it does not insert any operator, but it does label a certain point, and hence changes the moduli space one integrates over. This operator is called the 'puncture operator'.

All of this seems to lead to an enormous amount of semi-topological target space invariants that can be calculated, but there are many recursion relations between the several correlators. This is similar to how we showed before that all correlators for the cohomological field theories follow from the 2–and 3–point functions on the sphere. Here, it turns out that the set of all correlators has a structure which is related to the theory of integrable hierarchies. Unfortunately, a discussion of this is outside the scope of both these lectures and the author's current knowledge.

The Holomorphic Anomaly

We have now defined the partition function and correlation functions of topological string theory, but even though the expressions we obtained are much simpler than the path integrals for ordinary quantum field or string theories, it would still be very hard to explicitly calculate them. Fortunately, it turns out that the $t-$dependent partition and correlation functions are actually 'nearly holomorphic' in t, and this is a great aid in exactly calculating these quantities.

Let us make this 'near holomorphy' more precise. As we have seen, calculating correlation functions of primary operators in topological string theories amounts to taking $t-$derivatives of the corresponding perturbed partition function $Z[t]$ and consequently setting $t = 0$. Recall that $Z[t]$ is defined through adding terms to the action of the form

$$t^a \int_\Sigma \mathcal{O}_a^{(2)}, \tag{6.168}$$

Let us for definiteness consider the $B-$twisted model. We want to show that the above term is Q_B-exact. For simplicity, we assume that $\mathcal{O}_a^{(2)}$ is a bosonic operator, but what we are about to say can by inserting a few signs straightforwardly be generalized to the fermionic case. From the descent equations we studied above, we know that

$$(\mathcal{O}_a^{(2)})_{+-} = -\{G_+, [G_-, \mathcal{O}_a^{(0)}]\}, \tag{6.169}$$

where G_+ is the charge corresponding to the current G_{zz}, and G_- the one corresponding to $G_{\bar{z}\bar{z}}$. We can in fact express $G_\pm$ in terms of the $N = (2,2)$ supercharges $\mathcal{Q}$. So, according to [Von05], we have

$$H = 2\pi \mathrm{i}(L_0 + \bar{L}_0) = \frac{1}{2}\{\mathcal{Q}_+, \bar{\mathcal{Q}}_+\} - \frac{1}{2}\{\mathcal{Q}_-, \bar{\mathcal{Q}}_-\}P$$
$$= 2\pi \mathrm{i}(L_0 - \bar{L}_0) = \frac{1}{2}\{\mathcal{Q}_+, \bar{\mathcal{Q}}_+\} + \frac{1}{2}\{\mathcal{Q}_-, \bar{\mathcal{Q}}_-\}.$$

Thus, we find that the left– and right–moving SEM charges satisfy

$$T_+ = 2\pi \mathrm{i} L_0 = \frac{1}{2}\{\mathcal{Q}_+, \bar{\mathcal{Q}}_+\}T_- = 2\pi \mathrm{i}\bar{L}_0 = -\frac{1}{2}\{\mathcal{Q}_-, \bar{\mathcal{Q}}_-\}.$$

To find G in the $B-$model, we should write these charges as commutators with respect to $Q_B = \bar{\mathcal{Q}}_+ + \bar{\mathcal{Q}}_-$, which gives

$$T_+ = \frac{1}{2}\{Q_B, \mathcal{Q}_+\}T_- = -\frac{1}{2}\{Q_B, \mathcal{Q}_-\},$$

so we arrive at the conclusion that for the $B-$model,

$$G_+ = \frac{1}{2}\mathcal{Q}_+ G_- = -\frac{1}{2}\mathcal{Q}_-.$$

Now, we can rewrite (6.169) as

$$(\mathcal{O}_a^{(2)})_{+-} = -\{G_+, [G_-, \mathcal{O}_a^{(0)}]\} = \frac{1}{4}\{\mathcal{Q}_+, [\mathcal{Q}_-, \mathcal{O}_a^{(0)}]\} \tag{6.170}$$
$$= \frac{1}{8}\{\bar{Q}_B, [(\mathcal{Q}_- - \mathcal{Q}_+), \mathcal{O}_a^{(0)}]\},$$

which proves our claim that $\mathcal{O}_a^{(2)}$ is Q_B-exact.

An $N = (2, 2)$ sigma model with a real action does, apart from the term (6.168), also contain a term

$$t^{\bar{a}} \int_{\Sigma} \bar{\mathcal{O}}_a^{(2)}, \tag{6.171}$$

where $t^{\bar{a}}$ is the complex conjugate of t^a. It is not immediately clear that $\bar{\mathcal{O}}_a^{(2)}$ is a physical operator: we have seen that physical operators in the $B-$model correspond to forms that are $\bar{\partial}-$closed, but the complex conjugate of such a form is $\partial-$closed. However, by taking the complex conjugate of (6.170), we see that

$$(\bar{\mathcal{O}}_a^{(2)})_{+-} = \frac{1}{8}\{Q_B, [(\bar{\mathcal{Q}}_- - \bar{\mathcal{Q}}_+), \bar{\mathcal{O}}_a^{(0)}]\},$$

so not only is the operator Q_B-closed, it is even Q_B-exact. This means that we can add terms of the form (6.171) to the action, and taking $t^{\bar{a}}-$derivatives inserts Q_B-exact terms in the correlation functions. Naively, we would expect this to give a zero result, so all the physical quantities seem to be $t-$independent, and thus holomorphic in t. We will see in a moment that this naive expectation turns out to be almost right, but not quite.

However, before doing so, let us comment briefly on the generalization of the above argument in the case of the $A-$model. It seems that a straightforward generalization of the argument fails, since Q_A is its own complex conjugate, and the complex conjugate of the de Rham operator is also the same operator. However, note that the $N = (2,2)-$theory has a different kind of 'conjugation symmetry': we can exchange the two supersymmetries, or in other words, exchange θ^+ with $\bar{\theta}^+$ and θ^- with $\bar{\theta}^-$. This exchanges Q_A with an operator which we might denote as $Q_{\bar{A}} \equiv \mathcal{Q}_+ + \bar{\mathcal{Q}}_-$. Using the above argument, we then find that the physical operators $\mathcal{O}_a^{(2)}$ are $Q_{\bar{A}}-$exact, and that their conjugates in the new sense are Q_A-exact. We can now add these conjugates to the action with parameters $t^{\bar{a}}$, and we again naively find independence of these parameters. In this case it is less natural to choose t^a and $t^{\bar{a}}$ to be complex conjugates, but we are free to choose this particular 'background point' and study how the theory behaves if we then vary t^a and $t^{\bar{a}}$ independently.

Now, let us see how the naive argument showing independence of the theory of $t^{\bar{a}}$ fails. In fact, the argument above would certainly hold for topological *field* theories. However, in topological string theories (see [Von05]), we have to worry about the insertions in the path integral of

$$G \cdot \mu_i \equiv \int d^2z\, G_{zz}\, (\mu_i)_{\bar{z}}^{z},$$

and their complex conjugates, when commuting the Q_B towards the vacuum and making sure it gives a zero answer. Indeed, the Q_B-commutator of the above factor is not zero, but it gives

$$\{Q_B, G \cdot \mu_i\} = T \cdot \mu_i.$$

Now recall that $T_{\alpha\beta} = \partial_{h^{\alpha\beta}} S$. We did not give a very precise definition of μ_i above, but we know that it parameterizes the change in the metric under an infinitesimal change of the coordinates m_i on the moduli space. One can make this intuition precise, and then finds the following 'chain rule': $T \cdot \mu_i = \partial_{m^i} S$. Inserting this into the partition function, we find that

$$\frac{\partial F_g}{\partial t^{\bar{a}}} =$$
$$\int_{M_g} \prod_{i=1}^{3g-3} dm^i d\bar{m}^i \sum_{j,k} \frac{\partial^2}{\partial m^j \partial \bar{m}^k} \left\langle (\prod_{l \neq j} \int \mu_l \cdot G)(\prod_{l \neq k} \int \bar{\mu}_l \cdot \bar{G}) \int \bar{\mathcal{O}}_a^{(2)} \right\rangle,$$

where $F_g = \ln Z_g$ is the free energy at genus g, and the reason F_g appears in the above equation instead of Z_g is, as usual in quantum field theory, that the expectation values in the r.h.s. are normalized such that $\langle 1 \rangle = 1$, and so the l.h.s. should be normalized accordingly and equal $Z_g^{-1} \partial_{\bar{a}} Z_g = \partial_{\bar{a}} F_g$ [Von05].

Thus, as we have claimed before, we are integrating a total derivative over the moduli space of genus g surfaces. If the moduli space did not have a boundary, this would indeed give zero, but in fact the moduli space does have a boundary. It consists of the moduli which make the genus g surface degenerate. This can happen in two ways: an internal cycle of the genus g

surface can be pinched, leaving a single surface of genus $g-1$, as in Figure 6.20 (a), or the surface can split up into two surfaces of genus g_1 and $g_2 = g - g_1$, as depicted in Figure 6.20 (b). By carefully considering the boundary contributions to the integral for these two types of boundaries, it was shown in [BCO94] that

$$\frac{\partial F_g}{\partial t^{\bar{a}}} = \frac{1}{2} c_{\bar{a}\bar{b}\bar{c}} e^{2K} G^{\bar{b}d} G^{\bar{c}e} \left(D_d D_e F_{g-1} + \sum_{r=1}^{g-1} D_d F_r D_e F_{g-r} \right),$$

where G is the so–called *Zamolodchikov metric* on the space parameterized by the coupling constants $t^a, t^{\bar{a}}$; K is its Kähler potential, and the D_a are covariant derivatives on this space. The coefficients $c_{\bar{a}\bar{b}\bar{c}}$ are the 3–point functions on the sphere of the operators $\bar{\mathcal{O}}_a^{(0)}$. We will not derive the above formula in detail, but the reader should notice that the contributions from the two types of boundary are quite clear.

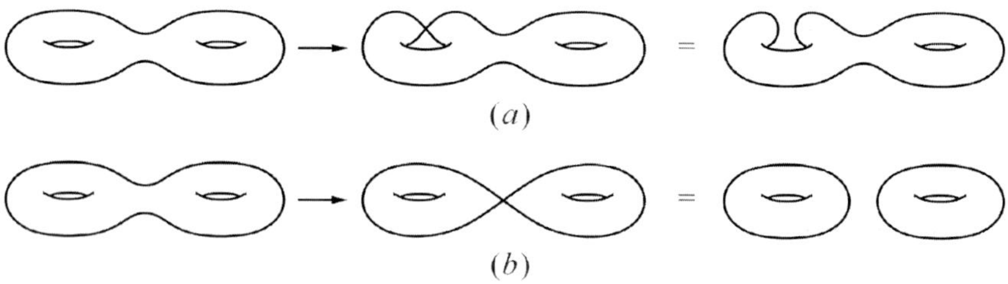

Fig. 6.20. At the boundary of the moduli space of genus g surfaces, the surfaces degenerate because certain cycles are pinched. This either lowers the genus of the surface (*a*) or breaks the surface into two lower genus ones (*b*) (see text for explanation).

Using this formula, one can inductively determine the $t^{\bar{a}}$ dependence on the partition functions if the holomorphic t^a–dependence is known. Holomorphic functions on complex spaces (or more generally holomorphic sections of complex vector bundles) are quite rare: usually, there is only a nD space of such functions. The same turns out to hold for our topological string partition functions: even though they are not quite holomorphic, their anti–holomorphic behavior is determined by the holomorphic dependence on the coordinates, and as a result there is a finite number of coefficients which determines them.

Thus, just from the above structure and without doing any path integrals, one can already determine the topological string partition functions up to a finite number of constants. This leads to a feasible program for completely determining the topological string partition function for a given target space and at given genus. From the holomorphic anomaly equation, one first has to find the general form of the partition function. Then, all one has left to do is to fix the unknown constants. Here, the fact that in the $A-$model the partition function counts a number of points becomes our help: by requiring that the $A-$model partition functions are integral, one can often fix the unknown constants and completely determine the $t-$dependent partition function. In practice, the procedure is still quite elaborate, so we will not describe any examples here, but several have been worked out in detail in the literature. Once again, the pioneering work for this can be found in the paper [BCO94].

6.3.8 Geometrical Transitions

Conifolds

Recall that a *conifold* is a generalization of the notion of a manifold. Unlike manifolds, a conifold can (or, should) contain conical singularities i.e., points whose neighborhood looks like a cone with a certain base. The base is usually a 5D manifold.

In string theory, a *conifold transition* represents such an evolution of the Calabi–Yau manifold in which its fabric rips and repairs itself, yet with mild and acceptable physical consequences in the context of string theory. However, the tears involved are more severe than those in an 'weaker' *flop transition* (see [Gre00]). The geometrically singular conifolds were shown to lead to completely smooth physics of strings. The divergences are 'smeared out' by D3–branes wrapped on the shrinking 3–sphere S^3, as originally pointed out by A. Strominger, who, together with D. Morrison and B. Greene have also found that the topology near the conifold singularity can undergo a topological phase–transition (see subsection 6.4.6). It is believed that nearly all Calabi–Yau manifolds can be connected via these 'critical transitions'.

More precisely, the conifold is the simplest example of a non–compact Calabi–Yau 3–fold: it is the set of solutions to the equation

$$x_1x_2 - x_3x_4 = 0$$

in $\mathbb{C}^4$. The resulting manifold is a cone, meaning in this case that any real multiple of a solution to this equation is again a solution. The point $(0, 0, 0, 0)$ is the 'tip' of this cone, and it is a singular point of the solution space. Note that by writing

$$x_1 = z_1 + iz_2, \qquad x_2 = z_1 - iz_2, \qquad x_3 = z_3 + iz_4, \qquad x_4 = -z_3 + iz_4,$$

where the z_i are still *complex* numbers, one can also write the equation as

$$z_1^2 + z_2^2 + z_3^2 + z_4^2 = 0.$$

Writing each z_i as $a_i + ib_i$, with a_i and b_i real, we get the two equations

$$|a|^2 - |b|^2 = 0, \qquad a \cdot b = 0. \tag{6.172}$$

Here $a \cdot b = \sum_i a_i b_i$ and $|a|^2 = a \cdot a$. Since the geometry is a cone, let us focus on a 'slice' of this cone given by

$$|a|^2 + |b|^2 = 2r^2,$$

for some $r \in \mathbb{R}$. On this slice, the first equation in (6.172) becomes

$$|a|^2 = r^2, \tag{6.173}$$

which is the equation defining a 3–sphere S^3 of radius r. The same holds for b, so both a and b lie on 3–spheres. However, we also have to take the second equation in (6.172) into account. Let us suppose that we fix an a satisfying (6.173). Then b has to lie on a 3–sphere, but also on the plane through the origin defined by $a \cdot b = 0$. That is, b lies on a 2–sphere. This holds for every a, so the slice we are considering is a fibration of 2–spheres over the 3–sphere. With a little more work, one can show that this fibration is trivial, so the conifold is a cone over $S^2 \times S^3$.

Since the conifold is a singular geometry, we would like to find geometries which approximate it, but which are non–singular. There are two interesting ways in which this can be done. The simplest way is to replace the defining equation by

$$x_1 x_2 - x_3 x_4 = \mu^2. \tag{6.174}$$

From the two equations constraining a and b, we now see that $|a|^2 \geq \mu^2$. In other words, the parameter r should be at least μ. At $r = \mu$, the $a-$sphere still has finite radius μ, but the $b-$sphere shrinks to zero size. This geometry is called the *deformed conifold.* Even though this is not clear from the picture, from the equation (6.174) one can straightforwardly show that it is nonsingular. One can also show that it is topologically equivalent to the cotangent bundle on the 3–sphere, T^*S^3. Here, the S^3 on which the cotangent bundle is defined is exactly the S^3 at the 'tip' of the deformed conifold.

The second way to change the conifold geometry arises from studying the two equations

$$x_1 A + x_3 B = 0, \qquad x_4 A + x_2 B = 0. \tag{6.175}$$

Here, we require A and B to be homogeneous complex coordinates on a $\mathbb{C}P^1$, i.e.,

$$(A, B) \neq (0, 0), \qquad (A, B) \sim (\lambda A, \lambda B)$$

where λ is any nonzero complex number. If one of the x_i is nonzero, say x_1, one can solve for A or B, e.g., $A = -\frac{x_3 B}{x_1}$, and insert this in the other equation to get

$$x_1 x_2 - x_3 x_4 = 0$$

which is the conifold equation. However, if all x_i are zero, any A and B solve the system of equations (6.175). In other words, we have constructed a geometry which away from the former singularity is completely the same as the conifold, but the singularity itself is replaced by a $\mathbb{C}P^1$, which topologically is the same as an S^2. From the defining equations one can again show that the resulting geometry is nonsingular, so we have now replaced our conifold geometry by the so–called *resolved conifold.*

Topological D–branes

Since topological string theories are in many ways similar to an ordinary (bosonic) string theories, one natural question which arises is: are there also open topological strings which can end on D–branes? To answer the above question rigorously, we would have to study boundary conditions on world–sheets with boundaries which preserve the $Q-$symmetry.

In the $A-$model, one can only construct 3D–branes wrapping so–called 'Lagrangian' submanifolds of M. Here, 'Lagrangian' means that the Kähler form ω vanishes on this submanifold. In the $B-$model, one can construct D–branes of any even dimension, as long as these branes wrap *holomorphic* submanifolds of M.

Just like in ordinary string theory, when we consider open topological strings ending on a D–brane, there should be a field theory on the brane world–volume describing the low–energy physics of the open strings. Moreover, since we are studying topological theories, one may expect such a theory to inherit the property that it only depends on a restricted amount of data of the manifolds involved. A key example is the case of the $A-$model on the deformed conifold, $M = T^*S^3$, where we wrap ND–branes on the S^3 in the base. (One can show that this is indeed a Lagrangian submanifold.) In ordinary string theory, the world–volume theory on ND–branes has a $U(N)$ gauge symmetry, so putting the ingredients together we can make the guess that the world–volume theory is a 3D topological field theory with $U(N)$ gauge symmetry. There is really only one candidate for such a theory: the *Chern–Simons gauge theory*. Recall that it consists of a single $U(N)$ gauge field, and has the action

$$S = \frac{k}{4\pi}\int_{S^3} \mathrm{Tr}\left(A\wedge dA + \frac{2}{3}A\wedge A\wedge A\right). \qquad (6.176)$$

Before the invention of D–branes, E. Witten showed that this is indeed the theory one gets. In fact, he showed even more: this theory actually describes the *full* topological string–field theory on the D–branes, even without going to a low–energy limit [Wit95].

Let us briefly outline the argument that gives this result. In his paper, Witten derived the open string–field theory action for the open $A-$model topological string; it reads

$$S = \int \mathrm{Tr}\left(\mathcal{A} * Q_A\mathcal{A} + \frac{2}{3}\mathcal{A} * \mathcal{A} * \mathcal{A}\right).$$

The form of this action is very similar to Chern–Simons theory, but its interpretation is completely different: $\mathcal{A}$ is a *string–field* (a wave function on the space of all maps from an open string to the space–time manifold), Q_A is the topological symmetry generator, which has a natural action on the string–field, and $*$ is a certain noncommutative product. Witten shows that the topological properties of the theory imply that only the constant maps contribute, so $\mathcal{A}$ becomes a field on M – and since open strings can only end on D–branes, it actually becomes a field on S^3. Moreover, recall that Q_A can be interpreted as a de Rham differential. Using these observations and the precise definition of the star product one can indeed show that the string–field theory action reduces to Chern–Simons theory on S^3.

6.3.9 Topological Strings and Black Hole Attractors

Topological string theory is naturally related to *black hole dynamics* (see subsection 7.3.3 below). Namely, critical string theory compactified on Calabi–Yau manifolds has played a central role in both the mathematical and physical development of modern string theory. The physical relevance of the data provided by the topological string $\hat{c} = 6$ (of A and B types) has been that it computes $F-$type terms in the corresponding four dimensional theory [BCO94, AGN94]. These higher–derivative $F-$type terms for Type II superstring on a Calabi–Yau manifold are of the general form

$$\int d^4x d^4\theta (W_{ab}W^{ab})^g F_g(X^\Lambda), \tag{6.177}$$

where W_{ab} is the *graviphoton super–field* of the $N = 2$ *super–gravity* and X^Λ are the *vector multiplet fields.* The lowest component of W is F the graviphoton field strength and the highest one is the Riemann tensor. The lowest components of X^Λ are the complex scalars parameterizing Calabi–Yau moduli and their highest components are the associated $U(1)$ vector–fields. These terms contribute to multiple graviphoton–graviton scattering. (6.177) includes (after θ integrations) an R^2F^{2g-2} term. The topological string partition function Z_{top} represents the canonical ensemble for multi–particle spinning five dimensional black holes [BMP97, KKV99].

Recently, [OSV04] proposed a simple and direct relationship between the second–quantized topological string partition function Z_{top} and black hole partition function Z_{BH} in four dimensions of the form

$$Z_{BH}(p^\Lambda, \phi^\Lambda) = |Z_{\text{top}}(X^\Lambda)|^2, \qquad \text{where} \qquad X^\Lambda = p^\Lambda + \frac{\mathrm{i}}{\pi}\phi^\Lambda$$

in a certain *Kähler gauge*. The l.h.s. here is evaluated as a function of integer magnetic charges p^Λ and continuous electric potentials ϕ^Λ, which are conjugate to integer electric charges q_Λ. The r.h.s. is the holomorphic square of the partition function for a gas of topological strings on a Calabi–Yau whose moduli are those associated to the charges/potentials $(p^\Lambda, \phi^\Lambda)$ via the attractor equations [OSV04]. Both sides of (6.178) are defined in a perturbation expansion in $1/Q$, where Q is the graviphoton charge carried by the black hole.[22] The non–perturbative completion of either side of (6.178) might in principle be defined as the partition function of the holographic CFT dual to the black hole, as in [SV96b]. Then we have the triple equality,

$$Z_{CFT} = Z_{BH} = |Z_{\mathrm{top}}|^2.$$

The existence of fundamental connection between 4D black holes and the topological string might have been anticipated from the following observation. Calabi–Yau spaces have two types of moduli: Kähler and complex structure. The world–sheet twisting which produces the A (B) model topological string from the critical superstring eliminates all dependence on the complex structure (Kähler) moduli at the perturbative level. Hence the perturbative topological string depends on only half the moduli. Black hole entropy on the other hand, insofar as it is an intrinsic property of the black hole, cannot depend on any externally specified moduli. What happens at leading order is that the moduli in vector multiplets are driven to attractor values at the horizon which depend only on the black hole charges and not on their asymptotically specified values. Hypermultiplet vevs on the other hand are not fixed by an attractor mechanism but simply drop out of the entropy formula. It is natural to assume this is valid to all orders in a $1/Q$ expansion. Hence the perturbative topological string and the large black hole partition functions depend on only half the Calabi–Yau moduli. It would be surprising if string theory produced two functions on the same space that were not simply related. Indeed [OSV04] argued that they were simply related as in (6.178).

Supergravity Area–Entropy Formula

Recall that a well–known hypothesis by J. Bekenstein and S. Hawking states that *the entropy of a black hole is proportional to the area of its horizon* (see [HE79]). This *area* is a function of the black hole mass, or in the extremal case, of its charges. Here we review the leading *semiclassical area–entropy formula* for a general $N = 2$, $d = 4$ extremal black hole characterized by magnetic and electric charges (p^Λ, q_Λ), recently reviewed in [OSV04]. The asymptotic values of the moduli in vector multiplets, parameterized by complex projective coordinates X^Λ, $(\Lambda = 0, 1, \dots, n_V)$ in the black hole solution, are arbitrary. These moduli couple to the electromagnetic fields and accordingly vary as a function of the radius. At the horizon they approach an attractor point whose location in the moduli space depends only on the charges. The locations of these attractor points can be found by looking for supersymmetric solutions with constant moduli. They are determined by the attractor equations,

$$p^\Lambda = \mathrm{Re}[CX^\Lambda], \qquad q_\Lambda = \mathrm{Re}[CF_{0\Lambda}], \tag{6.178}$$

[22] The string coupling g_s is in a hypermultiplet and decouples from the computation.

where $F_{0\Lambda} = \partial F_0/\partial X^\Lambda$ are the *holomorphic periods*, and the subscript 0 distinguishes these from the string loop corrected periods to appear in the next subsection. Both (p^Λ, q_Λ) and $(X^\Lambda, F_{0\Lambda})$ transform as vectors under the $Sp(2n+2; Z)$ duality group.

The $(2n_v+2)$ real equations (6.178) determine the (n_v+2) complex quantities (C, X^Λ) up to Kähler transformations, which act as

$$K \to K - f(X) - \bar{f}(\bar{X}), \quad X^\Lambda \to \mathrm{e}^f X^\Lambda, \qquad F_0 \to \mathrm{e}^{2f} F_0, \quad C \to \mathrm{e}^{-f} C,$$

where the *Kähler potential* K is given by

$$\mathrm{e}^{-K} = \mathrm{i}(\bar{X}^\Lambda F_{0\Lambda} - X^\Lambda \bar{F}_{0\Lambda}).$$

We could at this point set $C = 1$ and fix the Kähler gauge but later we shall find other gauges useful. It is easy to see that (as required) the charges (p^Λ, q_Λ) determined by the attractor equations (6.178) are invariant under Kähler transformations. Given the horizon attractor values of the moduli determined by (6.178) the *Bekenstein–Hawking entropy* S_{BH} may be written as

$$S_{BH} = \frac{1}{4} Area = \pi |Q|^2,$$

where $Q = Q_m + \mathrm{i}Q_\mathrm{e}$ is a complex combination of the magnetic and electric graviphoton charges and

$$|Q|^2 = \frac{\mathrm{i}}{2}\left(q_\Lambda \bar{C}\bar{X}^\Lambda - p^\Lambda \bar{C}\bar{F}_{0\Lambda}\right) = \frac{C\bar{C}}{4}\mathrm{e}^{-K}.$$

The normalization of Q here is chosen so that $|Q|$ equals the radius of the two sphere at the horizon.

It is useful to rephrase the above results in the context of type IIB superstrings in terms of geometry of Calabi–Yau. In this case the attractor equations fix the complex geometry of the Calabi–Yau. The electric/magentic charges correlate with three cycles of Calabi–Yau. Choosing a symplectic basis for the three cycles gives a choice of the splitting to electric and magnetic charges. Let A_Λ denote a basis for the electric three cycles, B^Σ the dual basis for the magnetic charges and Ω the holomorphic 3–form at the attractor point. Ω is fixed up to an overall multiplication by a complex number $\Omega \to \lambda\Omega$. There is a unique choice of λ such that the resulting Ω has the property that

$$p^\Lambda = \int_{A_\Lambda} \mathrm{Re}\,\Omega = \mathrm{Re}[CX^\Lambda], \qquad q_\Lambda = \int_{B^\Lambda} \mathrm{Re}\,\Omega = \mathrm{Re}[CF_{0\Lambda}],$$

$$\text{where} \quad \mathrm{Re}\,\Omega = \frac{1}{2}(\Omega + \overline{\Omega}).$$

In terms of this choice, the black hole entropy can be written as

$$S_{BH} = \frac{\pi}{4}\int_{CY} \Omega \wedge \overline{\Omega}.$$

Higher–Order Corrections

$F-$term corrections to the action are encoded in a string loop corrected holomorphic prepotential

$$F(X^\Lambda, W^2) = \sum_{h=0}^{\infty} F_h(X^\Lambda) W^{2h}, \tag{6.179}$$

where F_h can be computed by topological string amplitudes (as we review in the next section) and W^2 involves the square of the anti–self dual graviphoton field strength. This obeys the homogeneity equation

$$X^\Lambda \partial_\Lambda F(X^\Lambda, W^2) + W\partial_W F(X^\Lambda, W^2) = 2F(X^\Lambda, W^2). \tag{6.180}$$

Near the black hole horizon, the attractor value of W^2 obeys $C^2W^2 = 256$, and therefore the exact attractor equations read

$$p^\Lambda = \mathrm{Re}[CX^\Lambda], \qquad q_\Lambda = \mathrm{Re}\left[CF_\Lambda\left(X^\Lambda, \frac{256}{C^2}\right)\right]. \tag{6.181}$$

This is essentially the only possibility consistent with *symplectic invariance.* It has been then argued that the entropy as a function of the charges is

$$S_{BH} = \frac{\pi \mathrm{i}}{2}(q_\Lambda \bar{C}\bar{X}^\Lambda - p^\Lambda \bar{C}\bar{F}_\Lambda) + \frac{\pi}{2}\,\mathrm{Im}[C^3 \partial_C F], \tag{6.182}$$

where F_Λ, X^Λ and C are expressed in terms of the charges using (6.181).

Topological Strings

Partition Functions for Black Hole and Topological Strings. The notion of *topological string* was introduced in [Wit90]. Subsequently a connection between them and superstring was discovered: It was shown in [BCO94, AGN94], that the superstring loop corrected $F-$terms (6.179) can be computed as topological string amplitudes. The purpose of this subsection is to translate the super–gravity notation of the previous section to the topological string notation.

The second quantized partition function for the topological string may be written

$$\begin{aligned} Z_{\mathrm{top}}(t^A, g_{\mathrm{top}}) &= \exp\left[F_{\mathrm{top}}(t^A, g_{\mathrm{top}})\right], \qquad \text{where} \\ F_{\mathrm{top}}(t^A, g_{\mathrm{top}}) &= \sum_h g_{\mathrm{top}}^{2h-2}\, F_{\mathrm{top},h}(t^A), \end{aligned}$$

and $F_{\mathrm{top},h}$ is the $h-$loop *topological string amplitude.* The Kähler moduli are expressed in the flat coordinates

$$t^A = \frac{X^A}{X^0} = \theta^A + \mathrm{i}r^A,$$

where r^A are the Kähler classes of the Calabi–Yau M and θ^A are periodic $\theta^A \sim \theta^A + 1$.

We would like to determine relations between super–gravity quantities and topological string quantities. Using the homogeneity property (6.180) and the expansion (6.179), the holomorphic prepotential in super–gravity can be expressed as

$$\begin{aligned} F(CX^A, 256) &= (CX^0)^2 F\left(\frac{X^A}{X^0}, \frac{256}{(CX^0)^2}\right) \\ &= \sum_{h=0}^{\infty} (CX^0)^{2-2h} f_h(t^A), \end{aligned} \tag{6.183}$$

where $f_h(t^A)$ is related to $F_h(X^A)$ in (6.179) as

$$f_h(t^A) = 16^{2h} F_h\left(\frac{X^A}{X^0}\right).$$

This suggests an identification of the form $f_h(t^A) \sim F_{\mathrm{top},h}(t^A)$ and $g_{\mathrm{top}} \sim (CX^0)^{-1}$. For later purposes, we need precise relations between super–gravity and topological string quantities, including numerical coefficients. These can be determined by studying the limit of a large Calabi–Yau space.

In the *super–gravity notation*, the genus 0 and 1 terms in the large volume are given by

$$\begin{aligned} F\left(CX^A, 256\right) &= C^2 D_{ABC} \frac{X^A X^B X^C}{X^0} - \frac{1}{6} c_{2A} \frac{X^A}{X^0} + \cdots \\ &= (CX^0)^2 D_{ABC} t^A t^B t^C - \frac{1}{6} c_{2A} t^A + \cdots, \end{aligned}$$

where $\quad c_{2A} = \int_M c_2 \wedge \alpha_A,$

with c_2 being the *second Chern class* of M, and $C_{ABC} = -6D_{ABC}$ are the 4–cycle intersection numbers. These terms are normalized so that the mixed entropy S_{BH} is given by (6.182). On the other hand, the topological string amplitude in this limit is given by

$$F_{\mathrm{top}} = -\frac{(2\pi)^3 \mathrm{i}}{g_{\mathrm{top}}^2} D_{ABC} t^A t^B t^C - \frac{\pi \mathrm{i}}{12} c_{2A} t^A + \cdots \tag{6.184}$$

The normalization here is fixed by the holomorphic anomaly equations in [BCO94], which are nonlinear equations for $F_{\mathrm{top},h}$.

Comparing the one–loop terms in (6.183) and (6.184), which are independent of g_{top}, we find

$$F(CX^{\Lambda}, 256) = -\frac{2\mathrm{i}}{\pi} F_{\mathrm{top}}(t^{A}, g_{\mathrm{top}}).$$

Given this, we can compare the genus 0 terms to find

$$g_{\mathrm{top}} = \pm\frac{4\pi\mathrm{i}}{CX^{0}}.$$

This implies

$$\begin{aligned} \ln Z_{BH} &= -\pi \ \mathrm{Im}\left[F(CX^{\Lambda}, 256)\right] = F_{\mathrm{top}} + \bar{F}_{\mathrm{top}} \qquad \text{and} \\ Z_{BH}(\phi^{\Lambda}, p^{\Lambda}) &= |Z_{\mathrm{top}}(t^{A}, g_{\mathrm{top}})|^{2}, \qquad \text{with} \\ t^{A} &= \frac{p^{A} + \mathrm{i}\phi^{A}/\pi}{p^{0} + \mathrm{i}\phi^{0}/\pi}, \qquad g_{\mathrm{top}} = \pm\frac{4\pi\mathrm{i}}{p^{0} + \mathrm{i}\phi^{0}/\pi}. \end{aligned}$$

Supergravity Approach to Z_{BH}. The above relation

$$Z_{BH} = |Z_{\mathrm{top}}|^{2} \tag{6.185}$$

can have a simpler super–gravity derivation [OSV04].

A main ingredient in this derivation is the observation that the $N = 2$ super–gravity coupled to vector multiplets can be written as the action

$$S = \int d^{4}xd^{4}\theta \ (\text{super} - -\text{volume form}) + h.c. = \int d^{4}x\sqrt{-g}R + ..., \tag{6.186}$$

where the super–volume form in the above depends non–trivially on curvature of the fields. This reproduces the ordinary action after integrating over $d^{4}\theta$ and picking up the θ^{4} term in the super–volume. In the context of black holes the boundary terms accompanying (6.186) give the classical black hole entropy.

We now become the derivation of (6.185). As was observed in [BCO94, AGN94], topological string computes the terms

$$F = \sum_{h=0}^{\infty} \int d^{4}xd^{4}\theta F_{h}(X)(W^{2})^{g} + c.c. \tag{6.187}$$

There are various terms one can get from the above action after integrating over $d^{4}\theta$. Let us concentrate on one of the terms which turns out to be the relevant one for us: Take the top components of X^{Λ} and W^{2}, and absorb the $d^{4}\theta$ integral from the super–volume measure as in (6.186). We will work in the gauge $X^{0} \sim 1$ and thus $C \sim 1/g_{\mathrm{top}}$. As noted before in the near–horizon black hole geometry in this gauge the top component $W^{2} \sim 1/C^{2} \sim g_{\mathrm{top}}^{2}$ and the X^{Λ} are fixed by the attractor mechanism. Thus, we have the black hole free energy

$$\begin{aligned}\ln Z_{BH} &= \sum_{h=0}^{\infty} g_{\rm top}^{2h} F_{{\rm top},h}(X^{\Lambda}/X^0) \int d^4x d^4\theta + c.c. \\ &= \sum_{g=0}^{\infty} (g_{\rm top})^{2h-2} F_{{\rm top},h}(X^{\Lambda}/X^0) + c.c. \\ &= 2 \ {\rm Re} \ F_{\rm top}, \qquad \left(\text{using} \int d^4x d^4\theta \sim 1/g_{\rm top}^2\right).\end{aligned}$$

Upon exponentiation this leads to (6.185).

Here we have shown that if we consider one absorption of θ^4 term in (6.187) upon $d^4\theta$ integral we get the desired result. That there be no other terms is not obvious. For example another way to absorb the θ's would have given the familiar term R^2F^{2g-2} where F is the graviphoton field. However, such terms do not contribute in the black hole background. It would be nice to find a simple way to argue why these terms do not contribute and that we are left with this simple absorption of the θ integrals.

6.4 Other Applications of Path Integrals

6.4.1 Stochastic Optimal Control

A path–integral based optimal control model for nonlinear stochastic systems has recently been developed in [Kap05]. The author addressed the role of noise and the issue of efficient computation in stochastic optimal control problems. He considered a class of nonlinear control problems that can be formulated as a *path integral* and where the noise plays the role of temperature. The path integral displays symmetry breaking and there exist a critical noise value that separates regimes where optimal control yields qualitatively different solutions. The path integral can be computed efficiently by Monte Carlo integration or by Laplace approximation, and can therefore be used to solve high dimensional stochastic control problems.

Recall that *optimal control of nonlinear systems in the presence of noise* is a very general problem that occurs in many areas of science and engineering. It underlies *autonomous system behavior*, such as the control of movement and planning of actions of animals and robots, but also optimization of financial investment policies and control of chemical plants. The problem is stated as: *given that the system is in this configuration at this time, what is the optimal course of action to reach a goal state at some future time.* The *cost* of each time course of actions consists typically of a path contribution, that specifies the amount of work or other cost of the trajectory, and an end cost, that specifies to what extend the trajectory reaches the goal state.

Also recall that in the absence of noise, the optimal control problem can be solved in two ways: using (i) the *Pontryagin Maximum Principle* (PMP, see previous subsection), which represents a pair of ordinary differential equations that are similar to the Hamiltonian equations; or (ii) the *Hamilton–Jacobi–Bellman* (HJB) equation, which is a partial differential equation (PDE) [BK64].

In the presence of *Wiener noise*, the PMP formalism is replaced by a set of stochastic differential equations (SDEs), which become difficult to solve (compare with [YZ99]). The inclusion of noise in the HJB framework is mathematically quite straightforward, yielding the so–called *stochastic HJB equation* [Ste93]. However, its solution requires a *discretization of space and time* and the computation becomes intractable in both memory requirement and CPU time in high dimensions. As a result, deterministic control can be computed efficiently using the PMP approach, but stochastic control is intractable due to the curse of dimensionality.

For small noise, one expects that optimal stochastic control resembles optimal deterministic control, but for larger noise, the optimal stochastic control can be entirely different from the deterministic control [RN03]. However, there is currently no good understanding how *noise* affects optimal control.

In this subsection, we address both the issue of efficient computation and the role of noise in stochastic optimal control. We consider a class of nonlinear stochastic control problems, that can be formulated as a statistical mechanics problem. This class of control problems includes arbitrary dynamical systems, but with a limited control mechanism. It contains linear–quadratic [Ste93] control as a special case. We show that under certain conditions on the noise, the HJB equation can be written as a *linear* PDE

$$-\partial_t \psi = H\psi, \tag{6.188}$$

with H a (non–Hermitian) operator. Equation (6.188) must be solved subject to a boundary condition at the end time. As a result of the linearity of (6.188), the solution can be obtained in terms of a diffusion process evolving forward in time, and can be written as a path integral. The path–integral has a direct interpretation as a free energy, where noise plays the role of temperature.

This link between stochastic optimal control and a free energy has an immediate consequence that phenomena that allow for a free energy description, typically display phase transitions. [Kap05] has argued that for stochastic optimal control one can identify a critical noise value that separates regimes where the optimal control has been qualitatively different. He showed how the Laplace approximation can be combined with Monte Carlo sampling to efficiently calculate the optimal control.

Path–Integral Formalism

Let x^i be an nD stochastic variable that is subject to the SDE

$$dx^i = (b^i(x^i,t) + u^i)dt + d\xi^i \tag{6.189}$$

with $d\xi^i$ being an nD Wiener process with $\langle d\xi_i d\xi_j \rangle = \nu_{ij} dt$, and functions ν_{ij} independent of x^i, u^i and time t. The term $b^i(x^i,t)$ is an arbitrary nD function of x^i and t, and u^i represents an nD vector of control variables. Given the value of x^i at an initial time t, the *stochastic optimal control problem* is to find the control path $u^i(\cdot)$ that minimizes

$$C(x^i,t,u^i(\cdot)) = \left\langle \phi(x^i(t_f)) + \int_t^{t_f} d\tau (\frac{1}{2}u_i(\tau)Ru^i(\tau) + V(x^i(\tau),\tau)) \right\rangle_{x^i}, \tag{6.190}$$

with R a matrix, $V(x^i,t)$ a time–dependent potential, and $\phi(x^i)$ the *end cost.* The brackets $\langle\rangle_{x^i}$ denote expectation value with respect to the stochastic trajectories (6.189) that start at x^i.

One defines the *optimal cost–to–go function* from any time t and state x^i as

$$J(x^i,t) = \min_{u^i(\cdot)} C(x^i,t,u^i(\cdot)).$$

J satisfies the following stochastic HJB equation [Kap05]

$$\begin{aligned} -\partial_t J(x^i,t) &= \min_{u^i}\left(\frac{1}{2}u_i Ru^i + V + (b_i+u_i)\partial_{x^i}J(x^i,t) + \frac{1}{2}\nu_{ij}\partial_{x^ix^j}J(x^i,t)\right) \\ &= -\frac{1}{2}R^{-1}\partial_{x^i}J(x^i,t)\partial_{x^i}J + V + b_i\partial_{x^i}J(x^i,t) + \frac{1}{2}\nu_{ij}\partial_{x^ix^j}J(x^i,t), \end{aligned} \tag{6.191}$$

where $b_i = (b^i)^T$, and $u_i = (u^i)^T$, and

$$u^i = -R^{-1}\partial_{x^i}J(x^i,t) \tag{6.192}$$

is the optimal control at the point (x^i,t). The HJB equation is nonlinear in J and must be solved with end boundary condition $J(x^i,t_f) = \phi(x^i)$.

Let us define $\psi(x^i,t)$ through the Log Transform

$$J(x^i,t) = -\lambda \log \psi(x^i,t), \tag{6.193}$$

and assume that there exists a scalar λ such that

$$\lambda\delta_{ij} = (R\nu)_{ij}, \tag{6.194}$$

with δ_{ij} the Kronecker delta. In the one dimensional case, such a λ can always be found. In the higher dimensional case, this restricts the matrices $R \propto (\nu_{ij})^{-1}$. Equation (6.194) reduces the dependence of optimal control on the nD noise matrix to a scalar value λ that will play the role of temperature, while (6.191) reduces to the linear equation (6.188) with

$$H = -\frac{V}{\lambda} + b_i\partial_{x^i} + \frac{1}{2}\nu_{ij}\partial_{x^ix^j}J(x^i,t).$$

Let $\rho(y^i,\tau|x^i,t)$ with $\rho(y^i,t|x^i,t) = \delta(y^i - x^i)$ describe a *diffusion process* for $\tau > t$ defined by the *Fokker–Planck equation*

$$\partial_\tau\rho = H^\dagger\rho = -\frac{V}{\lambda}\rho - \partial_{x^i}(b_i\rho) + \frac{1}{2}\nu_{ij}\partial_{x^ix^j}J(x^i,t)\rho \tag{6.195}$$

with $H^\dagger$ the Hermitian–conjugate of H. Then $A(\tau) = \int dy^i \rho(y^i, \tau|x^i, t)\psi(y^i, \tau)$ is independent of τ and in particular $A(t) = A(t_f)$. It immediately follows that

$$\psi(x^i, t) = \int dy^i \rho(y^i, t_f|x^i, t) \exp(-\phi(y^i)/\lambda) \tag{6.196}$$

We arrive at the important conclusion that $\psi(x^i, t)$ can be computed either by backward integration using (6.188) or by forward integration of a diffusion process given by (6.195).

We can write the integral in (6.196) as a path integral. Following [Kap05] we can divide the time interval $t \to t_f$ in n_1 intervals and write $\rho(y^i, t_f|x^i, t) = \prod_{i=1}^{n_1} \rho(x_i^i, t_i|x_{i-1}^i, t_{i-1})$ and let $n_1 \to \infty$. The result is

$$\psi(x^i, t) = \int [dx^i]_{x^i} \exp\left(-\frac{1}{\lambda} S(x^i(t \to t_f))\right) \tag{6.197}$$

with $\int [dx^i]_{x^i}$ an integral over all paths $x^i(t \to t_f)$ that start at x^i and with

$$S(x^i(t \to t_f)) = \phi(x^i(t_f)+\int_t^{t_f} d\tau(\frac{1}{2}(\dot{x}^i - b_i(x^i, \tau))R(\dot{x}^i - b^i(x^i, \tau))+V(x^i, \tau)) \tag{6.198}$$

the Action associated with a path. From (6.193) and (6.197), the cost–to–go $J(x, t)$ becomes a log partition sum (i.e., a free energy) with temperature λ.

Monte Carlo Sampling

The path integral (6.197) can be estimated by stochastic integration from t to t_f of the diffusion process (6.195) in which particles get annihilated at a rate $V(x^i, t)/\lambda$ [Kap05]:

$$\begin{aligned} x^i &= x^i + b^i(x^i, t)dt + d\xi^i, \qquad \text{with probability} \quad 1 - Vdt/\lambda \\ x^i &= \dagger, \qquad \text{with probability} \quad Vdt/\lambda \end{aligned} \tag{6.199}$$

where $\dagger$ denotes that the particle is taken out of the simulation. Denote the trajectories by $x_\alpha^i(t \to t_f)$, $(\alpha = 1, \dots, N)$. Then, $\psi(x^i, t)$ and u^i are estimated as

$$\hat{\psi}(x^i, t) = \sum_{\alpha \in \text{alive}} w_\alpha, \qquad u^i dt = \frac{1}{\hat{\psi}(x^i, t)} \sum_{\alpha \in \text{alive}}^{N} w_\alpha d\xi_\alpha^i(t), \tag{6.200}$$

$$\text{with} \qquad w_\alpha = \frac{1}{N} \exp(-\phi(x_\alpha^i(t_f))/\lambda),$$

where 'alive' denotes the subset of trajectories that do not get killed along the way by the $\dagger$ operation. The normalization $1/N$ ensures that the annihilation process is properly taken into account. Equation (6.200) states that optimal control at time t is obtained by averaging the initial directions of the noise component of the trajectories $d\xi^i_\alpha(t)$, weighted by their success at t_f.

The above sampling procedure can be quite inefficient, when many trajectories get annihilated. One of the simplest procedures to improve it is by importance sampling. We replace the diffusion process that yields $\rho(y^i, t_f|x^i, t)$ by another diffusion process, that will yield $\rho'(y^i, t_f|x^i, t) = \exp(-S'/\lambda)$. Then (6.197) becomes,

$$\psi(x^i, t) = \int [dx^i]_{x^i} \exp\left(-S'/\lambda\right) \exp\left(-(S - S')/\lambda\right).$$

The idea is to chose ρ' such as to make the sampling of the path integral as efficient as possible. Following [Kap05], here we use the Laplace approximation, which is given by the k deterministic trajectories $x_\beta(t \to t_f)$ that minimize the Action

$$J(x^i, t) \approx -\lambda \log \sum_{\beta=1}^{k} \exp(-S(x^i_\beta(t \to t_f)/\lambda).$$

The Laplace approximation ignores all fluctuations around the modes and becomes exact in the limit $\lambda \to 0$. The Laplace approximation can be computed efficiently, requiring $\mathcal{O}(n^2 m^2)$ operations, where m is the number of time discretization.

For each Laplace trajectory, we can define a diffusion processes ρ'_β according to (6.199) with $b^i(x^i, t) = x^i_\beta(t)$. The estimators for ψ and u^i are given again by (6.200), but with weights

$$w_\alpha = \frac{1}{N} \exp\left(-\left(S(x^i_\alpha(t \to t_f)) - S'_\beta(x^i_\alpha(t \to t_f))\right)/\lambda\right).$$

S is the original Action (6.198) and S'_β is the new Action for the Laplace guided diffusion. When there are multiple Laplace trajectories one should include all of these in the sample.

6.4.2 Nonlinear Dynamics of Option Pricing

Classical theory of option pricing is based on the results found in 1973 by Black and Scholes [BS04] and, independently, Merton [Mer73]. Their pioneering work starts from the basic assumption that the asset prices follow the dynamics of a particular stochastic process (geometrical Brownian motion), so that they have a lognormal distribution [Hul00, PB99]. In the case of an efficient market with no arbitrage possibilities, no dividends and constant volatilities, they found that the price of each financial derivative is ruled by an ordinary partial differential equation, known as the (Nobel–Prize winning) *Black–Scholes–Merton* (BSM) formula. In the most simple case of a so–called European option, the BSM equation can be explicitly solved to get an analytical formula for the price of the option [Hul00, PB99]. When we consider other financial derivatives, which are commonly traded in real markets and allow anticipated exercise and/or depend on the history of the underlying asset, the BSM formula fails to give an analytical result. Appropriate numerical procedures have been developed in the literature to price exotic financial derivatives with path–dependent features, as discussed in detail in [Hul00, WDH93, PBS01]. The aim of this work is to give a contribution to the problem of efficient option pricing in financial analysis, showing how it is possible to use path integral methods to develop a fast and precise algorithm for the evaluation of option prices.

Following recent studies on the application of the path integral approach to the financial market as appeared in the econophysics literature (see [Mat02] for a comprehensive list of references), in [MNM02] the authors proposed an original, efficient path integral algorithm to price financial derivatives, including those with path–dependent and early exercise features, and to compare the results with those get with the standard procedures known in the literature.

Theory and Simulations of Option Pricing

Classical Theory and Path–Dependent Options

The basic ingredient for the development of a theory of option pricing is a suitable model for the time evolution of the asset prices. The assumption of the BSM model is that the price S of an asset is driven by a *Brownian motion* and verifies the stochastic differential equation (SDE) [Hul00, PB99]

$$dS = \mu S dt + \sigma S dw, \tag{6.201}$$

which, by means of the *Itô lemma*, can be cast in the form of an arithmetic Brownian motion for the logarithm of S

$$d(\ln S) = A dt + \sigma dw, \tag{6.202}$$

where σ is the *volatility*, $A = \left(\mu - \sigma^2/2\right)$, μ is the drift parameter and w is the realization of a Wiener process. Due to the properties of a Wiener process, (6.202) may be written as

$$d(\ln S) = A dt + \sigma \epsilon \sqrt{dt}, \tag{6.203}$$

where ϵ follows from a standardized normal distribution with mean 0 and variance 1. Thus, in terms of the logarithms of the asset prices $z' = \ln S'$, $z = \ln S$, the conditional transition probability $p(z'|z)$ to have at the time t' a price S' under the hypothesis that the price was S at the time $t < t'$ is given by [PB99, BRT99]

$$p(z'|z) = \frac{1}{\sqrt{2\pi(t'-t)\sigma^2}} \exp\left\{-\frac{[z'-(z+A(t'-t))]^2}{2\sigma^2(t'-t)}\right\}, \qquad (6.204)$$

which is a gaussian distribution with mean $z+A(t'-t)$ and variance $\sigma^2(t'-t)$. If we require the options to be exercised only at specific times $t_i, i = 1, \cdots, n$, the asset price, between two consequent times t_{i-1} and t_i, will follow (6.203) and the related transition probability will be

$$p(z_i|z_{i-1}) = \frac{1}{\sqrt{2\pi\Delta t\sigma^2}} \exp\left\{-\frac{[z_i-(z_{i-1}+A\Delta t)]^2}{2\sigma^2\Delta t}\right\}, \qquad (6.205)$$

with $\Delta t = t_i - t_{i-1}$.

A time–evolution model for the asset price is strictly necessary in a theory of option pricing because the fair price at time $t = 0$ of an option $\mathcal{O}$, without possibility of anticipated exercise before the expiration date or maturity T (a so–called *European option*), is given by the scaled expectation value [Hul00]

$$\mathcal{O}(0) = \mathrm{e}^{-rT} E[\mathcal{O}(T)], \qquad (6.206)$$

where r is the risk–free interest and $E[\cdot]$ indicates the mean value, which can be computed only if a model for the asset underlying the option is understood. For example, the value $\mathcal{O}$ of an European call option at the maturity T will be $\max\{S_T - X, 0\}$, where X is the strike price, while for an European put option the value $\mathcal{O}$ at the maturity will be $\max\{X - S_T, 0\}$. It is worth emphasizing, for what follows, that the case of an European option is particularly simple, since in such a situation the price of the option can be evaluated by means of analytical formulae, which are get by solving the BSM partial differential equation with the appropriate boundary conditions [Hul00, PB99]. On the other hand, many further kinds of options are present in the financial markets, such as American options (options which can be exercised at any time up to the expiration date) and exotic options [Hul00], i.e., derivatives with complicated payoffs or whose value depend on the whole time evolution of the underlying asset and not just on its value at the end. For such options with path-dependent and early exercise features no exact solutions are available and pricing them correctly is a great challenge.

In the case of options with possibility of anticipated exercise before the expiration date, the above discussion needs to be generalized, by introducing a slicing of the time interval T. Let us consider, for definiteness, the case of an option which can be exercised within the maturity but only at the times $t_1 = \Delta t, t_2 = 2\Delta t, \ldots, t_n = n\Delta t = T$. At each time slice t_{i-1} the value $\mathcal{O}_{i-1}$ of the option will be the maximum between its expectation value at the time t_i scaled with $\mathrm{e}^{-r\Delta t}$ and its value in the case of anticipated exercise $\mathcal{O}^Y_{i-1}$. If S_{i-1} denotes the price of the underlying asset at the time t_{i-1}, we can thus write for each $i = 1, \ldots, n$

$$\mathcal{O}_{i-1}(S_{i-1}) = \max\left\{\mathcal{O}^Y_{i-1}(S_{i-1}), \mathrm{e}^{-r\Delta t} E[\mathcal{O}_i|S_{i-1}]\right\}, \qquad (6.207)$$

where $E[\mathcal{O}_i|S_{i-1}]$ is the conditional expectation value of $\mathcal{O}_i$, i.e., its expectation value under the hypothesis of having the price S_{i-1} at the time t_{i-1}. In this way, to get the actual price $\mathcal{O}_0$, it is necessary to proceed backward in time and calculate $\mathcal{O}_{n-1}, \ldots, \mathcal{O}_1$, where the value $\mathcal{O}_n$ of the option at maturity is nothing but $\mathcal{O}_n^Y(S_n)$. It is therefore clear that evaluating the price of an option with early exercise features means to simulate the evolution of the underlying asset price (to get the $\mathcal{O}_i^Y$) and to calculate a (usually large) number of expectation conditional probabilities.

Standard Numerical Procedures

To value derivatives when analytical formulae are not available, appropriate numerical techniques have to be advocated. They involve the use of Monte Carlo (MC) simulation, binomial trees (and their improvements) and finite–difference methods [Hul00, WDH93].

A natural way to simulate price paths is to discretize (6.203) as

$$\ln S(t + \Delta t) - \ln S(t) = A\Delta t + \sigma\epsilon\sqrt{\Delta t},$$

or, equivalently,

$$S(t + \Delta t) = S(t) \exp\left[A\Delta t + \sigma\epsilon\sqrt{\Delta t}\right], \tag{6.208}$$

which is correct for any $\Delta t > 0$, even if finite. Given the spot price S_0, i.e., the price of the asset at time $t = 0$, one can extract from a standardized normal distribution a value ϵ_k, $(k = 1, \ldots, n)$ for the random variable ϵ to simulate one possible path followed by the price by means of (6.208):

$$S(k\Delta t) = S((k-1)\Delta t) \exp\left[A\Delta t + \sigma\epsilon_k\sqrt{\Delta t}\right].$$

Iterating the procedure m times, one can simulate m price paths $\{(S_0, S_1^{(j)}, S_2^{(j)}, \ldots, S_n^{(j)} \equiv S_T^{(j)}) : j = 1, \ldots, m\}$ and evaluate the price of the option. In such a MC simulation of the stochastic dynamics of asset price (Monte Carlo random walk) the mean values $E[\mathcal{O}_i|S_{i-1}], i = 1, \ldots, n$ are given by

$$E[\mathcal{O}_i|S_{i-1}] = \frac{\mathcal{O}_i^{(1)} + \mathcal{O}_i^{(2)} + \cdots + \mathcal{O}_i^{(m)}}{m},$$

with no need to calculate transition probabilities because, through the extraction of the possible ϵ values, the paths are automatically weighted according to the probability distribution function of (6.205). Unfortunately, this method leads to an estimated value whose numerical error is proportional to $m^{-1/2}$. Thus, even if it is powerful because of the possibility to control the paths and to impose additional constrains (as it is usually required by exotic and path-dependent options), the MC random walk is extremely time consuming when precise predictions are required and appropriate variance reduction procedures have to be used to save CPU time [Hul00]. This difficulty can be overcome by means of the method of the binomial trees and its extensions (see [Hul00] and references therein), whose main idea stands in a deterministic choice of the possible paths to limit the number of intermediate points. At each time step the price S_i is assumed to have only two choices: increase to the value $uS_i, u > 1$ or decrease to $dS_i, 0 < d < 1$, where the parameters u and d are given in terms of σ and Δt in such a way to give the correct values for the mean and variance of stock price changes over the time interval Δt. Also finite difference methods are known in the literature [Hul00] as an alternative to time-consuming MC simulations. They give the value of the derivative by solving the differential equation satisfied by the derivative, by converting it into a difference equation. Although tree approaches and finite difference methods are known to be faster than the MC random walk, they are difficult to apply when a detailed control of the history of the derivative is required and are also computationally time consuming when a number of stochastic variables is involved [Hul00]. It follows that the development of efficient and fast computational algorithms to price financial derivatives is still a key issue in financial analysis.

Option Pricing via Path Integrals

Recall that the path integral method is an integral formulation of the dynamics of a stochastic process. It is a suitable framework for the calculation of the transition probabilities associated to a given stochastic process, which is seen as the convolution of an infinite sequence of infinitesimal short-time steps [BRT99]. For the problem of option pricing, the path–integral method can be employed for the explicit calculation of the expectation values of the quantities of financial interest, given by integrals of the form [BRT99]

$$E[\mathcal{O}_i|S_{i-1}] = \int dz_i p(z_i|z_{i-1})\mathcal{O}_i(\mathrm{e}^{z_i}), \qquad (6.209)$$

where $z = \ln S$ and $p(z_i|z_{i-1})$ is the transition probability. $E[\mathcal{O}_i|S_{i-1}]$ is the conditional expectation value of some functional $\mathcal{O}_i$ of the stochastic process. For example, for an European call option at the maturity T the quantity of interest will be $\max\{S_T - X, 0\}$, X being the strike price. As already emphasized, and discussed in the literature [Hul00, WDH93, PBS01, RT02, Mat02], the computational complexity associated to this calculation is generally great: in the case of exotic options, with path-dependent and early exercise features, integrals of the type (6.209) cannot be analytically solved. As a consequence, we demand two things from a path integral framework: a very quick way to estimate the transition probability associated to a stochastic process (6.203) and a clever choice of the integration points with which evaluate the integrals (6.209). In particular, our aim is to develop an efficient calculation of the probability distribution without losing information on the path followed by the asset price during its time evolution.

Transition Probability

The probability distribution function related to a SDE verifies the *Chapman–Kolmogorov equation* [PB99]

$$p(z''|z') = \int dz p(z''|z)p(z|z'), \tag{6.210}$$

which states that the probability (density) of a transition from the value z' (at time t') to the value z'' (at time t'') is the 'summation' over all the possible intermediate values z of the probability of separate and consequent transitions $z' \to z$, $z \to z''$. As a consequence, if we consider a finite time interval $[t', t'']$ and we apply a time slicing, by considering $n+1$ subintervals of length $\Delta t = (t'' - t')/n + 1$, we can write, by iteration of (6.210)

$$p(z''|z') = \int_{-\infty}^{+\infty} \cdots \int_{-\infty}^{+\infty} dz_1 \cdots dz_n p(z''|z_n)p(z_n|z_{n-1}) \cdots p(z_1|z'),$$

which, thanks to (6.204), can be written as [MNM02]

$$\int_{-\infty}^{+\infty} \cdots \tag{6.211}$$

$$\cdots \int_{-\infty}^{+\infty} dz_1 \cdots dz_n \frac{1}{\sqrt{(2\pi\sigma^2 \Delta t)^{n+1}}} \exp\left\{ -\frac{1}{2\sigma^2 \Delta t} \sum_{k=1}^{n+1} \left[z_k - (z_{k-1} + A\Delta t)\right]^2 \right\}.$$

In the limit $n \to \infty$, $\Delta t \to 0$ such that $(n+1)\Delta t = (t'' - t')$ (infinite sequence of infinitesimal time steps), the expression (6.211), as explicitly shown in [BRT99], exhibits a Lagrangian structure and it is possible to express the transition probability in the path integral formalism as a convolution of the form [BRT99]

$$p(z'', t''|z', t') = \int_{\mathcal{C}} \mathcal{D}[\sigma^{-1}\tilde{z}] \exp\left\{ -\int_{t'}^{t''} L(\tilde{z}(\tau), \dot{\tilde{z}}(\tau); \tau) d\tau \right\},$$

where L is the Lagrangian, given by

$$L(\tilde{z}(\tau), \dot{\tilde{z}}(\tau); \tau) = \frac{1}{2\sigma^2} \left[\dot{\tilde{z}}(\tau) - A\right]^2,$$

and the integral is performed (with functional measure $\mathcal{D}[\cdot]$) over the paths $\tilde{z}(\cdot)$ belonging to $\mathcal{C}$, i.e., all the continuous functions with constrains $\tilde{z}(t') \equiv z'$, $\tilde{z}(t'') \equiv z''$. As carefully discussed in [BRT99], a path integral is well defined only if both a continuous formal expression and a discretization rule are given. As done in many applications, the Itô prescription is adopted here (see subsection 6.1.2 above).

A first, naïve evaluation of the transition probability (6.211) can be performed via Monte Carlo simulation, by writing (6.211) as

$$p(z'',t''|z',t') =$$
$$\int_{-\infty}^{+\infty} \cdots \int_{-\infty}^{+\infty} \prod_i^n dg_i \frac{1}{\sqrt{2\pi\sigma^2 \Delta t}} \exp\left\{-\frac{1}{2\sigma^2 \Delta t}\left[z'' - (z_n + A\Delta t)\right]^2\right\}, \quad (6.212)$$

in terms of the variables g_i defined by the relation

$$dg_k = \frac{dz_k}{\sqrt{2\pi\sigma^2 \Delta t}} \exp\left\{-\frac{1}{2\sigma^2 \Delta t}\left[z_k - (z_{k-1} + A\Delta t)\right]^2\right\}, \quad (6.213)$$

and extracting each g_i from a gaussian distribution of mean $z_{k-1} + A\Delta t$ and variance $\sigma^2 \Delta t$. However, as we will see, this method requires a large number of calls to get a good precision. This is due to the fact that each g_i is related to the previous g_{i-1}, so that this implementation of the path integral approach can be seen to be equivalent to a naïve MC simulation of random walks, with no variance reduction.

By means of appropriate manipulations [Sch81] of the integrand entering (6.211), it is possible, as shown in the following, to get a path integral expression which will contain a factorized integral with a constant kernel and a consequent variance reduction. If we define $z'' = z_{n+1}$ and $y_k = z_k - kA\Delta t$, $k = 1, \ldots, n$, we can express the *transition probability distribution* as

$$\int_{-\infty}^{+\infty} \cdots \int_{-\infty}^{+\infty} dy_1 \cdots dy_n \frac{1}{\sqrt{(2\pi\sigma^2 \Delta t)^{n+1}}} \cdot \exp\left\{-\frac{1}{2\sigma^2 \Delta t}\sum_{k=1}^{n+1}\left[y_k - y_{k-1}\right]^2\right\}, \quad (6.214)$$

in order to get rid of the contribution of the drift parameter. Now let us extract from the argument of the exponential function a quadratic form

$$\sum_{k=1}^{n+1}\left[y_k - y_{k-1}\right]^2 = y_0^2 - 2y_1y_0 + y_1^2 + y_1^2 - 2y_1y_2 + \ldots + y_{n+1}^2$$
$$= y^t M y + \left[y_0^2 - 2y_1y_0 + y_{n+1}^2 - 2y_ny_{n+1}\right], \quad (6.215)$$

by introducing the nD array y and the nxn matrix M defined as [MNM02]

$$y = \begin{pmatrix} y_1 \\ y_2 \\ \vdots \\ \vdots \\ y_n \end{pmatrix}, \qquad M = \begin{pmatrix} 2 & -1 & 0 & \cdots & \cdots & 0 \\ -1 & 2 & -1 & 0 & \cdots & 0 \\ 0 & -1 & 2 & -1 & \cdots & 0 \\ 0 & \cdots & -1 & 2 & -1 & 0 \\ 0 & \cdots & \cdots & -1 & 2 & -1 \\ 0 & \cdots & \cdots & \cdots & -1 & 2 \end{pmatrix}, \quad (6.216)$$

where M is a real, symmetric, non singular and tridiagonal matrix. In terms of the eigenvalues m_i of the matrix M, the contribution in (6.215) can be written as

$$y^t M y = w^t O^t M O w = w^t M_d w = \sum_{i=1}^{n} m_i w_i^2, \quad (6.217)$$

by introducing the orthogonal matrix O which diagonalizes M, with $w_i = O_{ij}y_j$. Because of the orthogonality of O, the Jacobian

$$J = \det \left| \frac{dw_i}{dy_k} \right| = \det |O_{ki}|,$$

of the transformation $y_k \to w_k$ equals 1, so that $\prod_{i=1}^n dw_i = \prod_{i=1}^n dy_i$. After some algebra, (6.215) can be written as

$$\sum_{k=1}^{n+1} [y_k - y_{k-1}]^2 = \sum_{i=1}^{n} m_i w_i^2 + y_0^2 - 2y_1 y_0 + y_{n+1}^2 - 2y_n y_{n+1} =$$

$$\sum_{i=1}^{n} m_i \left[w_i - \frac{(y_0 O_{1i} + y_{n+1} O_{ni})}{m_i} \right]^2 + y_0^2 + y_{n+1}^2 - \sum_{i=1}^{n} \frac{(y_0 O_{1i} + y_{n+1} O_{ni})^2}{m_i}. \tag{6.218}$$

Now, if we introduce new variables h_i obeying the relation

$$dh_i = \sqrt{\frac{m_i}{2\pi\sigma^2 \Delta t}} \exp\left\{ -\frac{m_i}{2\sigma^2 \Delta t} \left[w_i - \frac{(y_0 O_{1i} + y_{n+1} O_{ni})}{m_i} \right]^2 \right\} dw_i, \tag{6.219}$$

it is possible to express the *finite–time probability distribution* $p(z''|z')$ as [MNM02]

$$\begin{aligned}
&\int_{-\infty}^{+\infty} \cdots \int_{-\infty}^{+\infty} \prod_{i=1}^{n} dy_i \frac{1}{\sqrt{(2\pi\sigma^2 \Delta t)^{n+1}}} \exp\left\{ -\frac{1}{2\sigma^2 \Delta t} \sum_{k=1}^{n+1} [y_k - y_{k-1}]^2 \right\} \\
&= \int_{-\infty}^{+\infty} \cdots \int_{-\infty}^{+\infty} \prod_{i=1}^{n} dw_i \frac{1}{\sqrt{(2\pi\sigma^2 \Delta t)^{n+1}}} \mathrm{e}^{-(y_0^2 + y_{n+1}^2)/2\sigma^2 \Delta t} \\
&\times \exp\left\{ -\frac{1}{2\sigma^2 \Delta t} \sum_{i=1}^{n} \left[m_i \left(w_i - \frac{(y_0 O_{1i} + y_{n+1} O_{ni})}{m_i} \right)^2 - \frac{(y_0 O_{1i} + y_{n+1} O_{ni})^2}{m_i} \right] \right\} \\
&= \int_{-\infty}^{+\infty} \cdots \int_{-\infty}^{+\infty} \prod_{i=1}^{n} dh_i \frac{1}{\sqrt{2\pi\sigma^2 \Delta t \det(M)}} \qquad (6.220) \\
&\times \exp\left\{ -\frac{1}{2\sigma^2 \Delta t} \left[y_0^2 + y_{n+1}^2 + \sum_{i=1}^{n} \frac{(y_0 O_{1i} + y_{n+1} O_{ni})^2}{m_i} \right] \right\}.
\end{aligned}$$

The probability distribution function, as given by (6.220), is an integral whose kernel is a constant function (with respect to the integration variables) and which can be factorized into the n integrals

$$\int_{-\infty}^{+\infty} dh_i \exp\left\{ -\frac{1}{2\sigma^2 \Delta t} \frac{(y_0 O_{1i} + y_{n+1} O_{ni})^2}{m_i} \right\}, \tag{6.221}$$

given in terms of the h_i, which are gaussian variables that can be extracted from a normal distribution with mean $(y_0 O_{1i} + y_{n+1} O_{ni})^2/m_i$ and variance

$\sigma^2 \Delta t / m_i$. Differently to the first, naïve implementation of the path integral, now each h_i is no longer dependent on the previous h_{i-1}, and importance sampling over the paths is automatically accounted for.

It is worth noticing that, by means of the extraction of the random variables h_i, we are creating price paths, since at each intermediate time t_i the asset price is given by

$$S_i = \exp\{\sum_{k=1}^{n} O_{ik} h_k + iA\Delta t\}. \tag{6.222}$$

Therefore, this path integral algorithm can be easily adapted to the cases in which the derivative to be valued has, in the time interval $[0, T]$, additional constraints, as in the case of interesting path–dependent options, such as Asian and barrier options [Hul00].

Integration Points

The above illustrated method represents a powerful and fast tool to calculate the transition probability in the path integral framework and it can be employed if we need to value a generic option with maturity T and with possibility of anticipated exercise at times $t_i = i\Delta t$ $(n\Delta t = T)$ [MNM02]. As a consequence of this time slicing, one must numerically evaluate $n - 1$ mean values of the type (9), in order to check at any time t_i, and for any value of the stock price, whether early exercise is more convenient with respect to holding the option for a future time. To keep under control the computational complexity and the time of execution, it is mandatory to limit as far as possible the number of points for the integral evaluation. This means that we would like to have a linear growth of the number of integration points with the time. Let us suppose to evaluate each mean value

$$E[\mathcal{O}_i|S_{i-1}] = \int dz_i \, p(z_i|z_{i-1}) \mathcal{O}_i(\mathrm{e}^{z_i}),$$

with p integration points, i.e., considering only p fixed values for z_i. To this end, we can create a grid of possible prices, according to the dynamics of the stochastic process as given by (6.203)

$$z(t + \Delta t) - z(t) = \ln S(t + \Delta t) - \ln S(t) = A\Delta t + \epsilon\sigma\sqrt{\Delta t}. \tag{6.223}$$

Starting from z_0, we thus evaluate the expectation value $E[\mathcal{O}_1|S_0]$ with $p = 2m + 1, m \in \mathbb{N}$ values of z_1 centered on the mean value $E[z_1] = z_0 + A\Delta t$ and which differ from each other of a quantity of the order of $\sigma\sqrt{\Delta t}$

$$z_1^j = z_0 + A\Delta t + j\sigma\sqrt{\Delta t}, \qquad (j = -m, \ldots, +m).$$

Going on like this, we can evaluate each expectation value $E[\mathcal{O}_2|z_1^j]$ get from each one of the z_1's created above with p values for z_2 centered around the mean value

$$E[z_2|z_1^j] = z_1^j + A\Delta t = z_0 + 2A\Delta t + j\sigma\sqrt{\Delta t}.$$

Iterating the procedure until the maturity, we create a deterministic grid of points such that, at a given time t_i, there are $(p-1)i+1$ values of z_i, in agreement with the request of linear growth. This procedure of selection of integration points, together with the calculation of the transition probability previously described, is the basis of the path integral simulation of the price of a generic option.

By applying the results derived above, we have at disposal an efficient path integral algorithm both for the calculation of transition probabilities and the evaluation of option prices. In [MNM02] the application of the above path–integral method to European and American options in the BSM model was illustrated and comparisons with the results were get with the standard procedures known in the literature were shown. First, the path integral simulation of the probability distribution of the logarithm of the stock prices, $p(lnS)$, as a function of the logarithm of the stock price, for a BSM–like stochastic model, was given by (6.202). Once the transition probability has been computed, the price of an option could be computed in a path integral approach as the conditional expectation value of a given functional of the stochastic process. For example, the price of an European call option was given by

$$\mathcal{C} = \mathrm{e}^{-r(T-t)} \int_{-\infty}^{+\infty} dz_f \, p(z_f, T|z_i, t) \max[\mathrm{e}^{z_f} - X, 0], \tag{6.224}$$

while for an European put it will be

$$\mathcal{P} = \mathrm{e}^{-r(T-t)} \int_{-\infty}^{+\infty} dz_f \, p(z_f, T|z_i, t) \max[X - \mathrm{e}^{z_f}, 0], \tag{6.225}$$

where r is the risk–free interest rate. Therefore just 1D integrals need to be evaluated and they can be precisely computed with standard quadrature rules.

Continuum Limit and American Options

In the specific case of an *American option*, the possibility of exercise at any time up to the expiration date allows to develop, within the path integral formalism, a specific algorithm, which, as shown in the following, is precise and very quick [MNM02].

Given the time slicing considered above, the case of American options requires the limit $\Delta t \longrightarrow 0$ which, putting $\sigma \longrightarrow 0$, leads to a delta–like transition probability

$$p(z, t+\Delta t|z_t, t) \approx \delta(z - z_t - A\Delta t).$$

This means that, apart from volatility effects, the price z_i at time t_i will have a value remarkably close to the expected value $\bar{z} = z_{i-1} + A\Delta t$, given by the drift growth. In order to take care of the volatility effects, a possible solution is to estimate the integral of interest, i.e.,

$$E[\mathcal{O}_i|S_{i-1}] = \int_{-\infty}^{+\infty} dz\, p(z|z_{i-1})\mathcal{O}_i(\mathrm{e}^z), \tag{6.226}$$

by inserting in (6.226) the analytical expression for the $p(z|z_{i-1})$ transition probability

$$\begin{aligned} p(z|z_{i-1}) &= \frac{1}{\sqrt{2\pi \Delta t \sigma^2}} \exp\left\{-\frac{(z - z_{i-1} - A\Delta t)^2}{2\sigma^2 \Delta t}\right\} \\ &= \frac{1}{\sqrt{2\pi \Delta t \sigma^2}} \exp\left\{-\frac{(z - \bar{z})^2}{2\sigma^2 \Delta t}\right\}, \end{aligned}$$

together with a Taylor expansion of the kernel function $\mathcal{O}_i(\mathrm{e}^z) = f(z)$ around the expected value $\bar{z}$. Hence, up to the second–order in $z - \bar{z}$, the kernel function becomes

$$f(z) = f(\bar{z}) + (z - \bar{z})f'(\bar{z}) + \frac{1}{2}f''(\bar{z})(z - \bar{z})^2 + O((z - \bar{z})^3),$$

which induces

$$E[\mathcal{O}_i|S_{i-1}] = f(\bar{z}) + \frac{\sigma^2}{2}f''(\bar{z}), + \dots,$$

since the first derivative does not give contribution to (6.226), being the integral of an odd function over the whole z range. The second derivative can be numerically estimated as

$$f''(\bar{z}) = \frac{1}{\delta_\sigma^2}[f(\bar{z} + \delta_\sigma) - 2f(\bar{z}) + f(\bar{z} - \delta_\sigma)],$$

with $\delta_\sigma = O(\sigma\sqrt{\Delta t})$, as dictated by the dynamics of the stochastic process.

6.4.3 Nonlinear Dynamics of Complex Nets

Recall that many systems in nature, such as neural nets, food webs, metabolic systems, co–authorship of papers, the worldwide web, etc. can be represented as *complex networks*, or *small–world networks* (see, e.g., [WS98, DM03]). In particular, it has been recognized that many networks have scale–free topology; the distribution of the degree obeys the power law, $P(k) \sim k^{-\gamma}$. The study of the scale–free network now attracts the interests of many researchers in mathematics, physics, engineering and biology [Ich04].

Another important aspect of complex networks is their dynamics, describing e.g., the spreading of viruses in the Internet, change of populations in a food web, and synchronization of neurons in a brain. In particular, [Ich04] studied the synchronization of the random network of oscillators. His work follows the previous studies (see [Str00]) that showed that mean–field type synchronization, that Kuramoto observed in globally–coupled oscillators [Kur84], appeared also in the small–world networks.

Continuum Limit of the Kuramoto Net

Ichinomiya started with the standard network with N nodes, described by a variant of the *Kuramoto model.* Namely, at each node, there exists an oscillator and the phase of each oscillator θ_i is evolving according to

$$\dot{\theta}_i = \omega_i + K \sum_j a_{ij} \sin(\theta_j - \theta_i), \qquad (6.227)$$

where K is the coupling constant, a_{ij} is 1 if the nodes i and j are connected, and 0 otherwise; ω_i is a random number, whose distribution is given by the function $N(\omega)$.

For the analytic study, it is convenient to use the *continuum limit equation.* We define $P(k)$ as the distribution of nodes with degree k, and $\rho(k, \omega; t, \theta)$ the density of oscillators with phase θ at time t, for given ω and k. We assume that $\rho(k, \omega; t, \theta)$ is normalized as

$$\int_0^{2\pi} \rho(k, \omega; t, \theta) d\theta = 1.$$

For simplicity, we also assume $N(\omega) = N(-\omega)$. Thus, we suppose that the collective oscillation corresponds to the stable solution, $\dot{\rho} = 0$.

Now we construct the continuum limit equation for the network of oscillators. The evolution of ρ is determined by the *continuity equation* $\partial_t \rho = -\partial_\theta(\rho v)$, where v is defined by the continuum limit of the r.h.s of (6.227). Because one randomly selected edge connects to the node of degree k, frequency ω, phase θ with the probability $kP(k)N(\omega)\rho(k, \omega; t, \theta)/\int dk kP(k)$, $\rho(k, \omega; t, \theta)$ obeys the equation

$$\begin{aligned}\partial_t \rho(k, \omega; t, \theta) &= -\partial_\theta [\rho(k, \omega; t, \theta) \, (\omega \; + \\ &+ \frac{Kk \int d\omega' \int dk' \int d\theta' N(\omega') P(k') k' \rho(k', \omega'; t, \theta') \sin(\theta - \theta')}{\int dk' P(k') k'})].\end{aligned}$$

The mean–field solution of this equation was studied by [Ich04].

Path–Integral Approach to Complex Nets

Recently, [Ich05] introduced the *path–integral* (see subsection 4.4.6 above) *approach* in studying the dynamics of complex networks. He considered the stochastic generalization of the Kuramoto network (6.227), given by

$$\dot{x}_i = f_i(x_i) + \sum_{j=1}^{N} a_{ij} g(x_i, x_j) + \xi_i(t), \qquad (6.228)$$

where $f_i = f_i(x_i)$ and $g_{ij} = g(x_i, x_j)$ are functions of network activations x_i, $\xi_i(t)$ is a *random force* that satisfies $\langle \xi_i(t) = 0 \rangle$, $\langle \xi_i(t) \xi_j(t') \rangle = \delta_{ij} \delta(t - t') \sigma^2$.

He assumed $x_i = x_{i,0}$ at $t = 0$. In order to discuss the dynamics of this system, he introduced the so–called *Matrin–Siggia–Rose* (MSR) *generating functional* Z given by [Dom78]

$$Z[\{l_{ik}\},\{\bar{l}_{ik}\}] = \left(\frac{1}{\pi}\right)^{NN_t} \left\langle \int \prod_{i=1}^{N} \prod_{k=0}^{N_t} dx_{ik} d\bar{x}_{ik} \mathrm{e}^{-S} \exp(l_{ik} x_{ik} + \bar{l}_{ik} \bar{x}_{ik}) J \right\rangle,$$

where the *action* S is given by

$$S = \sum_{ik} [\frac{\sigma^2 \Delta t}{2} \bar{x}_{ik}^2 + i\bar{x}_{ik}\{x_{ik} - x_{i,k-1} - \Delta t(f_i(x_{i,k-1}) + \sum_j a_{ij} g(x_{i,k-1}, x_{j,k-1}))\}],$$

and $\langle \cdots \rangle$ represents the average over the ensemble of networks. J is the functional Jacobian term,

$$J = \exp\left(-\frac{\Delta t}{2} \sum_{ijk} \frac{\partial (f_i(x_{ik}) + a_{ij} g(x_{ik}, x_{jk}))}{\partial x_{ik}} \right).$$

Ichinomiya considered such a form of the network model in which

$$a_{ij} = \begin{cases} 1 \text{ with probability } p_{ij}, \\ 0 \text{ with probability } 1 - p_{ij}. \end{cases}$$

Note that p_{ij} can be a function of variables such as i or j. For example, in the 1D chain model, p_{ij} is 1 if $|i - j| = 1$, else it is 0. The average over all networks can be expressed as

$$\left\langle \exp\left[\sum_{ik} i\Delta t \bar{x}_{ik} \sum_j a_{ij} g(x_{i,k-1}, x_{j,k-1}) \right] \right\rangle =$$
$$\prod_{ij} \left[p_{ij} \exp\left\{ \sum_k i\Delta t \bar{x}_{ik} g(x_{i,k-1}, x_{j,k-1}) \right\} + 1 - p_{ij} \right],$$

so we get

$$\langle \mathrm{e}^{-S} \rangle = \exp(-S_0) \prod_{ij} \left[p_{ij} \exp\left\{ \sum_k i\Delta t \bar{x}_{ik} g(x_{i,k-1}, x_{j,k-1}) \right\} + 1 - p_{ij} \right],$$

where $S_0 = \sum_{ik} \frac{\sigma^2 \Delta t}{2} \bar{x}_{ik}^2 + i\bar{x}_{ik}\{x_{ik} - x_{i,k-1} - \Delta t f_i(x_{i,k-1})\}.$

This expression can be applied to the dynamics of any complex network model. [Ich05] applied this model to analysis of the *Kuramoto transition in random sparse networks*.

6.4.4 Dissipative Quantum Brain Model

The *conservative brain* model was originally formulated within the framework of the quantum field theory (QFT) by [RU67] and subsequently developed in [STU78, STU79, JY95, JPY96]. The conservative brain model has been recently extended to the *dissipative quantum dynamics* in the work of G. Vitiello and collaborators [Vit95, AV00, PV99, Vit01, PV03, PV04].

The canonical quantization procedure of a dissipative system requires to include in the formalism also the system representing the environment (usually the heat bath) in which the system is embedded. One possible way to do that is to depict the environment as the time–reversal image of the system [CRV92]: the environment is thus described as the *double* of the system in the time–reversed dynamics (the system image in the mirror of time).

Within the framework of dissipative QFT, the brain system is described in terms of an *infinite collection of damped harmonic oscillators* A_κ (the simplest prototype of a dissipative system) representing the DWQ [Vit95]. Now, the collection of damped harmonic oscillators is ruled by the Hamiltonian [Vit95, CRV92]

$$H = H_0 + H_I, \qquad \text{with}$$
$$H_0 = \hbar\Omega_\kappa(A_\kappa^\dagger A_\kappa - \tilde{A}_\kappa^\dagger \tilde{A}_\kappa), \qquad H_I = \mathrm{i}\hbar\Gamma_\kappa(A_\kappa^\dagger \tilde{A}_\kappa^\dagger - A_\kappa \tilde{A}_\kappa),$$

where Ω_κ is the frequency and Γ_κ is the damping constant. The $\tilde{A}_\kappa$ modes are the 'time–reversed mirror image' (i.e., the 'mirror modes') of the A_κ modes. They are the doubled modes, representing the environment modes, in such a way that κ generically labels their degrees of freedom. In particular, we consider the damped harmonic oscillator (DHO)

$$m\ddot{x} + \gamma\dot{x} + \kappa x = 0, \tag{6.229}$$

as a simple prototype for dissipative systems (with intention that thus get results also apply to more general systems). The damped oscillator (6.229) is a non–Hamiltonian system and therefore the customary canonical quantization procedure cannot be followed. However, one can face the problem by resorting to well known tools such as the *density matrix* ρ and the *Wigner function* W.

Let us start with the special case of a *conservative particle* in the absence of friction γ, with the standard Hamiltonian, $H = -(\hbar\partial_x)^2/2m + V(x)$.

Recall (from the previous subsection) that the *density matrix equation of motion*, i.e., *quantum Liouville equation*, is given by

$$\mathrm{i}\hbar\dot{\rho} = [H, \rho]. \tag{6.230}$$

The density matrix function ρ is defined by

$$\langle x + \frac{1}{2}y|\rho(t)|x - \frac{1}{2}y\rangle = \psi^*(x + \frac{1}{2}y, t)\psi(x - \frac{1}{2}y, t) \equiv W(x, y, t),$$

with the associated standard expression for the *Wigner function* (see [FH65]),

$$W(p,x,t) = \frac{1}{2\pi\hbar}\int W(x,y,t)\,\mathrm{e}^{\left(-\mathrm{i}\frac{py}{\hbar}\right)}dy.$$

Now, in the coordinate $x-$representation, by introducing the notation

$$x_\pm = x \pm \frac{1}{2}y, \tag{6.231}$$

the Liouville equation (6.230) can be expanded as

$$\begin{aligned} &\mathrm{i}\hbar\,\partial_t\langle x_+|\rho(t)|x_-\rangle = \\ &\left\{-\frac{\hbar^2}{2m}\left[\partial^2_{x_+} - \partial^2_{x_-}\right] + \left[V(x_+) - V(x_-)\right]\right\}\langle x_+|\rho(t)|x_-\rangle, \end{aligned} \tag{6.232}$$

while the Wigner function $W(p,x,t)$ is now given by

$$\begin{aligned} \mathrm{i}\hbar\,\partial_t W(x,y,t) &= H_o W(x,y,t), \qquad \text{with} \\ H_o &= \frac{1}{m}p_x p_y + V(x+\frac{1}{2}y) - V(x-\frac{1}{2}y), \\ \text{and} \qquad p_x &= -\mathrm{i}\hbar\partial_x, \qquad p_y = -\mathrm{i}\hbar\partial_y. \end{aligned} \tag{6.233}$$

The new Hamiltonian H_o (6.233) may be get from the corresponding Lagrangian

$$L_o = m\dot{x}\dot{y} - V(x+\frac{1}{2}y) + V(x-\frac{1}{2}y). \tag{6.234}$$

In this way, Vitiello concluded that the density matrix and the Wigner function formalism *required*, even in the conservative case (with zero mechanical resistance γ), the introduction of a 'doubled' set of coordinates, $x_\pm$, or, alternatively, x and y. One may understand this as related to the introduction of the 'couple' of indices *necessary* to label the density matrix elements (6.232).

Let us now consider the case of the *particle interacting* with a *thermal bath* at temperature T. Let f denote the *random force* on the particle at the position x due to the bath. The interaction Hamiltonian between the bath and the particle is written as

$$H_{int} = -fx. \tag{6.235}$$

Now, in the *Feynman–Vernon formalism* (see [Fey72]), the *effective action* $A[x,y]$ for the particle is given by

$$A[x,y] = \int_{t_i}^{t_f} L_o(\dot{x},\dot{y},x,y)\,dt + I[x,y],$$

with L_o defined by (6.234) and

$$\mathrm{e}^{\frac{\mathrm{i}}{\hbar}I[x,y]} = \langle(\mathrm{e}^{-\frac{\mathrm{i}}{\hbar}\int_{t_i}^{t_f} f(t)x_-(t)dt})_-(\mathrm{e}^{\frac{\mathrm{i}}{\hbar}\int_{t_i}^{t_f} f(t)x_+(t)dt})_+\rangle, \tag{6.236}$$

where the symbol $\langle . \rangle$ denotes the average with respect to the thermal bath; '$(.)_+$' and '$(.)_-$' denote time ordering and anti–time ordering, respectively; the coordinates $x_\pm$ are defined as in (6.231). If the interaction between the bath and the coordinate x (6.235) were turned off, then the operator f of the bath would develop in time according to $f(t) = \mathrm{e}^{\mathrm{i}H_\gamma t/\hbar} f \mathrm{e}^{-\mathrm{i}H_\gamma t/\hbar}$, where H_γ is the Hamiltonian of the isolated bath (decoupled from the coordinate x). $f(t)$ is then the force operator of the bath to be used in (6.236).

The interaction $I[x,y]$ between the bath and the particle has been evaluated in [SVW95] for a linear passive damping due to thermal bath by following Feynman–Vernon and Schwinger [FH65]. The final result from [SVW95] is:

$$I[x,y] = \frac{1}{2}\int_{t_i}^{t_f} dt\,[x(t)F_y^{ret}(t) + y(t)F_x^{adv}(t)] + \frac{\mathrm{i}}{2\hbar}\int_{t_i}^{t_f}\int_{t_i}^{t_f} dtds\, N(t-s)y(t)y(s),$$

where the retarded force on y, F_y^{ret}, and the advanced force on x, F_x^{adv}, are given in terms of the retarded and advanced Green functions $G_{ret}(t-s)$ and $G_{adv}(t-s)$ by

$$F_y^{ret}(t) = \int_{t_i}^{t_f} ds\, G_{ret}(t-s)y(s), \qquad F_x^{adv}(t) = \int_{t_i}^{t_f} ds\, G_{adv}(t-s)x(s),$$

respectively. In (6.237), $N(t-s)$ is the *quantum noise* in the fluctuating random force given by: $N(t-s) = \frac{1}{2}\langle f(t)f(s) + f(s)f(t)\rangle$.

The real and the imaginary part of the action are given respectively by

$$\mathrm{Re}\,(A[x,y]) = \int_{t_i}^{t_f} L\, dt, \tag{6.237}$$

$$L = m\dot{x}\dot{y} - \left[V(x + \frac{1}{2}y) - V(x - \frac{1}{2}y)\right] + \frac{1}{2}\left[xF_y^{ret} + yF_x^{adv}\right], \tag{6.238}$$

and $$\mathrm{Im}\,(A[x,y]) = \frac{1}{2\hbar}\int_{t_i}^{t_f}\int_{t_i}^{t_f} N(t-s)y(t)y(s)\, dtds. \tag{6.239}$$

Equations (6.237–6.239), are *exact* results for linear passive damping due to the bath. They show that in the classical limit '$\hbar \to 0$' nonzero y yields an 'unlikely process' in view of the large imaginary part of the action implicit in (6.239). Nonzero y, indeed, may lead to a negative real exponent in the evolution operator, which in the limit $\hbar \to 0$ may produce a negligible contribution to the probability amplitude. On the contrary, at quantum level nonzero y accounts for quantum noise effects in the fluctuating random force in the system–environment coupling arising from the imaginary part of the action (see [SVW95]).

When in (6.238) we use

$$F_y^{ret} = \gamma \dot{y} \qquad \text{and} \qquad F_x^{adv} = -\gamma \dot{x} \qquad \text{we get,}$$

$$L(\dot{x}, \dot{y}, x, y) = m\dot{x}\dot{y} - V\left(x + \frac{1}{2}y\right) + V\left(x - \frac{1}{2}y\right) + \frac{\gamma}{2}(x\dot{y} - y\dot{x}). \quad (6.240)$$

By using

$$V\left(x \pm \frac{1}{2}y\right) = \frac{1}{2}\kappa(x \pm \frac{1}{2}y)^2$$

in (6.240), the DHO equation (6.229) and its complementary equation for the y coordinate

$$m\ddot{y} - \gamma\dot{y} + \kappa y = 0. \quad (6.241)$$

are derived. The y–oscillator is the time–reversed image of the x–oscillator (6.229). From the manifolds of solutions to equations (6.229) and (6.241), we could choose those for which the y coordinate is constrained to be zero, they simplify to

$$m\ddot{x} + \gamma\dot{x} + \kappa x = 0, \qquad y = 0.$$

Thus we get the classical damped oscillator equation from a Lagrangian theory at the expense of introducing an 'extra' coordinate y, later constrained to vanish. Note that the constraint $y(t) = 0$ is *not* in violation of the equations of motion since it is a true solution to (6.229) and (6.241).

6.4.5 Cerebellum as a Neural Path–Integral

Recall that human motion is naturally driven by synergistic action of more than 600 skeletal muscles. While the muscles generate driving torques in the moving joints, subcortical neural system performs both local and global (loco)motion control: first reflexly controlling contractions of individual muscles, and then orchestrating all the muscles into synergetic actions in order to produce efficient movements. While the local reflex control of individual muscles is performed on the *spinal control level*, the global integration of all the muscles into coordinated movements is performed within the *cerebellum.*

All hierarchical subcortical neuro–muscular physiology, from the bottom level of a single muscle fiber, to the top level of cerebellar muscular synergy, acts as a *temporal* $< out|in >$ *reaction*, in such a way that the higher level acts as a command/control space for the lower level, itself representing an abstract image of the lower one:

1. At the *muscular level*, we have *excitation–contraction dynamics* [Hat77a, Hat78, Hat77b], in which $< out|in >$ is given by the following sequence of nonlinear diffusion processes: *neural-action-potential* $\rightsquigarrow$ *synaptic-potential* $\rightsquigarrow$ *muscular-action-potential* $\rightsquigarrow$ *excitation-contraction-coupling* $\rightsquigarrow$ *muscle-tension-generating* [Iva91, II06a]. Its purpose is the generation of muscular forces, to be transferred into driving torques within the joint anatomical geometry.

2. At the *spinal level*, $< out|in >$ is given by *autogenetic–reflex stimulus–response control* [Hou79]. Here we have a neural image of all individual muscles. The main purpose of the spinal control level is to give both positive and negative feedbacks to stabilize generated muscular forces within the 'homeostatic' (or, more appropriately, 'homeokinetic') limits. The individual muscular actions are combined into flexor–extensor (or agonist–antagonist) pairs, mutually controlling each other. This is the mechanism of *reciprocal innervation of agonists and inhibition of antagonists.* It has a purely mechanical purpose to form the so–called *equivalent muscular actuators* (EMAs), which would generate driving torques $T_i(t)$ for all movable joints.
3. At the *cerebellar level*, $< out|in >$ is given by *sensory–motor integration* [HBB96]. Here we have an abstracted image of all autogenetic reflexes. The main purpose of the cerebellar control level is integration and fine tuning of the action of all active EMAs into a synchronized movement, by *supervising* the individual autogenetic reflex circuits. At the same time, to be able to perform in new and unknown conditions, the cerebellum is continuously adapting its own neural circuitry by unsupervised (self–organizing) learning. Its action is subconscious and automatic, both in humans and in animals.

Naturally, we can ask the question: Can we assign a single $< out|in >$ measure to all these neuro–muscular stimulus–response reactions? We think that we can do it; so in this Letter, we propose the concept of *adaptive sensory–motor transition amplitude* as a unique measure for this temporal $< out|in >$ relation. Conceptually, this $< out|in >$ $-$*amplitude* can be formulated as the '*neural path integral*':

$$< out|in > \equiv \underset{amplitude}{\langle motor|sensory\rangle} = \int \mathcal{D}[w,x]\, \mathrm{e}^{\mathrm{i}\, S[x]}. \tag{6.242}$$

Here, the integral is taken over all *activated* (or, 'fired') *neural pathways* $x^i = x^i(t)$ of the cerebellum, connecting its input *sensory*$-$state with its output *motor*$-$state, symbolically described by *adaptive neural measure* $\mathcal{D}[w,x]$, defined by the weighted product (of discrete time steps)

$$\mathcal{D}[w,x] = \lim_{n\longrightarrow\infty} \prod_{t=1}^{n} w^i(t)\, dx^i(t),$$

in which the *synaptic weights* $w^i = w^i(t)$, included in all active neural pathways $x^i = x^i(t)$, are updated by the unsupervised *Hebbian–like learning* rule 6.7, namely

$$w^i(t+1) = w^i(t) + \frac{\sigma}{\eta}(w^i_d(t) - w^i_a(t)), \tag{6.243}$$

where $\sigma = \sigma(t)$, $\eta = \eta(t)$ represent local neural *signal* and *noise* amplitudes, respectively, while superscripts d and a denote *desired* and *achieved* neural states, respectively. Theoretically, equations (6.242–6.243) define an $\infty-$*dimensional neural network*. Practically, in a computer simulation we can use $10^7 \leq n \leq 10^8$, roughly corresponding to the number of neurons in the cerebellum.

The exponent term $S[x]$ in equation (6.242) represents the *autogenetic–reflex action*, describing reflexly–induced motion of all active EMAs, from their initial *stimulus*–state to their final *response*–state, along the family of extremal (i.e., Euler–Lagrangian) paths $x^i_{\min}(t)$. ($S[x]$ is properly derived in (6.246–6.247) below.)

Spinal Autogenetic Reflex Control

Recall (from Introduction) that at the spinal control level we have the autogenetic reflex *motor servo* [Hou79], providing the local, reflex feedback loops for individual muscular contractions. A voluntary contraction force F of human skeletal muscle is reflexly excited (positive feedback $+F^{-1}$) by the responses of its *spindle receptors* to stretch and is reflexly inhibited (negative feedback $-F^{-1}$) by the responses of its *Golgi tendon organs* to contraction. Stretch and unloading reflexes are mediated by combined actions of several autogenetic neural pathways, forming the *motor servo*.

In other words, branches of the afferent fibers also synapse with with interneurons that inhibit motor neurons controlling the antagonistic muscles – *reciprocal inhibition*. Consequently, the stretch stimulus causes the antagonists to relax so that they cannot resists the shortening of the stretched muscle caused by the main reflex arc. Similarly, firing of the Golgi tendon receptors causes inhibition of the muscle contracting too strong and simultaneous *reciprocal activation* of its antagonist. Both mechanisms of reciprocal inhibition and activation performed by the autogenetic circuits $+F^{-1}$ and $-F^{-1}$, serve to generate the well–tuned EMA–driving torques T_i.

Now, once we have properly defined the symplectic musculo–skeletal dynamics [Iva04] on the biomechanical (momentum) phase–space manifold T^*M^N, we can proceed in formalizing its hierarchical subcortical neural control. By introducing the *coupling Hamiltonians* $H^m = H^m(q,p)$, selectively corresponding *only* to the $M \leq N$ *active joints*, we define the *affine Hamiltonian control function* $H_{aff} : T^*M^N \to \mathbb{R}$, in local canonical coordinates on T^*M^N given by (adapted from [NS90] for the biomechanical purpose)

$$H_{aff}(q,p) = H_0(q,p) - H^m(q,p)\,T_m, \qquad (m = 1,\dots, M \leq N), \qquad (6.244)$$

where $T_m = T_m(t,q,p)$ are affine feedback torque one–forms, different from the initial driving torques T_i acting in all the joints. Using the affine Hamiltonian function (6.244), we get the *affine Hamiltonian servo–system* [Iva04],

$$\dot{q}^i = \frac{\partial H_0(q,p)}{\partial p_i} - \frac{\partial H^m(q,p)}{\partial p_i} T_m, \tag{6.245}$$
$$\dot{p}_i = -\frac{\partial H_0(q,p)}{\partial q^i} + \frac{\partial H^m(q,p)}{\partial q^i} T_m,$$
$$q^i(0) = q_0^i, \quad p_i(0) = p_i^0, \qquad (i = 1, \dots, N; \quad m = 1, \dots, M \leq N).$$

The affine Hamiltonian control system (6.245) gives our formal description for the autogenetic spinal motor–servo for all $M \leq N$ activated (i.e., working) EMAs.

Cerebellum – the Comparator

Having, thus, defined the spinal reflex control level, we proceed to model the top subcortical commander/controller, the *cerebellum*. It is a brain region anatomically located at the bottom rear of the head (the hindbrain), directly above the brainstem, which is important for a number of subconscious and automatic motor functions, including motor learning. It processes information received from the motor cortex, as well as from proprioceptors and visual and equilibrium pathways, and gives 'instructions' to the motor cortex and other subcortical motor centers (like the basal nuclei), which result in proper balance and posture, as well as smooth, coordinated skeletal movements, like walking, running, jumping, driving, typing, playing the piano, etc. Patients with cerebellar dysfunction have problems with precise movements, such as walking and balance, and hand and arm movements. The cerebellum looks *similar in all animals*, from fish to mice to humans. This has been taken as evidence that it performs a common function, such as regulating motor learning and the timing of movements, in all animals. Studies of simple forms of motor learning in the vestibulo–ocular reflex and eye–blink conditioning are demonstrating that timing and amplitude of learned movements are encoded by the cerebellum.

The cerebellum is responsible for coordinating precisely timed $< out|in >$ activity by integrating motor output with ongoing sensory feedback. It receives extensive projections from sensory–motor areas of the cortex and the periphery and directs it back to premotor and motor cortex [Ghe90, Ghe91]. This suggests a role in sensory–motor integration and the timing and execution of human movements. The cerebellum stores patterns of motor control for frequently performed movements, and therefore, its circuits are changed by experience and training. It was termed the *adjustable pattern generator* in the work of J. Houk and collaborators [HBB96]. Also, it has become the inspiring 'brain–model' in the recent robotic research [SA98, Sch98].

Comparing the number of its neurons (10^7–10^8), to the size of conventional neural networks, suggests that artificial neural nets *cannot* satisfactorily model the function of this sophisticated 'super–bio–computer', as its dimensionality is virtually infinite. Despite a lot of research dedicated to its structure

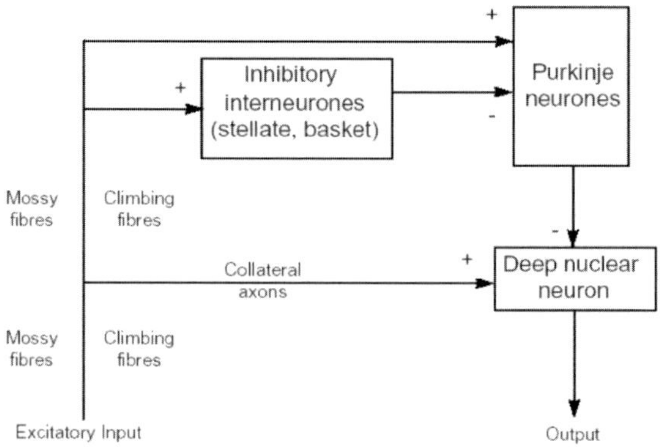

Fig. 6.21. Schematic $< out|in >$ organization of the primary cerebellar circuit. In essence, excitatory inputs, conveyed by collateral axons of Mossy and Climbing fibers activate directly neurones in the Deep cerebellar nuclei. The activity of these latter is also modulated by the inhibitory action of the cerebellar cortex, mediated by the Purkinje cells.

and function (see [HBB96] and references there cited), the real nature of the cerebellum still remains a 'mystery'.

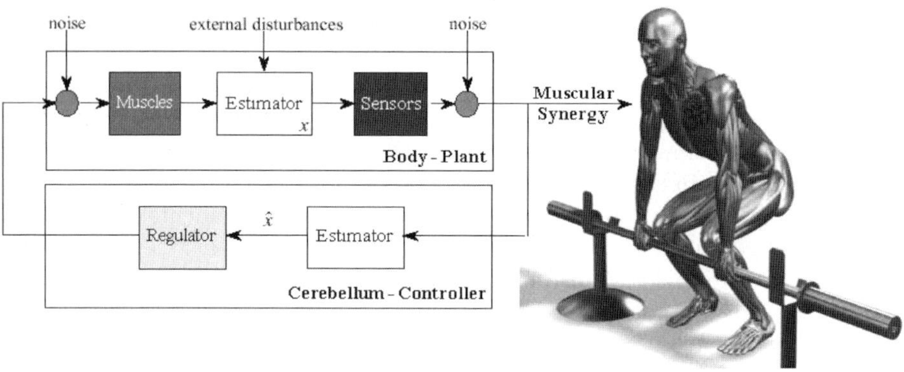

Fig. 6.22. The cerebellum as a motor controller.

The main function of the cerebellum as a motor controller is depicted in Figure 6.22. A coordinated movement is easy to recognize, but we know little about how it is achieved. In search of the neural basis of coordination, a model of spinocerebellar interactions was recently presented in [AG05], in which the structural and functional organizing principle is a division of the cerebellum into discrete micro–complexes. Each micro–complex is the recipient of a specific motor error signal - that is, a signal that conveys information about an inappropriate movement. These signals are encoded by spinal reflex circuits and conveyed to the cerebellar cortex through climbing fibre afferents. This organization reveals salient features of cerebellar information processing, but also highlights the importance of systems level analysis for a fuller understanding of the neural mechanisms that underlie behavior.

Hamiltonian Action and Neural Path Integral

Here, we propose a *quantum–like adaptive control* approach to modelling the 'cerebellar mystery'. Corresponding to the affine Hamiltonian control function (6.244) we define the *affine Hamiltonian control action*,

$$S_{aff}[q,p] = \int_{t_{in}}^{t_{out}} d\tau \left[p_i \dot{q}^i - H_{aff}(q,p) \right]. \tag{6.246}$$

From the affine Hamiltonian action (6.246) we further derive the associated expression for the *neural phase–space path integral* (in normal units), representing the *cerebellar sensory–motor amplitude* $< out|in >$,

$$\begin{aligned} \langle q^i_{out}, p_i^{out} | q^i_{in}, p_i^{in} \rangle &= \int \mathcal{D}[w,q,p] \, \mathrm{e}^{\mathrm{i}\, S_{aff}[q,p]} \qquad (6.247) \\ &= \int \mathcal{D}[w,q,p] \exp \left\{ \mathrm{i} \int_{t_{in}}^{t_{out}} d\tau \left[p_i \dot{q}^i - H_{aff}(q,p) \right] \right\}, \\ \text{with} \qquad \int \mathcal{D}[w,q,p] &= \int \prod_{\tau=1}^{n} \frac{w^i(\tau) dp_i(\tau) dq^i(\tau)}{2\pi}, \end{aligned}$$

where $w_i = w_i(t)$ denote the cerebellar synaptic weights positioned along its neural pathways, being continuously updated using the Hebbian–like self–organizing learning rule (6.243). Given the transition amplitude $< out|in >$ (6.247), the *cerebellar sensory–motor transition probability* is defined as its absolute square, $| < out|in > |^2$.

In (6.247), $q^i_{in} = q^i_{in}(t)$, $q^i_{out} = q^i_{out}(t)$; $p_i^{in} = p_i^{in}(t)$, $p_i^{out} = p_i^{out}(t)$; $t_{in} \leq t \leq t_{out}$, for all discrete time steps, $t = 1, ..., n \longrightarrow \infty$, and we are allowing for the affine Hamiltonian $H_{aff}(q,p)$ to depend upon all the $(M \leq N)$ EMA–angles and angular momenta collectively. Here, we actually systematically took a discretized differential time limit of the form $t_\sigma - t_{\sigma-1} \equiv d\tau$ (both σ and τ denote discrete time steps) and wrote $\frac{(q^i_\sigma - q^i_{\sigma-1})}{(t_\sigma - t_{\sigma-1})} \equiv \dot{q}^i$. For technical details regarding the path integral calculations on Riemannian and symplectic manifolds (including the standard regularization procedures), see [Kla97, Kla00].

Now, motor learning occurring in the cerebellum can be observed using functional MR imaging, showing changes in the cerebellar action potential, related to the motor tasks (see, e.g., [MA02]). To account for these electro–physiological currents, we need to add the *source* term $J_i(t) q^i(t)$ to the affine Hamiltonian action (6.246), (the current $J_i = J_i(t)$ acts as a source $J_i A^i$ of the *cerebellar electrical potential* $A^i = A^i(t)$),

$$S_{aff}[q,p,J] = \int_{t_{in}}^{t_{out}} d\tau \left[p_i \dot{q}^i - H_{aff}(q,p) + J_i q^i\right],$$

which, subsequently gives the cerebellar path integral with the action potential source, coming either from the motor cortex or from other subcortical areas.

Note that the standard *Wick rotation*: $t \mapsto it$ (see [Kla97, Kla00]), makes all our path integrals real, i.e.,

$$\int \mathcal{D}[w,q,p]\, \mathrm{e}^{\mathrm{i}\, S_{aff}[q,p]} \quad \underrightarrow{Wick} \quad \int \mathcal{D}[w,q,p]\, \mathrm{e}^{-\, S_{aff}[q,p]},$$

while their subsequent discretization gives the standard thermodynamic *partition functions*,

$$Z = \sum_j \mathrm{e}^{-w_j E^j/T}, \qquad (6.248)$$

where E^j is the energy eigenvalue corresponding to the affine Hamiltonian $H_{aff}(q,p)$, T is the temperature–like environmental control parameter, and the sum runs over all energy eigenstates (labelled by the index j). From (6.248), we can further calculate all statistical and thermodynamic system properties (see [Fey72]), as for example, *transition entropy* $S = k_B \ln Z$, etc.

6.4.6 Topological Phase Transitions and Hamiltonian Chaos

Phase Transitions in Hamiltonian Systems

Recall that *phase transitions* (PTs) are phenomena which bring about *qualitative* physical changes at the macroscopic level in presence of the same microscopic forces acting among the constituents of a system. Their mathematical description requires to translate into *quantitative* terms the mentioned qualitative changes. The standard way of doing this is to consider how the values of thermodynamic observables, get in laboratory experiments, vary with temperature, or volume, or an external field, and then to associate the experimentally observed discontinuities at a PT to the appearance of some kind of singularity entailing a loss of analyticity. Despite the smoothness of the statistical measures, after the *Yang–Lee Theorem* [YL52] we know that in the $N \to \infty$ limit non–analytic behaviors of thermodynamic functions are possible whenever the analyticity radius in the complex fugacity plane shrinks to zero, because this entails the loss of *uniform convergence* in N (number of degrees of freedom) of any sequence of real–valued thermodynamic functions, and all this depends on the distribution of the zeros of the grand canonical partition function. Also the other developments of the rigorous theory of PTs [Geo88, Rue78], identify PTs with the loss of analyticity.

In this subsection we will address a recently proposed geometric approach to thermodynamic phase transitions (see [CCC97, FCS99, FPS00, FP04]). Given any Hamiltonian system, the configuration space can be equipped with a metric, in order to get a Riemannian geometrization of the dynamics. At the beginning, several numerical and analytical studies of a variety of models showed that the fluctuation of the curvature becomes singular at the transition point. Then the following conjecture was proposed in [CCC97]: The phase transition is determined by a change in the topology of the configuration space, and the loss of analyticity in the thermodynamic observables is nothing but a consequence of such topological change. The latter conjecture is also known as the *topological hypothesis*.

The topological hypothesis states that suitable topology changes of equipotential submanifolds of the Hamiltonian system's configuration manifold can entail thermodynamic phase transitions [FPS00]. The authors of the topological hypothesis gave both a theoretical argument and numerical demonstration in case of $2D$ lattice φ^4 model. They considered classical many–particle (or many–subsystem) systems described by standard *mechanical Hamiltonians*

$$H(p,q)=\sum_{i=1}^{N}\frac{p_i^2}{2m}+V(q), \tag{6.249}$$

where the coordinates $q^i=q^i(t)$ and momenta $p_i=p_i(t)$, $(i=1,...,N)$, have continuous values and the system's potential energy $V(q)$ is bounded below.

Now, assuming a large number of subsystems N, the statistical behavior of physical systems described by Hamiltonians of the type (6.249) is usually encompassed, in the system's canonical ensemble, by the *partition function* in the system's phase–space

$$\begin{aligned} Z_N(\beta) &= \int\prod_{i=1}^{N} dp_i dq^i \mathrm{e}^{-\beta H(p,q)} = \left(\frac{\pi}{\beta}\right)^{\frac{N}{2}}\int\prod_{i=1}^{N} dq^i \mathrm{e}^{-\beta V(q)} \\ &= \left(\frac{\pi}{\beta}\right)^{\frac{N}{2}}\int_0^\infty dv\,\mathrm{e}^{-\beta v}\int_{M_v}\frac{d\sigma}{\|\nabla V\|}, \end{aligned} \tag{6.250}$$

where the last term is written using a *co–area formula* [Fed69], and v labels the *equipotential hypersurfaces* M_v of the system's configuration manifold M,

$$M_v=\{(q^1,\dots,q^N)\in\mathbb{R}^N|V(q^1,\dots,q^N)=v\}. \tag{6.251}$$

Equation (7.123) shows that for Hamiltonians (5.137) the relevant statistical information is contained in the *canonical configurational partition function*

$$Z_N^C=\int\prod_{i=1}^{N} dq^i\exp[-\beta V(q)].$$

Therefore, partition function Z_N^C is decomposed – in the last term of equation (7.123) – into an infinite summation of geometric integrals, $\int_{M_v} d\sigma\,/\|\nabla V\|$, defined on the $\{M_v\}_{v\in\mathbb{R}}$. Once the microscopic interaction potential $V(q)$ is given, the configuration space of the system is automatically foliated into the family $\{M_v\}_{v\in\mathbb{R}}$ of these equipotential hypersurfaces. Now, from standard statistical mechanical arguments we know that, at any given value of the inverse temperature β, the larger the number N of particles the closer to $M_v \equiv M_{u_\beta}$ are the microstates that significantly contribute to the averages – computed through $Z_N(\beta)$ – of thermodynamic observables. The hypersurface M_{u_β} is the one associated with the average potential energy computed at a given β,

$$u_\beta = (Z_N^C)^{-1} \int \prod_{i=1}^{N} dq^i V(q) \exp[-\beta V(q)].$$

Thus, at any β, if N is very large the effective support of the canonical measure shrinks very close to a single $M_v = M_{u_\beta}$.

Explicitly, the topological hypothesis reads: the basic origin of a phase transition lies in a suitable topology change of the $\{M_v\}$, occurring at some v_c. This topology change induces the singular behavior of the thermodynamic observables at a phase transition. By change of topology we mean that $\{M_v\}_{v<v_c}$ are not diffeomorphic to the $\{M_v\}_{v>v_c}$. In other words, canonical measure should 'feel' a big and sudden change of the topology of the equipotential hypersurfaces of its underlying support, the consequence being the appearance of the typical signals of a phase transition.

This point of view has the interesting consequence that – also at finite N – in principle *different* mathematical objects, i.e., manifolds of different cohomology type, could be associated to *different* thermodynamical phases, whereas from the point of view of measure theory [YL52] the only mathematical property available to signal the appearance of a phase transition is the loss of analyticity of the grand–canonical and canonical averages, a fact which is compatible with analytic statistical measures only in the mathematical $N \to \infty$ limit.

As it is conjectured that the counterpart of a phase transition is a breaking of diffeomorphicity among the surfaces M_v, it is appropriate to choose a *diffeomorphism invariant* to probe if and how the topology of the M_v changes as a function of v. This is a very challenging task because we have to deal with high dimensional manifolds. Fortunately a topological invariant exists whose computation is feasible, yet demands a big effort. Recall (from subsection 4.1.4 above) that this is the *Euler characteristic*, a diffeomorphism invariant of the system's configuration manifold, expressing its fundamental topological information.

Geometry of the Largest Lyapunov Exponent

Now, the topological hypothesis has recently been promoted into a *topological Theorem* [FP04]. The new Theorem says that non–analyticity is the 'shadow' of a more fundamental phenomenon occurring in the system's configuration manifold: a *topology change* within the family of equipotential hypersurfaces

(6.251). This topological approach to PTs stems from the numerical study of the Hamiltonian dynamical counterpart of phase transitions, and precisely from the observation of *discontinuous* or *cuspy patterns*, displayed by the *largest Lyapunov exponent* at the transition energy (or temperature).

Recall that the *Lyapunov exponents* measure the strength of *dynamical chaos* and cannot be measured in laboratory experiments, at variance with thermodynamic observables, thus, being genuine dynamical observables they are only measurable in numerical simulations of the microscopic dynamics. To get a hold of the reason why the largest Lyapunov exponent λ_1 should probe configuration space topology, let us first remember that for standard Hamiltonian systems, λ_1 is computed by solving the *tangent dynamics equation* for Hamiltonian systems (see Jacobi equation of geodesic deviation (4.34)),

$$\ddot{\xi}_i + \left(\frac{\partial^2 V}{\partial q^i \partial q^j}\right)_{q(t)} \xi^j = 0, \qquad (6.252)$$

which, for the nonlinear Hamiltonian system

$$\begin{aligned} \dot{q}^1 &= p_1, & \dot{p}_1 &= -\partial_{q^1} V, \\ &\dots & &\dots \\ \dot{q}^N &= p_N, & \dot{p}_N &= -\partial_{q^N} V, \end{aligned}$$

expands into *linearized Hamiltonian dynamics*

$$\begin{aligned} \dot{\xi}_1 &= \xi_{N+1}, & \dot{\xi}_{N+1} &= -\sum_{j=1}^{N} \left(\frac{\partial^2 V}{\partial q_1 \partial q_j}\right)_{q(t)} \xi_j, \\ &\dots & &\dots \\ \dot{\xi}_n &= \xi_{2N}, & \dot{\xi}_{2N} &= -\sum_{j=1}^{N} \left(\frac{\partial^2 V}{\partial q_N \partial q_j}\right)_{q(t)} \xi_j. \end{aligned} \qquad (6.253)$$

Using (6.252) we can get the analytical expression for the largest Lyapunov exponent

$$\lambda_1 = \lim_{t\to\infty} \frac{1}{t} \log \frac{\left[\xi_1^2(t) + \dots + \xi_N^2(t) + \dot{\xi}_1^2(t) + \dots + \dot{\xi}_N^2(t)\right]^{1/2}}{\left[\xi_1^2(0) + \dots + \xi_N^2(0) + \dot{\xi}_1^2(0) + \dots + \dot{\xi}_N^2(0)\right]^{1/2}}. \qquad (6.254)$$

If there are critical points of V in configuration space, that is points $q_c = [\overline{q}^1, \dots, \overline{q}^N]$ such that $\nabla V(q)|_{q=q_c} = 0$, according to the *Morse lemma* (see e.g., [Hir76]), in the neighborhood of any critical point q_c there always exists a coordinate system $\tilde{q}(t) = [\overline{q}^1(t), \dots, \overline{q}^N(t)]$ for which

$$V(\tilde{q}) = V(q_c) - \left(\overline{q}^1\right)^2 - \dots - \left(\overline{q}^k\right)^2 + \left(\overline{q}^{k+1}\right)^2 + \dots + \left(\overline{q}^N\right)^2, \qquad (6.255)$$

where k is the index of the critical point, i.e., the number of negative eigenvalues of the Hessian of V. In the neighborhood of a critical point, equation (6.255) yields

$$\partial^2 V/\partial q^i \partial q^j = \pm\delta_{ij},$$

which, substituted into equation (6.252), gives k *unstable directions* which contribute to the exponential growth of the norm of the tangent vector $\xi = \xi(t)$. This means that the strength of dynamical chaos, measured by the largest Lyapunov exponent λ_1, is affected by the existence of critical points of V. In particular, let us consider the possibility of a sudden variation, with the potential energy v, of the number of critical points (or of their indexes) in configuration space at some value v_c, it is then reasonable to expect that the pattern of $\lambda_1(v)$ – as well as that of $\lambda_1(E)$ since $v = v(E)$ – will be consequently affected, thus displaying *jumps* or *cusps* or other *singular patterns* at v_c.

On the other hand, recall that *Morse theory* teaches us that the existence of critical points of V is associated with topology changes of the hypersurfaces $\{M_v\}_{v\in\mathbb{R}}$, provided that V is a good *Morse function* (that is: bounded below, with no vanishing eigenvalues of its Hessian matrix). Thus the existence of critical points of the potential V makes possible a conceptual link between dynamics and configuration space topology, which, on the basis of both direct and indirect evidence for a few particular models, has been formulated as a topological hypothesis about the relevance of topology for PTs phenomena (see [FPS00, FP04, GM04]).

Here we give two simple examples of standard Hamiltonian systems of the form (6.249), namely Peyrard–Bishop system and mean–field XY model.

Peyrard–Bishop Hamiltonian System

The *Peyrard–Bishop system* [PB89][23] exhibits a *second–order phase transition*. It is defined by the following potential energy

$$V(q) = \sum_{i=1}^{N}\left[\frac{K}{2}(q^{i+1} - q^i)^2 + D(\mathrm{e}^{-aq^i} - 1)^2 + Dhaq^i\right], \qquad (6.256)$$

which represents the energy of a string of N base pairs of reduced mass m. Each hydrogen bond is characterized by the stretching q^i and its conjugate momentum $p_i = m\dot{q}^i$. The elastic transverse force between neighboring pairs is tuned by the constant K, while the energy D and the inverse length a determine, respectively, the plateau and the narrowness of the on–site potential well that mimics the interaction between bases in each pair. It is understood that K, D, and a are all positive parameters. The transverse, external stress $h \geq 0$ is a computational tool useful in the evaluation of the susceptibility.

[23] The Peyrard–Bishop system has been proposed as a simple model for describing the DNA thermally induced denaturation [GM04].

Our interest in it lies in the fact that a phase transition can occur only when $h = 0$. We assume periodic boundary conditions.

The transfer operator technique [DTP02] maps the problem of computing the classical partition function into the easier task of evaluating the lowest energy eigenvalues of a 'quantum' mechanical Morse oscillator (no real quantum mechanics is involved, since the temperature plays the role of $\hbar$). One can then observe that, as the temperature increases, the number of levels belonging to the discrete spectrum decreases, until for some critical temperature $T_c = 2\sqrt{2KD}/(ak_B)$ only the continuous spectrum survives. This passage from a localized ground state to an unnormalizable one corresponds to the second–order phase transition of the statistical model. Various critical exponents can be analytically computed and all applicable scaling laws can be checked. The simplicity of this model permits an analytical computation of the largest Lyapunov exponent by exploiting the geometric method proposed in [CCC97].

Mean–Field XY Hamiltonian System

The mean–field XY model describes a system of N equally coupled planar classical rotators (see [AR95, CCP99]). It is defined by a Hamiltonian of the class (6.249) where the potential energy is

$$V(\varphi) = \frac{J}{2N} \sum_{i,j=1}^{N} \left[1 - \cos(\varphi_i - \varphi_j)\right] - h \sum_{i=1}^{N} \cos\varphi_i. \tag{6.257}$$

Here $\varphi_i \in [0, 2\pi]$ is the rotation angle of the ith rotator and h is an external field. Defining at each site i a classical spin vector $\mathbf{s}_i = (\cos\varphi_i, \sin\varphi_i)$ the model describes a planar (XY) Heisenberg system with interactions of equal strength among all the spins. We consider only the ferromagnetic case $J > 0$; for the sake of simplicity, we set $J = 1$. The equilibrium statistical mechanics of this system is exactly described, in the thermodynamic limit, by the *mean–field theory* [AR95]. In the limit $h \to 0$, the system has a continuous phase transition, with classical critical exponents, at $T_c = 1/2$, or $\varepsilon_c = 3/4$, where $\varepsilon = E/N$ is the energy per particle.

The Lyapunov exponent λ_1 of this system is extremely sensitive to the phase transition. According to reported numerical simulations (see [CCP99]), $\lambda_1(\varepsilon)$ is positive for $0 < \varepsilon < \varepsilon_c$, shows a sharp maximum immediately below the critical energy, and drops to zero at ε_c in the thermodynamic limit, where it remains zero in the whole region $\varepsilon > \varepsilon_c$, which corresponds to the thermodynamic disordered phase. In fact in this phase the system is integrable, reducing to an assembly of uncoupled rotators.

Euler Characteristics of Hamiltonian Systems

Recall that *Euler characteristic* χ is a number that is a characterization of the various classes of geometric figures based only on the topological relationship between the numbers of vertices V, edges E, and faces F, of a geometric Figure. This number, $\chi = F - E + V$, is the same for all figures the boundaries of which are composed of the same number of connected pieces. Therefore, the Euler characteristic is a *topological invariant*, i.e., any two geometric figures that are homeomorphic to each other have the same Euler characteristic.

More specifically, a standard way to analyze a geometric Figure is to fragment it into other more familiar objects and then to examine how these pieces fit together. Take for example a surface M in the Euclidean 3D space. Slice M into pieces that are curved triangles (this is called a triangulation of the surface). Then count the number F of faces of the triangles, the number E of edges, and the number V of vertices on the tesselated surface. Now, no matter how we triangulate a compact surface Σ, its Euler characteristic, $\chi(\Sigma) = F - E + V$, will always equal a constant which is characteristic of the surface and which is invariant under diffeomorphisms $\phi : \Sigma \to \Sigma'$.

At higher dimensions this can be again defined by using higher dimensional generalizations of triangles (simplexes) and by defining the Euler characteristic $\chi(M)$ of the nD manifold M to be the alternating sum:

$$\{\text{number of points}\} - \{\text{number of 2-simplices}\} \quad + \\ \{\text{number of 3-simplices}\} - \{\text{number of 4-simplices}\} \quad + \quad \ldots$$

$$\text{i.e.,} \qquad \chi(M) = \sum_{k=0}^{n} (-1)^k (\text{number of faces of dimension } k).$$

and then define the Euler characteristic of a manifold as the Euler characteristic of any simplicial complex homeomorphic to it. With this definition, circles and squares have Euler characteristic 0 and solid balls have Euler characteristic 1.

The Euler characteristic χ of a manifold is closely related to its *genus* g as $\chi = 2 - 2g$.[24]

Recall that a more standard topological definition of $\chi(M)$ is

$$\chi(M) = \sum_{k=0}^{n} (-1)^k \, b_k(M), \tag{6.258}$$

where b_k are the kth *Betti numbers* of M.

In general, it would be hopeless to try to practically calculate $\chi(M)$ from (6.258) in the case of non–trivial physical models at large dimension. Fortunately, there is a possibility given by the *Gauss–Bonnet formula*, that relates

[24] Recall that the *genus* of a topological space such as a surface is a topologically invariant property defined as the largest number of nonintersecting simple closed curves that can be drawn on the surface without separating it, i.e., an integer representing the maximum number of cuts that can be made through it without rendering it disconnected. This is roughly equivalent to the number of holes in it, or handles on it. For instance: a point, line, and a sphere all have genus 0; a torus has genus 1, as does a coffee cup as a solid object (solid torus), a Möbius strip, and the symbol 0; the symbols 8 and B have genus 2; etc.

$\chi(M)$ with the total *Gauss–Kronecker curvature* of the manifold, (compare with (4.27) and (4.36))

$$\chi(M) = \gamma \int_M K_G \, d\sigma, \tag{6.259}$$

which is valid for even dimensional hypersurfaces of Euclidean spaces $\mathbb{R}^N$ [here $\dim(M) = n \equiv N - 1$], and where:

$$\gamma = 2/\operatorname{Vol}(S_1^n)$$

is twice the inverse of the volume of an $n-$dimensional sphere of unit radius S_1^n; K_G is the Gauss–Kronecker curvature of the manifold;

$$d\sigma = \sqrt{\det(g)}\, dx^1 dx^2 \cdots dx^n$$

is the invariant volume measure of M and g is its Riemannian metric (induced from $\mathbb{R}^N$). Let us briefly sketch the meaning and definition of the Gauss–Kronecker curvature. The study of the way in which an $n-$surface M curves around in $\mathbb{R}^N$ is measured by the way the normal direction changes as we move from point to point on the surface. The rate of change of the normal direction ξ at a point $x \in M$ in direction v is described by the *shape operator*

$$L_x(v) = -\mathcal{L}_v \xi = [v, \xi],$$

where v is a tangent vector at x and $\mathcal{L}_v$ is the Lie derivative, hence

$$L_x(v) = -(\nabla\xi_1 \cdot v, \dots, \nabla\xi_{n+1} \cdot v);$$

gradients and vectors are represented in $\mathbb{R}^N$. As L_x is an operator of the tangent space at x into itself, there are n independent eigenvalues $\kappa_1(x), \dots, \kappa_n(x)$ which are called the principal curvatures of M at x [Tho79]. Their product is the Gauss–Kronecker curvature:

$$K_G(x) = \prod_{i=1}^n \kappa_i(x) = \det(L_x).$$

Alternatively, recall that according to the *Morse theory*, it is possible to understand the topology of a given manifold by studying the regular critical points of a smooth *Morse function* defined on it. In our case, the manifold M is the configuration space $\mathbb{R}^N$ and the natural choice for the Morse function is the potential $V(q)$. Hence, one is lead to define the family M_v (6.251) of submanifolds of M.

A full characterization of the topological properties of M_v generally requires the critical points of $V(q)$, which means solving the equations

$$\partial_{q^i} V = 0, \qquad (i = 1, \dots, N). \tag{6.260}$$

Moreover, one has to calculate the indexes of all the critical points, that is the number of negative eigenvalues of the Hessian $\partial^2 V/(\partial q^i \partial q_j)$. Then the Euler characteristic $\chi(M_v)$ can be computed by means of the formula

$$\chi(M_v) = \sum_{k=0}^{N} (-1)^k \mu_k(M_v), \tag{6.261}$$

where $\mu_k(M_v)$ is the total number of critical points of $V(q)$ on M_v which have index k, i.e., the so–called *Morse numbers* of a manifold M, which happen to be upper bounds of the Betti numbers,

$$b_k(M) \leq \mu_k(M) \qquad (k = 0, \dots, n). \tag{6.262}$$

Among all the Morse functions on a manifold M, there is a special class, called *perfect Morse functions*, for which the Morse inequalities (6.262) hold as equalities. Perfect Morse functions characterize completely the topology of a manifold.

Now, we continue with our two examples started before.

Peyrard–Bishop System. If applied to any generic model, calculation of (6.261) turns out to be quite formidable, but the exceptional simplicity of the Peyrard–Bishop model (6.256) makes it possible to carry on completely the topological analysis without invoking equation (6.261).

For the potential in exam, equation (6.260) results in the nonlinear system

$$\frac{a}{R}(q^{i+1} - 2q^i + q^{i-1}) = h - 2(\mathrm{e}^{-2aq^i} - \mathrm{e}^{-aq^i}),$$

where $R = Da^2/K$ is a dimensionless ratio. It is easy to verify that a particular solution is given by

$$q^i = -\frac{1}{a} \ln \frac{1 + \sqrt{1 + 2h}}{2}, \qquad (i = 1, \dots, N).$$

The corresponding minimum of potential energy is

$$V_{\min} = ND \left(\frac{1 + h - \sqrt{1 + 2h}}{2} - h \ln \frac{1 + \sqrt{1 + 2h}}{2} \right).$$

Mean–Field XY Model. In the case of the mean–field XY model (6.257) it is possible to show analytically that a topological change in the configuration space exists and that it can be related to the thermodynamic phase transition. Consider again the family M_v of submanifolds of the configuration space defined in (6.251); now the potential energy per degree of freedom is that of the mean–field XY model, i.e.,

$$\mathcal{V}(\varphi) = \frac{V(\varphi)}{N} = \frac{J}{2N^2} \sum_{i,j=1}^{N} \left[1 - \cos(\varphi_i - \varphi_j)\right] - h \sum_{i=1}^{N} \cos \varphi_i,$$

where $\varphi_i \in [0, 2\pi]$. Such a function can be considered a Morse function on M, so that, according to Morse theory, all these manifolds have the same topology until a critical level $\mathcal{V}^{-1}(v_c)$ is crossed, where the topology of M_v changes.

A change in the topology of M_v can only occur when v passes through a critical value of $\mathcal{V}$. Thus in order to detect topological changes in M_v we have to find the critical values of $\mathcal{V}$, which means solving the equations

$$\partial_{\varphi_i}\mathcal{V}(\varphi) = 0, \qquad (i = 1, \dots, N). \tag{6.263}$$

For a general potential energy function $\mathcal{V}$, the solution of (6.263) would be a formidable task, but in the case of the mean–field XY model, the mean–field character of the interaction greatly simplifies the analysis, allowing an analytical treatment of (6.263); moreover, a projection of the configuration space onto a 2D plane is possible [CCP99, CPC03].

6.4.7 Force–Field Psychodynamics

In this section, which is written in the fashion of the above *quantum brain modelling* (see subsection 6.4.4 above), we present the top level of natural biodynamics, using geometrical generalization of the *Feynman path integral*. To formulate the basics of *force–field psychodynamics*, we use the *action–amplitude picture* of the $BODY \rightleftarrows MIND$ adjunction:

↓ Deterministic (causal) world of *Human BODY* ↓

$$Action: S[q^n] = \int_{t_{in}}^{t_{out}} (E_k - E_p + Wrk + Src^{\pm})\, dt$$

– –

$$Amplitude: \langle out|in\rangle = \sum\!\!\!\!\!\!\!\int \mathcal{D}[w_n q^n]\, \mathrm{e}^{\mathrm{i}S[q^n]}$$

↑ Probabilistic (fuzzy) world of *Human MIND* ↑

In the action integral, E_k, E_p, Wrk and $Src^{\pm}$ denote the kinetic end potential energies, work done by dissipative/driving forces and other energy sources/sinks, respectively. In the amplitude integral, the peculiar sign $\sum\!\!\!\!\!\!\int$ denotes integration along smooth paths and summation along discrete Markov chains; i is the imaginary unit, w_n are synaptic–like weights, while $\mathcal{D}$ is the Feynman path differential (defined below) calculated along the configuration trajectories q^n. The action $S[q^n]$, through the *least action principle* $\delta S = 0$, leads to all biodynamic equations considered so far (in generalized Lagrangian and Hamiltonian form). At the same time, the action $S[q^n]$ figures in the exponent of the path integral $\sum\!\!\!\!\!\!\int$, defining the probability transition amplitude $\langle out|in\rangle$. In this way, the whole body dynamics is incorporated in the mind dynamics. This *adaptive path integral* represents an *infinite–dimensional neural network*, suggesting an infinite capacity of human brain/mind.

For a long time the cortical systems for *language and actions* were believed to be independent modules. However, according to the recent research of [Pul05], as these systems are reciprocally connected with each other, information about language and actions might interact in distributed neuronal assemblies. A critical case is that of action words that are semantically related to different parts of the body (e.g. 'pick', 'kick', 'lick',...). The author suggests that the comprehension of these words might specifically, rapidly and automatically activate the motor system in a somatotopic manner, and that their comprehension rely on activity in the action system.

Motivational Cognition in the Life Space Foam

Applications of nonlinear dynamical systems (NDS) theory in psychology have been encouraging, if not universally productive/effective [Met97]. Its historical antecedents can be traced back to Piaget's [PHE92] and Vygotsky's [Vyg82] interpretations of the dynamic relations between action and thought, Lewinian theory of social dynamics and cognitive–affective development [Lew51, Gol99], and Bernstein's [Ber47] theory of self–adjusting, goal–driven motor action.

Now, both the original *Lewinian force–field theory* in psychology (see [Lew51, Gol99]) and modern decision–field dynamics (see [BT93, RBT01, BD02]) are based on the classical Lewinian concept of an individual's *life space*.[25] As a topological construct, Lewinian life space represents a person's psychological environment that contains *regions* separated by dynamical permeable *boundaries*. As a field construct, on the other hand, the life space is not empty: each of its regions is characterized by *valence* (ranging from positive or negative and resulting from an interaction between the person's *needs* and the dynamics of their *environment*). Need is an energy construct, according to Lewin. It creates *tension* in the person, which, in combination with other tensions, initiates and sustains behavior. Needs vary from the most primitive urges to the most idiosyncratic intentions and can be both internally generated (e.g., thirst or hunger) and stimulus–induced (e.g., an urge to buy something in response to a TV advertisement). Valences are, in essence, personal values dynamically derived from the person's needs and attached to various regions in their life space. As a field, the life space generates forces pulling the person towards positively–valenced regions and pushing them away from regions with negative valence. Lewin's term for these forces is *vectors*. Combinations of multiple vectors in the life space cause the person to move from one region towards another. This movement is termed *locomotion* and it may range from overt behavior to cognitive shifts (e.g., between alternatives in a decision–making process). Locomotion normally results in crossing the boundaries between regions. When their permeability is degraded, these

[25] The work presented in this subsection has been developed in collaboration with Dr. Eugene Aidman, Senior Research Scientist, Human Systems Integration, Land Operations Division, Defence Science & Technology Organisation, Australia.

boundaries become *barriers* that restrain locomotion. Life space model, thus, offers a meta–theoretical language to describe a wide range of behaviors, from goal–directed action to intrapersonal conflicts and multi–alternative decision–making.

In order to formalize the Lewinian life–space concept, a set of *action principles* need to be associated to Lewinian force–fields, (loco)motion paths (representing mental abstractions of biomechanical paths [Iva04]) and life space geometry. As an extension of the Lewinian concept, in this section we introduce a new concept of *life–space foam* (LSF, see Figure 6.23). According to this new concept, Lewin's life space can be represented as a *geometrical functor* with globally smooth macro–dynamics, which is at the same time underpinned by wildly fluctuating, non–smooth, local micro–dynamics, describable by *Feynman's*: (i) *sum–over–histories* $\oint_{paths}$, (ii) *sum–over–fields* $\oint_{fields}$, and (iii) *sum–over–geometries* $\oint_{geom}$.

LSF is thus a two–level *geometrodynamical functor*, representing these two distinct types of dynamics within the Lewinian life space. At its *macroscopic spatio–temporal level*, LSF appears as a 'nice & smooth' geometrical functor with globally predictable dynamics – formally, a smooth n–dimensional manifold M with local Riemannian metrics $g_{ij}(x)$, smooth force–fields and smooth (loco)motion paths, as conceptualized in the Lewinian theory. To model the global and smooth macro–level LSF–paths, fields and geometry, we use the general physics–like *principle of the least action.*

Now, the apparent smoothness of the macro–level LSF is achieved by the existence of another level underneath it. This *micro–level* LSF is actually a collection of wildly fluctuating force–fields, (loco)motion paths, curved regional geometries and topologies with holes. The micro–level LSF is proposed as an extension of the Lewinian concept: it is characterized by uncertainties and fluctuations, enabled by microscopic time–level, microscopic transition paths, microscopic force–fields, local geometries and varying topologies with holes. To model these fluctuating microscopic LSF–structures, we use three instances of *adaptive path integral*, defining a multi–phase and multi–path (also multi–field and multi–geometry) *transition* process from *intention* to the goal–driven *action.*

We use the new LSF concept to develop modelling framework for motivational dynamics (MD) and induced cognitive dynamics (CD).

According to Heckhausen (see [Hec77]), *motivation* can be thought of as a process of *energizing* and *directing the action.* The process of energizing can be represented by Lewin's *force–field analysis* and Vygotsky's *motive formation* (see [Vyg82, AL91]), while the process of directing can be represented by *hierarchical action control* (see [Ber47, Ber35, Kuh85]).

Motivation processes both precede and coincide with every goal–directed action. Usually these motivation processes include the sequence of the following four feedforward *phases* [Vyg82, AL91]: (*)

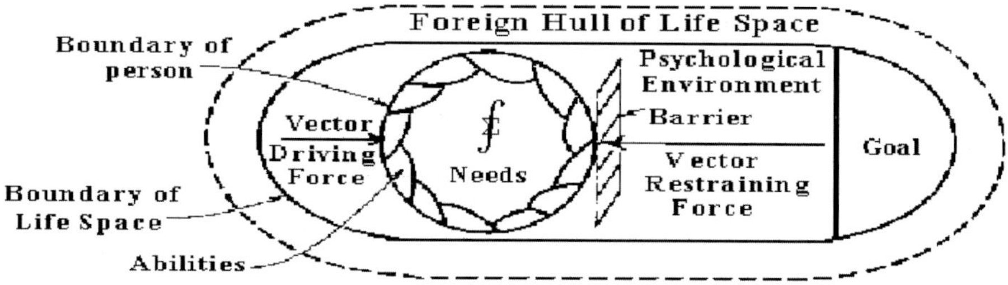

Fig. 6.23. Diagram of the *life space foam*: Lewinian life space with an adaptive path integral acting inside it and generating microscopic fluctuation dynamics.

1. *Intention Formation* $\mathcal{F}$, including: decision making, commitment building, etc.
2. *Action Initiation* $\mathcal{I}$, including: handling conflict of motives, resistance to alternatives, etc.
3. *Maintaining the Action* $\mathcal{M}$, including: resistance to fatigue, distractions, etc.
4. *Termination* $\mathcal{T}$, including parking and avoiding addiction, i.e., staying in control.

With each of the phases $\{\mathcal{F}, \mathcal{I}, \mathcal{M}, \mathcal{T}\}$ in (*), we can associate a *transition propagator* – an ensemble of (possibly crossing) feedforward paths propagating through the 'wood of obstacles' (including topological holes in the LSF, see Figure 6.24), so that the complete *transition functor* $\mathcal{T}A$ is a product of propagators (as well as sum over paths). All the phases–propagators are controlled by a unique *Monitor* feedback process.

In this subsection we propose an *adaptive path integral* formulation for the motivational–transition functor $\mathcal{T}A$. In essence, we sum/integrate over different paths and make a product (composition) of different phases–propagators. Recall that this is the most general description of the general *Markov stochastic process*.

We will also attempt to demonstrate the utility of the same LSF–formalisms in representing cognitive functions, such as memory, learning and decision making. For example, in the classical *Stimulus encoding* $\longrightarrow$ *Search* $\longrightarrow$ *Decision* $\longrightarrow$ *Response* sequence [Ste69, Ash94], the environmental input–triggered *sensory memory* and *working memory* (WM) can be interpreted as operating at the micro–level force–field under the executive control of the *Monitor* feedback, whereas *search* can be formalized as a *control* mechanism guiding retrieval from the long–term memory (LTM, itself shaped by learning) and filtering material relevant to decision making into the WM. The essential measure of these mental processes, the *processing speed* (essentially determined by Sternberg's reaction–time) can be represented by our (loco)motion speed $\dot{x}$.

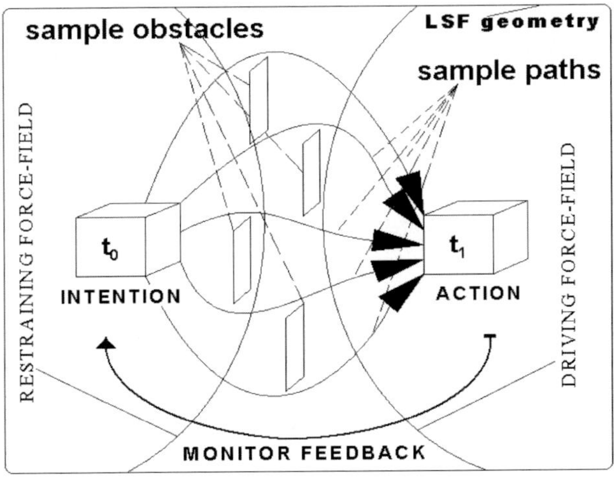

Fig. 6.24. *Transition–propagator* corresponding to each of the motivational phases $\{\mathcal{F}, \mathcal{I}, \mathcal{M}, \mathcal{T}\}$, consisting of an ensemble of feedforward paths propagating through the 'wood of obstacles'. The paths affected by driving and restraining force–fields, as well as by the local LSF–geometry. Transition goes from *Intention*, occurring at a sample time instant t_0, to *Action*, occurring at some later time t_1. Each propagator is controlled by its own *Monitor* feedback. All together they form the transition functor $\mathcal{T}A$.

Six Faces of the Life–Space Foam

The LSF has three forms of appearance: $paths + field + geometries$, acting on both macro–level and micro–level, which is six modes in total. In this section, we develop three least action principles for the macro–LSF–level and three adaptive path integrals for the micro–LSF–level. While developing our psycho–physical formalism, we will address the behavioral issues of motivational fatigue, learning, memory and decision making.

General Formalism

At both macro– and micro–levels, the total LSF represents a union of transition paths, force–fields and geometries, formally written as

$$\begin{aligned} LSF_{total} &:= LSF_{paths} \bigcup LSF_{fields} \bigcup LSF_{geom} \\ &\equiv \oint_{paths} + \oint_{fields} + \oint_{geom} . \end{aligned} \tag{6.264}$$

Corresponding to each of the three LSF–subspaces in (6.264) we formulate:

1. The *least action principle*, to model deterministic and predictive, macro–level MD & CD, giving a unique, global, causal and smooth path–field–geometry on the macroscopic spatio–temporal level; and

2. Associated *adaptive path integral* to model uncertain, fluctuating and probabilistic, micro–level MD & CD, as an ensemble of local paths–fields–geometries on the microscopic spatio–temporal level, to which the global macro–level MD & CD represents both time and ensemble *average* (which are equal according to the *ergodic hypothesis*).

In the proposed formalism, transition paths $x^i(t)$ are affected by the force–fields $\varphi^k(t)$, which are themselves affected by geometry with metric g_{ij}.

Global Macro–Level of LSF_{total}. In general, at the *macroscopic* LSF–level we first formulate the *total action* $S[\Phi]$, the central quantity in our formalism that has psycho–physical dimensions of $Energy \times Time = Effort$, with immediate cognitive and motivational applications: *the greater the action – the higher the speed of cognitive processes and the lower the macroscopic fatigue* (which includes all sources of physical, cognitive and emotional fatigue that influence motivational dynamics). The action $S[\Phi]$ depends on macroscopic paths, fields and geometries, commonly denoted by an abstract field symbol Φ^i. The action $S[\Phi]$ is formally defined as a temporal integral from the *initial* time instant t_{ini} to the *final* time instant t_{fin},

$$S[\Phi] = \int_{t_{ini}}^{t_{fin}} \mathfrak{L}[\Phi]\, dt, \tag{6.265}$$

with *Lagrangian density* given by

$$\mathfrak{L}[\Phi] = \int d^n x\, \mathcal{L}(\Phi_i, \partial_{x^j}\Phi^i),$$

where the integral is taken over all n coordinates $x^j = x^j(t)$ of the LSF, and $\partial_{x^j}\Phi^i$ are time and space partial derivatives of the Φ^i–variables over coordinates.

Second, we formulate the *least action principle* as a minimal variation δ of the action $S[\Phi]$

$$\delta S[\Phi] = 0, \tag{6.266}$$

which, using techniques from the calculus of variations gives, in the form of the so–called Euler–Lagrangian equations, a shortest (loco)motion path, an extreme force–field, and a life–space geometry of minimal curvature (and without holes). In this way, we effectively derive a *unique globally smooth transition functor*

$$\mathcal{TA} : INTENTION_{t_{ini}} \Rrightarrow ACTION_{t_{fin}}, \tag{6.267}$$

performed at a macroscopic (global) time–level from some initial time t_{ini} to the final time t_{fin}.

In this way, we get macro–objects in the global LSF: a single path described Newtonian–like equation of motion, a single force–field described by

Maxwellian–like field equations, and a single obstacle–free Riemannian geometry (with global topology without holes).

For example, recall that in the period 1945–1949 J. Wheeler and R. Feynman developed their *action-at-a-distance electrodynamics* [WF49], in complete experimental agreement with the classical Maxwell's electromagnetic theory, but at the same time avoiding the complications of divergent self–interaction of the Maxwell's theory as well as eliminating its infinite number of field degrees of freedom. In Wheeler–Feynman view, "Matter consists of electrically charged particles," so they found a form for the action directly involving the motions of the charges only, which upon variation would give the Newtonian–like equations of motion of these charges. Here is the expression for this action in the flat space–time, which is in the core of quantum electrodynamics:

$$
\begin{aligned}
S[x; t_i, t_j] &= \frac{1}{2} m_i \int (\dot{x}^i_\mu)^2 \, dt_i + \frac{1}{2} e_i e_j \int \int \delta(I^2_{ij}) \, \dot{x}^i_\mu(t_i) \dot{x}^j_\mu(t_j) \, dt_i dt_j \\
&\text{with} \qquad\qquad\qquad\qquad\qquad\qquad\qquad\qquad (6.268) \\
I^2_{ij} &= \left[x^i_\mu(t_i) - x^j_\mu(t_j)\right] \left[x^i_\mu(t_i) - x^j_\mu(t_j)\right],
\end{aligned}
$$

where $x^i_\mu = x^i_\mu(t_i)$ is the four–vector position of the ith particle as a function of the proper time t_i, while $\dot{x}^i_\mu(t_i) = dx^i_\mu/dt_i$ is the velocity four–vector. The first term in the action (6.268) is the ordinary mechanical action in Euclidean space, while the second term defines the electrical interaction of the charges, representing the Maxwell–like field (it is summed over each pair of charges; the factor $\frac{1}{2}$ is to count each pair once, while the term $i = j$ is omitted to avoid self–action; the interaction is a double integral over a delta function of the square of space–time interval I^2 between two points on the paths; thus, interaction occurs only when this interval vanishes, that is, along light cones [WF49]).

Now, from the point of view of Lewinian geometrical force–fields and (loco)motion paths, we can give the following life–space interpretation to the Wheeler–Feynman action (6.268). The mechanical–like locomotion term occurring at the single time t, needs a covariant generalization from the flat 4D Euclidean space to the nD smooth Riemannian manifold, so it becomes (see e.g., [Iva04])

$$
S[x] = \frac{1}{2} \int_{t_{ini}}^{t_{fin}} g_{ij} \, \dot{x}^i \dot{x}^j \, dt,
$$

where g_{ij} is the Riemannian metric tensor that generates the total 'kinetic energy' of (loco)motions in the life space.

The second term in (6.268) gives the sophisticated definition of Lewinian force–fields that drive the psychological (loco)motions, if we interpret electrical charges e_i occurring at different times t_i as motivational charges – needs.

Local Micro–Level of LSF_{total}. After having properly defined macro–level MD & CD, with a unique transition map F (including a unique motion path, driving field and smooth geometry), we move down to the *microscopic* LSF–level of rapidly fluctuating MD & CD, where we cannot define a unique and smooth path–field–geometry. The most we can do at this level of *fluctuating uncertainty*, is to formulate an adaptive path integral and calculate overall probability amplitudes for ensembles of local transitions from one LSF–point to the neighboring one. This *probabilistic transition micro–dynamics* functor is defined by a multi–path (field and geometry, respectively) and multi–phase *transition amplitude* $\langle Action|Intention\rangle$ of corresponding to the globally–smooth transition map (6.267). This absolute square of this probability amplitude gives the *transition probability* of occurring the final state of $Action$ given the initial state of $Intention$,

$$P(Action|Intention) = |\langle Action|Intention\rangle|^2.$$

The total transition amplitude from the state of $Intention$ to the state of $Action$ is defined on LSF_{total}

$$\mathcal{TA} \equiv \langle Action|Intention\rangle_{total} : INTENTION_{t_0} \Rrightarrow ACTION_{t_1}\,, \tag{6.269}$$

given by adaptive generalization of the Feynman's path integral [FH65, Fey72, Fey98]. The transition map (6.269) calculates the *overall probability amplitude* along a multitude of wildly fluctuating paths, fields and geometries, performing the *microscopic* transition from the micro–state $INTENTION_{t_0}$ occurring at initial micro–time instant t_0 to the micro–state $ACTION_{t_1}$ at some later micro–time instant t_1, such that all micro–time instants fit inside the global transition interval $t_0, t_1, ..., t_s \in [t_{ini}, t_{fin}]$. It is symbolically written as

$$\langle Action|Intention\rangle_{total} := \sum\!\!\!\!\!\!\int \mathcal{D}[w\Phi]\, \mathrm{e}^{\mathrm{i}S[\Phi]}, \tag{6.270}$$

where the Lebesgue integration is performed over all continuous $\Phi^i_{con} = paths + field + geometries$, while summation is performed over all discrete processes and regional topologies Φ^j_{dis}). The symbolic differential $\mathcal{D}[w\Phi]$ in the general path integral (6.270), represents an *adaptive path measure*, defined as a weighted product

$$\mathcal{D}[w\Phi] = \lim_{N\longrightarrow\infty} \prod_{s=1}^{N} w_s d\Phi^i_s, \qquad (i = 1, ..., n = con + dis), \tag{6.271}$$

which is in practice satisfied with a large N corresponding to infinitesimal temporal division of the four motivational phases (*). Technically, the path integral (6.270) calculates the *amplitude* for the transition functor $\mathcal{TA} : Intention \Rrightarrow Action$.

In the exponent of the path integral (6.270) we have the action $S[\Phi]$ and the imaginary unit $i = \sqrt{-1}$ (i can be converted into the real number –1 using the so–called *Wick rotation*, see next subsection).

In this way, we get a range of micro–objects in the local LSF at the short time–level: ensembles of rapidly fluctuating, noisy and crossing paths, force–fields, local geometries with obstacles and topologies with holes. However, by averaging process, both in time and along ensembles of paths, fields and geometries, we recover the corresponding global MD & CD variables.

Infinite–Dimensional Neural Network. The adaptive path integral (6.270) incorporates the *local learning process* according to the standard formula: *New Value = Old Value+Innovation.* The general *weights* $w_s = w_s(t)$ in (6.271) are updated by the *MONITOR* feedback during the transition process, according to one of the two standard neural learning schemes, in which the micro–time level is traversed in discrete steps, i.e., if $t = t_0, t_1, ..., t_s$ then $t + 1 = t_1, t_2, ..., t_{s+1}$:

1. A *self–organized, unsupervised* (e.g., Hebbian–like [Heb49]) learning rule:
$$w_s(t+1) = w_s(t) + \frac{\sigma}{\eta}(w_s^d(t) - w_s^a(t)), \qquad (6.272)$$
where $\sigma = \sigma(t)$, $\eta = \eta(t)$ denote *signal* and *noise*, respectively, while superscripts d and a denote *desired* and *achieved* micro–states, respectively; or
2. A certain form of a *supervised gradient descent learning*:
$$w_s(t+1) = w_s(t) - \eta \nabla J(t), \qquad (6.273)$$
where η is a small constant, called the *step size*, or the *learning rate* and $\nabla J(n)$ denotes the gradient of the 'performance hyper–surface' at the t–th iteration.

Both Hebbian and supervised learning are used for the local decision making process (see below) occurring at the intention formation faze $\mathcal{F}$.

In this way, local micro–level of LSF_{total} represents an infinite–dimensional neural network. In the cognitive psychology framework, our adaptive path integral (6.270) can be interpreted as *semantic integration* (see [BF71, Ash94]).

Motion and Decision Making in LSF_{paths}

On the macro–level in the subspace LSF_{paths} we have the (loco)*motion action principle*
$$\delta S[x] = 0,$$
with the *Newtonian–like action* $S[x]$ given by
$$S[x] = \int_{t_{ini}}^{t_{fin}} dt \, [\frac{1}{2} g_{ij}\, \dot{x}^i \dot{x}^j + \varphi^i(x^i)], \qquad (6.274)$$
where overdot denotes time derivative, so that $\dot{x}^i$ represents *processing speed*, or (loco)motion velocity vector. The first bracket term in (6.274) represents the kinetic energy T,

$$T = \frac{1}{2} g_{ij}\, \dot{x}^i \dot{x}^j,$$

generated by the *Riemannian metric tensor* g_{ij}, while the second bracket term, $\varphi^i(x^i)$, denotes the family of potential force–fields, driving the (loco)motions $x^i = x^i(t)$ (the *strengths* of the fields $\varphi^i(x^i)$ depend on their positions x^i in LSF, see LSF_{fields} below). The corresponding Euler–Lagrangian equation gives the Newtonian–like equation of motion

$$\frac{d}{dt} T_{\dot{x}^i} - T_{x^i} = -\varphi^i_{x^i}, \tag{6.275}$$

(subscripts denote the partial derivatives), which can be put into the standard Lagrangian form

$$\frac{d}{dt} L_{\dot{x}^i} = L_{x^i}, \qquad \text{with} \qquad L = T - \varphi^i(x^i).$$

In the next subsection we use the micro–level implications of the action S[x] as given by (6.274), for dynamical descriptions of the local decision–making process.

On the micro–level in the subspace LSF_{paths}, instead of a single path defined by the Newtonian–like equation of motion (6.275), we have an ensemble of fluctuating and crossing paths with weighted probabilities (of the unit total sum). This ensemble of micro–paths is defined by the simplest instance of our adaptive path integral (6.270), similar to the Feynman's original *sum over histories*,

$$\langle Action|Intention\rangle_{paths} = \oint \mathcal{D}[wx]\, \mathrm{e}^{\mathrm{i}S[x]}, \tag{6.276}$$

where $\mathcal{D}[wx]$ is a functional measure on the *space of all weighted paths*, and the exponential depends on the action $S[x]$ given by (6.274). This procedure can be redefined in a mathematically cleaner way if we Wick–rotate the time variable t to imaginary values $t \mapsto \tau = it$, thereby making all integrals real:

$$\oint \mathcal{D}[wx]\, \mathrm{e}^{\mathrm{i}S[x]} \xrightarrow{Wick} \oint \mathcal{D}[wx]\, \mathrm{e}^{-S[x]}. \tag{6.277}$$

Discretization of (6.277) gives the *thermodynamic–like partition function*

$$Z = \sum_j \mathrm{e}^{-w_j E^j/T}, \tag{6.278}$$

where E^j is the motion energy eigenvalue (reflecting each possible motivational energetic state), T is the temperature–like environmental control parameter, and the sum runs over all motion energy eigenstates (labelled by the index j). From (6.278), we can further calculate all thermodynamic–like and statistical properties of MD & CD (see e.g., [Fey72]), as for example, *transition entropy* $S = k_B \ln Z$, etc.

From cognitive perspective, our adaptive path integral (6.276) calculates all (alternative) pathways of information flow during the transition *Intention* $\longrightarrow$ *Action*.

In the language of transition–propagators, the integral over histories (6.276) can be decomposed into the product of propagators (i.e., Fredholm kernels or Green functions) corresponding to the cascade of the four motivational phases (*)

$$\langle Action|Intention\rangle_{paths} = \oint dx^{\mathcal{F}} dx^{\mathcal{I}} dx^{\mathcal{M}} dx^{\mathcal{T}} K(\mathcal{F},\mathcal{I})K(\mathcal{I},\mathcal{M})K(\mathcal{M},\mathcal{T}), \tag{6.279}$$

satisfying the Schrödinger–like equation (see e.g., [Dir49])

$$i\,\partial_t\langle Action|Intention\rangle_{paths} = H_{Action}\,\langle Action|Intention\rangle_{paths}, \tag{6.280}$$

where H_{Action} represents the Hamiltonian (total energy) function available at the state of *Action*. Here our 'golden rule' is: the higher the H_{Action}, the lower the microscopic fatigue.

In the connectionist language, our propagator expressions (6.279–6.280) represent *activation dynamics*, to which our $Monitor$ process gives a kind of *backpropagation* feedback, a version of the basic supervised learning (6.273).

Mechanisms of Decision–Making under Uncertainty. The basic question about our local decision making process, occurring under uncertainty at the intention formation faze $\mathcal{F}$, is: Which alternative to choose? (see [RBT01, Gro82, Gro99, Gro88, Ash94]). In our path–integral language this reads: Which path (alternative) should be given the highest probability weight w? Naturally, this problem is iteratively solved by the learning process (6.272–6.273), controlled by the $MONITOR$ feedback, which we term *algorithmic approach*.

In addition, here we analyze qualitative mechanics of the local decision making process under uncertainty, as a *heuristic approach*. This qualitative analysis is based on the micro–level interpretation of the Newtonian–like action $S[x]$, given by (6.274) and figuring both processing speed $\dot{x}$ and LTM (i.e., the force–field $\varphi(x)$, see next subsection). Here we consider three different cases:

1. If the potential $\varphi(x)$ is not very dependent upon position $x(t)$, then the more direct paths contribute the most, as longer paths, with higher mean square velocities $[\dot{x}(t)]^2$ make the exponent more negative (after Wick rotation (6.277)).
2. On the other hand, suppose that $\varphi(x)$ does indeed depend on position x. For simplicity, let the potential increase for the larger values of x. Then a direct path does not necessarily give the largest contribution to the overall transition probability, because the integrated value of the potential is higher than over another paths.

3. Finally, consider a path that deviates widely from the direct path. Then $\varphi(x)$ decreases over that path, but at the same time the velocity $\dot{x}$ increases. In this case, we expect that the increased velocity $\dot{x}$ would more than compensate for the decreased potential over the path.

Therefore, the most important path (i.e., the path with the highest weight w) would be one for which any smaller integrated value of the surrounding field potential $\varphi(x)$ is more than compensated for by an increase in kinetic–like energy $\frac{m}{2}\dot{x}^2$. In principle, this is neither the most direct path, nor the longest path, but rather a middle way between the two. Formally, it is the path along which the average Lagrangian is minimal,

$$< \frac{m}{2}\dot{x}^2 + \varphi(x) > \longrightarrow \min, \tag{6.281}$$

i.e., the *path that requires minimal memory* (both LTM and WM, see LSF_{fields} below) and *processing speed.* This mechanical result is consistent with the 'filter theory' of *selective attention* [Bro58], proposed in an attempt to explain a range of the existing experimental results. This theory postulates a low level filter that allows only a limited number of percepts to reach the brain at any time. In this theory, the importance of conscious, directed attention is minimized. The type of attention involving low level filtering corresponds to the concept of *early selection* [Bro58].

Although we termed this 'heuristic approach' in the sense that we can instantly feel both the processing speed $\dot{x}$ and the LTM field $\varphi(x)$ involved, there is clearly a psycho–physical rule in the background, namely the averaging minimum relation (6.281).

From the decision making point of view, all possible paths (alternatives) represent the *consequences* of decision making. They are, by default, *short–term consequences*, as they are modelled in the micro–time–level. However, the path integral formalism allows calculation of the *long–term consequences*, just by extending the integration time, $t_{fin} \longrightarrow \infty$. Besides, this *averaging decision mechanics* – choosing the optimal path – actually performs the 'averaging lift' in the LSF: from micro– to the macro–level.

Force–Fields and Memory in LSF_{fields}

At the macro–level in the subspace LSF_{fields} we formulate the *force–field action principle*

$$\delta S[\varphi] = 0, \tag{6.282}$$

with the action $S[\varphi]$ dependent on Lewinian force–fields $\varphi^i = \varphi^i(x)$ $(i = 1, ..., N)$, defined as a temporal integral

$$S[\varphi] = \int_{t_{ini}}^{t_{fin}} \mathfrak{L}[\varphi]\, dt, \tag{6.283}$$

with Lagrangian density given by

$$\mathfrak{L}[\varphi] = \int d^n x \, \mathcal{L}(\varphi_i, \partial_{x^j}\varphi^i),$$

where the integral is taken over all n coordinates $x^j = x^j(t)$ of the LSF, and $\partial_{x^j}\varphi^i$ are partial derivatives of the field variables over coordinates.

On the micro–level in the subspace LSF_{fields} we have the Feynman–type *sum over fields* φ^i $(i = 1, ..., N)$ given by the adaptive path integral

$$\langle Action|Intention\rangle_{fields} = \oint \mathcal{D}[w\varphi]\, \mathrm{e}^{\mathrm{i}S[\varphi]} \xrightarrow{Wick} \oint \mathcal{D}[w\varphi]\, \mathrm{e}^{-S[\varphi]}, \quad (6.284)$$

with action $S[\varphi]$ given by temporal integral (6.283). (Choosing special forms of the force–field action $S[\varphi]$ in (6.284) defines micro–level MD & CD, in the LSF_{fields} space, that is similar to standard quantum–field equations, see e.g., [Ram90].) The corresponding partition function has the form similar to (6.278), but with field energy levels.

Regarding topology of the force fields, we have in place $n-$*categorical Lagrangian–field structure* on the Riemannian LSF manifold M,

$$\Phi^i : [0,1] \to M, \; \Phi^i : \Phi^i_0 \mapsto \Phi^i_1,$$

generalized from the recursive homotopy dynamics (see [II06b]) above, using

$$\begin{array}{ccc} \frac{d}{dt} f_{\dot{x}^i} = f_{x^i} & \longrightarrow & \partial_\mu \left(\frac{\partial \mathcal{L}}{\partial_\mu \Phi^i} \right) = \frac{\partial \mathcal{L}}{\partial \Phi^i}, \\ \text{with} \quad [x_0, x_1] & \longrightarrow & [\Phi^i_0, \Phi^i_1]. \end{array}$$

Relationship between Memory and Force–Fields. As already mentioned, the subspace LSF_{fields} is related to our *memory storage* [Ash94]. Its global macro–level represents the *long–term memory* (LTM), defined by the least action principle (6.282), related to *cognitive economy* in the model of *semantic memory* [Rat78, CQ69]. Its local micro–level represents *working memory* (WM), a limited–capacity 'bottleneck' defined by the adaptive path integral (6.284). According to our formalism, each of Miller's 7 ± 2 units [Mil56] of the local WM are adaptively stored and averaged to give the global LTM capacity (similar to the physical notion of potential). This averaging memory lift, from WM to LTM represents *retroactive interference*, while the opposite direction, given by the path integral (6.284) itself, represents *proactive interference*. Both retroactive and proactive interferences are examples of the impact of cognitive contexts on memory. Motivational contexts can exert their influence, too. For example, a reduction in task–related recall following the completion of the task is one of the clearest examples of force–field influences on memory: the amount of details remembered of a task declines as the force–field tension to complete the task is reduced by actually completing it.

Once defined, the global LTM potential $\varphi = \varphi(x)$ is then affecting the locomotion transition paths through the path action principle (6.274), as well as general learning (6.272–6.273) and decision making process (6.281).

On the other hand, the two levels of LSF_{fields} fit nicely into the two levels of processing framework, as presented by [CL72], as an alternative to theories of separate stages for sensory, working and long–term memory. According to the *levels of processing framework*, stimulus information is processed at multiple levels simultaneously depending upon its characteristics. In this framework, our macro–level memory field, defined by the fields action principle (6.282), corresponds to the *shallow memory*, while our micro–level memory field, defined by the adaptive path integral (6.284), corresponds to the *deep memory*.

Geometries, Topologies and Noise in LSF_{geom}

On the macro–level in the subspace LSF_{geom} representing an $n-$dimensional smooth manifold M with the global Riemannian metric tensor g_{ij}, we formulate the *geometrical action principle*

$$\delta S[g_{ij}] = 0,$$

where $S = S[g_{ij}]$ is the $n-$dimensional *geodesic action* on M,

$$S[g_{ij}] = \int d^n x \sqrt{g_{ij}\, dx^i dx^j}. \tag{6.285}$$

The corresponding Euler–Lagrangian equation gives the *geodesic equation* of the *shortest path* in the manifold M,

$$\ddot{x}^i + \Gamma^i_{jk}\, \dot{x}^j\, \dot{x}^k = 0,$$

where the symbol Γ^i_{jk} denotes the so–called *affine connection* which is the source of *curvature*, which is geometrical description for *noise* (see [Ing97, Ing98]). The higher the local curvatures of the LSF–manifold M, the greater the noise in the life space. This noise is the source of our micro–level fluctuations. It can be internal or external; in both cases it curves our micro–LSF.

Otherwise, if instead we choose an $n-$dimensional Hilbert–like action (see [MTW73]),

$$S[g_{ij}] = \int d^n x \sqrt{\det |g_{ij}|} R, \tag{6.286}$$

where R is the scalar curvature (derived from Γ^i_{jk}), we get the $n-$dimensional Einstein–like equation: $G_{ij} = 8\pi T_{ij}$, where G_{ij} is the Einstein–like tensor representing geometry of the LSF manifold M (G_{ij} is the trace–reversed Ricci tensor R_{ij}, which is itself the trace of the *Riemann curvature tensor* of the manifold M), while T_{ij} is the $n-$dimensional *stress–energy–momentum* tensor. This equation explicitly states that *psycho–physics of the LSF is proportional to its geometry.* T_{ij} is important quantity, representing motivational *energy*, geometry–imposed *stress* and *momentum* of (loco)motion. As before, we have our 'golden rule': *the greater the $T_{ij}-$components, the higher the speed of cognitive processes and the lower the macroscopic fatigue.*

The choice between the geodesic action (6.285) and the Hilbert action (6.286) depends on our interpretation of time. If time is not included in the LSF manifold M (non–relativistic approach) then we choose the geodesic action. If time is included in the LSF manifold M (making it a relativistic–like $n-$dimensional space–time) then the Hilbert action is preferred. The first approach is more related to the information processing and the working memory. The later, space–time approach can be related to the long–term memory: we usually recall events closely associated with the times of their happening.

On the micro–level in the subspace LSF_{geom} we have the adaptive *sum over geometries*, represented by the path integral over all local (regional) Riemannian metrics $g_{ij} = g_{ij}(x)$ varying from point to point on M (modulo diffeomorphisms),

$$\langle Action|Intention\rangle_{geom} = \oint \mathcal{D}[wg_{ij}]\,\mathrm{e}^{\mathrm{i}S[g_{ij}]} \xrightarrow{Wick} \oint \mathcal{D}[wg_{ij}]\,\mathrm{e}^{-S[g_{ij}]}, \tag{6.287}$$

where $\mathcal{D}[g_{ij}]$ is diffeomorphism equivalence class of $g_{ij}(x) \in M$.

To include the topological structure (e.g., a number of holes) in M, we can extend (6.287) as

$$\langle Action|Intention\rangle_{geom/top} = \sum_{\text{topol.}} \oint \mathcal{D}[wg_{ij}]\,\mathrm{e}^{\mathrm{i}S[g_{ij}]}, \tag{6.288}$$

where the topological sum is taken over all connectedness–components of M determined by the *Euler characteristic* χ of M. This type of integral defines the *theory of fluctuating geometries*, a propagator between $(n-1)-$dimensional boundaries of the $n-$dimensional manifold M. One has to contribute a meaning to the integration over geometries. A key ingredient in doing so is to approximate (using simplicial approximation and Regge calculus [MTW73]) in a natural way the smooth structures of the manifold M by piecewise linear structures (mostly using topological simplices Δ). In this way, after the Wick–rotation (6.277), the integral (6.287–6.288) becomes a *simple statistical system*, given by partition function $Z = \sum_{\Delta} \frac{1}{C_{\Delta}} \mathrm{e}^{-S_{\Delta}}$, where the summation is over all triangulations Δ of the manifold M, while C_T is the order of the automorphism group of the performed triangulation.

Micro–Level Geometry: the source of noise and stress in LSF. The subspace LSF_{geom} is the source of noise, fluctuations and obstacles, as well as psycho–physical stress. Its micro–level is adaptive, reflecting the human ability to efficiently act within the noisy environment and under the stress conditions. By averaging it produces smooth geometry of certain curvature, which is at the same time the smooth psycho–physics. This macro–level geometry directly affects the memory fields and indirectly affects the (loco)motion transition paths.

The Mental Force Law. As an effective summary of this section, we state that the psychodynamic transition functor $\mathcal{TA} : INTENTION_{t_{ini}} \Rightarrow ACTION_{t_{fin}}$, defined by the generic path integral (6.270), can be interpreted as a *mental force law*, analogous to our musculo–skeletal *covariant force law*, $F_i = mg_{ij}a^j$, and its associated *covariant force functor* $\mathcal{F}_* : TT^*M \to TTM$ [II06a].

7

Quantum Gravity and Cosmological Dynamics

In this Chapter we apply the methods elaborated so far to present the 'Holy Grail' of modern physical and cosmological science, the search for the 'theory of everything' and the 'true' cosmological dynamics.

7.1 Search for Quantum Gravity

7.1.1 What Is Quantum Gravity?

The landscape of fundamental physics has changed substantially during the last few decades. Not long ago, our understanding of the weak and strong interactions was very confused, while general relativity was almost totally disconnected from the rest of physics and was empirically supported by little more than its three classical tests. Then two things have happened. The $SU(3) \times SU(2) \times U(1)$ *Standard Model* has found a dramatic empirical success, showing that *quantum field theory* (QFT) is capable of describing all accessible fundamental physics, or at least all non–gravitational physics. At the same time, *general relativity* (GR) has undergone an extraordinary 'renaissance', finding widespread application in astrophysics and cosmology, as well as novel vast experimental support – so that today GR is basic physics needed for describing a variety of physical systems we have access to, including advanced technological systems [Ash97].

These two parallel developments have moved fundamental physics to a position in which it has rarely been in the course of its history: We have today a group of fundamental laws, the Standard Model and GR, which – even if it cannot be regarded as a satisfactory global picture of Nature– is perhaps the best confirmed set of fundamental theories after Newton's universal gravitation and Maxwell's electromagnetism. More importantly, there aren't today experimental facts that openly challenge or escape this set of fundamental laws. In this unprecedented state of affairs, a large number of theoretical physicists from different backgrounds have begun to address *the* piece of the puzzle which is clearly missing: combining the two halves of the picture and understanding the quantum properties of the gravitational field. Equivalently, understanding the quantum properties of space–time. Interest and researches in quantum gravity have thus increased sharply in recent years. And the problem of understanding what is a quantum space–time is today at the core of fundamental physics.

Today we have some well developed and reasonably well defined tentative theories of quantum gravity. String theory and loop quantum gravity are the two major examples. Within these theories definite physical results have been obtained, such as the explicit computation of the 'quanta of geometry' and the derivation of the black hole entropy formula. Furthermore, a number of fresh new ideas, like noncommutative geometry, have entered quantum gravity. For an overview of the problem of quantum gravity, see [Ish97].

7.1.2 Main Approaches to Quantum Gravity

String Theory

Recall from the previous Chapter that string theory is by far the research direction which is presently most investigated. String theory presently exists at two levels. First, there is a well developed set of techniques that define the string perturbation expansion over a given metric background. Second, the understanding of the non–perturbative aspects of the theory has much increased in recent years [Pol95] and in the string community there is a widespread faith, supported by numerous indications, in the existence of a yet-to-be-found full non–perturbative theory, capable of generating the perturbation expansion. There are attempts of constructing this non–perturbative theory, generically denoted M theory. The currently popular one is Matrix–theory, of which it is far too early to judge the effectiveness [Mat02, IKK97].

The claim that string theory solves QG is based on two facts. First, the string perturbation expansion includes the graviton. More precisely, one of the string modes is a massless spin two, and helicity ± 2, particle. Such a particle necessarily couples to the energy–momentum tensor of the rest of the fields [Wei64, Wei80] and gives general relativity to a first approximation. Second, the perturbation expansion is consistent if the background geometry over which the theory is defined satisfies a certain consistency condition; this condition turns out to be a high energy modification of the Einstein's equations. The hope is that such a consistency condition for the perturbation expansion will emerge as a full–fledged dynamical equation from the yet–to–be–found non–perturbative theory.

From the point of view of the problem of quantum gravity, the relevant physical results from string theory are two [Rov97]:

Black hole entropy. The most remarkable physical results for quantum gravity is the derivation of the *Bekenstein–Hawking formula* for the entropy of a black hole as a function of the horizon area. This beautiful result has been obtained by [SV96b], and has then been extended in various directions. The result indicates that there is some unexpected internal consistency between string theory and QFT on curved space.

Microstructure of space–time. There are indications that in string theory the space–time continuum is meaningless below the Planck length. An old set of results on very high energy scattering amplitudes indicates that there is no way of probing the space–time geometry at very short distances. What happens is that in order to probe smaller distance one needs higher energy, but at high energy the string 'opens up from being a particle to being a true string' which is spread over space–time, and there is no way of focusing a string's collision within a small space–time region.

More recently, in the non–perturbative formulation of the Matrix–theory [Mat02], the space–time coordinates of the string x^i are replaced by matrices $(X^i)^n_m$. This can perhaps be viewed as a new interpretation of the space–time structure. The continuous space–time manifold emerges only in the long distance region, where these matrices are diagonal and commute; while the space–time appears to have a noncommutative discretized structure in the short distance regime. This features are still poorly understood, but they have intriguing resonances with noncommutative geometry [CDS98] and loop quantum gravity [Rov98].

A key difficulty in string theory is the lack of a complete non–perturbative formulation. During the last year, there has been excitement for some tentative non–perturbative formulations [Mat02]; but it is far too early to understand if these attempts will be successful. Many previously highly acclaimed ideas have been rapidly forgotten.

A distinct and even more serious difficulty of string theory is the lack of a *background independent* formulation of the theory. In the words of Ed Witten:

> 'Finding the right framework for an intrinsic, background independent formulation of string theory is one of the main problems in string theory, and so far has remained out of reach... This problem is fundamental because it is here that one really has to address the question of what kind of geometrical object the string represents.'

Most of string theory is conceived in terms of a theory describing excitations over this or that background, possibly with connections between different backgrounds. This is also true for (most) non–perturbative formulations such as Matrix theory. For instance, the (bosonic part of the) Lagrangian of Matrix–theory is

$$L \sim \frac{1}{2}\mathrm{Tr}\left(\dot{X}^2 + \frac{1}{2}[X^i, X^j]^2\right). \tag{7.1}$$

The indices that label the matrices X^i are raised and lowered with a Minkowski metric, and the theory is Lorentz invariant. In other words, the Lagrangian is really

$$L \sim \frac{1}{2}\mathrm{Tr}\left(g^{00}g^{ij}\dot{X}_i\dot{X}_j + \frac{1}{2}g^{ik}g^{jl}[X_i,X_j][X_k,X_l]\right), \tag{7.2}$$

where g is the flat metric of the background. This shows that there is a non–dynamical metric, and an implicit flat background in the action of the theory.

However, the world is not formed by a fixed background over which things happen. The background itself is dynamical. In particular, for instance, the theory should contain quantum states that are quantum superpositions of different backgrounds – and presumably these states play an essential role in the deep quantum gravitational regime, namely in situations such as the *Big–Bang*[1] or the final phase of black hole evaporation. The absence of a fixed background in nature (or active diffeomorphism invariance) is the key general lessons we have learned from gravitational theories [Rov97].

There has been a burst of recent activity in an outgrowth of string theory denoted *string cosmology* by [Ven91]. The aim of string cosmology is to extract physical consequences from string theory by applying it to the Big–Bang. The idea is to start from a Minkowski flat universe; show that this is unstable and therefore will run away from the flat (false–vacuum) state. The evolution then leads to a cosmological model that starts off in an inflationary phase. This scenario is described using mini–superspace technology, in the context of the low energy theory that emerge as limit of string theory. Thus, first one freezes all the massive modes of the string, then one freezes all massless modes except the zero modes (the spatially constant ones), obtaining a finite dimensional theory, which can be quantized non–perturbatively.

[1] Recall that the Big–Bang is the scientific theory that the universe emerged from a tremendously dense and hot state about 13.7 billion years ago. The theory is based on the observations indicating the expansion of space in accord with the Robertson-Walker model of general relativity, as indicated by the Hubble red–shift of distant galaxies taken together with the cosmological principle. Extrapolated into the past, these observations show that the universe has expanded from a state in which all the matter and energy in the universe was at an immense temperature and density. Physicists do not widely agree on what happened before this, although general relativity predicts a gravitational singularity.

The term Big–Bang is used both in a narrow sense to refer to a point in time when the observed expansion of the universe (Hubble's law) began calculated to be 13.7 billion (1.37×1010) years ago (2%) - and in a more general sense to refer to the prevailing cosmological paradigm explaining the origin and expansion of the universe, as well as the composition of primordial matter through nucleosynthesis as predicted by the Alpher–Bethe–Gamow theory [ABG48]. From this model, George Gamow was able to predict in 1948 the existence of cosmic microwave background radiation (CMB). The CMB was discovered in 1964 and corroborated the Big Bang theory, giving it more credence over its chief rival, the steady state theory. For details, see section on Cosmological Dynamics below.

Loop Quantum Gravity

The second most popular approach to quantum gravity, and the most popular among relativists, is loop quantum gravity [Rov98]. Loop quantum gravity is presently the best developed alternative to string theory. Like strings, it is not far from a complete and consistent theory and it yields a corpus of definite physical predictions, testable in principle, on quantum space–time.

Loop quantum gravity, however, attacks the problem from the opposite direction than string theory. It is a non-perturbative and background independent theory to start with. In other words, it is deeply rooted into the conceptual revolution generated by general relativity. In fact, successes and problems of loop quantum gravity are complementary to successes and problems of strings. Loop quantum gravity is successful in providing a consistent mathematical and physical picture of non perturbative quantum space–time; but the connection to the low energy dynamics is not yet completely clear.

The general idea on which loop quantum gravity is based is the following. The core of quantum mechanics is not identified with the structure of (conventional) QFT, because conventional QFT presupposes a background *metric* space–time, and is therefore immediately in conflict with GR. Rather, it is identified with the general structure common to all quantum systems. The core of GR is identified with the absence of a fixed observable background space–time structure, namely with active diffeomorphism invariance. Loop quantum gravity is thus a quantum theory in the conventional sense: a Hilbert space and a set of quantum (field) operators, with the requirement that its classical limit is GR with its conventional matter couplings. But it is not a QFT over a metric manifold. Rather, it is a 'quantum field theory on a differentiable manifold', respecting the manifold's invariances and where only coordinate independent quantities are physical.

Technically, loop quantum gravity is based on two inputs [Rov98, Rov97]:

- The formulation of classical GR based on the Ashtekar connection [Ash86, Ash87, Ash91]. The version of the connection now most popular is not the original complex one, but an evolution of the same, in which the connection is real.
- The choice of the holonomies of this connection, denoted *loop variables*, as basic variables for the quantum gravitational field [RS88].

This second choice determines the peculiar kind of quantum theory being built. Physically, it corresponds to the assumption that excitations with support on a loop are normalizable states. This is the key technical assumption on which everything relies.

It is important to notice that this assumption fails in conventional 4D *Yang–Mills theory*, because loop-like excitations on a metric manifold are too singular: the field needs to be smeared in more dimensions [Rov97]. Equivalently, the linear closure of the loop states is a 'far too big' non-separable state space. This fact is the major source of some particle physicists's suspicion at loop quantum gravity. What makes GR different from 4D Yang–Mills theory, however, is non–perturbative diffeomorphism invariance. The gauge invariant states, in fact, are not localized at all – they are, pictorially speaking, smeared by the (gauge) diffeomorphism group all over the coordinates manifold. More precisely, factoring away the diffeomorphism group takes us down from the state space of the loop excitations, which is 'too big', to a separable physical state space of the right size. Thus, the consistency of the loop construction relies heavily on diffeomorphism invariance. In other words, the diff–invariant invariant loop states (more precisely, the diff–invariant spin network states) are not physical excitations of a field *on* space–time. They are excitations *of* space–time itself.

Loop quantum gravity was briefly described by [Rov97] as follows:

Definition of theory. The mathematical structure of the theory has been put on a very solid basis. Early difficulties have been overcome. In particular, there were three major problems in the theory: the lack of a well defined scalar product, the overcompleteness of the loop basis, and the difficulty of treating the reality conditions.

- The problem of the lack of a scalar product on the Hilbert space has been solved with the definition of a diffeomorphism invariant measure on a space of connections [AL95]. Later, it has also became clear that the same scalar product can be defined in a purely algebraic manner [DR96]. The state space of the theory is therefore a genuine Hilbert space $\mathcal{H}$.
- The overcompleteness of the loop basis has been solved by the introduction of the spin network states [RS95]. A spin network is a graph carrying labels (related to $SU(2)$ representations and called 'colors') on its links and its nodes.
  Each spin network defines a spin network state, and the spin network states form a (genuine, non-overcomplete) orthonormal basis in $\mathcal{H}$.
- The difficulties with the reality conditions have been circumvented by the use of the real formulation [Bar94, Bar95a, Bar95b, Thi96].
  The kinematics of loop quantum gravity is now defined with a level of rigor characteristic of mathematical physics [AI92, ALM95] and the theory can be defined using various alternative techniques [DR96, DeP97].

Hamiltonian constraint. A rigorous definition version of the Hamiltonian constraint equation has been constructed. This is anomaly free, in the sense that the constraints algebra closes (but see later on). The Hamiltonian has the crucial properties of acting on nodes only, which implies that its action is naturally discrete and combinatorial [RS88, RS94]. This fact is at the roots of the existence of exact solutions [RS88], and of the possible finiteness of the theory

Matter. The old hope that QFT divergences could be cured by QG has recently received an interesting corroboration. The matter part of the Hamiltonian constraint is well–defined without need of renormalization. Thus, a main possible stumbling block is over: infinities did not appear in a place where they could very well have appeared [Rov97].

Black hole entropy. The first important physical result in loop quantum gravity is a computation of black hole entropy [Kra97, Rov96a, Rov96b].

Quanta of geometry. A very exciting development in quantum gravity in the last years has been by the computations of the quanta of geometry. That is, the computation of the discrete eigenvalues of area and volume.

In quantum gravity, any quantity that depends on the metric becomes an operator. In particular, so do the area A of a given (physically defined) surface, or the volume V of a given (physically defined) spatial region. In loop quantum gravity, these operators can be written explicitly. They are mathematically well defined self–adjoint operators in the Hilbert space $\mathcal{H}$. We know from quantum mechanics that certain physical quantities are quantized, and that we can compute their discrete values by computing the eigenvalues of the corresponding operator. Therefore, if we can compute the eigenvalues of the area and volume operators, we have a physical prediction on the possible quantized values that these quantities can take, at the Planck scale. These eigenvalues have been computed in loop quantum gravity. Here is for instance the main sequence of the spectrum of the area

$$A_{\mathbf{j}} = 8\pi\gamma\,\hbar G \sum_i \sqrt{j_i(j_i+1)}. \tag{7.3}$$

$\mathbf{j} = (j_1, \dots, j_n)$ is an $n-$tuplet of half–integers, labeling the eigenvalues, G and $\hbar$ are the Newton and Planck constants, and γ is a dimensionless free parameter, denoted the so–called *Immirzi parameter* [Imm97], not determined by the theory. A similar result holds for the volume. The spectrum (7.3) has been rederived and completed using various different techniques [DR96]. These spectra represent solid results of loop quantum gravity. Under certain additional assumptions on the behavior of area and volume operators in the presence of matter, these results can be interpreted as a corpus of detailed quantitative predictions on hypothetical Planck scale observations.

Besides its direct relevance, the quantization of the area and thee volume is of interest because it provides a physical picture of quantum space–time. The states of the spin network basis are eigenstates of some area and volume operators. We can say that a spin network carries quanta of area along its links, and quanta of volume at its nodes. The magnitude of these quanta is determined by the coloring. For instance, the half–integers $j_1 \dots j_n$ in (7.3) are the coloring of the spin network's links that cross the given surface. Thus, a quantum space–time can be decomposed in a basis of states that can be visualized as made by quanta of volume (the intersections) separated by quanta of area (the links). More precisely, we can view a spin network as sitting on the *dual* of a cellular decomposition of physical space. The nodes of the spin network sit in the center of the 3–cells, and their coloring determines the (quantized) 3–cell's volume. The links of the spin network cut the faces of the cellular decomposition, and their color **j** determine the (quantized) areas of these faces via equation (7.3).

7.1.3 Traditional Approaches to Quantum Gravity

Discrete Approaches

Discrete quantum gravity is the program of regularizing classical GR in terms of some lattice theory, quantize this lattice theory, and then study an appropriate continuum limit, as one may do in QCD. There are three main ways of discretizing GR.

Regge Calculus

Regge introduced the idea of triangulating space–time by means of a simplicial complex and using the lengths l_i of the links of the complex as gravitational variables [Reg61]. The theory can then be quantized by integrating over the lengths l_i of the links. For a recent review and extensive references see [WT92]. More recent work has focused in problems such as the geometry of Regge superspace [HWM97] and choice of the integration measure.

Dynamical Triangulations

Alternatively, one can keep the length of the links fixed, and capture the geometry by means of the way in which the simplices are glued together, namely by the triangulation. The *Einstein–Hilbert action* of Euclidean gravity is approximated by a simple function of the total number of simplices and links, and the theory can be quantized summing over distinct triangulations (for a detailed introduction, see [ACM98]). There are two coupling constants in the theory, roughly corresponding to the Newton and cosmological constants. These define a two dimensional space of theories. The theory has a nontrivial continuum limit if in this parameter space there is a critical point corresponding to a second order phase transition. The theory has phase transition and a critical point. The transition separates a phase with crumpled space–times from a phase with 'elongated' spaces which are effectively 2D, with characteristic of a *branched polymer* [BS95, AJL01c]. This polymer structure is surprisingly the same as the one that emerges from loop quantum gravity at short scale. Near the transition, the model appears to produce 'classical' S^4 space–times, and there is evidence for scaling, suggesting a continuum behavior.

State Sum Models

A third road for discretizing GR was opened by a celebrated paper by [PR68]. They started from a Regge discretization of 3D GR and introduced a second discretization, by posing the so–called *Ponzano–Regge ansatz* that the lengths l assigned to the links are discretized as well, in half–integers in Planck units

$$l = \hbar G\, j, \qquad j = 0, \qquad \frac{1}{2}, 1, \dots \tag{7.4}$$

(Planck length is $\hbar G$ in 3D.) The half integers j associated to the links are denoted 'coloring' of the triangulation. Coloring can be viewed as the assignment of a $SU(2)$ irreducible representation to each link of the Regge triangulation. The elementary cells of the triangulation are tetrahedra, which have six links, colored with six $SU(2)$ representations. $SU(2)$ representation theory naturally assigns a number to a sextuplet of representations: the Wigner $6-j$ symbol. Rather magically, the product over all tetrahedra of these $6-j$ symbols converges to (the real part of the exponent of) the Einstein–Hilbert action. Thus, Ponzano and Regge were led to propose a quantization of 3D GR based on the partition function

$$Z \sim \sum_{coloring} \prod_{tetrahedra} 6 - j(\text{color of the tetrahedron}), \tag{7.5}$$

where we have neglected some coefficients for simplicity. They also provided arguments indicating that this sum is independent from the triangulation of the manifold.

The formula (7.5) is simple and elegant, and the idea has recently had many surprising and interesting developments. 3D GR was quantized as a topological field theory by Ed Witten in [Wit88a] and using loop quantum gravity in [AHS89]. The *Ponzano–Regge quantization* based on equation (7.5) was shown to be essentially equivalent to the TQFT quantization in [Oog92], and to the loop quantum gravity in [Rov93] (for an extensive discussion of 3D quantum gravity, see [CN95]).

It turns out that the Ponzano–Regge ansatz (7.4) can be *derived* from loop quantum gravity [Rov93]. Indeed, (7.4) is the 2D version of the 3D formula (7.3), which gives the quantization of the area. Therefore, a key result of quantum gravity of the last years, namely the quantization of the geometry, derived in the loop formalism from a full fledged non–perturbative quantization of GR, was anticipated as an ansatz by the intuition of Ponzano and Regge.

Hawking's Euclidean Quantum Gravity

Hawking's Euclidean quantum gravity is the approach based on his formal sum over Euclidean geometries (i.e., an Euclidean path integral)

$$Z \sim \mathcal{N} \int \mathcal{D}[g] \, \mathrm{e}^{-\int d^4x \sqrt{g} R[g]}. \tag{7.6}$$

Firstly, Hawking's picture of quantum gravity as a sum–over–space–times, continues to provide a powerful intuitive reference point for most of the research related to quantum gravity. Indeed, many approaches can be seen as attempts to replace the ill–defined and non–renormalizable formal integral (7.6) with a well defined expression. The dynamical triangulation approach (see above) and the spin foam approach (see below) are examples of attempts to realize Hawking's intuition. Influence of Euclidean quantum gravity can also be found in the *Atiyah axioms* for TQFT.

Secondly, this approach can be used as an approximate method for describing certain regimes of non–perturbative quantum space–time physics, even if the fundamental dynamics is given by a more complete theory. In this spirit, Hawking and collaborators have continued the investigation of phenomena such as, for instance, pair creation of black holes in a background de Sitter space–time.

Effective Perturbative Quantum Gravity

If we expand classical GR around, say, the Minkowski metric,

$$g_{\mu\nu}(x) = \eta_{\mu\nu} + h_{\mu\nu}(x),$$

and construct a conventional QFT for the field $h_{\mu\nu}(x)$, we get, as it is well know, a non renormalizable theory. A small but intriguing group of papers has recently appeared, based on the proposal of treating this perturbative theory seriously, as a respectable low energy effective theory by its own. This cannot solve the deep problem of understanding the world in general relativistic quantum terms. But it can still be used for studying quantum properties of space–time in some regimes. This view has been advocated in a convincing way by John Donoghue, who has developed effective field theory methods for extracting physics from non renormalizable quantum GR [Don96].

QFT in Curved Space–Time

Quantum field theory in curved space–time is by now a reasonably established theory (see, e.g., [Wal94, BD82, Ful89], predicting physical phenomena of remarkable interest such as particle creation, vacuum polarization effects and Hawking's black-hole radiance [Haw75]. To be sure, there is no direct nor indirect experimental observation of any of these phenomena, but the theory is quite credible as an approximate theory, and many theorists in different fields would probably agree that these predicted phenomena are likely to be real.

The most natural and general formulation of the theory is within the algebraic approach [Haa92], in which the primary objects are the local observables and the states of interest may all be treated on equal footing (as positive linear functionals on the algebra of local observables), even if they do not belong to the same Hilbert space.

The great merit of QFT on curved space–time is that it has provided us with some very important lessons. The key lesson is that in general one loses the notion of a single preferred quantum state that could be regarded as *the* 'vacuum'; and that the concept of 'particle' becomes vague and/or observer-dependent in a gravitational context. In a gravitational context, *vacuum* and *particle* are necessarily ill defined or approximate concepts. It is perhaps regrettable that this important lesson has not been yet absorbed by many scientists working in fundamental theoretical physics [Rov97].

New Approaches to Quantum Gravity

Noncommutative Geometry

Noncommutative geometry is a research program in mathematics and physics which has recently received wide attention and raised much excitement. The program is based on the idea that space–time may have a noncommutative structure at the Planck scale. A main driving force of this program is the radical, volcanic and extraordinary sequence of ideas of A. Connes [Con94]. Connes observes that what we know about the structure of space–time derives from our knowledge of the fundamental interactions: special relativity derives from a careful analysis of Maxwell theory; Newtonian space–time and general relativity, derived both from a careful analysis of the gravitational interaction. Recently, we have learned to describe weak and strong interactions in terms of the $SU(3)\times SU(2)\times U(1)$ Standard Model. Connes suggests that the Standard Model might hide information on the minute structure of space–time as well. By making the hypothesis that the Standard Model symmetries reflect the symmetry of a noncommutative microstructure of space–time, Connes and Lott are able to construct an exceptionally simple and beautiful version of the Standard Model itself, with the impressive result that the Higgs field appears automatically, as the components of the Yang–Mills connection in the internal 'noncommutative' direction [CL90]. The theory admits a natural extension in which the space–time metric, or the gravitational field, is dynamical, leading to GR [CC96].

The key idea behind a non-commutative space–time is to use algebra instead of geometry in order to describe spaces. Consider a topological (Hausdorf) space M. Consider all continuous functions f on M. These form an algebra A, because they can be multiplied and summed, and the algebra is commutative. According to a celebrated result, due to Gel'fand, knowledge of the algebra A is equivalent to knowledge of the space M, i.e., M can be reconstructed from A. In particular, the points x of the manifold can be obtained as the 1D irreducible representations x of A, which are all of the form $x(f) = f(x)$. Thus, we can use the algebra of the functions, instead of using the space. In a sense, notices Connes, the algebra is more physical, because we never deal with space–time: we deal with fields, or coordinates, over space–time. But one can capture Riemannian geometry as well, algebraically. Consider the Hilbert space H formed by all the spinor fields on a given Riemannian (spin) manifold. Let D be the (curved) Dirac operator, acting on H. We can view A as an algebra of (multiplicative) operators on H. Now, from the triple (H, A, D), which Connes calls 'spectral triple', one can reconstruct the Riemannian manifold. In particular, it is not difficult to see that the distance between two points x and y can be obtained from these data by

$$d(x, y) = sup_{\{f \in A, ||D, f|| < 1\}} \, |x(f) - y(f)|, \tag{7.7}$$

a beautiful surprising algebraic definition of distance. A non-commutative space–time is the idea of describing space–time by a spectral triple in which the algebra A is a non-commutative algebra.

Remarkably, the gravitational field is captured, together with the Yang–Mills field, and the Higgs fields, by a suitable *Dirac operator* D [CC96], and the full action is given simply by the trace of a very simple function of the Dirac operator.

Even if we disregard noncommutativity and the Standard Model, the above construction represents an intriguing re–formulation of conventional GR, in which the geometry is described by the Dirac operator instead than the metric tensor. This formulation has been explored in [Lan98], where it is noticed that the eigenvalues of the Dirac operator are diffeomorphism invariant functions of the geometry, and therefore represent true observables in Euclidean GR. Their Poisson bracket algebra can be explicitly computed in terms of the energy–momentum eigenspinors. Surprisingly, the Einstein equations turn out to be captured by the requirement that the energy momentum of the eigen–spinors scale linearly with the eigenvalues.

Variants of Connes's version of the idea of non commutative geometry and noncommutative coordinates have been explored by many authors (see, e.g., [DFR94]) and intriguing connections with string theory have been suggested [CDS98, FG94].

Null–Surface Formulation

A second new set of ideas comes from [FKN95]. These authors have discovered that the (conformal) information about the geometry is captured by suitable families of null hypersurfaces in space–time, and have been able to reformulate GR as a theory of self–interacting families of surfaces. A remarkable aspect of the theory is that physical information about the space–time interior is transferred to null infinity, along null geodesics. Thus, the space–time interior is described in terms of how we would (literally) 'see it' from outside. This description is diffeomorphism invariant, and addresses directly the relational localization characteristic of GR: the space–time location of a region is determined dynamically by the gravitational field and is captured by when and where we see the space–time region from infinity. This idea may lead to interesting and physically relevant diffeomorphism invariant observables in quantum gravity. A discussion of the quantum gravitational fuzziness of the space–time points determined by this perspective can be found in [FKN97].

Spin Foam Models

From the mathematical point of view, the problem of quantum gravity is to understand what is QFT on a differentiable manifold without metric. A class of well understood QFT's on manifolds exists. These are the topological quantum field theories (TQFT). Topological field theories are particularly simple field theories. They have as many fields as gauges and therefore no local degree of freedom, but only a finite number of global degrees of freedom. An example is GR in 3D, say on a torus (the theory is equivalent to a Chern–Simons theory). In 3D, the Einstein equations require that the geometry is flat, so there are no gravitational waves. Nevertheless, a careful analysis reveals that the radii of the torus are dynamical variables, governed by the theory. Witten has noticed that theories of this kind give rise to interesting quantum models [Wit88b], and [Ati89] has provided a beautiful axiomatic definition of a TQFT. Concrete examples of TQFT have been constructed using Hamiltonian, combinatorial and path integral methods. The relevance of TQFT for quantum gravity has been suggested by many and the recent developments have confirmed these suggestions.

Recall that TQFT is a diffeomorphism invariant QFT. Sometimes, the expression TQFT is used to indicate all diffeomorphism invariant QFT's. This has lead to a widespread, but incorrect belief that any diffeomorphism invariant QFT has a finite number of degrees of freedom, unless the invariance is somehow broken, for instance dynamically. This belief is wrong. The problem of quantum gravity is precisely to define a diffeomorphism invariant QFT having an infinite number degrees of freedom and 'local' excitations. Locality in a gravity theory, however, is different from locality in conventional field theory. This point is often source of confusion. Here is Rovelli's clarification [Rov97]:

- In a conventional field theory on a metric space, the degrees of freedom are local in the sense that they can be localized on the metric manifold (an electromagnetic wave is here or there in Minkowski space).
- In a diffeomorphism invariant field theory such as general relativity, the degrees of freedom are still local (gravitational waves exist), but they are not localized with respect to the manifold. They are nevertheless localized with respect to each other (a gravity wave is three meters apart from another gravity wave, or from a black hole).
- In a topological field theory, the degrees of freedom are not localized at all: they are global, and in finite number (the radius of a torus is not in a particular position on the torus).

The first TQFT directly related to quantum gravity was defined by [TV92]. The Turaev–Viro model is a mathematically rigorous version of the 3D

Ponzano-Regge quantum gravity model described above. In the Turaev–Viro theory, the sum (7.5) is made finite by replacing $SU(2)$ with quantum $SU(2)_q$ (with a suitable q). Since $SU(2)_q$ has a finite number if irreducible representations, this trick, suggested by [Oog92, Oog92], makes the sum finite. The extension of this model to four dimensions has been actively searched for a while and has finally been constructed by [CY93], again following Ooguri's ideas. The Crane–Yetter (CY) model is the first example of 4D TQFT. It is defined on a simplicial decomposition of the manifold. The variables are spins ('colors') attached to faces and tetrahedra of the simplicial complex. Each 4–simplex contains 10 faces and 5 tetrahedra, and therefore there are 15 spins associated to it. The action is defined in terms of the quantum Wigner $15 - j$ symbols, in the same manner in which the Ponzano–Regge action is constructed in terms of products of $6 - j$ symbols.

$$Z \sim \sum_{coloring} \prod_{4-simplices} 15 - j(\text{color of the } 4 - \text{simplex}), \tag{7.8}$$

(where we have disregarded various factors for simplicity). Crane and Yetter introduced their model independently from loop quantum gravity. However, recall that loop quantum gravity suggests that in 4 dimensions the naturally discrete geometrical quantities are area and volume, and that it is natural to extend the Ponzano–Regge model to 4D by assigning colors to faces and tetrahedra.

The CY model is not a quantization of 4D GR, nor could it be, being a TQFT in strict sense. Rather, it can be formally derived as a quantization of $SU(2)$ BF theory. BF theory is a topological field theory with two fields, a connection A, with curvature F, and a 2–form B [Hor89], with action

$$S[A, B] = \int B \wedge F. \tag{7.9}$$

However, there is a strict relation between GR and BF. If we add to $SO(3,1)$ BF theory the constraint that the 2–form B is the product of two tetrad 1–forms

$$B = E \wedge E, \tag{7.10}$$

we get precisely GR. This observation has lead many to suggest that a quantum theory of gravity could be constructed by a suitable modification of quantum BF theory [Bae96c]. This suggestion has become very plausible, with the following construction of the *spin foam models*.

The key step in development of the spin foam models was taken by [Bar97], studying the 'quantum geometry' of the simplices that play a role in loop quantum gravity. Barbieri discovered a simple relation between the quantum operators representing the areas of the faces of the tetrahedra. This relation turns out to be the quantum version of the constraint (7.10), which turns BF theory into GR. Barret and Crane[BC97] added the Barbieri relation to (the $SO(3,1)$ version of) the CY model. This is equivalent to replacing the the 15-j

Wigner symbol, with a different function A_{BC} of the colors of the 4–simplex. This replacement defines a 'modified TQFT', which has a chance of having general relativity as its classical limit.

The Barret–Crane model is not a TQFT in strict sense. In particular, it is not independent from the triangulation. Thus, a continuum theory has to be formally defined by some suitable sum over triangulations

$$Z \sim \sum_{triang} \sum_{coloring} \prod_{4-simplices} A_{BC}(\text{color of the } 4 - \text{simplex}). \tag{7.11}$$

This essential aspect of the construction, however, is not yet understood.

The Barret Crane model can virtually be obtained also from loop quantum gravity. This is an unexpected convergence of two very different lines of research. Loop quantum gravity is formulated canonically in the frozen time formalism. While the frozen time formalism is in principle complete, in practice it is cumbersome, and anti-intuitive. Our intuition is four dimensional, not three dimensional. An old problem in loop quantum gravity has been to derive a space–time version of the theory. A space–time formulation of quantum mechanics is provided by the sum over histories. A sum over histories can be derived from the Hamiltonian formalism, as Feynman did originally. Loop quantum gravity provides a mathematically well defined Hamiltonian formalism, and one can therefore follow Feynman steps and construct a sum over histories quantum gravity starting from the loop formalism. This has been done in [RR97]. The sum over histories turns out to have the form of a sum over surfaces.

More precisely, the transition amplitude between two spin network states turns out to be given by a sum of terms, where each term can be represented by a (2D) branched 'colored' surface in space–time. A branched colored surface is formed by elementary surface elements carrying a label, that meet on edges, also carrying a labelled; edges, in turn meet in vertices (or branching points, see Figure 7.1). The contribution of one such surfaces to the sum over histories is the product of one term per each branching point of the surface. The branching points represent the 'vertices' of this theory, in the sense of Feynman. The contribution of each vertex can be computed algebraically from the 'colors' (half integers) of the adjacent surface elements and edges. Thus, space–time loop quantum gravity is defined by the partition function

$$Z \sim \sum_{surfaces} \sum_{colorings} \prod_{vertices} A_{loop}(\text{color of the vertex}) \tag{7.12}$$

The vertex A_{loop} is determined by a matrix elements of the Hamiltonian constraint. The fact that one gets a sum over surfaces is not too surprising, since the time evolution of a loop is a surface. Indeed, the time evolution of a spin network (with colors on links and nodes) is a surface (with colors on surface elements and edges) and the Hamiltonian constraint generates branching points in the same manner in which conventional Hamiltonians generate the vertices of the Feynman diagrams.

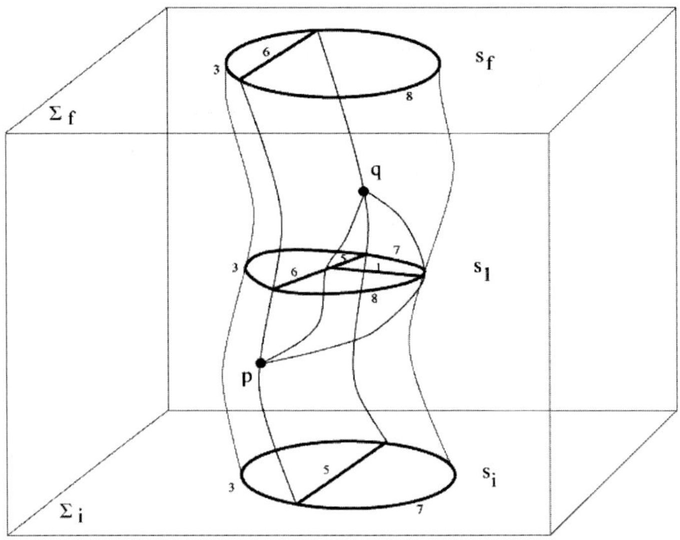

Fig. 7.1. A branched surface with two vertices.

Now, (7.12) has the same structure of the Barret–Crane model (7.8). To see this, simply notice that we can view each branched colored surface as located on the lattice *dual* to a triangulation. Then each vertex correspond to a 4-simplex; the coloring of the two models matches exactly (elementary surfaces → faces, edges → tetrahedra); and summing over surfaces corresponds to summing over triangulations. The main difference is the different weight at the vertices. The Barret–Crane vertex A_{BC} can be read as a covariant definition a Hamiltonian constraint in loop quantum gravity.

Thus, the space–time formulation of loop quantum GR is a simple modification of a TQFT. This approach provides a 4D pictorial intuition of quantum space–time, analogous to the Feynman graphs description of quantum field dynamics. John Baez has introduced the term 'spin foam' for the branched colored surfaces of the model, in honor of John Wheeler's intuitions on the quantum microstructure of space–time. Spin foams are a precise mathematical implementation of Wheeler's 'space–time foam' suggestions.

Hawking's Black Hole Entropy

A focal point of the research in quantum gravity in the last years has been the discussion of black hole (BH) entropy. This problem has been discussed from a large variety of perspectives and within many different research programs.

Let us very briefly recall the origin of the problem. In *classical* GR, future event horizons behave in a manner that has a peculiar thermodynamical flavor. This remark, together with a detailed physical analysis of the behavior of hot matter in the vicinity of horizons, prompted Bekenstein to suggest that there is entropy associated to every horizon. The suggestion was first consider ridicule, because it implies that a black hole is hot and radiates. But then S. Hawking, in a celebrated work [Haw75], showed that QFT in curved space–time predicts that a black hole emits thermal radiation, precisely at the temperature predicted by Bekenstein, and Bekenstein courageous suggestion was fully vindicated. Since then, the entropy of a BH has been indirectly computed in a surprising variety of manners, to the point that BH entropy and BH radiance are now considered almost an established fact by the community, although, of course, they were never observed nor, presumably, they are going to be observed soon. This confidence, perhaps a bit surprising to outsiders, is related to the fact thermodynamics is powerful in indicating general properties of systems, even if we do not control its microphysics. Many hope that the Bekenstein–Hawking radiation could play for quantum gravity a role analogous to the role played by the black body radiation for quantum mechanics. Thus, indirect arguments indicate that a Schwarzschild BH has an entropy

$$S = \frac{1}{4}\frac{A}{\hbar G} \tag{7.13}$$

The remaining challenge is to derive this formula from first principles [Rov97].

Later in the book we will continue our exposition of various approaches to quantum gravity.

7.2 Loop Quantum Gravity

7.2.1 Introduction to Loop Quantum Gravity

Recall (from subsection 7.1 above) that C. Rovelli developed (in the last decade of the 20th Century) the so–called loop approach to quantum gravity (see [Rov98] and references therein). The first announcement of this approach was given in [RS87]. Together with string theory, this approach provides another serious candidate theory of quantum gravity. It provides a physical picture of Planck scale quantum geometry, calculation techniques, definite quantitative predictions, and a tool for discussing classical problems such as black hole thermodynamics.

String theory and *loop quantum gravity* differ not only because they explore distinct physical hypotheses, but also because they are expressions of two separate communities of scientists, which have sharply distinct prejudices, and view the problem of quantum gravity in surprisingly different manners. As Rovelli says: "I heard the following criticism to loop quantum gravity: 'Loop quantum gravity is certainly physically wrong, because:
(1) it is not supersymmetric, and
(2) is formulated in four dimensions'.
But experimentally, the world still insists on looking four–dimensional and not supersymmetric. In my opinion, people should be careful of not being

blinded by their own speculation, and mistaken interesting hypotheses (such as supersymmetry and high–dimensions) for established truth. But string theory may claim extremely remarkable theoretical successes and is today the leading and most widely investigated candidate theory of quantum gravity" [Rov98].

High energy physics has obtained spectacular successes during this Century, culminated with the (far from linear) establishment of quantum field theory as the general form of dynamics and with the comprehensive success of the $SU(3) \times SU(2) \times U(1)$ *Standard Model*. Thanks to this success, now a few decades old, physics is in a condition in which it has been very rarely: there are no experimental results that clearly challenge, or clearly escape, the present fundamental theory of the world. The theory we have encompasses virtually everything – except gravitational phenomena. From the point of view of a particle physicist, gravity is then simply the last and weakest of the interactions. It is natural to try to understand its quantum properties using the strategy that has been so successful for the rest of microphysics, or variants of this strategy. The search for a conventional quantum field theory capable of embracing gravity has spanned several decades and, through an adventurous sequence of twists, moments of excitement and disappointments, has lead to string theory. The foundations of string theory are not yet well understood; and it is not yet entirely clear how a supersymmetric theory in 10 or 11 dimensions can be concretely used for deriving comprehensive univocal predictions about our world.

In string theory, gravity is just one of the excitations of a string (or other extended object) living over some background metric space. The existence of such background metric space, over which the theory is defined, is needed for the formulation and for the interpretation of the theory, not only in perturbative string theory, but in the recent attempts of a non-perturbative definition of the theory, such as M theory, as well, in my understanding. Thus, for a physicist with a high energy background, the problem of quantum gravity is now reduced to an aspect of the problem of understanding what is the mysterious non–perturbative theory that has perturbative string theory as its perturbation expansion, and how to extract information on Planck scale physics from it.

For a *relativist*, on the other hand, the idea of a fundamental description of gravity in terms of physical excitations over a background metric space sounds physically very wrong. The key lesson learned from general relativity is that there is no background metric *over* which physics happens. The world is more complicated than that. Indeed, for a relativist, general relativity is much more than the field theory of a particular force. Rather, it is the discovery that certain classical notions about space and time are inadequate at the fundamental level; they require modifications which are possibly as basics as the ones that quantum mechanics introduced. One of such inadequate notions is precisely the notion of a background metric space (flat or curved), over which physics happens. This profound conceptual shift has led to the understanding of relativistic gravity, to the discovery of black holes, to relativistic astrophysics and to modern cosmology.

From Newton to the beginning of this Century, physics has had a solid foundation in a small number of key notions such as space, time, causality and matter. In spite of substantial evolution, these notions remained rather stable and self-consistent. In the first quarter of this Century, quantum theory and general relativity have modified this foundation in depth. The two theories have obtained solid success and vast experimental corroboration, and can be now considered as established knowledge. Each of the two theories modifies the conceptual foundation of classical physics in a (more or less) internally consistent manner, but we do not have a novel conceptual foundation capable of supporting *both* theories. This is why we do not yet have a theory capable of predicting what happens in the physical regime in which both theories are relevant, the regime of Planck scale phenomena, 10^{-33} cm.

General relativity has taught us not only that space and time share the property of being dynamical with the rest of the physical entities, but also (more crucially) that space–time location is relational only. Quantum mechanics has taught us that any dynamical entity is subject to Heisenberg's uncertainty at small scale. Thus, we need *a relational notion of a quantum space–time*, in order to understand Planck scale physics. Thus, for a relativist, the problem of quantum gravity is the problem of bringing a vast conceptual revolution, started with quantum mechanics and with general relativity, to a conclusion and to a new synthesis (see [Rov97, Smo97b].) In this synthesis, the notions of space and time need to be deeply reshaped, in order to keep into account what we have learned with both our present 'fundamental' theories.

Unlike perturbative or non–perturbative string theory, loop quantum gravity is formulated without a background space–time, and is thus a genuine attempt to grasp what is quantum space–time at the fundamental level. Accordingly, the notion of space–time that emerges from the theory is profoundly different from the one on which conventional quantum field theory or string theory are based.

According to Rovelli, the main merit of string theory is that it provides a superbly elegant unification of known fundamental physics, and that it has a well defined perturbation expansion, finite order by order. Its main incompleteness is that its non–perturbative regime is poorly understood, and that we do not have a background–independent formulation of the string theory. In a sense, we do not really know what is the theory we are talking about. Because of this poor understanding of the non perturbative regime of the theory, Planck scale physics and genuine quantum gravitational phenomena are not easily controlled: except for a few computations, there is not much Planck scale physics derived from string theory so far. There are, however, two sets of remarkable physical results. The first is given by some very high energy scattering amplitudes that have been computed. An intriguing aspect of these results is that they indirectly suggest that geometry below the Planck scale cannot be probed – and thus in a sense does not exist – in string theory. The second physical achievement of string theory (which followed the D–branes revolution) is the derivation of the Bekenstein–Hawking black hole entropy formula for certain kinds of black holes.

On the other hand, the main merit of loop quantum gravity is that it provides a well–defined and mathematically rigorous formulation of a background–independent non–perturbative generally covariant quantum field theory. The theory provides a physical picture and quantitative predictions on the world at the Planck scale. The main incompleteness of the theory regards the dynamics, formulated in several variants. The theory has lead to two main sets of physical results. The first is the derivation of the (Planck scale) eigenvalues of geometrical quantities such as areas and volumes. The second is the derivation of black hole entropy for 'normal' black holes (but only up to the precise numerical factor).

The main physical hypotheses on which loop quantum gravity relies are only general relativity and quantum mechanics. In other words, loop quantum gravity is a rather conservative 'quantization' of general relativity, with its traditional matter couplings. In this sense, it is very different from string theory, which is based on a strong physical hypothesis with no direct experimental support '*that the world is made by strings*'.

Finally, strings and loop gravity, may not necessarily be competing theories: there might be a sort of complementarity, at least methodological, between the two. This is due to the fact that the open problems of string theory regard its background–independent formulation, and loop quantum gravity is precisely a set of techniques for dealing non–perturbatively with background independent theories. Perhaps the two approaches might even, to some extent, converge. Undoubtedly, there are similarities between the two theories: first of all the obvious fact that both theories start with the idea that the relevant excitations at the Planck scale are one dimensional objects – call them loops or strings. [Smo97a] also explored the possible relations between string theory and loop quantum gravity.

Loop quantum gravity is a quantum field theory on a differentiable 4–manifold. We have learned with general relativity that the space–time metric and the gravitational field are the same physical entity. Thus, a quantum theory of the gravitational field is a quantum theory of the space–time metric as well. It follows that quantum gravity cannot be formulated as a quantum field theory over a *metric manifold*, because there is no (classical) metric manifold whatsoever in a regime in which gravity (and therefore the metric) is a quantum variable [Rov98].

One could conventionally split the space–time metric into two terms: one to be consider as a background, which gives a metric structure to space–time; the other to be treated as a fluctuating quantum field. This, indeed, is the procedure on which old perturbative quantum gravity, perturbative strings, as well as current non-perturbative string theories (M–theory), are based. In following this path, one assumes, for instance, that the causal structure of space–time is determined by the underlying background metric alone, and not by the full metric. Contrary to this, in loop quantum gravity we assume that the identification between the gravitational field and the metric–causal structure of space–time holds, and must be taken into account, in the quantum regime as well. Thus, no split of the metric is made, and there is no background metric on space–time.

We can still describe space–time as a (differentiable) manifold (a space without metric structure), over which quantum fields are defined. A classical metric structure will then be defined by expectation values of the gravitational field operator. Thus, the problem of quantum gravity is the problem of understanding what is *a quantum field theory on a manifold*, as opposed to quantum field theory on a metric space. This is what gives quantum gravity its distinctive flavor, so different than ordinary quantum field theory. In all versions of ordinary quantum field theory, the metric of space–time plays an essential role in the construction of the basic theoretical tools (creation and annihilation operators, canonical commutation relations, gaussian measures, propagators); these tools cannot be used in quantum field over a manifold.

Technically, the difficulty due to the absence of a background metric is circumvented in loop quantum gravity by defining the quantum theory as a representation of a Poisson algebra of classical observables, which can be defined without using a background metric. The idea that the quantum algebra at the basis of quantum gravity is not the canonical commutation relation algebra, but the Poisson algebra of a different set of observables has long been advocated by [Ish84], whose ideas have been very influential in the birth of loop quantum gravity. The algebra on which loop gravity is the loop algebra [RS90].

In choosing the *loop algebra* as the basis for the quantization, we are essentially assuming that Wilson loop operators are well defined in the Hilbert space of the theory. In other words, that certain states concentrated on one dimensional structures (loops and graphs) have finite norm. This is a subtle non trivial assumptions entering the theory. It is the key assumption that characterizes loop gravity. If the approach turned out to be wrong, it will likely be because this assumption is wrong. The Hilbert space resulting from adopting this assumption is not a Fock space. Physically, the assumption corresponds to the idea that quantum states can be decomposed on a basis of *Faraday lines*–excitations (as Minkowski QFT states can be decomposed on a particle basis).

Furthermore, this is an assumption that *fails* in conventional quantum field theory, because in that context well defined operators and finite norm states need to be smeared in at least three dimensions, and 1D objects are too singular. The fact that at the basis of loop gravity there is a mathematical assumption that fails for *conventional Yang–Mills quantum field theory* is probably at the origin of some of the resistance that loop quantum gravity encounters among some high energy theorists. What distinguishes gravity from Yang–Mills (YM) theories, however, and makes this assumption viable in gravity even if it fails for YM theory is *diffeomorphism invariance*. The loop states are singular states that span a 'huge' non–separable state space. Non–perturbative diffeomorphism invariance plays two roles. First, it wipes away the infinite redundancy. Second, it 'smears' a loop state into a knot state, so that the physical states are not really concentrated in one dimension, but are, in a sense, smeared all over the entire manifold by the non–perturbative diffeomorphisms [Rov98].

Conventional field theories are not invariant under a diffeomorphism acting on the dynamical fields. Every field theory, suitably formulated, is trivially invariant under a diffeomorphism acting on *everything*. General relativity, on the contrary is invariant under such transformations. More precisely, every general relativistic theory has this property. Thus, diffeomorphism invariance is not a feature of just the gravitational field: it is a feature of physics, once the existence of relativistic gravity is taken into account. Thus, one can say that the gravitational field is not particularly 'special' in this regard, but that diffeomorphism invariance is a property of the physical world that can be disregarded only in the approximation in which the dynamics of gravity is neglected.

Now, *diffeomorphism invariance* is the technical implementation of a physical idea, due to Einstein. The idea is a deep modification of the pre–general–relativistic (pre–GR) notions of space and time. In pre–GR physics, we assume that physical objects can be localized in space and time with respect to a fixed non–dynamical background structure. Operationally, this background space–time can be defined by means of physical reference–system objects, but these objects are considered as dynamically decoupled from the physical system that one studies. This conceptual structure fails in a relativistic gravitational regime. In general relativistic physics, the physical objects are localized in space and time only with respect to each other. Therefore if we 'displace' all dynamical objects in space–time at once, we are not generating a different state, but an equivalent mathematical description of the same physical state. Hence, diffeomorphism invariance.

Accordingly, a physical state in GR is not 'located' somewhere. Pictorially, GR is not physics over a stage, it is the dynamical theory of (or including) the stage itself. Loop quantum gravity is an attempt to implement this subtle relational notion of space–time localization in quantum field theory. In particular, the basic quantum field theoretical excitations cannot be localized somewhere as, say, photons are. They are quantum excitations of the 'stage' itself, not excitations over a stage. Intuitively, one can understand from this discussion how knot theory plays a role in the theory. First, we define quantum states that correspond to loop–like excitations of the gravitational field, but then, when factoring away diffeomorphism invariance, the location of the loop becomes irrelevant. The only remaining information contained in the loop is then its knotting (a knot is a loop up to its location). Thus, diffeomorphism invariant physical states are labelled by knots. A knot represent an elementary quantum excitation of space. It is not here or there, since it *is* the space with respect to which here and there can be defined. A knot state is an elementary quantum of space. In this manner, loop quantum gravity ties the new notion of space and time introduced by general relativity with quantum mechanics.

7.2.2 Formalism of Loop Quantum Gravity

The starting point is classical general relativity formulated in terms of the *Ashtekar phase–space formalism* (see [Ash91]). Recall that classical general relativity can be formulated in the phase–space form as follows.

We fix a 3D manifold M (compact and without boundaries) and consider a smooth real $SU(2)-$connection $A^i_a(x)$ and a vector density $\tilde{E}^a_i(x)$ (transforming in the vector representation of $SU(2)$) on M. We use $a, b, \ldots = 1, 2, 3$ for spatial indices and $i, j, \ldots = 1, 2, 3$ for internal indices. The internal indices can be viewed as labelling a basis in the Lie algebra of $SU(2)$ or the three axis of a local triad. We indicate coordinates on M with x. The relation between these fields and conventional metric gravitational variables is as follows: $\tilde{E}^a_i(x)$ is the (densitized) inverse triad, related to the 3D metric $g_{ab}(x)$ of constant–time surfaces by

$$g\,g^{ab} = \tilde{E}^a_i \tilde{E}^b_i, \tag{7.14}$$

where g is the determinant of g_{ab}; and

$$A^i_a(x) = \Gamma^i_a(x) + \gamma k^i_a(x); \tag{7.15}$$

$\Gamma^i_a(x)$ is the *spin connection* associated to the triad,
(defined by $\partial_{[a}e^i_{b]} = \Gamma^i_{[a}e_{b]j}$, where e^i_a is the triad).
$k^i_a(x)$ is the extrinsic curvature of the constant time three surface.

In (7.15), γ is a constant, denoted the Immirzi parameter, that can be chosen arbitrarily (it will enter the Hamiltonian constraint). Different choices for γ yield different versions of the formalism, all equivalent in the classical domain. If we choose γ to be equal to the imaginary unit, $\gamma = \sqrt{-1}$, then A is the standard *Ashtekar connection*, which can be shown to be the projection of the self–dual part of the 4D spin connection on the constant time surface. If we choose $\gamma = 1$, we get the real Barbero connection. The Hamiltonian constraint of Lorentzian general relativity has a particularly simple form in the $\gamma = \sqrt{-1}$ formalism, while the Hamiltonian constraint of Euclidean general relativity has a simple form when expressed in terms of the $\gamma = 1$ real connection. Other choices of γ are viable as well. In particular, it has been argued that the quantum theory based on different choices of γ are genuinely physical inequivalent, because they yield 'geometrical quanta' of different magnitude [Rov98]. Apparently, there is a unique choice of γ yielding the correct 1/4 coefficient in the *Bekenstein–Hawking formula*.

The spinorial version of the Ashtekar variables is given in terms of the Pauli matrices $\sigma_i, i = 1, 2, 3$, or the $\mathfrak{su}(2)$ generators $\tau_i = -\frac{i}{2}\,\sigma_i$, by

$$\tilde{E}^a(x) = -\mathrm{i}\, \tilde{E}_i^a(x)\, \sigma_i = 2\tilde{E}_i^a(x)\, \tau_i, \tag{7.16}$$

$$A_a(x) = -\frac{\mathrm{i}}{2}\, A_a^i(x)\, \sigma_i = A_a^i(x)\, \tau_i \;. \tag{7.17}$$

Thus, $A_a(x)$ and $\tilde{E}^a(x)$ are 2×2 anti–Hermitian complex matrices.

The theory is invariant under local $SU(2)$ gauge transformations, three-dimensional diffeomorphisms of the manifold on which the fields are defined, as well as under (coordinate) time translations generated by the Hamiltonian constraint. The full dynamical content of general relativity is captured by the three constraints that generate these gauge invariances (see [Ash91]).

7.2.3 Loop Algebra

Certain classical quantities play a very important role in the quantum theory. These are: the trace of the holonomy of the connection, which is labelled by loops on the three manifold; and the higher order loop variables, obtained by inserting the E field (in n distinct points, or 'hands') into the holonomy trace. More precisely, given a loop α in M and the points $s_1, s_2, \ldots, s_n \in \alpha$ we define:

$$\mathcal{T}[\alpha] = -\mathrm{Tr}[U_\alpha], \tag{7.18}$$

$$\mathcal{T}^a[\alpha](s) = -\mathrm{Tr}[U_\alpha(s,s)\tilde{E}^a(s)] \tag{7.19}$$

and, in general

$$\mathcal{T}^{a_1 a_2}[\alpha](s_1, s_2) = -\mathrm{Tr}[U_\alpha(s_1,s_2)\tilde{E}^{a_2}(s_2)U_\alpha(s_2,s_1)\tilde{E}^{a_1}(s_1)], \tag{7.20}$$

$$\mathcal{T}^{a_1 \ldots a_N}[\alpha](s_1 \ldots s_N) = -\mathrm{Tr}[U_\alpha(s_1,s_N)\tilde{E}^{a_N}(s_N)U_\alpha(s_N,s_{N-1}) \ldots \tilde{E}^{a_1}(s_1)]$$

where $U_\alpha(s_1, s_2) \sim \mathcal{P}\exp\{\int_{s_1}^{s_2} A_a(\alpha(s))ds\}$ is the parallel propagator of A_a along α, defined by

$$\frac{d}{ds}U_\alpha(1,s) = \frac{d\alpha_a(s)}{ds}\; A_a(\alpha(s))\; U_\alpha(1,s). \tag{7.21}$$

These are the loop observables, previously introduced in YM theories.

The loop observables coordinate the phase space and have a closed Poisson algebra, denoted the loop algebra. This algebra has a remarkable geometrical flavor. For instance, the Poisson bracket between $\mathcal{T}[\alpha]$ and $\mathcal{T}^a[\beta](s)$ is non vanishing only if $\beta(s)$ lies over α; if it does, the result is proportional to the holonomy of the Wilson loops obtained by joining α and β at their intersection (by rerouting the 4 legs at the intersection). More precisely

$$\{\mathcal{T}[\alpha], \mathcal{T}^a[\beta](s)\} = \Delta^a[\alpha, \beta(s)]\; \left[\mathcal{T}[\alpha\#\beta] - \mathcal{T}[\alpha\#\beta^{-1}]\right]. \tag{7.22}$$

Here

$$\Delta^a[\alpha, x] = \int ds\; \frac{d\alpha^a(s)}{ds}\; \delta^3(\alpha(s), x) \tag{7.23}$$

is a vector distribution with support on α and $\alpha\#\beta$ is the loop obtained starting at the intersection between α and β, and following first α and then β. β^{-1} is β with reversed orientation.

A (non–$SU(2)$ gauge invariant) quantity that plays a role in certain aspects of the theory, particularly in the regularization of certain operators, is obtained by integrating the E field over a two dimensional surface S

$$E[S,f] = \int_S dS_a \tilde{E}^a_i f^i, \tag{7.24}$$

where f is a function on the surface S, taking values in the Lie algebra of $SU(2)$. In alternative to the full loop observables (7.18),(7.19),(7.20), one also can take the holonomies and $E[S,f]$ as elementary variables.

7.2.4 Loop Quantum Gravity

The kinematic of a quantum theory is defined by an algebra of 'elementary' operators (such as x and $i\hbar d/dx$, or creation and annihilation operators) on a Hilbert space $\mathcal{H}$. The physical interpretation of the theory is based on the connection between these operators and classical variables, and on the interpretation of $\mathcal{H}$ as the space of the quantum states. The dynamics is governed by a Hamiltonian, or, as in general relativity, by a set of quantum constraints, constructed in terms of the elementary operators. To assure that the quantum Heisenberg equations have the correct classical limit, the algebra of the elementary operator has to be isomorphic to the Poisson algebra of the elementary observables. This yields the *heuristic quantization rule*: 'promote Poisson brackets to commutators'. In other words, define the quantum theory as a linear representation of the Poisson algebra formed by the elementary observables. The kinematics of the quantum theory is defined by a unitary representation of the loop algebra.

We can start *à la Schrödinger,* by expressing quantum states by means of the amplitude of the connection, namely by means of functionals $\Psi(A)$ of the (smooth) connection. These functionals form a linear space, which we promote to a Hilbert space by defining a inner product. To define the inner product, we choose a particular set of states, which we denote 'cylindrical states' and begin by defining the scalar product between these.

Pick a graph Γ, say with n links, denoted $\gamma_1 \dots \gamma_n$, immersed in the manifold M. For technical reasons, we require the links to be analytic. Let $U_i(A) = U_{\gamma_i}$, $i = 1, \dots, n$ be the parallel transport operator of the connection A along γ_i. $U_i(A)$ is an element of $SU(2)$. Pick a function $f(g_1 \dots g_n)$ on $[SU(2)]^n$. The graph Γ and the function f determine a functional of the connection as follows

$$\psi_{\Gamma,f}(A) = f(U_1(A), \dots, U_n(A)), \tag{7.25}$$

(these states are called cylindrical states because they were previously introduced as cylindrical functions for the definition of a cylindrical measure).

Notice that we can always 'enlarge the graph', in the sense that if Γ is a subgraph of Γ' we can write

$$\psi_{\Gamma,f}(A) = \psi_{\Gamma',f'}(A), \tag{7.26}$$

by simply choosing f' independent from the U_i's of the links which are in Γ' but not in Γ. Thus, given any two cylindrical functions, we can always view them as having the same graph (formed by the union of the two graphs). Given this observation, we define the scalar product between any two cylindrical functions, by

$$(\psi_{\Gamma,f}, \psi_{\Gamma,h}) = \int_{SU(2)^n} dg_1 \dots dg_n \; \overline{f(g_1 \dots g_n)}\, h(g_1 \dots g_n). \tag{7.27}$$

where dg is the Haar measure on $SU(2)$. This scalar product extends by linearity to finite linear combinations of cylindrical functions. It is not difficult to show that (7.27) defines a well defined scalar product on the space of these linear combinations. Completing the space of these linear combinations in the Hilbert norm, we get a Hilbert space $\mathcal{H}$. This is the (unconstrained) quantum state space of loop gravity.[2] $\mathcal{H}$ carries a natural unitary representation of the diffeomorphism group and of the group of the local $SU(2)$ transformations, obtained transforming the argument of the functionals. An important property of the scalar product (7.27) is that it is invariant under both these transformations.

$\mathcal{H}$ is non-separable. At first sight, this may seem as a serious obstacle for its physical interpretation. But we will see below that after factoring away diffeomorphism invariance we may get a separable Hilbert space. Also, standard spectral theory holds on $\mathcal{H}$, and it turns out that using spin networks (discussed below) one can express $\mathcal{H}$ as a direct sum over finite dimensional subspaces which have the structure of Hilbert spaces of spin systems; this makes practical calculations very manageable.

Finally, we will use a Dirac notation and write

$$\Psi(A) = \langle A|\Psi\rangle, \tag{7.28}$$

in the same manner in which one may write $\psi(x) = \langle x|\Psi\rangle$ in ordinary quantum mechanics. As in that case, however, we should remember that $|A\rangle$ is not a normalizable state.

[2] This construction of $\mathcal{H}$ as the closure of the space of the cylindrical functions of smooth connections in the scalar product (7.27) shows that $\mathcal{H}$ can be defined without the need of recurring to C^* algebraic techniques, distributional connections or the Ashtekar-Lewandowski measure. The casual reader, however, should be warned that the resulting $\mathcal{H}$ topology is different than the natural topology on the space of connections: if a sequence Γ_n of graphs converges point–wise to a graph Γ, the corresponding cylindrical functions $\psi_{\Gamma_n,f}$ do not converge to $\psi_{\Gamma,f}$ in the $\mathcal{H}$ Hilbert space topology.

7.2.5 Loop States and Spin Network States

A subspace $\mathcal{H}_0$ of $\mathcal{H}$ is formed by states invariant under $SU(2)$ gauge transformations. We now define an orthonormal basis in $\mathcal{H}_0$. This basis represents a very important tool for using the theory. It was introduced in [RS95] and developed in [Bae96a, Bae96b]; it is denoted *spin network* basis.

First, given a loop α in M, there is a normalized state $\psi_\alpha(A)$ in $\mathcal{H}$, which is obtained by taking $\Gamma = \alpha$ and $f(g) = -\mathrm{Tr}(g)$. Namely

$$\psi_\alpha(A) = -\mathrm{Tr}(U_\alpha(A)). \tag{7.29}$$

We introduce a Dirac notation for the abstract states, and denote this state as $|\alpha\rangle$. These sates are called loop states. Using Dirac notation, we can write

$$\psi_\alpha(A) = \langle A|\alpha\rangle, \tag{7.30}$$

It is easy to show that loop states are normalizable. Products of loop states are normalizable as well. Following tradition, we denote with α also a multi–loop, namely a collection of (possibly overlapping) loops $\{\alpha_1, \dots, \alpha_n,\}$, and we call

$$\psi_\alpha(A) = \psi_{\alpha_1}(A) \times \dots \times \psi_{\alpha_n}(A) \tag{7.31}$$

– a multi–loop state. Multi–loop states represented the main tool for loop quantum gravity before the discovery of the spin network basis. Linear combinations of multi–loop states over–span $\mathcal{H}$, and therefore a generic state $\psi(A)$ is fully characterized by its projections on the multi–loop states, namely by

$$\psi(\alpha) = (\psi_\alpha, \psi). \tag{7.32}$$

The 'old' loop representation was based on representing quantum states in this manner, namely by means of the functionals $\psi(\alpha)$ over loop space defined in(7.32).

Next, consider a graph Γ. A 'coloring' of Γ is given by the following.

1. Associate an irreducible representation of $SU(2)$ to each link of Γ. Equivalently, we may associate to each link γ_i a half integer number s_i, the spin of the irreducible, or, equivalently, an integer number p_i, the 'color' $p_i = 2s_i$.
2. Associate an invariant tensor v in the tensor product of the representations $s_1 \dots s_n$, to each node of Γ in which links with spins $s_1 \dots s_n$ meet. An invariant tensor is an object with n indices in the representations $s_1 \dots s_n$ that transform covariantly. If $n = 3$, there is only one invariant tensor (up to a multiplicative factor), given by the Clebsh–Gordon coefficient. An invariant tensor is also called an *intertwining tensor*. All invariant tensors are given by the standard Clebsch–Gordon theory. More precisely, for fixed $s_1 \dots s_n$, the invariant tensors form a finite dimensional linear space. Pick a basis v_j is this space, and associate one of these basis elements to the

node. Notice that invariant tensors exist only if the tensor product of the representations $s_1 \dots s_n$ contains the trivial representation. This yields a condition on the coloring of the links. For $n = 3$, this is given by the well known Clebsh–Gordan condition: each color is not larger than the sum of the other two, and the sum of the three colors is even.

We indicate a colored graph by $\{\Gamma, \mathbf{s}, \mathbf{v}\}$, or simply $S = \{\Gamma, \mathbf{s}, \mathbf{v}\}$, and denote it a 'spin network'. (It was R. Penrose who first had the intuition that this mathematics could be relevant for describing the quantum properties of the geometry, and who gave the first version of *spin network theory* [Pen71a, Pen71b].)

Given a spin network S, we can construct a state $\Psi_S(A)$ as follows. We take the propagator of the connection along each link of the graph, in the representation associated to that link, and then, at each node, we contract the matrices of the representation with the invariant tensor. We get a state $\Psi_S(A)$, which we also write as

$$\psi_S(A) = \langle A|S\rangle. \tag{7.33}$$

One can then show the following.

- The spin network states are normalizable. The normalization factor is computed in [DR96].
- They are $SU(2)$ gauge invariant.
- Each spin network state can be decomposed into a finite linear combination of products of loop states.
- The (normalized) spin network states form an orthonormal basis for the gauge $SU(2)$ invariant states in $\mathcal{H}$ (choosing the basis of invariant tensors appropriately).
- The scalar product between two spin network states can be easily computed graphically and algebraically.

The spin network states provide a very convenient basis for the quantum theory.

The spin network states defined above are $SU(2)$ gauge invariant. There exists also an extension of the spin network basis to the full Hilbert space.

7.2.6 Diagrammatic Representation of the States

A diagrammatic representation for the states in $\mathcal{H}$ is very useful in concrete calculations. First, associate to a loop state $|\alpha\rangle$ a diagram in M, formed by the loop α itself. Next, notice that we can *multiply* two loop states, obtaining a normalizable state. We represent the product of n loop states by the diagram formed by the set of the n (possibly overlapping) corresponding loops (we denote this set 'multi–loop'). Thus, linear combinations of multi–loops diagrams represent states in $\mathcal{H}$. Representing states as linear combinations of multi–loops diagrams makes computation in $\mathcal{H}$ easy.

Now, the spin network state defined by the graph with no nodes α, with color 1, is clearly, by definition, the loop state $|\alpha\rangle$, and we represent it by the diagram α. The spin network state $|\alpha, n\rangle$ determined by the graph without nodes α, with color n can be obtained as follows. Draw n parallel lines along the loop α; cut all lines at an arbitrary point of α, and consider the $n!$ diagrams obtained by joining the legs after a permutation. The linear combination of these $n!$ diagrams, taken with alternate signs (namely with the sign determined by the parity of the permutation) is precisely the state $|\alpha, n\rangle$. The reason of this key result can be found in the fact that an irreducible representation of $SU(2)$ can be obtained as the totally symmetric tensor product of the fundamental representation with itself (for details, see [DR96]).

Next, consider a graph Γ with nodes. Draw n_i parallel lines along each link γ_i. Join pairwise the end points of these lines at each node (in an arbitrary manner), in such a way that each line is joined with a line from a different link. In this manner, one get a multi–loop diagram. Now antisymmetrize the n_i parallel lines along each link, obtaining a linear combination of diagrams representing a state in $\mathcal{H}$. One can show that this state is a spin network state, where n_i is the color of the links and the color of the nodes is determined by the pairwise joining of the legs chosen [DR96]. Again, simple $SU(2)$ representation theory is behind this result.

More in detail, if a node is trivalent (has 3 adjacent links), then we can join legs pairwise only if the total number of the legs is even, and if the number of the legs in each link is smaller or equal than the sum of the number of the other two. This can be immediately recognized as the *Clebsch–Gordan condition*. If these conditions are satisfied, there is only a single way of joining legs. This corresponds to the fact that there is only one invariant tensor in the product of three irreducible of $SU(2)$. Higher valence nodes can be decomposed in trivalent 'virtual' nodes, joined by 'virtual' links. Orthogonal independent invariant tensors are obtained by varying over all allowed colorings of these virtual links (compatible with the Clebsch–Gordan conditions at the virtual nodes). Different decompositions of the node give different orthogonal bases. Thus the total (links and nodes) coloring of a spin network can be represented by means of the coloring of the real and the virtual links (see Figure 7.2).

Viceversa, multi–loop states can be decomposed in spin network states by simply symmetrizing along (real and virtual) nodes. This can be done particularly easily diagrammatically, as illustrated by the graphical formulae in [RS95, DR96]. These are standard formulae. In fact, it is well known that the tensor algebra of the $SU(2)$ irreducible representations admits a completely graphical notation. This graphical notation has been widely used for instance in nuclear and atomic physics. The application of this diagrammatic calculus to quantum gravity is described in detail in [DR96].

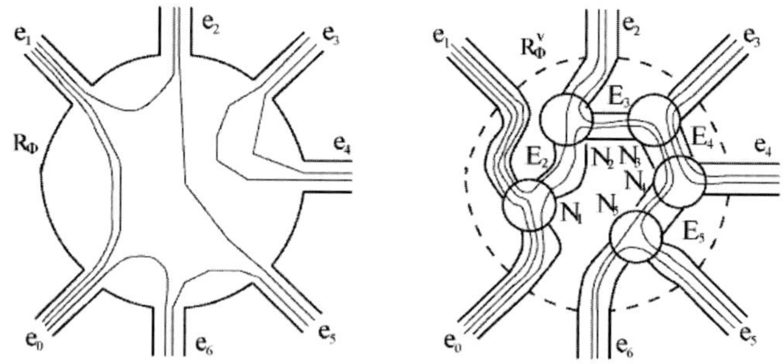

Fig. 7.2. Construction of 'virtual' nodes links over an $n-$valent node in a graph Γ.

7.2.7 Quantum Operators

Now, we define the *quantum operators*, corresponding to the $\mathcal{T}-$variables, as linear operators on $\mathcal{H}$. These form a representation of the loop variables Poisson algebra. The operator $\mathcal{T}[\alpha]$ acts diagonally

$$\mathcal{T}[\alpha]\Psi(A) = -\mathrm{Tr}\, U_\alpha(A)\, \Psi(A),$$

(recall that products of loop states and spin–network states are normalizable states). In diagrammatic notation, the operator simply adds a loop to a (linear combination of) multi–loops

$$\mathcal{T}[\alpha]\ |\Psi\rangle = |\alpha\rangle|\Psi\rangle.$$

Higher order loop operators are expressed in terms of the elementary 'grasp' operation. Consider first the operator $\mathcal{T}^a(s)[\alpha]$, with one hand in the point $\alpha(s)$. The operator annihilates all loop states that do not cross the point $\alpha(s)$. Acting on a loop state $|\beta\rangle$, it gives

$$\mathcal{T}^a(s)[\alpha]\ |\beta\rangle = l_0^2\ \Delta^a[\beta, \alpha(s)]\ \left[|\alpha\#\beta\rangle - |\alpha\#\beta^{-1}\rangle\right], \tag{7.34}$$

where we have introduced the elementary length l_0 by

$$l_0^2 = \hbar G = \frac{16\pi\hbar G_{\mathrm{Newton}}}{c^3} = 16\pi\ l_{Planck}^2 \tag{7.35}$$

and Δ^a and # were defined above. This action extends by linearity, continuity and by the Leibniz rule to products and linear combinations of loop states, and to the full $\mathcal{H}$. In particular, it is not difficult to compute its action on a spin network state [DR96]. Higher order loop operators act similarly. It is easy to verify that these operators provide a representation of the classical Poisson loop algebra.

All the operators in the theory are then constructed in terms of these basics loop operators, in the same way in which in conventional QFT one constructs all operators, including the Hamiltonian, in terms of creation and annihilation operators. The construction of the composite operators requires the development of regularization techniques that can be used in the absence of a background metric.

7.2.8 Loop v.s. Connection Representation

Imagine we want to quantize the one dimensional harmonic oscillator. We can consider the Hilbert space of square integrable functions $\psi(x)$ on the real line, and express the momentum and the Hamiltonian as differential operators. Denote the eigenstates of the Hamiltonian as $\psi_n(x) = \langle x|n\rangle$. It is well known that the theory can be expressed entirely in algebraic form in terms of the states $|n\rangle$. In doing so, all elementary operators are algebraic: $\hat{x}|n\rangle = \frac{1}{\sqrt{2}}(|n-1\rangle + (n+1)|n+1\rangle)$, $\hat{p}|n\rangle = \frac{-i}{\sqrt{2}}(|n-1\rangle - (n+1)|n+1\rangle)$. Similarly, in quantum gravity we can directly construct the quantum theory in the spin–network (or loop) basis, without ever mentioning functionals of the connections. This representation of the theory is denoted the *loop representation.*

A section of the first paper on loop quantum gravity by [RS90] was devoted to a detailed study of 'transformation theory' (in the sense of Dirac) on the state space of quantum gravity, and in particular on the relations between the loop states

$$\psi(\alpha) = \langle\alpha|\psi\rangle \tag{7.36}$$

and the states $\psi(A)$ giving the amplitude for a connection field configuration A, and defined by

$$\psi(A) = \langle A|\psi\rangle. \tag{7.37}$$

Here $|A\rangle$ are 'eigenstates of the connection operator', or, more precisely (since the operator corresponding to the connection is ill defined in the theory) the generalized states that satisfy

$$T[\alpha]\ |A\rangle = -\mathrm{Tr}[\mathcal{P}e^{\int_\alpha A}]\ |A\rangle. \tag{7.38}$$

However, at the time of [RS90] the lack of a scalar product made transformation theory quite involved.

On the other hand, the introduction of the scalar product (7.27) gives a rigorous meaning to the loop transform. In fact, we can write, for every spin network S, and every state $\psi(A)$

$$\psi(S) = \langle S|\psi\rangle = (\psi_S, \psi). \tag{7.39}$$

This equation defines a unitary mapping between the two presentations of $\mathcal{H}$: the 'loop representation', in which one works in terms of the basis $|S\rangle$; and the 'connection representation', in which one uses wave functionals $\psi(A)$.

The development of the connection representation followed a winding path through C^*-algebraic and measure theoretical methods. The work of [DeP97] has proven the unitary equivalence of the two formalisms.

7.3 Cosmological Dynamics

In this section we review the pinnacle of modern physical science, the cosmological dynamics, mainly following the strongest player int this exiting research field, Stephen Hawking.

7.3.1 Hawking's Cosmology in 'Plain English'

According to S. Hawking and his new collaborator T. Hertog, there is no one history of the universe. There is no immutable past, no 13.7 billion years of evolution for cosmologists to retrace. Instead, there are many possible histories, and the universe has lived them all [Gef06].

All this started in Hawking's early work with mathematical physicist R. Penrose in 1970s. They proved their *Singularity Theorems*, effectively showing that our expanding universe must have emerged from a black–hole like singularity, a point of infinite curvature, a place where gravity becomes so strong that space and time are curved beyond recognition, where general relativity (our best description of how space, time and matter interact) – no longer applied. Now, instead of gravitation theory, Hawking and Hertog suggest that the universe was so small at this time that quantum effects must have been important. Hertog claims, "The real lesson of these Singularity theorems is that the origin of the universe is a quantum event."

In 1983, Hawking and J. Hartle took the picture of the famous quantum double–slit experiment (see Introduction) and applied it to the evolution of the whole universe, using Feynman's *sum–over–histories* interpretation of quantum physics, i.e., path integral methodology. Recall that R. Feynman suggested that the way to interpret quantum phenomena, such as the *double–slit experiment*, was to assume that when a particle travels from point A to point B, it doesn't simply take one path; it rather takes every possible path simultaneously; e.g., the photon travels through both slits at the same time and interferes with itself. In this scheme, when a photon travels from a lam to our eye it moves in a straight line, but it also dances about in twists and swirls. The obvious question, then, is why do we ever see only one path, straight and simple? Feynman's answer was, because all the other paths cancel each other out. In the sum–over–histories interpretation, each path, or history, can be mapped out as a wave. Each wave has a different phase (effectively a starting time), and all the waves added together create an 'interference pattern', building upon one another where their phases align and cancelling each other out where their phases are mismatched. The sum of all the waves is one single wave, the so–called wave ψ−function, which describes the path we observe.

Applied to the universe, this idea has an obvious implication [Gef06]. Just as a particle travelling from point A to point B takes every possible path in between, so too must the history of the universe. Hertog says, "The universe doesn't have a single history, but every possible history, each with its own probability." This is Hawking's famous *wave function of the universe*.

The Wave Function of the Universe

This approach starts with the idea of not just our universe but all possible universes; then calculates the relative likelihoods of each of them, much like calculating the wave ψ–function of a particle. Recall that the wave function $\psi = \psi(t)$ of a particle is essentially the function that determines its most likely location at any given time. Wave–functions are largest where that particle is observed to be, but also extend throughout the known universe in accordance with the sum–over–histories method. Everything has a wave ψ–function: elementary particles possess wave–functions and make up all other matter in the universe, so this is a logical conclusion. Large and common objects such as rubber balls have wave–functions; for example, the wave–function of a ball sitting on a flat surface is largest where it is observed – say, a table – but also extends everywhere around us, or on the moon, or even in another galaxy. However, the likelihood of the ball suddenly appearing in any of these locations is infinitesimal. The likelihood of such changes in location depends on *Planck's constant* $\hbar$.

Hawking and Hartle have proposed to calculate the wave–function of the universe using the sum–over–histories method, which begins with the assumption that the universe has all possible histories. Moreover, they would calculate this sum in imaginary time, not ordinary time. This is because imaginary time travels at right angles to ordinary time and 'meets' with the three spatial dimensions to create a smooth surface similar to the surface of the earth. This eliminates the singularities (points of infinite curvature) present in ordinary time, allowing the history of the universe to be reliably calculated. Also unlike ordinary time, imaginary time has no beginning or end, so progression through it is determined entirely by physical laws.

For Hawking and Hartle's calculation, we must begin with a wave–function describing all possible universes – an infinite number of them. The wave–function is large near our own universe and infinitesimal near others in which life is impossible or the known laws of physics do not apply. Because of the wave–function's concentration in our own universe, it is the most likely of them all, but there is a chance (albeit vanishingly small) that an object from this universe would suddenly make a quantum leap into another one. Proving this conjecture mathematically is one of the primary goals of *quantum cosmology*, which applies quantum theory to the large structures of cosmology. The Hawking–Hartle theory also postulates the existence of wormholes connecting the different universes. According to them, the multitude of universes should be connected by wormholes, although these wormholes are not an efficient or readily available means of transportation. Some of these universes are very rich, and others are quite barren. Similarly, some are connected with many others, while others are isolated.

The wave–function of Hawking and Hartle raises two major controversies long debated among scientists. he first of these is an apparent return to the so–called *anthropic principle* (of S. Weinberg), which basically says that the universe is the way it is because we wouldn't be here if it was any other way. It has two basic forms: the 'weak' and 'strong' versions. The weak version states that the existence of intelligent life is experimental evidence to help us understand the universe's seemingly random physical constants. The strong version, much more controversial, states that these apparently random constants are not random but were instead chosen by some Supreme Being to make life in our universe possible. Hawking and Hartle's idea appeals to the weak anthropic principle, but apparently states that Supreme Being is not necessary to explain the existence of a universe uniquely suited to intelligent life. The second controversy relates to the famous *Schrödinger's cat paradox* and the *many–worlds theories*. The Schrödinger's cat paradox states that, by the *Heisenberg's Uncertainty Principle*, a particle is in a sum of all possible states until it is observed, a process called reducing the *wave–function reduction*. Schrödinger postulated a paradox that arises by this theory: suppose a cat is placed in a box connected to a gun which is in turn connected to a Geiger counter measuring a uranium atom. If the unstable atom decays, the Geiger counter will register it, the gun will go off, and the cat will be killed. If it doesn't decay, the process will not be initiated and the cat will live. Before observing the cat, quantum theory states that it is in a superimposed condition of both dead and live states. Most physicists either assume the wave–function is always being reduced by the cosmic observer, the Supreme Being, or simply ignore the problem. A third way of dealing with the paradox is the many–worlds theory, suggested by H. Everett, which states that the universe is constantly splitting off new offshoots, so that in one universe, the cat is dead, and in another it is alive. In the traditional many–worlds theory, contact between the worlds is mathematically impossible and therefore the idea cannot be tested. By the physics principle of *Occam's razor*, which states that the simpler of two competing explanations is generally correct, we should throw out the many–worlds theory because it is irrelevant to our universe and completely untestable. Hawking and Hartle's idea revives the many–worlds theory with one important twist: communication between the worlds in the form of wormholes discussed above is possible in their formulation. Thus, their idea is both testable and directly relevant to our universe.

Top–Down Cosmology with No–Boundary Proposal

However, there is a twist: the history that we see depends on the experimental setup. Recall that in the double–slit experiment, if we use a photon detector to find which of the two slits the photon went through, it no longer creates an interference patter, just a single spot on the film. In other words, the way we look at the photon changes the nature of its journey. The same thing happens in Hawking's universe: our observations of the cosmos today are determining the outcome, that is the entire history of the universe. A measurement made in the present is deciding what happened 13.7 billion years ago; by looking out at the universe, we assign ourselves a particular, concrete history [Gef06].

Although this might look as a violation of the *cause–and–effect laws*, Hawking says that it is all to do with perspective. If we could stand outside the world, we would be able to to see the present affecting the past, as when an observer affects a photon's path through the universe. From inside the universe, though, no observer sees causality violated. What we observe in the present, the 'final' state, is one entire, causally consistent history or another: from within any given history, cause and effect proceed in the usual manner. "Observations of final states determine histories of the universe. A worm's–eye view from inside the universe would have the normal causality. Backwards causality is an angel's view from outside the universe," says Hawking.

So the idea is that to unravel the past, we must sum together all possible histories of the universe. Hawking and Hertog equate the cosmic histories with how the geometry (and topology) of the universe evolves in each possible case of going from the *initial point* A (the beginning of time) to the *final point* B (now). We can specify the state of the universe at the final point B by making certain observations of the world around us (e.g., the universe has three large spatial dimensions, its geometry is close to flat, it is expanding, etc.). However, what about the initial point A? Mapping out the paths of a photon from a lamp to our eye is easy because we know the initial point A (the lamp) and the final point B (our eye). However, we know nothing about the universe at the beginning of time. And that is precisely what cosmology is supposed to tell us.

This is where the sum–over–histories interpretation comes into its own. Its mathematics contains an apparent oddity: the answers only come out right when the calculation is done in imaginary time. Hawking and Hartle's original work on the wave function of the universe (i.e., quantum properties of the cosmos), suggested that imaginary time, which previously seemed like a mathematical curiosity in the sum–over–histories, held the answer to understanding the origin of the universe. Add up the histories of the universe in imaginary time, and time is transformed into space. The result is that, when the universe was small enough to be governed by quantum mechanics, it had four spatial dimensions and no dimension of time: where time would usually come to an end at a singularity, a new dimension of space appears and the singularity vanishes [Gef06].

In terms of the universe's history, that means there is no initial point A; this is called the *no–boundary proposal.* "Like the surface of a sphere, our universe has no definable starting point." This has led Hawking to define a new kind of cosmology. The traditional approach, which he calls *bottom–up cosmology*, tries to specify the initial state of the universe and work from there. This is doomed to fail, Hawking says, because we know nothing about the initial conditions. Instead, he suggests, we should use the no–boundary proposal to do *top–down cosmology*, where the only input into our models of the universe comes from what we observe now. The result of this process, he says, solves a long-standing problem of cosmology, called 'fine-tuning'. For example, most cosmologists think that the universe went through an early burst of rapid expansion, the so–called *inflation*.[3] The problem here is that standard inflationary models require a very improbable initial state, one that

[3] Recall that in physical cosmology, the *inflation* is the idea that the nascent universe passed through a phase of exponential expansion that was driven by a negative-pressure vacuum energy density. As a direct consequence of this expansion, all of the observable universe originated in a small causally–connected region. Inflation answers the classic conundrums of the Big–Bang cosmology: why does the universe appear flat, homogeneous and isotropic in accordance with the cosmological principle when one would expect, on the basis of the physics of the Big–Bang, a highly curved, inhomogeneous universe. Inflation also explains the origin of the large-scale structure of the cosmos. Quantum fluctuations in the microscopic inflationary region, magnified to cosmic size, become the seeds for the growth of structure in the universe (see galaxy formation and evolution and structure formation). While the detailed particle physics mechanism responsible for inflation is not known, the basic picture makes a number of predictions that have been confirmed by observational tests. Inflation is thus now considered part of the standard hot Big–Bang cosmology. The hypothetical particle or field thought to be responsible for inflation is called the *inflaton field*.

Inflation suggests that there was a period of exponential expansion in the very early universe. The expansion is exponential because the distance between any two fixed observers is increasing exponentially, due to the metric expansion of space (a space–time with this property is called a *de Sitter space*, see below). The physical conditions from one moment to the next are stable: the rate of expansion, called the *Hubble parameter*, is nearly constant, which leads to high levels of symmetry. Inflation is often called a period of accelerated expansion because the distance between two fixed observers is increasing at an accelerating rate as they move apart (however, this does not mean that the Hubble parameter is increasing).

Cosmic inflation has the important effect of smoothing out inhomogeneities, anisotropies and the curvature of space. This pushes the universe into a very simple state, in which it is completely dominated by the inflaton field and the only significant inhomogeneities are the tiny quantum fluctuations in the inflaton. Inflation also dilutes exotic heavy particles, such as the magnetic monopoles predicted by many extensions to the Standard Model of particle physics. If the universe was only hot enough to form such particles before a period of inflation, they would not be observed in nature, as they would be so rare that it is quite likely that there are none in the observable universe. Together, these effects are called the inflationary 'no–hair Theorem' by analogy with Hawking's no hair Theorem for black holes. The 'no–hair' Theorem works essentially because the universe expands by an enormous factor during inflation. In an expanding universe, energy densities generally fall as the volume of the universe increases. For example, the density of ordinary 'cold' matter (dust) goes as the inverse of the volume: when linear dimensions double, the energy density goes down by a factor of eight. The energy density in radiation goes down even more rapidly as the universe expands: when linear dimensions are doubled, the energy density in radiation falls by a factor of sixteen. During inflation, the energy density in the inflaton field is roughly constant. However, the energy density in inhomogeneities, curvature, anisotropies and exotic particles is falling, and through sufficient infla-

must have 'finely tuned' values that cause inflation to start, then stop in a certain way after a certain time: complicated prescription whose only justification is to produce a flat universe without any strange topology, etc., a universe like ours. On the other hand, in the no–boundary proposal, there is simply no defined initial state. Hawking says, "In the usual approach it is difficult to explain how inflation began. But it occurs naturally in top–down approach with no–boundary condition. It doesn't need fine tuning."

To do top–down cosmology, Hawking and Hertog first take a whole raft of possible histories (i.e., evolving geometries), all of which would result in a universe with features familiar to us. "We then calculate the probability for other features of the universe, given the constraints. Top–down cosmology does not predict that all possible universes have to begin with a period of inflation, but that inflation occurs naturally within a certain subclass of universes," Hertog says. The process creates a probability for each scenario, and so Hertog can see which kind of history is most likely. "What we find is that the inflating histories generally have the largest probability."

String Theory Landscape

Hawking and Hertog's top–down/no–boundary cosmology adds an interesting twist to the ongoing debate in physics about the existence of multiple universes. At issue is the fact that string theory, physicists most popular candidate for a *theory of everything*, describes not just one universe but a near infinity of them. Some physicists are willing to accept that these theoretical universes actually exist, both because string theory does not seem to favour any particular universe over all others, and because their existence could explain the apparently fine-tuned features of our universe. Hawking's view is that the so–called *string theory landscape* is populated by the set of all possible histories. Rather than a branching set of individual universes, every possible version of a single universe exists simultaneously in a state of quantum superposition. When we choose to make a measurement, we select from this landscape a subset of histories that share the specific features measured. From

tion these become negligible. This leaves an empty, flat, and symmetric universe, which is filled with radiation when inflation ends.

A key requirement is that inflation must continue long enough to produce the present observable universe from a single, small inflationary *Hubble volume*. This is necessary to ensure that the universe appears flat, homogeneous and isotropic at the largest observable scales. This requirement is generally thought to be satisfied if the universe expanded by a factor of at least $10^{2}6$ during inflation. At the end of inflation, a process called *reheating* occurs, in which the inflaton particles decay into the radiation that starts the hot Big–Bang. It is not known how long inflation lasted but it is usually thought to be extremely short compared to the age of the universe. Assuming that the energy scale of inflation is between $10^{1}5$ and $10^{1}6$ GeV, as is suggested by the simplest models (like the *Friedmann equation*, see section on Cosmological Dynamics below), the period of inflation responsible for the observable universe probably lasted roughly 10^{-33} seconds.

observer's perspective, the history of the universe is derived from that subset of histories. In other words, we choose our past [Gef06].

7.3.2 Theories of Everything, Anthropic Principle and Wave Function of the Universe

If a cat, a cannonball, and an economics textbook are all dropped from the same height, they fall to the ground with exactly the same acceleration ($9.8\mathrm{m/s}^2$) under the influence of gravity [Har03]. This equality of the gravitational accelerations of different things is one of the most accurately tested laws of physics. The accuracy record holder at the moment is the lunar laser ranging demonstration that the Earth and the Moon fall with the same acceleration toward the Sun [AW01]. The accelerations are known to be equal to an accuracy of a few parts in a thousand billion. The equality of gravitational accelerations of different things is an example of a regularity of nature. Everything falls in exactly the same way. The regularity is universal. No exceptions!

Identifying and explaining the regularities of nature is the goal of science. Physics, like other sciences, is concerned with the regularities exhibited by particular systems. Stars, atoms, fluid flows, high temperature superconductors, black holes, and the elementary particles are just some of the many examples. Studies of these specific systems define the various subfields of physics – astrophysics, atomic physics, fluid mechanics, and so forth. But beyond the regularities exhibited by specific systems, physics has a special charge. This is to find the laws that govern the regularities that are exhibited by all physical systems, without exception, without qualification, and without approximation. The equality of gravitational accelerations of different things is an example. These are usually called the *fundamental laws* of physics. Taken together they are called informally a *theory of everything*. S. Hawking has been a leader in the quest for these universal laws [Haw84b].

Ideas for the nature of the fundamental laws of nature have changed as experiment and observation have revealed new realms of phenomena and reached new levels of precision in the last century. But since they have been studied, it has been thought that the fundamental laws consist of two parts:

- The *dynamical laws* that govern *regularities in time*. *Newton's laws of motion* governing the orderly progression of the planets, or the trajectory of a tennis ball are examples, as is the law that different things fall with the same acceleration in a gravitational field and the *Einstein equation* governing the evolution of the universe.
- The *initial conditions* that govern how things started out and therefore most often specify regularities in *space*.

This so–called *Newtonian determinism*, famously propagated by P.S. Laplace [Lap51], was the first theory of everything dated circa 1820.

Both parts of a theory of everything are needed to make any predictions. Newton's dynamical laws by themselves *do not* predict the trajectory of a tennis ball we might throw. To predict where it goes, we must also specify the position from which we throw it, the direction, and how fast. Technically, we must specify the ball's *initial condition*. One of Hawking's most famous achievements is such an initial condition [Haw84b], but not for tennis balls. Hawking's *no–boundary initial condition* is for the whole universe [Har03].

The search for a theory of the dynamical laws has been seriously under way since the time of Newton. Classical mechanics, Newtonian gravity, Maxwell's electrodynamics, special and general relativity, quantum mechanics, quantum field theory, and superstring theory are but some of the milestones in this search. But the search for a theory of the initial condition of the universe has been seriously under way for only the twenty years since Hawking's pioneering work on the subject. Why this difference? The examples used above to discuss regularities governed by the two parts of a theory of everything hint at the answer. The trajectory of a tennis ball was used to illustrate the regularities of dynamical laws, and the large scale distribution of galaxies was used to illustrate the regularities implied by the law of the initial condition. There is a difference in the *kind* and *scale* of regularities that the two laws predict.

Dynamical laws predict regularities in time. It is a fortunate empirical fact that the fundamental dynamical laws are *local* — both in space and time. As Einstein said, physics is simple only locally. The trajectory of a tennis ball depends only on conditions that are nearby both in space and time, and not, for example, either on what is going on in distant parts of the universe or a long time ago. This is fortunate because that means that dynamical laws can be discovered and studied in laboratories on Earth and extrapolated, assuming locality, to the rest of the universe. For example, because it is local, the law that different things fall with the same acceleration in a gravitational field can be discovered by experiments in laboratories here, and indeed all over the galaxy. Without that simplicity of the dynamical laws in the here and now it is possible that we would never have discovered them [Har03].

By contrast, the regularities governed by the law of the initial condition of the universe occur on large, cosmological scales. The universe isn't simple on small spatial scales. Look at the disorder or complexity in the room we're in right now for example. But *the universe is simple on large*, cosmological scales — more or less the same in one direction as in any other, more or less the same in one place as in any other.

Quantum Mechanics

There is another way in which our vision of the fundamental laws and the nature of a theory of everything has changed since the times of Newton and Laplace. That is quantum mechanics. We do not yet know the final form the fundamental laws will take. But the inference is inescapable from the physics of the last seventy–five years that they will conform to that subtle framework of prediction we call quantum mechanics.

Recall that in quantum mechanics, any system (the universe included) is described by a *wave function* $\Psi = \Psi(t)$. There is a dynamical law called the *Schrödinger equation* which governs how the wave function changes in time,

$$i\hbar \frac{d\Psi(t)}{dt} = H\Psi(t) \qquad \text{(quantum dynamical law)}.$$

Here the operator H, called the *Hamiltonian*, summarizes the dynamical theory. There are different forms of H for Maxwell's electrodynamics, for a theory of the strong interactions, etc. Like Newton's laws of motion, the Schrödinger equation does not make any predictions by itself, it requires an initial condition. This is

$$\Psi(0), \qquad \text{(initial condition)}.$$

When we consider the universe as a quantum–mechanical system, this initial condition is Hawking's *no–boundary quantum wave function of the universe* [Haw84b]

Probabilities are the key difference between classical and quantum mechanics. Let's first think about probabilities in classical physics. If I say that there is a 60% chance of hitting an audience member if I toss a ball in this room, I am not expressing a lack of confidence that its trajectory will be governed by the deterministic laws of Newtonian mechanics. Rather the 60% reflects my ignorance of the exact initial speed I'll impart to the ball, of the influence of air on its motion, and perhaps my ability to do an accurate calculation. If I practice to control the initial condition when it's thrown, the subsequent evolution of the tennis ball becomes more certain. Probabilities in classical physics result from ignorance.

But in quantum mechanics probabilities are fundamental and uncertainty inevitable. No amount of careful determination of the present state of the tennis ball will achieve certainty for its trajectory. In quantum mechanics there is some probability that a ball will take *any* trajectory as it leaves my hand. However, in classical situations one trajectory (the one obeying Newton's laws) is much more probable than all the others. The determinism of classical physics is an approximation, but an approximation on which we can rely in many practical circumstances [Har03].

Scientific Reduction

Where do all the other regularities in the universe come from, those particular to specific systems — those of the behavior of cats as they fall, those studied by the environmental sciences like biology, geology, economics, and psychology? They are the results of chance events that occur naturally over the history of a quantum mechanical universe. As M. Gell-Mann puts it in [GM94], they are *frozen accidents*. "Chance events of which particular outcomes have a multiplicity of long–term consequences, all related by their common ancestry".

The regularities of cats probably do depend a little bit on the fundamental physical laws, for example, an initial condition that is smooth across the universe, leads to three spatial dimensions, etc. But the origin of most of their regularities can be traced to the chance events of four billion years of biological evolution. Cats behave in similar ways because they have a common ancestry and develop in similar environments. The mechanisms which produce those chance events that led to cats are very much dependent on fundamental biochemistry and ultimately atomic physics. But the particular outcomes of those chances have little to do with the theory of everything [Har03].

Do psychology, economics, biology reduce to physics? The answer is YES, because everything considered in those subjects must obey the universal, fundamental laws of physics. Every one of the subjects of study in these sciences, humans, market tables, historical documents, bacteria, cats, etc., fall with the same acceleration in a gravitational field. The answer is NO, because the regularities of interest in these subjects are not predicted by the universal laws with near certainty *even in principle.* They are frozen quantum accidents that produce emergent regularities. The answer depends upon what we mean by reduce.

In summary:

- The fundamental laws of physics constituting a 'theory of everything' are those which specify the regularities exhibited by every physical system, without exception, without qualification, and without approximation.
- A theory of everything is not (and cannot be) a theory of everything in a quantum–mechanical universe.
- The regularities of human history, personal psychology, economics, biology, geology, etc. are consistent with the fundamental laws of physics, but do not follow from them.

But remember also, especially on this occasion, that all the beautiful regularities that we observe in the universe, certain or not, predictable or not, could be the result of quantum chances following from the fundamental dynamical theory and Hawking's no-boundary wave function of the universe. For more details on theories of everything, see [Har03].

Anthropic Reasoning and Quantum Cosmology

Prediction in quantum cosmology requires a specification of the universe's quantum dynamics and its (initial) quantum state. We expect only a few general features of the universe to be predicted with probabilities near unity conditioned on the dynamics and quantum state alone. Most useful predictions are of conditional probabilities that assume additional information beyond the dynamics and quantum state. Anthropic reasoning utilizes probabilities conditioned on 'us'. In this section, following [Har04b], we discuss anthropic reasoning and quantum cosmology.

If the universe is a quantum mechanical system, then it has a quantum state. This state provides the initial condition for cosmology. A theory of this state is an essential part of any final theory summarizing the regularities exhibited universally by all physical systems and is the objective of the subject of quantum cosmology. This section is concerned with the role the state of the universe plays in anthropic reasoning, that is the process of explaining features of our universe from our existence in it [BT86]. The main idea is that anthropic reasoning in a quantum mechanical context depends crucially on assumptions about the universe's quantum state.

A Model Quantum Universe

Every prediction in a quantum mechanical universe depends on its state if only very weakly. Quantum mechanics predicts probabilities for alternative possibilities, most generally the probabilities for alternative histories of the universe. The computation of these probabilities requires both a theory of the quantum state as well as the theory of the dynamics specifying its evolution.

To make this idea concrete while keeping the discussion manageable, we consider a model quantum universe. The details of this model are not essential to the subsequent discussion of anthropic reasoning but help to fix the notation for probabilities and provide a specific example of what they mean. Particles and fields move in a large, perhaps expanding box, say presently 20,000 Mpc on a side. Quantum gravity is neglected, which is an excellent approximation for accessible alternatives in our universe later than 10^{-43}s from the Big–Bang. Spacetime geometry is thus fixed with a well defined notion of time and the usual quantum apparatus of *Hilbert space*, states, and their unitary evolution governed by a *Hamiltonian* can be applied [Har04b].

The Hamiltonian H and the state $|\Psi\rangle$ in the *Heisenberg picture* are the assumed theoretical inputs to the prediction of quantum mechanical probabilities. Alternative possibilities at one moment of time t can be reduced to yes/no alternatives represented by an exhaustive set of orthogonal projection operators $\{P_\alpha(t)\}$, $\alpha = 1, 2, \cdots$ in this Heisenberg picture. The operators representing the same alternatives at different times are connected by

$$P_\alpha(t) = \mathrm{e}^{iHt/\hbar} P_\alpha(0)\, \mathrm{e}^{-iHt/\hbar} . \tag{7.40}$$

For instance, the P's could be projections onto an exhaustive set of exclusive ranges of the center-of-mass position of the Earth labelled by α. The probabilities $p(\alpha)$ that the Earth is located in one or another of these regions at time t is

$$p\,(\alpha|H, \Psi) = \|P_\alpha(t)\,|\Psi\rangle\|^2 . \tag{7.41}$$

The probabilities for the Earth's location at a different time is given by the same formula with different P's computed from the Hamiltonian by (7.40). The notation $p\,(\alpha|H, \Psi)$ departs from usual conventions (e.g., [Har93]) to indicate explicitly that all probabilities are conditioned on the theory of the Hamiltonian H and quantum state $|\Psi\rangle$.

Most generally quantum theory predicts the probabilities of sequences of alternatives at a series of times, that is *quantum histories*. An example is a sequence of ranges of center of mass position of the Earth at a series of times giving a coarse–grained description of its orbit. Sequences of sets of alternatives $\{P^k_{\alpha_k}(t_k)\}$ at a series of times t_k, $k = 1, \cdots, n$ specify a set of alternative histories of the model universe. An individual history α in the set corresponds to a particular sequence of alternatives $\alpha \equiv (\alpha_1, \alpha_2, \cdots, \alpha_n)$ and is represented by the corresponding chain of projection operators C_α [Har04b]

$$C_\alpha \equiv P^n_{\alpha_n}(t_n) \cdots P^1_{\alpha_1}(t_1) \quad , \quad \alpha \equiv (\alpha_1, \cdots, \alpha_n) . \tag{7.42}$$

The probabilities of the histories in the set are given by

$$p\,(\alpha|H, \Psi) \equiv p\,(\alpha_n, \cdots, \alpha_1|H, \Psi) = \|C_\alpha|\Psi\rangle\|^2, \tag{7.43}$$

provided the set decoheres, i.e., provided the branch state vectors $C_\alpha|\Psi\rangle$ are mutually orthogonal. Decoherence ensures the consistency of the probabilities (7.43) with the usual rules of probability theory[4].

To use either (7.41) or (7.43) to make predictions, a theory of *both* H and $|\Psi\rangle$ is needed. No state; no predictions.

What is Predicted?

Once M. Gell-Mann asked J. Hartle, "If you know the wave function of the universe, why aren't you rich?" (see [Har04b]). Hartle's answer was that there were unlikely to be any alternatives relevant to making money that were predicted as sure bets conditioned just on the Hamiltonian and quantum state alone. A probability $p(\text{rise}|H, \Psi)$ for the stock market to rise tomorrow could be predicted from H and $|\Psi\rangle$ through (7.41) in principle. But it seems likely that the result would be a useless $p(\text{rise}|H, \Psi) \approx 1/2$ conditioned just on the 'no boundary' wave function.

It iss plausible that this is the generic situation. To be manageable and discoverable, the theories of dynamics and the quantum state must be short, that is describable in terms of a few fundamental equations and the explanations of the symbols they contain. It's therefore unlikely that H and $|\Psi\rangle$ contain enough information to determine most of the interesting complexity of the present universe with significant probability [Har96, Har03]. We *hope* that the Hamiltonian and the quantum state are sufficient conditions to predict certain large scale features of the universe with significant probability. Approximately classical space–time, the number of large spatial dimensions, the approximate homogeneity and isotropy on scales above several hundred Mpc, and the spectrum of density fluctuations that were the input to inflaton are some examples of these. But even a simple feature like the time the Sun

[4] For a short introduction to decoherence see [Har93] or any of the classic expositions of decoherent (consistent) histories quantum theory [Gri02, Omn94, GM94].

will rise tomorrow will not be usefully predicted by our present theories of dynamics and the quantum state alone.

The time of sunrise *does* become predictable with high probability if a few previous positions and orientations of the Earth in its orbit are supplied in addition to H and $|\Psi\rangle$. That is a particular case of a *conditional probability* of the form

$$p(\alpha|\beta, H, \Psi) = \frac{p(\alpha, \beta|H, \Psi)}{p(\beta|H, \Psi)}, \tag{7.44}$$

for alternatives α (e.g., the times of sunrise) given H, $|\Psi\rangle$ and further alternatives β (e.g., a few earlier positions and orientations of the Earth). The joint probabilities on the right hand side of (7.62) are computed using (7.43) as described above.

Conditioning probabilities on specific information can weaken their dependence on H and $|\Psi\rangle$ but does not eliminate it. That is because any specific information available to us as human observers (like a few positions of the Earth) is but a small part of that needed to specify the state of the universe. The P_β used to define the joint probabilities in (7.62) by (7.43) therefore spans a very large subspace of Hilbert space. As a consequence $P_\beta|\Psi\rangle$ depends strongly on $|\Psi\rangle$. For example, to extrapolate present data on the Earth to its position 24 hours from now requires that the probability be high that it moves on a classical orbit in that time and that the probability be low that it is destroyed by a neutron star now racing across the galaxy at near light speed. Both of these probabilities depend crucially, if weakly, on the nature of the quantum state [Har94b].

Many useful predictions in physics are of conditional probabilities of the kind discussed in this section. We next turn to the question of whether we should be part of the conditions.

Anthropic Reasoning

In calculating the conditional probabilities for predicting some of *our* observations given others, there can be no objection of principle to including a description of 'us' as part of the conditions [Har04b],

$$p(\alpha|\beta, \text{`us'}, H, \Psi). \tag{7.45}$$

Drawing inferences using such probabilities is called *anthropic reasoning*. The motivation is the idea that probabilities for certain features of the universe might be sensitive to this inclusion.

The utility of anthropic reasoning depends on how sensitive probabilities like (7.64) are to the inclusion of 'us'. To make this concrete, consider the probabilities for a hypothetical cosmological parameter we will call Λ. We will assume that H and $|\Psi\rangle$ imply that Λ is constant over the visible universe, but only supply probabilities for the various constant values it might take through (7.43). We seek to compare $p(\Lambda|H, \Psi)$ with $p(\Lambda|\text{`us'}, H, \Psi)$. In principle, *both* are calculable from (7.43) and (7.62). There are three possible ways they might be related [Har04b]:

- $p(\Lambda|H,\Psi)$ is peaked around one value. The parameter Λ is determined either by H or $|\Psi\rangle$, or by both. Anthropic reasoning is not necessary; the parameter is already determined by fundamental physics.
- $p(\Lambda|H,\Psi)$ is distributed and $p(\Lambda|\text{'us'},H,\Psi)$ is also distributed. Anthropic reasoning is inconclusive. One might as well measure the value of Λ and use this as a condition for making further predictions, i.e., work with probabilities of the form $p(\alpha|\Lambda,H,\Psi)$.
- $p(\Lambda|H,\Psi)$ is distributed but $p(\Lambda|\text{'us'},H,\Psi)$ is peaked. Anthropic reasoning helps to explain the value of Λ.

The important point to emphasize is that a *theoretical hypothesis for H and $|\Psi\rangle$ is needed to carry out anthropic reasoning.* Put differently, a theoretical context is needed to decide whether a parameter like Λ *can* vary, and to find out *how* it varies, before using anthropic reasoning to restrict its range. The Hamiltonian and quantum state provide this context. Below we will consider the situation where the state is imperfectly known.

While there can be no objections of principle to including 'us' as a condition for the probabilities of our observations, there are formidable obstacles of practice [Har04b]:

- We are complex physical systems requiring an extensive environment and a long evolutionary history whose description in terms of the fundamental variables of H and $|\Psi\rangle$ may be uncertain, long, and complicated.
- The complexity of the description of a condition including 'us' may make the calculation of the probabilities long or impossible as a practical matter.

In practice, therefore, anthropic probabilities (7.64) can only be estimated or guessed. Theoretical uncertainty in the results is thereby introduced.

The objectivity striven for in physics consists, at least in part, in using probabilities that are not too sensitive to 'us'. We would not have science if anthropic probabilities for observation depended significantly on which individual human being was part of the conditions. The existence of schizophrenic delusions shows that this is possible so that the notion of 'us' should be restricted to exclude such cases.

For these reasons it is prudent to condition probabilities, not on a detailed description of 'us', but on the weakest condition consistent with 'us' that plausibly provides useful results. A short list of conditions of roughly decreasing complexity might include [Har04b]:

- human beings;
- carbon-based life;
- information gathering and utilizing systems (IGUSes);
- at least one galaxy;
- a universe older than 5 Gyr;

- no condition at all.

For example, the probabilities used to bound the cosmological constant Λ [BT86, Wei05] make use of the fourth and fifth on this list under the assumption that including earlier ones will not much affect the anthropically–allowed range for Λ. To move down in the above list of conditions is to move in the direction of increasing theoretical certainty and decreasing computational complexity. With anthropic reasoning, less is more.

Ignorance is NOT Bliss

The quantum state of a single isolated subsystem generally cannot be determined from a measurement carried out on it. That is because the outcomes of measurements are distributed probabilistically and the outcome of a single trial does not determine the distribution. Neither can the state be determined from a series of measurements because measurements disturb the state of the subsystem. The Hamiltonian can not be inferred from a sequence of measurements on one subsystem for similar reasons. In the same way, we can not generally determine either the Hamiltonian or the quantum state of the universe from our observations of it. Rather these two parts of a final theory are theoretical proposals, inferred from partial data to be sure, but incorporating theoretical assumptions of simplicity, beauty, coherence, mathematical precision, etc. To test these proposals we search among the conditional probabilities they imply for predictions of observations yet to be made with probabilities very near one. When such predictions occur we count it a success of the theory, when they do not we reject it and propose another. The main question is, do we need a theory of the quantum state? To analyze this question, let us consider various degrees of theoretical uncertainty about it [Har04b].

Total Ignorance. In the model cosmology in a box of Section II, theoretical uncertainty about the quantum state can be represented by a density matrix ρ that specifies probabilities for its eigenstates to be $|\Psi\rangle$. Total ignorance of the quantum state is represented by a ρ proportional to the unit matrix. To illustrate this and the subsequent discussion, assume for the moment that the dimension of the Hilbert space is very large but finite. Then total ignorance of the quantum state is represented by [Har04b]

$$\rho_{\text{tot. ign.}} = \frac{I}{Tr(I)}, \tag{7.46}$$

which assigns equal probability to any member of any complete set of orthogonal states.

The density matrix (7.70) predicts thermal equilibrium, infinite temperature, infinitely large field fluctuations, and maximum entropy [Har03]. In short, its predictions are inconsistent with observations. This is a more precise way of saying that every useful prediction depends in some way on a theory of the quantum state. Ignorance is never bliss.

What We Know. A more refined approach to avoiding theories of the quantum state is to assume that it is unknown except for reproducing our present observations of the universe. The relevant density matrix is [Har04b]

$$\rho_{\text{obs}} = \frac{P_{\text{obs}}}{Tr(P_{\text{obs}})}, \tag{7.47}$$

where P_{obs} is the projection on our current observations, that is "what we know". 'Observations' in this context mean what we directly observe and record here on Earth and not the inferences we draw from this data about the larger universe. That is because those inferences are based on assumptions about the very quantum state that (7.47) aims to ignore. For instance, we observed nebulae long before we understood what they were or where they are. The inference that the nebulae are distant clusters of stars and gas relies on assumptions about how the universe is structured on very large scales that are in effect weak assumptions on the quantum state.

Even if we made the overly generous assumption that we had somehow directly observed and recorded every detail of the volume 1 km above the surface of the Earth, say at a 1 mm resolution, that is still a tiny fraction ($\sim 10^{-60}$) of the volume inside the present cosmological horizon. The projection operator P_{obs} therefore defines a very large subspace of Hilbert space. We can expect that the entropy of the density matrix (7.47) will therefore be near maximal, close to that of (7.70), and its predictions similarly inconsistent with further observations.

In the context of anthropic reasoning, these results show that conditioning probabilities on 'us' alone is not enough to make useful predictions. Rather, a theory of H and $|\Psi\rangle$ are needed in addition as described in the previous section.

Let us hope that one day we will have a unified theory based on a principle that will specify both quantum dynamics (H) and a unique quantum state of the universe ($|\Psi\rangle$). That would truly be a final theory and a proper context for anthropic reasoning. For more details, see [Har04b].

Quantum State of the Universe

The central question of quantum cosmology is, What is the quantum state of the universe? To ask this question is to assume that the universe is a quantum mechanical system. We perhaps have little direct evidence of peculiarly quantum mechanical phenomena on large and even familiar scales, but there is no evidence that the phenomena that we do see cannot be described in quantum mechanical terms and explained by quantum mechanical laws. Further, every major candidate for a fundamental dynamical law from the standard model to M–theory conforms to the quantum mechanical framework for prediction. If this framework applies to the whole thing, there must be a quantum state of the universe [Har03].

Final Theories

The final theory predicts the regularities that are exhibited by all physical systems, without exception, without qualification, and without approximation. A possible view at present is that it consists of two parts [Har03]:

- A universal dynamical law such as string theory or its successors;
- A law for the initial quantum state of the universe such as *Hawking's no–boundary wave function of the universe.*

In a model universe in a box these two parts are represented by the Hamiltonian specifying the form of the Schrödinger equation

$$i\hbar \frac{d|\Psi(t)\rangle}{dt} = H\,|\Psi(t)\rangle \tag{7.48}$$

and the initial quantum state

$$|\Psi(0)\rangle \,. \tag{7.49}$$

Both of these pieces are necessary for prediction. The Schrödinger equation makes no predictions by itself. The probabilities p_α predicted by quantum mechanics for a set of alternatives represented by projection operators $\{P_\alpha\}$ are

$$p_\alpha = \| P_\alpha |\Psi(t)\rangle \|^2 \,. \tag{7.50}$$

To compute these the quantum state is needed at least at one time. No state, no predictions.

To put this in a different way, if the state is arbitrary, the predictions are arbitrary. Pick any probabilities p_α you like for the alternatives P_α. There is some state that will reproduce them. For example, we can take

$$|\Psi(t)\rangle = \sum\nolimits_\alpha p_\alpha^{\frac{1}{2}} |\Psi_\alpha\rangle \tag{7.51}$$

where the $|\Psi_\alpha\rangle$ are any set of eigenstates of the P_α's

$$P_\alpha |\Psi_\beta\rangle = \delta_{\alpha\beta} |\Psi_\beta\rangle. \tag{7.52}$$

The $|\Psi(t)\rangle$ constructed according to (7.51) will reproduce the pre-assigned probabilities p_α [Har03].

Neither is ignorance bliss. If we assume we know nothing about the state of the universe in a box then we should make predictions with a density matrix proportional to unity

$$\rho = \frac{I}{Tr(I)}, \tag{7.53}$$

reflecting that ignorance. But this density matrix corresponds to equilibrium at infinite temperature and its predictions are nothing like the universe we live in. In particular, there would be no evolution since $[H, \rho] = 0$. There would be no second law of thermodynamics since the entropy $-Tr[\rho \log \rho]$ is already at its maximum possible value. There would be no classical behavior since, although the expected value of a field averaged over a space–time volume R might be finite, its fluctuations, $\langle\phi(R)^2\rangle$, would be infinite [Har03].

The search for a unified fundamental dynamical law has been seriously under way at least since the time of Newton, with string theory or its generalizations being the most actively investigated direction today. By contrast, the search for a theory of the quantum state of the universe has only been actively under way since the time of Hawking. Why this difference?

Dynamical laws govern regularities in time and it is an empirical fact that the basic dynamical laws are local in space on scales above the Planck length. The laws that govern regularities in time across the whole universe are therefore discoverable and testable in laboratories on earth. By contrast many of the regularities predicted with near certainty by the quantum state of the universe are mostly in space on large cosmological scales. Only recently has there been enough data to confront theory with observation. That difference in the nature of the predicted regularities, or their difference in scales, should not obscure the fact that the state is just as much a part of the final theory as is its Hamiltonian.

Given these differences, what grounds do we have to hope that we can discover the quantum state of the universe? There are two: The first is the simplicity of the early universe revealed by observation — more homogeneous, more isotropic, more nearly in thermal equilibrium than the universe is today. It is therefore possible that the universe has a simple, discoverable initial quantum state and that all of the complex universe of galaxies, stars, planets, and life today arose from quantum accidents that have happened since and the action of gravitational attraction. The second reason is the idea that the quantum state and the dynamical theory may be naturally connected as in Hawking's no–boundary theory [Har03].

Effective Theories

We are used to the idea of effective dynamical theories that accurately describe limited ranges of phenomena. The Navier–Stokes equations, non–relativistic quantum mechanics, general relativity, quantum electrodynamics, and the standard model of particle physics are all familiar examples. To construct an effective theory we typically assume a coarse–grained description (restricting attention to energies below the Planck scale for instance) and assume some simple property that the state might predict there (classical space–time, for example).

Cosmology too has its effective theory and its standard model. This is summarized neatly by the following list (extended) from [Ree01]:

- space–time is classical, governed by the Einstein equation
- our universe is expanding,
- from a hot Big–Bang,
- in which light elements were synthesized,

- there was a period of inflation,
- which led to a flat universe today,
- structure was seeded by gaussian irregularities,
- which are relics of quantum fluctuations,
- the dominant matter is cold and dark,
- but a cosmological constant (or quintessence) is dynamically dominant.

Possibly all current observations in cosmology, at least the large scale ones, can be compressed into an effective 'Standard model' based on this list of ten assumptions and a few cosmological parameters. That is not unlike the situation in particle physics where most observations can be compressed into the Lagrangian of the Standard model and its eighteen or so parameters.

However, the success of such effective theories which operate in limited ranges of phenomena should not obscure the need to find fundamental ones which apply to all phenomena without qualification and without approximation. It would be inconsistent, we believe, to pursue a fundamental dynamical theory in the face of a successful effective, standard dynamical model, and not pursue a fundamental theory of the state of the universe because of the success of *its* effective, standard cosmological model. That not least because the fundamental theory could provide a unified explanation of its assumptions.

It must be said, however, that when the natural domains of fundamental theories are as far from controllable experiments as string theory and the quantum initial condition the possibility of definitive tests seems to recede. It could be that the predictions of string theory are limited to general relativity, gauge theories, supersymmetry, and the parameters of the standard particle model. In a similar way the predictions of the state of the universe could be limited to classical space–time, the initial conditions for inflation, and the quantum fluctuations that satisfy large scale structure. Perhaps that is prediction enough [Har03].

Directions

Continuing the search for a final theory incorporating dynamics and the initial quantum state is certainly one direction. But we would like to mention three questions that might lead to different approaches to the main one [Har03].

What's Environmental? Which features of the observed universe follow entirely from the dynamical theory (H) and which follow entirely from the initial condition ($|\Psi(0)\rangle$), and which are the result of quantum accidents that occurred over the course of the universe's history with probabilities specified by the combination of H and $|\Psi(0)\rangle$. Those that depend significantly on $|\Psi(0)\rangle$ are called 'environmental'. Some version of this question was number one on the list of top ten questions for the next millennium prepared by string theorists at the Strings 2000 conference [DLL01].

Take the coupling constants in effective dynamical theories for instance. The viscosities and equation–of–state in the Navier–Stokes equation are certainly environmental. They vary with system, place, and time. But at a given energy do the coupling constants of the standard model of the elementary particle interactions vary with place and time or with the possible history of the universe? If so then the initial quantum state is central to determining their probabilities.

Why Quantum Mechanics? The founders of quantum mechanics thought that the inherent indeterminacy of quantum theory "reflected the unavoidable interference in measurement dictated by the magnitude of the quantum of the action" (Bohr). But why then do we live in a quantum mechanical universe which, by definition, is never measured from the outside?

The most striking general feature of quantum mechanics is its exact linearity, the principle of superposition. But why should there be a principle of superposition in quantum cosmology which has only a single quantum state?

Why a Division into Dynamics and Initial Condition? The schema for a final theory which I have been describing posits a separate theory of dynamics and quantum state. Could they be connected? They already are in *Hawking's no–boundary wave function* [Haw84b]

$$\Psi = \int \mathcal{D}[g]\,\mathcal{D}[\phi]\;\mathrm{e}^{-I[g,\phi]}, \tag{7.54}$$

where the action for metric $g_{\alpha\beta}(x)$ coupled to matter $\phi(x)$ determines both the state and quantum dynamics. Is there a principle that determines both? Is there a connection between superstring theory and its successors and a unique quantum state?

A unified quantum theory of state and dynamics would be truly a final theory. Pursuing that vision is surely a direction for theoretical physics and cosmology [Har03].

From Quantum Mechanics to Quantum Gravity

How do our ideas about quantum mechanics affect our understanding of space–time? This familiar question leads to quantum gravity [Har06]. This subsection addresses a complementary question: How do our ideas about space–time affect our understanding of quantum mechanics?

Familiar non–relativistic quantum theory illustrates how quantum mechanics incorporates assumptions about space–time. The *Schrödinger equation* governs the evolution of the state in between measurements

$$i\hbar\frac{\partial\psi}{\partial t} = H\psi\,. \tag{7.55}$$

The state vector is 'reduced' at the time of a measurement according to the second law of evolution:

$$\psi \to P\psi/||P\psi||, \tag{7.56}$$

where P is the projection on the outcome of the measurement. Both of these laws of evolution assume a fixed background space–time. A fixed geometry of space–time is needed to define both the t in the Schrödinger equation and the space–like surface on which the state vector is reduced.

But, in quantum gravity, the geometry of space–time is not fixed. Rather geometry is a quantum variable, fluctuating and generally without a definite value. There is no fixed t. Quantum mechanics must therefore be generalized to deal with quantum space–time. This is sometimes called the 'problem of time' in quantum gravity.

Already our ideas about quantum theory have evolved as out ideas about space–time have changed. Milestones in the evolution of our concepts of space and time include: the separate space and absolute time of Newtonian physics, Minkowski space–time with different times in different Lorentz frames, the curved but fixed space–time of general relativity, the quantum fluctuations of space–time in quantum gravity, and the ideas of string/M-theory and loop quantum gravity that space–time is an approximation to something more fundamental. Changes in quantum theory have reflected this evolution. Non-relativistic quantum mechanics incorporates Newtonian time in the Schrödinger equation and the second law of evolution. Any one of the possible time–like directions in Minkowski space can be used to describe the unitary evolution of quantum fields and the results of different choices are unitarily equivalent. Quantum field theories in curved space–times based on different foliations by space–like surfaces are not generally unitarily equivalent. In quantum gravity there is no fixed space–time through which a state can unitarily evolve. Quantum mechanics therefore needs to be generalized for quantum gravity so that it does not require a fixed space–time foliable by space–like surfaces. And, if space–time is not fundamental, quantum mechanics will certainly need to be generalized for whatever replaces it.

However, familiar quantum mechanics also needs to be generalized for cosmology. This generalization is needed so that quantum theory can apply to closed systems such as the universe as a whole containing both observers and observed, measuring apparatus and measured subsystems (if any). These two generalizations can be connected in a common framework called *generalized quantum theory* which is abstracted from the consistent (or decoherent) histories formulation of the quantum mechanics of closed systems [Gri84].

The principles of generalized quantum mechanics were introduced in [Har91a] and developed more fully for example in [Har95a]. The principles have been axiomatized in a rigorous mathematical setting by [Ish94]. Three elements are needed to specify a generalized quantum theory [Har06]:

1. The sets of *fine-grained histories*. These are the most refined possible description of a closed system.
2. The allowed *coarse grainings*. A coarse graining of a set of histories is generally a partition of that set into mutually exclusive classes $\{c_\alpha\}$, (α discrete) called *coarse-grained histories*. The set of classes constitutes a set of coarse-grained histories with each history labelled by the discrete index α.

3. A *decoherence functional* defined for each allowed set of coarse-grained histories which measures the interference between pairs of histories in the set and incorporates a theory of the initial condition and dynamics of the closed system. A decoherence functional $D(\alpha', \alpha)$ must satisfy the following properties:

(i) *Hermiticity*:

$$D(\alpha', \alpha) = D^*(\alpha, \alpha')$$

(ii) *Positivity*:

$$D(\alpha, \alpha) \geq 0 \ .$$

(iii) *Normalization*:

$$\sum\nolimits_{\alpha'\alpha} D(\alpha', \alpha) = 1 \ .$$

(iv) *The Principle of Superposition*: If $\{\bar{c}_{\bar{\alpha}}\}$ is a coarse graining of a set of histories $\{c_\alpha\}$, that is, a further partition into classes $\{\bar{c}_{\bar{\alpha}}\}$, then

$$D(\bar{\alpha}', \bar{\alpha}) = \sum_{\alpha' \in \bar{\alpha}'} \sum_{\alpha \in \bar{\alpha}} D(\alpha', \alpha) \ .$$

Once these three elements are specified the process of prediction proceeds as follows: A set of histories is said to (medium) decohere if all the "off-diagonal" elements of $D(\alpha', \alpha)$ are sufficiently small. The diagonal elements are the probabilities $p(\alpha)$ of the individual histories in a decoherent set. These two definitions are summarized in the one relation

$$D(\alpha', \alpha) \approx \delta_{\alpha'\alpha} p(\alpha) \ . \tag{7.58}$$

As a consequence of (7.58) and properties (i)-(iv) above, the numbers $p(\alpha)$ lie between zero and one, sum to one, and satisfy the most general form of the probability sum rules

$$p(\bar{\alpha}) = \sum_{\alpha \in \bar{\alpha}} p(\alpha)$$

for any coarse graining $\{\bar{c}_{\bar{\alpha}}\}$ of the set $\{c_\alpha\}$. The $p(\alpha)$ are therefore probabilities. They are the predictions of generalized quantum mechanics for the possible coarse-grained histories of the closed system that arise from the theory of its initial condition and dynamics incorporated in the construction of D.

Feynman's 1948 space–time formulation of quantum mechanics [Fey48] supplies one route to constructing a fully 4D generalized quantum theory of space–time geometry. The quantum mechanics of a non–relativistic particle moving in one dimension ($x = x(t)$) between time $t = 0$ and time $t = T$ provides the simplest example. The particle's dynamics is assumed specified by an action functional $S[x(t)]$ and its initial quantum state is assumed to be a particular state vector $|\psi\rangle$.

1. *Fine–grained histories*: These are all paths $x(t)$ between $t = 0$ and $t = T$.

2. *Coarse–grainings*: An allowed coarse graining is any partition of the paths into an exhaustive set of exclusive classes c_α, (α discrete), each class being a *coarse-grained history*. For instance, the paths could be partitions by specifying a set of spatial intervals Δ_i, $i = 1, 2, \cdots$ and giving which two intervals $\alpha = (i, j)$ the particle passes through at two times. An example of a *space–time coarse graining* is provided by specifying a space–time region R and partitioning the paths into the class c_0 which never pass through R and the class c_1 that pass through R *some*time.
3. *Decoherence functional*: In a given set of coarse-grained histories $\{c_\alpha\}$ construct *branch state vector* $|\psi_\alpha\rangle$ for each coarse grained history by summing $\exp(iS)$ over all the paths in c_α and applying that to the initial state $|\psi\rangle$, viz.
$$|\psi_\alpha\rangle \equiv \int_{c_\alpha} \mathcal{D}[x] \exp\{iS[x(t)]/\hbar\}|\psi\rangle.$$
The decoherence functional is
$$D(\alpha', \alpha) = \langle\psi_{\alpha'}|\psi_\alpha\rangle.$$

This space–time formulation of non–relativistic quantum mechanics is easy to visualize, fully 4D, manifests Lagrangian symmetries, and has a close connection to the semiclassical approximation. It incorporates both unitary evolution and the reduction of the state vector in a unified way [Cav86].

The space–time formulation is equivalent to usual Hamiltonian quantum mechanics when the fine grained histories are *single valued in a time* as in non–relativistic quantum mechanics and Minkowski space quantum field theory. This fully 4D formulation generalizes usual quantum mechanics when the histories do not have this property, for instance if there is no fixed time or the histories are not single valued in time. But in those cases we cannot expect to find a notion of state of the system at a moment of time or its unitary evolution through time.

Here are several model situations [Har06]:

- Spacetime alternatives extended over time such as those defined by field averages over space–time regions with extent both in time and space [Har91b].
- Time-neutral quantum mechanics without a quantum mechanical arrow of time but with both initial and final conditions [HGM94].
- Quantum field theory in fixed background space–times that are not foliable by space–like surfaces such as space–times with closed time–like curves, spactimes exhibiting topology change, and evaporating black hole space–times [Har94c, Har98].
- Histories that move backward in time such as those of a single relativistic particle moving in 4D flat space–time [Har95a].

For each of these examples the three ingredients of generalized quantum theory were exhibited, fine grained histories, coarse graining, and decoherence functional.

Building on the lessons of these examples, a generalized quantum mechanics of quantum cosmological space–time geometry can be sketched. The fine grained histories are closed 4D cosmological metrics with 4D matter field configurations upon them. The allowed coarse grainings are partitions of these histories into 4D diffeomorphism invariant classes c_α. A decoherence functional $D(\alpha', \alpha)$ is constructed using amplitudes defined by sums over the histories in the classes c'_α and c_α, initial and final wave functions of the universe, and an inner product linking amplitudes and wave functions. The semiclassical limit for geometry is provided by the steepest descents approximations to the sums over metrics. What remains is a usual quantum field theory in the background space–time described by the metric which gives the biggest contribution to these sums. Thus, familiar familiar quantum mechanics is recovered for those initial conditions and those coarse-grainings in which space–time is fixed, classical, and can supply the necessary time for unitary evolution.

As a summary, we have [Har06]:

- Quantum mechanics can be generalized so that it is free from a fundamental notion of measurement, free of the need for a fixed background space–time, and free from the 'problem of time'.
- General relativity as a theory of four-dimensional space–time is more general than its 3+1 initial value problem. Similarly, a fully four-dimensional formulation of quantum theory is more general than its 3+1 formulation in terms of states evolving unitarily through space–like surfaces in a fixed background space–time.
- In a 4D generalized quantum mechanics of space–time geometry there is no 'problem of time', but there are also typically no states at a moment of time.
- In the context of a fully four-dimensional formulation of quantum theory, the familiar 3+1 quantum mechanics of states evolving unitarily through space–like surfaces is an approximation that is appropriate for those initial conditions and those coarse grained descriptions in which space–time geometry behaves classically and can supply the notion of time necessary to describe the evolution.

7.3.3 Quantum Gravity and Black Holes

In their search for quantum gravity, S. Hawking and R. Penrose use the *straightforward* application of quantum theory to general relativity [HE79, Pen67, HP96], rather than following the more fashioned string theory approach (described below).

According to Hawking, "Einstein's general relativity is a beautiful theory that agrees with every observation that has been made so far. It might require modifications on the Planck scale, and it might be only a *low energy approximation* to some more fundamental theory, like e.g., superstring theory, but it will not affect many of the predictions that can be get from gravity..." [HE79].

Space–Time Manifold, Gravity, Black Holes and Big–Bang

The crucial technique for investigating *Hawking–Penrose singularities and black holes*, has been the study of the global causal structure of space–time [HE79]. Define $I^+(p)$ to be the set of all points of the space–time manifold M that can be reached from the point p by future directed time like curves. One can think of $I^+(p)$ as the set of all events that can be influenced by what happens at p. One now considers the boundary $\dot{I}^+(S)$ of the future of a set S. It is easy to see that this boundary cannot be *time–like*. For in that case, a point q just outside the boundary would be to the future of a point p just inside. Nor can the boundary of the future be *space–like*, except at the set S itself. For in that case every past directed curve from a point q, just to the future of the boundary, would cross the boundary and leave the future of S. That would be a contradiction with the fact that q is in the future of S. Therefore, the boundary of the future is null apart from at S itself.

To show that each generator of the boundary of the future has a past end point on the set, one has to impose some global condition on the causal structure. The strongest and physically most important condition is that of *global space–time hyperbolicity*.[5] The significance of global hyperbolicity for singularity theorems stems from the following [HE79, HP96]. Let U be globally hyperbolic and let p and q be points of U that can be joined by a time like or null curve. Then there is a time–like or null–geodesic between p and q which maximizes the length of time like or null curves from p to q. The method of proof is to show the space of all time like or null curves from p to q is compact in a certain topology. One then shows that the length of the curve is an upper semi–continuous function on this space. It must therefore attain its maximum and the curve of maximum length will be a geodesic because otherwise a small variation will give a longer curve.

One can now consider the second variation of the length of a geodesic γ. One can show that γ can be varied to a longer curve if there is an infinitesimally neighboring geodesic from p which intersects γ again at a point r between p and q. The point r is said to be conjugate to p. One can illustrate this by considering two points p and q on the surface of the Earth. Without loss of generality one can take p to be at the north pole. Because the Earth has a positive definite metric rather than a Lorentzian one, there is a geodesic of minimal length, rather than a geodesic of maximum length. This minimal geodesic will be a line of longitude running from the north pole to the point

[5] Recall that an open set U is said to be globally hyperbolic if:

1. For every pair of points p and q in U the intersection of the future of p and the past of q has compact closure. In other words, it is a bounded diamond shaped region.
2. Strong causality holds on U. That is, there are no closed or almost closed time–like curves contained in U.

q. But there will be another geodesic from p to q which runs down the back from the north pole to the south pole and then up to q. This geodesic contains a point conjugate to p at the south pole where all the geodesics from p intersect. Both geodesics from p to q are stationary points of the length under a small variation. But now in a positive definite metric the second variation of a geodesic containing a conjugate point can give a shorter curve from p to q. Thus, on the Earth, the geodesic that goes down to the south pole and then comes up is not the shortest curve from p to q.

The reason one gets conjugate points in space–time is that gravity is an attractive force. It therefore curves space–time in such a way that neighboring geodesics are bent towards each other rather than away. One can see this from the *Newman–Penrose equation*[6]

$$\frac{d\rho}{dv} = \rho^2 + \sigma^{ij}\sigma_{ij} + \frac{1}{n}R_{\alpha\beta}l^\alpha l^\beta, \qquad (\alpha, \beta = 0, 1, 2, 3)$$

where $n = 2$ for null geodesics and $n = 3$ for time–like geodesics. Here v is an affine parameter along a congruence of geodesics, with tangent vector l^α which are hypersurface orthogonal. The quantity ρ is the average rate of convergence of the geodesics, while σ measures the shear. The term $R_{\alpha\beta}l^\alpha l^\beta$ gives the direct gravitational effect of the matter on the convergence of the geodesics. By the *Einstein equation*, $G_{\alpha\beta} = 8\pi T_{\alpha\beta}$ (here $G_{\alpha\beta}$ is the Einstein tensor), the term $R_{\alpha\beta}l^\alpha l^\beta$ will be non–negative for any null vector l^α if the matter obeys the so–called *weak energy condition*, which says that the energy density T_{00} is non–negative in any frame, i.e.,

$$T_{\alpha\beta}v^\alpha v^\beta \geq 0, \tag{7.59}$$

for any time–like vector v^α, is obeyed by the classical SEM–tensor of any reasonable matter [HE79, HP96].

Suppose the weak energy condition holds, and that the null geodesics from a point p begin to converge again and that ρ has the positive value ρ_0. Then the Newman–Penrose equation would imply that the convergence ρ would become infinite at a point q within an affine parameter distance $\frac{1}{\rho_0}$ if the null geodesic can be extended that far. If $\rho = \rho_0$ at $v = v_0$ then $\rho \geq \frac{1}{\rho^{-1} + v_0 - v}$. Thus there is a conjugate point before $v = v_0 + \rho^{-1}$.

Infinitesimally neighboring null geodesics from p will intersect at q. This means the point q will be conjugate to p along the null geodesic γ joining them. For points on γ beyond the conjugate point q there will be a variation of γ that gives a time like curve from p. Thus γ cannot lie in the boundary

[6] Recall that the *Newman–Penrose formalism* is a set of notation developed by E.T. Newman and R. Penrose [NP62, Teu73] for *general relativity* (GR). Their notation is an effort to treat GR in terms of *spinor notation*, which introduces complex forms of the usual variables used in GR. The most often–used variables in their formalism are the *Weyl scalars*, derived from the *Weyl tensor*.

of the future of p beyond the conjugate point q. So γ will have a future end point as a generator of the boundary of the future of p.

The situation with time–like geodesics is similar, except that the *strong energy condition* [HE79, HP96],

$$T_{\alpha\beta}v^{\alpha}v^{\beta} \geq \frac{1}{2}v^{\alpha}v_{\alpha}T, \tag{7.60}$$

that is required to make $R_{\alpha\beta}l^{\alpha}l^{\beta}$ non–negative for every time like vector l^{α}, is rather stronger than the weak energy condition (7.59). However, it is still physically reasonable, at least in an averaged sense, in classical theory. If the strong energy condition holds, and the time like geodesics from p begin converging again, then there will be a point q conjugate to p.

Finally there is the *generic energy condition*, which says:

1. The strong energy condition holds.
2. Every time–like or null geodesic has a point where $l_{[a}R_{b]cd[e}l_{f]}l^{c}l^{d} \neq 0$.

One normally thinks of a space–time singularity as a region in which the curvature becomes unboundedly large. However, the trouble with this definition is that one could simply leave out the singular points and say that the remaining manifold was the whole of space–time. It is therefore better to define space–time as the maximal manifold on which the metric is suitably smooth. One can then recognize the occurrence of singularities by the existence of incomplete geodesics that cannot be extended to infinite values of the affine parameter.

Hawking–Penrose Singularity Theorems

Hawking–Penrose Singularity is defined as follows [HE79, Pen67, HP96]:

A space–time manifold is singular if it is time–like or null geodesically incomplete but cannot be embedded in a larger space–time manifold.

This definition reflects the most objectionable feature of singularities, that there can be particles whose history has a beginning or end at a finite time. There are examples in which geodesic incompleteness can occur with the curvature remaining bounded, but it is thought that generically the curvature will diverge along incomplete geodesics. This is important if one is to appeal to quantum effects to solve the problems raised by singularities in classical general relativity.

Singularity Theorems include:

1. Energy condition (i.e., weak (7.59), strong (7.60), or generic (7.3.3)).
2. Condition on global structure (e.g., there should not be any closed time–like curves).
3. Gravity strong enough to trap a region (so that nothing could escape).

The various singularity theorems show that space–time must be time like or null geodesically incomplete if different combinations of the three kinds of conditions hold. One can weaken one condition if one assumes stronger versions of the other two. The Hawking–Penrose Singularity theorems have the generic energy condition, the strongest of the three energy conditions. The global condition is fairly weak, that there should be no closed time like curves. And the no escape condition is the most general, that there should be either a trapped surface or a closed space like three surface.

The theorems predict singularities in two situations. One is in the future in the gravitational collapse of stars and other massive bodies. Such singularities would be an end of time, at least for particles moving on the incomplete geodesics. The other situation in which singularities are predicted is in the past at the beginning of the present expansion of the universe.

The prediction of singularities means that classical general relativity is not a complete theory. Because the singular points have to be cut out of the space–time manifold one cannot define the field equations there and cannot predict what will come out of a singularity. With the singularity in the past the only way to deal with this problem seems to be to appeal to quantum gravity. But the singularities that are predicted in the future seem to have a property that Penrose has called, Cosmic Censorship. That is they conveniently occur in places like black holes that are hidden from external observers. So any break down of predictability that may occur at these singularities will not affect what happens in the outside world, at least not according to classical theory.

Hawking Cosmic Censorship Hypothesis says: "Nature abhors a naked singularity" [HE79, HP96]. However, there is unpredictability in the quantum theory. This is related to the fact that gravitational fields can have intrinsic entropy which is not just the result of coarse graining. *Gravitational entropy*, and the fact that time has a beginning and may have an end, are the two main themes of Hawking's research, because they are the ways in which gravity is distinctly different from other physical fields.

The fact that gravity has a quantity that behaves like entropy was first noticed in the purely classical theory. It depends on Penrose's Cosmic Censorship Conjecture. This is unproved but is believed to be true for suitably general initial data and state equations.

One makes the approximation of treating the region around a collapsing star as asymptotically flat. Then, as Penrose showed, one can conformally embed the space–time manifold M in a manifold with boundary $\bar{M}$. The boundary ∂M will be a null surface and will consist of two components, future and past null infinity, called $\mathcal{I}^+$ and $\mathcal{I}^-$. One says that *weak* Cosmic Censorship holds if two conditions are satisfied. First, it is assumed that the null geodesic generators of $\mathcal{I}^+$ are complete in a certain conformal metric. This implies that observers far from the collapse live to an old age and are not wiped out by a thunderbolt singularity sent out from the collapsing star. Second, it is assumed that the past of $\mathcal{I}^+$ is globally hyperbolic. This means there are no naked singularities that can be seen from large distances. Penrose has also a stronger form of Cosmic Censorship which assumes that the whole space–time is globally hyperbolic.

Weak Cosmic Censorship Hypothesis reads:

1. $\mathcal{I}^+$ and $\mathcal{I}^-$ are complete.
2. $I^-(\mathcal{I}^+)$ is globally hyperbolic.

If weak Cosmic Censorship holds, the singularities that are predicted to occur in gravitational collapse cannot be visible from $\mathcal{I}^+$. This means that there must be a region of space–time that is not in the past of $\mathcal{I}^+$. This region is said to be a **black hole** because no light or anything else can escape from it to infinity. The boundary of the black hole region is called the *event horizon.* Because it is also the boundary of the past of $\mathcal{I}^+$ the event horizon will be generated by null–geodesic segments that may have past end points but don't have any future end points. It then follows that if the weak energy condition holds the generators of the horizon cannot be converging. For if they were they would intersect each other within a finite distance [HE79, Pen67, HP96].

This implies that the area of a cross section of the event horizon can never decrease with time and in general will increase. Moreover if two black holes collide and merge together the area of the final black hole will be greater than the sum of the areas of the original black holes. This is very similar to the behavior of entropy according to the *Second Law of Thermodynamics:*[7]

Second Law of Black Hole Mechanics: $\delta\mathcal{A} \geq 0$.

Second Law of Thermodynamics: $\delta S \geq 0$.

The similarity with thermodynamics is increased by what is called the First Law of Black Hole Mechanics, which relates the change in mass of a black hole to the change in the area of the event horizon and the change in its angular momentum and electric charge. One can compare this to the First Law of Thermodynamics which gives the change in internal energy in terms of the change in entropy and the external work done on the system [HE79, HP96]:

First Law of Black Hole Mechanics: $\delta E = \frac{\kappa}{8\pi}\delta\mathcal{A} + \Omega\delta J + \Phi\delta Q$.

First Law of Thermodynamics: $\delta E = T\delta S + P\delta V$.

One sees that if the *area* $\mathcal{A}$ of the *event horizon* is *analogous to entropy* S then the quantity analogous to *temperature* is what is called the *surface gravity of the black hole* κ. This is a measure of the strength of the gravitational field on the event horizon. The similarity with thermodynamics is further increased by the so–called Zeroth Law of Black Hole Mechanics: the surface gravity is the same everywhere on the event horizon of a time independent black hole [HE79].

Zeroth Law of Black Hole Mechanics:
κ is the same everywhere on the horizon of a time independent black hole.

Zeroth Law of Thermodynamics:

---

[7] Recall that Second Law of Thermodynamics states: Entropy can never decrease and the entropy of a total system is greater than the sum of its constituent parts.

T is the same everywhere for a system in thermal equilibrium.

Encouraged by these similarities Bekenstein proposed that some multiple of the area of the event horizon actually was the entropy of a black hole. He suggested a generalized Second Law: the sum of this black hole entropy and the entropy of matter outside black holes would never decrease (see [SV96b]).

Generalized Second Law: $\delta(S + c\mathcal{A}) \geq 0$.

However, this proposal was not consistent. If black holes have an entropy proportional to horizon area $\mathcal{A}$ they should also have a non zero temperature proportional to surface gravity.

Path–Integral Model for Black Holes

Recall that the fact that gravity is attractive means that it will tend to draw the matter in the universe together to form objects like stars and galaxies. These can support themselves for a time against further contraction by thermal pressure, in the case of stars, or by rotation and internal motions, in the case of galaxies. However, eventually the heat or the angular momentum will be carried away and the object will begin to shrink. If the mass is less than about one and a half times that of the Sun the contraction can be stopped by the degeneracy pressure of electrons or neutrons. The object will settle down to be a white dwarf or a neutron star respectively. However, if the mass is greater than this limit there is nothing that can hold it up and stop it continuing to contract. Once it has shrunk to a certain critical size the gravitational field at its surface will be so strong that the light cones will be bent inward [HE79, HP96].

If the Cosmic Censorship Conjecture is correct the trapped surface and the singularity it predicts cannot be visible from far away. Thus there must be a region of space–time from which it is not possible to escape to infinity. This region is said to be a black hole. Its boundary is called the event horizon and it is a null surface formed by the light rays that just fail to get away to infinity. As we saw in the last subsection, the area $\mathcal{A}$ of a cross section of the event horizon can never decrease, at least in the classical theory. This, and perturbation calculations of spherical collapse, suggest that black holes will settle down to a stationary state.

Recall that the *Schwarzschild metric form*, given by

$$ds^2 = -(1 - \frac{2M}{r})dt^2 + (1 - \frac{2M}{r})^{-1}dr^2 \ + \ r^2(d\theta^2 + \sin^2\theta d\phi^2),$$

represents the gravitational field that a black hole would settle down to if it were non rotating. In the usual r and t coordinates there is an apparent singularity at the Schwarzschild radius $r = 2M$. However, this is just caused by a bad choice of coordinates. One can choose other coordinates in which the metric is regular there.

Now, if one performs the *Wick rotation*, $t = \mathrm{i}\tau$, one gets a positive definite metric, usually called Euclidean even though they may be curved. In the *Euclidean–Schwarzschild metric*

$$ds^2 = x^2\left(\frac{d\tau}{4M}\right)^2 + \left(\frac{r^2}{4M^2}\right)^2 dx^2 + r^2(d\theta^2 + \sin^2\theta d\phi^2)$$

there is again an apparent singularity at $r = 2M$. However, one can define a new radial coordinate x to be $4M(1 - 2Mr^{-1})^{\frac{1}{2}}$.

The metric in the $x - \tau$ plane then becomes like the origin of polar coordinates if one identifies the coordinate τ with period $8\pi M$. Similarly, other Euclidean black hole metrics will have apparent singularities on their horizons which can be removed by identifying the imaginary time coordinate with period $\frac{2\pi}{\kappa}$.

To see the significance of having imaginary time identified with some period β, let us consider the *amplitude* to go from some field configuration ϕ_1 on the surface t_1 to a configuration ϕ_2 on the surface t_2. This will be given by the matrix element of $\mathrm{e}^{\mathrm{i}H(t_2-t_1)}$. However, one can also represent this amplitude as a *path integral* over all fields ϕ between t_1 and t_2 which agree with the given fields ϕ_1 and ϕ_2 on the two surfaces,

$$< \phi_2, t_2|\phi_1, t_1 >=< \phi_2|\exp(-\mathrm{i}H(t_2 - t_1))|\phi_1 >= \int D[\phi]\exp(\mathrm{i}A[\phi]).$$

One now chooses the time separation $(t_2 - t_1)$ to be pure imaginary and equal to β. One also puts the initial field ϕ_1 equal to the final field ϕ_2 and sums over a complete basis of states ϕ_n. On the left one has the expectation value of $\mathrm{e}^{-\beta H}$ summed over all states. This is just the *thermodynamic partition function* Z at the temperature $T = \beta^{-1}$,

$$Z = \sum < \phi_n|\exp(-\beta H)|\phi_n >= \int D[\phi]\exp(-A[\phi]). \qquad (7.61)$$

On the r.h.s. of this equation one has a *path integral* (see chapter 6). One puts $\phi_1 = \phi_2$ and sums over all field configurations ϕ_n. This means that effectively one is doing the path integral over all fields ϕ on a space–time that is identified periodically in the imaginary time direction with period β. Thus the partition function for the field ϕ at temperature T is given by a path integral over all fields on a Euclidean space–time. This space–time is periodic in the imaginary time direction with period $\beta = T^{-1}$ [HE79, HP96].

If one calculates the path integral in flat space–time identified with period β in the imaginary time direction one gets the usual result for the partition function of black body radiation. However, as we have just seen, the Euclidean–Schwarzschild solution is also periodic in imaginary time with period $\frac{2\pi}{\kappa}$. This means that fields on the Schwarzschild background will behave as if they were in a thermal state with temperature $\frac{\kappa}{2\pi}$.

The periodicity in imaginary time explained why the messy calculation of frequency mixing led to radiation that was exactly thermal. However, this derivation avoided the problem of the very high frequencies that take part in the frequency mixing approach. It can also be applied when there are interactions between the quantum fields on the background. The fact that the path integral is on a periodic background implies that all physical quantities like expectation values will be thermal. This would have been very difficult to establish in the frequency mixing approach [HE79, HP96].

One can extend these interactions to include interactions with the gravitational field itself. One starts with a background metric g_0 such as the Euclidean–Schwarzschild metric that is a solution of the classical field equations. One can then expand the action A in a power series in the perturbations δg about g_0, as

$$A[g] = A[g_0] + A_2(\delta g)^2 + A_3(\delta g)^3 + ...$$

Here, the linear term vanishes because the background is a solution of the field equations. The quadratic term can be regarded as describing gravitons on the background while the cubic and higher terms describe interactions between the gravitons. The path integral over the quadratic terms are finite. There are non renormalizable divergences at two loops in pure gravity but these cancel with the fermions in super–gravity theories. It is not known whether super–gravity theories have divergences at three loops or higher because no one has been brave or foolhardy enough to try the calculation. Some recent work indicates that they may be finite to all orders. But even if there are higher loop divergences they will make very little difference except when the background is curved on the scale of the Planck length (10^{-33} cm).

More interesting than the higher order terms is the zeroth order term, the action of the background metric g_0 [HE79, HP96],

$$A = -\frac{1}{16\pi}\int R(-g)^{\frac{1}{2}}\, d^4x + \frac{1}{8\pi}\int K(\pm h)^{\frac{1}{2}}\, d^3x.$$

Recall that the usual *Einstein–Hilbert action* for general relativity is the volume integral of the scalar curvature R. This is zero for vacuum solutions so one might think that the action of the Euclidean-Schwarzschild solution was zero. However, there is also a surface term in the action proportional to the integral of K, the trace of the second fundamental form of the boundary surface. When one includes this and subtracts off the surface term for flat space one finds the action of the Euclidean–Schwarzschild metric is $\frac{\beta^2}{16\pi}$ where β is the period in imaginary time at infinity. Thus the dominant contribution to the path integral for the partition function Z given by (7.61), is $\mathrm{e}^{\frac{-\beta^2}{16\pi}}$,

$$Z = \sum \exp(-\beta E_n) = \exp\left(-\frac{\beta^2}{16\pi}\right).$$

If one differentiates $\log Z$ with respect to the period β one gets the expectation value of the energy, or in other words, the mass,

$$< E >= -\frac{d}{d\beta}(\log Z) = \frac{\beta}{8\pi}.$$

So this gives the mass $M = \frac{\beta}{8\pi}$. This confirms the relation between the mass and the period, or inverse temperature, that we already knew. However, one can go further. By standard thermodynamic arguments, the log of the partition function is equal to minus the free energy F divided by the temperature T, i.e., $\log Z = -\frac{F}{T}$. And the free energy is the mass or energy plus the temperature times the entropy S, i.e., $F = < E > + TS$. Putting all this together one sees that the action of the black hole gives an entropy of $4\pi M^2$,

$$S = \frac{\beta^2}{16\pi} = 4\pi M^2 = \frac{1}{4}\mathcal{A}.$$

This is exactly what is required to make the laws of black holes the same as the laws of thermodynamics [HE79, HP96]. The reason why does one get this intrinsic gravitational entropy which has no parallel in other quantum field theories, is that gravity allows different topologies for the space–time manifold.

In the case we are considering the Euclidean–Schwarzschild solution has a boundary at infinity that has topology $S^2 \times S^1$. The S^2 is a large space like two sphere at infinity and the S^1 corresponds to the imaginary time direction which is identified periodical. One can fill in this boundary with metrics of at least two different topologies. One is the Euclidean–Schwarzschild metric. This has topology $R^2 \times S^2$, that is the Euclidean two plane times a two sphere. The other is $R^3 \times S^1$, the topology of Euclidean flat space periodically identified in the imaginary time direction. These two topologies have different Euler numbers. The *Euler number* of periodically identified flat space is zero, while that of the Euclidean–Schwarzschild solution is two,

$$\text{Total action} = M(\tau_2 - \tau_1).$$

The significance of this is as follows: on the topology of periodically identified flat space one can find a periodic time function τ whose gradient is no where zero and which agrees with the imaginary time coordinate on the boundary at infinity. One can then work out the action of the region between two surfaces τ_1 and τ_2. There will be two contributions to the action, a volume integral over the matter Lagrangian, plus the Einstein–Hilbert Lagrangian and a surface term. If the solution is time independent the surface term over $\tau = \tau_1$ will cancel with the surface term over $\tau = \tau_2$. Thus the only net contribution to the surface term comes from the boundary at infinity. This gives half the mass times the imaginary time interval $(\tau_2 - \tau_1)$. If the mass is non–zero there must be non–zero matter fields to create the mass. One can show that the volume integral over the matter Lagrangian plus the Einstein–Hilbert Lagrangian also gives $\frac{1}{2}M(\tau_2 - \tau_1)$. Thus the total action is $M(\tau_2 - \tau_1)$. If one puts this contribution to the log of the partition function into the thermodynamic formulae one finds the expectation value of the energy to be the mass, as one would expect. However, the entropy contributed by the background field will be zero.

However, the situation is different with the Euclidean–Schwarzschild solution, which says:

$$\text{Total action including corner contribution} = M(\tau_2 - \tau_1)$$

$$\text{Total action without corner contribution} = \frac{1}{2}M(\tau_2 - \tau_1)$$

Because the Euler number is two rather than zero one cannot find a time function τ whose gradient is everywhere non–zero. The best one can do is choose the imaginary time coordinate of the Schwarzschild solution. This has a fixed two sphere at the horizon where τ behaves like an angular coordinate. If one now works out the action between two surfaces of constant τ the volume integral vanishes because there are no matter fields and the scalar curvature is zero. The trace K surface term at infinity again gives $\frac{1}{2}M(\tau_2 - \tau_1)$. However there is now another surface term at the horizon where the τ_1 and τ_2 surfaces meet in a corner. One can evaluate this surface term and find that it also is equal to $\frac{1}{2}M(\tau_2 - \tau_1)$. Thus the total action for the region between τ_1 and τ_2 is $M(\tau_2 - \tau_1)$. If one used this action with $\tau_2 - \tau_1 = \beta$ one would find that the entropy was zero. However, when one looks at the action of the Euclidean Schwarzschild solution from a 4–dimensional point of view rather than a 3+1, there is no reason to include a surface term on the horizon because the metric is regular there. Leaving out the surface term on the horizon reduces the action by one quarter the area of the horizon, which is just the intrinsic gravitational entropy of the black hole [HE79, HP96].

Quantum Cosmology

According to Hawking, *cosmology* used to be considered a *pseudo–science* and the preserve of physicists who may have done useful work in their earlier years but who had gone mystic in their dotage. There is a serious objection that cosmology cannot predict anything about the universe unless it makes some assumption about the *initial conditions*. Without such an assumption, all one can say is that things are as they are now because they were as they were at an earlier stage. Yet many people believe that science should be concerned only with the local laws which govern how the universe evolves in time. They would feel that the boundary conditions for the universe that determine how the universe began were a question for metaphysics or religion rather than science [HE79, HP96].

Hawking–Penrose theorems showed that according to general relativity there should be a singularity in our past. At this singularity the field equations could not be defined. Thus classical general relativity brings about its own downfall: it predicts that it cannot predict the universe. For Hawking this sounds rally disturbing: If the laws of physics could break down at the beginning of the universe, why couldn't they break down any where. In quantum theory it is a principle that anything can happen if it is not absolutely forbidden. Once one allows that singular histories could take part in the path integral they could occur any where and predictability would disappear completely. If the laws of physics break down at singularities, they could break down any where.

The only way to have a scientific theory is if the laws of physics hold everywhere including at the beginning of the universe. One can regard this as a triumph for the *Principle of Democracy*: Why should the beginning of the universe be exempt from the laws that apply to other points. If all points are equal one cannot allow some to be more equal than others.

To implement the idea that the laws of physics hold everywhere, one should take the path integral only over non–singular metrics. One knows in the ordinary path integral case that the measure is concentrated on non–differentiable paths. But these are the completion in some suitable topology of the set of smooth paths with well defined action. Similarly, one would expect that the path integral for quantum gravity should be taken over the completion of the space of smooth metrics. What the path integral cannot include is metrics with singularities whose action is not defined.

In the case of black holes we saw that the path integral should be taken over Euclidean, that is, positive definite metrics. This meant that the singularities of black holes, like the Schwarzschild solution, did not appear on the Euclidean metrics which did not go inside the horizon. Instead the horizon was like the origin of polar coordinates. The action of the Euclidean metric was therefore well defined. One could regard this as a quantum version of Cosmic Censorship: the break down of the structure at a singularity should not affect any physical measurement.

It seems, therefore, that the path integral for quantum gravity should be taken over non–singular Euclidean metrics. But what should the boundary conditions be on these metrics. There are two, and only two, natural choices. The first is metrics that approach the flat Euclidean metric outside a compact set. The second possibility is metrics on manifolds that are compact and without boundary. Therefore, the natural choices for path integral for quantum gravity are [HE79, HP96]: (i) asymptotically Euclidean metrics, and (ii) compact metrics without boundary. The first class of asymptotically Euclidean metrics is appropriate for *scattering* calculations. In these one sends particles in from infinity and observes what comes out again to infinity. All measurements are made at infinity where one has a flat background metric and one can interpret small fluctuations in the fields as particles in the usual way. One doesn't ask what happens in the interaction region in the middle. That is why one does a path integral over all possible histories for the interaction region, that is, over all asymptotically Euclidean metrics. However, in cosmology one is interested in measurements that are made in a finite region rather than at infinity. We are on the inside of the universe not looking in from the outside. To see what difference this makes let us first suppose that the path integral for cosmology is to be taken over all asymptotically Euclidean metrics.

The so–called No Boundary Proposal of Hartle and Hawking reads [HE79, HP96]: The path integral for quantum gravity should be taken over all compact Euclidean metrics. One can paraphrase this as: *the boundary condition of the universe is that it has no boundary.* According to Hawking, this no boundary proposal seems to account for the universe we live in. That is an isotropic and homogeneous expanding universe with small perturbations. We can observe the spectrum and statistics of these perturbations in the fluctuations in the microwave background. The results so far agree with the predictions of the no boundary proposal. It will be a real test of the proposal and the whole Euclidean quantum gravity program when the observations of the microwave background are extended to smaller angular scales.

In order to use the no boundary proposal to make predictions, it is useful to introduce a concept that can describe the state of the universe at one time:

$$\text{Probability of induced metric } h_{ij} \text{ on } \Sigma = \int_{\substack{\text{metrics on } M \text{ that} \\ \text{induce } h_{ij} \text{ on } \Sigma}} d[g]\ \exp(-A[g]).$$

Consider the probability that the space–time manifold M contains an embedded three dimensional manifold Σ with induced metric h_{ij}. This is given by a path integral over all metrics g_{ab} on M that induce h_{ij} on Σ. If M is simply–connected, which we will assume, the surface Σ will divide M into two parts M^+ and M^- [HE79, HP96],

$$\text{Probability of } h_{ij} = \Psi^+(h_{ij}) \times \Psi^-(h_{ij}), \qquad \text{where}$$
$$\Psi^+(h_{ij}) = \int_{\substack{\text{metrics on } M^+ \text{ that} \\ \text{induce } h_{ij} \text{ on } \Sigma}} d[g]\ \exp(-A[g]).$$

In this case, the probability for Σ to have the metric h_{ij} can be factorized. It is the product of two wave functions Ψ^+ and Ψ^-. These are given by path integrals over all metrics on M^+ and M^- respectively, that induce the given three metric h_{ij} on Σ. In most cases, the two wave functions will be equal and we will drop the superscripts + and +. Ψ is called the wave function of the universe. If there are matter fields ϕ, the wave function will also depend on their values ϕ_0 on Σ. But it will not depend explicitly on time because there is no preferred time coordinate in a closed universe. The no boundary proposal implies that the wave function of the universe is given by a path integral over fields on a compact manifold M^+ whose only boundary is the surface Σ. The path integral is taken over all metrics and matter fields on M^+ that agree with the metric h_{ij} and matter fields ϕ_0 on Σ.

One can describe the position of the surface Σ by a function τ of three coordinates x_i on Σ. But the wave function defined by the path integral cannot depend on τ or on the choice of the coordinates x_i. This implies that the wave function Ψ has to obey four functional differential equations. Three of these equations are called the momentum constraint One can describe the position of the surface Σ by a function τ of three coordinates x_i on Σ. But the wave function defined by the path integral cannot depend on τ or on the choice of the coordinates x_i. This implies that the wave function Ψ has to obey four functional differential equations. Three of these equations are called the *momentum constraint equation*: $\left(\frac{\partial \Psi}{\partial h_{ij}}\right)_{;j} = 0$. They express the fact that the wave function should be the same for different 3 metrics h_{ij} that can be get from each other by transformations of the coordinates x_i. The fourth equation is called the *Wheeler–DeWitt equation*

$$\left(G_{ijkl} \frac{\partial^2}{\partial h_{ij} \partial h_{kl}} - h^{\frac{1}{2}}\, {}^3R \right) \Psi = 0.$$

It corresponds to the independence of the wave function on τ. One can think of it as the *Schrödinger equation for the universe.* But there is no time derivative term because the wave function does not depend on time explicitly.

In order to estimate the wave function of the universe, one can use the *saddle point approximation* to the path integral as in the case of black holes. One finds a Euclidean metric g_0 on the manifold M^+ that satisfies the field equations and induces the metric h_{ij} on the boundary Σ. One can then expand the action A in a power series around the background metric g_0,

$$A[g] = A[g_0] + \frac{1}{2} \delta g A_2 \delta g + ...$$

As before, the term linear in the perturbations vanishes. The quadratic term can be regarded as giving the contribution of gravitons on the background and the higher order terms as interactions between the gravitons. These can be ignored when the radius of curvature of the background is large compared to the Planck scale. Therefore, according to [HE79, HP96] we have

$$\Psi \approx \frac{1}{(\det A_2)^{\frac{1}{2}}} \exp(-A[g_o]).$$

Consider now a situation in which there are no matter fields but there is a positive cosmological constant Λ. Let us take the surface Σ to be a three sphere and the metric h_{ij} to be the round three sphere metric of radius a. Then the manifold M^+ bounded by Σ can be taken to be the four ball. The metric that satisfies the field equations is part of a four sphere of radius $\frac{1}{H}$ where $H^2 = \frac{\Lambda}{3}$,

$$A = \frac{1}{16\pi} \int (R - 2\Lambda)(-g)^{\frac{1}{2}} d^4x + \frac{1}{8\pi} \int K(\pm h)^{\frac{1}{2}} d^3x.$$

For a 3–sphere Σ of radius less than $\frac{1}{H}$ there are two possible Euclidean solutions: either M^+ can be less than a hemisphere or it can be more. However there are arguments that show that one should pick the solution corresponding to less than a hemisphere.

One can interpret the wave function Ψ as follows. The real time solution of the Einstein equations with a Λ term and maximal symmetry is de Sitter space (see, e.g., [Wit98b]). This can be embedded as a hyperboloid in five dimensional Minkowski space. Here, we have two choices:

1. *Lorentzian–de Sitter metric,*
$$ds^2 = -dt^2 + \frac{1}{H^2}\cosh Ht(dr^2 + \sin^2 r(d\theta^2 + \sin^2\theta d\phi^2)).$$
One can think of it as a closed universe that shrinks down from infinite size to a minimum radius and then expands again exponentially. The metric can be written in the form of a Friedmann universe with scale factor $\cosh Ht$. Putting $\tau = it$ converts the cosh into cos giving the Euclidean metric on a four sphere of radius $\frac{1}{H}$.
2. *Euclidean metric,*
$$ds^2 = d\tau^2 + \frac{1}{H^2}\cos H\tau(dr^2 + \sin^2 r(d\theta^2 + \sin^2\theta d\phi^2)).$$
Thus one gets the idea that a wave function which varies exponentially with the three metric h_{ij} corresponds to an imaginary time Euclidean metric. On the other hand, a wave function which oscillates rapidly corresponds to a real time Lorentzian metric.

Hawking says: "The Euclidean path integral over all topologically trivial metrics can be done by time slicing and so is unitary when analytically continued to the Lorentzian. On the other hand, the path integral over all topologically non–trivial metrics is asymptotically independent of the initial state. Thus the total path integral is unitary and information is not lost in the formation and evaporation of black holes. The way the information gets out seems to be that a true event horizon never forms, just an apparent horizon."

Like in the case of the pair creation of black holes, one can describe the spontaneous creation of an exponentially expanding universe. One joins the lower half of the Euclidean four sphere to the upper half of the Lorentzian hyperboloid.

Unlike the black hole pair creation, one couldn't say that the de Sitter universe was created out of field energy in a pre–existing space. Instead, it would quite literally be created out of nothing: not just out of the vacuum but out of absolutely nothing at all because there is nothing outside the universe. In the Euclidean regime, the de Sitter universe is just a closed space like the surface of the Earth but with two more dimensions ([Wit98b]). If the cosmological constant is small compared to the Planck value, the curvature of the Euclidean four sphere should be small. This will mean that the saddle point approximation to the path integral should be good, and that the calculation of the wave function of the universe will not be affected by our ignorance of what happens in very high curvatures.

One can also solve the field equations for boundary metrics that aren't exactly the round three sphere metric. If the radius of the three sphere is less than $\frac{1}{H}$, the solution is a real Euclidean metric. The action will be real and the wave function will be exponentially damped compared to the round three sphere of the same volume. If the radius of the three sphere is greater than this critical radius there will be two complex conjugate solutions and the wave function will oscillate rapidly with small changes in h_{ij}.

Any measurement made in cosmology can be formulated in terms of the wave function. Thus the no boundary proposal makes cosmology into a science because one can predict the result of any observation. The case we have just been considering of no matter fields and just a cosmological constant does not correspond to the universe we live in. Nevertheless, it is a useful example, both because it is a simple model that can be solved fairly explicitly and because, as we shall see, it seems to correspond to the early stages of the universe.

Although it is not obvious from the wave function, a de Sitter universe has thermal properties rather like a black hole. One can see this by writing the de Sitter metric in a *static form* (rather like the Schwarzschild solution)

$$ds^2 = -(1 - H^2 r^2)dt^2 + (1 - H^2 r^2)^{-1} dr^2 + r^2(d\theta^2 + \sin^2\theta d\phi^2).$$

There is an apparent singularity at $r = \frac{1}{H}$. However, as in the Schwarzschild solution, one can remove it by a coordinate transformation and it corresponds to an event horizon.

If one returns to the static form of the de Sitter metric and put $\tau = it$ one gets a Euclidean metric. There is an apparent singularity on the horizon. However, by defining a new radial coordinate and identifying τ with period $\frac{2\pi}{H}$, one gets a regular Euclidean metric which is just the four sphere. Because the imaginary time coordinate is periodic, de Sitter space and all quantum fields in it will behave as if they were at a temperature $\frac{H}{2\pi}$. As we shall see, we can observe the consequences of this temperature in the fluctuations in the microwave background. One can also apply arguments similar to the black hole case to the action of the Euclidean–de Sitter solution [Wit98b]. One finds that it has an intrinsic entropy of $\frac{\pi}{H^2}$, which is a quarter of the area of the event horizon. Again this entropy arises for a topological reason: the Euler number of the four sphere is two. This means that there cannot be a global time coordinate on Euclidean–de Sitter space. One can interpret this *cosmological entropy* as reflecting an observers lack of knowledge of the universe beyond his event horizon [HE79, HP96]:

$$\text{Euclidean metric periodic with period } \frac{2\pi}{H} \Rightarrow \begin{cases} \text{Temperature } T = \frac{H}{2\pi}, \\ \text{Area } \mathcal{A} \text{ of event horizon} = \frac{4\pi}{H^2}, \\ \text{Entropy } S = \frac{\pi}{H^2}. \end{cases}$$

7.3.4 Generalized Quantum Mechanics

Recall that familiar textbook quantum mechanics assumes a fixed background space–time to define states on space–like surfaces and their unitary evolution between them. Quantum theory has changed as our conceptions of space and time have evolved. But quantum mechanics needs to be generalized further for quantum gravity where space–time geometry is fluctuating and without definite value. In this section, following [Har05b], we review a fully 4D, sum–over–histories, generalized quantum mechanics of cosmological space–time geometry. This generalization is constructed within the framework of generalized quantum theory. This is a minimal set of principles for quantum theory abstracted from the modern quantum mechanics of closed systems, most generally the universe. In this generalization, states of fields on space–like surfaces and their unitary evolution are emergent properties appropriate when space–time geometry behaves approximately classically. The principles of generalized quantum theory allow for the further generalization that would be necessary were space–time not fundamental. Emergent space–time phenomena are discussed in general and illustrated with the example of the classical space–time geometries with large space–like surfaces that emerge from the *no–boundary wave function of the universe*.

Does quantum mechanics apply to space–time? This is an old question. Belgian physicist L. Rosenfeld wrote one of the first papers on quantum gravity [Ros30], but late in his career came to the conclusion that the quantization of the gravitational field would be meaningless[8] [Ros63, Ros66]. Today, there are probably more colleagues of the opinion that quantum theory needs to be replaced than there are who think that it doesn't apply to space–time. But in the end this is an experimental question as Rosenfeld stressed.

The answer that J. Hartle proposes is: "Quantum mechanics can be applied to space–time provided that the usual textbook formulation of quantum theory is suitably generalized." A generalization is necessary because, in one way or another, the usual formulations rely on a fixed space–time geometry to define states on space–like surfaces and the time in which they evolve unitarily one surface to another. But in a quantum theory of gravity, space–time geometry is generally fluctuating and without definite value. The usual formulations are emergent from a more general perspective when geometry is approximately classical and can supply the requisite fixed notions of space and time.

A framework for investigating generalizations of usual quantum mechanics can be abstracted from the modern quantum mechanics of closed systems [Gri02, Omn94, GM94] which enables quantum mechanics to be applied to cosmology. The resulting framework, the so–called *generalized quantum the-*

[8] Rosenfeld considered the example of classical geometry curved by the expected value of the stress–energy of quantum fields. Some of the difficulties with this proposal, including experimental inconsistencies, are discussed in [PG81].

ory [Har91a, Har95b, Ish94], defines a broad class of generalizations of usual quantum mechanics.

A generalized quantum theory of a physical system (most generally the universe) is built on three elements which can be very crudely characterized as follows [Har05b]:

- The possible fine–grained descriptions of the system.
- The coarse–grained descriptions constructed from the fine–grained ones.
- A measure of the quantum interference between different coarse–grained descriptions incorporating the principle of superposition

The standard quantum two–slit experiment (see Figure 1.11 in Introduction) provides an immediate, concrete illustration. A set of possible fine–grained descriptions of an electron moving through the two–slit apparatus are its Feynman paths in time (histories) from the source to the detecting screen. One coarse–grained description is by which slit the electron went through on its way to detection in an interval Δ about a position y on the screen at a later time. Amplitudes $\psi_U(y)$ and $\psi_L(y)$ for the two coarse–grained histories where the electron goes through the upper or lower slit and arrives at a point y on the screen can be computed as a sum over paths in the usual way. The natural measure of interference between these two histories is the overlap of these two amplitudes integrated over the interval Δ in which the electron is detected. In this way usual quantum mechanics is a special case of generalized quantum theory.

Probabilities cannot be assigned to the two coarse–grained histories in the two–slit experiment because they interfere. The probability to arrive at the screen y should be the sum of the probabilities to go by way of the upper or lower slit. But in quantum theory, probabilities are squares of amplitudes and

$$|\psi_U(y) + \psi_L(y)|^2 \neq |\psi_U(y)|^2 + |\psi_L(y)|^2 \,.$$

Probabilities can only be predicted for sets of alternative coarse–grained histories for which the quantum interference is negligible between every pair of coarse–grained histories in the set (decoherence).

Usual quantum mechanics is not the only way of implementing the three elements of generalized quantum theory. Below we sketch a *sum–over–histories* generalized quantum theory of space–time. The fine–grained histories are the set of 4D cosmological space–times with matter fields on them. A coarse graining is a partition of this set into (diffeomorphism invariant) classes. A natural measure of interference is described. This is a fully 4D quantum theory without an equivalent 3+1 formulation in terms of states on space–like surfaces and their unitary evolution between them. Rather, the usual 3+1 formulation is emergent for those situations, and for those coarse–grainings, where space–time geometry behaves approximately classically. The intent of this development is not to propose a new quantum theory of gravity. This essentially low energy theory suffers from the usual ultraviolet difficulties. Rather, it is to employ this theory as a model to discuss how quantum mechanics can be generalized to deal with quantum geometry. A common expectation is that space–time is itself emergent from something more fundamental. In that case a generalization of usual quantum mechanics will surely be needed and generalized quantum theory can provide a framework for discovering it.

Quantum Mechanics Today

Three features of quantum theory are striking from the present perspective: its success, its rejection by some of the deepest thinkers, and the absence of compelling alternatives [Har05b]. Firstly, quantum mechanics must be counted as one of the most successful of all physical theories. Within the framework it provides, a truly vast range of phenomena can be understood and that understanding is confirmed by precision experiment. We perhaps have little evidence for peculiarly quantum phenomena on large and even familiar scales, but there is *no* evidence that all the phenomena we do see, from the smallest scales to the largest of the universe, cannot be described in quantum mechanical terms and explained by quantum mechanical laws. Indeed, the frontier to which quantum interference is confirmed experimentally is advancing to ever larger, more 'macroscopic' systems[9]. The textbook electron two–slit experiment has been realized in the laboratory [TEM89]. Interference has been confirmed for the biomolecule tetraphenylporphyrin ($C_{44}H_{30}N_4$) and the flurofullerine ($C_{60}F_{48}$) in analogous experiments [Hac03]. Experiments with superconducting squids have demonstrated the coherent superposition of macroscopic currents [Wal00, CNH03, Fri03]. In particular, the experiment of Friedman, et al. [Fri03] exhibited the coherent superposition of two circulating currents whose magnetic moments were of order $10^{10}\mu_B$ (where $\mu_B = e\hbar/2m_e c$ is the Bohr magneton). Experiments under development will extend the boundary further [Mar03]. Experiments of increasing ingenuity and sophistication have extended the regime in which quantum mechanics has been tested. No limit to its validity has yet emerged.

Secondly, even while acknowledging its undoubted empirical success, many of the greatest minds have rejected quantum mechanics as a framework for fundamental theory. Among the pioneers, the names of Einstein, Schrödinger, DeBroglie, and Bohm stand out in this regard. Among our distinguished contemporaries, Adler, Leggett, Penrose, and 't Hooft could probably be counted in this category. Much of this thought has in common the intuition that quantum mechanics is an effective approximation of a more fundamental theory built on a notion of reality closer to that classical physics.

Finally and remarkably, despite eighty years of unease with its basic premises, and despite having been tested only in a limited, largely microscopic, domain, no fully satisfactory alternative to quantum theory has emerged. By fully satisfactory we mean not only consistent with existing experiment, but

[9] For an insightful and lucid review see [Leg02].

also incorporating other seemingly secure parts of modern physics such as special relativity, field theory, and the standard model of elementary particle interactions. As S. Weinberg summarized the situation, "It is striking that it has not so far been possible to find a logically consistent theory that is close to quantum mechanics other than quantum mechanics itself" [Wei92]. Alternatives to quantum theory meeting the above criteria would be of great interest if only to guide experiment.

There are several directions under investigation today which aim at a theory from which quantum mechanics would be emergent. For an encyclopedic survey of different interpretations and alternatives to quantum mechanics, see [Aul00].

Bohmian mechanics [BH93] in its most representative form is a deterministic but highly non–classical theory of particle dynamics whose statistical predictions largely coincide with quantum theory [Har04]. Fundamental noise [Per98] or spontaneous dynamical collapse of the wave function [BG03, DH04] are the underlying ideas of another class of model theories whose predictions are distinguishable from those of quantum theory, in principle. S. Adler has proposed a statistical mechanics of deterministic matrix models from which quantum mechanics is emergent [Adl04]. G. 't Hooft has a different set of ideas for a determinism beneath quantum mechanism that are explained in [tHo06]. R. Penrose has championed a role for gravity in state vector reduction [Pen00, Pen04]. This has not yet developed into a detailed alternative theory, but has suggested experimental situations in which the decay of quantum superpositions could be observed [Pen04, Mar03].

In the face of an increasing domain of confirmed predictions of quantum theory and the absence as yet of compelling alternatives, it seems natural to extend quantum theory as far as it will go, to the largest scales of the universe and the smallest of quantum gravity [Har05b].

Spacetime and Quantum Theory

Usual, textbook quantum theory incorporates definite assumptions about the nature of space and time. These assumptions are readily evident in the two laws of evolution for the quantum state Ψ. The *Schrödinger equation* describes its unitary evolution between measurements.

$$i\hbar\, \partial_t \Psi = H\Psi \,. \tag{7.62}$$

At the time of an ideal measurement, the state is projected on the outcome and renormalized

$$\Psi \to \frac{P\Psi}{\|P\Psi\|} \,. \tag{7.63}$$

The Schrödinger equation (7.62) assumes a fixed notion of time. In the non–relativistic theory, t is the absolute time of Newtonian mechanics. In the flat space–time of special relativity, it is the time of any Lorentz frame. Thus, there are many times but results obtained in different Lorentz frames, are unitarily equivalent. The projection in the second law of evolution (7.63) is in Hilbert space. But in field theory or particle mechanics, the Hilbert space is constructed from configurations of fields or position in physical space. In that sense it is the state on a space–like surface that is projected (7.63).

Because quantum theory incorporates notions of space and time, it has changed as our ideas of space and time have evolved. The accompanying table briefly summarizes this co–evolution. It is possible to view this evolution as a process of increasing generalization of the concepts in the usual theory. Certainly the two laws of evolution (7.62) and (7.63) have to be generalized somehow if space–time geometry is not fixed. One such generalization is offered in this section, but there have been many other ideas [Kuc92]. And if space–time geometry is emergent from some yet more fundamental description, we can certainly expect that a further generalization, free of any reference to space–time, will be needed to describe that emergence [Har05b].

The Quantum Mechanics of Closed Systems

Here we briefly review the elements of the modern *quantum mechanics of closed systems* aimed at a quantum mechanics for cosmology. See, e.g., [Gri02, Omn94, GM94] for the classic expositions at length or [Har93] for a shorter summary for quantum mechanics of closed systems. To keep the present discussion manageable we focus on a simple model universe of particles moving in a very large box. Everything is contained within the box, in particular galaxies, stars, planets, observers and observed (if any), measured subsystems, and the apparatus that measures them. We assume a fixed background space–time supplying well–defined notions of time. The usual apparatus of Hilbert space, states, operators, Feynman paths, etc. can then be employed in a quantum description of the contents of the box. The essential theoretical inputs to the process of prediction are the Hamiltonian H and the initial quantum state $|\Psi\rangle$, the *wave function of the universe*. These are assumed to be fixed and given.

The most general objective of a quantum theory for the box is the prediction of the probabilities of exhaustive sets of coarse–grained alternative time histories of the particles in the closed system. For instance, we might be interested in the probabilities of an alternative set of histories describing the progress of the Earth around the Sun. Histories of interest here are typically very coarse–grained for at least three reasons: They deal with the position of the Earth's center–of–mass and not with the positions of all the particles in the universe. The center–of–mass position is not specified to arbitrary accuracy, but to the error we might observe it. The center–of–mass position is not specified at all times, but typically at a series of times.

But, as described in the Introduction, not every set of alternative histories that may be described can be assigned consistent probabilities because of quantum interference. Any quantum theory must therefore not only specify the sets of alternative coarse–grained histories, but also give a rule identifying which sets of histories can be consistently assigned probabilities as well as what those probabilities are. In the quantum mechanics of closed systems, that rule is simple: probabilities can be assigned to just those sets of histories for which the quantum interference between its members is negligible as a consequence of the Hamiltonian H and the initial state $|\Psi\rangle$. We now make this specific for our model universe of particles in a box.

Three elements specify this quantum theory; to facilitate later discussion, we give these in a space–time sum–over–histories formulation [Har05b]:

1. *Fine-grained histories*: The most refined description of the particles from the initial time $t = 0$ to a suitably large final time $t = T$ gives their position at all times in between, i.e., their Feynman paths. We denote these simply by $x(t)$.
2. *Coarse–graining*: The general notion of coarse-graining is a partition of the fine–grained paths into an exhaustive set of mutually exclusive classes $\{c_\alpha\}, \alpha = 1, 2, \cdots$. For instance, we might partition the fine–grained histories of the center–of–mass of the Earth by which of an exhaustive and exclusive set of position intervals $\{\Delta_\alpha\}, \alpha = 1, 2, \cdots$ the center–of–mass passes through at a series of times $t_1, \cdots t_n$. Each coarse–grained history consists of the bundle of fine–grained paths that pass through a specified sequence of intervals at the series of times. Each coarse–grained history specifies an orbit where the center–of–mass position is localized to a certain accuracy at a sequence of times.
3. *Measure of Interference*: Branch state vectors $|\Psi_\alpha\rangle$ can be defined for each coarse–grained history in a partition of the fine–grained histories into classes $\{c_\alpha\}$ as follows
$$\langle x|\Psi_\alpha\rangle = \int_{c_\alpha} \mathcal{D}[x] \exp(iS[x(t)]/\hbar) \, \langle x'|\Psi\rangle \, . \tag{7.64}$$
Here, $S[x(t)]$ is the action for the Hamiltonian H. The integral is over all paths starting at x' at $t = 0$, ending at x at $t = T$, and contained in the class c_α. This includes an integral over x'. (For those preferring the Heisenberg picture, this is equivalently
$$|\Psi_\alpha\rangle = \mathrm{e}^{-iHT/\hbar} P^n_{\alpha_n}(t_n) \cdots P^1_{\alpha_1}(t_1) \, |\Psi\rangle \tag{7.65}$$
when the class consists of restrictions to position intervals at a series of times and the P's are the projection operators representing them.)
The measure of quantum interference between two coarse–grained histories is the overlap of their branch state vectors
$$D(\alpha', \alpha) \equiv \langle \Psi_{\alpha'}|\Psi_\alpha\rangle \, . \tag{7.66}$$
This is called the *decoherence functional*.

When the interference between each pair of histories in a coarse–grained set is negligible

$$\langle \Psi_\alpha | \Psi_\beta \rangle \approx 0 \quad \text{all} \quad \alpha \neq \beta \,, \tag{7.67}$$

the set of histories is said to *decohere*[10]. The probability of an individual history in a decoherent set is

$$p(\alpha) = \| \, |\Psi_\alpha\rangle \|^2 \,. \tag{7.68}$$

The decoherence condition (7.66) is a sufficient condition for the probabilities (7.67) to be consistent with the rules of probability theory. Specifically, the p's obey the sum rules

$$p(\bar{\alpha}) \approx \sum_{\alpha \in \bar{\alpha}} p(\alpha) \tag{7.69}$$

where $\{\bar{c}_{\bar{\alpha}}\}$ is any coarse–graining of the set $\{c_\alpha\}$, i.e., a further partition into coarser classes. It was the failure of such a sum rule that prevented consistent probabilities from being assigned to the two histories previously discussed in the two-slit experiment (Figure 1). That set of histories does not decohere.

Decoherence of familiar quasi–classical variables is widespread in the universe. Imagine, for instance, a dust grain in a superposition of two positions, a multimeter apart, deep in intergalactic space. The 10^{11} cosmic background photons that scatter off the dust grain every second dissipate the phase coherence between the branches corresponding to the two locations on the time scale of about a nanosecond [JZ85].

Measurements and observers play no fundamental role in this generalization of usual quantum theory. The probabilities of measured outcomes can, of course, be computed and are given to an excellent approximation by the usual story. But, in a set of histories where they decohere, probabilities can be assigned to the position of the Moon when it is not being observed and to the values of density fluctuations in the early universe when there were neither measurements taking place nor observers to carry them out [Har05b].

Quantum Theory in 3+1 Form

The quantum theory of the model universe in a box in the previous section is in fully 4D space–time form. The fine–grained histories are paths in space–time, the coarse–grainings were partitions of these, and the measure of interference was constructed by space–time path integrals. No mention was made of states on space–like surfaces or their unitary evolution.

However, as originally shown by Feynman [Fey48, FH65], this space–time formulation is equivalent to the familiar 3+1 formulation in terms of states on space–like surfaces and their unitary evolution through a foliating family of such surfaces. This section briefly sketches that equivalence emphasizing

[10] This is the *medium* decoherence condition. For a discussion of other conditions, see, e.g., [GMH90, GMH95, Har04a].

properties of space–time and the fine–grained histories that are necessary for it to hold.

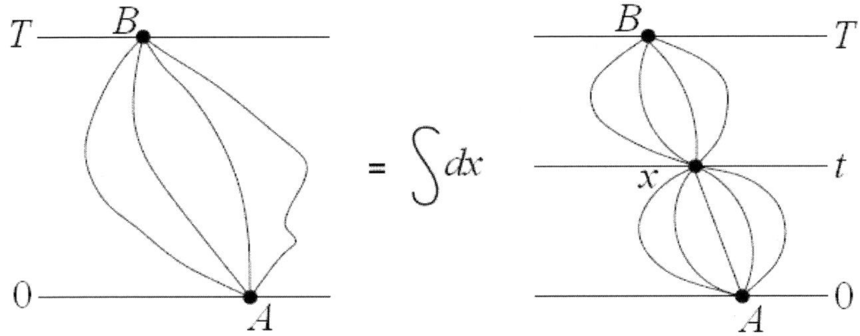

Fig. 7.3. The origin of states on a space–like surface. These space–time diagrams are a schematic representation of (7.70). The amplitude for a particle to pass from point A at time $t = 0$ to a point B at $t = T$ is a sum over all paths connecting them weighted by $\exp(iS[x(t)])$. That sum can be factored across an intermediate constant time surface as shown at right into product of a sum from A to x on the surface and a sum from x to B followed by a sum over all x. The sums in the product define states on the surface of constant time at t. The integral over x defines the inner product between such states, and the path integral construction guarantees their unitary evolution in t. Such factorization is possible only if the paths are single valued functions of time (modified and adapted from [Har05b]).

The key observation is illustrated in Figure 7.3. Sums–over–histories that are single–valued in time can be factored across constant time surfaces. A formula expressing this idea is [Har05b]:

$$\int_{[A,B]} \mathcal{D}[x]\, \mathrm{e}^{iS[x(t)]/\hbar} = \int \psi_B^*(x,t)\psi_A(x,t)\, dx. \tag{7.70}$$

The sum on the left is over all paths from A at $t = 0$ to B at $t = T$. The amplitude $\psi_A(x,t)$ is the sum of $\exp\{iS[x(t)]\}$ over all paths from A at $t = 0$ to x at a time t between 0 and T. The amplitude $\psi_B(x,t)$ is similarly constructed from the paths between x at t to B at T.

The wave function $\psi_A(x,t)$ defines a state on constant time surfaces. Unitary evolution by the Schrödinger equation follows from its path integral construction.[11] The inner product between states defining a Hilbert space is specified by (7.70). In this way, the familiar 3+1 formulation of quantum mechanics is recovered from its space–time form.

[11] Reduction of the state vector (7.63) also follows from the path integral construction [Cav86] when histories are coarse–grained by intervals of position at various times.

The equivalence represented in (7.70) relies on several special assumptions about the nature of space–time and the fine–grained histories. In particular, it requires[12]:

- A fixed Lorentzian space–time geometry to define time–like and space–like directions.
- A foliating family of space–like surfaces through which states can evolve.
- Fine–grained histories that are single-valued in the time labeling the space–like surfaces in the foliating family.

As an illustrative example where the equivalence does not hold, consider quantum field theory in a fixed background space–time with closed time–like curves (CTCs) such as those that can occur in wormhole space–times [MTY88]. The fine–grained histories are 4D field configurations that are single-valued on space–time. But there is no foliating family of space–like surfaces with which to define the Hamiltonian evolution of a quantum state. Thus, there is no usual 3+1 formulation of the quantum mechanics of fields in space–times with CTCs.

However, there is a 4D sum–over–histories formulation of field theory in space–times with CTCs [Har94a, FPS92, Ros98]. The resulting theory has some unattractive properties such as acausality and non–unitarity. But it does illustrate how closely usual quantum theory incorporates particular assumptions about space–time, and also how these requirements can be relaxed in a suitable generalization of the usual theory.

Generalized Quantum Theory

In generalizing usual quantum mechanics to deal with quantum space–time, some of its features will have to be left behind and others retained. What are the minimal essential features that characterize a quantum mechanical theory? The generalized quantum theory framework [Har91a, Har93, Ish94] provides one answer to this question. Just three elements abstracted from the quantum mechanics of closed systems above define a generalized quantum theory [Har05b]:

- *Fine–grained histories*: the sets of alternative fine–grained histories of the closed system which are the most refined descriptions of it physically possible.

[12] The usual 3+1 formulation is also restricted to coarse–grained histories specified by alternatives at definite moments of time. More general space–time coarse–grainings that are defined by quantities that extend over time can be used in the space–time formulation (see, e.g. [BH05] and references therein). Spacetime alternatives are the only ones available in a diffeomorphism invariant quantum gravity.

- *Coarse–grained histories*: these are partitions of a set of fine–grained histories into an exhaustive set of exclusive classes $\{c_\alpha\}, \alpha = 1, 2 \cdots$. Each class is a coarse–grained history.
- *Decoherence functional*: the measure of quantum interference $D(\alpha, \alpha')$ between pairs of histories in a coarse–grained set, meeting the following conditions:

(i) *Hermiticity*: $D(\alpha, \alpha') = D^*(\alpha', \alpha)$
(ii) *Positivity*: $D(\alpha, \alpha) \geq 0$
(iii) *Normalization*: $\Sigma_{\alpha\alpha'} D(\alpha, \alpha') = 1$
(iv) *Principle of superposition*: If $\{\bar{c}_{\bar{\alpha}}\}$ is a further coarse–graining of $\{c_\alpha\}$, then

$$\bar{D}(\bar{\alpha}, \bar{\alpha}') = \sum_{\substack{\alpha \in \bar{\alpha} \\ \alpha' \in \bar{\alpha}'}} D(\alpha, \alpha')$$

Probabilities $p(\alpha)$ are assigned to sets of coarse–grained histories when they decohere according to the basic relation

$$D(\alpha, \alpha') \approx \delta_{\alpha\alpha'}\, p(\alpha)\,. \tag{7.71}$$

These $p(\alpha)$ satisfy the basic requirements for probabilities as a consequence of (i)–(iv) above. In particular, they satisfy the sum rule

$$p(\bar{\alpha}) = \sum_{\alpha \in \bar{\alpha}} p(\alpha) \tag{7.72}$$

as a consequence of (i)–(iv) and decoherence. For instance, the probabilities of an exhaustive set of alternatives always sum to 1.

The sum–over–histories formulation of usual quantum mechanics given above is a particular example of a generalized quantum theory. The decoherence functional (7.64) satisfies the requirements (i)–(iv). But its particular form is not the only way of constructing a decoherence functional. Therein lies the possibility of generalization.

A Quantum Theory of Spacetime Geometry

The low energy, effective theory of quantum gravity is a quantum version of general relativity with a space–time metric $g_{\alpha\beta}(x)$ coupled to matter fields. Of course, the divergences of this effective theory have to be regulated to extract predictions from it (perhaps, most naturally by discrete approximations to geometry such as the *Regge calculus*, see, e.g. [Har85a, HW04]). These predictions can therefore be expected to be accurate only for limited coarse–grainings and certain states. But this effective theory does supply an instructive model for generalizations of quantum theory that can accommodate quantum space–time. This generalization is sketched in this section.

The key idea is that the fine–grained histories do not have to represent evolution *in* space–time. Rather they can be histories of space–time. For this discussion we take these histories to be spatially closed cosmological 4–geometries represented by metrics $g_{\alpha\beta}(x)$ on a fixed manifold $M = \mathbb{R} \times M^3$ where M^3 is a closed 3–manifold. For simplicity, we restrict attention to a single scalar matter field $\phi(x)$.

The three ingredients of a generalized quantum theory for space–time geometry are then as follows [Har05b]:

- *Fine–grained Histories*: A fine–grained history is defined by a 4D metric and matter field configuration on M.
- *Coarse–grainings*: The allowed coarse-grainings are partitions of the metrics and matter fields into 4D *diffeomorphism invariant* classes $\{c_\alpha\}$.
- *Decoherence Functional*: A decoherence functional constructed on sum–over–history principles analogous to that described above for usual quantum theory. Schematically, branch state vectors $|\Psi_\alpha\rangle$ can be constructed for each coarse–grained history by summing over the metrics and fields in the corresponding class c_α of fine–grained histories [Har05b]

$$|\Psi_\alpha\rangle = \int_{c_\alpha} \mathcal{D}[g]\mathcal{D}[\phi]\, \exp\{iS[g,\phi]/\hbar\}\, |\Psi\rangle\,. \tag{7.73}$$

A decoherence functional satisfying the requirements above is

$$D(\alpha', \alpha) = \langle \Psi_{\alpha'} | \Psi_\alpha \rangle\,. \tag{7.74}$$

Here, $S[g,\phi]$ is the action for general relativity coupled to the field $\phi(x)$, and $|\Psi\rangle$ is the initial cosmological state. The construction is only schematic because we did not spell out how the functional integrals are defined or regulated, nor did we specify the product between states that is implicit in both (7.73) and (7.74). These details can be made specific in models [Har95b, HM97, CH04], but they will not be needed for the subsequent discussion.

A few remarks about the coarse–grained histories may be helpful. To every physical assertion that can be made about the geometry of the universe and the fields within, there corresponds a diffeomorphism invariant partition of the fine–grained histories into the class where the assertion is true and the class where it is false. The notion of coarse–grained history described above therefore supplies the most general notion of alternative describable in space–time form. Among these we do not expect to find local alternatives because there is no diffeomorphism invariant notion of locality. In particular, we do not expect to find alternatives specified at a moment of time. We do expect to find alternatives referring to the kind of relational observables discussed in [GMH06] and the references therein. We also expect to find observables referring to global properties of the universe such as the maximum size achieved over the history of its expansion.

This generalized quantum mechanics of space–time geometry is in fully space–time form with alternatives described by partitions of 4D histories and a decoherence functional defined by sums over those histories. It is analogous to the space–time formulation of usual quantum theory reviewed above.

However, unlike the above theory, we cannot expect an equivalent 3+1 formulation, of the kind described above, expressed in terms of states on space–like surfaces and their unitary evolution between these surfaces. The fine–grained histories are not 'single–valued' in any geometrically defined variable labelling a space–like surface. They therefore cannot be factored across a space–like surface as in (7.70). More precisely, there is no geometrical variable that picks out a unique space–like surface in all geometries.[13]

Even without a unitary evolution of states the generalized quantum theory is fully predictive because it assigns probabilities to the most general sets of coarse–grained alternative histories described in space–time terms when these are decoherent.

How then is usual quantum theory used every day, with its unitarily evolving states, connected to this generalized quantum theory that is free from them? The answer is that usual quantum theory is an approximation to the more general framework that is appropriate for those coarse–grainings and initial state $|\Psi\rangle$ for which space–time behaves classically. One equation will show the origin of this relation. Suppose we have a coarse–graining that distinguishes between fine–grained geometries only by their behavior on scales well above the Planck scale. Then, for suitable states $|\Psi\rangle$ we expect that the integral over metrics in (7.74) can be well approximated semiclassically by the method of steepest descents. Suppose further for simplicity that only a single classical geometry with metric $\hat{g}_{\alpha\beta}$ dominates the semiclassical approximation. Then, (7.74) becomes [Har05b]

$$|\Psi_\alpha\rangle \approx \int_{\hat{c}_\alpha} \mathcal{D}[\phi] \exp\{iS[\hat{g},\phi]/\hbar\} |\Psi\rangle, \qquad (7.75)$$

where $\hat{c}_\alpha$ is the coarse–graining of $\phi(x)$ arising from c_α and the restriction of $g_{\alpha\beta}(x)$ to $\hat{g}_{\alpha\beta}(x)$. Equation (7.75) effectively defines a quantum theory of the field $\phi(x)$ in the fixed background space–time with the geometry specified by $\hat{g}_{\alpha\beta}(x)$. This is familiar territory. Field histories are single valued on space–time. This *sum–over–fields* can thus be factored across space–like surfaces in the geometry $\hat{g}$ as in (7.70) to define field states on space–like surfaces, their unitary evolution, and their Hilbert space product. Usual quantum theory is thus recovered when space–time behaves classically and provides the fixed space–time geometry on which usual quantum theory relies. From this perspective, familiar quantum theory and its unitary evolution of states is an

[13] Spacelike surfaces labelled by the trace of the extrinsic curvature K foliate certain classes of classical space–times obeying the Einstein equation [MT80]. However, there is no reason to require that non-classical histories be foliable in this way. It is easy to construct geometries where surfaces of a given K occur arbitrarily often.

effective approximation to a more general sum–over–histories formulation of quantum theory. The approximation is appropriate for those coarse-grainings and initial states in which space–time geometry behaves classically.

Beyond Spacetime

The generalized quantum theory of space–time sketched in the previous section assumed that geometry was a fundamental variable—part of the description of the fine–grained histories. But on almost every frontier in quantum gravity one finds the idea that continuum geometry is not fundamental, but will be replaced by something more fundamental. This is true for string theory [Sei06], loop quantum gravity [AL04], and the causal set program [Dow05, Hen06] although space does not permit a review of these speculations.

Can generalized quantum theory serve as a framework for theories where space–time is emergent rather than fundamental? Certainly we cannot expect to have a notion of 'history'. But we can expect some fine–grained description, or a family of equivalent ones, and that is enough. A generalized quantum theory needs [Har05b]:

- The possible fine–grained descriptions of the system.
- The coarse–grained descriptions constructed from the fine–grained ones.
- A measure of quantum interference between different coarse–grained descriptions respecting conditions (i)–(iv) above.

Generalized quantum theory requires neither space nor time and can therefore serve as the basis for a quantum theory in which space–time is emergent.

Emergence/Excess Baggage

The word 'emergent' appears in a number of places in the previous discussion. It probably has many meanings. This section aims at a more precise understanding of what is meant by the term in this essay.

Suppose we have a quantum theory defined by certain sets of fine–grained histories, coarse–grainings, and a decoherence functional. Let's call this the *fundamental theory*. It may happen that the decoherence and probabilities of limited kinds of sets of coarse–grained histories are given approximately by a second, *effective theory*. The two theories are related in the following way [Har05b]:

- Every fine–grained history of the effective theory is a coarse–grained history of the fundamental theory.
- The decoherence functionals approximately agree on a limited class of sets of coarse–grained histories.

$$D^{\mathrm{fund}}(\alpha', \alpha) \approx D^{\mathrm{eff}}(\alpha', \alpha)\,. \tag{7.76}$$

On the right, α' and α refer to the fine–grained histories of the effective theory. On the left, they refer to the corresponding coarse–grained histories of the fundamental theory.

When two theories are related in this way we can say that the effective theory is *emergent* from the fundamental theory. Loosely we can say that the restrictions, and the concepts that characterize them, are emergent. It should be emphasized that an approximate equality like (7.76) can be expected to hold, not just as a consequence of the particular dynamics incorporated into decoherence functionals, but also only for particular states.

Several examples of emergence in this sense have been considered in this essay: There is the possible emergence of a generalized quantum theory of space–time geometry from a theory in which space–time is not fundamental. There is the emergence of a 3+1 quantum theory of fields in a fixed background geometry from a 4D generalized quantum theory in which geometry is a quantum variable. There is the emergence of the approximate quantum mechanics of measured subsystems (textbook quantum theory) from the quantum mechanics of the universe. And there is the emergence of classical physics from quantum phyaics.

Instead of looking at an effective theory as a restriction of a more fundamental one, we may look at the fundamental theory as a generalization of the effective one. That perspective is important because generalization is a way of searching for more comprehensive theories of nature. In passing from the specific to the more general some ideas have to be discarded. They are often ideas that were once perceived to be general because of our special place in the universe and the limited range of our experience. But, in fact, they arise from special situations in a more general theory. They are 'excess baggage' that has to be discarded to reach a more comprehensive theory [Har90]. Emergence and excess baggage are two ways of looking at the same thing [Har05b].

Emergence of Signature

Classical space–time has Lorentz signature. At each point it is possible to choose one time–like direction and three orthogonal space–like ones. There are no physical space–times with zero time–like directions or with two time–like directions. But is such a seemingly basic property fundamental, or is it rather, emergent from a quantum theory of space–time which allows for all possible signatures? This section sketches a simple model where that happens.

Classical behavior requires particular states [Har94b]. Let's consider the possible classical behaviors of cosmological geometry assuming the *no–boundary quantum state of the universe* [HH83] in a theory with only gravity and a cosmological constant Λ. The *no–boundary wave function* is given by a *sum–over–geometries* of the schematic form [Har05b]

$$\Psi[h] = \int_e \mathcal{D}[g]\, \mathrm{e}^{-I[g]/\hbar} . \tag{7.77}$$

For simplicity, we consider a fixed manifold[14] M. The key requirement is that it be compact with one boundary for the argument of the wave function and no other boundary. The functional $I[g]$ is the Euclidean action for metric defining the geometry on M. The sum is over a complex contour $\mathcal{C}$ of g's that have finite action and match the three-metric h on the boundary that is the argument of Ψ.

Quantum theory predicts classical behavior when it predicts high probability for histories exhibiting the correlations in time implied by classical deterministic laws [GMH93, Har94b]. The state Ψ is an input to the process of predicting those probabilities as described above. However, plausibly the output for the predicted classical space–times in this model are the extrema of the action in (7.77). We will assume this (see [Har95b] for some justification). Further, to keep the discussion manageable, we will restrict it to the *real* extrema. These are the real tunneling geometries discussed in a much wider context in [GMH90].

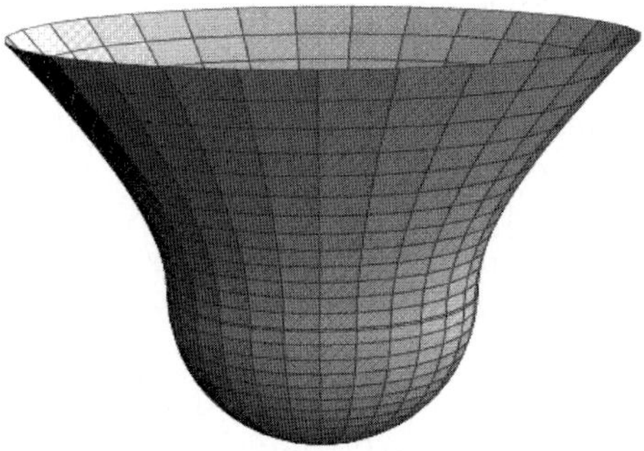

Fig. 7.4. The emergence of the Lorentz signature $(-,+,+,+)$ of space–time. The semiclassical geometry describing a classical space–time which becomes large according to the 'no-boundary' proposal for the universe's quantum state. The model is pure gravity and a cosmological constant. Purely Euclidean geometries $(+,+,+,+)$ or purely Lorentzian geometries are not allowed as described in the text. What is allowed is the real tunneling geometry illustrated above consisting of half a Euclidean four-sphere joined smoothly onto an expanding Lorentzian de Sitter space at the moment of maximum contraction. This can be described as the nucleation of classical Lorentz signatured space–time. There is no similar nucleation of a classical geometry with signature $(-,-,+,+)$ because it could not match the Euclidean one across a space–like surface (modified and adapted from [Har05b]).

Let us ask for the semiclassical geometries which become large, i.e., contain symmetric three surfaces with size much larger than $(1/\Lambda)^{1/2}$. There are none

[14] Even the notion of manifold may be emergent in a more general theory of certain complexes [Har85b, SW93].

with Euclidean signature. The purely Euclidean extremum is the round 4–sphere with linear size $(1/\Lambda)^{1/2}$ and contains no symmetric three surfaces with larger size. There are none with purely Lorentzian signature either because these cannot be regular on M. There are, however, tunneling solutions of the kind illustrated in Figure 7.4 in which half of a Euclidean 4–sphere is matched to expanding DeSitter space across a surface of vanishing extrinsic curvature.

Could a space–time with two time and two space directions be nucleated in this way? The answer is 'no' because the geometry on a surface could not have the three space–like directions necessary to match onto the half of a 4–sphere.

Thus, in this very simple model, with many assumptions, if we live in a large universe it must have one time and three space dimensions. The Lorentzian signature of classical space–time is an emergent property from an underlying theory not committed to this signature [Har05b].

Beyond Quantum Theory

The path of generalization in the previous sections began with the textbook quantum mechanics of measurement outcomes in a fixed space–time and ended in a quantum theory where neither measurements nor space–time are fundamental. In this journey, the principles of generalized quantum theory are preserved, in particular the idea of quantum interference and the linearity inherent in the principle of superposition. But the end of this path is strikingly different from its beginning.

The founders of quantum theory thought that the indeterminacy of quantum theory "reflected the unavoidable interference in measurement dictated by the magnitude of the quantum of the action" (Bohr). But what then is the origin of quantum indeterminacy in a closed quantum universe which is never measured? Why enforce the principle of superposition in a framework for prediction of the universe which has but a single quantum state? In short, the endpoint of this journey of generalization forces us to ask J. Wheeler's famous question, "How come the quantum?" [Whe86].

Could quantum theory itself be an emergent effective theory? Many have thought so. Extending quantum mechanics until it breaks could be one route to finding out. If space–time geometry is not fundamental, quantum mechanics will need further generalization and generalized quantum theory provides one framework for exploring that. For more details, see [Har05b].

7.3.5 Anthropic String Landscape

The *anthropic string landscape* refers to the large number of different *false vacua* in string theory. Recall that a *false vacuum* (see Figure 7.5) is a

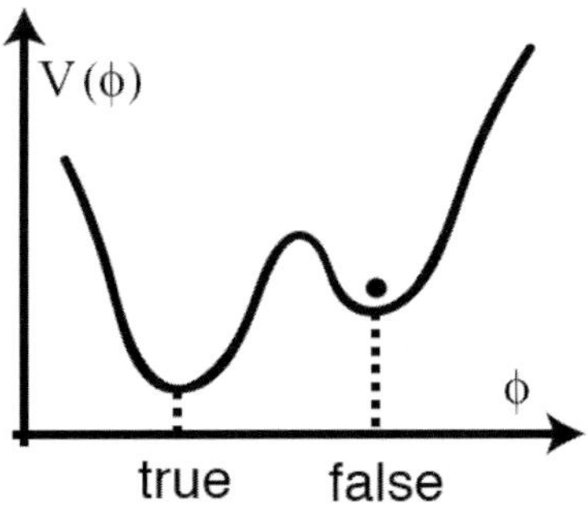

Fig. 7.5. A scalar field ϕ in a false vacuum. Note that the potential energy $V(\phi)$ is higher than that in the true vacuum or ground state, but there is a barrier preventing the field from classically rolling down to the true vacuum. Therefore, the transition to the true vacuum must be stimulated by the creation of high energy particles or through quantum mechanical tunnelling.

metastable sector of a *quantum field theory* (QFT) which appears to be a *perturbative vacuum* but is unstable to instanton effects which tunnel to a lower energy state. This *tunnelling*[15] can be caused by quantum fluctuations or the creation of high energy particles. Simply put, the false vacuum is a state

[15] Recall that tunneling is the quantum–mechanical effect of transitioning through a classically–forbidden energy state. It can be generalized to other types of classically–forbidden transitions as well. For example, consider rolling a ball up a hill. If the ball is not given enough velocity, then it will not roll over the hill. This scenario makes sense from the standpoint of classical mechanics, but is an inapplicable restriction in quantum mechanics simply because quantum mechanical objects do not behave like classical objects such as balls. On a quantum scale, objects exhibit wavelike behavior. For a quantum particle moving against a potential energy 'hill', the wave ψ–function describing the particle can extend to the other side of the hill. This wave represents the probability of finding the particle in a certain location, meaning that the particle has the possibility of being detected on the other side of the hill. This behavior is called tunnelling; it is as if the particle has 'dug' through the potential hill. As this is a quantum and non–classical effect, it can generally only be seen in nanoscopic phenomena, where the wave behavior of particles is more pronounced. Availability of states is necessary for tunnelling to occur. In the above example, the quantum mechanical ball will not appear inside the hill because there is no available 'space' for it to exist, but it can tunnel to the other side of the hill, where there is free space. Analogously, a particle can tunnel through the barrier, but unless there are states available within the barrier, the particle can only tunnel to the other side of the barrier. The wave–function describing a particle only expresses the probability of finding the particle at a location assuming a free state exists.

of a physical theory which is not the lowest energy state, but is nonetheless stable for some time. This is analogous to metastability for the *first order phase transitions*.[16]

---

[16] Recall that in thermodynamics, *phase transition* is the transformation of a thermodynamic system from one phase to another. The distinguishing characteristic of a phase transition is an abrupt sudden change in one or more physical properties, in particular the heat capacity, with a small change in a thermodynamic variable such as the temperature. The first–order phase transitions are those that involve a latent heat. During such a transition, a system either absorbs or releases a fixed (and typically large) amount of energy. Because energy cannot be instantaneously transferred between the system and its environment, first–order transitions are associated with 'mixed–phase regimes', in which some parts of the system have completed the transition and others have not. This phenomenon is familiar to anyone who has boiled a pot of water: the water does not instantly turn into gas, but forms a turbulent mixture of water and water vapor bubbles. Mixed–phase systems are difficult to study, because their dynamics are violent and hard to control. However, many important phase transitions fall in this category, including the solid/liquid/gas transitions and *Bose–Einstein condensation*.

On the other hand, the *second–order phase transitions* are the continuous phase transitions, such as the ferromagnetic transition and the superfluid transition. They have no associated latent heat.

In systems containing liquid and gaseous phases, there exist a special combination of pressure and temperature, known as the critical point, at which the transition between liquid and gas becomes a second–order transition. Near the critical point, the fluid is sufficiently hot and compressed that the distinction between the liquid and gaseous phases is almost non–existent. Phase transitions often (but not always) take place between phases with different symmetry. Consider, for example, the transition between a fluid (i.e., liquid or gas) and a crystalline solid. A fluid, which is composed of atoms arranged in a disordered but homogeneous manner, possesses continuous translational symmetry: each point inside the fluid has the same properties as any other point. A crystalline solid, on the other hand, is made up of atoms arranged in a regular lattice. Each point in the solid is not similar to other points, unless those points are displaced by an amount equal to some lattice spacing.

Generally, we may speak of one phase in a phase transition as being more symmetrical than the other. The transition from the more symmetrical phase to the less symmetrical one is a symmetry–breaking process. In the fluid–solid transition, for example, we say that continuous translation symmetry is broken.

The ferromagnetic transition is another example of a *symmetry–breaking transition*, in this case the symmetry under reversal of the direction of electric currents and magnetic field lines. This symmetry is referred to as *up–down symmetry* or *time–reversal symmetry*. It is broken in the ferromagnetic phase due to the formation of magnetic domains containing aligned magnetic moments. Inside each

In a physical theory in a *false vacuum*, the system moves to a lower energy state (either the true vacuum, or another, lower energy vacuum) through a process known as *bubble nucleation* [Col77]. In this, instanton effects cause a bubble to appear in which fields have their true vacuum values inside. Therefore, the interior of the bubble has a lower energy. The walls of the bubble (aka domain walls) have a surface tension, as energy is expended as the fields roll over the potential barrier to the lower energy vacuum. The most likely

---

domain, there is a magnetic field pointing in a fixed direction chosen spontaneously during the phase transition. The name 'time–reversal symmetry' comes from the fact that electric currents reverse direction when the time coordinate is reversed.

The presence of symmetry–breaking (or non–breaking) is important to the behavior of phase transitions. It was pointed out by Landau that, given any state of a system, one may unequivocally say whether or not it possesses a given symmetry. Therefore, it cannot be possible to analytically deform a state in one phase into a phase possessing a different symmetry. This means, for example, that it is impossible for the solid–liquid phase boundary to end in a critical point like the liquid–gas boundary. However, symmetry–breaking transitions can still be either first– or second–order.

Typically, the more symmetrical phase is on the high–temperature side of a phase transition, and the less symmetrical phase on the low–temperature side. This is certainly the case for the solid–fluid and ferromagnetic transitions. This happens because the *Hamiltonian* of a system usually exhibits all the possible symmetries of the system, whereas the low–energy states lack some of these symmetries (this phenomenon is known as *spontaneous symmetry breaking*). At low temperatures, the system tends to be confined to the low–energy states. At higher temperatures, thermal fluctuations allow the system to access states in a broader range of energy, and thus more of the symmetries of the Hamiltonian.

When symmetry is broken, one needs to introduce one or more extra variables to describe the state of the system. For example, in the ferromagnetic phase one must provide the net magnetization, whose direction was spontaneously chosen when the system cooled below the Curie point. Such variables are examples of *order parameters*.

Symmetry–breaking phase transitions play an important role in cosmology. It has been speculated that, in the hot early universe, the vacuum (i.e., the various quantum fields that fill space) possessed a large number of symmetries. As the universe expanded and cooled, the vacuum underwent a series of symmetry–breaking phase transitions. For example, the electro–weak transition broke the $SU(2) \times U(1)$ symmetry of the electro–weak field into the $U(1)$ symmetry of the present–day electromagnetic field. This transition is important to understanding the asymmetry between the amount of matter and antimatter in the present–day universe.

size of the bubble is determined in the *semiclassical approximation*[17] to be such that the bubble has zero total change in the energy: the decrease in energy by the true vacuum in the interior is compensated by the tension of the walls. Any increase in size of the bubble will decrease its potential energy, as the energy of the wall increases as the area of a sphere $4\pi r^2$ but the negative contribution of the interior increases more quickly, as the volume of a sphere $4/3\pi r^3$. Therefore, after the bubble is nucleated, it quickly begins expanding at very nearly the speed of light. The excess energy contributes to the very large kinetic energy of the walls. If two bubbles are nucleated and they eventually collide, it is thought that particle production occurs where the walls collide.

The *tunnelling rate* is increased by increasing the energy difference between the two vacua and decreased by increasing the height or width of the barrier.

The addition of gravity to the story leads to a considerably richer variety of phenomena. The key insight is that a false vacuum with positive potential energy density is a *de Sitter vacuum*, in which the potential energy acts as a cosmological constant and the universe is undergoing the exponential expansion of *de Sitter space*.[18] This leads to a number of interesting effects, first studied by Coleman and de Luccia [CL80]:

---

[17] Recall that the semiclassical approximation may refer to quantum-mechanical calculations that are obtained by considering a small perturbation of a classical calculation, for example the WKB approximation in non–relativistic quantum mechanics or the loop expansion or the instanton methods in quantum field theory. In quantum field theory, a semiclassical correction arises from *one–loop Feynman diagrams*. The *semiclassical effective action* is

$$\Gamma[\phi] = S[\phi] + \frac{1}{2}Tr\left[\ln S^{(2)}[\phi]\right] + ...$$

[18] Recall that nD *de Sitter space* (see, e.g., [Cox43]) is the maximally symmetric, simply–connected, *Lorentzian manifold* with constant positive curvature. It may be regarded as the Lorentzian analog of an $n-$sphere (with its canonical Riemannian metric). De Sitter space is most easily defined as a submanifold of Minkowski space in one higher dimension. Take Minkowski space $\mathbb{R}^{1,n}$ with the standard metric,

$$ds^2 = -dx_0^2 + \sum_{i=1}^{n} dx_i^2,$$

De Sitter space is the submanifold described by the hyperboloid,

$$-x_0^2 + \sum_{i=1}^{n} x_i^2 = \alpha^2,$$

where α is some positive constant with dimensions of length. The metric on de Sitter space is the metric induced from the ambient Minkowski metric. One

1. Tunnelling from a space with zero potential energy (e.g., Minkowski space) to negative potential energy: the walls of the bubble grow at the speed of light, as described above. However, the interior of the bubble rapidly collapses, as anti–de Sitter space and the universe ends (see ultimate fate of the universe and vacuum metastability event, below).
2. Tunnelling from a space of positive potential energy (de Sitter space) to one of vanishing potential energy (Minkowski space). In this case, the volume of the bubble continues to grow at the speed of light. Since the exterior of the bubble is expanding exponentially, however, and the Minkowski space is not, unlike the non–gravitational case, the whole of spa⌐e time need never be dominated by the lower energy vacuum. If the tu nelling

---

can check that the induced metric is nondegenerate and has Lorentzian signature. Topologically, a simply–connected de Sitter space is $\mathbb{R}\times S^{n-1}$. The isometry group of de Sitter space is the Lorentz group $O(1,n)$. The metric therefore then has $n(n+1)/2$ independent *Killing vectors* and is maximally symmetric. Every maximally symmetric space has constant curvature. The *Riemann curvature tensor* of de Sitter space is given by

$$R_{\rho\sigma\mu\nu} = \frac{1}{\alpha^2}(g_{\rho\mu}g_{\sigma\nu} - g_{\rho\nu}g_{\sigma\mu}).$$

De Sitter space is an *Einstein manifold* since the *Ricci tensor* is proportional to the metric

$$R_{\mu\nu} = \frac{n-1}{\alpha^2}g_{\mu\nu}.$$

In the language of general relativity, de Sitter space is the maximally symmetric, vacuum solution of *Einstein's field equation* with a positive cosmological constant given by:

$$\Lambda = (n-1)(n-2)/2\alpha^2.$$

The scalar curvature of de Sitter space is given by

$$R = \frac{n(n-1)}{\alpha^2} = \frac{2n}{n-2}\Lambda.$$

We can introduce static coordinates for de Sitter space as

$$x_0 = \sqrt{\alpha^2 - r^2}\sinh(t/\alpha), \qquad x_1 = \sqrt{\alpha^2 - r^2}\cosh(t/\alpha), \qquad x_i = rz_i \qquad (2 \le i \le n),$$

where z_i gives the standard embedding of the $(n-2)$–sphere in $\mathbb{R}^{n-1}$. In these coordinates the de Sitter metric takes the form

$$ds^2 = -\left(1 - \frac{r^2}{\alpha^2}\right)dt^2 + \left(1 - \frac{r^2}{\alpha^2}\right)^{-1}dr^2 + r^2 d\Omega_{n-2}^2.$$

When $n = 4$, de Sitter space is considered to be a cosmological model for the physical universe, called *de Sitter universe*. In this case, we have $\Lambda = 3/\alpha^2$ and $R = 4\Lambda = 12/\alpha^2$.

rate is slow enough, the exponentially expanding space in the false vacuum state can expand quickly enough that the bubbles of lower–energy space never begin to collide and convert all of space time to the lower energy state. That is, the tunnelling is competing with rapid expansion, and the exponential expansion can be rapid enough that the tunnelling effect is overwhelmed.
3. Tunnelling from positive potential energy to lower, positive potential energy. Just as for the above case, the more rapid exponential expansion of the higher energy false vacuum can continue to dominate.
4. Tunnelling from positive potential energy to negative potential energy. This effect is highly suppressed: the expansion of the positive energy vacuum dominates the contraction of the negative energy vacuum.

A final kind of tunnelling is the *Hawking–Moss instanton* [HM82]. This occurs when the size of the Coleman–de Luccia bubble is larger than the size of the universe, in a closed universe, or of the horizon. In this case, the entire universe tunnels from the false vacuum to the true vacuum at once.

In his original proposal for cosmic inflation [Gut81], A. Guth proposed that inflation could end through quantum mechanical bubble nucleation of the sort described above. It was soon understood that a homogeneous and isotropic universe could not be preserved through the violent tunnelling process. This led A. Linde [Lin82] and, independently, A. Albrecht and P. Steinhardt [AS82] to propose the so–called 'new inflation' or *slow–roll inflation*, in which no tunnelling occurs, and the inflationary scalar field instead rolls down a gentle slope. A more recent application of these tunnelling phenomena in cosmology and particle physics is the string landscape in which string theory is conjectured to be populated by an exponentially large 'discretuum' of false vacua, and the small observed value of the cosmological constant (see dark energy) can be explained by the anthropic principle and quantum mechanical tunnelling to the lowest positive energy vacuum.

In their paper, Coleman and de Luccia noted [CL80]: "The possibility that we are living in a false vacuum has never been a cheering one to contemplate. Vacuum decay is the ultimate ecological catastrophe; in the new vacuum there are new constants of nature; after vacuum decay, not only is life as we know it impossible, so is chemistry as we know it. However, one could always draw stoic comfort from the possibility that perhaps in the course of time the new vacuum

---

Many inflationary models are approximately de Sitter space and can be modelled by giving the Hubble parameter a mild time dependence. For simplicity, some calculations involving inflation in the early universe can be performed in de Sitter space rather than a more realistic inflationary universe. By using the de Sitter universe instead, where the expansion is truly exponential, there are many simplifications.

would sustain, if not life as we know it, at least some structures capable of knowing joy. This possibility has now been eliminated." The possibility that we are living in a false vacuum has been considered. If a bubble of lower energy vacuum were nucleated, it would approach at nearly the speed of light and destroy the Earth instantaneously, without any forewarning. Thus, this vacuum metastability event is a theoretical doomsday event. This was used in a science–fiction story by G.A.Landis in 1988.[19]

The *string landscape* arises from the idea that there are an extremely large number of metastable vacua (ground states) in string theory (see, e.g., [Dou]). The large number of possibilities arise from different choices of *Calabi–Yau manifolds* and different values of generalized magnetic fluxes over different homology cycles. This large number of de–Sitter like metastable vacua [BP00] is thought by some physicists to be large enough that the known laws of physics, the Standard Model and general relativity with a positive cosmological constant, occurs in at least one, although computing quantities such as masses of particles and Yukawa couplings for even a single vacuum is a technically difficult problem. The problem of enumerating all the vacua is thought to be NP complete [DD06].

The Landscape

In this subsection, following [Sus03], we now give a detailed description of the anthropic string landscape. Recall that the world–view shared by most physicists is that the laws of nature are uniquely described by some special action principle that completely determines the vacuum, the spectrum of elementary particles, the forces and the symmetries. Experience with quantum electrodynamics and quantum chromodynamics suggests a world with a small number of parameters and a unique ground state. For the most part, string theorists

[19] One scenario is that, rather than quantum tunnelling, a particle accelerator, which produces very high energies in a very small area, could create sufficiently high energy density as to penetrate the barrier and stimulate the decay of the false vacuum to the lower energy vacuum. Hut and Rees [HR83], however, have determined that because we had observed cosmic ray collisions at much higher energies than those produced in terrestrial particle accelerators, that these experiments will not, at least for the foreseeable future, pose a threat to our vacuum. Particle accelerations have reached energies of only approximately four thousand billion electron volts (4×10^3 GeV). Cosmic ray collisions have been observed at and beyond energies of $10^1 1$ GeV, the so–called *Greisen–Zatsepin–Kuzmin limit*. This event would be contingent on our living in a metastable vacuum, an issue which is far from resolved [TW82]. Worries about the vacuum metastability event are reminiscent of the controversy about turning the *Relativistic Heavy Ion Collider* on.

bought into this paradigm. At first it was hoped that string theory would be unique and explain the various parameters that quantum field theory left unexplained. When this turned out to be false, the belief developed that there were exactly five string theories with names like type–2a and Heterotic. This also turned out to be wrong. Instead, a continuum of theories were discovered that smoothly interpolated between the five and also included a theory called M–Theory. The language changed a little. One no longer spoke of different theories, but rather different solutions of some master theory. The space of these solutions is called the *moduli space of supersymmetric vacua*. Susskind calls it the *supermoduli–space*. Moving around on this supermoduli–space is accomplished by varying certain dynamical *moduli*. Examples of moduli are the size and shape parameters of the compact internal space that 4–dimensional string theory always needs. These moduli are not parameters in the theory but are more like fields. As you move around in ordinary space, the moduli can vary and have their own equations of motion. In a low energy approximation the moduli appear as massless scalar fields. The beauty of the supermoduli–space point of view is that there is only one theory but many solutions which are characterized by the values of the scalar field moduli. The mathematics of the string theory is so precise that it is hard to believe that there isn't a consistent mathematical framework underlying the supermoduli–space vacua.

However the continuum of solutions in the supermoduli–space are all supersymmetric with exact super–particle degeneracy and vanishing cosmological constant. Furthermore they all have massless scalar particles, the moduli themselves. Obviously none of these vacua can possibly be our world. Therefore the string theorist must believe that there are other discrete islands lying off the coast of the supermoduli–space . The hope now is that a single non–supersymmetric island or at most a small number of islands exist and that non–supersymmetric physics will prove to be approximately unique. This view is not inconsistent with present knowledge (indeed it is possible that there are no such islands) but we find it completely implausible. It is much more likely that the number of discrete vacua is astronomical [Sus03].

This change in viewpoint is demanded by two facts, one observational and one theoretical. The first is that the expansion of the universe is accelerating. The simplest explanation is a small but non–zero cosmological constant. Evidently we have to expand our thinking about vacua to include states with non–zero vacuum energy. The incredible smallness and apparent fine tuning of the cosmological constant makes it absurdly improbable to find a vacuum in the observed range unless there are an enormous number of solutions with almost every possible value of λ. It seems to me inevitable that if we find one such vacuum we will find a huge number of them. We will from now on call the space of all such string theory vacua the *landscape*.

The second fact is that some recent progress has been made in exploring the landscape [BP00, KKL03]. Before explaining the new ideas we need to define more completely what we mean by the landscape. The supermoduli–space is parameterized by the moduli which we can think of as a collection of scalar fields Φ_n. Unlike the case of Goldstone bosons, points in the moduli space are not related by a symmetry of the theory. Generically, in a quantum field theory, changing the value of a non–Goldstone scalar involves a change of potential energy. In other words there is a non–zero field potential $V(\Phi)$. Local minima of V are what we call vacua. If the local minimum is an absolute minimum the vacuum is stable. Otherwise it is only metastable. The value of the potential energy at the minimum is the cosmological constant for that vacuum.

To the extent that the low energy properties of string theory can be approximated by field theory, similar ideas apply. Bearing in mind that the low energy approximation may break down in some regions of the landscape, we will assume the existence of a set of fields and a potential. The space of these fields is the landscape.

The supermoduli–space is a special part of the landscape where the vacua are supersymmetric and the potential $V(\Phi)$ is exactly zero. These vacua are marginally stable and can be excited by giving the moduli arbitrarily small time derivatives. On the supermoduli–space the cosmological constant is also exactly zero. Roughly speaking, the supermoduli–space is a perfectly flat plain at exactly zero altitude (the value of V). Once we move off the plain, supersymmetry is broken and a non–zero potential developes, usually through some non–perturbative mechanism. Thus beyond the flat plain we encounter hills and valleys. We are particularly interested in the valleys where we find local minima of V. Each such minimum has its own vacuum energy. The typical value of the potential difference between neighboring valleys will be some fraction of M_p^4 where M_p is the Planck mass. The potential barriers between minima will also be of similar height. Thus if a vacuum is found with cosmological constant of order $10^{-120}M_p^4$, it will be surrounded by much higher hills and other valleys.

Next consider two large regions of space, each of which has the scalars in some local minimum, the two minima being different. If the local minima are landscape–neighbors then the two regions of space will be separated by a domain wall. Inside the domain wall the scalars go over a "mountain pass". The interior of the regions are vacuum like with cosmological constants. The domain wall which can also be called a membrane has additional energy in the form of a membrane tension. Thus there will be configurations of string theory which are not globally described by a single vacuum but instead consists of many domains separated by domain walls. Accordingly, the landscape in field space is reflected in a complicated terrain in real space.

There are scalar fields that are not usually thought of as moduli but once we leave the flat plain there is no any fundamental difference. These are the four–form field strengths first introduced in the context of the cosmological constant by Brown and Teitelboim [BT88]. A simple analogy exists to help visualize these fields and their potential. Think of 1+1 dimensional electrodynamics with electric fields[20] E and massive electrons. The electric field is constant in any region of space where there are no charges. The field energy is proportional to the square of the field strength. The electric field jumps by a quantized unit whenever an electron is passed. Going in one direction, say along the positive x axis, the field makes a positive unit jump when an electron is passed and a negative jump when a positron is passed. In this model different vacua are represented by different quantized values of the electric field while the electrons/positrons are the domain walls. The energy of a vacuum is proportional to E^2. This model is not fundamentally different than the case with scalar fields and a potential. In fact by bosonizing the theory it can be expressed as a scalar field theory with a potential: $V(\phi) = c\phi^2 + \mu \cos \phi$. If μ is not too small there are many minima representing the different possible 2–form field strengths, each with a different energy.

In $3+1$ dimensions the corresponding construction requires a 4–form field strength F whose energy is also proportional to F^2. This energy appears in the gravitational field equations as a positive contribution to the cosmological constant . The analogue of the charged electrons are membranes which appear in string theory and function as domain walls to separate vacua with different F. This theory can also be written in terms of a scalar field with a potential similar to $V(\phi)$.

Let's now consider a typical compactification of M–theory from eleven to 4 dimensions. The simplest example is gotten by choosing for the compact directions a 7–torus. The torus has a number of moduli representing the sizes and angles between the seven 1–cycles. The 4–form fields have as their origin a 7–form field strength, which is one of the fundamental fields of M–theory. The 7–form fields have 7 anti–symmetrized indecies. These non–vanishing 7–form can be configured so that three of the indecies are identified with compact dimensions and the remainder with uncompactified space–time. This can be done in thirty five $= (7\times6\times5)/(1\times2\times3)$ ways which means that there are that many distinct 4–form fields in the uncompactified non–compact space. More generally, in the kinds of compact manifolds used in string theory to try to reproduce standard model physics there can be hundreds of independent ways of "wrapping" three compact directions with flux, thus producing hundreds of 3+1 dimensional 4–form fields. As in the case of 1+1 dimensional quantum electrodynamics, the field strengths are quantized, each in integer multiples of a basic field unit. A vacuum is specified by a set of integers $n_1, n_2, ..., n_N$ where N can be as big as several hundred or more. The energy density of the energy of the 4–form fields has the form: $\epsilon = c_i N_i^2$, where the constants c_i depend on the details of the compact space.

[20] In 1+1 dimensions there is no magnetic field and the electric field is a two form, aka, a scalar density.

The analogue of the electrons and positrons of the 1+1 dimensional example are branes. The 11 dimensional M–theory has 5–branes which fill 5 spatial directions and time. By wrapping 5–branes the same way the fluxes of the 4–forms are wrapped on internal 3–cycles leaves 2–dimensional membranes in 3+1 dimensions. These are the domain walls which separate different values of field strength. There are N types of domain wall, each allowing a unit jump of one of the 4-forms.

Bousso and Polchinski [BP00] begin by assuming they have located some deep minimum of the field potential at some point Φ_0. The value of the potential is supposed to be very negative at this point, corresponding to a negative cosmological constant, λ_0 of order the Planck scale. Also the 4-forms are assumed to vanish at this point. They then ask what kind of vacua can they get by discretely increasing the 4–forms. The answer depends to some degree on the compactification radii on the internal space but with modest parameters it is not hard to get such a huge number of vacua that it is statistically likely to have one in the range $\lambda \sim 10^{-120} M_p^4$.

To see how this works we write the cosmological constant as the sum of two terms, is the cosmological constant for vanishing 4–form, and the contribution of the 4–forms: $\lambda = \lambda_0 + c_i N_i^2$. With a hundred terms and modestly small values for the c_n it is highly likely to find a value of λ in the observed range. Note that no fine tuning is required, only a very large number of ways to make the vacuum energy.

The problem with [BP00] was clearly recognized by the authors; The starting point is so far from the supermoduli–space that none of the usual tools of approximate supersymmetry are available to control the approximation. The example was intended only as a model of what might happen because of the large number of possibilities.

More recently Kachru, Kallosh, Linde and Trivedi [KKL03] have improved the situation by finding an example which is more under control. These authors subtly use the various ingredients of string theory including fluxes, branes, anti–branes and instantons to construct a rather tractable example with a small positive cosmological constant .

In addition to arguing that string theory does have many vacua with positive cosmological constant the argument in [KKL03] tends to dispel the idea that vacua, not on the supermoduli–space , must have vanishing cosmological constant . In other words there is no evidence in string theory that a hoped for but unknown mechanism will automatically force the cosmological constant to zero. It seems very likely that all of the non–supersymmetric vacua have finite λ.

The vacua in [KKL03] are not at all simple. They are jury–rigged, Rube Goldberg contraptions that could hardly have fundamental significance. But in an anthropic theory simplicity and elegance are not considerations. The only criteria for choosing a vacuum is utility, i.e., does it have the necessary elements such as galaxy formation and complex chemistry that are needed for life. That together with a cosmology that guarantees a high probability that at least one large patch of space will form with that vacuum structure is all we need [Sus03].

The Trouble with de Sitter Space

The classical vacuum solution of Einstein's equations with a positive cosmological constant is de Sitter space. It is doubtful that it has a precise meaning in a quantum theory such as string theory [GKS02, DLS02, DKS02]. We want to review some of the reasons for thinking that de Sitter space is at best a metastable state.

It is important to recognize that there are two very different ways to think about de Sitter space. The first is to take a global view of the space–time. The global geometry is described by the metric

$$ds^2 = R^2 \left\{dt^2 - (\cosh t)^2 d^2\Omega_3\right\},$$

where $d^2\Omega_d$ is the metric for a unit d–sphere and R is related to the cosmological constant by

$$R = (\lambda G)^{-1/2}.$$

Viewing de Sitter space globally would make sense if it were a system that could be studied from the outside by a 'meta–observer'. Naively, the meta–observer would make use of a (time dependent) Hamiltonian to evolve the system from one time to another. An alternate description would use a Wheeler de Witt formalism to define a wave function of the universe on global space–like slices.

The other way of describing the space is the *causal patch* description. The relevant metric is

$$ds^2 = R^2 \left\{(1 - r^2)\, dt^2 - \frac{1}{(1 - r^2)} dr^2 - r^2 d^2\Omega_2\right\}.$$

In this form the metric is static and has a form similar to that of a black hole. In fact the geometry has a horizon at $r = 1$. The static patch does not cover the entire global de Sitter space but is analogous to the region outside a black hole horizon. It is the region which can receive signals from, and send signals to, an observer located at $r = 0$. To such an observer de Sitter space appears to be a spherical cavity bounded by a horizon a finite distance away [Sus03].

Experience with black holes has taught us to be very wary of global descriptions when horizons are involved. In a black hole geometry there is no global conventional quantum description of both sides of the horizon. This suggests that a conventional quantum description of de Sitter space only makes sense within a given observer's causal patch. The descriptions in different causal patches are complementary [STU93, SHW94] but can not be put together into a global description without somehow modifying the rules of quantum mechanics.

As in the black hole case, a horizon implies a thermal behavior with a temperature and an entropy. These are given by: $T = 1/2\pi R$, $S = \pi R^2/G$. We will assume the causal patch description of some particular observer.

If the observed 'dark energy' in the universe really is a small positive cosmological constant the ultimate future of our universe will be eternal de Sitter space. This would mean not that the future is totally empty space but that the world will have all the features of an isolated finite thermal cavity with finite temperature and entropy. Thermal equilibrium for such a system is not completely featureless. On short time scales not much can be expected to happen but on very long time scales everything happens. A famous example involves a gas of molecules in a sealed room. Imagine that we start all the molecules in one corner of the room. In a relatively short time the gas will spread out to fill the room and come to thermal equilibrium. During the approach to equilibrium interesting dissipative structures such as droplets, eddies and vortices form and then dissipate. The usual assumption is that nothing happens after that. The entropy has reached its maximum value and the second law forbids any further interesting history. But on a sufficiently long time scale, large fluctuations will occur. In fact the phase point will return over and over to the neighborhood of any point in phase space including the original starting point. These *Poincare recurrences* generally occur on a time scale exponentially large in the thermal entropy of the system. Thus we define the *Poincare recurrence time* as: $T_r = \mathrm{e}^S$. On such long time scale the second law of thermodynamics will repeatedly be violated by large scale fluctuations.

Thus even a pure de Sitter space would have an interesting cosmology of sorts. The causal patch of any observer would undergo Poincare recurrences in which it would endlessly fluctuate back to a state similar to its starting point, but each time slightly different.

The trouble with such a cosmology is that it relies on very rare "miracles" to start it off each time. But there are other miracles which could occur and lead to anthropically acceptable worlds with a vastly larger probability than our world. Roughly speaking the relative probability of a fluctuation leading to a given configuration is proportional to the exponential of its entropy. An example of a configuration far more likely than our own would be a world in which everything would be just like our universe except the temperature of the cosmic microwave background was ten degrees instead of three. When we say everything is the same we are including such details as the abundance of the elements.

Ordinarily such a universe would be ruled out on the grounds that it would take a huge miracle for the helium and deuterium to survive the bombardment by the extra photons implied by the higher temperature. That is correct, a fantastic miracle would be required, but such miracles would occur far more frequently than the ultimate miracle of returning to the starting point. This can be argued just from the fact that a universe at 10 degrees K has a good deal more entropy than one at 3 degrees. In a world based on recurrences it would be overwhelming unlikely that cosmology could be traced back to something like the inflationary era without a miraculous reversal of the second law along the way. Thus we are forced to conclude that the sealed tin can model of the universe must be incorrect, at least for time scales as long as the recurrence time.

Another difficulty with an eternal de Sitter space involves a mathematical conflict between the symmetry of de Sitter space and the finiteness of the entropy [GKS02]. Basically the argument is that the finiteness of the de Sitter space entropy indicates that the spectrum of energy is discrete. It is possible to prove that the symmetry algebra of de Sitter space can not be realized in a way which is consistent with the discreteness of this spectrum. In fact this problem is not independent of the issues of recurrences. The discreteness of the spectrum means that there is a typical energy spacing of order: $\Delta E \sim \mathrm{e}^{-S}$. The discreteness of the spectrum can only manifest itself on time scales of order $(\Delta E)^{-1}$ which is just the recurrence time. Thus there are problems with realizing the full symmetries of de Sitter space for times as long as T_r.

Finally another difficulty for eternal de Sitter space is that it does not fit at all well with string theory. Generally the only objects in string theory which are rigorously defined are S–Matrix elements. Such an S matrix can not exist in a thermal background. Part of the problem is again the recurrences which undermine the existence of asymptotic states. Unfortunately there are no known observables in de Sitter space which can substitute for S–matrix elements. The unavoidable implication of the issues is that eternal de Sitter space is an impossibility in a properly defined quantum theory of gravity [Sus03].

De Sitter space is Unstable

In [KKL03] a particular string theory vacuum with positive λ was studied. One of the many interesting things that the authors found was that the vacuum is unstable with respect to tunnelling to other vacua. In particular the vacuum can tunnel back to the supermoduli–space with vanishing cosmological constant. Using instanton methods the authors calculated that the lifetime of the vacuum is less than the Poincare recurrence time. This is no accident. To see why it always must be so, let's consider the effective potential that the authors of [KKL03] derived. The only modulus which is relevant is the overall size of the compact manifold Φ. The de Sitter vacuum occurs at the point $\Phi = \Phi_0$. However, the absolute minimum of the potential occurs not at Φ_0 but at $\Phi = \infty$. At this point the vacuum energy is exactly zero and the vacuum one of the ten dimensional vacua of the supermoduli–space. There are always runaway solutions like this in string theory. The potential on the supermoduli–space is zero and so it is always possible to lower the energy by tunnelling to a point on the supermoduli–space.

Suppose we are stuck in the potential well at Φ_0. The vacuum of the causal patch has a finite entropy and fluctuates up and down the walls of the potential. One might think that fluctuations up the sides of the potential are

Bolzmann suppressed. In a usual thermal system there are two things that suppress fluctuations. The first is the Bolzmann suppression by factor $e^{-\beta E}$, and the second is entropy suppression by factor $e^{(S_f - S)}$, where S is the thermal entropy and S_f is the entropy characterizing the fluctuation which is generally smaller than S. However in a gravitational theory in which space is bounded (as in the static patch) the total energy is always zero, at least classically. Hence the only suppression is entropic. The phase point wanders around in phase space spending a time in each region proportional to its phase space volume, i.e., e^{-S_f}. Furthermore, the typical time scale for such a fluctuation to take place is of order $T_f \sim e^{(S-S_f)}$ [Sus03].

Now consider a fluctuation which brings the field ϕ to the top of the local maximum at $\phi = \phi_1$ in the entire causal patch. The entropy at the top of the potential is given in terms of the cosmological constant at the top. It is obviously positive and less than the entropy at ϕ_0. Thus the time for the field to fluctuate to ϕ_1 (over the whole causal patch) is strictly less than the recurrence time e^S. But once the field gets to the top there is no obstruction to it rolling down the other side to infinity. It follows that a de Sitter vacuum of string theory is never longer lived than T_r and furthermore we end up at a supersymmetric point of vanishing cosmological constant .

There are other possibilities. If the cosmological constant is not very small it may tunnel over the nearest mountain pass to a neighboring valley of smaller positive cosmological constant. This will also take place on a time scale which is too short to allow recurrences. By the same argument it will not stay in the new vacuum indefinitely. It may find a vacuum with yet smaller cosmological constant to tunnel to. Eventually it will have to make a transition out of the space of vacua with positive cosmological constants[21].

Bubble Cosmology

To make use of the enormous diversity of environments that string theory is likely to bring with it, we need a dynamical cosmology which, with high probability, will populate one or more regions of space with an anthropically favorable vacuum. There is a natural candidate for such a cosmology that we will explain from the global perspective [Sus03].

For simplicity let's temporarily assume that there are only two vacua, one with positive cosmological constant λ, and one with vanishing cosmological constant . Without worrying how it happened we suppose that some region of the universe has fallen into the minimum with positive cosmological constant . ¿From the global perspective it is inflating and new Hubble volumes are

[21] Tunnelling to vacua with negative cosmological constant may or may not be a possibility. However such a transition will eventually lead to a crunch–singularity. Whether the system survives the crunch is not known. It should be noted that transitions to negative cosmological constant are suppressed and can even be forbidden depending on magnitudes of the vacuum energies, and the domain wall tension. We will assume that such transitions do not occur [Sus03].

constantly being produced by the expansion. Pick a time–like observer who looks around and sees a static universe bounded by a horizon. The observer will eventually observe a transition in which his entire observable region slides over the mountain pass and settles to the region of vanishing λ. The observer sees the horizon–boundary quickly recede, leaving in its wake an infinite open Freedman, Robertson, Walker universe with negative spatial curvature. The final geometry has light–like and time like future infinities similar to flat space.

Now let us take the more global view. The bubble does not swallow the entire global space but leaves part of the space still inflating. Inevitably bubbles will form in this region. In fact if we follow the world line of any observer, it will eventually be swallowed by a bubble of $\lambda = 0$ vacuum.

The real landscape is not comprised of only two vacua. If an observer starts with a large value of the cosmological constant there will be many ways for the causal patch to descend to the supermoduli–space. From the global viewpoint a bubbles will form in neighboring valleys with somewhat smaller cosmological constant. Since each bubble has a positive cosmological constant it will be inflating but the space between bubbles is inflating faster so the bubbles go out of causal contact with one another. Each bubble evolves in isolation from all the others. Furthermore, in a time too short for recurrences, bubbles will nucleate within the bubbles. Following a single observer within his own causal patch, the cosmological constant decreases in a series of events until the causal patch finds itself in the supermoduli–space . Each observer will see a series of vacuums descending down to the supermoduli–space and the chances that he passes through an anthropically acceptable vacuum is most likely very small. But on the other hand the global space contains an infinite number of such histories and some of them will be acceptable.

The only problem with the cosmology that we just outlined is that it is formulated in global coordinates. From the viewpoint of any causal patch, all but one of the bubbles is outside the horizon. As I've emphasized, the application of the ordinary rules of quantum mechanics only makes sense within the horizon of an observer. We don't know the rules for putting together the various patches into one comprehensive global description and until we do there can not be any firm basis for the kind of anthropic cosmology I described. Nevertheless the picture is tempting [Sus03].

Cosmology as a Resonance

The idea of scalar fields and potentials is approximate once we leave the supermoduli–space. So is the notion of a stable de Sitter vacuum. The problem is familiar. How do we make precise sense of an unstable state in quantum mechanics. In ordinary quantum mechanics the clearest situation is when we can think of the unstable state as a resonance in a set of scattering amplitudes. The parameters of a resonance, i.e., its width and mass are well defined and don't depend on the exact way the resonance was formed. Thus even black holes have precise meaning as resonant poles in the S–matrix. Normally we can not compute the scattering amplitudes that describe the formation and evaporation of a black hole but it is comforting that an exact criterion exists.

In the case of a black hole the density of levels is enormous being proportional to the exponential of the entropy. The spacing between levels is therefore exponentially small. On the other hand the width of each level is not very small. The lifetime of a state is the time that it takes to emit a single quantum of radiation and this is proportional to the Schwarzschild radius. Therefore the levels are broadened by much more than their spacing. The usual resonance formulas are not applicable but the precise definition of the unstable state as a pole in the scattering amplitude is. We think the same things can be said about the unstable de Sitter vacua but it can only be understood by returning to the causal patch way of thinking [Sus03]. Therefore let's focus on the causal patch of one observer. We have discussed the observer's future history and found that it always ends in an infinite expanding supersymmetric open Freedman universe. Such a universe has the usual kind of asymptotic future consisting of time–like and light–like infinity. There is no temperature in the remote future and the geometry permits particles to separate and propagate as free particles just as in flat space–time.

Now let's consider the observer's past history. The same argument which says that the observer will eventually make a transition to $\lambda = 0$ in the far future can be run backward. The observer could only have gotten to the de Sitter vacuum by the time–reversed history so he must have originated from a collapsing open universe. The history may seem paradoxical since it requires the second law of thermodynamics to be violated in the past. A similar paradox arises in a more familiar setting. Let me return to the sealed room filled with gas molecules except that now one of the walls has a small hole that lets the gas escape to unbounded space. Suppose we find the gas filling the room in thermal equilibrium at some time. If we run the system forward we will eventually find all the molecules have escaped and are on their way out, never to return. But it is also true that if we run the equations of motion backwards we will eventually find all the molecules outside the room moving away. Thus the only way the starting configuration could have occurred is if the original molecules were converging from infinity toward the small hole in the wall.

If we are studying the system quantum mechanically, the metastable configuration with all the molecules in the room would be an *unstable resonance* in a scattering matrix describing the many body scattering of a system of molecules with the walls of the room. Indeed the energy levels describing the molecules trapped inside the room are complex due to the finite lifetime of the configuration.

This suggests a view of the intermediate de Sitter space as an unstable resonance in the scattering matrix connecting states in the asymptotic $\lambda = 0$ vacua. In fact we can estimate the width of the states. Since the lifetime of the de Sitter space is always longer than the recurrence time, generally by a huge factor, the width γ satisfies: $\gamma >> e^{-S}$.On the other hand the spacing between levels, ΔE, is of order e^{-S}. Therefore, $\gamma >> \Delta E$, so that the levels are very broad and overlapping as for the black hole [Sus03].

No perfectly precise definition exists in string theory for the moduli fields or their potential when we go away from the supermoduli–space. The only precise definition of the de Sitter vacua seems to be as complex poles in some new sector of the scattering matrix between states on the supermoduli–space.

Knowing that a black hole is a resonance in a scattering amplitude does not tell us much about the way real black holes form. Most of the possibilities for black hole formation are just the time reverse of the ways that it evaporate. In other words the overwhelming number of initial states that can lead to a black hole consist of thermal radiation. Real black holes in our universe form from stellar collapse which is just one channel in a huge collection of S–matrix 'in states'. In the same way the fact that cosmological states may be thought of in a scattering framework is in itself does not shed much light on the original creation process.

As seen above, vacua come in two varieties, supersymmetric and otherwise. Most likely the non–supersymmetric vacua do not have vanishing cosmological constant but it is plausible that there are so many of them that they practically form a continuum. Some tiny fraction have cosmological constant in the observed range. With nothing preferring one vacuum over another, the anthropic principle comes to the fore whether or not we like the idea. String theory provides a framework in which this can be studied in a rigorous way. Progress can certainly be made in exploring the landscape. The project is in its infancy but in time we should know just how rich it is. We can argue the philosophical merits of the anthropic principle but we can't argue with quantitative information about the number of vacua with each particular property such as the cosmological constant, Higgs mass or fine structure constant. That information is there for us to extract [Sus03].

Counting the vacua is important but not sufficient. More understanding of cosmological evolution is essential to determining if the large number of possibilities are realized as actualities. The vacua in string theory with $\lambda > 0$ are not stable and decay on a time scale smaller than the recurrence time. This is very general and also very fortunate since there are serious problems with stable de Sitter space. The instability also allows the universe to sample all or a large part of the landscape by means of bubble formation. In such a world the probability that some region of space has suitable conditions for life to exist can be large.

The bubble universe based on Linde's *eternal inflation* seems promising but it is unclear how to think about it with precision. There are real conceptual problems having to do with the global view of space–time. The main problem is to reconcile two pictures; the causal patch picture and the global picture. String theory has provided a testing ground for some important relevant ideas such as black hole complementarity [STU93, SHW94] and the Holographic principle [tHo93, Sus95]. Complementarity requires the observer's side of the horizon to have a self contained conventional quantum description. It also prohibits a conventional quantum description that covers the interior and exterior simultaneously. Any attempt to describe both sides as a single quantum system will come into conflict with one of three sacred principles [Sus02]. The first is the equivalence principle which says that a freely falling observer passes the horizon without incident. The second says that experiments performed outside a black hole should be consistent with the rules of quantum mechanics as set down by Dirac in his textbook. No loss of quantum information should take place and the time evolution should be unitary. Finally the rules of quantum mechanics forbid information duplication. This means that we can not resolve the so called information paradox by creating two copies (quantum Xeroxing) of every bit as it falls through the horizon; at least not within the formalism of conventional quantum mechanics. The complementarity and holographic principles have been convincingly confirmed by the modern methods of string theory [Mal01]. The inevitable conclusion is that a global description of geometries with horizons, if it exists at all, will not be based on the standard quantum rules.

Why is this important for cosmology? The point is that the eternal inflationary production of an infinity of bubbles takes place behind the horizon of any given observer. It is not something that has a description within one causal patch. If it makes sense, a global description is needed but if cosmic event horizons are at all like black hole horizons then any global description will involve wholly new elements. If we were to make a wild guess about which rule of quantum mechanics has to be given up in a global description of either black holes or cosmology we would guess it is the Quantum Xerox Principle [Sus02]. We would look for a theory which formally allowed quantum duplication but cleverly prevents any observer from witnessing it. Perhaps then the replication of bubbles can be sensibly described.

Progress may also be possible in sharpening the exact mathematical meaning of the de Sitter vacua. Away from the supermoduli–space , the concept of a local field and the effective potential is at best approximate in string theory. The fact that the vacua are false metastable states makes it even more problematic to be precise. In ordinary quantum mechanics the best mathematical definition of an unstable state is as a resonance is amplitudes for scattering between very precisely defined asymptotic states. Each metastable state corresponds to pole whose real and imaginary parts define the energy and inverse lifetime of the state.

We have argued that each causal patch begins and ends with an asymptotic 'roll' toward the supermoduli–space [Sus03]. The final state have the boundary conditions of an FRW open universe and the initial states are time reversals of these. This means we may be able to define some kind S–matrix connecting initial and final asymptotic states. The various intermediate metastable de Sitter phases would be exactly defined as resonant resonances in this amplitude.

At first this proposal sounds foolish. In general relativity initial and final states are very different. Black holes make sense. White holes do not.

Ordinary things fall into black holes and thermal radiation comes out. The opposite never happens. But this is deceiving. Our experience with string theory has made it clear that the fundamental micro–physical input is completely reversible and that black holes are most rigorously defined in terms of resonances in scattering amplitudes[22]. Knowing that a black hole is an intermediate state in a tremendously complicated scattering amplitude does not really tell us much about how real black holes form. For that we need to know about stellar collapse and the like. But it does provide an exact mathematical definition of the states that comprise the black hole ensemble.

From the causal patch viewpoint the evolutionary endpoints seems to be an approach to some point on the supermoduli–space. After the last tunnelling the universe enters an final open FRW expansion toward some flat supersymmetric solution. This is not to be thought of as a unique quantum state but as a large set of states with similar evolution. Running the argument backward (assuming microscopic reversibility) we expect the initial state to be the time reversal of one of the many future endpoints. We might even hope for a scattering matrix connecting initial and final states. de Sitter minima would be an enormously large density of complex poles in the amplitude. For more details, see [Sus03].

7.3.6 Top–Down Cosmology

In this section, following [HH02], we try to convince the readers that inflation actually starts at the 'top of the hill'.

Structure and complexity have developed in our universe, because it is out of equilibrium. This feature shows up in all known cosmological scenarios for the early universe, which rely on gravitational instability to generate local inhomogeneities from an almost homogeneous and isotropic state for the universe. Inflation seems the best explanation for this homogeneous and isotropic state because whatever drives the inflation will remove the local instability and iron out irregularities. However the inflationary expansion has to be globally unstable because otherwise it would continue forever and galaxies would never form.

The instability can be described as the evolution of an order parameter ϕ which can be treated as a scalar field with effective potential $V(\phi)$. If V'/V is small, ϕ will roll slowly down the potential and the universe will inflate by a large factor. However, this raises the question: Why did the universe start with a high value of the potential? Why didn't ϕ start at the global minimum of V?

There have been various attempts to explain why ϕ started high on the potential hill. In the old [Gut82] and new [Lin82, AS82] inflationary scenarios the universe was supposed to start with infinite temperature at a singularity. As the universe expanded and cooled, thermal corrections would make the

[22] The one exception is black holes in anti–de Sitter space which are stable.

effective potential time dependent. So even if ϕ started in the minimum of V, it could still end up in a metastable false vacuum state (in old inflation) or at a local maximum of V (in new inflation). The scalar field was then supposed to tunnel through the potential barrier or just fall off the top of the hill and slowly roll down. However both scenarios tended to predict a more inhomogeneous universe than we observe. They were also unsatisfactory because they assumed an initial singularity and a fairly homogeneous and isotropic pre–inflation hot Big–Bang phase. Why not just assume the singularity produced the standard hot Big–Bang, since we don't have a measure on the space of singular initial conditions for the universe?

In the *chaotic inflation scenario* [Lin83], quantum fluctuations of ϕ are supposed to drive the volume weighted average ϕ up the potential hill, leading to everlasting *eternal inflation*. However this effect is dependent on using the synchronous gauge: in other gauges the volume weighted average of the potential can go down. Looking from a 4 rather than 3+1 dimensional perspective, it is clear that the quantum fluctuations of a single scalar field are insufficient to drive de Sitter like eternal inflation, if the de Sitter space is larger than the Planck length. Eternal inflation may be possible at the Planck scale, but all presented methods would break down in this situation so it would mean that we could not analyze the origin of the universe.

The aim of this section however is to show that the universe can come into being and start inflating without the need for an initial hot Big–Bang phase or Planck curvature. It is required that the potential V has a local maximum which is below the Planck density and sufficiently flat on top, $V''/V > -4/3$. This last condition means only the homogeneous mode of the scalar field is tachyonic: the higher modes all have positive eigenvalues. It also means there is not a Coleman–De Luccia solution [Col80] describing quantum tunnelling from a false vacuum on one side of the maximum to the true vacuum on the other side. Instead there is only a homogeneous Hawking–Moss instanton [HM82] that sits on the top of the hill, at the local maximum of V.

It has long been a problem to understand how the universe could decay from a false vacuum in this situation. The Hawking–Moss instanton does not interpolate between the false and true vacua, because it is constant in space and time. Instead, what must happen is that the original universe can continue in the false vacuum state but that new completely disconnected universes can form at the top of the hill via Hawking–Moss instantons. For someone in one of these new universes, the universe in the false vacuum is irrelevant and can be ignored.

The top of the hill might seem the least likely place for the universe to start. However we shall show it is the most likely place for an inflationary universe to begin, if $V''/V > -4/3$. The reason is that although being at the top of the hill costs potential action, the saving of gradient action from having a constant scalar field is greater. Thus inflation will start at the top of the hill. In particular, this justifies Starobinsky's scenario of *trace anomaly inflation*, in which the universe starts in an unstable de Sitter state supported by the conformal anomaly of a large number of conformally coupled matter fields [Sta80].

The usual approach to the problem of initial conditions for inflation, is to assume some initial configuration for the universe, and evolve it forward in time. This could be described as the bottom up approach to cosmology. It is an essentially classical picture, because it assumes there is a single well defined metric for the universe. By contrast, here we adopt a quantum approach, based on the no boundary proposal [HH83], which states that the amplitude for an observable like the 3–metric on a space–like hypersurface Σ, is given by a path integral over all metrics whose only boundary is Σ. The quantum origin of our universe and the no boundary proposal naturally lead to a top down view of the universe, in which the histories that contribute to the path integral, depend on the observable being measured.

Following [HH02], we study the quantum cosmological origin of an expanding universe in theories like trace anomaly inflation, by investigating the semiclassical predictions of the no boundary proposal for the wave function of interest. One may argue that a clearer picture of the pre–inflationary conditions can only emerge from a deeper understanding of quantum gravity at the Planck scale. However, the amplitude of the cosmic microwave temperature anisotropies indicates that the universe may always have been much larger than the Planck scale. This suggests it might be possible to describe the origin of our universe within the semiclassical regime of quantum cosmology. Correspondingly, the effective potential must have a local maximum well below the Planck density, which is the case in the trace anomaly model.

Trace Anomaly Driven Inflation

Large N Cosmology

It has been argued that the theoretical foundations for inflation are weak, since it has proven difficult to realise inflation in classical M–theory. A large class of supergravity theories admit no warped de Sitter compactifications on a compact, static internal space [Gib85, MN01] and although some gauged $N = 8$ and $N = 4$ supergravities in $D = 4$ do permit de Sitter vacua [GZ83, Hul84], these vacua are too unstable for a significant period of inflation to occur. However, an appealing way to evade the no go theorems is to include higher derivative quantum corrections to the classical supergravity equations, such as the trace anomaly.

Since we observe a large number of matter fields in the universe, it is natural to consider the large N approximation [Tom77]. In the large N approximation, one performs the path integral over the matter fields in a given background to get an effective action that is a functional of the background metric [HH02],

$$\exp(-W[\mathbf{g}]) = \int \mathcal{D}[\phi] \exp(-S[\phi; \mathbf{g}]).$$

In the leading–order $1/N$ approximation, one can neglect graviton loops and look for a stationary point of the effective action for the matter fields combined with the gravitational action. This is equivalent to solving the Einstein equations with the source being the expectation value of the matter energy–momentum tensor derived from W,

$$R_{\mu\nu} - \frac{1}{2}Rg_{\mu\nu} = 8\pi G\langle T_{\mu\nu}\rangle.$$

The expectation value of the energy–momentum tensor is generally non–local and depends on the quantum state. However, during inflation, particle masses are small compared with the space–time curvature, $R >> m^2$, and in asymptotically free gauge theories, interactions become negligible in the same limit. Therefore, at the high curvatures during inflation, the energy–momentum tensor of a large class of grand unified theories is to a good approximation given by the expectation value $\langle T_{\mu\nu}\rangle$ of a large number of free, massless, conformally invariant fields[23]. The entire one–loop contribution to the trace of the energy–momentum tensor then comes from the conformal anomaly [Cap74], which is given for a general CFT by the following equation,

$$g^{\mu\nu}\langle T_{\mu\nu}\rangle = cF - aG + \alpha\prime\nabla^2 R, \tag{7.78}$$

where F is the square of the Weyl tensor, G is proportional to the Euler density and the constants a, c and α' are given in terms of the field content of the CFT by

$$a = \frac{1}{360(4\pi)^2}\left(N_S + 11N_F + 62N_V\right), \qquad c = \frac{1}{120(4\pi)^2}\left(N_S + 6N_F + 12N_V\right),$$
$$\alpha\prime = \frac{1}{180(4\pi)^2}\left(N_S + 6N_F - 18N_V\right),$$

with N_S the number of real scalar fields, N_F the number of Dirac fermions and N_V the number of vector fields.

The trace anomaly is entirely geometrical in origin and therefore independent of the quantum state. In a maximally symmetric space–time, the symmetry of the vacuum implies that the expectation value of the energy–momentum tensor is proportional to the metric,

$$\langle 0|T_{\mu\nu}|0\rangle = \frac{1}{4}g_{\mu\nu}g^{\rho\sigma}\langle 0|T_{\rho\sigma}|0\rangle.$$

Thus the trace anomaly acts just like a cosmological constant for these space–times, and a positive trace anomaly permits a de Sitter solution to the Einstein equations.

[23] For simplicity, it is assumed that scalar fields become conformally coupled at high energies, but the contribution of the interaction terms to $\langle T_{\mu\nu}\rangle$ is small at high curvature, as long as the couplings don't become very large [Par84].

The radius of the de Sitter solution is determined by the number of fields, N^2, in the CFT and is of order $\sim Nl_{pl}$. Therefore the one-loop contributions to the energy–momentum tensor are $\sim 1/N^2$, which means they are of the same order as the classical terms in the Einstein equations. On the other hand, the corrections due to graviton loops are $\sim 1/N^3$, so for large N quantum gravitational fluctuations are suppressed, confirming the consistency of the large N approximation.

For $\alpha' = 0$ in 7.78, the only $O(3,1)$ invariant solutions are de Sitter space and flat space, which are the initial and final stages of the simplest inflationary universe. In order for a solution to exist that interpolates between these two stages, one must have $\alpha' < 0$ in 7.78, as Starobinsky discovered [Sta80]. Starobinsky showed that if $\alpha' < 0$, the de Sitter solution is unstable, and decays into a matter dominated Friedman–Lemaitre Robertson–Walker universe, on a timescale determined by α'. The purpose of Starobinsky's work was to demonstrate that quantum effects of matter fields might resolve the Big–Bang singularity. From a modern perspective, it is more interesting that the conformal anomaly might have been the source of a finite, but significant period of inflation in the early universe. Rapid oscillations in the expansion rate at the end of inflation, would result in particle production and (p)reheating.

Starobinsky showed that the de Sitter solution is unstable both to the future and to the past, so it was not clear how the universe could have entered the de Sitter phase. This is the problem of initial conditions for trace anomaly driven inflation, which should be addressed within the framework of quantum cosmology, by combining inflation with a theory for the wave $\Psi--$function of the quantum universe. Hartle and Hawking suggested that the amplitude for the quantum state of the universe described by 3–metric $\mathbf{h}$ and matter fields $\phi(\mathbf{x})$ on a 3–surface Σ, should be given by [HH83]

$$\Psi[\Sigma, \mathbf{h}, \phi_\Sigma] = N \sum_M \int \mathcal{D}[\mathbf{g}]\mathcal{D}[\phi(\mathbf{x})]\mathrm{e}^{-S_E(\mathbf{g},\phi)}, \tag{7.79}$$

where the Euclidean path integral is taken over all compact four geometries bounded only by a 3–surface Σ, with induced metric $\mathbf{h}$ and matter fields ϕ_Σ. M denotes a diffeomorphism class of 4–manifolds and N is a normalization factor. The motivation to restrict the class of manifolds and metrics to geometries with only a single boundary is that in cosmology, in contrast with scattering calculations, one is interested in measurements in a finite region in the interior of space–time. The 'no boundary' proposal gives a definite ansatz for the wave function $\Psi[\Sigma, \mathbf{h}, \phi_\Sigma]$ of the universe and in principle removes the initial singularity in the hot Big–Bang model. At least within the semiclassical regime, this yields a well–defined probability measure on the space of initial conditions for cosmology.

One can appeal to quantum cosmology to explain how the de Sitter phase emerges in trace anomaly inflation, since the no boundary proposal can describe the creation of an inflationary universe from nothing. At the semiclassical level, this process is mediated by a compact instanton saddle–point of the Euclidean path integral, which extrapolates to a real Lorentzian universe at late times. To find the relative probability of different geometries in the no boundary path integral, one must compute their Euclidean action. Below we consider a model of anomaly–induced inflation consisting of gravity coupled to $\mathcal{N} = 4$, $U(N)$ super Yang–Mills theory, for which the AdS/CFT correspondence [Mal98] provides an attractive way to calculate the effective matter action on backgrounds without symmetry. The fact that we are using $\mathcal{N} = 4$, $U(N)$ super Yang–Mills theory is probably not significant since, as we shall describe, it is the large number of fields that matters here and not the Yang–Mills coupling. Therefore, we expect the presented results to be valid for any matter theory that is approximately massless during the de Sitter phase [HH02].

Effective Matter Action

We consider, in Euclidean signature, Einstein gravity coupled to a $\mathcal{N} = 4$, $U(N)$ super Yang–Mills theory with large N,

$$S = -\frac{1}{2\kappa}\int d^4x\sqrt{g}R - \frac{1}{\kappa}\int d^3x\sqrt{h}K + W, \qquad (7.80)$$

where W denotes the Yang–Mills effective action. The field content of the Yang–Mills theory is $N_S = 6N^2$, $N_F = 2N^2$ and $N_V = N^2$, yielding an anomalous trace [HH02]

$$g^{\mu\nu}\langle T_{\mu\nu}\rangle = \frac{N^2}{64\pi^2}(F - G).$$

The one–loop result for the conformal anomaly is exact, since it is protected by supersymmetry. Therefore, inflation supported by the trace anomaly of $\mathcal{N} = 4$, $U(N)$ super Yang–Mills would never end. The presence of non–conformally invariant fields in realistic matter theories, however, necessarily alters the value of α' in the anomaly 7.78. Since the coefficient of the $\nabla^2 R$ term plays such an important role in trace anomaly driven inflation, we ought to include this correction. As a first approximation, one can account for the non–conformally invariant fields by adding a local counterterm to the action,

$$S_{ct} = \frac{\alpha N^2}{192\pi^2}\int d^4x\sqrt{g}R^2.$$

This leads to an extra contribution to the conformal anomaly, which becomes

$$g^{\mu\nu}\langle T_{\mu\nu}\rangle = \frac{N^2}{64\pi^2}(F - G) + \frac{\alpha N^2}{16\pi^2}\nabla^2 R.$$

For $\alpha < 0$, the expansion now changes from exponential to the typical power law $\sim t^{2/3}$ of a matter dominated universe, on a time scale $\sim 12|\alpha|\log N$. One can construct more sophisticated models of anomaly driven inflation, by taking in account corrections from particle masses and interactions in a more precise way. One could, for instance, consider soft supersymmetry breaking during inflation. The coefficient α' could then vary in time, because the decoupling of massive particles at low energy [SS01] alters the number of degrees of freedom that contribute to the quantum effective action. For our purposes, however, it is sufficient to consider the theory above.

In no boundary cosmology, one is interested in solutions that describe a Lorentzian inflationary universe that emerges from a compact instanton solution of the Euclidean field equations. These geometries provide saddle–points of the Euclidean path integral (7.99) for the wave function of interest. Because our universe is Lorentzian at late times, it has been suggested that the relevant instanton saddle–points of the no boundary path integral are so–called 'real tunnelling' geometries [GH90, HH90]. Cosmological real tunnelling solutions are compact Riemannian geometries joined to an $O(3,1)$ invariant Lorentzian solution of Einstein's equations, across a hypersurface of vanishing extrinsic curvature $K_{\mu\nu}$. Such instanton solutions can then be used as background in a perturbative evaluation of the no boundary path integral, to find correlators of metric perturbations during inflation, which in turn determine the cosmic microwave anisotropies.

We now compute the effective matter action W on such perturbed instanton metrics. After eliminating the gauge freedom, the perturbed metric on the spaces of interest can be written as

$$ds^2 = B^2(\chi)\gamma_{\mu\nu}dx^\mu dx^\nu = B^2(\chi)((1+\psi)\hat{\gamma}_{\mu\nu} + \theta_{\mu\nu})\, dx^\mu dx^\nu, \tag{7.81}$$

where $\hat{\gamma}_{\mu\nu}$ is the metric on the unit S^4 and $\theta_{\mu\nu}$ is transverse and traceless with respect to the four sphere.

In order to evaluate the no boundary path integral, we must first compute the quantum effective action $W[B, \mathbf{h}]$ on the background (7.81). The effective action of the matter fields is computed as an expansion around the homogeneous background with metric

$$g_{\mu\nu} = B^2(\chi)\hat{\gamma}_{\mu\nu}.$$

To second order in the metric perturbation, $W[B, \mathbf{h}]$ is determined by the one and 2–point function of the energy–momentum tensor on the unperturbed $O(4)$ invariant background. The one-point function is given by the conformal anomaly. Since the FLRW background is conformal to the round four sphere, the 2–point function can be calculated by a conformal transformation from S^4. On S^4, the 2–point function is determined entirely by symmetry and the trace anomaly [HHR01]. Therefore, since the energy–momentum tensor transforms anomalously, the 2–point function on 7.81 should be fully determined by the 2–point function on S^4, the trace anomaly and the scale factor $B(\chi)$. For the matter theory we have in mind, all these quantities are independent of the coupling, so it follows that the effective action $W[B, \mathbf{h}]$ is independent of the coupling, to second order in the metric perturbation.

In [Rie84], it was found how the effective action that generates a conformal anomaly of the form 7.78, transforms under a conformal transformation. We can use this result to relate $W[B, \mathbf{h}]$ on the perturbed FLRW space to $W[r, \mathbf{h}]$ on the perturbed four sphere with radius r. Writing $B(\chi) = r\ \mathrm{e}^{\sigma(\chi)}$, where r is an arbitrary radius, the transformation is given by [HH02]

$$W[\sigma(\chi), h] = \tilde{W}[r, h] - \frac{N^2}{32\pi^2} \int d^4x \sqrt{\gamma} \Big[\sigma(R^{\mu\nu} R_{\mu\nu} - \frac{1}{3} R^2) + 2\nabla_\mu \sigma \nabla^\mu \sigma \nabla^2 \sigma$$
$$+ 2(R^{\mu\nu} - \frac{1}{2}\gamma^{\mu\nu} R) \nabla_\mu \sigma \nabla_\nu \sigma + (\nabla_\mu \sigma \nabla^\mu \sigma)^2 \Big].$$

Here $\tilde{W}$ denotes the effective action on the perturbed four sphere of radius r with metric $\gamma_{\mu\nu}$, and the Ricci scalar R and covariant derivative ∇_μ refer to the same space.

The generating functional $\tilde{W}[r, h]$ was computed in [HHR01], by using the AdS/CFT correspondence [Mal98],

$$Z[\mathbf{h}] \equiv \int \mathcal{D}[\mathbf{g}] \exp(-S_{grav}[\mathbf{g}]) = \int \mathcal{D}[\phi] \exp(-S_{CFT}[\phi; \mathbf{h}]) \equiv \exp(-W_{CFT}[\mathbf{h}]), \tag{7.82}$$

where $Z[\mathbf{h}]$ is the supergravity partition function on AdS_5. The AdS/CFT calculation is performed by introducing a fictional ball of (Euclidean) AdS that has the perturbed sphere as its boundary. In the classical gravity limit, the CFT generating functional can then be obtained by solving the IIB supergravity field equations, to find the bulk metric $\mathbf{g}$ that matches onto the boundary metric $\mathbf{h}$, and adding a number of counterterms that depend on the geometry of the boundary, in order to render the action finite as the boundary is moved off to infinity. To second order in the perturbation $\mathbf{h}$, the quantum effective action (including the R^2 counterterm) is given by

$$\tilde{W} = \tilde{W}^{(0)} + \tilde{W}^{(1)} + \tilde{W}^{(2)} + \dots \qquad \text{where} \tag{7.83}$$

$$\tilde{W}^{(0)} = -\frac{3\beta N^2 \Omega_4}{8\pi^2} + \frac{3\alpha N^2 \Omega_4}{4\pi^2} + \frac{3N^2 \Omega_4}{32\pi^2} (4\log 2 - 1), \tag{7.84}$$

$$\tilde{W}^{(1)} = \frac{3N^2}{16\pi^2 r^2} \int d^4x \sqrt{\hat{\gamma}}\, \psi, \tag{7.85}$$

$$\tilde{W}^{(2)} = -\frac{3N^2}{64\pi^2 r^4} \int d^4x \sqrt{\hat{\gamma}} \left[\psi \left(\hat{\nabla}^2 + 2 \right) \psi - \alpha\psi \left(\hat{\nabla}^4 + 4\hat{\nabla}^2 \right) \psi \right]$$
$$+ \frac{N^2}{256\pi^2 r^4} \sum_p \left(\int d^4x' \sqrt{\hat{\gamma}}\, \theta^{\mu\nu}(x') H^{(p)}_{\mu\nu}(x') \right)^2 (\Psi(p) - 4\alpha p(p+3)),$$

where p labels the eigenvalues of the Laplacian $\hat{\nabla}^2$ on the round four sphere and

$$\Psi(p) = p(p+1)(p+2)(p+3)\left[\psi(p/2+5/2)+\psi(p/2+2)-\psi(2)-\psi(1)\right] + p^4 + 2p^3 - 5p^2 - 10p - 6 + 2\beta p(p+1)(p+2)(p+3),$$

and we have allowed for a finite contribution, with coefficient β, of the third counterterm, which is necessary to cancel a logarithmic divergence of the tensor perturbation. One gets the quantum effective action of the Yang–Mills theory on a general, perturbed FLRW geometry. For completeness, we also give the Einstein–Hilbert action of the perturbed four sphere,

$$S_{EH} = -\frac{3\Omega_4 r^2}{4\pi G} - \frac{3}{4\pi G}\int d^4x\sqrt{\hat{\gamma}}\psi + \frac{1}{16\pi G r^2}\int d^4x\sqrt{\hat{\gamma}}\left(\frac{3}{2}\psi\hat{\nabla}^2\psi + 2\theta^{\mu\nu}\theta_{\mu\nu} - \frac{1}{4}\theta^{\mu\nu}\hat{\nabla}^2\theta_{\mu\nu}\right).$$

We shall use these results in section III, where we discuss the instability of anomaly–induced inflation. But first, we return to the background evolution. In the next paragraph, we discuss a class of $O(4)$ invariant 'real tunnelling' instanton solutions of the Starobinsky model (7.80) and study their role in the no boundary path integral for the wave function of an inflationary universe.

Real Tunnelling Geometries

It is easily seen that the total action is stationary under all perturbations $h_{\mu\nu}$, if the background is a round four sphere with radius

$$r_s^2 = \frac{N^2 G}{4\pi}. \tag{7.86}$$

By slicing the four sphere at the equator $\chi = \pi/2$ and writing $\chi = \frac{\pi}{2} - it$, it analytically continues into the Lorentzian to the de Sitter solution mentioned above, with the cosmological constant provided by the trace anomaly of the large N Yang–Mills theory.

Other compact, real instanton solutions of the form

$$ds^2 = d\tau^2 + b^2(\tau)d\Omega_3^2 \tag{7.87}$$

were found in [HHR01], by numerically integrating the Einstein equations, which can be obtained directly from the trace anomaly by using energy–momentum conservation. Imposing regularity at the North Pole (at $\tau = 0$) of the instanton leaves only the third derivative of the scale factor at the North Pole as an adjustable parameter. It is convenient to define dimensionless variables $\tilde{\tau} = \tau/r_s$ and $f(\tilde{\tau}) = b(\tau)/r_s$. For $\alpha < 0$, there exists a second regular, compact 'double bubble' instanton, with $f'''(0) = -2.05$, together with a one-parameter family of instantons with an irregular South Pole. For $f'''(0) < -1$, the scale factor of the latter has two peaks. For $-1 < f'''(0) < 0$ on the other hand, they are similar to the singular Hawking–Turok instantons that have been considered in the context of scalar field inflaton [HT98].

The Lorentzian part of the real tunnelling saddle–points is obtained by analytically continuing the instanton metric across a hypersurface of vanishing extrinsic curvature. The double bubble instanton can be continued across its 'equator' to give a closed FLRW universe, or into an open universe by a double continuation across the South Pole. Our numerical studies show that the closed universe rapidly collapses and that the open space–time hyper-inflates, with the scale factor blowing up at a finite time. Similarly, the singular instantons can be continued into an open FLRW universe across $\tau = 0$, by setting $\tau = it$ and $\Omega_3 = i\phi$. For $f'''(0) < -1$ this again gives hyper–inflation, but for $-1 < f'''(0) < 0$ one gets a realistic inflationary universe. The four sphere solution as well as the singular instantons that are small perturbations of S^4 at the regular pole, are most interesting for cosmology, since they yield long periods of inflation.

Using the expression (7.84) for $W[\sigma(\chi)]$ and the relations

$$\chi(\tau) = 2\lim_{\epsilon\to 0}\tan^{-1}\left[\tan(\epsilon/2)\exp\left(\int_\epsilon^\tau \frac{d\tau\prime}{b(\tau\prime)}\right)\right], \qquad B(\tau) = \frac{b(\tau)}{\sin(\chi)},$$

one can numerically compute the action of the real tunnelling geometries [Her02]. On an unperturbed FLRW background, conformal to the round four sphere, the total Euclidean action becomes [HH02]

$$S^{(0)} = \frac{3N^2\Omega_3}{32\pi^2}\int d\chi \sin^3\chi\left[\frac{1}{3}(12(\log 2 + \sigma - \beta) - 3 + 6\sigma'^2 - \sigma'^4 - 4\sigma'^3\cot\chi)\right.$$
$$\left. - e^{2\sigma}(\sigma'^2 + 2) + 2\alpha(\sigma'' + 3\sigma'\cot\chi + \sigma'^2 - 2)^2\right],$$

where $\sigma = \log(B/r)$. On the round four sphere, $\sigma \to 0$, so the action reduces to

$$S^{(0)} = \frac{3N^2\Omega_4}{32\pi^2}(8\alpha - 3 + 4(\log 2 - \beta)). \tag{7.88}$$

We find that for all $\alpha < 0$ the regular double bubble instanton has much lower action than the four sphere. The singular double bubble instantons have divergent action, but the Hawking–Turok type instantons have finite action. For given α, the action of the latter class depends on the third derivative of the scale factor at $\tau = 0$. This is the analogue of the situation in scalar field inflaton, where the action of the Hawking–Turok instantons depends on the value of the inflaton field at the North Pole. The action of the singular instantons tends smoothly to the S^4 action 7.88 as $f'''(0) \to -1$ and it decreases monotonically with increasing $f'''(0)$.

To summarize, we found a one–parameter family of finite–action, compact solutions of the Euclidean field equations that can be analytically continued across a space–like surface Σ of vanishing curvature, to Lorentzian geometries that describe realistic inflationary universes. The condition on Σ guarantees that a real solution of the Euclidean field equations is continued to a real Lorentzian space–time. The Euclidean region is essential, since there is no way to round off a Lorentzian geometry without introducing a boundary. What is the relevance then, in the context of the no boundary proposal, of these real tunnelling geometries with regard to the problem of initial conditions in cosmology?

At least at the semiclassical level, the no boundary proposal gives a measure on the space of initial conditions for cosmology. The weight of each classical trajectory is approximately $|\Psi|^2 \sim e^{-2S_R}$, where S_R is the real part of the Euclidean action of the solution. For real tunnelling solutions this comes entirely from the part of the manifold on which the geometry is Riemannian. The simplicity of this situation has led to the interpretation of the no boundary proposal as a bottom up theory of initial conditions. In particular, it has been argued that if a given theory allows different instantons, the no boundary proposal predicts our universe to be created through the lowest–action solution, since this would give the dominant contribution to the path integral. Applying this interpretation to trace anomaly driven inflation, one must conclude that the no boundary proposal predicts the creation of a hyper–inflating universe emerging from the double bubble instanton, or a nearly empty open universe that occurs by semiclassical tunnelling via a singular instanton with $|f'''(0)|$ small.

The situation is similar in many theories of scalar field inflaton. Restricting attention to real tunnelling geometries, a bottom up interpretation of the no boundary proposal generally favors the creation of large space–times. One typically gets a probability distribution that is peaked around instantons in which the field at the surface of continuation is near the minimum of its potential, yielding very little inflation. Hence, the most probable universes are nearly empty open universes or collapsing closed universes, depending on the analytic continuation one considers. Weak anthropic arguments have been invoked to try to rescue the situation [HT98], by weighing the a priori no boundary probability with the probability of the formation of galaxies. However, for the most natural inflaton potentials, this still predicts a value of Ω_0 that is far too low to be compatible with observations. Another attempt [Tur00], based on introducing a volume factor that represents the projection onto the subset of states containing a particular observer, leads to eternal inflation at the Planck density, where the theory breaks down. In fact, invoking conditional probabilities is contrary to the whole idea of the no boundary proposal, which by itself specifies the quantum state of the universe.

Clearly the predictions of a bottom up interpretation of the no boundary proposal do not agree with observation. This is because it is an essentially classical interpretation, which is neither relevant nor correct for cosmology. The quantum origin of the universe implies its quantum state is given by a path integral. Therefore, one must adopt a quantum approach to the problem of initial conditions, in which one considers the no boundary path integral 7.99 for a given quantum state of the universe. We shall apply such a quantum approach in section IV, to describe the origin of an inflationary universe, in theories like trace anomaly inflation. It turns out that the relevant saddle–points are not exactly real tunnelling geometries. Instead, one must consider complex saddle points, in which the geometry becomes gradually Lorentzian at late times [HH02].

Instability of Anomaly–Induced Inflation

Metric Perturbations

Two-point functions of metric perturbations can be computed directly from the no boundary path integral. One perturbatively evaluates the path integral around an $O(4)$ invariant instanton background to get the real–space Euclidean correlator, which is then analytically continued into the Lorentzian universe, where it describes the quantum fluctuations of the graviton field in the primordial de Sitter phase [GT99, HT00]. The quantum state of the Lorentzian fluctuations is uniquely determined by the condition of regularity on the instanton [HHR01]. Both scalar and tensor perturbations are given by a path integral of the form [HH02]

$$\langle h_{\mu\nu}(x) h_{\mu\prime\nu\prime}(x\prime)\rangle \sim \int \mathcal{D}[\mathbf{h}] \exp(-S^{(2)}) h_{\mu\nu}(x) h_{\mu\prime\nu\prime}(x\prime), \qquad (7.89)$$

where $S^{(2)}$ denotes the second order perturbation of the action

$$S = S_{EH} + S_{GH} + S_{R^2} + \tilde{W}, \qquad (7.90)$$

with $\tilde{W}$ given by (7.83). For the scalars, eliminating the remaining gauge freedom introduces Faddeev–Popov ghosts. These ghosts supply a determinant $(\hat{\nabla}^2 + 4)^{-1}$, which cancels a similar factor in the scalar action, rendering it second order[24]. The action for the tensors $\theta_{\mu\nu}$ on the other hand is non-local and fourth order. Nevertheless, the metric perturbation and its first derivative should not be regarded as two independent variables, since this would lead to meaningless probability distributions in the Lorentzian [HT01]. Instead the path integral should be taken over the fields $\theta_{\mu\nu}$ only[25], to compute correlators of the form (7.89). The Euclidean action for $\theta_{\mu\nu}$ is positive definite, so the path integral over all $\theta_{\mu\nu}$ converges and determines a well-defined Euclidean quantum field theory. One might worry that the higher derivatives would lead to instabilities in the Lorentzian. This is not the case, however, since the no boundary prescription to compute Lorentzian propagators by Wick rotation from the Euclidean, implicitly imposes the final boundary condition that the fields remain bounded, which eliminates the runaways [HHR01, HT01].

[24] The gauge freedom also leads to closed loops of Faddeev–Popov ghosts but they can be neglected in the large N approximation.

[25] This means one loses unitarity. However, probabilities for observations tend towards those of the second order theory, as the coefficients of the fourth order terms in the action tend to zero. Hence unitarity is restored at the low energies that now occur in the universe.

The path integral (7.89) is Gaussian, so the correlation functions can be read off from the perturbed action and equation (7.86):

$$\langle\psi(x)\psi(x\prime)\rangle = \frac{32\pi^2 r_s^4}{3|\alpha|N^2}\left(-\hat{\nabla}^2 + 1/2\alpha\right)^{-1}, \qquad \text{and} \tag{7.91}$$

$$\langle\theta_{\mu\nu}(x)\theta_{\mu\prime\nu\prime}(x\prime)\rangle = \frac{128\pi^2 r_s^4}{N^2}\sum_{p=2}^{\infty}\frac{W^{(p)}_{\mu\nu\mu\prime\nu\prime}(x,x\prime)}{p^2+3p+6+\Psi(p)-4\alpha p(p+3)}, \tag{7.92}$$

where the bitensor $W^{(p)}_{\mu\nu\mu'\nu'}(x,x')$ is defined as the usual sum over degenerate rank–2 harmonics on the four sphere.

The scalar 2–point function 7.91 is just the propagator of a particle with physical mass $m^2 = (2\alpha r_s^2)^{-1}$. Since we are assuming $\alpha < 0$, we have $m^2 < 0$ so this particle is a tachyon, which is the perturbative manifestation of the Starobinsky instability. Making α more negative, makes the tachyon mass squared less negative, and therefore weakens the instability. Indeed, the number of efoldings in the primordial de Sitter phase emerging from the four sphere instanton is given by $N_{efolds} \sim 12|\alpha|(\log N - 1)$. Therefore, in the interesting regime, we have $-m^2 << m_{pl}^2$, so semiclassical gravity should be a good approximation.

This result sheds light on the problem of initial conditions in trace anomaly inflation. One can think of the non derivative term in the scalar correlator as a potential $V(\psi)$, with the unperturbed de Sitter solution at $\psi = 0$ at the maximum. If $|\alpha|$ is not too small, then the top of the potential is sufficiently flat, so that the lowest–action regular instanton is a homogeneous Hawking–Moss instanton [HM82], with ψ constant at the top. Since the instability of the de Sitter phase is characterized entirely by the coefficient α of the R^2 counterterm. this means the problem of initial conditions in anomaly–induced inflation is similar to the corresponding problem in many theories of scalar field inflaton, where one ought to explain why the inflation starts initially at the top of the hill.

Homogeneous Fluctuations

The most interesting instantons in both trace anomaly driven inflation as well as most theories of scalar field inflaton possess a homogeneous fluctuation mode which decreases their action [HHR01, GT01]. The presence of such a negative mode is the perturbative manifestation of the conformal factor problem. Indeed, since the conformal factor problem is closely related to the instability of gravity under gravitational collapse, one expects instantons that are appealing from a cosmological perspective, to possess a negative mode.

Writing the scalar propagator 7.91 on the four sphere instanton in momentum space gives [HH02]

$$\langle\psi(x)\psi(x\prime)\rangle = \frac{32\pi^2 r_s^4}{3|\alpha|N^2}\sum_{p=0}^{\infty}\frac{W^{(p)}(\mu(x,x\prime))}{p(p+3)+m^2}, \tag{7.93}$$

where the biscalar $W^{(p)}$ equals the usual sum over degenerate scalar harmonics on the four sphere with eigenvalue $\lambda_p = -p(p+3)$ of the Laplacian.

There are many negative modes if $-1/8 < \alpha < 0$. This is usually the perturbative indication of the existence of a lower-action instanton solution. For instance, in scalar field inflaton with a double well potential, the Hawking–Moss instanton possesses several negative modes if $V_{,\phi\phi}/H^2 < -4$, which is precisely the condition for the existence of a lower–action Coleman–De Luccia instanton that straddles the maximum. On the other hand, if $\alpha < -1/8$ in 7.93 then only the homogeneous ($p = 0$) negative mode remains, which is again similar to the well–known negative homogeneous mode of the Hawking–Moss instanton in theories with a scalar potential that is sufficiently flat.

The presence of a physical negative mode supports the interpretation of an instanton as describing the decay of an unstable state through semiclassical tunnelling [Col80]. On the other hand, it has been argued that it questions its use in the no boundary path integral to define the initial quantum state of the universe[26] [GT01]. Within the semiclassical approximation, however, it is more appropriate to project out the negative mode, since the semiclassical approach is based on the *assumption* that the path integral can be expanded around solutions of the classical field equations.

The conclusions of [GT01] are based on a perturbation calculation around compact, real instanton backgrounds, that does not take in account the wave function of interest. One expects, however, the configuration specifying the quantum state of the Lorentzian universe to project out the negative mode from the perturbation spectrum. Consider for example the wave function of a universe described by a 3–sphere with radius $R^2 = V_0/3$ and field $\phi = 0$, in a theory of gravity coupled to a single scalar field with potential $V_0(1-\phi^2)^2$. In the semiclassical approximation, this is given by half of a Hawking–Moss instanton with the field constant at the top of the potential. Obviously, this solution has no negative mode, since the boundary condition on the 3–sphere Σ removes the lowest eigenvalue solution of the Schrödinger equation for the perturbations. Since the negative mode corresponds to a homogeneous fluctuation, this is probably true also for large 3–spheres in the Lorentzian regime. Therefore, one expects that in the top down approach to cosmology, where the quantum state of the universe is taken in account, the negative mode is automatically projected out.

[26] In scalar field inflaton, one can view the singular Hawking–Turok instantons as constrained instantons, with additional data specified on an internal boundary. For some theories, the constraint introduced in [KTW00] to resolve the singularity, also removes the negative mode, at least perturbatively [GT01]. However, it does not remove the instability non–perturbatively and for the most obvious potentials, the lowest–action constrained instanton gives very little inflation.

Quantum Matter and the Microwave Background

Before discussing the top down approach in more detail, we pause to briefly comment on some of the characteristic predictions for observations of trace anomaly inflation.

To extract accurate predictions for the cosmic microwave anisotropies, one must evolve the perturbations through the Starobinsky instability, to get initial conditions for the inhomogeneities during the radiation and matter eras. Details of this calculation will be presented elsewhere [HH02], but some interesting features of the microwave temperature anisotropies predicted by anomaly–induced inflation, can be extracted from the correlators (7.91) and (7.92) in the primordial de Sitter era. As can be seen from (7.92), the quantum matter couples to the tensors. Starobinsky [Sta83] and Vilenkin [Vil85] assumed that the amplitude of primordial gravity waves was not significantly altered by the quantum matter loops. This assumption can now be examined using AdS/CFT, which has allowed us to include the effect of the Weyl^2 counterterm and the non–local part of the matter effective action. We find that at small scales, matter fields dominate the tensor propagator and make it decay like $p^4 \log p$. In other words, the CFT appears to give space–time a rigidity on small scales, an example of how quantum loops of matter can change gravity at short distances. In fact, this suppression should occur even if inflation is not driven by the trace anomaly, since we observe a large number of matter fields, whose effective action is expected to dominate the propagator on small scales.

Secondly, both the higher derivative counterterms and the matter fields introduce anisotropic stress, which is an important difference with scalar field inflaton. This can be seen from decomposing the tensors $\theta_{\mu\nu}$ into a scalar ϕ and tensor t_{ij} under $O(4)$. The former is the difference between the two potentials in the Newtonian gauge and corresponds to anisotropic stress. Typically reheating at the end of anomaly–induced inflation leads to creation of particles that are not in thermal equilibrium with the photon-baryon fluid, so one expects some anisotropic stress to survive during the radiation era. To make more precise predictions, however, a better understanding is required of the (probably time–dependent) values of the coefficients α and β of the higher derivative counterterms in the theory.

Finally, we should mention that for the tensor propagator the higher derivative terms also give rise to poles in the complex p–plane. These are harmless, however, since the contour obtained from the Euclidean goes around the complex poles [HHR01]. In other words, defining the theory in the Euclidean, implicitly removes the instabilities associated with the complex poles, like a final boundary condition removes the runaway solution of the classical radiation reaction force [HT01].

Origin of Inflation

We have seen that the predictions of the bottom up approach to the problem of initial conditions in inflation do not agree with observation. This is because it is based on an essentially classical picture, in which one assumes some initial condition for the universe and evolves it forward in time. The quantum origin of our universe, however, means that its wave function is determined by a path integral, in which one sums over all possible histories that lead to a given quantum state, together with some suitable boundary conditions on the paths. This naturally leads to a top down view of the universe. In a top down context, rather than comparing the relative probabilities of different semiclassical geometries, one looks for the most probable evolution that leads to a certain outcome.

We now apply the quantum top down interpretation of the no boundary proposal to study the origin of an inflationary universe, in theories where the instability of the inflationary phase can be described in terms of a single scalar field with an effective potential that has a local maximum. As shown above, this includes trace anomaly driven inflation, since the emergence of an anomaly driven inflationary universe is very similar to the creation of an exponentially expanding universe in theories of new inflation.

We consider a model consisting of gravity coupled to a single scalar field, with a double well potential $V(\phi) = A(1 - \frac{C}{2}\phi^2)^2$ (with $A, C > 0$). For $C < 2/3$, the potential has a maximum at $\phi = 0$ with $V_{,\phi\phi}/V$ sufficiently low so that there exists no Coleman–De Luccia instanton, but only a Hawking–Moss instanton with $\phi = 0$ everywhere on top of the hill. Implementing a top down approach, we consider the quantum amplitudes $\Phi[\Sigma, \tilde{\mathbf{h}}, K, \phi_\Sigma]$ for different conformal 3-geometries $\tilde{\mathbf{h}}$ with trace K of the second fundamental form, on an expanding surface Σ during inflation. According to the no boundary proposal, the defining path integral should be taken over all compact Riemannian geometries that induce the prescribed configuration on Σ.

In the K–representation, the Euclidean action is given by [HH02]

$$S = -\frac{1}{2\kappa}\int d^4x\sqrt{g}R - \frac{1}{3\kappa}\int d^3x\sqrt{h}K + \int d^4x\sqrt{g}\left(\frac{1}{2}g^{\mu\nu}\partial_\mu\phi\partial_\nu\phi + V(\phi)\right), \tag{7.94}$$

The usual wave function $\Psi[\mathbf{h}, \phi_\Sigma]$ is obtained from $\Phi[\tilde{\mathbf{h}}, K, \phi_\Sigma]$ by an inverse Laplace transform,

$$\Psi[\mathbf{h}, \phi_\Sigma] = \int_\Gamma \mathcal{D}\left[\frac{K}{4i\kappa}\right] \exp\left[\frac{2}{3\kappa}\int d^3x\sqrt{h}K\right] \Phi[\tilde{\mathbf{h}}, K, \phi_\Sigma]$$

where the contour Γ runs from $-i\infty$ to $+i\infty$.

Within the semiclassical approximation, the no boundary wave function is approximately given by the saddle–point contributions. Restricting attention to saddle–points that are invariant under the action of an $O(4)$ isometry group, the instanton metric can be written as

$$ds^2 = d\tau^2 + b^2(\tau)d\Omega_3^2, \tag{7.95}$$

and the Euclidean field equations read

$$\phi" = -K\phi\prime + V_{,\phi}, \tag{7.96}$$

$$K\prime + K^2 = -(\phi_{,\tau}^2 + V), \tag{7.97}$$

where $\phi' = \phi_{,\tau}$ and $K = 3b_{,\tau}/b$. The Lorentzian trace $K_L = -3\dot{a}/a$ is obtained by analytic continuation. We first calculate the wave function for real K, and then analytically continue to imaginary, or Lorentzian $K_L = -iK$.

At the semiclassical level, there are two contributions to the given amplitude. For small ϕ_Σ and any Euclidean K, there always exists a non–singular, Euclidean $O(4)$ invariant solution of the field equations, with the prescribed boundary conditions. This solution is part of a deformed sphere, or Hawking–Turok instanton. In the approximation $K = 3H\cot(H\tau)$, with $H^2 = A/3$, and $V(\phi) \sim A(1 - C\phi^2)$, the solution of (7.107) is given by

$$\phi = \phi_\Sigma \frac{{}_2F_1(3/2+q, 3/2-q, 2, z(K))}{{}_2F_1(3/2+q, 3/2-q, 2, z(K_\Sigma))}, \qquad \text{where}$$
$$q = \sqrt{9/4 + 6C}, \qquad z(K) = \frac{1}{2}\left[1 - \frac{K}{(A^2+K^2)^{1/2}}\right].$$

At the South Pole $K \to +\infty$, so in the instanton the scalar field slowly rolls up the hill from its value at the regular South Pole to the prescribed value ϕ_Σ on the 3-sphere with trace K_Σ. The weight of the Hawking–Turok geometry in the no boundary path integral for the wave function $\Phi[K, \phi_\Sigma]$ is approximately given by [HH02]

$$\begin{aligned} S[K_\Sigma, \phi_\Sigma] &= -\frac{1}{3\kappa}\int d^3x\sqrt{h}K - \int d^4x\sqrt{g}V(\phi) \\ &= -\frac{12\pi^2}{A}\left[1 - \frac{K}{(A^2+K^2)^{1/2}}\right] - \frac{24\pi^2 C}{A^2(1-C)}\phi_\Sigma^2\, z^2(K_\Sigma)\times \\ &\left[1 - 2z(K_\Sigma) + 3C(1 - z(K_\Sigma))\left(z(K_\Sigma) + \frac{2-3C}{1+3C}\right)\frac{F'}{F}[(z(K_\Sigma))]\right]. \end{aligned}$$

For small ϕ_Σ, there is a second semiclassical contribution to the wave function, coming from universes that are created via an $O(5)$ symmetric Hawking–Moss instanton with ϕ constant at the top of the hill, but in which a quantum fluctuation disturbs the field, causing it to run down to its prescribed value ϕ_Σ at the 3–sphere boundary with trace K_Σ. It follows that for $K_\Sigma = 0$, the action of the Hawking–Turok geometry is more negative than the action of the Hawking–Moss instanton. This would seem to suggest that the universe is least likely to start at the top of the hill. However, we are not interested in the amplitude for a Euclidean space–time, but in the no boundary wave function of a Lorentzian expanding universe.

Within the regime in which ϕ remains small over the whole geometry, one can derive the amplitude in the Lorentzian from the result for $S[K_\Sigma, \phi_\Sigma]$, by analytic continuation into the complex $K-$plane. In a Lorentzian universe, Euclidean K is pure imaginary, $K_L = -iK$. Since the action is invariant under diffeomorphically related contours in the complex $\tau-$plane, we may deform the contour into one with straight sections, along the real and imaginary $K-$axis. The real part of the action for the Hawking–Moss instanton is constant on the imaginary $K-$axis, unlike the action for the Hawking–Turok geometry. According to the no boundary proposal, the relative probability of both geometries is given by

$$P[K_L, \phi_\Sigma] = \frac{A_{HM}^2}{A_{HT}^2} \mathrm{e}^{-2\Re[\Delta S]},$$

where $\Delta S = S_{HM} - S_{HT}$. The prefactors account for small fluctuations around the classical solutions and can be neglected for small ϕ.

As mentioned above, this scenario is realised in trace anomaly driven inflation. The unperturbed de Sitter solution (7.86) in anomaly–induced inflation emerges from the Hawking–Moss geometry, while the inhomogeneous Hawking–Turok evolution corresponds to one of the singular instantons discussed above. The field configuration on Σ determines the third derivative of the scale factor at the regular South Pole, or equivalently the initial value of the order parameter ϕ governing the instability. For $\alpha < -1/8$, the instability of the de Sitter phase is sufficiently weak, so that the universe is most likely to start at the top, in an unstable de Sitter state. This result also justifies presented calculation of metric perturbations, which were based on a perturbative expansion of the path integral about the round four sphere.

Finally, we should mention that because we are interested in real matter fields on Σ, the analytic continuation into the complex $K-$plane means ϕ must be complex in the bulk of the instanton. More precisely, at the South Pole, we must have

$$\mathrm{Im}[\phi] = \phi_\Sigma \, \mathrm{Im}[F(z_\Sigma)] / \, \mathrm{Re}[F(z_\Sigma)].$$

This has no physical meaning though, since the stationary phase approximation is just a mathematical construction to evaluate the path integral over real ϕ.

For more details on top–down inflation, see [HH02].

7.3.7 Cosmology in the String Landscape

In this final section on cosmological dynamics, we present the seminal paper by S. Hawking and T. Hertog [HH06], proposing a top–down/no–boundary approach to quantum cosmology in the string landscape. They put forward a framework for cosmology that combines the string landscape with no boundary initial conditions. In this framework, amplitudes for alternative histories for the universe are calculated with final boundary conditions only. This leads to a top down approach to cosmology, in which the histories of the universe depend on the precise question asked. The authors study the observational consequences of no boundary initial conditions on the landscape, and outline a scheme to test the theory. This is illustrated in a simple model landscape that admits several alternative inflationary histories for the universe. Only a few of the possible vacua in the landscape will be populated. They also discuss in what respect the top down approach differs from other approaches to cosmology in the string landscape, like eternal inflation.

Now, it seems likely that string theory contains a vast ensemble of stable and meta–stable vacua, including some with a small positive effective cosmological constant [Bp00] and the low energy effective field theory of the *Standard Model*. Recent progress on the construction of meta–stable *de Sitter vacua* [KRL03] lends further support to the notion of a *string landscape* [Sus03], and a statistical analysis gives an idea of the distribution of some properties among the vacua [Dou]. But it has remained unclear what is the correct framework for cosmology in the string landscape. There are good reasons to believe, however, that a proper understanding of the cosmological dynamics will be essential for the landscape to be predictive [BDG04].

Recall that in particle physics, one usually computes *S–matrix* elements. This is useful to predict the outcome of laboratory experiments, where one prepares the initial state and measures the final state. It could be viewed as a bottom–up approach to physics, in which one evolves forward in time a particular initial state of the system. The predictability of this approach arises from and relies upon the fact that one has control over the initial state, and that experiments can be repeated many times to gain statistically significant results.

But cosmology poses questions of a very different character. In our past there is an epoch of the early universe when quantum gravity was important. The remnants of this early phase are all around us. The central problem in cosmology is to understand why these remnants are what they are, and how the distinctive features of our universe emerged from the Big–Bang. Clearly it is not an S–matrix that is the relevant observable[27] for these predictions, since we live in the middle of this particular experiment. Furthermore, we have no control over the initial state of the universe, and there is certainly no opportunity for observing multiple copies of the universe.

In fact if one does adopt a bottom–up approach to cosmology, one is immediately led to an essentially classical framework, in which one loses all ability to the central question of cosmology: why our universe is the way it is. In particular a bottom–up approach to cosmology either requires one to postulate an initial state of the universe that is carefully fine–tuned [GV93], as if prescribed by an outside agency, or it requires one to invoke the notion of

[27] See [BF01, Wit01, Bou05, GMH05] for recent work on the existence and the construction of observables in cosmological space–times.

eternal inflation [Vil83], which prevents one from predicting what a typical observer would see.

Here we put forward a different approach to cosmology in the string landscape, based not on the classical idea of a single history for the universe but on the quantum sum over histories [Har95a]. We argue that the quantum origin of the universe naturally leads to a framework for cosmology where amplitudes for alternative histories of the universe are computed with boundary conditions at late times only. We thus envision a set of alternative universes in the landscape, with amplitudes given by the no–boundary path integral [HH83].

The measure on the landscape provided by no–boundary initial conditions allows one to derive predictions for observations. This is done by evaluating probabilities for alternative histories that obey a set of constraints at late times. The constraints provide information that is supplementary to the fundamental laws and act as a selection principle. In particular, they select the subclass of histories that contribute to the amplitude of interest. One then identifies alternatives within this subclass that have probabilities near one. These include, in particular, predictions of future observations. The framework proposed here is thus more like a top–down approach to cosmology, where the histories of the universe depend on the precise question asked [HH06].

Quantum State

In cosmology one is generally not concerned with observables at infinity or with properties of the entire 4–geometry, but with alternatives in some finite region in the interior of the space–time. The amplitudes for these more restricted sets of observables are obtained from the amplitudes of four dimensional metric and matter field configurations, by integrating over the unobserved quantities. A particularly important case is the amplitude of finding a compact space–like surface S with induced 3–metric g^3_{ij} and matter field configuration ϕ [HH06],

$$\Psi[g^3,\phi] \sim \int_C [\mathcal{D}g][\mathcal{D}\phi]\ \mathrm{e}^{iS[g,\phi]}. \qquad (7.98)$$

Here the path integral is taken over the class C of space–times which agree with g^3_{ij} and ϕ on a compact boundary S. The quantum state of the universe is determined by the remaining specification of the class C.

Usually one sums over histories that have an initial and a final boundary. This is useful for the computation of S–matrix elements to predict the outcome of laboratory experiments, where one prepares the initial state and measures the final state. It is far from clear, however, that this is the appropriate setup for cosmology, where one has no control over the initial state, and no opportunity for observing multiple copies of the universe. In fact, if one does apply this approach to cosmology one is naturally led to an essentially classical picture, in which one simply assumes the universe began and evolved in a way that is well defined and unique.

The so–called *pre–Big–Bang cosmologies* [GV93] are examples of models that are based on a bottom–up approach. In these models one specifies an initial state on a surface in the infinite past and evolves this forward in time. A natural choice for the initial state would be flat space, but that would obviously remain flat space. Thus one instead starts with an unstable state in the infinite past, tuned carefully in order for the *Big–Crunch/Big–Bang transition* to be smooth and the path integral to be peaked around a single *semi–classical history*. Several explicit solutions of such bouncing cosmologies have been found in various mini–superspace approximations [KS04]. It has been shown, however, using several different techniques, that solutions of this kind are unstable [LMS02, HH04]. In particular, one finds that generic small perturbations at early times (or merely taking in account the remaining degrees of freedom) dramatically change the evolution near the transition. Rather than evolving towards an expanding semi–classical universe at late times, one generically produces a strong curvature singularity. Hence the evolution of pre–Big–Bang cosmologies always includes a genuinely quantum gravitational phase, unless the initial state is extremely fine–tuned. It is therefore more appropriate to describe these cosmologies by a path integral in quantum cosmology, and not in terms of a single semi–classical trajectory. The universe will not have a single history but every possible history, each with its own probability.

In fact, the quantum state of the universe at late times is likely to be independent of the state on the initial surface. This is because there are geometries in which the initial surface is in one universe and the final surface in a separate disconnected universe. Such metrics exist in the Euclidean regime, and correspond to the quantum annihilation of one universe and the quantum creation of another. Moreover, because there are so many different possible universes, these geometries dominate the path integral. Therefore even if the path integral had an initial boundary in the infinite past, the state on a surface S at late times would be independent of the state on the initial surface. It would be given by a path integral over all metric and matter field configurations whose only boundary is the final surface S. But this is precisely the *no–boundary quantum state* [HH83]

$$\Psi[g^3, \phi] \sim \int_C [\mathcal{D}g][\mathcal{D}\phi] \; \mathrm{e}^{-S_E[g,\phi]}, \tag{7.99}$$

where the integral is taken over all regular geometries bounded only by the compact 3–geometry S with induced metric g^3_{ij} and matter field configuration ϕ. The *Euclidean action* S_E is given by[28]

$$S_E = -\frac{1}{2}\int d^4x\sqrt{g}\,(R + L(g,\phi)) - \int_S d^3x\sqrt{g^3}K, \tag{7.100}$$

[28] We have set $8\pi G = 1$.

where $L(g, \phi)$ is the matter Lagrangian.

One expects that the dominant contributions to the path integral will come from saddle points in the action. These correspond to solutions of the Einstein equations with the prescribed final boundary condition. If their curvature is bounded away from the Planck value, the saddle point metric will be in the semi–classical regime and can be regarded as the most probable history of the universe. Saddle point geometries of particular interest include geometries where a Lorentzian metric is rounded off smoothly in the past on a compact *Euclidean instanton*. Well known examples of such geometries are the *Hawking–Moss (HM) instanton* [HM82] which matches to *Lorentzian de Sitter space*, and the Coleman–De Luccia (CdL) instanton [Col80], which continues to an open FLRW universe. The former occurs generically in models of gravity coupled to scalar fields, while the latter requires a rather fine-tuned potential.

The usual interpretation of these geometries is that they describe the decay of a false vacuum in de Sitter space. However, they have a different interpretation in the no–boundary proposal [GT99]. Here they describe the beginning of a new, independent universe with a completely self–contained 'no–boundary' description[29]. By this we mean, in particular, that the expectation values of observables that are relevant to local observers within the universe can be unambiguously computed from the no–boundary path integral, without the need for assumptions regarding the pre–bubble era. The original de Sitter universe may continue to exist, but it is irrelevant for observers inside the new universe. The no–boundary proposal indicates, therefore, that the pre-bubble inflating universe is a redundant theoretical construction.

It is appealing that the no–boundary quantum state (7.99) is computed directly from the action governing the dynamical laws. There is thus essentially a single theory of dynamics and of the quantum state. It should be emphasized however that this remains a *proposal* for the wave function of the universe. We have argued it is a natural choice, but the ultimate test is whether its predictions agree with observations.

Prediction in Quantum Cosmology

Quantum cosmology aims to identify which features of the observed universe follow directly from the fundamental laws, and which features can be understood as consequences of quantum accidents or late time selection effects. In no–boundary cosmology, where one specifies boundary conditions at late times only, this program is carried out by evaluating probabilities for alternative histories that obey certain constraints at the present time. The final boundary conditions provide information that is supplementary to the fundamental laws, which selects a subclass of histories and enables one to identify

[29] The interpretation of these saddle point geometries is in line with their interpretation that follows from holographic reasoning, as described e.g., in [DKS02]. Some of our conclusions, however, differ from [DKS02].

alternatives that (within this subclass) have probabilities near one. In general the probability for an alternative α, given H, Ψ and a set of constraints β, is given by [HH06]

$$p(\alpha|\beta, H, \Psi) = \frac{p(\alpha, \beta|H, \Psi)}{p(\beta|H, \Psi)}. \tag{7.101}$$

The conditions β in (7.101) generally contain environmental selection effects, but they also include features that follow from quantum accidents in the early universe[30].

A typical example of a condition β is the dimension D of space. For good reasons, one usually considers string compactifications down to three space dimensions. However, there appears to be no dynamical reason for the universe to have precisely four large dimensions. Instead, the no–boundary proposal provides a framework to calculate the quantum amplitude for every number of spatial dimensions consistent with string theory. The probability distribution of various dimensions for the universe is of little significance, however, because we have already measured we live in four dimensions. Our observation only gives us a single number, so we cannot tell from this whether the universe was likely to be four dimensional, or whether it was just a lucky chance. Hence as long as the no–boundary amplitude for three large spatial dimensions is not exactly zero, the observation that $D = 3$ does not help to prove or disprove the theory. Instead of asking for the probabilities of various dimensions for the universe, therefore, we might as well use our observation as a final boundary condition and consider only amplitudes for surfaces S with three large dimensions. The number of dimensions is thus best used as a constraint to restrict the class of histories that contribute to the path integral for a universe like ours. This restriction allows one to identify definite predictions for future observations.

The situation with the low energy effective theory of particle interactions may well be similar. In string theory this is the effective field theory for the modular parameters that describe the internal space. It is well known that string theory has solutions with many different compact manifolds. The corresponding effective field theories are determined by the topology and the geometry of the internal space, as well as the set of fluxes that wind the 3–cycles. Furthermore, for each effective field theory the potential for the moduli typically has a large number of local minima. Each local minimum of the potential is presumably a valid vacuum of the theory. These form a landscape [Sus03] of possible stable or meta–stable states for the universe at the present time, each with a different theory of low energy particle physics.

In the bottom–up picture it is thought that the universe begins with a grand unified symmetry, such as $E_8 \times E_8$. As the universe expands and cools the symmetry breaks to the Standard Model, perhaps through intermediate stages. The idea is that string theory predicts the pattern of breaking, and

[30] These are quantum accidents that became 'frozen', leaving an imprint on the universe at late times.

the masses, couplings and mixing angles of the Standard Model. However, as with the dimension of space, there seems to be no particular reason why the universe should evolve precisely to the internal space that gives the Standard Model[31]. It is therefore more useful to compute no–boundary amplitudes for a space–like surface S with a given internal space. This is the top–down approach, where one sums only over the subclass of histories which end up on S with the internal space for the Standard Model [HH06].

We now turn to the predictions α we can expect to derive from amplitudes like (7.101). We have seen that the relative amplitudes for radically different geometries are often irrelevant. By contrast, the probabilities for neighboring geometries are important. The most powerful predictions are obtained from the relative amplitudes of nearby geometries, conditioned on various discrete features of the universe. This is because these amplitudes are not determined by the selection effects of the final boundary conditions. Rather, they depend on the quantum state $|\Psi\rangle$ itself.

Neighboring geometries correspond to small quantum fluctuations of continuous quantities, like the temperature of the cosmic microwave background (CMB) radiation or the expectation values of the string theory moduli in a given vacuum. In inflationary universes these fluctuations are amplified and stretched, generating a pattern of spatial variations on cosmological scales in those directions of moduli space that are relatively flat[32]. The spectra depend on the quantum state of the universe. Correlators of fluctuations in the no–boundary state can be calculated by perturbatively evaluating the path integral around instanton saddle points [GT99]. In general if $\mathcal{P}(x_1)$ and $\mathcal{Q}(x_2)$ are two observables at x_1 and x_2 on a final surface S, then their correlator is formally given by the following integral over a complete set of observables $\mathcal{O}(x)$ on S [GT99],

$$\langle \mathcal{P}(x_1)\mathcal{Q}(x_2)\rangle \sim \sum_B \int [\mathcal{D}\ \mathcal{O}(S)]\Psi_B[\mathcal{O}]^*\Psi_B[\mathcal{O}]\mathcal{P}(x_1)\mathcal{Q}(x_2). \tag{7.102}$$

Here the sum is taken over backgrounds B that satisfy the prescribed conditions on S. The amplitude Ψ_B for fluctuations about a particular background geometry $(\bar{g},\bar{\phi})$ is given by

$$\Psi_B[g^3,\phi] \sim \mathrm{e}^{-S_0(\bar{g},\bar{\phi})} \int [\mathcal{D}\delta g][\mathcal{D}\delta\phi]\ \mathrm{e}^{-S_2[\delta g,\delta\phi]}, \tag{7.103}$$

[31] An extension of the bottom-up approach invokes the notion of eternal inflation to accommodate the possibility that the position in the moduli space falls into different minima in different places in space, leading to a mosaic structure for the universe. The problem with this approach is that one cannot predict what a typical local observer within such a universe would see.

[32] Spatial variations of coupling constants from scalar moduli field fluctuations generate large scale iso–curvature fluctuations in the matter and radiation components [Kof03].

where the metric $g = \bar{g} + \delta g$ and the fields $\phi = \bar{\phi} + \delta\phi$. The C_l's of the CMB temperature anisotropies are classic examples of observables that can be calculated from correlators like this. Whilst the full correlator (7.102) generally involves a sum over several saddle points, for most practical purposes only the lowest action instanton matters.

In no–boundary backgrounds like the HM geometry, where a real Euclidean instanton is matched onto a real Lorentzian metric, one can find the correlators by first calculating the 2–point functions in the *Euclidean region*. The Euclidean correlators are then analytically continued into the *Lorentzian region*, where they describe the quantum mechanical vacuum fluctuations of the various fields in the state determined by no–boundary initial conditions. The path integral unambiguously specifies boundary conditions on the Euclidean fluctuation modes. This essentially determines a reflection amplitude $R(k)$, where k is the wave–number, which depends on the instanton geometry. The spectra in the Lorentzian, and in particular the primordial gravitational wave spectrum [GHT00], depend on the instanton background through $R(k)$.

The relative amplitudes of neighboring geometries can thus be used to predict, from first principles, the precise shape of the primordial fluctuation spectra that we observe. This provides a test of the no–boundary proposal and, more generally, an observational discriminant between different proposals for the state of the universe, because the spectra contain a signature of the initial conditions [HH06].

Anthropic Reasoning

Recall that in general *anthropic reasoning* [BT86] aims to explain certain features of our universe from our existence in it. One possible motivation for this line of reasoning is that the observed values and correlations of certain parameters in particle physics and cosmology appear necessary to ensure life emerges in our universe. If this is indeed the case it seems reasonable to suppose that certain environmental selection effects need to be taken in account in the calculation of probabilities for observations.

It has been pointed out many times, however, that anthropic reasoning is meaningless if it is not implemented in a theoretical framework that determines which parameters can vary and how they vary. Top down cosmology, by combining the string landscape with the no–boundary proposal, provides such a framework. The anthropic principle is implemented in the top–down approach by specifying a set of conditions β in (7.101) that select the subclass of histories where life is likely to emerge. More specifically, anthropic reasoning in the context of top–down cosmology amounts to the evaluation of conditional probabilities like [HH06]

$$p(\alpha|O, H, \Psi), \qquad (7.104)$$

where O represents a set of conditions that are required for the appearance of complex life. The utility and predictability of anthropic reasoning depends on how sensitive the probabilities (7.104) are to the inclusion of O. Anthropic reasoning is useful and predictive only if (7.104) is sharply peaked around the observed value of α, and if the *a priori* theoretical probability $p(\alpha|H, \Psi)$ itself is broadly distributed [Har05a].

Anthropic reasoning, therefore, can be naturally incorporated in the top–down approach. In particular it may provide a qualitative understanding for the origin of certain conditions β that one finds are useful in top–down cosmology. Consider the number of dimensions of space, for example. We have argued that this is best used as a final constraint, but the top–down approach itself does not explain why this particular property of the universe cannot be predicted from first principles. In particular, the top–down argument does not depend on whether four dimensions is the only arena for life. Rather, it is that the probability distribution over dimensions is irrelevant, because we cannot use our observation that $D = 3$ to falsify the theory. But it may turn out that anthropically weighted probabilities (7.104) are always sharply peaked around $D = 3$. In this case one can essentially interpret the number of dimensions as an anthropic requirement, and it would be an example where anthropic reasoning is useful to understand why one needs to condition on the number of dimensions in top–down cosmology.

We emphasize, however, that the top–down approach developed here goes well beyond conventional anthropic reasoning. Firstly, the top–down approach gives *a priori* probabilities that are more sharply peaked, because it adopts a concrete prescription for the quantum state of the universe – as opposed to the usual assumption that predictions are independent of Ψ. Hence the framework we propose is more predictive than conventional anthropic reasoning[33].

Top down cosmology is also more general than anthropic reasoning, because there is a wider range of selection effects that can be quantitatively taken in account. In particular the conditions β that are supplied in (7.101) need not depend on whether they are necessary for life to emerge. The set of conditions generally includes environmental selection effects similar to anthropic requirements, but it also includes chance outcomes of quantum accidents in the early universe that became frozen. The latter need not be relevant to the emergence of life. Furthermore, they cannot be taken in account by simply adding an *a posteriori* selection factor proportional to the number density of some reference object, because they change the entire history of the universe!

Models of Inflation

How can one get a nonzero amplitude for the present state of the universe if, as we claim, the metrics in the sum over histories have no–boundary apart from

[33] Anthropic selection effects have been used to constrain the value of the cosmological constant [Wei87], and the dark matter density [Lin88]. In these studies it is assumed, however, that the a priori probability distributions are independent of the state of the universe. This reduces the predictability of the calculations, and could in fact be misleading.

the surface S at the present time? We do not have a definitive answer, but one possibility would be if the four dimensional part of the saddle point metric was an inflating universe at early times. Hartle and Hawking [HH83] have shown that such metrics can be rounded off in the past, without a singular beginning and with curvature bounded well away the Planck value. They give a nonzero value of the no–boundary amplitude for almost any universe that arises from an early period of inflation. Thus to illustrate the top–down approach described above, we consider a simple model with a few positive extrema of the effective potential.

We assume the instability of the inflationary phase can be described as the evolution of a scalar order parameter ϕ moving in a double well potential $V(\phi)$. We take the potential to have a broad flat–topped maximum V_0 at $\phi = 0$ and a minimum at ϕ_1. The value at the bottom is the present small cosmological constant Λ. A concrete example would be gravity coupled to a large number of light matter fields [Sta80]. The trace anomaly generates a potential which has unstable de Sitter space as a self–consistent solution.

We are interested in calculating the no–boundary amplitude of an expanding non–empty region of space–time similar to the one we observe today. In the semi–classical approximation, this will come from one or more saddle points in the action. These correspond to solutions of the Einstein equations. One solution is de Sitter space with the field ϕ sitting at the minimum of the potential $V(\phi)$. This will have a very large amplitude, but will be complete empty and therefore does not contribute to the top–down amplitude for a universe like ours. To get an expanding universe with $\Omega_m \sim \mathcal{O}(1)$ and with small perturbations that lead to galaxies, it seems necessary to have a period of inflation[34].

We therefore consider the no–boundary amplitude[35] $\Phi[\tilde{g}^3, K, \phi]$ for a closed inflating universe bounded by a 3–surface S with a large approximately constant *Hubble parameter* $H = \dot{a}/a$ (and corresponding trace $K = -3\dot{a}/a = -3H$), and a nearly constant field ϕ near the top of V. The value of ϕ on S is chosen sufficiently far away from the minimum of V to ensure there are at least enough efoldings of inflation for the universe at the present time to be approximately flat.

[34] One might think it would be more likely for a universe like ours to arise from a fluctuation of the big de Sitter space directly into a hot Big–Bang, rather than from a homogeneous fluctuation up the potential hill that leads to a period of inflation. The amplitude of a hot Big–Bang fluctuation is much smaller, however, than the amplitude of the inflationary saddle points we discuss below (see also [AS04]). The latter do not directly connect to the large de Sitter space, but they could be connected with very little cost in action by a thin bridge [Haw84b].

[35] We work in the K representation of the wave function (see e.g., [Haw84b]), where one replaces g_{ij}^3 on the three-surface S by $\tilde{g}_{ij}^3$, the three-metric up to a conformal factor, and K, the trace of the second fundamental form. The action S_E^k differs from (7.100) in that the surface term has a coefficient 1/3.

We first calculate the wave function for imaginary K, or real Euclidean $K_e = iK$, and then analytically continue the result to real Lorentzian K. There are two distinct saddle point contributions to the amplitude for an inflating universe in this model [HH02]. In the first case, the universe is created by the HM instanton with constant $\phi = 0$. Then quantum fluctuations disturb the field, causing it to classically roll down the potential to its prescribed value on S. Histories of this kind thus have a long period of inflation, and lead to a perfectly flat universe today. The action of the HM geometry is given by [HH06]

$$S_{HM}^k(K) = -\frac{12\pi^2}{V_0}\left(1 - \frac{K_e}{(V_0^2 + K_e^2)^{1/2}}\right), \tag{7.105}$$

where $K_e = 3b_{,\tau}/b$.

There is, however, a second saddle point contribution which comes from a deformed four sphere, with line element

$$ds^2 = d\tau^2 + b^2(\tau)d\Omega_3^2, \tag{7.106}$$

where $\phi(\tau)$ varies across the instanton. The Euclidean field equations for $O(4)$–invariant instantons are

$$\phi" = -K_e\phi\prime + V_{,\phi}, \qquad K_e\prime + K_e^2 = -(\phi_{,\tau}^2 + V), \tag{7.107}$$

where $\phi' = \phi_{,\tau}$. These equations admit a solution, which is part of a *Hawking–Turok instanton* [HT98], where ϕ slowly rolls up the potential from some value ϕ_0 at the (regular) South Pole to its prescribed value on the 3–surface S. Hence this solution represents a class of histories where the scalar starts as far down the potential as the condition that the present universe be approximately flat allows it to. This naturally leads to fewer efoldings of inflation, and hence a universe that is only approximately flat today. The Euclidean action $S_{HT}^k(K)$ of the deformed four sphere was given in [HH02], in the approximation that ϕ is reasonably small everywhere.

A comparison of the action of both saddle points shows that the deformed four sphere dominates the path integral for amplitudes with real Euclidean K_e on S. This would seem to suggest that the universe is least likely to start with ϕ at the top of the hill. However, we are interested in the amplitude for an expanding Lorentzian universe, with real Lorentzian K on S. If one analytically continues the action into the complex K_e–plane, one finds the action of the deformed four sphere rapidly increases along the imaginary K_e– axis whereas the real part of S_{HM}^K remains constant, and the dominant contribution to amplitudes for larger K on S actually comes from the HM geometry. The reason for this is that a constant scalar field saves more in gradient energy, than it pays in potential energy for being at the top of the hill. Hence a Lorentzian, expanding universe with large Hubble parameter H is most likely to emerge in an inflationary state, with ϕ constant at the maximum of the potential.

Top down cosmology thus predicts that in models like trace anomaly inflation, expanding universes with small perturbations that lead to galaxies, start with a long period of inflation, and are perfectly flat today. Furthermore, as discussed earlier, the precise shape of the primordial fluctuation spectra can be computed from the Euclidean path integral, by perturbatively evaluating around the HM saddle point.

Prediction in a Potential Landscape

The predictions we obtained in the previous section extend in a rather obvious way to models where one has a potential landscape. A generic potential landscape admits a large class of alternative inflationary histories with no–boundary initial conditions. There will be HM geometries at all positive saddle points of the potential. For saddle points with more than one descent direction, there will generally be a lower saddle point with only one descent direction, and with lower action. If this descent direction is sharply curved $|V''(0)|/H^2 > 1$, one would not expect a significant top–down amplitude to come from the saddle point. Thus only broad saddle points with a single descent direction will give rise to amplitudes for universes like our own. The requirement[36] that the primordial fluctuations be sufficiently large to form galaxies, however, sets a lower bound on the value of V_0 [HH06].

Only a few of the saddle points will satisfy the demanding condition that they be broad, because it requires that the scalar field varies by order the Planck value across them. Because the dominant saddle points are in the semi–classical regime, the solutions will evolve from the saddle points to the neighboring minima of V. Thus top–down cosmology predicts that only a few of the possible vacua in the landscape will have significant amplitudes.

Alternative Proposals

To conclude, we briefly comment on a number of different approaches to the problem of initial conditions in cosmology, and we clarify in what respect they differ from the top–down approach we have put forward.

We have already discussed the pre–Big–Bang cosmologies [GV93], where one specifies initial conditions in the infinite past and follows forward in time a single semi–classical history of the universe. Pre–Big–Bang cosmology is thus based on a bottom–up approach to cosmology. It requires one to postulate a fine–tuned initial state, in order to have a smooth deterministic transition through the Big–Crunch singularity.

We have also discussed the anthropic principle [BT86]. This can be implemented in top–down cosmology, through the specification of final boundary conditions that select histories where life emerges. Anthropic reasoning within the top–down approach is reasonably well–defined, and useful to the extent that it provides a qualitative understanding for the origin of certain late time conditions that one finds are needed in top–down cosmology.

---

[36] Extra constraints from particle physics, when combined with the cosmological constraints discussed here, will probably further raise the value of V_0.

Eternal Inflation

A different approach to string cosmology has been to invoke the phenomenon of eternal inflation [Vil83] to populate the landscape. There are two different mechanisms to drive eternal inflation, which operate in different moduli space regions of the landscape. In regions where the moduli potential monotonically increases away from its minimum, it is argued that inflation can be sustained forever by quantum fluctuations up the potential hill. Other regions of the landscape are said to be populated by the nucleation of bubbles in meta–stable de Sitter regions. The interior of these bubbles may or may not exit inflation, depending on the shape of the potential across the barrier.

Both mechanisms of eternal inflation lead to a mosaic structure for the universe, where causally disconnected thermalized regions with different values for various effective coupling constants are separated from each other by a variety of inflating patches. It has proven difficult, however, to calculate the probability distributions for the values of the constants that a local observer in an eternally inflating universe would measure. This is because there are typically an infinite number of thermalized regions.

One could also consider the no–boundary amplitude for universes with a mosaic structure. However, these amplitudes would be much lower than the amplitudes for final states that are homogeneous and lie entirely within a single minimum, because the gradient energy in a mosaic universe contributes positively to the Euclidean action. Histories in which the universe eternally inflates, therefore, hardly contribute to the no–boundary amplitudes we measure. Thus the global structure of the universe that eternal inflation predicts, differs from the global structure predicted by top–down cosmology. Essentially this is because eternal inflation is again based on the classical idea of a unique history of the universe, whereas the top–down approach is based on the quantum sum over histories. The key difference between both cosmologies is that in the proposal based on eternal inflation there is thought to be only one universe with a fractal structure at late times, whereas in top–down cosmology one envisions a set of alternative universes, which are more likely to be homogeneous, but with different values for various effective coupling constants.

It nevertheless remains a challenge to identify predictions that would provide a clear observational discriminant between both proposals. We emphasize, however, that even a precise calculation of conditional probabilities in no–boundary cosmology, which takes in account the back–reaction of quantum fluctuations, will make no reference to the exterior of our past light cone. Indeed, the top–down framework we have put forward indicates that the mosaic structure of an eternally inflating universe is a redundant theoretical construction, which should be excised by *Ockham's razor*. It appears unlikely, therefore, that something like a *volume–weighted probability distribution*, which underlies the idea of eternal inflation, can arise from calculations in top–down cosmology. The implementation of selection effects in both approaches is fundamentally different, and this should ultimately translate into distinct predictions for observations.

Interpretations

In summary, the bottom up approach to cosmology would be appropriate, if one knew that the universe was set going in a particular way in either the finite or infinite past. However, in the absence of such knowledge one is required to work from the top–down [HH06].

In a top–down approach one computes amplitudes for alternative histories of the universe with final boundary conditions only. The boundary conditions act as late time constraints on the alternatives and select the subclass of histories that contribute to the amplitude of interest. This enables one to test the proposal, by searching among the conditional probabilities for predictions of future observations with probabilities near one. In top–down cosmology the histories of the universe thus depend on the precise question asked, i.e., on the set of constraints that one imposes. There are histories in which the universe eternally inflates, or is eleven dimensional, but we have seen they hardly contribute to the amplitudes we measure.

A central idea that underlies the top–down approach is the interplay between the fundamental laws of nature and the operation of chance in a quantum universe. In top–down cosmology, the structure and complexity of alternative universes in the landscape is predictable from first principles to some extent, but also determined by the outcome of quantum accidents over the course of their histories.

We have illustrated our framework in a simple model of gravity coupled to a scalar with a double well potential, and a small fundamental cosmological constant Λ. Imposing constraints that select the subclass of histories that are three dimensional and approximately flat at late times, with sufficiently large primordial perturbations for structure formation to occur, we made several predictions in this model.

In particular we have shown that universes within this class are likely to emerge in an inflationary state. Furthermore, we were able to identify the dominant inflationary path as the history where the scalar starts all the way at the maximum of its potential, leading to a long period of inflation and a perfectly flat universe today. Moreover, one can calculate the relative amplitudes of neighboring geometries by perturbatively evaluating the path integral around the dominant saddle point. Neighboring geometries correspond to small quantum fluctuations of various continuous quantities, like the temperature of the CMB radiation or the expectation values of moduli fields. In inflationary universes these fluctuations are amplified and stretched, generating a pattern of spatial variations on cosmological scales in those directions of moduli space that are relatively flat. The shape of these primordial spectra depends on the (no) boundary conditions on the dominant geometry and provides a strong test of the no–boundary proposal.

When one extends these considerations to a potential that depends on a multi–dimensional moduli space, one finds that only a few of the minima of the potential will be populated, i.e., will have significant amplitudes.

The top–down approach we have described leads to a profoundly different view of cosmology, and the relation between cause and effect. Top down cosmology is a framework in which one essentially traces the histories backwards, from a space–like surface at the present time. The no–boundary histories of the universe thus depend on what is being observed, contrary to the usual idea that the universe has a unique, observer independent history. In some sense no–boundary initial conditions represent a sum over all possible initial states. This is in sharp contrast with the bottom–up approach, where one assumes there is a single history with a well defined starting point and evolution. Our comparison with eternal inflation provides a clear illustration of this. In a cosmology based on eternal inflation there is only one universe with a fractal structure at late times, whereas in top–down cosmology one envisions a set of alternative universes, which are more likely to be homogeneous, but with different values for various effective coupling constants [HH06].

7.3.8 Brane Cosmology

In this section, mainly following [BB03] and [Lan02], we give a review of cosmological consequences of the brane world scenario, that is cosmological behavior of a brane–universe, i.e., a 3D space, where ordinary matter is confined, embedded in a higher dimensional space–time.

It has been recently suggested that there might exist some extra spatial dimensions, not in the traditional Kaluza–Klein sense where the extra–dimensions are compactified on a small enough radius to evade detection in the form of Kaluza–Klein modes, but in a setting where the extra dimensions could be large, under the assumption that ordinary matter is confined onto a 3D subspace, called *brane* (more precisely '3–brane', referring to the three spatial dimensions) embedded in a larger space, called *bulk* [Lan02].

Recall that the idea of extra dimensions was proposed in the early twentieth century by Nordstrom and a few years later by Kaluza and Klein [Kal21]. It has reemerged over the years in theories combining the principles of quantum mechanics and relativity. In particular theories based on supersymmetry, especially superstring theories, are naturally expressed in more than four dimensions [Pol99]. Four dimensional physics is retrieved by Kaluza–Klein reduction, i.e., compactifying on a manifold of small size, typically much smaller than the size of an atomic nucleus.

Recent developments in string theory and its extension M–theory have suggested another approach to compactify extra spatial dimensions. According to these developments, the standard model particles are confined on a brane–hypersurface embedded in a higher dimensional bulk. Only gravity and other exotic matter such as the dilaton can propagate in the bulk. Our universe may be such a brane–like object. This idea was originally motivated phenomenologically (see [Aka82, GW82]) and later revived in string theory. Within the brane world scenario, constraints on the size of extra dimensions become weaker, because the standard model particles propagate only in three spatial dimensions. Newton's law of gravity, however, is sensitive to the presence of extra–dimensions. Gravity is being tested only on scales larger than a tenth of a millimeter and possible deviations below that scale can be envisaged.

From the string theory point of view, brane worlds of the kind discussed in this review spring from a model suggested by Horava and Witten [HW96]. The strong coupling limit of the $E_8 \times E_8$ heterotic string theory at low energy is described by eleven dimensional supergravity with the eleventh dimension compactified on an orbifold with Z_2 symmetry, i.e., an interval. The two boundaries of space–time (i.e., the orbifold fixed points) are 10–dimensional planes, on which gauge theories (with the E_8 gauge groups) are confined. Later Witten argued that 6 of the 11 dimensions can be consistently compactified on a *Calabi–Yau threefold* and that the size of the Calabi–Yau manifold can be substantially smaller than the space between the two boundary branes [Wit98b]. Thus, in that limit space–time looks five–dimensonal with four dimensional boundary branes [LOS99]. This provides the underlying picture for many brane world models proposed so far.

Another important ingredient was put forward by Arkani-Hamed, Dimopoulos and Dvali (ADD), [HDD98], who suggested that by confining the standard model particle on a brane the extra dimensions can be larger than previously anticipated. They considered a flat bulk geometry in $(4+n)$–dimensions, in which n dimensions are compact with radius R (toroidal topology). The 4D Planck mass M_P and the $(4+n)$D Planck mass $M_{\rm fund}$, the gravitational scale of the extra dimensional theory, are related by $M_P^2 = M_{\rm fund}^{2+n} R^n$. Gravity deviates from Newton's law only on scales smaller than R. Since gravity is tested only down to sizes of around a millimeter, R could be as large as a fraction of a millimeter.

ADD assumed that the bulk geometry is flat. Considerable progress was made by Randall and Sundrum, who considered *non–flat*, i.e., warped bulk geometries [RS99a, RS99b]. In their models, the bulk space–time is a slice of Anti–de Sitter space–time, i.e., a space–time with a negative cosmological constant. Their discovery was that, due to the curvature of the bulk space time, Newton's law of gravity can be obtained on the brane of positive tension embedded in an infinite extra–dimension. Small corrections to Newton's law are generated and constrain the possible scales in the model to be smaller than a millimetre.

They also proposed a two–brane model in which the hierarchy problem, i.e., the large discrepancy between the Planck scale at 10^{19} GeV and the electroweak scale at 100 GeV, can be addressed. The large hierarchy is due to the highly curved AdS background which implies a large gravitational red shift between energy scale on the two branes. In this scenario, the standard model particles are confined on a brane with negative tension sitting at $y = r_c$, whereas a positive tension brane is located at $y = 0$. The large hierarchy is generated by the appropriate inter–brane distance, i.e., the radion. It can be shown that the Planck mass $M_{\rm Pl}$ measured on the negative tension brane is (for $k = \sqrt{-\Lambda_5 \kappa_5^2/6}$) given by $M_{\rm Pl}^2 \approx {\rm e}^{2kr_c} M_5^3/k$, where M_5 is the 5D Planck mass and Λ_5 the (negative) cosmological constant in the bulk. Thus, we see that, if M_5 is not very far from the electroweak scale $M_W \approx$TeV, we need $kr_c \approx 50$, in order to generate a large Planck mass on our brane. Hence, by tuning the radius r_c of the extra dimension to a reasonable value, one can get a very large hierarchy between the weak and the Planck scale. Clearly a complete realization of this mechanism requires an explanation for such a value of the radion. In other words, the radion needs to be stabilized at a certain value. The stabilization mechanism is not thoroughly understood, though models with a bulk scalar field have been proposed and have the required properties [GW99].

Another puzzle which might be addressed with brane models is the cosmological constant problem. One may invoke an extra dimensional origin for the apparent (almost) vanishing of the cosmological constant. The self-tuning idea [HDK00] advocates that the energy density on our brane does not lead to a large curvature of our universe. On the contrary, the extra dimension becomes highly curved, preserving a flat Minkowski brane with apparent vanishing cosmological constant. Unfortunately, the simplest realization of this mechanism with a bulk scalar field fails due to the presence of a naked singularity in the bulk. This singularity can be shielded by a second brane whose tension has to be fine-tuned with the original brane tension [FLL00]. In a sense, the fine tuning problem of the cosmological constant reappears through the extra dimensional back-door.

Finally, we will later discuss in some detail another spectacular consequence of brane cosmology, namely the possible modification to the Friedmann equation at very high energy. This effect was first recognised in [LOS00] in the context of inflatonary solutions. As we will see, *Friedmann's equation* for the Randall–Sundrum model has the form [CGS99, CGK99]

$$H^2 = \frac{\kappa_5^4}{36}\rho^2 + \frac{8\pi G_N}{3}\rho + \Lambda,$$

relating the expansion rate of the brane H to the (brane) matter density ρ and the (effective) cosmological constant Λ. The cosmological constant can be tuned to zero by an appropriate choice of the brane tension and bulk cosmological constant, as in the Randall–Sundrum case. Notice that at high energies, for which $\rho \gg 96\pi G_N/\kappa_5^4$, where κ_5^2 is the five dimensional gravitational constant, the Hubble rate becomes $H \propto \rho$, while in ordinary cosmology $H \propto \sqrt{\rho}$. The latter case is retrieved at low energy, i.e., $\rho \ll 96\pi G_N/\kappa_5^4$. Clearly, modifications to the Hubble rate can only be significant before nucleosynthesis. They may have drastic consequences on early universe phenomena such as inflation.

The Randall–Sundrum Brane World

Randall–Sundrum Model of the Universe

The so–called Randall–Sundrum (RS) models in particle physics propose that the real world is a higher–dimensional universe described by *warped geometry*, that is a Lorentzian manifold whose metric tensor can be written in form

$$ds^2 = g_{ab}(y)dy^a dy^b + f(y)g_{ij}dx^i dx^j.$$

Note that the geometry almost decomposes into a Cartesian product of the y–geometry and the x–geometry, except that the x–part is warped, i.e., it is rescaled by a scalar function of the other coordinates y. For this reason, the metric of a warped geometry is often called a *warped product metric*. Warped geometries are useful in that separation of variables can be used when solving PDEs over them.

While studying the so–called *technicolor models*,[37] L. Randall and R. Sundrum have proposed in [RS99b] that our universe is a 5D *Anti de Sitter space*[38]

[37] The technicolor models are theories beyond the Standard Model (sometimes, but not always, GUTs) which do not have a scalar Higgs field. Instead, they have a larger number of fermion fields than the Standard Model and involve a larger gauge group. This larger gauge group is spontaneously broken down to the Standard Model group as fermion condensates form. The idea of technicolor is to build a model in which the sort of dynamics we see in *quantum chromodynamics* (QCD) can be used to explain the masses of the W and Z bosons. In QCD, there are quarks that feel both the weak interaction and the strong interaction. The strong interaction binds them together in condensates which spontaneously break electroweak symmetry. In fact, QCD itself gives masses to the W and Z bosons, but these masses are tiny compared to the observed masses. Technicolor uses a QCD–like theory at a higher energy scale to give the observed masses to the W and Z bosons. Unfortunately the simplest models are already experimentally ruled out by precision tests of the electroweak interactions. There is currently no fully satisfactory model of technicolor, but it is possible that some form of technicolor will be experimentally discovered at the *Large Hadron Collider*.

[38] Recall that nD anti de Sitter space represents the maximally symmetric, simply–connected, *Lorentzian manifold* with constant negative curvature. It may be regarded as the Lorentzian analog of nD hyperbolic space. In the language of general relativity, anti de Sitter space is the maximally symmetric, vacuum solution of Einstein's field equation with a negative cosmological constant Λ.A coordinate patch covering part of the space gives the half–space coordinatization of anti de Sitter space. The metric for this patch is

$$ds^2 = \frac{1}{y^2}\left(dt^2 - dy^2 - \sum_i dx_i^2\right).$$

In the limit as $y = 0$, this reduces to a *Minkowski metric* and the elementary particles except for the graviton are localized on a (3+1)D *branes*.[39]

There are two popular RS models. The first, called RSI, has a finite size for the extra dimension with two branes, one at each end. The second, RS2, is similar to the first, but one brane has been placed infinitely far away, so that there is only one brane left in the model.

The RSI model attempts to address the hierarchy problem. The warping of the extra dimension is analogous to the warping of space–time in the vicinity of

---

$$dy^2 = \left(dt^2 - \sum_i dx_i^2 \right).$$

Thus, the anti–de Sitter space contains a conformal Minkowski space at infinity, the so–called *conformal infinity* ('infinity' having y–coordinate zero in this patch). The constant time slices of this coordinate patch are hyperbolic spaces in the Poincaré half–plane metric.

There are two types of AdS space: one where time is periodic, and the universal cover with non–periodic time. The coordinate patch above covers half of a single period of the space–time. Because the conformal infinity of AdS is timelike, specifying the initial data on a spacelike hypersurface would not determine the future evolution uniquely (i.e., deterministically) unless there are boundary conditions associated with the conformal infinity.

[39] Recall that branes or p–branes are spatially extended objects that appear in string theory and its relatives (M–theory and brane cosmology). The variable p refers to the spatial dimension of the brane. That is, a 0–brane is a zero–dimensional particle, a 1–brane is a string, a 2–brane is a 'membrane', etc. Every p–brane sweeps out a $(p+1)$D world volume as it propagates through space–time.

The so–called *brane cosmology* refers to several theories in particle physics and cosmology motivated by, but not rigorously derived from, superstring theory and M–theory. The central idea is that our visible, 4D universe is entirely restricted to a brane inside a higher–dimensional space, called the *bulk*. The additional dimensions may be taken to be *compact*, in which case the observed universe contains the extra dimensions, and then no reference to the bulk is appropriate in this context. In the bulk model, other branes may be moving through this bulk. Interactions with the bulk, and possibly with other branes, can influence our brane and thus introduce effects not seen in more standard cosmological models. As one of its attractive features, the model can explain the weakness of gravity relative to the other fundamental forces of nature, thus solving the so–called *hierarchy problem*. In the brane picture, the other three forces (electromagnetism and the weak and strong nuclear forces) are localised on the brane, but gravity has no such constraint and so much of its attractive power 'leaks' into the bulk. As a consequence, the force of gravity should appear significantly stronger on small (sub–millimetre) scales, where less gravitational force has 'leaked'. The Randall–Sundrum, pre–Big–Bang, ekpyrotic and cyclic scenarios are particular models of brane cosmology which have attracted a considerable amount of attention. The theory hypothesises that the origin of the Big–Bang could have occurred when two parallel branes touched.

a massive object, such as a black hole. This warping, or red–shifting, generates a large ratio of energy scales so that the natural energy scale at one end of the extra dimension is much larger than at the other end,

$$ds^2 = \frac{1}{k^2 y^2}(dy^2 + \eta_{\mu\nu}\, dx^\mu\, dx^\nu),$$

where k is some constant and η has $(-+++)-$metric signature. This space has boundaries at $y = 1/k$ and $y = 1/Wk$, with $0 \leq \frac{1}{k} \leq \frac{1}{Wk}$, where k is around the Planck scale, W is the *warp factor* and Wk is around a TeV. The boundary at $y = 1/k$ is called the *Planck brane* and the boundary at $y = 1/Wk$ is called the *TeV–brane*. The particles of the Standard Model reside on the TeV brane. The distance between both branes is only $-ln(W)/k$.

In another coordinate system,

$$\varphi = -\frac{\pi \ln(ky)}{\ln(W)},$$

so that $0 \leq \varphi \leq \pi$ and

$$ds^2 = \left(\frac{\ln(W)}{\pi k}\right)^2 d\varphi^2 + \mathrm{e}^{\frac{2\ln(W)\varphi}{\pi}} \eta_{\mu\nu}\, dx^\mu\, dx^\nu.$$

The RSII model uses the same geometry as RSI, but there is no TeV brane. The particles of the standard model are presumed to be on the Planck brane. This model was originally of interest because it represented an infinite 5D model which, in many respects, behaved as a 4D model. This setup may also be of interest for studies of the *AdS/CFT* conjecture.[40]

RS Scenario

Originally, Randall and Sundrum suggested a two–brane scenario in five dimensions with a highly curved bulk geometry as an explanation for the large hierarchy between the Planck scale and the electroweak energy–scale [RS99a]. In this scenario, the standard model particles live on a brane with (constant) negative tension, whereas the bulk is a slice of Anti–de Sitter (AdS) space–time, i.e., a space–time with a negative cosmological constant. In the bulk there is another brane with positive tension. This is the so–called Randall–Sundrum I (RSI) model. Analysing the solution of Einstein's equation on

[40] The *AdS/CFT correspondence* (anti-De-Sitter space/conformal field theory correspondence), sometimes called the *Maldacena duality*, is the conjectured equivalence between a string theory defined on one space, and a quantum field theory without gravity defined on the conformal boundary of this space, whose dimension is lower by at least one. The name suggests that the first space is the product of anti de Sitter space (AdS) with some closed manifold like sphere, orbifold, or noncommutative space, and that the quantum field theory is a conformal field theory (CFT).

the positive tension brane and sending the negative tension brane to infinity, an observer confined to the positive tension brane recovers Newton's law if the curvature scale of the AdS is smaller than a millimeter [RS99b]. The higher–dimensional space is *non–compact*, which must be contrasted with the Kaluza–Klein mechanism, where all extra–dimensional degrees of freedom are compact. This one–brane model, on which we will concentrate here, is the so–called Randall–Sundrum II (RSII) model. It was shown, there is a *continuum of Kaluza–Klein modes* for the gravitational field, contrasting with the discrete spectrum if the extra dimension is periodic. This leads to a correction to the force between two static masses on the brane. To be specific, it was shown that the potential energy between two point masses confined on the brane is given by

$$V(r) = \frac{G_N m_1 m_2}{r}\left(1 + \frac{l^2}{r^2} + O(r^{-3})\right).$$

In this equation, l is related to the 5D bulk cosmological constant Λ_5 by $l^2 = -6/(\kappa_5^2 \Lambda_5)$ and is therefore a measure of the curvature scale of the bulk space–time. Gravitational experiments show no deviation from Newton's law of gravity on length scales larger than a millimeter. Thus, l has to be smaller than that length scale.

The static solution of the Randall–Sundrum model can be obtained as follows: The total action consists of the Einstein–Hilbert action and the brane action, which in the RS model have the form

$$S_{\mathrm{EH}} = -\int dx^5 \sqrt{-g^{(5)}}\left(\frac{R}{2\kappa_5^2} + \Lambda_5\right),$$
$$S_{\mathrm{brane}} = \int dx^4 \sqrt{-g^{(4)}}\,(-\sigma).$$

The parameter Λ_5 (the bulk cosmological constant) and σ (the brane tension) are constant and κ_5 is the 5D gravitational coupling constant. The brane is located at $y = 0$ and we assume a Z_2 symmetry, i.e., we identify y with $-y$. The ansatz for the metric is

$$ds^2 = \mathrm{e}^{-2K(y)}\eta_{\mu\nu}dx^\mu dx^\nu + dy^2.$$

Einstein's equations, derived from the action above, give two independent equations:

$$6K'^2 = -\kappa_5^2\Lambda_5, \qquad 3K'' = \kappa_5^2\sigma\delta(y).$$

The first equation can be easily solved:

$$K = K(y) = \sqrt{-\frac{\kappa_5^2}{6}\Lambda_5}\; y \equiv ky, \tag{7.108}$$

which tells us that Λ_5 must be negative. If we integrate the second equation from $-\epsilon$ to $+\epsilon$, take the limit $\epsilon \to 0$ and make use of the Z_2–symmetry, we get $6K'|_0 = \kappa_5^2\sigma$.Together with (7.108) this tells us that

$$\Lambda_5 = -\frac{\kappa_5^2}{6}\sigma^2. \tag{7.109}$$

Thus, there must be a fine–tuning between the brane tension and the bulk cosmological constant for static solutions to exist. Here we will discuss the cosmology of this model in detail.

Einstein's equations on the brane

There are two ways of deriving the cosmological equations and we will describe both of them below. The first one is rather simple and makes use of the bulk equations only. The second method uses the geometrical relationship between 4D and 5D quantities. We begin with the simpler method.

Friedmann's Equation from 5D Einstein Equations

In the following subsection we will set $\kappa_5 \equiv 1$. We write the bulk metric as follows:

$$ds^2 = a^2b^2(dt^2 - dy^2) - a^2\delta_{ij}dx^i dx^j. \tag{7.110}$$

This metric is consistent with homogeneity and isotropy on the brane located at $y = 0$. The functions a and b are functions of t and y only. Furthermore, we have assumed flat spatial sections, it is straightforward to include a spatial curvature. For this metric, Einstein equations in the bulk read:

$$a^2b^2G^0{}_0 \equiv 3\left[2\frac{\dot{a}^2}{a^2} + \frac{\dot{a}\dot{b}}{ab} - \frac{a''}{a} + \frac{a'b'}{ab} + kb^2\right] = a^2b^2\left[\rho_B + \rho\bar{\delta}(y - y_b)\right], \tag{7.111}$$

$$a^2b^2G^5{}_5 \equiv 3\left[\frac{\ddot{a}}{a} - \frac{\dot{a}\dot{b}}{ab} - 2\frac{a'^2}{a^2} - \frac{a'b'}{ab} + kb^2\right] = -a^2b^2T^5_5, \tag{7.112}$$

$$a^2b^2G^0{}_5 \equiv 3\left[-\frac{\dot{a}'}{a} + 2\frac{\dot{a}a'}{a^2} + \frac{\dot{a}b'}{ab} + \frac{a'\dot{b}}{ab}\right] = -a^2b^2T^0_5, \tag{7.113}$$

$$a^2b^2G^i{}_j \equiv \left[3\frac{\ddot{a}}{a} + \frac{\ddot{b}}{b} - \frac{\dot{b}^2}{b^2} - 3\frac{a''}{a} - \frac{b''}{b} + \frac{b'^2}{b^2} + kb^2\right]\delta^i{}_j$$
$$= -a^2b^2\left[p_B + p\bar{\delta}(y - y_b)\right]\delta^i{}_j, \tag{7.114}$$

where the bulk energy–momentum tensor T^a_b has been kept general here. For the RS model we will now take $\rho_B = -p_B = \Lambda_5$ and $T^0_5 = 0$. Later we will use these equations to derive Friedmann's equation with a bulk scalar field. In the equations above, a dot represents the derivative with respect to t and a prime a derivative with respect to y.

Let us integrate the 00–component over y from $-\epsilon$ to ϵ and use the fact that $a(y) = a(-y)$, $b(y) = b(-y)$, $a'(y) = -a(-y)$ and $b'(y) = -b(-y)$ (i.e., Z_2–symmetry). Then, taking the limit $\epsilon \to 0$, we get

$$\frac{a'}{a}\,|_{y=0} = \frac{1}{6}ab\rho. \tag{7.115}$$

Integrating the ij–component in the same way and using the last equation gives

$$\frac{b'}{b}\,|_{y=0} = -\frac{1}{2}ab(\rho + p). \tag{7.116}$$

These two conditions are called the junction conditions. The other components of the Einstein equations should be compatible with these conditions. It is not difficult to show that the restriction of the 05 component to $y = 0$ leads to

$$\dot{\rho} + 3\frac{\dot{a}}{a}\,(\rho + p) = 0, \tag{7.117}$$

where we have made use of the junction conditions (7.115) and (7.116). This is nothing but matter conservation on the brane.

Proceeding in the same way with the 55–component gives

$$\frac{\ddot{a}}{a} - \frac{\dot{a}\dot{b}}{ab} + kb^2 = -\frac{a^2b^2}{3}\left[\frac{1}{12}\rho\,(\rho\ + 3p) + q_B\right].$$

Changing to cosmic time $d\tau = abdt$, writing $a = \exp(\alpha(t))$ and using the energy conservation gives ([FTW00], [BDM00])

$$\frac{n(H^2 e^{4\alpha})}{d\alpha} = \frac{2}{3}\Lambda_5 e^{4\alpha} + \frac{n}{d\alpha}\left(e^{4\alpha}\frac{\rho^2}{36}\right).$$

In this equation $aH = da/d\tau$. This equation can easily be integrated to give

$$H^2 = \frac{\rho^2}{36} + \frac{\Lambda_5}{6} + \frac{\mu}{a^4}.$$

The final step is to split the total energy–density and pressure into parts coming from matter and brane tension, i.e., to write $\rho = \rho_M + \sigma$ and $p = p_M - \sigma$. Then we find Friedmann's equation

$$H^2 = \frac{8\pi G}{3}\rho_M\left[1 + \frac{\rho_M}{2\sigma}\right] + \frac{\Lambda_4}{3} + \frac{\mu}{a^4}, \qquad \text{using}$$
$$\frac{8\pi G}{3} = \frac{\sigma}{18}, \qquad \frac{\Lambda_4}{3} = \frac{\sigma^2}{36} + \frac{\Lambda_5}{6}.$$

Comparing the last equation with the fine–tuning relation (7.109) in the static RS solution, we see that $\Lambda_4 = 0$ in this case. If there is a small mismatch between the brane tension and the 5D cosmological constant, then an effective 4D cosmological constant is generated. Another important point is that the 4D Newton constant is directly related to the brane tension. The constant μ appears in the derivation above as an integration constant. The term including μ is called the *dark radiation* term (see e.g., [Muk00, IJK02]). The parameter

μ can be obtained from a full analysis of the bulk equations [MSM00]. An extended version of *Birkhoff's Theorem* tells us that if the bulk space–time is AdS, this constant is zero [BCG00]. If the bulk is AdS–Schwarzschild instead, μ is non–zero but a measure of the mass of the bulk black hole. In the following we will assume that $\mu = 0$ and $\Lambda_4 = 0$.

The most important change in Friedmann's equation compared to the usual 4D form is the appearance of a term proportional to ρ^2. It tells us that if the matter energy density is much larger than the brane tension, i.e., $\rho_M \gg \sigma$, the expansion rate H is proportional ρ_M, instead of $\sqrt{\rho_M}$. The expansion rate is, in this regime, larger in this brane world scenario. Only in the limit where the brane tension is much larger than the matter energy density, the usual behavior $H \propto \sqrt{\rho_M}$ is recovered. This is the most important change in brane world scenarios. It is quite generic and not restricted to the Randall–Sundrum brane world model. From Friedmann's equation and from the energy–conservation equation we can derive *Raychaudhuri's equation*:

$$\frac{dH}{d\tau} = -4\pi G(\rho_M + p_M)\left[1 + \frac{\rho_M}{\sigma}\right].$$

We will use these equations later in order to investigate inflation driven by a scalar field confined on the brane.

Notice that at the time of nucleosythesis the brane world corrections in Friedmann's equation must be negligible, otherwise the expansion rate is modified and the results for the abundances of the light elements are unacceptably changed. This implies that $\sigma \geq (1\text{MeV})^4$. Note, however, that a much stronger constraint arises from current tests for deviation from Newton's law [Maa01] (assuming the Randall–Sundrum fine–tuning relation (7.109)): $\kappa_5^{-3} > 10^5$ TeV and $\sigma \geq (100\text{GeV})^4$. Similarily, cosmology constrains the amount of dark radiation. It has been shown that the energy density in dark radiation can at most be 10 percent of the energy density in photons [Ida00].

Another derivation of Einstein's equation

There is a more powerful way of deriving Einstein's equation on the brane [SMS00]. Consider an arbitrary (3+1)D hypersurface $\mathcal{M}$ with unit normal vector n_a embedded in a 5D space–time. The induced metric and the extrinsic curvature of the hypersurface are defined as

$$h^a{}_b = \delta^a{}_b - n^a n_b, \qquad K_{ab} = h_a{}^c h_b{}^n \nabla_c n_n.$$

For the derivation we need three equations, two of them relate 4D quantities constructed from h_{ab} to full 5D quantities constructed from g_{ab}. We just state these equations here and refer to [Wal84] for the derivation of these equations. The first equation is the Gauss equation, which reads

$$R^{(4)}_{abcd} = h_a{}^j h_b{}^k h_c{}^l h_n{}^m R_{jklm} - 2K_{a[c}K_{n]b}.$$

This equation relates the 4D curvature tensor $R^{(4)}_{abcd}$, constructed from h_{ab}, to the 5D one and K_{ab}. The next equation is the Codazzi equation, which relates K_{ab}, n_a and the 5D *Ricci tensor*:

$$\nabla^{(4)}_b K^b{}_a - \nabla^{(4)}_a K = n^c h^b{}_a R_{bc}.$$

One decomposes the 5D curvature tensor R_{abcd} into the *Weyl tensor* C_{abcd} and the Ricci tensor:

$$R_{abcd} = \frac{2}{3}\left(g_{a[c}R_{n]b} - g_{b[c}R_{n]a}\right) - \frac{1}{6}Rg_{a[c}g_{b]n} + C_{abcd}.$$

If one substitutes the last equation into the Gauss equation and constructs the 4D Einstein tensor, one gets

$$G^{(4)}_{ab} = \frac{2}{3}\left(G_{cd}h^c{}_a h^n{}_b + \left(G_{cd}n^c n^n - \frac{1}{4}G\right)h_{ab}\right) \tag{7.118}$$

$$+ KK_{ab} - K_a{}^c K_{bc} - \frac{1}{2}\left(K^2 - K^{cd}K_{cd}\right)h_{ab} - E_{ab},$$

where $\quad E_{ab} = C_{acbd}n^c n^n.$

We would like to emphasize that this equation holds for any hypersurface. If one considers a hypersurface with energy momentum tensor T_{ab}, then there exists a relationship between K_{ab} and T_{ab} (T is the trace of T_{ab}) [Isr66]:

$$[K_{ab}] = -\kappa_5^2\left(T_{ab} - \frac{1}{3}h_{ab}T\right),$$

where [...] denotes the *jump*:

$$[f](y) = \lim_{\epsilon\to 0}\left(f(y+\epsilon) - f(y-\epsilon)\right).$$

These equations are called junction conditions and are equivalent in the cosmological context to the junction conditions (7.115) and (7.116). Splitting $T_{ab} = \tau_{ab} - \sigma h_{ab}$ and inserting the junction condition into equation (7.118), we get Einstein's equation on the brane:

$$G^{(4)}_{ab} = 8\pi G\tau_{ab} - \Lambda_4 h_{ab} + \kappa_5^4\pi_{ab} - E_{ab}. \tag{7.119}$$

The tensor π_{ab} is defined as

$$\pi_{ab} = \frac{1}{12}\tau\tau_{ab} - \frac{1}{4}\tau_{ac}\tau_b{}^c + \frac{1}{8}h_{ab}\tau_{cd}\tau^{cd} - \frac{1}{24}\tau^2 h_{ab}, \qquad \text{whereas}$$

$$8\pi G = \frac{\kappa_5^4}{6}\sigma, \qquad \Lambda_4 = \frac{\kappa_5^2}{2}\left(\Lambda_5 + \frac{\kappa_5^2}{6}\sigma^2\right).$$

Note that in the Randall–Sundrum case we have $\Lambda_4 = 0$ due to the fine–tuning between the brane tension and the bulk cosmological constant. Moreover $E_{ab} = 0$ as the Weyl–tensor vanishes for an AdS space–time. In general, the energy conservation and the Bianchi identities imply that

$$\kappa_5^4 \nabla^a \pi_{ab} = \nabla^a E_{ab} \qquad (7.120)$$

on the brane.

Clearly, this method is powerful, as it does not assume homogeneity and isotropy nor does it assume the bulk to be AdS. In the case of an AdS bulk and a Friedmann–Robertson walker brane, the previous equations reduce to the Friedmann equation and Raychaudhuri equation derived earlier. However, the set of equations on the brane *are not closed in general* [Maa00], as we will see below.

Slow–Roll Inflation on the Brane

We have seen that the Friedmann equation on a brane is drastically modified at high energy where the ρ^2 terms dominate. As a result the early universe cosmology on branes tends to be different from standard 4d cosmology. In that vein it seems natural to look for brane effects on early universe phenomena such as inflation [MWB00, CLL01] and on phase–transitions [BB03].

The energy density and the pressure of a scalar field are given by

$$\rho_\phi = \frac{1}{2}\phi_{,\mu}\phi^{,\mu} + V(\phi), \qquad p_\phi = \frac{1}{2}\phi_{,\mu}\phi^{,\mu} - V(\phi),$$

where $V(\phi)$ is the potential energy of the scalar field. The full evolution of the scalar field is described by the (modified) Friedmann equation, the Klein–Gordon equation and the Raychaudhuri equation.

We will assume that the field is in a slow–roll regime, the evolution of the fields is governed by (from now on a dot stands for a derivative with respect to cosmic time)

$$3H\dot{\phi} \approx -\frac{\partial V}{\partial \phi}, \qquad H^2 \approx \frac{8\pi G}{3} V(\phi)\left(1 + \frac{V(\phi)}{2\sigma}\right).$$

It is not difficult to show that these equations imply that the slow–roll parameter are given by

$$\epsilon \equiv -\frac{\dot{H}}{H^2} = \frac{1}{16\pi G}\left(\frac{V'}{V}\right)^2 \left[\frac{4\sigma(\sigma + V)}{(2\sigma + V)^2}\right] \qquad (7.121)$$

$$\eta \equiv \frac{V''}{3H^2} = \frac{1}{8\pi G}\left(\frac{V''}{V}\right)\left[\frac{2\sigma}{2\sigma + V}\right]. \qquad (7.122)$$

The modifications to General Relativity are contained in the square brackets of these expressions. They imply that for a given potential and given initial conditions for the scalar field the slow–roll parameters are suppressed compared to the predictions made in General Relativity. In other words, brane world effects ease slow–roll inflation [MWB00]. In the limit $\sigma \ll V$ the parameter are heavily suppressed. It implies that steeper potentials can be used to drive slow–roll inflation [CLL01]. Let us discuss the implications for cosmological perturbations.

According to Einstein's equation (7.119), perturbations in the metric are sourced not only by matter perturbations but also by perturbations of the bulk geometry, encoded in the perturbation of E_{ab}. This term can be seen as an external source for perturbations, absent in General Relativity. If one regards E_{ab} as an energy–momentum tensor of an additional fluid (called the *Weyl–fluid*), its evolution is connected to the energy density of matter on the brane, as one can see from (7.120). If one neglects the anisotropic stress of the Weyl-fluid, then at low energy and superhorizon scales, it decays as radiation, i.e., $\delta E_{ab} \propto a^{-4}$. However, the bulk gravitational field exerts an anisotropic stress onto the brane, whose time–evolution cannot be obtained by considering the projected equations on the brane alone [Maa00]. Rather, the full 5D equations have to be solved, together with the junction conditions. The full evolution of E_{ab} in the different cosmological eras is currently not understood. However, as we will discuss below, partial results have been obtained for the case of a de Sitter brane, which suggest that E_{ab} does not change the spectrum of scalar perturbations. It should be noted however, that the issue is not settled and that it is also not clear if the subsequent cosmological evolution during radiation and matter era leaves an imprint of the bulk gravitational field in the anisotropies of the microwave background radiation [LMS01]. With this in mind, we will, for scalar perturbations, neglect the gravitational backreaction described by the projected Weyl tensor.

Considering scalar perturbations for the moment, the perturbed line element on the brane has the form

$$ds^2 = -(1+2A)dt^2 + 2\partial_i B dt dx^i + ((1-2\psi)\delta_{ij} + D_{ij}E)dx^i dx^j,$$

where the functions A, B, E and ψ depend on t and x^i.

An elegant way of discussing scalar perturbations is to make use of of the gauge invariant quantity [BB03]

$$\zeta = \psi + H\frac{\delta\rho}{\dot{\rho}}. \tag{7.123}$$

In General Relativity, the evolution equation for ζ can be obtained from the energy–conservation equation. It reads, on large scales [BB03],

$$\dot{\zeta} = -\frac{H}{\rho + p}\delta p_{\mathrm{nad}}, \qquad \text{where} \tag{7.124}$$

$$\delta p_{\mathrm{nad}} = \delta p_{tot} - c_s^2 \delta\rho$$

is the non–adiabatic pressure perturbation. The energy conservation equation, however, holds for the Randall–Sundrum model as well. Therefore, (7.124) is still valid for the brane world model we consider. For inflation driven by a single scalar field δp_{nad} vanishes and therefore ζ is constant on superhorizon scales during inflation. Its amplitude is given in terms of the fluctuations in the scalar field on spatially flat hypersurfaces:

$$\zeta = \frac{H\delta\phi}{\dot{\phi}}. \tag{7.125}$$

The quantum fluctuation in the (slow–rolling) scalar field obey $\langle(\delta\phi)^2\rangle \approx (H/2\pi)^2$, as the Klein–Gordon equation is not modified in the brane world model we consider. The amplitude of scalar perturbations is [LL00] $A_S^2 = 4\langle\zeta^2\rangle/25$. Using the slow–roll equations and (7.125) one gets [MWB00]

$$A_S^2 \approx \left(\frac{512\pi}{75M_{pl}^6}\right)\frac{V^3}{V'^2}\left[\frac{2\sigma+V}{2\sigma}\right]^3|_{k=aH} \tag{7.126}$$

Again, the corrections are contained in the terms in the square brackets. For a given potential the amplitude of scalar perturbations is enhanced compared to the prediction of General Relativity.

The arguments presented so far suggest that, at least for scalar perturbations, perturbations in the bulk space–time are not important during inflation. This, however, might not be true for tensor perturbations, as gravitational waves can propagate into the bulk. For tensor perturbations, a wave equations for a single variable can be derived [LMW00]. The wave equation can be separated into a 4D and a 5D part, so that the solution has the form $h_{ij} = A(y)h(x^\mu)e_{ij}$, where e_{ij} is a (constant) polarization tensor. One finds that the amplitude for the zero mode of tensor perturbation is given by [LMW00]

$$A_T^2 = \frac{4}{25\pi M_{pl}^4}H^2F^2(H/\mu)|_{k=aH}, \qquad \text{with} \tag{7.127}$$

$$F(x) = \left[\sqrt{1+x^2} - x^2\sinh^{-1}\left(\frac{1}{x}\right)\right]^{-1/2}, \qquad \text{using}$$

$$\frac{H}{\mu} = \left(\frac{3}{4\pi\sigma}\right)^{1/2} HM_{pl}.$$

It can be shown that modes with $m > 3H/2$ are generated but they decay during inflation. Thus, one expects in this scenario only the massless modes to survive until the end of inflation [LMW00], [GRS01].

From equation (7.127) and (7.126) one sees that the amplitudes of scalar and tensor perturbations are enhanced at high energies. However, scalar perturbations are more enhanced than tensors. Thus, the relative contribution of tensor perturbations will be suppressed, if inflation is driven at high energies.

Finally, we would like to mention that there are also differences between General Relativity and the brane world model we consider for the prediction of two–field brane inflation. Usually correlations are separated in adiabatic and isocurvature modes for two–field inflation. In the RS model, this correlation is suppressed if inflation is driven at high energies. This implies that isocurvature and adiabatic perturbations are uncorrelated, if inflation is driven at energies much larger than the brane tension [BB03].

Coming back to cosmological perturbations, the biggest problem is that the evaluation of the projected Weyl tensor is only possible for the background cosmology. As soon as one tries to analyze the brane cosmological perturbations, one faces the possibility that the E_{0i} terms might not vanish. In particular this means that the equation for the density contrast $\delta = \delta\rho/\rho$, which is given by ($w_m = p/\rho$, k is the wavenumber)

$$\ddot{\delta} + (2 - 3\omega_m)H\dot{\delta} - 6\omega_m(H^2 + \dot{H})\delta = (1 + \omega_m)\delta R_{00} - \omega_m \frac{k^2}{a^2}\delta,$$

cannot be solved as δR_{00} involves δE_{00} and can therefore not be deduced solely from the brane dynamics [Maa00].

The Randall–Sundrum model discussed here is the simplest brane world model. We have not discussed other important conclusions one can draw from the modifications of Friedmann's equation, such as the evolution of primordial black holes, its connection to the AdS/CFT correspondence and inflation driven by the trace anomaly of the conformal field theory living on the brane. These developments are important in many respects, because they give not only insights about the early universe but gravity itself.

Including a Bulk Scalar Field

Here we are going to generalize the previous results obtained with an empty bulk. To be specific, we will consider the inclusion of a scalar field in the bulk. As we will see, one can extend the projective approach wherein one focuses on the dynamics of the brane, i.e., one studies the projected Einstein and the Klein–Gordon equation [MW00, MB01]. As in the RS setting, the dynamics do not closed, as bulk effects do not decouple. We will see that there are now two objects representing the bulk back-reaction: the projected Weyl tensor $E_{\mu\nu}$ and the loss parameter $\Delta\Phi_2$. In the case of homogeneous and isotropic cosmology on the brane, the projected Weyl tensor is determined entirely up to a dark radiation term. Unfortunately, no information on the loss parameter is available. This prevents a rigorous treatment of brane cosmology in the projective approach.

Another route amounts to studying the motion of a brane in a bulk space–time. This approach is successful in the RS case thanks to Birkhoff's Theorem which dictates a unique form for the metric in the bulk [BCG00]. In the case of a bulk scalar field, no such Theorem is available. One has to resort to various ansatze for particular classes of bulk and brane scalar potentials [BB03]).

BPS Backgrounds

Properties of BPS Backgrounds

As the physics of branes with bulk scalar fields is pretty complicated, we will start with a particular example where both the bulk and the brane dynamics are fully under control [CEG00, You00]. We specify the bulk Lagrangian as

$$S = \frac{1}{2\kappa_5^2}\int d^5x\sqrt{-g_5}\left(R - \frac{3}{4}\left((\partial\phi)^2 + V(\phi)\right)\right),$$

where $V(\phi)$ is the bulk potential. The boundary action depends on a brane potential $U_B(\phi)$

$$S_B = -\frac{3}{2\kappa_5^2}\int d^4x\sqrt{-g_4}U_B(\phi_0),$$

where $U_B(\phi_0)$ is evaluated on the brane. The BPS backgrounds arise as particular case of this general setting with a particular relationship between the bulk and brane potentials. This relation appears in the study of $N = 2$ supergravity with vector multiplets in the bulk. The bulk potential is given by

$$V = \left(\frac{\partial W}{\partial\phi}\right)^2 - W^2,$$

where $W(\phi)$ is the superpotential. The brane potential is simply given by the superpotential $U_B = W$. We would like to mention, that the last two relations have been also used in order to generate bulk solutions without necessarily imposing supersymmetry. Notice that the RS case can be retrieved by putting $W = cst$. Supergravity puts further constraints on the superpotential which turns out to be of the exponential type [BB03]

$$W = 4k\mathrm{e}^{\alpha\phi},$$

with $\alpha = -1/\sqrt{12}, 1/\sqrt{3}$. In the following we will often choose this exponential potential with an arbitrary α as an example. The actual value of α does not play any role and will be considered generic.

The bulk equations of motion comprise the Einstein equations and the Klein–Gordon equation. In the BPS case and using the following ansatz for the metric

$$ds^2 = a(y)^2\eta_{\mu\nu}dx^\mu dx^\nu + dy^2, \tag{7.128}$$

these second order differential equations reduce to a system of two first order differential equations

$$\frac{a'}{a} = -\frac{W}{4}, \qquad \phi' = \frac{\partial W}{\partial\phi}.$$

Notice that when $W = cst$ one easily retrieves the exponential profile of the RS model.

An interesting property of BPS systems can be deduced from the study of the boundary conditions. The Israel junction conditions reduce to

$$\frac{a'}{a}|_B = -\frac{W}{4}|_B$$

and for the scalar field

$$\phi'|_B = \frac{\partial W}{\partial \phi}|_B$$

This is the main property of BPS systems: the boundary conditions coincide with the bulk equations, i.e., as soon as the bulk equations are solved one can locate the BPS branes anywhere in this background, there is no obstruction due to the boundary conditions. In particular two-brane systems with two boundary BPS branes admit *moduli* corresponding to massless deformations of the background. They are identified with the positions of the branes in the BPS background.

Let us treat the example of the exponential superpotential. The solution for the scale factor reads

$$a = (1 - 4k\alpha^2 x_5)^{1/4\alpha^2}, \tag{7.129}$$

and the scalar field is given by

$$\phi = -\frac{1}{\alpha}\ln(1 - 4k\alpha^2 x_5). \tag{7.130}$$

For $\alpha \to 0$, the bulk scalar field decouples and these expressions reduce to the RS case. Notice a new feature here, namely the existence of singularities in the bulk, corresponding to

$$a(x_5)|_{x_*} = 0.$$

In order to analyse singularities it is convenient to use conformal coordinates: $du = dx_5/a(x_5)$. In these coordinates light follows straight lines $u = \pm t$. It is easy to see that the singularities fall in two categories depending on α. For $\alpha^2 < 1/4$ the singularity is at infinity $u_* = \infty$. This singularity is null and absorbs incoming gravitons. For $\alpha^2 > 1/4$ the singularity is at finite distance. It is time–like and not wave–regular, i.e., the propagation of wave packets is not uniquely defined in the vicinity of the singularity. For all these reasons these naked singularities in the bulk are a major drawback of brane models with bulk scalar fields [BB03]. In the two-brane case the second brane has to sit in front of the naked singularity.

De Sitter and anti de Sitter Branes

Let us modify slightly the BPS setting by detuning the tension of the BPS brane. This corresponds to adding or substracting some tension compared to the BPS case, $U_B = TW$, where T is real number. Notice that this modification only affects the boundary conditions, the bulk geometry and scalar field are still solutions of the BPS equations of motion. In this sort of situation, one can show that the brane does not stay static. For the detuned case, the result is either a boosted brane or a rotated brane. We will soon generalize these results so we postpone the detailed explanation to later. Defining by $u(x^\mu)$ the position of the brane in conformal coordinates, one gets

$$(\partial u)^2 = \frac{1-T^2}{T^2}.$$

The brane velocity vector $\partial_\mu u$ is of constant norm. For $T > 1$, the brane velocity vector is time–like and the brane moves at constant speed. For $T < 1$ the brane velocity vector is space–like and the brane is rotated. Going back to a static brane, we see that the bulk geometry and scalar field become x^μ dependent.

Let us consider the brane geometry when $T > 1$. In particular one can study the Friedmann equation for the induced bulk factor

$$H^2 = \frac{T^2-1}{16} W^2,$$

where W is evaluated on the brane. Of course we get the fact that cosmological solutions are only valid for $T > 1$. Now in the RS case $W = 4k$ leading to

$$H^2 = (T^2-1)k^2.$$

In the case $T > 1$ the brane geometry is driven by a positive cosmological constant. This is a de Sitter brane. When $T < 1$ the cosmological constant is negative, corresponding to an AdS brane. We are going to generalize these results by considering the projective approach to the brane dynamics.

Bulk Scalar Fields and the Projective Approach

The Friedmann Equation

We will first follow the traditional coordinate dependent path. This will allow us to derive the matter conservation equation, the Klein–Gordon and the Friedmann equations on the brane. Then we will concentrate on the more geometric formulation where the role of the projected Weyl tensor will become transparent [BBD01, BB03]. Again, in this subsection we will put $\kappa_5 \equiv 1$.

We consider a static brane that we choose to put at the origin $x_5 = 0$. and impose $b(0,t) = 1$. This guarantees that the brane and bulk expansion rates

$$4H = \partial_\tau \sqrt{-g}|_0, \ 3H_B = \partial_\tau \sqrt{-g_4}|_0$$

coincide. We have identified the brane cosmic time $d\tau = ab|_0 dt$. We will denote by prime the normal derivative $\partial_n = \frac{1}{ab}|_0 \partial_{x_5}$. Moreover we now allow for some matter to be present on the brane

$$T^{\mu\ matter}_{\nu} = (-\rho_m, p_m, p_m, p_m).$$

The bulk energy-momentum tensor reads

$$T_{ab} = \frac{3}{4}\left(\partial_a\phi\partial_b\phi\right) - \frac{3}{8}g_{ab}\left((\partial\phi)^2 + V\right).$$

The total matter density and pressure on the brane are given by

$$\rho = \rho_m + \frac{3}{2}U_B,\ p = p_m - \frac{3}{2}U_B.$$

The boundary condition for the scalar field is unchanged by the pre nce of matter on the brane.

The (05) Einstein equation leads to matter conservation

$$\dot{\rho}_m = -3H(\rho_m + p_m).$$

By restricting the (55) component of the Einstein equations we get

$$H^2 = \frac{\rho^2}{36} - \frac{2}{3}Q - \frac{1}{9}E + \frac{\mu}{a^4}$$

in units of κ_5^2. The last term is the dark radiation, whose origin is similar to the RS case. The quantity Q and E satisfy the differential equations [BDM00]

$$\dot{Q} + 4HQ = HT^5_5, \qquad \dot{E} + 4HE = -\rho T^0_5.$$

These equations can be easily integrated to give

$$H^2 = \frac{\rho_m^2}{36} + \frac{U_B\rho_m}{12} - \frac{1}{16a^4}\int d\tau\frac{da^4}{d\tau}(\dot{\phi}^2 - 2U) - \frac{1}{12a^4}\int d\tau a^4\rho_m\frac{dU_B}{d\tau},$$

up to a dark radiation term and we have identified

$$U = \frac{1}{2}\left(U_B^2 - \left(\frac{\partial U_B}{\partial\phi}\right)^2 + V\right).$$

This is the Friedmann equation for a brane coupled to a bulk scalar field. Notice that retarded effects springing from the whole history of the brane and scalar field dynamics are present. Below we will see that these retarded effects come from the projected Weyl tensor. They result from the exchange between the brane and the bulk. Notice, that Newton's constant depends on the value of the bulk scalar field evaluated on the brane ($\phi_0 = \phi(t, y = 0)$):

$$\frac{8\pi G_N(\phi_0)}{3} = \frac{\kappa_5^2 U_B(\phi_0)}{12}.$$

On cosmological scale, time variation of the scalar field induce a time variations of Newton's constant. This is highly constrained experimentally, leading to tight restrictions on the time dependence of the scalar field [BB03].

To get a feeling of the physics involved in the Friedmann equation, it is convenient to assume that the scalar field is evolving slowly on the scale of the variation of the scale factor. Neglecting the evolution of Newton's constant, the Friedmann equation reduces to

$$H^2 = \frac{8\pi G_N(\phi)}{3}\rho_m + \frac{U}{8} - \frac{\dot{\phi}^2}{16}.$$

Several comments are in order. First of all we have neglected the contribution due to the ρ_m^2 term as we are considering energy scales below the brane tension. Then the main effect of the scalar field dynamics is to involve the potential energy U and the kinetic energy $\dot{\phi}^2$. Although the potential energy appears with a positive sign we find that the kinetic energy has a negative sign. For an observer on the brane this looks like a violation of unitarity. The minus sign for the kinetic energy is due to the fact that one does not work in the Einstein frame where Newton's constant does not vary, a similar minus sign appears also in the effective 4D theory when working in the brane frame.

The time dependence of the scalar field is determined by the Klein–Gordon equation. The dynamics is completely specified by

$$\ddot{\phi} + 4H\dot{\phi} + \frac{1}{2}(\frac{1}{3} - \omega_m)\rho_m\frac{\partial U_B}{\partial \phi} = -\frac{\partial U}{\partial \phi} + \Delta\Phi_2,$$

where $p_m = \omega_m\rho_m$. We have identified

$$\Delta\Phi_2 = \phi''|_0 - \frac{\partial U_B}{\partial \phi}\frac{\partial^2 U_B}{\partial \phi^2}|_0.$$

This cannot be set to zero and requires the knowledge of the scalar field in the vicinity of the brane. When we discuss cosmological solutions below, we will assume that this term is negligible.

The evolution of the scalar field is driven by two effects. First of all, the scalar field couples to the trace of the energy momentum tensor via the gradient of U_B. Secondly, the field is driven by the gradient of the potential U, which might not necessarily vanish.

The Friedmann equation vs the projected Weyl tensor

We are now coming back to the origin of the non–trivial Friedmann equation. Using the *Gauss–Codazzi equation* one can get the Einstein equation on the brane [MW00, MB01]

$$\bar{G}_{ab} = -\frac{3}{8}Uh_{ab} + \frac{U_B}{4}\tau_{ab} + \pi_{ab} + \frac{1}{2}\partial_a\phi\partial_b\phi - \frac{5}{16}(\partial\phi)^2 h_{ab} - E_{ab}.$$

Now the projected Weyl tensor can be determined in the homogeneous and isotropic cosmology case. Indeed only the E_{00} component is independent. Using the Bianchi identity $\bar{D}^a\bar{G}_{ab} = 0$ where $\bar{D}_a$ is the brane covariant derivative, one gets that

$$\dot{E}_{00} + 4HE_{00} = \partial_\tau \left(\frac{3}{16}\dot{\phi}^2 + \frac{3}{8}U \right) + \frac{3}{2}H\dot{\phi}^2 + \frac{\dot{U}_B}{4}\rho_m, \qquad \text{leading to}$$

$$E_{00} = \frac{1}{a^4} \int d\tau a^4 \left(\partial_\tau \left(\frac{3}{16}\dot{\phi}^2 + \frac{3}{8}U \right) + \frac{3}{2}H\dot{\phi}^2 + \frac{\dot{U}_B}{4}\rho_m \right).$$

Upon using

$$\bar{G}_{00} = 3H^2,$$

one gets the Friedmann equation. It is remarkable that the retarded effects in the Friedmann equation all spring from the projected Weyl tensor. Hence the projected Weyl tensor proves to be much richer in the case of a bulk scalar field than in the empty bulk case.

Self–Tuning and Accelerated Cosmology

Dynamics of the brane is not closed, it is an open system continuously exchanging energy with the bulk. This exchange is characterized by the dark radiation term and the loss parameter. Both require a detailed knowledge of the bulk dynamics. This is of course beyond the projective approach where only quantities on the brane are evaluated. In the following we will assume that the dark radiation term is absent and that the loss parameter is negligible. Furthermore, we will be interested in the effects of a bulk scalar field for late–time cosmology (i.e., well after nucleosynthesis) and not in the case for inflation driven by a bulk scalar field [BB03].

Let us consider the self–tuned scenario as a solution to the cosmological constant problem. It corresponds to the BPS superpotential with $\alpha = 1$. In that case the potential $U = 0$ for any value of the brane tension. The potential $U = 0$ can be interpreted as a vanishing of the brane cosmological constant. The physical interpretation of the vanishing of the cosmological constant is that the brane tension curves the fifth dimensional space–time leaving a flat brane intact. Unfortunately, the description of the bulk geometry in that case has shown that there was a bulk singularity which needs to be hidden by a second brane whose tension is fine–tuned with the first brane tension. This reintroduces a fine-tuning in the putative solution to the cosmological constant problem [FLL00].

Let us generalize the selftuned case to $\alpha \neq 1$, i.e., $U_B = TW$, $T > 1$ and W is the exponential superpotential. The resulting induced metric on the brane is of the FRW type with a scale factor

$$a(t) = a_0 \left(\frac{t}{t_0} \right)^{1/3+1/6\alpha^2},$$

leading to an acceleration parameter

$$q_0 = \frac{6\alpha^2}{1+2\alpha^2} - 1.$$

For the supergravity value $\alpha = -\frac{1}{\sqrt{12}}$ this leads to $q_0 = -4/7$. This is in coincidental agreement with the supernovae results. This model can serve as a *brane quintessence model* [BB03].

The brane cosmological eras

Let us now consider the possible cosmological scenarios with a bulk scalar field [BBD01, BB03]. We assume that the potential energy of the scalar field U is negligible throughout the radiation and matter eras before serving as quintessence in the recent past.

At very high energy above the tension of the brane the non–conventional cosmology driven by the ρ_m^2 term in the Friedmann equation is obtained. Assuming radiation domination, the scale factor behaves like

$$a = a_0 \left(\frac{t}{t_0}\right)^{1/4},$$

and the scalar field

$$\phi = \phi_i + \beta \ln \left(\frac{t}{t_0}\right).$$

In the radiation dominated era, no modification is present, provided $\phi = \phi_i$, which is a solution of the Klein–Gordon equation as the trace of the energy–momentum of radiation vanishes (together with a decaying solution, which we have neglected). In the matter dominated era the scalar field evolves due to the coupling to the trace of the energy–momentum tensor. This has two consequences. Firstly, the kinetic energy of the scalar field starts contributing in the Friedmann equation. Secondly, the effective Newton constant does not remain constant. The cosmological evolution of Newton's constant is severely constrained since nucleosynthesis. This restricts the possible time variation of ϕ [BB03].

In order to be more quantitative let us come back to the exponential superpotential case with a detuning parameter T. The time dependence of the scalar field and scale factor become

$$\phi = \phi_1 - \frac{8}{15}\alpha \ln \left(\frac{t}{t_e}\right), \qquad a = a_e \left(\frac{t}{t_e}\right)^{\frac{2}{3}-\frac{8}{45}\alpha^2},$$

where t_e and a_e are the time and scale factor at matter-radiation equality. Notice the slight discrepancy of the scale factor exponent with the standard model value of 2/3. The redshift dependence of the Newton constant is

$$\frac{G_N(z)}{G_N(z_e)} = \left(\frac{z+1}{z_e+1}\right)^{4\alpha^2/5}.$$

For the supergravity model with $\alpha = -\frac{1}{\sqrt{12}}$ and $z_e \sim 10^3$ this leads to a decrease by (roughly) 37% since nucleosynthesis. This is marginally compatible with experiments [BB03].

Finally let us analyse the possibility of using the brane potential energy of the scalar field U as the source of acceleration now. We have seen that when matter is negligible on the brane, one can build brane quintessence models. We now require that this occurs only in the recent past. As can be expected, this leads to a fine-tuning problem as $M^4 \sim \rho_c$,where $M^4 = (T-1)\frac{3W}{2\kappa_5^2}$ is the amount of detuned tension on the brane. Of course this is nothing but a reformulation of the usual cosmological constant problem. Provided one accepts this fine–tuning, as in most quintessence models, the exponential model with $\alpha = -\frac{1}{\sqrt{12}}$is a cosmological consistent quintessence model with a five dimensional origin.

Therefore, the main difference between a brane world model with a bulk scalar field and the Randall–Sundrum model is that the gravitational constant becomes time–dependent. As such it has much in common with scalar–tensor theories [FM03], but there are important differences due to the projected Weyl tensor $E_{\mu\nu}$ and its time–evolution. The bulk scalar field can play the role of the quintessence field, as discussed above, but it could also play a role in an inflationary era in the very early universe. In any case, the cosmology of such a system is much richer and, because of the variation of the gravitational constant, more constrained. It remains to be seen if the bulk scalar field can leave a trace in the CMB anisotropies and Large Scale Structures (for first results see [BBD01]).

Moving Branes in a Static Bulk

So far, we were mostly concerned with the evolution of the brane, without referring to the bulk itself. In fact, the coordinates introduced in (7.110) are a convenient choice for studying the brane itself, but when it comes to analysing the bulk dynamics and its geometry, these coordinates are not the best choice. We have already mentioned the extended Birkhoff Theorem above. It states that for the case of a vacuum bulk space–time, the bulk is necessarily static, in certain coordinates. A cosmological evolving brane is then moving in that space–time, whereas for an observer confined on the brane the motion of the brane will be seen as an expanding (or contracting) universe. In the case of a scalar field in the bulk, a similar Theorem is unfortunately not available, which makes the study of such systems much more complicated. We will now discuss these issues in some detail, following in particular [Dav02a] and [Dav02b].

Motion in AdS–Schwarzschild Bulk

We have already discussed the static background associated with BPS configurations (including the Randall–Sundrum case) above. Here we will focus on other backgrounds for which one can integrate the bulk equations of motion. Let us write the following ansatz for the metric

$$ds^2 = -A^2(r)dt^2 + B^2(r)dr^2 + R^2(r)d\Sigma^2, \qquad (7.131)$$

where $d\Sigma^2$ is the metric on the 3d symmetric space of curvature $q = 0, \pm 1$. In general, the function A, B and R depend on the type of scalar field potential. This is to be contrasted with the case of a negative bulk cosmological constant where Birkhoff's Theorem states that the most general solution of the (bulk) Einstein equations is given by $A^2 = f$, $B^2 = 1/f$ and $R = r$ where $f(r) = q + r^2/l^2 - \mu/r^2$. We have denoted by $l = 1/k = \sqrt{-6/(\Lambda_5 \kappa_5^2)}$ the AdS scale and μ the black hole mass. This solution is the so–called *AdS–Schwarzschild solution*.

Let us now study the motion of a brane of tension T/l in such a background. The equation of motion is determined by the junction conditions. The method will be reviewed later when a scalar field is present in the bulk. The resulting equation of motion for a boundary brane with a Z_2 symmetry is

$$\left(\dot{r}^2 + f(r)\right)^{1/2} = \frac{T}{l} r,$$

for a brane located at r [KLS99]. Here $\dot{r}$ is the velocity of the brane measured with the proper time on the brane. This leads to the following Friedmann equation

$$H^2 \equiv \left(\frac{\dot{r}}{r}\right)^2 = \frac{T^2 - 1}{l^2} - \frac{q}{r^2} + \frac{\mu}{r^4}.$$

So the brane tension leads to an effective cosmological constant $(T^2 - 1)/l^2$. The curvature gives the usual term familiar from standard cosmology while the last term is the dark radiation term whose origin springs from the presence of a black–hole in the bulk. At late time the dark radiation term is negligible for an expanding universe, we retrieve the cosmology of a FRW universe with a non–vanishing cosmological constant. The case $T = 1$ corresponds of course to the RS case.

Moving Branes

Let us now describe the general formalism, which covers the case of the AdS–Schwarzschild space–time mentioned above. Consider a brane embedded in a static background. It is parametrized by the coordinates $X^A(x^\mu)$ where $A = 0 \ldots 4$ and the x^μ are world volume coordinates. Locally the brane is characterized by the local frame

$$e^A_\mu = \frac{\partial X^A}{\partial x^\mu},$$

which are tangent to the brane. The induced metric is given by

$$h_{\mu\nu} = g_{AB} e^A_\mu e^B_\nu,$$

and the extrinsic curvature

$$K_{\mu\nu} = e^A_\mu e^B_\nu D_A n_B,$$

where n^A is the unit vector normal to the brane defined by (up to a sign ambiguity)

$$g_{AB}n^A n^B = 1, \qquad n_A e^A_\mu = 0.$$

For a homogeneous brane embedded in the space–time described by the metric (7.131), we have $T = T(\tau)$, $r = r(\tau)$ where τ is the proper time on the brane. The induced metric is

$$ds_B^2 = -d\tau^2 + R^2(\tau)d\Sigma^2.$$

The local frame becomes

$$e^A_\tau = (\dot{T}, \dot{r}, 0, 0, 0), \qquad e^A_i = (0, 0, \delta^A_i),$$

while the normal vector reads

$$n_A = (AB\dot{r}, -B\sqrt{1+\dot{r}^2}, 0, 0, 0).$$

The components of the extrinsic curvature tensor can found to be

$$K_{ij} = -\frac{\sqrt{1+B^2\dot{r}^2}}{B}RR'\delta_{ij}, \qquad K_{\tau\tau} = \frac{1}{AB}\frac{n}{dr}(A\sqrt{1+B^2\dot{r}^2}).$$

The junction conditions are given by

$$K_{\mu\nu} = -\frac{\kappa_5^2}{2}\left(\tau_{\mu\nu} - \frac{1}{3}\tau h_{\mu\nu}\right).$$

This implies that the brane dynamics are specified by the equations of motion

$$\frac{\sqrt{1+B^2\dot{r}^2}}{B}\frac{R'}{R} = \frac{\kappa_5^2}{6}\rho, \qquad \text{and}$$
$$\frac{1}{AB}\frac{n}{dr}(A\sqrt{1+B^2\dot{r}^2}) = -\frac{\kappa_5^2}{6}(2\rho + 3p),$$

where we have assumed a fluid description for the matter on the brane. These two equations determine the dynamics of any brane in a static background.

Let us now close the system of equations by stating the scalar field boundary condition [BB03]

$$n^A\partial_A\phi = \frac{\kappa_5^2}{2}\frac{d\xi}{d\phi}(\rho - 3p),$$

where the coupling to the brane is defined by the Lagrangian

$$S_{brane} = \int d^4x\mathcal{L}[\psi_m, \tilde{h}_{\mu\nu}],$$

where ψ_m represents the matter fields and

$$\tilde{h}_{\mu\nu} = \mathrm{e}^{2\xi(\phi)}h_{\mu\nu}.$$

This reduces to

$$\phi' = \frac{\kappa_5^2}{2}\frac{B}{\sqrt{1+B^2\dot{r}^2}}\frac{d\xi}{d\phi}(-\rho+3p).$$

Combining the junction conditions leads to the conservation equation

$$\dot{\rho} + 3H(\rho+p) = (\rho-3p)\dot{\xi}.$$

This is nothing but the conservation of matter in the Jordan frame defined by $\tilde{h}_{\mu\nu}$.

We now turn to a general analysis of the brane motion in a static bulk. To do that it is convenient to parametrized the bulk metric slightly differently

$$ds^2 = -f^2(r)h(r)dt^2 + \frac{dr^2}{h(r)} + r^2 d\Sigma^2.$$

Now, the Einstein equations lead to (redefining $\phi \to \frac{\sqrt{3}}{2\kappa_5}\phi$ and $V \to \frac{3}{8\kappa_5^2}V$)

$$\frac{3}{r^2}\left(h + \frac{rh'}{2} - q\right) = -\kappa_5^2\left(\frac{h\phi'^2}{2} + V\right), \tag{7.132}$$

$$\frac{3}{r^2}\left(h + \frac{rh'}{2} - q + \frac{hrf'}{f}\right) = \kappa_5^2\left(\frac{h\phi'^2}{2} - V\right), \tag{7.133}$$

and the Klein–Gordon equation

$$h\phi'' + \left(\frac{3h}{r} + \frac{hf'}{f} + h'\right)\phi' = \frac{dV}{d\phi}.$$

Subtracting (7.132) from (7.133) and solving the resulting differential equation, we get

$$f = \exp\left(\frac{\kappa_5^2}{3}\int dr\, r\phi'^2\right).$$

It is convenient to evaluate the spatial trace of the projected Weyl tensor. This is obtained by computing both the bulk Weyl tensor and the vector normal to the moving brane. With $A = \sqrt{h}f,\ R = r,\ B = 1/\sqrt{h}$, this gives

$$\frac{\mu}{r^4} = -\frac{E^i_i}{3} = \frac{r}{4f^2}\left(\frac{hf^2}{r^2}\right)' + \frac{q}{2r^2}.$$

This is the analogue of the dark radiation term for a general background. The equations of motion can be cast in the form

$$\mu' = -\frac{\kappa_5^2}{3}(\mu - \frac{kr^2}{2})r\phi'^2,$$

$$\mathcal{H}' + 4\frac{\mu}{r^5} = -\frac{2\kappa_5^2}{3}(\mathcal{H} - \frac{q}{r^2})r\phi'^2,$$

$$\kappa_5^2 V = 6\mathcal{H} + \frac{3}{4}r\mathcal{H}' - 3\frac{\mu}{r^4}, \qquad \text{using} \qquad \mathcal{H} = \frac{q-h}{r^2}.$$

This allows to retrieve easily some of the previous solutions. Choosing ϕ to be constant leads to $f = 1$, μ is constant and

$$\mathcal{H} = -\frac{1}{l^2} + \frac{\mu}{r^4}.$$

This is the AdS–Schwarzschild solution.

For $q = 0$ the equations of motion simplify to

$$\frac{\kappa_5^2}{3} r\phi' = -\frac{n \ln \mu}{d\phi}, \tag{7.134}$$

$$n\left(\frac{\mathcal{H}}{\mu^2}\right) = \frac{1}{\mu} n\left(r^{-4}\right),$$

$$\frac{\kappa_5^2}{6} V = -\frac{3}{4\kappa_5^2} \frac{d\mu}{d\phi} \frac{n}{d\phi}\left(\frac{\mathcal{H}}{\mu}\right) + \mathcal{H} \tag{7.135}$$

In this form it is easy to see that the dynamics of the bulk are completely integrable. First of all the solutions depend on an arbitrary function $\mu(\phi)$ which determines the dynamics. Notice that $f = \mu_0/\mu$, where μ_0 is an arbitrary constant. The radial coordinate r is obtained by simple integration of (7.134)

$$r = r_0 \mathrm{e}^{-\frac{\kappa_5^2}{3} \int \frac{d\phi}{n \ln \mu} d\phi}.$$

Finally the rest of the metric follows from

$$h = -\frac{4\kappa_5^2}{3} r^2 \mu^2 \int d\phi \frac{d\phi}{d\mu} \mathrm{e}^{4\frac{\kappa_5^2}{3} \int \frac{d\phi}{n \ln \mu} d\phi}.$$

The potential V then follows (7.135). This is remarkable and shows why Birkhoff's Theorem is not valid in the presence of a bulk scalar field. Moreover, it is intriguing that the generalization of the dark energy term dictates the bulk dynamics completely.

It is interesting to recast the Friedmann equation in the form

$$H^2 = \mathcal{H} + \frac{\kappa_5^4}{36} \mu^2 \rho^2,$$

where H is the Hubble parameter on the brane in cosmic time. One can retrieve standard cosmology by studying the dynamics in the vicinity of a critical point $\frac{d\mu}{d\phi} = 0$. Parametrizing

$$\mu = \frac{6A}{\kappa_5^2} + B\phi^2,$$

leads to the Friedmann equation

$$H^2 = \frac{\kappa_5^4}{36}(\rho^2 - \theta)\mu^2 + \frac{\mu}{a^4} + o(a^{-4}).$$

Here θ is an arbitrary integration constant. Notice that this is a small deviation from the RS case as $\phi = r^{-B/A}$ goes to zero at large distances. Hence, standard cosmology is retrieved at low energy and long distance.

Cosmology of a Two–Brane System

Here we will once more include an ingredient suggested by particle physics theories, in particular M–theory. So far we have assumed that there is only one brane in the whole space–time. According to string theory, there should be at least another brane in the bulk. Indeed, in heterotic M–theory these branes are the boundaries of the bulk space–time [HW96]. However, even from a purely phenomenological point of view there is a reason to include a second brane: the bulk singularity (or the AdS horizon). As we have seen above, the inclusion of a bulk scalar field often implies the presence of a naked singularity located away from the positive tension brane. The second brane which we include now should shield this singularity, so that the physical space–time stretches between the two branes. Another motivation is the hierarchy problem. Randall and Sundrum proposed a two brane model (one with positive and one with negative tension), embedded in a 5D AdS space–time. In their scenario the standard model particles would be confined on the *negative tension brane*. As they have shown, in this case gravity is weak due to the warping of the bulk space–time. However, as will become clear from the above results, in order for this model to be consistent with gravitational experiments, the interbrane distance has to be *fixed* [GT00]. This can be achieved, for example, with a bulk scalar field. As shown in [GT00], gravity in the two–brane model of RS is described by a scalar–tensor theory, in which the interbrane–distance, called radion, plays the role of a scalar field. The bulk scalar field will modify the Brans–Dicke parameter of the scalar field and will introduce a second scalar field in the low–energy effective theory, so that the resulting theory at low energy in the case of two branes and a bulk scalar field is a bi–scalar–tensor theory [BB03]. In the following we will investigate the cosmological consequences when the distance between the branes is *not* fixed (for some aspects not covered here see e.g., [CGR00, LS02]).

The Low–Energy Effective Action

In order to understand the cosmology of the two–brane system, we derive the low-energy effective action by utilizing the moduli space approximation. From the above discussion, it becomes clear, that the *general* solution of the bulk Einstein equations for a given potential is difficult to find. The moduli space approximation gives the *low–energy–limit* effective action for the two brane system, i.e., for energies much smaller than the brane tensions.

In the static BPS solutions described above, the brane positions can be chosen arbitrarily. In other words, they are *moduli fields*. It is expected that by putting some matter on the branes, these moduli field become time-dependent, or, if the matter is inhomogeneously distributed, space–time dependent. Thus, the first approximation is to replace the brane–positions with space–time dependent functions. Furthermore, in order to allow for the gravitational zero–mode, we will replace the flat space–time metric $\eta_{\mu\nu}$ with $g_{\mu\nu}(x^\alpha)$. We do assume that the evolution of these fields is slow, which means that we neglect terms like $(\partial\phi)^3$ when constructing the low-energy effective action.

As already mentioned, the moduli space approximation is only a good approximation at energies much less than the brane tension. Thus, we do not recover the quadratic term in the moduli space approximation. We are interested in the *late time* effects after nucleosynthesis, where the corrections have to be small.

Replacing $\eta_{\mu\nu}$ with $g_{\mu\nu}(x^\alpha)$ in (7.128) and collecting all the terms one finds from the 5D action after an integration over y:

$$\begin{aligned} S_{\rm MSA} =& \int d^4x\sqrt{-g_4}\left[f(\phi,\sigma)\mathcal{R}^{(4)} + \frac{3}{4}a^2(\phi)\frac{U_B(\phi)}{\kappa_5^2}(\partial\phi)^2\right.\\ &\left. - \frac{3}{4}a^2(\sigma)\frac{U_B}{\kappa_5^2}(\sigma)(\partial\sigma)^2\right], \qquad \text{with} \\ f(\phi,\sigma) =& \frac{1}{\kappa_5^2}\int_\phi^\sigma dya^2(y), \end{aligned}$$

and $a(y)$ given by (7.129). The moduli ϕ and σ represent the location of the two branes. Note that the kinetic term of the field ϕ has the wrong sign. This is an artifact of the frame we use here. As we will see below, it is possible to go to the Einstein frame with a simple conformal transformation, in which the sign in front of the kinetic term is correct for both fields.

In the following we will concentrate on the BPS system with above exponential superpotential. Let us redefine the fields according to

$$\tilde{\phi}^2 = \left(1 - 4k\alpha^2\phi\right)^{2\beta}, \qquad \tilde{\sigma}^2 = \left(1 - 4k\alpha^2\sigma\right)^{2\beta}, \tag{7.136}$$

with $\beta = \frac{2\alpha^2+1}{4\alpha^2}$; and then

$$\tilde{\phi} = Q\cosh R, \qquad \tilde{\sigma} = Q\sinh R. \tag{7.137}$$

A conformal transformation $\tilde{g}_{\mu\nu} = Q^2 g_{\mu\nu}$ leads to the Einstein frame action:

$$\begin{aligned} S_{\rm EF} =& \frac{1}{2k\kappa_5^2(2\alpha^2+1)}\int d^4x\sqrt{-g}\left[\mathcal{R} - \frac{12\alpha^2}{1+2\alpha^2}\frac{(\partial Q)^2}{Q^2}\right.\\ &\left. - \frac{6}{2\alpha^2+1}(\partial R)^2\right]. \end{aligned}$$

Note that in this frame both fields have the correct sign in front of the kinetic terms. For $\alpha \to 0$ (i.e., the RS case) the Q–field decouples. This reflects the fact, that the bulk scalar field decouples, and the only scalar degree of freedom is the distance between the branes. One can read off the gravitational constant to be

$$16\pi G = 2k\kappa_5^2(1+2\alpha^2).$$

The matter sector of the action can be found easily: if matter lives on the branes, it 'feels' the induced metric. That is, the action has the form

$$S_m^{(1)} = S_m^{(1)}(\Psi_1, g_{\mu\nu}^{B(1)}) \quad \text{and} \quad S_m^{(2)} = S_m^{(2)}(\Psi_2, g_{\mu\nu}^{B(2)}),$$

where $g_{\mu\nu}^{B(i)}$ denotes the induced metric on each branes. In going to the Einstein frame one gets

$$S_m^{(1)} = S_m^{(1)}(\Psi_1, A^2(Q,R)g_{\mu\nu}) \quad \text{and} \quad S_m^{(2)} = S_m^{(2)}(\Psi_2, B^2(Q,R)g_{\mu\nu}),$$

where matter now couples explicitely to the fields via the functions A and B, which we will give below (neglecting derivative interactions).

The theory derived with the help of the moduli space approximation has the form of a *multi–scalar–tensor theory*, in which matter on both branes couple differently to the moduli fields. We note, that methods different from the moduli–space approximation have been used in the literature in order to get the low–energy effective action or the resulting field equations for a two–brane system. Qualitatively, the features of the resulting theories agree with the moduli–space approximation discussed above.

In the following we will discuss observational constraints imposed on the parameter of the theory.

Observational Constraints

In order to constrain the theory, it is convenient to write the moduli Lagrangian in the form of a non-linear sigma model with kinetic terms

$$\gamma_{ij}\partial\phi^i\partial\phi^j,$$

where $i = 1,2$ labels the moduli $\phi^1 = Q$ and $\phi^2 = R$. The sigma model couplings are here

$$\gamma_{QQ} = \frac{12\alpha^2}{1+2\alpha^2}\frac{1}{Q^2}, \qquad \gamma_{RR} = \frac{6}{1+2\alpha^2}.$$

Notice the potential danger of the $\alpha \to 0$ limit, the RS model, where the coupling to Q becomes very small. In an ordinary Brans–Dicke theory with a single field, this would correspond to a vanishing Brans-Dicke parameter which is ruled out experimentally. Here we will see that the coupling to matter is such that this is not the case. Indeed we can write the action expressing the coupling to ordinary matter on our brane as

$$A = a(\phi)f^{-1/2}(\phi,\sigma), \qquad B = a(\sigma)f^{-1/2}(\phi,\sigma),$$

where we have neglected the derivative interaction.

Let us introduce the parameters

$$\alpha_Q = \partial_Q \ln A, \qquad \alpha_R = \partial_R \ln A.$$

We find that ($\lambda = 4/(1+2\alpha^2)$)

$$A = Q^{-\frac{\alpha^2\lambda}{2}} (\cosh R)^{\frac{\lambda}{4}},$$

leading to

$$\alpha_Q = -\frac{\alpha^2\lambda}{2}\frac{1}{Q}, \qquad \alpha_R = \frac{\lambda \tanh R}{4}.$$

Observations constrain the parameter

$$\theta = \gamma^{ij}\alpha_i\alpha_j$$

to be less than 10^{-3} [BB03]. We get therefore a bound on

$$\theta = \frac{4}{3}\frac{\alpha^2}{1+2\alpha^2} + \frac{\tanh^2 R}{6(1+2\alpha^2)}.$$

The bound implies that

$$\alpha \leq 10^{-2}, \qquad R \leq 0.2.$$

The smallness of α indicates a strongly warped bulk geometry such as an Anti–de Sitter space–time. In the case $\alpha = 0$, we can easily interpret the bound on R. Indeed in that case

$$\tanh R = \mathrm{e}^{-k(\sigma-\phi)},$$

i.e., this is nothing but the exponential of the radion field measuring the distance between the branes. We get that gravity experiments require the branes to be sufficiently far apart. When $\alpha \neq 0$ but small, one way of obtaining a small value of R is for the hidden brane to become close from the would-be singularity where $a(\sigma) = 0$.

We would like to mention that the parameter θ can be calculated also for matter on the negative tension brane. Then, following the same calculations as above, it can be seen that the observational constraint for θ cannot be satisfied. Thus, if the standard model particles are confined on the negative tension brane, the moduli have necessarily to be stabilized. In the following we will assume that the standard model particles are confined on the positive tension brane and study the cosmological evolution of the moduli fields.

Cosmological Implications

The discussion in the last subsection raises an important question: the parameter α has to be choosen rather small, in order for the theory to be consistent with observations. Similarly the field R has to be small too. The field R is dynamical and one would like to know if the cosmological evolution drives the field R to small values such that it is consistent with the observations today. Otherwise are there natural initial conditions for the field R? In the following we study the cosmological evolution of the system in order to answer these questions.

The field equations for a homogenous and isotropic universe can be obtained from the action. The Friedmann equation reads

$$H^2 = \frac{8\pi G}{3}(\rho_1 + \rho_2 + V_{\rm eff} + W_{\rm eff}) + \frac{2\alpha^2}{1+2\alpha^2}\dot{\phi}^2 + \frac{1}{1+2\alpha^2}\dot{R}^2. \qquad (7.138)$$

where we have defined $Q = \exp\phi$. The field equations for R and ϕ read

$$\ddot{R} + 3H\dot{R} = -8\pi G\frac{1+2\alpha^2}{6}\left[\frac{\partial V_{\rm eff}}{\partial R} + \frac{\partial W_{\rm eff}}{\partial R} + \alpha_R^{(1)}(\rho_1 - 3p_1) + \alpha_R^{(2)}(\rho_2 - 3p_2)\right] \qquad (7.139)$$

$$\ddot{\phi} + 3H\dot{\phi} = -8\pi G\frac{1+2\alpha^2}{12\alpha^2}\left[\frac{\partial V_{\rm eff}}{\partial \phi} + \frac{\partial W_{\rm eff}}{\partial \phi} + \alpha_\phi^{(1)}(\rho_1 - 3p_1) + \alpha_\phi^{(2)}(\rho_2 - 3p_2)\right]. \qquad (7.140)$$

The coupling parameter are given by

$$\alpha_\phi^{(1)} = -\frac{2\alpha^2}{1+2\alpha^2}, \qquad \alpha_\phi^{(2)} = -\frac{2\alpha^2}{1+2\alpha^2}, \qquad (7.141)$$

$$\alpha_R^{(1)} = \frac{\tanh R}{1+2\alpha^2}, \qquad \alpha_R^{(2)} = \frac{(\tanh R)^{-1}}{1+2\alpha^2}. \qquad (7.142)$$

We have included matter on both branes as well as potentials $V_{\rm eff}$ and $W_{\rm eff}$ on each branes. We now concentrate on the case where matter is only on our brane. In the radiation dominated epoch the trace of the energy–momentum tensor vanishes, so that Q and ϕ quickly become constant. The scale factor scales like $a(t) \propto t^{1/2}$.

In the matter–dominated era, the solution to these equations is given by

$$\rho_1 = \rho_e\left(\frac{a}{a_e}\right)^{-3-2\alpha^2/3}, \qquad a = a_e\left(\frac{t}{t_e}\right)^{2/3-4\alpha^2/27},$$

together with

$$\phi = \phi_e + \frac{1}{3}\ln\frac{a}{a_e}, \qquad R = R_0\left(\frac{t}{t_e}\right)^{-1/3} + R_1\left(\frac{t}{t_e}\right)^{-2/3},$$

as soon as $t \gg t_e$. Note that R indeed decays. This implies that small values of R compatible with gravitational experiments are favoured by the cosmological evolution. Note, however, that the size of R in the early universe is constrained by nucleosynthesis as well as by the CMB anisotropies. A large discrepancy between the values of R during nucleosynthesis and now induces a variation of the particle masses, or equivalently Newton's constant, which is excluded experimentally. One can show that by putting matter on the negative tension brane as well, the field R evolves even faster to zero. This behavior is reminiscent of the attractor solution in scalar–tensor theories [BB03].

In the 5D picture the fact that R is driven to small values means that the negative tension brane is driven towards the bulk singularity. In fact, solving the equations numerically for more general cases suggest that R can even by negative, which is, in the 5D description meaningless, as the negative tension brane would move through the bulk singularity. Thus, in order to make any further progress, one has to understand the bulk singularity better[41]. Of course, one could simply assume that the negative tension brane is destroyed when it hits the singularity. A more interesting alternative would be if the brane is repelled instead. It was speculated that this could be described by some effective potential in the low-energy effective action [BB03].

Brane Collision

We have seen that brane world models are plagued with a singularity problem: the negative tension brane might hit a bulk singularity. In that case our description of the physics on the brane requires techniques beyond the field theory approach that we have followed in this review. It is only within a unified theory encompassing general relativity and quantum mechanics that such questions might be addressed. String theory may be such a theory. The problem of the nature of the resolution of cosmological singularities in string theory is still a vastly unchartered territory. There is a second kind of singularity which arises when two branes collide. In such a case there is also a singularity in the low energy effective action as one of the extra dimensions shrink to zero size. It was speculated that brane collisions play an important role in cosmology, especially in order to understand the Big–Bang itself.

As soon as the space–time contains several branes and that these branes move with respect to each other, they might collide. A fascinating possibility, which has been actively explored by [KOS01b, Buc02, GIT02], is that the Big–Bang is such a brane collision. Rather than entering into the details of these various models, let us point out here a simple and general analysis [LMW02] of the collision of (parallel or concentric) branes separated by vacuum, i.e., branes separated by patches of AdS–Schwarzschild space–times (allowing for different Schwarschild–type mass and cosmological constant in each region) with the 5D AdS–Schwarzschild metric (AdS for $\Lambda < 0$, which is the case we are interested in; for $\Lambda > 0$, this the Schwarzschild–de Sitter metric)

$$ds^2 = -f(R)\,dT^2 + \frac{dR^2}{f(R)} + R^2\,\gamma_{ij}\,dx^i dx^j, \qquad \text{where} \tag{7.143}$$
$$f(R) \equiv k - \frac{\Lambda}{6}R^2 - \frac{C}{R^2}$$

[41] For $\alpha = 0$ the theory is equivalent to the Randall–Sundrum model. In this case the bulk singularity is shifted towards the Anti–de Sitter boundary.

(where C is an arbitrary integration constant). Although we are interested here by 3-branes embedded in a 5D space–time, this analysis is immediately applicable to the case of n-branes moving in a $(n+2)$-dimensional space–time, with the analogous symmetries.

To analyze the collision, it is convenient to introduce an angle α, which characterizes the motion of the brane with respect to the coordinate system (7.143), defined by

$$\alpha = \sinh^{-1}(\epsilon \dot{R}/\sqrt{f}),$$

where $\epsilon = +1$ if R decreases from 'left' to 'right', $\epsilon = -1$ otherwise. Considering a collision involving a total number of N branes, both ingoing and outgoing, thus separated by N space–time regions, one can label alternately branes and regions by integers, starting from the leftmost ingoing brane and going anticlockwise around the point of collision. The branes will thus be denoted by odd integers, $2k-1$ $(1 \leq k \leq N)$, and the regions by even integers, $2k$ $(1 \leq k \leq N)$. Let us introduce the angle $\alpha_{2k-1|2k}$ which characterizes the motion of the brane $\mathcal{B}_{2k-1}$ with respect to the region $\mathcal{R}_{2k}$, and which is defined by [Lan02]

$$\sinh \alpha_{2k-1|2k} = \frac{\epsilon_{2k} \dot{R}_{2k-1}}{\sqrt{f_{2k}}}. \tag{7.144}$$

Conversely, the motion of the region $\mathcal{R}_{2k}$ with respect to the brane by the Lorentz angle

$$\alpha_{2k|2k-1} = -\alpha_{2k-1|2k}.$$

It can be shown that the junction conditions for the branes can be written in the form

$$\tilde{\rho}_{2k-1} \equiv \pm \frac{\kappa^2}{3} \rho_{2k-1} R = \epsilon_{2k} \sqrt{f_{2k}} \exp\left(\pm\alpha_{2k-1|2k}\right) - \epsilon_{2k-2} \sqrt{f_{2k-2}} \exp\left(\mp\alpha_{2k-2|2k-1}\right), \tag{7.145}$$

with the plus sign for ingoing branes $(1 \leq k \leq N_{in})$, the minus sign for outgoing branes $(N_{in}+1 \leq k \leq N)$. An outgoing positive energy density brane thus has the same sign as an ingoing negative energy density brane.

The advantage of this formalism becomes obvious when one writes the *geometrical consistency relation* that expresses the matching of all branes and space–time regions around the collision point. In terms of the angles defined above, it reads simply

$$\sum_{i=1}^{2N} \alpha_{i|i+1} = 0. \tag{7.146}$$

Moreover, introducing the generalized angles

$$\alpha_{j|j'} = \sum_{i=j}^{j'-1} \alpha_{i|i+1}, \qquad \text{if} \quad j < j', \qquad \text{and} \quad \alpha_{j'|j} = -\alpha_{j|j'},$$

the sum rule for angles (7.146) combined with the junction conditions (7.145) leads to the laws of energy conservation and momentum conservation. The energy conservation law reads [Lan02]

$$\sum_{k=1}^{N} \tilde{\rho}_{2k-1}\gamma_{j|2k-1} = 0,$$

where $\gamma_{j|j'} \equiv \cosh \alpha_{j|j'}$ corresponds to the Lorentz factor between the brane/region j and the brane/region j' and can be obtained, if j and j' are not adjacent, by combining all intermediary Lorentz factors (this is simply using the velocity addition rule of special relativity). The index j corresponds to the reference frame with respect to which the conservation rule is written. Similarly, the *momentum conservation law* in the jth reference frame can be expressed in the form

$$\sum_{k=1}^{N} \tilde{\rho}_{2k-1}\gamma_{2k-1|j}\beta_{2k-1|j} = 0,$$

with $\gamma_{j|j'}\beta_{j|j'} \equiv \sinh \alpha_{j|j'}$. One thus gets, just from geometrical considerations, conservation laws relating the brane energies densities and velocities before and after the collision point. These results apply to any collision of branes in vacuum, with the appropriate symmetries of homogeneity and isotropy. An interesting development would be to extend the analysis to branes with small perturbations and investigate whether one can find scenarios which can produce quasi–scale invariant adiabatic spectra, as seems required by current observations.

On the other hand, in heteroric M-theory the regime where the distance between the branes becomes small corresponds to the regime where the string coupling constant becomes small and therefore a perturbative heterotic treatment may be available. In particular for adiabatic processes the resulting small instanton transition has been thoroughly studied. Here we would like to present an analysis of such a collision and of the possible outcome of such a collision. A natural and intuitive phenomenon which may occur during a collision is the existence of a cosmological bounce. Such objects are not avalaible in 4d under mild assumptions, and therefore can be exhibited as a purely extra dimensional signature [BB03].

We will describe a nD theory with a scalar field and gravity whose solutions present a cosmological singularity at $t = 0$. It turns out that this model is the low energy approximation of a purely $(n + 1)$D model where the extra dimension is an interval with two boundary branes. The singularity corresponds to the brane collision. In the $(n + 1)$D picture, one can extend the motion of the branes past each other, hence providing a continuation of the brane motion after the collision. The $(n + 1)$D space–time is equivalent to an orbifold where the identification between space–time points is provided by a Lorentz boost. These spaces are the simplest possible space–times with a singularity. As with ordinary spatial orbifolds, one may try to define string theory in such backgrounds and analyse the stringy resolution of the singularity. Unfortunately, these orbifolds are not stable in general relativity ruling them out as candidate stringy backgrounds. Let us now briefly outline some of the arguments.

We have already investigated the moduli space approximation for models with a bulk scalar field. Here we will consider that at low energy the moduli space consists of a single scalar field ϕ coupled to gravity

$$S = \int d^n x \sqrt{-g} \left(R - \frac{1}{2} (\partial \phi)^2 \right) . \tag{7.147}$$

Cosmological solutions with

$$ds^2 = a^2(t)[-dt^2 + dx^i dx_i]$$

can be easily obtained

$$a = a(1)|t|^{\frac{1}{n-2}}, \ \phi = \phi(1) + \epsilon \sqrt{\frac{2(n-1)}{n-2}} \ln |t|, \tag{7.148}$$

where $\epsilon = \pm 1$. There are two branches corresponding to $t < 0$ and $t > 0$ connected by a singularity at $t = 0$.

So what is the extra dimensional origin of a such a model? One can uplift the previous system to $(n+1)$ dimensions by defining

$$\psi = \mathrm{e}^{\gamma \phi} \qquad \text{and} \qquad \bar{g}_{\mu\nu} = \psi^{-4/(n-2)} g_{\mu\nu},$$

where $\gamma = \sqrt{(n-2)/8(n-1)}$. Consider now the purely gravitational $(n+1)$ dimensional theory with the metric

$$ds^2_{n+1} = \psi^4 dw^2 + \bar{g}_{\mu\nu} dx^\mu dx^\nu,$$

where $w \in [0, 1]$. The two boundaries at $w = 0$ and $w = 1$ are boundary branes. The dimensional reduction on the interval $w \in [0, 1]$, i.e., integrating over the extra dimension, yields the effective action (7.147) provided one restricts the two fields $\psi(x^\mu)$ and $\bar{g}_{\mu\nu}(x^\mu)$ to be dependent on nD only.

Let us now consider the nature of $(n+1)$ D space–time obtained from the solutions (7.148). The $(n+1)$D metric becomes

$$ds^2_{n+1} = B^2 t^2 dw^2 + \eta_{\mu\nu} dx^\mu dx^\nu,$$

for a given B depending on the integrations constants $\phi(1)$ and $a(1)$. The geometry of space–time is remarkably simple. It is a direct product $R^{n-1} \times M$ where M is the two dimensional compactified Milne space whose metric is

$$ds^2_M = -dt^2 + B^4 t^2 dw^2.$$

Using the light cone coordinates

$$x^{\pm} = \pm t \mathrm{e}^{\pm B^2 w},$$

the metric of Milne space reads

$$ds_M^2 = dx_+ dx_-,$$

coinciding with the two dimensional Minkowski metric. There is one subtlety here, the original identification of the extra–dimensional interval is here transcribed in the fact that Milne space is modded out by the boost

$$x^{\pm} \to \mathrm{e}^{\pm 2B^2} x^{\pm},$$

as we have identified the interval with S^1/Z_2 and the boundary branes are the fixed points of the Z_2 action as in the RS model.

The two boundary branes collide at $x^{\pm} = 0$, their trajectories are given by

$$x_0^{\pm} = \pm t, \qquad x_1^{\pm} = \pm t \mathrm{e}^{\pm B^2}.$$

At the singularity one can hope that the branes go past each other and evolve henceforth. Unfortunatly, Horowitz and Polchinski have shown that the structure of the orbifold space–time is unstable [HP02]. By considering a particle in this geometry, they showed that space–time collapses to a space–like singularity. Indeed one can focus on a particle and its nth image under the orbifold action. In terms of collision the impact parameter b becomes constant as n grows while the center of mass energy $\sqrt{s}$ grows like $\cosh nB^2$. As soon as n is large enough,

$$G\sqrt{s} > b^{n-2},$$

the two particle approach each other within their Schwarzschild radii therefore forming a black hole through gravitational collapse. So the orbifold space–time does not make sense in general relativity, i.e., not defining a time–dependent background for string theory.

Hence, it seems that the most simple example of brane collision needs to be modified in order to provide a working example of singularity with a meaningful string theoretic resolution. It would be extremely relevant if one could find examples of stable backgrounds of string theory where a cosmological singularity can be resolved using stringy arguments. A particularly promising avenue is provided by S–branes where a cosmological singularity is shielded by a horizon [BB03]. Time will certainly tell which of these approaches could lead to a proper understanding of cosmological singularities and their resolutions, an issue highly relevant to brane cosmology both in the early universe and the recent past.

Open Questions

In this section we have reviewed different aspects of brane cosmology in a hopefully pedagogical manner reflecting our own biased point of view. Let us finally summarize some of the open questions:

- In the case of the single brane model by Randall & Sundrum, the homogeneous cosmological evolution is well understood. An unsolved issue in this model, however, is a complete understanding of the evolution of cosmological perturbations. The effects of the bulk gravitational field, encoded in the projected Weyl–tensor, on CMB physics and Large Scale Structures are not known. The problem is twofold: first, the bulk equation are partial, nonlinear differential equations and second, boundary conditions on the brane have to be imposed. The current formalism have not yet been used in order to tackle these problems (for perturbations in brane world theories, see [RVS02]).
- Models with bulk scalar fields: Although we have presented some results on the cosmological evolution of a homogeneous brane, we assumed that the bulk scalar field does not strongly vary around the brane. Clearly, this needs to be investigated in some detail through a detailed investigation of the bulk equations, presumably with the help of numerical methods. Furthermore, for models with two branes, the cosmology has to be explored also in the high energy regime, in which the moduli–space approximation is not valid. Some exact cosmological solutions have been found in [LOS00].
- Both the bulk scalar field as well as the inter–brane distance in two brane models could play an important role at least during some part of the cosmological history. Maybe one of the fields plays the role of dark energy. In that case, it is only natural that masses of particles vary, as well as other parameter, such as the fine structure constant α_{em} [BBD01]. Details of this interesting proposal have yet to be worked out.
- The bulk singularity, which was thought to be shielded away with the help of a second brane, seems to play an important role in a cosmological setting. We have seen that the negative tension brane moves towards the bulk singularity and eventually hits it. Therefore, cosmology forces us to think about this singularity, even if it was shielded with a second brane. Cosmologically, the brane might be repelled, which might be described by a potential. Alternatively, one might take quantum corrections into account in form of a Gauss–Bonnett term in the bulk.
- Brane collisions provide a different singularity problem in brane cosmology. String theory has to make progress in order to understand this singularity as well. From the cosmological point of view, the question is, if a transition between the brane collision can provide a new cosmological era and how cosmological perturbations evolve before and after the bounce.

For more details, see [BB03, Lan02].

7.3.9 Hawking's Brane New World

In this subsection, following [HHR00], we present S. Hawking's Brane New World.

As seen above, Randall and Sundrum have suggested [RS99b] that 4D gravity may be recovered in the presence of an infinite fifth dimension provided that we live on a domain wall embedded in anti–de Sitter space (AdS). Their linearized analysis showed that there is a massless bound state of the graviton associated with such a wall as well as a continuum of massive Kaluza–Klein modes.[42]

[42] Recall that *Kaluza–Klein theory* (KK) is a model that seeks to unify the two fundamental forces of gravitation and electromagnetism. The theory was first published in 1921 and was discovered by the mathematician T. Kaluza who extended general relativity to a 5D space–time. The resulting equations can be separated out into further sets of equations, one of which is equivalent to Einstein field equations, another set equivalent to Maxwell's equations for the electromagnetic field and the final part an extra scalar field now termed the *radion*.

In modern geometry, the extra fifth dimension can be understood to be the circle group $U(1)$, as electromagnetism can essentially be formulated as a gauge theory on a fiber bundle, the circle bundle, with gauge group $U(1)$. Once this geometrical interpretation is understood, it is relatively straightforward to replace $U(1)$ by a general Lie group, to get *Yang–Mills theories*. If a distinction is drawn, then it is that Yang–Mills theories occur on a flat space–time, whereas KK treats the more general case of curved space–time. The base space of KK need not be 4D space–time; it can be any (pseudo)Riemannian manifold, or even a supersymmetric manifold or *orbifold* or even a *noncommutative space*.

As an approach to the unification of the forces, it is straightforward to apply the KK in an attempt to unify gravity with the strong and electroweak forces by using the symmetry group of the Standard Model, $SU(3)\times SU(2)\times U(1)$. However, a naive attempt to convert this interesting geometrical construction into a bona-fide model of reality founders on a number of issues, including the fact that the fermions must be introduced in an artificial way (in nonsupersymmetric models). A less problematic approach to the unification of the forces is taken by modern string theory and M–theory. Nonetheless, KK remains an important touchstone in theoretical physics and is often embedded in more sophisticated theories. It is studied in its own right as an object of geometric interest in K–theory.

Even in the absence of a completely satisfying theoretical physics framework, the idea of exploring extra, compactified, dimensions is of considerable interest in the experimental physics and astrophysics communities. A variety of predictions, with real experimental consequences, can be made (in the case of large extra dimensions/warped models). For example, on the simplest of principles, one might expect to have standing waves in the extra compactified dimension(s). If an extra dimension is of radius R, the energy of such a standing wave would (naively) be $E = nhc/R$ with n an integer, h being Planck's constant and c the speed of light. This set of possible energy values is often called the *Kaluza–Klein tower*.

To build the Kaluza–Klein theory, one picks an invariant metric on the circle S^1 that is the fiber of the $U(1)-$bundle of electromagnetism. Suppose this metric gives the circle a total length of Λ. One then considers metrics $\hat{g}$ on the bundle P

that are consistent with both the fiber metric, and the metric on the underlying manifold M. The consistency conditions are:

(i) the projection of $\hat{g}$ to the vertical subspace $\mathrm{Vert}_p P \subset T_p P$ needs to agree with metric on the fiber over a point in the manifold M, and

(ii) the projection of $\hat{g}$ to the horizontal subspace $\mathrm{Hor}_p P \subset T_p P$ of the tangent space at point $p \in P$ must be isomorphic to the metric g on M at $\pi(p)$.

The *Kaluza–Klein action* for such a metric is given by .

$$S(\widehat{g}) = \int_P R(\widehat{g}) \ \mathrm{vol}(\widehat{g}).$$

The scalar curvature, written in components, then expands to

$$R(\widehat{g}) = \pi^* \left(R(g) - \frac{\Lambda^2}{2} |F|^2 \right),$$

where π^* is the *pull–back* of the fiber–bundle projection $\pi : P \to M$. The connection A on the fiber bundle is related to the electromagnetic field strength as

$$\pi^* F = dA.$$

That there always exists such a connection, even for fiber bundles of arbitrarily complex topology, is a result from homology and specifically, *K–theory*. Applying *Fubini's Theorem* and integrating on the fiber, one gets

$$S(\widehat{g}) = \Lambda \int_M \left(R(g) - \frac{1}{\Lambda^2} |F|^2 \right) \ \mathrm{vol}(g).$$

Varying the action with respect to the component A, one regains the Maxwell equations. Applying the variational principle to the base metric g, one gets the *Einstein equation*

$$R_{ij} - \frac{1}{2} g_{ij} R = \frac{1}{\Lambda^2} T_{ij},$$

with the *Maxwell stress–energy tensor* being given by

$$T^{ij} = F^{ik} F^{jl} g_{kl} - \frac{1}{4} g^{ij} |F|^2.$$

The original theory identifies Λ with the fiber metric g_{55}, and allows Λ to vary from fiber to fiber. In this case, the coupling between gravity and the electromagnetic field is not constant, but has its own dynamical field, the radion.

In the above, the size of the loop Λ acts as a coupling constant between the gravitational field and the electromagnetic field. If the base manifold is 4D, the *Kaluza–Klein manifold* P is 5D. The fifth dimension is a *compact space*, and is called the compact dimension. The phenomenon of having a higher-dimensional manifold where some of the dimensions are compact is referred to as *compactification*.

The above development generalizes in a more-or-less straightforward fashion to general principal $G-$bundles for some arbitrary Lie group G taking the place of $U(1)$. In such a case, the theory is often referred to as a *Yang–Mills theory*, and is sometimes taken to be synonymous. If the underlying manifold is supersymmetric, the resulting theory is a *supersymmetric Yang–Mills theory*.

RS used *horospherical coordinates* based on slicing AdS into flat hypersurfaces. An issue that has not received much attention so far is the role of boundary conditions at these Cauchy horizons in AdS. With stationary perturbations, one can impose the boundary conditions that the horizons remain regular. Indeed, without this boundary condition the solution for stationary perturbations is not well defined. Even for non-perturbative departures from the RS solution, like black holes, one can impose the boundary condition that the AdS horizons remain regular [CHR00, EHM00]. Non–stationary perturbations on the domain wall, however, will give rise to gravitational waves that cross the horizons. This will tend to focus the null geodesic generators of the horizon, which will mean that they will intersect each other on some caustic. Beyond the caustic, the null geodesics will not lie in the horizon. However, null geodesic generators of the future event horizon cannot have a future endpoint [HE73] and so the endpoint must lie to the past. We conclude that if the past and future horizons remain non–singular when perturbed (as required for a well–defined boundary condition) then they must intersect at a finite distance from the wall. By contrast, the past and future horizons don't intersect in the RS ground state but go off to infinity in AdS.

The RS horizons are like the horizons of extreme black holes. When considering perturbations of black holes, one normally assumes that radiation can flow across the future horizon but that nothing comes out of the past horizon. This is because the past horizon isn't really there, and should be replaced by the collapse that formed the black hole. To justify a similar boundary condition on the Randall-Sundrum past horizon, one needs to consider the initial conditions of the universe.

The main contender for a theory of initial conditions is the *no boundary proposal* [HH83] that the quantum state of the universe is given by a Euclidean path integral over compact metrics. The simplest way to implement this proposal for the Randall Sundrum idea is to take the Euclidean version of the wall to be a four sphere at which two balls of AdS_5 are joined together. In other words, take two balls in AdS_5, and glue them together along their four sphere boundaries. The result is topologically a five sphere, with a delta function of curvature on a 4D domain wall separating the two hemispheres. If one analytically continues to Lorentzian signature, one gets a 4D de Sitter hyperboloid, embedded in Lorentzian anti de Sitter space.

The past and future RS horizons, are replaced by the past and future light cones of the points at the centers of the two balls. Note that the past and future horizons now intersect each other and are non extreme, which means they are stable to small perturbations. A perfectly spherical Euclidean domain wall will give rise to a 4D Lorentzian universe that expands forever in an inflationary manner.[43]

[43] Such inflationary brane-world solutions have been studied in [CR99, Kal99, Nih99].

In order for a spherical domain wall solution to exist, the tension of the wall must be larger than the value assumed by RS, who had a flat domain wall. We shall assume that matter on the wall increases its effective tension, permitting a spherical solution. Below we consider a strongly coupled large N CFT on the domain wall. On a spherical domain wall, the conformal anomaly of the CFT increases the effective tension of the domain wall, making the spherical solution possible. The Lorentzian geometry is a de Sitter universe with the conformal anomaly driving inflation.

The no boundary proposal allows one to calculate unambiguously the graviton correlator on the domain wall. In particular, the Euclidean path integral itself uniquely specifies the allowed fluctuation modes, because perturbations that have infinite Euclidean action are suppressed in the path integral. Therefore, in this framework, there is no need to impose by hand an additional, external prescription for the vacuum state for each perturbation mode. In addition, the AdS/CFT correspondence allows a fully quantum mechanical treatment of the CFT, in contrast with the usual classical treatment of matter fields in inflationary cosmology.

Finally, we analytically continue the Euclidean correlator into the Lorentzian region, where it describes the spectrum of quantum mechanical vacuum fluctuations of the graviton field on an inflating domain wall with conformally invariant matter living on it. We find that the quantum loops of the large N CFT give space–time a rigidity that strongly suppresses metric fluctuations on small scales. Since any matter would be expected to behave like a CFT at small scales, this result probably extends to any inflationary model with sufficiently many matter fields. It has long been known that matter loops lead to short distance modifications of gravity. Our work shows that these modifications can lead to observable consequences in an inflationary scenario.

Although we have carried out our calculations for the RS model, we shall show that results for 4D Einstein gravity coupled to the CFT can be recovered by taking the domain wall to be large compared with the AdS scale. Thus our conclusion that metric fluctuations are suppressed holds independently of the RS scenario.

The spherical domain wall considered in this section analytically continues to a Lorentzian de Sitter universe that inflates forever. However, Starobinsky [Sta80] showed that the conformal anomaly driven de Sitter phase is unstable to evolution into a matter dominated universe. If such a solution could be obtained from a Euclidean instanton then it would have an $O(4)-$symmetry group, rather than the $O(5)-$symmetry of a spherical instanton.

The AdS/CFT correspondence [Mal98, GKP98, Wit98b] provides an explanation of the RS behavior.[44] It relates the RS model to an equivalent 4D theory consisting of general relativity coupled to a strongly interacting conformal field theory and a logarithmic correction. Under certain circumstances,

[44] This was first pointed out in unpublished remarks of Maldacena and Witten.

the effects of the CFT and logarithmic term are negligible and pure gravity is recovered.

RS Scenario from AdS/CFT

The AdS/CFT correspondence [Mal98, GKP98, Wit98b] relates IIB supergravity theory in $AdS_5 \times S^5$ to a $\mathcal{N} = 4$ $U(N)$ superconformal field theory. If g_{YM} is the coupling constant of this theory then the *'t Hooft parameter* is defined to be $\lambda = g_{YM}^2 N$. The CFT parameters are related to the supergravity parameters by [Mal98]

$$l = \lambda^{1/4} l_s, \qquad \frac{l^3}{G} = \frac{2N^2}{\pi}, \tag{7.149}$$

where l_s is the string length, l the AdS radius and G the 5D Newton constant. Note that λ and N must be large in order for stringy effects to be small. The CFT lives on the conformal boundary of AdS_5. The correspondence takes the following form [HHR00]:

$$Z[\mathbf{h}] \equiv \int \mathcal{D}[\mathbf{g}] \exp(-S_{grav}[\mathbf{g}]) = \int \mathcal{D}[\phi] \exp(-S_{CFT}[\phi; \mathbf{h}]) \equiv \exp(-W_{CFT}[\mathbf{h}]), \tag{7.150}$$

here $Z[\mathbf{h}]$ denotes the supergravity partition function in AdS_5. This is given by a path integral over all metrics in AdS_5 which induce a given conformal equivalence class of metrics $\mathbf{h}$ on the conformal boundary of AdS_5. The correspondence relates this to the generating functional W_{CFT} of connected Green's functions for the CFT on this boundary. This functional is given by a path integral over the fields of the CFT, denoted schematically by ϕ. Other fields of the supergravity theory can be included on the left hand side; these act as sources for operators of the CFT on the right hand side.

A problem with equation (7.150) as it stands is that the usual gravitational action in AdS is divergent, rendering the path integral ill–defined. A procedure for solving this problem was developed in [Wit98b, TL98, HS98, BK99, EJM99, KLS99]. First one brings the boundary into a finite radius. Next one adds a finite number of counterterms to the action in order to render it finite as the boundary is moved back off to infinity. These counterterms can be expressed solely in terms of the geometry of the boundary. The total gravitational action for AdS_{n+1} becomes

$$S_{grav} = S_{EH} + S_{GH} + S_1 + S_2 + \dots .$$

The first term is the usual Einstein–Hilbert action[45] with a negative cosmological constant:

---

[45] We use a positive signature metric and a curvature convention for which a sphere has positive Ricci scalar.

$$S_{EH} = -\frac{1}{16\pi G}\int d^{n+1}x\sqrt{g}\left(R + \frac{n(n-1)}{l^2}\right)$$

the overall minus sign arises because we are considering a Euclidean theory. The second term in the action is the *Gibbons–Hawking boundary*term, which is necessary for a well–defined variational problem [GH77]:

$$S_{GH} = -\frac{1}{8\pi G}\int d^n x\sqrt{h}K,$$

where K is the trace of the extrinsic curvature of the boundary[46] and h the determinant of the induced metric. The first two counterterms are given by the following [BK99, EJM99, KLS99] (we use the results of [KLS99] rotated to Euclidean signature)

$$S_1 = \frac{n-1}{8\pi Gl}\int d^n x\sqrt{h}, \qquad \text{and} \qquad S_2 = \frac{l}{16\pi G(n-2)}\int d^n x\sqrt{h}R,$$

where R now refers to the Ricci scalar of the boundary metric. The third counterterm is

$$S_3 = \frac{l^3}{16\pi G(n-2)^2(n-4)}\int d^n x\sqrt{h}\left(R_{ij}R^{ij} - \frac{n}{4(n-1)}R^2\right), \qquad (7.151)$$

where R_{ij} is the Ricci tensor of the boundary metric and boundary indices i, j are raised and lowered with the boundary metric h_{ij}. This expression is ill–defined for $n = 4$, which is the case of most interest to us. With just the first two counter–terms, the gravitational action exhibits logarithmic divergences [TL98, HS98] so a third term is needed. This term cannot be written solely in terms of a polynomial in scalar invariants of the induced metric and curvature tensors; it makes explicit reference to the cut–off (i.e.,the finite radius to which the boundary is brought before taking the limit in which it tends to infinity). The form of this term is the same as (7.151) with the divergent factor of $1/(n-4)$ replaced by $\log(R/\rho)$, where R measure the boundary radius and ρ is some finite renormalization length scale.

Following [GKP98], we can now use the AdS/CFT correspondence to explain the behavior discovered by Randall and Sundrum. The (Euclidean) RS model has the following action:

$$S_{RS} = S_{EH} + S_{GH} + 2S_1 + S_m.$$

Here $2S_1$ is the action of a domain wall with tension $(n-1)/(4\pi Gl)$. The final term is the action for any matter present on the domain wall. The domain wall tension can cancel the effect of the bulk cosmological constant to produce a

[46] Our convention is the following. Let n denotes the outward unit normal to the boundary. The extrinsic curvature is defined as $K_{\mu\nu} = h^\rho_\mu h^\sigma_\nu \nabla_\rho n_\sigma$, where $h^\nu_\mu = \delta^\nu_\mu - n_\mu n^\nu$ projects quantities onto the boundary.

flat domain wall. However, we are interested in a spherical domain wall so we assume that the matter on the wall gives an extra contribution to the effective tension. We shall discuss a specific candidate for the matter on the wall later on. The wall separates two balls B_1 and B_2 of AdS.

We want to study quantum fluctuations of the metric on the domain wall. Let $\mathbf{g}_0$ denote the 5D background metric we have just described and $\mathbf{h}_0$ the metric it induces on the wall. Let $\mathbf{h}$ denote a metric perturbation on the wall. If we wish to calculate correlators of $\mathbf{h}$ on the domain wall then we are interested in a path integral of the form[47] [HHR00]

$$\langle h_{ij}(x)h_{i'j'}(x')\rangle = \int \mathcal{D}[\mathbf{h}]Z[\mathbf{h}]\, h_{ij}(x)h_{i'j'}(x'), \qquad \text{where}$$

$$\begin{aligned} Z[\mathbf{h}] &= \int_{B_1\cup B_2} \mathcal{D}[\delta\mathbf{g}]\mathcal{D}[\phi]\exp(-S_{RS}[\mathbf{g}_0+\delta\mathbf{g}]) \\ &= \exp(-2S_1[\mathbf{h}_0+\mathbf{h}]) \\ &\times \int_{B_1\cup B_2} \mathcal{D}[\delta\mathbf{g}]\mathcal{D}[\phi]\exp(-S_{EH}[\mathbf{g}_0+\delta\mathbf{g}] - S_{GH}[\mathbf{g}_0+\delta\mathbf{g}] - S_m[\phi;\mathbf{h}_0+\mathbf{h}]), \end{aligned}$$

$\delta\mathbf{g}$ denotes a metric perturbation in the bulk that approaches $\mathbf{h}$ on the boundary and ϕ denotes the matter fields on the domain wall. The integrals in the two balls are independent so we can replace the path integral by

$$\begin{aligned} Z[\mathbf{h}] &= \exp(-2S_1[\mathbf{h}_0+\mathbf{h}])\left(\int_B \mathcal{D}[\delta\mathbf{g}]\exp(-S_{EH}[\mathbf{g}_0+\delta\mathbf{g}] - S_{GH}[\mathbf{g}_0+\delta\mathbf{g}])\right)^2 \\ &\times \int \mathcal{D}[\phi]\exp(-S_m[\phi;\mathbf{h}_0+\mathbf{h}]), \end{aligned}$$

where B denotes either ball. We now take $n = 4$ and use the AdS/CFT correspondence (7.150) to replace the path integral over $\delta\mathbf{g}$ by the generating functional for a conformal field theory:

$$\begin{aligned} &\int_B \mathcal{D}[\delta\mathbf{g}]\exp(-S_{EH}[\mathbf{g}_0+\delta\mathbf{g}] - S_{GH}[\mathbf{g}_0+\delta\mathbf{g}]) \\ &\quad = \exp(-W_{RS}[\mathbf{h}_0+\mathbf{h}] + S_1[\mathbf{h}_0+\mathbf{h}] + S_2[\mathbf{h}_0+\mathbf{h}] + S_3[\mathbf{h}_0+\mathbf{h}]), \end{aligned}$$

we shall refer to this CFT as the RS–CFT since it arises as the dual of the RS geometry. It has gauge group $U(N_{RS})$, where N_{RS} is given by equation (7.149). Strictly speaking, we are using an extended form of the AdS/CFT conjecture, which asserts that supergravity theory in a finite region of AdS

[47] In principle, we should worry about gauge fixing and ghost contributions to the gravitational action. A convenient gauge to use in the bulk is transverse traceless gauge. We shall only deal with metric perturbations that also appear transverse and traceless on the domain wall. The gauge fixing terms vanish for such perturbations and the ghosts only couple to these perturbations at higher orders.

is dual to a CFT on the boundary of that region with an ultraviolet cut-off related to the radius of the boundary. The path integral for the metric perturbation becomes

$$Z[\mathbf{h}] = \exp(-2W_{RS}[\mathbf{h}_0+\mathbf{h}]+2S_2[\mathbf{h}_0+\mathbf{h}]+2S_3[\mathbf{h}_0+\mathbf{h}]) \int \mathcal{D}[\phi] \exp(-S_m[\phi;\mathbf{h}_0+\mathbf{h}]).$$

The RS model has been replaced by a CFT and a coupling to matter fields and the domain wall metric given by the action

$$-2S_2[\mathbf{h}_0 + \mathbf{h}] - 2S_3[\mathbf{h}_0 + \mathbf{h}] + S_m[\phi; \mathbf{h}_0 + \mathbf{h}].$$

The remarkable feature of this expression is that the term $-2S_2$ is precisely the (Euclidean) Einstein–Hilbert action for 4D gravity with a Newton constant given by the RS value $G_4 = G/l$. Therefore the RS model is equivalent to 4D gravity coupled to a CFT with corrections to gravity coming from the third counter term. This explains why gravity is trapped to the domain wall.

At first sight this appears rather amazing. We started off with a quite complicated 5D system and have argued that it is dual to 4D Einstein gravity with some corrections and matter fields. However in order to use this description, we have to know how to calculate with the RS–CFT. At present, the only way we know of doing this is via AdS/CFT, i.e., going back to the 5D description. The point of the AdS/CFT argument is to explain why the RS 'alternative to compactification' works and also to explain the origin of the corrections to Einstein gravity in the RS model. Note that if the matter on the domain wall dominates the RS–CFT and the third counterterm then these can be neglected and a purely 4D description is adequate.

CFT on the Domain Wall

Long ago, Starobinsky studied the cosmology of a universe containing conformally coupled matter [Sta80]. CFTs generally exhibit a conformal anomaly when coupled to gravity (for a review, see [Duf94]). Starobinsky gave a de Sitter solution in which the anomaly provides the cosmological constant. By analyzing homogeneous perturbations of this model, he showed that the de Sitter phase is unstable but could be long lived, eventually decaying to a FRW cosmology.

Here we will consider the RS analogue of Starobinsky's model by putting a CFT on the domain wall. On a spherical domain wall, the conformal anomaly provides the extra tension required to satisfy the Israel equations. It is appealing to choose the new CFT to be a $\mathcal{N} = 4$ superconformal field theory because then the AdS/CFT correspondence makes calculations relatively easy[48]. This

[48] We emphasize that this use of the AdS/CFT correspondence is independent of the use described above because this new CFT is unrelated to the RS CFT [HHR00].

requires that the CFT is strongly coupled, in contrast with Starobinsky's analysis[49].

Our 5D Euclidean action is the following [HHR00]:

$$S = S_{EH} + S_{GH} + 2S_1 + W_{CFT}. \qquad (7.152)$$

We seek a solution in which two balls of AdS_5 are separted by a spherical domain wall. Inside each ball, the metric can be written

$$ds^2 = l^2(dy^2 + \sinh^2 y d\Omega_n^2), \qquad \text{with} \quad 0 \leq y \leq y_0.$$

The domain wall is at $y = y_0$ and has radius: $R = l \sinh y_0$. The effective tension of the domain wall is given by the Israel equations as

$$\sigma_{eff} = \frac{3}{4\pi G l} \coth y_0.$$

The actual tension of the domain wall is $\sigma = 3/(4\pi G l)$. We therefore need a contribution to the effective tension from the CFT. This is provided by the conformal anomaly, which takes the value [TL98, HS98]

$$\langle T \rangle = -\frac{3N^2}{8\pi^2 R^4},$$

This contributes an effective tension $-\langle T \rangle/4$. We can now get an equation for the radius of the domain wall:

$$\frac{R^3}{l^3}\sqrt{\frac{R^2}{l^2} + 1} = \frac{N^2 G}{8\pi l^3} + \frac{R^4}{l^4}. \qquad (7.153)$$

It is easy to see that this has a unique positive solution for R.

We are particularly interested in how perturbations of this model would appear to inhabitants of the domain wall. Thus we are interested in metric perturbations on the sphere

$$ds^2 = (R^2 \hat{\gamma}_{ij} + h_{ij})\, dx^i dx^j.$$

Here $\hat{\gamma}_{ij}$ is the metric on a unit $n-$sphere. We shall only consider *tensor perturbations*, for which h_{ij} is transverse and traceless with respect to $\hat{\gamma}_{ij}$. In order to calculate correlators of the metric perturbation, we need to know the action to second order in the perturbation. The most difficult part here is obtaining W_{CFT} to second order.

[49] Note that the conformal anomaly is the same at strong and weak coupling [HS98] so any differences arising from strong coupling can only show up when we perturb the system.

CFT Generating Function

We want to work out the effect of the perturbation on the CFT on the sphere. To do this we use AdS/CFT. Introduce a fictional AdS region that fills in the sphere. Let $\bar{l}, \bar{G}$ be the AdS radius and Newton constant of this region. We emphasize that this region has nothing to do with the regions of AdS that 'really' lie inside the sphere in the RS scenario. This new AdS region is bounded by the sphere. If we take $\bar{l}$ to zero then the sphere is effectively at infinity in AdS so we can use AdS/CFT to calculate the generating functional of the CFT on the sphere. In other words, $\bar{l}$ is acting like a cut–off in the CFT and taking it to zero corresponds to removing the cut–off. However the relation

$$\frac{\bar{l}^3}{\bar{G}} = \frac{2N^2}{\pi}, \tag{7.154}$$

implies that if $\bar{l}$ is taken to zero then we must also take $\bar{G}$ to zero since N is fixed (and large).

For the unperturbed sphere, the metric in the new AdS region is

$$ds^2 = \bar{l}^2(dy^2 + \sinh^2 y \, \hat{\gamma}_{ij} \, dx^i dx^j),$$

and the sphere is at $y = y_0$ given by $R = \bar{l} \sinh y_0$. Note that $y_0 \to \infty$ as $\bar{l} \to 0$ since R is fixed. In order to use AdS/CFT for the perturbed sphere, we need to know how the perturbation extends into the bulk. This is done by solving the linearized Einstein equations. It is always possible to choose a gauge in which the bulk metric perturbation takes the form

$$h_{ij}(y, x) \, dx^i dx^j,$$

where h_{ij} is transverse and traceless with respect to the metric on the spherical spatial sections:

$$\hat{\gamma}^{ij}(x) h_{ij}(y, x) = \hat{\nabla}^i h_{ij}(y, x) = 0,$$

with $\hat{\nabla}$ denoting the covariant derivative defined by the metric $\hat{\gamma}_{ij}$. Since we are only dealing with tensor perturbations, this choice of gauge is consistent with the boundary sitting at constant y. If scalar metric perturbations were included then we would have to take account of a perturbation in the position of the boundary.

The linearized Einstein equations in the bulk are (for any dimension) [HHR00]

$$\nabla^2 h_{\mu\nu} = -\frac{2}{\bar{l}^2} h_{\mu\nu}, \tag{7.155}$$

where μ, ν are $n + 1$D indices. It is convenient to expand the metric perturbation in terms of tensor spherical harmonics $H_{ij}^{(p)}(x)$. These obey

$$\hat{\gamma}^{ij} H_{ij}^{(p)}(x) = \hat{\nabla}^i H_{ij}^{(p)}(x) = 0,$$

and they are tensor eigenfunctions of the Laplacian:

$$\hat{\nabla}^2 H_{ij}^{(p)} = (2 - p(p+n-1))\, H_{ij}^{(p)},$$

where $p = 2, 3, \ldots$. We have suppressed extra labels $k, l, m, \ldots$ on these harmonics. The harmonics are orthonormal with respect to the obvious inner product. See [Hig87] for more details of their properties. The metric perturbation can be written as a sum of separable perturbations of the form

$$h_{ij}(y, x) = f_p(y) H_{ij}^{(p)}(x).$$

Substituting this into equation (7.155) gives

$$f_p''(y) + (n-4)\coth y f_p'(y) - (2(n-2) + (p(p+n-1) + 2(n-3))\mathrm{cosech}^2 y) f_p(y) = 0. \tag{7.156}$$

The roots of the indicial equation are $p+2$ and $-p-n+3$, yielding two linearly independent solutions for each p. In order to compute the generating functional W_{CFT} we have to calculate the Euclidean action of these solutions. However, because the latter solution goes as $y^{-(p+n-3)}$ at the origin $y = 0$ of the instanton, the corresponding fluctuation modes have infinite Euclidean action[50]. Hence they are suppressed in the path integral. Therefore, in contrast to the methods where one requires a (rather *ad hoc*) prescription for the vacuum state of each perturbation mode, there is no need to impose boundary conditions by hand in our approach: the Euclidean path integral defines its own boundary conditions, which automatically gives a unique Green function. The path integral unambiguously specifies the allowed fluctuation modes as those which vanish at $y = 0$. Note that boundary conditions at the origin in Euclidean space replace the need for boundary conditions at the horizon in Lorentzian space. The solution regular at $y = 0$ is given by

$$f_p(y) = \frac{\sinh^{p+2} y}{\cosh^p y} F(p/2, (p+1)/2, p + (n+1)/2, \tanh^2 y).$$

This solution can also be written in terms of associated Legendre functions:

$$f_p(y) \propto (\sinh y)^{(5-n)/2} P_{-(n+1)/2}^{-(p+(n-1)/2)}(\cosh y) \propto (\sinh y)^{(4-n)/2} Q_{p+(n-2)/2}^{n/2}(\coth y), \tag{7.157}$$

and the latter can be related to Legendre functions if $n/2$ is an integer, using

$$Q_\nu^m(z) = (z^2 - 1)^{m/2} \frac{n^m Q_\nu}{dz^m}.$$

The full solution for the metric perturbation is

---

[50] This can be seen by surrounding the origin by a small sphere $y = \epsilon$ and calculating the surface terms in the action that arise on this sphere. They are the same as the surface terms in equations (7.160) and (7.161) below, which are obviously divergent for the modes in question.

$$h_{ij}(y,x) = \sum_p \frac{f_p(y)}{f_p(y_0)} H_{ij}^{(p)}(x) \int d^n x' \sqrt{\hat{\gamma}} h^{kl}(x') H_{kl}^{(p)}(x'). \tag{7.158}$$

We have a solution for the metric perturbation throughout the bulk region. The AdS/CFT correspondence can now be used to give the generating functional of the CFT on the perturbed sphere:

$$W_{CFT} = S_{EH} + S_{GH} + S_1 + S_2 + \dots. \tag{7.159}$$

We shall give the terms on the right hand side for $n = 4$.

The Einstein–Hilbert action with cosmological constant is

$$S_{EH} = -\frac{1}{16\pi \bar{G}} \int d^5 x \sqrt{g} \left(R + \frac{12}{\bar{l}^2} \right),$$

and perturbing this gives

$$\begin{aligned} S_{bulk} = &-\frac{1}{16\pi \bar{G}} \int d^5 x \sqrt{g} \left(-\frac{8}{\bar{l}^2} + \frac{1}{4} h^{\mu\nu} \nabla^2 h_{\mu\nu} + \frac{1}{2\bar{l}^2} h^{\mu\nu} h_{\mu\nu} \right) \\ &- \frac{1}{16\pi \bar{G}} \int d^4 x \sqrt{\gamma} \left(-\frac{1}{2} n^\mu h^{\nu\rho} \nabla_\nu h_{\mu\rho} + \frac{3}{4} h_{\nu\rho} n^\mu \nabla_\mu h^{\nu\rho} \right), \end{aligned}$$

where Greek indices are 5D and we are raising and lowering with the unperturbed 5D metric. $n = ldy$ is the unit normal to the boundary and ∇ is the covariant derivative defined with the unperturbed bulk metric. $\gamma_{ij} = R^2 \hat{\gamma} n_{ij}$ is the unperturbed boundary metric. It is important to keep track of all the boundary terms arising from integration by parts. Evaluating on shell gives [HHR00]

$$\begin{aligned} &S_{EH} = \\ &\frac{\bar{l}^3}{2\pi \bar{G}} \int d^4 x \sqrt{\hat{\gamma}} \int_0^{y_0} dy \sinh^4 y - \frac{\bar{l}^3}{16\pi \bar{G}} \int d^4 x \sqrt{\hat{\gamma}} \left(\frac{3}{4\bar{l}^4} h^{ij} \partial_y h_{ij} - \frac{\coth y_0}{\bar{l}^4} h^{ij} h_{ij} \right). \end{aligned} \tag{7.160}$$

where we are now raising and lowering with $\hat{\gamma}_{ij}$. The Gibbons–Hawking term is

$$S_{GH} = -\frac{\bar{l}^3}{2\pi \bar{G}} \int d^4 x \sqrt{\hat{\gamma}} \left(\sinh^3 y_0 \cosh y_0 - \frac{1}{8\bar{l}^4} h^{ij} \partial_y h_{ij} \right). \tag{7.161}$$

The first counter term is

$$\begin{aligned} S_1 &= \frac{3}{8\pi \bar{G} \bar{l}} \int d^4 x \sqrt{\gamma} \\ &= \frac{3\bar{l}^3}{8\pi \bar{G}} \int d^4 x \sqrt{\hat{\gamma}} \left(\sinh^4 y_0 - \frac{1}{4\bar{l}^4} h^{ij} h_{ij} \right). \end{aligned} \tag{7.162}$$

The second counter term is

$$
\begin{aligned}
S_2 &= \frac{\bar{l}}{32\pi\bar{G}}\int d^4x\sqrt{\gamma}R \\
&= \frac{\bar{l}^3}{32\pi\bar{G}}\int d^4x\sqrt{\hat{\gamma}}\left(12\sinh^2 y_0 - \frac{2}{\bar{l}^4\sinh^2 y_0}h^{ij}h_{ij} + \frac{1}{4\bar{l}^4\sinh^2 y_0}h^{ij}\hat{\nabla}^2 h_{ij}\right).
\end{aligned}
$$

Thus with only two counter terms we would have

$$
\begin{aligned}
W_{CFT} &= \frac{3N^2\Omega_4}{8\pi^2}\log\frac{R}{\bar{l}} \\
&- \frac{\bar{l}^3}{16\pi\bar{G}}\int d^4x\sqrt{\hat{\gamma}}\left(-\frac{1}{4\bar{l}^4}h^{ij}\partial_y h_{ij} + \frac{1}{\bar{l}^4}h^{ij}h_{ij}\left(\frac{3}{2} - \sqrt{1+\frac{\bar{l}^2}{R^2}}\right)\right. \\
&\left. + \frac{1}{\bar{l}^2R^2}h^{ij}h_{ij} - \frac{1}{8\bar{l}^2R^2}h^{ij}\hat{\nabla}^2 h_{ij}\right).
\end{aligned}
$$

Ω_4 is the area of a unit 4–sphere and we have used equation (7.154). The expansion of $\partial_y h_{ij}$ at $y = y_0$ is obtained from

$$
\partial_y h_{ij} = \sum_p \frac{f_p'(y_0)}{f_p(y_0)}H_{ij}^{(p)}(x)\int d^4x'\sqrt{\hat{\gamma}}h^{kl}(x')H_{kl}^{(p)}(x') \qquad \text{and}
$$

$$
\begin{aligned}
\frac{f_p'(y_0)}{f_p(y_0)} &= 2 + \frac{\bar{l}^2}{2R^2}(p+1)(p+2) + p(p+1)(p+2)(p+3)\frac{\bar{l}^4}{4R^4}\log(\bar{l}/R) \\
&+ \frac{\bar{l}^4}{8R^4}\left[p^4 + 2p^3 - 5p^2 - 10p - 2 - p(p+1)(p+2)(p+3)(\psi(1)+\psi(2)\right. \\
&\left. - \psi(p/2+2) - \psi(p/2+5/2))\right] + \mathcal{O}\left(\frac{\bar{l}^6}{R^6}\log(\bar{l}/R)\right).
\end{aligned}
$$

The psi function is defined by $\psi(z) = \Gamma'(z)/\Gamma(z)$. Substituting into the action we find that the divergences as $\bar{l} \to 0$ cancel at order $R^4/\bar{l}^4$ and $R^2/\bar{l}^2$. The term of order $\bar{l}^4/R^4$ in the above expansion makes a contribution to the finite part of the action:

$$
\begin{aligned}
W_{CFT} &= \frac{3N^2\Omega_4}{8\pi^2}\log\frac{R}{\bar{l}} + \\
\frac{N^2}{256\pi^2R^4}&\sum_p\left(\int d^4x'\sqrt{\hat{\gamma}}h^{kl}(x')H_{kl}^{(p)}(x')\right)^2\left(2p(p+1)(p+2)(p+3)\log(\bar{l}/R) + \Psi(p)\right),
\end{aligned}
$$

where

$$
\begin{aligned}
\Psi(p) &= p(p+1)(p+2)(p+3)\left[\psi(p/2+5/2) + \psi(p/2+2) - \psi(2) - \psi(1)\right] \\
&+ p^4 + 2p^3 - 5p^2 - 10p - 6.
\end{aligned}
$$

To cancel the logarithmic divergences as $\bar{l} \to 0$, we have to introduce a length scale ρ defined by $\bar{l} = \epsilon\rho$ and add a counter term proportional to $\log\epsilon$ to cancel the divergence as ϵ tends to zero. The counter term is

$$S_3 = -\frac{\bar{l}^3}{64\pi\bar{G}} \log\epsilon \int d^4x\sqrt{\gamma}\left(\gamma^{ik}\gamma^{jl}R_{ij}R_{kl} - \frac{1}{3}R^2\right)$$
$$= -\frac{\bar{l}^3}{64\pi\bar{G}} \log\epsilon \int d^4x\sqrt{\hat{\gamma}}\left(-12 + \frac{1}{R^4}\left[2h^{ij}h_{ij} - \frac{3}{2}h^{ij}\hat{\nabla}^2 h_{ij} + \frac{1}{4}h^{ij}\hat{\nabla}^4 h_{ij}\right]\right).$$

This term does indeed cancel the logarithmic divergence, leaving us with

$$W_{CFT} = \frac{3N^2\Omega_4}{8\pi^2}\log\frac{R}{\rho} + \tag{7.163}$$
$$\frac{N^2}{256\pi^2 R^4}\sum_p \left(\int d^4x'\sqrt{\hat{\gamma}}h^{kl}(x')H^{(p)}_{kl}(x')\right)^2 (2p(p+1)(p+2)(p+3)\log(\rho/R) + \Psi(p))$$

Note that varying W_{CFT} twice with respect to h_{ij} yields the expression for the transverse traceless part of the correlator $\langle T_{ij}(x)T_{i'j'}(x')\rangle$ on a round four sphere. At large p, this behaves like $p^4\log p$, as expected from the flat space result [GKP98]. In fact this correlator can be determined in closed form solely from the trace anomaly and symmetry considerations. However, we shall be be interested in calculating cosmologically observable effects, for which our mode expansion is more useful.

The Total Action

Recall that our 5D action is

$$S = S_{EH} + S_{GH} + 2S_1 + W_{CFT}.$$

In order to calculate correlators of the metric, we need to evaluate the path integral [HHR00]

$$Z[\mathbf{h}] = \int_{B_1\cup B_2} \mathcal{D}[\delta\mathbf{g}]\exp(-S) =$$
$$\exp(-2S_1[\mathbf{h}_0+\mathbf{h}] - W_{CFT}[\mathbf{h}_0+\mathbf{h}])\left(\int_B \mathcal{D}[\delta\mathbf{g}]\exp(-S_{EH}[\mathbf{g}_0+\delta\mathbf{g}] - S_{GH}[\mathbf{g}_0+\delta\mathbf{g}])\right)^2.$$

Here $\mathbf{g}_0$ and $\mathbf{h}_0$ refer to the unperturbed background metrics in the bulk and on the wall respectively and $\mathbf{h}$ denotes the metric perturbation on the wall. Many of the terms required here can be obtained from above results by simply replacing $\bar{l}$ and $\bar{G}$ with l and G. For example, from equation (7.162) we get

$$S_1[\mathbf{h}_0+\mathbf{h}] = \frac{3l^3}{8\pi G}\int d^4x\sqrt{\hat{g}}\left(\sinh^4 y_0 - \frac{1}{4l^4}\right),$$

where y_0 is defined by $R = l\sinh y_0$. The path integral over $\delta\mathbf{g}$ is performed by splitting it into a classical and quantum part:

$$\delta\mathbf{g} = \mathbf{h} + \mathbf{h}',$$

where the boundary perturbation $\mathbf{h}$ is extended into the bulk using the linearized Einstein equations and the requirement of finite Euclidean action, i.e., $\mathbf{h}$ is given in the bulk by equation (7.158). $\mathbf{h}'$ denotes a quantum fluctuation that vanishes at the domain wall. The gravitational action splits into separate contributions from the classical and quantum parts:

$$S_{EH} + S_{GH} = S_0[\mathbf{h}] + S'[\mathbf{h}'],$$

where S_0 can be read off from equations (7.160) and (7.161) as

$$S_0 = -\frac{3l^3\Omega_4}{2\pi G}\int_0^{y_0} dy \sinh^2 y_0 \cosh^2 y_0 + \frac{l^3}{16\pi G}\int d^4x\sqrt{\hat{\gamma}}\left(\frac{1}{4l^4}h^{ij}\partial_y h_{ij} + \frac{\coth y_0}{l^4}h^{ij}h_{ij}\right),$$

Note that S' cannot be converted to a surface term since $\mathbf{h}'$ does not satisfy the Einstein equations. We shall not need the explicit form for S' since the path integral over $\mathbf{h}'$ just contributes a factor of some determinant Z_0 to $Z[\mathbf{h}]$. We get

$$Z[\mathbf{h}] = Z_0 \exp(-2S_0[\mathbf{h}_0 + \mathbf{h}] - 2S_1[\mathbf{h}_0 + \mathbf{h}] - W_{CFT}[\mathbf{h}_0 + \mathbf{h}]).$$

The exponent is given by

$$\begin{aligned} 2S_0 + 2S_1 + W_{CFT} &= -\frac{3l^3\Omega_4}{\pi G}\int_0^{y_0} dy \sinh^2 y \cosh^2 y + \frac{3\Omega_4 R^4}{4\pi G l} + \frac{3N^2\Omega_4}{8\pi^2}\log\frac{R}{\rho} \\ &+ \frac{1}{l^4}\sum_p \left(\int d^4x'\sqrt{\hat{\gamma}}h^{kl}(x')H^{(p)}_{kl}(x')\right)^2 \left[\frac{l^3}{32\pi G}\left(\frac{f_p'(y_0)}{f_p(y_0)} + 4\coth y_0 - 6\right)\right. \\ &+ \left.\frac{N^2}{256\pi^2\sinh^4 y_0}\left(2p(p+1)(p+2)(p+3)\log(\rho/R) + \Psi(p)\right)\right]. \end{aligned}$$

We have kept the unperturbed action in order to demonstrate how the conformal anomaly arises: it is simply the coefficient of the $\log(R/\rho)$ term divided by the area $\Omega_4 R^4$ of the sphere. If we set the metric perturbation to zero and vary R (using $R = l \sinh y_0$) then we reproduce equation (7.153).

Having calculated R, we can now choose a convenient value for the renormalization scale ρ. If we were dealing purely with the CFT then we could keep ρ arbitrary. However, since the third counter term involves the square of the Weyl tensor (the integrand is proportional to the difference of the Euler density and the square of the Weyl tensor), we can expect pathologies to arise if this term is present when we couple the CFT to gravity. In other words, when coupled to gravity, different choices of ρ lead to different theories. We shall choose the value $\rho = R$ so that the third counter term exactly cancels the divergence in the CFT, with no finite remainder and hence no residual curvature squared terms in the action.

The (Euclidean) graviton correlator can be read off from the action as [HHR00]

$$\langle h_{ij}(x)h_{i'j'}(x')\rangle = \frac{128\pi^2 R^4}{N^2}\sum_{p=2}^{\infty} W^{(p)}_{iji'j'}(x,x')F(p,y_0)^{-1} \tag{7.164}$$

where we have eliminated l^3/G using equation (7.153). The function $F(p,y_0)$ is given by

$$F(p,y_0) = \mathrm{e}^{y_0}\sinh y_0\left(\frac{f_p'(y_0)}{f_p(y_0)} + 4\coth y_0 - 6\right) + \Psi(p),$$

and the bitensor $W^{(p)}_{iji'j'}(x,x')$ is defined as

$$W^{(p)}_{iji'j'}(x,x') = \sum_{k,l,m,\ldots} H^{(p)}_{ij}(x)H^{(p)}_{i'j'}(x'),$$

with the sum running over all the suppressed labels $k,l,m,\ldots$ of the tensor harmonics.

The appearance of N^2 in the denominator in equation (7.164) suggests that the CFT suppresses metric perturbations on all scales. This is misleading because R also depends on N. The function $F(p,y_0)$ has the following limiting forms for large and small radius:

$$\lim_{y_0\to\infty} F(p,y_0) = \Psi(p) + p^2 + 3p + 6, \tag{7.165}$$

$$\lim_{y_0\to 0} F(p,y_0) = \Psi(p) + p + 6. \tag{7.166}$$

$F(p,y_0)$ has poles at $p = -4,-5,-6,\ldots$ with zeros between each pair of negative integers starting at $-3,-4$. When we analytically continue to Lorentzian signature, we shall be particularly interested in zeros lying in the range $p \geq -3/2$. There is one such zero exactly at $p = 0$, another near $p = 0$ and a third near $p = -3/2$. For large radius, these extra zeros are at $p \approx -0.054$ and $p \approx -1.48$ while for small radius they are at $p \approx 0.094$ and $p \approx -1.60$. For intermediate radius they lie between these values, with the zeros crossing through $-3/2$ and 0 at $y_0 \approx 0.632$ and $y_0 \approx 1.32$ respectively.

Comparison with 4D Gravity

Above we have discussed how the RS scenario reprodues the predictions of 4D gravity when the effects of matter on the domain wall dominates the effects of the RS–CFT. In our case we have a CFT on the domain wall. This has action proportional to N^2. The RS–CFT is a similar CFT (but with a cut–off) and therefore has action proportional to N_{RS}^2. Hence we can neglect it when $N \gg N_{RS}$. The logarithmic counterterm is also proportional to N_{RS}^2 and therefore also negligible. We therefore expect the predictions of 4D gravity to be recovered when $N \gg N_{RS}$. We shall now demonstrate this explicitly.

First consider the radius R of the domain wall given by equation (7.153). It is convenient to write this in terms of the rank N_{RS} of the RS–CFT (given by $l^3/G = 2N_{RS}^2/\pi$)

$$\frac{R^3}{l^3}\sqrt{\frac{R^2}{l^2}+1} = \frac{N^2}{16N_{RS}^2} + \frac{R^4}{l^4}.$$

If we assume $N \gg N_{RS} \gg 1$ then the solution is

$$\frac{R}{l} = \frac{N}{2\sqrt{2}N_{RS}}\left[1 + \frac{N_{RS}^2}{N^2} + \mathcal{O}(N_{RS}^4/N^4)\right].$$

Note that this implies $R \gg l$, i.e., the domain wall is large compared with the anti–de Sitter length scale.

Now let's turn to a 4D description in which we are considering a four sphere with no interior. The only matter present is the CFT. The metric is simply

$$ds^2 = R_4^2\hat{\gamma}_{ij}dx^i dx^j,$$

where R_4 remains to be determined. The action is the 4D Einstein–Hilbert action (without cosmological constant) together with W_{CFT}. There is no Gibbons–Hawking term because there is no boundary. Without a metric perturbation, the action is simply [HHR00]

$$S = -\frac{1}{16\pi G_4}\int d^4x\sqrt{\gamma}R + W_{CFT} = -\frac{3\Omega_4 R_4^2}{4\pi G_4} + \frac{3N^2\Omega_4}{8\pi^2}\log\frac{R_4}{\rho}.$$

where G_4 is the 4D Newton constant. We want to calculate the value of R_4 so we can't choose $\rho = R_4$ yet. Varying R_4 gives

$$R_4^2 = \frac{N^2 G_4}{4\pi},$$

and N is large hence R_4 is much greater than the 4D Planck length. Substituting $G_4 = G_5/l$, this reproduces the leading order value for R found above from the 5D calculation.

We can now go further and include the metric perturbation. The perturbed 4D Einstein–Hilbert action is

$$S_{EH}^{(4)} = -\frac{1}{16\pi G_4}\int d^4x\sqrt{\hat{\gamma}}\left(12R_4^2 - \frac{2}{R_4^2}h^{ij}h_{ij} + \frac{1}{4R_4^2}h^{ij}\hat{\nabla}^2 h_{ij}\right). \qquad (7.167)$$

Adding the perturbed CFT gives

$$S = -\frac{3N^2\Omega_4}{16\pi^2} + \frac{3N^2\Omega_4}{8\pi^2}\log\frac{R_4}{\rho} + \sum_p\left(\int d^4x'\sqrt{\hat{\gamma}}h^{kl}(x')H_{kl}^{(p)}(x')\right)^2$$

$$\left[\frac{1}{64\pi G_4 R_4^2}(p^2+3p+6) + \frac{N^2}{256\pi^2 R_4^4}\left(2p(p+1)(p+2)(p+3)\log(\rho/R_4) + \Psi(p)\right)\right].$$

Setting $\rho = R_4$, we find that the graviton correlator for a 4D universe containing the CFT is

$$\langle h_{ij}(x)h_{i'j'}(x')\rangle = 8N^2G_4^2 \sum_{p=2}^{\infty} W^{(p)}_{iji'j'}(x,x') \left[p^2 + 3p + 6 + \Psi(p)\right]^{-1}. \quad (7.168)$$

This can be compared with the expression obtained from the 5D calculation, which can be written

$$\langle h_{ij}(x)h_{i'j'}(x')\rangle = \frac{8N^2G^2}{l^2}\left[1+\mathcal{O}(N_{RS}^2/N^2)\right] \sum_{p=2}^{\infty} W^{(p)}_{iji'j'}(x,x') \left[p^2 + 3p + 6 + \Psi(p)\right.$$
$$\left. + \, 4p(p+1)(p+2)(p+3)(N_{RS}^2/N^2)\log(N_{RS}/N) + \mathcal{O}(N_{RS}^2/N^2)\right]^{-1}.$$

We have expanded in terms of

$$\frac{N_{RS}^2}{N^2} = \frac{\pi l^3}{2N^2 G}.$$

The four and 5D expressions clearly agree (for $G_4 = G/l$) when $N \gg N_{RS}$, i.e., $R \gg l$. There are corrections of order $(N_{RS}^2/N^2)\log(N_{RS}/N)$ coming from the RS–CFT and the logarithmic counter term. In fact, these corrections can be absorbed into the renormalization of the CFT on the domain wall if, instead of choosing $\rho = R$, we choose

$$\rho = R\left(1 - \frac{2N_{RS}^2}{N^2}\log(N_{RS}/N)\right).$$

The corrections to the 4D expression are then of order N_{RS}^2/N^2. We shall not give these correction terms explicitly although they are easily obtained from the exact result (7.164).

Lorentzian Correlator

Here we will show how the Euclidean correlator calculated above is analytically continued to give a correlator for Lorentzian signature [HHT00]. Let us first introduce a new label $p' = i(p + 3/2)$, so that on the four sphere [HHR00]

$$\hat{\nabla}^2 H^{(p')}_{ij} = \lambda_{p'} H^{(p')}_{ij}, \qquad \text{where} \quad p' = 7i/2, 9i/2, \ldots \qquad \text{and}$$
$$\lambda_{p'} = (p'^2 + 17/4).$$

Recall that there are extra labels on the tensor harmonics that we have suppressed. The set of rank–two tensor eigenmodes on S^4 forms a representation of the symmetry group of the manifold. Hence the sum of the degenerate eigenfunctions with eigenvalue $\lambda_{p'}$ defines a maximally symmetric bitensor $W_{(p')}{}^{ij}{}_{i'j'}(\mu(\Omega,\Omega'))$, where $\mu(\Omega,\Omega')$ is the distance along the shortest geodesic between the points with polar angles Ω and Ω'.

The motivation for the unusual labelling is that in terms of the label p' the bitensor on S^4 has exactly the same formal expression as the corresponding bitensor on Lorentzian de Sitter space. This property will enable us to analytically continue the Euclidean correlator into the Lorentzian region without Fourier decomposing it. In other words, instead of imposing by hand a prescription for the vacuum state of the graviton on each mode separately and propagating the individual modes into the Lorentzian region, we compute the two–point tensor correlator in real space, directly from the no boundary path integral. Since the path integral unambiguously specifies the allowed fluctuation modes as those which vanish at the origin of the instanton, this automatically gives a unique Euclidean correlator. The technical advantage of our method is that dealing directly with the real space correlator makes the derivation independent of the gauge ambiguities involved in the mode decomposition [HHT00].

We begin by continuing the graviton correlator (equation (7.164)) obtained via the 5D calculation. The analytic continuation of the correlator for 4D gravity (equation (7.168)) is completely analogous. In terms of the new label p', the Euclidean correlator (7.164) between two points on the wall is given by [HHR00]

$$\langle h_{ij}(\Omega)h_{i'j'}(\Omega')\rangle = \frac{128\pi^2R^4}{N^2}\sum_{p'=7i/2}^{i\infty} W^{(p')}_{iji'j'}(\mu)G(p',y_0)^{-1}, \qquad \text{where} \tag{7.169}$$

$$\begin{aligned} G(p',y_0) &= F(-ip'-3/2,y_0)\\ &= \mathrm{e}^{y_0}\sinh y_0\left(\frac{g'_{p'}(y_0)}{g_{p'}(y_0)}+4\coth y_0-6\right)+\big(p'^4-4ip'^3+p'^2/2-5ip'-63/16\\ &+(p'^2+1/4)(p'^2+9/4)[\psi(-ip'/2+5/4)+\psi(-ip'/2+7/4)-\psi(1)-\psi(2)]\big),\\ &\text{with}\quad g_{p'}(y)=Q^2_{-ip'-1/2}(\coth y), \end{aligned}$$

which follows from (7.157). The function $G(p',y_0)$ is real and positive for all values of p' in the sum and for arbitrary $y_0\geq 0$.

We have the Euclidean correlator defined as an infinite sum. However, the eigenspace of the Laplacian on de Sitter space suggests that the Lorentzian propagator is most naturally expressed as an integral over real p'. We must therefore first analytically continue our result from imaginary to real p'. The coefficient $G(p',y_0)^{-1}$ of the bitensor is analytic in the upper half complex p'-plane, apart from three simple poles on the imaginary axis. One of them is always at $p'=3i/2$, regardless of the radius of the sphere. Let the position of the remaining two poles be written $p'_k=i\Lambda_k(y_0)$. If we take the radius of the domain wall to be large compared with the AdS scale (which is necessary for corrections to 4D Einstein gravity to be small) then[51] $0<\Lambda_k\leq 3/2$,

[51] If we decrease the radius of the domain wall, then the poles move away from each other. Their behavior follows from the discussion below equations 7.165 and

with $\Lambda_1 \sim 0$ and $\Lambda_2 \sim 3/2$. Since $G(p', y_0)$ is real on the imaginary p'-axis, the residues at these poles are purely imaginary. In order to extend the correlator into the complex p'-plane, we must also understand the continuation of the bitensor itself. The condition of regularity at opposite points on the four sphere imposed by the completeness relation is sufficient to uniquely specify the analytic continuation of $W^{(p')}_{iji'j'}(\mu)$ into the complex p'–plane.

Now we are able to write the sum in equation (7.169) as an integral along a contour $\mathcal{C}_1$ encircling the points $p' = 7i/2, 9i/2, ..ni/2$, where n tends to infinity. This yields

$$\langle h_{ij}(\Omega)h_{i'j'}(\Omega')\rangle = \frac{-i64\pi^2 R^4}{N^2}\int_{\mathcal{C}_1} dp' \tanh p'\pi W^{(p')}_{iji'j'}(\mu) G(p', y_0)^{-1}. \tag{7.170}$$

Since we know the analytic properties of the integrand in the upper half complex p'–plane, we can distort the contour for the p' integral to run along the real axis. At large imaginary p' the integrand decays and the contribution vanishes in the large n limit. However as we deform the contour towards the real axis, we encounter three extra poles in the $\cosh p'\pi$ factor, the pole at $p' = 3i/2$ becoming a double pole due to the simple zero of $G(p', y_0)$. In addition, we have to take in account the two poles of $G(p', y_0)^{-1}$ at $p' = i\Lambda_k$.

For the $p' = 5i/2$ pole, it follows from the normalization of the tensor harmonics that $W^{(5i/2)}_{iji'j'} = 0$. Indirectly, this is a consequence of the fact that spin–2 perturbations do not have a dipole or monopole component. The meaning of the remaining two poles of the $\tanh p'\pi$ factor has been extensively discussed in [HHT00], where the continuation is described of the two–point tensor fluctuation correlator from a 4D–$O(5)$ instanton into open de Sitter space. They represent non–physical contributions to the graviton propagator, arising from the different nature of tensor harmonics on S^4 and on Lorentzian de Sitter space. In fact, a degeneracy appears between $p'_t = 3i/2$ and $p'_t = i/2$ tensor harmonics and respectively $p'_v = 5i/2$ vector harmonics and $p'_s = 5i/2$ scalar harmonics on S^4. More precisely, the tensor harmonics that constitute the bitensors $W^{iji'j'}_{(3i/2)}$ and $W^{iji'j'}_{(i/2)}$ can be constructed from a vector (scalar) quantity. Consequently, the contribution to the correlator from the former pole is pure gauge, while the latter eigenmode should really be treated as a scalar perturbation, using the perturbed scalar action. Henceforth we shall exclude them from the tensor spectrum. This leaves us with the poles of $G(p', y_0)$ at $p' = i\Lambda_k$. If we deform the contour towards the real axis, we must compensate for them by subtracting their residues from the integral. We will see that these residues correspond to discrete 'super–curvature' modes in the Lorentzian tensor correlator.

The contribution from the closing of the contour in the upper half p'–plane vanishes. Hence our final result for the Euclidean correlator reads [HHR00]

7.166. For $y_0 \leq 0.632$, Λ_1 becomes slightly smaller than zero while for $y_0 \leq 1.32$, Λ_2 becomes slightly greater than 3/2.

$$\langle h_{ij}(\Omega)h_{i'j'}(\Omega')\rangle = \frac{-i64\pi^2R^4}{N^2}\Bigg[\int_{-\infty}^{+\infty} dp' \tanh p'\pi W^{(p')}_{iji'j'}(\mu)G(p',y_0)^{-1} + 2\pi\sum_{k=1}^{2}\tan\Lambda_k\pi W^{(i\Lambda_k)}_{iji'j'}(\mu)\mathbf{Res}(G(p',y_0)^{-1};i\Lambda_k)\Bigg].$$

The analytic continuation from a four sphere into Lorentzian closed de Sitter space is given by setting the polar angle $\Omega = \pi/2 - it$. Without loss of generality we may take $\mu = \Omega$, and μ then continues to $\pi/2 - it$. We then get the correlator in de Sitter space where one point has been chosen as the origin of the time coordinate.

The extra factor $ie^{p\pi}/\sinh p'\pi$ combines with the factor $-i\tanh p'\pi$ in the integrand to $e^{p'\pi}/\cosh p'\pi$. Furthermore, since $G(-p',y_0) = \bar{G}(p',y_0)$, we can rewrite the correlator as an integral from 0 to ∞. We finally get the Lorentzian tensor Feynman (time–ordered) correlator,

$$\begin{aligned}\langle h_{ij}(x)h_{i'j'}(x')\rangle = {}& \frac{128\pi^2R^4}{N^2}\Bigg[\int_{0}^{+\infty} dp' \tanh p'\pi W^{L(p')}_{iji'j'}(\mu)\Re(G(p',y_0)^{-1}) \\ & + \pi\sum_{k=1}^{2}\tan\Lambda_k\pi W^{L(i\Lambda_k)}_{iji'j'}(\mu)\mathbf{Res}(G(p',y_0)^{-1};i\Lambda_k)\Bigg] \\ & + i\frac{128\pi^2R^4}{N^2}\Bigg[\int_{0}^{+\infty} dp' W^{L(p')}_{iji'j'}(\mu)\Re(G(p',y_0)^{-1}) \\ & - \pi\sum_{k=1}^{2} W^{L(i\Lambda_k)}_{iji'j'}(\mu)\mathbf{Res}(G(p',y_0)^{-1};i\Lambda_k)\Bigg].\end{aligned}$$

In this integral the bitensor $W^{L(p')}_{iji'j'}(\mu(x,x'))$ may be written as the sum of the degenerate rank–two tensor harmonics on closed de Sitter space with eigenvalue $\lambda_{p'} = (p'^2 + 17/4)$ of the Laplacian. Note that the normalization factor $\tilde{Q}_{p'} = p'(4p'^2 + 25)/48\pi^2$ of the bitensor is imaginary at $p' = i\Lambda_k$ and the residues of G^{-1} are also imaginary, so the quantities in square brackets are all real. Both integrands in this equation vanish as $p' \to 0$, so the correlator is well-behaved in the infrared.

For cosmological applications, one is usually interested in the expectation of some quantity squared, like the microwave background multipole moments. For this purpose, all that matters is the symmetrized correlator, which is just the real part of the Feynman correlator.

Gravitational waves provide an extra source of time–dependence in the background in which the cosmic microwave background photons propagate. In particular, the contribution of gravitational waves to the CMB anisotropy is given by the integral in the *Sachs–Wolfe formula*, which is basically the integral along the photon trajectory of the time derivative of the tensor perturbation. Hence the resulting microwave multipole moments C_l can be directly determined from the graviton correlator.

We can therefore understand the effect of the strongly coupled CFT on the microwave fluctuation spectrum. On the 4–sphere, this is easily obtained by varying the Einstein–Hilbert action with a cosmological constant. In terms of the bitensor, this yields [HHR00]

$$\langle h_{ij}(\Omega) h_{i'j'}(\Omega')\rangle = 32\pi G_4 R^2 \sum_{p'=7i/2}^{i\infty} \frac{W^{(p')}_{iji'j'}(\mu(\Omega,\Omega'))}{\lambda_{p'}-2},$$

which continues to

$$\langle h_{ij}(x) h_{i'j'}(x')\rangle = 32\pi G_4 R^2 \int_0^{+\infty} \frac{dp'}{\lambda_{p'}-2} W^{L(p')}_{iji'j'}(\mu(x,x')).$$

Note that (apart from the pole at $p' = 3i/2$ corresponding to the gauge mode mentioned before) there are no supercurvature modes.

Summary on Hawking's Brane New World

We have studied a Randall–Sundrum cosmological scenario consisting of a domain wall in anti–de Sitter space with a large N conformal field theory living on the wall. The conformal anomaly of the CFT provides an effective tension which leads to a de Sitter geometry for the domain wall. We have computed the spectrum of quantum mechanical vacuum fluctuations of the graviton field on the domain wall, according to Euclidean no boundary initial conditions. The Euclidean path integral unambiguously specifies the tensor correlator with no additional assumptions. This is the first calculation of quantum fluctuations for RS cosmology.

In the usual inflationary models, one considers the classical action for a single scalar field. In that context, it is consistent to neglect quantum matter loops, on the grounds that they are small. On the other hand, in this section we have studied a strongly coupled large N CFT living on the domain wall, for which quantum loops of matter are important. By using the AdS/CFT correspondence, we have performed a fully quantum mechanical treatment of this CFT. The most notable effect of the large N CFT on the tensor spectrum is that it suppresses small scale fluctuations on the microwave sky. It can be seen that the CFT yields a $(p'^4 \ln p')^{-1}$ behavior for the graviton propagator at large p' (in agreement with the flat space results of [Tom77]), instead of the usual p'^{-2} falloff. In other words, quantum loops of the CFT give space–time a rigidity that strongly suppresses metric fluctuations on small scales. Note that this is true independently of how the de Sitter geometry arises, i.e.,it is also true for 4D Einstein gravity. In addition, the coupling of the CFT to tensor perturbatons gives rise to two additional discrete modes in the tensor spectrum. Although this is a novel feature in the context of inflationary tensor perturbations, it is not surprising. In conventional open inflationary scenarios for instance, the coupling of scalar field fluctuations with scalar metric perturbations introduces a supercurvature mode with an eigenvalue of the Laplacian close to the discrete de Sitter gauge mode [YST96]. The former discrete mode at $p' = i\Lambda_1 \sim 3i/2$ is the analogue of this well known supercurvature mode in the scalar fluctuation spectrum. The second mode has an eigenvalue $p' = i\Lambda_2 \sim 0$. Its interpretation is less clear, but it is clearly an effect of the matter on the domain wall. However it hardly contributes to the correlator because $\tan \Lambda_2 \pi$ is very small.

The effect of the CFT on large scales is more difficult to quantify because of the complicated p'–dependence of the tensor correlator in the low–p' regime. Generally speaking, however, long–wavelength tensor correlations in closed (or open) models for inflation are very sensitive to the details of the underlying theory, as well as to the boundary conditions at the instanton. Since tensor fluctuations do give a substantial contribution to the large scale CMB anisotropies, this may provide an additional way to observationally distinguish different inflationary scenarios.

Most matter fields can be expected to behave like a CFT at small scales. Furthermore, fundamental theories such as string theory predict the existence of a large number of matter fields. Therefore, our results based on a quantum treatment of a large N CFT may be accurate at small scales for any matter. If this is the case then our result shows that tensor perturbations at small angular scales are much smaller than predicted by calculations that neglect quantum effects of matter fields.

Cosmic Inner Flight

We will finish this Chapter on cosmic dynamics by quoting the words of Paramhansa Yogananda [Yog46], depicting the echo of the Big–Bang occurring 13.7 billion years ago:

> "My body became immovably rooted; breath was drawn out of my lungs as if by some huge magnet. Soul and mind instantly lost their physical bondage, and streamed out like a fluid piercing light from my every pore. The flesh was as though dead, yet in my intense awareness I knew that never before had I been fully alive. My sense of identity was no longer narrowly confined to a body, but embraced the circumambient atoms. People on distant streets seemed to be moving gently over my own remote periphery... The whole vicinity lay bare before me. My ordinary frontal vision was now changed to a vast spherical sight, simultaneously all-perceptive... All objects within my panoramic gaze trembled and vibrated like quick motion pictures... The unifying light alternated with materializations of form, the metamorphoses revealing the law of cause and effect in creation... An oceanic joy broke upon calm endless shores of my soul. The Spirit of God, I realized, is exhaustless Bliss; His body is countless tissues of light. A swelling glory within me began to envelop towns, continents, the earth, solar and stellar systems, tenuous nebulae, and floating

universes. The entire cosmos, gently luminous, like a city seen afar at night, glimmered within the infinitude of my being. The sharply etched global outlines faded somewhat at the farthest edges; there I could see a mellow radiance, ever–undiminished. It was indescribably subtle; the planetary pictures were formed of a grosser light... The divine dispersion of rays poured from an Eternal Source, blazing into galaxies, transfigured with ineffable auras. Again and again I saw the creative beams condense into constellations, then resolve into sheets of transparent flame. By rhythmic reversion, sextillion worlds passed into diaphanous luster; fire became firmament. I cognized the center of the empyrean as a point of intuitive perception in my heart. Irradiating splendor issued from my nucleus to every part of the universal structure. Blissful amrita, the nectar of immortality, pulsed through me with a quick silver–like fluidity. The creative voice of God I heard resounding as AUM, the vibration of the Cosmic Motor."

References

AA67. Atiyah, M.F., Anderson, D.W.: $K-$Theory. Benjamin, New York, (1967)

AAM76. Anderson, B.D., Arbib, M.A., Manes, E.G.: Foundations of System Theory: Finitary and Infinitary Conditions. Lecture Notes in Economics and Mathematical Systems Theory, Springer, New York, (1976)

AB84. Atiyah, M.F., Bott, R.: The moment map and equivariant cohomology. Topology, **23**, 1–28, (1984)

ABG48. Alpher, R.A., Bethe, H., Gamow, G.: The Origin of Chemical Elements. Phys. Rev. **73**, 803, (1948)

AC91. Ablowitz, M.J., Clarkson, P.A.: Solitons, nonlinear evolution equations and inverse scattering. London Math. Soc., **149**, CUP, Cambridge, UK, (1991)

AC94. Connes, A.: Noncommutative Geometry, Academic Press, New York, (1994)

ACM98. Ambjørn. J., Carfora, M., Marzuoli, A.: The Geometry of Dynamical Triangulations. Springer, Berlin, (1998)

ADJ97. Ambjørn. J., Durhuus, B., Jonsson, T.: Quantum geometry. Cambridge Monographs on Mathematical Physics, Cambridge Univ. Press, Cambridge, UK, (1997)

AEH05. Ahmed, E., Elgazzar, A.S., Hegazi, A.S.: An Overview of Complex Adaptive Systems. Mansoura J. Math. (to appear)

AFH86. Albeverio, S., Fenstat. J., Hoegh-Krohn, R., Lindstrom, T.: Nonstandard Methods in Stochastic Analysis and Mathematical Physics. Academic Press, New York, (1986)

AG05. Apps, R., Garwicz, M.: Anatomical and physiological foundations of cerebellar information processing. Nature Rev. Neurosci., **6**, 297–311, (2005)

AGM94. Alekseevsky, D.V., Grabowski. J., Marmo, G., Michor, P.W.: Poisson structures on the cotangent bundle of a Lie group or a principle bundle and their reductions. J. Math. Phys., **35**, 4909–4928, (1994)

AGM97. Alekseevsky, D., Grabowksi. J., Marmo, G., Michor, P.W.: Completely integrable systems: a generalization. Mod. Phys. Let. A, **12**(22), 1637–1648, (1997)

AGN94. Antoniadis, I., Gava, E., Narain, K.S., Taylor, T.R.: Topological amplitudes in string theory. Nucl. Phys. B **413**, 162, (1994)

AH61. Atiyah, M.F., Hirzebruch, F.: Vector bundles and homogeneous spaces. Proc. Symp. Pure Math. **3**, 7–38, (1961)

AH88. Atiyah, M.F., Hitchin, N.J.: The geometry and dynamics of magnetic monopoles, Princeton Univ. Press, Princeton, NJ, (1988)

AHC01. Chamseddine, A.H.: Complexified gravity in noncommutative spaces. Comm. Math. Phys. **218**, 283, (2001)

AHS89. Ashtekar, A., Husain, V., Samuel. J., Rovelli, C., Smolin, L.: 2+1 quantum gravity as a toy model for the 3+1 theory, Classical and Quantum Gravity **6**, L185, (1989)

AI92. Ashtekar, A., Isham, C.J.: Representations of the holonomy algebras of gravity and non-Abelian gauge theories. Class. Quant. Grav. **9**, 1433–85, (1992)

AJ90. Atiyah, M.F., Jeffrey, L.: Topological Lagrangians and Cohomology. J. Geom. Phys.**7**, 119, (1990)

AJL00a. Ambjørn. J., Jurkiewicz. J., Loll, R.: Lorentzian and Euclidean quantum gravity – analytical and numerical results. In M-Theory and Quantum Geometry, eds. L. Thorlacius and T. Jonsson, NATO Science Series, Kluwer, 382–449, (2000)

AJL00b. Ambjørn. J., Jurkiewicz. J., Loll, R.: A nonperturbative Lorentzian path integral for gravity. Phys. Rev. Lett. **85**, 924–927, (2000)

AJL01a. Ambjørn. J., Jurkiewicz. J., Loll, R.: Dynamically triangulating Lorentzian quantum gravity. Nucl. Phys. B **610**, 347–382, (2001)

AJL01b. Ambjørn. J., Jurkiewicz. J., Loll, R.: Nonperturbative 3d Lorentzian quantum gravity. Phys. Rev. D **64**, 044–011, (2001)

AJL01c. Ambjørn. J., Jurkiewicz. J., Loll, R.: Dynamically triangulating Lorentzian quantum gravity. Nucl. Phys. B **610**, 347–382, (2001)

AJL01d. Ambjørn. J., Jurkiewicz. J., Loll, R.: Computer simulations of 3d Lorentzian quantum gravity. Nucl. Phys. B **94**, 689–692, (2001)

AK93. Ambjørn. J., Kristjansen, C.F.: Nonperturbative 2D quantum gravity and Hamiltonians unbounded from below. Int. J. Mod. Phys. A **8**, 1259–1282, (1993)

AL04. Ashtekar, A., Lewandowski. J.: Background Independent Gravity: A status report, Class. Quant. Grav. **21**, R53, (2004)

AL05. Achimescu, S., Lipan, O.: Signal Propagation in Nonlinear Stochastic Gene Regulatory Networks. 3rd Int. Conf. Path. Netw., Sys. Rhodes, Greece 2005.

AL91. Aidman, E.V., Leontiev, D.A.: From being motivated to motivating oneself: a Vygotskian perspective. Stud. Sov. Thought, **42**, 137–151, (1991)

AL95. Ashtekar, A., Lewandowski. J.: Projective techniques and functional integration. J. Math. Phys. **36**, 2170, (1995)

AL98. Ambjørn. J., Loll, R.: Non-perturbative Lorentzian quantum gravity, causality and topology change. Nucl. Phys. B **536**, 407–434, (1998)

ALM95. Ashtekar, A., Lewandowski. J., Marolf, D., Mourao. J., Thiemann, T.: Quantization of diffeomorphism invariant theories of connections with local degrees of freedom. J. Math. Phys. **36**, 6456–6493, (1995)

AM78. Abraham, R., Marsden. J.: Foundations of Mechanics. Benjamin, Reading, (1978)

AM90. Akbulut, S., McCarthy. J.D.: Casson's invariant for homological 3−spheres- An exposition, Mathematical Notes **36**, Princeton Univ. Press, Princeton, (1990)

AM91. Aringazin, A., Mikhailov, A.: Matter fields in space–time with vector non–metricity. Clas. Quant. Grav. **8**, 1685, (1991)

AMR88. Abraham, R., Marsden. J., Ratiu, T.: Manifolds, Tensor Analysis and Applications. Springer, New York, (1988)

AN00. Amari, Nagaoka, H.: Methods of Information Geometry, Oxford Univ. Press and Amer. Math. Soc., (2000)

AN99. Aoyagi, T., Nomura, M.: Oscillator Neural Network Retrieving Sparsely Coded Phase Patterns. Phys. Rev. Lett. **83**, 1062–1065, (1999)

AR95. Antoni, M., Ruffo, S.: Clustering and relaxation in long-range Hamiltonian dynamics. Phys. Rev. E, **52**, 2361–2374, (1995)

AS04. Albrecht, A., Sorbo, L.: Can the Universe Afford Inflation? Phys. Rev. D **70**, 063528, (2004)

AS63. Atiyah, M.F., Singer, I.M.: The Index of Elliptic Operators on Compact Manifolds, Bull. Amer. Math. Soc. **69**, 322-433, (1963)

AS68. Atiyah, M.F., Singer, I.M.: The Index of Elliptic Operators I, II, III. Ann. Math. **87**, 484–604, (1968)

AS71. Atiyah, M.F., Segal, G.B.: Exponential isomorphisms for $\lambda-$rings. Quart. J. Math. Oxford Ser. **22**, 371–378, (1971)

AS72. Abramowitz, M., Stegun, I. (Eeds.): Handbook of Mathematical Functions. National Bureau of Standards, (1972)

AS82. Albrecht, A., Steinhardt, P.J.: Cosmology For Grand Unified Theories With Radiatively Induced Symmetry Breaking. Phys. Rev. Lett. **48**, 1220, (1982)

AS92. Abraham, R., Shaw, C.: Dynamics: the Geometry of Behavior. Addison–Wesley, Reading, MA, (1992)

AT05. Aguirre, A., Tegmark, M.: Multiple Universes, Cosmic Coincidences, and Other dark Matters, JCAP **0501**, 003, (2005)

AV00. Alfinito, E., Vitiello, G.: Formation and life–time of memory domains in the dissipative quantum model of brain, Int. J. Mod. Phys. B, **14**, 853–868, (2000)

AW01. Anderson. J.D., Williams. J.G.: Long Range Tests of the Equivalence Principle, Class. Quant. Grav. **18**, 2447, (2001)

AW91. Axelrod, S., della Pietra, S., Witten, E.: Geometric Quantization of Chern-Simons Gauge Theory. J. Diff. Geom. **33**, 787, (1991)

Ada62. Adams. J.F.: Vector fields on spheres. Ann. Math. **75**, 603-632, (1962)

Ada78. Adams. J.F.: Infinite Loop Spaces. Princeton Univ. Press, Princeton, NJ, (1978)

Adl04. Adler, S.L.: Quantum Theory as an Emergent Phenomenon. Cambridge University Press, Cambridge, UK, (2004)

Aka82. Akama, K.: An Early Proposal of Brane World. Lect. Not. Phys. **176**, 267, (1982)

Ale68. Alekseev, V.: Quasi-random dynamical systems I, II, III. Math. USSR Sbor. **5**, 73–128, (1968)

Ale94. Alexander, D.S.: A history of complex dynamics. From Schröder to Fatou and Julia. Aspects of Mathematics. Vieweg E24, (1994)

Alo03. Alon, U.: Biological Networks: The tinkerer as an engineer. Science **301**, 1866-1867, (2003)

Ama85. Amari, S.I.: Differential Geometrical Methods in Statistics. Springer, New York, (1985)

Ame93. Amemiya, Y.: On nonlinear factor analysis. Proc. Social Stat. Section. Ann. Meet. Ame. Stat. Assoc. 290–294, (1993)

And01. Andrecut, M.: Biomorphs, program for $Mathcad^{TM}$, Mathcad Application Files, Mathsoft, (2001)

And64. Anderson, P.W.: Lectures on the Many Body Problem. E.R. Caianiello (ed), Academic Press, New York, (1964)

Arb98. Arbib, M. (ed.): Handbook of Brain Theory and Neural Networks (2nd ed.). MIT Press, Cambridge, MA, (1998)

Ark01. Arkin, A.P.: Synthetic cell biology. Curr. Opin. Biotech., **12**, 638-644, (2001)

Arn78. Arnold, V.I.: Ordinary Differential Equations. MIT Press, Cambridge, MA, (1978)

Arn88. Arnold, V.I.: Geometrical Methods in the Theory of Ordinary differential equations. Springer, New York, (1988)

Arn89. Arnold, V.I.: Mathematical Methods of Classical Mechanics (2nd ed). Springer, New York, (1989)

Arn92. Arnold, V.I.: Catastrophe Theory. Springer, Berlin, (1992)

Arn93. Arnold, V.I.: Dynamical systems. Encyclopaedia of Mathematical Sciences, Springer, Berlin, (1993)

Ash86. Ashtekar, A.: New variables for classical and quantum gravity. Phys. Rev. Lett. **57**, 2244–2247, (1986)

Ash87. Ashtekar, A.: New Hamiltonian formulation of general relativity. Phys. Rev. D **36**, 1587–1602, (1987)

Ash88. Ashtekar, A.: New perspectives in canonical gravity. Bibliopolis, (1988)

Ash91. Ashtekar, A.: Lecture notes on non-perturbative canonical gravity. Advaced Series in Astrophysics and Cosmology, Vol. 6, World Scientific, Singapore, (1991)

Ash94. Ashcraft, M.H.: Human Memory and Cognition (2nd ed). HarperCollins, New York, (1994)

Ash97. Ashby, N.: Relativistic effects in the global positioning system. Plenary lecture on quantum gravity at the GR15 conference, Puna, India, (1997)

Ati87. Atiyah, M.F.: Magnetic Monopoles in hyperbolic spaces. In Proceedings of Bombay Colloquium 1984 on vector bundles in algebraic varieties. Oxford Univ. Press, pp. 1–34, (1987)

Ati88a. Atiyah, M.F.: Topological quantum field theory. Publ. Math. IHES **68**, 175–186, (1988)

Ati88b. Atiyah, M.F.: New invariants of three and four dimensional manifols. In The Mathematical Heritage of Hermann Weyl, eds. R. Well et al., Proc. Symp. Pure. Math. 48, Am. Math. Soc., Providence, (1988)

Ati89. Atiyah, M.F.: The Geometry and Physics of Knots. Cambridge Univ. Press, Cambridge, (1989)

Ati00. Atiyah, M.F.: K–Theory Past and Present. arXiv:math/0012213, (2000)

Aul00. Auletta, G.: Foundations and Interpretations of Quantum Mechanics. World Scientific, Singapore, (2000)

B-Y97. Bar-Yam, Y.: Dynamics of Complex Systems. Perseus Books, Reading, (1997)

BA90. Braam, P.J. Austin, D.M.: Boundary values of hyperbolic monopoles, Nonlinearity 3(3), 809–823, (1990)

BB04. Ben-Bassat, O., Boyarchenko, M.: Submanifolds of generalized complex manifolds. J. Sympl. Geom. **2**(3), 309–355, (2004)

BB98. C. M. Bender and S. Boettcher. Real Spectra in Non-Hermitian Hamiltonians Having PT Symmetry. Phys. Rev. Lett. 80, 5243, (1998)

BBD01. Brax, P., van de Bruck, C., Davis, A.C.: Brane-world cosmology, bulk scalars and perturbations. JHEP **0110**, 026, (2001)

BBJ02. Bender, C.M., Brody, D.C., Jones, H.F.: Complex Extension of Quantum Mechanics. Phys. Rev. Lett.**89**, 270401, (2002)

BBM99. Bender, C.M., Boettcher, S., Meisinger, P.N.: $\mathcal{PT}$-Symmetric Quantum Mechanics. J. Math. Phys. **40**, 2201, (1999)

BBR91. Birmingham, D., Blau, M., Rakowski, M., Thompson, G. Topological field theory. Phys. Rep. **209**, 129, (1991)

BC97. Barret. J., Crane, L.: Relativistic spin networks and quantum gravity. arXiv:gr-qc/9709028, (1997)

BCD06. Bender, C.M, Chen J-H., Daniel W. Darg, D.W., Milton, K.A.: Classical Trajectories for Complex Hamiltonians. arXiv:math-ph/0602040, (2006)

BCG00. Bowcock, P., Charmousis, C., Gregory, R.: General brane cosmologies and their global space–time structure, Class. Quant. Grav. **17**, 4745, (2000)

BB03. P. Brax, C. van de Bruck, Cosmology and Brane Worlds: A Review. arXiv: hep-th/0303095, (2003)

BCG91. Bryant, R., Chern, S., Gardner, R., Goldscmidt, H., Griffiths, P.: Exterior Differential Systems. Springer, Berlin, (1991)

BCO94. Bershadsky, M., Cecotti, S., Ooguri, H., Vafa, C.: Kodaira-Spencer theory of gravity and exact results for quantum string amplitudes. Commun. Math. Phys. **165**, 311, (1994)

BD02. Busemeyer. J.R., Diederich, A.: Survey of decision field theory. Math. Soc. Sci., **43**, 345–370, (2002)

BD82. Birrel, N.D., Davies, P.C.W.: Quantum fields in curved space. Cambridge Univ. Press, Cambridge, UK, (1982)

BD95. Baez. J., Dolan. J.: Higher dimensional algebra and topological quantum field theory. J. Math. Phys. **36**, 6073–6105, (1995)

BD98. Baez. J., Dolan. J.: Higher–Dimensional Algebra III: $n-$categories and the Algebra of Opetopes. Adv. Math. **135**(2), 145–206, (1998)

BDB00. Van de Bruck, C., Dorca, M., Brandenberger, R.H., Lukas, A.: Cosmological perturbations in brane-world theories: Formalism. Phys. Rev. D **62**, 123515, (2000)

BDG04. Banks, T., Dine, M., Gorbatov, E.: Is There a String Theory Landscape? JHEP **0408**, 058, (2004)

BDM00. Van de Bruck, C., Dorca, M., Martins, C.J., Parry, M.: Cosmological consequences of the brane/bulk interaction. Phys. Lett. B **495**, 183, (2000)

BF01. Banks, T., Fischler, W.: M-Theory Observables for Cosmological Space-times. arXiv:hep-th/0102077, (2001)

BF71. Bransford. J.D., Franks. J.J.: The Abstraction of Linguistic Ideas. Cogn. Psych., **2**, 331–350, (1971)

BFM00. Bertoldi, G., Faraggi, A., Matone, M.: Equivalence principle, higher dimensional Moebius group and the hidden antisymmetric tensor of quantum mechanics. Class. Quant. Grav. **17**, 3965, (2000)

BFR04. Bagnoli, F., Franci, F., Rechtman, R.: Chaos in a simple cellular automata model of a uniform society. In Lec. Not. Comp. Sci., Vol. 3305, 513–522, Springer, London, (2004)

BG03. Bassi, A., Ghirardi, G.C.: Dynamical Reduction Models. Phys. Rep. **379**, 257–426, (2003)

BG88. Baulieu, L., Grossman, B.: Monopoles and topological field theory. Phys. Lett. B **214**(2), 223, (1988)

BGG03. Bryant, R., Griffiths, P., Grossman, D.: Exterior Differential Systems and Euler–Lagrange partial differential equations. Univ. Chicago Press, Chicago, (2003)

BGG89. Batlle, C., Gomis. J., Gràcia, X., Pons. J.M.: Noether's Theorem and gauge transformations: application to the bosonic string and CP_2^{n-1} model. J. Math. Phys. **30**, 1345, (1989)

BGM02. Bruzzo, U., Gorini, V., Moschella, U. (eds.): Geometry and Physics of Branes. Institute of Physics, Bristol, (2002)

BGT95. Bucher, M., Goldhaber, A.S., Turok, N.: An Open Universe from Inflation. Phys. Rev. D **52**, 3314, (1995)

BH01. Brody, D., Hughston, L.: Geometric quantum mechanics. J. Geom. Phys. **38**, 19, (2001)

BH05. Bosse, A.W., Hartle, J.B.: Representations of Spacetime Alternatives and Their Classical Limits. Phys. Rev. **A72**, 022105, (2005)

BH93. Bohm, D., Hiley, B.J.: The Undivided Universe. Routledge, London, (1993)

BH96. Banaszuk, A., Hauser. J.: Approximate feedback linearization: a homotopy operator approach. SIAM J. Cont. & Optim., **34**(5), 1533–1554, (1996)

BK64. Bellman, R., Kalaba, R.: Selected papers on mathematical trends in control theory. Dover, New York, (1964)

BK92. Berleant, D., Kuipers, B.: Qualitative–Numeric Simulation with Q3, in Recent Advances in Qualitative Physics. eds. Boi Faltings and Peter Struss, MIT Press, Cambridge, (1992)

BK99. Balasubramanian, V., Kraus, P.: A Stress Tensor for Anti-de Sitter Gravity. Commun. Math. Phys. 208, 413, (1999)

BL92. Blackmore, D.L., Leu, M.C.: Analysis of swept volumes via Lie group and differential equations. Int. J. Rob. Res., **11**(6), 516–537, (1992)

BM00. Choquet-Bruhat, Y., DeWitt-Morete, C.: Analysis, Manifolds and Physics, Part II: 92 Applications (rev. ed). North-Holland, Amsterdam, (2000)

BM82. Choquet-Bruhat, Y., DeWitt-Morete, C.: Analysis, Manifolds and Physics (2nd ed). North-Holland, Amsterdam, (1982)

BMP97. Breckenridge. J.C., Myers, R.C., Peet, A.W., Vafa, C.: D–branes and spinning black holes. Phys. Lett. B **391**, 93, (1997)

BMT04. Basu, S., Mehreja, R., Thiberge, S., Chen, M.T., Weiss, R.: Spatiotemporal control of gene expression with pulse generating networks. Proc Natl. Acad. Sci. USA, **101**, 6355–6360, (2004)

BMW01. Bridgman, H.A., Malik, K.A., Wands, D.: Cosmic vorticity on the brane. Phys. Rev. D **63**, 084012, (2001)

BMW02. Bridgman, H.A., Malik, K.A., Wands, D.: Cosmological perturbations in the bulk and on the brane. Phys. Rev. D **65**, 043502, (2002)

BO95. Basar, T., Olsder, G.J.: Dynamic Noncooperative Game Theory (2nd ed.), Academic Press, New York, (1995)

BP00. Bousso, R., Polchinski. J.: Quantization of four-form fluxes and dynamical neutralization of the cosmological constant. JHEP **06**, 006, (2000)

BP02. Benvenuto, N., Piazza, F.: On the complex backpropagation algorithm. IEEE Trans. Sig. Proc., **40**(4), 967–969, (1992)

BPS75. Belavin, A.A., Polyakov, A.M., Swartz, A.S., Tyupkin, Yu.S.: SU(2) instantpons discovered. Phys. Lett. B **59**, 85, (1975)

BPS98. Blackmore, D.L., Prykarpatsky, Y.A., Samulyak, R.V.: The Integrability of Lie-invariant Geometric Objects Generated by Ideals in the Grassmann Algebra. J. Nonlin. Math. Phys., **5**(1), 54–67, (1998)

BRT89. Birmingham, D., Rakowski, M., Thompson, G. BRST quantization of topological field theories. Nucl. Phys. B **315**, 577, (1989)

BRT99. Bennati, E., Rosa-Clot, M., Taddei, S.: A Path Integral Approach to Derivative Security Pricing: I. Formalism and Analytical Results, Int. Journ. Theor. Appl. Finance **2**, 381, (1999)

BS00. Becskei, A., Serrano, L.: Engineering stability in gene networks by autoregulation. Nature, **405**, 590–593, (2000)

BS04. Banos, B., Swann, A.: Potentials for hyper-Kähler metrics with torsion, Class. Quant. Grav. **21**, 3127–3135, (2004)

BS04. Bashkirov, D., Sardanashvily, G.: Covariant Hamiltonian Field Theory. Path Integral Quantization. Int. J. Theor. Phys. **43**, 1317–1333, (2004)

BS85. Bamber, D., van Santen. J.P.H.: How many parameters can a model have and still be testable? J. Math. Psych., **29**, 443–473, (1985)

BS95. Bakker, B.V., Smit. J.: Curvature and scaling in 4D dynamical triangulation. Nucl. Phys. **B439**, 239, (1995)

BSM97. Breitenbach, G., Schiller, S., Mlynek. J.: Measurement of the quantum states of squeezed light. Nature, **387**, 471–475 (1997)

BT03. Bauchau, O.A., Trainelli, L.: The Vectorial Parameterization of Rotation. Non. Dyn., **31**(1), 71–92, (2003)

BT86. Barrow. J., Tipler, F.: The Anthropic Cosmological Principle. Oxford University Press, Oxford, (1986)

BT88. Brown. J.D., Teitelboim, C.: Neutralization of the Cosmological Constant by Membrane Creation. Nucl. Phys. **B297**, 787, (1988)

BT93. Busemeyer. J.R., Townsend. J.T.: Decision field theory: A dynamic–cognitive approach to decision making in an uncertain environment. Psych. Rev., **100**, 432–459, (1993)

Bae01. Baez. J.: The Meaning of Einstein's Equation. arXiv: gr-qc/0103044.

Bae02. Baez. J.: Categorified gauge theory. Lecture in the Joint Spring Meeting of the Pacific Northwest Geometry Seminar and Cascade Topology Seminar, (2002)

Bae96a. Baez. J.C.: Spin Network States in Gauge Theory. Adv. Math. **117**, 53, (1996)

Bae96b. Baez. J.C.: Spin Networks in Nonperturbative Quantum Gravity. In Kauffman, LH, (ed.) The Interface of Knots and Physics. Am. Math. Soc., Providence, Rhode Island, (1996)

Bae96c. Baez. J.C.: 4-Dimensional BF Theory as a Topological Quantum Field Theory, Lett. Math. Phys. **38**. 128, (1996)

Bae97. Baez. J.: An introduction to $n-$categories. $7th$ Conference on Category Theory and Computer Science, E. Moggi and G. Rosolini (eds), Lecture Notes in Computer Science, Springer, Berlin, (1997)

Ban07. Banos, B.: Monge-Ampere equations and generalized complex geometry. J. Geom. Phys. **57**, 841, (2007)

Ban84. Bando, S.: On the three dimensional compact Kähler manifolds of nonnegative bisectional curvature. J. Diff. Geom. **19**, 283–297, (1984)

Ban. Banks, T., Johnson, M: Regulating Eternal Inflation. arXiv:hep-th/0512141, (2005)

Bar93. Barry Jay, C. : Matrices, Monads and the Fast Fourier Transform. Univ. Tech., Sidney, (1993)

Bar94. Barbero, F.: Real-polynomial formulation of general relativity in terms of connections. Phys. Rev. **D49**, 6935–6938, (1994)

Bar95a. Barbero, F.: Real Ashtekar Variables for Lorentzian Signature Space–Times. Phys. Rev. **D 51**, 5507–5510, (1995)
Bar95b. Barbero, F.: Reality Conditions and Ashtekar Variables: a Different Perspective. Phys. Rev. **D 51**, 5498–5506, (1995)
Bar97. Barbieri, A.: Quantum tetrahedra and simplicial spin networks. arXiv:gr-qc/9707010, (1997)
Bau00. Baum, H.: Twistor and Killing spinors on Lorentzian manifolds and their relations to CR and Kaehler geometry. Int. Congr. Diff. Geom. in memory of Alfred Gray, Bilbao, Spain, (2000)
Bax82. Baxter, R.J.: Exactly solved models in statistical mechanics. Academic Press, (1982)
Ben67. Bénabou. J.: Introduction to bicategories. In: Lecture Notes in Mathematics. Springer, New York, (1967)
Ben96. Bennett, C. et al.: 4-Year COBE DMR Cosmic Microwave Background Observations:Maps and Basic Results, Ap. J. 464, L1, (1996)
Ber00. De Bernardis, P. et. al.: A Flat Universe from High-Resolution Maps of the Cosmic Microwave Background Radiation, Nature **404**, 955, (2000)
Ber35. Bernstein, N.A.: Investigations in Biodynamics of Locomotion (in Russian). WIEM, Moscow, (1935)
Ber47. Bernstein, N.A.: On the structure of motion (in Russian). Medgiz, Moscow, (1947)
Ber65. Berger, M.: Sur les *variétés* d'Einstein compactes. C. R. *IIIe Réunion* Math. Expression latine, Namur, 35–55, (1965)
Ber74. Berezin, F.: Sov. Math. Izv. 38, 1116, (1974); Sov. Math. Izv. 39, 363, (1975); Comm. Math. Phys. 40, 153, (1975); Comm. Math. Phys. 63, 131, (1978)
Bir82. Birrell, N.D., Davies, P.C.W.: Quantum Fields in Curved Space. Cambridge Univ. Press, Cambridge, UK, (1982)
Bla83. Blattner, R.: Nonlinear Partial Differential Operators and Quantization Procedure, in Proceedings, Clausthall 1981, Springer-Verlag, New York, 209–241, (1983)
Bla84. Blanchard, P.: Complex analytic dynamics on the Riemann sphere. Bull. AMS **11**, 85–141, (1984)
Boc04. Boccara, N.: Modeling complex systems. Springer, Berlin, (2004)
Boh61. Bohr, N.: Atomic Theory and the Description of Nature. Cambridge Univ. Press, Cambridge, (1961)
Bon95. Bontempi, G.: Modelling with uncertainty in continuous dynamical systems: the probability and possibility approach. IRIDIA – ULB Technical Report, 95–16, (1995)
Boo86. Boothby, W.M.: An Introduction to Differentiable Manifolds and Riemannian Geometry, Academic Press, New York, (1986)
Bot59. Bott, R.: The Stable Homotopy of the Classical Groups, Ann. Math. **70**, 313–337, (1959)
Bou05. Bousso, R.: Cosmology and the S-matrix. Phys. Rev. D **71**, 064024, (2005)
Bp00. Bousso, R., Polchinski. J.: Quantization of Four Form Fluxes and Dynamical Neutralization of the Cosmological Constant. JHEP **0006**, 006, (2000)
Bre04. Brent, R.: A partnership between biology and engineering. Nature, **22**, 1211–1214, (2004)
Bro58. Broadbent, D.E.: Perception and communications. Pergamon Press, London, (1958)

Bry04. Bryant, R.: Gradient Kähler Ricci Solitons. arXiv:math/0407453, (2004)
Buc02. Bucher, M.: A braneworld universe from colliding bubbles. Phys. Lett. B **530**, 1, (2002)
CC77. Callan, C., Coleman, S.: Fate of the false vacuum. II. First quantum corrections. Phys. Rev. D **16**, 1762, (1977)
CC96. Cheeger. J., Colding, T.H.: Lower bounds on Ricci curvature and almost rigidity of wraped products. Ann. Math. **144**, 189–237, (1996)
CC96. Chamseddine, A.H., Connes, A.: The Spectral Action Principle. Phys. Rev. Lett. **24**, 4868–4871, (1996)
CC99. Cao, H.D., Chow, B.: Recent Developments on the Ricci Flow. Bull. Amer. Math. Soc. **36**, 59–74, (1999)
CCC97. Caiani, L., Casetti, L., Clementi, C., Pettini, M.: Geometry of dynamics, Lyapunov exponents and phase transitions. Phys. Rev. Lett. **79**, 4361, (1997)
CCI91. Cariñena. J., Crampin, M., Ibort, L.: On the multisymplectic formalism for first order field theories. Diff. Geom. Appl. **1**, 345, (1991)
CCP96. Christiansen, F., Cvitanovic, P., Putkaradze, V.: Hopf's last hope: spatiotemporal chaos in terms of unstable recurrent patterns. Nonlinearity, **10**, 1, (1997)
CCP99. Casetti, L., Cohen, E.G.D., Pettini, M.: Origin of the Phase Transition in a Mean–Field Model. Phys. Rev. Lett., **82**, 4160, (1999)
CD98. Chen, G., Dong, X.: From Chaos to Order. Methodologies, Perspectives and Application. World Scientific, Singapore, (1998)
CDM01. Colless, M., Dalton, G., Maddox, S., Sutherland, W.: The 2dF Galaxy Redshift Survey: Spectra and Redshifts, MNRAS 328, 1039–1063, (2001)
CDS98. Connes, A., Douglas, M., Schwarz, A.: Noncommutative Geometry and Matrix Theory: Compactification on Tori. JHEP **9802**, 003, (1998)
CEG00. Csaki, C., Erlich. J., Grojean, C., Hollowood, T.J.: General properties of the self-tuning domain wall approach to the cosmological constant problem, Nucl. Phys. B **584**, 359, (2000)
CF93. Cariñena. J., Fernández-Núñez. J.: Geometric theory of time-dependent singular Lagrangians, Fortschr. Phys. **41**, 517, (1993)
CF94. Crane, L., Frenkel, I.: Four dimensional topological quantum field theory, Hopf categories, and the canonical bases. Jour. Math. Phys. **35**, 5136–5154, (1994)
CG83. Cohen, M.A., Grossberg, S.: Absolute stability of global pattern formation and parallel memory storage by competitive neural networks. IEEE Trans. Syst., Man, Cybern., **13**(5), 815–826, (1983)
CGI95. Cariñena. J., Gomis. J., Ibort, L. and Román, N.: Canonical transformation theory for presymplectic systems. J. Math. Phys. **26**, 1961, (1985)
CGK99. Csaki, C., Graesser, M., Kolda, C.F., Terning. J.: Cosmology of one extra dimension with localized gravity. Phys. Lett. B **462**, 34, (1999)
CGP88. Cvitanovic, P., Gunaratne, G., Procaccia, I.: Topological and metric properties of Hénon-type strange attractors. Phys. Rev. A **38**, 1503-1520, (1988)
CGR00. Csaki, C., Graesser, M., Randall, L., Terning. J.: Cosmology of brane models with radion stabilization. Phys. Rev. D **62**, 045015, (2000)
CGR00. Charmousis, C., Gregory, R., Rubakov, V.A., Wave function of the radion in a brane world.. Phys. Rev. D **62**, 067505, (2000)
CGS99. Cline. J.M., Grojean, C., Servant, G.: Cosmological expansion in the presence of extra dimensions. Phys. Rev. Lett. **83**, 4245, (1999)

CH04. Craig, D., Hartle, J.B.: Generalized Quantum Theories of Recollapsing, Homogeneous Cosmologies. Phys. Rev. D **69**, 123525–123547, (2004)

CH64. Conway, E.D., Hopf, E.: Hamilton's Theory and Generalized Solutions of the Hamilton–Jacobi Equation. J. Math. Mech. **13**, 939–986, (1964)

CHR00. Chamblin, A., Hawking, S.W., Reall, H.S.: Brane-world black holes. Phys. Rev. D **61**, 065007, (2000)

CHS85. Candelas, P., Horowitz, G.T., Strominger, A., Witten, E.: Vacuum configurations for superstrings, Nucl. Phys. B **258**, 46, (1985)

CL72. Craik, F., Lockhart, R.: Levels of processing: A framework for memory research. J. Verb. Learn. & Verb. Behav., **11**, 671–684, (1972)

CL80. Coleman, S., De Luccia, F.: Gravitational effects on and of vacuum decay. Phys. Rev. D **21**, 3314, (1980)

CL85. Caldeira, A.O., Leggett, A.J.: Influence of damping on quantum interference: An exactly soluble model. Phys. Rev. A **31**, 1059–1066, (1985)

CL84. Cheng, T.-P., Li, L.-F.: Gauge Theory of Elementary Particle Physics. Clarendon Press, Oxford, (1984)

CL90. Connes, A., Lott. J.: Particle models and non-commutative geometry. Nucl. Phys. **B18**, 29–47, (1990)

CLL01. Copeland, E.J., Liddle, A.R., Lidsey. J.E.: Steep inflation: Ending braneworld inflation by gravitational particle production. Phys. Rev. D **64**, 023509, (2001)

CLM94. Chinea, D., de León, M., Marrero. J.: The constraint algorithm for time-dependent Lagrangians. J. Math. Phys. **35**, 3410, (1994)

CN95. Carlip, S., Nelson. J.E.: Comparative Quantizations of (2+1)-Dimensional Gravity. Phys. Rev. D **51**, 5643, (1995)

CNH03. Chiorescu, I., Nakamura, Y., Harmans, C., Mooij. J.: Coherent Quantum Dynamics of a Superconducting Flux Orbit, Science, **299**, 1869, (2003)

CP02. Clementi, C. Pettini, M.: A geometric interpretation of integrable motions. Celest. Mech. & Dyn. Astr., **84**, 263–281, (2002)

CPC03. Casetti, L., Pettini, M., Cohen, E.G.D.: Phase transitions and topology changes in configuration space. J. Stat. Phys., **111**, 1091, (2003)

CPP02. García–Compeán, H., Plebański. J., Przanowski, M., Turrubiates, F.: Deformation Quantization of Geometric Quantum Mechanics. J. Phys. **A35**, 4301, (2002)

CQ69. Collins, A.M., Quillian, M.R.: Retrieval Time From Semantic Memory. J. Verb. Learn. & Verb. Behav., **8**, 240–248, (1969)

CR00. Cao, H.D., Hamilton, R.S.: Gradient Kähler-Ricci solitons and periodic orbits, Comm. Anal. Geom. **8**, 517–529, (2000)

CR89. Cariñena. J., Rañada, M.: Poisson maps and canonical transformations for time-dependent Hamiltonian systems. J. Math. Phys. **30**, 2258, (1989)

CR93. Cariñena. J., Rañada, M.: Lagrangian systems with constraints. A geometric approach to the method of Lagrange multipliers. J. Phys. A. **26**, 1335, (1993)

CR99. Chamblin, H.A., Reall, H.S.: Dynamic dilatonic domain walls, Nucl. Phys. B **562**, 133, (1999)

CRV92. Celeghini, E., Rasetti, M., Vitiello, G.: Quantum Dissipation. Annals Phys., **215**, 156, (1992)

CS96. Constantine, G.M., Savits, T.H.: A multivariate Faà di Bruno formula with applications. Trans. Amer. Math. Soc. **348**(2), 503–520, (1996)

CST86. Cremmer, E., Schwimmer, A., Thorn, C.: The Vertex Function in Witten's Formulation of String Field Theory. Phys. Lett. **B179**, 57, (1986)

CT01. X. X. Chen and G. Tian. Ricci Flow on Kähler–Einstein manifolds. arXiv:math/0108179, (2001)

CT02. Chen, X., Tian, G.: Ricci flow on Kahler-Einstein surfaces. Invent. Math. **147**, 487-544, (2002)

CT03. Chau, A., Tam, L.F.: Gradient Kähler-Ricci solitons and a uniformization conjecture. arXiv:math.DG/0310198, (2003)

CT04. Chau, A., Tam, L.F.: A note on the uniformization of gradient Kähler-Ricci solitons. arXiv:math.DG/0404449, (2004)

CV91. Cecotti, S., Vafa, C.: Topological – antitopological fusion, Nucl. Phys. B **367**, 359, (1991)

CY89. Crutchfield, J.P., Young, K.: Computation at the onset of chaos. In *Complexity, Entropy and the Physics of Information*, p. 223, SFI Studies in the Sciences Complexity, Vol. VIII, W.H. Zurek (Ed.), Addison-Wesley, Reading, MA, (1989)

CY93. Crane, L., Yetter, D.: A categorical construction of 4D topological quantum field theories. In Quantum Topology, R. Baadhio and L.H. Kauffman ed. World Scientific, Singapore, (1993)

Cal57. Calabi, E.: On Kähler manifolds with vanishing canonical class. In Algebraic geometry and topology: a symposium in honor of S. Lefschetz, eds. R.H. Fox, D.C. Spencer and A.W. Tucker, Princeton Univ. Press, (1957)

Can86. Canarutto, D.: Bundle splittings, connections and locally principle fibred manifolds. Bull. U.M.I. Algebra e Geometria, Seria VI **V-D**, 18, (1986)

Cao85. Cao, H.D.: Deformation of Kähler metrics to Kähler-Einstein metrics on compact Kähler manifolds. Invent. Math., 81, 359–372, (1985)

Cao94. Cao, H.D.: Existence of gradient Kähler-Ricci solitons, Elliptic and parabolic methods in geometry. Minneapolis, MN, (1994)

Cao97. Cao, H.D.: Limits of solutions to the Kähler-Ricci flow. J. Diff. Geom. **45**, 257–272, (1997)

Cap74. Capper, D.M., Duff, M.J.: Trace Anomalies in Dimensional Regularization, Nuovo Cimento **23A**, 173, (1974)

Cas92. Cassidy, D.: Uncertainty: The Life and Science of Werner Heisenberg. Freeman, New York, (1992)

Cav05. Cavalcanti, G.R.: New aspects of the dd^c–lemma. arXiv:math.DG/0501406, (2005)

Cav86. Caves, C.: Quantum Mechanics and Measurements Distributed in Time I: A Path Integral Approach. Phys. Rev. D **33**, 1643, (1986); Quantum Mechanics and Measurements Distributed in Time II: Connections among Formalisms, ibid. D **35**, 1815, (1987)

Ch99. Chu, C.-S., Ho, P.-M.: Noncommutative Open String and D–brane, Nucl. Phys. **B550**, 151, (1999)

Cha02. Charmousis, C.: Dilaton space–times with a Liouville potential. Class. Quant. Grav. **19**, 83, (2002)

Cha05. Chamseddine, A.H.: Hermitian Geometry and Complex Space-Time. arXiv:hep-th/0503048, (2005)

Cha48. Chandra, H.: Relativistic Equations for Elementary Particles. Proc. Roy. Soc. London A, **192**(1029), 195–218, (1948)

Che02. Chen, X.: Recent progress in Kähler geometry, in Proceedings of the International Congress of Mathematicians, Vol. II (Beijing, 2002), Higher Ed. Press, Beijing, 273–282, (2002)

Che46. Chern, S.S.: Characteristic classes of Hermitian manifolds, Ann. of Math. **47**, 85–121, (1946)

Che55. Chevalley, C.: Theorie differential equations groupes de Lie. Vol. 1–3. Hermann D.C., Paris, (1955)

Che96. Chern, S.S.: Riemannian Geometry as a Special Case of Finsler Geometry, Cont. Math., 51–58, Vol. 196, Amer. Math. Soc. Providence, RI, (1996)

Cho91. Chow, B.: The Ricci flow on the 2-sphere. J. Diff. Geom. **33**, 325–334, (1991)

Col69. Coleman, S.: Acausality in Theory and Phenomenology in Particle Physics, ed. A. Zichichi, New York, (1969)

Col77. Coleman, S.: Fate of the false vacuum: Semiclassical theory. Phys. Rev. D15, 292936, (1977)

Col80. Coleman, S., De Luccia, F.: Gravitational Effects on and of Vacuum Decay. Phys. Rev. D **21**, 3305, (1980)

Con94. Connes, A.: Noncommutative Geometry. Academic Press, New York, (1994)

CCN85. Conway, J.H., Curtis, R.T., Norton, S.P., Parker, R.A., Wilson, R.A: Atlas of Finite Groups: Maximal Subgroups and Ordinary Characters for Simple Groups. Clarendon Press, Oxford, (1985)

Cop35. Copson, E.T.: Theory of Functions of a Complex Variable. Oxford Univ. Press, London, (1935)

Cox43. Coxeter, H.S.M.: A geometrical background for de Sitter's world. Am. Math. Mont. **50**, 217–228, (1943)

Cox92. Cox, E.: Fuzzy Fundamentals, IEEE Spectrum, 58–61, (1992)

Cox94. Cox, E.: The Fuzzy Systems Handbook. AP Professional, (1994)

Cra04. Crainic, M.: Generalized complex structures and Lie brackets. math.DG/0412097, (2004)

Cra91. Crawford, J.: Clifford algebra: Notes on the spinor metric and Lorentz, Poincaré and conformal groups. J. Math. Phys. **32**, 576, (1991)

Cro80. Croke, C.: Some Isoperimetric Inequalities and Consequences. Ann. Sci. E. N. S., Paris, **13**, 419–435, (1980)

Cve92. Cveticanin, L.: Approximate analytical solutions to a class of nonlinear equations with complex functions. J. Sound Vibr. **157**, 289–302, (1992)

Cve92. Cveticanin, L.: An approximate solution for a system of two coupled differential equations. J. Sound Vibr. **152**, 375–380, (1992)

Cve93. Cveticanin, L.: An asymptotic solution for weak nonlinear vibrations of the rotor. Mech. Mach. The. **28**, 495–505, (1993)

Cve98. Cveticanin, L.: Analytical methods for solving strongly nonlinear differential equations. J. Sound Vibr. **214**, 325–328, (1998)

Cve01. Cveticanin, L.: Analytic approach for the solution of the complex valued strong nonlinear differential equation. Physica A **297**, 348–360, (2001)

Cve05. Cveticanin, L.: Approximate solution of a strongly nonlinear complex differential equation. J. Sound Vibr. **284**, 503–512, (2005)

Cvi00. Cvitanovic, P.: Chaotic field theory: a sketch. Physica A **288**, 61–80, (2000)

DC76. Dowker, J.S., Critchley, R.: Effective Lagrangian and Energy–Momentum Tensor in de Sitter Space. Phys. Rev. D **13**, 3224, (1976)

DD06. Denef, F., Douglas, M.R.: Computational complexity of the landscape. arXiv:hep-th/0602072, (2006)

DD04. Denef, F., Douglas, M.R.: Distributions of Flux Vacua. arXiv:hep-th/0404116, (2005)

DDK01. Deruelle, N., Dolezel, T., Katz. J.: Perturbations of brane worlds. Phys. Rev. D **63**, 083513, (2001)

DDT01. Dorey, P., Dunning, C. Tateo, R.: Supersymmetry and the spontaneous breakdown of PT-symmetry. J. Phys. A: Math. Gen. **34**, L391, (2001)

DEF99. Deligne, P., Etingof, P., Freed, D.S., Jeffrey, L.C., Kazhdan, D., Morgan. J.W., Morrison, D.R., Witten, E.: Quantum Fields and Strings: A Course for Mathematicians, Am. Math. Soc., (1999)

DFR94. Doplicher, S., Fredenhagen, K., Roberts. J.E.: Space-time quantization inducedby classical gravity. Phys. Lett. **B331**, 39–44, (1994)

DG03. Dragovic, V., Gajic, B.: The Wagner Curvature Tensor in Nonholonomic Mechanics. Reg. Chaot. Dyn., **8**(1), 105–124, (2003)

DH04. Dowker, F., Henson. J.: A Spontaneous Collapse Model on a Lattice. J. Stat. Phys. **115**, 1349, (2004)

DH85. Douady, A., Hubbard. J.: On the dynamics of polynomial-like mappings. Ann. scient. Éc. Norm. Sup., 4e ser. **18**, 287–343, (1985)

DHS91. Domany, E., van Hemmen. J.L., Schulten, K. (eds.): Models of Neural Networks. Springer, Berlin, (1991)

DKS02. Dyson, L., Kleban, M., Susskind, L.: Disturbing Implications of a Cosmological Constant. JHEP **0210**, 011, (2002)

DKS02. Dyson, L., Kleban, M., Susskind, L.: Disturbing Implications of a Cosmological Constant. JHEP **0210**, 011, (2002)

DL01. Dasgupta, A., Loll, R.: A proper-time cure for the conformal sickness in quantum gravity. Nucl. Phys. B **606**, 357–379, (2001)

DLL01. Duff, M.J., Liu. J.T., Lu. J. (eds.): Strings: Proceedings of the 2000 International Superstrings Conference, World Scientific, Singapore, (2001)

DLS02. Dyson, L., Lindesay. J., Susskind, L.: Is There Really a de Sitter/CFT Duality? arXiv:hep-th/0202163, (2002)

DM03. Dorogovtsev, S.N., Mendes. J.F.F.: Evolution of Networks. Oxford Univ. Press, (2003)

DN79. Devaney, R., Nitecki, Z.: Shift automorphisms in the Hénon mapping. Comm. math. Phys. **67**, 137–48, (1979)

DP80. Dubois, D., Prade, H.: Fuzzy Sets and Systems. Academic Press, New York, (1980)

DP97. Dodson, C.T.J., Parker, P.E.: A User's Guide to Algebraic Topology. Kluwer, Dordrecht, (1997)

DR94. Dittrich, W., Reuter, M.: Classical and Quantum Dynamics. Springer Verlag, Berlin, (1994)

DR96. DePietri, R., Rovelli, C.: Geometry Eigenvalues and Scalar Product from Recoupling Theory in Loop Quantum Gravity. Phys. Rev. **D54**, 2664–2690, (1996)

DT92. Ding, W., Tian, G.: Kähler–Eistein metric and the generalized Futaki invariant. Inv. Math. **110**, 315–335, (1992)

DTP02. Dauxois, T., Theodorakopoulos, N., Peyrard, M.: Thermodynamic instabilities in one dimension: correlations, scaling and solitons. J. Stat. Phys. **107**, 869, (2002)

DVV91. Dijkgraaf, R., Verlinde, H., Verlinde, E.: Notes On Topological String Theory And 2D Quantum Gravity, in String Theory and Quantum Gravity, Proceedings of the Trieste Spring School 1990, (eds.) M. Green et al., World Scientific, 91–156, (1991)

Das02. Dasgupta, A.: The real Wick rotations in quantum gravity. JHEP, **0207**, (2002)

Dav02a. Davis, S.C.: Cosmological brane world solutions with bulk scalar fields. JHEP **0203**, 054, (2002a)

Dav02b. Davis, S.C.: Brane cosmology solutions with bulk scalar fields. JHEP **0203**, 058, (2002b)

Dav81. Davydov, A.S.: Biology and Quantum Mechanics, Pergamon Press, New York, (1981)

Dav89. Davies, E.B.: Heat Kernels and Spectral Theory. Cambridge Univ. Press, (1989)

Dav91. Davydov, A.S.: Solitons in Molecular Systems. (2nd ed), Kluwer, Dordrecht, Ger, (1991)

DeP97. DePietri, R.: On the relation between the connection and the loop representation of quantum gravity, Class. and Quantum Grav. **14**, 53–69, (1997)

Dea90. Dearnaley, R.: The Zero-Slope Limit of Witten's String Field Theory with Chan-Paton Factors. Nucl. Phys. **B334**, 217, (1990)

Die69. Dieudonne. J.A.: Foundations of Modern Analysis (in four volumes). Academic Press, New York, (1969)

Die88. Dieudonne. J.A.: A History of Algebraic and Differential Topology 1900–1960. Birkháuser, Basel, (1988)

Dim59. Dimentberg, F.M.: Izgibnije Kolabanija Vrashchajushihsja Valov. Izd. Akad. Nauk SSSR, Moscow, (1959)

Dir25. Dirac, P.A.M.: The Fundamental Equations of Quantum Mechanics. Proc. Roy. Soc. London A, **109**(752), 642–653, (1925)

Dir26a. Dirac, P.A.M.: Quantum Mechanics, a Preliminary Investigation of the Hydrogen Atom. Proc. Roy. Soc. London A, **110**(755), 561–579, (1926)

Dir26b. Dirac, P.A.M.: The Elimination of the Nodes in Quantum Mechanics. Proc. Roy. Soc. London A, **111**(757), 281–305, (1926)

Dir26c. Dirac, P.A.M.: Relativity Quantum Mechanics with an Application to Compton Scattering. Proc. Roy. Soc. London A, **111**(758), 281–305, (1926)

Dir26d. Dirac, P.A.M.: On the Theory of Quantum Mechanics. Proc. Roy. Soc. London A, **112**(762), 661–677, (1926)

Dir26e. Dirac, P.A.M.: The Physical Interpretation of the Quantum Dynamics. Proc. Roy. Soc. London A, **113**(765), 1–40, (1927)

Dir28a. Dirac, P.A.M.: The Quantum Theory of the Electron. Proc. Roy. Soc. London A, **117**(778), 610–624, (1928)

Dir28b. Dirac, P.A.M.: The Quantum Theory of the Electron. Part II. Proc. Roy. Soc. London A, **118**(779), 351–361, (1928)

Dir29. Dirac, P.A.M.: Quantum Mechanics of Many-Electron Systems. Proc. Roy. Soc. London A, **123**(792), 714–733, (1929)

Dir32. Dirac, P.A.M.: Relativistic Quantum Mechanics. Proc. Roy. Soc. London A, **136**(829), 453–464, (1932)

Dir36. Dirac, P.A.M.: Relativistic Wave Equations. Proc. Roy. Soc. London A, **155**(886), 447–459, (1936)

Dir49. Dirac, P.A.M.: The Principles of Quantum Mechanics. Oxford Univ Press, Oxford, (1949)

Dir58. Dirac, P.A.M.: Generalized Hamiltonian Dynamics. Proc. Roy. Soc. London A, **246**(1246), 326–332, (1958)

Dom78. de Dominicis, C.: Dynamics as a substitute for replicas in systems with quenched random impurities. Phys. Rev. B **18**, 4913–4919, (1978)

Don84. Donaldson, S.K.: The Nahm equations and the classification of monopoles, Commun. Math. Phys. 96, 387–407, (1984)

Don87. Donaldson, S.: The Orientation of Yang–Mills Moduli Spaces and Four–Manifold Topology. J. Diff. Geom. **26**, 397, (1987)

Don96. Donoghue. J.F.: The Quantum Theory of General Relativity at Low Energies. Helv. Phys. Acta. **69**, 269–275, (1996)

Dou. Douglas, M.: "The statistics of string / M theory vacua. JHEP **0305**, 46, (2003)

Dow02. Dowker, F.: Topology change in quantum gravity. In Proceedings of Stephen Hawking's 60th birthday conference. Cambridge, UK, (2002)

Dow05. Dowker, F.: Causal Sets and the Deep Structure of Spacetime. In 100 Years of Relativity, ed. by A. Ashtekar, World Scientific, Singapore, (2005)

Dri86. Drinfeld, V.G.: Quantum groups. Proc. ICM, Berkeley, 798, (1985)

Duf94. Duff, M.: Twenty years of the Weyl anomaly. Class. Quant. Grav. **11**, 1387, (1994)

Dun99. Dunne, G.V.: Aspects of Chern–Simons Theory. arXiv:hep-th/9902115, (1999)

Dus84. Dustin, P.: Microtubules. Springer, Berlin, (1984)

EB01. Endy, D., Brent, R.: Modelling cellular behaviuor. Nature **409**, 391-395, (2001)

EG91. Eastwood, M.G., Graham, C.R.: Invariants of conformal densities. Duke Math. Jour. **63**, 633–671, (1991)

EHM00. Emparan, R., Horowitz, G.T., Myers, R.C.: Exact Description of Black Holes on Branes II: Comparison with BTZ Black Holes and Black Strings. JHEP 0001, 007, (2000)

EIS67. Eccles, J.C., Ito M., Szentagothai J.: The Cerebellum as a Neuronal Machine. Springer, Berlin, (1967)

EJM99. Emparan, R. Johnson C.V. and Myers R.C.: Surface terms as counterterms in the AdS-CFT correspondence. Phys. Rev. D **60**, 104001, (1999)

EL00. Elowitz, M.B., Leibler, S.: A synthetic oscillatory network of transcriptional regulators. Nature, **403**, 335–338, (2000)

EL78. Eells, J., Lemaire, L.: A report on harmonic maps. Bull. London Math. Soc. **10**, 1–68, (1978)

EMN92. Ellis, J., Mavromatos, N., Nanopoulos, D.V.: String theory modifies quantum mechanics. CERN-TH/6595, (1992)

EMN99. Ellis, J., Mavromatos, N., Nanopoulos, D.V.: A microscopic Liouville arrow of time. Chaos, Solit. Fract., **10**(2–3), 345–363, (1999)

EMR91. Echeverría-Enríquez, A., Muñoz-Lecanda, M., Román-Roy, N.: Geometrical setting of time-dependent regular systems. Alternative models, Rev. Math. Phys. **3**, 301, (1991)

EMR95. Echeverría-Enríquez, A., Muñoz-Lecanda, M., Román-Roy, N.: Non-standard connections in classical mechanics. J. Phys. A **28**, 5553, (1995)

EMR98. Echeverría-Enríquez, A., Muñoz-Lecanda, M., Román-Roy, N.: Multivector fields and connections. Setting Lagrangian equations in field theories. J. Math. Phys. **39**, 4578, (1998)

EP00. Eastwood, M., Penrose, R.: Drawing with Complex Numbers. arXiv math.MG/0001097, (2000)
ES46. Einstein, A., Strauss, E.: Ann. Math. **47**, 731, (1946)
Eas02. Eastwood, M.: Higher symmetries of the Laplacian. arXiv:hep-th/0206233, (2002)
Ecc64. Eccles, J.C.: The Physiology of Synapses. Springer, Berlin, (1964)
Elk99. Elkin, V.I.: Reduction of Nonlinear Control Systems. A Differential Geometric Approach, Kluwer, Dordrecht, (1999)
Elw82. Elworthy, K.D.: Stochastic Differential Equations on Manifolds. Cambridge Univ. Press, Cambridge, (1982)
Erm96. Ermentrout, G.B.: Type I membranes, phase resetting curves, and synchrony. Neural Computation, **8**(5), 979–1001, (1996)
FA04. Field, T., Anandan. J.: Geometric phases and coherent states. J. Geom. Phys. **50**, 56–78, (2004)
FCS99. Franzosi, R., Casetti, L., Spinelli, L., Pettini, M.: Topological aspects of geometrical signatures of phase transitions. Phys. Rev. E, **60**, 5009–5012, (1999)
FFF94. Fatibene, L., Ferraris, M., Francaviglia, M.: Nöther formalism for conserved quantities in classical gauge field theories II. J. Math. Phys., **35**, 1644, (1994)
FG94. Fröhlich, J., Gawedzki, K.: Conformal field theory and the geometry of strings. CRM Proceedings and Lecture Notes **7**, 57–97, (1994)
FGR98. Fajstrup, L., Goubault, E., Raussen, M.: Detecting Deadlocks in Concurrent Systems, In D. Sangiorgi R. de Simone, editor, CONCUR '98; Concurrency Theory, number 1466 in Lecture Notes in Computer Science, 332–347. Springer, Berlin, (1998)
FH65. Feynman, R.P., Hibbs, A.: Quantum Mechanics and Path Integrals. McGraw-Hill, New York, (1965)
FHH79. Fischetti, M.V., Hartle, J.B., Hu, B.L.: Quantum Effects in the Early Universe I. Influence of Trace Anomalies on Homogeneous, Isotropic, Classical Geometries. Phys. Rev. D **20**, 1757, (1979)
FHT03. Freed, D., Hopkins, M. Teleman, C.: Twisted K–theory and loop group representations. arXiv:math.AT/0312155, (2003)
FK82. Ferraris, M., Kijowski. J.: On the equivalence of the relativistic theories of gravitation. GRG **14**, 165, (1982)
FK83. Frölich, H., Kremer, F.: Coherent Excitations in Biological Systems. Springer, New York, (1983)
FKL05. Freivogel, B., Kleban, M., Susskind, L.: Observational Consequences of a Landscape. arXiv:hep-th/0505232, (2005)
FKN95. Frittelli, S., Kozameh, C., Newman, E.T.: General Relativity via Characteristic Surfaces. J. Math. Phys. **36**, 4984, (1995)
FKN97. Frittelli, S., Kozameh, C., Newman, E.T., Rovelli, C., Tate, R.S.: Fuzzy space–time points from the null–surface formulation of general relativity, Class. Quantum Gravity, **14**, A143, (1997)
FLL00. Forste, S., Lalak, Z., Lavignac, S., Nilles, H.P.: A comment on self-tuning and vanishing cosmological constant in the brane world. Phys. Lett. B **481**, 360, (2000)
FM03. Fujii, Y., Maeda, K-I.: The Scalar-Tensor Theory of Gravitation. Cambridge Univ. Press, (2003)

FP04. Franzosi, R., Pettini, M.: Theorem on the origin of phase transitions. Phys. Rev. Lett., **92**(6), 60601, (2004)

FP04. Franzosi, R., Pettini, M.: Theorem on the origin of phase transitions. Phys. Rev. Lett., **92**(6), 60601, (2004)

FPR04. Forger, M., Paufler, C., Römer, H.: Hamiltonian Multivector Fields and Poisson Forms in Multisymplectic Field Theory. arXiv:math-ph/0407057.

FPS00. Franzosi, R., Pettini, M. Spinelli, L.: Topology and phase transitions: a paradigmatic evidence. Phys. Rev. Lett. 84(13), 2774–2777, (2000)

FPS92. Friedman, J.L., Papastamatiou, N.J., Simon, J.Z.: Unitarity of Interacting Fields in Curved Spacetime. Phys. Rev. D **46**, 4441, (1992)

FR04. Forger, M., Römer, H.: Currents and the energy–momentum tensor in classical field theory: a fresh look at an old problem. Ann. Phys. (N.Y.) **309**, 306–389, (2004)

FS92. Freeman, J.A., Skapura, D.M.: Neural Networks: Algorithms, Applications, and Programming Techniques. Addison-Wesley, Reading, MA, (1992)

FT82. Fradkin, E.S., Tseytlin, A.A.: Higher Derivative Quantum Gravity: One Loop Counterterms and Asymptotic Freedom, Nucl. Phys. **B201**, 469, (1982)

FT84. Fradkin, E.S., Tseytlin, A.A.: Conformal Anomaly in Weyl Theory and Anomaly Free Superconformal Theories. Phys. Lett. **134B**, 187, (1984)

FTW00. Flanagan, E.E., Tye, S.H., Wasserman, I.: Cosmological expansion in the RS brane world scenario. Phys. Rev. D **62**, 044039, (2000)

FU84. Freed, D., Uhlenbeck, K.K.: Instantons and four manifolds, Springer, New York, (1984)

Fat19. Fatou, P.: Sur les équations fonctionnelles. Bull. Soc. math. France **47**, 161–271, (1919)

Fat22. Fatou, P.: Sur les fonctions méromorphes de deux variables and Sur certaines fonctions uniformes de deux variables. C.R. Acad. Sc. Paris **175**, 862-65 and 1030–33, (1922)

Fed69. Federer, H.: Geometric Measure Theory. Springer, New York, (1969)

Fei80. Feigenbaum, M.: Universal Behavior in Nonlinear Systems. Los Alamos Science. **1**, 4, (1980)

Fer99. Ferber, J.: Multi-Agent Systems. An Introduction to Distributed Artificial Intelligence. Addison-Wesley, Reading, MA, (1999)

Fey48. Feynman, R.P.: Space-time Approach to Non-Relativistic Quantum Mechanics. Rev. Mod. Phys. **20**, 267, (1948)

Fey51. Feynman, R.P.: An Operator Calculus Having Applications in Quantum Electrodynamics. Phys. Rev. **84**, 108–128, (1951)

Fey72. Feynman, R.P.: Statistical Mechanics, A Set of Lectures. WA Benjamin, Inc., Reading, Massachusetts, (1972)

Fey98. Feynman, R.P.: Quantum Electrodynamics. Advanced Book Classics, Perseus Publishing, (1998)

Fla63. Flanders, H.: Differential Forms with Applications to the Physical Sciences. Acad. Press, (1963)

Fla76. Flaherty, E.J.: Hermitian and Kählerian Geometry in General Relativity, Lecture Notes in Physics, Vol. 46, Springer, Heidelberg, (1976)

Fla80. Flaherty, E.J.: Complex Variables in Relativity. In General Relativity and Gravitation, One Hundred Years after the Birth of Albert Einstein, ed A. Held Plenum, New York, (1980)

Flo87. Floer, A.: Morse theory for fixed points of symplectic diffeomorphisms. Bull. AMS. **16**, 279, (1987)

Flo88. Floer, A.: Morse theory for Lagrangian intersections. J. Diff. Geom., **28**(9), 513–517, (1988)

Fok29. Fokker, A.D.: Z. Phys., **58**, 386–393, (1929)

For90. Fordy, A.P. (ed.): Soliton Theory: A Survey of Results. MUP, Manchester, UK, (1990)

Fre01. Freed, D.S.: The Verlinde Algebra is Twisted Equivariant K–Theory. Turk. J. Math **25**, 159-167, (2001)

Fri88. Friberg, O.: A Set of Parameters for Finite Rotations and Translations. Comp. Met. App. Mech. Eng., **66**, 163–171, (1988)

Fri03. Friedman, J. et. al.: Quantum Superposition of Distinct Macroscopic States, Nature, **406**, 43–46, (2000)

Ful89. Fulling, S.: Aspects of quantum field theory in curved space–time. Cambridge Univ. Press, (1989)

Fut83. Futaki, A.: An obstruction to the existence of Einstein Kähler metrics. Inv. Math. **73**(3), 437–443, (1983)

GBL03. Gardner, T.S., di Bernardo, D., Lorenz, D., Collins. J.J.: Inferring genetic networks and identifying compound mode of action via expression profiling. Science, **301**, 102–105, (2003)

GEH02. Guet, C., Elowitz, M.B., Hsing, W., Leibler, S.: Combinatorial synthesis of genetic networks. Science, **296**(5572), 1466–1470, (2002)

GG03. Gaucher, P., Goubault, E.: Topological Deformation of Higher Dimensional Automata. Homology, Homotopy and Applications, **5**(2), 39–82, (2003)

GH77. Gibbons, G.W., Hawking, S.W.: Action integrals and partition functions in quantum gravity. Phys. Rev. D **15**, 2752, (1977)

GH90. Gibbons, G.W., Hartle, J.B.: Real Tunneling Geometries and the Large-Scale Topology of the Universe. Phys. Rev. D **42**, 2458, (1990)

GH94. Griffiths, P., Harris. J.: Principles of Algebraic Geometry, Wiley Interscience, New York, (1994)

GH96. Gérard, R., Tahara, H.: Singular nonlinear partial differential equations, Aspects of Mathematics, Friedr. Vieweg & Sohn, Braunschweig, (1996)

GHP78. Gibbons, G.W., Hawking, S.W., Perry, M.J.: Path Integrals and the Indefiniteness of the Gravitational Action, Nucl. Phys. **B138**, 141, (1978)

GHT00. Gratton, S., Hertog, T., Turok, N.: An Observational Test of Quantum Cosmology. Phys. Rev. D **62**, 063501 (2000)

GIM04. Gotay, M.J., Isenberg. J., Marsden. J.E., Montgomery, R.: Momentum Maps and Classical Relativistic Fields II: Canonical Analysis of Field Theories. arXiv:math-ph/0411032.

GIM98. Gotay, M.J., Isenberg. J., Marsden. J.E.: Momentum Maps and the Hamiltonian Structure of Classical Relativistic Fields. arXiv:hep/9801019.

GIT02. Gen, U., Ishibashi, A., Tanaka, T.: Brane big-bang brought by bulk bubble. Phys. Rev. D **66**, 023519, (2002)

GJ87a. Gross, D.J., Jevicki, A.: Operator Formulation of Interacting String Field Theory (I). Nucl. Phys. **B283**, 1, (1987)

GJ87b. Gross, D.J., Jevicki, A.: Operator Formulation of Interacting String Field Theory (II). Nucl. Phys. **B287**, 225, (1987)

GJ94. van Groesen E., De Jager E.M.: (Editors) Mathematical Structures in Continuous Dynamical Systems. Studies in Mathematical Physics, vol. 6 North-Holland, Amsterdam, (1994)

GK92. Georgiou, G.M., Koutsougeras, C.: Complex domain backpropagation, IEEE Trans. Circ. Sys., **39**(5), 330–334, (1992)

GKP98. Gubser, S.S., Klebanov, I.R., Polyakov, A.M.: Gauge Theory Correlators from Noncritical String Theory. Phys. Lett. **B428**, 105, (1998)

GKS02. Goheer, N., Kleban, M., Susskind, L.: The Trouble with de Sitter space. arXiv:hep-th/0212209, (2002)

GM01. Gordon, C., Maartens, R.: Density perturbations in the brane world. Phys. Rev. D **63**, 044022, (2001)

GM04. Grinza, P., Mossa, A.: Topological origin of the phase transition in a model of DNA denaturation. Phys. Rev. Lett. **92**(15), 158102, (2004)

GM87. Gross, D., Mende, P.: The High-Energy Behavior Of String Scattering Amplitudes. Phys. Lett **B197**, 129, (1987)

GM88. Garland, H., Murray, M.K.: Kac-Moody monopoles and periodic instantons, Commun. Math. Phys. 120, 335–351, (1988)

GM90. Giachetta, G., Mangiarotti, L.: Gauge-invariant and covariant operators in gauge theories. Int. J. Theor. Phys., **29**, 789, (1990)

GM92. Gotay, M.J., Marsden. J.E.: Stress-energy–momentum tensors and the Belinfante-Rosenfeld formula. Contemp. Math. **132**, AMS, Providence, 367–392, (1992)

GM94. Gell-Mann, M.: The Quark and the Jaguar. W. Freeman, San Francisco, (1994)

GMH05. Giddings, S.B., Marolf, D., Hartle, J.B.: Observables in Effective Gravity. arXiv:hep-th/0512200, (2005)

GMH06. Giddings, S., Marolf, D., Hartle. J.: Observables in Effective Gravity. arXiv:hep-th/0512200, (2005)

GMH90. Gell-Mann, M., Hartle, J.B.: Alternative Decohering Histories in Quantum Mechanics, in the Proceedings of the 25th International Conference on High Energy Physics, Singapore, August, 2-8, 1990, ed. by K.K. Phua and Y. Yamaguchi, South East Asia Theoretical Physics Association and Physical Society of Japan, distributed by World Scientific, Singapore, (1990)

GMH90. Gibbons, G.W., Hartle, J.B.: Real Tunneling Geometries and the Large-scale Topology of the Universe. Phys. Rev. D **42**, 2458, (1990)

GMH93. Gell-Mann, M., Hartle, J.B.: Classical Equations for Quantum Systems. Phys. Rev. D **47**, 3345, (1993)

GMH95. Gell-Mann, M., Hartle, J.B.: Strong Decoherence. In the Proceedings of the 4th Drexel Symposium on Quantum Non-Integrability — The Quantum-Classical Correspondence, Drexel University, September 8-11, 1994, ed. by D.-H. Feng and B.-L. Hu, International Press, Boston/Hong-Kong, (1995)

GMP05. Grunwald, P., Myung, I.J., Pitt, M.A.(eds.): Advances in Minimum Description Length: Theory and Applications. MIT Press, Cambridge, MA (2005)

GMS02a. Giachetta, G., Mangiarotti, L., Sardanashvily, G.: Covariant geometric quantization of non-relativistic Hamiltonian mechanics. J. Math. Phys. **43**, 56, (2002)

GMS02b. Giachetta, G., Mangiarotti, L., Sardanashvily, G.: Geometric quantization of mechanical systems with time-dependent parameters. J. Math. Phys., **43**, 2882, (2002)

GMS05. Giachetta, G., Mangiarotti, L., Sardanashvily, G.: Lagrangian supersymmetries depending on derivatives. Global analysis and cohomology. Commun. Math. Phys. **259**, 103–128, (2005)

GMS97. Giachetta, G., Mangiarotti, L., Sardanashvily, G.: New Lagrangian and Hamiltonian Methods in Field Theory, World Scientific, Singapore, (1997)

GMS99. Giachetta, G., Mangiarotti, L., Sardanashvily, G.: Covariant Hamilton equations for field theory. J. Phys. A **32**, 6629, (1999)

GN80. Gotay, M.J., Nester, J.M.: Generalized constraint algorithm and special presymplectic manifolds. In G. E. Kaiser. J. E. Marsden, Geometric methods in mathematical physics, Proc. NSF-CBMS Conf., Lowell/Mass. 1979, Berlin: Springer-Verlag, Lect. Notes Math. **775**, 78–80, (1980)

GNH78. Gotay, M., Nester, J., Hinds, G.: Presymplectic manifolds and the Dirac-Bergmann theory of constraints. J. Math. Phys. **19**, 2388, (1978)

GP00. Grantcharov, G., Poon, Y.S.: Geometry of hyper-Kähler connections with torsion Comm. Math. Phys. **213**(1), 19–37, (2000)

GP78. Gibbons, G.W., Perry, M.J.: Quantizing Gravitational Instantons, Nucl. Phys. **B146**, 90, (1978)

GPS95. Gordon, R., Power, A.J., Street, R.: Coherence for tricategories. Memoirs Amer. Math. Soc. **117**(558), (1995)

GRS01. Gorbunov, D.S., Rubakov, V.A., Sibiryakov, S.M.: Gravity waves from inflating brane or mirrors moving in adS(5).. JHEP **0110**, 015, (2001)

GRS96. Gilks, W.R., Richardson, S., Spiegelhalter, D.J.: Markov Chain Monte Carlo in Practice. Chapman & Hall, (1996)

GS02. Gen, U., Sasaki, M.: Quantum radion on de Sitter branes. arXiv:gr-qc/0201031, (2002)

GS95. Giachetta, G., Sardanashvily, G.: Stress-energy–momentum of affine–metric gravity. Generalized Komar superpotential. arXiv: gr-qc/9511008.

GS96. Giachetta, G., Sardanashvily, G.: Stress-Energy-Momentum of Affine-Metric Gravity. Class.Quant.Grav. **13** L, 67–72, (1996)

GS97. Giachetta, G., Sardanashvily, G.: Dirac Equation in Gauge and Affine-Metric Gravitation Theories. Int. J. Theor. Phys. **36**, 125–142, (1997)

GS98. Grosche, C., Steiner, F.: Handbook of Feynman path integrals. Springer Tracts in Modern Physics 145, Springer, Berlin, (1998)

GSV05. Garriga, J., Schwartz-Perlov, D., Vilenkin, A., Winitzki, S.: Probabilities in the Inflationary Multiverse. arXiv:hep-th/0509184, (2005)

GSW87. Green, M.B., Schwarz, J.H., Witten, E.: Superstring Theory, 2 Vols., Cambridge Univ. Press, Cambridge, (1987)

GT00. Garriga. J., Tanaka, T.: Gravity in the brane-world. Phys. Rev. Lett. **84**, 2778, (2000)

GT99. Gratton, S., Turok, N.: Cosmological Perturbations from the No Boundary Path Integral. Phys. Rev. D **60**, 123507, (1999)

GT01. Gratton, S., Turok, N.: Homogeneous Modes of Cosmological Instantons. Phys. Rev. D **63**, 123514, (2001)

GV93. Gasperini, M., Veneziano, G.: Pre-Big–Bang in String Cosmology, Astropart. Phys. **1**, 317, (1993)

GW82. Gibbons, G.W., Wiltshire, D.L.: Spacetime as a membrane in higher dimensions. Nucl. Phys. B **717**, (1987)

GW99. Goldberger, W.D., Wise, M.B.: Modulus stabilization with bulk fields. Phys. Rev. Lett. **83**, 4922, (1999)

GZ83. Gates, S.J., Zwiebach, B.: Gauged $N = 4$ Supergravity Theory with a New Scalar Potential. Phys. Lett. **B123**, 200, (1983)

Gal83. Gallavotti, G.: The Elements of Mechanics. Springer-Verlag, Berlin, (1983)

Gar72. García, P.: Connections and 1-jet fibre bundles. Rend. Sem. Univ. Padova, **47**, 227, (1972)

Gar77. García, P.: Gauge algebras, curvature and symplectic structure. J. Diff. Geom., **12**, 209, (1977)

Gar85. Gardiner, C.W.: Handbook of Stochastic Methods for Physics, Chemistry and Natural Sciences, (2nd ed). Springer-Verlag, New York, (1985)

Gau84. Gauduchon, P.: La 1-forme de torsion d'une varit hermitienne compacte. Math. Ann. **267**, 495, (1984)

Gef06. Gefter, A.: Stephen Hawking's Strange New Universe. New Sci. 22 April, (2006)

Geo88. Georgii, H.O.: Gibbs Measures and Phase Transitions. Walter de Gruyter, Berlin, (1988)

Ghe90. Ghez, C.: Introduction to motor system. In: Kandel, E.K. and Schwarz. J.H. (eds.) Principles of neural science. 2nd ed. Elsevier, Amsterdam, 429–442, (1990)

Ghe91. Ghez, C.: Muscles: Effectors of the Motor Systems. In: Principles of Neural Science. 3rd Ed. (Eds. E.R. Kandel. J.H. Schwartz, T.M. Jessell), Appleton and Lange, Elsevier, 548–563, (1991)

Gia92. Giachetta G.: Jet manifolds in nonholonomic mechanics. J. Math. Phys. **33**, 1652, (1992)

Gib85. Gibbons, G.W.: Aspects of Supergravity Theories. In Supersymmetry, Supergravity, and Related Topics, eds. F. del Aguila. J.A. de Azcarraga and L.E. Ibanez. World Scientific, Singapore, 346–351, (1985)

Gla63a. Glauber, R.J.: The Quantum Theory of Optical Coherence. Phys. Rev. **130**, 2529–2539, (1963)

Gla63b. Glauber, R.J.: Coherent and Incoherent States of the Radiation Field. Phys. Rev. **131**, 2766–2788, (1963)

Gle87. Gleick, J.: Chaos: Making a New Science. Penguin–Viking, New York, (1987)

Gol56. Goldberg, S.I.: Construction of universal bundles. Ann. Math. **63**, 64, (1956)

Gol99. Gold, M.: A Kurt Lewin Reader, the Complete Social Scientist. Am. Psych. Assoc., Washington, (1999)

Gom94. Gómez, J.C.: Using symbolic computation for the computer aided design of nonlinear (adaptive) control systems. Tech. Rep. EE9454, Dept. Electr. and Comput. Eng., Univ. Newcastle, Callaghan, NSW, AUS, (1994)

Goo98. Goodwine, J.W.: Control of Stratified Systems with Robotic Applications. PhD thesis, California Institute of Technology, Pasadena, Cal, (1998)

Gor26. Gordon, W.: Z. Phys., **40**, 117–133, (1926)

Got71. Goto, T.: Relativistic quantum mechanics of one-dimensional mechanical continuum and subsidary condition of dual resonance model. Prog. Theor. Phys, **46**, 1560, (1971)

Got82. Gotay, M.: On coisotropic imbeddings of presymplectic manifolds. Proc. Amer. Math. Soc. **84**, 111, (1982)

Got91a. Gotay, M.J.: A multisymplectic framework for classical field theory and the calculus of variations. I: Covariant Hamiltonian formalism. In M. Francaviglia (ed.), Mechanics, analysis and geometry: 200 years after Lagrangian. North-Holland, Amsterdam, 203–235, (1991)

Got91b. Gotay, M.J.: A multisymplectic framework for classical field theory and the calculus of variations. II: Space + time decomposition. Differ. Geom. Appl. **1**(4), 375–390, (1991)

Got91c. Gotay, M.: A multisymplectic framework for classical field theory and the calculus of variations. I. Covariant Hamiltonian formalism. In: M.Francaviglia (ed.) Mechanics, Analysis and Geometry: 200 Years after Lagrange. 203–235. North-Holland, Amsterdam, (1991)

Gou95. Goubault, E.: The Geometry of Concurrency. PhD thesis, Ecole Normale Supérieure, (1995)

Gre00. Greene, B.R.: The Elegant Universe: Superstrings, Hidden Dimensions, and the Quest for the Ultimate Theory. Random House, (2000)

Gre96. Greene, B.R.: String Theory on Calabi-Yau Manifolds. Lectures given at the TASI-96 summer school on Strings, Fields and Duality, (1996)

Gri02. Griffiths, R.B.: Consistent Quantum Theory. Cambridge University Press, Cambridge, (2002)

Gri83a. Griffiths, P.A.: Exterior Differential Systems and the Calculus of Variations, Birkhauser, Boston, (1983)

Gri83b. Griffiths, P.A.: Infinitesimal variations of Hodge structure. III. Determinantal varieties and the infinitesimal invariant of normal functions. Compositio Math., **50**(2-3), 267–324, (1983)

Gri84. Griffiths, R.B.: Consistent Histories and the Interpretation of Quantum Mechanics. J. Stat. Phys. **36**, 219, (1984)

Gro69. Grossberg, S.: Embedding fields: A theory of learning with physiological implications. J. Math. Psych. **6**, 209–239, (1969)

Gro82. Grossberg, S.: Studies of Mind and Brain. Dordrecht, Holland, (1982)

Gro83. Grothendieck, A.: Pursuing stacks (Unpublished manuscript, distributed from UCNW). Bangor, UK, (1983)

Gro88. Grossberg, S.: Neural Networks and Natural Intelligence. MIT Press, Cambridge, MA, (1988)

Gro99. Grossberg, S.: How does the cerebral cortex work? Learning, attention and grouping by the laminar circuits of visual cortex. Spatial Vision **12**, 163–186, (1999)

Gru00. Grunwald, P.: The minimum description length principle. J. Math. Psych., **44**, 133–152, (2000)

Gru99. Grunwald, P.: Viewing all models as 'probabilistic'. Proceedings of the Twelfth Annual Conference on Computational Learning Theory (COLT' 99), Santa Cruz, CA, (1999)

Gua04a. Gualtieri, M.: Generalized complex geometry. arXiv:math.DG/0401221, (2004a)

Gua04b. Gualtieri, M.: Generalized geometry and the Hodge decomposition. arXiv:math.DG/04090903, (2004b)

Gun03. Gunion, J.F.: Class Notes on Path-Integral Methods. U.C. Davis, 230B, (2003)

Gut81. Guth, A.H.: The Inflationary Universe: A Possible Solution to the Horizon and Flatness Problems. Phys. Rev. D **23**, 347, (1981)

Gut82. Guth, A.H.: The Inflationary Universe: A Possible Solution to the Horizon and Flatness Problems. Phys. Rev. D **23**, 347, (1981)

Gut90. Gutzwiller, M.C.: Chaos in Classical and Quantum Mechanics. Springer, New York, (1990)

Gut98. Gutkin, B.S., Ermentrout, B.: Dynamics of membrane excitability determine interspike interval variability: A link between spike generation mechanisms and cortical spike train statistics. Neural Comput., **10**(5), 1047–1065, (1998)

HBB96. Houk, J.C., Buckingham. J.T., Barto, A.G.: Models of the cerebellum and motor learning. Behavioral and Brain Sciences, **19**(3), 368–383, (1996)

HC99. Hilbert, D., Cohn-Vossen, S.: Geometry and the Imagination. (Reprint ed.), Amer. Math. Soc, (1999)

HDD98. Arkani-Hamed, N., Dimopoulos, S., Dvali, G.: Phenomenology, Astrophysics and Cosmology of theories with sub-millimeter dimensions and TeV scale quantum gravity. Phys. Lett. **B429**, 263, (1998)

HDK00. Arkani-Hamed, N., Dimopoulos, S., Kaloper, N., Sundrum, R.: A small cosmological constant from a large extra dimension.. Phys. Lett. B **480**, 193, (2000)

HDR02. Hasty, J., Dolnik, M., Rottschafer, V., Collins. J.J.: A synthetic gene network for entraining and amplifying cellular oscillations. Phys. Rev. Lett. **88**(14), 1–4, (2002)

HE73. Hawking, S.W., Ellis, G.F.R.: The large scale structure of space-time. Cambridge Univ. Press, (1973)

HE79. Hawking, S.W., Israel, W. (ed.): General relativity: an Einstein centenary survey. Cambridge Univ. Press, Cambridge, (1979)

HGM94. Hartle, J.B., Gell-Mann, M.: Time Symmetry and Asymmetry in Quantum Mechanics and Quantum Cosmology in Physical Origins of Time Asymmetry ed. by J. Halliwell. J. Perez-Mercader, and W. Zurek. Cambridge Univ. Press, Cambridge, (1994)

HH02. Hawking, S.W., Hertog, T.: Why Does Inflation Start at the Top of the Hill? Phys. Rev. D **66**, 123509, (2002)

HH04. Hertog, T., Horowitz, G.T.: Towards a Big–Crunch Dual. JHEP **0407**, 073, (2004)

HH06. Hawking, S.W., Hertog, T.: Populating the Landscape: A Top Down Approach. Phys.Rev. D **73**, 123527, (2006)

HH52. Hodgkin, A.L., Huxley, A.F.: A quantitative description of membrane current and application to conduction and excitation in nerve. J. Physiol., **117**, 500–544, (1952)

HH81. Hartle, J.B., Horowitz, G.T.: Ground-State Expectation value of the Metric in the $1/N$ or Semiclassical Approximation to Quantum Gravity. Phys. Rev. D **24**, 257, (1981)

HH83. Hartle, J.B., Hawking, S.W.: The Wave Function of the Universe. Phys. Rev. D **28**, 2960, (1983)

HH90. Halliwell, J.J., Hartle, J.B.: Integration Contours for the No-Boundary Wave Function of the Universe. Phys. Rev. D **41**, 1815, (1990)

HHR00. Hawking, S.W., Hertog, T., Reall, H.S.: Brane New World. Phys. Rev. D **62**, 043501, (2000)

HHR01. Hawking, S.W., Hertog, T., Reall, H.S.: Trace Anomaly Driven Inflation. Phys. Rev. D **63**, 083504, (2001)

HHT00. Hawking, S.W., Hertog, T., Turok, N.: Gravitational Waves in Open de Sitter Space. arXiv:hep-th/0003016, (2000)

HI97. Hoppensteadt, F.C., Izhikevich, E.M.: Weakly Connected Neural Networks. Springer, New York, (1997)

HKK03. Hori, K., Katz, S., Klemm, A., Pandharipande, R., Thomas, R., Vafa, C., Vakil, R., Zaslow, E.: Mirror symmetry. Clay Mathematics Monographs 1, American Mathematical Society, Providence, Clay Mathematics Institute, Cambridge, MA, (2003)

HKS01. Hellerman, S., Kaloper, N., Susskind, L.: String Theory and Quintessence. arXiv:hep-th/0104180, (2001)

HL84. Hamoui, A., Lichnerowicz, A.: Geometry of dynamical systems with time–dependent constraints and time-dependent Hamiltonians: An approach towards quantization. J. Math. Phys. **25**, 923, (1984)

HL90. Halliwell, J.J., Louko. J.: Steepest-Descent Contours in the Path Integral Approach to Quantum Cosmology, III. A General Method with Applications to Anisotropic Minisuperspace Models. Phys. Rev. D **42**, 3997, (1990)

HLR86. Horowitz, G.T., Lykken. J., Rohm, R., Strominger, A.: Purely Cubic Action for String Field Theory. Phys. Rev. Lett. **57**, 283, (1986)

HM82. Hawking, S.W., Moss, I.G.: Supercooled phase transitions in the very early universe. Phys. Lett. **B110**, 358, (1982)

HM88. Hitchin, N.J., Murray, M.K.: Spectral curves and the ADHMN method, Commun. Math. Phys. 114, 463–474, (1988)

HM89. Hurtubise, J., Murray, M.K.: On the construction of monopoles for the classical groups, Commun. Math. Phys. 122, 35–89, (1989).

HM90. Hurtubise, J., Murray, M.K.: Monopoles and their spectral data, Commun. Math. Phys. 133, 487–508, (1990)

HM97. Hartle, J.B. and D. Marolf, Comparing Formulations of Generalized Quantum Mechanics for Reparametrization Invariant Systems. Phys. Rev. D **56**, 6247-6257, (1997)

HMM95. Hitchin, N.J., Manton, N.S., Murray, M.K.: Symmetric Monopoles, Nonlinearity 8, 661–692, (1995)

HMM95. Hehl, F., McCrea. J., Mielke, E., Ne'eman, Y.: Metric–affine gauge theory of gravity. Phys. Rep. **258**, 1, (1995)

HMS82. Hawking, S.W., Moss, I.G., Stewart, J.M.: Bubble Collisions in the Very Early Universe. Phys. Rev. D **26**, 2681, (1982)

HN54. Huxley, A.F., Niedergerke, R.: Changes in the cross–striations of muscle during contraction and stretch and their structural interpretation. Nature, **173**, 973–976, (1954)

HP02. Horowitz, G.T., Polchinski, J.: Instability of spacelike and null orbifold singularities. Phys. Rev. D **66**, 103512, (2002)

HP29a. Heisenberg, W., Pauli, W.: Z. Phys., **56**, 1–61, (1929)

HP29b. Heisenberg, W., Pauli, W.: Z. Phys., **59**, 168–190, (1929)

HP96. Hameroff, S.R., Penrose, R.: Orchestrated reduction of quantum coherence in brain microtubules: A model for consciousness. In: Hameroff, S. R., Kaszniak, A.W. and Scott, A.C. Eds: Toward a Science of Consciousness: the First Tucson Discussion and Debates, 507–539. MIT Press, Cambridge, MA, (1996)

HP96. Hawking, S.W., Penrose, R.: The Nature of Space and Time. Princetone Univ. Press, Princetone, NJ, (1996)

HR83. Hut, P., Rees, M.J.: How stable is our vacuum? Nature, 508509, (1983)

HS01. Himemoto, Y., Sasaki, M.: Brane-world inflation without inflaton on the brane. Phys. Rev. D **63**, 044015, (2001)

HS74. Hirsch, M. Smale, S.: Differential Equations, Dynamical Systems, and Linear Algebra, Academic Press, New York, (1974)

HS98. Hofbauer, J., Sigmund, K.: Evolutionary games and population dynamics, Cambridge Univ. Press, U.K., (1998)

HS98. Henningson, M., Skenderis, K.: The Holographic Weyl Anomaly. JHEP **9807**, 023, (1998)

HSK92. Hauser, J., Sastry, S., Kokotovic, P.: Nonlinear control via approximate input–output linearization: The ball and beam example, IEEE Trans. Aut. Con., AC–37, 392–398, (1992)

HT85. Hopfield, J.J., Tank, D.W.: Neural computation of decisions in optimisation problems. Biol. Cybern., **52**, 114–152, (1985)

HT92. Henneaux, M., Teitelboim, C.: Quantization of Gauge systems. Princeton Univ. Press, (1992)

HT93. Hunt, L, Turi. J.: A new algorithm for constructing approximate transformations for nonlinear systems. IEEE Trans. Aut. Con., AC–38,1553–1556, (1993)

HT98. Hawking, S.W., Turok, N.: Open Inflation without False Vacua. Phys. Lett. **B425**, 25, (1998)

HTS02. Himemoto, Y., Tanaka, T., Sasaki, M.: A bulk scalar in the braneworld can mimic the 4d inflaton dynamics. Phys. Rev. D **65**, 104020 (2002)

HT00. Hertog, T., Turok, N.: Gravity Waves from Instantons. Phys. Rev. D **62**, 083514, (2000)

HT01. Hawking, S.W., Hertog, T.: Living with ghosts. arXiv:hep-th/0107088, (2001)

HW04. Hamber, H., Williams, R.M.: Non-perturbative Gravity and the Spin of the Lattice Gravition. Phys. Rev. D **70**, 124007, (2004)

HW96. Horava, P., Witten, E.: Heterotic and type I string dynamics from eleven dimensions, Nucl. Phys. B **460**, 506, (1996); ibid Eleven-Dimensional Supergravity on a Manifold with Boundary, Nucl. Phys. B **475**, 94, (1996)

HWM97. Hartle, J.B., Williams, R.M., Miller, W.A., Williams, R.: Signature of the Simplicial Supermetric, Class. Quant. Grav. **14**, 2137–2155, (1997)

HYW03. Hao, N., Yildirim, N., Wang, Y., Elston, T.C., Dohlman, H.G.: Regulators of G protein signaling and transient activation of signaling: experimental and computational analysis reveals negative and positive feedback controls on G protein activity. J. Biol. Chem., 278, 46506–46515, (2003)

Haa92. Haag, R.: Local Quantum Physics. Springer, Berlin, (1992)

Hac03. Hackermüller, L. et al.: The Wave Nature of Biomolecules and Flurofullerenes. Phys. Rev. Lett. **91**, 090408, (2003)

Hak83. Haken, H.: Synergetics: An Introduction (3rd ed). Springer, Berlin, (1983)

Hak93. Haken, H.: Advanced Synergetics: Instability Hierarchies of Self-Organizing Systems and Devices (3nd ed.). Springer, Berlin. (1993)

Ham82. Hamilton, R.: Three-manifolds with positive Ricci curvature. J. Diff. Geom. **17**, 255–306, (1982)

Ham86. Hamilton, R.: Four-manifolds with positive curvature operator. J. Diff. Geom. **24**, 153–179, (1986)

Ham87. Hameroff, S.R.: Ultimate Computing: Biomolecular Consciousness and Nanotechnology. North-Holland, Amsterdam, (1987)

Ham88. Hamilton, R.: The Ricci flow on surfaces. Contem. Math. **71**, 237–261, (1988)

Ham88. Hamilton, R.S.: The Ricci flow on surfaces. Contem. Math. **71**, 237–261, (1988)

Ham93. Hamilton, R.: The formation of singularities in the Ricci flow, volume II. Internat. Press, (1993)

Har03. Hartle, J.B.: The State of the Universe, in The Future of Theoretical Physics and Cosmology: Stephen Hawking 60th Birthday Symposium, ed. by G.W. Gibbons, E.P.S. Shellard, and S.J. Ranken. Cambridge Univ. Press, UK, (2003)

Har03. Hartle, J.B.: Theories of Everything and Hawking's Wave Function of the Universe. In The Future of Theoretical Physics and Cosmology, ed. by G.W. Gibons, E.P.S. Shellard and S.J. Rankin. Cambridge Univ. Press, Cambridge, (2003)

Har04. Hartle, J.B.: Bohmian Histories and Decoherent Histories. Phys. Rev. **A69**, 042111, (2004)

Har04a. Hartle, J.B.: Linear Positivity and Virtual Probability. Phys. Rev. **A70**, 022104, (2004)

Har04b. Hartle, J.B.: Anthropic Reasoning and Quantum Cosmology. The New Cosmology, ed. R. Allen et al., AIP, (2004)

Har05a. Hartle, J.B.: Anthropic Reasoning and Quantum Cosmology, Proceedings of Strings and Cosmology Conference, Texas A&M, AIP Conf. Proc. 743, 298, (2005)

Har05b. Hartle, J.B.: Generalizing Quantum Mechanics for Quantum Spacetime. Proc. of the 23rd Solvay Conference, The Quantum Structure of Space and Time, Brussels, (2005)

Har06. Hartle, J.B.: Generalizing Quantum Mechanics for Quantum Gravity. Int. J. Theor. Phys. **45**, 1390-1396, (2006)

Har85a. Hartle, J.B.: Simplicial Minisuperspace I. General Discussion J. Math. Phys. **26**, 804, (1985); Simplicial Minisuperspace III: Integration Contours in a Five-Simplex Model, ibid. **30**, 452, (1989)

Har85b. Hartle, J.B.: Unruly Topologies in Two Dimensional Quantum Gravity, Class. Quant. Grav. **2**, 707, (1985)

Har90. Hartle, J.B.: Excess Baggage. In Elementary Particles and the Universe: Essays in Honor of Murray Gell-Mann ed. by J. Schwarz. Cambridge Univ. Press, Cambridge, (1990)

Har91a. Hartle, J.B., The Quantum Mechanics of Cosmology. In Quantum Cosmology and Baby Universes: Proceedings of the 1989 Jerusalem Winter School for Theoretical Physics, ed. by S. Coleman, Hartle, J.B., T. Piran, and S. Weinberg, World Scientific, Singapore, 65–157, (1991)

Har91b. Hartle, J.B.: Spacetime Coarse Grainings in Non-Relativistic Quantum Mechanics. Phys. Rev. D **44**, 3173–3196, (1991)

Har93. Hartle, J.B.: The Quantum Mechanics of Closed Systems. In Directions in General Relativity, Volume 1: A Symposium and Collection of Essays in honor of Professor Charles W. Misner's 60th Birthday, ed. by B.-L. Hu, M.P. Ryan, and C.V. Vishveshwara. Cambridge University Press, Cambridge, (1993)

Har94a. Hartle, J.B.: Unitarity and Causality in Generalized Quantum Mechanics for Non-Chronal Spacetimes. Phys. Rev. D **49**, 6543, (1994)

Har94b. Hartle, J.B.: Quasiclassical Domains In A Quantum Universe, in Proceedings of the Cornelius Lanczos International Centenary Conference, North Carolina State University, December 1992, ed. by J.D. Brown, M.T. Chu, D.C. Ellison, R.J. Plemmons, SIAM, Philadelphia, (1994)

Har94c. Hartle, J.B.: Unitarity and Causality in Generalized Quantum Mechanics for Non-Chronal Spacetimes, Phys Rev D **49**, 6543, (1994)

Har95a. Hartle, J.B.: Spacetime Quantum Mechanics and the Quantum Mechanics of Spacetime in Gravitation and Quantizations, Proceedings of the 1992 Les Houches Summer School, edited by B. Julia and J. Zinn-Justin, Les Houches Summer School Proceedings Vol. LVII, North Holland, Amsterdam, (1995)

Har95b. Hartle, J.B., Spacetime Quantum Mechanics and the Quantum Mechanics of Spacetime in Gravitation and Quantizations, Proceedings of the 1992 Les Houches Summer School, ed. by B. Julia and J. Zinn-Justin, Les Houches Summer School Proceedings Vol. LVII, North Holland, Amsterdam, (1995)

Har96. Hartle, J.B.: Scientific Knowledge from the Perspective of Quantum Cosmology in Boundaries and Barriers : On the Limits to Scientific Knowledge, ed. by John L. Casti and Anders Karlqvist, Addison-Wesley, Reading, MA, (1996)

Har98. Hartle, J.B.: Generalized Quantum Theory in Evaporating Black Hole Spacetimes, in Black Holes and Relativistic Stars: A Symposium in Honor of S. Chandrasekhar, ed. by R.M. Wald, University of Chicago Press, Chicago, (1998)

Hat02. Hatcher, A.: Algebraic Topology. Cambridge Univ. Press, Cambridge, (2002)

Hat77a. Hatze, H.: A myocybernetic control model of skeletal muscle. Biol. Cyber. **25**, 103–119, (1977)

Hat77b. Hatze, H.: A complete set of control equations for the human musculoskeletal system. J. Biomech. **10**, 799–805, (1977)

Hat78. Hatze, H.: A general myocybernetic control model of skeletal muscle. Biol. Cyber., **28**, 143–157, (1978)

Haw75. Hawking, S.W.: Particle creation by black holes. Commun. Math. Phys, **43**, 199-220, (1975)

Haw84a. Hawking, S.W.: The Cosmological Constant in Probably Zero. Phys. Lett. **B134**, 403, (1984)

Haw84b. Hawking, S.W.: The Quantum State of the Universe, Nucl. Phys. **B239**, 257, (1984)

Haw87. Hawking, S.W. In Quantum Field Theory and Quantum Statistics: Essays in Honor of the 60th Birthday of E.S. Fradkin, eds. A. Batalin, C.J. Isham and C.A. Vilkovisky, Hilger, Bristol, UK, (1987)

Hay94. Haykin, S. Neural Networks: A Comprehensive Foundation. Macmillan, (1994)

Heb49. Hebb, D.O.: The Organization of Behavior, Wiley, New York, (1949)

Hec77. Heckhausen, H.: Achievement motivation and its constructs: a cognitive model. Motiv. Emot, **1**, 283–329, (1977)

Hec87. Hecht-Nielsen, R.: Counterpropagation networks. Applied Optics, **26**(23), 4979–4984, (1987)

Hee90. Heermann, D.W.: Computer Simulation Methods in Theoretical Physics. (2nd ed), Springer, Berlin, (1990)

Hel01. Helgason, S.: Differential Geometry, Lie Groups and Symmetric Spaces. Am. Math. Soc., Providence, (2001)

Hen06. Henson, J.: The Causal Set Approach to Quantum Gravity. arXiv:gr-qc/0601121, (1980)

Hen66. Hénon, M.: Sur la topologie des lignes de courant dans un cas particulier. C. R. Acad. Sci. Paris A, **262**, 312–314, (1966).

Hen69. Hénon, M.: Numerical study of quadratic area preserving mappings. Q. Appl. Math. **27**, (1969)

Hen76. Hénon, M.: A two-dimensional mapping with a strange attractor. Com. Math. Phys. **50**, 6977, (1976)

Her02. Hertog, T.: The Origin of Inflation, Ph.D. dissertation, University of Cambridge, (2002)

Hig87. Higuchi, A.: Symmetric tensor spherical harmonics on the N-sphere and their application to the de Sitter group $SO(N, 1)$. J. Math. Phys. **28** (7), 1553, (1987)

Hil38. Hill, A.V.: The heat of shortening and the dynamic constants of muscle, Proc. R. Soc. B, **76**, 136–195, (1938)

Hir66. Hirzebruch, F.: Topological Methods in Algebraic Geometry, Springer-Verlag, Berlin, (1966)

Hir76. Hirsch, M.W.: Differential Topology. Springer, New York, (1976)

Hit03. Hitchin, N.J.: Generalized Calabi–Yau manifolds. Quat. J. Math. **54**, 281-308, (2003)

Hit05. Hitchin, N.J.: Instantons, Poisson structures and generalized Kähler geometry. arXiv:math.DG/0503432, (2005)

Hit82. Hitchin, N.J.: Monopoles and Geodesics, Commun. Math. Phys. 83, (1982), 579–602.

Hit83. Hitchin, N.J.: On the construction of monopoles, Commun. Math. Phys. 89, (1983), 145–190.

Hod64. Hodgkin, A.L.: The Conduction of the Nervous Impulse. Liverpool Univ. Press, Liverpool, (1964)

Hop82. Hopfield, J.J.: Neural networks and physical systems with emergent collective computational activity. Proc. Natl. Acad. Sci. USA., **79**, 2554–2558, (1982)

Hop84. Hopfield, J.J.: Neurons with graded response have collective computational properties like those of two–state neurons. Proc. Natl. Acad. Sci. USA, **81**, 3088–3092, (1984)

Hor89. Horowitz, G.T.: Exactly soluble diffeomorphism invariant theories. Comm. Math. Phys. **125**(3), 417, (1989)

Hou79. Houk, J.C.: Regulation of stiffness by skeletomotor reflexes. Ann. Rev. Physiol., **41**, 99–114, (1979)

Hul00. Hull, J.C.: Options, Futures, and Other Derivatives. (4th ed.), Prentice Hall, New Jersey, (2000)

Hul84. Hull, C.M.: New Gauging of $N = 8$ Supergravity. Phys. Rev. D **30**, 760, (1984)

Hur93. Hurmuzlu, Y.: Dynamics of bipedal gait. J. Appl. Mech., **60**, 331–343, (1993)

Hux57. Huxley, A.F.: Muscle structure and theories of contraction. Progr. Biophys. Chem., **7**, 255–328, (1957)

IB05. Ivancevic, V., Beagley, N.: Brain-like functor control machine for general humanoid biodynamics. Int. J. Math. Math. Sci. **11**, 1759–1779, (2005)

II05. Ivancevic, V., Ivancevic, T.: Human–Like Biomechanics. Springer, Mechanical Engineering Ser., (2005)

II06a. Ivancevic, V., Ivancevic, T.: Natural Biodynamics. World Scientific, Series: Mathematical Biology, (2006)

II06b. Ivancevic, V., Ivancevic, T.: Geometrical Dynamics of Complex Systems. Springer, Series: Microprocessor-Based and Intelligent Systems Engineering, Vol. 31, (2006)

II07. Ivancevic, V., Ivancevic, T.: High-Dimensional Chaotic and Attractor Systems. Springer, Series: Springer, Intelligent Systems, Control and Automation: Science and Engineering, Vol. 32, (2007)

IJB99a. Ivancevic, T., Jain, L.C., Bottema, M.: New Two–feature GBAM–Neurodynamical Classifier for Breast Cancer Diagnosis. Proc. KES'99, IEEE Press, (1999)

IJB99b. Ivancevic, T., Jain, L.C., Bottema, M.: A New Two–Feature FAM–Matrix Classifier for Breast Cancer Diagnosis. Proc. KES'99, IEEE Press (1999)

IJK02. Ichiki, K., Yahiro, M., Kajino, T., Orito, M., Mathews, G.J.: Observational constraints on dark radiation in brane cosmology. Phys. Rev. D **66**, 043521, (2002)

IK80. Iyanaga, S., Kawada, Y. (eds.): Pontryagin's Maximum Principle. In Encyclopedic Dictionary of Mathematics. MIT Press, Cambridge, MA, 295-296, (1980)

IKK97. Ishibashi, N., Kawai, H., Kitazawa, Y., Tsuchiya, A.: A Large-N Reduced Model as Superstring. Nucl. Phys. **B498**, 467, (1997)

ILI95. Ivancevic, V., Lukman, L., Ivancevic, T.: Selected Chapters in Human Biomechanics. Textbook (in Serbian). Univ. Novi Sad Press, Novi Sad, (1995)

IN92. Igarashi, E., Nogai, T.: Study of lower level adaptive walking in the saggital plane by a biped locomotion robot. Advanced Robotics, **6**, 441–459, (1992)

IP01a. Ivancevic, V., Pearce, C.E.M.: Poisson manifolds in generalised Hamiltonian biomechanics. Bull. Austral. Math. Soc. **64**, 515–526, (2001)

IP01b. Ivancevic, V., Pearce, C.E.M.: Topological duality in humanoid robot dynamics. ANZIAM J. **43**, 183–194, (2001)

IP05b. Ivancevic, V., Pearce, C.E.M.: Hamiltonian dynamics and Morse topology of humanoid robots. Gl. J. Mat. Math. Sci. (to appear)

IS01. Ivancevic, V., Snoswell, M.: Fuzzy–stochastic functor machine for general humanoid–robot dynamics. IEEE Trans. on Sys, Man, Cyber. B, **31**(3), 319–330, (2001)

Ich04. Ichinomiya, T.: Frequency synchronization in random oscillator network. Phys. Rev. E **70**, 026116, (2004)

Ich05. Ichinomiya, T.: Path-integral approach to the dynamics in sparse random network. Phys. Rev. E **72**, 016109, (2005)

Ida00. Ida, D.: Brane-world cosmology. JHEP **0009**, 014, (2000)

Ike90. Ikeda, S.: Some remarks on the Lagrangian Theory of Electromagnetism. Tensor, N.S. **49**, (1990)

Ila01. Ilachinski, A.: Cellular automata. World Scientific, Singapore, (2001)

Imm97. Immirzi, G.: Quantum Gravity and Regge Calculus. Nucl. Phys. Proc. Suppl. **57**, 65–72, (1997)

Ing97. Ingber, L.: Statistical mechanics of neocortical interactions: Applications of canonical momenta indicators to electroencephalography, Phys. Rev. E, **55**(4), 4578–4593, (1997)

Ing98. Ingber, L.: Statistical mechanics of neocortical interactions: Training and testing canonical momenta indicators of EEG, Mathl. Computer Modelling **27**(3), 33–64, (1998)

Ish84. Isham, C.J.: In B. DeWitt, R.Stora, (ed.), Relativity, Groups and Topology, Les Houches Session XL. North Holland, Amsterdam, (1984)

Ish94. Isham, C.J.: Quantum Logic and the Histories Approach to Quantum Theory. J. Math. Phys. **35**, 2157, (1994)

Ish97. Isham, C.J.: Structural issues in quantum gravity, in General Relativity and Gravitation: World Scientific, Singapore, (1997)

Isi03. Isidro, J.M.: Duality, Quantum Mechanics and (Almost) Complex Manifolds. Mod. Phys. Lett. **A18**(28), 1975, (2003)

Isi04a. Isidro, J.M.: Quantum Mechanics in Infinite Symplectic Volume. Mod. Phys. Lett. **A19**(5), 349, (2004)

Isi04b. Isidro, J.M.: Quantum-Mechanical Dualities on the Torus. Mod. Phys. Lett. **A19**, 1733, (2004)

Isi89. Isidori, A.: Nonlinear Control Systems. An Introduction, (2nd ed) Springer, Berlin, (1989)

Isr66. Israel, W.: Singular Hypersurfaces and Thin Shells in General Relativity, Nuovo Cim. B **44S10**, 1, (1966)

Ito60. Ito, K.: Wiener Integral and Feynman Integral. Proc. Fourth Berkeley Symp. Math., Stat., Prob., **2**, 227–238, (1960)

Iva91. Ivancevic, V.: Introduction to Biomechanical Systems: Modelling, Control and Learning (in Serbian). Scientific Book, Belgrade, (1991)

Iva95. Ivancevic, T.: Some Possibilities of Multilayered Neural Networks Application in Biomechanics of Muscular Contractions, Human Motion and Sport Training. Master Thesis (in Serbian), University of Novi Sad, YU, (1995)

Iva02. Ivancevic, V.: Generalized Hamiltonian biodynamics and topology invariants of humanoid robots. Int. J. Mat. Mat. Sci., **31**(9), 555–565, (2002)

Iva04. Ivancevic, V.: Symplectic Rotational Geometry in Human Biomechanics. SIAM Rev., **46**(3), 455–474, (2004)

Iva05a. Ivancevic, V.: Dynamics of Humanoid Robots: Geometrical and Topological Duality. In Biomathematics: Modelling and simulation (ed. J.C. Misra), World Scientific, Singapore (to appear)

Iva05b. Ivancevic, V.: A Lagrangian model in human biomechanics and its Hodge–de Rham cohomology. Int. J. Appl. Math. Mech., **3**, 35–51, (2005)

Iva05c. Ivancevic, V.: Lie–Lagrangian model for realistic human bio-dynamics. Int. J. Humanoid Robotics 3(2), 205–218, (2006)

Izh01. Izhikevich, E.M.: Resonate-and-fire neurons. Neu. Net., **14**, 883–894, (2001)

Izh04. Izhikevich, E.M.: Which model to use for cortical spiking neurons? IEEE Trans. Neu. Net., **15**, 1063–1070, (2004)

Izh99. Izhikevich, E.M.: Class 1 neural excitability, conventional synapses, weakly connected networks, and mathematical foundations of pulse-coupled models. IEEE Trans. Neu. Net., **10**, 499–507, (1999)

JPY96. Jibu, M., Pribram, K.H., Yasue, K.: From conscious experience to memory storage and retrieval: the role of quantum brain dynamics and boson condensation of evanescent photons, Int. J. Mod. Phys. B, **10**, 1735, (1996)

JT80. Jaffe, A., Taubes, C.: Vortices and monopoles, Birkhäuser, Boston, MA, (1980)

JY95. Jibu, M., Yasue, K.: Quantum brain dynamics and consciousness. John Benjamins, Amsterdam, (1995)

JZ85. Joos, E., Zeh, H.D.: Emergence of Classical Properties through Interaction with the Environment, Zeit. Phys. B **59**, 223, (1985)

Jar98a. Jarvis, S.: Euclidean monopoles and rational maps, Proc. London Math. Soc. 3(77), 170–192, (1998a)

Jar98b. Jarvis, S.: Construction of Euclidean monopoles, Proc. London Math. Soc. 3(77), 193–214, (1998b)

Jon85. Jones, V.F.R.: A new polynomial invariant of knots and links. Bull. Amer. Math. Soc. **12**, 103, (1985)

Jul18. Julia, G.: Mémoires sur l'itération des fonctions rationelles. J. Math. **8**, 47–245, (1918)

KF86. Klein, M.V., Furtak, T.E.: Optics (2nd ed), Wiley, New York, (1986)

KFA69. Kalman, R.E., Falb, P., Arbib, M.A.: Topics in Mathematical System Theory. McGraw Hill, New York, (1969)

KHS93. Koruga, D.L., Hameroff, S.I., Sundareshan, M.K., Withers. J., Loutfy, R.: Fullerence C60: History, Physics, Nanobiology and Nanotechnology. Elsevier Science Pub, (1993)

KK01. Kawahara, G., Kida, S.: Periodic motion embedded in plane Couette turbulence: regeneration cycle and burst. J. Fluid Mech. **449**, 291–300, (2001)

KKL01. Kallosh, R., Kofman, L., Linde, A.D.: Pyrotechnic universe. Phys. Rev. D **64**, 123523, (2001)

KKL03. Kachru, S., Kallosh, R., Linde, A., Trivedi, S.P.: De Sitter Vacua in String Theory. arXiv:hep-th/0301240, (2003)

KKV99. Katz, S., Klemm, A., Vafa, C.: M–theory, topological strings and spinning black holes. Adv. Theor. Math. Phys. **3**, 1445, (1999)

KLR03. Kushner, A., Lychagin, V., Roubtsov, V.: Contact geometry and Non-linear Differential Equations. Cambridge Univ. Press, Cambridge, (2003)

KLS99. Kraus, P., Larsen, F., Siebelink, R.: The gravitational action in asymptotically AdS and flat space–times. Nucl. Phys. B **563**, 259, (1999)

KLV85. Krasil'shchik, I., Lychagin, V., Vinogradov, A.: Geometry of Jet Spaces and Nonlinear Partial Differential Equations, Gordon and Breach, Glasgow, (1985)

KMS93. Kolar, I., Michor, P.W., Slovak. J.: Natural Operations in Differential Geometry. Springer, Berlin, (1993)

KN00. Kotz, S., Nadarajah, S.: Extreme Value Distributions. Imperial College Press, London, (2000)

KN63/9. Kobayashi, S., Nomizu, K.: Foundations of Differential Geometry, Vols. 1,2., Interscience Publ., New York, (1963/1969).

KN96. Kobayashi, S., Nomizu, K.: Foundations of Differential Geometry, Wiley Interscience, New York, (1996)

KO89. Kamron, N., Olver, P.J.: Le probléme d'equivalence à une divergence prés dans le calcul des variations des intégrales multiples. C. R. Acad. Sci. Paris,t. **308**, Sèrie I, 249–252, (1989)

KOS01. Khoury, J., Ovrut, B.A., Seiberg, N., Steinhardt, P.J., Turok, N.: From Big Crunch to Big–Bang. arXiv:hep-th/0108187, (2001)

KOS01a. Khoury, J., Ovrut, B.A., Steinhardt, P.J., Turok, N.: The ekpyrotic universe: Colliding branes and the origin of the hot Big–Bang. Phys. Rev. D **64** 123522, (2001)

KOS01b. Khoury, J., Ovrut, B.A., Steinhardt, P.J., Turok, N.: The ekpyrotic universe: Colliding branes and the origin of the hot Big–Bang. Phys. Rev. D **64**, 123522, (2001)

KR03. Kock, A., Reyes, G.E.: Some calculus with extensive quantities: wave equation. Theory and Applications of Categories, **11**(14), 321–336, (2003)

KR95. Kauffman, L.H., Radford, D.E.: Invariants of 3

KR99. Kauffman, L.H., Radford, D.E.: Quantum algebra structures on $n \times n$ matrices. J. Algebra **213**, 405–436, (1999)

KRL03. Kachru, S., Kallosh, R., Linde, A., Trivedi, S.P.: De Sitter Vacua in String Theory. Phys. Rev. D **68**, 046005, (2003)

KS00. Koyama, K., Soda, J.: Evolution of cosmological perturbations in the brane world. Phys. Rev. D **62**, 123502, (2000)

KS02. Koyama, K., Soda, J.: Bulk gravitational field and cosmological perturbations on the brane. Phys. Rev. D **65**, 023514, (2002)

KS04. Karczmarek, J., Strominger, A.: Matrix Cosmology. JHEP **0404**, 055, (2004)

KS76. Kijowski, J., Szczyrba, W.: A Canonical Structure for Classical Field Theories. Commun. Math. Phys. **46**, 183, (1976)

KT75. Kamber, F., Tondeur, P.: Foliated Bundles and Characteristic Classes. Lecture Notes in Math. 493, Springer, Berlin, (1975)

KT79. Kijowski. J., Tulczyjew, W.: A Symplectic Framework for Field Theories, Springer-Verlag, Berlin, (1979)

KTW00. Kirklin, K., Turok, N., Wiseman, T.: Singular Instantons Made Regular. Phys. Rev. D **63**, 083509, (2000)

KV03a. Katic, D., Vukobratovic, M.: Advances in Intelligent Control of Robotic Systems, Book series: Microprocessor-Based and Intelligent Systems Engineering, Kluwer Acad. Pub., Dordrecht, (2003)

KV03b. Katic, D., Vukobratovic, M.: Survey of intelligent control techniques for humanoid robots. J. Int. Rob. Sys, **37**, 117–141, (2003)

KV98. Katic, D., Vukobratovic, M.: A neural network-based classification of environment dynamics models for compliant control of manipulation robots. IEEE Trans. Sys., Man, Cyb., B, **28**(1), 58–69, (1998)

Kac51. Kac, M.: On Some Connection between Probability Theory and Differential and Integral Equations. Proc. 2nd Berkeley Sympos. Math. Stat. and Prob., 189–215, (1951)

Kal21. Kaluza, T.: Sitzungsber. Preuss. Akad. Wiss. Berlin (Math. Phys.) K1, 966, (1921)

Kal60. Kalman, R.E.: A new approach to linear filtering and prediction problems. Transactions of the ASME, Ser. D. J. Bas. Eng., **82**, 34–45, (1960)

Kal99. N. Kaloper. Bent domain walls as braneworlds. Phys. Rev. D **60**, 123506, (1999)

Kan58. Kan, D.M.: Adjoint Functors. Trans. Am. Math. Soc. **89**, 294–329, (1958)

Kan98. Kanatchikov, I.: Canonical structure of classical field theory in the polymomentum phase space. Rep. Math. Physics, **41**(1), 49–90, (1998)

Kap05. Kappen, H.J.: A linear theory for control of nonlinear stochastic systems. Phys. Rev. Let. (to appear)

Kau94. Kauffman, L.H.: Knots and Physics, (2nd ed.) World Scientific, Singapore (1994)

Kla00. Klauder. J.R.: Beyond Conventional Quantization, Cambridge Univ. Press, Cambridge, (2000)

Kla97. Klauder. J.R.: Understanding Quantization. Found. Phys. **27**, 1467–1483, (1997)

Kle27. Klein, O.: Z. Phys., **41**, 407–442, (1927)

Koc81. Kock, A.: Synthetic Differential Geometry, London Math.Soc. Lecture Notes Series No. 51, Cambridge Univ. Press, Cambridge, (1981)

Kof03. Kofman, L.: Probing String Theory with Modulated Cosmological Fluctuations. astro-ph/0303641, (2003)

Kos04. Kosmann-Schwarzbach, Y.: Derived brackets. Lett. Math. Phys. **69**, 61–87, (2004)

Kos92. Kosko, B.: Neural Networks and Fuzzy Systems, A Dynamical Systems Approach to Machine Intelligence. Prentice–Hall, New York, (1992)

Kos96. Kosko, B.: Fuzzy Engineering. Prentice Hall, New York, (1996)

Kra97. Krasnov, K.: Geometrical entropy from loop quantum gravity. Phys. Rev. D **55**, 3505, (1997)

Kre84. Krener, A.: Approximate linearization by state feedback and coordinate change, Systems Control Lett., 5, 181–185, (1984)

Kru97. Krupkova, O.: The Geometry of Ordinary Variational Equations. Springer, Berlin, (1997)

Kuc92. Kuchař, K.: Time and Interpretations of Quantum Gravity. In Proceedings of the 4th Canadian Conference on General Relativity and Relativistic Astrophysics, ed. by G. Kunstatter, D. Vincent, and J. Williams, World Scientific, Singapore, (1992)

Kuh85. Kuhl, J.: Volitional Mediator of cognition-Behaviour consistency: Self-regulatory Processes and action versus state orientation (pp. 101–122). In: J. Kuhl & S. Beckman (Eds.) Action control: From Cognition to Behaviour. Springer, Berlin, (1985)

Kur76. Kuramoto, Y., Tsuzuki, T.: Persistent propagation of concentration waves in dissipative media far from thermal equilibrium. Progr. Theor. Physics **55**, 365, (1976)

Kur84. Kuramoto, Y.: Chemical Oscillations. Waves and Turbulence. Springer, New York, (1984)

LD05. Levine, M., Davidson, E.H.: Gene regulatory networks for development. Proc Natl. Acad. Sci. USA, **102**(14), 4936–4942, (2005)

LDC02. Leong, B., Dunsby, P., Challinor, A., Lasenby, A.: 1+3 covariant dynamics of scalar perturbations in braneworlds. Phys. Rev. D **65**, 104012, (2002)

LGS98. Lygeros, J., Godbole, D.N., Sastry, S.: Verified hybrid controllers for automated vehicles, IIEEE Trans. Aut. Con., **43**, 522–539, (1998)

LL00. Liddle, A.R., Lyth, D.: Cosmological Inflation and Large Scale Structure. Cambridge Univ. Press, (2000)

LL59. Landau, L.D., Lifshitz, E.M.: Fluid Mechanics. Pergamon Press, (1959)

LL98. Labastida, J.M.F., Lozano, C.: Lectures on topological quantum field theory, in Proceedings of the CERN–Santiago de Compostela–La Plata Meeting on 'Trends in Theoretical Physics', eds. H. Falomir, R. Gamboa, F. Schaposnik, Amer. Inst. Physics, New York, (1998)

LM03. Lopez, M.C., Marsden, J.E.: Some remarks on Lagrangian and Poisson reduction for field theories. J. Geom. Phys., **48**, 52–83, (2003)

LM87. Libermann, P., Marle, C.M.: Symplectic Geometry and Analytical Mechanics, Reidel, Dordrecht, (1987)

LM93. de León, M., Marrero, J.: Constrained time-dependent Lagrangian systems and Lagrangian submanifolds. J. Math. Phys. **34**, 622, (1993)

LM96. de León, M., Martín de Diego, D.: On the geometry of nonholonomic Lagrangian systems. J. Math. Phys. **37**, 3389, (1996)

LM97. Lewis, A.D., Murray, R.M.: Controllability of simple mechanical control systems, SIAM J. Con. Opt., **35**(3), 766–790, (1997)

LM99. Lewis, A.D., Murray, R.M.: Configuration controllability of simple mechanical control systems, SIAM Review, **41**(3), 555–574, (1999)

LMD04. de León, M., Martin, D., de Diego, A.: Santamaria-Merino: Symmetries in Classical Field Theory. arXiv:math-ph/0404013.

LMM97. de León, M., Marrero, J., Martín de Diego D.: Nonholonomic Lagrangian systems in jet manifolds. J. Phys. A. **30**, 1167, (1997)

LMS01. Langlois, D., Maartens, R., Sasaki, M., Wands, D.: Large-scale cosmological perturbations on the brane. Phys. Rev. D **63**, 084009, (2001)

LMS02. Liu, H., Moore, G.W., Seiberg, N.: Strings in a Time-Dependent Orbifold. JHEP **0206**, 045, (2002)

LMW00. Langlois, D., Maartens, Wands, D.: Gravitational waves from inflation on the brane. Phys. Lett. **B489**, 259, (2000)

LMW02. Langlois, D., Maeda, K.I., Wands, D.: Conservation laws for collisions of branes (or shells) in general relativity. Phys. Rev. Lett. **88**, 181301, (2002)

LOS00. Lukas, A., Ovrut, B.A., Stelle, K.S., Waldram, D.: Boundary inflation. Phys. Rev. D **61**, 023506, (2000)

LOS99. Lukas, A., Ovrut, B.A., Stelle, K.S., Waldram, D.: The universe as a domain wall. Phys. Rev. D **59**, 086001, (1999); ibid Heterotic M-theory in five dimensions, Nucl. Phys. B **552**, 246, (1999)

LP94. Langer. J., Perline, R.: Local geometric invariants of integrable evolution equations. J. Math. Phys., **35**(4), 1732–1737, (1994)

LR01. Langlois, D., Rodriguez-Martinez, M.: Brane cosmology with a bulk scalar field. Phys. Rev. D **64**, 123507, (2001)

LR93. Lychagin, V.V., Roubtsov, V.N., Chekalov, I.V.: A classification of Monge–Ampère equations. Ann. Scient. Ec. Norm. Sup. **4** ème série, 26, 281-308, (1993)

LRR02. Lee, T.I., Rinaldi, N.J., Robert, F., Odom, D.T., Bar-Joseph, Z., et al.: Transcriptional regulatory networks in Saccharomyces cerevisiae. Science, **298**, 799–804, (2002)

LS02. Langlois, D., Sorbo, L.: Effective action for the homogeneous radion in brane cosmology.. Phys. Lett. B **543**, 155, (2002)

LS97. Louko. J., Sorkin, R.D.: Complex actions in two-dimensional topology change. Class. Quant. Grav. **14**, 179–204, (1997)

LSA03. Louzoun, Y., Solomon, S., Atlan, H., Cohen, I.R.: Proliferation and competition in discrete biological systems. Bull. Math. Biol. **65**, 375, (2003)

LV97. Li, M., Vitanyi, P.: An Introduction to Kolmogorov Complexity and its Applications. Springer, New York, (1997)

LVW89. Lerche, W., Vafa, C., Warner, N.P.: Chiral Rings In N=2 Superconformal Theories, Nucl. Phys. B **324**, 427, (1989)

LW05. Lipan, O., Wong, W.H.: The use of oscillatory signals in the study of genetic networks. Proc. Natl. Acad. Sci. USA, **102**, 7063-7068, (2005)

LWX97. Liu, Z.J., Weinstein, A., Xu, P.: Manin triples for Lie bialgebroids. J. Diff. Geom. **45**(3), 547–574, (1997)

LY79. Li, P., Yau, S.T. Estimates of eigenvalues of a compact riemannian manifold. In Proceedings of Symposia in Pure Mathematics **36**, 205–239, (1979)

Lam94. Lamport, L.: The temporal logic of actions. ACM Transactions on Programming Languages and Systems, **16**(3), 872–923, (1994)

Lan00. Langlois, D.: Brane cosmological perturbations. Phys. Rev. D **62**, 126012, (2000)

Lan01. Langlois, D.: Evolution of cosmological perturbations in a brane-universe. Phys. Rev. Lett. **86**, 2212, (2001)

Lan02. Langlois, D.: Brane cosmology: an introduction. In the proceedings of YITP Workshop: Braneworld: Dynamics of Space-time Boundary. (2002)

Lan95. Landsman, N.P.: Against the Wheeler–DeWitt equation. Class. Quan. Grav. L **12**, 119–124, (1995)

Lan98. Landi, G.: An introduction to Noncommutative Spaces and Their Geometries, Springer, Berlin, (1998)

Lap51. Laplace, P.S.: A Philosophical Essay on Probabilities, translated from the 6th French edition, Dover publications, New York, (1951)

Leg02. Leggett, A.J.: Testing the Limits of Quantum Mechanics: Motivation, State-of-Play, Prospects. J. Phys. Cond. Matter, **14**, R415, (2002)

Lei02. Leinster, T.: A survey of definitions of $n-$category. Theor. Appl. Categ. **10**, 1–70, (2002)

Lei03. Leinster, T.: Higher Operads, Higher Categories, London Mathematical Society Lecture Notes Series, Cambridge Univ. Press, (2003)

Lei04. Leinster, T.: Operads in higher-dimensional category theory. Theor. Appl. Categ. **12**, 73–194, (2004)

Lew00a. Lewis, A.D.: Simple mechanical control systems with constraints, IEEE Trans. Aut. Con., **45**(8), 1420–1436, (2000)

Lew00b. Lewis, A.D.: Affine connection control systems. Proceedings of the IFAC Workshop on Lagrangian and Hamiltonian Methods for Nonlinear Control 128–133, Princeton, (2000)

Lew51. Lewin, K.: Field Theory in Social Science. Univ. Chicago Press, Chicago, (1951)

Lew95. Lewis, A.D.: Aspects of Geometric Mechanics and Control of Mechanical Systems. Technical Report CIT-CDS 95-017 for the Control and Dynamical Systems Option, California Institute of Technology, Pasadena, CA, (1995)

Lew97. Lewin, K.: Resolving Social Conflicts: Field Theory in Social Science, American Psych. Assoc., New York, (1997)

Lew98. Lewis, A.D.: Affine connections and distributions with applications to nonholonomic mechanics, Reports on Mathematical Physics, **42**(1/2), 135–164, (1998)

Lew99. Lewis, A.D.: When is a mechanical control system kinematic?, in Proceedings of the 38th IEEE Conf. Decis. Con., 1162–1167, IEEE, Phoenix, AZ, (1999)

Li04. Li, Y.: Chaos in Partial Differential Equations. Int. Press, Sommerville, MA, (2004)

Li04a. Li, Y.: Persistent homoclinic orbits for nonlinear Schrödinger equation under singular perturbation. Dyn. PDE, **1**(1), 87–123, (2004)

Li04b. Li, Y.: Existence of chaos for nonlinear Schrödinger equation under singular perturbation. Dyn. PDE. **1**(2), 225-237, (2004)

Li04c. Li, Y.: Homoclinic tubes and chaos in perturbed sine-Gordon equation. Cha. Sol. Fra., **20**(4), 791–798, 2004)

Li04d. Li, Y.: Chaos in Miles' equations. Cha. Sol. Fra., **22**(4), 965–974, (2004)

Li80. Li, P.: On the sobolev constant and the $p-$ spectrum of a compact Riemannian manifold. Ann. Sci. E.N.S., Paris, **13**, 451–468, (1980)

Lin82. Linde, A.D.: A New Inflationary Universe Scenario: A Possible Solution of the Horizon, Flatness, Homogeneity, Isotropy and Primordial Monopole Problems. Phys. Lett. **B108**, 389, (1982)

Lin82. Linde, A.: A New Inflationary Universe Scenario: A Possible Solution Of The Horizon, Flatness, Homogeneity, Isotropy And Primordial Monopole Problems. Phys. Lett. **B108**, 389, (1982)

Lin83. Linde, A.D.: Chaotic Inflation. Phys. Lett. **B129**, 177, (1983)

Lin86a. Linde, A.D.: Eternal Chaotic Inflation. Mod. Phys. Lett. **A1**, 81, (1986)

Lin86b. Linde, A.D.: Eternally Existing Selfreproducing Chaotic Inflationary Universe. Phys. Lett. **175B**, 395, (1986)

Lin88. Linde, A.D.: Inflation and Axion Cosmology. Phys. Lett. B201, 437, (1988)

Lol01. Loll, R.: Discrete Lorentzian quantum gravity. Nucl. Phys. B **94**, 96–107, (2001)

Lol98. Loll, R.: Discrete approaches to quantum gravity in four dimensions. Living Reviews in Relativity, **13**, (1998)

Loo00. Loo, K.: A Rigorous Real Time Feynman Path Integral and Propagator. J. Phys. A: Math. Gen, **33**, 9215–9239, (2000)

Loo99. Loo, K.: A Rigorous Real Time Feynman Path Integral. J. Math. Phys., **40**(1), 64–70, (1999)

Lor63. Lorenz, E.N.: Deterministic Nonperiodic Flow. J. Atmos. Sci., **20**, 130–141, (1963)

Lu90. Lu. J.-H.: Multiplicative and affine Poisson structures on Lie groups. PhD Thesis, Berkeley Univ., Berkeley, (1990)

Lu91. Lu. J.-H.: Momentum mappings and reduction of Poisson actions. in Symplectic Geometry, Groupoids, and Integrable Systems, eds.: P. Dazord and A. Weinstein, 209–225, Springer, New York, (1991)

Lyc79. Lychagin, V.V.: Contact geometry and nonlinear second order differential equations. Uspèkhi Mat. Nauk, **34**, 137–165 (in Russian), (1979)

Lyo92. Lyons, G.W.: Complex Solutions for the Scalar Field Model of the Universe. Phys. Rev. D **46**, 1546, (1992)

MA02. Mascalchi, M. *et al.*: Proton MR Spectroscopy of the Cerebellum and Pons in Patients with Degenerative Ataxia, Radiology, **223**, 371, (2002)

MA94. Miron, R., Anastasiei, M.: The Geometry of Lagrangian Spaces: Theory and Applications, Kluwer Academic Publishers, (1994)

MB01. Mennim, A., Battye, R.A.: Cosmological expansion on a dilatonic braneworld. Class. Quant. Grav. **18**, 2171, (2001)

MBP00. Myung, I.J., Balasubramanian, V., Pitt, M.A.: Counting probability distributions: differential geometry and model selection. Proceedings of the National Academy of Science, USA, **97**, 11170–11175, (2000)

MFB00. Myung, I.J., Forster, M., Browne, M.W.: Special issue on model selection. J. Math. Psych., **44**, 1–2, (2000)

MFV90. Morandi, G., Ferrario, C., Lo Vecchio, G., Marmo, G., Rubano, C.: The inverse problem in the calculus of variations and the geometry of the tangent bundle, Phys. Rep. **188**, 147, (1990)

MH92. Meyer, K.R., Hall, G.R.: Introduction to Hamiltonian Dynamical Systems and the N–body Problem. Springer, New York, (1992)

MK05. Moon, S.J., Kevrekidis, I.G.: An equation-free approach to coupled oscillator dynamics: the Kuramoto model example. Submitted to Int. J. Bifur. Chaos, (2005)

MKA88. Miron, R., Kirkovits, M.S., Anastasiei, M.: A Geometrical Model for Variational Problems of Multiple Integrals. Proc. Conf. Diff. Geom Appl. Dubrovnik, Yugoslavia, (1988)

ML81. Morris, C., Lecar, H.: Voltage oscillations in the barnacle giant muscle fiber. Biophys. J., **35**, 193–213, (1981)

MLS94. Murray, R.M., Li, X., Sastry, S.: Robotic Manipulation, CRC Press, Boco Raton, Fl, (1994)

MM92. Marathe, K., Martucci, G.: The Mathematical Foundations of Gauge Theories. North-Holland, Amsterdam, (1992)

MMZ99. Manasevich, R., Mawhin, J., Zanolin, F.: Periodic solutions of some complex–valued Lienard and Rayleigh equations. Nonl. Anal. **36**, 997–1014, (1999)

MN01. Maldacena, J., Nunez, C.: Supergravity Description of Field Theories on Curved Manifolds and No Go Theorem, Int. J. Mod. Phys. **A16**, 822, (2001)

MN95a. Mavromatos, N.E., Nanopoulos, D.V.: A Non-critical String (Liouville) Approach to Brain Microtubules: State Vector reduction, Memory coding and Capacity. ACT-19/95, CTP-TAMU-55/95, OUTP-95-52P, (1995)

MN95b. Mavromatos, N.E., Nanopoulos, D.V.: Non-Critical String Theory Formulation of Microtubule Dynamics and Quantum Aspects of Brain Function. ENSLAPP-A-524/95, (1995)

MNM02. Montagna, G., Nicrosini, O., Moreni, N.: A path integral way to option pricing. Physica A **310**, 450–466, (2002)

MOS99. Mangiarotti, L., Obukhov, Yu., Sardanashvily, G.: Connections in Classical and Quantum Field Theory. World Scientific, Singapore, (1999)

MP03. Myung, I.J., Pitt, M.A.: Model Evaluation, Testing and Selection. To appear in K. Lambert and R. Goldstone (eds.), Handbook of Cognition, Sage Publ., (2003)

MP94. Massa, E., Pagani, E.: Jet bundle geometry, dynamical connections and the inverse problem of Lagrangian mechanics. Ann. Inst. Henri Poincaré **61**, 17, (1994)

MPS98. Marsden. J.E., Patrick, G.W., Shkoller, S.: Multisymplectic Geometry, Variational Integrators, and Nonlinear PDEs. Comm. Math. Phys., **199**, 351–395, (1998)

MQ86. Mathai V., Quillen, D.: Super-connections, Thom classes and equivariant differential forms. Topology **25**, 85–110, (1986)

MR91. Mathur, V.S., Rajeev, S.G.: What Are the Anti-Particles of $K(L, S)$? Mod. Phys. Lett. A **6**, 2741, (1991)

MR92. Muñoz-Lecanda, M. and Román-Roy, N.: Lagrangian theory for presymplectic systems. Ann. Inst. Henrí Poincaré **57**, 27, (1992)

MR99. Marsden. J.E., Ratiu, T.S.: Introduction to Mechanics and Symmetry: A Basic Exposition of Classical Mechanical Systems. (2nd ed), Springer, New York, (1999)

MRS04. Munteanu, F., Rey, A.M., Salgado, M.: The Günther's formalism in classical field theory: momentum map and reduction. J. Math. Phys. **45**(5), 1730–1750, (2004)

MS00. Meyer, K.R., Schmidt, D.S.: From the restricted to the full three-body problem. Trans. Amer. Math. Soc., **352**, 2283–2299, (2000)

MS00. Murray, M.K., Singer, M.A.: On the complete integrability of the discrete Nahm equations Communications in Mathematical Physics, 210(2) 497–519, (2000)

MS00a. Mangiarotti, L., Sardanashvily, G.: Connections in Classical and Quantum Field Theory, World Scientific, Singapore, (2000)

MS00b. Mangiarotti, L., Sardanashvily, G.: Constraints in Hamiltonian time-dependent mechanics. J. Math. Phys. **41**, 2858, (2000)

MS03. Moroianu, A., Semmelmann, U.: Twistor forms on Kähler manifolds. Ann. Scuola Norm. Sup. Pisa Cl. Sci. **2**(4), 823–845, (2003)

MS74). Milnor. J.W., Stasheff. J.D.: Characteristic Classes, Princeton Univ. Press, Princeton, NJ, (1974)

MS78. Modugno, M., Stefani, G.: Some results on second tangent and cotangent spaces. Quadernidell' Instituto di Matematica dell' Universit a di Lecce Q., **16**, (1978)

MS95. Marmo, G., Simoni, A., Stern, A.: Poisson Lie group symmetries for the isotropic rotator. Int. J. Mod. Phys. A **10**, 99–114, (1995)

MS96. Murray, M.K., Singer, M.A.: Spectral curves of non-integral hyperbolic monopoles, Nonlinearity 9, 973–997, (1996)

MS98. Mangiarotti, L., Sardanashvily, G.: Gauge Mechanics. World Scientific, Singapore, (1998)

MS99. Mangiarotti, L., Sardanashvily, G.: On the geodesic form of non-relativistic dynamic equations. arXiv: math-ph/9906001.

MSM00. Mukohyama, S., Shiromizu, T., Maeda, K.I.: Global structure of exact cosmological solutions in the brane world. Phys. Rev. D **62**, 024028, (2000)

MT03. Minic, D., Tze, C.: Background independent quantum mechanics and gravity. Phys. Rev. D **68**, 061501, (2003)

MT80. Marsden, J.E., Tipler, F.: Maximal Hypersurfaces and Foliations of Constant Mean Curvature in General Relativity, Physics Reports **66**, 109, (1980)

MTW73. Misner, C.W., Thorne, K.S., Wheeler. J.A.: Gravitation. Freeman, San Francisco, (1973)

MTY88. Morris, M., Thorne, K.S., Yurtsver, U.: Wormholes, Time Machines, and the Weak Energy Condition. Phys. Rev. Lett. **61**, 1446, (1988)

MUW85. Mazenko, G.F., Unruh, W.G., Wald, R.M.: Does a Phase Transition in the Early Universe Produce the Conditions Needed for Inflation? Phys. Rev. D **31**, 273, (1985)

MW00. Maeda, K.I., Wands, D.: Dilaton-gravity on the brane. Phys. Rev. D **62**, 124009, (2000)

MW74. Marsden. J.E., Weinstein, A.: Reduction of symplectic manifolds with symmetry. Rept. Math. Phys. **5**, 121–130, (1974)

MWB00. Maartens, R., Wands, D., Bassett, B.A., Heard, I.: Chaotic inflation on the brane. Phys. Rev. D **62**, 041301, (2000)

MZA03. Mangan, S., Zaslaver, A., Alon, U.: The coherent feedforward loop serves as a sign-sensitive delay element in transcription networks. JMB, **334**(2), 197–204, (2003)

Maa00. Maartens, R.: Cosmological dynamics on the brane. Phys. Rev. D **62**, 084023, (2000)

Maa01. Maartens, R.: Geometry and dynamics of the brane-world. arXiv:gr-qc/0101059, (2001)

MacL71. MacLane, S.: Categories for the Working Mathematician. Springer, New York, (1971)

Mah98. Mahmoud, G.M.: Approximate solutions of a class of complex nonlinear dynamical systems. Physica A **253**, 211–222, (1998)

Mal01. Maldacena, J.M.: Eternal Black Holes in AdS. arXiv:hep-th/0106112, (2001)

Mal98. Maldacena, J.: The Large N Limit of Superconformal Field Theories and Gravity, Adv. Theor. Math. Phys. **2**, 231, (1998)

Man77. Manton, N.S.: The force between 't Hooft-Polyakov monopoles, Nucl. Phys. B126, 525–541, (1977)

Man80a. Mandelbrot, B.: Fractal aspects of the iteration of $z \mapsto \lambda z(1-z)$ for complex λ, z, Annals NY Acad. Sci. **357**, 249–259, (1980)

Man80b. Mandelbrot, B.: The Fractal Geometry of Nature. WH Freeman and Co., New York, (1980)

Man82. Manton, N.S.: A remark on the scattering of BPS monopoles, Phys. Lett. B110, 54–56, (1982)

Man98. Manikonda, V.: Control and Stabilization of a Class of Nonlinear Systems with Symmetry. PhD Thesis, Center for Dynamics and Control of Smart Structures, Harvard Univ., Cambridge, MA, (1998)

Mar03. Marshall, W., Simon, C., Penrose, R., Bouwmeester, D.: Quantum Superposition of a Mirror. Phys. Rev. Lett. **91**, 13, (2003)

Mar86. Margein, C: Positive Pinched manifolds are space forms. Proceedings of Symp. Pure Math. **44**, 307–328, (1986)

Mat02. Matacz, A.: Path Dependent Option Pricing: the Path Integral Partial Averaging Method. J. Comp. Finance, **6**(2), (2002)

Mat82. Matsumoto, M.: Foundations of Finsler Geometry and Special Finsler Spaces. Kaisheisha Press, Kyoto, (1982)

May81. Mayer, P.A.: A Differential Geometric Formalism for the Ito Calculus. Lecture Notes in Mathematics, Vol. 851, Springer, New York, (1981)

McCul87. McCullagh, P.: Tensor methods in statistics. Monographs in statistics and applied probability. Chapman & Hall, Cambridge, UK, (1987)

Men98. Mendes, R.V.: Conditional exponents, entropies and a measure of dynamical self-organization. Phys. Let. A **248**, 167–1973, (1998)

Mer73. Merton, R.: Theory of Rational Option Pricing. Bell J. Econom. Managem. Sci. **4**, 141–183, (1973)

Mes00. Messiah, A.: Quantum Mechanics (two volumes bound as one). Dover Pubs, (2000)

Met97. Metzger, M.A.: Applications of nonlinear dynamical systems theory in developmental psychology: Motor and cognitive development. Nonlinear Dynamics, Psychology, and Life Sciences, **1**, 55–68, (1997)

Mic01. Michor, P.W.: Topics in Differential Geometry. Lecture notes of a course in Vienna, (2001)

Mil56. Miller, G.A.: The magical number seven, plus or minus two: Some limits on our capacity for processing information, Psych. Rev., **63**, 81–97, (1956)

Mil63. Milnor, J.: Morse Theory. Princeton Univ. Press, Princeton, (1963)

Mil65. Milnor, J.: Lectures on the H–Cobordism Theorem. Math. Notes. Princeton Univ. Press, Princeton, (1965)

Mil99. Milinković, D.: Morse homology for generating functions of Lagrangian submanifolds. Trans. Amer. Math. Soc. **351**(10), 3953–3974, (1999)

Mla91. Mladenova, C.: Mathematical Modelling and Control of Manipulator Systems. Int. J. Rob. Comp. Int. Man., **8**(4), 233–242, (1991)

Mla99. Mladenova, C.: Applications of Lie Group Theory to the Modelling and Control of Multibody Systems. Mult. Sys. Dyn., **3**(4), 367–380, (1999)

Moh91. Mohler, R.R.: Nonlinear systems, Vol. 2. Applications to bilinear control. Prentice Hall, Inc. (1991)

Mok88. Mok, N.: The uniformization Theorem for compact Kähler manifolds of non-negative holomorphic bisectional curvature. J. Diff. Geom. **27**, 179–214, (1988)

Moo89. Moore, W.: Schr'odinger: Life and Thought. Cambridge Univ. Press, Cambridge, (1989)

Mor34. Morse, M.: The Calculus of Variations in the Large. Amer. Math. Soc. Coll. Publ. No. 18, Providence, RI, (1934)

Mor79. Mori, S.: Projective manifolds with ample tangent bundles. Ann. Math., **76**(2), 213–234, (1979)

Mos73. Moser, J.: Stable and Random Motions in Dynamical Systems. Princeton Univ. Press, Princeton, (1973)

Muk00. Mukohyama, S.: Gauge-invariant gravitational perturbations of maximally symmetric space–times. Phys. Rev. D **62**, 084015, (2000)

Mur01. Murray, M.K.: Monopoles. arXiv:math-ph/0101035, (2001)

Mur84. Murray, M.K.: Non-Abelian Magnetic Monopoles, Commun. Math. Phys. 96, 539–565, (1984)

Mus99. Mustafa, M.T.: Restrictions on harmonic morphisms. Confomal Geometry and Dynamics (AMS), **3**, 102–115, (1999)

NHP89. Noakes, L., Heinzinger, G., Paden, B.: Cubic splines on curved spaces, IMA J. Math. Con. Inf. , **6**(4), 465–473, (1989)

NMP84. Novikov, S.P., Manakov, S.V., Pitaevskii, L.P., Zakharov, V.E.: Theory of Solitons, Plenum/Kluwer, Dordrecht, (1984)

NP62. Newman, E.T., Penrose, R.: An Approach to Gravitational Radiation by a Method of Spin Coefficients. J. Math. Phys. **3**(3), 566–768, (1962)

NS90. Nijmeijer, H., van der Schaft, A.J.: Nonlinear Dynamical Control Systems. Springer, New York, (1990)

NS98. Nelson, D.R., Shnerb, N.M.: Non-Hermitian localization and population biology. Phys. Rev. E **58**, 1383, (1998)

NU00a. Neagu, M., Udrişte, C.: Multi-Time Dependent Sprays and Harmonic Maps on $J^1(T,M)$, Third Conference of Balkan Society of Geometers, Politehnica University of Bucarest, Romania, (2000)

NU00b. Neagu, M. Udrişte, C.: Torsions and Curvatures on Jet Fibre Bundle $J^1(T,M)$. arXiv:math.DG/0009069.

Nah82. Nahm, W.: The construction of all self-dual monopoles by the ADHM method, in Monopoles in Quantum Field Theory, Proceedings of the monopole meeting in Trieste 1981, World Scientific, Singapore, (1982)

Nan04. Nanayakkara, A.: Classical trajectories of 1D complex non-Hermitian Hamiltonian systems. J. Phys. A: Math. Gen. **37**, 4321, (2004)

Nan95. Nanopoulos, D.V.: Theory of Brain Function, Quantum Mechanics and Superstrings. CERN-TH/95128, (1995)

Nas83. Nash, C., Sen, S.: Topology and Geometry for Physicists. Academic Press, London, (1983)

Nay73. Nayfeh, A.H.: Perturbation Methods. Wiley, New York, (1973)

Nea00. Neagu, M.: Upon $h-$normal $\Gamma-$linear connection on $J^1(T,M)$. arXiv:math.DG/0009070, (2000)

Nea02. Neagu, M.: The Geometry of Autonomous Metrical Multi-Time Lagrange Space of Electrodynamics. Int. J. Math. Math. Sci. **29**, 7–16, (2002)

New61. Newman, E.: J. Math. Phys. **2**, 324, (1961)

New80. Newhouse, S.: Lectures on dynamical systems. In Dynamical Systems, C.I.M.E. Lectures, 1–114, Birkhauser, Boston, MA, (1980)

Nih99. Nihei, T.: Dynamics of Scalar field in a Brane World. Phys. Lett. B **465**, 81, (1999)

Nik95. Nikitin, I.N.: Quantum string theory in the space of states in an indefinite metric. Theor. Math. Phys. **107**(2), 589601, (1995)

Nis86. Nishikawa, S.: Deformation of Riemannian metrics and manifolds with bounded curvature ratios. Proceedings of Symp. Pure Math. **44**, 345–352, (1986)

OSV04. Ooguri, H., Strominger, A., Vafa, C.: Black hole attractors and the topological string. Phys. Rev. D **70**, 106007, (2004)

OTK02. Ozbudak, E.M., Thattai, M., Kurtser, I., Grossman, A.D., van Oudenaarden, A.: Regulation of noise in the expression of a single gene. Nature Gen., **31**,69–73, (2002)

OTL04. Ozbudak, E.M., Thattai, M., Lim, H.N., Shraiman, B.I., van Oudenaarden, A.: Multistability in the lactose utilization network of Escherichia coli. Nature, **427**(6976), 737–40, (2004)

Obe00. Oberste-Vorth, R.: Horseshoes among Hénon mappings. In Mastorakis, N. (ed.) Recent Advances in Applied and Theoretical Mathematics, WSES Press, 116–121, (2000)

Obe01. Oberlack, M.: A unified approach for symmetries in plane parallel turbulent shear flows. J. Fluid Mech. **427**, 299–328, (2001)

Obe87. Oberste-Vorth, R.: Complex Horseshoes and the Dynamics of Mappings of Two Complex Variables. PhD thesis, Cornell Univ., (1987)

Oht98. Ohta, Y.: Topological Field Theories associated with Three Dimensional Seiberg-Witten monopoles. Int.J.Theor.Phys. **37**, 925–956, (1998)

Oku81. Okubo, S.: Canonical quantization of some dissipative systems and nonuniqueness of Lagrangians. Phys. Rev. A **23**, 2776, (1981)

Olv86. Olver, P.J.: Applications of Lie Groups to Differential Equations (2nd ed.) Graduate Texts in Mathematics, Vol. 107, Springer, New York, (1986)

Omn94. Omnès, R.: Interpretation of Quantum Mechanics, Princeton University Press, Princeton, (1994)

Omo86. Omohundro, S.M.: Geometric Perturbation Theory in Physics. World Scientific, Singapore, (1986)

Oog92. Ooguri, H.: Topological Lattice Models in Four Dimensions. Mod. Phys. Lett. **A7**, 2799–2810, (1992)

Oog92. Ooguri, H.: Partition Functions and Topology-Changing Amplitudes in the 3D Lattice Gravity of Ponzano and Regge. Nucl. Phys. **B382**, 276, (1992)

Ori01. Oriti, D.: Spacetime geometry from algebra: Spin foam models for non-perturbative quantum gravity. Rept. Prog. Phys. **64**, 1489–1544, (2001)

PB89. Peyrard, M., Bishop, A.R.: Statistical mechanics of a nonlinear model for DNA denaturation. Phys. Rev. Lett. **62**, 2755, (1989)

PB99. Paul, W., Baschnagel. J.: Stochastic Processes: from Physics to Finance. Springer, Berlin, (1999)

PBK62. Pontryagin, L., Boltyanskii, V., Gamkrelidze, R., Mishchenko, E.: The mathematical theory of optimal processes. Interscience, (1962)

PBS01. Potters, M., Bouchaud. J.P., Sestovic, D.: Hedged Monte-Carlo: low variance derivative pricing with objective probabilities. Physica A **289**, 517–25, (2001)

PC05. Park. J., Chung, W.-K.: Geometric Integration on Euclidean Group With Application to Articulated Multibody Systems. IEEE Trans. Rob. **21**(5), 850–863, (2005)

PC91. Pecora, L.M., Carroll, T.L.: Driving systems with chaotic signals. Phys. Rev. A, **44**, 2374–2383, (1991)

PC98. Pecora, L.M., Carroll, T.L.: Master stability functions for synchronized coupled systems. Phys. Rev. Lett. **80**, 2109–2112, (1998)

PE34. Pauli, W., Weisskopf, V.: ´Uber die Quantisierung der skalaren relativistischen. Helv. Phys. Acta, **7**, 708–731, (1934)

PG81. Page, D.N., Geilker, C.D.: Indirect Evidence for Quantum Gravity. Phys. Rev. Lett. **47**, 979–982, (1981)

PHE92. Piaget, J., Henriques, G., Ascher, E.: Morphisms and categories. Erlbaum Associates, Hillsdale, NJ, (1992)

PI03. Pearce, C.E.M., Ivancevic, V.: A generalised Hamiltonian model for the dynamics of human motion. In Differential Equations and Applications, Vol. 2, Eds. Y.J. Cho. J.K. Kim and K.S. Ha, Nova Science, New York, (2003)

PI04. Pearce, C.E.M., Ivancevic, V.: A qualitative Hamiltonian model for the dynamics of human motion. In Differential Equations and Applications, Vol. 3, Eds. Y.J. Cho. J.K. Kim and K.S. Ha, Nova Science, New York, (2004)

PLS00. Pappas, G.J., Lafferriere, G., Sastry, S.: Hierarchically consistent control systems. IEEE Trans. Aut. Con., **45**(6), 1144–1160, (2000)

PMZ02. Pitt, M.A., Myung, I.J., Zhang, S.: Toward a method of selecting among computational models of cognition Psychological Review, **109**(3), 472–491, (2002)

PO82. Ponomarev, V., Obukhov, Yu.: Generalized Einstein-Maxwell theory. GRG **14**, 309, (1982)

PR02. Paufler, C., Römer, H.: The geometry of Hamiltonian n–vector–fields in multisymplectic field theory. J. Geom. Phys. **44**(1), 52–69, (2002)

PR68. Ponzano, G., Regge, T.: Spectroscopy and Group Theoretical Methods. In F. Block (ed.), North-Holland, Amsterdam, (1968)

PR86. Penrose, R., Rindler, W.: Spinors and Space–Time. Cambridge University Press, Cambridge, (1986)

PS02. Pappas, G.J., Simic, S.: Consistent hierarchies of affine nonlinear systems. IEEE Trans. Aut. Con., **47**(5), 745–756, (2002)

PS75. Prasad, M.K., Sommerfield, C.M.: Exact classical solutions for the ‘t Hooft monopole and Julia-Zee dyon. Phys. Rev. Lett. **35**, 760–762, (1975)

PSS96. Pons, J.M., Salisbury, D.C., Shepley, L.C.: Gauge transformations in the Lagrangian and Hamiltonian formalisms of generally covariant theories. arXiv gr-qc/9612037, (1996)

PST86. Park, J.K., Steiglitz, K. and Thurston, W.P.: Soliton–like Behavior in Automata. Physica D, **19**, 423–432, (1986)

PV03. Pessa, E., Vitiello, G.: Quantum noise, entanglement and chaos in the quantum field theory of mind/brain states. Mind and Matter, **1**, 59–79, (2003)

PV04. Pessa, E., Vitiello, G.: Quantum noise induced entanglement and chaos in the dissipative quantum model of brain. Int. J. Mod. Phys. B, **18** 841–858, (2004)

PV99. Pessa, E., Vitiello, G.: Quantum dissipation and neural net dynamics, Biolectrochem. Bioener., **48**, 339–342, (1999)

Par84. Parker, L., Toms, D.J.: Renormalization-group Analysis of Grand Unified Theories in Curved Spacetime. Phys. Rev. D **29**, 1584, (1984)

Pen00. Penrose, R.: Wavefunction Collapse as a Real Gravitational Effect, in Mathematical Physics 2000, ed. by A. Fokas, T.W.B. Kibble, A. Grigourion, and B. Zegarlinski, Imperial College Press, London, 266–282, (2000)

Pen04. Penrose, R.: The Road to Reality. Jonathan Cape, London, (2004)

Pen67. Penrose, R.: Twistor algebra. J. Math. Phys., **8**, 345–366, (1967)

Pen67. Penrose, R.: J. Math. Phys. **8**, 345, (1967)

Pen71a. Penrose, R.: Angular momentum: an approach to combinatorial space–time, in Bastin, T. (ed.), Quantum Theory and Beyond, 151–180. Cambridge Univ. Press, Cambridge, UK, (1971)

Pen71b. Penrose, R.: Applications of negative dimensional tensors, in Welsh, D. (ed.) Combinatorial Mathematics and its Application, 221–243, (1971)

Pen94. Penrose, R.: Shadows of the Mind. Oxford Univ. Press, Oxford, (1994)

Pen97. Penrose, R.: The Large, the Small and the Human Mind. Cambridge Univ. Press, (1997)

Per86. Perelomov, A.: Generalized Coherent States and their Applications. Springer, Berlin, (1986)

Per98. Percival, I.: Quantum State Diffusion. Cambridge Univ. Press, Cambridge, UK, (1998)

Pes77. Pesin, Ya.B.: Lyapunov Characteristic Exponents and Smooth Ergodic Theory. Russ. Math. Surveys, **32**(4), 55–114, (1977)

Pet98. Petersen, P.: Riemannian Geometry. Springer, New York, (1998)

Pet99. Petersen, P.: Aspects of Global Riemannian Geometry. Bull. Amer. Math. Soc., **36**(3), 297–344, (1999)

Pic86. Pickover, C.A.: Computer Displays of Biological Forms Generated From Mathematical Feedback Loops. Computer Graphics Forum, **5**, 313, (1986)

Pic87. Pickover, C.A.: Mathematics and Beauty: Time–Discrete Phase Planes Associated with the Cyclic System. Computer Graphics Forum, **11**, 217, (1987)

Pol95. Polchinski, J.: Dirichlet-branes and Ramond-Ramond charges. Phys. Rev. Lett. **75**, 4724, (1995)

Pol98. Polchinski, J.: String theory. 2 Vols. Cambridge Univ. Press, (1998)

Pol99. Polchinski, J.: String Theory, Two Volumes. Cambridge Univ. Press, (1999)

Pom78. Pommaret, J.: Systems of Partial Differential Equations and Lie Pseudogroups, Gordon and Breach, Glasgow, (1978)

Pop59. Popper, K.R: The Logic of Scientific Discovery. New York, NY: Basic Books, (1959)

Pos86. Postnikov, M.M.: Lectures in Geometry V, Lie Groups and Lie Algebras, Mir Publ., Moscow, (1986)

Pra91. Pratt, V.: Modelling concurrency with geometry. in: Proc. of the 18th ACM Symposium on Principles of Programming Languages, (1991)

Pry96. Prykarpatsky, A.K.: Geometric models of the Blackmore's swept volume dynamical systems and their integrability, In: Proc. of the IMACS-95, Hamburg 1995, ZAMP, **247**(5), 720–724, (1996)

Pul05. Pulvermüller, F.: Brain mechanicsms kinking language and action. Nature Rev. Neurosci. **6**, 576–582, (2005)

Put93. Puta, M.: Hamiltonian Mechanical Systems and Geometric Quantization, Kluwer, Dordrecht, (1993)

RBT01. Roe, R.M., Busemeyer, J.R., Townsend, J.T.: Multi-alternative decision field theory: A dynamic connectionist model of decision making. Psych. Rev., **108**, 370–392, (2001)

REB03. Rani, R., Edgar, S.B., Barnes, A.: Killing Tensors and Conformal Killing Tensors from Conformal Killing Vectors. Class.Quant.Grav., **20**, 1929–1942, (2003)

RG98. Rao, A.S., Georgeff, M.P.: Decision Procedures for BDI Logics. Journal of Logic and Computation, **8**(3), 292–343, (1998)

RH89. Rose, R.M., Hindmarsh. J.L.: The assembly of ionic currents in a thalamic neuron. I The three-dimensional model. Proc. R. Soc. Lond. B, **237**, 267–288, (1989)

RMR04. Bou-Rabee, N.M., Marsden, J.E., Romero, L.A.: Tippe Top Inversion as a Dissipation-Induced Instability. SIAM J. Appl. Dyn. Sys. **3**, 352, (2004)

RN03. Russell, Norvig: Artificial Intelligence. A modern Approach. Prentice Hall, (2003)

RR97. Reisenberger, M., Rovelli, C.: 'Sum over surfaces' form of loop quantum gravity. Phys. Rev. D **56**, 3490–3508, (1997)

RS75. Reed, M., Simon, B.: Methods of modern mathematical physics, Vol. 2: Fourier analysis, self-adjointness. Academic Press, San Diego, (1975)

RS87. Rovelli, C., Smolin, L.: A new approach to quantum gravity based on loop variables. Int. Conf. Gravitation and Cosmology, Goa, Dec 14-19 India, (1987)

RS88. Rovelli, C., Smolin, L.: Knot theory and quantum gravity. Phys. Rev. Lett. **61**, 1155, (1988)

RS90. Rovelli, C., Smolin, L.: Loop representation of quantum general relativity, Nucl. Phys., **B331**:(1), 80–152, (1990)

RS94. Rovelli, C., Smolin, L.: The physical Hamiltonian in non–perturbative quantum gravity. Phys. Rev. Lett. **72**, 446, (1994)

RS95. Rovelli, C., Smolin, L.: Spin Networks and Quantum Gravity. Phys. Rev. **D52**, 5743–5759, (1995)

RS99a. Randall, L., Sundrum, R.: A large mass hierarchy from a small extra dimension. Phys. Rev. Lett. **83**, 3370, (1999)

RS99b. Randall, L., Sundrum, R.: An alternative to compactification. Phys. Rev. Lett. **83**, 4690, (1999)

RT02. Rosa-Clot, M., Taddei, S.: A Path Integral Approach to Derivative Security Pricing. Int. J. Theor. Appl. Finance, **5**(2), 123–146, (2002)

RT71. Ruelle, D., Takens, F.: On the nature of turbulence. Comm. Math. Phys., **20**, 167–192, (1971)

RT90. Reshetikhin, N., Turaev, V.: Ribbon graphs and their invariants derived from quantum groups. Comm. Math. Phys. **127**, 1–26, (1990)

RT91. Reshetikhin, N.Yu, Turaev, V.G.: Invariants of four-manifolds via link polynomials and quantum groups. Invent. Math. **10**, 547, (1991)

RT99. Reshetikhin, N., Takhtajan, L.: Deformation Quantization of Kahler Manifolds. arXiv:math.QA/9907171, (1999)

RU67. Ricciardi, L.M., Umezawa, H.: Brain physics and many-body problems, Kibernetik, **4**, 44, (1967)

RVS02. Riazuelo, A., Vernizzi, F., Steer, D., Durrer, R.: Gauge invariant cosmological perturbation theory for braneworlds. arXiv:hep-th/0205220, (2002)

Raj07. Rajeev, S.G.: Dissipative Mechanics Using Complex-Valued Hamiltonians. arXiv:quant-ph/0701141, (2007)

Ram90. Ramond, P.: Field Theory: a Modern Primer. Addison–Wesley, Reading, MA, (1990)

Rat78. Ratcliff, R.: A theory of memory retrieval. Psych. Rev., **85**, 59–108, (1978)

Ree01. Rees, M.J.: The State of Modern Cosmology. In N.G. Turok, (ed.), Critical Dialogues in Cosmology, World Scientific, Singapore, (1997)

Reg61. Regge, T.: General relativity without coordinates. Nuovo Cim. A **19**, 558–571, (1961)

Rei48. Reidemeister, K.: Knotentheorie, Chelsea Pub., New York, (1948)

Rha84. de Rham, G.: Differentiable Manifolds. Springer, Berlin, (1984)

Ric93. Ricca, R.L.: Torus knots and polynomial invariants for a class of soliton equations, Chaos, **3**(1), 83–91, (1993)

Rie84. Riegert, R.J.: Non-Local Action for the Trace Anomaly. Phys. Lett. **134B**, 56, (1984)

Ros30. Rosenfeld, L.: Über die Gravitationswirkungen des Lichtes, Z. Phys. **65**, 589–599, (1930)

Ros63. Rosenfeld, L.: On Quantization of Fields. Nucl. Phys. **40**, 353–356, (1963)

Ros66. Rosenfeld, L.: Quantentheorie und Gravitation in Entstehung, Entwicklung, und Perspektiven der Einsteinschen Gravitationstheorie, Akademie-Verlag, Berlin, 185–197, (1966) [English translation in Selected Papers of Léon Rosenfeld, ed. by R.S. Cohen and J. Stachel, D. Reidel, Dordrecht, (1979]

Ros98. Rosenberg, S.: Testing Causality Violation on Spacetimes with Closed Timelike Curves. Phys. Rev. D **57**, 3365, (1998)

Rov93. Rovelli, C.: The basis of the Ponzano-Regge-Turaev-Viro-Ooguri model is the loop representation basis. Phys. Rev. D **48**, 2702–2707, (1993)

Rov96a. Rovelli, C.: Black Hole Entropy from Loop Quantum Gravity. Phys. Rev. Lett. **14**, 3288–3291, (1996)

Rov96b. Rovelli, C.: Loop Quantum Gravity and Black Hole Physics, Helv. Phys. Acta. **69**, 582–611, (1996)

Rov97. Rovelli, C. Half way through the woods, in Earman, J, and Norton, J, (eds.) The Cosmos of Science, 180–223, Univ. Pittsburgh Press, Konstanz, (1997)

Rov97. Rovelli, C.: Strings, loops and others: a critical survey of the present approaches to quantum gravity. Plenary lecture on quantum gravity at the GR15 conference, Puna, India, (1997)

Rov98. Rovelli, C.: Loop quantum gravity. Living Rev. Rel. **1**, (1998)

Rub01. Rubakov, V.A.: Large and infinite extra dimensions: An introduction. Phys. Usp. **44**, 871, (2001)

Rue78. Ruelle, D.: Thermodynamic formalism. Encyclopaedia of Mathematics and its Applications. Addison–Wesley, Reading, MA, (1978)

Ryd96. Ryder, L.: Quantum Field Theory. Cambridge Univ. Press, (1996)

SA98. Schaal, S., Atkeson, C.G.: Constructive incremental learning from only local information. Neural Comput., **10**, 2047–2084, (1998)

SB98. Sutton, R.S., Barto, A.G.: Reinforcement Learning: An Introduction. MIT Press, Cambridge, MA, (1998)

SC01. Segel, L.A., Cohen, I.R. (eds.): Design principles for the immune system and other distributed autonomous systems, Oxford Univ. Press, (2001)

SFM98. Skiffington, S., Fernandez, E., McFarland, K.: Towards a validation of multiple features in the assessment of emotions. Eur. J. Psych. Asses. **14**(3), (1998)

SGL93. Shih, C.L., Gruver, W. Lee, T.: Inverse kinematics and inverse dynamics for control of a biped walking machine. J. Robot. Syst., **10**, 531–555, (1993)

SHT02. Shimizu-Sato, S., Huq, E., Tepperman. J.M., Quail, P.H.: A light-switchable gene promoter system. Nature, Biotech., **20**, 1041–1044, (2002)

SHW94. Stephens, C.R., 't Hooft, G., Whiting, B.F.: Black Hole Evaporation without Information Loss. Class. Quant. Grav. **11**, 621–648, (1994)

SI89. Sastry, S.S., Isidori, A.: Adaptive control of linearizable systems. IEEE Trans. Aut. Con., **34**(11), 1123–1131, (1989)

SK93. Shih, C.L., Klein, C.A.: An adaptive gait for legged walking machines over rough terrain. IEEE Trans. Syst. Man, Cyber. A, **23**, 1150–1154, (1993)

SK98a. Shabanov, S.V., Klauder, J.R.: Path Integral Quantization and Riemannian-Symplectic Manifolds. Phys. Lett. B4 **35**, 343–349, (1998)

Sch96. Schafer, R.D.: An Introduction to Nonassociative Algebras. Dover, New York, (1996)

SK98b. Shabanov, S.V., Klauder, J.R.: Path Integral Quantization and Riemannian-Symplectic Manifolds. Phys. Lett. B **435**, 343–349, (1998)

SMS00. Shiromizu, T., Maeda, K.I., Sasaki, M.: The Einstein equations on the 3-brane world. Phys. Rev. D **62**, 024012, (2000)

SS01. Shapiro, I.L., Sola, J.: Massive Fields Temper Anomaly-Induced Inflation: The Clue to Graceful Exit? arXiv:hep-th/0104182, (2001)

STP95. Schaub, H., Tsiotras, P, Junkins., J.: Principal Rotation Representations of Proper N x N Orthogonal Matrices. Int. J. Eng. Sci., **33**,(15), 2277–2295, (1995)

STU78. Stuart, C.I.J., Takahashi, Y., Umezawa, H.: On the stability and non-local properties of memory. J. Theor. Biol. **71**, 605–618, (1978)

STU79. Stuart, C.I.J., Takahashi, Y., Umezawa, H.: Mixed system brain dynamics: neural memory as a macroscopic ordered state, Found. Phys. **9**, 301, (1979)

STU93. Susskind, L., Thorlacius, L., Uglum. J.: The Stretched Horizon and Black Hole Complementarity. Phys. Rev. D **48**, 3743–3761, (1993)

STZ93. Satarić, M.V., Tuszynski, J.A., Zakula, R.B.: Kinklike excitations as an energy-transfer mechanism in microtubules. Phys. Rev. E, **48**, 589–597, (1993)

SV96a. Strominger, A., Vafa, C.: On the microscopic origin of the Bekenstein–Hawking entropy. Phys. Lett. B3 **79**, 99, (1996)

SV96b. Strominger, A., Vafa, C.: Microscopic Origin of the Bekenstein-Hawking Entropy. Phys. Lett. **B379**, 99–104, (1996)

SVW95. Srivastava, Y.N., Vitiello, G., Widom, A.: Quantum dissipation and quantum noise, Annals Phys., **238**, 200, (1995)

SW72. Sulanke R., Wintgen, P.: Differential geometry und faser-bundel; bound 75, Veb. Deutscher Verlag der Wissenschaften, Berlin, (1972)

SW93. Schleich, K., Witt, D.: Generalized Sums over Histories for Quantum Gravity: I. Smooth Conifolds, Nucl. Phys. **402**, 411, (1993); II. Simplicial Conifolds, ibid. **402**, 469, (1993)

SW94a. Seiberg, N., Witten, E.: Monopoles, duality and chiral symmetry breaking in $N = 2$ supersymmetric QCD. Nucl. Phys. **B426**, 19–52, (1994)

SW94b. Seiberg, N., Witten, E.: Electric-magnetic duality, monopole condensation, and confinement in $N = 2$ supersymmetric Yang–Mills theory. Nucl. Phys. **431**, 484–550, (1994)

SW99. Seiberg, N., Witten, E.: String Theory and Noncommutative Geometry. JHEP **9909**, 032, (1999)

SWV99. Srivastava, Y.N., Widom, A., Vitiello, G.: Quantum measurements, information and entropy production. Int. J. Mod. Phys. B13, 3369–3382, (1999)

SY94. Schoen, R., Yau, S.T.: Lectures on differential geometry, in Conference Proceedings and Lecture Notes in Geometry and Topology, 1, International Press, Cambridge, MA, (1994)

SZ97. Scully, M.O., Zubairy, M.S.: Quantum Optics. Cambridge Univ. Press, (1997)

SZT98. Satarić, M.V., Zeković, S., Tuszynski, J.A., Pokorny, J.: Mössbauer effect as a possible tool in detecting nonlinear excitations in microtubules. Phys. Rev. E **58**, 6333–6339, (1998)

Sar98. Sardanashvily, G.: Hamiltonian time-dependent mechanics. J. Math. Phys. **39**, 2714, (1998)

Sar02b. Sardanashvily, G.: The Lyapunov stability of first order dynamic equations with respect to time-dependent Riemannian metrics. arXiv nlin.CD/0201060.

Sar02c. Sardanashvily, G.: The bracket and the evolution operator in covariant Hamiltonian field theory, arXiv: math-ph/0209001.

Sar03. Sardanashvily, G.: Geometric quantization of relativistic Hamiltonian mechanics. Int. J. Theor. Phys. **42**, 697–704, (2003)

Sar92. Sardanashvily, G., Zakharov, O.: Gauge Gravitation Theory. World Scientific, Singapore, (1992)

Sar92. Sardanashvily, G.: On the geometry of spontaneous symmetry breaking. J. Math. Phys. **33**, 1546, (1992)

Sar93. Sardanashvily, G.: Gauge Theory in Jet Manifolds. Hadronic Press, Palm Harbor, FL, (1993)

Sar94. Sardanashvily, G.: Constraint field systems in multimomentum canonical variables. J. Math. Phys. **35**, 6584, (1994)

Sau89. Saunders, D.J.: The Geometry of Jet Bundles. Lond. Math. Soc. Lect. Notes Ser. **142**, Cambr. Univ. Pr., (1989)

Sch01. Schlichenmaier, M.: Berezin–Toeplitz Quantization and Berezin's Symbols for Arbitrary Compact Kähler Manifolds, in Coherent States, Quantization and Gravity, M. Schlichenmaier et al. (eds.) Polish Scientific Publishers PWN, Warsaw, (2001)

Sch01. Schleich, W.: Quantum Optics in Phase Space. Wiley, New York, (2001).

Sch78). Schwarz, A.S.: New topological invariants in the theory of quantised fields. Lett. Math. Phys. **2**, 247, (1978)

Sch81. Schulman, L.S.: Techniques and Applications of Path Integration, John Wiley & Sons, New York, (1981)

Sch93. Schwarz, M.: Morse Homology, Birkháuser, Basel, (1993)

Sch96. Schellekens, A.N.: Introduction to conformal field theory. Fortsch. Phys., **44**, 605, (1996)

Sch98. Schaal, S.: Robot learning. In M. Arbib (ed). Handbook of Brain Theory and Neural Networks (2nd ed.), MIT Press, Cambridge, (1998)

Sch99. Schaal, S.: Is imitation learning the route to humanoid robots?. Trends Cogn. Sci., **3**, 233–242, (1999)

Sei06. Seiberg, N.: Emergent Spacetime. arXiv:hep-th/0601234, (2006)

Sei95. Seiler, W.M.: Involution and constrained dynamics II: The Faddeev-Jackiw approach. J. Phys. A, **28**, 7315–7331, (1995)

Sem02. Semmelmann, U.: Conformal Killing forms on Riemannian manifolds. arXiv math.DG/0206117, (2002)

She05. Shen, Z.: Riemann–Finsler geometry with applications to information geometry, unpublished material, (2005)

Shi98. Shishikura, M.: The Hausdorff dimension of the boundary of the Mandelbrot set and Julia sets. Ann. of Math. **147**, 225–267, (1998)

Shu93. Shuster, M.D.: A Survey of Attitude Representations. J. Astr. Sci., **41**(4), 316–321, (1993)

Siu80. Siu, Y.T., Yau, S.T.: Compact Kähler manifolds of positive bisectional curvature. Invent. Math. **59**, 189–204, (1980)

Siv77. Sivashinsky, G.I.: Nonlinear analysis of hydrodynamical instability in laminar flames – I. Derivation of basic equations. Acta Astr. **4**, 1177, (1977)

Sma60. Smale, S.: The generalized Poincaré conjecture in higher dimensions, Bull. Amer. Math. Soc., **66**, 373–375, (1960)

Sma67. Smale, S.: Differentiable dynamical systems, Bull. Amer. Math. Soc., **73**, 747–817, (1967)

Sma99. van der Smagt, P.: (ed.) Self–Learning Robots. Workshop: Brainstyle Robotics, IEE, London, (1999)

Smo97a. Smolin, L.: Loops and Strings. Living Reviews, September, (1997)

Smo97b. Smolin, L.: The Life of the Cosmos. Oxford Univ. Press, Oxford, (1997)

Sni80. S̀niatycki, J.: Geometric Quantization and Quantum Mechanics, Springer-Verlag, Berlin, (1980)

Sni80. Sniatycki, J.: Geometric Quantization and Quantum Mechanics. Springer-Verlag, Berlin, (1980)

Soc91. Socolovsky, M.: Gauge transformations in fibre bundle theory. J. Math. Phys., **32**, 2522, (1991)

Spa82. Sparrow, C.: The Lorenz Equations: Bifurcations, Chaos, and Strange Attractors. Springer, New York, (1982)

Sta00. Stanislavsky, A.A.: Memory effects and macroscopic manifestation of randomness. Phys. Rev. E **61**, 4752, (2000)

Sta63. Stasheff. J.D.: Homotopy associativity of $H-$spaces I & II. Trans. Amer. Math. Soc., **108**, 275–292, 293–312, (1963)

Sta80. Starobinsky, A.A.: A New Type of Isotropic Cosmological Models without Singularity. Phys. Lett. **B91**, 99, (1980)

Sta83. Starobinsky, A.A.: The Perturbation Spectrum Evolving from a Nonsingular, Initially de Sitter cosmology, and the Microwave Background Anisotropy. Sov. Astron. Lett. **9**, 302, (1983)

Ste69. Sternberg, S.: Memory-scanning: Mental processes revealed by reaction-time experiments. Am. Sci., **57**(4), 421–457, (1969)

Ste72. Steenrod, N.: The Topology of Fibre Bundles, Princeton Univ. Press, Princeton, (1972)

Ste77. Stelle, K.S.: Renormalization of Higher Derivative Quantum Gravity. Phys. Rev. D **16**, 953, (1977)

Ste90. Stewart, J.M.: Perturbations of Friedmann-Robertson-Walker cosmological models. Class. Quant. Grav. **7**, 1169, (1990)

Ste93. Stengel, R.: Optimal control and estimation. Dover, New York, (1993)

Sto68. Stong, R.E.: Notes on Cobordism Theory. Princeton Univ. Press, Princeton, (1968)

Str00. Strogatz, S.: From Kuramoto to Crawford: exploring the onset of synchronization in populations of coupled oscillators. Physica D, **143**, 1–20, (2000)

Str90. Strominger, A.: Special Geometry. Commun. Math. Phys. **133**, 163, (1990)

Str91. Strominger, A.: Baby Universes. In Quantum Cosmology and Baby Universes: Proceedings of the 1989 Jerusalem Winter School for Theoretical Physics, ed. by S. Coleman, Hartle, J.B., T. Piran and S. Weinberg, World Scientific, Singapore, (1991)

Stu99. Stuart, J.: Calculus (4th ed.). Brooks/Cole Publ., Pacific Grove, CA, (1999)

Sug00. Sugino, F.: Witten's open string field theory in constant B-field background. J. High Energy Phys. JHEP03, 017, (2000)

Sus02. Susskind, L.: Twenty Years of Debate with Stephen. arXiv:hep-th/0204027, (2002)

Sus03. Susskind, L.: The Anthropic Landscape of String Theory. arXiv:hep-th/0302219, (2003)

Sus83. Sussmann, H.J.: Lie brackets and local controllability: a sufficient condition for scalar–input systems, SIAM J. Con. Opt., **21**(5), 686–713, (1983)

Sus87. Sussmann, H.J.: A general theorem on local controllability, SIAM J. Con. Opt., **25**(1), 158–194, (1987)

Sus95. Susskind, L.: The World as a Hologram. J. Math. Phys. **36**, 6377–6396, (1995)

Sut97. Sutcliffe, P.: BPS monopoles, Int. J. Mod. Phys. A12, 4663–4706, (1997)

Swi75. Switzer, R.K.: Algebraic Topology – Homology and Homotopy. (in Classics in Mathematics), Springer, New York, (1975)

Syn61. Synge, J.L.: On a certain nonlinear differential equation. Proc. R. Irish. Acad. **62**, 1, (1961)

TEM89. Tonomura, A., Endo. J., Matsuda, T., Kawasaki, T.: Demonstration of Single-electron Build-up of an Interference Pattern, Am. J. Phys. **57**, 117–120, (1989)

TF91. Tsue, Y., Fujiwara, Y.: Time-Dependent Variational Approach to (1+1)-Dimensional Scalar-Field Solitons, Progr. Theor. Phys. **86**(2), 469–489, (1991)

tHo93. 't Hooft, G.: Dimensional Reduction in Quantum Gravity. gr-qc/9310026, (1993)

tHo06. 't Hooft, G.: Determinism Beneath Quantum Mechanics. In Quo Vadis Quantum Mechanics, ed. by A. Elitzur, S. Dolen, and N. Kolenda, Springer Verlag, Heidelburg, (2005)

TH98. Turok, N., Hawking, S.W.: Open Inflation, the Four Form and the Cosmological Constant. Phys. Lett. **B432**, 271, (1998)

TL98. Tseytlin, A., Liu, H.: $D = 4$ Super Yang-Mills, $D = 5$ gauged supergravity and $D = 4$ conformal supergravity. Nucl. Phys. B **533**, 88, (1998)

TP01. Tabuada, P., Pappas, G.J.: Abstractions of Hamiltonian Control Systems. Proceedings of the 40th IEEE Conf. Decis. Con., Orlando, FL, (2001)

TPS98. Tomlin, C., Pappas, G.J., Sastry, S.: Conflict resolution for air traffic management: A case study in multi-agent hybrid systems, IEEE Trans. Aut. Con., **43**, 509–521, (1998)

TS01. Thompson. J.M.T., Stewart, H.B.: Nonlinear Dynamics and Chaos: Geometrical Methods for Engineers and Scientists. Wiley, New York, (2001)

TV92. Turaev, V., Viro, O.: State Sum Invariants of 3-Manifolds and Quantum 6-J Symbols. Topology **31**, 865–902, (1992)

TVP99. Tabony, J., Vuillard, L., Papaseit, C.: Biological self-organisation and pattern formation by way of microtubule reaction-diffusion processes. Adv. Complex Syst. **2**(3), 221–276, (1999)

TW82. Turner, M.S., Wilczek, F.: Is our vacuum metastable? Nature, 298, 633634, (1982)

TW95. Tucker, R., Wang, C.: Black holes with Weyl charge and non-Riemannian waves. Class. Quant. Grav. **12**, 2587, (1995)

TZ00. Tian, G., Zhu, X.: Uniqueness of Kähler-Ricci solitons, Acta Math. **184**, 271–305, (2000)

Tau90. Taubes, C.H.: Casson's invariant and gauge theory. J. Diff. Geom. **31**, 547–599, (1990)

Tau94. Taubes, C.H.: The Seiberg-Witten invariants and symplectic forms, Math. Res. Lett. **1**, 809–822, (1994)

Tau95a. Taubes, C.H.: More constraints on symplectic manifolds from Seiberg-Witten invariants, Math. Research Letters, **2**, 9–14, (1995)

Tau95b. Taubes, C.H.: The Seiberg-Witten invariants and The Gromov invariants, Math. Research Letters **2**, 221–238, (1995)

Tei83. Teitelboim, C.: Causality versus gauge invariance in quantum gravity and supergravity. Phys. Rev. Lett. **50**, 705–708, (1983)

Teu73. Teukolsky, S.: Perturbations of a rotating black hole. Astrophys. J. 185, 635–647, (1973)

Thi79. Thirring, W.: A Course in Mathematical Physics (in four volumes). Springer, New York, (1979)

Thi96. Thiemann, T.: Anomaly-Free Formulation of Nonperturbative Four–Dimensional Lorentzian Quantum Gravity. Phys. Lett. **B380**, 257–264, (1996)

Tho79. Thorpe, J.A.: Elementary Topics in Differential Geometry. Springer, New York, (1979)

Tho93. Thompson, G.: Non-uniqueness of metrics compatible with a symmetric connection. Class. Quant. Grav. **10**, 2035, (1993)

Tia97. Tian, G.: Kähler-Einstein metrics with positive scalar curvature. Invent. Math. **130**, 1–39, (1997)

Tia98. Tian, G.: Some aspects of Kähler Geometry. Lecture note taken by M. Akeveld, (1997)

Tom77. Tomboulis, E.: 1/N Expansion and Renormalization in Quantum Gravity. Phys. Lett. **B70**, 361, (1977)

Tur00. Turok, N.: Before Inflation. Lecture at CAPP2000. arXiv:astro-ph/0011195, (2000)

UJ98. Unruh, W.G., Jheeta, M.: Complex Paths and the Hartle-Hawking Wave Function for Slow Roll Cosmologies. arXiv:gr-qc/9812017, (1998)

UN99. Udriste, C., Neagu, M.: Extrema of p-energy functional on a Finsler manifold. Diff. Geom. Dyn. Sys. 1(1), 10–19, (1999)

Udr00. Udriste, C.: Geometric dynamics. Kluwer Academic Publishers, (2000)

VW94. Vafa, C., Witten, E.: A strong coupling test of S duality. arXiv:hep-th/9408074, (1994)

Vaf97. Vafa, C.: Lectures on Strings and Dualities. arXiv:hep-th/9702201, (1997)

Vai94. Vaisman, I.: Lectures on the Geometry of Poisson Manifolds. Birkhäuser Verlag, Basel, (1994)

Van88. Vance, J.M.: Rotor Dynamics of Turbomachinery. Wiley, New York, (1988)

Ven91. Veneziano, G.: Scale factor duality for classical and quantum strings. Phys. Lett. **B265**, 287, (1991)

Ver88. Verlinde, E.: Fusion rules and modular transformations in 2D conformal field theory. Nucl. Phys. **B300**, 360–376, (1988)

Vil83. Vilenkin, A.: Birth of Inflationary Universes. Phys. Rev. D **27**, 2848, (1983)

Vil85. Vilenkin, A.: Classical and Quantum Cosmology of the Starobinsky Inflationary Model. Phys. Rev. D **32**, 2511, (1985)

Vio83. Dubois–Violette, M.: Structures Complexes au-dessus des Variétés, Applications, in Mathématique et Physique, Séminaire de l'Ecole Normale Supérieure 1979–1982, L. Boutet de Monvel et al. (eds.), Birkhäuser, Boston, (1983)

Vit01. Vitiello, G.: My Double Unveiled. John Benjamins, Amsterdam, (2001)

Vit95. Vitiello, G.: Dissipation and memory capacity in the quantum brain model, Int. J. Mod. Phys. B, **9**, 973–989, (1995)

Voi02. Voisin, C.: Hodge Theory and Complex Algebraic Geometry I. Cambridge Univ. Press, Cambridge, (2002)

Von05. Vonk, M.: A mini-course on topological strings. arXiv: hep-th/0504147.

Vyg82. Vygotsky, L.S.: Historical meaning of the Psychological crisis. Collected works. Vol. 1. Pedag. Publ., Moscow, (1982)

WA00. Wall, M.M., Amemiya, Y.: Estimation for polynomial structural equation models. Journal of American Statistical Association, **95**, 929–940, (2000)

WA98. Wall, M.M., Amemiya, Y.: Fitting nonlinear structural equation models. Proc. Social Stat. Section. Ann. Meet. Ame. Stat. Assoc. 180–185, (1998)

WDH93. Wilmott, P., Dewynne, J., Howinson, S.: Option Pricing: Mathematical Models and Computation. Oxford Financial Press, (1993)

WF49. Wheeler. J.A., Feynman, R.P.: Classical Electrodynamics in Terms of Direct Interparticle Action. Rev. Mod. Phys. **21**, 425–433, (1949)

WS98. Watts, D.J., Strogatz, S.H.: Collective dynamics of 'small-world' networks. Nature, **393**, 440–442, (1998)

WT92. Williams, R.M., Tuckey, P.: Regge Calculus: A bibliography and brief review. Class. Quant. Grav. **9**, 1409, (1992)

WW83a. Wehner, M.F., Wolfer, W.G.: Numerical evaluation of path–integral solutions to Fokker–Planck equations. I., Phys. Rev. A **27**, 2663–2670, (1983)

WW83b. Wehner, M.F., Wolfer, W.G.: Numerical evaluation of path–integral solutions to Fokker–Planck equations. II. Restricted stochastic processes, Phys. Rev. A, **28**, 3003–3011, (1983)

Wal00. Van der Wal, C. et al.: Quantum Superposition of Macroscopic Persistent-current States, Science, **290**, 773–777, (2000)

Wal84. Wald, R.: General Relativity. University of Chicago Press, (1984)

Wal94. Wald, R.: Quantum field theory in curved space–time and black hole thermodynamics, Chicago Univ. Press, (1994)

Wei36. Weiss, P.: Proc. Roy. Soc., A **156**, 192–220, (1936)

Wei64. Weinberg, S.: Photons and Gravitons in S-Matrix Theory: Derivation of Charge Conservation and Equality of Gravitational and Inertial Mass. Phys. Rev. **B135**, 1049, (1964)

Wei79. Weinberg, E.: Parameter counting for multimonopole solutions. Phys. Rev. D **20**, 936–944, (1979)

Wei80. Weinberg, S.: Conceptual foundations of the unified theory of weak and electromagnetic interactions. Rev. Mod. Phys. **52**, 515–523, (1980)

Wei87. Weinberg, S.: Anthropic Bound on the Cosmological Constant. Phys. Rev. Lett. **59**, 2607, (1987)

Wei90. Weinstein, A.: Affine Poisson structures. Internat. J. Math., **1**, 343–360, (1990)

Wei92. Weinberg, S.: Dreams of a Final Theory, Pantheon Books, New York, (1992)

Wei99. Weiss, U.: Quantum Dissipative Systems. World Scientific, Singapore, (1999)

Wei05. Weinberg, S.: The Cosmological Constant Problems. astro-ph/0005265, (2005)

Whe61. Wheeler, J.A.: Geometrodynamics and the Problem of Motion. Rev. Mod. Phys. **33**, 63–78, (1961)

Whe62. Wheeler. J.A.: Geometrodynamics. Academic Press, New York, (1962)

Whe86. Wheeler. J.A.: How Come the Quantum? In New Techniques and Ideas in Quantum Measurement Theory, ed. by D. Greenberger, Ann. N.Y. Acad. Sci 480, 304–316, (1986)

Whi87. Whitney, D.E.: Historical perspective and state of the art in robot force control. Int. J. Robot. Res., **6**(1), 3–14, (1987)

Whi84. Whitt, B.: Fourth-Order Gravity as General Relativity Plus Matter. Phys. Lett. **B145**, 176, (1984)

Wie61. Wiener, N.: Cybernetics. Wiley, New York, (1961)

Wig90. Wiggins, S.: Introduction to Applied Dynamical Systems and Chaos. Springer, New York, (1990)

Wik05. Wikipedia, the free encyclopedia. (2005) http://wikipedia.org.

Wil00. Wilson, D.: Nonlinear Control, Advanced Control Course (Student Version), Karlstad Univ., (2000)

Wil93. Willmore, T.J.: Riemannian Geometry. Oxford Univ. Press, Oxford, (1993)

Wil97. Williams, R.M.: Recent progress in Regge calculus. Nucl. Phys. B **57**, 73–81, (1997)

Wit01. Witten, E. Quantum Gravity in de Sitter Space. arXiv:hep-th/0106109, (2001)

Wit02. Witten, E.: The Universe on a String. Astronomy magazine, June, (2002)

Wit82. Witten, E.: Supersymmetry and Morse theory. J. Diff. Geom., **17**, 661–692, (1982)

Wit86a. Witten, E.: Interacting Field Theory of Open Superstrings. Nucl. Phys. **B276**, 291, (1986)

Wit88a. Witten, E.: 2+1 Gravity as an Exactly Soluble Model. Nucl. Phys. **B311**, 46–78, (1988)

Wit88b. Witten, E: Topological quantum field theory, Commun. Math. Phys., **117**, 353, (1988)

Wit88c. Witten, E.: Space-Time and Topological Orbifolds. Phys. Rev. Lett. **61**, 670–673, (1988)

Wit88d. Witten, E.: Topological Sigma Models. Commun. Math. Phys. **118**, 411, (1988)

Wit88e. Witten, E.: Topological Gravity. Phys. Lett. B **206**, 601, (1988)

Wit89. Witten, E.: Quantum field theory and the Jones polynomial. Commun. Math. Phys. **121**, 351, (1989)

Wit90. Witten, E.: On the structure of the topological phase of two–dimensional gravity. Nucl. Phys. B **340**, 281, (1990)

Wit91. Witten, E.: Introduction To Cohomological Field Theories. Int. J. Mod. Phys. A **6**, 2775, (1991)

Wit92. Witten, E.: Mirror manifolds and topological field theory. In Essays on mirror manifolds, ed. S.-T. Yau, International Press, 120–158, (1992)

Wit94. Witten, E.: Monopoles and four manifolds. Math. Res. Lett. **1**, 769–796, (1994)

Wit95. Witten, E.: Chern–Simons gauge theory as a string theory. Prog. Math. **133**, 637, (1995)

Wit95. Witten, E.: String Theory Dynamics in Various Dimensions. Nucl. Phys. B **443**, 85, (1995)

Wit96. Witten, E.: Bound States of Strings and p-Branes. Nucl. Phys. **B460**, 335, (1996)

Wit98a. Witten, E.: Magic, mystery, and matrix. Notices AMS, **45**(9), 1124–1129, (1998)

Wit98b. Witten, E.: Anti-de Sitter space and holography. Adv. Theor. Math. Phys., **2**, 253, (1998)

Woo92. Woodhouse, N.: Geometric Quantization. Clarendon Press, Oxford, (1992)

XH94. Xu, Z., Hauser, J.: Higher order approximate feedback linearization about a manifold. J. Math. Sys. Est. Con., **4**, 451–465, (1994)

XH95. Xu, Z., Hauser. J.: Higher order approximate feedback linearization about a manifold for multi-input systems. IEEE Trans. Aut. Con, AC-40, 833–840, (1995)

YA01. Yalcin, I., Amemiya, Y.: Nonlinear factor analysis as a statistical method. Statistical Science, **16**, 275–294, (2001)

YL52. Yang, C.N., Lee, T.D.: Statistical theory of equation of state and phase transitions I: Theory of condensation. Phys. Rev. **87**, 404–409, (1952)

YST96. Yamamoto, K., Sasaki, M., Tanaka, T.: Quantum fluctuations and CMB anisotropies in one-bubble open inflation models. Phys. Rev. D54, 5031-5048, (1996)

YZ99. Yong, J., Zhou, X.: Stochastic controls. Hamiltonian Systems and HJB Equations. Springer, New York, (1999)

Yag87. Yager, R.R.: Fuzzy Sets and Applications: Selected Papers by L.A. Zadeh, Wiley, New York, (1987)

Yan52. Yano, K.: Some remarks on tensor fields and curvature. Ann. of Math. **55**(2), 328–347, (1952)

Yan65. Yano, K.: Differential Geometry on Complex and Almost Complex Manifolds, Pergamon Press, New York, (1965)

Yau78. Yau, S.T.: On the Ricci curvature of a compact Kähler manifold and the complex Monge-Ampere equation, I^*. Comm. Pure Appl. Math. **31**, 339–441, (1978)

Yog46. Yogananda, P.: An Experience in Cosmic Consciousness. In Autobiography of a Yogi. Self-Realization Felowship, LA, CA, (1946)

You00. Youm, D.: Bulk fields in dilatonic and self-tuning flat domain walls, Nucl. Phys. B **589**, 315, (2000)

Zak92. Zakharov, O.: Hamiltonian formalism for nonregular Lagrangian theories in fibred manifolds. J. Math. Phys. **33**, 607, (1992)

ZGD91. Zhang, R.B., Gould, M.D., Bracken, A.J.: Quantum Group Invariants and Link Polynomials. Commun. Math. Phys, **137**, 13, (1991)

ZOC95. Zhang, R.B., Wang, B.L., Carey, A.L., McCarthy, J.: Topological Quantum Field Theory and Seiberg-Witten Monopoles. arXiv:hep-th/9504005, (1995)

Index

International Series on
INTELLIGENT SYSTEMS, CONTROL AND AUTOMATION: SCIENCE AND ENGINEERING

Editor: Professor S. G. Tzafestas, *National Technical University, Athens, Greece*

1. S.G. Tzafestas (ed.): *Microprocessors in Signal Processing, Measurement and Control*. 1983 ISBN 90-277-1497-5
2. G. Conte and D. Del Corso (eds.): *Multi-Microprocessor Systems for Real-Time Applications*. 1985 ISBN 90-277-2054-1
3. C.J. Georgopoulos: *Interface Fundamentals in Microprocessor-Controlled Systems*. 1985 ISBN 90-277-2127-0
4. N.K. Sinha (ed.): *Microprocessor-Based Control Systems*. 1986 ISBN 90-277-2287-0
5. S.G. Tzafestas and J.K. Pal (eds.): *Real Time Microcomputer Control of Industrial Processes*. 1990 ISBN 0-7923-0779-8
6. S.G. Tzafestas (ed.): *Microprocessors in Robotic and Manufacturing Systems*. 1991 ISBN 0-7923-0780-1
7. N.K. Sinha and G.P. Rao (eds.): *Identification of Continuous-Time Systems*. Methodology and Computer Implementation. 1991 ISBN 0-7923-1336-4
8. G.A. Perdikaris: *Computer Controlled Systems*. Theory and Applications. 1991 ISBN 0-7923-1422-0
9. S.G. Tzafestas (ed.): *Engineering Systems with Intelligence*. Concepts, Tools and Applications. 1991 ISBN 0-7923-1500-6
10. S.G. Tzafestas (ed.): *Robotic Systems*. Advanced Techniques and Applications. 1992 ISBN 0-7923-1749-1
11. S.G. Tzafestas and A.N. Venetsanopoulos (eds.): *Fuzzy Reasoning in Information, Decision and Control Systems*. 1994 ISBN 0-7923-2643-1
12. A.D. Pouliezos and G.S. Stavrakakis: *Real Time Fault Monitoring of Industrial Processes*. 1994 ISBN 0-7923-2737-3
13. S.H. Kim: *Learning and Coordination*. Enhancing Agent Performance through Distributed Decision Making. 1994 ISBN 0-7923-3046-3
14. S.G. Tzafestas and H.B. Verbruggen (eds.): *Artificial Intelligence in Industrial Decision Making, Control and Automation*. 1995 ISBN 0-7923-3320-9
15. Y.-H. Song, A. Johns and R. Aggarwal: *Computational Intelligence Applications to Power Systems*. 1996 ISBN 0-7923-4075-2
16. S.G. Tzafestas (ed.): *Methods and Applications of Intelligent Control*. 1997 ISBN 0-7923-4624-6
17. L.I. Slutski: *Remote Manipulation Systems*. Quality Evaluation and Improvement. 1998 ISBN 0-7932-4822-2
8. S.G. Tzafestas (ed.): *Advances in Intelligent Autonomous Systems*. 1999 ISBN 0-7932-5580-6
9. M. Teshnehlab and K. Watanabe: *Intelligent Control Based on Flexible Neural Networks*. 1999 ISBN 0-7923-5683-7
0. Y.-H. Song (ed.): *Modern Optimisation Techniques in Power Systems*. 1999 ISBN 0-7923-5697-7
1. S.G. Tzafestas (ed.): *Advances in Intelligent Systems*. Concepts, Tools and Applications. 2000 ISBN 0-7923-5966-6
2. S.G. Tzafestas (ed.): *Computational Intelligence in Systems and Control Design and Applications*. 2000 ISBN 0-7923-5993-3
3. J. Harris: *An Introduction to Fuzzy Logic Applications*. 2000 ISBN 0-7923-6325-6

International Series on

INTELLIGENT SYSTEMS, CONTROL AND AUTOMATION: SCIENCE AND ENGINEERING

24. J.A. Fernández and J. González: *Multi-Hierarchical Representation of Large-Scale Space*. 2001 ISBN 1-4020-0105-3
25. D. Katic and M. Vukobratovic: *Intelligent Control of Robotic Systems*. 2003 ISBN 1-4020-1630-1
26. M. Vukobratovic, V. Potkonjak and V. Matijevic: *Dynamics of Robots with Contact Tasks*. 2003 ISBN 1-4020-1809-6
27. M. Ceccarelli: *Fundamentals of Mechanics of Robotic Manipulation*. 2004 ISBN 1-4020-1810-X
28. V.G. Ivancevic and T.T. Ivancevic: *Human-Like Biomechanics*. A Unified Mathematical Approach to Human Biomechanics and Humanoid Robotics. 2005 ISBN 1-4020-4116-0
29. J. Harris: *Fuzzy Logic Applications in Engineering Science*. 2005 ISBN 1-4020-4077-6
30. M.D. Zivanovic and M. Vukobratovic: *Multi-Arm Cooperating Robots*. Dynamics and Control. 2006 ISBN 1-4020-4268-X
31. V.G. Ivancevic and T. Ivancevic: *Geometrical Dynamics of Complex Systems*. A Unified Modelling Approach to Physics, Control, Biomechanics, Neurodynamics and Psycho-Socio-Economical Dynamics. 2006 ISBN 1-4020-4544-1
32. V.G. Ivancevic and T.T. Ivancevic: *High-Dimensional Chaotic and Attractor Systems*. A Comprehensive Introduction. 2007 ISBN 1-4020-5455-6
33. K.P. Valavanis: *Advances in Unmanned Aerial Vehicles*. State of the Art and the Road to Autonomy. 2007 ISBN 978-1-4020-6113-4
34. V.G. Ivancevic and T.T. Ivancevic: *Complex Dynamics*. Advanced System Dynamics in Complex Variables. 2007 ISBN 978-1-4020-6411-1